12TH INTERNATIONAL CONFERENCE ON MOLTEN SLAGS, FLUXES AND SALTS

Supporting the Transition to Sustainable Technologies

17–19 June 2024
BRISBANE, AUSTRALIA

Edited by:
P Hayes and E Jak

The Australasian Institute of Mining and Metallurgy
Publication Series No 3/2024

AusIMM

Published by:
The Australasian Institute of Mining and Metallurgy
Ground Floor, 204 Lygon Street, Carlton Victoria 3053, Australia

ISBN 978-1-922395-39-9 (2 volume set)

ISBN 978-1-922395-42-9 (volume 1)

ISBN 978-1-922395-43-6 (volume 2)

ADVISORY COMMITTEE

Emer Prof Peter Hayes
FAusIMM(CP)
University of Queensland – Conference Chair

Prof Evgueni Jak
University of Queensland – Conference Chair

Dr Denis Shishin
AAusIMM
University of Queensland – Conference Co-Chair

Dr Maksym Shevchenko
University of Queensland – Conference Co-Chair

Prof Akbar Rhamdhani
Swinburne University of Technology

Prof Brian Monaghan
University of Wollongong

Dr Chunlin Chen
CSIRO

Prof Geoff Brooks
Swinburne University of Technology

Dr Jeff Chen
University of Queensland

Dr John Rankin
Swinburne University of Technology

Prof Oleg Ostrovski
UNSW

Dr Sharif Jahanshahi
Metalogical Solutions

Dr Stuart Nicol
MAusIMM
Glencore Technology

Dr Svetlana Sineva
University of Queensland

INTERNATIONAL ADVISORY COMMITTEE

AFRICA
Buhle Xakalashe, Mintek, South Africa

ASIA AND PACIFIC
Adj Prof Gerardo Alvear, University of Queensland, Australia

Prof Chenguang Bai, Chongqing University, China

Prof Jung Wook Cho, POSTECH, Korea

Prof Yongsug Chung, Korea Polytechnic University, Korea

Prof Xiaohui Fan, Central South University, China

Dr Taufiq Hidayat, Institut Teknologi Bandung, Indonesia

Prof Shuqiang Jiao, University of Science and Technology Beijing, China

Prof In-Ho Jung, Seoul National University, Korea

Prof Youn-Bae Kang, POSTECH, Korea

Prof Jong Hyeon Le, Chungnam National University, Korea

Prof Joonho Lee, Korea University, Korea

Prof Chenjun Liu, Northeaster University, China

Assoc Prof Hiroyuki Matsuura, Tokyo University, Japan

Prof Kazuki Morita, University of Tokyo, Japan

Prof Toru Okabe, University of Tokyo, Japan

Prof Joo Hyun Park, Hanyang University, Korea

Prof Hiroyuki Shibata, Tohoku University, Japan

Prof Il Sohn, Yonsei University, Korea

Prof Toshihiro Tanaka, Osaka University, Japan

Prof Jilai Xue, University of Science and Technology Beijing, China

Prof Baijun Yan, University of Science and Technology Beijing, China

EUROPE
Dr Moritz to Baben, GTT Technologies, Germany

Prof Bo Bjorkman, Luleå University of Technology, Sweden

Prof Bart Blanpain, KU Leuven, Belgium

Prof Timo Fabritus, University of Oulu, Finland

Prof Bernd Friedrich, RWTH Aachen University, Germany

Dr Muxing Guo, Katholieke Universiteit Leuven, Belgium

Prof Miroslaw Karbowniczek, AGH University of Science and Technology, Poland

Prof Zushu Li, University of Warwick, United Kingdom

Dr Alexander Pisch, French National Centre for Scientific Research, France

Prof Markus Reuter, SMS Group GmbH, Germany

Prof Johannes Schenk, Montanuniversitaet Leoben, Austria

Prof Merete Tangstad, NTNU, Norway

Prof Emer Pekka Taskinen, Aalto University, Finland

NORTH AMERICA

Prof Antoine Allanore, Massachusetts Institute of Technology, United States of America

Prof Mansoor Barati, University of Toronto, Canada

Prof Patrice Chartrand, Polytechnique Montréal, Canada

Dr Sina Mostachel, SNC-Lavalin, Canada

Prof Chris Pistorius, Carnegie Mellon University, United States of America

Prof Ramana Reddy, University of Alabama, United States of America

Assoc Prof Leili Tafagodhi, McMaster University, Canada

SOUTH AMERICA

Prof Roberto Parra, University of Concepción, Chile

AUSIMM

Julie Allen
Head of Events

Kathryn Laslett
Conference Program Manager

REVIEWERS

We would like to thank the following people for their contribution towards enhancing the quality of the papers included in this volume:

Dr Alejandro Abadias Llamas

Dr Hamed Abdeyazdan

Adj Prof Gerardo Alvear

Prof Antoine Allanore

Dr Anton Andersson

Prof Chenguang Bai

Prof Mansoor Barati

Prof Bo Bjorkman

Prof Geoff Brooks

Dr Chunlin Chen

Dr Jiang (Jeff) Chen

Prof Jung Wook Cho

Prof Yongsug Chung

Prof Timo Fabritius

Prof Xiaohui Fan

Dr Jonah Gamutan

Dr Muxing Guo

Dr Joseph Hamuyuni

Dr Taufiq Hidayat

Dr Sharif Jahanshahi

Prof In-Ho Jung

Prof Youn-Bae Kang

Prof Miroslaw Karbowniczek

Georgii Khartcyzov

Dr Anna Klemettinen

Dr Lassi Klemettinen

Dr Alex Kondratiev

Prof Jong Hyeon Lee

Prof Joonho Lee

Prof Zushu Li

Prof Chengjun Liu

Assoc Prof Hiroyuki Matsuura

Prof Brian Monaghan

Dr Sina Mostaghel

Dr Evgenii Nekhoroshev

Dr Stuart Nicol

Prof Oleg Ostrovski

Prof Joohyun Park

Dr Alexander Pisch

Dr Chris Pistorius

Dr John Rankin

Prof Ramana Reddy

Prof Markus Reuter

Prof Akbar Rhamdhani

Prof Johannes Schenk

Dr Maksym Shevchenko

Prof Hiroyuki Shibata

Dr Denis Shishin

Dr Svetlana Sineva

Prof Il Sohn

Prof Emeritus Pekka Taskinen

Dr Moritz to Baben

Xi Rui (Jason) Wen

Prof Jilai Xue

Prof Baijun Yan

Dr Johan Zietsman

FOREWORD

On behalf of the organising committee, welcome to the 12th International Conference on Slags, Molten Salts and Fluxes (MOLTEN2024). This important series of MOLTEN conferences brings together the leading researchers and engineers from around the world to exchange information and discuss the latest ideas, developments and concepts in the field. For those interested in developing and optimising high temperature processing systems, this conference, held once every four years, is our version of the Olympics!

Our society faces major and immediate challenges in the form of the sustainability of our natural ecosystems and environment. This has resulted in commitments to transition globally to circular economy and corresponding recycling, to renewable energy, electrical power storage, the use of electricity in transportation and reduction in the use of fossil fuels. The new technologies required to achieve these objectives all rely on the use of metals and metal compounds.

The majority of metals undergo some form of high temperature processing in the molten state whether in production, refining or recycling. Critical for the design, development and optimisation of efficient, environmentally benign processes is the availability of fundamental scientific information on the physical and chemical properties of these multi-phase systems. It is the aim of this conference to make a positive contribution to the exchange and dissemination of knowledge on this important class of processing systems, and thereby enhance the rate of progress to more sustainable industrial processes.

The conference proceedings contains both peer reviewed articles, based on the oral presentations made at the conference, and summaries of poster presentations. These articles cover topics in the following themes.

1. Experimental measurements:
   - Thermodynamic properties, multi-phase equilibria, minor element distributions.
   - Physico-chemical properties (viscosity, surface tension, conductivity…).
   - Reaction / Process kinetics.

2. Mathematical descriptions:
   - Thermodynamic databases and models.
   - Physico-chemical property models.
   - Process simulations.

3. Research:
   - Industrial slag / flux / molten salt design and optimisation.
   - Refractory-melt interactions.
   - New energy and metal production technologies.
   - Recycling and environment sustainability.

4. Implementation of research outcomes by industry:
   - Application of fundamental research to industrial practice.

We trust you will find this record of discussions at MOLTEN2024 useful and stimulating.

Yours faithfully,

Peter Hayes and Evgueni Jak

on behalf of the Advisory Committee of MOLTEN2024

x

SPONSORS

Gold Sponsor

HATCH

Conference Dinner Sponsor

GLENCORE TECHNOLOGY

Host City Partner

brisbane
australia

Supporting Partners

MET SOC
CIM ICM
Metallurgy & Materials Society
Société de la métallurgie et des matériaux

GDMB
Gesellschaft der
Metallurgen und Bergleute e.V.

ASOSIASI PROFESI METALURGI INDONESIA
PROMETINDO

I•M3 Institute of Materials, Minerals & Mining

JIM

KIM⁺
THE KOREAN INSTITUTE OF METALS AND MATERIALS

SOC

SME
Society for
Mining, Metallurgy
& Exploration

SAIMM
THE SOUTHERN AFRICAN INSTITUTE
OF MINING AND METALLURGY

CONTENTS

Volume 1

History of MOLTEN

Experimental measurements of – Physico-chemical properties (viscosity, surface tension, conductivity...)

Experimental measurements of – Reaction/Process kinetics

Experimental measurements of – Thermodynamic properties, multi-phase equilibria, minor element distributions

Implementation of research outcomes by industry – Application of fundamental research to industrial practice

Mathematical descriptions of – Physico-chemical property models

Mathematical descriptions of – Process simulations

Volume 2

Mathematical descriptions of – Thermodynamic databases and models

Research on – Industrial slag/flux/molten salt design and optimisation

Research on – New energy and metal production technologies

Research on – Recycling and environment sustainability

Research on – Refractory-melt interactions

Mathematical descriptions of – Thermodynamic databases and models

Thermodynamic modelling of species distribution in AlCl$_3$:BMIC salts

M K Nahian[1] and R G Reddy[2]

1. Graduate student, The University of Alabama, Tuscaloosa AL 35487, USA.
   Email: mknahian@crimson.ua.edu
2. Professor, The University of Alabama, Tuscaloosa AL 35487, USA. Email: rreddy@eng.ua.edu

ABSTRACT

The thermodynamic model for species distribution in aluminium chloride (AlCl$_3$) and 1-butyl-3-methylimidazolium chloride (BMIC) system was developed. The experimental thermodynamic data, considering a range of aluminium chloride species, such as AlCl$_4^-$, Al$_2$Cl$_7^-$, Al$_3$Cl$_{10}^-$, Al$_4$Cl$_{13}^-$, and Al$_2$Cl$_6$ were used in developing the model. When the X$_{AlCl3}$ was in the range of 0 to 0.50, only anion species Cl$^-$ and AlCl$_4^-$ were existing in the solution. As the X$_{AlCl3}$ value increased more than 0.50, the available anions were Al$_2$Cl$_7^-$, Al$_3$Cl$_{10}^-$, Al$_4$Cl$_{13}^-$, and Al$_2$Cl$_6$. Specifically, the concentration of Al$_2$Cl$_7^-$ increased in the composition range from 0.50 to 0.67 X$_{AlCl3}$, after which it decreased. This change in the concentration of Al$_2$Cl$_7^-$ was due to its reaction with excess AlCl$_3$, leading to the formation of Al$_3$Cl$_{10}^-$. The average cathode current density was determined by the electrochemical experiments varying X$_{AlCl3}$ from 0.50 to 0.71 in the AlCl$_3$:BMIC solutions. The average cathode current density increased as the X$_{AlCl3}$ increased from 0.50 to 0.67. However, it decreased when the X$_{AlCl3}$ > 0.67. This trend can be attributed to the electroactive species Al$_2$Cl$_7^-$. The Al$_2$Cl$_7^-$ promotes the reduction rate at the cathode, resulting in an increase in the average cathode current density.

INTRODUCTION

Ionic liquids (ILs) exhibit distinctive chemical and physical characteristics such as extensive liquidus temperature range, high thermal stability, low melting point, and large electrochemical window (Pradhan and Reddy, 2014). Typically, ILs contain organic cations and either organic or inorganic anions with melting points less than 100°C. The adjustable nature of ILs enables the tailoring of specific properties to suit diverse fields, including batteries, fuel cells, and solvents in chemical processes. They are regarded as green electrolytes for metal deposition due to their low vapor pressure and less pollutant emission (Pradhan and Reddy, 2012, 2014). Among them, the 1-butyl-3-methyl-imidazolium chloride (BMIC) ionic liquid, when combined with aluminium chloride (AlCl$_3$), shows great promise in fields like batteries and low-temperature aluminium electrodeposition. In this IL, different important properties like density, viscosity, electrical conductivity, and diffusion coefficient are dictated by the species and their interaction with each other (Fannin *et al*, 1984). With changing AlCl$_3$ mole fraction (X$_{AlCl3}$) in the AlCl$_3$:BMIC system, species concentration changes which results in the change of the chloro-acidity. Depending on the content of X$_{AlCl3}$, AlCl$_3$:BMIC can be basic (X$_{AlCl3}$ < 0.5), neutral, and acidic (X$_{AlCl3}$ > 0.5) (Karpinski and Osteryoung, 1984). The molecular structures of AlCl$_3$ and BMIC are presented in the Figure 1.

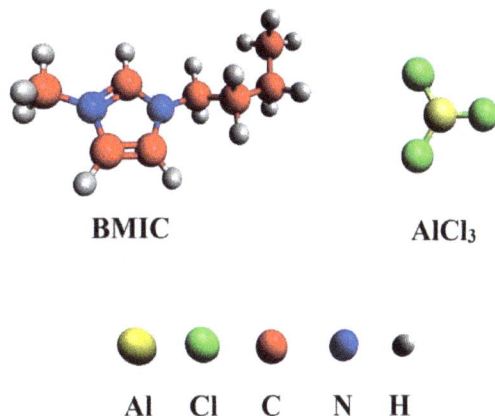

BMIC   AlCl$_3$

Al   Cl   C   N   H

FIG 1 – Molecular structure of BMIC and AlCl$_3$ (drawn by Avogadro program (Hanwell *et al*, 2012).

Although there is a growing interest in low-temperature refining of aluminium using $AlCl_3$:BMIC, production of aluminium on a pilot scale using ILs is in progress (Pradhan, Mantha and Reddy, 2009). The average cathode current density during aluminium electrodeposition refers to the amount of electric current per unit area that flows to the cathode during the electrodeposition process. This parameter is a critical factor in controlling the rate at which metal ions are reduced and deposited onto the cathode surface. Parameters such as electrolyte composition, temperature, and applied voltage are varied to achieve the desired cathode current density for successful electrodeposition. To obtain maximum aluminium electrodeposition efficiently, it is crucial to understand the optimal composition of the $AlCl_3$:BMIC electrolyte.

This study discusses the species concentration profile of the $AlCl_3$:BMIC ionic liquid electrolyte with average cathode current density data during the aluminium electrochemical process. The species concentration profile was constructed across the X_{AlCl3} range from 0 to 1 (Nahian and Reddy, 2023). Average cathode current density in several literature with respect to X_{AlCl3} (Wang, 2020; Kamavaram, Mantha and Reddy, 2003, 2005). Understanding this correlation is vital as the species concentration profile influences the electrochemical behaviour of the electrolyte, while average cathode current density data provides insights into the rate of reduction reactions during aluminium electrodeposition.

MODELLING AND EXPERIMENTAL PROCEDURE

Thermodynamic modelling of $AlCl_3$:BMIC species

In $AlCl_3$:BMIC, BMIC dissociates and produces BMI^+ cation and Cl^-:

$$BMIC = BMI^+ + Cl^- \qquad (1)$$

$AlCl_3$ prefers to form $0.5Al_2Cl_6$. This $0.5Al_2Cl_6$ reacts with Cl^- to produce $AlCl_4^-$:

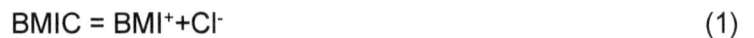
$$Cl^- + 0.5Al_2Cl_6 = AlCl_4^- \qquad (2)$$

$AlCl_4^-$ reacts with additional Al_2Cl_6 and produces $Al_2Cl_7^-$:

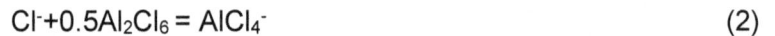
$$AlCl_4^- + 0.5Al_2Cl_6 = Al_2Cl_7^- \qquad (3)$$

In a similar way higher order polymeric species form:

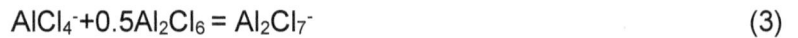
$$Al_2Cl_7^- + 0.5Al_2Cl_6 = Al_3Cl_{10}^- \qquad (4)$$

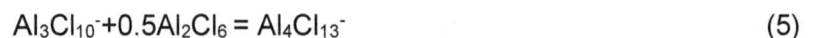
$$Al_3Cl_{10}^- + 0.5Al_2Cl_6 = Al_4Cl_{13}^- \qquad (5)$$

From above reaction the only cation is BMI^+ and the anions are Cl^-, $AlCl_4^-$, $Al_2Cl_7^-$, $Al_3Cl_{10}^-$, $Al_4Cl_{13}^-$. Al_2Cl_6 is the neutral species. Based on thermodynamic data, mass and charge balance anionic species concentration profile was developed for 25°C considering the anionic and the neutral species (Nahian and Reddy, 2023).

Electrodeposition of aluminium

The experimental procedure for aluminium electrodeposition using $AlCl_3$:BMIC is described somewhere else (Wang, 2020; Ahmed and Reddy, 2021; Pradhan and Reddy, 2014; Kamavaram, Mantha and Reddy, 2005). Electrolytes were prepared by mixing the required amount of $AlCl_3$ and BMIC. All experiments conducted in a 50 mL beaker. Cathode and anode were copper and aluminium alloys. Average cathode current density during this electrodeposition process was recorded for different X_{AlCl3}. As the electrolyte was sensitive to moisture, all experiments conducted under an argon atmosphere. The potential difference between two electrodes and the current of the electrochemical process measured using a Multimeter (Keithley Instruments Inc). Kepco power source (Kepco Programmable Power Supply) or Gamry potentiostat (Interface 1010E full-featured potentiostat) was used to apply the required potential. The temperature of the cell measured using thermometer by inserting into the electrolyte. The current was measured as a function of temperature. Electrochemical deposition time of aluminium was 2 hrs. The electrolyte was stirred at a constant speed. The current density was determined by dividing the current by the deposited aluminium area on the copper electrode. The schematic diagram of the electrochemical cell is represented in Figure 2.

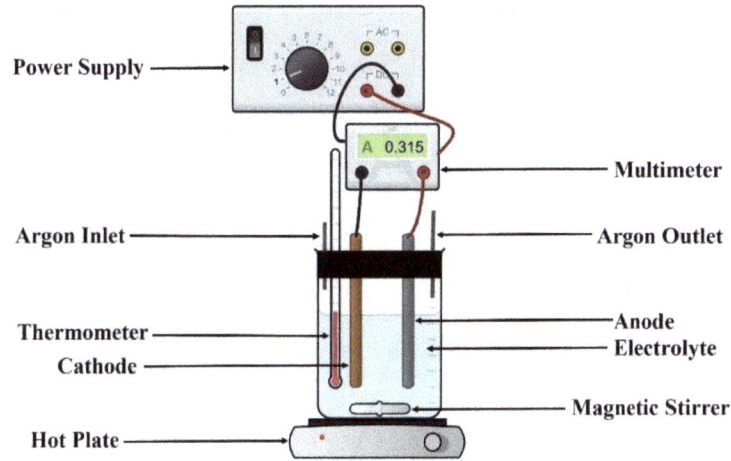

FIG 2 – Schematic of a typical electrochemical cell for aluminium deposition using IL electrolyte.

The average cathode current density for different experimental conditions is presented in Table 1. For experiments 2 and 3, the average cathode current density data were taken from the plots.

TABLE 1

Experimental conditions for aluminium deposition using $AlCl_3$:BMIC electrolyte.

Experiments	Anode	Cathode	Applied potential (V)	Temperature (°C)	X_{AlCl3}	Average cathode current density (A/m²)	Reference
1	Al2020	Cu	1.5	100	0.50	6.4	Wang, 2020
					0.55	21.1	
					0.58	169.3	
					0.62	180.5	
					0.64	282.6	
					0.67	396.7	
					0.69	317.2	
					0.71	231.0	
2	Al MMC	Cu	1.0	103 ± 2	0.57	211	Kamavaram, Mantha and Reddy, 2005
					0.60	242	
					0.64	241	
					0.67	261	
3	A360	Cu/Al	1.3	100	0.60	169	Kamavaram, Mantha and Reddy, 2003
					0.62	183	
					0.64	183	

RESULTS AND DISCUSSIONS

The species concentration of $AlCl_3$:BMIC at 25°C is presented in Figure 3. X_{Cl^-}, X_{AlCl4^-}, X_{Al2Cl7^-}, $X_{Al3Cl10^-}$, $X_{Al4Cl13^-}$, and X_{Al2Cl6} represents the mole fraction of Cl^-, $AlCl_4^-$, $Al_2Cl_7^-$, $Al_3Cl_{10}^-$, $Al_4Cl_{13}^-$ and Al_2Cl_6 respectively.

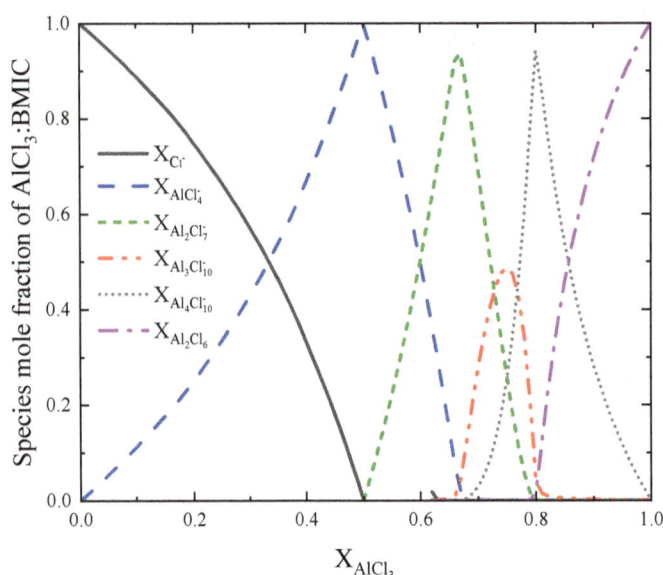

FIG 3 – Different species mole fraction of $AlCl_3$:BMIC as a function of X_{AlCl3} at 25°C (reproduced from (Nahian and Reddy, 2023).

When X_{AlCl3} varies from 0 to 0.5, the only anionic species are Cl^- and $AlCl_4^-$. Other anionic species $Al_2Cl_7^-$, $Al_3Cl_{10}^-$, $Al_4Cl_{13}^-$ exist when X_{AlCl3} > 0.5. The $AlCl_4^-$, $Al_2Cl_7^-$, $Al_3Cl_{10}^-$, $Al_4Cl_{13}^-$ exhibit peaks at X_{AlCl3} values of 0.50, 0.67, 0.75, and 0.80, respectively.

The average cathode current density as a function of X_{AlCl3} (varying from 0.50 to 0.71) in $AlCl_3$:BMIC electrolyte is given in Figure 4.

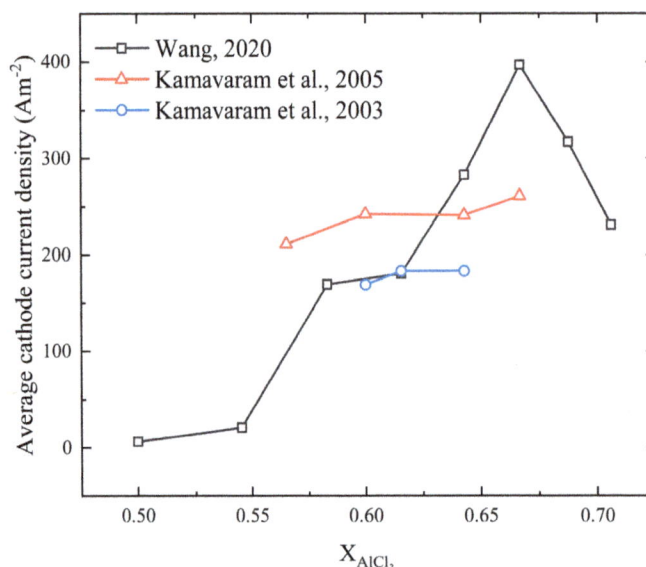

FIG 4 – Average cathode current density as a function of X_{AlCl3} during electrodeposition at 100°C, 120 rev/min, 1.5 V, and 2 hrs (Wang, 2020; Kamavaram, Mantha and Reddy, 2003, 2005).

In Figure 4, it is observed that the average cathode current density increases significantly with increasing X_{AlCl3} till X_{AlCl3} = 0.67. When X_{AlCl3} > 0.67, the average cathode current density starts to decrease. For experiment 1, the increase in average current density curve is steep X_{AlCl3} compared to experiments 2 and 3. Experiment 1 has also the highest average cathode current density at X_{AlCl3} = 0.67. This may be due to the composition of the anode. In experiment 1, Al2020 anode was used which has 93.43 wt per cent aluminium. On the other hand, Al MMC and A360 were used in experiments 2 and 3 which contain 66.22 and 76.2 wt per cent aluminium. Comparatively higher average current density was observed for experiment 2 at $X_{AlCl3} \geqslant 0.60$ because of the slightly high deposition temperature.

Species concentration profile is developed for 25°C and the average cathode current density data presented here is for 100°C. The effect of temperature on the species distribution of chloroaluminate IL, such as the mixture of 1-ethyl-3-methylimidazolium chloride (EMIC) or 1-butyl-3-methylimidazolium chloride with $AlCl_3$, is not significant, as evidenced in previous studies (Ahmed, Nahian and Reddy, 2023; Øye et al, 1991; Shi, Jiang and Zhao, 2021; Nahian and Reddy, 2023). Therefore, average cathode current density data obtained at 100°C can be effectively compared with the species concentration profile obtained at room temperature. Figure 5, presents a comparative analysis of the mole fractions of different species at 25°C alongside the average cathode current density for aluminium electrodeposition at 100°C using the $AlCl_3$:BMIC electrolyte system.

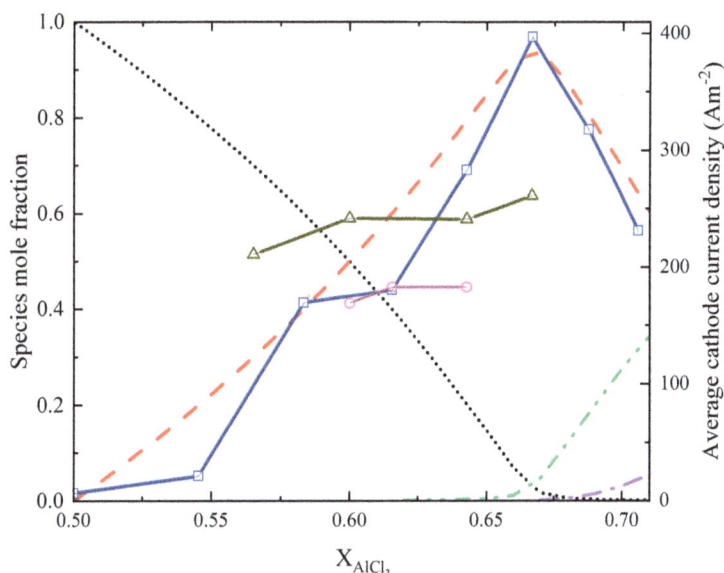

FIG 5 – Species mole fraction (X_{AlCl4^-}, X_{Al2Cl7^-}, $X_{Al3Cl10^-}$, and $X_{Al4Cl13^-}$) at 25°C and average cathode current density at 100 and 103 ± 2°C for Al electrodeposition in $AlCl_3$:BMIC.

($\cdots\cdots$ X_{AlCl4^-}, $---$ X_{Al2Cl7^-}, $-\cdot\cdot-$ $X_{Al3Cl10^-}$, $-\cdot-$ $X_{Al4Cl13^-}$, □ current density (Wang, 2020),
△ current density (Kamavaram, Mantha and Reddy, 2005), ○ current density (Kamavaram, Mantha and Reddy, 2003).

In the aluminium electrodeposition using $AlCl_3$:BMIC electrolyte, the average cathode current density shows a good correlation with the concentration of electrolyte species. At X_{AlCl3} = 0.50, where the only available anionic species is X_{AlCl4^-}, it has been found that the deposition of aluminium is not possible due to the non-reducibility of $AlCl_4^-$ anion (Carlin, Crawford and Bersch, 1992; Kamavaram, Mantha and Reddy, 2003). Consequently, the average cathode current density is negligible at X_{AlCl3} = 0.50. Beyond this composition (X_{AlCl3} > 0.50), the $Al_2Cl_7^-$ anion starts to appear, concurrently with a decrease in $AlCl_4^-$. It is reported that $Al_2Cl_7^-$ is the electroactive species that is responsible for the deposition of aluminium (Kamavaram, Mantha and Reddy, 2005). The aluminium electrodeposition process involves the reduction of $Al_2Cl_7^-$ to $AlCl_4^-$, as shown in the following reaction:

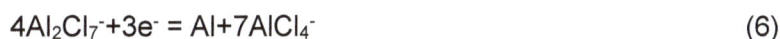

$$4Al_2Cl_7^- + 3e^- = Al + 7AlCl_4^- \tag{6}$$

At X_{AlCl3} = 0.50, the average cathode current density increases due to the increase in X_{Al2Cl7^-}. It reaches the maximum when X_{Al2Cl7^-} is highest (at X_{AlCl3} = 0.67). However, as X_{AlCl3} exceeds 0.67, there is a decline in X_{Al2Cl7^-} concentration resulting in a subsequent decrease in average cathode current density. It can be deduced that $Al_2Cl_7^-$ anionic species are determining species that control the overall average cathode current density.

CONCLUSIONS

A species concentration profile for the $AlCl_3$:BMIC electrolyte system based on thermodynamic modelling was discussed. The model was constructed incorporating species like Cl^-, $AlCl_4^-$, $Al_2Cl_7^-$, $Al_3Cl_{10}^-$, $Al_4Cl_{13}^-$ and Al_2Cl_6. The average cathode current density during the aluminium electrodeposition process using $AlCl_3$:BMIC electrolyte as a function of $AlCl_3$ mole fraction (X_{AlCl3})

was also correlated with this species concentration profile calculated using a thermodynamic model to get insight into the aluminium deposition process. At X_{AlCl3} = 0.50, the average cathode current density was negligible. This is attributed to anion $AlCl_4^-$ which cannot be reduced during the electrochemical process. Hence there is no significant cathode current density was observed. Above X_{AlCl3} = 0.50, the average cathode current density increases and when X_{AlCl3} > 0.67, the average cathode current density starts to decrease. This trend is due to the $Al_2Cl_7^-$ which is responsible for the aluminium deposition. These findings contribute to the understanding of the intricate interplay between the electrolyte composition, species concentration, and average cathode current density in the aluminium electrodeposition process, providing valuable insight for optimising such electrochemical processes for the production of aluminium for practical applications.

ACKNOWLEDGEMENTS

The authors acknowledge the financial support from the National Science Foundation (NSF) award number 1762522 and ACIPCO for this research project. Authors also thank the Department of Metallurgical and Materials Engineering, the University of Alabama for providing the experimental and analytical facilities.

REFERENCES

Ahmed, A N and Reddy, R, 2021. Electrochemical separation of aluminum from mixed scrap using ionic liquids, *The Minerals, Metals and Materials Series*, pp 1036–1044.

Ahmed, A N, Nahian, M K and Reddy, R G, 2023. Chloro-aluminate species distribution correlation with electrical conductivity of 1-ethyl-3-methyl imidazolium chloride (EMIC)-aluminum chloride ($AlCl_3$) system, *The Minerals, Metals and Materials Series*, pp 1115–1120.

Carlin, R T, Crawford, W and Bersch, M, 1992. Nucleation and morphology studies of aluminum deposited from an ambient-temperature chloroaluminate molten salt, *Journal of The Electrochemical Society*, 139(10): 2720–2727.

Fannin, A A, Floreani, D A, King, L A, Landers, J S, Piersma, B J, Stech, D J, Vaughn, R L, Wilkes, J S and Williams, J L, 1984. Properties of 1,3-dialkylimidazolium chloride-aluminium chloride ionic liquids, part 2, phase transitions, densities, electrical conductivities and viscosities, *The Journal of Physical Chemistry*, 88(12):2614–2621.

Hanwell, M D, Curtis, D E, Lonie, D C, Vandermeersch, T, Zurek, E and Hutchison, G R, 2012. Avogadro: An advanced semantic chemical editor, visualization and analysis platform, *Journal of Cheminformatics*, 4(1).

Kamavaram, V, Mantha, D and Reddy, R G, 2003. Electrorefining of aluminum alloy in ionic liquids at low temperatures, *Journal of Mining and Metallurgy, Section B: Metallurgy*, 39(1–2):43–58.

Kamavaram, V, Mantha, D and Reddy, R G, 2005. Recycling of aluminum metal matrix composite using ionic liquids, *Electrochimica Acta*, 50(16–17):3286–3295.

Karpinski, Z J and Osteryoung, R A, 1984. Determination of equilibrium constants for the tetra chloroaluminate ion dissociation in ambient-temperature ionic liquids, *Inorganic Chemistry*, 23(10):1491–1494.

Nahian, M K and Reddy, R G, 2023. Electrical conductivity and species distribution of aluminium chloride and 1-butyl-3-methylimidazolium chloride Ionic Liquid Electrolytes, *Journal of Physical Organic Chemistry*, 36(10).

Øye, H A, Jagtoyen, M, Oksefjell, T and Wilkes, J S, 1991. Vapour pressure and thermodynamics of the system 1-methyl-3-ethyl-imidazolium chloride-aluminium chloride, *Materials Science Forum*, 73–75:183–190.

Pradhan, D and Reddy, R G, 2012. Dendrite-free aluminium electrodeposition from $AlCl_3$-1-ethyl-3-methyl-imidazolium chloride ionic liquid electrolytes, *Metallurgical and Materials Transactions B*, 43(3):519–531.

Pradhan, D and Reddy, R G, 2014. Mechanistic study of al electrodeposition from EMIC–$AlCl_3$ and BMIC–$AlCl_3$ electrolytes at low temperature, *Materials Chemistry and Physics*, 143(2):564–569.

Pradhan, D, Mantha, D and Reddy, R G, 2009. The effect of electrode surface modification and cathode overpotential on deposit characteristics in aluminium electrorefining using EMIC–$AlCl_3$ ionic liquid electrolyte, *Electrochimica Acta*, 54(26):6661–6667.

Shi, M, Jiang, J and Zhao, H, 2021. Electrodeposition of aluminium in the 1-ethyl-3-methylimidazolium tetra chloroaluminate ionic liquid, *Electrochem*, 2(2):185–196.

Wang, Y, 2020. Development of aluminium electrorefining in ionic liquids: The effects of experimental conditions on the deposition behaviour and microstructure, thesis, The University of Alabama.

Continuous method of thermodynamic optimisation using first-derivative matrices for large multicomponent systems

E Nekhoroshev[1], D Shishin[2], M Shevchenko[3] and E Jak[4]

1. Theme Leader in Computational Thermodynamics, Pyrosearch group, University of Queensland, Brisbane Qld 4850. Email: e.nekhoroshev@uq.edu.au
2. Theme Leader in Thermodynamic Modelling, Pyrosearch group, University of Queensland, Brisbane Qld 4850. Email: d.shishin@uq.edu.au
3. Theme Leader in Experiments and Thermodynamic Optimization, Pyrosearch group, University of Queensland, Brisbane Qld 4850. Email: m.shevchenko@uq.edu.au
4. Professor of Chemical Engineering, Pyrosearch group, University of Queensland, Brisbane Qld 4850. Email: e.jak@uq.edu.au

ABSTRACT

Real-life applications require thermodynamic databases consisting of many elements. Currently, Pyrosearch supports a 20-component thermodynamic FactSage™, ver 8.2 (by Thermfact and GTT technologies) database for Cu-, Pb- and Fe-production provided to a group of industrial sponsors. The database development follows an agile cycle of experiment planning, modelling/update of database subsystems, quality tests and release. A unique challenge is development and maintenance of a database self-consistent across all binary/ternary and multicomponent systems. Practice shows that binary and ternary system data do not provide the necessary accuracy for predictions within multicomponent systems. Commercial databases are often developed using only available literature data and it is hard to estimate the prediction quality for industry-relevant multicomponent compositions. Therefore, to sustain a high reliability there is a need of targeted experiments including multicomponent data.

To support the continuous development process, a new optimisation methodology has been developed. After long years of optimisation experience, it is an authors' belief that for optimisation, it is essential to pinpoint a few trusted target points instead of using a whole cloud of available experimental data. In case if the optimisation procedure is automated, using all data, even with trust weights, still leads to inferior optimisation quality. This is caused by the fact that available optimisation algorithms usually assume a normal error distribution of experimental data which can be false in cases when experimental results are affected by sample contamination, evaporation, incomplete reactions, etc.

Typically, iterative optimisation involves multiple re-calculations of targets values which can take significant time and the whole procedure is not interactive. It is hard to find an exact combination of target weights to achieve desired outcomes. This problem has been solved by the authors by replacing slow FactSage™/ChemApp™ calculations by parameter value/target value relations using first-order derivative matrices. Precalculated derivative matrices allow fast analysis of parameter influence across multiple systems, selection of interaction combinations with superior performance, and analytical matrix solution to the optimisation problems using weighted linear least squares. The current optimisation system supports in-sheet ChemApp™ (by GTT-technologies, version 614) calculations for Microsoft® Excel® using custom Python® binding code, graphical and statistical representation of experimental data and target points, and real-time first-derivative matrix analysis of parameter sensitivity.

INTRODUCTION

This paper discusses results of development of a continuous optimisation methodology and a new efficient method of semi-automatic optimisation based on use of first derivative matrices of target points by model parameters, abbreviated JET (Jacobian of Experimental Targets).

Pyrosearch, UQ (University of Queensland) specialises in development of a 20-component thermodynamic database covering the Pb-Zn-Cu-Fe-Ca-Si-S-O-Al-Mg-Cr-As-Sn-Sb-Bi-Ag-Au-Ni-Co-Na system for various non-ferrous industrial applications. While development of most of the open and commercial thermodynamic databases relies heavily relies on the availability of experimental data in the literature, Pyrosearch has a strong team of researchers specialising in high-precision measurements of phase equilibria including gas phase, slags, matte, speiss, metal, solids using electron probe microanalysis (EPMA) technique (Llovet *et al*, 2021). Rather than relying on availability of experimental data in the literature sources, the database development follows an agile cycle of experimental investigation of new systems, the initial optimisation, further experimental and modelling refining of these systems by cross-checking database prediction quality with multicomponent experimental data. Besides consistent experiment planning, systematic use of high-quality EPMA data for optimisation highlighted certain legacy inconsistencies and limitations in the thermodynamic database which eventually had to be resolved. For example, compared to FToxid database of FactSage™ software, all liquid slag end members now have consistent physical properties with proper extrapolation down to low temperatures.

The 20-component database includes 728 oxide systems, 123 Na_2O-containing systems, 345 metal-matte-speiss systems, with the total of 1196 2-, 3- and 4-metal component systems. There are over 450 stoichiometric compounds, around 100 large and small solution phases and more than 3000 excess parameters in solution phases. Development and maintenance of such a database is a huge effort from both experimental and modelling points of view. Higher order systems in a database depend on all lower-order subsystems which causes a pyramid dependence effect. Re-optimisation of a single binary system can result in a significant work of re-optimisation of all ternary and cross-checking quaternary systems including that binary system. For this reason, many systems, such as $CaO-SiO_2$, $Al_2O_3-SiO_2$ etc has not been updated in FToxid since 1990s (Eriksson and Pelton, 1993).

One of the goals of the work described in this paper was to, first, develop an overall methodology of continuous optimisation which allows to significantly reduce the effort and time of accessing new experimental data and second, develop a practical optimisation method suitable for dealing with large bodies of multicomponent data and able to simultaneously assess multiple subsystems and analyse parameter influence across multicomponent data sets.

OVERALL OPTIMISATION METHODOLOGY

The continuous optimisation methodology is based on the overall accepted CALPHAD methodology. A thermodynamic database consists of many individual stoichiometric and solution phases; solution models are chosen depending on chemistry, structure and type of interactions. Simple solid solutions are typically modelled using simple polynomial models, solid solutions with multiple sublattices and complex substitution mechanisms are expanded within the compound energy formalism (CEF) (Hillert, 2021). Slag and matte/liquid metal phase are modelled using Modified Quasichemical Formalism (Pelton *et al*, 2000), which is suitable to describe strong interactions such as metal oxide/silica in slag and metal/sulfur, metal/oxygen in matte and metals. Each model provides expressions for Gibbs free energy of a solution as a function of composition, temperature, and, optionally, pressure. These expressions include properties of pure solution end members and various interaction parameters which together comprise *model parameters* of a solution. Model parameters of all phases in a system, limited to the elements of that system, make up *model parameters* of that system.

Model parameters need to be evaluated during an optimisation procedure based on various experimental data available for a system. According to thermodynamics, it is possible to derive all thermodynamic properties of a system from its Gibbs free energy. Optimisation solves the inverse problem of finding parameters of Gibbs energy functions from existing experimental data.

Generally, there can be numerous combinations of model parameters based on the same thermodynamic data. To limit the uncertainty, it is highly desirable to consider multiple types of experimental data together: phase equilibria, calorimetry, activity measurements, etc. For example, phase equilibria alone should not be used to derive Gibbs energy parameters: there might be multiple combinations of phases with different individual stabilities which cancel each other at equilibrium resulting in similar phase diagrams.

In systems pertaining to pyrometallurgy, there is a limited number of solid phases with extensive cover of liquidus area: spinel, tridymite, monoxide, olivine, hematite/corundum, etc. There are referred to as *major* phases. The thermodynamic properties of major phases usually have been extensively studied and their known thermodynamic properties can be used to derive the parameters in a liquid phase. There is also a much larger number of miscellaneous stoichiometric and slightly non-stoichiometric *minor* phases. While for some of them the thermodynamic properties are known, most of them are defined by the phase equilibria with a liquid and/or major solid phases.

Experimental data can have different quality depending on their source and the measurement method. The uncertainty in experimental data can be divided into *bias* and *scatter*. Biases can appear due to multiple reasons: incomplete chemical reactions, material losses, etc. Scatter is a random deviation due to individual history of a sample and its measurement. Because of the reasons explained in more detail later, for continuous optimisation it is preferred to use *target points* instead of raw experimental data.

Each target point can have one or more *target values*. It can be temperature, composition, partial pressure, or something else. For each target value, there should be an unambiguous way of calculating it. Similar target values are grouped together into *target groups*.

Ideally, all target points should be free from experimental biases, which are not handled well by optimisation algorithms, and scatter, which bloats the number of points necessary to derive reliable model parameters. For high-precision data, like EPMA measurements, target points can be the same as the experimental points. In case of highly scattered data (data clouds) and if there are several data groups with distinct biases, the best judgement and expertise of the researcher is required to carefully define custom target points (they do not have to be a subset of experimental points).

For composition areas where measurements are missing, *virtual target points*, introduced as a best estimate of thermodynamic properties, can be used to avoid unrealistic model parameters.

Ideally, for each subsystem there must be developed a set of target points that define its optimisation and can be reused multiple times to consistently update model parameters for that system.

Since target points, contrary to raw experimental data, are supposed to be free from bias and scatter, with target points it becomes possible to evaluate a quality of optimisation or compare the predictive ability of different versions of a database solely using predictions in target points: compact RMS (residual mean squares) tables with ΔT and Δx errors, detailed visualisations of individual deviations, scatter plots, etc.

JET (JACOBIAN OF EXPERIMENTAL TARGETS) OPTIMISATION ALGORITHM

Cost functions in thermodynamic optimisation

Let us take a simplified example of the CaO-SiO$_2$ system with just a few target points and a few variable parameter values. The complete optimisation of the system (Nekhoroshev *et al*, in prep) contains more target points and model parameters, but for illustrative purposes we consider only four target points in the phase diagram (Figure 1) and three parameters in Table 1.

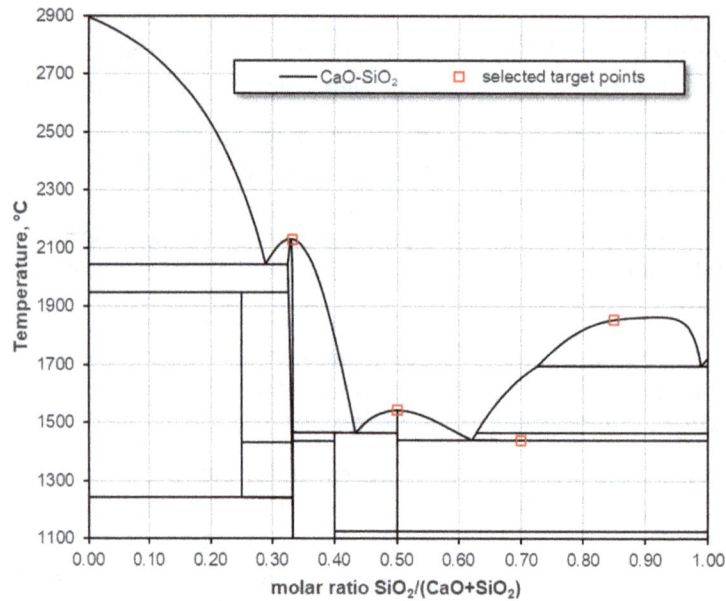

FIG 1 – Optimised phase diagram of the CaO-SiO$_2$ system (Shevchenko *et al*, in prep) together with a subset of target points: α-Ca$_2$SiO$_4$ and CaSiO$_3$ melting points, CaSiO$_3$-SiO$_2$ eutectic and a point on the miscibility gap.

TABLE 1

Optimal values of three selected parameters in slag phase of the CaO-SiO$_2$ system.

Parameter	Optimal value, J·mol^{-1}
$g^{00}_{Ca,Si}(T^0)$	-125520.8
$g^{01}_{Ca,Si}(T^0)$	+23544.7
$g^{07}_{Ca,Si}(T^0)$	-191208.8

The influence of parameters of the Modified Quasichemical Model in pair fraction expansion is mostly localised in composition domain. The main parameter $g^{00}_{Ca,Si}(T^0)$ mostly affects the area around the maximum short-range ordering composition at 33.33 mol per cent of SiO$_2$ (calcium disilicate liquidus). The auxiliary parameters $g^{01}_{Ca,Si}(T^0)$ and $g^{07}_{Ca,Si}(T^0)$ mostly affect the middle part of the diagram (pseudowollastonite liquidus) and the miscibility gap, correspondingly.

Figure 2 illustrates the dependencies of calculated temperatures in the target points obtained if the parameters are varied independently from each other starting from the optimal values in Table 1 as baseline references.

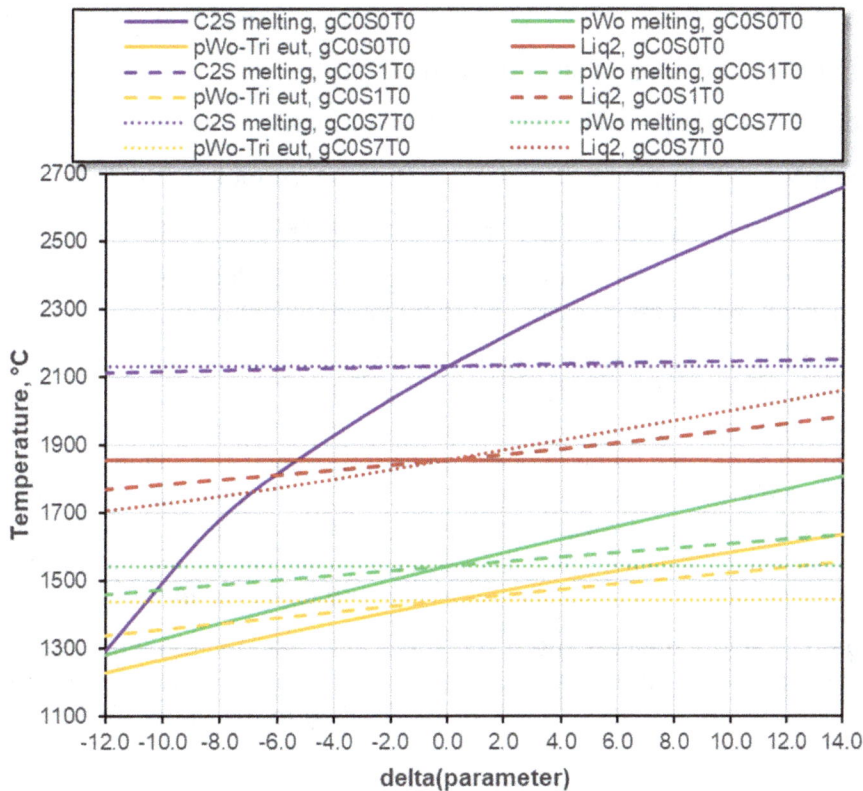

FIG 2 – Calculated formation target temperatures in selected target points on the CaO-SiO$_2$ phase diagram as functions of parameter deviations from the optimal values.

First, one should note the lines in Figure 2 are smooth and do not exhibit any breaks or kinks. For the most of them, the dependence is also approximately linear. It is always possible to make parameter dependencies exhibit smooth behaviour using a correct calculation set-up of target values.

The area of mathematical optimisation is extremely broad and covers various problems in engineering and physics. There are also multiple types of optimisation algorithms, each one of them is best suited for a certain type of task. Optimisation algorithms with certain assumptions about the cost function usually require less function evaluations to converge while algorithms which treat a cost function as a black box usually need more function evaluations. In thermodynamic optimisation, each function evaluation is performed using thermodynamic software such as FactSage™ or ChemApp™ and can be time-consuming. Phase target calculations generally take longer to calculate, especially in case of miscibility gaps. Isothermal calculations are usually much faster but in case of multicomponent systems can have limited applicability. Even if a single calculation is quite fast, multiple function evaluations can significantly slow down the whole optimisation procedure. It is important to find an appropriate trade-off between generality and efficiency.

To ensure the smoothness of the cost function one should pay attention to phase selection. As a rule, calculated thermodynamic properties exhibit breaks when a phase appears or disappears in a consequent iteration but generally, this situation can be avoided using a proper calculation set-up. For example, precipitation targets with multiple possible forming solids should be avoided and replaced with formation targets of a specific solid phase. Bulk compositions for isothermal targets should be selected in such a phase that a stable phase assemblage does not change when model parameters are updated.

Second, if set-up correctly, target values are smooth functions of model parameters and although the derivatives cannot be evaluated directly, they can be calculated using a finite difference method with an appropriate step size for each parameter. Numerical evaluation of second derivatives is usually too computationally expensive, though.

In general, the following information is required to choose an appropriate algorithm:

- Does the cost function have to be a scalar, or it can be written down as a vector of residuals?
- Is the cost function smooth and differentiable?
- If the first derivative exists, can it be evaluated analytically?
- Can we cheaply evaluate the second derivative of the target function?
- Does the target function have one global minimum or several local minima?

A scalar cost function, for example, a sum of squared residuals, loses some information compared to a vector cost function, which can result in more function evaluations.

Existence of first order derivatives greatly simplifies the optimisation task, giving rise to a group of optimisation methods related to the *gradient descent* algorithm. The gradient descent algorithm uses the direction of the maximum change in function value (the gradient) to update the solution. The extent of a step can be evaluated using information on the second derivative (Hessian matrix), giving rise to the Newton's method. If the Hessian matrix is not known, some information about its properties can be implicitly evaluated based on the change of gradient direction with consecutive iterations (quasi-Newton methods). In the limiting case when the second-order derivative of the cost function is zero (a linear system), the optimal solution can be found with one iteration.

The disadvantage of gradient descent methods is that they rely only on local information about the cost function and if multiple minima exist, the exact outcome of the optimisation routine will depend on the initial estimate. Usually, a set of initial points leading to the same minimum form a *basin of attraction* of this minimum.

To overcome this limitation, global optimisation methods can be used, such as Direct Search (Kolda, Lewis and Torczon, 2003), or genetic algorithms (Katoch, Chauhan and Kumar, 2020).

A typical scalar cost function in thermodynamic optimisation is illustrated in Figures 3 and 4. Since target value dependence on model parameters is mostly linear, as shown in Figure 2, the sum of squared residuals is supposed to have a near-parabolic shape (Figure 3). In case of two and more parameters, a parabolic valley can be more or less elongated indicating how fast residuals grow depending on parameters (Figure 4).

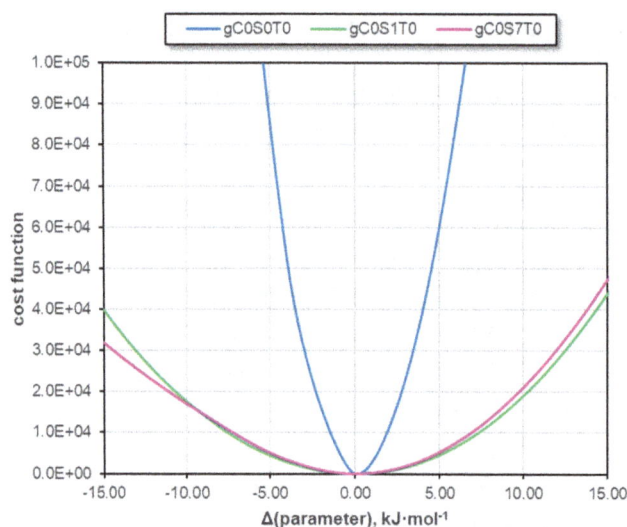

FIG 3 – The cost function (sum of squared residuals) as a function of deviation of a parameter from an optimal value with other parameters fixed at optimal values.

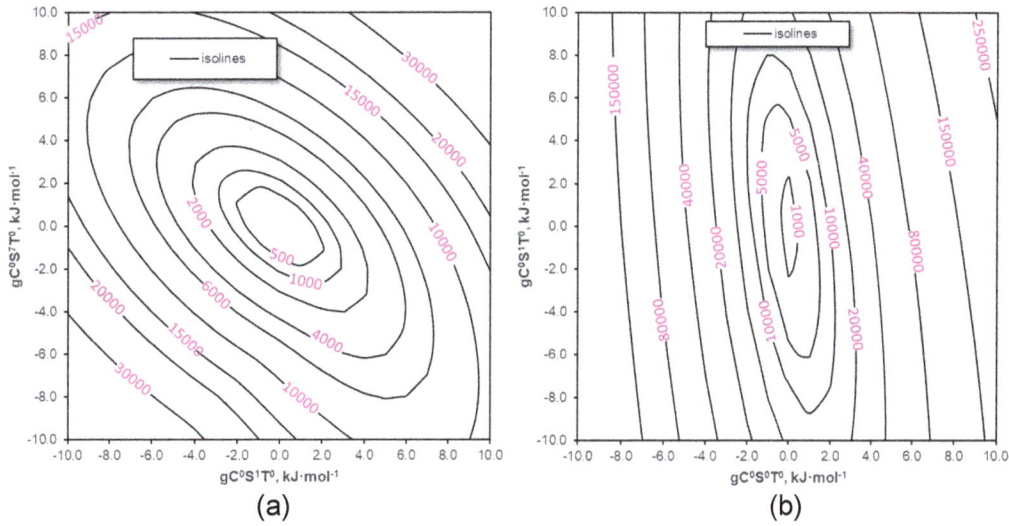

FIG 4 – Minimum valleys for the sum of squared residuals as a function of two parameters with the third parameter fixed: $S\left(g_{Ca,Si}^{01}(T^0), g_{Ca,Si}^{07}(T^0)\right)$ (a), $S\left(g_{Ca,Si}^{00}(T^0), g_{Ca,Si}^{01}(T^0)\right)$ (b).

Non-linear least squares method is an algorithm which, in our previous experience, works well for thermodynamic optimisation. It is available, for example, as *lsqnonlin* function in MATLAB or *least_squares* function in SciPy package for Python®. To use those functions, a rough initial guess for model parameters must be known which is usually found manually. The initial guess does not have to describe target points well, the only requirement is to be able to perform all thermodynamic calculations without errors and with a proper set of stable phases which ensures the existence of a smooth derivative in the initial point.

Non-linear least square optimisation with direct thermodynamic calculations is an acceptable solution for systems of small size (binary and ternary). However, it quickly becomes unfeasible for large multicomponent systems with thousands of target points. Also, maintenance of a multicomponent thermodynamic database implies that most subsystems are only slightly or moderately modified when new experimental data become available without a need of a total re-optimisation from scratch.

Also, it is very important to analyse parameter sensitivity and how model parameters influence target points across multiple subsystems and use this information to produce large self-consistent blocks of a thermodynamic database.

Derivation of Jacobian matrix and its use in optimisation

It was proposed by Professor Evgueni Jak that calculated target values can be efficiently approximated using linear relations between calculated values and model parameters. Let us suppose that any calculated values (temperatures, compositions, enthalpies, activities etc) are implicit functions of model parameters:

$$y_1 = f_1(p_1, p_2, \dots, p_m)$$
$$y_2 = f_2(p_1, p_2, \dots, p_m)$$
$$\dots$$
$$y_n = f_n(p_1, p_2, \dots, p_m) \tag{1}$$

In a proximity of an *m*-dimensional model parameter vector $\bar{p}^0 = (p_1^0, p_2^0, \dots, p_m^0)$, *n f*-functions can be decomposed into linear parts of Taylor series of parameter deviations $\Delta\bar{p} = (\Delta p_1, \Delta p_2, \dots, \Delta p_m)$:

$$y_1 \approx y_1^0 + \frac{\partial f_1}{\partial p_1}(\bar{p}^0) \cdot (p_1 - p_1^0) + \frac{\partial f_1}{\partial p_2}(\bar{p}^0) \cdot (p_2 - p_2^0) + \dots + \frac{\partial f_1}{\partial p_m}(\bar{p}^0) \cdot (p_m - p_m^0)$$

$$y_2 \approx y_2^0 + \frac{\partial f_2}{\partial p_1}(\bar{p}^0) \cdot (p_1 - p_1^0) + \frac{\partial f_2}{\partial p_2}(\bar{p}^0) \cdot (p_2 - p_2^0) + \dots + \frac{\partial f_2}{\partial p_m}(\bar{p}^0) \cdot (p_m - p_m^0)$$

...

$$y_n \approx y_n^0 + \frac{\partial f_n}{\partial p_1}(\bar{p}^0) \cdot (p_1 - p_1^0) + \frac{\partial f_n}{\partial p_2}(\bar{p}^0) \cdot (p_2 - p_2^0) + \cdots + \frac{\partial f_n}{\partial p_m}(\bar{p}^0) \cdot (p_m - p_m^0) \tag{2}$$

The same relations can be written in a compact matrix form:

$$\bar{y} = \bar{y}^0 + A(\bar{p} - \bar{p}^0) \tag{3}$$

where the elements of matrix A are composed of partial derivatives in Equation 2.

Matrix A can be used for fast real-time updates of target values under conditions that parameter differences are not very large. It is important to note that once a derivative matrix is calculated, no consequent calls to thermodynamic software are needed. It can also be used for the inverse task of finding a new optimal set of parameters $\bar{p}^{opt}$ which satisfies the Linear Least Squares condition. To do that, we need to solve the following matrix equation:

$$(\bar{y} - \bar{y}^0) = A(\bar{p} - \bar{p}^0) \tag{4}$$

In general, matrix A is not square and cannot be inverted directly. To avoid this obstacle, Equation 4 is multiplied by a transposed matrix A on both sides:

$$A^T(\bar{y} - \bar{y}^0) = A^T A(\bar{p} - \bar{p}^0) \tag{5}$$

Now, matrix $D = A^T A$ is a symmetric square matrix (which can easily be shown) and we can take its proper inverse to find $(\bar{p} - \bar{p}^0)$:

$$\bar{p} - \bar{p}^0 = (A^T A)^{-1} A^T (\bar{y} - \bar{y}^0) = D^{-1} A^T (\bar{y} - \bar{y}^0) \tag{6}$$

A modification of Equation 6 accepts individual weights for target points as well:

$$\bar{p} - \bar{p}^0 = (A^T W A)^{-1} A^T W (\bar{y} - \bar{y}^0) \tag{7}$$

where W is a diagonal square matrix, whose diagonal elements are the corresponding target point weights.

From the practical point of view, direct matrix regression Equation 6 works well not only for a case when the initial guess is quite close to the optimal solution, but also in case when it is not – the only requirement is that all target values can be properly calculated initially (TD software converges). In this case, it is usually required to recalculate the derivative matrix two to three times before full convergence is achieved.

Derivation of the Linear Least Squares method assuming normal error distribution

This section covers in detail the mathematical reasons behind inefficiency of using raw experimental data for optimisation and why it is preferrable to replace it by target points.

Linear Least Squares (LLS) are so far one of the most popular methods of statistical analysis of experimental data. The usual way of deriving the formulas estimating the parameters a_0 and a_1 is writing down the expression of the sum of error squares and minimising it:

$$S(a_0, a_1) = \sum_{i=1}^{n} (y_i^{exp} - y_i^{calc})^2 = \sum_{i=1}^{n} (y_i^{exp} - (a_1 x_i + a_0))^2 \tag{8}$$

The condition of a minimum of Equation 8 is equivalent to the condition of partial derivatives of S by a_1 and a_0 being equal to zero:

$$\frac{\partial S}{\partial a_1} = -2 \sum_{i=1}^{n} x_i^{exp} (y_i^{exp} - a_1 x_i - a_0) = 0$$

$$\frac{\partial S}{\partial a_0} = -2 \sum_{i=1}^{n} (y_i^{exp} - a_1 x_i - a_0) = -2n(\bar{y} - a_1 \bar{x} - a_0) = 0 \tag{9}$$

Solving for a_0 in the second equation of Equation 9 and substituting it in the first equation, after consequent rearrangement the final equations for a_1 and a_0 can derived:

$$a_1 = \frac{\sum_{i=1}^{n}(y_i - \bar{y})(x_i - \bar{x})}{\sum_{i=1}^{n}(x_i - \bar{x})^2}$$

$$a_0 = \bar{y} - a_1\bar{x} \tag{10}$$

From the standard definition it is unclear why, for example, the cost function must be expressed as a sum of squares of residuals. The reason behind this is that LLS is a method assuming *normal* error distribution. There is an alternative and more general approach to derive the LLS Equation 10 using maximum likelihood estimation (MLE) which explicitly uses error distributions and results in the same formulas for LLS.

Assuming a normal error distribution, we can write that the likelihood (estimated expectation) of an experimental point y^{exp} to correspond to a calculated point y^{calc} with added error value following a normal error distribution is:

$$L(y^{calc}, \sigma | y^{exp}, x^{exp}) = L(a_1, a_0, \sigma | y^{exp}, x^{exp}) = \frac{1}{\sigma\sqrt{2\pi}} \exp\left[-\frac{(y^{exp}-y^{calc})^2}{2\sigma^2}\right] =$$

$$\frac{1}{\sigma\sqrt{2\pi}} \exp\left[-\frac{(y^{exp}-a_1 x^{exp}-a_0)^2}{2\sigma^2}\right] \tag{11}$$

If for different points errors are independent, then for *n* points, the total likelihood is expressed as a product of individual likelihoods:

$$L(a_1, a_0, \sigma | y_1^{exp}, x_1^{exp}, y_2^{exp}, x_2^{exp}, \dots y_n^{exp}, x_n^{exp}) = \prod_{i=1}^{n} L(a_1, a_0, \sigma | y_i^{exp}, x_i^{exp}) \tag{12}$$

For the maximum likelihood of a_1 and a_0, partial derivatives of the Equation 12 must be equal to zero; however, is it more convenient to use a logarithm of the likelihood Equation 13:

$$\ln L = \sum_{i=1}^{n}\left[-\ln(\sigma\sqrt{2\pi}) - \frac{(y^{exp}-a_1 x^{exp}-a_0)^2}{2\sigma^2}\right] \tag{13}$$

$$\frac{\partial \ln L}{\partial a_1} = \frac{1}{\sigma^2}\sum_{i=1}^{n} x^{exp}(y^{exp} - a_1 x^{exp} - a_0) = 0$$

$$\frac{\partial \ln L}{\partial a_0} = \frac{1}{\sigma^2}\sum_{i=1}^{n}(y^{exp} - a_1 x^{exp} - a_0) = 0 \tag{14}$$

Equation 14, obtained by partial differentiation of Equation 13 are essentially the same as Equation 9 obtained independently except for a multiplier and they lead to the same working formulas as Equation 10.

The second derivation method explicitly states the implied error distribution of experimental data. It can be proven that only error distributions from so called *exponential family* (normal, Bernoulli, Poisson, chi-squared and some others) give rise to the same Equations 10.

Unfortunately, LLS cannot be effectively used for experimental data with non-normal distributions, such as data with experimental biases.

The most common source of errors in thermochemical measurements can be sample contamination, incomplete reactions, mass losses, kinetic hurdles, undesirable recrystallisation, etc. These kinds of uncertainties add a bias to the data which can be temperature and composition dependent. Since an experimental bias does not follow a normal error distribution, its presence quickly degrades the quality of optimisation predictions.

Although only shown for the case of ordinary least squares with one independent variable *x*, the same can be stated for the case of multivariate linear least squares as well as non-linear least squares and other optimisation methods.

In scientific community, it is usually assumed that the reliability of different data sets must be represented using *group weights*. However, we consider that even the weighted version of LLS cannot handle experimental biases well. To get reliable optimisation results, the optimisation algorithm should work on *unbiased* data, which can only be achieved by careful selection of *target points*.

CHEMAPP™ CALCULATIONS IN EXCEL® USING PYTHON® BINDING

Automated optimisation algorithms need to be able to read/write model parameters in a thermodynamic database, run thermodynamic calculations and read calculated values. Our optimisation process uses ChemApp™ library as a backend (https://gtt-technologies.de). ChemApp™ is preferred to FactSage™ because of much greater flexibility. However, it is harder to set-up and use for an average engineer because ChemApp™ use implies good skills in at least one of programming languages.

To overcome this problem and make ChemApp™ available for a user who is not familiar with programming, we developed a proprietary Excel® application called PyroApp which exports ChemApp™-related functionality together with rich additional functionality for optimisation purposes in form of ordinary Excel® functions.

From the programming point of view, there are three layers in PyroApp structure. In the lowest layer, a ChemApp™ library is hooked up to Python® environment using a foreign function interface (FFI) implemented in a third-party library called *ctypes*. Since the library version and name are not hard coded into the program during compilation, it is possible to customise ChemApp™ library selection and even load several copies of the same ChemApp™ library into a process memory for multithreaded calculations. The lowest layer library is called chemapp™ (not to confuse with a similar library from Ex Mente (https://ex-mente.com)). It exports all native calling routines from the original ChemApp™ documentation into Python® environment and also provides useful object-oriented abstractions, such as Component, Phase, Constituent and other classes. An analogous Python® library from Ex Mente has similar functionality, however, we use our own implementation which can be freely modified for our purposes.

The second layer Python® library is called *pyroapp* and it contains a rich collection of functions grouped according to their functionality. A non-exhaustive list of main functions is given in Table 2.

The life cycle of an Excel® function is different from a function life cycle in C++ or Python®.

First, it is impossible to make Excel® function execute in a particular order because of the way cell update mechanism in Excel® works. Also, it is impossible to pass any non-primitive parameters as function arguments (objects, structures, unions). Each Excel® function call must be self-contained: if ChemApp™ library is called, the whole cycle of initialisation, loading datafile into memory and execution of useful routines must be performed before the function returns. To comply with this requirement, every function in Table 2, starting with *ca_* prefix takes a datafile name as its first argument. To avoid repetitive datafile loading, similar operations should be organised into tables which are passed as an array input to a function, so ChemApp™ initialisation and datafile reading is done only once. Consequently, such a function usually outputs an array of results. For convenience, it is preferrable to work with dynamic Excel® arrays with variable length instead of the old-fashioned Ctrl + Shift + Enter fixed-size arrays. They have been first introduced into MS Office 365 and have a special syntax for cell addresses like A1# or A1#.

Second, Excel® runtime manages calls to Excel® functions in an automated way. Inside Excel®, cells with functions are mapped into a tree-like structure which keeps track of cell changes. If any cell in Excel® is updated, the runtime mechanism checks if this cell has dependencies and updates them if needed. This mechanism does not allow a user-defined function execution. Since PyroApp functions work with external objects (thermodynamic datafiles), Excel® runtime is not able to track down any changes that happen there. To enforce a proper update mechanism, most of ca_ functions have an optional last argument caller a *trigger*. A trigger can point to any cell or cell range and there are no restrictions on what values they contain. The only function of a trigger cell is to initiate function execution by changing its content to another arbitrary value. Properly set-up trigger cells allow a user to update thermodynamic calculations after model parameters have been updated in a datafile.

The third layer of PyroApp is Excel® binding created using third-party xlwings module. This module works both as an Excel® add-in and a Python® library. It creates a glue VBA code which allows to run Python® environment as a server and imports marked Python® functions as Excel® functions. Each Excel® book with activated PyroApp has a configuration sheet called xlwings.conf with settings for both xlwings (Python® installation and executable, imported Python® modules) and for PyroApp.

TABLE 2

Main functionality groups of PyroApp application.

Function group	Functions	Purpose
info	*ca_version*, *ca_islite*, *ca_dimensions_max*, *ca_user_id*, etc.	Functions related to retrieving general information from a ChemApp™ library: version, calculation size limitations, licence and dongle info, etc.
opt	*condition_number*, *linear_regression*, *best_combination*, *target_scores*, *pd_error_table*, etc.	Functions for Jacobian of experimental targets (JET) optimisation and analysis of a derivative matrix; also functions for custom pivot tables and residual mean squares (RMS) summaries
tdcalc	*ca_list_*... (*components, phases, solutions, compounds, species*) *ca_get_*... (*constituent_H298, constituent_S298, constituent_Cp, compound_H298, compound_S298, compound_Cp, interaction_parameters_g, interaction_parameters_m*), *ca_set_*... (the same as above), *ca_list_interactions* *ca_calculate*	The majority of ChemApp™ functionality, routines for reading and writing to open text datafiles and thermodynamic calculations adapted to the Excel® environment
utils	*data_change_basis*, *data_generate_mesh*, etc.	Utility functions which help generate multicomponent composition meshes and transform compositions from one chemical formula basis to another
factsage™	*factsage™_get_compound_H298*, *factsage™_get_compound_S298*, *factsage™_set_compound_H298*, *factsage™_set_compound_S298*, *factsage™_write_parameters*, etc.	Utility functions to write final optimised parameters to either a FactSage™ compound database or FactSage™ solution database

One of the remarkable options in PyroApp settings is the possibility to split a table calculation between a few copies of ChemApp™ library running concurrently in different threads. At optimal conditions, it can raise CPU occupancy up to 100 per cent compared to 15–20 per cent for a single threaded ChemApp™ or FactSage™. Modern operating systems kernels are capable of loading several copies of the same binary library into the computer memory. Usually, read-only code-containing parts (also called text sections) are loaded once for all processes and threads to lower memory usage while sections reserved for data (program variables etc) are loaded separately for each user process/thread.

In PyroApp, multithreading is done strictly inside the Python® code and for Excel® functions work the same way as if they were single threaded. Currently, PyroApp has a very simple dispatch mechanism which splits a calculation table into equal parts according to the number of threads indicated in xlwings.conf sheet. Possibly a more advanced mechanism with a task queue will be implemented in future to handle cases when some threads exit earlier and remain idle.

The main function which performs Gibbs Energy Minimization (GEM) calculations, called *ca_calculate*, requires special attention. A lot of thought and effort were spent to fit miscellaneous ChemApp™ calculations under the same hood and develop a unified set of function arguments for all of them.

Broadly, ChemApp™ calculations can be split into normal calculations and stream calculations. Stream calculations allow to pre-calculate some properties of input mass flows to use them later to output property differences. In Excel®, the same effect can be achieved by joining an output of one calculation to the input of another, so stream calculations have been deliberately left out in the Excel® interface, although it is still possible to use them in the lower level chemapp™ library.

Each ca_calculate call has four obligatory and two optional arguments. The obligatory arguments are *datafile* (absolute or relative path to a datafile), *input_header*, *input_values* and *output_header*. The optional arguments are *entered* (indicates default phase selection for the whole table, can be overridden by table entries) and *update_token*, whose purpose was explained earlier.

Input header is a 3xn cell array indicating n input conditions used for calculation. The first cell of a condition entry indicates its type. All native condition types from *tqsetc* routine of ChemApp™ are accepted; there are also multiple custom types, such as FORMATION (name of formation target phase), PRECIPITATION (name of a precipitation target phase), ENTERED, DORMANT, ELIMINATED (custom phase selection for each calculation in a table), THIGH, TLOW (values passed to *tqclim* function), etc. The second cell of a condition entry can contain a phase name, the third cell can contain a constituent/component name, if applicable.

Input_values is a table of different values of conditions defined in *input_header*. *Output_header* is a 3xm cell array with a structure similar to that of *input_header*. It accepts all native ChemApp™ output options for *tqgetr* routine and has custom-defined output options as well. It is worth noting that *ca_calculate* function can output calculated bond fractions for quasichemical models, a feature which, to the best of our knowledge, is missing from Equilib module of FactSage™ package. To output a bond fraction, BOND option is used together with a name of the phase and the name of the quadruplet (for example, Ca,Si//O,O).

DERIVATIVE MATRIX SET-UP FOR JET OPTIMISATION

A few things are required to calculate a matrix of first-order derivatives of target values by model parameters: First, corresponding *ca_get_* and *ca_set_* functions must be set-up with all variable parameters in one column; second, thermodynamic calculations have to be set-up as well with all calculated target values in one column. A derivative matrix is calculated by a special macro code called *derivative_matrix* which can be either called directly from Excel® as a VBA macro or from command line Python® environment (for example, ipython). The latter option is more convenient because it allows to track calculation progress for large calculations. A macro call is initialised with settings stored in the upper left corner of the sheet used for matrix calculations. It requires range addresses of model parameters, parameter steps, a residual vector, upper left coordinate for matrix output, a trigger cell to initiate recalculations as well as a binary mask to turn on/off recalculation of derivatives for certain parameters.

Once a derivative matrix is calculated, it can be used for multiple purposes. There is a PyroApp function called *linear_regression* which manages finding an optimal solution for model parameters based on a precalculated derivative matrix. As arguments, this function takes a matrix of derivatives, a vector of differences between target values and calculated values, a vector of combined weights (on/off target masks + group and individual weights), a binary array indicating which parameters are allowed to change and an array of manual increments for parameters. The output parameters (expressed as deltas and not absolute values) are multiplied by the derivative matrix using MMULT Excel® function and added to the vector of calculated target values.

The resulting vector is called a vector of predicted target values; predicted target values are then plotted and/or used in statistical analysis. This set-up allows to observe changes in predicted target values caused by changes in model parameters in a real-time mode. Arguments to *linear_regression* function are flexible and allow both manual experimenting with parameter selection as well as automatic solution within a chosen subset of parameters. If a suitable combination of parameters has been found, the parameter deltas are added to the initial parameters for the datafile and everything is recalculated again. Non-linearity might cause slight disagreements between previously predicted and current calculated values. However, the cycle can be repeated one to two times until predicted deltas reach negligible values. Real time updates are indispensable for fine tuning of model behaviour across large data sets using target weights and selection/deselection of specific model parameters; it is a unique feature of JET optimisation. In conventional optimisers, the final result is not known until a complete cycle of iterations is performed and fine-tuning with those optimisers is not an easy task.

If the initial parameter set was far from optimal, the derivative matrix should be recalculated for consequent model refinement. In this case, JET is used similarly to conventional non-linear solvers. The number of matrix recalculations is usually not large, three to four times are enough to optimise a system from scratch. It is worth noting that if an initial estimate was close enough to the desired values in target points, the same derivative matrix allows to have significant changes in parameters if in total they do not result in large changes in target values. For example, parameters with certain powers can be replaced with parameters with other powers, which can prove to be very convenient.

For most of the systems in the 20-component database, a continuous refining is needed to keep the model updated with new experimental data coming in. Usually, the number of parameters and their powers are kept the same. Currently, it is possible to calculate matrices for many target points and a large number of components (five to six). Parameters for a subsystem can be change consistently with data sets for multicomponent system to ensure that the overall prediction quality does not deteriorate.

For a fewer number of newly investigated systems, an optimisation from scratch is required. If in the absence of model parameters some target values cannot be calculated (a phase is not stable etc), a very rough manual fit with two to three parameters for a binary or one to two parameters for a ternary system is required. For convenience, it can be done using FactSage™ Phase Diagram module. Once all target values can be calculated without errors, the quality of the initial estimate is not important (it is fine even if formation targets are 300–500°C off). In the first round, parameter selection is activated only for the existing non-zero parameters to further refine the system within the initial parameter set. Once a couple of iterations have passed, more parameters with different powers can be included. Usually, datafiles are created with many zero-valued parameters of different powers to avoid changing formulas; a zero-valued parameter is equivalent to not having the parameter at all. It is recommended to add parameters one by one; there is a special function of PyroApp, called *best_combination* which highlights a parameter which reduces the total cost function the most. Since regression with a calculated matrix is non-iterative, different parameters can be ranked quite fast. This method has recommended itself in course of many optimisations.

CONCLUSIONS

The current paper explains the principles of the continuous optimisation methodology, suitable for hybrid experimental and modelling development of a large multicomponent database. One of the key points of the continuous optimisation methodology is the use of target points instead of raw experimental data which allow to denoise input to an optimisation algorithm and make possible transparent automated assessment of optimisation quality of multiple subsystems across a database. Benefits and drawbacks of different optimisation algorithms are discussed. It is shown that thermodynamic calculations in optimisation process can be set-up in a way that first-order derivatives of target values by model parameters always exist.

The paper also explains the principles of using a first-order derivative matrix (jacobian) for real-time parameter sensitivity analysis and JET (Jacobian of Experimental Targets) optimisation. The main functionality of a proprietary application for Excel® with ChemApp™ backend, called PyroApp, is demonstrated together with data set-up and optimisation procedure.

REFERENCES

Eriksson, G and Pelton, A D, 1993. Critical Evaluation and Optimisation of the Thermodynamic Properties of the CaO-Al$_2$O$_3$, Al$_2$O$_3$-SiO$_2$, and CaO-Al$_2$O$_3$-SiO$_2$ Systems, *Metallurgy Transactions*, 24:807–816.

Hillert, M, 2001. The Compound Energy Formalism, *Journal of Alloys and Compounds*, 320(2):161–176.

Katoch, S, Chauhan, S S and Kumar, V, 2020. A review on genetic algorithm: past, present, and future, *Multimedia Tools and Applications*, 80:8091–8126.

Kolda, T G, Lewis, R M and Torczon, V, 2003. Optimisation by Direct Search: New Perspectives on Some Classical and Modern Methods, *SIAM Review*, 45(3):385–482.

Llovet, X, Moy, A, Pinard, P T and Fournelle, J H, 2021. Electron probe microanalysis: A review of recent developments and applications in materials science and engineering, *Progress in Materials Science*, 116:100673.

Pelton, A D, Decterov, S A, Eriksson, G, Robelin, C and Dessureault, Y, 2000. The Modified Quasichemical Model. I – Binary Solutions, *Metallurgical and Materials Transactions B*, 31(4):651–659.

Nekhoroshev, E, Shevchenko, M, Cheng, S, Shishin, D and, Jak, E, *in preparation*. Third generation CALPHAD re-optimisation of the $CaO\text{-}SiO_2$ system integrated with experimental phase equilibria studies, *Journal of Ceramics International*.

Thermodynamic modelling of the Fe-Al-Ti-O system and evolution of Al-Ti complex inclusions during Ti-added ultra-low carbon steel production

Y-J Park[1] and Y-B Kang[2]

1. PhD candidate, Pohang University of Science and Technology, Pohang, Gyeongbuk 37673, Republic of Korea. Email: yjpark92@postech.ac.kr
2. Professor, Pohang University of Science and Technology, Pohang, Gyeongbuk 37673, Republic of Korea. Email: ybkang@postech.ac.kr

ABSTRACT

Ti-added ultra-low carbon (ULC) steel is produced by adding Al and Ti during the secondary refining process to deoxidise molten steel and to bind C and N which are detrimental to the deep drawing quality of the steel product. However, the production of Ti-added ULC steel faces challenges, including submerged entry nozzle clogging and surface defects on cold rolled coils. The troubles were often attributed to 'Al-Ti complex inclusions' in the molten steel. However, the evolution of oxide inclusions including their stability has been still unclear. In the present study, the evolution of these inclusions in Ti-added ULC steel was investigated by phase diagram measurement, CALPHAD thermodynamic modelling, high-temperature stability test for the oxides, and observation of transient behaviour of oxide inclusions in the steel. The phase diagram of the Al-Ti-O system was elucidated with an emphasis on oxygen potential, which significantly controls the stability of the oxides. CALPHAD thermodynamic models and a self-consistent database were developed. The transient behaviour of Al-Ti complex inclusions in Ti-added ULC steel was observed *in situ* using a confocal scanning laser microscope. This was additionally validated by the new thermodynamic model. The equilibrium phases of the oxide inclusions were investigated using a high-temperature resistance furnace, and were analysed using a secondary electron microscope, energy dispersive spectrometer, and electron back-scattered diffraction. By considering the morphology and crystal structure of these inclusions, the evolution of Al-Ti oxide inclusions was elucidated. Various behaviour of Al-Ti inclusions and their influence on steel cleanliness are discussed.

INTRODUCTION

Interstitial-free (IF) steel, also known as ultra-low carbon (ULC) steel, is widely used in automotive outer panels due to its excellent deep drawing quality (DDQ). In order to further enhance the DDQ, IF steel requires minimising interstitial elements such as C and N to the utmost extent. In particular, Ti is one of the most commonly used elements for improving the properties of IF steel. The addition of Ti has several beneficial effects, such as an increase in the plastic strain ratio (*r*-value), a decrease in the recrystallisation temperature, and reduced sensitivity to coiling temperatures (Gupta and Bhattacharya, 1989; Fukuda, 1973). As a result, the demand and supply of Ti-added ULC steel have been consistently increasing due to its excellent performance in maximising DDQ properties.

During the production of Ti-added ULC steel, the steel undergoes deoxidation in the RH (Ruhrstahl-Heraeus) process by adding Al: $2\underline{Al} + 3\underline{O} = Al_2O_3(s)$. Subsequently, Ti is added as Ti sponge or ferro-Ti alloy to achieve the target composition. While Ti-free ULC steel only considers Al_2O_3 oxide as the non-metallic inclusion, in the case of Ti-added ULC steel, Ti may also react with dissolved O. This could happen if the Al deoxidation was not enough to kill the O or reoxidation occurs during the secondary refining or continuous casting processes. This may induce Ti-containing inclusions and further, Al-Ti complex inclusions. It has been reported that Ti-added ULC steel encounters more severe submerged entry nozzle (SEN) clogging during continuous casting, compared Ti-free ULC steel (Basu, Choudhary and Girase, 2004; Cui *et al*, 2010; Lee *et al*, 2019). The accumulation of clogging deposits becomes more significant with increasing Ti content, resulting in poor productivity (Dorrer *et al*, 2019; Bernhard *et al*, 2019).

The origin of the severe SEN clogging due to Ti addition includes refractory-steel reaction, reoxidation from various sources (air aspiration, Ar gas quality, refractory carbothermic reaction etc), enhanced wettability, etc. In addition to this, the evolution of inclusions by Ti addition may play a significant role, however, the phase stability of Al-Ti complex inclusions is yet to be understood clearly. Frequently mentioned oxide phase as the Al-Ti complex inclusion has been 'Al$_2$TiO$_5$' and there have been cases where Al$_2$TiO$_5$ was observed in Ti-added ULC steel melts (Wang, Nuhfer and Sridhar, 2009; Bai, Sun and Zhang, 2019). Several oxide stability diagrams of the Fe-Al-Ti-O system predicted by thermodynamic calculation included Al$_2$TiO$_5$, as a probable equilibrium phase in molten steel. However, there has been an experimental result showing the instability of Al$_2$TiO$_5$ in Ti-added ULC steel melts (Jo, 2011). Although the oxide stability diagram has been used to understand the inclusion evolution, published diagrams do not agree with each other and do not support experimental results on inclusion evolution. Therefore, the oxide stability of the Al-Ti complex inclusions has been a long-standing debate.

This is because the phase equilibria of the Fe-Al-Ti-O oxide system have not been fully investigated. The present authors recently carried out an experimental study on the phase equilibria in the Al-Ti-O system with an emphasis on the impact of oxygen partial pressure (Park, Kim and Kang, 2021). Al$_2$TiO$_5$, which was shown to form a pseudobrookite solid solution with Ti$_3$O$_5$ should decompose under low oxygen partial pressure relevant to the molten steel during the secondary refining process. This indicates that it is no longer a stoichiometric oxide. The decomposition behaviour of Al$_2$TiO$_5$ may provide insights into the understanding of inclusion evolution in Ti-added ULC steel. However, for further application to Ti-added ULC steel, it is necessary to complete the modelling of the Fe-Al-Ti-O system, including both molten metal and several oxide phases in the system, in order to study the behaviour of complex inclusions in molten steel.

In the present study, thermodynamic modelling of the Fe-Al-Ti-O system was carried out to evaluate the stability of Al-Ti complex inclusions in Fe-Al-Ti-O melts. The evolution of Al-Ti complex inclusions in Ti-added ULC steel was also investigated using various methodologies including Scanning-Electron Microscopy with Energy-Dispersive X-ray Spectroscopy (SEM-EDS), Electron Back-Scattered Diffraction (EBSD), and Confocal Scanning Laser Microscopy (CSLM).

THERMODYNAMIC MODELLING OF THE FE-AL-TI-O SYSTEM

Thermodynamic calculations and optimisation for the Fe-Al-Ti-O system were conducted using FactSage™ thermodynamic software (Bale *et al*, 2016). The Gibbs energies of pure substances were sourced from the SGTE compilation of Dinsdale (1991). The Fe-Al-Ti-O system was modelled by considering solution phases such as liquid alloy, liquid oxide, pseudobrookite, cubic spinel, ilmenite, rutile, monoxide, some metallic solutions, and stoichiometric compounds. The modelling assessed various sub-binary systems and extended to multi-component systems to cover the entire subsystem of the Fe-Al-Ti-O system. In certain instances, model parameters were adjusted slightly to meet the requirements of the phase diagrams and thermodynamic properties of solutions. Details of the models and parameter optimisation can be found elsewhere (Park and Kang, 2024).

Thermodynamic models

Liquid metal and liquid oxide

The Modified Quasichemical Model (Pelton *et al*, 2000; Pelton and Chartrand, 2001) was employed in modelling the liquid metal and the liquid oxide separately, aiming to consider short-range ordering (SRO) observed in the liquid solutions across a broader concentration range. For example, in the case of liquid metal, components A, B, C,... are distributed over a quasi-lattice as atoms. In the pair approximation, the pair-exchange reaction is considered as:

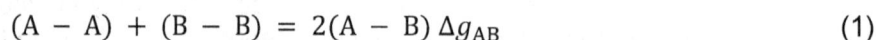

$$(A - A) + (B - B) = 2(A - B) \quad \Delta g_{AB} \tag{1}$$

where (A − A), (B − B), and (A − B) represent first-nearest-neighbour (FNN) pairs. The non-configurational Gibbs energy change for the formation of two moles of (A − B) pairs is Δg_{AB} which is used as a model parameter in each binary system. Gibbs energies of ternary and higher-order systems are then estimated using the appropriate interpolation method (Pelton and Chartrand, 2001), and a few of adjustable ternary model parameters can be added. The Gibbs energy of liquid oxide phase was modelled similarly.

Solid solutions including pseudobrookite, ilmenite, cubic spinel, etc

Solid phases having more than two sublattices such as pseudobrookite $((Ti,Al,Fe)_3O_5)$, cubic spinel $(Fe(Al,Ti)_2O_4)$, and ilmenite $((Ti,Al,Fe)_2O_3)$ etc were modelled using the compound energy formalism (CEF) (Hillert, 2001) to described the Gibbs energies of the solutions in the Fe-Al-Ti-O system. For example, the model for the Fe-Al-Ti-O pseudobrookite is described below:

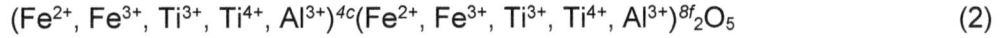

$$(Fe^{2+}, Fe^{3+}, Ti^{3+}, Ti^{4+}, Al^{3+})^{4c}(Fe^{2+}, Fe^{3+}, Ti^{3+}, Ti^{4+}, Al^{3+})^{8f}{}_2O_5 \qquad (2)$$

where ions enclosed in parentheses occupy the same octahedral (*4c* and *8f*) sublattice.

The Gibbs energy of the pseudobrookite solution is expressed as:

$$G^m = \sum_i \sum_j Y_i^{4c} Y_j^{8f} G_{ij} - TS^{\text{config}} + G^{\text{excess}} \qquad (3)$$

where Y_i^{4c} and Y_j^{8f} represent the site fractions of constituents '*i*' and '*j*' on the *4c* and *8f* octahedral sublattices, respectively; G_{ij} is the Gibbs energy of an 'end member $[i]^{4c}[j]^{8f}{}_2O_5$' in which *4c* and *8f* sites are occupied only by *i* and *j* cations, respectively; S^{config} is the configurational entropy which takes into account random mixing of cations on each sublattice. The Gibbs energies (G_{ij}) of end-members are the primary model parameters.

However, due to the presence of several non-neutral charged end-members, it is not feasible to ascertain the Gibbs energies of all end-members based solely on experimental data. To appropriately determine the values of these Gibbs energies, physically meaningful linear combinations of the end-members' Gibbs energies (G_{ij}) involved in specific site exchange reactions among cations are utilised as parameters in the model.

Taking the example of a Ti_3O_5 pseudobrookite oxide, the measurable Gibbs energy, $G°(Ti_3O_5)$ (Gibbs energy of pure Ti_3O_5), is directly employed in the CEF modelling. The model parameters also encompass site exchange reactions between cations at the *4c* and *8f* sites, denoted by Δ and *I* parameters. These parameters are crucial in determining the Gibbs energies of end-members. Generally, these exchange energies are expected to be small and are frequently set to zero. This practice allows for a more controlled determination of the Gibbs energies of end-members, as opposed to arbitrary assignments without a rationale. Adopting this approach enhances the predictive capability and stability of the model, particularly for extensive solid solutions like pseudobrookite, ilmenite, and spinel phases.

The Gibbs energies of other solid phases without sublattice was described by a simple Bragg-Williams random mixing model.

Model calculation

Using the reported thermodynamic modelling of subsystems of the Fe-Al-Ti-O system, the present study extended the previous modelling to construct a model database for the Fe-Al-Ti-O system. Through model calculations of the Al-Ti-O oxide system, a definition of Al-Ti complex inclusions has been achieved. Considering the characteristics of Al_2TiO_5, formerly considered as an Al-Ti complex inclusion, the present study aimed to interpret the evolution and decomposition behaviours of Al-Ti complex inclusions under various oxygen pressure conditions (see section on 'Al-Ti-O oxide phase diagram at 1600°C').

Additionally, deoxidation equilibria in Ti-added ULC steel were described by modelling the deoxidation equilibria by Al and by Ti. While the previous research conducted by Paek *et al* (2016), using MQM for the liquid metal phase was adopted, additional modelling of Ti-deoxidation equilibria was performed using MQM to provide a more refined interpretation of equilibrium relationships in Ti-added ULC steel production (see section on 'Ti deoxidation equilibria at 1600°C').

At the final stage, the equilibrium [O] contents and corresponding inclusion phases in molten steel at 1600°C were interpreted based on the Al and Ti contents in the liquid steel. Unlike the model calculations investigated previously, the model database optimised in the present study allows for the calculation of not only the oxide phases in equilibrium with ULC liquid steel but also the Fe, Al and Ti solubilities within the oxide phase. This capability is expected to facilitate tracking the generation and transition behaviours of inclusions under reoxidation conditions.

Al-Ti-O oxide phase diagram at 1600°C

The calculated result of the Al-Ti-O_2 phase diagram (PO_2-R_{Ti}) is shown in Figure 1. It should be stressed that there is a wide solid solution encompassing 'Al$_2$TiO$_5$' (R_{Ti} = 0.33) and 'Ti$_3$O$_5$' (R_{Ti} = 1.0) over a wide range of PO_2 (1–10^{-14} atm), according to the present authors' experimental investigation (Park, Kim and Kang, 2021). These two oxides have the same crystal structure, therefore, these were treated in a single solid solution (pseudobrookite s.s.). However, this solid solution separates into two phases at lower temperatures (T < 1500°C), which was first revealed by the present authors (Park, Kim and Kang, 2021). This kind of phase equilibrium has not been known previously. According to this phase equilibria, Al$_2$TiO$_5$ should undergo the following phase transformation at lower PO_2 relevant to the steel refining process:

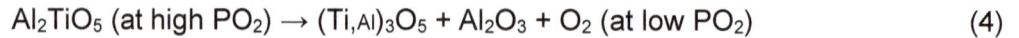

$$\text{Al}_2\text{TiO}_5 \text{ (at high PO}_2) \rightarrow \text{(Ti,Al)}_3\text{O}_5 + \text{Al}_2\text{O}_3 + \text{O}_2 \text{ (at low PO}_2) \qquad (4)$$

where 'Al$_2$TiO$_5$ (at high PO_2)' and '(Ti,Al)$_3$O$_5$' represent Al-rich pseudobrookite s.s. and Ti-rich pseudobrookite s.s., respectively. This reaction would represent a situation that Al-Ti complex inclusions (near Al$_2$TiO$_5$) generated at high oxygen potential due to reoxidation in Ti-added ULC steel could generate Ti-rich oxide inclusion, Al$_2$O$_3$, and the additional O in the liquid steel. This suggests that the oxide inclusion reoxidised in Ti-added ULC steel may play as an O-carrier in the liquid steel.

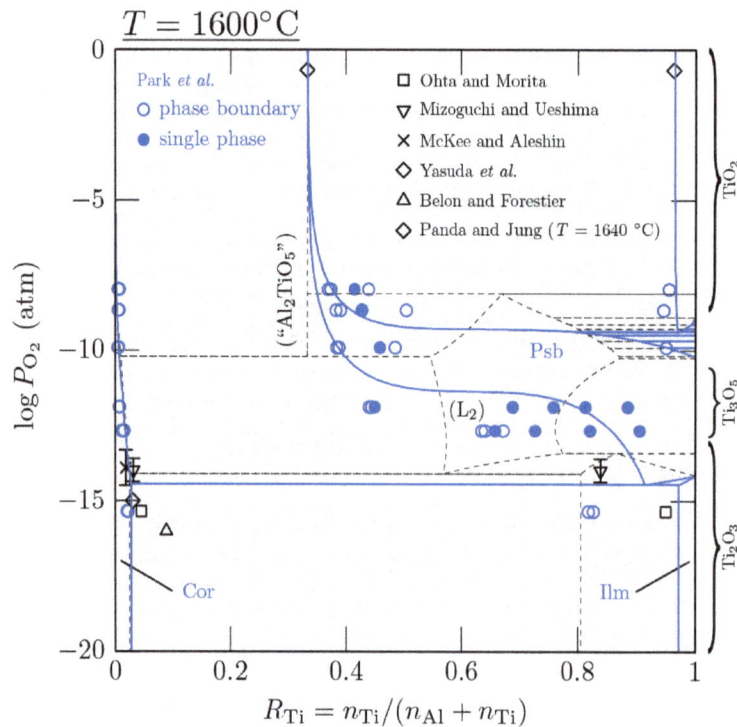

FIG 1 – Calculated phase diagram of the Al-Ti-O_2 system at 1600°C (Park and Kang, 2024). Blue solid lines were calculated using the present thermodynamic model, and black dotted lines were calculated using FactSage FTOxid database (ver. 8.2). Symbols are the reported experimental data (see Park, Kim and Kang, 2021).

These models were extended to incorporate Fe as a component. Detail modelling results can be found elsewhere (Park and Kang, 2024).

Ti deoxidation equilibria at 1600°C

By combining the existing thermodynamic model of the Fe-Al-O liquid metal, with that of the Fe-Ti-O liquid metal in the present study, the thermodynamic model for the deoxidation equilibria in the Fe-Al-Ti-O liquid metal could be completed. Figure 2 shows the Ti-deoxidation equilibria in the liquid Fe-Ti-O at 1600°C. Long-dashed, full, and short-dashed lines represent the equilibrium compositions of liquid metal in equilibrium with ilmenite, pseudobrookite, and liquid oxide, respectively. The present study describes the Ti-deoxidation equilibria over a wide composition range. The developed thermodynamic model is now used to predict high-temperature equilibria between Ti-added ULC steel and oxide inclusions.

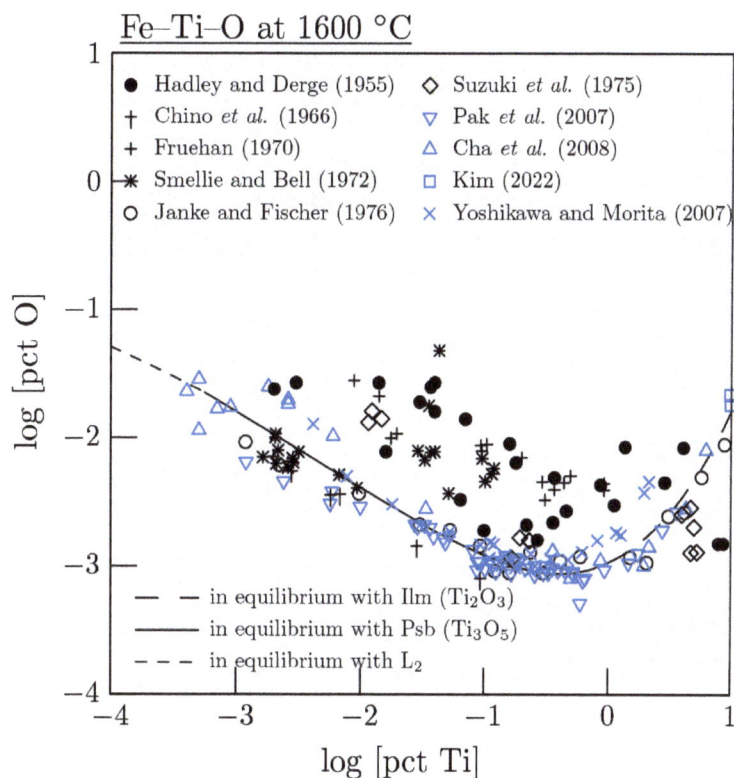

FIG 2 – Ti-deoxidation equilibria in liquid iron at 1600°C (Park and Kang, 2024). Solid lines were calculated using the present thermodynamic model.

Oxide stability diagram of the Fe-Al-Ti-O system

The calculated oxide stability diagram of the Fe-Al-Ti-O system at 1600°C using the thermodynamic models developed in the present study is shown in Figure 3. Thin black solid lines and blue solid lines are the iso-O (ppm) lines, and the phase boundaries between two different oxide phases, respectively. The most probable oxide phase formed during the Ti-added ULC steel production is obviously corundum $((Al,Ti,Fe)_2O_3)$ by typical compositions of this steel grade. Compared to the previous stability diagram by Kang and Lee (2017), the phase boundaries between corundum and pseudobrookite oxide are similar to each other: no separate Al_2TiO_5 phase region is seen. However, the incorporation of Fe in the pseudobrookite s.s. resulted in a somewhat wider stability region of the pseudobrookite s.s. This subsequently resulted in the stability region of the liquid oxide and ilmenite (Ti_2O_3) phases being reduced.

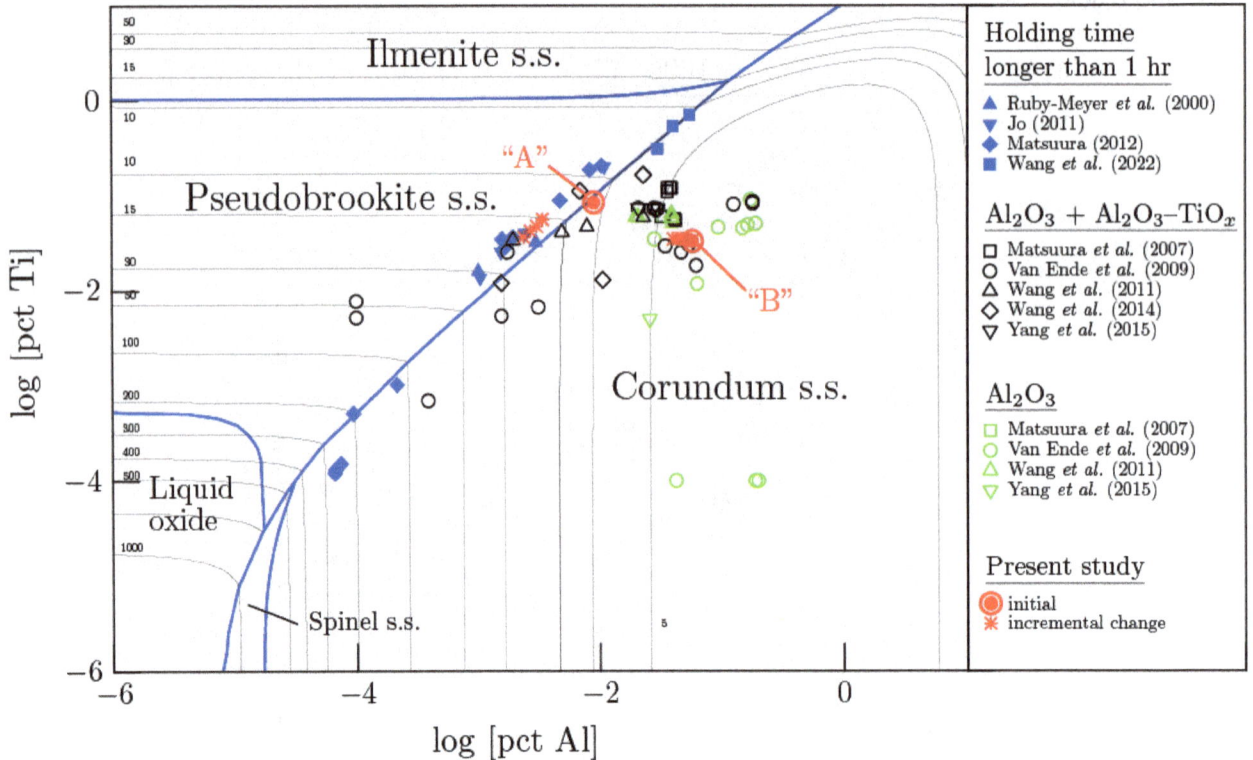

FIG 3 – Calculated oxide stability diagram of the Fe-Al-Ti-O system at 1600°C in the present study, modified from a figure from Park and Kang (2024). Symbols marked by 'A' and 'B' were obtained in the experimental work of the present authors. Other symbols were taken from literature (see Park and Kang, 2024).

EVOLUTION OF AL-TI COMPLEX INCLUSION IN LIQUID STEEL

Through the thermodynamic modelling of the Al-Ti-O system, the decomposition reaction of the Al-Ti complex oxide, as expressed in Reaction 4, was elucidated. Reaction 4 can be extended to the Fe-containing system, the Fe-Al-Ti-O system. In the reaction involving Fe, such as expressed in Reaction 5, the solubility of Fe in the pseudobrookite inclusion and the incorporation of the decomposition product O_2 into the molten steel can contribute to an increase in $\underline{O}$ contents.

$$\text{(Fe,Al,Ti)}_3O_5 \text{ (at high PO}_2) \rightarrow \text{(Ti,}_\text{Al})_3O_5 + Al_2O_3 + \underline{O} + \text{Fe (at low PO}_2) \tag{5}$$

where '(Fe,Al,Ti)_3O_5 (at high PO_2)' represents a reoxidised pseudobrookite s.s. The O generated in Reaction 5 can react with $\underline{Al}$ and/or $\underline{Ti}$ in the liquid steel to produce additional inclusions. The Reaction 5 may be manipulated in the reverse order, which may represent the reoxidation of liquid Ti-added ULC steel:

$$Al_2O_3 + \text{Fe} + \underline{Ti} + \underline{O} \rightarrow \text{(Fe,Al,Ti)}_3O_5 \tag{6}$$

This reversed reaction would result in the formation of Fe-containing pseudobrookite s.s. from alumina inclusions. This scenario was experimentally investigated in the present study by employing CSLM. A Ti-added ULC steel sample that contains Al_2O_3 inclusions and dissolved Ti and O was observed. Figure 4a shows only alumina inclusion. On the other hand, the same steel sample was intentionally but mildly oxidised and the resultant oxide inclusions were shown in Figure 4b. Certainly, shape of the inclusions in the two figures was distinguished. Those in Figure 4b were identified as pseudobrookite phase. Details of the procedure of the experiment and post-analysis are presented elsewhere (Park and Kang, 2024).

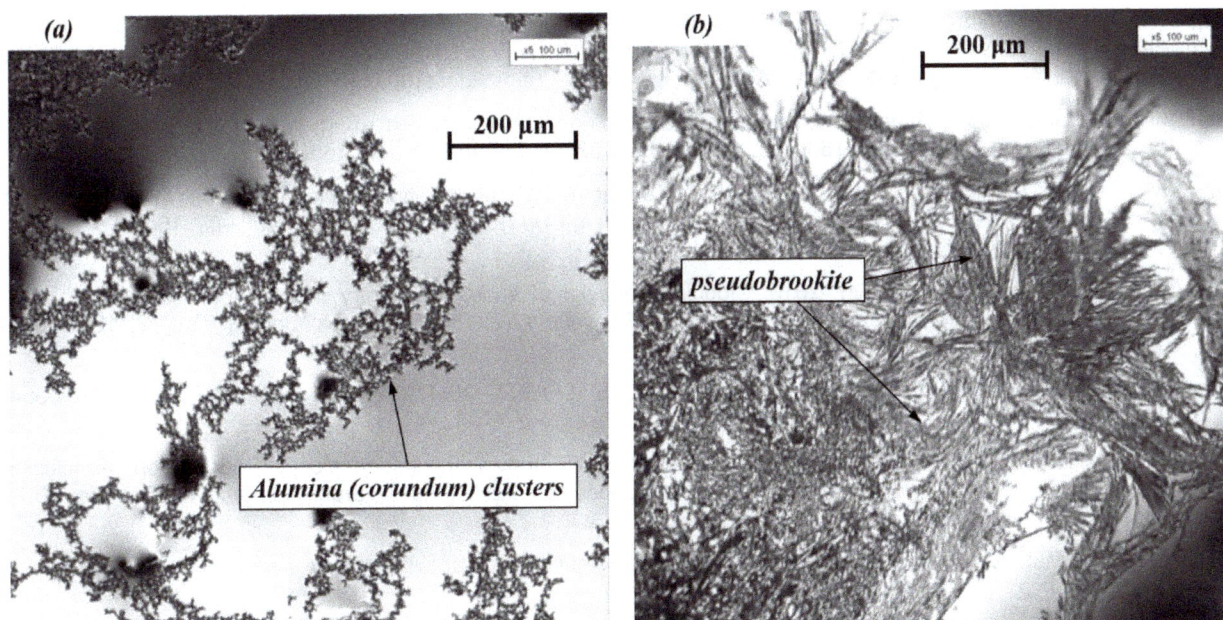

FIG 4 – Inclusions morphologies floating on a Ti-added ULC steel observed by CSLM: (a) without reoxidation and (b) with reoxidation.

The inclusions observed in Figure 4b, whose morphology seems to be a whisker-shape, were analysed using SEM-EDS after cooling the steel sample. The inclusions were composed of Al, Ti, Fe, and O Figure 5. According to the list of phases in the present system and the result of thermodynamic model calculation, this is most likely pseudobrookite phase.

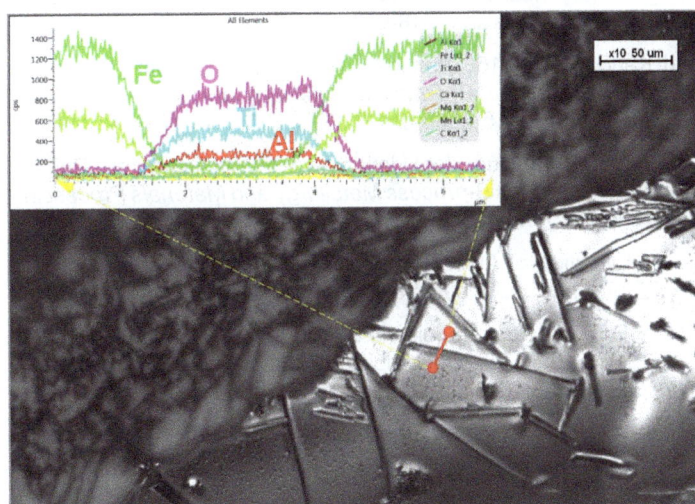

FIG 5 – Post analysis of inclusions seen in Figure 4b using SEM-EDS.

CONCLUSIONS

A self-consistent thermodynamic model of the Fe-Al-Ti-O system was developed, and the evolution of Al-Ti complex inclusions and their transition behaviours were investigated using various experimental techniques. It was shown that the Al-Ti complex inclusion should be defined as a part of pseudobrookite s.s. instead of the stoichiometric 'Al_2TiO_5', which is not stable in liquid steel. The predicted transition behaviours of the Al-Ti complex inclusions were postulated, and it was demonstrated that the complex inclusions can act as a reoxidation source, potentially deteriorating steel cleanliness.

REFERENCES

Bai, X-F, Sun, Y and Zhang, Y, 2019. Transient evolution of inclusions during Al and Ti additions in Fe-20 Mass pct Cr alloy, *Metals*, 9(6):702. https://doi.org/10.3390/met9060702

Bale, C W, Bélisle, E, Chartrand, P, Decterov, S A, Eriksson, G, Gheribi, A E, Hack, K, Jung, I-H, Kang, Y-B, Melançon, J, Pelton, A D, Petersen, S, Robelin, C, Sangster, J, Spencer, P and Van Ende, M-A, 2016. FactSage Thermochemical Software and Databases, 2010–2016, *CALPHAD*, 54:35–53. https://doi.org/10.1016/j.calphad.2016.05.002

Basu, S, Choudhary, S K and Girase, N U, 2004. Nozzle clogging behaviour of Ti-bearing Al-killed ultra low carbon steel, *ISIJ International*, 44(10):1653–1660. https://doi.org/10.2355/isijinternational.44.1653

Bernhard, C, Dorrer, P, Michelic, S K and Rössler, R, 2019. The role of FeTi addition to micro-inclusions in the production of ULC steel grades via the RH Process route, *BHM Berg-Und Hüttenmännische Monatshefte*, 164(11):475–478. https://doi.org/10.1007/s00501-019-00900-2

Cui, H, Bao, Y, Wang, M and Wu, W, 2010. Clogging behavior of submerged entry nozzles for Ti-bearing IF steel, *International Journal of Minerals, Metallurgy and Materials*, 17(2):154–158. https://doi.org/10.1007/s12613-010-0206-y

Dinsdale, A T, 1991. SGTE data for pure elements, *CALPHAD*, 15(4):317–425. https://doi.org/10.1016/0364-5916(91)90030-n

Dorrer, P, Michelic, S K, Bernhard, C, Penz, A and Rössler, R, 2019. Study on the influence of FeTi-addition on the inclusion population in Ti-stabilized ULC steels and its consequences for SEN-clogging, *Steel Research International*, 90(7). https://doi.org/10.1002/srin.201800635

Fukuda, N, 1973. Manufacture of Ti-stabilized extra low carbon steel for deep drawing quality, *Tetsu-to-Hagane*, 59(2):231–240.

Gupta, I and D, Bhattacharya. 1989. Metallurgy of formable vacuum degassed interstitial-free steels.

Hillert, M, 2001. The compound energy formalism, *Journal of Alloys and Compounds*, 320(2):161–176. https://doi.org/10.1016/s0925-8388(00)01481-x

Jo, J O, 2011. Thermodynamics of Titanium and Aluminum in Liquid Alloy Steel, PhD dissertation, Hanyang University.

Kang, Y-B and Lee, J-H, 2017. Reassessment of oxide stability diagram in the Fe–Al–Ti–O system, *ISIJ International*, 57(9):1665–1667. https://doi.org/10.2355/isijinternational.isijint-2017-182

Lee, J, Kang, M, Kim, S, Kim, J, Kim, M and Kang, Y, 2019. Influence of Al/Ti ratio in Ti-ULC steel and refractory components of submerged entry nozzle on formation of clogging deposits, *ISIJ International*, 59(5):749–758. https://doi.org/10.2355/isijinternational.isijint-2018-672

Paek, M, Jang, J, Pak, J and Kang, Y-B, 2015. Aluminum Deoxidation Equilibria in liquid iron: Part, I, Experimental, *Metallurgical and Materials Transactions B, Process Metallurgy and Materials Processing Science*, 46(4):1826–1836. https://doi.org/10.1007/s11663-015-0368-0

Paek, M, Pak, J and Kang, Y-B, 2015. Aluminum deoxidation equilibria in liquid iron: Part II, Thermodynamic Modeling, *Metallurgical and Materials Transactions B-process Metallurgy and Materials Processing Science*, 46(5):2224–2233. https://doi.org/10.1007/s11663-015-0369-z

Park, Y-J and Kang, Y-B, 2024. Evolution of Al–Ti Complex Oxide Inclusions in Interstitial-Free Steel Analyzed Using CALPHAD, SEM, EDS, EBSD and CSLM, submitted to *Metall Mater Trans B*.

Park, Y-J, Kim, W-Y and Kang, Y-B, 2021. Phase equilibria of Al_2O_3–TiO_x system under various oxygen partial pressure: Emphasis on stability of Al_2TiO_5–Ti_3O_5 pseudobrookite solid solution, *Journal of the European Ceramic Society*, 41(14):7362–7374. https://doi.org/10.1016/j.jeurceramsoc.2021.06.052

Pelton, A D and Chartrand, P, 2001. The modified quasi-chemical model: Part II, Multicomponent solutions, *Metallurgical and Materials Transactions A*, 32(6):1355–1360. https://doi.org/10.1007/s11661-001-0226-3

Pelton, A D, Degterov, S A, Eriksson, G, Robelin, C and Dessureault, Y, 2000. The modified quasichemical model I—Binary solutions, *Metallurgical and Materials Transactions B-process Metallurgy and Materials Processing Science*, 31(4):651–659. https://doi.org/10.1007/s11663-000-0103-2

Wang, C, Nuhfer, N T and Sridhar, S, 2009. Transient behavior of inclusion chemistry, shape and structure in Fe-Al-Ti-O melts: effect of titanium source and laboratory deoxidation simulation, *Metallurgical and Materials Transactions B-process Metallurgy Materials Processing Science*, 40(6):1005–1021. https://doi.org/10.1007/s11663-009-9267-6

Sulfur distribution ratio in iron and steelmaking slags

R G Reddy[1] and A Yahya[2]

1. Professor, The University of Alabama, Tuscaloosa, 35487 Alabama, USA.
   Email: rreddy@eng.ua.edu
2. Graduate student, The University of Alabama, Tuscaloosa, 35487 Alabama, USA.
   Email: ayahya@crimson.ua.edu

ABSTRACT

The sulfur distribution ratio (Ls) is an expression of the amount of sulfur in slag to the amount of sulfur in molten metal. The sulfide capacities calculated from Reddy-Blander (RB) model were used to calculate the sulfur distribution ratio (Ls). A new Reddy model for sulfur distribution ratio (Ls) was developed for modelling Ls for ironmaking, steelmaking, and secondary steelmaking conditions. Besides calcium oxide, the major component in the ladle furnace (LF) slags is alumina. The Ls was calculated for typical LF slags as a function of temperature and composition. The model calculated Ls values are in good agreement with the experimental and industrial slags data. The extremely low oxygen potential in these furnaces after fully-killed steel is favourable for desulfurisation.

For any given slag's composition and temperature, its sulfide capacity can be calculated *a priori*. The equilibrium sulfur distribution ratio between this slag and liquid metal in a particular vessel can also be calculated *a priori*. The Ls for ladle furnace (LF) slags is about twice those for blast furnaces (BF) slags and ten times those of basic oxygen furnace (BOF) slags. The desulfurisation of pig iron in BF and desulfurisation of steel in LF during secondary steelmaking are recommended. Steel desulfurisation in steelmaking furnaces is not effective. Since the model used to calculate the sulfur distribution ratio (Ls) is *a priori*, it is a useful tool for the iron and steel makers in improving the product quality and optimisation of the industrial furnace operations.

INTRODUCTION

The sulfur distribution ratio (Ls) can be expressed as the amount of sulfur in slag to the amount of sulfur in molten metal. Sulfide capacity is of crucial importance to iron and steelmakers who are continuously trying to improve product quality. Sulfur drastically decreases the ductility and strength of steel, and it is undesirable except for free-cutting steels. Sulfur also causes hot-shortness and helps propagate hydrogen-induced cracks. In certain grades of clean steel, eg armour plates, plates for offshore oil installations and pipelines, the sulfur content should be less than 20 ppm (0.002 wt per cent S) to prevent hydrogen-induced cracking. The sulfide capacities calculated from the Reddy-Blander (RB) model were used to calculate the sulfur distribution ratio.

Reddy-Blander (RB) model for sulfide capacity

Reddy and Blander (RB) developed a model (Reddy and Blander, 1987, 1989; Chen, Reddy and Blander, 1989; Reddy, Hu and Blander, 1992; Reddy, 2003b), for calculating sulfide capacities of binary and multicomponent aluminate and silicate slags. They showed that sulfide capacities could be calculated *a priori* based on the knowledge of the chemical and solution properties of oxides and sulfides. In another publication, Zhao and Reddy (1995) extended the model to binary aluminate melts. Derin, Yucel and Reddy (2010) applied the model to ternary silicate melts containing FeO. RB model was extended to other slag systems, titania slags (Derin, Yucel and Reddy, 2004), lead industrial slags (Derin, Yucel and Reddy, 2005, 2006), multicomponent aluminosilicates (Yahya and Reddy, 2011; Pelton, Eriksson and Romero-Serrano, 1993) and phosphates (Yang *et al*, 2014; Pelton, 2000) and sulfates (Pelton, 1999), arsenates (Reddy and Font, 2003), antimonate (Font and Reddy, 2005), and prediction of sulfur and oxygen partial pressures in copper flash smelting slags (Derin and Reddy, 2003). In all systems, excellent agreement between the model and experimental data was observed. The RB model calculated the sulfide capacities of binary, ternary and higher order aluminate slags of the system $CaO-FeO-MgO-MnO-Al_2O_3-SiO_2$. In this work, the RB model predicted sulfide capacities were incorporated in developing a new Reddy model for sulfur distribution ratios in iron and steelmaking slags.

Basic oxides like CaO, FeO, MgO and MnO dissociate into cations and oxygen ions and on the other hand, acidic oxides such as SiO_2 and Al_2O_3 consume oxygen ions and form complex silicate and aluminate ions. Using CaO-$AlO_{1.5}$ as an example, the binary system is divided into two regions; one basic and one acidic. The model equations used to calculate the sulfide capacity are:

Basic binary ($0 \leq X_{AlO_{1.5}} \leq 0.33$)

$$C_S = 100 \cdot M_S \cdot K_{Ca} \cdot a_{CaO} \cdot \left(\frac{1 - 2X_{AlO_{1.5}}}{\overline{M}} \right) \tag{1}$$

Where M_S is the atomic weight of sulfur and K_{Ca} is the equilibrium constant for the reaction, $a_{Ca}O$ is the activity of the calcium oxide in the binary system, and $\overline{M}$ is the average molecular weight of solution. For a given slag's temperature and composition, K_{Ca} and $a_{Ca}O$ were obtained from thermodynamic data (Bale, Pelton and Thompson, 2002; Gokcen and Reddy, 1996).

Acidic binary ($0.33 \leq X_{AlO_{1.5}} < 1$)

Acidic binary ($0.33 \leq X_{AlO_{1.5}} \leq 0.5$)

Acidic binaries are divided into two regions. For intermediate acidic binary ($0.33 \leq X_{AlO1.5} \leq 0.5$), the solution is a mixture of polymeric species and monomer ions (Flory, 1953). The model equation for this region is:

$$C_S = 100 \cdot M_S \cdot K_{Ca} \cdot a_{CaO} \cdot exp\left(-1.25\left(1 - \frac{1}{m}\right)\right) \cdot \frac{X_{AlO_{1.5}}}{\overline{M}} \tag{2}$$

Where m is the polymer chain length given by:

$$\frac{1}{m} = (1 - a_{CaO}) \cdot \left(\frac{1}{X_{AlO_{1.5}}} - 2 \right) \tag{3}$$

Acidic binary ($0.5 \leq X_{AlO1.5} < 1$)

For the remaining portion of the acidic binary, the solution is mostly polymeric. The model equation for this region is:

$$C_S = 100 \cdot M_S \cdot K_{Ca} \cdot a_{CaO} \cdot exp(-1.25) \cdot \frac{X_{AlO_{1.5}}}{\overline{M}} \tag{4}$$

The most complicated multicomponent system investigated in this work has six components. In this system, four oxides are basic; CaO, FeO, MgO and MnO, while two oxides are acidic; SiO_2 and Al_2O_3. To calculate the sulfide capacity of this system, it is broken down into four ternaries: CaO-SiO_2-Al_2O_3, FeO-SiO_2-Al_2O_3, MgO-SiO_2-Al_2O_3 and MnO-SiO_2-Al_2O_3. Equations 1–4 were used to calculate the sulfide capacities of each ternary system. The activity of the basic component was obtained using the same mole fractions of the acidic components as in the multicomponent system. The average molecular weight of solution is obtained for the ternary system. Instead of using $X_{AlO1.5}$, $X_{AlO1.5}$ + X_{SiO2} was used, where these acidic mole fractions were the same as in the multicomponent system. Once the sulfide capacity of each ternary was calculated, the following equation was used to calculate the sulfide capacity of the multicomponent system:

$$\log C_S = y_{Ca} \log C_{S(Ca)} + y_{Fe} \log C_{S(Fe)} + y_{Mg} \log C_{S(Mg)} + y_{Mn} \log C_{S(Mn)} \tag{5}$$

Where y_i is the equivalent cationic fraction of species i. The comparison of C_s for available experimental data to RB model results of CaO-FeO-MgO-MnO-$AlO_{1.5}$-SiO_2 and several other multicomponent slag systems at various temperatures (Reddy, 2003a) were discussed. An excellent agreement was obtained between the experimental sulfide capacities to that of RB model *a priori* calculated sulfide capacities of the metallurgical process important slag systems.

Reddy model for sulfur distribution ratio

At normal iron and steelmaking conditions, sulfur is present in molten metal in a dissolved form. The gas/metal reaction is thus simply:

$$\frac{1}{2}S_{2(gas)} = [S] in\ iron \quad \Delta G_6^\circ = -135\ 060 + 23.43T\ (J/mol) \tag{6}$$

The equilibrium constant of Reaction 6 is written as:

$$K_6 = exp\left(\frac{-\Delta G_6^\circ}{RT}\right) = \frac{a_S}{p_{S_2}^{1/2}} \tag{7}$$

The sulfur in equilibrium between slag and molten metal is written as the reaction:

$$[S] + (O^{2-}) = \frac{1}{2}O_{2(gas)} + (S^{2-}) \tag{8}$$

The equilibrium constant for the Reaction 8 is written as:

$$K_8 = \left(\frac{p_{O_2}^{1/2}}{a_{[S]}}\right) \bullet \left(\frac{a_{S^{2-}}}{a_{O^{2-}}}\right) \tag{9}$$

The gas/slag reaction for sulfur can then be written as:

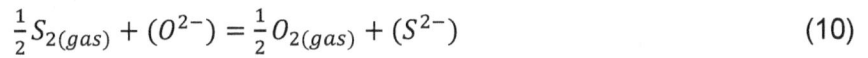

$$\frac{1}{2}S_{2(gas)} + (O^{2-}) = \frac{1}{2}O_{2(gas)} + (S^{2-}) \tag{10}$$

The equilibrium constant for this reaction is:

$$K_{10} = \left(\frac{a_{S^{2-}}}{a_{O^{2-}}}\right) \bullet \left(\frac{p_{O_2}}{p_{S_2}}\right)^{\frac{1}{2}} \tag{11}$$

The expression for sulfide capacity (Fincham and Richardson, 1954) is:

$$C_S = (wt.\%\ S\ in\ slag) \bullet \left(\frac{p_{O_2}}{p_{S_2}}\right)^{\frac{1}{2}} \tag{12}$$

Combining Equations 11 and 12:

$$K_{10} = \left(\frac{a_{S^{2-}}}{a_{O^{2-}}}\right) \bullet \frac{C_S}{(wt.\%\ S\ in\ slag)} \tag{13}$$

Since sulfur is present in molten metal in small amounts (<0.2 wt per cent), dilute solution approximation can be used. Henry's law can then be applied to obtain the following expression for the activity of sulfur dissolved in liquid metal:

$$a_{[S]} = f_S \bullet [wt.\%\ S\ in\ metal] \tag{14}$$

The value of the activity coefficient (f_S) can be calculated from the molten metal composition and interaction parameters. Using Equations 6, 8, 10, 12 and 14 and rearranging them, an expression for the distribution ratio is obtained:

$$L_S = \frac{(wt.\%\ S\ in\ slag)}{[wt.\%\ S\ in\ metal]} = \frac{(S)}{[S]} = \frac{f_S \bullet C_S}{K_6 \bullet p_{O_2}^{1/2}} \tag{15}$$

For a given slag composition and temperature, the Cs can be calculated using the RB model as described above. The oxygen partial pressure depends on the process and operating conditions. It can be determined by considering each process separately.

RESULTS AND DISCUSSION

Sulfur distribution ratio in ironmaking

In the process of ironmaking, iron oxides are reduced by coke to produce pig iron. The oxygen partial pressure in such reducing conditions is low and can be determined by the following reaction:

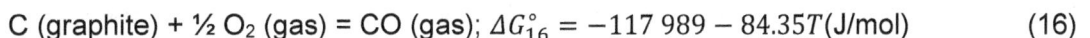

$$C\ (graphite) + \tfrac{1}{2}\ O_2\ (gas) = CO\ (gas);\ \Delta G_{16}^\circ = -117\ 989 - 84.35T\ (J/mol) \tag{16}$$

The equilibrium constant of Reaction 16 is:

$$K_{16} = \frac{p_{CO}}{a_C \cdot p_{O_2}^{1/2}} = exp\left(\frac{-\Delta G_{16}^\circ}{RT}\right) \tag{17}$$

The partial pressure of carbon monoxide can be calculated considering the different gases present in the blast furnace. The gases such as hydrocarbons (ie methane) and air, liquids (ie oils), and solids (ie pulverised coal) are also injected through the tuyeres in the BF. Some plants may use certain hydrocarbons such as methane, but most gas input into the furnace is air. Air contains about 0.79 and 0.21 mole fractions of nitrogen and oxygen, respectively. Assuming that all the oxygen input into the furnace is consumed by Reaction 16, two moles of carbon monoxide will be produced for each mole of oxygen. Assuming that the total furnace pressure is 1 atm, the partial pressure of carbon monoxide can be calculated as: $p_{CO} = X_{CO} \cdot p_{Tot.} = 0.35 \cdot 1 = 0.35$ atm. For activity of carbon is unity, substituting Equation 17 into Equation 15, the Ls is given as:

$$L_S = \frac{K_{16} \cdot f_S \cdot C_S}{0.35 \cdot K_6} \tag{18}$$

The activity coefficient of sulfur in liquid iron (f_S) is determined from the metal composition using the following equation:

$$log(f_S) = \sum e_S^j \cdot [wt.\% j]_{in\ iron} \tag{19}$$

The term e is called interaction coefficient. Interaction coefficients slightly change with changes in temperature (Deo and Boom, 1993a, 1993b). A typical pig iron composition in weight percentages would be: 3 C, 1 Mn, 1 Si, 0.1 P and 0.1 S. Using Equation 19, the activity coefficient of sulfur is calculated as: $f_S = 2.54$. The Equation 18 then becomes:

$$L_S = \frac{2.54 \cdot K_{16} \cdot C_S}{0.35 \cdot K_6} \tag{20}$$

Equation 20 was used to calculate the sulfur distribution ratio of blast furnace slags in equilibrium with molten iron. The model calculated distribution ratios for liquidus slag/metal at 1400 and 1500°C are plotted against the experimental data (Filer and Darken, 1952; Hatch and Chipman, 1949) in Figure 1.

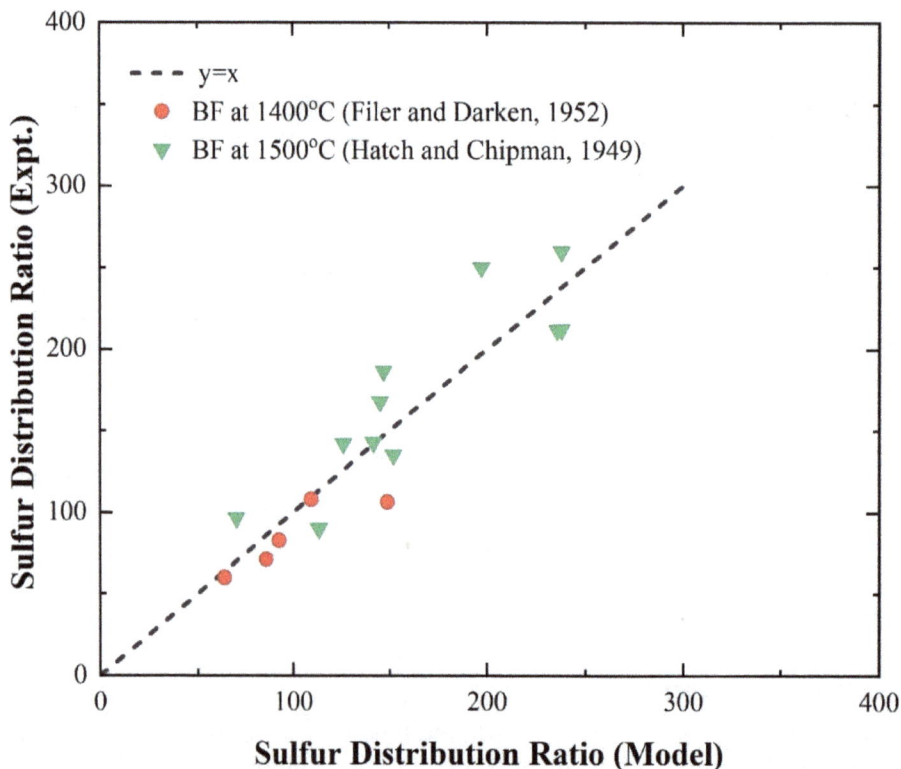

FIG 1 – Comparison of experimental data and model calculated data for sulfur distribution ratio (Ls) of BF ironmaking slags and iron.

The model calculated sulfur distribution ratio data are compared with the experimental data in the BF slags and liquid iron system over a wide compositions and two temperatures summarised in Figure 1. The present model reproduces most of the experimental data within experimental uncertainties, the average of relative value for this model was found to be 14.5 per cent. There are some points where the difference is noticeable. The method used here assumes equilibrium between slag and metal. It also assumes good mixing of each phase. These conditions are difficult to control and verify in laboratory experiments. Turkdogan (1996) reported that distribution ratios at turndown are about one-third to one-half of the slag/metal equilibrium values. There are two main observations that can be made from Figure 1. Increasing the temperature from 1400 to 1500°C nearly doubled the sulfur distribution ratio. The other observation is the large values of the distribution ratios, which were up to 300 at 1500°C.

Sulfur distribution ratio in steelmaking

Steelmaking is an oxidation process to remove the oxidisable elements from hot metal and scrap to the furnace slag as the steel is decarburised with oxygen blowing. Aluminium, silicon, phosphorous and manganese are all readily oxidised into the slag phase. The BOF charge normally contains only small amounts of Al, P and Mn. The oxidised silicon makes an important component of steelmaking slags. The oxidising conditions needed to decarburise the steel will also oxidise some of the iron. Lime is universally used as a fluxing agent to form steelmaking slags. Magnesia saturation of slag is needed to protect the furnace lining. The iron oxide goes into the slag phase and becomes an important component of slag. It follows that the main components of steelmaking slags are CaO, SiO_2, FeO and MgO. For a typical steel and steelmaking slag compositions used to determine the sulfur distribution ratio are Steel (wt per cent): 0.5 C, 0.3 Mn, 0.01 Si, 0.06 S and 0.01 P; and Slag (wt per cent): 47 CaO, 24 SiO_2, 15 FeO, 9 MgO, 3 MnO and 2 Al_2O_3. At or near turndown, among all other oxidation reactions, the iron oxidation reaction is what determines the oxygen potential:

$$\text{Fe (l)} + \tfrac{1}{2}\,O_2\,(g) = (FeO); \quad \Delta G^\circ_{21} = -238\,070 + 49.45T\,(J/mol) \tag{21}$$

Since the calculations performed here are for slag/metal equilibrium, Reaction 21 is used to calculate the oxygen partial pressure. The equilibrium constant for Reaction 21 is then obtained from the equation:

$$K_{21} = \frac{a_{FeO}}{a_{Fe} \cdot p_{O_2}^{1/2}} = exp\left(\frac{-\Delta G^\circ_{21}}{RT}\right) \tag{22}$$

Equation 22 for the pressure of oxygen, activity of Fe equal to 1 and substituting into Equation 15, then Equation 23 can be obtained:

$$L_S = \frac{f_S \cdot C_S \cdot K_{21}}{K_6 \cdot a_{FeO}} \tag{23}$$

All parameters in Equation 23 can be calculated from the temperature and compositions of slag and metal. The calculation will be performed here at 1600°C for the above slag and metal compositions. The sulfur activity coefficient was calculated as: f_S = 1.122. The activity of iron oxide was obtained (Bale, Pelton and Thompson, 2002) and interaction parameters (Ghosh, 2001a) and Gibbs energy (Ghosh, 2001b; Gokcen and Reddy, 1996). The sulfide capacity of the slag was calculated using the RB model as described above. The sulfur distribution ratio was calculated as:

$$L_S = \frac{f_S \cdot C_S \cdot K_{21}}{K_6 \cdot a_{FeO}} = \frac{(1.122) \cdot (5.3430 \times 10^{-2}) \cdot (1.1383 \times 10^4)}{(3.4887 \times 10^2) \cdot (2.4105 \times 10^{-1})} = 8.11 \tag{24}$$

The equilibrium sulfur distribution value for steelmaking is 8.11. The distribution ratio reported (Turkdogan, 1983) for similar system is 9.2. The model calculated data is in good agreement with the experimental data.

Sulfur distribution ratio in secondary steelmaking

In secondary steelmaking, the steel produced at the BOF or EAF is treated for impurity reduction and composition control in the ladle furnace (LF). The first step taken is usually to deoxidise the steel. When the aluminium bars with steel core are added at the top of the furnace, they sink to the bottom as their density increases with aluminium melting. This process achieves very low contents of dissolved oxygen, and the product is called fully-killed steel.

The added aluminium will dissolve in steel and react with dissolved oxygen. This oxygen partial pressure is determined by the reaction:

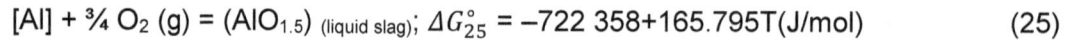

$$[Al] + \tfrac{3}{4} O_2 \text{ (g)} = (AlO_{1.5}) \text{ (liquid slag)}; \quad \Delta G^{\circ}_{25} = -722\,358 + 165.795T (J/mol) \tag{25}$$

From Reaction 25, The equilibrium constant is written as:

$$K_{25} = \frac{a_{AlO_{1.5}}}{a_{[Al]} \cdot p_{O_2}^{3/4}} \tag{26}$$

Rearranging Equation 26 for the partial pressure of oxygen and substituting into Equation 15:

$$L_S = \left(\frac{f_S \cdot C_S}{K_6}\right) \cdot \left(\frac{K_{25} \cdot a_{[Al]}}{a_{AlO_{1.5}}}\right)^{\frac{2}{3}} \tag{27}$$

The activity of $AlO_{1.5}$ can be obtained for a given slag composition and temperature. The activity of aluminium dissolved in steel can be calculated as:

$$a_{[Al]} = f_{Al} \cdot [\%Al] \tag{28}$$

The activity coefficient is calculated form steel composition using:

$$log(f_{Al}) = \sum e^j_{Al} \cdot [wt.\% j]_{in\ iron} \tag{29}$$

A typical of steel composition (wt per cent) in the ladle furnace is: 0.4 C, 0.1 Mn, 0.1 Al, 0.01 Si, 0.01 P and 0.05 S. Substituting the interaction coefficients and the composition, the activity coefficient of aluminium f_{Al} = 1.1 and activity coefficient of sulfur f_S = 1.2 were calculated. A typical LF slag composition in (wt per cent) is: 47 CaO, 1 FeO, 9 MgO, 1 MnO, 32 Al_2O_3 and 10 SiO_2. By knowing the activity of $AlO_{1.5}$ in slag and Cs for slag was calculated using RB model at the temperature of steel in the ladle furnace, the distribution ratio calculated using Equation 27 as follows:

$$L_S = \left(\frac{f_S\, C_S}{K_6}\right) \cdot \left(\frac{K_{25} \cdot f_{Al}\, [\%Al]}{a_{AlO_{1.5}}}\right)^{\frac{2}{3}} = \left(\frac{1.2 \times 2.5247 \times 10^{-3}}{2.7844 \times 10^2}\right) \cdot \left(\frac{9.1268 \times 10^{10}\ x\ 1.1\ x\ 0.1}{2.8731 \times 10^{-2}}\right)^{\frac{2}{3}} = 670 \tag{30}$$

The equilibrium sulfur distribution value for secondary steelmaking is 670. The experimental distribution ratio reported by Inoue and Suito (1994) for similar system is 663. The experimental and model data are in good agreement.

The large Ls value for LF indicates that steel desulfurisation in the ladle furnace is very effective. Fully-killed steel can be desulfurised in the ladle furnace with lime-alumina slags to achieve very low sulfur contents (below 10 ppm). The sulfur distribution ratios for LF are about twice those for BF and about ten times those of BOF. It is clear that steel desulfurisation in steelmaking furnaces is not effective.

In industrial operations, many factors influence the sulfur distribution ratio. These factors include slag volume, gas bubbling practice, slag skimming practice, sampling procedure, analysis techniques, sample exposure to the environment and the location and time at which the sample was taken. The values calculated using Reddy model here are to give a general idea about the actual distribution ratio. The true distribution ratio can be higher or lower, but for the most part it should be close to the model calculated value. Calculation of Ls for all three systems for an extended compositions and temperatures are in progress. Since the method used here is *a priori*, it is a useful tool for the furnace operators. They can adjust the slag composition towards one with high distribution ratio and thus have high desulfurising power.

CONCLUSIONS

The sulfide capacities predicted using the Reddy-Blander model were used in calculation of sulfur distribution ratios for ironmaking, steelmaking and secondary steelmaking processes. The Reddy model calculated data shows good agreement with the experimental data. For a given slag's composition and temperature, its sulfide capacity can be calculated *a priori* using the RB model. Further, the sulfur distribution ratio between this slag and liquid steel in a particular vessel can also be calculated *a priori* using the Reddy model. The sulfur distribution ratios for LF are about twice those for BF and about ten times those of BOF. The desulfurisation of pig iron in BF or desulfurisation of steel in LF during secondary steelmaking are recommended. Since the model used to calculate the sulfur distribution ratio (Ls) is *a priori*, it is a useful tool for the iron and steel makers in optimisation of the industrial furnace operations.

ACKNOWLEDGEMENTS

The authors acknowledge the financial support from the ACIPCO for this research project. Authors also thank the Department of Metallurgical and Materials Engineering, the University of Alabama for providing the experimental and analytical facilities.

REFERENCES

Bale, C W, Pelton, A D and Thompson, W T, 2002. FactSage 5.1: Thermochemical Software for Windows. TM Montreal, Quebec: Thermfact Ltd.

Chen, B, Reddy, R G and Blander, M, 1989. Sulfide Capacities of CaO-FeO-SiO$_2$ Slags, *Proceedings of the Third International Conference on Molten Slags and Fluxes*, pp 270–272.

Deo, B and Boom, R, 1993a. *Fundamentals of Steelmaking Metallurgy*, pp 205–206 (Prentice Hall: New York).

Deo, B and Boom, R, 1993b. *Fundamentals of Steelmaking Metallurgy*, pp 54–55 (Prentice Hall: New York).

Derin, B and Reddy, R G, 2003. Sulfur and Oxygen Partial Pressure Ratios Prediction in Copper Flash Smelting Plants Using Reddy-Blander Model, *TMS*, 1:625–632.

Derin, B, Yucel, O and Reddy, R G, 2005. Sulfide Capacities of PbO-SiO$_2$ and PbO-SiO$_2$-AlO$_{1.5}$ (sat.) Slags, *Mining and Materials Processing Institute of Japan (MMIJ)*, pp 1279–1287.

Derin, B, Yucel, O and Reddy, R G, 2006. Predicting of Sulfide Capacities of Industrial Lead Smelting Slags, *Advanced Processing of Metals and Materials, TMS*, 1:237–244.

Derin, B, Yucel, O and Reddy, R G, 2010. Sulfide Capacity Modeling of FeOx-MO-SiO$_2$ (MO=CaO, MnO, MgO) Melts, *Minerals and Metallurgical Processing*, 28(1):33–36.

Derin, B, Yucel, O and Reddy, R G, 2004. Modeling of Sulfide Capacities of Binary Titanate Slags, *EPD Congress 2004*, TMS, pp 155–160.

Filer, E W and Darken, L S, 1952. Equilibrium between blast-furnace metal and slag as determined by remelting, *Transactions of AIME*, 194:253–257.

Fincham, C J B and Richardson, F D, 1954. The behaviour of sulphur in silicate and aluminate melts, *Proceedings of the Royal Society of London*, 223A, pp 40–62.

Flory, P J, 1953. *Principles of Polymer Chemistry* (Cornell University Press: New York).

Font, J M and Reddy, R G, 2005. Modelling of Antimonate Capacity in copper and Nickel Smelting Slags, *Trans Inst Min Metall C*, 114:C160–C164.

Ghosh, A, 2001a. *Secondary Steelmaking: Principles and Applications*, appendix 2.1 (CRC Press: New York).

Ghosh, A, 2001b. *Secondary Steelmaking: Principles and Applications*, appendix 2.2 (CRC Press: New York).

Gokcen, N A and Reddy, R G, 1996. *Thermodynamics, Thermodynamic simulator (TSIM) for thermodynamic calculations*, 371 p (Plenum Press: NY).

Hatch, G G and Chipman, J, 1949. Sulphur Equilibria between Iron Blast Furnace Slags and Metal, *JOM*, 1:185:274–284.

Inoue, R and Suito, H, 1994. Thermodynamics of Fe Equilibrated with O, N, and S in Liquid CaO-Al203-MgO Slags, *Metallurgical and Materials Transactions*, 25B:235–244.

Pelton, A D, 1999. Thermodynamic Calculation of gas solubilities in oxide melts and glasses, *Glastechnische Berichte*, 72:40–62.

Pelton, A D, 2000. Thermodynamic Modelling of Complex Solutions, *The Brimacombe Memorial Symposium*, ISS, CIM, TMS, pp 763–780.

Pelton, A D, Eriksson, G and Romero-Serrano, A, 1993. Calculation of Sulfide Capacities of Multi-component Slags, *Metall Trans B*, 24:817–825.

Reddy, R G and Font, J M, 2003. Arsenic Capacities of Copper Smelting Slags, *Metallurgical and Materials Transactions B*, 34B:565–571.

Reddy, R G and Blander, M, 1987. Modeling of Sulfide Capacities of Silicate Melts, *Metallurgical Transactions*, 18B:591–596.

Reddy, R G and Blander, M, 1989. Sulfide Capacities of MnO-SiO_2 Melts, *Metallurgical Transactions*, 20B:137–140.

Reddy, R G, 2003a. Impurity Capacities in Metallurgical Slags, *Materials Processing Fundamentals and New Technologies*, TMS, 1:25–48.

Reddy, R G, 2003b. Emerging Technologies in Extraction and Processing of Metals, *Metallurgical and Materials Transactions B*, 34B:137–152.

Reddy, R G, Hu, H and Blander, M, 1992. Sulfide Capacities of Silicate Slags, *Proceedings of the Fourth International Conference on Molten Slags and Fluxes*, pp 144–148.

Turkdogan, E T, 1983. *Physicochemical properties of molten slags and glasses*, The Metals Society, London, pp 298–301.

Turkdogan, E T, 1996. *Fundamentals of Steelmaking*, pp 237–244 (The Institute of Materials: London).

Yahya, A and Reddy, R G, 2011. Sulfide Capacities of CaO-MgO-$AlO_{1.5}$, MgO-MnO-$AlO_{1.5}$ and CaO-MgO-MnO-$AlO_{1.5}$ Slags, *Trans Inst Min Metall C*, 120(1):45–48.

Yang, X-M, Li, J-Y, Zhang, M, Chai, G-M and Zhang, J, 2014. Prediction model of sulfide capacity for CaO-FeO-Fe_2O_3-Al_3O_3-P_2O_5 slags in a large variation range of oxygen potential based on the ion and molecule coexistence theory, *Metallurgical and Materials Transactions*, 45B:2118–2137.

Zhao, W and Reddy, R G, 1995. Sulfide Capacities of CaO-$AlO_{1.5}$ Melts, *EPD Congress*, TMS, pp 39–47.

Challenges and limitations in development of large thermodynamic databases for multiple molten phases using the Modified Quasichemical Formalism

D Shishin[1], M Shevchenko[2], E Nekhoroshev[3] and E Jak[4]

1. Research Fellow, PYROSEARCH, University of Queensland, Brisbane Qld 4072.
   Email: d.shishin@uq.edu.au
2. Research Fellow, PYROSEARCH, University of Queensland, Brisbane Qld 4072.
   Email: m.shevchenko@uq.edu.au
3. Postdoctoral Research Fellow, PYROSEARCH, University of Queensland, Brisbane Qld 4072.
   Email: e.nekhoroshev@uq.edu.au
4. Professor, Centre Director, PYROSEARCH, University of Queensland, Brisbane Qld 4072.
   Email: e.jak@uq.edu.au

ABSTRACT

In the field of pyrometallurgy, the presence of multiple molten phases with distinct chemical compositions, such as slags, mattes, speiss liquids, metals and molten salts, is a well-known phenomenon. These phases exhibit strongly non-ideal solution behaviour and are mutually miscible to varying degrees. To describe these complex molten liquids, the Modified Quasichemical Model (MQM) has proven effective and has been applied to numerous binary, ternary and several higher-order multicomponent systems related to pyrometallurgy. In recent decades, the complexity of high-temperature processes has escalated due to increased impurity concentrations in primary ores, the incorporation of recycled consumer products and the integration of metallurgical plants through by-product and waste exchanges. To address these issues, PYROSEARCH laboratory has been developing large 20-component thermodynamic database using a generalised CALPHAD (CALculation of PHAse Diagrams) methodology integrated with experimental investigations. Over the past decade, strategic decisions were made balancing the ability to generate experimental data, prediction power, accuracy, stability of calculations, as well as computational time, which are discussed in the paper. Examples provided for issues of >3 liquid immiscibility, non-ionic behaviour and exaggerated 'sharp' enthalpies of mixing in silicate slags, dealing with multiple oxidation states, as well as comparative analysis of pair and quadruplet approximation in MQM.

INTRODUCTION

Pyrometallurgical production of metals employs selective separation of metals and non-metals using their preference to one of the liquid phases, a gas phase, or rarely a solid phase. Commonly distinguished liquid types are molten salts, slags (oxide-based liquids), mattes (sulfide-based liquids), metallic liquids and speiss. Speiss liquids are formed between transition metals Fe, Co, Ni, Cu and Zn on the one side and metalloids As, Sb, (Sn) on the other side. Molten salts are based on pairing between alkali and alkali-earth metals on the one side and halogens, sulfates, phosphates, or carbonates on the other side. Molten salts are less common in primary pyrometallurgical production of metals from ores but applied in many electrometallurgical applications. They are also becoming increasingly important due to higher rates of recycling of e-waste, attempts to process tailings, incineration of waste, as well as in new emerging technologies for energy storage and generation.

In terms of the bond polarity these types of liquids can be ordered as follows:

$$Salts > Slags > Mattes > Speiss > Metals$$

Molten salts, slags and most mattes at stochiometric metal-to-sulfur compositions exhibit an ionic behaviour. Covalent bonds are observed within the anions of molten salts, ie SO_4^{2-} or CO_3^{2-}, and in slags with high proportion of silica or boron oxide. Incomplete dissociation into ions, with the portion of preserved covalent bonds is expected for matte and speiss. Also, when matte (Sundström, Eksteen and Georgalli, 2008) and speiss solutions deviate towards excess metal in terms of stoichiometry, metallic type of bonds give important contribution to the observed physico-chemical properties.

The division of liquid into types is relative, they all are mutually miscible to a certain extent and bond types change with changing composition. The basis for distinction is the ability to form miscibility gaps and the formation of distinct layers. Separation of the liquid layers through skimming or taping is at the foundation of many pyrometallurgical processes. Examples are shown in Table 1. The microstructures of samples, containing all five types of liquids are shown in Figure 1. These come from lead production. Molten sulfate salt is not typical for lead smelting and appears due to the introduction of zinc tailings in primary lead production. Speiss forms due to introduction of recycled materials.

TABLE 1
Examples of pyrometallurgical processes bases on separation of liquid (molten) phases.

Process	Target production	Common design of furnaces	Liquid (molten) phases to be separated
Smelting from primary ores	Cu, Ni	Flash furnace, top-submerged lance, side and bottom-blown furnaces	Slag, matte
E-scrap smelting	Cu, Pb, Ni, Au, Ag, ...	Top-submerged lance, top blown rotary furnace	Slag, metal
Smelting	Pb	Same as above	(Salt), slag, metal
Fuming	Zn	Side and bottom-blown, plasma, electric arc	Slag, metal (or speiss)
Converting	Cu, Ni	Peirce-Smith, side and bottom-blown	Slag, matte
Fire refining	Cu	Cylindrical anode furnace, tilting furnace	Slag, metal
Slag reduction	Pb	Lead blast furnace	Slag, matte, metal
Drossing	Pb	Kettles	(Liquid speiss), matte, metal
Doré process	Au, Ag	Bottom-blown oxygen cupel	Slag, metal
Smelting	Fe	Blast furnace	Slag, metal
Iron recycling	Fe	Electric arc furnace	Slag, metal
Steel production	Fe	Basic oxygen furnace	Slag, metal
Electrolysis (Hall-Héroult)	Al	Electrolytic cell	Salt, metal

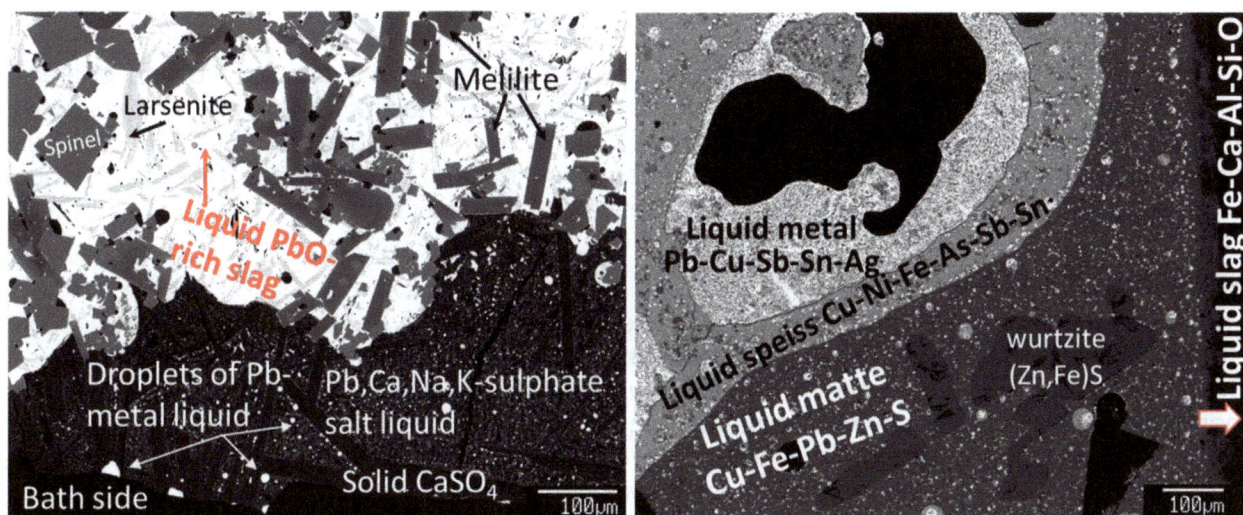

FIG 1 – Example of images obtained using backscattered electron microscopy of samples containing co-existing liquids with miscibility gaps. Left: liquid salt/slag/metal/solids equilibrium at conditions of lead sulfide smelting. Right: liquid slag/matte/speiss/metal/solids equilibrium at conditions of lead drossing. Slag phase is present, but not shown because it is too dark at the selected contrast.

Modern metallurgical production plants deal with rising impurity levels in primary ores. The recycled materials and wastes are also introduced in the process. The exchange of the by-products is used to increase the overall recovery, create new marketable products and reduce the environmental impact. These changes make the chemistry more complex. Any step-changes in existing operations can be performed more efficiently using predictive computational tools. But these tools should be applicable for a wide range of conditions in terms of composition, temperature, oxidation / reduction potential. They cannot be developed exclusively on the data from existing operations. Calculations of thermodynamic equilibrium based on Gibbs energy minimisations can provide such predictions. When combined with kinetic factors observed in real process, they may significantly reduce the need for costly pilot tests and trials (Castillo-Sanchez *et al*, 2023). Powerful thermodynamic software has been developed over the last decades. For practical predictions, thermodynamic software requires mathematical models describing the Gibbs energy of all potentially existing phases within the chemical system, as a function of temperature, pressure and composition.

Most mathematical models, including thermodynamic models, contain adjustable parameters. A combination of thermodynamic models together with the set of optimised adjustable parameters is often referred to as a thermodynamic database. High-quality models potentially have predictive power in complex systems with small number of adjustable parameters. But high-quality thermodynamic database is a product which must provide high accuracy. It is achieved by generating sufficiently abundant data used to fix the values of model parameters. The data used to optimise model parameters should be accurate and represent thermodynamic equilibrium. To achieve the accuracy and reduce the number of potential uncertainties, experimental studies are typically performed in less complex systems, which often means systems of smaller number of components. But the final goal is multi-component calculations. The aim of this publication is to demonstrate challenges we encountered during the development of thermodynamic models of liquid phases for the 20-component system directly relevant to non-ferrous pyrometallurgy applications. The first challenge was that the development had to be conducted in steps, with elements added in time:

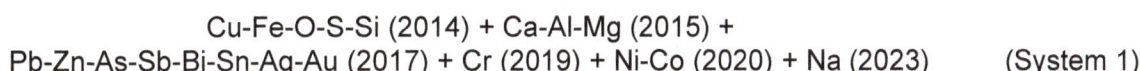

Cu-Fe-O-S-Si (2014) + Ca-Al-Mg (2015) +
Pb-Zn-As-Sb-Bi-Sn-Ag-Au (2017) + Cr (2019) + Ni-Co (2020) + Na (2023)   (System 1)

The decisions made in early stages of the project largely determined the direction of development of the models. As the size of the database grew, the larger resources required to introduce.

METHODOLOGY

Development of large multicomponent solution databases require the appropriate and strategic selection of the thermodynamic models and a method of obtaining model parameters while maintaining accuracy, predictive power and self-consistency. The models should be structurally based, so that model parameters are more than empirical mathematical expressions. For instance, they can correlate to the energy of interactions between the atoms in the phase.

Modified Quasichemical Formalism

The Modified Quasichemical Model (MQM) (Pelton *et al*, 2000; Pelton and Chartrand, 2001; Chartrand and Pelton, 2001; Pelton, Chartrand and Eriksson, 2001), a collaborative effort between CRCT (Montreal, Canada) and GTT (Germany), has demonstrated its ability to describe the thermodynamics of many chemical systems, binary, ternary and of more components. Over the years, the model has undergone several improvements, as detailed in recent publications (Jung and Van Ende, 2020). These enhancements now offer greater flexibility in choosing constituents, formulating expressions for excess Gibbs energy and interpolating binary parameters into ternary and higher-order systems. Notably, other established models, such as the Bragg-Williams ideal solution and the associate solution model can be reproduced within the MQM framework. These advancements warrant consideration of renaming MQM to the Modified Quasichemical Formalism (MQF), highlighting its ability to accommodate various models within its structure, as shown in Figure 2. Present publication will outline recent experience in model selection options within the MQF.

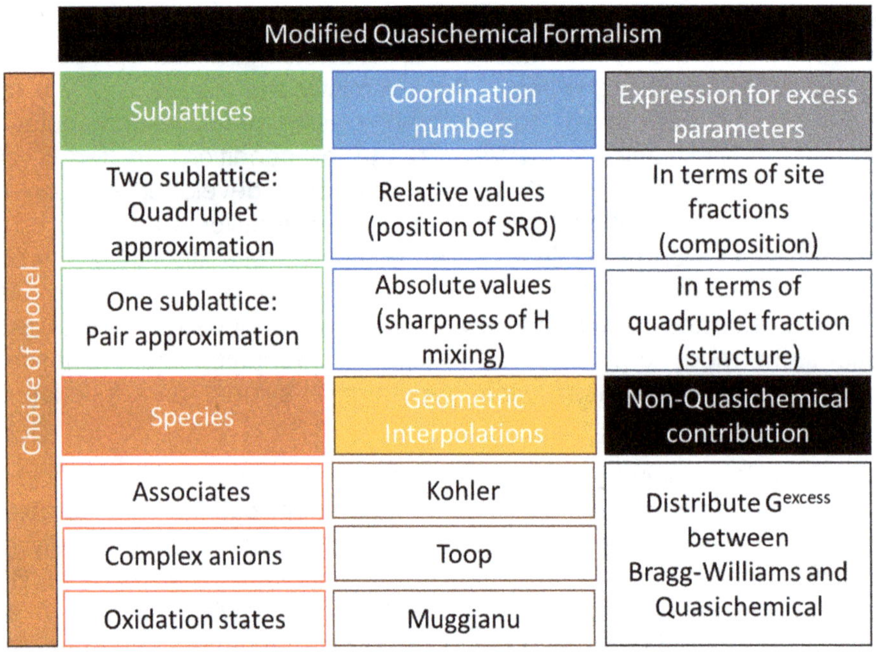

Modified Quasichemical Formalism			
Choice of model	**Sublattices**	**Coordination numbers**	**Expression for excess parameters**
	Two sublattice: Quadruplet approximation	Relative values (position of SRO)	In terms of site fractions (composition)
	One sublattice: Pair approximation	Absolute values (sharpness of H mixing)	In terms of quadruplet fraction (structure)
	Species	**Geometric Interpolations**	**Non-Quasichemical contribution**
	Associates	Kohler	Distribute G^excess between Bragg-Williams and Quasichemical
	Complex anions	Toop	
	Oxidation states	Muggianu	

FIG 2 – Choice of model options within the Modified Quasichemical Formalism.

In many liquid solutions, strong attractive interactions between certain constituents at the atomic level result in the significantly higher than random probability of finding them next to each other. This phenomenon, known as Short Range Ordering (SRO), alters the probability distribution, leading to a deviation from ideal entropy of mixing (Pelton and Kang, 2007). The main purpose of the MQF is to take into account this deviation. By definition, the ideal Bragg-Williams model neglects the SRO effect. The Associate solution model, which is an alternative way to treat SRO, assumes that strongly interacting constituents are bonded to each other. As a result, it cannot reproduce the random mixing entropy once the molecular associate is introduced (Pelton and Kang, 2007). The full implementation of MQF is only available in FactSage™ software, versions 7.0 to 8.3 (Bale *et al*, 2016). Thermocalc™ software, version 2024a (Andersson *et al*, 2002), relies on the Ionic Two-Sublattice Model (ITSM) (Sundman, 1991). Unlike MQF, ITSM enables charged and uncharged ions within the same solution, allowing the description of metallic and ionic solutions within the same model.

Ideas how to further improve the flexibility of MQF include the introduction of multiple compositions of SRO (Wang and Chartrand, 2021) and variable geometric interpolation technique, beyond the choice Kohler-, Toop- and Muggianu-types (Wang *et al*, 2023). But these ideas have not been implemented in thermodynamic databases developed by the authors.

Extended CALPHAD approach

Thermodynamic software requires expressions for the Gibbs energies of complex solution phases in the form of $G = f(T, P, x_1..x_n)$. where T is temperature, P is total pressure and $x_1...x_n$ are compositional variables. The adjustable parameters, contained in these expressions, must be derived from experimental data, or in rare cases, from the first-principal methods. Enthalpy of formation, entropy and heat capacity expressions ($\Delta H_f°$, $S°$, CP) of solution components, as well as energy parameters for the interactions between them are all examples of model parameters. CALPHAD principles have become a standard for the thermodynamic description of a chemical system (Jung and Van Ende, 2020). The principles for a given system are as follows: a single set of model parameters for all potentially stable phases within the system should be obtained, which provides good description of all types of available experimental data, such as phase equilibria, activity measurements, the distribution of elements among phases, calorimetry and crystallography (Shishin *et al*, 2013).

Mathematical models of complex systems, in general, can include the simultaneous contributions of four and more factors. In contrast, in the thermodynamics of multicomponent liquids, decreasing probability of complex interactions among more than three atoms warrants the use of only binary and ternary interaction parameters (Gorsse and Senkov, 2018). This imposes a temptation to study and access only binary and ternary systems, relying on model extrapolation into multicomponent systems. While this may be applicable to metallic liquids, our experience showed that the values of ternary parameters in oxide systems must be verified inside several related quaternary, or even higher order systems. Extrapolations often did not provide the results which were accurate enough. The experimental methodology had to be improved to produce high quantity of experimental data in 4-, 5-components systems.

Thus, CALPHAD principles must be further extended when developing a large self-consistent multicomponent thermodynamic database. A larger single set of parameters need to be used and it should describe all data in all related systems. When performing predictions in multicomponent systems, it is necessary that model parameters are defined (fixed) for all the binary and eventually, for all ternary sub-systems and all solutions. For the database of 20-components, the theoretical number is 20!/2!(20−2)! = 190 binary subsystems and 20!/3!(20−3)! = 1140 ternary subsystems. Such huge numbers, especially for ternary systems, show the advantage of the model with high predictive power. Thermodynamic calculations using the current version of the database are used to determine the conditions of experiments and compositional areas are selected to improve the confidence of a given parameter. This helps to increase the number of successful experiments and reduce the need to mesh-type planning of experiments. In cases, when experiments cannot be performed due to challenging chemical nature, eg Au-O liquids not existing at ambient conditions, the values of parameters must be obtained indirectly, from the data of the multicomponent systems. High number of experiments and parameters means the process of development is inevitably iterative, requiring re-optimisation of previously assessed systems. Thus, the integrated experimental and modelling approach, is critical for the development of large thermodynamic databases.

Solid phases

At present, there are 95 solid solutions and 630 solid compounds with fixed stoichiometry in the database developed in PYROSEARCH. Models for liquids cannot be developed independently from solid phases. Phase equilibrium between the liquid and solid phase is one of the most important sources of thermodynamic information and typically can be measured more accurately, compared to activity or calorimetric information for molten solutions. For notoriously large and complex solid solutions, such as spinel, the models are typically developed within the Compound Energy Formalism (CEF) (Hillert, 2001), which assumes the distinct crystal sublattices and random mixing within each sublattice. The main model parameters are the Gibbs energy functions of all solution end-members, as well as excess parameters for the interaction of atoms within the same sublattice. As in the case of MQF, different models can be developed within CEF. In CEF, more than two sublattices can be used, which MQF does not allow. This feature is used for pyroxenes, melilite, feldspar. The same components, even charged ions, can be present on several sublattices. This approach creates many 'virtual' end-members, necessitating the development of careful strategies to reduce the number of adjustable parameters though linear combinations and reciprocal reactions (Hillert *et al*, 2009). Still, for solutions like monoxide in the Fe-O system, simpler Bragg-Williams approach was preferred to describe the entropy of mixing (Hidayat *et al*, 2015). For non-stoichiometric solid speiss solutions, MQF with single sublattice demonstrated similar results as CEF with less parameters (Shishin and Jak, 2018) and more reasonable extrapolation of mixing properties outside of stoichiometry. This may be potentially beneficial for ternary and multicomponent systems.

OVERVIEW OF RESULTS AND ISSUES

Let us consider the selection of the models for important molten phases observed within the non-ferrous pyrometallurgy using the matrix in Figure 2. The summary is provided in Table 2 and the following sections discuss in more detail the selected models, their advantages and disadvantages.

TABLE 2

Selection of sublattice and species within the MQF, chosen for the molten phases within the 20-component System (1) targeted for non-ferrous pyrometallurgy.

Solution name	Selected model	Advantages	Limitations
Slag	$(Cu^{+1}, Fe^{+2}, Fe^{+3}, Si^{+4}, Al^{+3}, Ca^{+2}, Mg^{+2}, Pb^{+2}, Zn^{+2}, Ni^{+2}, Co^{2+}, Sn^{+2}, Sn^{4+}, Sb^{+3}, As^{+3}, Bi^{+3}, Ag^{+1}, Au^{+1}, Cr^{2+}, Cr^{3+}, Na^{+}, NaFe^{4+}, NaAl^{4+})\,(O^{-2}, S^{-2})$	Takes into account Second Nearest Neighbour SRO between basic and acidic components	Cannot describe deviation from MeO_x and MeS_x stoichiometry, where x = 0.5, 1.0, 1.5, 2.0
		Compared to purely associate models, less ill-defined parameters	Does not contain Cu^{2+}, Cr^{6+}, Pb^{4+}
		Associate still can be introduced for particularly difficult systems (ie $NaFe^{4+}$)	Assumption of fully ionic liquid does not work near pure SiO_2
		Separate from liquid matte/metal/speiss and salt overcomes 3-phase immiscibility limit in FactSage™	
liquid matte/ metal/ speiss	$(Cu^{I}, Cu^{II}, Fe^{II}, Fe^{III}, Pb^{II}, Zn^{II}, Ni^{II}, Sn^{II}, Sb^{III}, As^{III}, Bi^{III}, Ag^{I}, Au^{I}, Cr^{II}, Co^{II}, Ca^{II}, Mg^{II}, Na^{I}, O^{II}, S^{II})$	Possible to describe complete stoichiometry range (even for oxides)	Cannot describe Second Nearest Neighbour SRO
		Easy to add new component no ill-defined end-members	Lack of flexibility of ternary parameters and geometric interpolations
		Easy to introduce Bragg-Williams parameters from literature	Separate from slag, creates conflicts with oxide liquids
Salt	$(Ca^{+2}, Mg^{+2}, Pb^{+2}, Zn^{+2}, Na^{+1}, Cu^{2+}, Fe^{3+})(SO_4^{-2}, O^{-2})$	Relatively easy to expand using existing molten salt assessments	Separate from slag, which is not the case for some systems

Slag

The slag solution was the original reason for the development of the MQM (Pelton and Blander, 1986; Wu, Eriksson and Pelton, 1993; Wu *et al*, 1993; Jak *et al*, 1997; Jak, Zhao and Hayes, 1997). Many binary and ternary oxide systems were reasonably well described at the time.

Sublattices

One sublattice model was chosen with O^{2-} being the only anion. Since the end-members of the solution were pure oxides, no deviation towards excess metal, or towards oxygen was allowed. For the systems containing Fe-O and Cr-O, species with several oxidation states were introduced (Decterov and Pelton, 1996), ie Fe^{2+} and Fe^{3+}, Cr^{2+} and Cr^{3+}, which partially solved the oxygen/metal non-stoichiometry within these sub-subsystems for a limited range of composition. Wide liquid-liquid miscibility gaps between metallic liquid and oxide liquids for most metal-oxygen systems justified the practicality of the model. This eventually grew into FToxid public database of FactSage™ (Bale *et al*, 2016), which is currently on of the largest self-consistent oxide database (Jung and Van Ende, 2020). The PYROSEARCH database (Jak *et al*, 2019) uses the same approach. It is not as big in terms of number of elements, but is continuously re-assessed iteratively, supported by new experimental results. For instance, new data have been generated even for foundational system of Ca-Fe-O-Si in 2022, resulting in major revision. At this point, the properties of all oxide end-members and all excess parameters in FToxid and the database of the PYROSEARCH database are different.

Alternatives for slags have been developed by Selleby and Sundman (1996), Selleby (1997) and Hillert, Sundman and Wang (1990), who used ITSM, which become the foundation of the Metal Oxide Solutions (TCOX) thermodynamic database from Thermo-Calc Software. Yet another approach was used by (Jantzen *et al*, 2021) who started the development of the independent GTOx (GTT oxide) database in 2000s, citing the need to 'give a second opinion to our customers who have licensed the FToxid (FACT oxide) database … and a better description for … high Na_2O- and K_2O- and vanadium-containing systems'. They used an associate model. For comparison, the models for the slag within the Ca-Fe-Si-O system, which is foundational for many non-ferrous pyrometallurgical processes, are given below:

MQF (FToxid and PYROSEARCH database): $(Fe^{+2}, Fe^{+3}, Si^{+4}, Ca^{+2})(O^{-2})$,

Ionic Two-Sublattice Model (TCOx): $(Fe^{+2}, Si^{+4}, Ca^{+2})_P(O^{-2}, SiO_4^{-4}, FeO_{1.5}, SiO_2, Vacancy^{-Q})_Q$,

Associate model (GTOx): $(Ca_2O_2, Fe_2O_2, Fe_2O_3, Fe_3O_4/1.5, Si_2O_4, CaSiO_3, Ca_2SiO_4/1.5, FeSiO_3, Fe_2SiO_4/1.5, CaFe_2O_4/1.5, CaFeSi_2O_6/2)$

All these models have been further expanded to include more metals. The TCOx model can describe metallic liquid and slag using the same solution. The current model for slag in UQpy (Uncertainty Quantification with Python) is provided in Table 2.

Selection of species – oxidation states

Table 2 shows the selection of 'Species' for the PYROSEARCH slag model. A decision was made in 2013 to only include Cu^{+1}, but not Cu^{+2} in the slag model for non-ferrous applications. It was demonstrated (Shishin and Decterov, 2012) that the formation of Cu^{+2} is significant at $p(O_2)$ above 10^{-3} atm for temperature > 1100°C, as shown in Figure 3. These highly oxidising conditions are not encountered for most process steps in the pyrometallurgy of Cu, Ni, Pb, Zn, Fe, Sn or others. The decision significantly reduced the number parameters between Cu^{+2} and other components of the slag, which are difficult to define. Very little reliable experimental data existed for these high $p(O_2)$ conditions. The same logic applies to Pb^{+4}, Cr^{+6}, which are even less stable at high temperatures and are not expected in liquid slags, but are introduced as solid compounds, when necessary. Some applications, such as using molten copper oxide for high temperature for solar energy storage and oxygen production (Jafarian, Arjomandi and Nathan, 2017) do require correct predictions of Cu^{+2}/Cu^{+1} in liquid oxide. These are covered by the liquid matte/metal/speiss solution model (Shishin *et al*, 2013), which is discussed further. Some evidence exists that the formation of accretions close to the uptake shaft of the copper flash smelting furnaces may involve liquid slag phase equilibria within the Cu-Fe-O-Si system at oxygen partial pressures exceeding 10^{-2} atm. If confirmed, the accurate description would require the introduction of Cu^{+2}. This in turn, would shift the stoichiometry of slags equilibrated with metallic copper further away from Cu^0, contrary to diagram in Figure 3. Introduction of both Cu^0 and Cu^{+2} is not possible within the selected formalism, since charges and uncharged species together are not allowed. Even if was possible, that would create more calculation conflicts with liquid matte/metal/speiss solution model. The partial solution could be the possibility to turn on and off the Cu^{+2} species in FactSage™ software, but this makes phase selection complicated for metallurgical engineers.

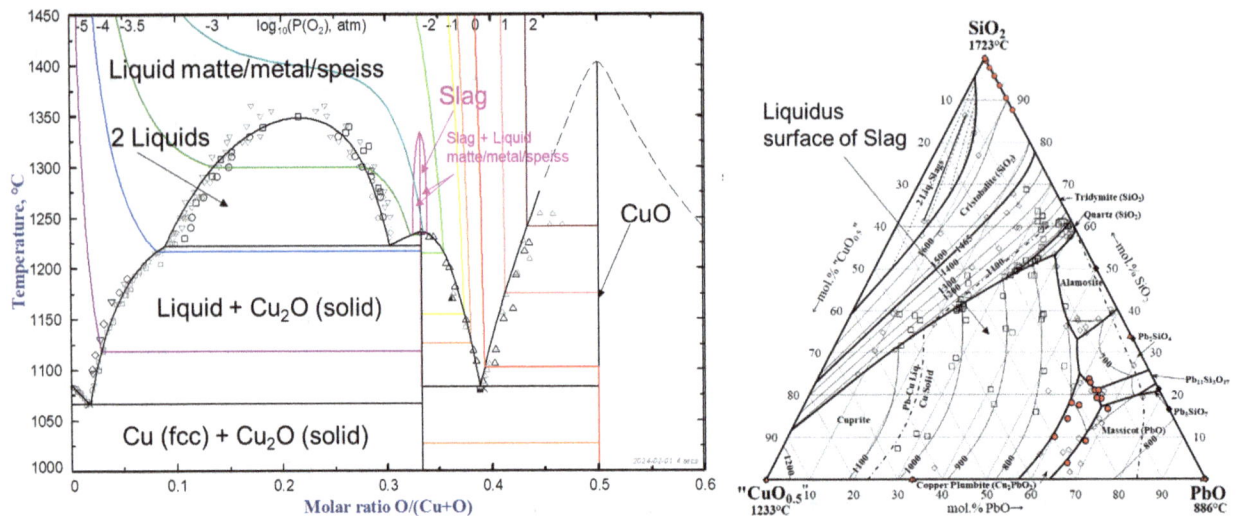

FIG 3 – Left: Phase diagram of the Cu-O system (Shishin and Decterov, 2012). The compositional domain of slag solution is much smaller than liquid/matte/metal solution. Right: Example of slag liquidus projection in the Cu-Pb-Si-O system (Wen *et al*, 2023) demonstrates the practicality of this approach for the description of multicomponent systems in non-ferrous pyrometallurgy.

Selection of species – associates

The presence of Na_2O in slags which also contain Al_2O_3 and SiO_2 result in the so called 'charge compensation' effect. Cations of Al^{+3} in the presence of Na^{+1}, assume the position of Si^{+4} and act as a network former (Decterov, 2018). This strong ternary effect cannot be satisfactory modelled without the introduction of $NaAl^{+4}$ associate in the MQF (Lambotte and Chartrand, 2013). The introduction did not require specific changes to the formalism and good results were achieved for systems exhibiting strong 'charge compensation'. Still, model parameters for the $NaAl^{+4}$ species are not well-defined. Thermodynamic properties of virtual liquid $NaAlO_2$ cannot be obtained from the Na_2O-Al_2O_3 data. They can be better fixed, when considering binary and ternary data within the Na_2O-Al_2O_3-SiO_2 simultaneously, but even so, the excess Gibbs energy parameters between $NaAl^{+4}$ and cations other than Si^{+4} are ill-defined. Predictions in multicomponent systems become less reliable and can result in spurious miscibility gaps, such as one observed by (Nekhoroshev, 2019) in the CaO-Na_2O-Al_2O_3-SiO_2 system. The decision to introduce $NaAl^{+4}$, (and $NaFe^{+4}$ for the same reasons), created a challenge. Significant resources are required to 'scan' the multicomponent systems for possible problems, generate experimental data and make sure the parameters involving these associates are fixed. This direction of work has not fully started yet, since Na^{+1} was the last element to be introduced in the chemical System (1).

Further evolution of the slag model within the MQF was largely driven by the need to describe the solubility of sulfur (Kang and Pelton, 2009), giving the rise of the two-sublattice model. The anion S^{-2} was introduced on the second sublattice as shown in Table 2 and used the mathematical expressions originally developed for molten salts. In addition to sulfide capacity, the resulting model could be used to predict phase equilibria involving sulfides (mattes) and oxysulfides (Jo, Lee and Kang, 2013), but limited to fixed sulfur-to-metal and oxygen-to-metal stoichiometry. For instance, within the Fe-O-S-Si system, no deviation of stoichiometry is possible near the composition of FeS, but the model was capable of describing miscibility between slag and matte. This approach is suitable to understand the formation of steel inclusions in the slag, but proved not applicable for slag/matte equilibria in copper, lead and nickel smelting. In the PYROSEARCH database, mattes are described using a separate solution. Metal/sulfur non-stoichiometry in mattes is a significant factor for accurate predictions for non-ferrous processes. The introduction of sulfur-containing quadruplets, such as $Cu^{+1}Fe^{2+}O^{-2}S^{-2}$, allowed quantitative modelling solubility of copper in slags (Shishin *et al*, 2018b), as well as slag/matte equilibria involving Pb, Ni and other metals (Hidayat *et al*, 2023; Shishin *et al*, 2020; Sineva *et al*, 2023). Still, the selected model does not allow the description of complete miscibility between slag and matte, which is observed in Cu-free Fe-O-S-Si and Ca-Fe-O-S systems. As in the case of Cu^{2+}, the selection of the model was driven by multicomponent practical predictions, at the expense of lower order systems with small practical value. Example of alternative approach, describing slag, matte and metal with using single solution is available $(Fe^{+2}, Ca^{+2})P(O^{-2}, S^{-2}, S^0, FeO_{1.5}, Vacancy^{-Q})_Q$ (Dilner and Selleby, 2017) in the TCOx database. Large compositional areas of the system in question were not experimentally studied and the agreement with existing experimental data is semi-quantitative. The FToxid slag was further expanded towards F^-, SO_4^{-2}, PO_4^{-3}, CO_3^{-2} and other anions, with limited concentrations. Certain challenges in modelling oxyfluoride systems required formalism improvements (Lambotte and Chartrand, 2011), which later were generalised (Wang and Chartrand, 2021). The two-sublattice MQM in Quadruplet Approximation assumes the ionic nature of oxides and salts. It uses the Gibbs energy functions for solution end-members, ie Cu_2O, Cu_2S, FeO and FeS, to express the magnitude of First Nearest Neighbour SRO, while Second Nearest Neighbour SRO is taken into account by excess parameters.

Coordination numbers

The absolute values of coordination numbers were originally selected to provide the entropy of mixing to be zero at the composition of maximum short-range ordering for the system with infinitely negative enthalpy of mixing. The numerical values were calculated by (Pelton and Blander, 1984) and these values are still used for the coordination numbers of most species in present-day slag solution: 0.68872188 for +1 cations, 1.37744375 for +2 cations, 2.06616563 for +3 cations and 2.75488750 for +4 cations. The absolute values are not as important as the ratios of coordination numbers. Possibly due to the low importance, these absolute values are the only model parameters that have been preserved since the early versions of the model. The ratios of coordination numbers determine the composition of maximum short-range ordering. In later publication (Pelton *et al*, 2000), it was recommended that for the systems with small negative or positive values of enthalpy of mixing, the coordination numbers should be closer to their physical values. In the case of Cu^{+1}, this recommendation was originally not applied, so it was assigned the 'default' value for +1 cations. This worked relative well but required the introduction of the positive Bragg-Williams parameters to describe the tridymite liquidus related to the Cu_2O-SiO_2 system (Hidayat and Jak, 2014; Hidayat *et al*, 2017a). The latest systematic experimental work in the high-SiO_2 region studying the miscibility gas in slags for the systems like Cu-Fe-O-Si (Wen, Shevchenko and Jak, 2021), Cu-Pb-O-Si, Cu-Zn-O-Si demonstrated (Wen *et al*, 2023) the need for correction. Example is shown in Figure 3 (right) and labelled as 2 Liq. Slags. After the coordination number for Cu^{+1} was doubled, it was possible to improve the description of the miscibility gaps.

Sharp enthalpy and non-Quasichemical contribution

A large data set of new experimental data have been obtained for the foundational system Ca-Fe-O-Si, including all subsystems, in 2020–2024 (Cheng *et al*, 2019, 2021). With higher accuracy and large density of experimental points, it was quite difficult to achieve the desirable agreement by using the similar set of model parameters, as in the earlier assessment (Hidayat *et al*, 2016a, 2016b, 2017b). The major re-vision was initiated, which included the update in the properties of all liquid oxide end-members. Properties of many solid compounds also required updates. The principles are described in the parallel presentation (Shevchenko, Shishin and Jak, 2024). Furthermore, it became clear the liquidus of Ca_2SiO_4 in the Ca-Fe-O-Si and other systems could not be described even after the changes in thermodynamic properties of solids. The reasons for that were difficult to isolate, but in the end it turned out that for the slag phase, the Gibbs energy of mixing was too 'sharp' at the composition of Ca_2SiO_4. The phenomenon of sharp enthalpy of mixing was previously discussed for the metallic systems (Pelton and Kang, 2007), but not applied in slags. The solution was proposed by the present authors to combine the Bragg-Williams and Quasichemical contributions in slags, when required. After the Gibbs energy of mixing was corrected by the combination of the Quasichemical and Bragg-Williams parameters for the CaO-SiO₂ system, as shown in Figure 4, better description in the area of Ca_2SiO_4 was obtained for the several systems containing $CaO-SiO_2$, such as $ZnO-CaO-SiO_2$, $Al_2O_3-CaO-SiO_2$ and $MgO-CaO-SiO_2$. These results are not yet published.

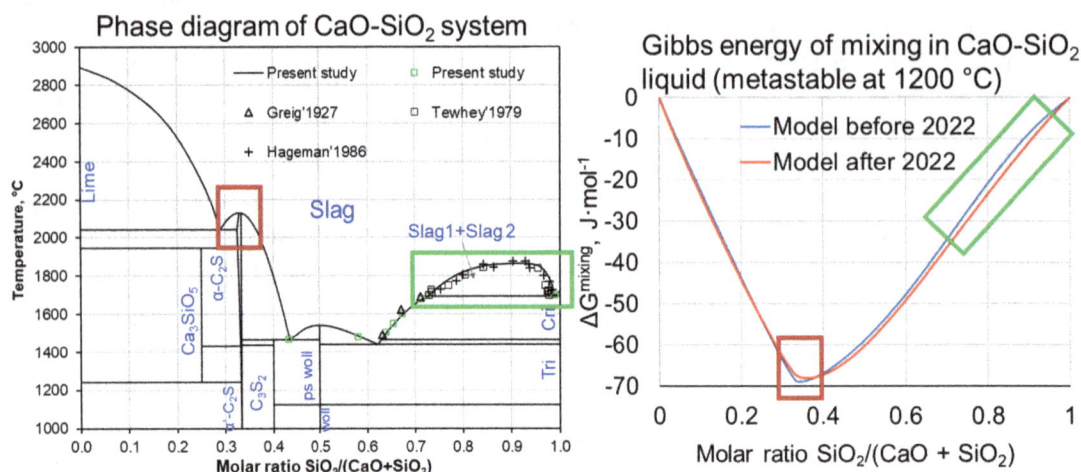

FIG 4 – Two regions in the CaO-SiO₂ system which required significant revision of model within the MQF, not just the revision of model parameters – enthalpy of mixing is too sharp near Ca_2SiO_4 and excess curvature of Gibbs energy of mixing near SiO_2.

Unusual excess parameters – non-ionic slags

The models for slags developed using the MQF typically require few model parameters to describe the main features of thermodynamic properties and phase diagrams. The use of multiple parameters to describe experimental data, especially series of negatively correlated parameters, is discouraged and may indicate internal inconsistencies in the experimental data series. Also, it is unusual to use excess parameters with high powers, ie $g_{AB}^{i,j}$ where i or $j > 6$, except for the Me-Si-O systems (Me = Li^+, Ca^{+2}, Mg^{+2}, Fe^{2+}, Fe^{3+}, Zn^{+2}, Cu, Ni^{+2}, Co^{+2}, Cr^{+2}...). Parameters with high power on SiO_2 have been commonly used for the to introduce the miscibility gaps in slag close to SiO_2 (Konar, Van Ende and Jung, 2017; Prostakova *et al*, 2012, 2013; Hidayat *et al*, 2017b; Wu *et al*, 1993; Jung, Decterov and Pelton, 2007). An example of miscibility gap in slag is shown in Figure 4 (left, green area). Furthermore, these parameters often contained large entropy terms to reproduce the closing the miscibility gaps at high temperature. For many systems the experimental data on the miscibility gaps were rare and controversial, so little attention was given to the values of these parameters. Recent Electron Probe microanalysis (EPMA) measurements in high-SiO₂ areas of the ternary and higher-order systems (Shevchenko, Shishin and Jak, 2022; Khartcyzov *et al*, 2022; Cheng *et al*, 2021) revealed that predicted miscibility gaps extended systematically too far into the multicomponent compositional space compared to experimental data and could not be described well enough using ternary parameters. This indicated that Gibbs energy of mixing was too concave in many Me-Si-O systems in the region shown using green area in Figure 4 (right). Less concave Gibbs energy of mixing, but similar shape of the miscibility gap could be achieved using excess parameters with higher power on SiO_2, more than 9. These parameters were never used before. It is believed the need to use unconventional parameter series is attributed to the fact that the model for slags assumes fully ionic behaviour. The high-SiO₂ liquids are not fully ionic (weak electrolytes) and exhibit long-range electrostatic interactions between the uncompensated charged Me^{x+} cations and O^{2-} anions, an effect similar to the Debye-Hückel model and clustering well known to be present in the diluted aqueous solutions. The Gibbs energy of mixing of the components can be better represented as a polynomial sequence, to simulate the long-distance ionic interactions (Figure 4). After the introduction of the polynomial sequences in some systems, unrealistic entropy contributions to excess parameters were no longer necessary, resulting in better description of tridymite liquidus at temperatures below the miscibility gap for many systems, particularly CaO-FeO-Fe₂O₃-SiO₂. Systematic replacement of parameters for high-SiO₂ liquids is expected to provide step-like improvement in many other systems.

An alternative approach was used in a PhD thesis by Nekhoroshev (2019), who introduced dimers of $(Na_2)^{+2}$ and $(K_2)^{+2}$ to better describe liquidus in high-SiO$_2$ region of the Na$_2$O-SiO$_2$ and K$_2$O-SiO$_2$ systems. Still, this approach, if applied to the multicomponent database, would result in many undefined excess parameters among dimers and other components of the solutions. In the PYROSEARCH model, a combination of $g_{Na^{+1}Si^{+4}}^{0,8}$ and $g_{Na^{+1}Si^{+4}}^{0,15}$ was tested for Na$_2$O-SiO$_2$ and good description of SiO$_2$-rich was achieved, as shown in Figure 5.

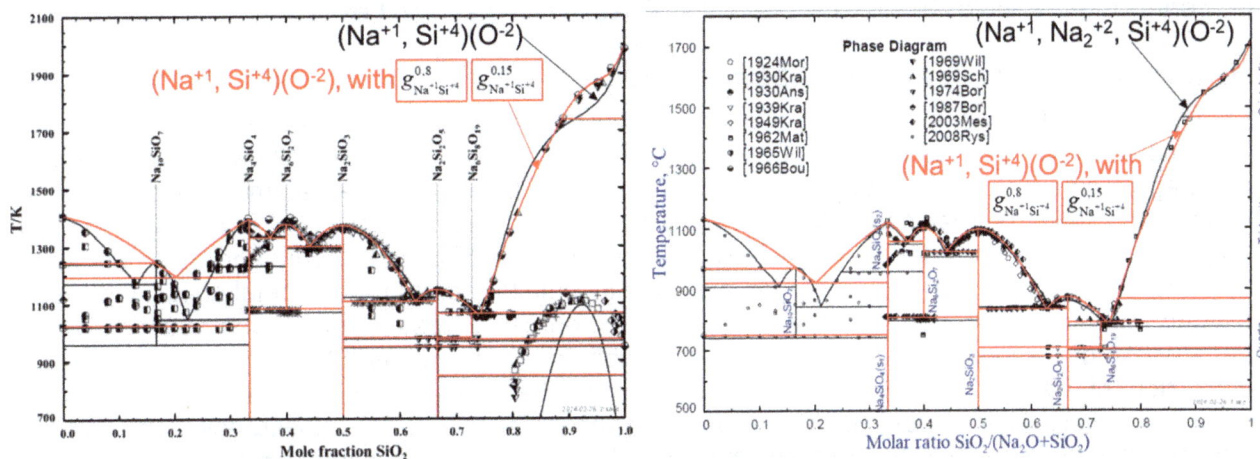

FIG 5 – Phase diagrams of the Na$_2$O-SiO$_2$ assessed using different models within the MQF – black lines are earlier works (Lambotte and Chartrand, 2011) and (Nekhoroshev, 2019), red lines are PYROSEARCH model results.

liquid matte/metal/speiss

Apart from differences in chemistry, the choice to treat liquid matte/metal/speiss as a separate model from slag and salt comes from the technical limitation of FactSage™. For the equilibrium calculations, a maximum of three immiscibilities within the same solution model is allowed, which is indicated by using the J-option in the user interface. In theory, more immiscible liquids of the same solution can be included in the calculation by creating copies of the same solution with different names and slightly different properties, but stability and speed of such calculations is questionable. Even three-phase equilibria in the multicomponent system pushes FactSage™ to the limit and sometimes gives incorrect results. A generalised model for slags, mattes, metals and speiss within the MQF is theoretically possible but not practical. We observe slag/matte/metal/speiss equilibrium in experimental results (see Figure 1, right). It would not be possible to reproduce this in the calculation, if all these phases were described using a single solution.

Sublattices and species

The first model for mattes within the MQF was developed in 1990s (Dessureault and Pelton, 1993; Decterov, Dessureault and Pelton, 2000; Kongoli, Dessureault and Pelton, 1998). One-sublattice approach was used, with the main goal of describing strong SRO between metals and sulfur. The species were not charged. In theory, this approach allows the description of liquids within the complete range of composition from metals to sulfides and to elemental sulfur. Still, in these publications, mattes were artificially separated from molten copper and from molten lead metal solutions to benefit from existing models for these metallic solutions. The one-sublattice approach was expanded by Waldner and Pelton (2004a, 2004b, 2005) and Waldner and Sitte (2011). Metallic copper and matte were merged and included in FactSage™ software as FTmisc database (FTsulf since version 8.2), but the results of this work were not published until much later (Waldner, 2020, 2022).

Multiple coordination numbers for elements

The concept of treating metals and mattes within the same solution was accepted by the developers of the PYROSEARCH model. When expanding the database towards non-ferrous applications, it was also necessary to describe the solubility of oxygen in metals and mattes. The selected model within the MQF was also applicable to oxide liquids (see Figure 3 left) in cases when Second Nearest Neighbour SRO was not that important, ie for oxide liquids without SiO_2. The Cu-O and Fe-O systems had two compositions of SRO each, which was solved by introducing extra species intro the model. They had different thermodynamic properties and coordination numbers, ie Cu^I and Cu^{II}, Fe^{II} and Fe^{III} (Shishin and Decterov, 2012; Hidayat *et al*, 2015; Shishin *et al*, 2013). The notations ^{I}, ^{II} and ^{III} have the meaning of valency with O^{II} and S^{II} and determined by the ratio of coordination numbers. A drawback of this approach was that thermodynamic properties of pure liquid O^{II}, Cu^{II}, Fe^{III} were ill-defined, since they don't exist.

Heat capacity term in excess Gibbs energy

Within the selected model, the heat capacity of liquids at the compositions of maximum SRO, for instance at Cu_2S, is an additive function of endmembers, ie liquid Cu^I and S^{II}, which is not a very good approximation from the physical point of view. The bonds between atoms give a significant contribution towards heat capacity. Liquid copper has metallic bonds, Cu_2S is believed to have large proportion of ionic bonds, while S should retain much of covalent bonds in the liquid state. Still, uncertainties in heat capacities are significantly smaller compared to metal-sulfur and metal-oxygen interactions, so very good description of experimental results have been obtained by the introduction of composition-dependent excess Gibbs energy functions. Still, in the case of Cu_2S, a correction of heat capacity may be required, since the only experimental work on this topic indicates lower value (Figure 6). The problem of heat capacity was not addressed in a recent publication (Waldner, 2020). Typically, excess Gibbs energy dependence on temperature (T) is expressed as a + bT, but additional $cT\ln T$ term should be used in this case. Of course, such a correction would initiate the re-assessment of all parameters in systems related to the Cu-S, ie Cu-As-S (Prostakova, Shishin and Jak, 2021), Pb-Cu-S (Shishin, Chen and Jak, 2020) and dozens more. Optimistically this may help resolving some of the accuracy issues explained in sections below.

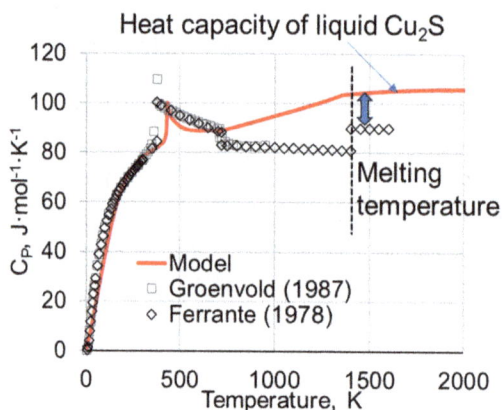

FIG 6 – Heat capacity of Cu_2S. Symbols are experimental data (Groenvold and Westrum, 1987; Ferrante *et al*, 1978). Line is calculated using the PYROSEARCH model for the liquid phase at the composition of Cu_2S. Double arrow indicates that heat capacity of liquid should be corrected using the $cT\ln T$ term in excess Gibbs energy.

Figure 6 shows another important concept. The reasonable function is used for liquid Cu_2S for temperatures below melting and down to 0 K. Large portions of FactSage™ public database do not have reasonable heat capacity functions far below melting temperatures, which is compensated by excess Gibbs energy functions. Further improvements in the accuracy of predictions demonstrated the need to revise these functions. In the CALPHAD community, the need to invest resources into re-assessments of the systems due to corrections in heat capacity below 298 K is often justified by using the term 'third generation' CALPHAD database, with publications appearing for single-component (He *et al*, 2022) and binary-system re-assessments. In some cases the use of this highly publicized term results in the publication of low-quality assessments (Abdul *et al*, 2023).

Lack of direct parameters between sulfides

The analysis of slag/matte equilibria with the Cu-Fe-O-S-Si and higher order systems (Shishin *et al*, 2018a, 2018b) indicates possible need for the re-assessment of the Cu-Fe-S system. The predicted $P(O_2)$ versus wt per cent Cu in matte, a systematic deviation of about +0.15 in $Log_{10}[P(O_2)$, atm] is observed, when compared to experimental data obtained using the equilibration with the flow of $CO-CO_2-SO_2-Ar$ gas. No apparent reasons for this deviation can be found from the analysis of existing data within the Cu-Fe-S (matte) (Waldner, 2022) or Cu-Fe-O-Si (slag) (Hidayat *et al*, 2017a) subsystems. The data in the Cu-Fe-O-S system (Shishin, Jak and Decterov, 2015) is somewhat scattered and no systematic deviation is observed. So, the slag/matte data indirectly indicates the systematic deviation of the activity 'FeS' according to the reaction:

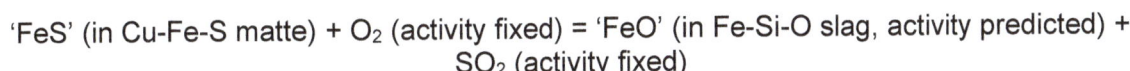

'FeS' (in Cu-Fe-S matte) + O_2 (activity fixed) = 'FeO' (in Fe-Si-O slag, activity predicted) + SO_2 (activity fixed)

This assumes the achievement of equilibrium and correct calibration of the gas flow in these complicated experiments.

Another case is the Cu-Ni-S system, where the data on the tie-lines between metallic phase and matte could not be successfully re-conciliated with the matte-digenite phase diagram data by Walder or in the PYROSEARCH database.

The current working theory is that accumulated deviation of Gibbs energy exist for the Cu_2S (as in Figure 6), or a misbalance in Gibbs energy between Cu_2S and other sulfides, ie FeS or NiS. In the case of Slag, these types of inconsistencies can be compensated by binary parameters between end-members. In one-sublattice model for mattes, ternary parameters must be used. As shown in Figure 7, main ternary parameters do not act along the line of maximum SRO, which makes it very hard to compensate issues in certain areas of the diagram, without affecting other areas. It is tempting to introduce the associates inside the MQF, such as Cu_2S, FeS, possibly NiS to get access more excess parameters. It would be possible to describe the existing binary system data and very likely, ternary data due to larger and more flexible set of available binary parameters. It is hard though, to predict how such model would behave in a system with many elements. Logically, Cu_2O, FeO etc associates would need to be introduced as well. Computational time will increase significantly, since FactSage™ still needs to calculate the bond fractions between all species on the sublattice, such as $Cu-Cu_2S$, Cu_2S-FeO etc.

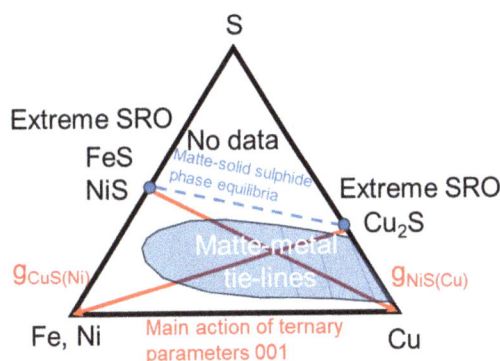

FIG 7 – The problem of ternary parameters in one-sublattice model for liquid matte/metal/speiss. No direct parameters along the FeS-Cu_2S and NiS-Cu_2S lines.

Geometric interpolation

Within the current model, the obvious interpolation method in binary parameters into higher-order space is Kohler-Toop with sulfur being an asymmetric component. With the expansion of the database towards As, Sb and Sn, the strong interaction between copper, iron and these metalloids (Shishin *et al*, 2019, 2023) was described using the same model within the QMF. Same model, but slightly different parameters were used in parallel studies by (Kidari and Chartrand, 2023b, 2023a). Unlike the sulfide systems, where barely any data exist for the sulfur-rich region beyond the composition of maximum SRO, the regions AsS-Cu_2S-S and AsS-Ag_2S-S were relatively well studied experimentally. In these systems, the choice of geometric interpolation is not obvious. For the Ag-As-S, silver was selected as an asymmetric component (Kidari and Chartrand, 2023a). It is worth mentioning, that no liquid immiscibility experimental data within the Ag-Ag$_2$S-Ag$_3$As area of the diagram were available, even though the gap was predicted by the model. In the case of Cu-As-S, Muggianu-type of interpolation provided reasonable results, even though the mathematical expressions of the Muggianu-type interpolation of excess parameters written in terms of pair fraction may be incorrect in FactSage™ software. It was very challenging to describe experimental data on the immiscibility within the Cu-Cu$_2$S-Cu$_3$As together with other areas of the diagram, indicating that different types of interpolation techniques may be required for different areas of the diagram. This can only be achieved by introducing the associates within the existing MQF. While helping some systems, choice of geometric interpolations between associates and all other components can be a daunting task.

Salt

The choice of the model for salt was largely inspired by the works of Coursol et al (2005a, 2005b). It was decided not to merge slag and salt within the same solution but introduce the O^{-2} anions into the salt model (see Table 2). Antimony and arsenic in the +5 oxidation state will in all probability dissolve in sulfate liquids, forming antimonates and arsenates. Sulfate liquids have been demonstrated to remove As and Sb from copper metal (Coursol et al, 2004) and these studies are of interest for lead refining. Formation of liquid sulfates should be controlled, as it accelerates the hot corrosion of furnace components during the lead smelting and may also cause the formation of dust accretions in off-gas pathways. A possible further expansion is the introduction of SbO$_4^{3-}$, AsO$_4^{3-}$ anions.

CONCLUSIONS

This review demonstrates the versatility of the Modified Quasichemical Formalism (MQF) for developing various molten phase models. The choice of the model is a balance between the predicting power of and the flexibility to provide the accurate description of reality through the optimisation of model parameters. When the focus is on multicomponent solutions, the developments in experimental techniques and generation of abundant and accurate experimental become critical. The data generated in 2014–2024 pushed the existing models to the limit in terms of accuracy and challenged some decisions that were made at the early stages of the development.

Examples of long standing issues that have been resolved recently by introducing unconventional model choices into MQF, including:

- combining Bragg-Williams and Quasichemical negative contributions into the slag model for the CaO-SiO$_2$, which made the tip at the enthalpy of mixing less sharp
- introducing the excess Gibbs energy parameters with powers of > 9 to describe non-ionic behaviour of slags near pure SiO$_2$.

An example of issue that has not been resolved yet is the inability to reconcile the activity, distribution of elements and phase equilibria for several Cu-S-containing systems. It is possible that the choice of one-sublattice approach without associates for liquid matte/metal/speiss limits the model flexibility and results in:

- incorrect heat capacity at the composition of maximum SRO requiring systematic introduction of the $T\ln T$ term in excess Gibbs energies
- inability to introduce direct excess binary parameters between two sulfides at the compositions of maximum SRO
- inability to create different geometric interpolation techniques for different areas of the ternary system X-Cu-S.

The issues and potential solutions often originate in systems of two or three components, but become evident only in multicomponent systems. For the established multicomponent thermodynamic database, any changes in low-order systems require significant efforts and iterative re-assessment of model parameters in hundreds of related higher-order systems. This phenomenon is often compared in the CALPHAD community to re-building the foundation of the inverted pyramid. To perform re-assessments effectively, large data sets of hundreds of thousands experimental points and tools for their effective management must be developed. Newly obtained experimental information should trigger the semi-automated re-assessment. In this way, alternative model approaches potentially providing further improvements can be tested in timely manner.

ACKNOWLEDGEMENTS

The research funding and technical support is provided by the consortium of copper and lead producing companies: Anglo American Platinum (South Africa), Aurubis AG (Germany), BHP Billiton Olympic Dam Operation (Australia), Boliden (Sweden), Glencore Technology (Australia), Metso Oy (Finland), Peñoles (Mexico), RHI Magnesita (Austria), Rio Tinto Kennecott (USA) and Umicore NV (Belgium), as well Australian Research Council Linkage program LP190101020 'Future copper metallurgy for the age of e-mobility and the circular economy'. Present study would not be possible without the facilities and technical assistance of the staff of the Australian Microscopy and Microanalysis Research Facility at the Centre for Microscopy and Microanalysis in The University of Queensland.

REFERENCES

Abdul, W, Mawalala, C, Pisch, A and Bannerman, M N, 2023. CaO-SiO$_2$ assessment using 3rd generation CALPHAD models, *Cem Concr Res*, 173:107309.

Andersson, J, Helander, T, Höglund, L, Shi, P and Sundman, B, 2002. Thermo-Calc and DICTRA, Computational tools for materials science, *CALPHAD*, 26:273–312.

Bale, C W, Belisle, E, Chartrand, P, Decterov, S A, Eriksson, G, Gheribi, A E, Hack, K, Jung, I H, Kang, Y B, Melancon, J, Pelton, A D, Petersen, S, Robelin, C, Sangster, J, Spencer, P and Van Ende, M A, 2016. FactSage thermochemical software and databases, 2010–2016. *CALPHAD*, 54:35–53.

Castillo-Sanchez, J-R, Oishi, K, St-Germain, L, Ait-Amer, D and Harvey, J-P, 2023. The power of computational thermochemistry in high-temperature process design and optimization: Part 1 – Unit operations, *CALPHAD*, 82:102593.

Chartrand, P and Pelton, A D, 2001. The Modified Quasichemical Model III – Two Sublattices, *Metall Mater Trans A*, 32:1397–1407.

Cheng, S, Shevchenko, M, Hayes, P C and Jak, E, 2019. Experimental Phase Equilibria Studies in the FeO-Fe$_2$O$_3$-CaO-SiO$_2$ System in Air: Results for the Iron-Rich Region, *Metall Mater Trans B*, 51:1587–1602.

Cheng, S, Shevchenko, M, Hayes, P C and Jak, E, 2021. Experimental Phase Equilibria Studies in the FeO-Fe$_2$O$_3$-CaO-SiO$_2$ System and the Subsystems CaO-SiO$_2$, FeO-Fe$_2$O$_3$-SiO$_2$ in Air, *Metall Mater Trans B*, 52:1891–1914.

Coursol, P, Pelton, A D, Chartrand, P and Zamalloa, M, 2005a. The CaSO$_4$-CaO-Ca$_3$(AsO$_4$)$_2$ phase diagram, *Can Metall Q*, 44:547–553.

Coursol, P, Pelton, A D, Chartrand, P and Zamalloa, M, 2005b. The CaSO$_4$-Na$_2$SO$_4$-CaO phase diagram, *Can Metall Q*, 44:537–545.

Coursol, P, Stubina, N, Carissimi, E, Zamalloa, M and Mackey, P J, 2004. Industrial development of a novel sulfate system for copper pyro-refining, *JOM*, 56:41–45.

Decterov, S A and Pelton, A D, 1996. Critical Evaluation and Optimization of the Thermodynamic Properties and Phase Diagrams of the CrO-Cr$_2$O$_3$-SiO$_2$ and CrO-Cr$_2$O$_3$-SiO$_2$-Al$_2$O$_3$ Systems, *J Phase Equilib*, 17:488–494.

Decterov, S A, 2018. Thermodynamic database for multicomponent oxide systems, *Chimica Techno Acta*, 5:16–48.

Decterov, S A, Dessureault, Y and Pelton, A D, 2000. Thermodynamic Modeling of Zinc Distribution among Matte, Slag and Liquid Copper, *Can Metall Q*, 39:43–54.

Dessureault, Y and Pelton, A D, 1993. An optimized thermodynamic database for matte, slag, speiss and metal phases in lead smelting, *Can Inst Min Metall Pet*, 143–151.

Dilner, D and Selleby, M, 2017. Thermodynamic description of the Fe-Ca-O-S system, *CALPHAD*, 57:118–125.

Ferrante, M J, Stuve, J M, Daut, G E and Pankratz, L B, 1978. Low-temperature heat capacities and high-temperature enthalpies of cuprous and cupric sulfides, Albany Metallurgy Research Centre, Bureau of Mines, Albany, Oregon, USA.

Gorsse, S and Senkov, O N, 2018. About the reliability of CALPHAD predictions in multicomponent systems, *Entropy*, 20:899.

Groenvold, F and Westrum Jr, E F, 1987. Thermodynamics of copper sulfides, I, Heat capacity and thermodynamic properties of copper(I) sulfide, Cu_2S, from 5 to 950 K, *J Chem Thermodyn*, 19:1183–1198.

He, Z, Haglöf, F, Chen, Q, Blomqvist, A and Selleby, M, 2022. A Third Generation Calphad Description of Fe: Revisions of Fcc, Hcp and Liquid, *Journal of Phase Equilibria and Diffusion*, 43:287–303.

Hidayat, T and Jak, E, 2014. Thermodynamic modeling of the "Cu_2O"-SiO_2, "Cu_2O"-CaO and "Cu_2O"-CaO-SiO_2 systems in equilibrium with metallic copper, *Int J Mater Res*, 105:249–257.

Hidayat, T, Fallah-Mehrjardi, A, Abdeyazdan, H, Shishin, D, Shevchenko, M, Hayes, P C and Jak, E, 2023. Integrated experimental and thermodynamic modelling study in the Pb-Fe-O-S-Si system: Effect of Temperature and p(SO2) on Slag-Matte-Metal-Tridymite Equilibria, *Metall Mater Trans B*, 54:536–549.

Hidayat, T, Shishin, D, Decterov, S A and Jak, E, 2016a. Critical thermodynamic re-evaluation and re-optimization of the CaO-FeO-Fe_2O_3-SiO_2 system, *CALPHAD*, 56:58–71.

Hidayat, T, Shishin, D, Decterov, S A and Jak, E, 2016b. Thermodynamic optimization of the Ca-Fe-O system, *Metall Trans B*, 47:256–281.

Hidayat, T, Shishin, D, Decterov, S A and Jak, E, 2017b. Experimental Study and Thermodynamic Re-evaluation of the FeO-Fe_2O_3-SiO_2 System, *J Phase Equilib Diffus*, 38:477–492.

Hidayat, T, Shishin, D, Decterov, S and Jak, E, 2017a. Critical assessment and thermodynamic modeling of the Cu-Fe-O-Si system, *CALPHAD*, 58:101–114.

Hidayat, T, Shishin, D, Jak, E and Decterov, S, 2015. Thermodynamic Reevaluation of the Fe-O System, *CALPHAD*, 48:131–144.

Hillert, M, 2001. The Compound Energy Formalism, *J Alloys Compd*, 320:161–176.

Hillert, M, Kjellqvist, L, Mao, H, Selleby, M and Sundman, B, 2009. Parameters in the compound energy formalism for ionic systems, *CALPHAD*, 33:227–232.

Hillert, M, Sundman, B and Wang, X, 1990. An assessment of the calcia-silica system, *Metall Trans B*, 21:303–312.

Jafarian, M, Arjomandi, M and Nathan, G J, 2017. Thermodynamic potential of molten copper oxide for high temperature solar energy storage and oxygen production, *Appl Energy*, 201:69–83.

Jak, E, Decterov, S A, Wu, P, Hayes, P C and Pelton, A D, 1997. Thermodynamic Optimisation of the Systems PbO-SiO_2, PbO-ZnO, ZnO-SiO_2 and PbO-ZnO-SiO_2, *Metall Mater Trans B*, 28B:1011–1018.

Jak, E, Hidayat, T, Prostakova, V, Shishin, D, Shevchenko, M and Hayes, P C, 2019. Integrated experimental and thermodynamic modelling research for primary and recycling pyrometallurgy, *EMC* Düsseldorf, Germany: GDMB Verlag.

Jak, E, Zhao, B and Hayes, P C, 1997. Experimental determination of phase equilibria in lead/zinc smelting slags and sinters, in *Proceedings of the 5th International Conference on Molten Slags Fluxes and Salts (MOLTEN '97)*, pp 719–726.

Jantzen, T, Yazhenskikh, E, Hack, K, Baben, M T, Wu, G and Mueller, M, 2021. Addition of V_2O_5 and V_2O_3 to the CaO-FeO-Fe_2O_3-MgO-SiO_2 database for vanadium distribution and viscosity calculations, *CALPHAD*, 74:102284.

Jo, Y, Lee, H-G and Kang, Y-B, 2013. Thermodynamics of the MnO-FeO-MnS-FeS-SiO_2 system at SiO_2 saturation under reducing condition: immiscibility in the liquid phase, *ISIJ Int*, 53:751–760.

Jung, I-H and Van Ende, M-A, 2020. Computational Thermodynamic Calculations: FactSage from CALPHAD Thermodynamic Database to Virtual Process Simulation, *Metall Mater Trans B*, 51:1851–1874.

Jung, I-H, Decterov, S A and Pelton, A D, 2007. Thermodynamic modeling of the CoO-SiO_2 and CoO-FeO-Fe_2O_3-SiO_2 systems, *Int J Mater Res*, 98:816–825.

Kang, Y-B and Pelton, A, 2009. Thermodynamic Model and Database for Sulfides Dissolved in Molten Oxide Slags, *Metall Mater Trans B*, 40:979–994.

Khartcyzov, G, Shevchenko, M, Cheng, S, Hayes, P C and Jak, E, 2022. Experimental phase equilibria studies in the "$CuO_{0.5}$"-CaO-SiO_2 ternary system in equilibrium with metallic copper, *Ceram Int*, 48:9927–9938.

Kidari, O and Chartrand, P, 2023a. Thermodynamic Evaluation and Optimization of the Ag-As-S system, *J Phase Equilib Diffus*, 44:269–299.

Kidari, O and Chartrand, P, 2023b. Thermodynamic evaluation and optimization of the As-Co, As-Fe and As-Fe-S systems, *CALPHAD*, 82:102589.

Konar, B, Van Ende, M-A and Jung, I-H, 2017. Critical evaluation and thermodynamic optimization of the Li-O and Li_2O-SiO_2 systems, *J Eur Ceram Soc*, 37:2189–2207.

Kongoli, F, Dessureault, Y and Pelton, A D, 1998. Thermodynamic Modeling of Liquid Fe-Ni-Cu-Co-S Mattes, *Metall Mater Trans B*, 29B:591–601.

Lambotte, G and Chartrand, P, 2011. Thermodynamic optimization of the (Na_2O + SiO_2 + NaF + SiF_4) reciprocal system using the Modified Quasichemical Model in the Quadruplet Approximation, *J Chem Thermodyn*, 43:1678–1699.

Lambotte, G and Chartrand, P, 2013. Thermodynamic modeling of the (Al_2O_3 + Na_2O), (Al_2O_3 + Na_2O + SiO_2) and (Al_2O_3 + Na_2O + AlF_3 + NaF) systems, *J Chem Thermodyn*, 57:306–334.

Nekhoroshev, E, 2019. Thermodynamic Optimization Of The Na_2O-K_2O-Al_2O_3-CaO-MgO-B_2O_3-SiO_2 System, PhD thesis, Polytechnique Montréal.

Pelton, A D and Blander, M, 1984. Computer-Assisted Analysis of the Thermodynamic Properties and Phase Diagrams of Slags, in *Proceedings of the Second International Symposium on Metallurgical Slags and Fluxes* (eds: H A Fine and D R Gaskell), pp 281–294 (TMS-AIME: Warrendale).

Pelton, A D and Blander, M, 1986. Thermodynamic Analysis of Ordered Liquid Solutions by a Modified Quasi-Chemical Approach, Application to Silicate Slags, *Metall Trans B*, 17B:805–815.

Pelton, A D and Chartrand, P, 2001. The Modified Quasichemical Model, II – Multicomponent Solutions, *Metall Mater Trans A*, 32:1355–1360.

Pelton, A D and Kang, Y-B, 2007. Modeling short-range ordering in solutions, *Int J Mater Res*, 98:907–917.

Pelton, A D, Chartrand, P and Eriksson, G, 2001. The modified Quasichemical Model, IV – Two Sublattice Quadruplet Approximation, *Metall Mater Trans A*, 32:1409–1415.

Pelton, A D, Decterov, S A, Eriksson, G, Robelin, C and Dessureault, Y, 2000. The Modified Quasichemical Model, I – Binary Solutions, *Metall Mater Trans B*, 31:651–659.

Prostakova, V, Chen, J, Jak, E and Decterov, S A, 2012. Experimental study and thermodynamic optimization of the CaO-NiO, MgO-NiO and NiO-SiO_2 systems, *CALPHAD*, 37:1–10.

Prostakova, V, Chen, J, Jak, E and Decterov, S, 2013. Experimental study and thermodynamic optimization of the MgO-NiO-SiO_2 system, *J Chem Thermodyn*, 62:43–55.

Prostakova, V, Shishin, D and Jak, E, 2021. Thermodynamic optimization of the Cu-As-S system, *CALPHAD*, 72:102247.

Selleby, M and Sundman, B, 1996. A Reassessment of the Ca-Fe-O System, *CALPHAD*, 20:381–392.

Selleby, M, 1997. An Assessment of the Ca-Fe-O-Si System, *Metall Trans B*, 28B:577–596.

Shevchenko, M, Shishin, D and Jak, E, 2022. Integrated phase equilibria experimental study and thermodynamic modeling of the Cr-Si-O, Fe-Cr-O and Fe-Cr-Si-O systems, *Ceram Int*, 48(22):33418–33439. https://doi.org/10.1016/j.ceramint.2022.07.286

Shevchenko, M, Shishin, D and Jak, E, 2024. Advancement in experimental methodologies to produce phase equilibria and thermodynamic data in multicomponent systems, in *Proceedings of the 12th International Conference on Molten Slags Fluxes and Salts (MOLTEN 2024)*, pp 531–558 (The Australasian Institute of Mining and Metallurgy: Melbourne).

Shishin, D and Decterov, S A, 2012. Critical assessment and thermodynamic modeling of Cu-O and Cu-O-S systems, *CALPHAD*, 38:59–70.

Shishin, D and Jak, E, 2018. Critical assessment and thermodynamic modeling of the Cu-As system, *CALPHAD*, 60:134–143.

Shishin, D, Chen, J and Jak, E, 2020. Thermodynamic Modeling of the Pb-S and Cu-Pb-S Systems with Focus on Lead Refining Conditions, *J Phase Equilib Diffus*, 41:218–233.

Shishin, D, Chen, J, Hidayat, T and Jak, E, 2019. Thermodynamic modelling of the Pb-As and Cu-Pb-As systems supported by experimental study, *J Phase Equilib Diff*, 40:758–767.

Shishin, D, Fallah Mehrjardi, A, Shevchenko, M, Hidayat, T and Jak, E, 2020. Experimental study, thermodynamic calculations and industrial implications of slag/matte/metal equilibria in the Cu-Pb-Fe-O-S-Si system, *J Mater Res Technol*, 19:899–912. https://doi.org/10.1016/j.jmrt.2022.05.058

Shishin, D, Hidayat, T, Chen, J, Hayes, P C and Jak, E, 2018a. Experimental Investigation and Thermodynamic Modelling of the Distributions of Ag and Au between Slag, Matte and Metal in the Cu-Fe-O-S-Si System, *Journal of Sustainable Metallurgy*, 5:240–249.

Shishin, D, Hidayat, T, Chen, J, Hayes, P C and Jak, E, 2018b. Integrated Experimental Study and Thermodynamic Modelling of the Distribution of Arsenic between Phases in the Cu-Fe-O-S-Si System, *J Chem Thermodyn*, 135:175–182.

Shishin, D, Hidayat, T, Jak, E and Decterov, S, 2013. Critical assessment and thermodynamic modeling of Cu-Fe-O system, *CALPHAD*, 41:160–179.

Shishin, D, Jak, E and Decterov, S A, 2015. Thermodynamic assessment and database for the Cu–Fe–O–S system, *CALPHAD*, 50:144–160.

Shishin, D, Shevchenko, M, Starykh, R, Sineva, S, Prostakova, V and Jak, E, 2023. Development and application of matte/speiss/metal thermodynamic database for optimization of processing of drosses, dusts and reverts from lead, zinc and copper production, in *Proceedings of the 10th International Conference on Lead and Zinc Processing (Lead-Zinc 2023)*, Journal of Physics: Conference Series, 2738:012032. https://doi.org/10.1088/1742-6596/2738/1/012032.

Sineva, S, Shishin, D, Prostakova, V, Fallah-Mehrjardi, A and Jak, E, 2023. Experimental Study and Thermodynamic Analysis of High-temperature Processing of mixed Ni-Cu sulfide feed, in *Proceedings of the 62nd Annual Conference of Metallurgists (COM 2023)*, pp 779–787. https://doi.org/10.1007/978-3-031-38141-6_102

Sundman, B, 1991. Modification of the Two-sublattice Model for Liquids, *CALPHAD*, 15:109–119.

Sundström, A, Eksteen, J J and Georgalli, G A, 2008. A review of the physical properties of base metal mattes, *Journal of The South African Institute of Mining and Metallurgy*, 108:431–448.

Waldner, P and Pelton, A D, 2004a. Critical Thermodynamic Assessment and Modeling of the Fe-Ni-S System, *Metall Mater Trans B*, 35:897–907.

Waldner, P and Pelton, A D, 2004b. Thermodynamic modeling of the Ni-S system, *Z Metallkd*, 95:672–681.

Waldner, P and Pelton, A D, 2005. Thermodynamic modeling of the Fe-S system, *J Phase Equilib Diff*, 26:23–28.

Waldner, P and Sitte, W, 2011. Thermodynamic modeling of the Cr-S system, *Int J Mater Res*, 102:1216–1225.

Waldner, P, 2020. Gibbs Energy Modeling of the Cu-S Liquid Phase: Completion of the Thermodynamic Calculation of the Cu-S System, *Metall Mater Trans B*, 51:805–817. https://doi.org/10.1007/s11663-020-01796-x

Waldner, P, 2022. The High-Temperature Cu-Fe-S System: Thermodynamic Analysis and Prediction of the Liquid–Solid Phase Range, *J Phase Equilib Diff*, 43:495–510. https://doi.org/10.1007/s11669-022-00988-z

Wang, K and Chartrand, P, 2021. Generic energy formalism for reciprocal quadruplets within the two-sublattice quasichemical model, *CALPHAD*, 74:102293.

Wang, K, Li, D, Zou, X, Cheng, H, Li, C, Lu, X and Chou, K, 2023. Generic bond energy formalism within the modified quasichemical model for ternary solutions, *J Mol Liq*, 370:120932.

Wen, X, Shevchenko, M and Jak, E, 2021. Experimental study of "$CuO_{0.5}$"-"FeO"-SiO_2 and "FeO"-SiO_2 systems in equilibrium with metal at 1400–1680°C, *J Alloys Compd*, 885:160853.

Wen, X, Shevchenko, M, Nekhoroshev, E and Jak, E, 2023. Phase equilibria and thermodynamic modelling of the PbO–ZnO-"$CuO_{0.5}$"-SiO_2 system, *Ceram Int*, 49(14)A:23817–23834. https://doi.org/10.1016/j.ceramint.2023.04.223

Wu, P, Eriksson, G and Pelton, A D, 1993. Critical Evaluation and Optimization of the Thermodynamic Properties and Phase Diagrams of the Calcia-Iron(II) Oxide, Calcia-Magnesia, Calcia-Manganese(II) Oxide, Iron(II) Oxide-Magnesia, Iron(II) Oxide-Manganese(II) Oxide and Magnesia-Manganese(II) Oxide Systems, *J Am Ceram Soc*, 76:2065–2075.

Wu, P, Eriksson, G, Pelton, A D and Blander, M, 1993. Prediction of the Thermodynamic Properties and Phase Diagrams of Silicate Systems – Evaluation of the FeO-MgO-SiO_2 System, *ISIJ Int*, 33:26–35.

Prognostic models for electroslag remelting process and slag engineering

G Stovpchenko[1], L Medovar[2], L Lisova[3], D Stepanenko[4] and D Togobitskaya[5]

1. Chief Foreign Scientist, Tianjin Heavy Industries Research and Development Co, Ltd, Tianjin 300457, China; EO Paton Electric Welding Institute of National Academy of Sciences of Ukraine (NASU), Kyiv 03150, Ukraine. Email: anna_stovpchenko@ukr.net; stovpchenko@cfhi.com
2. Chief Foreign Scientist, Tianjin Heavy Industries Research and Development Co, Ltd, Tianjin 300457, China; EO Paton Electric Welding Institute of NASU, Kyiv 03150, Ukraine. Email: medovar@ukr.net; medovar@cfhi.com
3. Research Associate, Chair of Metal Forming and Casting, School of Engineering and Design, Technical University of Munich, Munich 80333, Germany; EO Paton Electric Welding Institute of NASU, Kyiv 03150, Ukraine. Email: llisova@ukr.net
4. Head of the Department of Physicochemical Problems of Metallurgical Processes, ZI Nekrasov Iron and Steel Institute NASU, Dnipro 49107, Ukraine. Email: d.gorodenskiy@gmail.com
5. Leading Researcher of Physicochemical Problems of Metallurgical Processes, ZI Nekrasov Iron and Steel Institute NASU, Dnipro 49107, Ukraine. Email: dntog@ukr.net

ABSTRACT

Classical electroslag remelting of consumable electrodes is a drop-by-drop melt feeding through a slag layer into a renewing molten metal bath, slowly solidifying into a homogeneous, dense, defect-free ingot in a copper water-cooled mould. The new look at the additive nature of the electroslag remelting (ESR) process in a protective atmosphere allowed us to formulate the principles of a comprehensive thermodynamic-based model that can predict dynamic changes in slag and metal composition at certain ingot remelting. The model considers drastically different slag-to-metal mass ratios at the beginning and end of remelting and predicts gas, slag and metal composition in a chain of thermodynamic subsystems. Despite consisting of calcium fluoride and stable oxides, the ESR slag can oxidise main and active elements from steel and alloys (primarily aluminium, titanium and silicon) due to chemical reactions between slag and metal. ESR is not an electrochemical process in its nature. However, slags are ionic melts and deviation in their composition causes a change in their properties, affecting both operation mode and ingot quality.

Another important understanding derived from the modern metallurgy technological route is that the ability to refine metal from impurities is not a priority for the ESR because consumable electrode has already passed all stages of refining and deep degassing at ladle treatment that changes ESR slags engineering principles. The critical importance became slag's ability to generate process heat and keep the melting composition in the metal bath unchanged (except for non-metallic inclusions assimilation). Slag engineering for ESR required a compromise between chemical inertness to a metal composition and desired physical properties deriving from technological reasons. The Directed Chemical Bonds Concept (DCBC) in a multicomponent oxide system was used to build predictive models of electric conductivity and the melting temperature of fluoride-oxide slag based on their chemical composition. Both models help to design a customised composition of effective slags for steel and alloy groups or individual grades, and they are significant steps in the development of a comprehensive model of electroslag remelting.

INTRODUCTION

In the technological chain of high-quality steel product manufacturing, electroslag remelting (ESR) takes one of the last but not least places. It refines the melt from impurities and inclusion and refines the ingot structure, increasing yield. That is exactly due to the homogeneity of the chemical composition and the dense dendritic structure over the cross-section and height of the ingot ESR metal, which is in demand for the most critical applications in modern industry (Medovar and Boyko, 2013; Hoyle, 1983).

The low rate of metal supply through the slag layer, which simultaneously performs the functions of an electric heater and thermal buffer for the process and refining media, guarantees a relatively small depth of the liquid metal bath. The progressively replaced composition of the liquid metal bath provides significantly less segregation of elements than in a conventional same-diameter ingot. The second important feature of the ESR ingot is its dense dendritic structure, no shrinkage and smooth side surface, which ensures a high product yield. As a result, the low rate of ESR ingot formation in copper water-cooled mould (tens to hundreds of times lower than traditional ingots and continuously cast billets) is a prerequisite of well-recognised worldwide 'ESR quality'.

The electrical conductivity of the slag bath determines the performance and efficiency of the melting process, while the slag melting temperature and interval determine metal overheat, liquid metal bath depth, and ingot surface formation pattern, which in turn are efficiency factors for metal refining and homogeneity of ESR ingot. Thus, ESR slag is an indispensable component of the process, and its composition and properties are crucial for metal quality, process efficiency and cost. Knowledge of slag's chemical and physical properties and their impact on the remelting process is key to achieving high quality and efficiency. This article aims to provide a clear understanding of the changes that occur in the physicochemical system during the remelting of consumable electrodes in a protective gas atmosphere using a single slag charged at the start of the process. It emphasises the significant differences in slag-metal interaction conditions between ESR ingot formation's start and end points. The article asserts that slag is a critical component for ensuring consistency and stability in the ESR process and demonstrates the efficacy of an atomistic approach in predicting its crucial physical properties.

CONCEPT OF GAS-SLAG-METAL SYSTEM CHARACTERISATION AT ESR REMELTING IN INERT GAS

Electroslag remelting is an arc-less process powered by the heat generated by the passage of electric current through the electric conductive slag bath (Figure 1(3)). When slag temperature exceeds the melting point of a metal, the consumable electrode begins to melt. Metal drops go down from the electrode tip through a slag layer and accumulate in the metal pool (Figure 1(4)), which gradually solidifies, forming an ingot (Figure 1(6)). The reacting system at the ESR in each instant moment consists of a film on the edge of the consumable electrode, a liquid slag bath with falling metal drops inside, and a liquid metal pool whose bottom end solidifies, forming an ingot (Medovar and Boyko, 2013, Stovpchenko et al, 2020a).

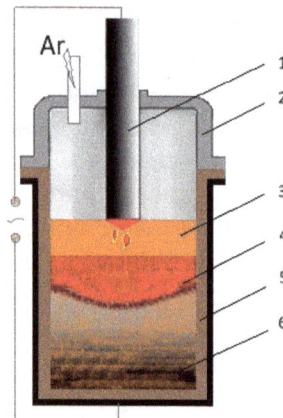

FIG 1 – ESR in argon protective atmosphere: (1) consumable electrode; (2) protective gas hood; (3) slag bath; (4) liquid metal pool; (5) mould; (6) solid ingot.

High temperatures of the melts and the developed surface of slag-metal interactions bring reactions closer to equilibrium (Mills, 2021; Hou et al, 2021; Duan et al, 2019). Nevertheless, specific features of gas-slag-metal system interactions at electroslag remelting must be outlined to clearly understand characteristic conditions that are a basis for further technological improvement.

Gas atmosphere contacts with slag only. Gaseous products of slag-metal reactions and evaporations from slag are removed in the furnace atmosphere (which is inert and slowly renewed). The gas atmosphere over the slag bath has no direct contact with liquid metal and does not participate in the mentioned interactions, but in the case when it contains oxygen, the electrode surface can be oxidised. Using an oxygen-free atmosphere (mostly argon or nitrogen, depending on steel grade) solves this problem by preventing the growth of oxygen activity in a slag and oxygen transfer to the metal caused by variable valence oxides (primarily iron oxides) involving atmospheric oxygen. For this reason, the closed hood that covers a whole electrode to keep the protective gas atmosphere became a benchmark in today's ESR plants. Incrementally solidifying in the bottom of a liquid metal pool, the ingot leaves the reaction zone (after interacting with slag). On the contrary, molten slag continues to participate in refining processes and assimilates slag-metal interaction products and non-metallic inclusions from the consumable electrode.

The most common ESR practice must envisage charging the entire slag mass at the start of the remelting process and no additions of premelted slag (or slag components) are used. Thus, it is crucial to understand that the slag-to-metal ratio changes throughout the entire remelting process, as all the products of the interactions between the slag and metal and most of the nonmetal inclusions from consumable electrodes are accumulated in the slag bath.

From the formal logic, it is understandable that a permanent change and drastic difference in the slag-metal mass ratio (Figure 2) causes radically different conditions for slag-metal interaction during electroslag remelting.

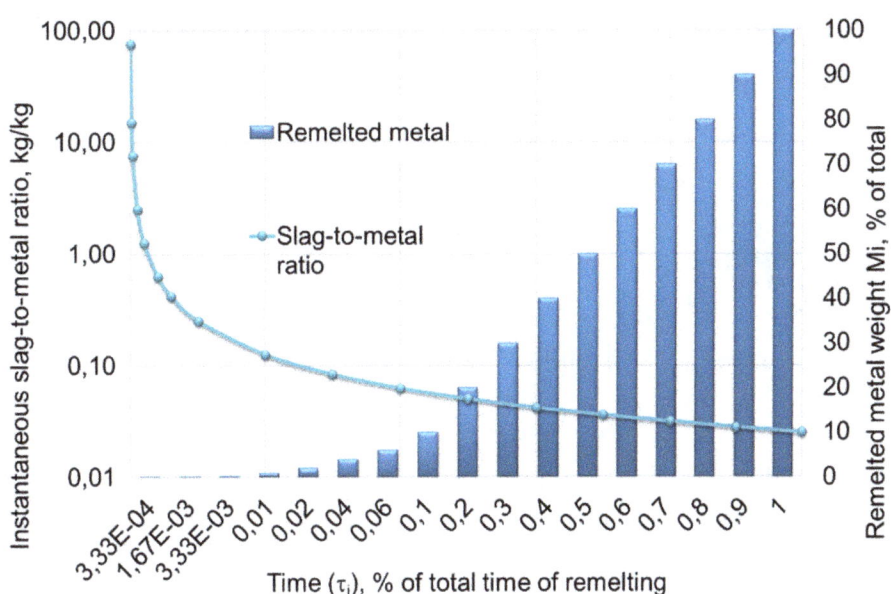

FIG 2 – Graphical representation of slag-metal interaction mass conditions at electroslag remelting (ESR) ingot formation (in an inert atmosphere) when a whole slag consumption makes 25 kg/ton of metal charged at the start.

The diagram summarises and demonstrates the characteristics of the physical and chemical processes that occur during ESR. The slag-to-metal ratio changes are shown for specific slag consumption of 25 kg/t of metal. However, the trend uncovered is general, regardless of its given value. Specific slag consumption usually makes 20–40 kg per ton, and exactly the last figure is usually used for thermodynamic calculations of the ESR, shoving just a very average result. However, due to the enormous slag-to-metal ratio in the first minutes of the remelting, the oxygen-bearing slag oxidates a metal the most at the beginning of the process. This effect cannot be neglected because the oxidative effect of slag remains tangible for quite a long time and can be a reason for ingot inhomogeneity along its height.

The physical properties of a slag itself are not less significant and they must be tested for their relevance to the requirements for an efficient ESR process. It is also important to find deeper dependencies between experimentally found values and chemical composition using atomistic approaches, which allows the prediction of molten slag physical properties and makes the choice of slags more deterministic. Directed Chemical Bonds Concept (DCBC) that was created by Professor E Prikhodko and continues to be developed by his followers at the ZI Nekrasov Institute of Ferrous

Metallurgy of the National Academy of Sciences of Ukraine (NASU) (Prikhodko, 1995a, 1995b; Prikhodko *et al*, 2013; Stepanenko, 2023; Stovpchenko *et al*, 2022) was chosen for these purposes.

METHODOLOGY OF RESEARCH

The first stage (Figure 3) of the evaluation procedure of slag-metal interaction is to construct the proposed 'quasi-dynamic' description (Figure 2) of the ESP process – a dynamic model of ESR in an inert atmosphere – in the form of a line of subsystems. This line of subsystems (masses of gas, slag and metal phases) for the predictive model of physicochemical interaction at ESR while consumable electrode remelting should be compiled using industrial data at certain ingot formation: typical slag consumption and rate of remelting. The input data is the mass of metal (mi) that is remelted to a certain point of time (τ_i), considering experimental data about the melting rate, the whole mass of slag and the volume of argon in the close chamber of the ESR plant. The proper time intervals (*i*), depending on the chosen purpose and number of calculations, should be determined, and for each conventional time, a subsystem of the whole masses of slag, gas and instantaneous mass of molten metal should be composed. For all subsystems, the remelted metal mass is only a variable (m_i); the primary mass of components of slag and gas phases are the same.

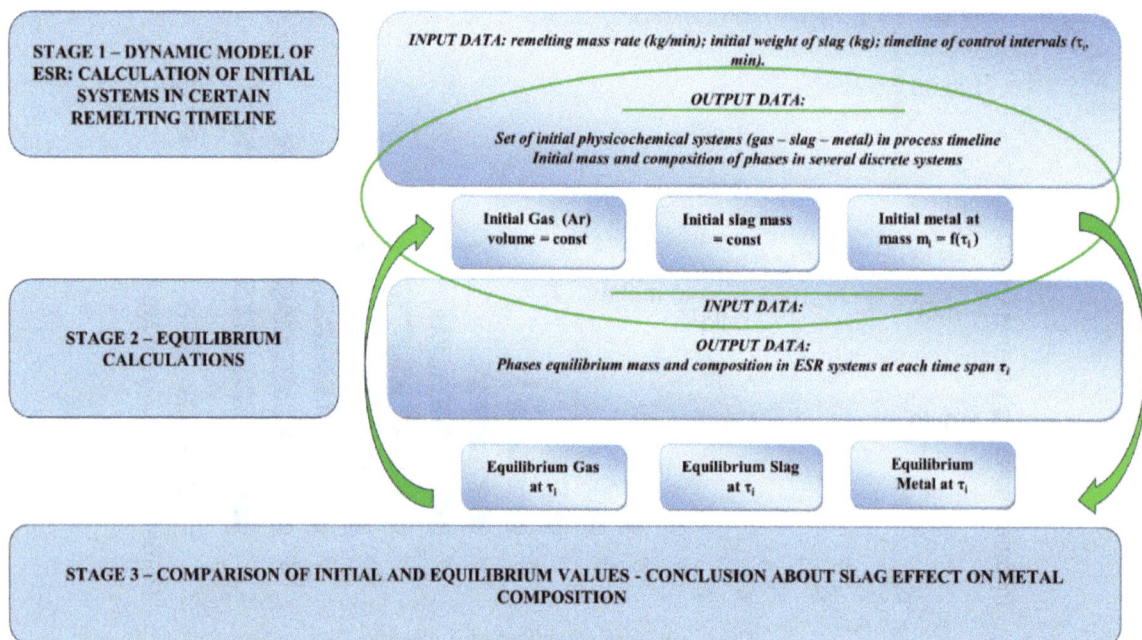

FIG 3 – The procedure of physicochemical modelling of the gas-slag-metal interaction in the inert atmosphere at the electroslag remelting (ESR) process course.

For each physicochemical subsystem, the masses and compositions of gas, slag, and metal phases at the chosen time of electroslag remelting should be calculated and used as initial data in software for equilibrium composition calculations. In the second modelling stage, thermodynamic calculations of the equilibrium state of the 'gas-slag-metal' system using CALPHAD Methodology (Lukas, Fries and Sundman, 2008; Sundman, 2016) implemented in HSC Chemistry, Pandat, ThermoCalc and similar possibility software can be used for this purpose. In the presented examples, thermodynamic calculations for the 'gas-slag-metal' system were made using HSC Chemistry software.

In the third stage, primary mass data of subsystems and equilibrium calculations are compared to recognise changes in phases' masses and composition. The overall analysis of received results allows for the finding of prospective slag compositions from a chemical point of view.

Changes in metal composition were calculated according to the proposed dynamic physicochemical model of ESR for comparative slag $34CaF_2/30Al_2O_3/27CaO/0.5SiO_2/2.5MgO/6TiO_2$ and Inconel 718 (Table 1) at 1873 K.

TABLE 1

The chemical composition of the Inconel 718 used in the calculation.

Elements content, wt%								
Ni	Cr	Fe	Nb	Mo	Ti	Al	C	B
53.7	18.2	18.2	5.4	3	1	0.5	0.03	0.004

The total system for dynamic ESR model demonstration comprised 300 kg of slag, 6.47 kg of argon atmosphere and a part of the total ingot mass (19 000 kg) of 1000 mm in diameter, corresponding to a certain time interval. This instantaneous metal mass and slag-to-metal ratio in a certain conditional time of remelting were found from the melting rate (14.2 kg/min), which was taken from the ESR practice of producing similar ingots. The main advantage of the proposed dynamic model is that it allows for predicting probable changes in ingot composition along its height caused by interaction conditions changes in gas-slag-metal. The main limitation is that it is very time-consuming (for ten times points and two slags, it would require creating and calculating 20 subsystems), making it impractical for cases when process dynamic is the same or ingot inhomogeneity is not a problem.

For example, a comparison of the effects of different slags on a metal composition could be done by traditional single-system thermodynamic estimation, slightly modifying it. Our approach involves using a higher slag-to-metal ratio typical for the stationary stage of the ESR. The suggested reacting system includes the entire slag mass and the mass of a liquid metal pool in a volume equal to a half-sphere with a depth equal to a certain ingot radius. Such a system gives more pronounced changes in phase composition than using a specific consumption value. The estimation presented here was made for an ingot 1000 mm in diameter. Such a single reacting system consisted of 2200 kg of liquid metal, 300 kg of a slag bath and 6.47 kg of argon. In the following calculations, the slag-to-metal ratio makes 0.14 (instead of a 0.016 value for the specific consumption at the above-described 19 000 kg ESR ingot). The single-system comparison was made for slags (#1–2, 1–3, 2–5 and 4–3) with the wider melting range whose compositions are given in Table 2.

TABLE 2

Composition and experimentally determined properties of slags system CaF_2-Al_2O_3-TiO_2-MgO.

Slag #	Components content, wt%				Melting temperature / melting range, K	Electrical conductivity at 1873 K
	CaF_2	Al_2O_3	TiO_2	MgO		
1–1	10	60	21	9	No data	625
1–2	20	54	18	8	1623 / 330	375
1–3	30	47	16	7	1653 / 190	279
2–1	70	0	30	0	No data	1064
2–2	63	10	27	0	1513 / 90	967
2–3	56	20	24	0	1543 / 80	760
2–4	49	30	21	0	1603 /80	449
2–5	42	40	18	0	1693 / 130	322
3–3	56	24	20	0	1533 / 60	541
4–1	50	22	18	10	1603 / 70	523
4–2	45	19	16	20	1473 /100	967
4–3	40	17	13	30	1583 / 110	625

Slags' physical properties should also be relevant to the ESR process. The molten slags' electrical conductivity and melting temperature range were measured experimentally (Table 2). The electrical conductivity of slags was measured using a bridge-type three-electrode measuring cell (Kolisnyk *et al*, 1980) with KCL solution as the reference media. The melting interval was determined visually using an optical microscope to monitor slag particle behaviour on a heated molybdenum plate in argon. A tungsten rhenium thermocouple WRE5/20 was used to control the temperature.

The dependency between experimentally determined values and chemical composition was searched using the model of the ordered structure of oxide melts. The basics of Professor Prikhodko's physicochemical model (Prikhodko, 1995a, 1995b; Prikhodko *et al*, 2013) is the description of the act of elementary interaction of each pair of elements with unpolarised atomic radii R_{ui}, R_{uj} at a distance d between them in a multicomponent system by calculating a set of partial parameters in Equations 1–4:

$$\begin{cases} R_{u_i} + R_{u_j} = d \\ lgR_{u_i} = lg\dot{R}_{u_i} - Z_{ij} \cdot tg\alpha_i \\ lgR_{u_j} = lg\dot{R}_{u_j} - Z_{ij} \cdot tg\alpha_A \end{cases} \tag{1}$$

where Z_{ij} is calculated using the formula:

$$Z_{ij} = \frac{lg\frac{\dot{R}_{u_i} \cdot tg\alpha_i}{\dot{R}_{u_j} \cdot tg\alpha_j}}{(tg\alpha_i + tg\alpha_j)} + \frac{\Delta e_{ij}}{2} \tag{2}$$

The slags' chemical composition effect on their properties was assessed by the index of the cationic sublattice ΔZm non-equilibrium state, which is expressed by the equation:

$$\Delta Zm = \left(Z_{K(K-A)} - Z_{K(K-K)}\right) - \left(\frac{\frac{R_{u_A}}{R_{u_K}} - 0.53}{15{,}43(tg\alpha_K)^{1{,}5075}} + 0{,}51\right) \tag{3}$$

where R_{uK}, R_{uA} are the ionic radii of cations and anions; $Z_{K(K-A)}$ and $Z_{K(K-K)}$ are the average statistical charges of atoms in the cation-anion and cation-cation bonds, respectively, and $tg\alpha_K$ is a weighted average parameter characterising the change in the radii of cations when their charge changes:

$$tg\alpha_K = \sum_{i=1}^{n} tg\alpha_i \tag{4}$$

This semi-empirical approach considers slag a chemically united system whose melt properties depend on structure. Based on this concept, an oxide-fluoride melt composition can be transformed into integral criteria. Derived criteria are used to predict certain slag properties.

RESEARCH RESULTS

Outcomes of slag-metal interaction prediction in the course of electroslag remelting

Figure 4 presents the equilibrium content of elements in the metal phase that are most oxidised by sags' oxides and the slag-to-metal ratio corresponding to a certain conventional time of remelting in seven certain subsystems (shown as points on built curves).

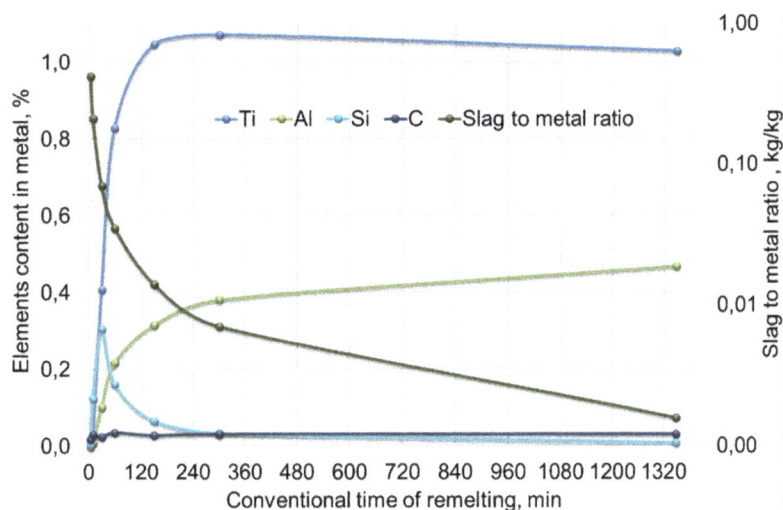

FIG 4 – Equilibrium content of active to oxygen elements in metal phase and slag to metal ratio at Inconel 718 remelting with slag $34CaF_2$-$30Al_2O_3$-$27CaO$-$0.5SiO_2$-$2.5MgO$-$6TiO_2$.

The slag-metal mass ratio at the fifth minute of remelting is nearly 42 kg/kg and drops throughout the process to 0.016 kg/kg at the completion of the melt. It is well visible that the system is most reactive when the slag-to-metal ratio is very high at the beginning of the remelting. Due to that, slag's oxidation action sufficiently affects the content of elements, which react with oxygen at this period. The content in the metal of active elements whose presence is critical for alloy properties (Ti, Al) at the process beginning (first 30–60 minutes) is close to zero. The sufficient changes in the content of both titanium and aluminium occur for around the first two hours, while the mass losses of the metal phase and gains of the slag phase mainly occur (Figure 5).

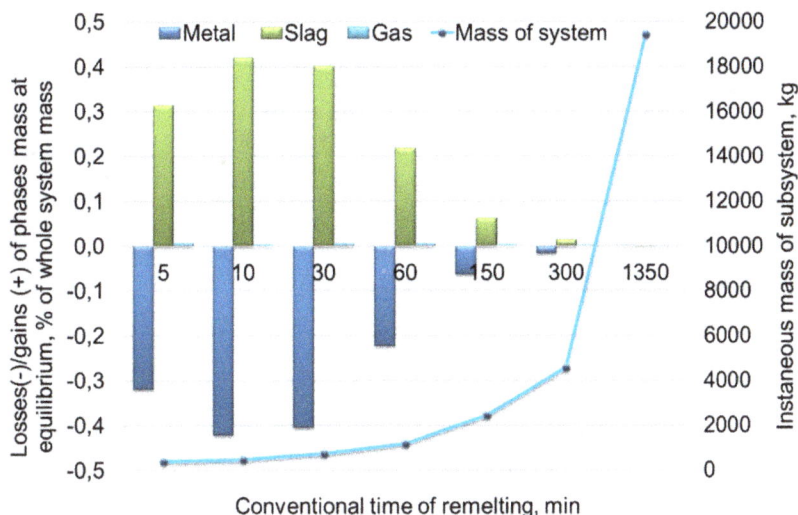

FIG 5 – Phases losses/gains in the reactive subsystems at a course of ESR in argon atmosphere (line is plotted by additional vertical axis).

After that period, when metal phase amounts start to be bigger (and the slag-to-metal ratio – smaller) than some critical values, the relative losses of equilibrium metal phase mass go down. This critical time can be supposed to be the time when the oxidative ability of slag is exhausted. Nevertheless, it does not mean that exchange reactions between slag and metal phases are stopped. It is visible (Figure 4) that titanium content grows during up to four hours. Despite a slight permanent reduction of its content during the lasting time, it is still higher than the initial value.

Aluminium content grows to the end of the remelting but doesn't reach the initial content due to silicon reduction from the slag, which goes most actively in the first minutes. At the start of the process, the silicon content goes to maximum and reduces during more than two hours of remelting and it is still in metal until the end of the melt. Carbon in metal shifts by 0.01–0.02 per cent from the initial value in the period of metal losses. Thus, the calculations under the developed dynamic model of the gas-slag-metal interaction during the ESR process under a protective atmosphere show that even in an inert gas atmosphere, changes in oxygen-active elements content in the metal phase take place, which cause inhomogeneity of their content along ingot height.

When the dynamic of the ESR process is the same (same ingot, melting rate) or precise evaluation is not important, the assessment of interactions in the gas-slag-metal system can be done using a single system approach – the results of such a way for comparing four ESR slag actions (Figure 6).

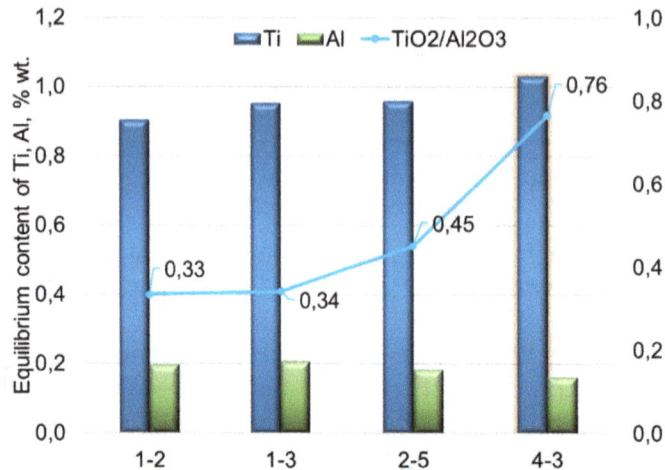

FIG 6 – Al and Ti equilibrium content in the metal phase interacted with slags having different TiO_2/Al_2O_3 ratios in their compositions (Table 2).

Inconel 718 has an extensive solidification range (1483–1617 K), so slags with the widest melting range (1–2, 1–3, 2–5 and 4–3) were compared. For slag with the highest TiO_2/Al_2O_3 ratio, the titanium content in the metal is slightly higher than the initial one (1 wt per cent), possibly due to the reduction of titanium from the slag. This version is also supported by the greatest loss in aluminium content in metal compared to other slags used. However, the cause may also be a metal loss, as was found for the comparative slag $34CaF_2$-$30Al_2O_3$-$27CaO$-$0.5SiO_2$-$2.5MgO$-$6TiO_2$. The titanium content close to the original was calculated using 1–3 and 2–5 slags. Aluminium losses in the metal are minimal for slags 1–2 and 1–3. Slag 1–3 is the most inert from the point of view of preserving both elements.

Outcomes of slag's physical properties modelling using Directed Chemical Bonds Concept

The atomistic approach was used to interpret slag's measured properties from the Directed Chemical Bonds Concept (DCBC) standpoint to build predictive models for slag engineering considering the physical properties of ESR slag melts. Table 3 lists the unpolarised radii of atoms $(\dot{R}_{u_i})$ and the gradient of changes in the radii of atoms from their charge $(tg\alpha_i)$ that belongs to elements in the studied slag system CaF_2-Al_2O_3-TiO_2-MgO (Table 2).

TABLE 3

Radii of elements $(\dot{R}_{u_i})$ and gradients of their change from their charge ($tg\alpha_i$) to interaction.

Element	$\dot{R}_{u_i}$, $(10^{-1}nm)$	$tg\alpha_i$
Ca	2.02	0.151
Ti	1.45	0.085
Al	1.468	0.156
Mg	1.6	0.196
O	0.73	0.136
F	0.96	0.142

Table 4 presents the interatomic interaction parameters calculated for experimental slag compositions of the CaF_2-Al_2O_3-TiO_2-MgO system.

TABLE 4

The parameters of interatomic interaction in the studied slags' melts.

Slag #	d	Δe	tgα	ΔZm	Zka	Zkk	Zaa	Zak
1–1	3.2513	-2.7444	0.1502	0.0108	-0.23031	-1.32999	-3.57101	-2.5141
1–2	3.2274	-2.5349	0.1506	0.0062	-0.12664	-1.22344	-3.45441	-2.40823
1–3	3.2209	-2.3552	0.1505	0.0031	-0.04222	-1.13722	-3.3533	-2.3131
2–1	3.8161	-2.7419	0.1315	0.0269	-0.42814	-1.56892	-3.45975	-2.31482
2–2	3.6791	-2.6378	0.1351	0.0194	-0.33979	-1.46586	-3.42745	-2.2989
2–3	3.5347	-2.5097	0.1383	0.0125	-0.24118	-1.35587	-3.38104	-2.26926
2–4	3.3836	-2.3573	0.1413	0.0062	-0.13198	-1.2385	-3.32003	-2.22589
2–5	3.2256	-2.1782	0.1439	0.0007	-0.01087	-1.1124	-3.24295	-2.16775
3–3	3.4051	-2.2597	0.1412	0.0051	-0.08413	-1.18915	-3.26822	-2.17643
4–1	3.5018	-2.5083	0.15	0.0108	-0.14606	-1.2391	-3.41106	-2.3627
4–2	3.5744	-2.6832	0.1577	0.0111	-0.17787	-1.25373	-3.52106	-2.50572
4–3	3.6024	-2.7749	0.1649	0.0058	-0.16662	-1.2217	-3.59282	-2.60856

These parameters characterise the chemical and structural state of the slag systems: d (10^{-1} nm) is the average interatomic distance of interacting ions in the melt; Δe represents the number of electrons localised in the direction of their connection, e; $tg\alpha$ means the weighted average value of the gradients of changes in ion radii with changing its charge; ρ expresses the ratio of the number of cations to the number of anions; ΔZm, as was mentioned above, is the indicator of non-equilibrium state of cationic sublattice.

The analysis showed that the parameter of the non-equilibrium state of cationic sublattice ΔZm demonstrates a close relationship with the chemical composition of the slags under study (Figure 7). The physicochemical essence of the parameter ΔZm characterises the stability of the cationic sublattice: the stability condition is that ΔZm tends to zero.

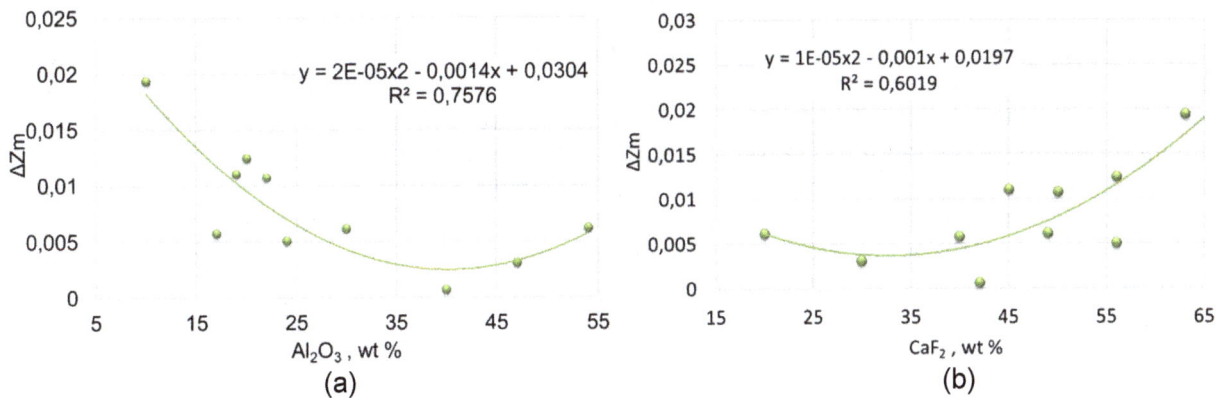

(a) $y = 2E\text{-}05x^2 - 0,0014x + 0,0304$, $R^2 = 0,7576$

(b) $y = 1E\text{-}05x^2 - 0,001x + 0,0197$, $R^2 = 0,6019$

FIG 7 – Relationship between the content of Al_2O_3 (a) and CaF_2 (b) with the non-equilibrium index of the cationic sublattice ΔZm.

Combined with the established relationship ΔZm – Al_2O_3, the minimal value of integrated criteria is reached at Al_2O_3 content, approaching its maximum values from the investigated range (Figure 7a) and at CaF_2 content is close to the minimal level and TiO_2 13–18 per cent (for the system under study).

The established dependences between ΔZm parameter and composition of the slags' system under study and its properties (Figure 7) make it possible to consider slags with minimal content (wt%) of CaF_2-30 (Figure 7b) and TiO_2-13 (Figure 7c) having not very high melting temperatures range and also satisfactory melting range according to Table 2.

As shown in Figure 8, the parameter ΔZm has an acceptable relationship with slags' physical properties.

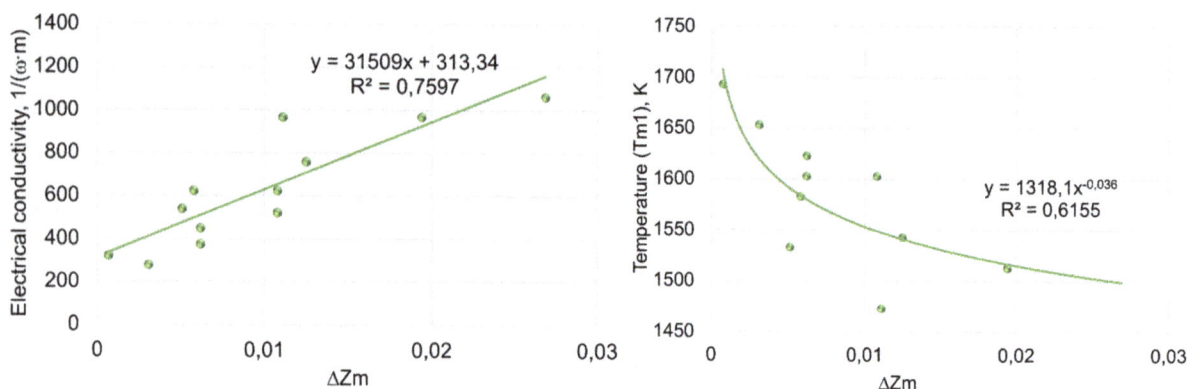

FIG 8 – Relationship between the non-equilibrium parameter of the cationic sublattice ΔZm and the properties of experimental slags: (a) electrical conductivity; (b) liquidus temperature of slag melting.

The involvement of the DCBC *ab initio* approach allows building models of electrical conductivity and temperature of melting from the integral parameter of the non-equilibrium state of cationic sublattice, ΔZm. Found dependencies are grounds for using this integral criterion when predicting these and other important properties of oxide-fluoride and other metallurgical slags.

DISCUSSION

Possibilities and limitations of the dynamic model of electrosalg remelting process

The current level of development of numerical methods for thermodynamic systems simulation has opened up new opportunities for improving existing and developing new industrial processes. However, digitalisation alone is insufficient since the simulation results must accurately reflect the real system. The accuracy of the system description and the reliability of thermodynamic programs play a decisive role in assessing complex processes, especially in the metallurgical industry, where sampling and analysis of samples are quite complex and expensive.

The provided view on ESR ingot composition formation while ESR differs from the traditional approach to the physicochemical calculation, which is usually used for ESR ingot composition estimation. Representation of this process as a line of subsystems allows for a step ahead in understanding the ESR ingot formation pattern and makes it possible to predict changes in slag-metal interaction during the process. The simpler single-system approach of thermodynamic calculations for the stable stage of electroslag remelting in argon is also relevant for comparing different slags under the same process conditions (same ingot, melting rate and temperature) or as a preliminary estimation. Calculations carried out using both the dynamic physicochemical model ESR process with several subsystems, and a single-system approach show that even with neutral gas and oxide-fluoride slags, gas-slag-metal interaction causes changes in the content of active-to-oxygen elements such as titanium and aluminium, silicon etc. The intensity of chemical composition changes is biggest at the higher slag-to-metal ratio. The highest slag-to-metal ratio is at the beginning of the process and permanently reduces during the process to the value of specific slag consumption usually used for thermodynamics estimation of the ESR.

The chemical composition of the phases changes as a result of oxidation-reduction reactions. Besides, slag's ability to desulfurise metal and assimilate non-metallic inclusions gradually reduces during the ESR process. Moreover, a slag bath accumulates all products of chemical interactions between metal and slag, changing its chemical and, accordingly, physical properties, which are crucial for the stability of the process and the resulting quality of the ingot.

The slag is inert to metal when no or very minor oxidation occurs and it is usually good for ingot quality and technology operation. For such slags (with low content of stable oxides or fluorides only), the thermodynamic calculations show no changes outside the permissible range in the chemical composition of phases. Otherwise, when changes are significant, the composition of a consumable electrode should be adjusted, or slag should be changed for a less reactive one, or, in quite rare cases, metal alloying through the slag should be envisaged, requiring additional calculations.

The proposed more precise description of the electroslag remelting process as a line of subsystems can be used to predict changes in ingot composition along its height. The most important advantage of the dynamic model is its possibility to predict and exclude factors affecting the metal composition along the ESR ingot height or to respond to them properly. Such analysis can be done in different contexts depending on the purpose of the research (eg changes in the content of active components in the system). Summarising, the predictive physicochemical model of ESR ingot formation in dynamic can be used to achieve several goals:

- Adjust existing composition or search for new slags. For example, new slag composition could replace high CaF_2 composition at ESR of alloyed and high-alloyed steels in both stationary and short-collar moulds. The industrial tests conducted on the developed slag have shown the following benefits: oxide inclusion content was reduced by 0.5 points and power consumption – by 17 per cent. The more environmentally friendly with more than two times less calcium fluoride composition at a lower cost of 23–25 per cent was developed (Stovpchenko *et al*, 2020b; Lisova *et al*, 2020).

- Maintaining the active components content (Ti, Al etc) at high-alloyed steels and alloys while remelting (Stovpchenko *et al*, 2023) and a low level of oxygen in steel during ESR (Medovar *et al*, 2023a, 2023b; Stovpchenko *et al*, 2018a).

- Development of slags with specific properties, for example, self-disintegrating slag (Stovpchenko *et al*, 2018b).

- Development of slags for new steels and alloys (AHSS, superalloys etc) remelting (Stovpchenko, Gusiev and Medovar, 2014; Davidchenko *et al*, 2017; Medovar *et al*, 2023a).

In general, the proposed physicochemical dynamic model of the ESR process in the new paradigm could be a good tool for predicting how a slag composition affects an ingot's homogeneity on its height and the yield of a suitable metal. Such prediction is crucial in manufacturing high-alloyed steels and alloys by electroslag remelting. Uncontrolled changes in slag composition while remelting can cause ingot scrappage. However, it is even more dangerous when minor changes occur unheeded and are ignored, posing a significant risk of responsible parts damage at exploitation.

Possibilities and limitations of the Directed Chemical Bonds Concept in the prediction of electrical conductivity and melting temperature of ESR slags

Based on DCBC principles, atomistic models were developed to predict the physical properties of Ti-bearing ESR slags. Although these models were built using a limited amount of experimental data, the results demonstrate that the parameter of the non-equilibrium state of cationic sublattice ΔZm can be used to predict the properties of molten slags.

ΔZm parameter's validity explains that the crystal lattice restructuring of solid oxides at high temperatures occurs with a disordered liquid slag melt formation. The non-equilibrium state of the cationic sublattice of slag (Figure 9) can be caused by structural differentiation depending on the geometric sizes ratio of neighbouring ions and the relative arrangement of cation-anion clusters formed in complex composition melts.

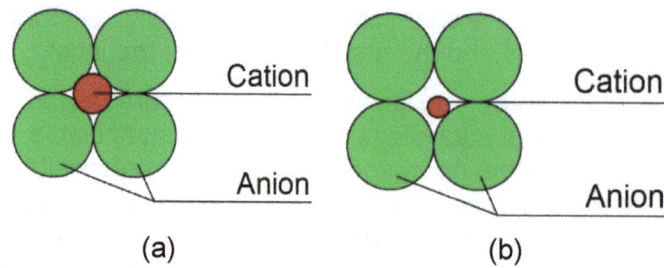

FIG 9 – Scheme of the cationic sublattice displacement inside the anion sublattice: (a) equilibrium state; (b) non-equilibrium state.

In the centre of such clusters are cations (calcium, aluminium, titanium, magnesium) surrounded by oxygen anions (fluorine and or complex anions). In a homogeneous melt, electron density distribution between cations and anions can be relatively stable. When an element is ionised, its radius changes significantly (Table 3): cations decrease in size due to transferring part of their electronic field to the anions in their immediate environment. In this case, the multicomponent slag system tends to the most favourable energy state, which can be described using the integral parameters of interatomic interaction (the model parameter ΔZm tends to zero). The bonds between cations and anions in a liquid melt at a given chemical composition and temperature retain short-range order, which determines the structure and complex properties of the liquid melt.

Atomistic models constructed using the integral parameter of the non-equilibrium state of cationic sublattice, ΔZm, offer promising possibilities for evaluating the physical properties of slags. ΔZm and other integral criteria offer promising possibilities for properties evaluation for other slags' and metals' melts (Togobitskaya, Bel'kova and Stepanenko, 2023; Muravyova *et al*, 2021; Stepanenko *et al*, 2017; Babachenko *et al*, 2020). Integral criteria are particularly effective when studying the multivariate compositions of melts, where more than one component is altered.

A proposed DCBC predictive model using *ab initio* principles helps develop a fundamental understanding of the phenomena occurring during slag melting and their interaction with metal melts in various metallurgical processes. It should be mentioned that slag engineering is now one of the trends because of the urgent necessity to be prepared for changes in the green transition of steelmaking, which is already ongoing. The issue is that an increased share of direct reduced iron (DRI) in electric arc furnaces increases the forming slag's volume and changes its composition. To keep efficient steel production, it became very important to have the possibility to make theoretical predictions of different slag (and alloys as well) characteristics. The melting temperature and electric conductivity were considered in this article, and viscosity, sulfide capacity, surface tension, and others can be predicted for various multicomponent molten slags whose composition changes in a wide range. The proposed concept makes it possible to reduce experimental measurements and tests, which are time-consuming and resource-consuming. This methodology is particularly relevant in metallurgy and materials science, where a detailed understanding of the underlying principles of chemical composition's effect on properties is essential for developing new materials and technologies.

Future works include experimental proof of dynamic models using slag and metal sampling at heavy ESR ingot producing and improvement of accuracy of DCBC models for physical properties of slag prediction using a sufficiently larger amount of experimental data for Ti-bearing and other metallurgical slags.

CONCLUSIONS

Both the dynamic physicochemical model ESR process in neutral gas presented as a chain of subsystems and a single-system estimation revealed that despite using neutral gas and oxide-fluoride slags participate in gas-slag-metal interaction, causing changes in the content of active-to-oxygen elements such as titanium, aluminium and silicon in a metal. The highest intensity of chemical composition changes occurs at the higher slag-to-metal ratio, which is at the start of the process and permanently reduces at its end to the value of a specific slag consumption.

The dynamic model is promising to study the composition changes along ingot height caused by the oxidative action of slag during the process, making it possible to predict ready ESR ingot's homogeneity and increase the yield of a suitable metal.

DCBC presenting a semi-empiric system providing effective concentration and convolution of the ab initio information on the type of bonds in multicomponent systems was employed to find integral criteria for predicting physicochemical and thermophysical properties of slags from their composition. The parameter ΔZm characterising the non-equilibrium state of the cation sublattice in the oxide melt is the suitable predictor of electric conductivity and melting temperature of studied slags.

The developed physicochemical dynamic model of the ESR combined with an ab initio approach of the DCBC that can predict slag properties becomes a step towards a comprehensive model of the ESR ingot formation and tailored functional slags development for specific compositions of alloys whose homogeneity is crucial for quality.

REFERENCES

Babachenko, A I, Togobitskaya, D N, Kononenko, A A, Snigura, I R and Kuksa, O V, 2020. Justification for choosing alloying and Micro-Alloying elements to improve the mechanical properties of railway wheels, *Steel in Translation*, 50(11):815–821. https://doi.org/10.3103/s0967091220110029

Davidchenko, S, Stovpchenko, G, Salnikov, A, Medovar, L, Logozinskii, I and Kuzmenko, A, 2017. Potentials of the Acid Slag for ESR, in *Proceedings of the Liquid Metal Processing and Casting Conference 2017*, pp 145–148.

Duan, S-C, Shi, X, Wang, F, Zhang, M-C, Sun, Y, Guo, H-J and Guo, J, 2019. A review of methodology development for controlling loss of alloying elements during the electroslag remelting process, *Metallurgical and Materials Transactions B*, 50(6):3055–3071. https://doi.org/10.1007/s11663-019-01665-2

Hou, D, Wang, D-Y, Jiang, Z-H, Qu, T-P, Wang, H-H and Dong, J-W, 2021. Investigation on Slag–Metal-Inclusion multiphase reactions during electroslag remelting of die steel, *Metallurgical and Materials Transactions B*, 52(1):478–493. https://doi.org/10.1007/s11663-020-02032-2

Hoyle, G, 1983. *Electroslag Processes, Principles and Practice*, 512 p. (Applied Science Publishers: London/New York).

Kolisnyk, V, et al. 1980. Patent #957081 USSR, G01 N27/02, Device for measuring the 07.09.82 electrical conductivity of molten slag, Bulletin #33:07.09.82.

Lisova, L O, Stovpchenko, G P, Goncharov, I O, Gusiev, Ia V and Medovar, L B, 2020. Thermodynamics of interactions and physical properties of slags of 30CaF2/30CaO/30Al2O3 (SiO2, MgO) system at electroslag remelting, *Sovremennaja Èlektrometallurgija*, 2020(1):8–13. https://doi.org/10.37434/sem2020.01.01

Lukas, H L, Fries, S G and Sundman, B, 2008. Computational thermodynamics: the CALPHAD method, *Choice Reviews Online*, 45(11):45–6195. https://doi.org/10.5860/choice.45-6195

Medovar, B I and Boyko, G A, 2013. *Electroslag Technology* (Springer Science and Business Media).

Medovar, L, Stovpchenko, G, Lisova, L, Jiang, Z, Dong, Y and Kang, C, 2023a. Features and Restrictions of Electroslag Remelting with Silica-Bearing Slags for Lightweight High Manganese Steel, *Steel Research International*, 94(11). https://doi.org/10.1002/srin.20230016

Medovar, L, Stovpchenko, G, Sybir, A, Gao, J, Ren, L and Kolomiets, D, 2023b. Electroslag hollow ingots for nuclear and petrochemical pressure vessels and pipes, *Metals*, 13(7):1290. https://doi.org/10.3390/met13071290

Mills, K C, 2021. Metallurgical slags, in *Encyclopedia of Glass Science, Technology, History and Culture* (eds: P Richet, R Conradt, A Takada and J Dyon), pp 843–855. https://doi.org/10.1002/9781118801017.ch7.4

Muravyova, I G, Togobitskaya, D N, Bel'kova, A I, Ivancha, N G and Nesterov, A S, 2021. Prediction of composition and properties of final smelting products based on integral indices of the blast furnace burden and temperature blasting mode, *Steel in Translation*, 51(8):531–537. https://doi.org/10.3103/s096709122108009x

Prikhodko, E V, 1995a. Metallokhimiya mnogokomponentnykh system (Metal chemistry of multicomponent systems), Moscow: *Metallurgiya*, 1995.

Prikhodko, E V, 1995b. Methodology for determining the parameters of directed interatomic interaction in molecular and crystalline compounds, *Metallophysics and Newest Technologies*, 17:(1)54–60.

Prikhodko, E, et al. 2013. Prediction of physical and chemical properties of oxide systems, Dnepropetrovsk, Porohi (*Thresholds*), 339 p.

Stepanenko, D, 2023. Calculation of activation energy of viscosity for evaluation of metallurgical slag melts structure, *Lithuanian Journal of Physics*, 63(1). https://doi.org/10.3952/physics.2023.63.1.6

Stepanenko, D A, Volkova, O, Heller, H-P, Otorvin, P I and Chebykin, D A, 2017. Selecting optimal slag conditions in the blast furnace, *Steel in Translation*, 47(9):610–613. https://doi.org/10.3103/s0967091217090133

Stovpchenko, G, Davidchenko, S V, Lisova, L O, Gusiev, Ya V and Medovar, L B, 2020b. Investigation of manufacturability and effectiveness of the new slag for electroslag remelting, *Sovremennaâ Èlektrometallurgiâ*, 2020(3):11–17. https://doi.org/10.37434/sem2020.03.01

Stovpchenko, G, Lisova, L, Medovar, L, Polishko, G, Brun, N, Bourson, P, Strelchuk, V and Nasieka, I, 2018a. Electroslag remelting for low oxygen metal manufacturing, in *Proceedings of the 7th International Congress on science and technology of steelmaking, The challenge of industry*, pp 1–10.

Stovpchenko, G P, Lisova, L O, Goncharov, I O and Gusiev, I V, 2018b. Physico-chemical properties of the ESR slags system CaF2-Al2O3-(MgO, TiO2), *Journal of Achievements in Materials and Manufacturing Engineering*, 2(89):64–72. https://doi.org/10.5604/01.3001.0012.7110

Stovpchenko, G P, Sybir, A V, Polishko, G O, Medovar, L B and Gusiev, Ya V, 2020a. Mass Transfer in Electroslag Processes with Consumable Electrode and Liquid Metal, *Uspehi Fiziki Metallov*, 21(4):481–498. https://doi.org/10.15407/ufm.21.04.481

Stovpchenko, G, Togobitskaya, D, Lisova, L, Stepanenko, D and Medovar, L, 2022. Predictive models for molten slags viscosity and electrical conductivity based on directed chemical bonds concept, *Ironmaking and Steelmaking*, 49(6):572–580. https://doi.org/10.1080/03019233.2022.2026043

Stovpchenko, G P, Lisova, L O, Medovar, L B and Goncharov, I O, 2023. Thermodynamic and physical properties of CAF2–(AL2O3–TIO2–MGO) system slags for electroslag remelting of Inconel 718 alloy, *Materials Science*, 58(4):494–504. https://doi.org/10.1007/s11003-023-00690-6

Stovpchenko, G, Gusiev, I and Medovar, L, 2014. Features of Slag-Metal interaction at electroslag remelting of superalloys, in *Proceedings of the 8th International Symposium on Superalloy 718 and Derivatives*, pp 47–56. https://doi.org/10.1002/9781119016854.ch4

Sundman, B, Kattner, U R, Sigli, C, Stratmann, M, Tellier, R L, Palumbo, M and Fries, S G, 2016. The Open CALPHAD thermodynamic software interface, *Computational Materials Science*, 125:188–196. https://doi.org/10.1016/j.commatsci.2016.08.045

Togobitskaya, D N, Bel'kova, A I and Stepanenko, D O, 2023. Model decision-making system in the task of choosing the optimal composition of the blast furnace burden under specific operating conditions of BF, *Acta Metallurgica Slovaca*, 29(2):67–74. https://doi.org/10.36547/ams.29.2.1764

Research on –
Industrial slag / flux / molten salt design and optimisation

Recent advances in understanding phosphorus in oxygen steelmaking

G Brooks[1]

1. Professor, Swinburne University of Technology, Melbourne Vic 3160.
   Email: gbrooks@swin.edu.au

ABSTRACT

The control of phosphorus in steelmaking is one of the famous challenges of pyrometallurgy. It is well established that highly basic slags, the presence of FeO in the slag and lower temperatures (less than 1650°C) are favourable for phosphorous removal. The thermodynamics of this reaction have been extensively studied under laboratory conditions and the general trends confirmed through industrial trials. However, in general, phosphorus slag/metal distributions measured in industry are far from equilibrium and quite specific to the oxygen steelmaking technology being used. This variability was quantified by Urban, Weinberg and Cappel (2015), who found that the phosphorus partition ratio measured in plants varied with blowing practice, stirring intensity and even converter life. Some recent kinetic studies by Gu, Dogan and Coley (2017) have shown how the kinetics of these reactions vary with FeO content and dissolved sulfur at the droplet scale. A global kinetic model by Rout *et al* (2018) provides meaningful predictions of where dephosphorisation occurs in an industrial basic oxygen furnace (BOF) and some recent heat transfer modelling by Madhavan *et al* (2021, 2022) implies that temperature difference between droplets and slag has a role to play in explaining why equilibrium is not achieved. This paper will review the current state of knowledge and provide suggestions on predicting phosphorous removal for industrial operations and how greater removal could be achieved.

INTRODUCTION

The control of phosphorus is one of the central challenges facing steelmakers. In oxygen steelmaking, it is well established that maintaining a slag with high basicity (high CaO and MgO), low tapping temperatures (<1650°C) and high FeO will promote the removal of phosphorus from steel. High basicity also means high flux additions and added expense, low tapping temperatures may necessitate reheating the steel between steelmaking and casting and high FeO contents in the slag will adversely affect yield. As a result, steelmakers are always looking for ways to minimise phosphorus coming into their feed materials and the adverse effects associated with removing phosphorous.

Phosphorus in even low quantities (above 0.1 wt per cent) can adversely affect ductility, fracture toughness and cause hot shortness. Ironmaking conditions are not ideal for removing phosphorous, that is, reducing conditions encourage both Fe and P formation from their oxide states. In general, phosphorus is either removed from hot metal from a post blast furnace operation before entering the oxygen steelmaking or in the oxygen steelmaking furnace, typically below 0.010 wt per cent in the tapped steel depending on the requirement of the grade being produced (Turkdogan, 1996). Some steelmakers use a two staged oxygen steelmaking process to remove phosphorus from hot metal, as a means of minimising flux addition, though this is also likely to lower overall productivity.

The thermodynamics of phosphorus removal has been extensively studied (Assis *et al*, 2015). A recent review identified 24 different equilibrium relationships proposed in the literature (Urban, Weinberg and Cappel, 2015), based on different models for dealing with the complex solution behaviour of PO_4^{3-} anion in slags. It is well beyond the scope of this paper to critically review this vast literature but to note that many of the equilibrium relationships take the following form:

$$\log (\%P)/[\%P] = A/T + 2.5 \log (\%Fe) + B. (\%CaO) + C. (\%MgO) - D \qquad (1)$$

Where () indicates slag, [] metal, T is temperature in Kelvin, concentrates are in wt per cent and A, B, C and D are constants.

This equation shows:

- The importance of temperature on phosphorus partition in slags – phosphorous oxide becoming less stable with temperature – with the ratio (%P)/[P] decreasing from approximately

50 at 1620°C for a commercial basic oxygen furnace (BOF) operation to below 35 at 1690°C (Urban, Weinberg and Cappel, 2015). High tapping temperatures (>1650°C) reflect processing requirements post steelmaking and the superheat required at the casting machine.

- The importance of FeO in the slag to provide dissolved oxygen into the metal for reaction with dissolved P (ie the least stable oxide in the slag will be accompanied by significant dissolved oxygen), other relatively unstable oxides (SiO_2 and MnO) will also provide dissolved oxygen but FeO content in the slag is major variable. Increasing the FeO content of the slag from 14 wt per cent to 24 wt per cent can typically double the partition ratio in industrial furnaces (Urban, Weinberg and Cappel, 2015). Increasing FeO in the slag effects yield from the operation and is itself controlled by the carbon content of the steel, so in many cases not a variable that can be easily manipulated ie the FeO content of the slag is set by the grade of the steel being tapped (Turkdogan, 1996).

- The importance of providing basic oxide fluxing agents to stabilise the phosphorus in the slag phase. Lime/silica weight ratios (an approximate measure of 'basicity' commonly used in industry) range from two to four depending on the grade of the steel and the operating temperature of the steelmaking furnace. Increasing the lime/silica ratio from 2.6 to 4.0 in BOFs can typically double the partitioning ratio of phosphorus but at the cost of extra flux and the risk of creating a high liquidus temperature slag that may not be practical to melt or easily tap from the vessel (Urban, Weinberg and Cappel, 2015).

There are variations to the relationship shown in Equation 1 (eg a temperature term is introduced to reflect the impact temperature on the activity of CaO and MgO) and significant debate about which is the most accurate formulae but these debates, in part reflect the variation in experimental data and difficulties in controlling and measuring all variables accurately in high temperature experiments (Assis *et al*, 2015). An important complication is the tendency of phosphorus oxide to dissolve into a solid solution with solid calcium silicate phases that can exist in steelmaking conditions, particularly early in the blow when temperatures in the vessel can be below 1600°C and various researchers have attempted to measure this behaviour in experiments and pilot plant testing (Millam *et al*, 2013).

The large body of work in quantifying the behaviour of phosphorus in steelmaking systems whilst providing an excellent basis by which to qualitatively understand the general trends of oxygen steelmaking furnaces, they do not allow direct control of the process because these thermodynamic relationships over predict phosphorus partitioning significantly in industrial operation. Most industrial furnaces are not at equilibrium regarding phosphorus and this paper explores the reasons for this and recent work in trying to provide a better predictive basis for phosphorous behaviour in oxygen steelmaking.

NON-EQUILIBRIUM CONDITIONS IN OXYGEN STEELMAKING

There is clear evidence that industrial oxygen steelmaking operations are not at equilibrium regarding phosphorus at the completion of the blow. An extensive review of plant data from around the world (including data from 30 plants) by Urban, Weinberg and Cappel (2015) found that tapped phosphorus partitioning ratios (P)/[P]) for oxygen steelmaking operations are far from equilibrium. For top blown (BOF) operations the (P)/[P] ratio may vary from 50 to 100, whilst combined blowing operations vary between from 80 to 140 and bottom blown between 100 to 200. Urban, Weinberg and Cappel reported that calculated equilibrium ratios over the range of conditions considered for these industrial systems ranged from 200 to 700. Turkdogan (1996) also recognised these differences had plant data showing that the greatest deviation from equilibrium could be found for top blown BOFs producing steel below 0.1 wt per cent C, where-as, bottom blowing operators were tapping close to equilibrium regarding phosphorous.

Urban, Weinberg and Cappel (2015) provided industrial data show that tapped phosphorus partition ratios in top blown BOFs were influenced by slag splashing, introduction of inert gas bottom stirring and converter life. They concluded that phosphorus partitioning in most oxygen steelmaking vessels is kinetically controlled, that bottom blown vessels are closest to equilibrium (which is consistent with Turkdogan (1996)) and that increased stirring in the vessel is important to improving the kinetics. They also concluded that equilibrium-based relationships could not be used to control industrial operations and that each plant needed to develop empirical models based on statistical techniques to predict their phosphorous partitioning ratios at tapping.

In addition to plant data, which is dominated by at 'turn down' (a few minutes before tapping) and/or 'at tap' data, there has an extensive pilot plant study carried at Luleå using a 6 t top blown converter in which samples from different positions within the furnace were removed at regular intervals during the blow (Millam *et al*, 2013). This is the only known study in the open literature which quantified how phosphorus distributes itself in different phases at different stages of an oxygen steelmaking blow. The data generated from the study is detailed and difficult to interpret but the major findings regarding how dephosphorisation proceeded in this pilot plant study can be summarised:

- Early in the blow (under ten minutes), the majority of phosphorus removed from the metal was present in a solid solution phase containing Ca_2SiO_4 and $Ca_3(PO_4)_2$.

- After ten minutes, the liquid slag phase dominates the make of the slag layer and phosphorus previously in the solid solution phase is dissolved into the liquid phase.

- That amount of phosphorus in the slag phase is maximised below ten minutes and then is stable in the molten slag layer until tapping.

- The amount of phosphorus found in metal droplets in the slag emulsion reaches a maximum total quantity at the ten minute mark in the blow and then decreases to zero after 15 minutes.

This analysis implies that the formation of a solid solution phase early in the blow (when the temperature in the slag layer well below 1600°C and solid phases are thermodynamically stable) is the dominant mechanism in controlling the overall removal of phosphorous. This conclusion does not consider the possibility that the solid solution samples measured from the samples were formed during the cooling of the samples post-sampling, as these are the very phases expected to form upon cooling of a molten steelmaking slag. Also, the recognition of the potential importance of the solid solution in removing phosphorus is difficult to relate to the overall observations of Urban, Weinberg and Cappel (2015) that increased mixing in the vessel is key to the kinetics of dephosphorisation. It is also hard to understand why the total phosphorous removal is stable after ten minutes given that gas injection is still mixing the overall system beyond ten minutes. The indication that droplets found in the slag emulsion have less and less phosphorous further suggests that the role of droplets in dephosphorisation may be crucial in the slowdown of dephosphorisation observed in the pilot plant study.

DROPLET KINETICS AND DEPHOSPHORISATION

The role of droplet formation and the chemical behaviour of these droplets in steelmaking slags has been the subject of extensive study in the last 20 years, with groups at McMaster University and Swinburne University of Technology working on both modelling and fundamental experimental aspects of these systems (Gu, Dogan and Coley, 2017; Rout *et al*, 2018). Fully reviewing this literature is beyond the scope of this paper but a few key points from the most recent published studies from these groups can be summarised, as follows:

- Droplets of iron containing carbon and phosphorus readily 'bloat' when reacted within steelmaking slag, with nucleating CO gas bubbles being entrapped with the metal droplets, effectively lowering the density of droplets and promoting their movement upwards in a slag. The buoyant droplets spend between 30 to 60 seconds in a bloated state depending on the amount of carbon available, the FeO content of the slag and other parameters (Gu, Dogan and Coley, 2017). These observations have been made using X-ray fluoroscopy in laboratories and there is also plant sample that support this observation (ie iron droplets in the slag emulsion 'bloat').

- Small droplets (1 g) of iron containing carbon and phosphorus react very quickly in a steelmaking slag, with phosphorous in the metal being reduced from 0.088 wt per cent to below 0.04 wt per cent within 20 seconds.

- After an initial rapid dephosphorisation, the formation of CO starts consuming dissolved oxygen available for phosphorus oxidation and some reversion of phosphorus can occur within the droplet. Both the nucleation of gas bubbles and the blocking of reaction sites at the gas/metal interface is influenced by slight changes in sulfur content in the metal droplets.

- Models of the bloating behaviour extended to the industrial BOF scenario by Rout *et al* (2018) calculated the rate of different refining reactions in three different regions of the furnace, namely: (i) the jet impact region, (ii) the slag-bulk metal zone and (iii) the slag metal droplet emulsion zone and predicted that the reaction between droplets and the slag dominated the dephosphorisation kinetics. The model further predicted that the rate of dephosphorisation slowed down after ten minutes in the blow because there was insufficient carbon in the metal to 'bloat' droplets and this resulted in significant less residence time of the droplets and thus slower kinetics. When validating the models developed against industrial data, there was some evidence to suggest that variations in the FeO content in the emulsion slag layer could influence the final phosphorus level of the product steel through both changes in the thermodynamic drive and kinetics of the droplet reaction kinetics. In the modelling of Rout *et al* (2018), the temperature at the droplets and emulsion was estimated using simplistic models and the droplets were assumed to be uniform in temperature during their time in the emulsion.

The modelling work of Rout *et al* (2018) did successfully capture several key aspects of steelmaking kinetics but the models were limited by a lack of reliable kinetic model for predicting FeO formation in steelmaking conditions – the kinetics of FeO were estimated by difference (ie calculate the kinetics of the other reactions and assume the remaining oxygen reacts with iron to form FeO) – and a simplistic treatment of heat transfer during the steelmaking process, particularly, in relation to the temperature of droplets generated from the impact region of the furnace and the gradients in the droplets as they react in the slag emulsion.

The modelling work of Madhavan *et al* (2021, 2022) attempted to combine the insights around droplet kinetics from Rout *et al* (2018) and recent experimental work by Gu, Dogan and Coley (2017), with a rigorous treatment of heat transfer within the system. From available plant data and modelling of heat transfer in the impact region, Madhavan *et al* (2022) estimated that the hot spot reaches a maximum of 2300°C in the middle of a blow (approximately ten minutes) and that droplets generated from the impact region heat the slag in the emulsion mainly by radiation, particularly in the first ten minutes of the blow. The model further predicts that the slag layer is consistently hotter than the hot metal bath, notably in the middle of the blow (50 to 100°C) because of the heat transfer from the droplets to the slag emulsion region. The model further predicts that the slag and hot metal get closer in temperature in the last five minutes of the blow. There is limited industrial data available to confirm this prediction but these results do suggest that some of the slowing of the overall kinetics observed in dephosphorisation could be explained by the slag region near the reacting droplets in the emulsion becoming hotter, resulting in lowering of the thermodynamic drive to remove phosphorus, which in turn could expected to slow the kinetics. Certainly, there is a complex interplay between droplet generation, reaction within the droplets whilst in the emulsion and the transfer of heat between droplets and the surrounding slag and this is likely to be important to understanding the overall kinetics of dephosphorisation in oxygen steelmaking.

CONCLUSIONS

There is significant evidence that top blown oxygen steelmaking vessels are far from equilibrium regarding phosphorus at the end of the blow. There is pilot plant data that suggests the majority of phosphorous removed from the metal is present in a solid solution phase containing Ca_2SiO_4 and $Ca_3(PO_4)_2$ and that metals droplets in the emulsion play a significant role in kinetic processes. Fundamental studies of droplet kinetics suggest that the 'bloating' of droplets is significant in understanding the behaviour of phosphorus in steelmaking and that the low residence time of droplets towards the end of the blow may explain why the system is far from equilibrium. Heat transfer modelling of droplets in the slag emulsion show that the droplets transfer significant heat to that layer and this is likely to affect the local kinetics of dephosphorisation. Models of oxygen steelmaking kinetics are limited by a lack of understanding of the kinetics of FeO formation, which is important as key phenomena such as decarburisation, 'bloating' and phosphorus removal are strongly related to the FeO content in the slag.

In general, there is a lack of plant data particularly relating to behaviour of droplets in the emulsion, the temperature gradients within the process and how they change with time. The recent developments in understanding droplet behaviour have the potential to accurately model key aspects of the process but improved industrial data will assist in refining and improving these models. Insights into the kinetics of this reaction would allow more efficient use of fluxes and reduced environmental impact of the process.

ACKNOWLEDGEMENTS

The author would like to acknowledge of the contribution of collaborators from Tata Steel, Swinburne University of Technology and McMaster University to my understanding of steelmaking kinetics, particularly, Akbar Rhamdhani, Ken Coley, Neslihan Dogan, Bapin Rout, Nirmal Madhavan and Aart Overbosch. The author is particularly grateful to the long-term support of the Tata Steel Ijmuiden for much of the work described in this paper.

REFERENCES

Assis, A, Tyler, M, Seetharman, S and Fruehan, R, 2015. Phosphorous Equilibrium Between Liquid Iron and CaO-SiO$_2$-MgO-Al$_2$O$_3$-FeO-P2O5 Slag Part 1, *Met Mat Trans B*, 46B:2255–2263.

Gu, K, Dogan, N and Coley, K, 2017. The Influence of Sulfur on Dephosphorization Kinetics Between Bloated Metal Droplets and Slag Containing FeO, *Met Mat Trans B*, 48B:2343–2353.

Madhavan, N, Brooks, G, Rhamdhani, M, Rout, B and Overbosch, A, 2022. Global Heat Transfer in Oxygen Steelmaking Process, *Metals*, 12:992.

Madhavan, N, Brooks, G, Rhamdhani, M, Rout, B, Overbosch, A, Gu, K, Kadrolkar, A and Dogan, N, 2021. Droplet Heat Transfer in Oxygen Steelmaking, *Met Mat Trans B*, 52B:4141–4155.

Millam, M S, Overbosch, A, Kapilashrami, Malmberg, D and Bramming, M, 2013. Some Observations and Insights on BOS refining, *Ironmaking and Steelmaking*, 40(6):460–469.

Rout, B, Brooks, G, Rhamdhani, A, Li, Z, Schrama, F and Sun, J, 2018. Dynamic Model of Basic Oxygen Steelmaking Process Based on Multi-Zone Reaction Kinetics: Model Derivation and Validation, *Met Mat Trans B*, 49B:537–557.

Turkdogan, E, 1996. *Fundamentals of Steelmaking* (Institute of Materials: London).

Urban, W, Weinberg, M and Cappel, J, 2015. Dephosphorisation Strategies and Modelling in Oxygen Steelmaking, *Iron and Steel Technology*, April 2015, pp 91–102.

Reoxidation of Al-killed ultra-low C steel by Fe$_t$O in CaO-Al$_2$O$_3$-MgO$_{sat.}$-Fe$_t$O slag representing RH slag by experiment and kinetic modelling

Y-M Cho[1], W-Y Cha[2] and Y-B Kang[3]

1. Graduate Student, Pohang University of Science and Technology, Pohang, Gyeongbuk 37673, Republic of Korea. Email: minn907@postech.ac.kr
2. Senior Researcher, POSCO, Pohang, Gyeongbuk 37673, Republic of Korea. Email: chawoo@posco.com
3. Professor, Pohang University of Science and Technology, Pohang, Gyeongbuk 37673, Republic of Korea. Email: ybkang@postech.ac.kr

ABSTRACT

Molten slag used in the secondary refining process (RH process) for the production of Ultra-Low C (ULC) steel is typically composed of CaO-Al$_2$O$_3$-MgO$_{sat.}$-Fe$_t$O with minor constituents. Fe$_t$O in the RH slag is the main source of reoxidation of the molten ULC steel in the ladle by 2Al + 3(Fe$_t$O) = (Al$_2$O$_3$) + 3Fe, thereby serving as the source of the alumina inclusions. On the other hand, Fe$_t$O also enhances the fluidity of the slag, thereby increasing inclusion absorption capacity. Optimum slag chemistry design is therefore required to produce clear ULC steel. The reoxidation kinetics was investigated in the present study by employing high-temperature experiments and developing the reaction rate model. Initial compositions of slag ((% CaO)$_0$/(% Al$_2$O$_3$)$_0$ and (% Fe$_t$O)$_0$)) and reaction temperature were varied and change of [% Al] in the molten steel was measured. The rate-controlling step was analysed. It was found that the rate-controlling step in some cases changed during the reoxidation: from a mass transport of Al in the molten steel to a mixed transport including Al$_2$O$_3$ in the molten slag. The experimental data validated the reaction rate model based on the elucidated reaction mechanism and FactSage thermodynamic database. The model suggested that high (% CaO)$_0$/(% Al$_2$O$_3$)$_0$ suppresses the reoxidation only when (% Fe$_t$O)$_0$ is low (5 or lower in the present study).

INTRODUCTION

Ultra-Low carbon (ULC) steel is distinguished as clean steel with exceptional formability and ductility, finding application in the production of various structural components for automobiles, necessitating intricate shaping and forming processes. The low carbon content (typically within the range of 10 to 30 mass ppm) in this steel grade makes it susceptible to oxidation during steelmaking and casting procedures. The production of ULC steel involves successive steps, including Blast Furnace (BF), Basic Oxygen Furnace (BOF), Ruhrstahl Heraeus (RH), and Continuous Casting (CC), with reoxidation posing a significant challenge during the RH and CC processes, resulting in the generation of non-metal inclusions (NMI).

ULC steel undergoes deoxidisation by Al alloys during the RH process. Once the liquid steel is adequately refined, it proceeds to a casting machine via a tundish. Reoxidation during the RH and casting processes can occur due to various factors, such as reactions with the RH slag/tundish flux, open-eye formation in the tundish, reactions with tundish and submerged entry nozzle refractories and aspiration through the sliding gate (Park and Kang, 2023). Consequently, Al in the liquid steel is depleted, resulting in the production of NMIs, such as alumina. This compromise not only hampers process efficiency but also adversely impacts the quality of the final products.

Narrowing the focus of the present study to reoxidation caused by RH slag reveals that Fe$_t$O, SiO$_2$ and MnO may be accountable for this phenomenon. Higher concentrations of these weak oxides can accelerate the reoxidation reaction. Conversely, since the slag absorbs non-metallic inclusions (NMI) at the steel–slag interface, a faster dissolution rate of NMI is necessary. Previous reports have summarised the dissolution rate of alumina inclusion in various slags (Park et al, 2020). Specifically, the dissolution rate of alumina inclusion in Fe$_t$O-containing RH slag was investigated using a high-temperature dissolution experimental technique coupled with the modified invariant interface approximation (Feichtinger et al, 2014). Findings indicate that a decrease in slag viscosity is a key factor in increasing the dissolution rate, aligning with proposals for various slag types. Given that the primary components of RH slag are CaO, Al_2O_3 and Fe_tO, an increase in Fe_tO content ($(\% \ Fe_tO)$) would enhance the dissolution rate. However, simultaneously, increasing the $(\% \ Fe_tO)$ would compromise the cleanliness of the liquid steel. Therefore, a comprehensive understanding of both phenomena – the dissolution of alumina into the RH slag and the reoxidation of the liquid steel by the RH slag – is crucial concerning $(\% \ Fe_tO)$ in the RH slag. The role of $(\% \ Fe_tO)$ from a dissolution perspective was recently discussed by the authors (Park *et al*, 2020).

The present article reports the authors' results that explore the reaction kinetics between ULC-Al killed liquid steel and Fe_tO-containing slag (representing RH slag) through high-temperature chemical reaction experiments and rate model analysis (Cho, Cha and Kang, 2021). Various rate-controlling steps were considered, compared with experimental data and a gradual change in the rate-controlling step was highlighted. The study concludes with practical suggestions for RH operation based on these findings.

EXPERIMENTAL PROCEDURE

A series of high-temperature reactions between Fe–Al alloy and CaO-Al_2O_3-$MgO_{sat.}$-Fe_tO slag were carried out, mostly at 1550°C under a well-controlled atmosphere. The reaction represents an interfacial reaction between Al-killed ULC steel and RH slag. The initial Al content in the liquid steel ($[\% \ Al]_0$) was set to about 0.1. Nine slag samples were synthesized by melting reagent grade of CaO (calcined from $CaCO_3$), Al_2O_3 and MgO in a graphite crucible. The melt was cooled, crushed and then burned under air to remove any residual C. Appropriate amount of reagent grade of FeO was then mixed to have various initial compositions of the slag samples: C/A ratio ($(\% \ CaO)_0/(\% \ Al_2O_3)_0$ and $(\% \ Fe_tO)_0$ were varied, where $_0$ means the initial state. The initial MgO content ($(\% \ MgO)_0$) was set to its saturation content, estimated by FactSage FTOxid database (Bale *et al*, 2016). The initial compositions of the slag samples are listed in Table 1. The compositions of the slag and the steel were given in mass per cent.

TABLE 1
Slag compositions used in the present study.

Name	C/A ratio	(% Fe_tO)$_0$	(% MgO)$_0$
A5		5	
A10	1.2	10	
A15		15	
B5		5	
B10	1.0	10	saturated
B15		15	
C5		5	
C10	0.8	10	
C15		15	

Some 500 g of Fe-0.01 per cent Al steel was melted in an induction furnace at 1550°C, under purified Ar atmosphere. The initial O content in the alloy was approximately 0.002 per cent. Then, 40 g of the synthesized slag prepared as above was charged onto the molten steel surface using an alumina guide tube through a hole available in the upper endcap. This moment was set to the beginning of the reaction ($t = 0$). After the pre-determined time, a small portion of the liquid steel was periodically sampled using a quartz tube, followed by quenching in water. The composition of the liquid steel was analysed by Inductively Coupled Plasma – Atomic Emission Spectroscopy (ICP-AES) for soluble Al content ($[\% \ S \ Al]$) and total Al content ($[\% \ T \ Al]$) by inert gas fusion infrared absorptiometry for total O content ($[\% \ T \ O]$). The distribution of non-metallic inclusions in the alloy was analysed using field-emission scanning electron microscopy with an energy-dispersive X-ray spectroscopy. All the data were reported previously (Cho, Cha and Kang, 2021) and were reproduced in this paper.

EXPERIMENTAL RESULTS

The time evolution of the alloy's composition is shown in Figure 1 for the case of high C/A ratio (= 1.2): [% T Al], [% S Al], [% T O] versus reaction time t (seconds) of liquid steel reacted with slags (A5, A10 and A15) at 1550°C. Both decreased continuously and those decreased faster when (% $Fe_tO)_0$ was higher. The following chemical reaction occurred:

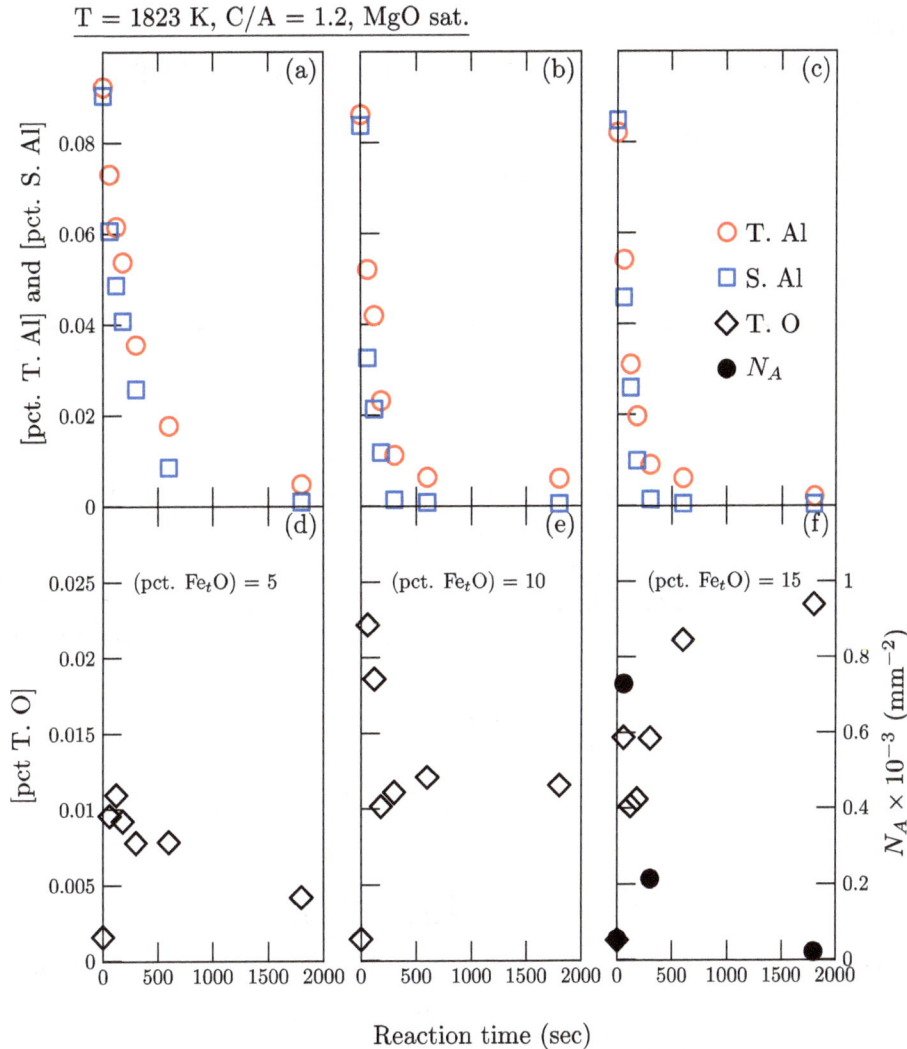

$$2\underline{Al} + 3(FeO) = (Al_2O_3) + 3Fe \tag{1}$$

FIG 1 – Composition change of the liquid steel ([% Al]$_0$ = 0.1 reacted with CaO-Al$_2$O$_3$-MgO$_{sat.}$-Fe$_t$O slag): (a) to (c) [% T Al] and [% S Al], (d) to (f) [% T O] of the liquid steel reacted with the slag A5, A10 and A15, respectively, measured in the present study. Closed circles in (f) are the measured inclusion number density N$_A$, mm^{-2}) (Cho, Cha and Kang, 2021).

[% T Al] was slightly higher than [% S Al] and the difference should correspond to the insoluble Al content relevant to alumina inclusion in the steel. [% T O] is a sum of soluble O content ([% S O]) and insoluble O content ([% I O]) where the latter corresponds to the O content relevant to alumina inclusion in the steel. Although a simultaneous analysis technique of [% S O] and [% I O] in Al-killed steel was recently developed (Hong and Kang, 2021), it was not attempted in the present study. Therefore, the measured [% T O] should be treated carefully. When (% $Fe_tO)_0$ = 5, [% T O] first increased, showed a maximum, then decreased. It was thought that (Al$_2$O$_3$) in Equation 1 was mostly alumina inclusion, which gradually disappeared in the alloy due to buoyancy force. On the other hand, when (% $Fe_tO)_0$ = 15, [% T O] first increased, showed a maximum, then decreased. This was followed by a minimum in the [% T O], then [% T O] increased again. There should have been the following reaction:

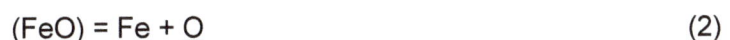

$$(FeO) = Fe + \underline{O} \tag{2}$$

and the $\underline{O}$ can be attributed to the increase of [% T O]. Indeed, N_A, the measured inclusion number density by FE-SEM, showed different behaviour to that of [% T O]. Therefore, the measured [T O] after passing the minimum in the [% T O] is likely to correspond to soluble O content. It is evident that increasing (% $Fe_tO)_0$ increased O content in the liquid alloy and this results in 'reoxidation' in the liquid steel. Other cases (B5, B10, B15, C5, C10, C15) were also similar (Cho, Cha and Kang, 2021). Now, the task is to understand how the slag composition (C/A, (% $Fe_tO)_0$) influences the reoxidation reaction rate.

REOXIDATION KINETICS MODEL

The reoxidation reaction rate was represented by a temporal change of [% S Al], d[% S Al]/dt, as the soluble Al is the reactant of the reoxidation reaction (Equation 1). Assuming that: (1) local equilibrium at the slag-alloy interface is valid for Equation 1; and (2) mass balance in the system is kept thereby the change of the mass of slag components (FeO, Al_2O_3) can be calculated by the change of the mass of alloy components ([% S Al]); the reaction rate could be formulated in two ways:

$$\frac{d[\% \text{ S Al}]}{dt} = -\frac{A\rho_{\text{steel}}}{W_{\text{steel}}} k_{\text{Al}}^{\text{M}} \left([\% \text{ S Al}] - [\% \text{ S Al}]^i\right) \tag{3}$$

if the reoxidation reaction was only controlled by mass transport of Al (from bulk to the slag-alloy interface), or

$$\frac{d[\% \text{ S Al}]}{dt} = -\frac{AM_{\text{Al}}}{W_{\text{steel}}} k_{\text{Al}-\text{Al}_2\text{O}_3}^{\text{app}} \left(\frac{\rho_{\text{steel}}}{M_{\text{Al}}}[\% \text{ S Al}] - \frac{1}{L_{\text{Al}-\text{Al}_2\text{O}_3}} \frac{\rho_{\text{slag}}}{M_{\text{Al}_2\text{O}_3}}(\% \text{ Al}_2\text{O}_3)\right) \tag{4}$$

if the reoxidation reaction was only controlled by mass transport of Al and Al_2O_3 (from bulk to the slag-steel interface), simultaneously, where t, A, ρ_j, W_j, M_j, k_{Al}^{M} and $L_{\text{Al}-\text{Al}_2\text{O}_3}$ are the reaction time, the reaction area (interfacial area between the slag and the alloy), the density of the phase j, the mass of the phase j, the atomic or molecular mass of j, the mass transport coefficient of Al in the alloy and the distribution coefficient of Al between the alloy and the slag which is expressed as (Cho, Cha and Kang, 2021):

$$L_{\text{Al}-\text{Al}_2\text{O}_3} = \frac{(\% \text{ Al}_2\text{O}_3)^i}{[\% \text{ S Al}]^i} \frac{\rho_{\text{slag}}}{\rho_{\text{steel}}} \frac{M_{\text{Al}}}{M_{\text{Al}_2\text{O}_3}} \tag{5}$$

The superscript i means the 'interface'. The apparent rate constant for the mixed rate model in Equation 4 is:

$$k_{\text{Al}-\text{Al}_2\text{O}_3}^{\text{app}} \equiv \frac{1}{k_{\text{Al}}^{\text{M}}} + \left(\frac{1}{2L_{\text{Al}-\text{Al}_2\text{O}_3}}\right) \frac{1}{k_{\text{Al}_2\text{O}_3}^{\text{S}}} \tag{6}$$

where $k_{\text{Al}_2\text{O}_3}^{\text{S}}$ is the mass transport coefficient of Al_2O_3 in the slag. As was proposed by Kim and Kang (2018), the mass transport coefficient of Al_2O_3 in slag phase was assumed to be inversely proportional to viscosity of the slag:

$$k_{\text{Al}_2\text{O}_3}^{\text{S}} = k_{\text{Al}_2\text{O}_3}^{\text{S},\circ} \times \frac{\eta}{\eta^\circ} \tag{7}$$

where η is the viscosity of the slag, and $k_{\text{Al}_2\text{O}_3}^{\text{S},\circ}$, and η° are the mass transport coefficient of Al_2O_3 and the viscosity of a reference slag. k_{Al}^{M} and $k_{\text{Al}_2\text{O}_3}^{\text{S},\circ}$ were optimised by fitting the experimental data to the model equation. It should be noted that the Equation 4 contains the interfacial concentration terms via $L_{\text{Al}-\text{Al}_2\text{O}_3}$, which varies during the reaction. The interface compositions ([% S Al]^i and (% $Al_2O_3)^i$) also vary during the reaction and those can be calculated using CALPHAD approach without assuming the volume or thickness of the interface reaction zone. Detailed procedure to obtain the interface compositions and subsequent steps to calculate $L_{\text{Al}-\text{Al}_2\text{O}_3}$, $k_{\text{Al}-\text{Al}_2\text{O}_3}^{\text{app}}$ and d[% S Al]/dt can be found in the present authors' article (Cho, Cha and Kang, 2021).

REOXIDATION KINETICS ANALYSIS

Figure 2 shows the extent of the reoxidation by temporal change of the normalised soluble Al content in the logarithmic scale (log [% S Al]/[% S Al]$_0$) (Cho, Cha and Kang, 2021). Closed symbols are the measured data. Increasing (% Fe$_t$O)$_0$ from 5 to 15 (from bottom panel to top panel) decreased [% S Al] faster, therefor the reoxidation occurred faster.

T = 1823 K, C/A = 0.8 – 1.2, MgO sat., (pct. FeO)$_0$ = 5-15

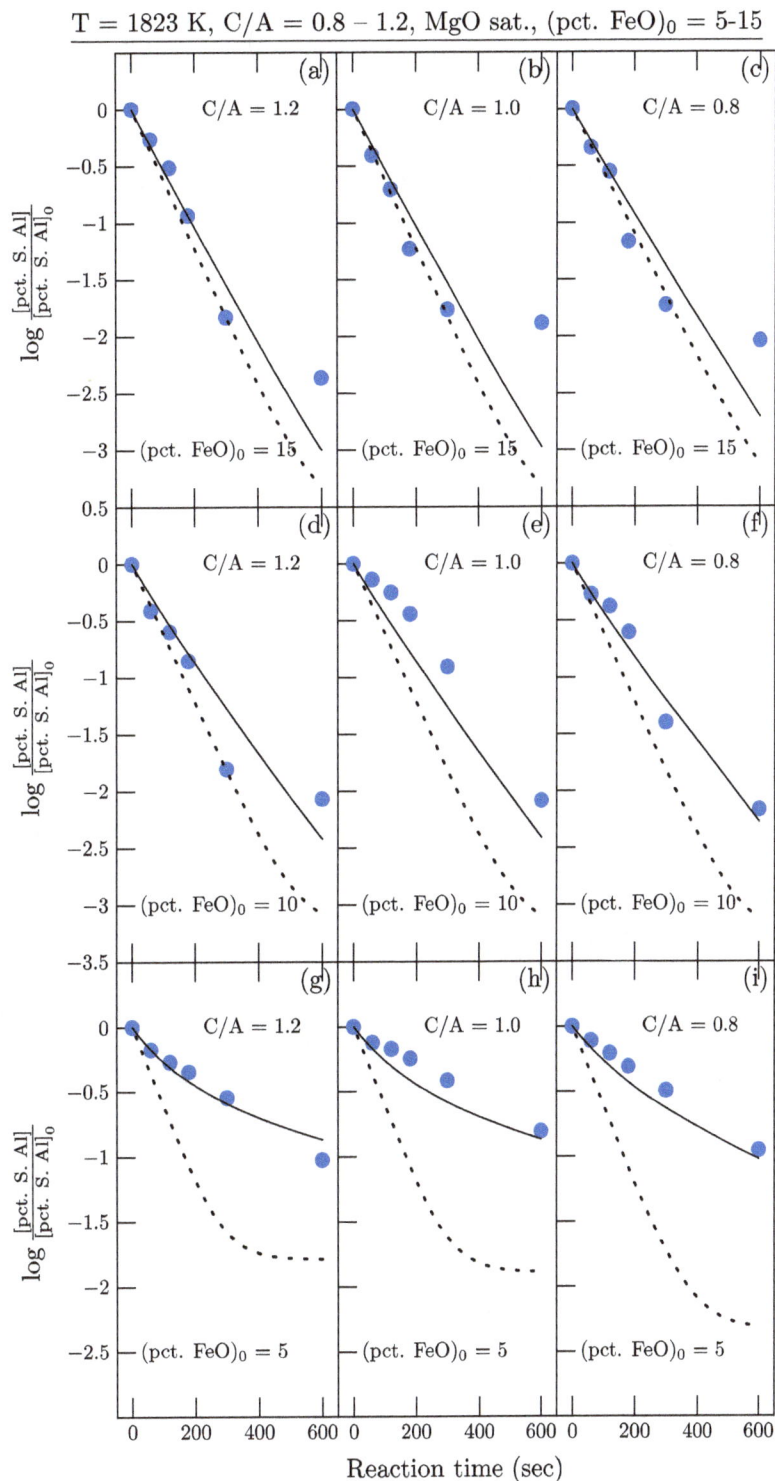

FIG 2 – Comparison between the experimental data and the model calculations for log [% S Al]. Symbols are the experimental data, the dotted lines from Equation 3 assuming the single rate controlling mechanism and the solid lines from Equation 4 assuming the mixed rate controlling mechanism (Cho, Cha and Kang, 2021).

The dotted lines were the calculated extent of the reoxidation by Equation 3, assuming the reoxidation was exclusively controlled by the mass transport of Al in the steel only. It was reasonable when (% Fe$_t$O)$_0$ was high (15), but deviated gradually from the experimental data as (% Fe$_t$O)$_0$ decreased. The lower (% Fe$_t$O)$_0$ means the higher viscosity of the slag. This suggests that the mixed controlling concept formulated by Equation 4 would be adequate. After a series of calculations, k_{Al}^M

and $k^{S,\circ}_{Al_2O_3}$ were optimised by fitting the experimental data to the model equation. The solid lines are the calculated extent of the reoxidation by Equation 4 in the context of the mixed controlling rate. A good agreement was achieved. It should be stressed that the extent of the reoxidation in view of the experiment and the model calculations does not clearly suggest the role of C/A on the reoxidation.

The developed model was manipulated to extract the reaction mechanism. The extent of reoxidation by Equation 4 requires $k^{app}_{Al-Al_2O_3}$ and $L_{Al-Al_2O_3}$ and these two properties require the property at the interface $[\% \text{ S Al}]^i$. Figure 3 shows the calculated $[\% \text{ S Al}]^i$, $L_{Al-Al_2O_3}$ and $k^{app}_{Al-Al_2O_3}$ for the nine slag samples (numbers near the curly bracket means $(\% \text{ Fe}_tO)_0$).

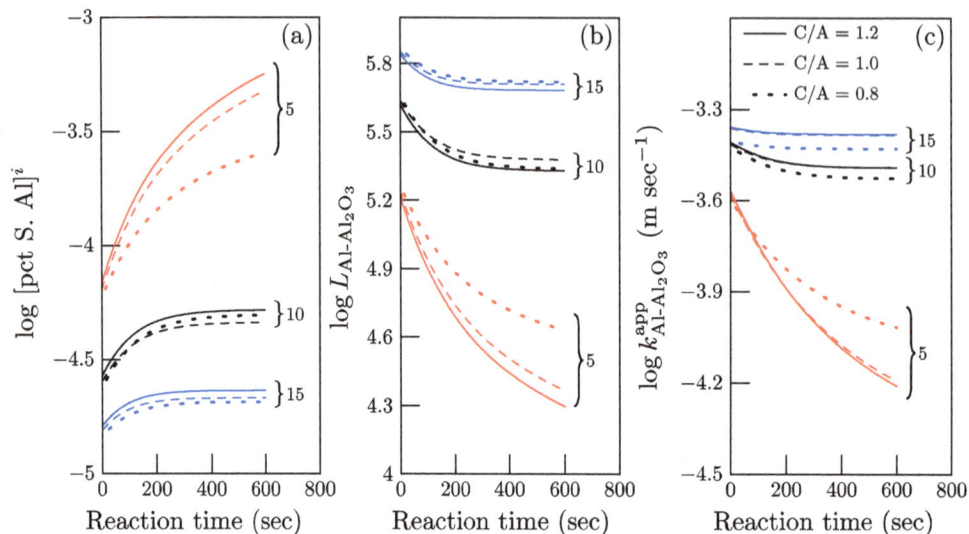

FIG 3 – Calculated various properties from during the reoxidation reaction at 1550°C: (a) $[\% \text{ S Al}]^i$, (b) $L_{Al-Al_2O_3}$ and (c) $k^{app}_{Al-Al_2O_3}$ (Cho, Cha and Kang, 2021).

The three properties depend on t, $(\% \text{ Fe}_tO)_0$ and C/A. In general, as the reoxidation proceeded, soluble Al at the interface increased, thereby decreasing the driving force of the reaction. This is reflected in the decrease in the Al distribution ratio, inducing a lower driving force. This lowers the apparent rate constant and finally, the reoxidation reaction ceased to proceed further. It should be stressed that the apparent rate constant is indeed not a constant, but varies during the reaction.

$(\% \text{ Fe}_tO)_0$ is the most dominant factor in the control of the reoxidation rate. When $(\% \text{ Fe}_tO)_0$ is high, the C/A ratio is not a critical factor in the three properties in Figure 2. However, at the lower $(\% \text{ Fe}_tO)_0$ = 5, C/A ratio does influence the three properties: increasing C/A ratio increased $[\% \text{ S Al}]^i$ and decreased $L_{Al-Al_2O_3}$ and $k^{app}_{Al-Al_2O_3}$. Therefore, increasing C/A ratio would decrease the reoxidation rate. High C/A ration decreased the reoxidation rate because $L_{Al-Al_2O_3}$ is low, which is due to high $[\% \text{ S Al}]^i$. At the slag-steel interface, $[\% \text{ S Al}]^i$ is determined by the equilibrium of Equation 1. It is known that increasing C/A in the CaO-Al$_2$O$_3$-Fe$_t$O slag lowers the activity of Fe$_t$O at the same $(\% \text{ Fe}_tO)$. This results in a higher soluble Al content. Although the slag used in the present study and RH slags in practical operations contain some other minor components such as MgO, increasing C/A ratio in RH slag of low $(\% \text{ FeO})_0$ (ie ~5) can lower the reoxidation rate. This conclusion is in agreement with recent reports (Ji *et al*, 2018; Ji *et al*, 2020).

In practical RH slag operation, Fe$_t$O originates from BOF slag, introduced due to partial entrapment during tapping. Fe$_t$O is a major cause of reoxidation during RH processing. Deoxidising the slag by eliminating Fe$_t$O using Al or C sources is a potential method to control reoxidation. However, reducing Fe$_t$O is challenging due to slag solidification and fume generation, impacting workability. Yet, higher Fe$_t$O aids in absorbing alumina inclusions, enhancing viscosity (Park *et al*, 2020). Balancing Fe$_t$O content is crucial, considering reoxidation prevention versus viscosity benefits. Alternatively, killing Fe$_t$O and adding a non-reducible flux can achieve the same goal, but cost and refractory life become additional considerations.

CONCLUSIONS

In the investigation of ULC steel reoxidation during the RH process, an analysis of the reaction rate and mechanism was conducted through a series of laboratory-scale experiments and the utilisation of a reaction rate model that integrates CALPHAD thermodynamics and mass transport theory (Cho, Cha and Kang, 2021). The observed increase in reoxidation rate with higher initial percentages of Fe_tO $((\% \, Fe_tO)_0)$ and the significance of variations in the C/A ratio were particularly notable when $(\% \, Fe_tO)_0$ was low (5 in this study). The predominant influence on the reoxidation rate was found to be the mass transport of Al in liquid steel. However, as $(\% \, Fe_tO)_0$ decreased, the rate-controlling step shifted towards mixed transport control, involving Al_2O_3 transport. A new reaction rate model was developed, incorporating local equilibrium through the CALPHAD method, mixed transport rate in both phases and the viscosity effect on mass transport coefficients of Al_2O_3 or Fe_tO in the slag. This model demonstrated a good alignment with the experimental data. At high $(\% \, Fe_tO)_0$, the C/A ratio in the slag had minimal impact on the reoxidation rate. Conversely, at low $(\% \, Fe_tO)_0$, increasing C/A effectively reduced the reoxidation rate, showcasing its potential to suppress reoxidation.

ACKNOWLEDGEMENTS

This work was financially supported by POSCO, Republic of Korea.

REFERENCES

Bale, C W, Belisle, E, Chartrand, P, Decterov, S A, Eriksson, G, Cheribi, A E, Hack, K, Jung, I H, Kang, Y B, Melancon, J, Pelton, A D, Petersen, S, Robelin, C, Sangster, J, Spencer, P and Van Ende, M A, 2016. FactSage thermochemical software and databases, 2010–2016, *Calphad*, 54:35–53.

Cho, Y-M, Cha, W-Y and Kang, Y-B, 2021. Reoxidation of Al-killed ultra-low C steel by Fe_tO in RH Slag: experiment, reaction rate model development, and mechanism analysis, *Metallurgical and Materials Transactions B*, 52(5):3032–3044.

Feichtinger, S, Michelic, S K, Kang, Y B and Bernhard C, 2014. In situ observation of the dissolution of SiO_2 particles in $CaO-Al_2O_3-SiO_2$ slags and mathematical analysis of its dissolution pattern, *Journal of the American Ceramic Society*, 97(1):316–325.

Hong, H-M and Kang, Y-B, 2021. Simultaneous analysis of soluble and insoluble oxygen contents in steel specimens using inert gas fusion infrared absorptiometry, *ISIJ International*, 61(9):2464–2473.

Ji, Y, Liu, C, Lu, Y, Yu, H, Huang, F and Wang, X, 2018. Effects of FeO and CaO/Al_2O_3 ratio in slag on the cleanliness of Al-killed steel, *Metallurgical and Materials Transactions B*, 49(6):3127–3136.

Ji, Y, Liu, C, Lu, Y, Yu, H, Huang, F and Wang, X, 2020. Oxygen transfer phenomenon between slag and molten steel for production of IF steel, *Journal of Iron and Steel Research International*, 27:402–408.

Kim, M-S and Kang, Y-B, 2018. Development of a multicomponent reaction rate model coupling thermodynamics and kinetics for reaction between high Mn-high Al steel and $CaO-SiO_2$-type molten mold flux, *Calphad*, 61:105–115.

Park, J-H and Kang, Y-B, 2023. Reoxidation phenomena of liquid steel in secondary refining and continuous casting Processes – A Review, *Steel Research International*. https://doi.org/10.1002/srin.202300598.

Park, Y, Cho, Y M, Cha, W Y and Kang, Y B, 2020. Dissolution kinetics of alumina in molten $CaO-Al_2O_3-Fe_tO-MgO-SiO_2$ oxide representing the RH slag in steelmaking process, *Journal of the American Ceramic Society*, 103(3):2210–2224.

Effect of liquefaction controlling components in carbon-free mould powder for the continuous casting of ultra-low carbon steels

N Gruber[1]

1. Senior Scientist, Montanuniversitaet Leoben – Chair of Ceramics, Leoben 8700, Austria. Email: nathalie.gruber@unileoben.ac.at

ABSTRACT

In the continuous casting of ultralow carbon (ULC) steels, free carbon is used to control the melting behaviour of mould powders. If the carbon is not completely removed during melting, it is enriched at the top of the slag pool. Liquid steel may come into contact with this layer because of the turbulence of the molten metal, resulting in its recarburisation, which negatively affects the desired product quality. Thus, a reduction in carbon input is desirable. For this purpose, SiC and/or Si_3N_4 with and without antioxidants were selected as melt-control additives to replace carbon in the mould powders. Thermodynamic calculations were performed to quantify their effect on the melting behaviour based on the chemical composition of a flux already applied to ULC steels. To experimentally assess the liquefaction behaviour, laboratory mould powders were prepared and annealed in steel crucibles closed with a lid. Crucibles were inserted into a furnace that was already preheated to selected temperatures between 900–1200°C for 10 mins and quenched to room temperature. Subsequently, the samples were mineralogically investigated. The results confirmed those obtained from the thermodynamic calculations. Si_3N_4, and SiC in particular, are suitable raw materials for delaying the solid-solid reactions of raw material components during melting. Owing to their stability at high temperatures, the necessary SiO_2 content to form a liquid phase is not available, resulting in lower amounts of the liquid phase. The addition of antioxidants to delay the oxidation of SiC further reduces this positive effect. Attempts to decrease the SiC content without negatively affecting the melting behaviour resulted in a reduction in the CO_2 emission by at least 27 g CO_2/kg of mould powder when compared to the carbon-containing standard mould powder. These investigations revealed differences in the melting behaviours of granules and loose powders, which are related to their respective production processes.

INTRODUCTION

In the continuous casting of steel mould powders added to the liquid steel pool form different horizontal layers during melting, a homogeneous liquid layer exists in contact with the steel and the original mould powder is still present on the upper surface. In between a sintered layer owing to the formation of liquid and new solid phases exists. The slag infiltrates the gap between the steel and the mould. A liquid layer is in contact with the strand, a glassy layer is close to the mould and a crystalline layer lies in between. In particular, the structure of the crystalline layer affects heat transfer from the strand to the mould (Mills *et al*, 2004).

A common approach for controlling the melting behaviour of mould powders is the addition of carbon particles of different types and sizes, for example, graphite or carbon black, to a mixture of raw-material components. At lower temperatures, they reduce the contact between raw material particles and delay solid-solid reactions. A liquid phase is formed with increasing temperature. Because of the nonwetting carbon particles, the liquid droplets remain separated and a continuous liquid is prohibited. This phenomenon is known as the skeleton effect (Mills, 1990; Kawamoto *et al*, 1994; Kölbl, Marschall and Harmuth, 2009; Kromhout, 2013). After the carbon is burned off, which depends on the oxygen supply in the high-temperature areas of the mould powder layers, a homogeneous liquid is formed. If the carbon is not completely oxidised, it accumulates on top of the liquid slag layer, resulting in a carbon-rich layer with a carbon content of more than 20 wt per cent for mould powders with carbon contents of solely 3.5 wt per cent when delivered (Supradist, Cramb and Schwerdtfeger, 2004). This effect has also been confirmed by laboratory investigations (Yan *et al*, 2015). Owing to the oscillation of the mould and the turbulence of the steel flow from the submerged entry nozzle, liquid steel may come into contact with the carbon-enriched layer, resulting in its recarburisation. In particular, for ultralow-carbon (ULC) steels, this reaction changes the steel properties, which reduces the strand quality. Various approaches have been proposed to inhibit this reaction. A thicker slag layer increases the distance between the carbon-rich layer and steel bath, making contact more difficult. Furthermore, different efforts have been made to improve the raw material composition of the powders. Graphite is replaced by quick-burning or activated carbons to support its oxidation, and oxidising agents such as MnO_2 or catalysts such as Fe_2O_3 for reactions at low oxygen activity and high temperatures are added (Yi, Song and Peng 2013; Terada *et al*, 1991; Han *et al*, 2023; Nakato *et al*, 1991). Another approach is the replacement of free carbon in the mould powder by using alternative raw material components showing similar properties as carbon (Takeuchi *et al*, 1976, 1978; Debiesme *et al*, 1996, 1998). With respect to the skeleton effect, the non-wettability of slags is in the focus of these studies. In the first attempt, nitrides were added to the mixture instead of carbon and tested. For this purpose, BN, Si_3N_4, MnN, CrN, FeN, AlN, TiN, or ZrN were suggested (Terada *et al*, 1991; Takeuchi *et al*, 1976, 1978). Boron nitride (BN) is the most appropriate option owing to its similarity to carbon in terms of crystalline structure and wetting behaviour. Mould powder investigations in the laboratory showed that the sintering tendency of fluxes containing BN was reduced. The minimum amount required to effectively control the melting behaviour of the mould powders depends on the sizes of the BN particles and the oxidic and fluoridic base materials. A decrease in the raw material particle size resulted in decreased melting rates, which also accounted for the BN. The fusion rates were determined for the original mould powder with carbon and for mixtures containing different amounts of BN in the laboratory. Those containing 2.10 wt per cent of BN showed similar behaviour to the original sample. Furthermore, compared with carbon black, BN exhibited a reduced sintering tendency and reduced gas formation. In contrast, during the oxidation of boron nitride, B_2O_3 is formed, acting as a fluxing agent and causing the formation of the first liquid phase. Thus, antioxidants such as Al or CaSi were added to the samples to impede its oxidation at lower temperatures. However, boron oxide not only improves the sintering tendency but also lowers the slag viscosity. Therefore, solely for samples showing a CaO/SiO_2 (C/S) ratio of 0.9, proper viscosities for the continuous casting process were obtained. Based on the laboratory results, the mould powder was selected for testing with a defined steel grade in a continuous caster. After the trial, the surface of the strand was analysed with respect to surface quality. The results revealed that an increase in nitrogen content on the strand surface being as detrimental as the effect of recarburisation. Therefore, this amount must be reduced. The realisation was achieved by deliberating retaining some carbon from the mould powder to ensure controlled melting behaviour (Terada *et al*, 1991; Takeuchi *et al*, 1976, 1978).

Additionally, Si_3N_4 was considered to effectively control the melting of mould powders. The benefit compared with BN is that SiO_2, which is already included in the original mould powder composition, is formed during oxidation. Thus, the viscosity of the slag does not change after liquification. Investigations in the laboratory suggested a particle size of 5 µm and a specific surface of 2.5–3.5 m^2g^{-1} to meet the required demands. Subsequently, a mould powder was prepared for industrial trials. During operation, suitable melting rates were observed, but the surface of the steel strand also showed considerable nitride absorption (Debiesme *et al*, 1996, 1998; Lefebvre *et al*, 1996; Sun *et al*, 2019).

The concept of associating specially coloured mould powders with steel grades to facilitate on-site identification during operation requires the elimination of carbon from the mould powder composition. Carbon-free samples were investigated using a heating microscope. The results showed that the softening, melting and flow points agreed with those of the carbon-containing products. In contrast, the melting rate increased considerably. The authors (Macho *et al*, 2005) assumed sufficient thermal insulation of the steel in the mould, as long as a sufficiently thick mould powder layer rests on the mould slag.

Efforts have also been made by various research groups to replace free carbon with SiC (Lefebvre *et al*, 1996; Maillart, Chaumat and Hodaj, 2010; Ning and Jinghao, 1998; Safarian and Tangstad, 2009; Park *et al*, 2016). Similar melting rates were achieved for mould powders containing 2–2.5 wt per cent of carbon or 5–6 per cent SiC. A detailed investigation of the layers formed during the experiment revealed three layers. Contrary to the standard, solely carbon-containing product, the sintered layer increased considerably, whereas the original sample layer remained very thin. This was explained by the oxidation of SiC, resulting in a decrease in the basicity of the mould powder, thus accelerating liquid-phase formation. Consequently, carbon was partially replaced by SiC. These products, which contained different ratios of free carbon to SiC, were then tested in contact with liquid steel. Analysis of the chemical compositions of the steel samples indicated an increasing tendency for recarburisation when the carbon content exceeded 1 wt per cent.

In the investigations quoted above, phenomenological methods were primarily used to compare the melting behaviour of mould powder compositions where carbon was partially or totally replaced as the melt-control additive. However, the effect of these additives during melting has not yet been described. Thus, in this study, SiC and/or Si_3N_4, with and without the addition of antioxidants, were selected to replace carbon in a mould powder already used for the continuous casting of steels for investigation in laboratory experiments.

EXPERIMENTAL

Mould powder, already used in the continuous casting of ultralow-carbon steels, was selected (hereafter denoted as MP0). Based on its chemical composition, this mould powder was reproduced as a mixture of raw material components that are generally used for the production of mould powders (MP2). As a reference, a sample without melt-controlling additives was prepared (MP1). Subsequently, to substitute the carbon content of MP2, SiC (MP3–6) and/or Si_3N_4 (MP9/MP8) were added to control the melting behaviour of the mixture. For this purpose, the composition was modified to achieve the chemical composition of MP1 after total oxidation of the carbide and nitride phases. An amount of 25 wt per cent of the total SiO_2 content in the mould powder was substituted by SiC and/or Si_3N_4. For refractory materials, the addition of antioxidants is the most common technology to prevent carbon oxidation and increase service lifetime owing to enhanced corrosion resistance. Additionally, the desired properties of the product, such as the mechanical properties, are improved (Ghosh, Jagannathan and Ghosh, 2001; Wang and Yamaguchi, 2001; Dai, Xiao and Ding, 2021). Laboratory mould powders MP4–MP6 contain Si and/or Al in addition to SiC, which prohibits the premature oxidation of SiC. A similar addition of Si and Al serves to protect carbon in refractories; so-called antioxidants (Ghosh, Jagannathan and Ghosh, 2001; Wang and Yamaguchi, 2001; Dai, Xiao and Ding, 2021).

Prior to laboratory investigations, thermodynamic calculations of the phase distribution with respect to temperature were performed using FactSage™, version 8.2 (by CRCT-ThermFact Inc. and GTT-Technologies). The mineral composition and possible liquid phase of each sample were calculated in equilibrium state up to 1550°C, which corresponds to the steel bath temperature in mould, while maintaining the total initial oxygen amount constant (no oxidation). As a result, the quantities of the phases formed in dependence on temperature are depicted in diagrams, and the amount of liquid phase at 1000°C was calculated, indicating the melting tendency of each sample for comparison. The experimental investigations of MP1–MP3 aimed to assess the effect of the additives used on the liquefaction behaviour, and it was revealed that, contrary to the calculations performed here, partial oxidation reactions occurred. To minimise the discrepancy between the chemical composition of the as-received commercially fabricated mould powder and the mixtures prepared in this study, the quantity of each raw material component was calculated based on its chemical composition. Quartz, feldspar, calcite, magnesite, fluorite, carbon, wollastonite, soda, corundum, fly ash, glass, blast furnace slag, and hematite were used as base materials, and SiC, Si, Al, and Si_3N_4 were added to control the melting behaviour of the specimen. The mixtures were prepared as listed in Tables 1 and 2. After homogenising, 20 g of each loose mould powder mixture were filled into steel crucibles (30·30·40 mm^3) closed by a steel lid and inserted into the furnace preheated to a selected temperature between 900 and 1200°C. Melt-control additives are expected to be partially oxidised in the crucible atmosphere. At least at lower temperatures, this reaction was incomplete. The oxygen supply was hindered by the lid, simulating the oxygen support under service conditions by diffusion. The crucibles were kept in the furnace for 10 mins before being removed and quenched at room temperature. Mineralogical investigations were conducted on polished sections using reflected light microscopy and scanning electron microscopy coupled with energy-dispersive X-ray microanalysis. The latter enables the identification of phases that are stable in dependence on temperature. The samples were ground for X-ray analysis. Finally, the results of the thermodynamic and laboratory investigations were compared.

TABLE 1

Raw material compositions of MP1–MP9 (in wt%).

	MP1	MP2	MP3	MP4	MP5	MP6	MP7	MP8	MP9
	no melt-control additive	with carbon	with SiC	SiC, antiox. Si	SiC, antiox. Al	SiC, antiox. Si and Al	antiox. Al	with Si$_3$N$_4$	with Si$_3$N$_4$ and SiC
Quartz	6.8	6.9	6.3		4.9	0.3			
Feldspar	4.3	4.9	3.8			3.6			
Calcite	2.1	3.3	17.8	7.6		12.0		3.0	15.3
Fly ash	13.3	11.1	13.0	7.4				16.0	7.7
Glass						8.1	22.7		15.3
Magnesite	9.4	9.1	8.1	9.5	10.9	9.3	9.9	9.3	7.7
Fluorite	19.5	18.7	18.7	19.5	20.4	19.2	20.3	19.8	19.1
Carbon		4.4							
Wollastonite	38.1	35.7	20.0	34.0	45.5	28.6	39.9	37.1	19.7
Soda	5.3	4.9	5.1	6.1	6.3	4.0	2.5	6.2	3.4
Corundum				2.4		0.7		0.5	1.7
Blast furnace slag				2.7	0.5	2.8			2.5
SiC			6.0	6.5	6.8	6.4			4.8
Hämatite	1.3	1.2	1.2	1.7	2.4	2.3	2.4	1.2	1.6
Si metallic				2.60		1.3			
Al metallic					2.3	1.3	2.3		
Si$_3$N$_4$								6.9	1.7

Owing to the suitable application of SiC to delay the reactions of the raw materials (Results), further attempts were made to reduce the SiC content of MP3 without negatively affecting its melting rate. For MP3, 25 wt per cent of the total SiO$_2$ content in MP0 was replaced with SiC. For MP10–MP13, this amount was continuously reduced. Mixtures were prepared according to Table 2, homogenised and granulated using a bottle roller to compare their melting behaviours. Subsequently, they were annealed in a preheated furnace according to the procedure described above.

TABLE 2
Raw material compositions of MP10–MP13 (in wt%).

	MP10	MP11	MP12	MP13
	20 wt% SiO₂ replaced by SiC	15 wt% SiO₂ replaced by SiC	10 wt% SiO₂ replaced by SiC	5 wt% SiO₂ replaced by SiC
Quartz			1.6	3.7
Feldspar	4.1	7.9	5.0	4.7
Calcite	4.14	3.5		
Fly ash	16.6	13.4	15.3	15.3
Magnesite	10.1	10.3	10.3	10.3
Fluorite	21.7	21.5	21.5	21.2
Wollastonite	40.1	41.2	44.1	43.9
Soda	6.0	5.3	5.8	5.8
SiC	5.8	4.3	2.8	1.4
Hämatite	1.3	1.5	1.3	1.4

RESULTS AND DISCUSSION

Thermodynamic calculations

Figure 1 summarises the phase compositions of MP0=MP2, MP3 and MP9 with increasing temperature. As expected, the phases differed only slightly among the calculated compositions; however, their stability ranges varied considerably. In all specimens, cuspidine ($Ca_4(Si_2O_7)(F,OH)_2$), formed by the reaction of the raw materials during heating, is the last solid phase that forms before a homogeneous liquid. If the iron oxide in the mould powder is reduced by another component, solid iron and iron silicide with diverse Fe/Si ratios are also formed. At low temperatures, nepheline ($Na_3KAl_4Si_4O_{16}$) is observed in all mixtures and combeite ($Na_2Ca_2Si_3O_9$) and fluorite (CaF_2) are observed in MP0–MP3 and MP8–MP9. MP3, MP8 and MP9 represent magnesium-fluoride silicates ($Mg_5F_2Si_2O_8$ or $Mg_9F_2Si_4O_8$).

In comparison to the carbon-containing mould powder, the replacement of SiO_2 by SiC and/or Si_3N_4 caused an increase in the quantity of cuspidine (MP3, MP8 and MP9). Owing to the incomplete oxidation of the melt-control components, the availability of SiO_2 was reduced in these samples. Thus, the formation of phases with a higher SiO_2 content was prohibited. Although diopside ($CaMgSi_2O_6$) can be found in MP0 and MP1, it is not stable in the samples containing SiC and/or Si_3N_4. The same applies to mixtures of SiC and antioxidants. Cuspidine and merwinite ($Ca_3Mg(SiO_4)_2$), olivine (($Mg,Fe)_2SiO_4$) and melilite (($Ca,Na)_2(Al,Mg,Fe^{2+})(Al,Si)SiO_7$) were present instead of clinopyroxene. Compared with the carbon-containing samples, the quantity of cuspidine did not increase significantly. Furthermore, because of the stability of SiC, particularly at lower temperatures, the CaO/SiO_2 ratio is low, and the formation of the liquid phase shifts to higher temperatures, where SiC oxidation is promoted, thus increasing the C/S ratio. This behaviour was not observed in the samples containing free carbon.

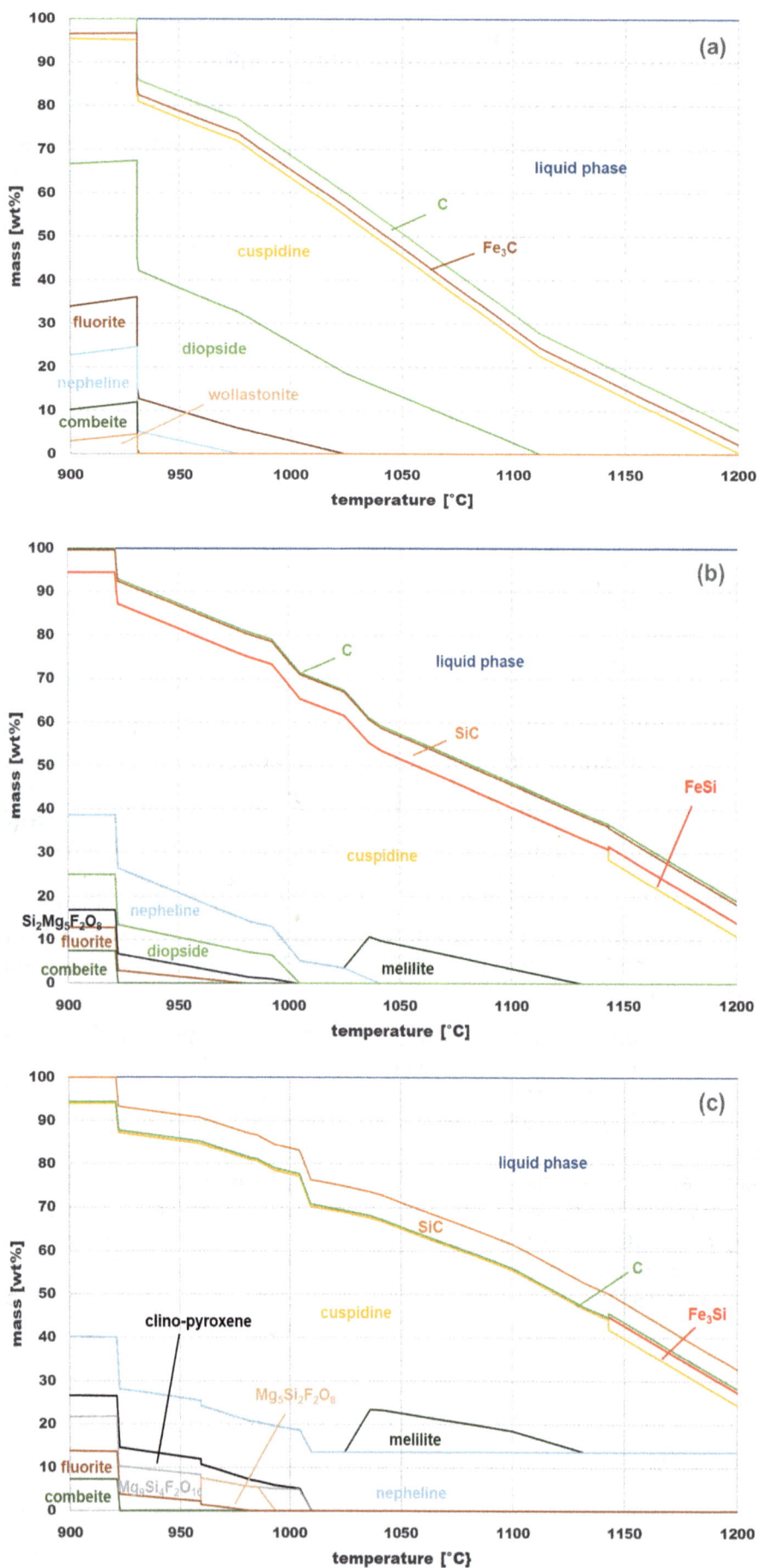

FIG 1 – Mineral compositions as a function of temperature for: (a) MP0 and MP2, (b) MP3, and (c) MP9.

For all the mould powders, the solidus (T_s) and liquidus (T_l) temperatures were determined using calculations. Additionally, the calculated amounts of liquid phase present at 1000°C (I_{1000}) were compared to enable the evaluation of the melting tendencies. The results are summarised in Table 3. In the equilibrium state, the temperature of the first liquid phase is not affected by the melting-controlling components. The amount of liquid phase at 1000°C is not significantly different for all mixtures, with the exception of MP5, which shows the lowest value. A more significant difference was observed in the liquidus temperature. The sample containing SiC (MP5) as the melt-control additive exhibits the highest T_l. By contrast, further addition of antioxidants results in an opposite effect and decreases the liquidus temperature by approximately 80°C again. In these samples, liquid phase formation is promoted, but its amount at 1000°C is only slightly higher and still lower than that of MP0 and MP2. The thermodynamic calculations do not end in a total liquefaction of the mould powders up to 1550°C because the melt-control additives (carbon, SiC, Si_3N_4) are not totally oxidised. Therefore, they are still stable at 1550°C. In Table 3, they are indicated by the subscript 'residual'. Additionally, their initial amount is given, which is marked by 'initial'.

TABLE 3
Melting tendency based on thermodynamic calculations.

	$I_{1000°C}$ [wt.%]	T_s [°C]	T_l [°C]	C residual [wt.%]	SiC residual [wt.%]	SiC initial [wt.%]	Si_3N_4 residual [wt.%]	Si_3N_4 initial [wt.%]
MP0	28.3	931	1202	3.2				
MP1	26.7	923	1190					
MP2	28.3	931	1202	3.2				
MP3	24.2	922	1235	0.8	4.3	6.0		
MP4	26.8	847	1209		7.1	7.1		
MP5	18.8	852	1225		7.0	7.0		
MP6	24.3	852	1210		7.0	7.0		
MP7	27.5	927	1216					
MP8	25.1	922	1235				4.9	6.9
MP9	23.8	921	1235	3.2	4.4	4.8	1.2	1.7

$I_{1000°C}$ amount of liquid phase at 1000°C

T_s solidus temperature

T_l liquidus temperature

residual quantity of the respective melt-control additive at 1550°C

initial initial quantity of the respective melt-control additive in the sample

Laboratory investigations of different mould powder compositions

The microscopical investigations of the standard mould powder MP0 after heat treatment at 900°C revealed the beginning reactions of the raw material components to form new phases. In this sample, the formation of cuspidine on the surface of the wollastonite particles is observed. The diffusion of Na_2O into the wollastonite results in the formation of sodium calcium silicates that separate the wollastonite in the centre from the cuspidine rim (Figure 2a). Fluorine does not only diffuse into wollastonite but also into albite to form fluorpectolite ($NaCa_2Si_3O_8F$) and into the partly decarburised magnesite (Figure 2a). Similar reactions were observed in MP1 and MP2. In addition, diopside ($CaMgSi_2O_6$) was formed. In contrast, no reaction between the raw material particles was observed in the sample containing SiC as the melting-control component (Figure 2b). However, this beneficial effect was reduced when antioxidants were added to the mixture. Al and/or Si are oxidised even at low temperatures. Furthermore, their oxides contribute to the melting of the raw materials, particularly fines. Thus, for MP4–MP6 liquid phase was already present owing to the reactions of the fines (Figure 2c). The cuspidine and sodium calcium silicate phases had already formed. The replacement of carbon with Si_3N_4 did not produce the desired effect. Due to the diffusion of Na_2O

into glassy phases, their melting temperature is reduced, which results in the formation of small amounts of liquid phase even at 900°C. Diffusion of F into the dissociated magnesite was also observed. When silicon nitride and silicon carbide were added together (MP9), the reactivity of the raw material components was marginally reduced. However, the diffusion of F into the magnesite and the diffusion of CaO, MgO and Na_2O into Si_3N_4 still occurred.

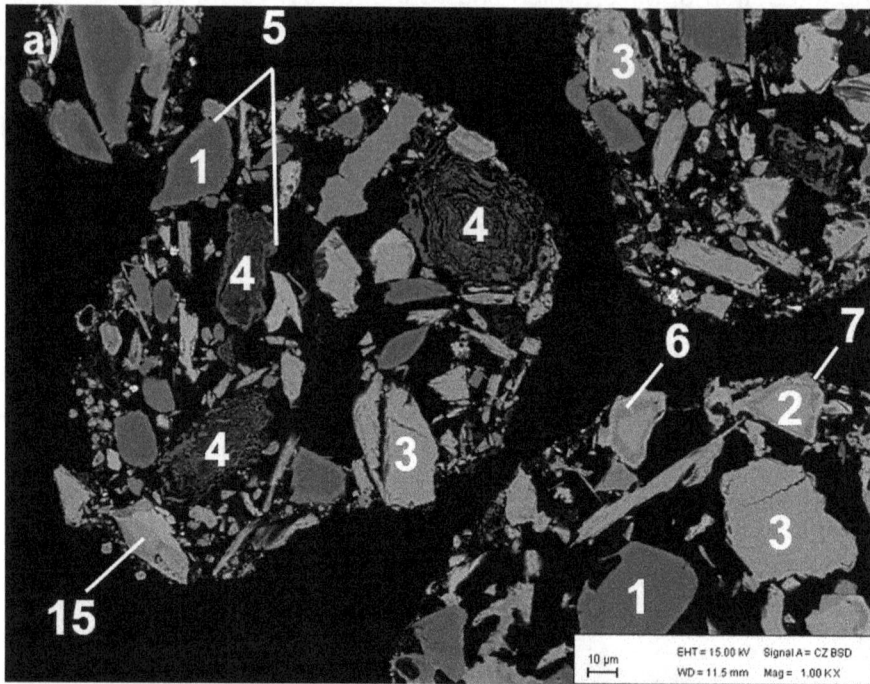

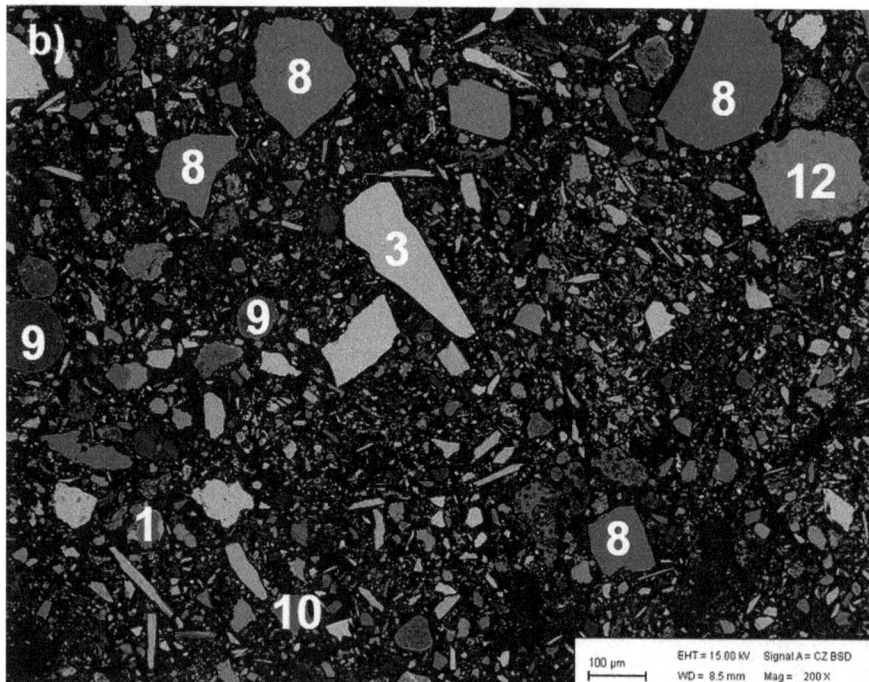

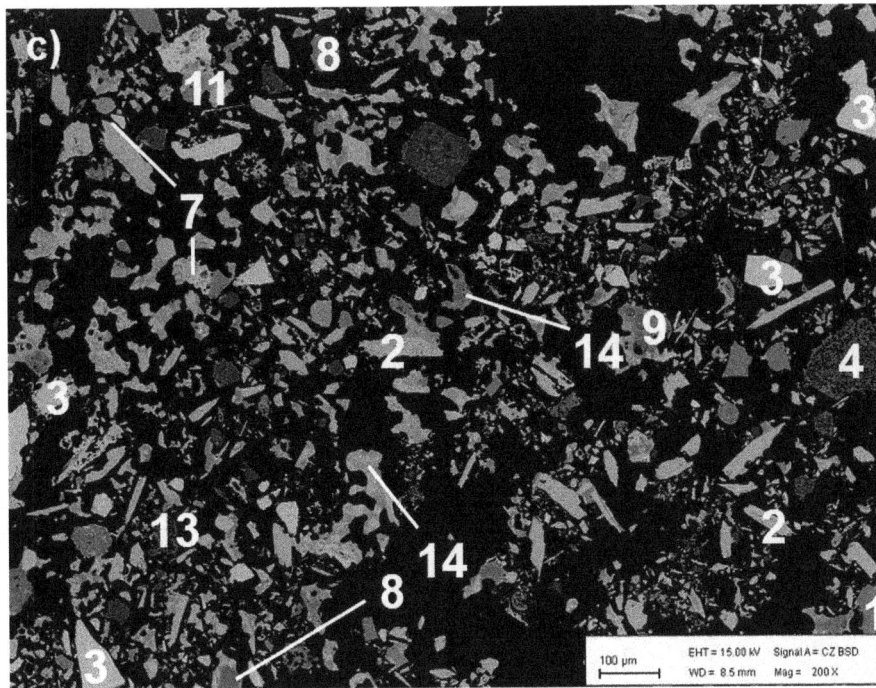

FIG 2 – Backscattered electron image of (a) MP0, (b) MP3, and (c) MP4 after temperature treatment at 900°C for 10 mins. (1) albite, (2) wollastonite, (3) fluorite, (4) dissociated magnesite, (5) fluorpectolite, (6) sodium calcium silicate, (7) cuspidine, (8) SiC, (9) fly ash, (10) quartz, (11) slag, (12) calcite, (13) corundum, (14) liquid phase, and (15) diopside.

Comparing the samples after the temperature treatment at 1000°C, considerable differences in the appearance of the respective mixtures could be observed. The mould powder MP0 used in the continuous casting process contained only the residuals of the initially larger raw material particles (Figure 3a). These were wollastonite, fluorite and albite. The cuspidine rim at the boundaries of the wollastonite and the area of the sodium calcium silicate separating the wollastonite in the centre increased. Furthermore, owing to the reaction of the fines, a considerable amount of liquid phase in contact with the newly formed cuspidine is observed. MP1, which contained no melt-control additives, exhibited the diffusion of primarily Na_2O and F into the other raw materials. The residuals of the decarburised magnesite are enriched in F, and silica-containing particles, especially at their surfaces, in Na_2O, but the liquid phase is hardly present. A similar behaviour was observed for MP2. In the SiC-containing sample (MP3), only small amounts of cuspidine were formed (Figure 3b). SiC inhibits the solid-solid reaction between the components. Thus, the formation of the new phases shifted to higher temperatures. The appearance of MP4 confirmed the previously described effect of Si addition, which resulted in an increased sintering tendency compared with MP3. New solid- and liquid-phase formations are observed. In contrast, if Si is replaced totally or at least partly by Al (MP5 and MP6), the amount of liquid phase is reduced, but is still higher than in MP3. Furthermore, the quantity of raw material particles increased at this temperature, as shown in Figure 3c. The use of Al alone as an additive resulted in the highest amount of the liquid phase, which is in accordance with the highest sintering tendency (Table 3). In MP8, silicon nitride also delays the solid-solid reactions of the raw materials but is less efficient than SiC. Hence, raw materials and new phases, such as cuspidine, are present. The MP9 showed a similar appearance (Figure 3d). For both MP8 and MP9 specimens, the oxidation of a substantial amount of Si_3N_4 was detected.

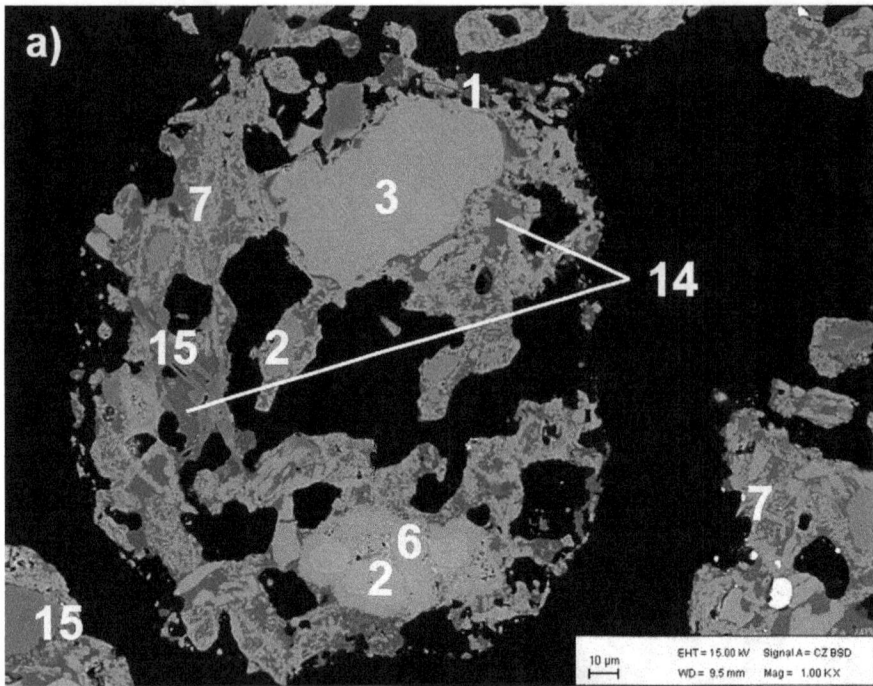

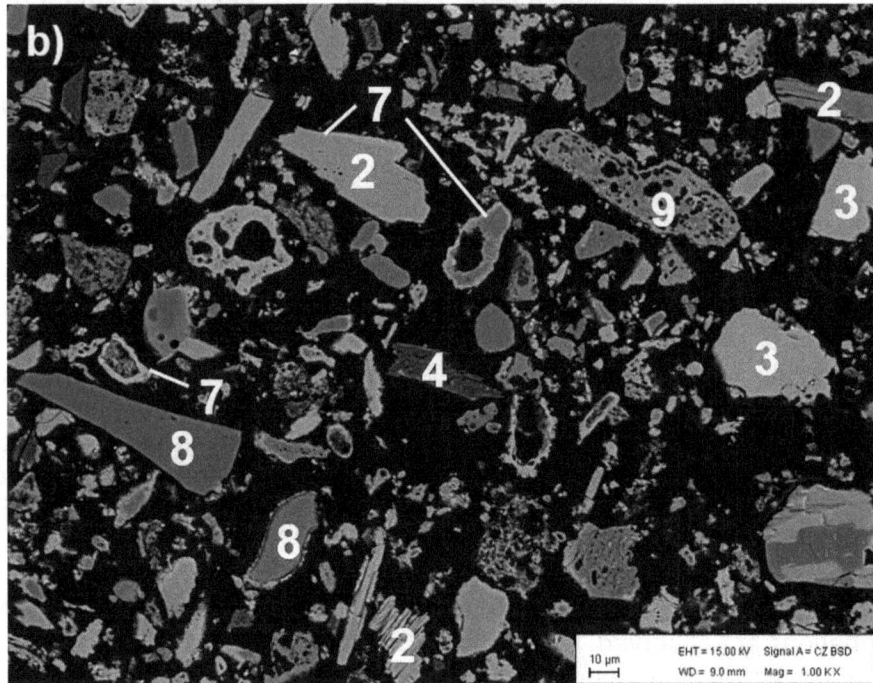

FIG 3 – Backscattered electron image of (a) MP0, (b) MP3, (c) MP6, and (d) MP9 after temperature treatment at 1000°C for 10 mins. (1) albite, (2) wollastonite, (3) fluorite, (4) dissociated magnesite, (6) sodium calcium silicate, (7) cuspidine, (8) SiC, (9) fly ash, (14) liquid phase, and (15) diopside.

After heat treatment at 1100°C, the samples contained mainly liquid phase and cuspidine (Figure 4). Residuals of large raw material particles were also observed. In MP1, additional fluoride dendrites were observed, which formed during quenching of the samples. A different appearance was observed for MP3 containing SiC. Large raw material particles showed small reaction boundaries on their surfaces, indicating a slow reaction rate. A liquid phase was present but was not coherent (Figure 3b). SiC was also detected in the samples containing additional antioxidants or silicon nitride (MP4–MP6 and MP9). In these specimens, SiC remained in direct contact with the already formed continuous liquid phase (Figure 4c). Locally, the SiC particles accumulated. Through oxidation, the melt-control additives caused the formation of pores in the liquid slag in all samples. This was particularly observed in MP8 and MP9, in which Si_3N_4 was used (Figure 4d).

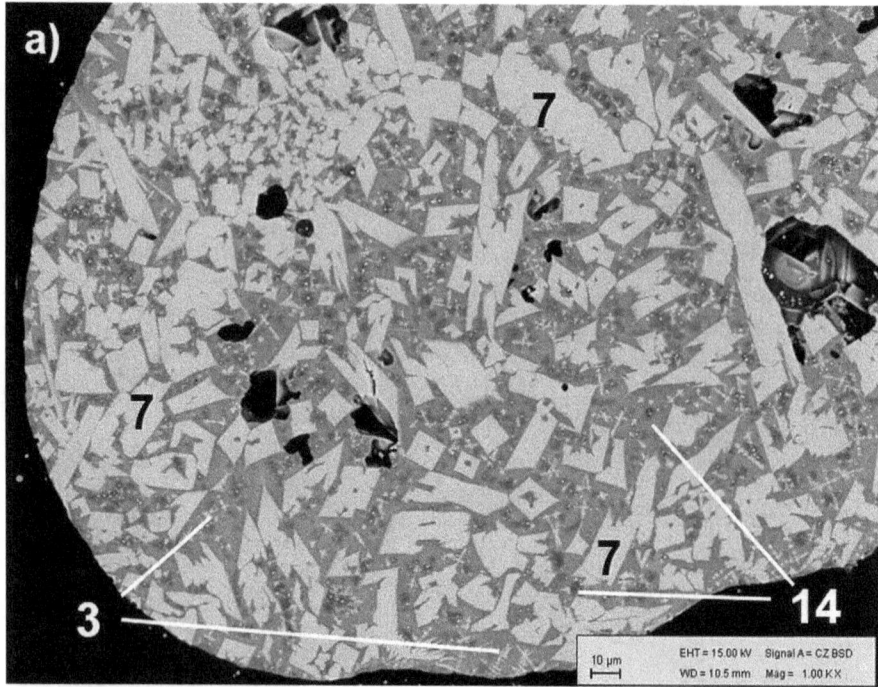

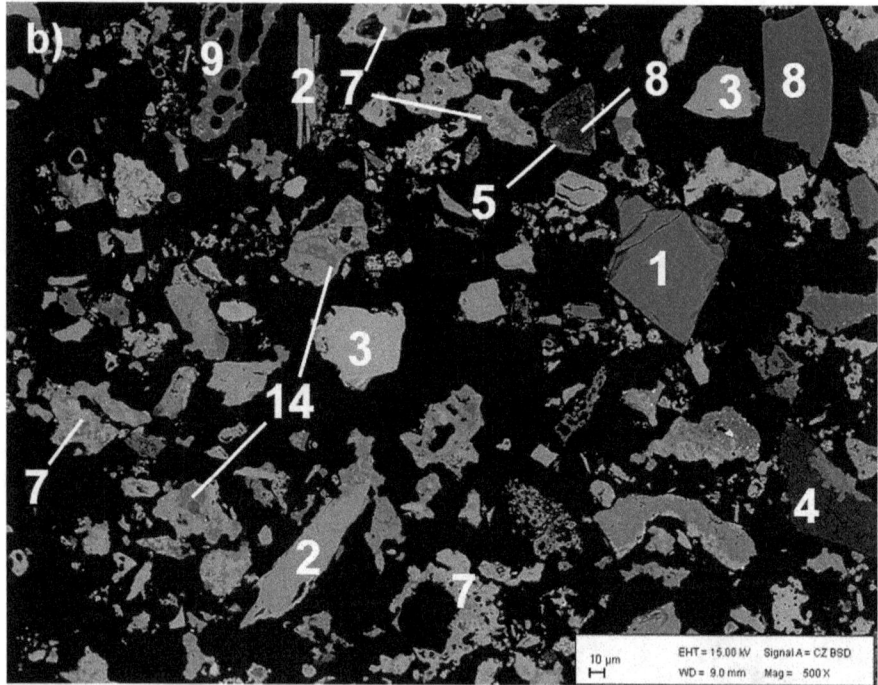

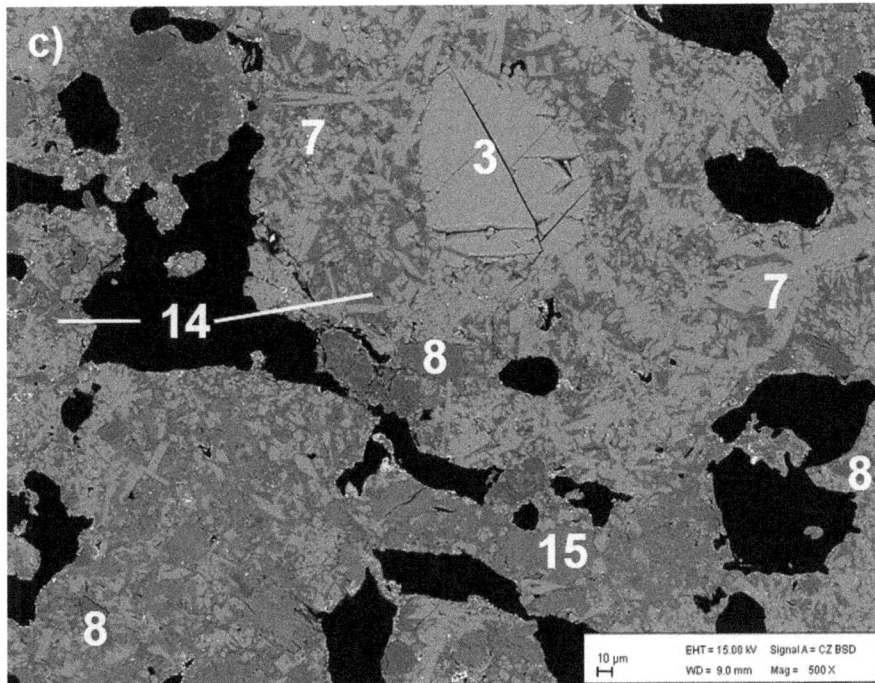

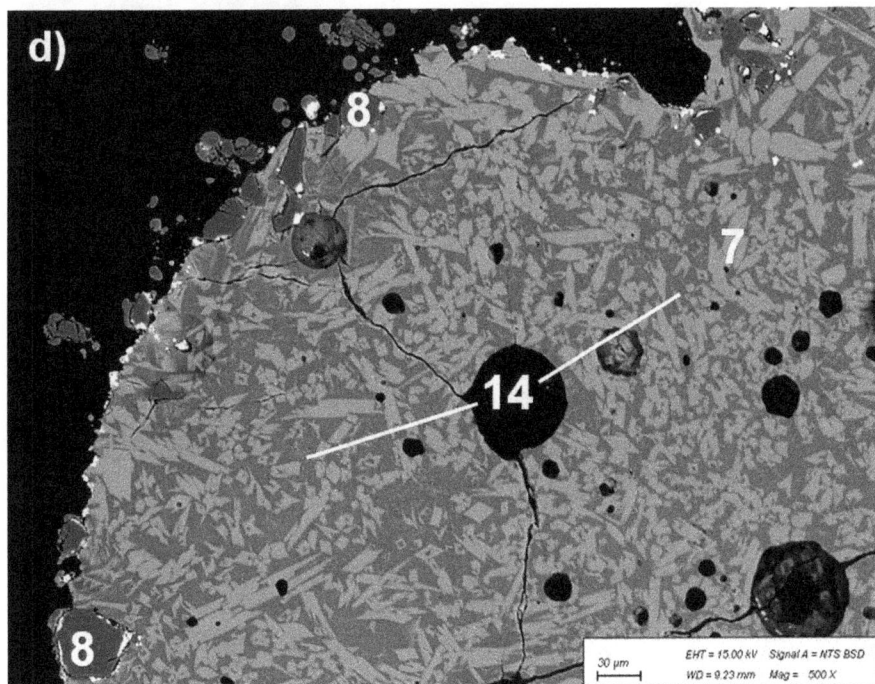

FIG 4 – Backscattered electron image of (a) MP0, (b) MP3, (c) MP4, and (d) MP9 after temperature treatment at 1100°C for 10 mins. (1) albite, (2) wollastonite, (3) fluorite, (4) dissociated magnesite, (5) fluoropectolite, (7) cuspidine, (8) SiC, (9) fly ash, (14) liquid phase, and (15) diopside.

After the treatment of the mould powder MP0 as received at 1200°C, the mineralogical investigations revealed the independent melting of each granule (Figure 5a). No continuous liquid phase was formed, even though all granules were at least partly liquefied. Some still showed cuspidine in the solid state, whereas others consisted only of the liquid phase. This can be explained by the slightly different chemical compositions of the granules with respect to their F and/or Na_2O contents, which influenced the melting temperature of the granules.

In general, the appearance of the mixtures was very similar. Except for MP0, MP3, and partly MP5, where small quantities of cuspidine in a stocky shape were still present (Figure 5a, 5b), a homogeneous liquid phase was formed (Figure 5d). Nevertheless, cuspidine was found in all other samples. But due to its dendritic shape, it is indicated that it crystallised during cooling and was not present at 1200°C in the solid state (Figure 5a, 5b, 5c). When SiC was used, some residuals still existed within the liquid slag at this temperature. It was also present in samples containing antioxidants, but its particle size was smaller. This confirms the previously suggested assumption that the addition of Al and especially Si to SiC counteracts the beneficial properties of SiC by promoting reactions between the components. This was also observed when silicon carbide and silicon nitride are used together (MP9).

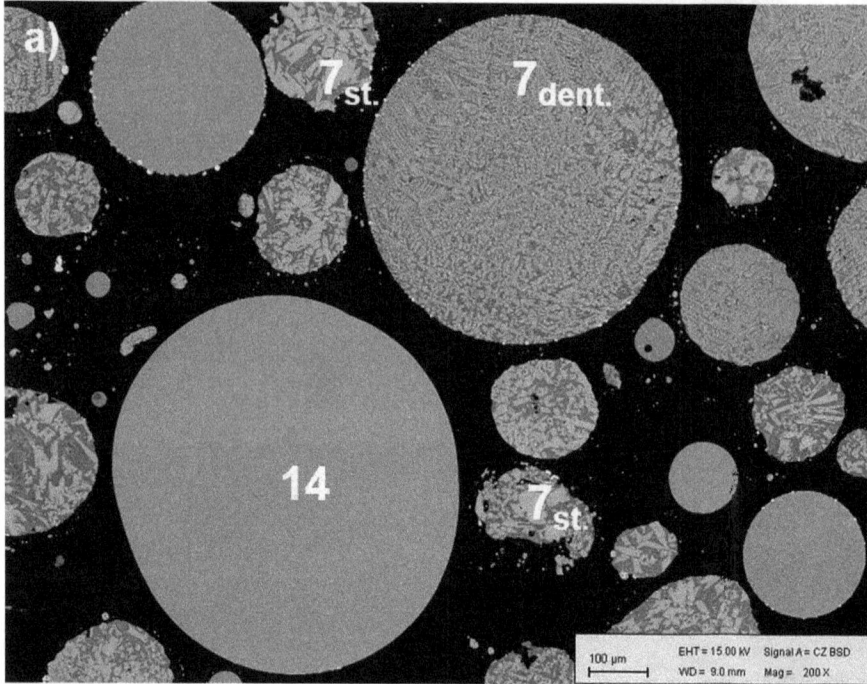

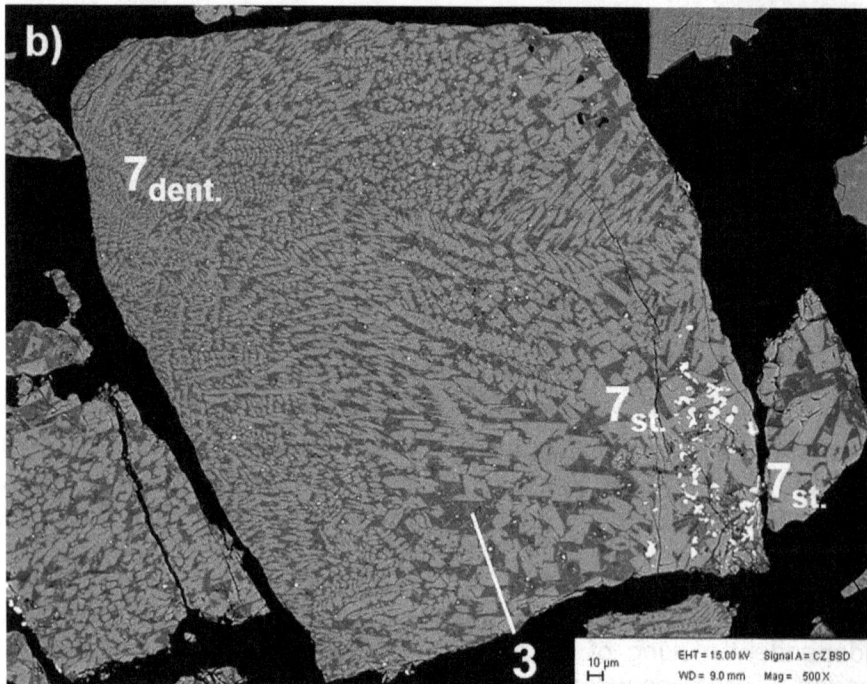

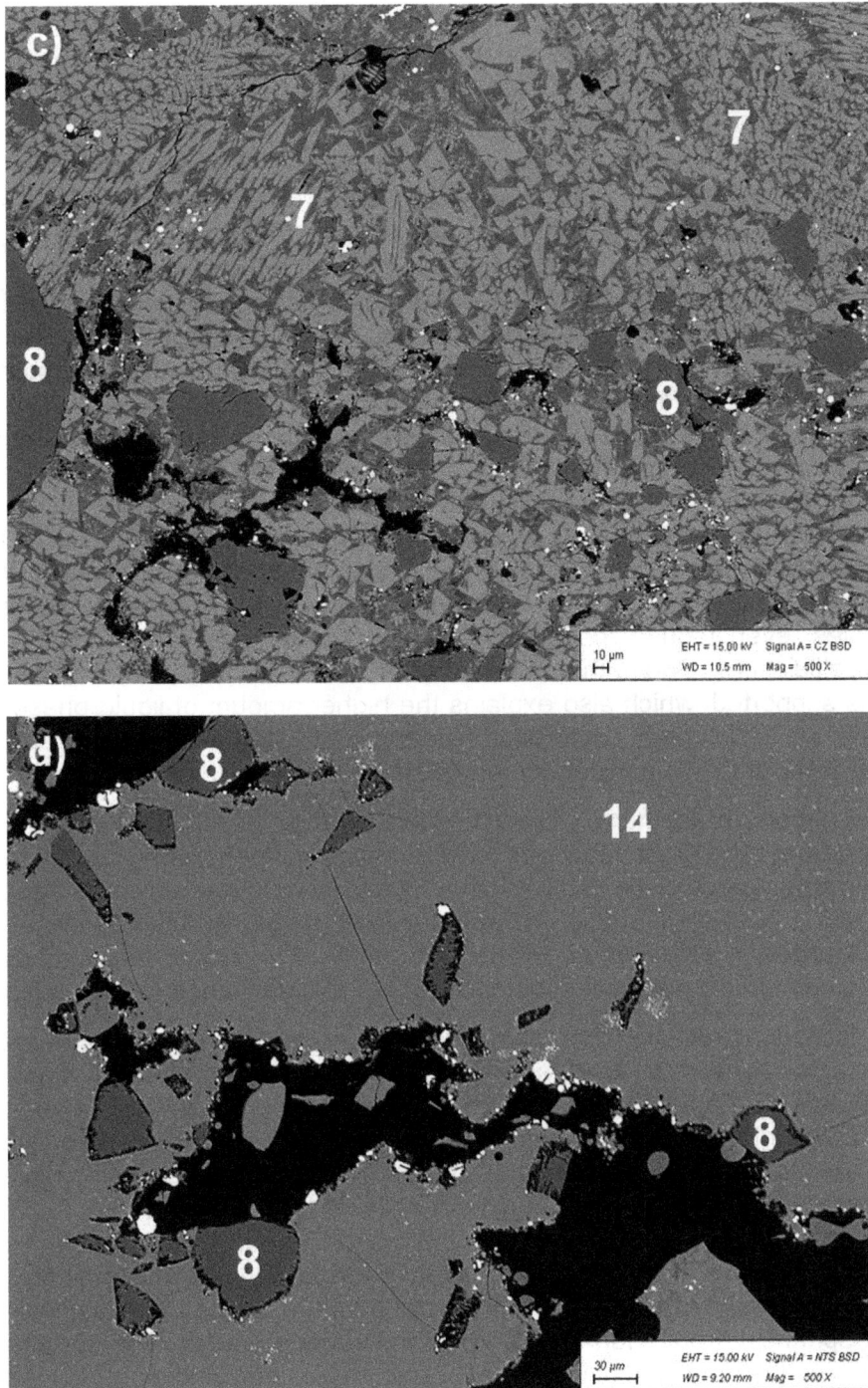

FIG 5 – Backscattered electron image of (a) MP0, (b) MP3, (c) MP5, and (d) MP9 after temperature treatment at 1200°C for 10 mins. (3) fluorite (7) cuspidine: stocky (st.) and dendritic (dent.), (8) SiC, and (14) liquid phase.

Reduction of the SiC content

To investigate the effect of the SiC content on the melting behaviour of the mould powder, mixtures were prepared according to the raw material composition given in Table 2. The investigations revealed that for this mould powder composition, SiC content >2.8 wt per cent did not have a significant further effect on reducing the reactions of the raw material particles. SiC particles were still present in the samples after the heat treatment at 1200°C. In contrast, the sample containing solely 1.4 wt per cent SiC (MP13) showed the same melting tendency as the mould powder without melt-control additives. On comparing the results of MP12 with those of the standard carbon-containing mould powder, it is evident that the residuals of the raw materials are detected at only higher temperatures and the amount of the liquid phase is reduced. This indicates that even small amounts of SiC partially inhibited the reaction between the raw material particles and shifted the formation of new solid and liquid phases to higher temperatures. At 1200°C, SiC is the only solid phase, but its amount is considerably reduced.

CONCLUSIONS

Different melt-control components used to replace carbon in mould powders were investigated. Thermodynamic calculations and laboratory experiments at selected temperatures, together with mineralogical investigations, were conducted to compare their effects. The melting behaviours of the different mixtures were compared with those of the mould powder already used in the continuous casting of steel. These investigations revealed that SiC is a promising substitute for controlling the melting rate of mould powders. For temperatures below 1100°C, a reduced solid-solid diffusion was observed. This shifted the formation of new solid and liquid phases to higher temperatures and increased the stability of the raw material particles. Owing to its high oxidation resistance, SiC initially remains in a solid state. Thus, the amount of SiO_2 required to form the liquid phase is not available. With increasing temperature, SiC is partially oxidised and the availability of SiO_2 increases. This influences the formation of new phases. At lower temperatures, where SiC is unoxidised, phases with lower SiO_2 contents are formed. Comparing the amounts of liquid phase with those of the standard mould powder at temperatures below 1000°C, a lower value was obtained for the SiC-containing samples, even though their solidus temperatures were lower. This agrees with the thermodynamic calculations. Due to the oxidation of Al and Si at lower temperatures, the formation of new phases is supported, which also explains the higher amount of liquid phase at 900°C. In a subsequent study, the SiC content could be considerably reduced, which could lead to a reduction of at least 27 g CO_2/kg of mould powder compared to the carbon-containing product.

Another possible carbon replacement is silicon nitride. However, its effect is weaker than that of SiC. At temperatures below 1100°C the reactions between particles were prohibited, but with increasing temperatures, the appearance of the samples after heat treatment was similar to that of the standard mould powder (MP0), because it shows a lower stability than SiC and the necessary SiO_2 amount to form the liquid phase is available at lower temperatures. In addition, the combination of silicon nitride and silicon carbide did not meet expectations with respect to controlling melting behaviour.

The present investigations revealed that the melting rate was lower for loose powders. This can be explained by the close contact between the raw material particles in the case of granules and their production processes. During the production of granulated mould powders, raw material particles are added to a water/binder mixture and homogenised. Sodium carbonate is dissolved and distributed over the entire volume (Han et al, 2022). Thus, the formation of new phases during heating is promoted. Furthermore, the independent melting of the granules during temperature treatment at 1100°C and 1200°C is explained by the diffusion of Na_2O to the surface of the granules during spray drying. Owing to the increased sodium oxide content, the melting temperature is reduced and the granules start sintering from the boundaries. With an increasing amount of liquid phase, a nearly spherical drop is formed with minimal surface energy. The surface energy would further be reduced by the formation of a coherent liquid phase from single drops. However, owing to the lack of residence time at the annealing temperature, this process was only partially observed during the experiment.

The possible pick-up of carbon or nitrogen into the liquid steel during casting, if SiC or Si_3N_4 is used to control the melting behaviour is discussed below. In previous studies, carbon was only partly substituted by nitrides to prevent renitriding of the liquid metal (Debiesme et al, 1998). Thus, if Si_3N_4 is used to totally substitute carbon, nitriding will most probably take place during casting. Nevertheless, it was shown that its effect to control the melting rate of a mould powder is weaker than that of SiC being the preferred substitute for carbon. Even for this carbide, only partial substitution of carbon was suggested to ensure the same melting behaviour. If 3 wt per cent SiC was used, the carbon content was defined to be <1 wt per cent. Experiments in contact to the steel revealed a lower effect on recarburisation of the liquid metal (Ning and Jinghao, 1998). As shown in the present paper, similar melting rates were achieved by using solely 2.8 wt per cent SiC, which is even lower than in the samples mentioned before. Consequently, recarburisation is not expected to take place. Nevertheless, this assumption has to be confirmed by industrial trials.

ACKNOWLEDGEMENTS

The authors gratefully acknowledge the funding support of K1-MET GmbH, a metallurgical competence centre. The research program of the K1-MET competence centre is supported by the Competence Centre for Excellent Technologies (COMET), an Austrian program for competence centres. COMET is funded by the Federal Ministry for Climate Action, Environment, Energy, Mobility, Innovation, and Technology; the Federal Ministry for Labour and Economy; the Federal States of Upper Austria, Tyrol, and Styria; as well as the Styrian Business Promotion Agency (SFG) and the Standortagentur Tyrol. The Upper Austrian Research GmbH continuously supports K1-MET. In addition to public funding from COMET, this research project was partially financed by the scientific partner Montanuniversitaet Leoben and the industrial partners RHI Magnesita, voestalpine Stahl, and voestalpine Stahl Donawitz.

REFERENCES

Dai, L, Xiao, G and Ding, D, 2021. Review of Oxidation Resistance Technology of Carbon-containing Refractories. *Materials Reports*, 35(3):3057–3066.

Debiesme, B, Radot, J P, Coulombet, D, Lefebvre, C, Pontoire, J N, Roux, Y and Damerval, C, 1996. Mould cover for continuous casting of steel, especially very-low-carbon steels. US Patent 5876482A.

Debiesme, B, Radot, J P, Coulombet, D, Lefebvre, C, Pontoire, J N, Roux, Y and Damerval, C, 1998. Mould cover powder for continuous casting of steel, especially very-low-carbon steels. US Patent 6328781B1.

Ghosh, N K, Jagannathan, K P and Ghosh, D N, 2001. Oxidation of magnesia-carbon refractories with addition of aluminium and silicon in air, *Interceram - International Ceramic Review*, 50(3):196–202.

Han, F, Wen, G, Tang, P and Chen, F, 2023. Catalytic Effect of Iron Oxide on the Combustion of Carbonaceous Materials in Mold Flux for Continuous Casting, *Metallurgical and Materials Transactions B*, 54(5):2605–2613.

Han, F, Yu, L, Wen, G, Wang, X, Zhang, F, Jia, J and Gu, S, 2022. The effect of composition segregation of mold powder produced by spray granulation on the sintering performance, *Journal of Materials Research and Technology*, 20:448–458.

Kawamoto, M, Nakajima, K, Kanazawa, T and Nakai, K, 1994. Design Principles of Mold Powder for High Speed Continuous Casting, *ISIJ International*, 34(7):593–598.

Kölbl, N, Marschall, I and Harmuth, H, 2009. Investigation of the melting behavior of mould powders, in *Proceedings VIII International Conference on Molten Slags, Fluxes and Salts* (ed: M Sanchez), pp 1031–1040 (GECAMIN: Santiago).

Kromhout, J A, 2013. Mould powders for high speed continuous casting of steel, PhD Thesis, Technical University Delft, Delft. Available from: <http://resolver.tudelft.nl/uuid:4b52997d-9fc9-41b3-a4e8-70ffab1b3fdf> [Accessed: 20 November 2023].

Lefebvre, C, Radot, J P, Pontoir, J N and Roux, Y, 1996. Development of continuous casting mould powder for the limitation of recarburization in ULC steels, La Revue de Métallurgie-CIT, 93(4):489–496.

Macho, J J, Hecko, G, Golinmowski, B and Frazee, M, 2005. The development and evaluation of a new generation of no free carbon continuous casting fluxes, in *Preprints 33rd McMaster Symposium on Iron and Steelmaking*, pp 131–146 (McMaster University: Hamilton).

Maillart, O, Chaumat, V and Hodaj, F, 2010. Wetting and interfacial interactions in the $CaO-Al_2O_3-SiO_2$/silicon carbide system, *Journal of Material Science*, 45(8):2126–2132.

Mills, K C, 1990. NPLRep, DMM(A), ECSC Contract No. 7210.22/451, Report EUR 13177 EN, Teddington, UK.

Mills, K C, Fox, A B, Thackray, R P and Li, Z, 2004. The performance and properties of mould fluxes, in *Proceedings of the VII International Conference on Molten Slags, Fluxes and Salts*, pp 713–722 (The South African Institute of Mining and Metallurgy: Johannesburg).

Nakato, H, Takeuchi, S, Fujii, T, Nozaki, T and Washio, M, 1991. Characteristics of new mold fluxes for strand casting of low and ultra-low carbon slabs, in *Proceedings of the 75th Steelmaking Conference*, pp 639–646.

Ning, L and Jinghao, C, 1998. Study on SiC as the substitute of the mold fluxes in continuous casting of ULC steel, *Journal of Shaoguan University (Social Science)*, 19:112–120.

Park, J, Jeon, J, Lee, K, Park, J H and Chung, Y, 2016. Initial wetting an spreading rates between SiC and $CaO-SiO_2-MnO$ slag, *Metallurgical and Materials Transactions B*, 47(3):1832–1838.

Safarian, J and Tangstad, M, 2009. Wettability of silicon carbide by $CaO-SiO_2$ slags, *Metallurgical and Materials Transactions B*, 40(6):920–928.

Sun, L, Liu, C, Fang, J, Zhang, J and Lu, C, 2019. Crystallization behavior and thermal properties of B_2O_3-containing $MgO-Al_2O_3-SiO_2-Li_2O$ glass ceramic and its wettability on Si3N4 ceramic, *Journal of the European Ceramic Society*, 39(4):1532–1539.

Supradist, M, Cramb, A W and Schwerdtfeger, K, 2004. Combustion of Carbon in Casting Powder in a Temperature Gradient, *ISIJ International*, 44(5):817–826.

Takeuchi, H, Mori, H, Nishida, T, Yanai, T and Mukunashi, K, 1978. Development of a Carbon-free Casting Powder for Continuous Casting of Steels, *Tetsu-to-Hagané*, 64(10):1548–1557.

Takeuchi, H, Nishida, T, Ohno, T, Kataoko, N and Hikara, Y, 1976. Kohlenstofffreies Gießpulver für Kokillen- und Strangguß von Stahl (Carbon-free casting powder for chill and continuous casting of steel), Deutsches Patent (26 26 354), München, Deutschland: eutsches Patentamt (DPMA: German Patent and Trade Mark Office).

Terada, S, Kaneko, S, Ishikawa, T and Yoshida, Y, 1991. Development of mold fluxes for ultra-low-carbon steels, *Iron and Steelmaker*, 18(9):41–44.

Wang, T and Yamaguchi, A, 2001. Oxidation Protection of MgO–C Refractories by Means of Al8B4C7, *Journal of the American Ceramic Society*, 84(3):577–582.

Yan, W, Yang, Y D, Chen, W Q, Barati, M and McLean, A, 2015. Design of mould fluxes for continuous casting of special steels, *Canadian Metallurgical Quarterly*, 54(4):467–476.

Yi, Z M, Song, J L and Peng, Q C, 2013. Analysis of Carbonization Mechanism in ULCS Continuous Casting and Control Measures, *Advanced Materials Research*, 739:214–217.

Fluxing options and slag operating window for Metso's DRI smelting furnace

J Hamuyuni[1], K Vallo[2], T Haimi[3], F Tesfaye[4], J Pihlasalo[5] and M Lindgren[6]

1. Product Manager, Metso, Espoo 28100, Finland. Email: joseph.hamuyuni@metso.com
2. Product Manager, Metso, Espoo 02230, Finland. Email: kimmo.vallo@metso.com
3. Senior Sales Manager, Metso, Espoo 02230, Finland. Email: timo.haimi@metso.com
4. Process Metallurgist, Metso, Espoo 02230, Finland. Email: fiseha.tesfaye@metso.com
5. Senior Metallurgist, Metso, Pori 28100, Finland. Email: jouni.pihlasalo@metso.com
6. Director R&D, Metso, Pori 28100, Finland. Email: mari.lindgren.@metso.com

ABSTRACT

Decarbonisation of the steel industry has recently attracted a lot of research and technological development. Most of the technological developments are centred around the replacement of blast furnace (BF), which has been the main primary smelting furnace in the ironmaking step of steel production. Replacement of the BF is not an easy task because of BF's many established capabilities over the years, especially its ability to use low-grade iron feed.

In this regard, Metso has developed its own technology, Metso's Outotec direct reduced iron (DRI) smelting furnace, to tackle this problem. The recently developed six-in line electric furnace for DRI smelting can be used for primary smelting of DRI from blast furnace grade iron ore. Unlike the electric arc furnace (EAF), the DRI smelting furnace can handle larger slag volumes that emanate from the high gangue in BF grade DRI. Therefore, fluxing plays an important role in operating the DRI smelting furnace. This paper presents fluxing options and proposes slag quality and operating windows for the DRI smelting furnace. Also, the effect of slag on refractory durability will be discussed.

INTRODUCTION

Decarbonising steel industry is one of the practical pathways to mitigate global warming. Carbon dioxide (CO_2) emissions from steel industry contribute a large share of greenhouse gases that leads to global warming. Carbon dioxide is responsible for about 73 per cent of all greenhouse gases, while the balance is mostly from methane (18 per cent), N_2O (6 per cent) and F-gases (2.5 per cent) (Olivier, Schure and Peters, 2017). Currently and depending on geographical location, it is estimated that between 5–14 per cent of all CO_2 emissions come from the steel industry alone (Gielen, 2003; Kuramochi, 2016). The Paris agreement puts an ambitious goal of limiting global temperature increase to 1.5°C. And while this is a daunting task, it also presents process technology development opportunities.

Currently, blast furnace which is the dominant technology for ironmaking step of the steel production, is the main source of CO_2 emissions. This is because of the consumption of large amount of fossil reductants. For every ton of pig iron, there is ca 0.5 tons of fossil reductants consumed (Fick *et al*, 2014). Opportunities for decarbonisation of steelmaking therefore, depends on finding a low CO_2 emission alternative process to the blast furnace. This can be built around existing secondary smelting processes or completely stand-alone process.

Metso's alternative to the blast furnace, is a high capacity 6-in line direct reduced iron (DRI) smelting furnace. When implemented together with a direct reduction process and current existing downstream processes, it is estimated to cut down CO_2 emissions by up to 60–80 per cent, depending on electricity sources. Metso's DRI smelting furnace is based on Outotec proven proprietary technologies and should offer competitive advantages towards greener steel industry. It offers capacities of hot metal more than 1.2 million tons per annum (Mt/a) and can handle larger than electric arc furnace (EAF) slag volumes, making it possible to use DRI from blast furnace grade iron ores. Also, various types of DRI from both natural gas and hydrogen based reducing processes can be used as feed.

DRI smelting furnace fluxing philosophy is built on using burnt dolomite and lime, in such proportion that regardless of the ore, the slag basicity stays within a narrow window. The fluxing philosophy also ensures that resultant slag can take impurities but stays within manageable viscosity.

Selection of suitable refractory is important, and Metso has conducted its own test work on various refractories on the market to determine the best candidate and operation mode for increased campaign life of furnace.

This paper presents data on fluxing philosophy and refractory selection for DRI smelting furnace. Operating window for resultant slag is assessed using thermodynamic software FactSage™, ver 8.2 (by GTT-Technologies, Germany) and the results are presented. Additionally, few observations made in ongoing refractory studies are discussed.

FEED FOR THE DRI SMELTING FURNACE AND FLUXING PHILOSOPHY

The design basis of the DRI smelting furnace is to target blast furnace grade iron ores, which have typically <67 per cent iron in the concentrate and much higher gangue content of >5 per cent. These types of ores are not suitable for smelting in EAF because of the low gangue tolerance of the EAF and globally only about 3 per cent of iron ores can be used to produce EAF grade DRI. Metso's ideal green steel flow sheet is by using hydrogen reduced DRI as in the Circored process, but the DRI smelting furnace can also handle DRI from shaft furnace and coal (Rotary kiln) based technologies. Table 1 shows typical feeds from the three DRI technologies. The numbers in the table are representative and not to be taken as standard for DRI available on the market today.

TABLE 1

Typical direct reduced iron (DRI) specification from different technologies.

	Natural gas reduced DRI	Solid carbon reduced DRI	Hydrogen reduced DRI
Total Fe (%)	>85	85–93	>85
Metallisation degree (%)	90–94	88–90	85–92
Carbon content (%)	0.5–3	0.1–2	0
Gangue (%)	2.8–6	3–8	3–8

The metal product quality targets of the DRI smelting furnace include carbon of 2–3 per cent and low in phosphorus and sulfur. Depending on the target carbon content, the metal and slag temperature are typically 1500–1600°C and 1600–1700°C, respectively. The DRI smelting furnace uses burnt dolomite and lime as flux, with the proportions of these varying according to the composition of types of feed. The main determining components are the amount of CaO, MgO, Al_2O_3 and SiO_2 contained in feed. By comparison, typical blast furnace flux is limestone. The main components of the resultant slag are therefore the above four oxides. The basicity (CaO/SiO_2) is typically targeted somewhere between 1.2–1.4 for most feeds. Additionally, it is important to keep the FeO content of slag at low levels (<5 per cent) that enables it acceptable as raw material for the cement industry.

Thermodynamic calculations for assessment of operating window

The thermodynamic calculations in this study were carried out with FactSage 8.1. Thermodynamic parameters for the chemical system (Al_2O_3-CaO-FeO-MgO-SiO_2) were obtained from FactPS and FToxid databases. Impurities MnO, P_2O_5 and TiO_2 were left out of the calculated chemical system, because their contents are so small and not considered to be significant factors regarding the thermodynamic properties of the slag.

Liquidus temperature projections were calculated with FactSage for different compositions. Four projections with a base diagram of CaO-SiO_2-MgO were calculated with constant content of FeO (2.00 per cent). Al_2O_3 content of the slag was 5.00, 10.00, 15.00 or 20.00 per cent in different projections. Calculated temperature range for all the projections was from 1200 to 1800°C (1473.15–2073.15 K). Interval between liquidus temperature isotherms was 20°C. The final calculation was done with only those stable phases to make the final calculation faster and to produce smoother liquidus temperature isotherms.

Stable phases – solutions and compounds (pure substances PS) for the FactSage liquidus projections are listed in Appendix A.

Calculations with FactSage software package

The four liquidus projections calculated with FactSage are presented in Figures 1–4.

At 5.00 per cent Al_2O_3, the global liquidus temperature minimum of 1270–1280°C is located near the four-phase intersection point [Clinopyroxene#1 / SiO_2_Tridymite(h)(s4) / Wollastonite] with the liquid slag. The composition of this four-phase intersection point at 1270°C is 65.25 per cent SiO_2, 28.10 per cent CaO and 6.65 per cent MgO.

At 10.00 per cent Al_2O_3, the global liquidus temperature minimum of between 1192–1220°C is located near the four-phase intersection point Clinopyroxene#1 / SiO_2_Tridymite(h)(s4) / Wollastonite with the liquid slag. The composition of this four-phase intersection point at 1192.7°C is 68.46 per cent SiO_2, 27.84 per cent CaO and 3.70 per cent MgO.

At 15.00 per cent Al_2O_3, the global liquidus temperature minimum of between 1228–1240°C is located near the four-phase intersection points [$CaAl_2Si_2O_8$_Anorthite(s2) / Cordierite / Orthopyroxene] and [$CaAl_2Si_2O_8$_Anorthite(s2) / Cordierite / SiO_2_Tridymite(h)(s4)] with the liquid slag. The composition of these four-phase intersection points at 1233.5 and 1227.8°C are 70.33 per cent SiO_2, 16.97 per cent CaO and 12.70 per cent MgO and 73.45 per cent SiO_2, 16.61 per cent CaO and 9.94 per cent MgO, respectively.

At 20.00 per cent Al_2O_3, the global liquidus temperature minimum of between 1280–1300°C is located near the four-phase intersection point [$CaAl_2Si_2O_8$_Anorthite(s2) / Melilite / Spinel#1] with the liquid slag. The composition of this four-phase intersection points at 1280°C is 52.41 per cent SiO_2, 37.70 per cent CaO and 9.89 per cent MgO.

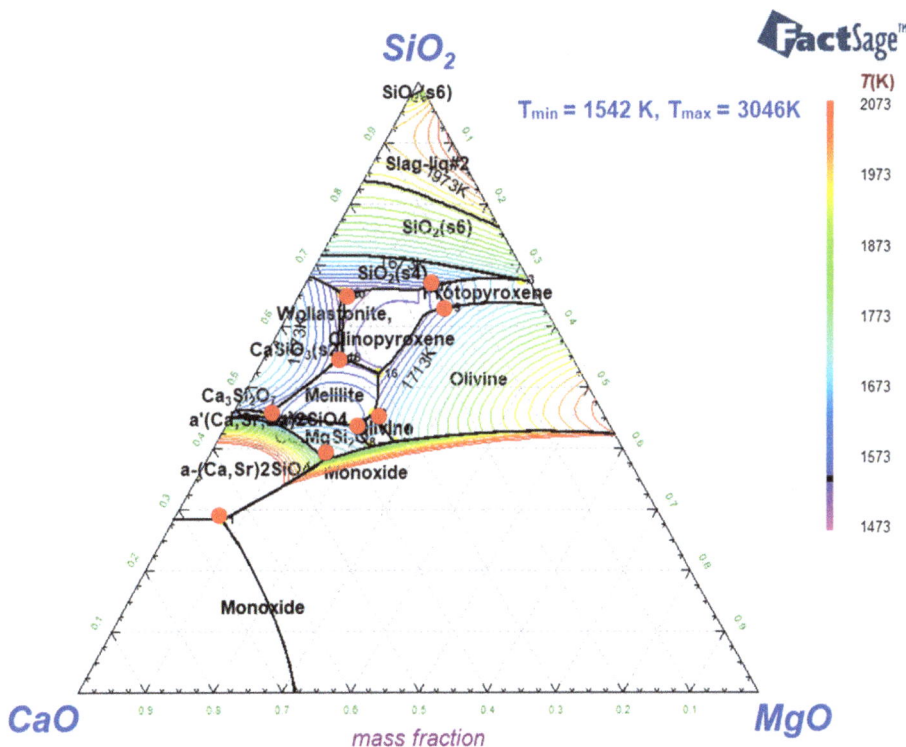

FIG 1 – Liquidus projection in system CaO-SiO_2-MgO with 2.00 per cent FeO and 5.00 per cent Al_2O_3 calculated with FactSage 8.1. The red dots indicate selected four-phase interaction points with the liquid slag.

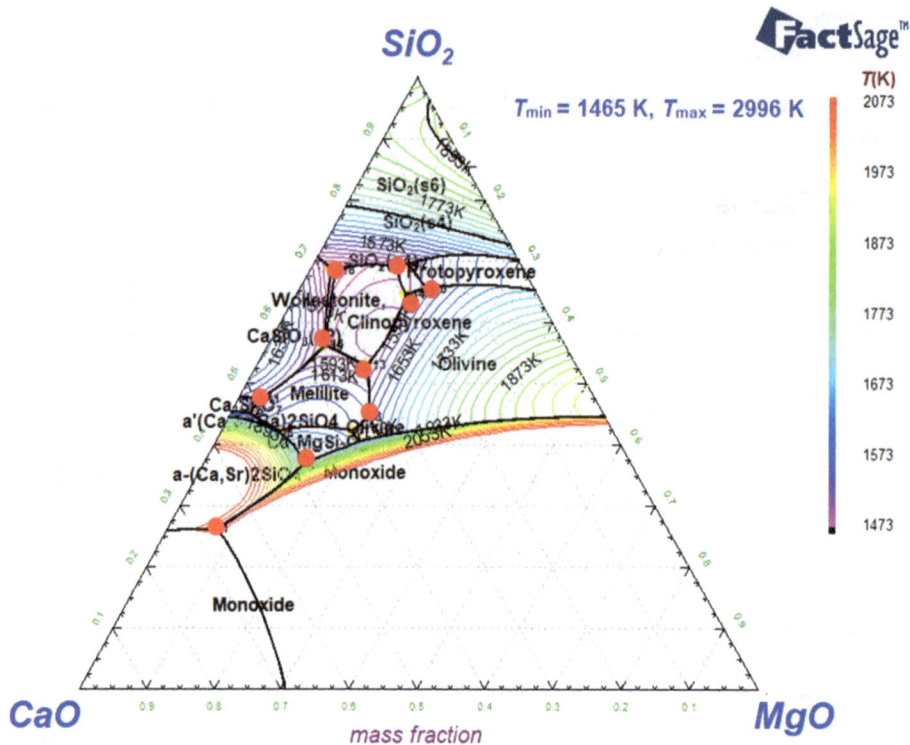

FIG 2 – Liquidus projection in system CaO-SiO₂-MgO with 2.00 per cent FeO and 10.00 per cent Al₂O₃ calculated with FactSage 8.1. The red dots indicate selected four-phase interaction points with the liquid slag.

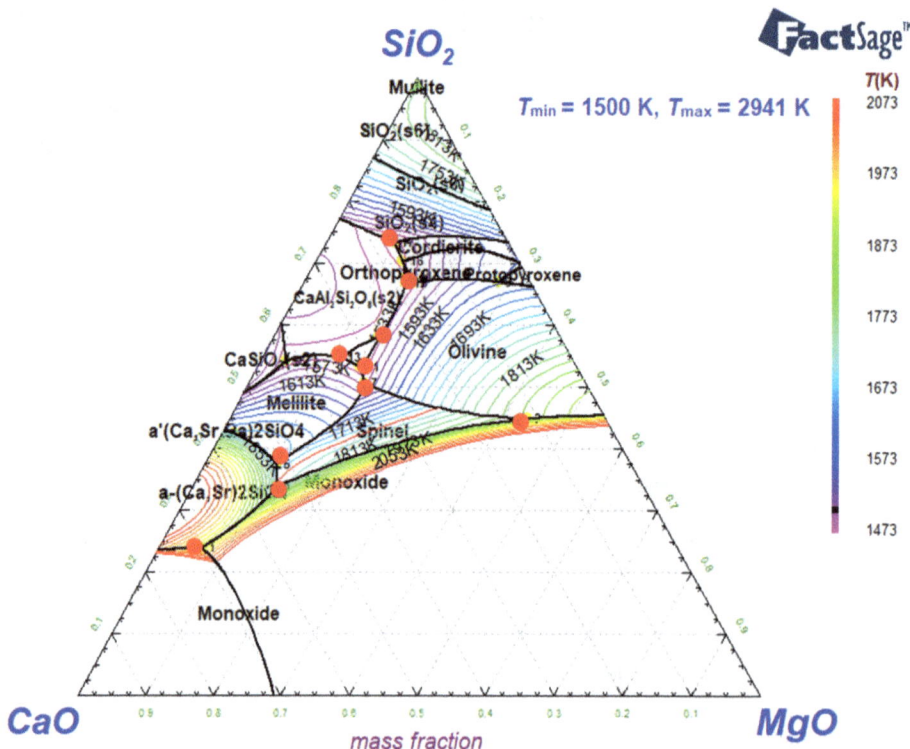

FIG 3 – Liquidus projection in system CaO-SiO₂-MgO with 2.00 per cent FeO and 15.00 per cent Al₂O₃ calculated with FactSage 8.1. The red dots indicate selected four-phase interaction points with the liquid slag.

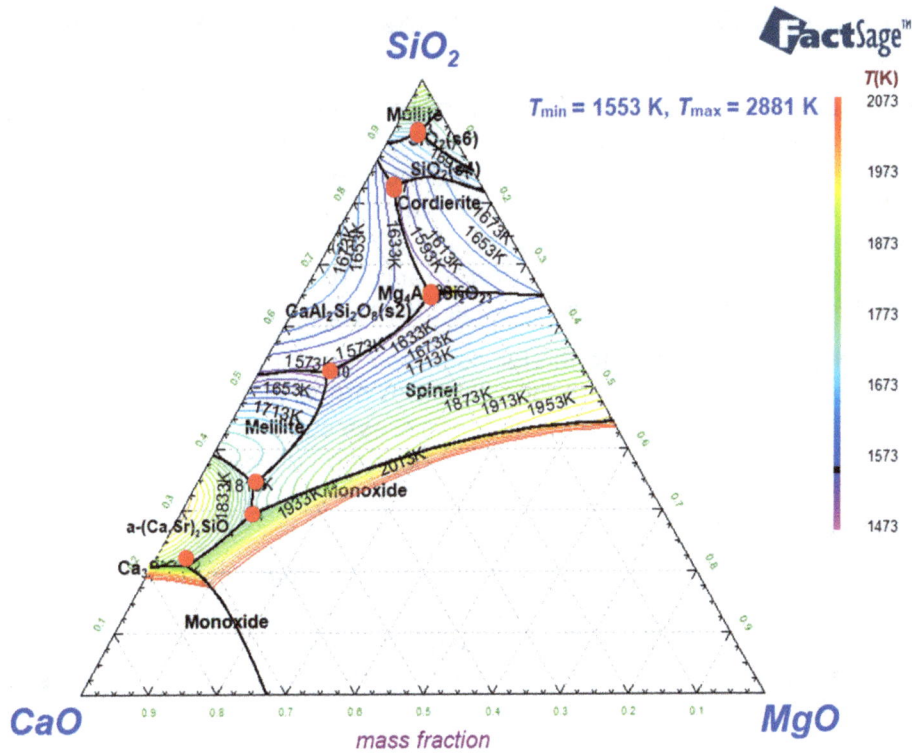

FIG 4 – Liquidus projection in system CaO-SiO$_2$-MgO with 2.00 per cent FeO and 20.00 per cent Al$_2$O$_3$ calculated with FactSage 8.1. The red dots indicate selected four-phase interaction points with the liquid slag.

Viscosity calculations with FactSage

Viscosity values have been calculated for certain slag compositions. Slag #1 has a CaO/SiO$_2$ ratio of 1.143 and slag #2 has a CaO/SiO$_2$ ratio of 1.5. For both slags, only the amounts of Al$_2$O$_3$ and MgO were varied in a proportional amount.

The results obtained are illustrated in Figure 5. In general, at any given composition of Al$_2$O$_3$ or MgO, slag #1 with lower CaO/SiO$_2$ ratio (dashed lines in Figure 5) has higher viscosity than that of slag #2. In all cases, substitution of Al$_2$O$_3$ with MgO resulted in lower viscosity. For the calculated slag compositions (Figure 5), viscosity at 1400°C remains below 0.65 Pa.s. Results are consistent with an earlier thermodynamic study conducted by Metso (internal report) (Pihlasalo and Tesfaye, 2022).

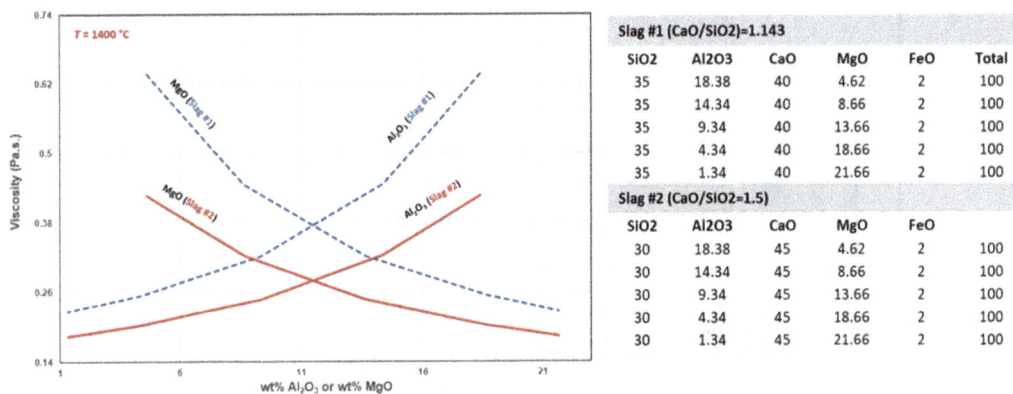

Slag #1 (CaO/SiO2)=1.143

SiO2	Al2O3	CaO	MgO	FeO	Total
35	18.38	40	4.62	2	100
35	14.34	40	8.66	2	100
35	9.34	40	13.66	2	100
35	4.34	40	18.66	2	100
35	1.34	40	21.66	2	100

Slag #2 (CaO/SiO2=1.5)

SiO2	Al2O3	CaO	MgO	FeO	
30	18.38	45	4.62	2	100
30	14.34	45	8.66	2	100
30	9.34	45	13.66	2	100
30	4.34	45	18.66	2	100
30	1.34	45	21.66	2	100

FIG 5 – Viscosities of two different slags (CaO/SiO$_2$ = 1.143 (dashed lines) and CaO/SiO$_2$ = 1.5 (solid lines)) with varying amounts of Al$_2$O$_3$ and MgO. Calculated with FactSage 8.1 software package.

Recommendation for DRI slag compositions at different operating temperatures

The operating temperature for the DRI smelting furnace is dependent on the liquidus temperature of the metal phase, which decreases as a function of carbon content of iron. Lowest liquidus temperature of Fe-C is 1147°C at 4.3 per cent C and melting temperature of pure Fe is 1538°C. The metal temperature is required to be 50°C higher than its liquidus temperature. The temperature difference between metal and slag is usually about 150°C. This means that the slag temperature (= operating temperature) is required to be in the range of 1350–1750°C. The slag needs 50°C superheat, which then corresponds to slag liquidus of temperature range of 1300–1700°C.

The suitable operating windows were found for operation temperatures between 1300 and 1500°C. At 1600°C, some operating windows were found but they are all high in MgO. At 1700°C, no suitable operating windows were found, because inside those areas suitable based on liquidus temperature, the isotherms are close to each other, which makes operation difficult.

REFRACTORY TESTING

Refractories play a crucial role in smelting furnaces by containing molten slag and metal for long periods extending to several years usually referred to as a campaign life. They must withstand chemical attack, must be thermal resistant and must maintain a certain minimum mechanical strength during the campaign life. Choosing the correct refractory for a process is therefore mandatory.

Metso has been conducting its own test work to understand the best refractory for the DRI smelting furnace. Finger tests and crucible tests were carried out on two types of refractories. The choice of refractory was based on preliminary assessment of potential candidate refractories.

A finger test is performed by dipping a piece of refractory that has been cut into a finger like shape, in hot slag at temperatures resembling typical operating conditions, Figure 6. In this study, the slag was heated to 1500°C and the samples were held in liquid slag for 6 hrs while stirring during the test was helped by induction.

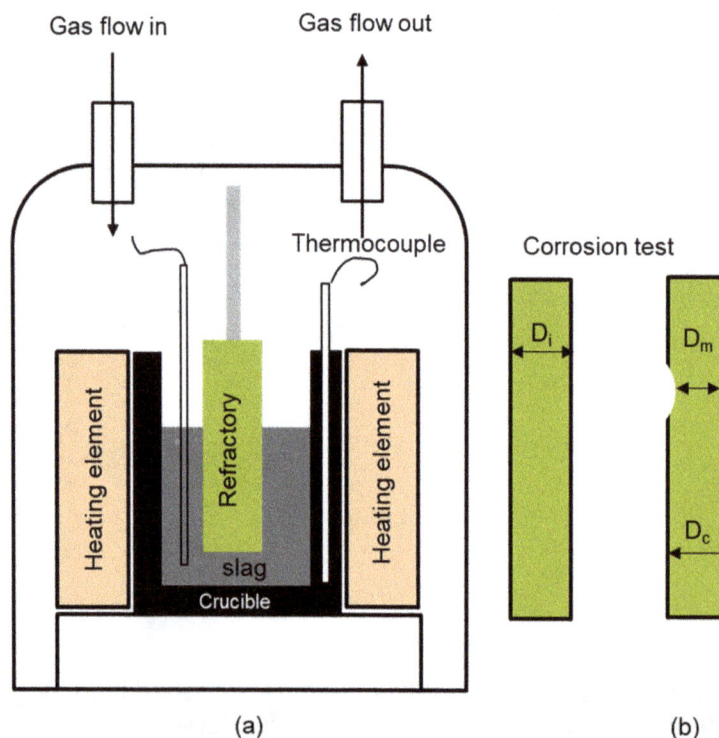

FIG 6 – Typical experimental set-up for finger tests: a) controlled inert atmosphere, sample evolution during tests in b), D_i is original dimensions of test sample, D_m, corrosion neck and D_c characteristic dimension of corroded piece in transverse direction sufficiently far off the corrosion neck, adapted from reference (Reynaert, Sniezek and Szczerba, 2020).

Crucible test (or cup test) is done by placing molten slag a in a cup or crucible shaped refractory. Although similar temperature for slag as in the finger test is used, the holding time for this test is much longer. In this study, the slag was held at 1500°C for about a week. Figure 7 shows the schematic of a crucible test set-up. Table 2 shows the conditions for the finger and crucible tests. The refractories tested are magnesia-chromite and pure magnesia grades.

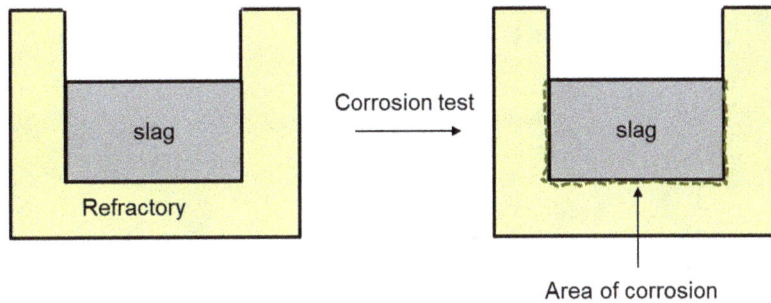

FIG 7 – Schematic of crucible test showing slag infiltration area. Adapted from reference (Reynaert, Sniezek and Szczerba, 2020).

TABLE 2

Test parameters for the refractory.

Type of test	Temperature (°C)	Holding time (hours)
Finger	1500	6
Crucible	1500	168

Results of finger tests are shown by the different forms of refractory degradation, especially in Figure 8. Figure 9 shows after test crucibles used for cup test. Microstructures of refractory after tests (not included here) were evaluated using scanning electron microscope - energy dispersive X-ray spectroscopy (SEM-EDS). Overall, the magnesia-chromite brick showed resistance to dissolution and cracking compared to the pure magnesia brick, which indicates that from the tested grades, magnesia-chromite is the recommended type of brick. In Figure 8, there is more degradation in pure magnesia brick. The worst case is observed when using this type of refractory in slag of higher basicity, as shown in Figure 8, the thin end of finger fell off during test.

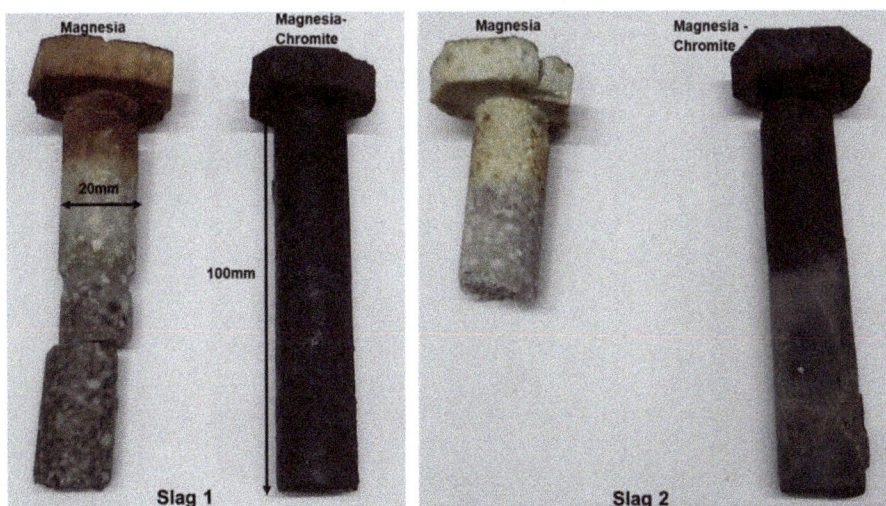

FIG 8 – Results of finger tests for two selected refractory types. Slag of two different basicities 1.2 and 1.4 have been used and are labelled slag 1 and 2, respectively.

However, this is an ongoing investigation and more information will be generated to fine-tune the selection and to understand the various types of loads caused by the industrial operation.

FIG 9 – Samples of crucible tests showing drilled hole for slag containment.

SUMMARY

Purpose of this paper was to present the fluxing philosophy of Metso's Outotec DRI smelting furnace. One of the main tasks was to identify operation window for smelting process, by using thermodynamic tools. The other purpose was to compare typical refractories for the DRI smelting furnace.

Suitable operation window in terms of liquidus temperature was found for production of iron metal with high carbon content. This operation window is large in terms of CaO/SiO_2 ratio, Al_2O_3 and MgO contents of the slag. However, operation area optimal in terms of liquidus temperature for iron metal with low carbon content is difficult to operate, because small changes in composition change operation temperature a lot. The slag obtained from Metso's DRI smelting furnace is of specification required for cement industry. For refractory selection, Magnesia chromite brick showed superior qualities for the DRI smelting furnace type of slag.

REFERENCES

Fick, G, Mirgaux, O, Neau, P and Patisson, F, 2014. Using biomass for pig iron production: A technical, environmental, and economical assessment, *Waste and Biomass Valorization*, 5:43–55.

Gielen, D, 2003. CO_2 removal in the iron and steel industry, *Energy Conversion and Management*, 44(7):1027–1037.

Kuramochi, T, 2016. Assessment of midterm CO_2 emissions reduction potential in the iron and steel industry: a case of Japan, *Journal of Cleaner Production*, 132:81–97.

Olivier, J G, Schure, K M and Peters, J A H W, 2017. Trends in global CO_2 and total greenhouse gas emissions, *PBL Netherlands Environmental Assessment Agency*, 5:1–11.

Pihlasalo, J and Tesfaye, F, 2022. Thermodynamic evaluation of the slag chemistry for DRI smelting, R2022–10677, confidential report.

Reynaert, C, Sniezek, E and Szczerba, J, 2020. Corrosion tests for refractory materials intended for the steel industry—A review, *Ceram–Silikáty*, 64:278–288.

APPENDIX A

Stable phases – solutions and compounds (pure substances PS) for the FactSage liquidus projections:

1   Slag-liq oxide [Al,Ca,Fe,Mg, (Mis.gap at high SiO_2)].

2   Spinel (Cubic): Al- Fe-Mg-O or Al-Fe-Mg-O. Replaces AlSp.Mis. gap if Al,Fe(3).

3   Monoxide Rocksalt-structure: Fe(2),Ca,Mg; dilute Al,Fe(3). Miscibility gap if CaO is present.

4   Clinopyroxene: (Ca,Mg,Fe) [Mg,Fe,Fe^{3+},Al]{Al,Fe^{3+},Si}SiO_6. Possible mis. gap when Ca is present.

5   Orthopyroxene: (Ca,Mg,Fe) [Mg,Fe,Fe^{3+},Al]{Al,Fe^{3+},Si}SiO_6. Possible miscibility gap when Ca is present.

6   Protopyroxene: (Ca,Mg,Fe) [Mg,Fe,Fe^{3+},Al]{Al,Fe^{3+},Si}SiO_6. Possible mis.gap when Ca is present.

7   Low-Clinopyroxene: {Ca,Mg}1{Mg}1$(Si)_2(O)_6$. Possible miscibility gap.

8   Wollastonite: $CaSiO_3$, dilute $MgSiO_3$, $FeSiO_3$.

9   Bredigite: $Ca_3(Ca,Mg)_4Mg(SiO_4)_4$.

10 a'$(Ca)_2SiO_4$: (alpha-prime $(Ca)_2SiO_4$); dilute Mg,Fe.

11 a-$(Ca)_2SiO_4$: (alpha-$(Ca)_2SiO_4$) – dilute Fe,Mg.

12 Melilite: $(Ca)_2$ [Mg,Fe^{2+},Fe^{3+},Al]{Fe^{3+},Al,Si}$_2O_7$.

13 Olivine [Ca,Mg,Fe]1[Ca,Mg,Fe]1[Si]1$[O]_4$. Possible mis. gap if Ca is present.

14 Cordierite: $Al_4(Mg,Fe)_2Si_5O_{18}$.

15 Mullite: [Al,Fe]2[Al,Si,Fe]$[O]_5$, accounts for non-stoichiometry.

16 $Ca_2(Al,Fe)_8SiO_{16}$: X-phase – $(CaO)_2[(Al,Fe)_2O_3]_4SiO_2$.

17 $Ca(Al,Fe)_{12}O_{19}$: $CaAl_{12}O_{19}$, dilute $CaFe_{12}O_{19}$.

18 $Ca(Al,Fe)_6O_{10}$: T-phase (pure end-members are not stable).

19 $Ca(Al,Fe)_4O_7$: $CaAl_4O_7$, dilute $CaFe_4O_7$.

20 Solution Phase $Ca(Al,Fe)_2O_4$. Possible miscibility gap (use [I] option).

21 Solution Phase $Ca_2(Al,Fe)_2O_5$: $Ca_2Fe_2O_5$, dilute $Ca_2Al_2O_5$.

22 Solution Phase $Ca_3(Al,Fe)_2O_6$: $Ca_3Al_2O_6$, dilute $Ca_3Fe_2O_6$.

23 Solution Phase M_2O_3(Corundum): Al_2O_3-Fe_2O_3 solid solution. Possible miscibility gap.

Mixed alkali effect on structure of Al$_2$O$_3$-based slags

S H Hyun[1] and J W Cho[2]

1. Senior Researcher, POSCO Steelmaking Research Group, Pohang Gyeongbuk 37859, Republic of Korea. Email: sh.hyun@posco.com
2. Professor, POSTECH GIFT, Pohang Gyeongbuk 37673, Republic of Korea. Email: jungwook@postech.ac.kr

INTRODUCTION

Designing and optimising industrial slag, flux, and salt based on a detailed understanding of the atomic-scale structure of supercooled liquids is necessary to effectively control their physicochemical properties. However, the structural analysis of the industrial multi-oxide system is challenging due to the interaction between multiple network formers, intermediate oxides, and network modifiers. In particular, the Al$_2$O$_3$-based multi-oxide system presents additional complications due to the amphoteric nature of aluminate and the easy conversion of boron between the three- and four-coordinated states depending on the alkali composition. Furthermore, maintaining charge compensation between AlO$_4$ and alkali cation is very difficult at high temperatures.

Baek *et al* (2018) have reported that the AlO$_4$ structure preferentially bonded with Na ion is drastically decreased by increasing temperature in the sodium aluminoborosilicate system. If the alkali does not play a charge-compensating role for Al, the aluminium coordination increases to AlO$_5$ or AlO$_6$, which can act as network modifiers. In addition, the released alkali ions can reduce network connectivity by generating non-bridging oxygen.

However, establishing a highly stable network structure in the melt state at high temperatures is crucial for designing chemically stable slag that does not react with high aluminium molten steel and prevents the formation of undesirable crystal phases. The chemical reaction between Al of molten steel and SiO$_2$ of mold flux occurs very rapidly during the continuous casting of high Al steel. The chemical reaction of $3(SiO_2) + 4[Al] = 3[Si] + 2(Al_2O_3)$ changes the composition of the designed lime silica-mold flux and varies physical and chemical properties, eventually yielding problems with the essential functions. Permanent changes in the chemical composition of the mold flux will cause unexpected problems in the casting process, resulting in inadequate lubrication and severe product defects. Thus, Al$_2$O$_3$-based mold flux is an excellent alternative to prevent chemical reactions. However, controlling the size or morphology of crystals of an Al$_2$O$_3$-based system is difficult due to their thermodynamic characteristics.

Therefore, an innovative design method of Al$_2$O$_3$-based glass-forming liquid with a very stable network structure at high temperatures to retard chemical reactions with proper lubrication is essential. This work proposes exploiting the mixed alkali effect (MAE), which exhibits non-linear changes in chemical and physical properties such as ionic conductivity, viscosity, chemical durability, and glass transition temperature (T_g), as a key to enhancing the charge compensation stability at high temperatures. The mechanism of MAE has yet to be well understood, despite various physical and phenomenological models proposed over the past several decades. The main aim of this study is to develop the industrial composition with enhanced structural stability by MAE. To determine the MAE, we analysed the structure by changing the chemical composition and examined the relative bond stability between AlO$_4$ and alkali cation by increasing the temperature to 1550°C from 1350°C. The structure analysis was performed using solid-state magic angle spinning (MAS) nuclear magnetic resonance (NMR) and Raman spectroscopy to assess the structural changes of aluminate and borate according to temperature and composition.

EXPERIMENTAL

Table 1 gives the sample names and alkali oxide ratios of six Al$_2$O$_3$-rich glassy slag samples prepared for the present study. The relative proportions of the alkali oxides are changed to find the composition that maximises the stability of the AlO$_4$ structure by MAE. The molar ratio of the total alkali oxides to Al$_2$O$_3$ is designed to be 1:1 for all samples. The samples were prepared from regent-grade SiO$_2$, CaCO$_3$, Al$_2$O$_3$, Li$_2$CO$_3$, Na$_2$CO$_3$, K$_2$CO$_3$, B$_2$O$_3$, and CaF$_2$. The starting materials were well mixed by a mixing machine for an hour, followed by homogenisation and melting in a graphite crucible using a box furnace at 1350°C and an induction furnace at 1550°C, respectively, with an air atmosphere. Then the melts were poured into a steel plate at room temperature to obtain 12 glass samples. After the conventional melting-quenching method, all glass sample was crushed to make powder for the NMR spectroscopy.

A solid-state ^{27}Al MAS NMR (Advance III HD, Bruker, Germany) analysis was performed using a 4 mm CP MAS (Triple Resonance) probe operating at 130.32 MHz. The ^{27}Al spectra were recorded at a 10 kHz spinning rate. $AlCl_3$, as a reference, was set at 0 ppm. ^{11}B MAS NMR spectra were obtained with the same solid-state MAS NMR operated at 160.46 MHz and 10 kHz spinning rate with a 4 mm CP MAS probe. Boric acid in H_2O, as a reference, was set at 0 ppm.

A Raman spectrometer (LabRaman High Resolution, Horiba Jobin-Yvon, France) to investigate the structure of aluminate and silicate units analysed the Raman spectra of glass samples quenched at 1350°C and 1550°C. The Raman spectra were obtained at room temperature in 200–1600 cm^{-1} with a wavelength of 514 nm Ar ions as the excitation laser source for 300 secs.

TABLE 1
Alkali oxide ratios of the Al_2O_3-rich glassy slag samples.

Name	Alkali oxide ratio
LN	$Li_2O:Na_2O = 1:1$
L3N	$Li_2O:Na_2O = 1:3$
3LN	$Li_2O:Na_2O = 3:1$
LNK	$Li_2O:Na_2O:K_2O = 1:1:1$
L2N3K	$Li_2O:Na_2O:K_2O = 3:2:1$
3L2NK	$Li_2O:Na_2O:K_2O = 1:2:3$

RESULTS AND DISCUSSION

To assess the structure of aluminate and borate according to each alkali oxide ratio and temperature, the solid-state ^{27}Al and ^{11}B MAS NMR chemical shifts were analysed. Figure 1 show the ^{27}Al NMR data of 1350°C and 1550°C samples, respectively. It is well known that the ^{27}Al chemical shift depends on the coordination number of Al-O. Although this data has non-symmetric line shapes, there is almost a network former of AlO_4 structure rather than modifiers such as AlO_5 or AlO_6 units. This observation suggests that most of the alkali cations maintain a strong bonding with AlO_4 as a charge compensator even at the high temperature of 1550°C.

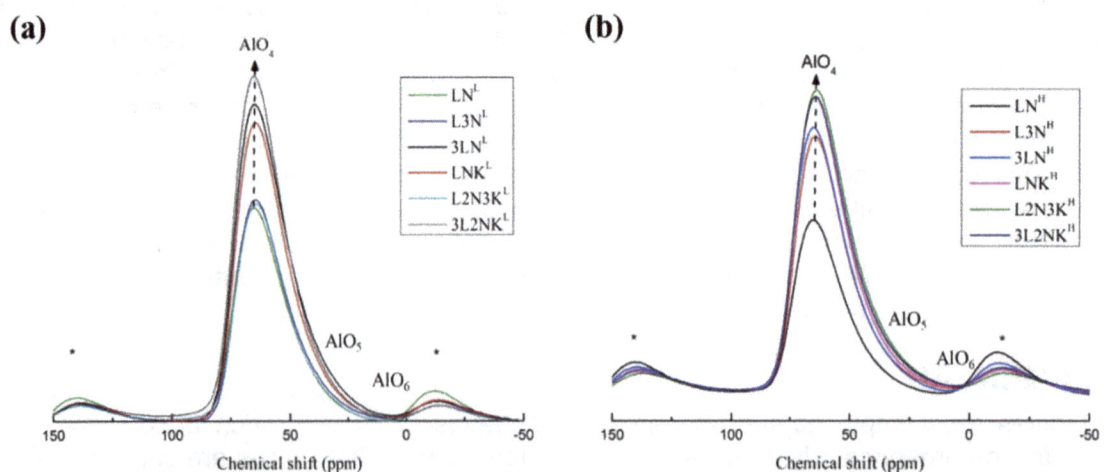

FIG 1 – ^{27}Al MAS NMR spectra of samples obtained at (a) 1350°C and (b) 1550°C.

Figure 2 show the ^{11}B MAS NMR chemical shift for the glass samples. The borate in the multi-oxide system generally exists as a BO_3 triangular ring or non-ring structure with a 3-coordinate boron and a BO_4 tetrahedron having a 4-coordinate boron. Among them, the tetrahedral borate structure can be formed only by charge compensation with alkali or alkaline earth cations, like the mechanism of AlO_4 formation. Moreover, both 3-coordinate boron and 4-coordinate boron can act as a glass former and do not significantly affect the glass-forming ability. Figure 2 shows a weak BO_4 peak compared to the BO_3 for all samples, which implies that most of the alkali cations are trying to achieve charge compensation with aluminate structure and not with borate. In the case of sample 3LN, it has slightly higher BO_4 peaks in both the 1350°C and 1550°C.

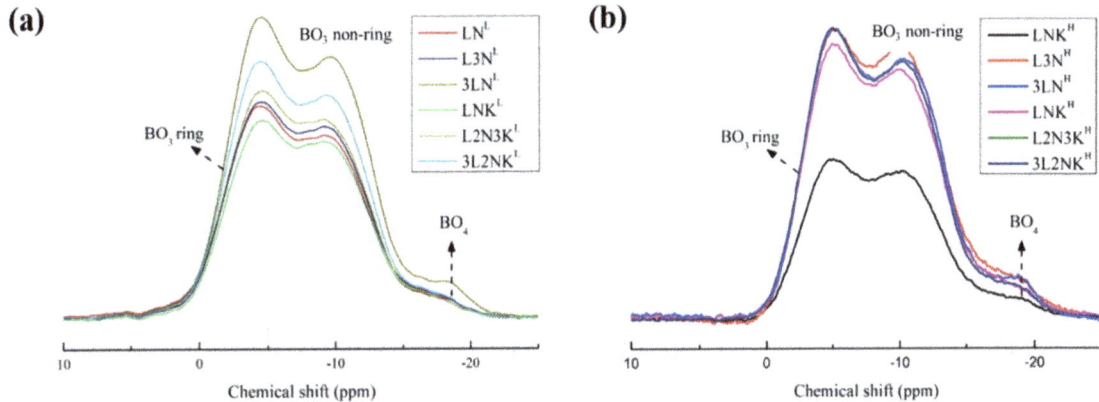

FIG 2 – ^{11}B MAS NMR spectra of samples obtained at (a) 1350°C and (b) 1550°C.

Alkali ions have good charge compensation ability with AlO_4 in the order of K>Na>Li. Therefore, due to the relatively large amount of Li_2O for the 3LN sample, Li cations will actively form a little more tetrahedral BO_4 without charge compensation with aluminate. Finally, these ^{11}B NMR results show the bonding tendency between the network former and alkali cation, consistent with the predominant consequence of AlO_4 bonding from ^{27}Al spectra.

The ^{27}Al NMR data show a featureless peak which indicates almost AlO_4 nature, but slight differences in line width were detected. The generation of different coordination environments contributes to an increase in full width at half maximum (FWHM). To determine structure variation depending on temperature and alkali mixing ratio, FWHM values of ^{27}Al NMR spectra were plotted in Figure 3. In Figure 3a, the FWHM was non-linearly decreased as the content of the stronger charge compensator Na increased due to the MAE. If the charge compensation for AlO_4 is not maintained, it will transition to the five- or six-coordinated Al. And then, the ^{27}Al spectrum broadens asymmetrically towards low chemical shift. In our previous work (Hyun et al, 2022), we demonstrated the new characteristic of the MAE, ie an increase in charge compensation stability of a small alkali when mixed with a large cation. Therefore, the non-linear variation of FWHM is due to the charge compensation stability of Li-Al being increased by the MAE. In Figure 3c, the FWHM values were nearly constant regardless of compositional changes.

However, it should be emphasised that the temperature affects the aluminium coordination environment. In Figures 3b and 3d, the minimum FWHM value was shown when the alkali molar ratio was equal (1:1 or 1:1:1) in the glasses quenched at a high temperature of 1550°C. The association between alkali and aluminium weakens with increasing temperature, and finally, the alkali cation easily dissociates from aluminium. Although the Na-Al charge compensation is stronger than that of Li-Al, Na inevitably tends to release from Al at high temperatures. However, the MAE enhances the charge compensation and is finally maximised in the 1:1 composition. This observation indicates that the MAE strengthens charge compensation even from K or Na.

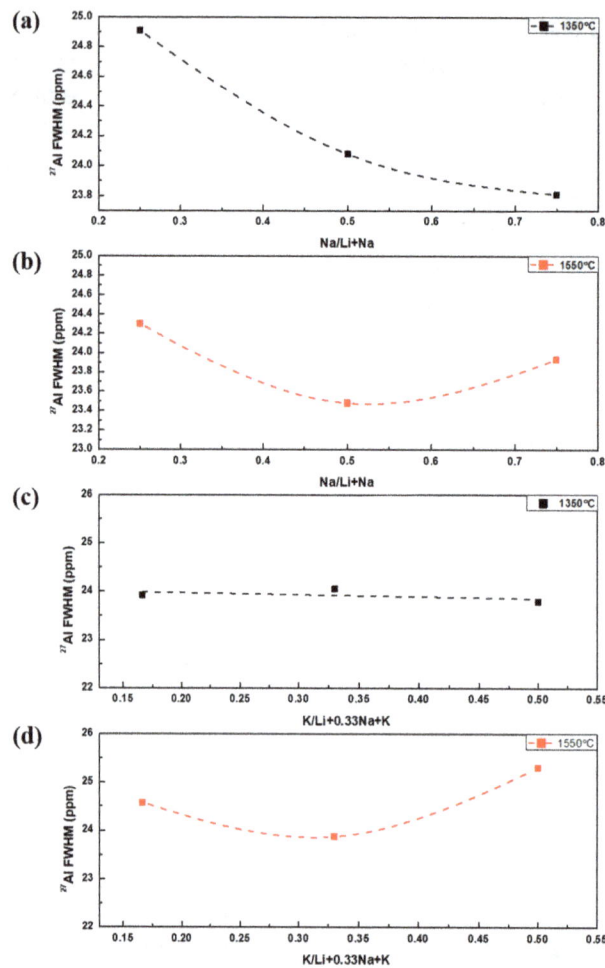

FIG 3 – FWHM values of ^{27}Al MAS NMR for glasses with different homogenisation holding temperatures (1350°C and 1550°C) and composition. (a) and (b): two types of alkalis were mixed. (c) and (d): three types of alkalis were mixed.

In Figure 4, The Raman peak at 770 cm^{-1} due to the presence of the AlO$_4$ revalidates this new knowledge. The relative intensities of the 770 cm^{-1} peaks from glass quenched at two different temperatures indicate precisely the opposite trend of the NMR results. The AlO$_4$ intensity was increased with the substitution of Na. However, the Raman feature related to AlO$_4$ is maximised in Li: Na = 1:1 glass-forming liquid at a high temperature of 1550°C. Therefore, the MAE is applicable to designing chemically stable Al$_2$O$_3$-rich slag systems for the continuous casting process.

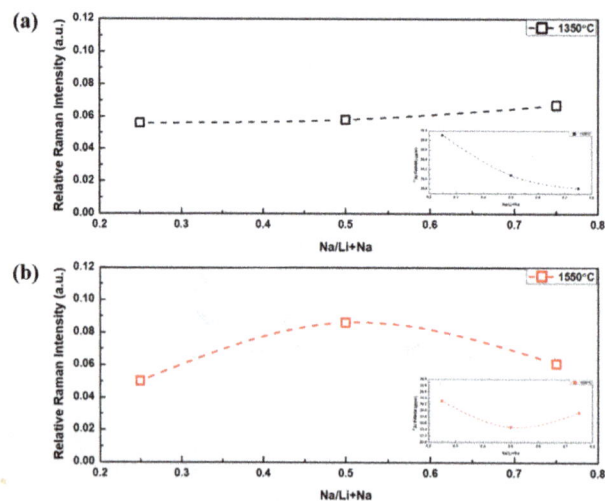

FIG 4 – Relative intensity of AlO$_4$ peak at 770 cm^{-1} in the Raman spectra for glassy slag samples obtained at (a) 1350°C and (b) 1550°C. Inset: ^{27}Al NMR results from Figure 3.

CONCLUSIONS

This study presents a structural study of multi-oxide systems for industrial aluminoborosilicate glassy slag systems. We provide insights into understanding complex structural changes of mixed alkali Al_2O_3-rich slag systems. To evaluate the compositional and temperature effect of the structural stability of Al, Al_2O_3-rich slags with various molar ratios of two or three alkali oxides were obtained at different temperatures of 1350°C and 1550°C. The stability of charge compensation between Al and alkali was significantly changed at high temperatures. Among them, Li:Na = 1:1 and Li:Na:K = 1:1:1 samples show the best charge compensation stability even at a high temperature which means the mixed alkali effect works most.

REFERENCES

Baek, J, Shin, S, Kim, S and Cho, J, 2018. Thermal history driven molecular structure transitions in alumino-borosilicate glass, *Journal of the American Ceramic Society*, 101(8):3271–3275.

Hyun, S, Yeo, T, Ha, H and Cho, J, 2022. Structural evidence of mixed alkali effect for aluminoborosilicate glasses, *Journal of Molecular Liquids*, 347:118319.

Effect of solid-solved FeO and MnO on hydration of free MgO in steelmaking slag

R Inoue[1], N Kado[2], T Iwama[3] and S Ueda[4]

1. Researcher, Tohoku University, Sendai Miyagi 980-8577, Japan.
   Email: ryo.inoue.e2@tohoku.ac.jp
2. Researcher, JFE Mineral Company, Chiba Chiba 260-0835, Japan.
   Email: n-kado@jfe-mineral.co.jp
3. Assistant Professor, Tohoku University, Sendai Miyagi 980-8577, Japan.
   Email: takayuki.iwama.a6@tohoku.ac.jp
4. Professor, Tohoku University, Sendai Miyagi 980-8577, Japan. Email: tie@tohoku.ac.jp

ABSTRACT

Steelmaking slags are usually used as roadbed and civil engineering materials. However, the expansion phenomenon, which is caused by the volume expansion during hydration of free CaO and free MgO contained in steelmaking slag, is indispensable. Since the hydration of free MgO in steelmaking slag is much slower than that of free CaO, it is qualitatively considered that free MgO contributes to the hydration expansion of steelmaking slag over a long period of time. Similar to free CaO, free MgO consists of undissolved MgO and crystallised MgO in steelmaking slag. The latter precipitates during slag cooling. When divalent metal oxides (FeO, MnO, CaO) are dissolved in the crystallised MgO to form a MgO-based solid solution, it is expected to inhibit the hydration reaction similar to precipitated CaO. From the previous hydration test of Magnesioferrite, which was calcined after mixing MgO and Fe_2O_3 reagents, it was found that the higher the Fe_2O_3 concentration, the slower the hydration reaction progressed. In this study, by clarifying the influence of calcination conditions of MgO-FeO and MgO-FeO-MnO solid solutions, which were prepared by the calcination of mixture of MgO, FeO and MnO reagents, on their hydration reactivity, the effect of composition and thermal treatment on the formation of free MgO that is not hydrated (or easily hydrated) were discussed. To investigate the influence of MgO grain size, coarse MgO particles, which were obtained by crushing MgO crucible, were also used for heat treatment. With increasing the amount of FeO and MnO in the MgO-FeO-MnO solid solution, the hydration reaction became more suppressed. The expansion suppression effect was higher with solid-solved FeO than with solid-solved MnO. When the calcination temperature of the MgO-FeO solid solution was increased, the Fe^{3+}/Fe^{2+} concentration ratio increased, resulting in a higher hydration rate.

INTRODUCTION

To achieve a better global environment, a recycling-oriented society has to be built. From the perspective of effective resource utilisation, it is necessary to establish actively a reuse system for by-products and waste from industries.

With regard to the reuse of iron and steel slags, various technologies have been conducted. Slow-cooled blast furnace slag is mainly used as road construction material, and granulated blast furnace slag is used as a cement admixture by taking advantage of its latent hydraulic property. Therefore, the utilisation rate of blast furnace slag has reached almost 100 per cent in a report of Nippon Slag Association (2023a). On the other hand, steelmaking slag, whose annual amount is approximately 14 Mt in Japan in another report of Nippon Slag Association (2023b), can be reused as road construction materials, civil engineering materials, cement, concrete aggregate etc, due to its hard and dense properties, as described in the reports of Ichihara (2010) and Nippon Slag Association (2023c). When steelmaking slag is used as roadbed and landfill materials, lime phase (free CaO) and magnesia phase (free MgO) isolated in the slag react with water to form calcium hydroxide and magnesium hydroxide. From a report of Hori *et al* (2012), because the volume after the hydration expands to approximately double, it causes road upheavals and ground cracks. Michikawa, Ono and Eba (2016) investigated the hydration reactivity of steelmaking slag by estimating the expansion ratio of slag after hydration from the free CaO content in the slag.

Takayama *et al* (1996) and Hori *et al* (2012) said that aging treatments, which are performed to advance the slag expansion caused by free CaO before burial, include atmospheric aging and accelerated aging. In the former the slag is left in a slag yard for more than six months to promote hydration through rainfall, and in the latter the hydration reaction progresses by using steam or hot water. Sasaki and Hamasaki (2014) found that the higher the ambient temperature, the faster the slag expansion becomes stable, and in the case of steam aging, expansion stability is achieved in 48 hours. Namely, accelerated aging can be expected to provide stable hydration treatment in a short time. Inoue and Suito (1995) found that free CaO in steelmaking slag includes undissolved CaO and crystallised (precipitated during cooling) CaO, and in the latter divalent metal oxides (FeO, MnO, MgO) are solid-solved. The larger the amount of the solid-solved components, the more the hydration reaction of CaO was suppressed, and when the slag was slowly cooled below 1200°C, calcium ferrite precipitates in spots in the crystallised CaO, which significantly reduces the FeO concentration in the CaO matrix phase, showed that crystallised CaO becomes more easily hydrated.

On the other hand, regarding the hydration reaction of MgO, Ohira and Obata (2009) clarified the effect of particle size on the hydration of lightly calcined magnesia. Inoue *et al* (2021) noted that since the hydration reaction of pure MgO reagents proceeds slowly at room temperature and pressure, it is desirable to accelerate the hydration by thermal treatment under high temperature and pressure using an autoclave. Because of slow hydration rate of free MgO in steelmaking slag compared with that of free CaO, it has been qualitatively thought by Ichihara (2010) that MgO hydration contributes to the slow volume expansion of steelmaking slag over a long period of time. Similar to free CaO in steelmaking slag, free MgO includes undissolved MgO and crystallised (precipitated during cooling) MgO, and divalent metal oxides (FeO, MnO, CaO) are solid-solved in the crystallised MgO and affect the hydration reaction, as reported by Inoue *et al* (2021). They found that the higher the iron concentration, the slower the hydration reaction progressed, when Magnesiowüstite synthesised by the calcination of the mixture of MgO and Fe_2O_3 reagents was hydrated. On the other hand, Amita *et al* (1989) said that the size and morphology of crystals in MgO particle are influenced by heating.

In this study, the effect of calcination conditions on the hydration reactivity of MgO-FeO solid solution, which was prepared by calcination the mixture of MgO and FeO reagents, was clarified. Furthermore, by hydrating the MgO-FeO-MnO solid solution, the difference in hydration reactivity between MnO and FeO in solid solution was investigated. In addition, by comparing those results with the hydration properties of MgO reagent and heat-treated product of a commercially available dense MgO crucibles, the condition for producing free MgO that is not hydrated (or easily hydrated) was discussed.

EXPERIMENTAL

Sample preparation

Three types of MgO-FeO solid solutions with MgO/FeO mass ratio = 9/1, 8/2, and 7/3 were prepared. Their compositions are plotted with solid circle marks in the MgO-FeO phase diagram, which was presented by Wu *et al* (1993), in Figure 1. It can be seen that all of them are solid solutions at 1400 to 1600°C.

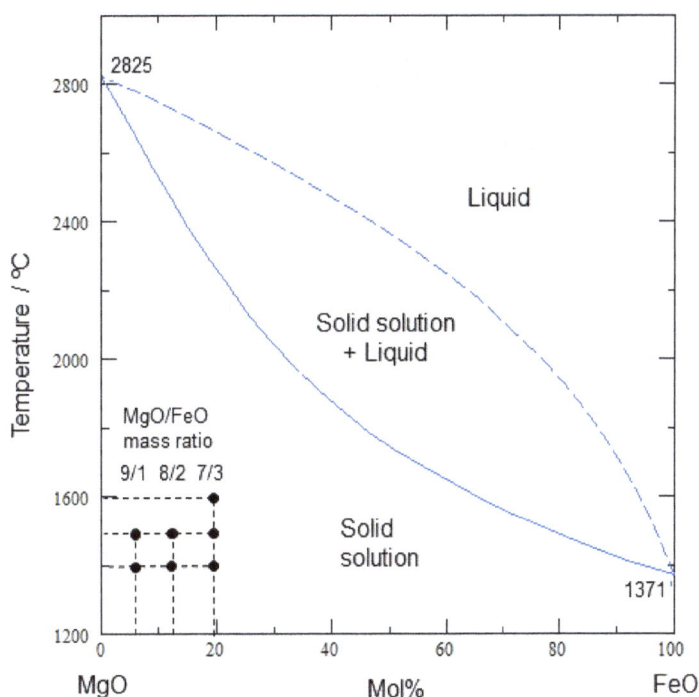

FIG 1 – Synthesised MgO-FeO solid solutions plotted in MgO-FeO phase diagram drawn by Wu *et al* (1993).

Figure 2 shows the synthesis of MgO-FeO solid solution. Appropriate amounts of special grade MgO reagent (manufactured by Wako Pure Chemical Industries, Ltd.) and FeO reagent (manufactured by Japan Pure Chemical Co., Ltd.) were weighed, mixed and pressed to a cylindrical shape at 8 ton/cm^2. To prevent direct reaction between the MgO calcination crucible and the solid solutions, pre-sintered MgO-FeO powder was placed on the inner bottom of the MgO crucible. The cylindrical sample was placed on the powder layer and slowly lowered to the soaking zone inside the vertical resistance furnace for calcination in an Ar flow (200 cm^3/min), which was deoxidised by Ti sponge at 1073 K. Heating temperature and holding time are summarised in Table 1. In this table, a pure Al$_2$O$_3$ crucible with thin wall (SSA-S, type-T3, manufactured by Nikkato) was used instead of pure MgO crucible with thick wall to promote temperature drop by reducing the heat accumulation on the crucible wall during rapid cooling. After a given time, the crucible was taken out of the furnace and quenched by He gas blowing (hereinafter abbreviated as 'He quenching') or dipping into water (hereinafter abbreviated as 'water-quenching'). In addition, a MgO-FeO−MnO solid solution in which a part of FeO was replaced with MnO was also prepared using the same method. The obtained solid solution sample was ground to 75–150 μm for hydration treatment.

Two types of pure MgO were employed for hydration treatment: special grade MgO reagent (manufactured by Wako Pure Chemical Industries, Ltd.) and a dense MgO crucible (MG-12, manufactured by Nikkato) pulverised to 75 to 150 μm.

MgO, FeO, MnO reagents

```
          ┌──────────┐
          │ Weighing │
          └──────────┘
               ↓
  ┌→ ┌─────────────────────────┐
  │  │ Mix and press to cylinder│
  │  └─────────────────────────┘
  │            ↓
  │  ┌─────────────────────────┐
  │  │ Sintering in deoxidized Ar│
  │  └─────────────────────────┘
  │       1400-1600°C,12 or 24h
  │       ↓                  ↓
  │  ┌───────────┐  ┌────────────────┐
  │  │ He blowing│  │ Water quenching│
  │  └───────────┘  └────────────────┘
  │                         ↓
  │                 ┌──────────────────┐
  │                 │ Drying (350°C,1h)│
  │                 └──────────────────┘
  │            ↓
  │       ┌───────┐
  │       │ Crush │  75-150µm
  │       └───────┘
  │            ↓
  └──────┌──────┐
         │ XRD  │
         └──────┘
               ↓
       ┌──────────────────┐
       │ Chemical analysis│
       └──────────────────┘
```

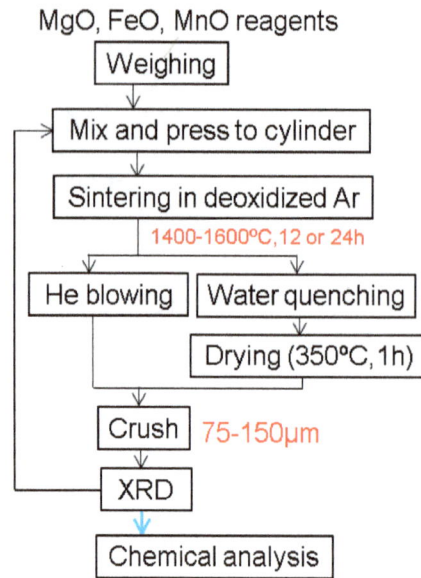

FIG 2 – Flow chart of preparation of MgO-FeO solid solution.

TABLE 1

Calcination conditions for preparation of solid solution samples.

MgO/FeO/MnO mass ratio	Temp. (°C)	Time (hrs)	Crucible
	1400	12	
		24	Dense MgO with thick wall
7 / 3 / 0	1500	12	
	1600	12	Dense MgO with thick wall
			Dense Al_2O_3 with thin wall
7 / 2 / 1	1400		
8 / 2 / 0	1400		
	1500	12	Dense MgO with thick wall
9 / 1 / 0	1400		
	1500		

Hydration treatment

An autoclave was used as the hydration reactor because hydration process is accelerated at high temperature and high pressure, as written in Dictionary of Research (2023).

At first, a Teflon test tube was weighed (w_1). After an appropriate amount of the solid solution powder was charged into the test tube, its weight was measured (w_2). The difference between w_1 and w_2 corresponds to the initial sample weight before the hydration reaction. Several drops of distilled water were added into the test tube to moisten the solid solution powder. The test tube was then placed in a pressure vessel of an autoclave and subjected to hydration treatment at 120°C for 12 hours. After the test tube removed from autoclave was heated at approximately 110°C for 2 hours in a drying oven to remove excess moisture, the weight of the test tube was measured to determine the sample weight after hydration. This autoclave treatment was repeated two more times in the same manner. For comparison, similar experiments were also conducted using a MgO reagent and a powder obtained by pulverising a dense MgO crucible.

Analysis

The mineral phases in solid solution samples were identified by powder X-ray diffraction (XRD) with the Crystallography Open Database (COD; <http://www.crystallography.net/cod/>; Downs and Hall-Wallace, 2003; Gražulis *et al*, 2009, 2012, 2015; Merkys *et al*, 2016, 2023; Quirós *et al*, 2018; Vaitkus, Merkys and Gražulis, 2021; Vaitkus *et al*, 2023).

Since it was found in preliminary experiments that the FeO amount in MgO-FeO solid solution cannot be evaluated from the relationships between MgO-FeO peak angle in XRD pattern and FeO concentration of MgO-FeO solid solution, which were given in Powder Diffraction File™ 4/Minerals supplied by JCPDS-International Centre for Diffraction Data (2020), it was determined by chemical analysis. The total iron and Fe^{2+} concentrations in the MgO-FeO and MgO-FeO-MnO solid solution powders were determined according to the potassium dichromate titration method of Japanese Industrial Standard JIS-M-8212 (1983) and JIS-M-8213 (1983), respectively. The MgO and MnO concentrations in the powder were determined by ICP emission spectrophotometry after dissolving 0.1 g of each powder with 10 mL of (1+1)HCl solution.

RESULTS AND DISCUSSION

Synthesis conditions of MgO-FeO solid solution

Number of calcinations

To investigate the change in the amount of solid-solved FeO with the number of calcination times of MgO-FeO solid solution, the following experiment was conducted. A mixture with MgO/FeO mass ratio = 7/3 was pressed to cylindrical shape and calcined in deoxidised Ar flow at 1400°C for 12 hours, then immediately quenched by He quenching. After pulverising this to less than 150 μm, it was pressed again and sintered using the same procedure as the first heating. Figure 3 shows the XRD results for those two samples. In this figure, only the (Mg, Fe)O peaks are observed in both samples despite the different number of calcination, indicating that one calcination is sufficient for the synthesis of MgO-FeO solid solution.

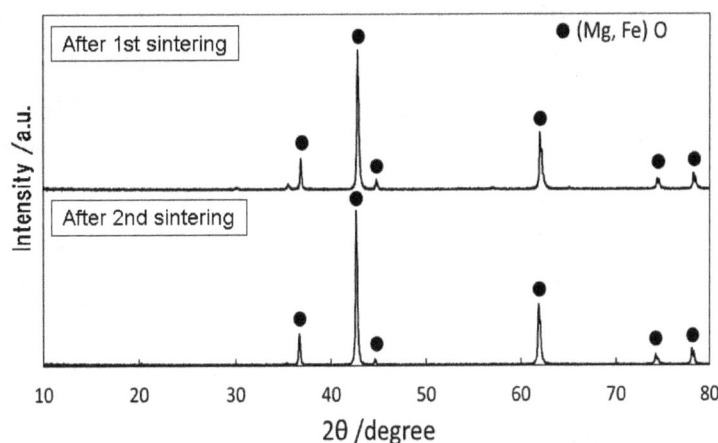

FIG 3 – Results of X-ray diffraction (XRD) analysis of MgO-FeO solid solution (MgO/FeO mass ratio = 7/3) calcined once and twice.

Table 2 shows the total iron and FeO concentrations determined by chemical analysis. In this table, the Fe^{3+} concentration is the difference between the total iron concentration and the Fe^{2+} concentration, and Fe_tO is the FeO concentration calculated assuming that all iron is present as FeO. It can be seen from the values listed in this table that the Fe^{3+} concentration is almost zero after second calcination, and that each solid solution contains approximately 30 mass per cent FeO. Therefore, one calcination is considered sufficient to synthesise MgO-FeO solid solution. In the subsequent solid solution synthesis, the calcination operation was performed only once. It should be noted that Fe^{3+} (Fe_2O_3) can dissolve in MgO according to MgO-Fe_2O_3 phase diagram reported by Phillips, Somiya and Muan (1961). The amount of solid-solved FeO and Fe_2O_3 will be explained later.

TABLE 2

Variations of chemical composition of MgO-FeO solid solutions (MgO/FeO mass ratio = 7/3) depending on calcination number and quenching method.

Quenching	Calcination number	T.Fe	Fe^{2+}	Fe^{3+}	Fe^{3+}/Fe^{2+} ratio	FeO	MgO	Fe_tO
		(mass%)				(mass%)		
He blowing	1	23.75	23.49	0.26	0.011	30.22	69.4	30.55
	2	22.64	22.63	0.01	0.0004	29.11	70.8	29.13
Water	1	24.13	21.24	2.89	0.136	27.33	68.5	31.05

Quenching method

In the previous section, the heated sample with MgO/FeO mass ratio = 7/3 was rapidly cooled by He quenching. To investigate the influence of the quenching method on the synthesis of MgO-FeO solid solution, the MgO crucible containing cylindrical sample subjected to the same calcination operation was quenched in water. The total iron and FeO concentrations in this MgO-FeO solid solution are listed in Table 2. According to XRD analysis, the sample obtained by water-quenching was also pure (Mg, Fe)O.

To clarify the influence of the quenching method on hydration behaviour, a sample with MgO/FeO mass ratio = 7/3, which was calcined once at 1400°C, rapidly cooled by He quenching or water-quenching, and crushed to a particle size of 75 to 150 μm, was subjected to the hydration treatment using an autoclave three times. Table 3 gives the sample weight changes after each hydration treatment.

TABLE 3

Mass change after hydration of MgO-FeO solid solutions (MgO/FeO mass ratio = 7/3) prepared by different quenching methods.

Hydration treatment	He quenching (69.4 mass%MgO)			Water-quenching (68.5 mass%MgO)		
	Weight (g)		Hydration rate (%)	Weight (g)		Hydration rate (%)
	Before hydration	After hydration		Before hydration	After hydration	
1st	0.5899	0.6021	6.67	0.7644	0.7924	12.0
2nd	0.6021	0.6095	10.7	0.7929	0.8135	21.0
3rd	0.6095	0.6118	12.0	0.8136	0.8328	29.2

The hydration rate of MgO can be determined by dividing the mass change after hydration by the mass when all MgO becomes $Mg(OH)_2$. For example, a sample of 0.5899 g (W_i) became 0.6021 g after the first autoclave treatment, so the mass change ΔW is expressed by Equation 1.

$$\Delta W = 0.6021 - 0.5899 = 0.0122 \text{ [g]} \tag{1}$$

When assuming that this ΔW is entirely due to the MgO hydration reaction, $MgO + H_2O = Mg(OH)_2$,

Reacted H_2O moles = ΔW / (H_2O molecular weight)

$$= 0.0122 / 18.00 = 6.78 \times 10^{-4}$$

$$= \text{Reacted MgO moles}$$

Mass of hydrated MgO = (Reacted MgO moles) × (MgO molecular weight)

$$= 6.78 \times 10^{-4} \times 40.31 = 0.0273 \text{ [g]}$$

Mass of MgO in solid solution = W_i × (mass% MgO) / 100

$$= 0.5899 \times 69.4 / 100 = 0.4094 \text{ [g]}$$

Hydration rate = (Mass of hydrated MgO) / (Mass of MgO in solid solution) × 100

$$= (\Delta W / 18.00 \times 40.31) / \{W_i \times (\text{mass\% MgO}) / 100\} \times 100$$

$$= 223.9 \, \Delta W / \{W_i \times (\text{mass\% MgO}) / 100\} \tag{2}$$

$$= 22390 \times 0.0122 / (0.5899 \times 69.4) = 6.67 \, [\%]$$

Table 3 shows the hydration rate of two types of MgO-FeO solid solutions with MgO/FeO mass ratio = 7/3, which were obtained by He quenching and water-quenching after the calcination at 1400°C, calculated by Equation 2 after each autoclave treatment. In those two MgO-FeO solid solutions, the hydration rate increases with the number of hydration treatment.

From the relationship between the hydration rate after third hydration treatment in Table 3 and the FeO concentration of solid solution (Table 2), the hydration rate is high in the water-quenched sample with low FeO concentration and high Fe^{3+}/Fe^{2+} concentration ratio. This is the same trend as the previous reports that the hydration of CaO (by Inou and Suito, 1995) and MgO (by Inoue et al, 2021) was suppressed by the solid solution of divalent metal oxides.

To clarify the reason for the difference in hydration rate depending on the quenching method, the presence of microcracks in the samples after hydration was examined by SEM observation. However, no microcracks were observed on the surface of any sample. Furthermore, when the specific surface area of the sample after third hydration was determined by the gas adsorption method (BET method), it was 0.7282 and 0.7352 $[m^2/g]$ for the sample obtained by He quenching and water-quenching, respectively. Although the latter is slightly larger, it is difficult to conclude that the difference in specific surface area due to the quenching method would affect the hydration reactivity. Table 2 shows that the sample after He quenching is not oxidised and contains a higher amount of FeO compared with that after water-quenching. Therefore, He quenching is used as the cooling method in the following experiments.

Effect of calcination conditions on hydration reactivity of MgO-FeO solid solution

MgO-FeO solid solutions were calcined by varying the mixing ratio of MgO and FeO reagents, heating temperature and duration and the thickness of the crucible for calcination. The effects of those factors on the hydration reactivity of the solid solution were studied.

When XRD analysis was performed on the MgO-FeO solid solutions obtained under each calcination condition given in Table 1, only the (Mg, Fe)O peak was identified in all samples before hydration.

The mineral phases in the solid solution after third autoclave treatment were also identified using XRD. Figure 4 shows the results for the sample with MgO/FeO mass ratio = 7/3, which was calcined at 1400, 1500 or 1600°C for 12 hours and hydrated. Figures 5 and 6 show those with MgO/FeO mass ratio = 8/2 and 9/1, respectively, which were calcined at 1400 or 1500°C for 12 hours and hydrated.

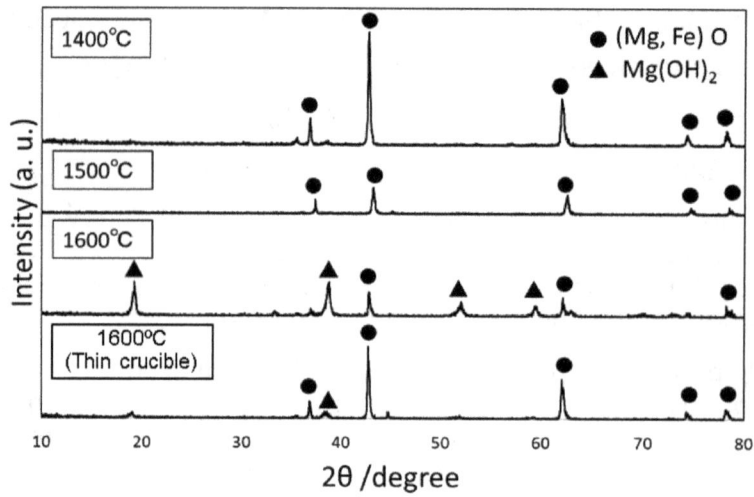

FIG 4 – XRD results after hydration of MgO-FeO solid solution (MgO/FeO mass ratio = 7/3) as a function of calcination temperature.

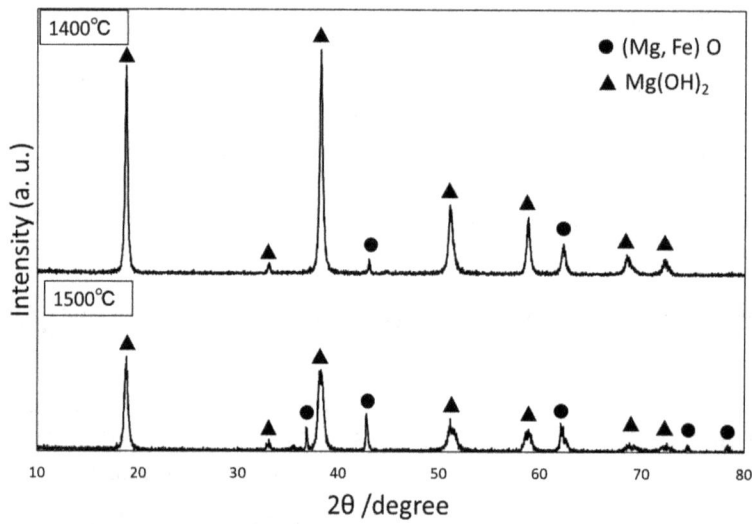

FIG 5 – XRD results after hydration of MgO-FeO solid solution (MgO/FeO mass ratio = 8/2) as a function of calcination temperature.

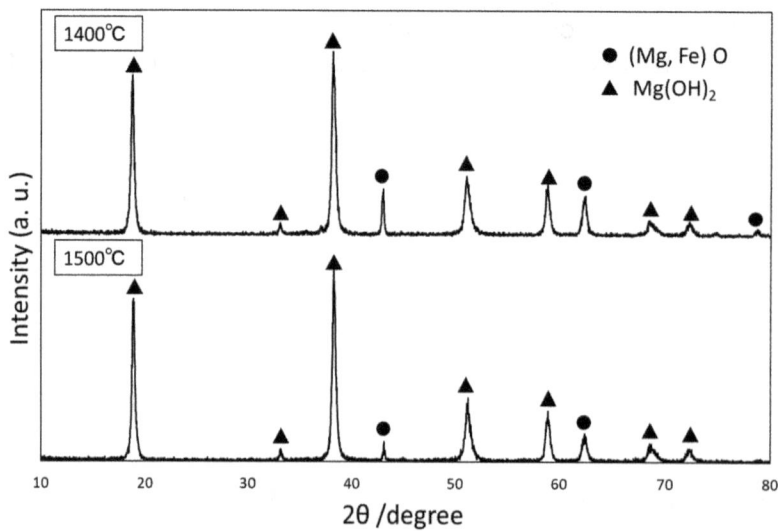

FIG 6 – XRD results after hydration of MgO-FeO solid solution (MgO/FeO mass ratio = 9/1) as a function of calcination temperature.

In all samples, only XRD peaks of (Mg, Fe)O and $Mg(OH)_2$ are observed. If only MgO in (Mg, Fe)O solid solution changes to $Mg(OH)_2$ during hydration, excess FeO should be generated as oxides or hydrates in terms of mass balance, but the XRD peaks of FeO, Fe_2O_3, $Fe(OH)_2$, FeO(OH), $Fe(OH)_3$, etc were not observed in Figures 4 to 6. Therefore, it is considered that amorphous Fe-based hydrates were generated. From the comparison of these figures, the $Mg(OH)_2$ peak becomes higher with increasing the MgO/FeO mass ratio, but clear correlation between the calcination temperature and the XRD peak height of $Mg(OH)_2$ was not found.

Table 4 shows the chemical compositions of solid solutions determined by chemical analysis. Figure 7 shows the relationship between the FeO concentration and calcination temperature, and Figure 8 shows that between the Fe^{3+}/Fe^{2+} concentration ratio and calcination temperature. In Figure 7, the effect of calcination temperature on the solid-solved FeO amount is small, while it can be seen from Figure 8 that the higher the calcination temperature and the lower the FeO concentration, the larger the Fe^{3+}/Fe^{2+} concentration ratio becomes. This might be thought to be because only a small portion of the sample surface is oxidised by the atmospheric oxygen caught in the He gas blow during He quenching.

TABLE 4

Chemical compositions of solid solutions prepared under various conditions.

MgO/FeO/MnO mass ratio	Temp. (°C)	Time (hrs)	T.Fe	Fe^{2+} (mass%)	Fe^{3+}	Fe^{3+}/Fe^{2+} ratio	FeO	MgO	MnO (mass%)	Fe_tO
7 / 3 / 0	1400	12	23.75	23.49	0.26	0.011	30.22	69.4	–	30.55
	1500		23.29	22.64	0.65	0.029	29.12	69.9	–	29.96
	1600		23.44	22.43	1.01	0.045	28.86	69.7	–	30.16
	1600*		23.62	22.14	1.48	0.067	28.48	69.4	–	30.39
	1400	24	23.97	23.26	0.71	0.031	29.92	69.0	–	30.84
7 / 2 / 1	1400	12	15.22	14.74	0.48	0.033	18.97	69.8	9.94	20.10
8 / 2 / 0	1400	12	15.55	14.50	1.05	0.072	18.65	79.8	–	20.00
	1500		15.80	13.89	1.91	0.14	17.86	79.3	–	20.33
9 / 1 / 0	1400	12	8.00	7.09	0.91	0.13	9.12	89.5	–	10.29
	1500		7.89	6.72	1.17	0.17	8.64	89.7	–	10.15

* Dense Al_2O_3 crucible with thin wall.

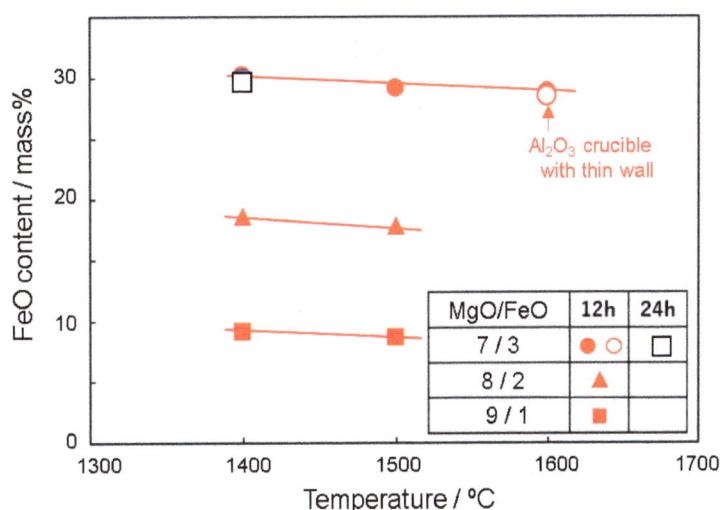

FIG 7 – Relationship between FeO content in MgO-FeO solid solution and calcination temperature.

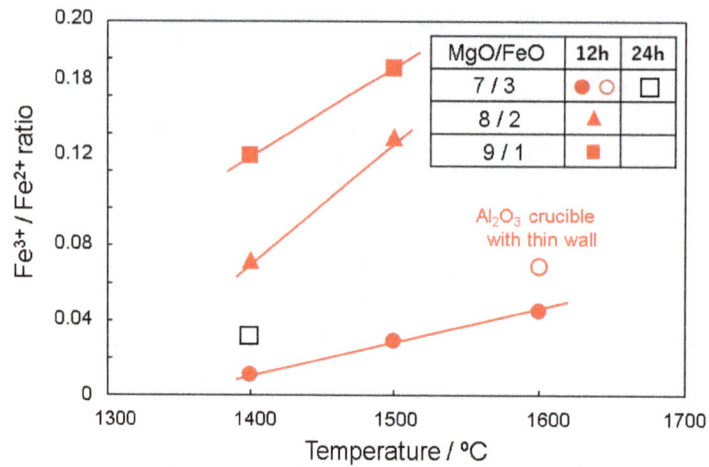

FIG 8 – Relationship between Fe^{3+}/Fe^{2+} ratio in MgO-FeO solid solution and calcination temperature.

Each sample was subjected to autoclave hydration three times. The results are listed in Table 5. The specific surface areas of MgO-FeO solid solutions (MgO/FeO mass ratio = 7/3, 8/2, 9/1) heated at 1400°C for 12 hours were determined by the BET method. The measured results are also shown in the same table. It can be seen from this table that the specific surface area increases with decreasing the amount of solid-solved FeO and the hydration rate increases accordingly.

TABLE 5

Mass change after hydration of MgO-FeO(-MnO) solid solutions prepared under various calcination conditions.

MgO/FeO/MnO mass ratio	Temp. (°C)	Time (hrs)	Weight (g)		Hydration rate (%)	Specific surface area (m²/g)
			Before hydration	After third hydration		
7 / 3	1400	12	0.5899	0.6118	12.0	0.73
	1500		0.9009	0.9369	12.8	
	1600		1.0002	1.2408	77.3	
	1600*		1.0738	1.1565	24.8	
	1400	24	1.0642	1.1149	15.5	
7 / 2 / 1	1400	12	1.0542	1.1455	27.8	
8 / 2	1400	12	1.0080	1.2505	67.5	9.17
	1500		1.0631	1.3675	86.1	
9 / 1	1400	12	1.0958	1.4730	77.0	10.01
	1500		1.0043	1.4044	99.4	

* Dense Al_2O_3 crucible with thin wall.

Based on the values given in Table 5, the hydration rates plotted against FeO concentration and Fe^{3+}/Fe^{2+} ratio of sample are shown in Figures 9 and 10, respectively, where the hydration rate is the value after third autoclave treatment. From Figure 9, the hydration rate increases with decreasing FeO concentration. At almost the same FeO concentration, the hydration rate increases with the calcination temperature. In Figure 10, the larger the Fe^{3+}/Fe^{2+} concentration ratio, the higher the hydration rate becomes. This is thought to be because, as mentioned above, the higher the calcination temperature, the easier the sample is oxidised during He quenching, and the Fe^{3+}/Fe^{2+} ratio becomes larger, resulting in easier hydration. The effect of solid-solved MnO on hydration reactivity of MgO-FeO will be described later.

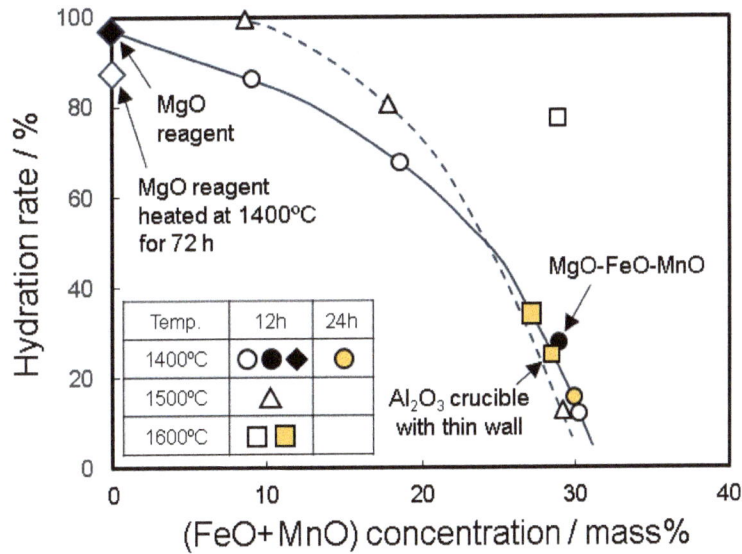

FIG 9 – Relationship between hydration rate and FeO (or FeO+MnO) content of MgO-FeO (MgO-FeO-MnO) solid solution as functions calcination temperature and time.

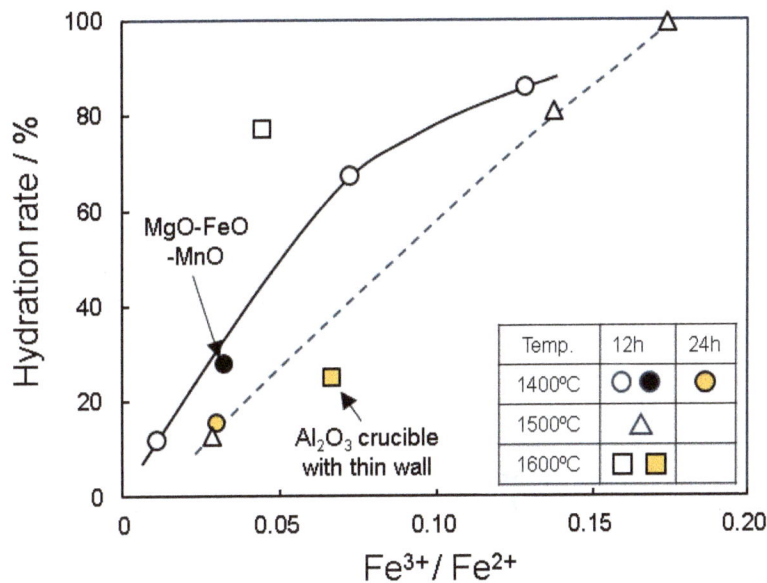

FIG 10 – Relationship between hydration rate and Fe^{3+}/Fe^{2+} ratio of MgO-FeO and MgO-FeO-MnO solid solutions as functions calcination temperature and time.

It is understood in the $MgO-Fe_2O_3$ system phase diagram, which was published by Phillips, Somiya and Muan (1961), shown in Figure 11 that the MgO-FeO sample with MgO/FeO = 7/3 calcined at 1400°C remains in the Magnesiowüstite region at rapid cooling, even if all FeO in this solid solution changes to Fe_2O_3 through oxidation. In contrast, at slow cooling, it will be in the coexistence region of Magnesiowüstite and Magnesioferrite at temperatures lower than point a (1374°C). In this region, when the iron concentration in Magnesiowüstite phase decreases toward point b with lowering temperature, the hydration reactivity of Magnesiowüstite increases with decreasing its iron oxide concentration, and the Magnesiowüstite/Magnesioferrite mass ratio of decreases. Therefore, the hydration rate increases by increasing Fe^{3+}/Fe^{2+} ratio due to the effect of hydration of the $MgO-Fe_2O_3$ system. It is concluded that by not only increasing the amount of solid-solved iron oxide but also rapidly cooling, the hydration of MgO–Fe oxide solid solution can be suppressed. In the samples obtained by calcination of MgO-FeO solid solution with MgO/FeO mass ratio = 7/3 at 1600°C for 12 hours using two types of crucibles with different wall thicknesses, it is found in Figure 9 that the hydration rate is lower when a thin-walled Al_2O_3 crucible is used, even though the Fe^{3+}/Fe^{2+} ratio is higher. The reason for this is not clear from this study, but it is presumed that the amount of heat stored in the crucible was so small that the cooling was fast during He quenching. it is also possible that a small amount of Al_2O_3 was dissolved in the MgO-FeO solid solution and suppressed hydration.

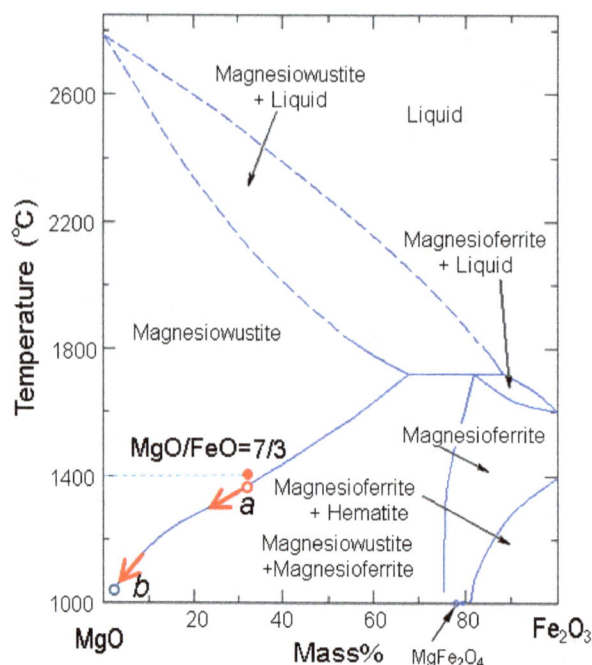

FIG 11 – Composition change of Magnesiowüstite during slow cooling in MgO-Fe_2O_3 phase diagram drawn by Phillips, Somiya and Muan (1961).

Effect of solid-solved MnO on hydration reactivity of MgO-FeO

The chemical composition and hydration rate of MgO-FeO-MnO solid solution are listed in Tables 4 and 5, respectively. Relationship between hydration rate and FeO+MnO content of MgO-FeO-MnO solid solution is shown in Figure 9, in which the hydration rate is not significantly different when solid-solving MnO instead of FeO. The relationship between hydration rate and Fe^{3+}/Fe^{2+} ratio for MgO-FeO-MnO solid solution shown in Figure 10 is near the line of that for the MgO-FeO solid solution. The relationship between Fe^{3+}/Fe^{2+} ratio and (FeO+MnO) content in the MgO-FeO-MnO solid solution was almost similar to that between Fe^{3+}/Fe^{2+} ratio and FeO content in the MgO-FeO solid solution, indicating that MnO did not affect the Fe^{3+}/Fe^{2+} ratio. It is said from Figures 9 and 10 that the ability of solid-solved MnO to the hydration of MgO-FeO solid solution is slightly inferior to that of FeO.

Effect of thermal history on hydration reactivity of MgO

Since Amita *et al* (1989) reported that the crystal size and morphology of MgO particles change with heating, it is expected that the hydration reactivity of pure MgO changes depending on the thermal history. In this study, two types of MgO powder, special grade MgO reagent powder and MgO crucible (MG-12 manufactured by Nikkato) pulverised to 75 to 150 µm, were heated at 1400°C for 12 or 72 hours in an Ar-1vol%H_2 flow, and then rapidly quenched with He blowing. The effect of thermal history on MgO hydration rate was investigated by performing hydration treatment three times in an autoclave. The obtained results are listed in Table 6.

TABLE 6

Changes in weight and hydration rate after hydration of MgO heated for various times.

Sample	Heating duration (h)	Before hydration		After third hydration	
		Weight (g)	Specific surface area (m²/g)	Weight (g)	Hydration rate (%)
MgO reagent	0	1.0881		1.5631	97.8
	12	1.0374		1.4842	96.5
	72	1.0589		1.4758	88.2
Crushed MgO crucible	0	1.5541	4.34	2.0882	77.0
	12	0.9963	1.80	1.1177	27.3
	72	1.0571	0.68	1.0863	6.2

Figure 12 shows the relationship between heating time at 1400°C and hydration rate for MgO reagent and MgO crucible-pulverised powder. In Table 6 and Figure 12, even when the MgO reagent is heated at 1400°C for 72 hours, the hydration rate decreases by only about 10 per cent compared to the untreated one. However, when the MgO crucible powder is heated at 1400°C for 12 and 72 hours, the hydration rate is approximately 50 per cent and more than 70 per cent lower than that of the untreated one, respectively. The SEM observation revealed that calcination of the MgO reagent did not progress even after long-term heating, whereas calcination of the MgO crucible-pulverised particles powder easily progressed. This is thought to be because the MgO crucible contains a small amount of SiO_2 and Al_2O_3 as binders, which promotes fusion between the powders, making hydration difficult.

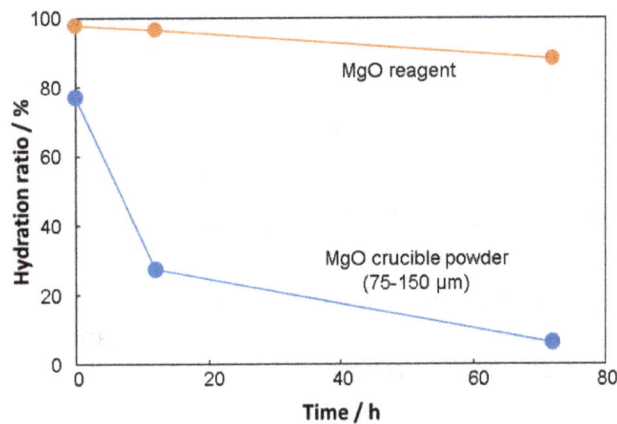

FIG 12 – Relationship between heating time at 1400°C and hydration rate for MgO reagent and MgO crucible powder.

To clarify the reason for the decrease in hydration reactivity due to calcination, the specific surface area of the MgO crucible powder (before hydration treatment) was measured using the BET method. The results are shown in Table 6. The specific surface area of the MgO crucible powder heated for a long time is significantly smaller than that of the unheated one. Since the specific surface area becomes smaller by calcination, hydration is suppressed. For the comparison of the hydration reactivity between MgO powders (Table 6) and MgO-FeO solid solutions prepared under various calcination conditions (Table 5), the hydration rates of MgO reagent powders heated for 12 and 72 hours are shown in Figure 9. The experimental point for MgO reagent heated for 12 hours is on the extrapolated curve of the relationship between FeO concentration and hydration rate of MgO-FeO solid solution heated for 12 hours. The relationship between hydration rate and the specific surface area is shown in Figure 13. In this figure, when comparing at the same specific surface area, the hydration rate of the MgO-FeO solid solution is lower than that of the unheated MgO crucible powder. This is also considered to be the effect of FeO solid solution.

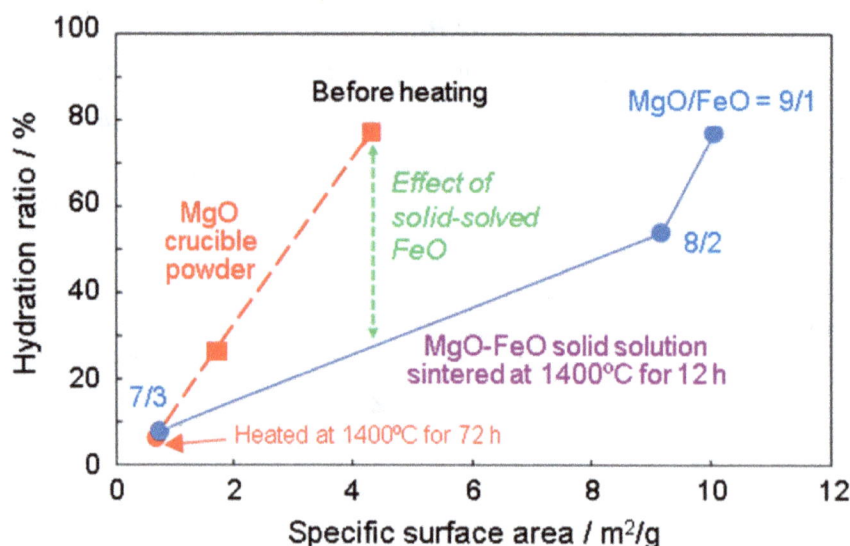

FIG 13 – Relationship between hydration rate and specific surface area for MgO-FeO solid solutions and MgO crucible powders.

CONCLUSIONS

To investigate the effect of solid-solved FeO on the hydration reactivity of free MgO in steelmaking slag, the preparation method of MgO-FeO solid solution was investigated, and conducted hydration experiments were performed on the synthesised MgO-FeO and MgO-FeO-MnO solid solutions using an autoclave. As a result, the following conclusions were obtained.

When the solid solution was rapidly cooled after calcination, a pure solid solution was obtained in one calcination process.

When MgO-FeO solid solution was hydrated under high temperature and pressure using an autoclave, the higher the amount of FeO dissolved in the MgO-FeO solid solution, the more the hydration reaction was suppressed.

When the calcination temperature of the MgO-FeO solid solution was increased, the Fe^{3+}/Fe^{2+} concentration ratio increased, resulting in a higher hydration rate.

Dense MgO crucible-pulverised powder had a lower hydration rate than commercially available MgO reagent.

The effect of heat treatment on the hydration rate of the MgO reagent was small. However, the longer the MgO crucible-pulverised powder was heated, the more the particles calcined and the specific surface area value decreased, resulting in a lower hydration rate.

ACKNOWLEDGEMENTS

We would like to express our gratitude by noting that a part of this research was conducted with the support of the Steel Environment Fund in 2018–2019.

REFERENCES

Amita, K, Nakagawa, K, Ikegami, Y, Ishihara, T and Hashizume, G, 1989. Effect of Additives on the shape of magnesium hydroxide obtained from hydration of magnesium oxide, *Gypsum and Lime*, 218:29–34.

Dictionary of Research, 2023. Autoclave. Available from:<https://www.wdb.com/kenq/dictionary/autoclabe> [Accessed: 6 January 2024].

Downs, R T and Hall-Wallace, M, 2003. The American Mineralogist Crystal Structure Database, *American Mineralogist*, 88:247–250.

Gražulis, S, Chateigner, D, Downs, R T, Yokochi, A T, Quiros, M, Lutterotti, L, Manakova, E, Butkus, J, Moeck, P and Le Bail, A, 2009. Crystallography Open Database – an open-access collection of crystal structures, *Journal of Applied Crystallography*, 42:726–729. https://doi.org/10.1107/S0021889809016690

Gražulis, S, Daškevič, A, Merkys, A, Chateigner, D, Lutterotti, L, Quirós, M, Serebryanaya, N R, Moeck, P, Downs, R T and Le Bail, A, 2012. Crystallography Open Database (COD): an open-access collection of crystal structures and platform for world-wide collaboration, *Nucleic Acids Research*, 40:D420–D427. https://doi.org/10.1093/nar/gkr900

Gražulis, S, Merkys, A, Vaitkus, A and Okulič-Kazarinas, M, 2015. Computing stoichiometric molecular composition from crystal structures, *Journal of Applied Crystallography*, 48(1):85–91. https://doi.org/10.1107/S1600576714025904

Horii, K, Tsutsumi, N, Kitano, Y and Kato, T, 2012. Processing and reusing technologies for steelmaking slag, *Shinnittetsu Giho*, 394:125–131.

Ichihara, A, 2010. Pressurized steam aging equipment for steelmaking slag, *Sanyo Technical Report*, 17(1):54–57.

Inoue, R and Suito, H, 1995. Hydration of crystallised lime in BOF slags, *ISIJ International*, 35(3):272–279.

Inoue, R, Uchidate, M, Kusukawa, S, Kado, N, Takasaki, Y and Ueda, S, 2021. Control of Hydration of Free Magnesia in Steelmaking Slag, *Journal of Sustainable Metallurgy*, 7:818–830.

Japanese Industrial Standard, 1983. JIS-M-8212. As 622.341.1-4:543.062:546.72 — Methods for Determination of Total Iron in Iron Ores, December 1983.

Japanese Industrial Standard, 1983. JIS-M-8213. As 622.341.1:543.242.5:546.722-31.062 — Methods for Determination of Ferrous Oxide in Iron Ores, December 1983.

JCPDS-International Centre for Diffraction Data, 2020. Powder Diffraction FileTM (PDF®) Search. Available from: <https://www.icdd.com/pdf-4-minerals/>

Merkys, A, Vaitkus, A, Butkus, J, Okulič-Kazarinas, M, Kairys, V and Gražulis, S, 2016. COD::CIF::Parser: an error-correcting CIF parser for the Perl language, *Journal of Applied Crystallography*, 49(1):292–301. https://doi.org/10.1107/S1600576715022396

Merkys, A, Vaitkus, A, Grybauskas, A, Konovalovas, A, Quirós, M and Gražulis, S, 2023. Graph isomorphism-based algorithm for cross-checking chemical and crystallographic descriptions, *Journal of Cheminformatics*, 15. https://doi.org/10.1186/s13321-023-0692-1

Michikawa, S, Ono, A and Eba, H, 2016. Determination of solid solubility and quantitative analysis of free-lime in steel slag by powder X-ray diffraction, *Tetsu-to-Hagané*, 102(11):623–629.

Nippon Slag Association, 2023a. Amounts of blast furnace slag produced and Used in FY 2021 [online]. Available from: <https://www.slg.jp/e/Statistics/Amounts of Blast Furnace Slag 2021FY.pdf> [Accessed: 6 January 2024].

Nippon Slag Association, 2023b. Amounts of steel slag produced and Used in FY 2021 [online]. Available from: <https://www.slg.jp/e/Statistics/Amounts of Steel Slag 2021FY.pdf> [Accessed: 6 January 2024].

Nippon Slag Association, 2023c. Characteristics and applications of iron and steel slag [online]. Available from: <https://www.slg.jp/e/slag/usage.html> [Accessed: 6 January 2024].

Ohira, Y and Obata, E, 2009. Effect of particle size on hydration rate of magnesium oxide, *Kagaku Kogaku Ronbunshu*, 35(5):543–547.

Phillips, B, Somiya, S and Muan, A, 1961. Melting relations of magnesium oxide-iron oxide mixtures in air, *J Am Ceram Soc*, 44(4):167–169.

Quirós, M, Gražulis, S, Girdzijauskaitė, S, Merkys, A and Vaitkus, A, 2018. Using SMILES strings for the description of chemical connectivity in the Crystallography Open Database, *Journal of Cheminformatics*, 10. https://doi.org/10.1186/s13321-018-0279-6

Sasaki, T and Hamazaki, T, 2014. Development of steam-aging process for steel slag, *Shinnittetsu-Sumikin Giho*, 399:21–25.

Takayama, S, Idemitsu, T, Aida N, Sugi M and Tokuhara H, 1996. The utilization of steel making slag improved by the method of steam aging as upper base materials, *Doboku-Gakkai Ronbunshu*, 544:177–186.

Vaitkus, A, Merkys, A and Gražulis, S, 2021. Validation of the Crystallography Open Database using the Crystallographic Information Framework, *Journal of Applied Crystallography*, 54(2):661–672. https://doi.org/10.1107/S1600576720016532

Vaitkus, A, Merkys, A, Sander, T, Quirós, M, Thiessen, P A, Bolton, E E and Gražulis, S, 2023. A workflow for deriving chemical entities from crystallographic data and its application to the Crystallography Open Database, *Journal of Cheminformatics*, 15. https://doi.org/10.1186/s13321-023-00780-2

Wu, P, Eriksson, G, Pelton, A D and Blander, M, 1993. Prediction of the thermodynamic properties and phase diagrams of silicate systems–evaluation of the $FeO-MgO-SiO_2$ system, *ISIJ International*, 33(1):26–35.

University research on molten slags, matte, speiss and metal systems for high temperature processing – challenges, opportunities and solutions

E Jak[1], M Shevchenko[2], D Shishin[3], E Nekhoroshev[4], J Chen[5] and P Hayes[6]

1. Professor, Director, Pyrometallurgy Innovation Laboratory (Pyrosearch), The University of Queensland, Brisbane Qld 4072. Email: e.jak@uq.edu.au
2. Theme Leader, Pyrometallurgy Innovation Laboratory (Pyrosearch), The University of Queensland, Brisbane Qld 4072. Email: m.shevchenko@uq.edu.au
3. Theme Leader, Pyrometallurgy Innovation Laboratory (Pyrosearch), The University of Queensland, Brisbane Qld 4072. Email: d.shishin@uq.edu.au
4. Theme Leader, Pyrometallurgy Innovation Laboratory (Pyrosearch), The University of Queensland, Brisbane Qld 4072. Email: e.nekhoroshev@uq.edu.au
5. Theme Leader, Pyrometallurgy Innovation Laboratory (Pyrosearch), The University of Queensland, Brisbane Qld 4072. Email: j.chen5@uq.edu.au
6. Professor Emeritus, Pyrometallurgy Innovation Laboratory (Pyrosearch), The University of Queensland, Brisbane Qld 4072. Email: p.hayes@uq.edu.au

ABSTRACT

Recent decades have seen significant advancements in analytical and experimental techniques, thermodynamic theory, and computational capabilities in high temperature research which are particularly timely to address the challenges of increasingly complex and variable feedstocks in both primary and secondary pyrometallurgical processes, the need to optimise and improve them. Currently a research program is in progress on the development of the 20-component multi-phase thermodynamic database for pyrometallurgical smelting, refining and recycling systems describing thermodynamics and phase equilibria of molten slag, matte, speiss, salts, alloys and associated solids. The integrated experimental and modelling program has been supported for nearly two decades by several consortia of 14 major international companies with over 30 operations around the world, integrates many components, incorporates many in-house developments, and brings together many different groups of professionals of different levels from junior and mid-career to senior from both industry and university sectors. The aims of the paper are:

- to outline the many components and issues of the overall program including analytical, experimental and thermodynamic modelling research as well as implementation of the results in industrial practice, professional education, planning and organisation issues.

- to highlight the opportunities, challenges, and possible solutions.

- to present these to all different groups of professionals involved in the current program.

- and thus, importantly, to facilitate possible future further developments and collaborations in the field of phase equilibria and thermodynamics of complex high-temperature systems.

INTRODUCTION

Purpose of the paper

Significant technological society changes associated with decarbonisation, developments of renewable energy sources, electric vehicles, increasingly complex electrical and electronic devices have led to increasingly complex and variable pyrometallurgical processing streams. To address these challenges, optimise existing processes, and develop new pyrometallurgical technologies knowledge-based approaches are required. Fundamental information and powerful computer-based predictive tools can provide accurate and reliable descriptions of phase equilibria and thermodynamics in multi-phase multi-component systems. The Pyrosearch Laboratory (Pyrosearch) at The University of Queensland (UQ) developed and uses integrated experimental and thermodynamic modelling methodology to create computational tools used by Australian and international metal production and recycling industrial companies. The aim of the overall research program is to support the industry in transition towards higher sustainability and efficiency. The current objectives of the program are to: i) develop a 20-component thermodynamic multiphase, multicomponent database for non-ferrous smelting, refining and recycling systems describing molten slag, matte, speiss, salts, alloys and associated solids using the integrated experimental and thermodynamic modelling approach; and ii) implement the research outcomes into industrial practice. The research program has been continuously supported over the last nearly two decades by the consortia of over a dozen of major international metallurgical and recycling companies with over three dozen industrial operations around the world, as well as by Australian and USA government grants. Significant progress has been achieved in fundamental and applied research, and in implementation of research outcomes into industrial practice.

The aims of this paper are to outline the developed research approach, different components integrated into one program, the issues encountered by the authors while working on this program, possible solutions and opportunities, and to present these to different groups of professionals to facilitate possible further developments and collaborations in the field.

The combination of many components in one overall research program is an important basis of the program.

The integrated experimental and modelling approach to develop thermodynamic databases presented by authors in this paper incorporates a *complex combination of many components* of different nature including but not limited to the: i) analytical, ii) experimental, iii) thermodynamic modelling, iv) computational science, v) organisational, vi) financial, vii) industrial implementation and viii) educational. This combination is an important basis of the program. Every component itself is complex and challenging, requires specific in-depth discussion and is of interest to a specific professional audience. The next level challenge is to bring together all these components into one integrated research program undertaken by a group of professionals with different but closely related complementary expertise. Each component therefore will be described at a high, generic, relatively simplified and superficial level to allow: a) to keep the focus on the combination of them; and b) to make it accessible to all different audiences of professionals of different spheres of expertise and levels. References will be given to the more detailed discussions on each component of interest to each specific group of professionals.

The combination of different groups of professionals in one team is another important factor, these include:

- Young undergraduate students – future industrial metallurgists and researchers learning and making their career choices.

- Junior, middle career and senior researchers involved in scientific developments, and assisting in implementation of the research outcomes into industrial practice and in education.

- Senior academics involved in the development of the research programs and in university education.

- University and scientific organisations senior management involved in major organisations decision-making.

- Young and middle career industrial research and development (R&D) metallurgists implementing the scientific and research advances into practice and improving metallurgical operations.

- Young and middle career industrial process metallurgists running the industrial processes, identifying and implementing improvements along with the R&D staff.

- Senior industrial executive R&D management leading industrial research and development programs.

- Senior industrial executive management making decisions on the companies' major developments and finance.

- Government officials looking after legislation and policies, as well as government funding.

Each of these groups is important for the sustainable development and operation of strategic and long-term research programs. The authors attempt to bring together all these groups in understanding of: a) the overall combination; and b) each component in order to facilitate further collaborative research and implementation developments in the field.

More on current challenges and opportunities for pyrometallurgy

As indicated above, significant technological changes are taking place in our societies right now and in the coming decades associated with decarbonisation and the developments of renewable energy sources, electric vehicles, and new increasingly complex electrical and electronic devices. The growth in the production of advanced materials used in these sophisticated technological devices has led to significantly increased demands in both volumes and variety of primary metals production and in metals recycling. As metal scarcity grows, primary concentrates are becoming increasingly complex. This is compounded by the rising number and diversity of metals found in recycled technological devices, which significantly increases the complexity and variability of metallurgical feed streams. Both short-term (daily-weekly-monthly) and long-term (multi-year) variations are observed. To address these challenges and minimise environmental impact, modifications to existing and development of new metallurgical technologies are crucial to enhance process efficiency and productivity. Adaptation of pyrometallurgical operations to these challenges requires implementation of a knowledge-based approaches using the fundamental information and powerful predictive computer tools for the accurate and reliable predictions of separation of the multi-component multi-phase process streams directly related to the phase equilibria and thermodynamics in multi-phase multi-component systems.

Significant gaps in knowledge on high temperature properties exist – these are due to difficulties associated with high-temperature research. *The demand* for the accurate fundamental information on phase equilibria, thermodynamic and physicochemical properties of the complex multi-component systems from metallurgical and recycling industries is growing due to the need to address: a) stricter environmental regulations, b) stronger economic competition, c) better equipment and options in process control and data management, and, importantly, d) increased complexity and variability of the process chemistry. *The supply* of the needed fundamental data on the chemistry of the processes is becoming possible since: a) new experimental techniques are becoming available due to b) the developments of modern advanced analytical techniques, dramatic improvement of their capabilities and availability, c) new theoretical modelling approaches, and d) the significantly increased computer capabilities. This demand/supply combination is the basis for *the renaissance in research on the high temperature chemistry of metallurgical systems.*

Phase equilibria and thermodynamics are critical for the high temperature processes modelling and optimisation

As illustrated in Figure 1, the optimisation and improvement of the existing processes and the development of new processes requires good control of the process parameters, that in turn requires adequate characterisation and modelling of all factors influencing the process output as a function of the input parameters. For pyrometallurgical operations the factors that require accurate characterisation for the stable operation include:

1. Thermodynamics and phase equilibria.

2. Physicochemical properties (eg viscosity etc).

3. Micro-kinetic (20–1000 µm).

4. Macro-kinetic (0.1–10 m) factors.

5. Plant data accuracy.

6. Plant control accuracy, where the thermodynamics and phase equilibria are particularly critical and important.

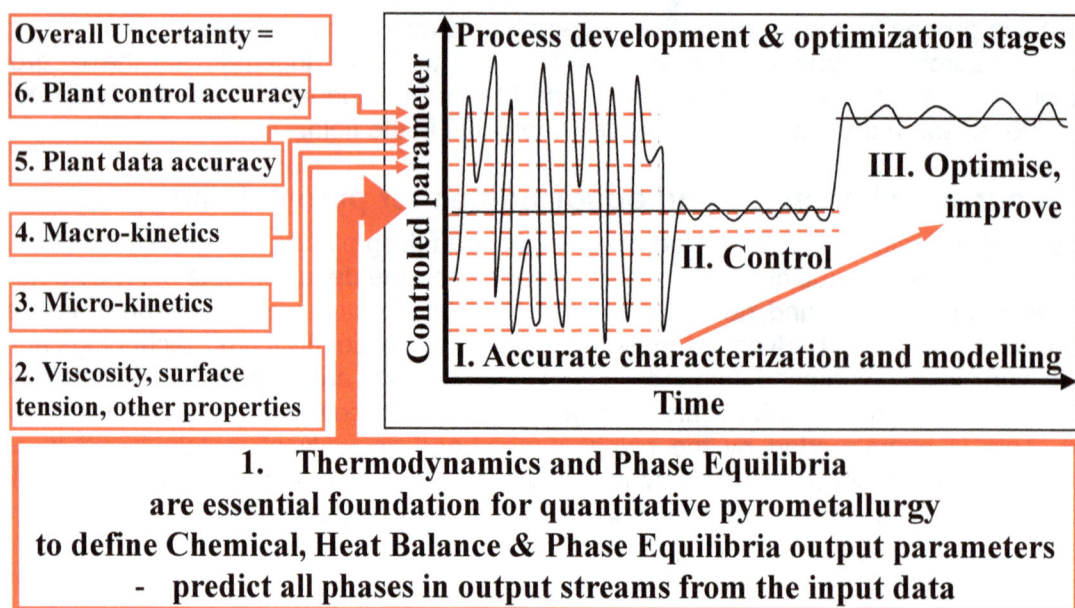

FIG 1 – Some factors influencing process uncertainty and the key stages in achieving process optimisation – characterisation/modelling and control.

In any pyrometallurgical process involving molten phases, adequate and simultaneous control of at least three key output parameters is essential (see Table 1):

1. *Chemical target* (product composition).

2. *Heat balance* (to control process temperature).

3. *Phase equilibria* (liquidus temperatures or proportion of solids) (Jak, 2018).

These three output parameters are directly controlled by the amounts and compositions of the materials introduced into the reactor. All these input and output parameters are interrelated- each output parameter depends on all input parameters with different sensitivities and is strongly determined by the thermodynamics and phase equilibria of the system. *Reliable models accurately describing thermodynamics and phase equilibria of the whole chemical system over a wide range of compositions, temperatures and pressures are therefore critical for the optimisation and control of existing and for the development of future pyrometallurgical processes.*

TABLE 1

Typical input, target and calibration parameters used in pyrometallurgical processing.

Examples of input parameters	Kinetic "calibration" parameters	Target output parameters
Oxygen coefficient (O_2/feed ratio) Duration of blow and rate	Oxygen efficiency	*Chemical* - e.g. matte grade, Cu or Pb in slag, %S in blister
O_2 enrichment, fuel (coal, oil), dust and reverts, recycling, feed rate, composition (e.g. Cu/S/Fe), mineralogy, electric power	Heat loss	*Heat Balance* - temperature
SiO_2 and CaO fluxes, slagging impurities levels (e.g. Al_2O_3, CaO, MgO)	Flux utilisation	*Phase equilibria* - Liquidus, %solids, freeze-lining thickness, e.g. Fe/SiO_2, CaO/SiO_2 in slag

In industrial practice, a range of process models and predictive tools are used depending on the needs, availability of necessary data and other factors (Jak, 2018). For example, simplified thermodynamically-based models with the kinetic 'calibration' parameters valid for limited range of conditions specific for a given reactor, eg Nikolic *et al* (2018), can be very effective as process advisers for improved process control, preparation of blends, short- and long-term production planning, limited scale-up predictions, staff training, and, importantly, for the development of the feed-forward control. These thermodynamic predictive tools can further be used to develop process predictive tools (virtual reactors) and be incorporated into 'Pyro-GPS' systems (similar to the GPS systems) used for optimisation of complex plant flow sheets (see Figure 2).

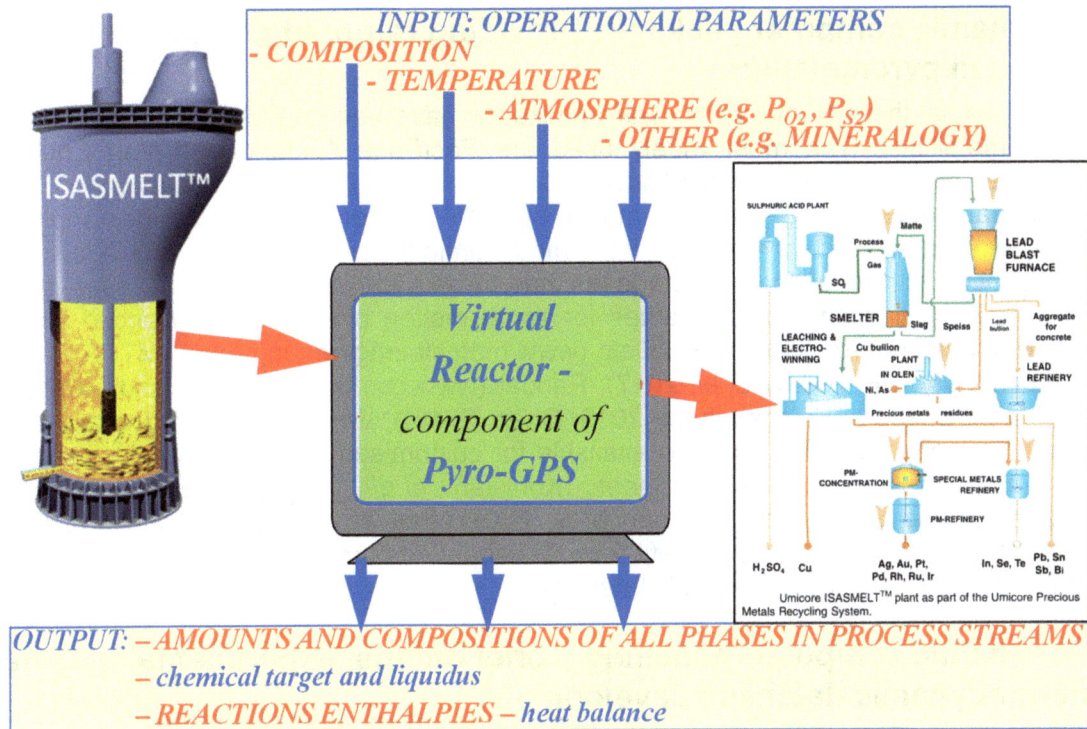

FIG 2 – Illustration of the key role of fundamentally based thermodynamic tools in predicting process outcomes and assisting in process optimisation (Jak, 2018).

The research program on the development of the multi-component multi-phase thermodynamic predictive tool outlined in this paper aims to provide the required strategic foundation for further developments in pyrometallurgy.

Research requirements for the development of computer predictive tool for pyrometallurgical thermodynamics and phase equilibria

Pyrometallurgical processing systems generally involve large number of phases with complex properties. The key liquid phases include slag, matte, speiss, salts (eg sulfates) and alloys. The majority of the key solid phases are solutions, many with wide ranges of compositions (eg spinels, melilites, olivines, pyroxenes, dicalcium silicate). Due to strong atomic-level interactions and intricate internal structures, most of these phases exhibit complex relationships between their thermodynamic properties, composition, and temperature. Accurately predicting the thermodynamic properties of key liquid and solid solution phases encountered in pyrometallurgy necessitates the use of suitable *theoretical thermodynamic models*. These models must be capable of capturing the complex dependence of thermodynamic functions on composition, temperature and pressure, and performing reliable extrapolations and interpolations beyond the range of available experimental data. Thermodynamic predictions for the multi-component multi-phase systems also require appropriate *computational capabilities* – the computer packages capable of performing complex and multiple calculations at practically acceptable speed and incorporating: i) adequate thermodynamic models, ii) Gibbs free energy minimisation module, iii) input/output interface and, importantly, iv) the databases containing thermodynamic model parameters on all possible phases including stoichiometric compounds as well as solutions (slags, mattes, salts, speiss, alloys, solids etc). Prediction of the thermodynamic properties of the actual phases requires *experimental data* to fix model parameters. In summary, the critical research components necessary to predict equilibrium chemical behaviour of the pyrometallurgical processing systems include: a) appropriate theoretical thermodynamic models, b) adequate computational capabilities, and c) sufficient and accurate experimental data. The following sections will provide some comments relevant to these critical research components.

DEVELOPMENT OF THE TOOLS

Thermodynamic computer packages providing required computational capabilities in pyrometallurgy

Several commercial thermodynamic software packages have been developed for predictions in a field of pyrometallurgy. The most widely used are FactSage™ (Jung and Van Ende, 2020) (www.factsage.com) and Thermo-Calc (Andersson *et al*, 2002) (www.thermocalc.com). FactSage contains the databases for wide range of molten systems, including oxides, sulfides, molten salts, and metals, which are based on high-quality thermodynamic assessments. Thermo-Calc, in addition to extensive metallic thermodynamic databases, has oxide and molten salts databases, as well as capabilities in modelling physical properties for some molten phases and advanced kinetics for alloys. Other available products are MatCalc (www.matcalc-engineering.com), MTDATA National Physical Laboratory, UK, (Gisby *et al*, 2017), Pandat (www.computherm.com) and MPE (CSIRO, Australia (Chen, Zhang and Jahanshahi, 2013; Chen, 2015)). When customised user interface is needed, related software tools are often available for customised applications such as ChemApp (Petersen and Hack, 2007), ChemSheet (Koukkari *et al*, 2000), ChemAppPy (www.gtt-technologies.de) and SimuSage (Petersen *et al*, 2007), that are compatible with the FactSage databases. Thermo-Calc users can benefit from developer kits for Python, Matlab, and a collection of subroutines in a form of DLL called TQ-Interface.

Thermodynamic computer modelling – brief outline, experimental data needs and thermodynamic database development

A number of key points related to the thermodynamic modelling are listed below:

- The Gibbs energies of stoichiometric compounds including pure end members are calculated from the entropy at 298 K (S_{298}), the enthalpy at 298 K (ΔH_{298}) and heat capacities ($Cp(T)$) – these are generally considered as the most accurate and self-consistent data – important foundation of the solution models in thermodynamic databases (see Figure 3).

- A two-component thermodynamic solution model consists of: i) the expression for end-members (S_{298}, ΔH_{298} and $Cp(T)$), ii) adequate solution model expression related to the internal atomic-scale structure of the solution phase that determines an ideal entropy of mixing, and iii) binary excess parameters describing deviations from the corresponding ideal behaviour as a function of composition, temperature and pressure determined by the trends of internal interactions between the two components at atomic scale. The solution model expression must be sophisticated enough to adequately describe complex Gibbs free energy of mixing determined by the internal structure of the phase, eg M-shaped entropy of mixing in the solutions with strong internal atomic-scale interactions and ordering in phases such as silicate slag (molten oxide solution) (Pelton and Blander, 1986). For example: a) the Modified Quasichemical Model (Pelton *et al*, 2000; Pelton and Chartrand, 2001; Pelton, Chartrand and Eriksson, 2001) is used for the liquid slag and matte phases, b) the Compound Energy Formalism (or sublattice model) (Chartrand and Pelton, 2001; Hillert, 2001) is used for the solid solutions with complex crystal structure, and c) the Bragg-Williams random mixing model is used for solutions with weak or positive deviation from ideal in the present study within the FactSage computer package (Bale *et al*, 2016) or custom tools developed using ChemApp (Petersen and Hack, 2007).

- A three-component thermodynamic solution model incorporates mathematical expressions for the extrapolation of the Gibbs energy of binary systems into ternary space, as well as mathematical formalism for the ternary excess thermodynamic parameters. The latter describes the effect of the third component on the binary interactions of the given two components. Generally, only binary and ternary thermodynamic parameters in the solution models of slag and other condensed phases are used since the effect of the next order interactions is negligible or does not exist at all. The extrapolation of the Gibbs energy of binary systems into ternary and higher order composition space is different for different components and is specific to a particular ternary system (Chartrand and Pelton, 2000; Pelton, 2001). The Gibbs free energies of the solutions with the number of components higher than three are generally described by binary and ternary parameters.

- Condensed phases in pyrometallurgy such as slag, matte, salts, speiss, solid oxide solutions are complex, non-ideal solutions, with strong interactions, so that the binary and ternary thermodynamic parameters for accurate description of solution properties cannot be derived without experimental data. Experiments over the whole range of conditions with adequate accuracy are necessary for the description of the thermodynamic properties of the multi-component solutions – so the *critical conditions for the accurate multi-component multi-phase model are: a) sufficiently abundant, and b) sufficiently accurate experimental data of all types,* including:

  o Heat capacities (Cp) as a function of temperature, S_{298}, ΔH_{298} for all stoichiometric compounds and all end-members of the solutions, calorimetric data.

  o Gibbs free energies of phases G_i and of reactions ΔG_R, thermodynamic activities.

  o Phase equilibria (liquidus/melting points, miscibility gaps, solid solubilities, distribution coefficients).

  o Structural information (such as, solid solutions, crystal structures, site occupancies etc).

- The lack of experimental data severely limits the development of accurate model parameters. This requirement can be summarised and will be further referred to in the paper by the slogan: *'No experiment, no model parameter'*.

- Thermodynamic databases of multi-component multi-phase systems are developed through thermodynamic 'optimisations' when all available thermodynamic and phase equilibrium experimental data for the system are evaluated simultaneously in order to obtain one set of model equations for the Gibbs energies of all phases as functions of temperature and composition. From these equations, all of the thermodynamic properties and the phase diagrams can be back-calculated. In this way, all the data are rendered self-consistent and consistent with thermodynamic principles. Thermodynamic property data, such as Cp(T), S_{298}, ΔH_{298}, G_i, ΔG_R and activity data, can aid in the evaluation of the phase diagram, and phase diagram measurements can be used to deduce thermodynamic properties. Discrepancies in the available data can be identified. These discrepancies can then be resolved through new experimental studies. The gaps in experimental data can in identified so that further experimental investigations are undertaken in areas essential for further thermodynamic optimisations.

- The self-consistent thermodynamic databases developed in this way are used for interpolations and extrapolations in a thermodynamically correct manner and for predictions of equilibrium phase compositions and phase assemblages in multi-component, multi-phase systems.

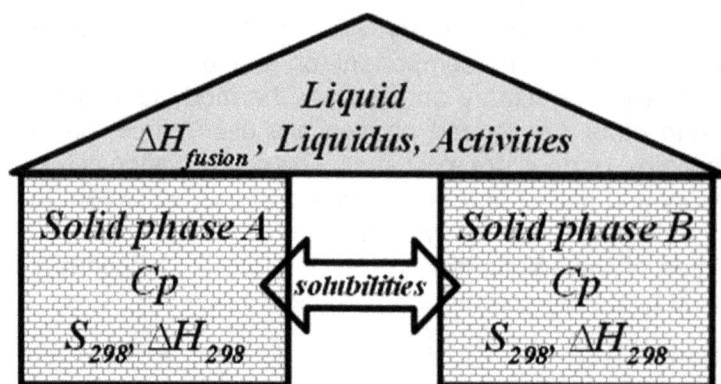

FIG 3 – Thermodynamic database structure.

Thermodynamic solution models used in the present program

Balance between predictive capability, complexity and accuracy of model predictions and available experimental data

The balance between predictive capability of a model, model complexity and accuracy of model predictions, extrapolations and interpolations depends on the available experimental data. Generally, the simpler the model (the less parameters it has) – the more predictive power it has, but at the same time the lesser capacity to describe complex trends and therefore the lesser accuracy the model has. The more complex model is—the more parameters generally it has—the more chances those parameters are not accurate enough when extrapolated, therefore significant additional experimental data are needed to determine parameters of more complex models. It is common that abundant and accurate experimental data set cannot be described with too simple and predictive model, therefore: a) sufficiently abundant, and b) accurate experimental data set requires more sophisticated model to be described, and therefore can fix parameters for such more sophisticated thermodynamic model, which in turn eventually results in more accurate and reliable description and predictions of the thermodynamics and phase equilibria of the multi-component systems. This can be re-formulated in an alternative way – the accurate and reliable sophisticated models with good predictive power require: a) sufficiently abundant, and b) sufficiently accurate experimental data set—supporting the statement indicated above 'no experiments—no model parameter'.

Brief outline of the thermodynamic models used in the present program

For the complex molten solutions with strong interactions and short range ordering at atomic scale, and corresponding complex trends in Gibbs free energy of mixing several approaches were developed, these are:

1. Introduction of associates at the composition of maximum ordering within the single sublattice.

2. Two-sublattice ionic models with complex ions corresponding to the composition of short-range ordering.

3. Quasichemical models in which the entropy of mixing takes into account the effect of short-range ordering.

The Modified Quasichemical Model (MQM) available in FactSage (Bale *et al*, 2016) used for the molten oxides (slags), sulfides and salts solutions in the present study incorporates many important additional complex functionalities, such as:

1. flexibility in the choice of pair (quadruplet) fractions or site fractions for the polynomial expansion for the Gibbs energy of quasichemical reactions (Pelton *et al*, 2000).

2. Freedom in the choice of coordination numbers, allowing the control in position and shape of short-range ordering for different systems sharing the same component (Pelton *et al*, 2000).

3. Combining quasichemical and Bragg-Williams random mixing contributions to the excess Gibbs energy (Pelton and Chartrand, 2001).

4. The opportunity to introduce associate species in cases when binary quasichemical interaction cannot describe the observed phenomenon, ie charge compensation effect.

5. Different interpolation methods of binary model parameters into multicomponent space (Pelton and Chartrand, 2001; Decterov, 2018).

6. Possible quadruplet formalism for ionic liquids with several cations and anions (Pelton, Chartrand and Eriksson, 2001).

All these modifications constitute a framework, which allows the description of the complex trends in the real solutions using Modified Quasichemical Formalism (MQF), which, in combination with sufficiently abundant and accurate experimental data, can achieve high levels of accuracy in multi-component compositional space.

For complex solid solutions such as spinels, melilites, pyroxenes, olivines and many other, the Compound Energy Formalism (CEF) (Hillert, 2001) is a widely accepted modelling method (Frisk and Selleby, 2001). The main two features of CEF are the assumption of distinct sublattices within the crystal structure, and the random mixing within each sublattice. The main model parameters are in the Gibbs energy functions of all solution endmembers, as well as excess parameters for the interaction of atoms (species) within the same sublattice. As in the case of MQF, different models can be developed within CEF. Large flexibility of CEF comes from the option of using more than two sublattices, which MQF does not allow. The same components, including charged ions, can be present on several sublattices. This approach creates many 'virtual' endmembers, necessitating the development of careful strategies to reduce the number of adjustable parameters through linear combinations and reciprocal reactions (Hillert *et al*, 2009). In multicomponent solutions, the introduction of a new solution component results in the dramatic increase in undefined endmembers, which makes expansion of solutions, such as Spinel, extremely challenging. In certain applications, such as non-stoichiometric solid speiss solutions, MQF with single sublattice demonstrated similar results to CEF with less parameters (Shishin and Jak, 2018). The MQF-based models for solid solutions are now being tested in ternary and multicomponent systems.

As indicated above, the Bragg-Williams random mixing model is used for simple solutions in the present study.

Further more detailed discussions on thermodynamic models are given in Shishin, Shevchenko and Jak (2024).

Overall experimental data requirements for constructing multi-component multi-phase thermodynamic database

As discussed earlier, a sufficiently abundant and accurate set of experimental data is crucial for a reliable thermodynamic description of pyrometallurgical systems. Some additional, more detailed points regarding data requirements are given below:

- All potential phases across the *entire composition and temperature range* need characterisation. This ensures the model's versatility for use by a diverse range of professionals to predict behaviour: i) at various stages throughout a pyrometallurgical process, and ii) within different reactor zones. Imagine a GPS system – if a specific area is not mapped, the route guidance might be inaccurate if your journey takes you through that unmapped region.

- At a minimum, *all binary systems* should be investigated. For simple binary systems with no intermediate compounds around 10–20 liquidus data points may be sufficient. However, complex and important binary systems containing multiple binary compounds and solid solutions may require 40–50 experiments to determine both the liquid slag's binary thermodynamic parameters and the solids' thermodynamic properties. As an example, the simple Sb_2O_3-SiO_2 binary system (no intermediate compounds) only required ten experimental points, while the more complex PbO-SiO_2 system (with four intermediate compounds at the liquidus and more at lower temperatures) needed 40 experiments (Wen *et al*, 2023).

- All ternary systems have to be investigated at least to a minimum level of ~20–30 liquidus points for a simple ternary and up to 200–300 experiments for a complex and important ternary with binary and ternary compounds and solid solutions to determine ternary thermodynamic parameters for the liquid slag as well as thermodynamic properties of solids. For example, the simple $PbO-Cu_2O-ZnO$ ternary system required only 12 experimental points (Wen *et al*, 2023) whereas a complex $PbO-'Fe_2O_3'-SiO_2$ ternary system required 146 experiments (Shevchenko and Jak, 2019c).

- Selected 4-component subsystems should be experimentally investigated to validate the extrapolation of ternary thermodynamic parameters.

- To ensure reliable and accurate model predictions for industrially relevant conditions, selected 4-component and higher-order subsystems should be experimentally characterised.

- The current 20-component, multi-phase gas-slag-matte-speiss-metal-solids systems with the $PbO-ZnO-Cu_2O'-FeO-Fe_2O_3-CaO-Al_2O_3-MgO-SiO_2-S$ major and As-Sn-Sb-Bi-Ag-Au-Ni-Co-Cr-Na minor elements include:

  o $22!/(20!\cdot2!) = 231$ binaries and $22!/(19!\cdot3!) = 1540$ ternaries in the 22-component $PbO-ZnO-'Cu_2O'-FeO-Fe_2O_3-CaO-Al_2O_3-MgO-SiO_2-As_2O_3-As_2O_5-SnO-SnO_2-Sb_2O_3-Bi_2O_3-Ag_2O-Au_2O-NiO-CoO-CrO-Cr_2O_3-Na_2O$ oxide system, and

  o $15!/(13!\cdot2!) = 105$ binaries and $15!/(12!\cdot3!) = 455$ ternaries in the 15-component Pb-Zn-Cu-Fe-S-As-Sn-Sb-Bi-Ag-Au-Ni-Co-Cr-Na metal/matte/speiss system.

- Complete experimental characterisation of all these binaries and ternaries would require many years of work of a laboratory such as Pyrosearch (these numbers do not even include 4- and higher order sub-systems nor combined slag-matte-metal systems). *Careful planning of the experimental work therefore is a critical factor* to ensure the continuing practical impact from this research program is delivered – this is discussed in the following sections.

Requirements for the overall experimental research program

As indicated above – adequate experimental characterisation of the current 20-component gas-slag-sulfate-matte-speiss-metal-solids systems is a very big task requiring significant efforts over extended period of time. Such significant research program, to be realistic, sustainable and practically executable requires a systematic planned gradual approach that would meet:

- the financial and organisational criteria would attract: i) continuing financial support with continuously returned value to the industrial and government sponsors, and ii) high-expertise continuous research expert team.

- the scientific or technical criteria.

These criteria are discussed as follows:

- Experimental and analytical methods should be available to generate: a) sufficient, and b) accurate data – this is the first necessary criterion.

- Intermediate, not fully optimised but, at the same time, accurate and reliable enough for some practical calculations thermodynamic database should be made available to the sponsors to provide immediate value to sponsors to justify continuing financial support – the *iterative approach* adopted in this program provides a practical solution of *bringing continuing value to sponsors* and thus maintaining the continuing financial support of the program.

- *Continuous support by research team of the implementation* of the intermediate results into the industrial practice is a critical factor to ensure this study is bringing the value to sponsors and thus maintaining the continuing financial support of the program.

- Criteria to produce the minimum number of experiments sufficient for model development is important given the large number of the overall experiments needed.

- Integration of experimental and modelling is critical to ensure: i) efficient progress of both experimental and modelling components, ii) high accuracy of both experiments and models are achieved, and iii) minimum but sufficient number of experiments is undertaken to develop the model of the required accuracy.

- Effective planning of experimental and modelling components is critical for the program with the large number of sub-systems being investigated, large number of experiments performed, large number of research staff involved and with the large number of sponsors and users of the database.

- Adequate computational method and computer tool for the experimental and modelling data management and manipulation in the process of the continuing iterative re-optimisation of the thermodynamic parameters is critical given the large number of experiments and large number of sub-systems continuously studied and re-optimised.

- Expert research team with adequate combined skills capable of running different components of the overall research program is a critical necessary factor.

As a summary – the following components are identified as critical combination to respond to the challenges:

- Financial and organisational criteria:
  - iterative approach in the thermodynamic database development
  - ongoing implementation into industrial practice
  - effective planning
  - minimum sufficient experimental data for the thermodynamic model development.

- Scientific or technical criteria:
  - experimental data: a) sufficiently abundant, and b) sufficiently accurate data for the thermodynamic model development
  - integration of experimental and modelling studies
  - adequate experimental and modelling data for computational management/ manipulation during re-optimisation
  - expert research team with necessary combined skills.

The following chapters will provide further comments by the authors on the above critical components.

METHODOLOGY

Integration of experiments and thermodynamic modelling is critical

The integrated experimental/thermodynamic modelling approach has been implemented at Pyrosearch and has enabled the rapid and efficient development of complex thermodynamic databases for the multicomponent systems: thermodynamic parameters are fixed by the new targeted experimental data points rather than by fitting into the pre-existing from literature and other sources experimental data. The continuously improved database is used for systematic assessment of completed experiments and identification of further new experiments that need to be carried out to accurately determine the binary and ternary model parameters. This systematic iterative procedure involves: i) the identification of priorities for experiments from thermodynamic assessment, ii) the development and application of advanced experimental and analytical techniques to characterise the required sub-systems, and performing experiments, and iii) the ongoing revision of thermodynamic parameters through re-optimisation of the required low- and corresponding high-order systems.

Steps i), ii) and iii) are repeated progressively improving predictions and database accuracy. The interval between planning and completing experiments in case of the Pyrosearch program is reduced to few days or weeks, and all tasks planning and undertaking of experiments and thermodynamic optimisations, are undertaken in many cases by the same researcher.

The integration of experimental and modelling components of the research ensures modelling is used: i) to analyse the agreement between different types of data, ii) to identify discrepancies in previous and, importantly, recent experimental results, and iii) to suggest required further experiments if needed or to conclude the work on a given sub-system, as well as iv) to plan further experiments as need (bulk composition, proportion of solids, equilibration path). Experiments in turn are critical to continuously test and gradually improve model accuracy. It is critical that the interval between request for and execution of experiments is minimum to allow dynamic model development and dynamic continuous planning of further experiments. The model-focused and model-based experiments planning ensures minimum but sufficient number of experiments are undertaken for a given system and at the same time ensures consistency and accuracy of experiments as well as of the model are tested during the research program progress.

Experimental phase equilibria methodology – critical for the high temperature phase equilibria and thermodynamics

As discussed above, the thermodynamic model development requires different types of experimental data, phase equilibria are one of them but a critical type that is described in the present paper. The requirement for: a) the sufficiently abundant, and b) the sufficiently accurate high-temperature phase equilibria measurements poses significant experimental difficulties due to a number of issues including but not limited to the highly reactive/corrosive properties of liquid phases in these systems, issues with containment, the high vapour pressures of some metal species, changing composition during equilibration due to interaction with substrate, establishing gas/slag/matte/metal equilibria, evaporation etc. Previous experimental techniques cannot provide wide enough range of applicability and the accuracy required for thermodynamic database development at high temperatures, particularly for systems containing complex solutions and multicomponent systems.

The equilibration/quenching/microanalysis methodology developed by Pyrosearch (Jak, Hayes and Lee, 1995; Jak, 2012) has overcome all of these limitations and it is now the preferred approach to phase equilibria determination in these complex systems (Figure 4). The methodology involves equilibration of small synthetic samples (typically less than 0.5 g prepared by mixing pure powders and pelletising) at accurately controlled temperature and gas atmosphere conditions. The sample is then rapidly quenched to low temperature thus retaining the phase assemblage and phase compositions present at the equilibration temperature. The microstructures in the samples prepared using standard metallographic techniques are analysed with optical and scanning electron microscopy (SEM). The phase compositions present in the samples are measured using advanced microanalytical techniques including: i) electron probe X-ray microanalysis (EPMA), and ii) laser ablation inductively coupled plasma mass spectrometry (LA-ICPMS). The compositions and process conditions are deliberately selected so that multi-phase materials are formed in the equilibrated samples. Any change to the sample bulk composition only changes the proportions of the phases, not their compositions that are measured with microanalysis after the experiment. The methodology then has the important advantage that the results do not depend on the small bulk composition changes (within certain limits) that may take place during equilibration. Recent developments in experimental techniques now allow the phase equilibrium measurements and elemental distributions between phases to be obtained in slags/sulfates/mattes/speisses/metal alloys/solid solutions in closed and open systems over a wide and continuously extended range of conditions thus meeting the conditions of being: a) sufficiently abundant, and b) sufficiently accurate. Contamination from crucible materials is avoided by using synthetic substrates of the primary phase material. A range of metal substrates (Pt, Pd, Ir, Re, Rh/Pt, Au, Mo, W, Fe, Co, Ni, Cu) have also been used for specific systems and process conditions. Successful equilibrium experiments have been carried at temperatures up to 1740°C and now being extended to higher temperatures. In all cases, proof of equilibrium is established through the '4-point' test that includes the following:

1. Changing the equilibration time to confirm that no further changes take place as the time is increased.

2. Confirming the chemical homogeneity of each of the phases.

3. Approaching equilibrium from different directions followed by analysis of the results.

4. Analysing possible reactions taking place during equilibration using available analytical techniques looking for possible signs of incomplete reaction pathways during equilibration.

SEM imaging and EPMA analysis of the trends of the compositions across the phases are particularly effective in this analysis.

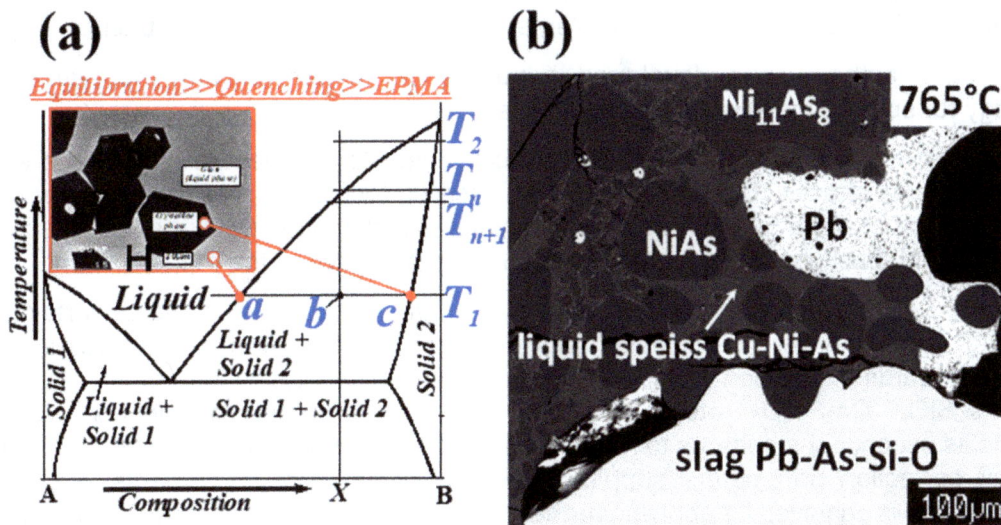

FIG 4 – (a) Illustration of the Pyrosearch approach to high temperature phase equilibria determination using equilibration-rapid quenching-microanalysis, enabling accurate measurement of phase compositions in multi-phase systems under defined process conditions, (b) An example of a complex multiphase, multicomponent system at equilibrium (Jak, 2018).

Overviews of the latest advances in the integrated experimental and modelling research approach to the thermodynamic database development for these pyrometallurgical systems are provided in Shevchenko, Shishin and Jak (2024). The use of these microanalysis techniques has provided breakthrough capabilities in phase equilibrium studies for pyrometallurgical applications, greatly extending the range of elements, bulk compositions and conditions that can be characterised and the accuracy achieved. Development and application of this experimental approach is the critical foundation of the overall research program.

Microanalytical techniques further developments

The availability of suitable analytical techniques is an important pre-requisite for obtaining: a) sufficiently abundant, and b) sufficiently accurate experimental data. The implementation of EPMA and LA-ICP-MS microanalytical techniques in the high-temperature phase equilibria has significantly expanded the range of measurable conditions and improved the accuracy of experimental data for phase equilibria studies. Therefore, the continuous development and refinement of these microanalytical methods is a crucial component of this research program. As an example of such advancements, the authors' experience with adapting and further improving these techniques is detailed below.

EPMA

The EPMA method involves the use of the electron microscope to both image the prepared cross-sections of the samples and simultaneously measure the compositions of the individual phases present. The impact of the electron beam on the samples results in the emission of X-rays characteristic of each of the elements present. Wavelength Dispersive Spectrometry (WDS) is used to select the wave lengths for particular elements for analysis. The measurement of intensities of the X-rays reflects the concentrations of the element present under the electron beam. Final analysis is obtained following corrections based on the overall matrix composition, for example, the ZAF correction where Z stands for the average atomic number, A – for absorption, and F – for fluorescence. The technique allows quantitative measurement of chemical compositions of both crystalline and amorphous phases on objects down to ~1 µm diameter. The region of interest is selected and navigated by scanning electron microscopy. EPMA also allows microanalysis of microcrystalline materials with non-zero diameter beam. Notwithstanding its superior spatial resolution, EPMA has often been perceived as reliable only for major elements with concentrations

>1 wt per cent. Recent advances in the EPMA instrumentation such as the field-emission electron gun, large-size diffraction crystals, and automation of aggregated intensity from multiple spectrometers (Donovan, Singer and Armstrong, 2016) have made it feasible to use EPMA to measure at least the more abundant trace elements (>100 ppm) on a regular basis.

Over the years, authors have developed a number of further custom improvements to this methodology for specific metallurgical applications.

One of the custom improvements are the analytical protocols to measure minor/trace elements in chemically complex phases (Chen *et al*, 2021). These protocols address major issues associated with trace element analysis by EPMA such as minimal detection limit (MDL), spectral interferences, beam damage and background correction to ensure the accuracy of the measurement.

The standard ZAF correction procedure was further improved. Systematic uncertainties of the standard ZAF correction were identified in many binary and ternary systems by measurement of the secondary standards – stoichiometric compounds existing within the system of interest, eg $PbSiO_3$, Zn_2SiO_4, Fe_2SiO_4, $Ca_2Fe_2O_5$, etc. The authors routinely synthesize and use stoichiometric compounds as secondary standards to monitor analytical uncertainties associated with the standard ZAF matrix correction parameters supplied with the commercial instruments. Improvement in analysis accuracy is achieved by applying further polynomial corrections to the measurement results (Shevchenko and Jak, 2017, 2018, 2019a, 2019b, 2021; Cheng *et al*, 2021; Khartcyzov *et al*, 2023). These corrections are specific for the same machine and standard procedures used, and allow back-correction of the data obtained in previous years before this ZAF correction bias was known. Alternative matrix correction parameters such as updated mass absorption coefficients (MACs) are also tested and applied where appropriate (Abdeyazdan *et al*, 2024).

Another potential issue affecting the analysis accuracy in multiphase systems is the Boundary secondary Fluorescence effect (BSF) – the element of interest was measured at a higher level than the true value when analysing a phase close to a boundary with another phase that contains a significantly higher level of the same element (Llovet *et al*, 2012). It was observed for transition elements (Cr, Mn, Fe, Co, Ni, Cu, Zn) in light element-rich phases (SiO_2, CaO, Al_2O_3) (Hidayat, Hayes and Jak, 2012; Hidayat *et al*, 2012; Xia, Liu and Taskinen, 2016; Hamuyuni, Klemettinen and Taskinen, 2016; Shevchenko and Jak, 2018), or when two neighbouring transition element-rich phases are adjacent to each other, such as over-estimated per cent Fe in metallic Cu surrounded by FeO-rich slag. The authors developed methods to first experimentally estimate the effect using cold-pressed minerals/materials then to apply correction to the unknown measurement results (Shevchenko and Jak, 2019d); in addition to that, authors demonstrated the effect could be minimised by measuring low energy < 5 keV characteristic lines (L lines of some of the transition metals, eg Cu, Zn) of element of interest (Shevchenko and Jak, 2021; Khartcyzov *et al*, 2023).

Focused Ion Beam – Scanning Electron Microscope (FIB-SEM)

An alternative approach to eliminate the secondary fluorescence effect at boundaries in quantitative X-ray microanalysis has recently been developed by the authors Chen *et al* (2024). Utilising FIB-SEM dual-beam system, the phase/features of interest with less than 10 µm size is physically extracted from the matrix and remount onto a suitable non-reactive substrate for subsequent EPMA analysis.

LA-ICP-MS

Originally developed for earth science research, Laser Ablation Inductively Coupled Plasma Mass Spectrometry has recently been adapted for use on the metallurgical systems for the quantitative analysis of trace/minor elements in solid materials (Avarmaa *et al*, 2015). This technique employs a focused laser beam to ablate a small portion of the sample, which is then ionised in an inductively coupled plasma and subjected to mass spectrometry for elemental identification and quantification. LA-ICP-MS offers significant advantages over EPMA, notably superior detection limits (down to ppb), enabling precise measurement of trace elements within targeted phases. Previous studies by the authors have showcased LA-ICP-MS's efficacy in accurately measuring trace elements in slag phases, contributing novel fundamental data and resolving discrepancies in literature regarding minor element distributions in non-ferrous melt smelting and converting processes. However, several challenges have constrained the utility of LA-ICP-MS in metallurgy research. Primarily, its analytical precision heavily depends on the availability of matrix-matched standards, which are presently limited for sulfides and metals. Additionally, existing glass standards (eg NIST SRMs) lack certain elements crucial for metallurgical applications, notably the platinum group elements. Consequently, authors have dedicated ongoing efforts focused on developing in-house standards. The authors pioneered the creation of silicate glass standards encompassing all six platinum group elements, utilising an innovative high-temperature and high-pressure apparatus. Furthermore, the authors engineered copper alloy standards doped with various minor elements, specifically tailored for trace element analysis in copper metal via LA-ICP-MS (Chen, Jak and Misztela, 2023). Significant improvement in analysis accuracy has been demonstrated by Chen, Hayes and Jak (2024). In comparison to EPMA, LA-ICP-MS also suffers from inferior spatial resolution (>15 µm), posing challenges for measuring small features, particularly solids. To overcome this limitation, the authors are actively investigating new laser systems, such as femto-second laser systems, with the potential to enable analysis of features smaller than 5 µm. The downside to the use of LA-ICP-MS is that the material used for analysis is physically removed from the sample – the same area cannot be reanalysed as is the case with non-destructive EPMA analysis.

Speciation of multi-valent elements (eg Fe^{2+}/Fe^{3+})

The quantitative determination of multi-valent elements, particularly Fe^{2+}/Fe^{3+} in the slag phase, has long been crucial for thermodynamic modelling in pyrometallurgy. Traditional analysis methods involve wet-chemical titration, necessitating large quantities (over 1 g) of homogeneous slag material. Mossbauer Spectroscopy offers an alternative approach, requiring less material (~200 mg), but accurate quantification often entails complex spectra deconvolution and fitting, leading to potential ambiguities depending on the fitting methods utilised. Microanalysis techniques such as X-ray Absorption Near Edge Spectroscopy (commonly employing a synchrotron beam line) and the 'Flank Method' (utilising EPMA) have been developed to quantitatively measure Fe^{2+}/Fe^{3+} in specific geological materials, demonstrating good accuracy with the use of matrix-matching standards (Borfecchia et al, 2012; Hoefer and Brey, 2007). Currently, the authors are actively exploring these options for Fe^{2+}/Fe^{3+} measurement in slag and other oxide phases.

Further development of the microanalytical techniques is required: i) to extend the ranges of compositions that can be investigated, and ii) to further improve the accuracy of the measurements to meet for thermodynamic database development requirement for: a) sufficiently abundant, and b) sufficiently accurate data. More details are given in Chen, Hayes and Jak (2024).

Application of microanalytical techniques – new opportunities for high-temperature industrial R&D

As discussed above – implementation of the advanced thermodynamic modelling tools into industrial pyrometallurgical process control requires significantly more advanced characterisation of the actual processes to incorporate adequate calibration parameters to account for kinetic and other factors. The microanalytical techniques (EPMA, LA-ICP-MS) that became available relatively recently provide the capabilities to characterising heterogeneous processes responding to the new demands and created fresh possibilities to improve the characterisation of the industrial systems and thereby make further improvements to industrial practice and support implementation of the advanced models. Most of pyrometallurgical processes involve heterogeneous reactions taking place at microscopic scale at and around the phase interfaces. The microanalytical techniques EPMA and LA-ICP-MS enable advanced microstructural characterisation and accurate composition measurements minimum elemental detection limit down to ~300 ppm at spatial resolutions of approximately 0.5 to ~1–3 micron and down to ppb at spatial resolution of approximately 20 to 80 micron respectively. Microanalysis of industrial quenched samples provides an exceptional opportunity to characterise processes taking place in the reactors. Detailed analysis of the composition trends as microscopy scale of 10–30 micron from the phase boundaries and at larger scale across the quenched samples provides important information on the kinetic reactions taking place in the industrial process and therefore a basis for the incorporation of thermodynamic models to describe real industrial processes.

The examples of applications of microstructural analysis of industrial samples include: i) phase equilibria (melting etc), ii) reactions extents (eg Cu losses), iii) specific troubleshooting (eg freeze-lining, refractories, deposits).

The recent examples include the measurement of distribution of Co, Ni, Zn, Pb, Bi, Sn, As, Te, Se, Ga, In, and Ge in complex copper converting slags (Chen *et al*, 2021), analysis of iron sinter micro-structure and their link to breakage characteristics (Cheng, Hayes and Jak, 2022), understanding the mechanism of refractory degradation in copper smelting (Fallah Mehrjardi *et al*, 2016), freeze-lining formation in non-ferrous applications (Fallah-Mehrjardi *et al*, 2014a, 2014b), investigation of reasons for the corrosion of steel walls in lead smelting (Watt *et al*, 2018), distribution of arsenic in industrial samples of nickel sulfide converting (Hidayat *et al*, 2017), conditions for the formation of spinels in nickel smelting and their effect on matte droplet settlement (Sineva *et al*, 2023).

Experimental data requirements and the importance of planning experiments

Effective planning of experimental and modelling components is a critical component for the program with large number of systems being investigated, large number of experiments performed, large number of research staff involved and with the large number of sponsors and users of the database. The following section presents further details on the planning derived by the authors from experience.

In case of the current Pyrosearch experience following up of the principle 'no experiments – no model parameters', the experimental component of the work takes approximately 75 per cent of resources and is a foundation of the overall program. The planning starts from the evaluation of the approximate number of possible experiments to be undertaken for a coming year for a given level of funding-each lead researcher in the team evaluates capabilities depending on experience as well as on the complexity of the system and commits to a particular number of successful experiments per week – this gives an approximate indication of a number of possible experiments and corresponding timing.

In order to prioritise the work to deliver more valuable outcomes, all elements in the overall multi-component system under investigation are divided into *major* elements (Me^{major}), *minor slagging* elements ($Me^{minor\_slagging}$) and *minor other* elements ($Me^{minor\ other}$). The current program includes systematic phase equilibria studies of all low order 2- and 3- metal sub-systems from the major elements list, selected key multicomponent 4- and higher number of metals sub-systems from the major elements and slagging minor elements lists, all low order 2-metal sub-systems $Me^{minor\ other}$-Me^{major}, $Me^{minor\ slagging\ or\ other}$-$Me^{minor\ slagging}$ and 3-metal sub-systems $Me^{minor\ slagging\ or\ other}$-$Me_1^{major}$-$Me_2^{major}$, $Me^{minor\ other}$-Me^{major}-$Me^{minor\_slagging}$, and minor elements distributions of elements between key phases (eg slag-metal, slag-matte, matte-metal). Transition from minor to major element list is done gradually after the all necessary major elements sub-systems have been characterised. For example, the current 20-component system at Pyrosearch has Me^{major} = Cu, Pb, Zn, Fe, Ca, Si, Al, Mg, S, O; $Me^{minor\_slagging}$ =Cr, Na and Me^{minor} = As, Sn, Sb, Bi, Ag, Au, Ni, Co. Nickel Ni is included as a major element in matte/speiss/metal and selected oxide solid solution phases.

The following *key directions* are identified for the purpose of systematic planning: i) slag-solids (S-free) phase equilibria (with and without metal phase), ii) slag-matte or-sulfate (S-containing) equilibria (with and without metal phase), iii) matte-speiss-metal equilibria, and iv) and minor elements distributions of elements between key phases (eg slag-metal, slag-matte, matte-metal). The gaseous species equilibria are not currently in focus of the Pyrosearch program. Two types of experiments are performed – closed experiments undertaken in sealed ampoules for the slag/matte/metal, slag/metal and matte/metal systems; and open or semi-open experiments with the PO_2 and PsO_2 in the gas/slag/matte and gas/slag/metal systems controlled by the $CO/CO_2/SO_2$ gas mixtures.

The techniques developed during this program for the first time enable the systematic accurate measurements of this kind to be undertaken, and these measurements provide an important foundation for the development of the thermodynamic database as well as for the overall quantitative description of the high-temperature thermochemistry. The total number of experiments needed to completely and quantitatively characterise the whole chemical system as functions of key operational parameters is very large and is increasing exponentially with further addition of new elements. The experimental needs therefore are carefully, critically and continuously reviewed.

The current criteria for priorities in selecting systems for further improvement are:

1. Importance for industrial sponsors.
2. Incorporation of all directions and elements.
3. Resolution of identified discrepancies in database predictions.
4. Finalising incomplete diagrams.
5. Prioritising possible, easy and high-success rate experiments.
6. Higher thermodynamic model parameters sensitivity/importance relative to the experimental accuracy.

Selection of experiments to be undertaken is facilitated by analysis of the sensitivity of the predicted values to the thermodynamic parameters relative to the achievable accuracy of experiments.

The issues indicated by the sponsors during the meetings are regularly summarised in a specific table, continuously reviewed and extended. In addition, to assess the accuracies of the databases at particular conditions relevant to the industrial practices, *systematic targeted programs on the laboratory-scale characterisation and modelling* of complex thermochemistry *of the key industrial processes of sponsors* are undertaken. This involves identifying/reconfirming chemistry of the main process streams and conditions, accurately characterising/reproducing complex industrial chemistry by undertaking series of experiments in the well-controlled laboratory conditions and analysing specially quenched industrial plant samples. The results are then used to analyse systematic uncertainties of the predictions against accurate and reliable data obtained in laboratory study and then to identify the needs and the ways for improvements of the models through the new targeted cycle of low order experiments and thermodynamic optimisations.

The integration of experimental and thermodynamic modelling studies helps to optimise the experimental program – to focus on what's needed and to minimise the overall number of experiments undertaken for the required accuracy of the thermodynamic predictions.

The actual program for the coming year is then developed based on the selected priorities and on the indication of a number of possible experiments indicated by the leading researchers.

The current focus of the research program of Pyrosearch is the 20-component development of the database for the system with the 'Cu_2O'-PbO-ZnO-FeO-Fe_2O_3-CaO-SiO_2-Al_2O_3-MgO-S major, Cr and Na slagging and As, Bi, Sn, Sb, Au, Ag, Ni, Co other minor elements containing gas, slag, sulfate, matte, speiss and metal alloy molten solution phases, 26 major oxide, sulfide and metallic solution phases such as spinel, melilites, olivines, pyroxenes and other, 79 solid solutions with limited ranges of solubility and over 380 oxide stoichiometric compounds.

In order to prepare an accurate description for all compositions and process conditions with the ten major elements following the criteria outlined above, the numbers of subsystems to be experimentally characterised is illustrated in Figure 5. The application of the priorities outlined above for the current list of major and minor elements, total 176 of binary and 694 ternary oxide sub-systems and several hundred more for the sulfates and matte/speiss/metal sub-systems are considered for experimental and further thermodynamic modelling characterisation.

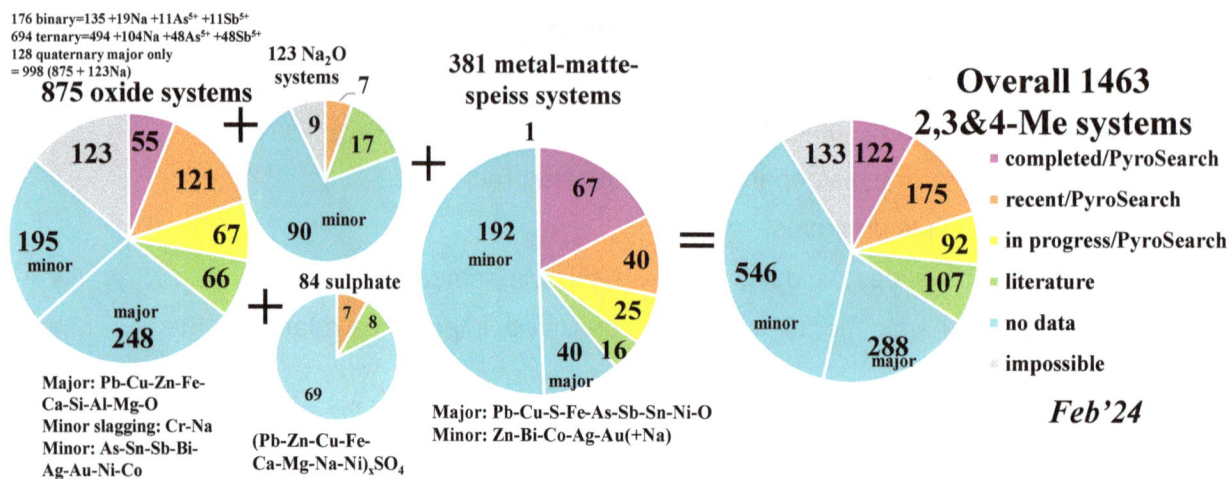

FIG 5 – Summary of chemical systems in 20-component system with 'Cu$_2$O'-PbO-ZnO-FeO-Fe$_2$O$_3$-CaO-SiO$_2$-Al$_2$O$_3$-MgO-S major, Cr and Na slagging and As, Bi, Sn, Sb, Au, Ag, Ni, Co, other minor elements and current status of database development.

Current experimental phase equilibria data status

Since early 1990s to date the Pyrosearch team has conducted ~20 000 experiments in 389 subsystems (122 completed, 175 studied at advanced level, and 92 in progress at initial level). Data from 107 systems are available in the literature, 133 systems are impossible to experimentally investigate using the current methodologies or they do not exist practically (eg Cu$_2$O-Fe$_2$O$_3$, FeO-SnO$_2$, Fe$_2$O$_3$-SnO). There are no measured phase equilibrium data on 288 major + 546 minor sub-systems. Pyrosearch team is currently undertaking approximately 1100–1300 successful experiments per annum in over >100 sub-systems.

Advanced thermodynamic optimisation methodology requirements and outline

Continuing re-optimisation of thermodynamic models using such significant number of experiments, management of the thermodynamic parameters and of experimental data is an important issue.

A major issue emerging in the *development* of multicomponent databases is the exponential increase in chemical interactions to be described with the increasing number of components. Adding a single component to a 20-component system results in the significant increase of the number of the sub-systems that require experimental and thermodynamic modelling characterisation from 171 to 190 binary and from 969 to 1140 ternary sub-systems. All types of experimental data from Cp's, enthalpies and entropies of endmembers and stoichiometric compounds to the multi-component phase equilibria data have to be described by the model simultaneously. Thus, any improvement in a given experimental value require fast and significant re-optimisation. Incorporating a new experimental data on a given binary sub-system into the database requires the iterative re-optimisation of all corresponding ~20 ternary and ~120 quaternary sub-systems – the so-called 'pyramid effect'. Experimental – thermodynamic modelling integration requires fast and frequent iterative re-optimisations, that in turn need continuous accumulation of optimisation results at each given stage for the use during the next iteration.

A semi-automated thermodynamic database development methodology has been developed and implemented to tackle these issues. The key points in this methodology are as follows. The sets of target experimental points to be described and corresponding weights are selected based on the experimental information available (note – not all available experimental data are included, and at the same time so called 'virtual' target points are included where no experiments are available). A matrix of first derivatives showing the sensitivity of each target point to each possible model parameter are calculated using traditional Gibbs energy minimisation calculations, which is a relatively slow step. Using the revised approach, the initial slow Gibbs energy minimisation calculations, are replaced by a fast analytical approach with linear extrapolation of the existing values through matrix multiplications of the form $\Delta \bar{y} = A \cdot \Delta \bar{x}$, where A is the matrix of first-order derivatives (n target values by k model parameters), $\Delta \bar{x} = \bar{x} - \bar{x}_0$ is the difference between the final set of model parameters and their initial approximation and $\Delta \bar{y} = \bar{y} - \bar{y}_0$ is the difference between the final model predictions and model predictions at $\bar{x}_0$. The optimum values of the model parameters $\bar{x}_{optimal}$ are then obtained by solving the relationship: $\bar{x}_{optimal} = \bar{x}_0 + \left(\mathbf{A}^T \cdot \mathbf{A}\right)^{-1} \cdot \mathbf{A}^T \cdot \left(\bar{y}_{target} - \bar{y}_0\right)$ for the condition $\bar{y} \rightarrow \bar{y}_{target}$. This non-iterative analytical (rather than numerical) optimisation approach is orders of magnitude faster than the combination of the thermodynamic calculations using Gibbs energy minimisation and numerical non-linear minimisation. The first derivative-based linear extrapolation approach enables immediate re-optimisation of model parameters for any single target or weight change. It also enables: a) the real-time graphical presentation of predicted and target points, as well as b) the real-time systematic tabular statistical analysis presentation of agreement between predictions and target values, which makes the optimisation process truly interactive. More details are given in Nekhoroshev et al (2024).

The accuracy of the database is analysed using a systematic set of graphical representations together with compact pivot tables across different primary phases/subsystems, including such information as average, min, max, temperature RMS + composition RMS, where RMS (root mean squares) are defined as a square root of average squared differences between target and predicted values.

The new formalised and semi-automated methodology makes it possible to increase the efficiency and flexibility of collaborative work between researchers by organising parallel simultaneous optimisations by several researchers, thus *distributing the database development intellectual efforts* between the research team members. Once formalised and semi-automated, the procedure can be used to optimise the model parameters by enabling each member of a group of researchers to contribute to the thermodynamic parameters optimisation by planning and undertaking new experiments, adding corresponding target points and correcting weights, rather than 'manually' optimising model parameters.

Thus, the discrepancies and conflicts within the system are resolved by the formalised semi-automated system significantly more efficiently and with less oversight. Only periodically the re-calculations using Gibbs energy minimisations are needed to update the matrix of derivatives. The new methodology enables researchers undertaking experimental work on a particular sub-system to personally contribute to the thermodynamic optimisation of that system and to select further experimental target points further increasing productivity. Optimisation is undertaken in iterative cycles, the major discrepancies are identified at each step, and new experiments are conducted to resolve discrepancies within time intervals from several days to several weeks rather than months and years using the more traditional approaches.

Outline of the thermodynamic database

The current 20-component database in focus includes the 'Cu_2O'-PbO-ZnO-FeO-Fe_2O_3-CaO-SiO_2-Al_2O_3-MgO-S major, Cr and Na slagging and As, Bi, Sn, Sb, Au, Ag, Ni, Co other minor elements in the gas, slag, sulfate, matte, speiss and metal alloy molten solution phases and many solid solution and stoichiometric phases. The database theoretically consists of 1540 binary and ternary oxide, 455 binary and ternary metal-matte-speiss, and several thousand quaternary systems as well as the systems including oxygen and sulfur together (oxygen in matte, sulfur in slag). Excluding systems of low importance (eg containing several minor elements together), there are currently considered 998 oxide binary, ternary and selected 4-component sub-systems (including 123 Na_2O-containing systems recently added), 381 metal-matte-speiss systems, 84 sulfate systems with the total of 1463 2-, 3-, and selected 4-metal component systems. The database includes over 450 stoichiometric compounds, >100 gaseous species, around 130 solution phases (among which there are 29 large solutions) with more than 3000 excess parameters. Most of compound and gaseous species have self-consistent CP functions over a wide range of temperatures, starting from 0 K and optimised S_{298} and ΔH_{298} values,. The largest liquid solutions in the system are Slag and Liquid Metal/Matte/Speiss, the latter being modelled as one thermodynamic solution with miscibility gaps. The Slag phase has 50 endmembers made up of 25 free metal cations (including Fe^{2+} and Fe^{3+}, Sn^{2+} and Sn^{4+}, As^{3+} and As^{5+}, Sb^{3+} and potentially Sb^{5+}, Cr^{2+} and Cr^{3+}, and two associates ($NaAl^{4+}$, $NaFe^{4+}$) multiplied by [O^{2-}

,S^{2-}] anions. There are 276 binary and 2024 ternary oxide systems in slag phase with around 990 excess parameters. Liquid Metal/Matte/Speiss solution consists of 20 endmembers on a single sublattice with over 730 interactions between them. A new liquid solution, Salt, has recently been introduced into the database, describing the interactions between molten sulfates, oxides, and arsenates. Currently, there are eight cations in the first sublattice and three anions in the second sublattice, making up 24 different end members with 33 interactions among them in the Salt solution model.

Most extensive solid solutions in the database are Spinel with 162 endmembers and 611 interaction parameters (some of them are identical for different combinations of elements in sublattices, though), Monoxide (12 endmembers, 74 interaction parameters), Melilite (63 endmembers, 23 interaction parameters), ortho- and clino-pyroxenes (36 and 75 endmembers, respectfully). Figure 6 provides a summary of the current Pyrosearch database.

Slag: $(Cu^{+1}, Fe^{+2}, Fe^{+3}, Si^{+4}, Al^{+3}, Ca^{+2}, Mg^{+2}, Pb^{+2}, Zn^{+2}, Ni^{+2}, Co^{2+}, Sn^{+2}, Sn^{+4}, Sb^{+1}, As^{+1}, As^{+5}, Bi^{+1}, Ag^{+1}, Au^{+1}, Cr^{2+}, Cr^{3+}, Na^{+}, AlNa^{+4}, FeNa^{+4})(O^{-2}, S^{-2})$, MQM

Liquid matte/metal/speiss: $(Cu^{I}, Cu^{II}, Fe^{II}, Fe^{III}, Pb^{II}, Zn^{II}, Ni^{II}, Sn^{II}, Sb^{III}, As^{III}, Bi^{II}, Ag^{I}, Au^{I}, Cr^{II}, Co^{II}, Ca^{II}, Mg^{I}, Na^{1+}, O^{II}, S^{II})$, MQM

Liquid salt $(Ca, Mg, Pb, Na, Zn, Fe, Cu, Ni)(SO_4, O, AsO_4, SiO_4)$, MQM

Spinel: $[Cu^{+2}, Fe^{+2}, Fe^{+3}, Ni^{2+}, Al^{+3}, Mg^{+2}, Zn^{+2}, Cr^{2+}, Cr^{3+}, Co^{+2}, Co^{3+}]^{tet}$ $[Cu^{+2}, Fe^{+2}, Fe^{+3}, Ni^{2+}, Cr^{3+}, Al^{+3}, Ca^{+2}, Mg^{+2}, Zn^{+2}, Sn^{4+}, Co^{+2}, Co^{3+}, Vacancy^0]_2^{oct}O_4$, CEF

Monoxide: $(FeO, FeO_{1.5}, NiO, CoO, CrO_{1.5}, AlO_{1.5}, CaO, MgO, CuO, ZnO, SnO_2, Na_2O)$, B-W

Corundum: $(FeO_{1.5}, AlO_{1.5}, CrO_{1.5}, SnO_{1.5})$, B-W

Mullite: $(Al^{3+}, Fe^{3+}, Cr^{3+})_2(Al^{3+}, Sn^{4+}, Sn^{2+}, Si^{4+})(O^{-2}, Vacancy^0)_5$, CEF

Olivine: $[Fe^{2+}, Ni^{2+}, Co^{2+}, Cr^{2+}, Ca^{2+}, Mg^{2+}, Zn^{2+}]^{M2}[Fe^{2+}, Ni^{2+}, Cr^{2+}, Co^{2+}, Ca^{2+}, Mg^{2+}, Zn^{2+}]^{M1}SiO_4$, CEF

Melilite: $[Ca^{2+}, Pb^{2+}, Na^{+}]_2[Al^{3+}, Zn^{2+}, Mg^{2+}, Fe^{2+}, Fe^{3+}, Ni^{2+}, Co^{2+}][Fe^{3+}, Al^{3+}, Si^{4+}]_2O_7$, CEF

Pyroxenes (proto-, clino-, ortho-): $[Fe^{2+}, Ca^{2+}, Mg^{2+}, Ni^{2+}, Zn^{2+}]^{M2}[Fe^{2+}, Fe^{3+}, Mg^{2+}, Al^{3+}, Ni^{2+}, Zn^{2+}]^{M1}[Fe^{3+}, Al^{3+}, Si^{4+}]^B SiAlO_6$, CEF

Dicalcium silicates: $(Ca_2SiO_4, Fe_2SiO_4, Mg_2SiO_4, Pb_2SiO_4, Zn_2SiO_4, Ni_2SiO_4, Co_2SiO_4, Sn_2SiO_4, CaO, Ca_{1.7}AlO_3, Ca_{1.5}AsO_4)$, B-W

Wollastonite, pseudowollastonite: $(CaSiO_3, FeSiO_3, MgSiO_3, ZnSiO_3, PbSiO_3)$, B-W

SFCA: $[CaO, FeO][Fe_2O_3, CaSiO_3, Va][Fe_2O_3, Al_2O_3]_2$, CEF

Willemite: $[Zn^{2+}, Fe^{2+}, Mg^{2+}, Ni^{2+}, Cu^{2+}][Zn^{2+}, Fe^{2+}, Mg^{2+}, Ni^{2+}, Cu^{2+}]SiO_4$, CEF

Feldspar: $(Ca, Pb, Na, Va)[Al, Zn][Al, Si]_3O_8$, CEF

Zincite: $(ZnO, FeO, Fe_2O_3, Al_2O_3, CaO, MgO, CuO, NiO, SnO_2, Ca_2Fe_6Zn_6O_{17})$, B-W

Melanotekite: $Pb_2(Fe,Al)_2Si_2O_9$, B-W

Larsenite: $Pb(Zn, Mg, Fe, Ni)SiO_4$, CEF

Magnetoplumbite: $(PbO, CaO)[Fe_2O_3, PbFeO_2, PbZnO_2, Al_2O_3][Fe_2O_3, Al_2O_3]_5$, CEF

Delafossite: $[Cu^{+}][Al^{3+}, Fe^{3+}, Cr^{3+}, Sn^{3+}]O_2$, B-W

Fcc and bcc solids alloys: $(Fe, Cu, Ni, Co, Cr, O, S, Pb, Zn, As, Sn, Sb, Bi, Ag, Au)$, B-W

Digenite-bornite: $(Cu_2S, FeS, PbS, ZnS, Ni_2S, Ag_2S, Vacancy_2S)$, B-W

Villamaninite: $(Cu, Ni, Fe)S_2$, **Millerite** (NiS, CuS), B-W

Sphalerite, wurtzite $(Zn, Fe)S$

Pb-Ag-Au-Zn compounds and solutions

Pb-Ca-Mg-Bi compounds and solutions

MeS cubic: $(FeS, PbS, CaS, MgS, Cu_2S, Vacancy, S)$, B-W

$(Cu, Ni)_2As_3$, B-W;

Cu_2As (solid speiss): $(Cu^{I}, As^{III}, Ag^{I}, Au^{I}, Fe^{I}, Ni^{I})$, MQMQA

ORTH $[Fe, Ni][As, Sb, S]_2$, CEF;

GAMM Cu_3X (Cu, Ni, Sb, Sn) MQMQA;

HEX $MeX_{-}(Fe, Ni, As, Sb, Sn, Cu)$, MQMQA

~450 solid Compounds and small solutions: sulfate, sulfide $(FeAsS, Cu_3AsS_4)$, oxide, intermetallic, e.g. $Ca_2Sb, Ca_5Sb_3, ZnSb, Sn_4Sb_3$

Ideal gas: >150 species, including $N_2, H_2O, CO, CO_2, SO_2, SO, As_2, AsS, AsO, Zn, ZnS, ZnO, AgS, Pb, PbS, PbO, Bi, Bi_2, BiO, BiS, Sn, SnO, SnS, Ag, AgO, AgS, Sb, Sb_2, SbO, SbS$ and more.

FIG 6 – Summary of phases in the current Pyrosearch database for the pyrometallurgical processing and the thermodynamic models used to describe these. MQM = Modified Quasichemical Model (Pelton *et al*, 2000; Pelton and Chartrand, 2001), CEF = Compound Energy Formalism (Hillert, 2001; Hidayat *et al*, 2015), B-W = Bragg-Williams ideal mixing model.

Current status of the thermodynamic database development

Recent accurate and abundant experimental measurements by Pyrosearch have provided sufficient data to identify and then rectify many significant systematic uncertainties in the previous thermodynamic database. These systematic uncertainties include thermodynamic descriptions of all liquid end-members as well as in the description of the silicate slag systems with the Modified Quasichemical model. The whole database therefore had to be fully revised starting from the properties of pure liquid endmembers such as SiO_2, CaO, MgO etc, key sub-systems such as CaO-SiO_2, FeO-Fe_2O_3-SiO_2 etc, and all other binary, ternary and higher order sub-systems. The thermodynamic optimisation methodology developed at Pyrosearch had been critical to enable this significant task to be completed.

Brief outline of the major developments is given below.

Correcting the heat capacities of liquid end-members and of solid phases

Previous databases contained oversimplified descriptions of heat capacities of elements and compounds in which usually a single function of temperature is assigned to all phases of the same composition, ie all polymorphs and corresponding liquid (eg quartz, tridymite, cristobalite, and liquid/amorphous SiO_2). This approach does not accurately define the real experimental data on heat capacities of most phases. These over-simplified descriptions resulted in the step-like changes in heat capacities (and therefore entropies, enthalpies and Gibbs free energies) at melting points of the endmembers and corresponding uncertainties in the multicomponent solutions. Most importantly, the heat capacities of liquid solutions (slags) calculated from these liquid endmembers were typically 20–40 per cent higher than the real values for the range of temperatures relevant to industrial practices (700–1700°C). To compensate for these systematic uncertainties, artificial distortions to the interaction parameters between the liquid species were previously introduced thus significantly limiting the predictive power of the database; this becomes increasingly problematic with the increased solution complexity. These uncertainties have recently been corrected as one of the components of the major database revision, as shown in Figure 7. Also, heat capacities of all solid phases were revised to obtain physically relevant descriptions below the room temperature down to 0 K and at very high temperatures (above the melting point of solid endmembers) thus eliminating a number of erroneous predictions of liquid phase present at very low temperatures, solids at very high temperatures, and liquid miscibility gaps in the ranges of compositions and conditions where they are not actually observed.

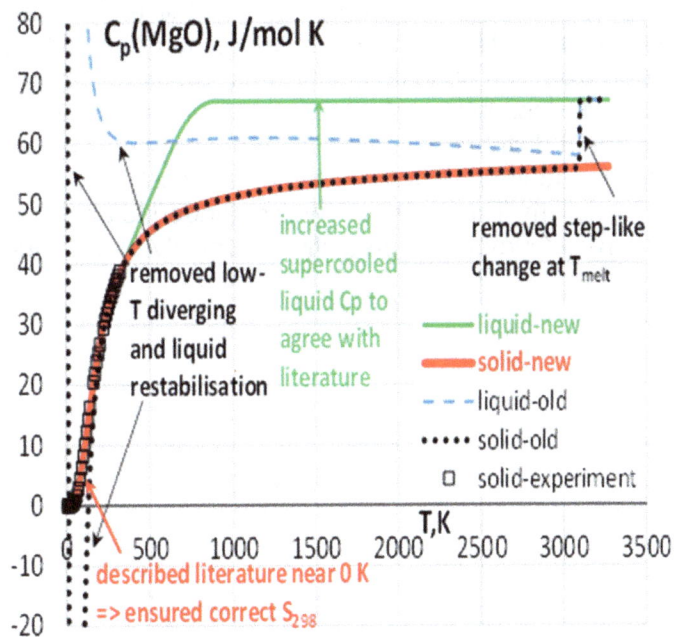

FIG 7 – Example of heat capacity versus temperature.

Gibbs free energies of solids at high temperatures

Extensive and accurate experimental results obtained by the Pyrosearch team at temperatures up to 1750°C have revealed significant uncertainties in the description of Gibbs free energies of high melting temperature of pure and binary compounds at higher temperatures (above 1750°C). The melting temperatures and enthalpies of several compounds (CaO, Ca_2SiO_4, Al_2O_3, MgO, SnO_2, Mg_2SiO_4, NiO etc) were corrected significantly. This correction, after overall major revision of the database, resulted in significant improvement of the description of phase equilibria in the temperature range important for industrial operations. For example, the melting temperature of SnO_2 in air was corrected from 1625°C to 2059°C (Shevchenko *et al*, 2021; Shevchenko, Shishin and Jak,

2024) that resulted in the improvement of the predictions of the cassiterite SnO_2 liquidus of 50–150°C for the range of temperatures relevant to industrial practice (700–1700°C).

Revising high-SiO₂ liquid

Accurate EPMA measurements of phase compositions and systematic high-temperature experiments in the high-SiO_2 areas recently performed by the Pyrosearch team have demonstrated that the currently used Quasi-chemical thermodynamic model of the SiO_2-containing liquids cannot accurately describe the experimental data in the areas of the tridymite/cristobalite liquidus, the monotectic and the miscibility gaps in a number of the Me-Si-O systems (Me = Ca, Mg, Fe^{2+}, Fe^{3+}, Zn, Cu, Ni, Co, Cr...). The reason for these systematic difficulties has been attributed to the fact that the Modified Quasichemical Formalism (MQM) only accounts for nearest neighbour interactions in the slag, assumed to be 100 per cent ionic liquid. The high-SiO_2 liquids, however, are not fully ionic (weak electrolytes), thus resulting in the long-range electrostatic interactions between the uncompensated charged M^{x+} cations and O^{2-} anions, an effect similar to the Debye-Hückel model and clustering well known to be present in dilute aqueous solutions. Identification and understanding of the underlying reasons for these behaviours provided the opportunity to adjust the models by: i) introducing a correction to the Gibbs energy of mixing (Figure 8) of the components in the form of a polynomial sequence so as to more closely simulate the long-distance ionic interactions (Debye-Hückel/clustering) in dilute MO_x solutions in SiO_2, that in turn enables ii) the minimisation or removal of the unrealistic temperature-dependent, excess entropy terms with high power on SiO_2 that were commonly used in previous optimisations for the binary systems and resulted in inaccurate extrapolation to the ternaries, and iii) making the entropy of mixing of the slag phase much closer to the ideal mixing limit in the experimentally determined immiscible regions. The introduction of these enthalpy and entropy of mixing corrections enabled the tridymite/cristobalite liquidus to be more accurately described in the key binary systems, such as 'FeO'-SiO_2, CaO-SiO_2, MgO-SiO_2, $CuO_{0.5}$-SiO_2, NiO-SiO_2. For example, these improvements enabled the elimination of the 2–5 wt per cent (~50–100°C) uncertainty in the prediction of the tridymite/cristobalite liquidus in the FeO-Fe_2O_3-CaO-SiO_2 system in reducing and oxidising conditions and thus significant improvement of the accuracy of the database. Introducing description of these long-range interactions also improved the description of the minor element distributions in multicomponent systems; it also allowed the use of smaller ternary parameters in the MO_x-$M'O_y$-SiO_2 systems, essentially treating them as near-ideal within the Quasichemical formalism. This approach was extended to all SiO_2-containing subsystems. This significantly improved the accuracy and predictive capability of the liquid slag model.

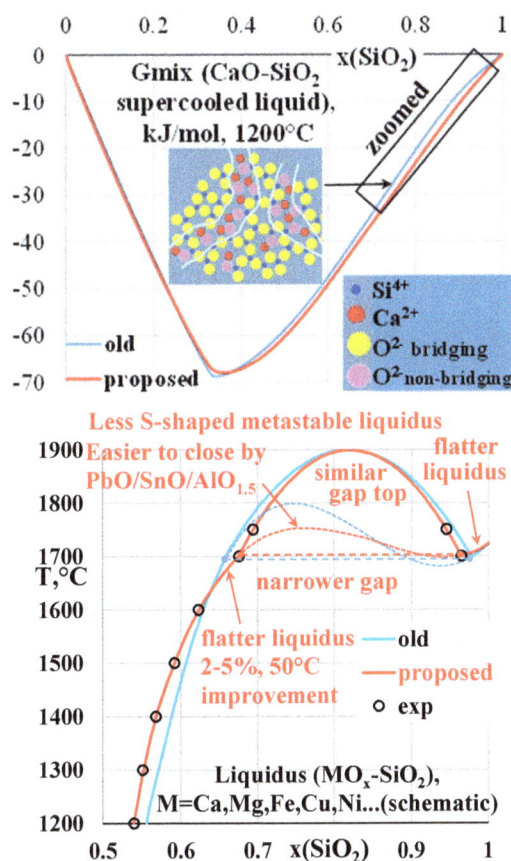

FIG 8 – Example of the correction of the high-SiO$_2$ slag model taking into account the different structure of the slags in high and low silica composition ranges.

Correcting the slag thermodynamic description around maximum ordering

The Modified Quasichemical Model (MQM) formalism adopts a one-dimensional solution (Ising model) as an approximate description of the correlation between the entropy of mixing and the strength of the second nearest bonds (the 3D solution of the Ising Model is not available). This 1D approximation works quite well for most systems, but in the cases of strong interactions the limitations of the 1D approximation can become significant, eg resulting in a sharp reversal in gradient at the minimum of the enthalpy/entropy of mixing and consequently of the Gibbs free energy of the slag (see Figure 8). For example, this sharp change means it was not possible using the MQM model alone to describe experimental data in the Ca$_2$SiO$_4$ primary phase field in the Fe$_x$O-CaO-SiO$_2$ system (Figure 9). The following approach was developed and implemented to rectify this issue: i) introduce a special polynomial sequence describing the enthalpy of mixing to reduce the 'sharpness' of the tip at the composition of maximum ordering, and ii) limit the configurational entropy of mixing to within ±2 J·mol^{-1} K^{-1} based on the available statistical physics theory. Systematic corrections of the entropies of mixing for all relevant binary systems have been undertaken to improve the accuracy of the descriptions and predictions of the database (eg Figure 9).

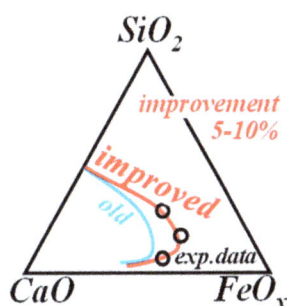

FIG 9 – Schematically shown Ca$_2$SiO$_4$ primary phase field in the CaO-FeO-Fe$_2$O$_3$ system at constant temperature.

Major revision of the whole thermodynamic database

Major revision of the whole (20 component) oxide database has been carried out a significant milestone that took substantial team efforts over several years; it included re-optimisation of ~250 subsystems using ~12 000 experimental phase equilibria points (including approximately ~10 000 experimental points by Pyrosearch) and significant number of other experimental thermodynamic data (Cp's etc), introduction of the new values for ~2500 thermodynamic parameters for 87 liquid and solid solution phases and ~700 stoichiometric solids. The new completely revised thermodynamic database forms an important basis for further developments.

Further more detailed discussions on thermodynamic optimisations are given in Shishin, Shevchenko and Jak (2024).

DEVELOPING THE TEAM

Multi-skilled team – critical factor for the strategic research program

Necessary components of the overall program of development of the thermodynamic database include application of the advanced theoretical models, development and application of appropriate sophisticated computational methodology, experimental methodology and analytical techniques as well as implementation of research results into industrial practice. Such program therefore requires a team with: i) high level expertise, ii) in different areas, and iii) of different levels (ie senior, intermediate and junior research academics, PhD students, qualified research assistants). The Pyrosearch current structure incorporates: i) five Theme leaders (advanced experts in high temperature experimentation, laboratory equipment and methods, analytical methods, thermodynamic optimisations and process modelling, and computational thermodynamics), ii) six Research Fellows; iii) seven to eight PhD and Masters students; iv) six well-trained continuing research assistants; v) seven casual undergraduate research assistants – current University metallurgy undergraduate students, and a few visitors. This is not a usual University group (many University groups have the majority of the research undertaken by PhD students). A strong continuing senior research academics core with the high level of expertise in different interrelated areas is the important factor. It is therefore critical to address the issues of continuity of the senior research academics within existing University system.

Industry support is critical for continuing research program

The long-term strategic research program requires continuing financial support, and industry therefore is a critical component providing long-term stability. *Continuous* implementation of the intermediate results into industrial practice is therefore a critical factor to ensure this study is *continuously* bringing the value to sponsors and thus helps to maintain the *continuing* financial support of the program. In the present research program this is achieved by: i) iterative, regular and frequent transfer of the intermediate thermodynamic database to sponsors, and ii) proactive implementation activity undertaken by the Pyrosearch staff.

The continuing focus on the industrial impact in turn guarantees the research is continuously directed at the practically important, priority industrial needs

The overall research program in case of Pyrosearch is organised through three consortia of industrial sponsors with common interests – Cu, Pb and Fe and slag recycling, where industry contributions are complemented by competitive government funds. The size of the overall program is beyond a capacity of a single industrial sponsor to support it. Pyrosearch important functions in this regard therefore are: i) to bring various industrial companies together into consortia and with common interests, and ii) obtain additional government and University support thus providing value to each individual industrial company.

Implementation of research outcomes into industrial practice is an important factor

The collaboration between the industry partners and Pyrosearch has enabled the development of a set of powerful predictive tools based on extensive experimental fundamental information, which can be used to analyse a broad range of industrial systems independent of the technologies used. The collaborative program does not stop there – the important next step is implementation of these advanced tools and information into industrial practice. The tools can be used by companies to improve process efficiencies and productivities, optimise the utilisation of existing plant, predict changes to plant practice to adjust for changes in process feed compositions, design new processes or operations. It has been recognised and acknowledged that the successful implementation of the advanced tools and information into industrial practice requires adequate professional education and training in their use. This has been achieved through a variety of ways in collaboration with sponsor companies.

Upskilling of engineering professionals is achieved through providing:

- Dedicated courses in chemical thermodynamic theory, pyrometallurgy fundamentals and the application of FactSage tools to industrial problems, delivered by distance and in person.

- Extended visits, placements and secondments of engineers with the research team.

- Undertaking Research Masters and PhD studies at Pyrosearch.

- Undertaking one-on-one projects with the research team on industrial problems.

Significant progress has been achieved in this direction. For example, over the last few years over 160 engineers successfully completed an on-line Pyrometallurgy course, 182 completed in-person and on-line FactSage course, seven metallurgical engineers visited Pyrosearch for training or collaborative projects, ten Masters and PhD students from other local sponsor Universities did collaborative research in-person at Pyrosearch, four full time Pyrosearch UQ PhD students are now working in sponsor companies, 56 one-on-one focused industrial research projects were undertaken by Pyrosearch staff in collaboration with the sponsor's metallurgists.

The active collaborations between industry and the research team provide important opportunities to obtain high value from the capabilities of the predictive tools. Testing the tools against industrial practice provides: i) confidence in their potential for extended use in an industrial context – a critical factor in technology transfer, and ii) guidance in further research directions.

Metallurgical research-focused academics roles

Universities are viewed traditionally as institutions where knowledge is stored and passed on through teaching to the undergraduate and postgraduate students. Much less understood are other important roles played by the University academics; those with specialist knowledge in the metallurgical engineering can actively support the metallurgical industry in a number of other ways, by:

- Undertaking knowledge generation through fundamental and applied research.

- Developing and maintaining collaborative research links with industry.

- Attracting additional financial support for research from Government sources,.

- Developing research infrastructure within institutions, facilities that can be utilised to support industry related research.

- Facilitating communication and support relationships between the metallurgy industry and senior university management.

- Promoting the important role of the metallurgical industry to young people and the broader community.

In addition, by insuring the sustainability of strong research teams with specialist knowledge and expertise, University academics can:

- Develop and maintain industry consortia for research and education of common interest to sponsors.

- Actively assist in the optimisation of existing and development of new metallurgical processes.

- Ensure the successful implementation of research outcomes into industrial practice.

- Ensure the availability of expert capabilities and advanced technological for application to individual confidential R&D support.

- Facilitate active technological exchange and links between sponsors metallurgists and between sponsor-nominated Universities.

- Minimise the adverse impacts of periodic economic cycles on the retention of knowledge and expertise for the metallurgical industry.

Metallurgical undergraduate education – critical for future progress

Despite of the importance of metallurgical industry, the number of undergraduate students selecting studies in metallurgical engineering and the number of university programs offering educational opportunities in metallurgical engineering have been in decline for many years – this is a worldwide trend. There appear to be a number of contributing factors to this trend:

- Unfavourable public perceptions of the industry and lack of information on the role of metallurgy in our society.

- The lack of industry engagement and support in promoting studies and careers in the field.

- The funding formulae used by Government to support Universities, which are in the main based on the number of enrolments rather than industry or societal needs.

- University and Journal Ranking systems do not favour small specialist fields.

The lack of professionals in the field poses a significant threat to successfully implementing process improvements in existing operations and hinders the major changes needed for future economies. The public and students are increasingly aware of the impacts of climate change and the necessity of transforming our industrial practices to address these challenges. This shift in attitude presents an opportunity to reframe the narrative and public perception of the industry, shifting it from negative to positive. The dramatic shift towards electrification across all sectors – renewable energy, transportation, and electronic devices – has heightened awareness of the critical metals needed to enable these transitions. Additionally, there is a growing need for material conservation and increased metal recycling to minimise environmental impact, reduce energy consumption, and achieve circularity in the use of valuable, non-renewable resources. These are powerful messages the industry can leverage to explain how lead metallurgy plays a vital role in facilitating the recycling of critical, strategic, and precious metals.

Increasing public awareness of the key role of metallurgy is necessary but is not necessarily sufficient to attract students and graduates. The industry needs to be proactive in improving workplace environments and practices so that they are attractive to prospective students and engineers. Students are looking for opportunities where they feel they can make a positive contribution to the environment and are able to use the latest technologies to achieve this. Industry can take positive steps in attracting undergraduates in their early years at university when they have an opportunity to redirect their careers into the discipline, through for example, providing on-site experiences in the form of vacation work, plant visits and internships, financial support scholarships, professional development and social activities.

The allocation of university funding is for the most part determined by class size rather than strategic needs of industry or the country. Small classes for specialist courses are more likely to be removed from the curriculum, small programs are closed in the name of efficient use of limited financial resources. This is the pattern around the world – the only way to change the current trend and provide educational resources is for industry to be more proactive. Industry leaders need to inform university management and governments of the importance of the discipline and actively engage with and provide support to those institutions providing programs in metallurgical engineering.

We strongly believe there are opportunities to increase the number of metallurgy graduates if Industry, Universities and Governments are proactive and work collaboratively together to solve this problem to their mutual benefits.

University teaching- and research-focused academics potential important central role in triple academia-industry-government helix

The role of University academics in the research, education and implementation of the scientific advancements into industrial practice thus making real impact to the society is critical. University academics are in the centre of and are inherently connected to all stake-holders. University academics:

- by definition, are actual experts in a particular field

- have active links and knowledge of industry through research: a) at the applied R&D level, as well as b) at the industrial R&D management

- are in direct contact with the University management

- are in direct continuing contact and aware of the issues of the changing next generations of specialists

- and at the same time educate and directly influence the next generation of specialists

- and are in contact with the government policies and bodies through the research funding applications and schemes.

There is a great potential for more engagement of the University academics into the policy development and decision-making of the industry R&D and University management as well as government bodies.

EXTENDING THE METHODOLOGY

Thermodynamics and other physicochemical properties

An additional advantage of the use of structurally-based thermodynamic models, such as, the quasichemical formalism (QCF), is that they also provide important fundamental information about the behaviour of atoms within the materials. This information can be used to describe the physicochemical properties of the materials, for example, the viscosities of fluids (Kondratiev and Jak, 2005; Jak, 2009). Thermodynamic models are also critical to describe the viscosities of heterogeneous partly crystallised systems by predicting the proportion of solids and the composition of the remaining liquid phase (Kondratiev and Jak, 2001; Kondratiev, Jak and Hayes, 2002; Jak et al, 2003). Extending the thermodynamic computer database predictive capabilities to other slag physicochemical properties is an important development (Thibodeau, Gheribi and Jung, 2016a, 2016b; Kang and Chartrand, 2016; Smith et al, 2020; Kang, 2015).

Future needs in the high-temperature molten phases area

As outlined in the paper above, thermodynamic modelling: a) is the central and critical component in the mathematical description and further optimisation of the high-temperature industrial processes, b) has the required methodologies for the development of the accurate description of thermodynamic properties and of phase equilibria, c) has already progressed to the useful level of adequate prediction of high temperature processes, and at the same time d) has great further potential p close the gaps in knowledge. It is therefore an opinion of the authors that more efforts and resources should be invested into further experimental work and thermodynamic modelling, similar to the continuing program undertaken by Pyrosearch. International collaboration of various groups can facilitate faster overall progress.

Further potential exists in the wider use of the first principles (ab-initio) and molecular dynamic simulations in phenomenological modelling of the type presented by authors. Currently, these models do not allow accurate representation of complex solutions at high temperatures. However, they allow to find important reference points for the thermodynamic model, such as enthalpies of formation at 0 K and sometimes heat capacities as functions of T for individual pure compounds and endmembers, including those which were not studied experimentally due to their metastability. These theoretical predictions could become a valuable complement to the experimental techniques described above.

The incorporation of the thermodynamic modelling into description of the industrial processes with kinetics playing significant role is another challenge. Relatively long times (in some cases up to several seconds) are currently required for a thermodynamic calculation at a single condition in a multi-component multi-phase system. Kinetics modelling usually would require multiple thermodynamic calculations making the overall modelling critically slow for practical use. This issue may be and should be gradually solved using significantly more powerful computer capabilities as well as special computational approaches. Linear extrapolation using first derivatives outlined above in the thermodynamic optimisation section can be a useful approach in relation to this issue.

The incorporation of thermodynamic modelling into predictions and optimisations of complex multi-unit industrial flow sheets is another great opportunity and at the same significant challenge. Similarly to the kinetic modelling, the relatively long time (in some cases up to several seconds) required for a thermodynamic calculation at a single condition in a multi-component multi-phase system is a barrier for the implementation of thermodynamic modelling into flow-sheet packages such as ASPEN, SYSCAD, METSIM where multiple calculations are needed for numerical minimisation methods for flow sheets with recycling streams. Similar to the kinetic modelling, incorporation of thermodynamic models into the flow sheet modelling can be facilitated with the significantly more powerful computer capabilities as well as special computational approaches. Again, linear extrapolation using first derivatives outlined above in the thermodynamic optimisation section can be a useful approach in relation to this issue.

CONCLUDING STATEMENTS

The key messages to all groups of professionals is that the metallurgical and recycling industry needs complex computerised optimisation tools similar to the GPS systems now used widely. The development of such GPS-like systems in pyrometallurgy is now possible although the task is challenging. To achieve that- the following components are essential, the availability of: i) sufficiently abundant and sufficiently accurate experimental data (that in turn require analytical capabilities), ii) appropriate theoretical thermodynamic models, and iii) adequate computational capabilities. The development and implementation of the advanced fundamental information and advanced predictive tools requires long-term efforts of multi-skilled expert teams. To support such teams, the industry financial support and engagement as well as industrial implementation are essential. We believe that historically we are at the start of new digitalisation Pyro-GPS period. The tasks are significant, and the opportunities are significant. The risks not to progress in the direction of digitalisation are also significant.

The messages:

- to young researchers and metallurgists – the metallurgy field is important to the society.

- to industry R&D – more advanced information and tools are becoming available bringing more opportunities.

- to industry management – there are significant opportunities for improving efficiencies and profitability of the operations, pro-active research support is important, educated metallurgists are needed to convert opportunities into profits.

- to University management, government, broader community –metallurgy is important, and the risks not to act now and to miss those opportunities are high; it is important to maintain and expand metallurgy research and education.

- to metallurgy research academics – you can play a central role in the academia-industry-government triple helix.

Collaborative work of many specialists, many industrial companies and many research groups is critical – we are all part of the solution and we all can contribute to the solution.

ACKNOWLEDGEMENTS

The Pyrosearch team acknowledges the financial and technical support from:

- *Pb research industrial consortia* including Aurubis (Germany), Boliden (Sweden), Kazzinc Ltd, Glencore (Kazakhstan), Nyrstar (Australia), Peñoles (Mexico) and Umicore (Belgium).

- *Cu research industrial consortia* including Anglo American Platinum (South Africa), Aurubis (Germany), BHP Billiton Olympic Dam (Australia), Boliden (Sweden), Glencore Technology (Australia), Metso Outotec Oy (Finland), Peñoles (Mexico), RHI Magnesita (Austria), Rio Tinto Kennecott (USA) and Umicore (Belgium).

- *Fe and slag recycling research industrial consortia* including Aurubis AG (Germany), BHP Fe ore (Australia), Rio Tinto ore (Australia), SWERIM (Sweden) and Umicore (Belgium).

- Australian Research Council (ARC) Linkage program.

- US Defence program 'Materials Recovery Technology for Defense Supply Resiliency' and Gopher (USA).

Special acknowledgement to *Maurits Van Camp* (formerly from Umicore) for his immense input in this program – his legacy stays in this continuing program, in the significant impact it makes to the research and industry communities, and the overall society and in our memories.

REFERENCES

Abdeyazdan, H, Shevchenko, M, Chen, J, Hayes, P C and Jak, E, 2024. Phase equilibria in the ZnO-MgO-SiO_2 and PbO-ZnO-MgO-SiO_2 systems for characterizing MgO-based refractory - slag interactions, *J Eur Ceram Soc*, 44:510–531.

Andersson, J O, Helander, T, Hoglund, L, Shi, P and Sundman, B, 2002. Thermo-Calc and DICTRA, computational tools for materials science, *CALPHAD*, 26:273–312.

Avarmaa, K, O'Brien, H, Johto, H and Taskinen, P, 2015. Equilibrium Distribution of Precious Metals Between Slag and Copper Matte at 1250–1350°C, *J Sustain Metall*, 1:216–228.

Bale, C W, Belisle, E, Chartrand, P, Decterov, S A, Eriksson, G, Gheribi, A E, Hack, K, Jung, I H, Kang, Y B, Melancon, J, Pelton, A D, Petersen, S, Robelin, C, Sangster, J, Spencer, P and Van Ende, M A, 2016. FactSage thermochemical software and databases, 2010–2016, *CALPHAD*, 54:35–53.

Borfecchia, E, Mino, L, Gianolio, D, Groppo, C, Malaspina, N, Martinez-Criado, G, Sans, J A, Poli, S, Castelli, D and Lamberti, C, 2012. Iron oxidation state in garnet from a subduction setting: a micro-XANES and electron microprobe (flank method) comparative study, *J Anal At Spectrom*, 27:1725–1733.

Chartrand, P and Pelton, A D, 2000. On the Choice of Geometric Thermodynamic Models, *J Phase Equilib*, 21:141–147.

Chartrand, P and Pelton, A D, 2001. The Modified Quasichemical Model, III - Two Sublattices, *Metall Mater Trans A*, 32:1397–1407.

Chen, C, 2015. Application of MPE Model to Iron Ore Sintering, Ironmaking and Steelmaking Processes, *Steel Res Int*, 86:612–618.

Chen, C, Zhang, L and Jahanshahi, S, 2013. Application of MPE model to direct-to-blister flash smelting and deportment of minor elements, in *Proceedings of Copper 2013*, pp 857–871.

Chen, J, Diao, H, O'Neil, Hugh and Jak, E, 2024. A novel method for minimizing the boundary secondary fluorescence in X-ray microanalysis by FIB-SEM, Unpublished material.

Chen, J, Fallah-Mehrjardi, A, Specht, A and O'Neill, H S C, 2021. Measurement of Minor Element Distributions in Complex Copper Converting Slags Using Quantitative Microanalysis Techniques, *JOM*.

Chen, J, Hayes, P C and Jak, E, 2024. Chemical- and micro-analytical techniques for molten slags, mattes, speisses and alloys, in *Proceedings of the 12th International Conference on Molten Slags, Fluxes and Salts (MOLTEN 2024)*, pp 321–334 (The Australasian Institute of Mining and Metallurgy: Melbourne).

Chen, J, Jak, E and Misztela, M, 2023. Development of Cu-based Reference Materials for Laser Ablation ICP-MS, The 27th Australian Conference on Microscopy and Microanalysis, Perth, Australia.

Cheng, S, Hayes, P C and Jak, E, 2022. Iron Ore Sinter Macro- and Micro-Structures and Their Relationships to Breakage Characteristics, *Minerals*, 12.

Cheng, S, Shevchenko, M, Hayes, P C and Jak, E, 2021. Experimental Phase Equilibria Studies in the FeO-Fe_2O_3-CaO-SiO_2 System and the Subsystems CaO-SiO_2, FeO-Fe_2O_3-SiO_2 in Air, *Metall Mater Trans B*, 52:1891–1914.

Decterov, S A, 2018. Thermodynamic database for multicomponent oxide systems, *Chimica Techno Acta*, 5:16–48.

Donovan, J, Singer, J and Armstrong, J, 2016. A new EPMA method for fast trace element analysis in simple matrices, *American Mineralogist*, 101:1839–1853.

Fallah Mehrjardi, A, Hayes, P C, Azekenov, T, Ushkov, L and Jak, E, 2016. Phase Chemistry Study of The Interactions Between Slag and Refractory in Coppermaking Processes, in *Advances in Molten Slags, Fluxes and Salts: Proceedings of the 10th Int Conf on Molten Slags, Fluxes and Salts 2016* (eds: R G Reddy, P Chaubal, P C Pistorius and U Pal), pp 1071–1076 (Springer International Publishing).

Fallah-Mehrjardi, A, Hayes, P C, Vervynckt, S and Jak, E, 2014a. Investigation of Freeze-Linings in a Nonferrous Industrial Slag, *Metall Mater Trans B*, 45:850–863.

Fallah-Mehrjardi, A, Jansson, J, Taskinen, P, Hayes, P C and Jak, E, 2014b. Investigation of the Freeze-Lining Formed in an Industrial Copper Converting Calcium Ferrite Slag, *Metall Mater Trans B*, 45:864–874.

Frisk, K and Selleby, M, 2001. The Compound Energy Formalism: Applications, *J Alloys Compd*, 320:177–188.

Gisby, J, Taskinen, P, Pihlasalo, J, Li, Z, Tyrer, M, Pearce, J, Avarmaa, K, Bjorklund, P, Davies, H, Korpi, M, Martin, S, Pesonen, L and Robinson, J, 2017. MTDATA and the Prediction of Phase Equilibria in Oxide Systems: 30 Years of Industrial Collaboration, *Metall Mater Trans B*, 48:91–98.

Hamuyuni, J, Klemettinen, L and Taskinen, P, 2016. Experimental phase equilibrium data of the system $Cu-O-CaO-Al_2O_3$ at copper saturation, *CALPHAD*, 55:199–207.

Hidayat, T, Hayes, P C and Jak, E, 2012. Experimental Study of Ferrous Calcium Silicate Slags: Phase Equilibria at Between 10^{-8} atm and 10^{-9} atm, *Metall Mater Trans B*, 43:27–38.

Hidayat, T, Henao, H M, Hayes, P C and Jak, E, 2012. Phase Equilibria Studies of the Cu-Fe-O-Si System in Equilibrium with Air and with Metallic Copper, *Metall Mater Trans B*, 43:1034–1045.

Hidayat, T, Shishin, D, Grimsey, D, Hayes, P and Jak, E, 2017. The Integration of Plant Sample Analysis, Laboratory Studies and Thermodynamic Modeling to Predict Slag-Matte Equilibria in Nickel Sulfide Converting, *Metall Mater Trans B*, 49:132–145.

Hidayat, T, Shishin, D, Jak, E and Decterov, S, 2015. Thermodynamic Reevaluation of the Fe-O System, *CALPHAD*, 48:131–144.

Hillert, M, 2001. The Compound Energy Formalism, *J Alloys Compd*, 320:161–176.

Hillert, M, Kjellqvist, L, Mao, H, Selleby, M and Sundman, B, 2009. Parameters in the compound energy formalism for ionic systems, *CALPHAD*, 33:227–232.

Hoefer, H E and Brey, G P, 2007. The iron oxidation state of garnet by electron microprobe: its determination with the flank method combined with major-element analysis, *Am Mineral*, 92:873–885.

Jak, E, 2009. Viscosity model for slags in the $Al_2O_3-CaO-'FeO'-K_2O-Na_2O-MgO-SiO_2$ system. in *Molten slags and salts 2009*, pp 434–448.

Jak, E, 2012. Integrated experimental and thermodynamic modelling research methodology for metallurgical slags with examples in the copper production field, in *Ninth Intl Conf on Molten Slags, Fluxes and Salts* (The Chinese Society for Metals).

Jak, E, 2018. Modelling Metallurgical Furnaces—Making the Most of Modern Research and Development Techniques, in *Extraction 2018* (eds: B Davis, M Moats and S Wang), pp 103–125 (Springer).

Jak, E, Hayes, P C and Lee, H-G, 1995. Improved methodologies for the determination of high temperature phase equilibria, *Korean Journal of Minerals and Materials Institute (Seoul)*, 1:1–8.

Jak, E, Kondratiev, A, Christie, S and Hayes, P C, 2003. The prediction and representation of phase equilibria and physicochemical properties in complex slag systems, *Metallurgical and Materials Transactions B*, 34:595–603.

Jung, I-H and Van Ende, M-A, 2020. Computational Thermodynamic Calculations: FactSage from CALPHAD Thermodynamic Database to Virtual Process Simulation, *Metall Mater Trans B*, 51:1851–1874.

Kang, Y-B and Chartrand, P, 2016. Calculation of property diagram as a Zero-Phase Fraction line of auxiliary phase, *CALPHAD*, 55:69–75.

Kang, Y-B, 2015. Relationship between surface tension and Gibbs energy and application of Constrained Gibbs Energy Minimization, *CALPHAD*, 50:23–31.

Khartcyzov, G, Shevchenko, M, Cheng, S, Hayes, P C and Jak, E, 2023. Experimental phase equilibria study and thermodynamic modelling of the $CuO_{0.5}-AlO_{1.5}-SiO_2$ ternary system in equilibrium with metallic copper, *Ceram Int*, 49:11513–11528.

Kondratiev, A and Jak, E, 2001. Modeling of Viscosities of the Partly Crystallized Slags in the $Al_2O_3-CaO-"FeO"-SiO_2$ System, *Metall Mater Trans B*, 32B:1027–1032.

Kondratiev, A and Jak, E, 2005. A Quasi-Chemical Viscosity Model for Fully Liquid Slags in the $Al_2O_3-CaO-'FeO'-SiO_2$ System, *Metall Trans B*, 36B:623–638.

Kondratiev, A, Jak, E and Hayes, P C, 2002. Predicting slag viscosities in metallurgical systems, *JOM*, 54:41–45.

Koukkari, P, Penttila, K, Hack, K and Petersen, S, 2000. ChemSheet - an efficient worksheet tool for thermodynamic process simulation, *EUROMAT 99, Biann Meet Fed Eur Mater Soc (FEMS)*, 3:323–330.

Llovet, X, Pinard, P T, Donovan, J J and Salvat, F, 2012. Secondary fluorescence in electron probe microanalysis of material couples, *J Phys D Appl Phys*, 45:225301/1–225301/12.

Nekhoroshev, E, Shishin, D, Shevchenko, M and Jak, E, 2024. Continuous method of thermodynamic optimisation using first-derivative matrices for large multicomponent systems, in *Proceedings of the 12th International Conference on Molten Slags, Fluxes and Salts (MOLTEN 2024)*, pp 901–914 (The Australasian Institute of Mining and Metallurgy: Melbourne).

Nikolic, S, Shishin, D, Hayes, P C and Jak, E, 2018. Case study on the application of research to operations - calcium ferrite slags, in *Extraction 2018* (ed: B Davis) (Springer: Cham).

Pelton, A D and Blander, M, 1986. Thermodynamic Analysis of Ordered Liquid Solutions by a Modified Quasi-Chemical Approach, Application to Silicate Slags, *Metall Trans B*, 17B:805–815.

Pelton, A D and Chartrand, P, 2001. The Modified Quasichemical Model, II - Multicomponent Solutions, *Metall Mater Trans A*, 32:1355–1360.

Pelton, A D, 2001. A General Geometric Thermodynamic Model for Multicomponent Solutions, *CALPHAD*, 25:319–328.

Pelton, A D, Chartrand, P and Eriksson, G, 2001. The modified Quasichemical Model, IV - Two Sublattice Quadruplet Approximation, *Metall Mater Trans A*, 32:1409–1415.

Pelton, A D, Decterov, S A, Eriksson, G, Robelin, C and Dessureault, Y, 2000. The Modified Quasichemical Model, I - Binary Solutions, *Metall Mater Trans B*, 31:651–659.

Petersen, S and Hack, K, 2007. The thermochemistry library ChemApp and its applications, *Int J Mater Res*, 98:935–945.

Petersen, S, Hack, K, Monheim, P and Pickartz, U, 2007. SimuSage - the component library for rapid process modeling and its applications, *Int J Mater Res*, 98:946–953.

Shevchenko, M and Jak, E, 2017. Experimental Phase Equilibria Studies of the PbO-SiO$_2$ System, *J Am Ceram Soc*, 101:458–471.

Shevchenko, M and Jak, E, 2018. Experimental Liquidus Studies of the Pb-Fe-Si-O System in Equilibrium with Metallic Pb, *Metall Mater Trans B*, 49:159–180.

Shevchenko, M and Jak, E, 2019a. Experimental Liquidus Studies of the Binary Pb-Cu-O and Ternary Pb-Cu-Si-O Systems in Equilibrium with Metallic Pb-Cu Alloys, *J Phase Equilib Diff*, 40:671–685.

Shevchenko, M and Jak, E, 2019b. Experimental Liquidus Studies of the Pb-Fe-Ca-O System in air, *J Phase Equilib Diff*, 40:128–137.

Shevchenko, M and Jak, E, 2019c. Experimental Liquidus Studies of the Pb-Fe-Si-O System in air, *J Phase Equilib Diff*, 40:319–355.

Shevchenko, M and Jak, E, 2019d. Experimental Liquidus Studies of the Zn-Fe-Si-O System in air, *IJMR (International Journal of Materials Research)*, 110:600–607.

Shevchenko, M and Jak, E, 2021. Integrated experimental phase equilibria study and thermodynamic modelling of the binary ZnO-Al$_2$O$_3$, ZnO-SiO$_2$, Al$_2$O$_3$-SiO$_2$ and ternary ZnO-Al$_2$O$_3$-SiO$_2$ systems, *Ceramics International*, 47:20974–20991.

Shevchenko, M, Ilyushechkin, A, Abdeyazdan, H and Jak, E, 2021. Integrated experimental phase equilibria study and thermodynamic modelling of the PbO-SnO-SnO$_2$-SiO$_2$ system in air and in equilibrium with Pb-Sn metal, *J Alloys Compd*, 888:161402.

Shevchenko, M, Shishin, D and Jak, E, 2024. Advancement in experimental methodologies to produce phase equilibria and thermodynamic data in multicomponent systems, in *Proceedings of the 12th International Conference on Molten Slags, Fluxes and Salts (MOLTEN 2024)*, pp 531–558 (The Australasian Institute of Mining and Metallurgy: Melbourne).

Shishin, D and Jak, E, 2018. Critical assessment and thermodynamic modeling of the Cu-As system, *CALPHAD*, 60:134–143.

Shishin, D, Shevchenko, M and Jak, E, 2024. Challenges and limitations in development of large thermodynamic databases for multiple molten phases using the Modified Quasichemical Formalism, in *Proceedings of the 12th International Conference on Molten Slags, Fluxes and Salts (MOLTEN 2024)*, pp 931–948 (The Australasian Institute of Mining and Metallurgy: Melbourne).

Sineva, S, Shishin, D, Prostakova, V, Fallah-Mehrjardi, A and Jak, E, 2023. Experimental Study and Thermodynamic Analysis of High-temperature Processing of mixed Ni-Cu sulfide feed, The 62nd Annual Conference of Metallurgists: COM 2023 (MetSoc of CIM).

Smith, A L, Capelli, E, Konings, R J M and Gheribi, A E, 2020. A new approach for coupled modelling of the structural and thermo-physical properties of molten salts, Case of a polymeric liquid LiF-BeF$_2$, *J Mol Liq*, 299:112165.

Thibodeau, E, Gheribi, A E and Jung, I-H, 2016a. A Structural Molar Volume Model for Oxide Melts Part I: Li$_2$O-Na$_2$O-K$_2$O-MgO-CaO-MnO-PbO-Al$_2$O$_3$-SiO$_2$ Melts-Binary Systems, *Metall Mater Trans B*, 47:1147–1164.

Thibodeau, E, Gheribi, A E and Jung, I-H, 2016b. A Structural Molar Volume Model for Oxide Melts Part II: Li_2O-Na_2O-K_2O-MgO-CaO-MnO-PbO-Al_2O_3-SiO_2 Melts-Ternary and Multicomponent Systems, *Metall Mater Trans B*, 47:1165–1186.

Watt, W, Hidayat, T, Shishin, D and Jak, E, 2018. Advanced Thermochemical Fundamental and Applied Research to Improve Integrity of the Steel Water Jacketed Furnace at Port Pirie, in *Extraction 2018*.

Wen, X, Shevchenko, M, Nekhoroshev, E and Jak, E, 2023. Phase equilibria and thermodynamic modelling of the PbO–ZnO-$CuO_{0.5}$-SiO_2 system, *Ceram Int.*

Xia, L, Liu, Z and Taskinen, P A, 2016. Equilibrium study of the Cu-O-SiO_2 system at various oxygen partial pressures, *J Chem Thermodyn*, 98:126–134.

Genetic design of personalised slag for manufacturing die steel via electroslag remelting method and an industrial application case

Z H Jiang[1], Y M Li[2], J L Tian[3] and J W Dong[4]

1. Professor, Northeastern University, Shenyang 110819, China. Email: jiangzh@smm.neu.edu.cn
2. PhD candidate, Northeastern University, Shenyang 110819, China.
   Email: 15984327096@163.com
3. Associate Professor, Northeastern University, Shenyang 110819, China.
   Email: neujialong@163.com
4. PhD candidate, Northeastern University, Shenyang 110819, China. Email: neuemta@163.com

ABSTRACT

Electroslag remelting is widely used to produce various special steels, mainly because of its ability to provide high cleanliness and excellent homogeneity of ingot. Undoubtedly an optimal slag plays a key role in removing inclusions and enhancing the mechanical properties of a specific special steel. In order to improve the metallurgical performance of the slag in electroslag refining of die steel, a variety of slags have already been developed using traditional trial-and-error methods. Nonetheless, these slags may not be inherently optimal for the specific requirements of actual die steel refining processes. In this work, we have designed a CaF_2-CaO-Al_2O_3-SiO_2-MgO quinary slag system using genetic algorithms method aiming to remove the existing large size inclusions (D and Ds type) in high quality die steel. The comprehensive effects of five components (CaF_2, CaO, Al_2O_3, SiO_2 and MgO) were studied based on existing fundamental theory. The slag composition was screened considering the parameters such as melting point, electrical conductivity, calcium ion concentration, and appropriate viscosity etc. Furthermore, the candidate optimal slag was applied in the industrial production of die steel by electroslag remelting method. The detection results indicate that majority large inclusions (D and Ds type) have been successfully removed, showing significant advantage to traditional slags. This result has demonstrated the genetic method to be efficient and reliable. In addition, depending on the personalised demand of other steel and alloys, this method allows for the adjustment of the screening frameworks and processes to develop the candidate optimal slags. This method has not been reported previously in the field of electroslag composition design, and compared with the traditional trial-and-error method, the new method will save a lot of time and experimental costs.

INTRODUCTION

In steel production, the presence of non-metallic inclusions often results in the formation of micropores and cracks, initiating fatigue fracture and structural defects (Sabih, Wanjara and Nemes, 2005). To counteract this, the electroslag remelting (ESR) process is employed to achieve a uniform composition, structural densification, and the removal of inclusions, thereby increasing steel purity and quality (Jiang et al, 2023). In the ESR process, the role of slag is critically important, acting as a decisive factor in determining the quality of electroslag ingots (Sebastian and Bernd, 2015; Shi et al, 2015; Shi, 2020). Consequently, precise design and control of the slag composition are essential.

Traditionally, the design of slag composition relies on an empirical trial-and-error approach. While practical, this method can be time-consuming and may not yield optimal results, necessitating a more systematic and scientifically grounded strategy (Dong, Jang and Yu, 2016; Dong et al, 2014). In response to the diverse metallurgical requirements in the ESR process, modern slags have transcended traditional constituents such as CaF_2 and Al_2O_3 (Schneider et al, 2019; Sebastian and Bernd, 2015) and specific quantities of CaO, MgO, and SiO_2 have been incorporated. The concentrations of slag components could vary widely to meet the specific requirements of different steel categories (Sebastian and Bernd, 2015; Duan, 2020; Ju et al, 2020). The typical composition of CaF_2–CaO–Al_2O_3-based ESR slag includes 40 per cent–70 per cent CaF_2, 0 per cent–40 per cent CaO, and 20 per cent–40 per cent Al_2O_3, with small amounts of SiO_2 and MgO (Ju, Gu and Zhang, 2021; Wan et al, 2022).

In material design field, genetic algorithms have been extensively utilised (Campbell and Olson, 2000; Xu, del Castillo and van der Zwaag, 2009). This study pioneers genetic algorithms for optimising slag composition, a previously unexplored technique in slag composition design. Both considering the ESR ingots' metallurgical quality and surface quality, the slag's physical and chemical characteristics including melting point, viscosity, electrical conductivity, basicity, density, and ion/molecule concentrations have been identified as the slag system's crucial parameters ('genes'), leveraging genetic algorithms to select superior slag compositions.

In the conventional trial-and-error approach, optimising slag compositions is restricted to minor adjustments, typically limited to a few percentage points from the original composition. This method limits the number of experiments and frequently yields suboptimal outcomes. However, the application of genetic algorithms allows for a significantly broader range of variation for each component in the slag based on years of production experience. This approach, generating an extensive data set, considerably enhances the likelihood of identifying the optimal composition window in the new generation of slag compositions. Moreover, this approach can save time and material costs compared to conventional trial-and-error methods.

This novel slag design approach has been applied in the die steel production. When using traditional ANF-6 slag, excessive D and Ds inclusions in 1.2343 die steel were usually detected. The D and Ds inclusions are mainly composed of $CaO–Al_2O_3$ or $CaO–MgO–Al_2O_3$. These inclusions are challenging to remove during electroslag remelting and lead to minute point-like formations in ingots (Guo *et al*, 2021), detrimentally affecting the steel's plasticity, toughness, fatigue resistance, workability, and specific physical properties (Ragnarsson and Sichen, 2009). This work employs genetic algorithms for slag design, successfully developing a new slag used for manufacturing high cleanliness 1.2343 die steel. The newly developed slag preserves the surface quality of the ESR ingot and effectively mitigates the issue of excessive D and Ds inclusions.

METHODS

First, we created a comprehensive database incorporating all possible combinations of slag compositions. This study identified six physicochemical properties—density, melting point, electrical conductivity, optical basicity, viscosity, and ion/molecule concentration—as key parameters ('genes') of the slag system. Subsequently, genetic algorithms are utilised to determine slag compositions exhibiting superior performance. For instance, the calcium ion 'gene' activity is adjusted to mitigate excessive D and Ds inclusions in 1.2343 die steel. Figure 1 shows the process of the genetic algorithm employed to screen the slag composition.

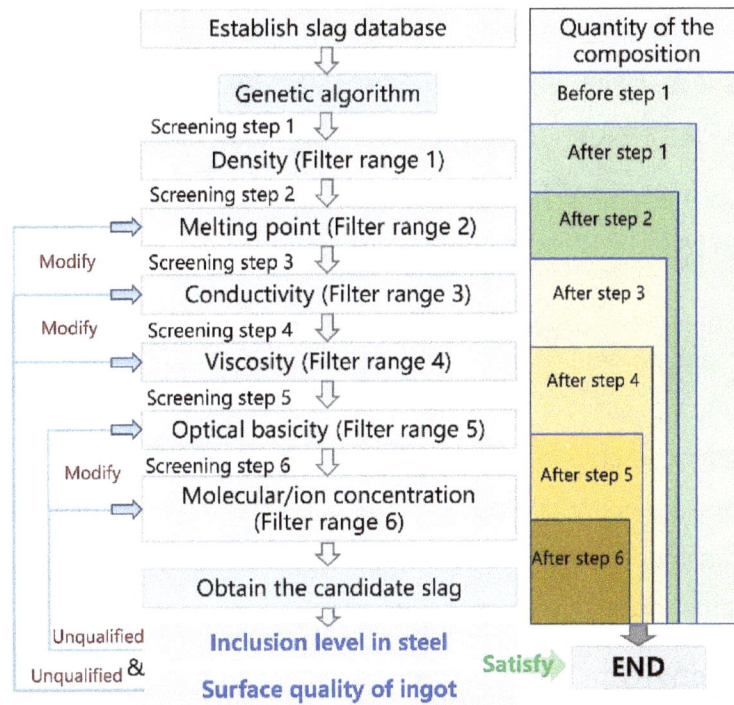

FIG 1 – Genetic algorithm employed in the selection of slag composition.

Establishment of slag composition combinations

Relying on extensive theoretical and practical experience, this study sets specific variation ranges for every component in the quinary slag system, as detailed in Table 1, to streamline calculations (Ju, Gu and Zhang, 2021; Duan, 2020; Wan *et al*, 2022; Dong, 2007). Employing a 1 per cent step increment for component variation, Python programming generated a total of 138 347 unique combinations.

TABLE 1

Concentration ranges for all components utilised in the optimisation (in weight fraction).

	CaO	SiO$_2$	Al$_2$O$_3$	MgO	CaF$_2$
Min	0	0	20	0	30
Max	40	15	40	15	70

Prediction of physical parameters of slag

Prediction of density of slag

The density of various slag compositions could be obtained based on the following empirical formula (Jang *et al*, 2015):

$$\frac{100}{\rho}=0.416m\left(SiO_2\right)+0.303m\left(CaO\right)+0.372m\left(MgO\right)+0.328m\left(Al_2O_3\right)+0.389m\left(CaF_2\right) \quad (1)$$

where m(i) represents the mass of substance i in 100 g of slag, expressed in grams, and the unit for density is grams per cubic centimetre (g/cm^3).

Prediction of melting point of slag

To calculate the melting point of various slags, the computational model has been proposed in previous work (Zhao, Zhang and Ju, 2013):

$$T = 1682.9399+164.479X_1-162.6886X_2+250.209X_3-1415.624X_4$$
$$-819.2188X_5-7937.5X_1X_2+8687.5X_1X_3-8262.5X_1X_4$$
$$+146.875X_1X_5-12725X_2X_3+5075X_2X_4+1068.75X_2X_5 \tag{2}$$
$$-1575X_3X_4+181.25X_3X_5+3906.25X_4X_5+1751.042X_1^2$$
$$+15.88542X_2^2-5495.84X_3^2+504.1668X_4^2+244.0104X_5^2$$

where:

X_1	is the mass percent of Al_2O_3
X_2	is the mass percent of MgO
X_3	is the mass percent of SiO_2
X_4	is the mass percent of CaO
X_5	is the mass percent of CaF_2
T	represents the temperature in °Celsius

Prediction of conductivity of slag

The relationship between conductivity and slag composition could be quantified by the following formula (Ju, Lv and Jiao, 2012).

$$K\left(\Omega^{-1}\cdot cm^{-1}\right)=100exp\left(1.911-1.38x_x-5.69x_x^{\;2}\right)+0.39\left(T-1973\right) \tag{3}$$

$$x_x=x(Al_2O_3)+0.2x(CaO)+0.8x(MgO)+0.75x(SiO_2)+0.2x(CaF_2) \tag{4}$$

where X (i) represents the molar percentage of component i.

Prediction of optical basicity of slag

The optical basicity of slag containing calcium fluoride could be calculated using the following formula (Zhang, Chou and Pal, 2013):

$$\Lambda=\Sigma\Lambda_i\times X_i \tag{5}$$

$$X_i=\frac{A_i}{2}\times n_i\times\frac{N_i}{\Sigma\left[\frac{A_i}{2}\times n_i\times N_i\right]} \tag{6}$$

where:

A_i	denotes the anionic charge of component i
n_i	is the anionic number of component i
N_i	denotes the mole fraction of component i
Λ_i	denotes the optical basicity of component i

The optical basicities of each constituent element are specified in Table 2, offering comprehensive quantitative data (Zhang, Chou and Pal, 2013; Hao and Wang, 2016).

TABLE 2
Optical basicity value for each component.

Component	Optical basicity
CaO	1
MgO	0.92
Al_2O_3	0.66
SiO_2	0.47
CaF_2	0.67

Prediction of molecule and ion activity in the slag

To determine the activity of molecule and ions in CaF_2–Al_2O_3–CaO–SiO_2–MgO slag, this work employs the molecule ion coexistence theory (MICT), examining ternary phase diagrams such as CaO–CaF_2–SiO_2, CaO–CaF_2–MgO, CaO–CaF_2–Al_2O_3, CaO–SiO_2–MgO, CaO–SiO_2–Al_2O_3, CaO–MgO–Al_2O_3, CaF_2–SiO_2–MgO, CaF_2–SiO_2–Al_2O_3, CaF_2–MgO–Al_2O_3, and SiO_2–MgO–Al_2O_3 (Guillot and Guissani, 1996; Sun et al, 2021; Zhao, Li and He, 2022; Duan, 2020; Dong, 2007) Finally, the structural units in CaO–CaF_2–SiO_2–MgO–Al_2O_3 slag system are identified as four simple ions (Ca^{2+}, Mg^{2+}, O^{2-}, F^-), two simple molecules (Al_2O_3, SiO_2), along with 23 complex molecules. Using thermodynamic data provided in Table 3 and Python programming, the concentration of molecule and ions in the molten slags could be calculated.

TABLE 3
Chemical reaction formulas of complex molecules that may be formed, their standard Gibbs free energy changes, and mass action concentrations of structural units in the CaF_2–Al_2O_3–CaO–SiO_2–MgO system.

Reactions	ΔG_f^0(J/mol)	Mass action concentration
$Ca^{2+} + O^{2-} = CaO$	/	$N_1 = N_{Ca^{2+}} + N_{O^{2-}} = \dfrac{2x_1}{\Sigma X}$
$Mg^{2+} + O^{2-} = MgO$	/	$N_2 = N_{Mg^{2+}} + N_{O^{2-}} = \dfrac{2x_1}{\Sigma X}$
$Ca^{2+} + 2F^- = CaF_2$	/	$N_3 = N_{Ca^{2+}} + 2N_{F^-} = \dfrac{3x_1}{\Sigma X}$
SiO_2	/	$N_4 = \dfrac{x_4}{\Sigma X}$
MgO	/	$N_5 = \dfrac{x_5}{\Sigma X}$
$3\left(Ca^{2+}+O^{2-}\right)+Al_2O_3=\left(3CaO \cdot Al_2O_3\right)$	$\Delta G_1^0 = -21757 - 29.288T$	$N_6 = K_1 N_1^3 N_5$
$12\left(Ca^{2+}+O^{2-}\right)+7Al_2O_3=\left(12CaO \cdot 7Al_2O_3\right)$	$\Delta G_2^0 = 617977 - 612.119T$	$N_7 = K_2 N_1^{12} N_5^7$
$\left(Ca^{2+}+O^{2-}\right)+Al_2O_3=\left(CaO\square Al_2O_3\right)$	$\Delta G_3^0 = 59413 - 59.413T$	$N_8 = K_3 N_1 N_5$
$\left(Ca^{2+}+O^{2-}\right)+2Al_2O_3=\left(CaO \cdot 2Al_2O_3\right)$	$\Delta G_4^0 = -16736 - 25.522T$	$N_9 = K_4 N_1 N_5^2$
$\left(Ca^{2+}+O^{2-}\right)+6Al_2O_3=\left(CaO \cdot 6Al_2O_3\right)$	$\Delta G_5^0 = -22594 - 31.798T$	$N_{10} = K_5 N_1 N_5^6$
$\left(Mg^{2+}+O^{2-}\right)+Al_2O_3=\left(MgO \cdot Al_2O_3\right)$	$\Delta G_6^0 = -18828 - 6.276T$	$N_{11} = K_6 N_2 N_5$

Table 3 – Continued ...

Reactions	ΔG_f^0 (J/mol)	Mass action concentration
$3(Ca^{2+}+O^{2-})+2Al_2O_3+(Ca^{2+}+2F^-)$ $=(3CaO \cdot 2Al_2O_3 \cdot CaF_2)$	$\Delta G_7^0 = -44492-73.15T$	$N_{12}=K_7 N_1^3 N_5^2 N_3$
$11(Ca^{2+}+O^{2-})+7Al_2O_3+(Ca^{2+}+2F^-)$ $=(11CaO \cdot 7Al_2O_3 \cdot CaF_2)$	$\Delta G_8^0 = -228760-155.8T$	$N_{13}=K_8 N_1^{11} N_5^7 N_3$
$3(Ca^{2+}+O^{2-})+SiO_2=(3CaO \cdot SiO_2)$	$\Delta G_9^0 = -118826-6.694T$	$N_{14}=K_9 N_1^3 N_4$
$3(Ca^{2+}+O^{2-})+2SiO_2=(3CaO \cdot 2SiO_2)$	$\Delta G_{10}^0 = -236814+9.623T$	$N_{15}=K_{10} N_1^3 N_4^2$
$2(Ca^{2+}+O^{2-})+SiO_2=(2CaO \cdot SiO_2)$	$\Delta G_{11}^0 = -102090-24.26T$	$N_{16}=K_{11} N_1^2 N_4$
$(Ca^{2+}+O^{2-})+SiO_2=(CaO \cdot SiO_2)$	$\Delta G_{12}^0 = -21757-36.819T$	$N_{17}=K_{12} N_1 N_4$
$2(Mg^{2+}+O^{2-})+(SiO_2)=(2MgO \cdot SiO_2)$	$\Delta G_{13}^0 = -56902-3.347T$	$N_{18}=K_{13} N_2^2 N_4$
$(Mg^{2+}+O^{2-})+(SiO_2)=(MgO \cdot SiO_2)$	$\Delta G_{14}^0 = -23849-29.706T$	$N_{19}=K_{14} N_2 N_4$
$3Al_2O_3+2SiO_2=(3Al_2O_3 \cdot 2SiO_2)$	$\Delta G_{15}^0 = -4354-10.467T$	$N_{20}=K_{15} N_5^3 N_4^2$
$2(Ca^{2+}+O^{2-})+Al_2O_3+SiO_2$ $=(2CaO \cdot Al_2O_3 \cdot SiO_2)$	$\Delta G_{16}^0 = -116315-38.911T$	$N_{21}=K_{16} N_1^2 N_5 N_4$
$(Ca^{2+}+O^{2-})+Al_2O_3+2SiO_2$ $=(CaO \cdot Al_2O_3 \cdot 2SiO_2)$	$\Delta G_{17}^0 = -4184-73.638T$	$N_{22}=K_{17} N_1 N_5 N_4^2$
$(Ca^{2+}+O^{2-})+(Mg^{2+}+O^{2-})+SiO_2$ $=(CaO \cdot MgO \cdot SiO_2)$	$\Delta G_{18}^0 = -124683+3.766T$	$N_{23}=K_{18} N_1 N_2 N_4$
$(Ca^{2+}+O^{2-})+(Mg^{2+}+O^{2-})+2SiO_2$ $=(CaO \cdot MgO \cdot 2SiO_2)$	$\Delta G_{19}^0 = -80333+51.882T$	$N_{24}=K_{19} N_1 N_2 N_4^2$
$2(Ca^{2+}+O^{2-})+(Mg^{2+}+O^{2-})+2SiO_2$ $=(2CaO \cdot MgO \cdot 2SiO_2)$	$\Delta G_{20}^0 = -73638+63.597T$	$N_{25}=K_{20} N_1^2 N_2 N_4^2$
$3(Ca^{2+}+O^{2-})+(Mg^{2+}+O^{2-})+2SiO_2$ $=(3CaO \cdot MgO \cdot 2SiO_2)$	$\Delta G_{21}^0 = -205016-31.798T$	$N_{26}=K_{21} N_1^3 N_2 N_4^2$
$2(Mg^{2+}+O^{2-})+2Al_2O_3+5SiO_2$ $=(2MgO \cdot 2Al_2O_3 \cdot 5SiO_2)$	$\Delta G_{22}^0 = -14422-14.808T$	$N_{27}=K_{21} N_2^2 N_5^2 N_4^5$
$3(Ca^{2+}+O^{2-})+2SiO_2+(Ca^{2+}+2F^-)$ $=(3CaO \cdot 2SiO_2 \cdot CaF_2)$	$\Delta G_{23}^0 = -255180-8.20T$	$N_{28}=K_{22} N_1^3 N_4^2 N_3$

Prediction of viscosity of slag

This work determined the viscosity of slag at 1550°C using FactSage™ 8.2, incorporating variations in slag components such as CaF_2 (30 per cent–70 per cent), Al_2O_3 (20 per cent–40 per cent), CaO (0 per cent–40 per cent), MgO (0 per cent–15 per cent), and SiO_2 (0 per cent–15 per cent) with 2 per cent step increments. This approach reduced the potential combinations to 9717, decreasing computational requirements. A quadratic polynomial fitting analysis of these combinations, executed using Python, demonstrated high accuracy with an average absolute error of 1.7708 per cent, thereby validating its applicability to the larger data set of 138 347 combinations. The viscosity fitting results at 1550°C are as follows:

$$\eta = 43342344120.47134 + 2260652.27797391 \cdot A$$
$$-492659.33981768 \cdot B - 247519.83942283 \cdot C$$
$$-442894.24850764 \cdot D - 465902.84576523 \cdot E$$
$$-4356840.93482216 \cdot A^2 - 8686148.7534862 \cdot A \cdot B$$
$$-8688600.14848205 \cdot A \cdot C - 8686646.40438652 \cdot A \cdot D$$
$$-8686416.31844023 \cdot A \cdot E - 4329307.81862237 \cdot B^2 \quad (7)$$
$$-8661067.03228913 \cdot B \cdot C - 8659113.28819614 \cdot B \cdot D$$
$$-8658883.20218896 \cdot B \cdot E - 4331759.21364703 \cdot C^2$$
$$-8661564.683198 \cdot C \cdot D - 8661334.59722407 \cdot C \cdot E$$
$$-4329805.46955192 \cdot D^2 - 8659380.8531375 \cdot D \cdot E$$
$$-4329575.38355188 \cdot E^2$$

where A, B, C, D, and E denote the mass percentages of CaF_2, Al_2O_3, CaO, MgO, and SiO_2, respectively, while η represents the slag system's viscosity at 1550°C, measured in Pa·s.

Screening method

The ESR product in this work is 1.2343 die steel and the chemical composition of this steel is presented in Table 4. Initially the 1.2343 die steel was remelted with ANF-6 slag, primarily composed of 70 per cent CaF_2 and 30 per cent Al_2O_3. Thus this work focused on optimising the slag composition based on ANF-6 slag.

TABLE 4

Chemical composition of 1.2343 die steel.

Element	C	Mn	Si	Cr	Mo	P	S
wt%	0.38–0.45	0.30–0.50	0.9–1.0	4.8–5.5	1.2–1.5	0.03 max	0.03 max

The initial step involves calculating various physical properties of the original slag (ANF-6 slag with 70 per cent CaF_2 and 30 per cent Al_2O_3), with values presented in Table 5. Upon calculating the physical property parameters of the original slag, we have chosen it as the reference slag system.

TABLE 5

Physical parameters of ANF-6 slag.

Physical parameters	Numerical value
Densities (g/cm³)	2.7
Melting point (°C)	1466.8
Conductivity ($\Omega^{-1} \cdot cm^{-1}$)	2.0
Optical basicity	0.67
Calcium ion activity (mol/g)	0.0067
Viscosity (Pa·s)	0.027

During the process of ESR, metal droplets melting from the consumable electrode tip are subject to the buoyant forces of the slag, which can be quantitatively described by the buoyancy formula $F_{bouyangcy} = \rho_{slag} \cdot V_{metal\ drop}$, where $F_{bouyangcy}$ represents the buoyant force, ρ_{slag} denotes the density of the slag, and $V_{metal\ drop}$ is the volume of the metal (Duchesne and Hughes, 2017). The buoyancy force prolongs the slag-metal reaction time, significantly influencing the metal's purification and alloy composition, and impacts the final product's chemical and microstructural qualities. Thus, the objective is to select a slag composition with a higher density than the ANF-6 slag or the average of all combinations.

Regarding melting point, it is commonly accepted that the slag's melting point should be 100°C–200°C lower than the metal's melting point (Dong, 2007; Jang et al, 2023).

In terms of electrical conductivity, optimal conductivity is crucial for electroslag remelting efficiency and ingot quality. Lower conductivity decreases the electrode gap, increasing heat generation. However, lower conductivity may result in short-circuiting and unstable melting. Conversely, high conductivity enlarges the electrode gap, lowering the temperature and increasing heat loss, potentially causing an open arc (Birol, Polat and Saridede, 2015; Liu et al, 2016). Therefore, the slag's conductivity should be slightly lower than that of the ANF-6 slag or just below the median of all slag compositions to achieve a balance between efficiency, stability, and quality.

Concerning optical basicity, higher basicity slags, with elevated CaO content, enhance dephosphorization and desulfurisation but may increase gas content in steel; On the contrary, acidic slags are less effective in desulfurisation and facilitate gas content reduction, primarily hydrogen. Under protective or vacuum conditions, the advantages of acidic slag diminish (Birol, Polat and Saridede, 2015; Zhang, Chou and Pal, 2013; Hao and Wang, 2016; Dong, 2007). Thus, higher basicity slags are preferable, and the goal is to select a slag composition with optical basicity surpassing that of ANF-6 slag or exceeding the average of all slag combinations, ensuring that the selected basicity is the highest among these values.

Regarding inclusions in 1.2343 die steel, reducing the concentration of free calcium ions in the molten slag is crucial to addressing the excessive D and Ds inclusions. The aim is to lower calcium ion activity in the slag. The objective is to select slag compositions with calcium ion activity lower than that in ANF-6 slag and the average across all slag combinations.

In terms of viscosity, the slag needs to maintain an appropriate viscosity with minimal fluctuation under temperature changes. Optimal viscosity ensures good slag fluidity and is crucial for maintaining the surface quality of the ESR ingots (Sebastian, Johannes and Bernd, 2012). Generally, a lower viscosity is preferred (Dong, 2007; Li and Zhang, 2000). Therefore, the objective is to select slag compositions with a viscosity close to that of the ANF-6 slag or within the median viscosity value of all slag combinations.

Screening results of the new slag

Based on our analysis and calculations, the criteria for selecting slag compositions are as follows: density of 2.80–2.97 g/cm³, melting point of 1150–1350°C, electrical conductivity of 1–2 $\Omega^{-1} \cdot cm^{-1}$, optical basicity of 0.70–0.82, free calcium ion concentration of 0.0035–0.0067 mol/g, and viscosity of 0.030–0.050 Pa·s.

In Figure 2, the sections highlighted with light blue shading represent the retained results after screening for each physicochemical parameter. After the screening, from the original 138 347 combinations, only 3961 remain, reducing the data set to 2.86 per cent of its initial data volume. The average, median, and model values for the filtered residue series are presented in Table 6.

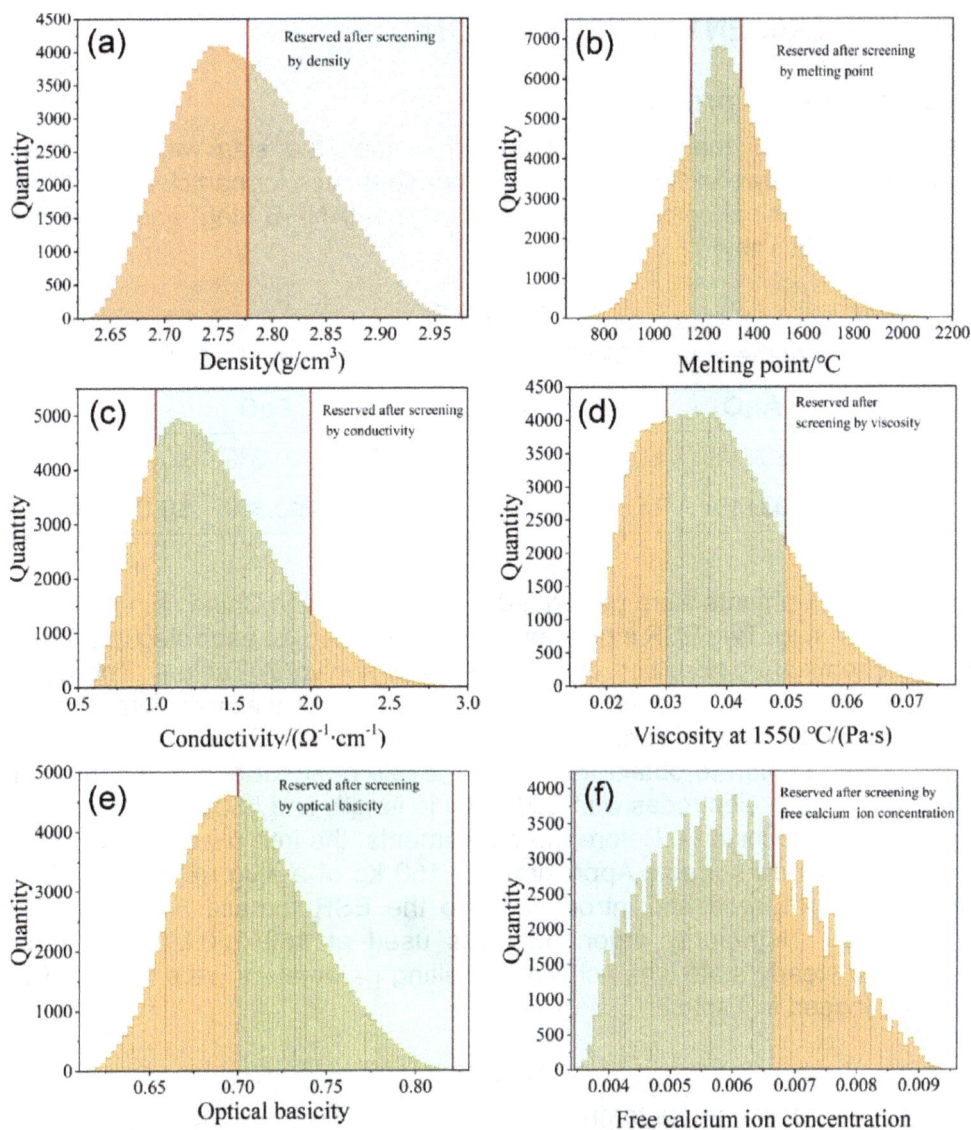

FIG 2 – Screening for each physical property parameter.

TABLE 6

The average, median, and model values for the filtered slag composition.

	CaF$_2$	Al$_2$O$_3$	CaO	MgO	SiO$_2$
Mean value/wt%	43.4	31.8	15.9	4.4	4.5
Median/wt%	43	32	15	4	4
Model/wt%	43	33	15	4	3

Despite narrowing down the viable slag combinations to 2.86 per cent of the original after screening, a wide range of options remains. Considering the limited experimental capacity, it's critical to choose a representative slag composition. To avoid extreme results and ensure experimental reliability, this study calculated the average content of each component, rounding to the nearest whole number for the target slag composition. This method balances the diverse slag characteristics, accurately representing average slag behaviour.

Based on the results presented in Table 6, the mass fraction composition of the slag system can be specified as follows: 43 per cent CaF$_2$, 32 per cent Al$_2$O$_3$, 16 per cent CaO, 4 per cent MgO, and 5 per cent SiO$_2$. This particular slag composition is denoted as the model slag.

INDUSTRIAL EXPERIMENTAL VALIDATION

Industrial experimental procedure

For slag used for industrial trial, it's essential to pre-melt the slag with specified granularity: ≥80 per cent of particles between 1-8 mm, and ≤10 per cent each for particles smaller than 1 mm or larger than 8 mm. The nominal compositions of the original (ANF-6 Slag) and experimental (Model slag) slags are outlined in Table 7.

TABLE 7
Compositions of the original (ANF-6 slag) and experimental (Model slag) slags.

Slag	CaF_2	Al_2O_3	CaO	MgO	SiO_2	FeO	P	S
Original	70±1%	30±1%	≤0.3%	≤0.3%	≤0.3%	≤0.3%	≤0.0005%	≤0.03%
Model	43±1%	32±1%	15±1%	4±1%	5±1%	≤0.3%	≤0.0005%	≤0.03%

Four ESR industrial experiments were performed at a steel plant in China using two different slags: ANF-6 slag and Model slag. Two ESR ingots have been remelted for each slag in a 5000 kg capacity furnace, and the chemical composition of the steel was shown in Table 4. The electrodes were melted in an electric arc furnace. After tapping of the melt, the ladle was transferred to a ladle furnace station to reach objective chemical composition and oxygen level. Thereafter, the ladle was sent to the vacuum degassing station to obtain the desired levels of sulfur and hydrogen contents in the melt. At casting station, the electrodes with 2600 mm in length and 550 mm in diameter have been cast using uphill casting technique. Before the experiments, the iron oxide scale was mechanically removed from the electrode surface. Approximately 150 kg of a slag mixture (CaF_2, Al_2O_3, CaO, SiO_2, and MgO) was prepared and introduced into the ESR furnace, initiating arcing with the consumable electrode. High-purity argon gas was used at 150–200 L/min to create an inert atmosphere. Once a steady state was achieved, melting parameters were recorded every minute and systematically logged in Table 8.

TABLE 8
Parameters for the electroslag remelting process.

Parameter	Value
Mold height/mm	3000
Mold diameter/mm	800
Slag mass/kg	150
Ar flow rate/(L·min^{-1})	150–200
Secondary voltage/V	42–52
Alternating current/A	10000–13000
Average melting rate/(kg·min^{-1})	10.0–10.5

Industrial experimental results

Figure 3 illustrates a smooth surface of the ESR ingot with an easily detachable thin slag skin during deslagging. This observation indicates that the physical properties of the model slag, such as melting point and viscosity, are well-suited for electroslag remelting, confirming its superior processability.

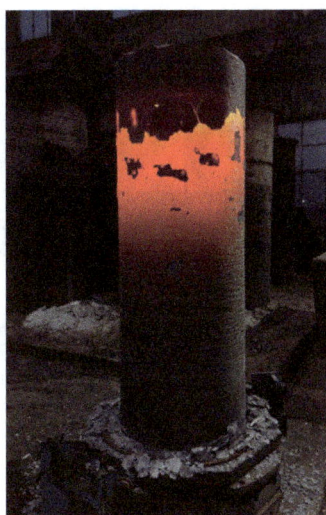

FIG 3 – The ESR ingot remelted by model slag.

After forging and heat treatment, samples with dimensions of 10 mm × 20 mm × 20 mm was cut from the axial centre of the slab with 180 mm × 1200 mm cross-section, used for inclusion detection. Figure 4 shows the calcium content of the electrode and ESR ingot, along with the inclusion rating results. Regarding inclusion rating, a notable reduction in the number and size of D/Ds inclusions was observed after adopting the model slag. In terms of elemental content, it was noted that the calcium concentration in ESR ingots significantly decreased with the introduction of the Model slag.

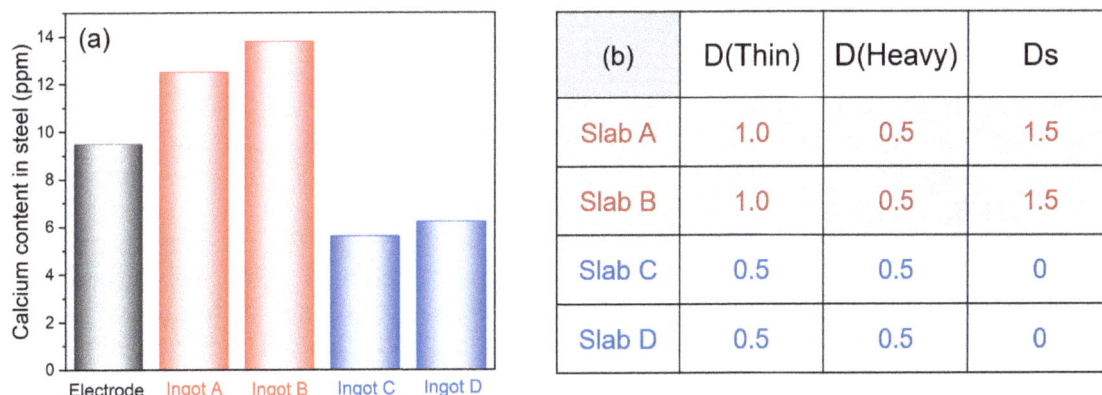

(b)	D(Thin)	D(Heavy)	Ds
Slab A	1.0	0.5	1.5
Slab B	1.0	0.5	1.5
Slab C	0.5	0.5	0
Slab D	0.5	0.5	0

FIG 4 – (a) Calcium content in electrode and ingots; (b) D and Ds inclusion rating results. Ingots A and B (Slabs A and B) were smelted using ANF-6 slag, while ingots C and D (Slabs A and B) were smelted using the model slag.

Data from Figure 4a indicates that, passing through the ANF-6 slag pool, the calcium content in Ingot A and Ingot B shows an increase trend relative to the raw consuming electrode. Conversely, when metal droplets traverse the Model slag pool to form Ingot C and Ingot D, there is a noticeable decrease in calcium content compared to the electrode. Given that calcium is not a required element in electrode alloys and primarily exists in the form of inclusions in the electrode, the reduction of calcium content during electroslag remelting is not attributable to the loss of dissociative calcium elements in the electrode matrix. This phenomenon suggests that the Model slag pool could effectively reduce the calcium content in the ingots by absorbing calcium-containing inclusions in the electrode. This is consistent with the results in Figure 5, which depicts electron microscopy scans of typical D and Ds inclusions in ESR ingots remelted using ANF-6 slag and Model slag. The results show a reduction in inclusion size and dramatically decreased concentrations of Ca in ingots remelted with the Model slag.

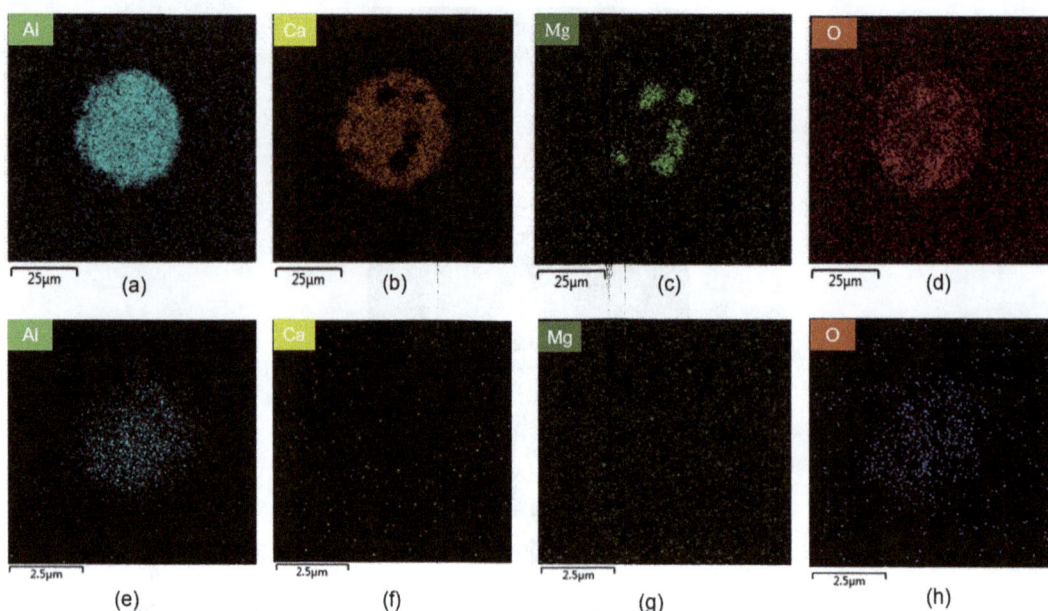

FIG 5 – (a), (b), (c), (d) SEM images of inclusions in Slab A remelted using ANF-6 slag; (e), (f), (g), (h) SEM images of inclusions in Slab C remelted using the model slag.

The industrial experimental results demonstrated that the utilisation of the Model slag not only preserved the excellent surface quality of the ingots but also effectively removed the excessive D and Ds inclusions in 1.2343 die steel.

Discussion

Based on MICT theory, the free calcium ions concentrations in ANF-6 slag and Model slag have been calculated. The results reveal a positive correlation between free calcium ion concentrations in molten slag and calcium content in ingots, as depicted in Figure 6. When remelting the steel with ANF-6 slag, due to sufficient slag-metal reaction, high free calcium ions in molten slag could lead to 'calcium treatment' on the original Al_2O_3 type inclusion in electrode. Thus, higher calcium content and excessive $CaO–Al_2O_3$ (D and Ds inclusions) have been detected in the slab. While, remelted with the Model slag with low free calcium ions concentration, the 'calcium treatment' effect have been effectively suppressed. Furthermore, calcium ions in liquid steel could diffuse into the molten slag driven by the concentration gradient. This could explain well that fewer calcium content and less D and Ds inclusions have been detected in the 1.2343 steel remelted with model slag.

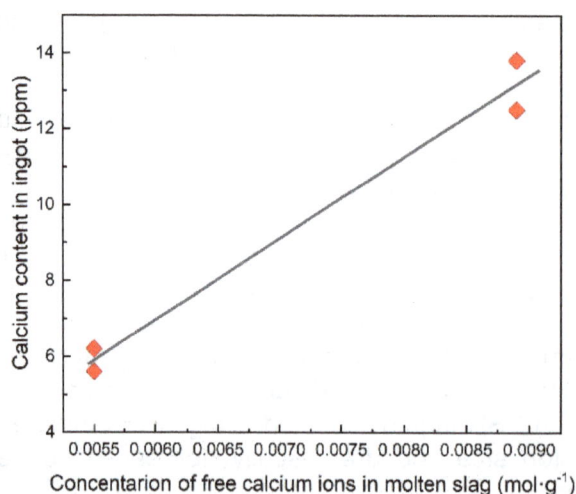

FIG 6 – Correlation between free calcium ion concentration in the slag and calcium content in the ESR ingot.

This work has utilised the molecule ion coexistence theory to calculate the effects of varying mass percentages of CaF_2, Al_2O_3, CaO, SiO_2, and MgO on the activity of calcium ions at 1550°C. Mass percentages of these components will range from 30 per cent to 70 per cent for CaF_2, 20 per cent to 40 per cent for Al_2O_3, 0 per cent to 40 per cent for CaO, 0 per cent to 15 per cent for SiO_2, and 0 per cent to 15 per cent for MgO, with each component varying in 1 per cent increments across all possible combinations. The calculated results presented in Figure 7a reveal that the correlation coefficient for CaF_2 is the highest, indicating that the concentration of CaF_2 in the slag directly affects the concentration of free calcium ions in the molten state. Furthermore, with the mass fraction of CaF_2 held constant, the impact of other components on the concentration of free calcium ions was calculated, shown in Figure 7b.

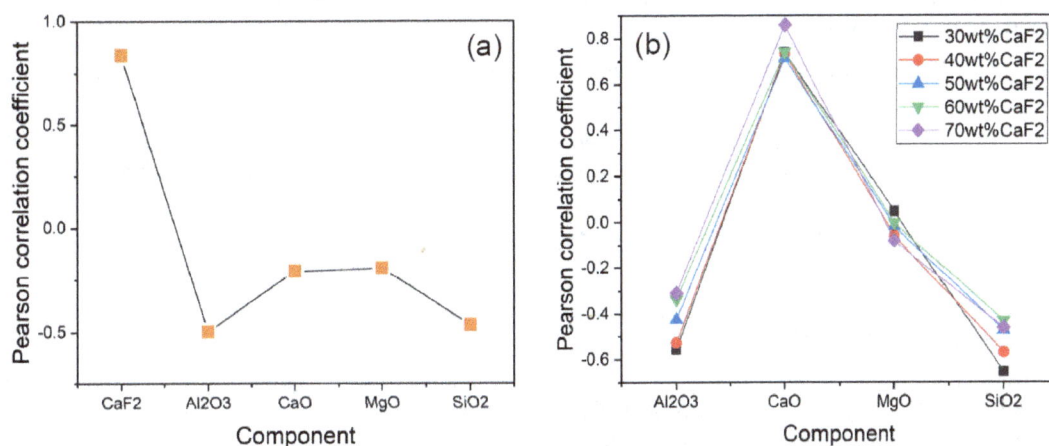

FIG 7 – Correlation coefficient for the relationship between each component and calcium ion concentration based on the molecule–ion coexistence theory: (a) Pearson correlation coefficients of each component with free calcium ion concentration; (b) Influence of other components on free calcium ion concentration when the CaF_2 mass fraction is fixed.

The Pearson correlation coefficients in Figure 7 show that CaF_2 significantly affects the concentration of free calcium ions in the slag. Additionally, the concentration of CaO also exerts an influence. Conversely, the concentration of MgO in the slag system has minimal impact on the concentration of free calcium ions. Moreover, Al_2O_3 and SiO_2 in the slag system moderately suppress the concentration of free calcium ions.

CONCLUSION

- Employing well-defined screening criteria, the slag composition exhibiting superior performance was theoretically selected. Subsequent experimental verification confirmed the exceptional performance of this new slag composition, thereby demonstrating the feasibility and efficiency of using genetic algorithms for slag system optimisation.

- CaF_2 and CaO were identified to increase the activity of calcium ions in the slag, with CaF_2 exerting a greater effect than CaO. SiO_2 and Al_2O_3 were observed to decrease calcium ion activity, with SiO_2 being slightly more effective than Al_2O_3 in this aspect. Meanwhile, MgO demonstrated no significant impact on calcium ion activity.

- To further reduce the concentration of free calcium ions and free alumina activity in the slag, it is recommended to decrease the content of CaF_2 and increase the contents of CaO and SiO_2.

ACKNOWLEDGEMENTS

This work was supported by Program of Introducing Talents of Discipline to Universities (No. B21001), Science and Technology Research Projects of Liaoning Province (2023JH1/10400051).

REFERENCES

Birol, B, Polat, G and Saridede, M N, 2015. Estimation Model for Electrical Conductivity of Molten CaF_2-Al_2O_3-CaO Slags Based on Optical Basicity, *JOM*, 67:427–435.

Campbell, C and Olson, G, 2000. Systems design of high performance stainless steels, I, Conceptual and computational design, *Journal of Computer-Aided Materials Design,* 7:145–170.

Dong, Y W, 2007. Mathematical Modeling of Solidification during Electroslag Remelting Process and Development of New Slags, PhD thesis, Northeastern University, Shenyang.

Dong, Y W, Jang, Z H and Yu, A, 2016. Dissolution Behavior of Alumina-Based Inclusions in CaF_2-Al_2O_3-CaO-MgO-SiO_2 Slag Used for the Electroslag Metallurgy Process, *Metals,* 6:272–279.

Dong, Y W, Jiang, Z H, Cao, Y L, Yu, A and Hou, D, 2014. Effect of Slag on Inclusions During Electroslag Remelting Process of Die Steel, *Metallurgical and Materials Transactions B,* 45:1315–1324.

Duan, S C, 2020. Fundamental Study on the Controlling Loss of Alloying Elements of Large IN718 Electroslag Remelting Ingots, PhD thesis, University of Science and Technology, Beijing.

Duchesne, M A and Hughes, R W, 2017. Slag density and surface tension measurements by the constrained sessile drop method, *Fuel,* 188:173–181.

Guillot, B and Guissani, Y, 1996. Towards a theory of coexistence and criticality in real molten salts, *Molecular Physics,* 87:37–86.

Guo, Y, Xia, Z, Shen, Z, Li, Q, Sun, C, Zheng, T and Ren, W, 2021. Enhancement of removing inclusions from liquid melt film during the ESR process assisted by a static magnetic field, *Journal of Materials Processing Technology,* 290:1–9.

Hao, X and Wang, X H, 2016. A New Sulfide Capacity Model for CaO-Al_2O_3-SiO_2-MgO Slags Based on Corrected Optical Basicity, *Steel Research International,* 87:359–363.

Jang, W F, Tian, L, Bo, H and Ma, T F, 2023. A new method for evaluating the melting performance of the slag, *Ironmaking and Steelmaking,* 50:775–781.

Jang, Z H, Dong, Y W, Geng, X and Liu, F B, 2015. *Physical Chemistry and Transport Phenomena During Electroslag Metallurgy,* pp 68–75 (Northeastern University Press: Shenyang).

Jiang, Z H, Dong, Y W, Geng, X and Liu, F B, 2023. Development and application of electroslag remelting technology for high quality special steels, *Iron and Steel,* 58:15–25.

Ju, J T, Gu, Y and Zhang, Q, 2021. Effect of CaF_2 and CaO/Al_2O_3 ratio on evaporation and melting characteristics of low-fluoride CaF_2–CaO–Al_2O_3–MgO–TiO_2 slag for electroslag remelting, *Ironmaking and Steelmaking,* 50:8.

Ju, J, Ji, G, Tang, C, Yang, K and Zhu, Z, 2020. Investigation of fluoride evaporation from CaF_2–CaO–Al_2O_3–MgO–TiO_2–(Li_2O) slag for electroslag remelting, *Scientific Reports,* 10:1–12.

Ju, J, Lv, Z L and Jiao, Z Y, 2012. Experimental Study on the Electrical Conductivity of CaF_2-SiO_2-Al_2O_3-CaO-MgO Slag System, *Journal of Iron and Steel Research International,* 24:27–31.

Li, J X and Zhang, J, 2000. CaO-MgO-CaF_2-Al_2O_3-SiO_2 Five-element slag system Calculation model of viscosity, *Journal of University of Science and Technology Beijing,* 22:316–319.

Liu, J H, Zhang, G H, Wu, Y D and Chou, K C, 2016. Study on electrical conductivity of FexO-CaOSiO2-Al_2O_3 slags, *Canadian Metallurgical Quarterly,* 55:221–225.

Ragnarsson, L and Sichen, D, 2009. Inclusions Generated during Ingot Casting of Tool Steel, *Steel Research International,* 81:40–47.

Sabih, A, Wanjara, P and Nemes, J, 2005. Characterization of internal voids and cracks in cold heading of dual phase steel, *ISIJ International,* 45:1179–1186.

Schneider, R S E, Molnar, M, Kloesch, G and Schueller, C, 2019. Effect of the Al_2O_3 Content in the Slag on the Chemical Reactions and Nonmetallic Inclusions during Electroslag Remelting, *Metallurgical and Materials Transactions B,* 51:1904–1911.

Sebastian, R and Bernd, F, 2015a. Influencing the Electroslag Remelting Process by varying Fluorine Content of the utilized Slag, *The 8th European Metallurgical Conference (EMC).* Düsseldorf.

Sebastian, R and Bernd, F, 2015b. Process and Refining Characteristics of ESR using MgO containing Slag Systems, *International Symposium on Liquid Metal Processing and Casting,* Leoben/Germany.

Sebastian, R, Johannes, M and Bernd, F, 2012. The Influence of Selected Slag Properties and Process Variables on the Solidification Structure during ESR, *1st International Conference on Ingot Casting, Rolling and Forging (ICRF).* Aachen/Germany.

Shi, C B, Li, J, Cho, J W, Jiang, F and Jung, I H, 2015. Effect of SiO_2 on the Crystallization Behaviors and In-Mold Performance of CaF_2-CaO-Al_2O_3 Slags for Drawing-Ingot-Type Electroslag Remelting, *Metallurgical and Materials Transactions B,* 46:2110–2120.

Shi, C, 2020. Deoxidation of Electroslag Remelting (ESR) – A Review, *ISIJ International,* 60:1083–1096.

Sun, H, Yang, J, Zhang, R H and Yang, W K, 2021. Effect of Slag Basicity on Dephosphorization at Lower Basicity and Lower Temperature Based on Industrial Experiments and Ion-Molecular Coexistence Theory, *Metallurgical and Materials Transactions B,* 52:3403–3422.

Wan, X X, Shi, C B, Yu, Z and Li, J, 2022. Effect of CaF_2 and Li_2O on structure and viscosity of low-fluoride slag for electroslag remelting of rotor steel, *Journal of Non-Crystalline Solids,* 597:121914.

Xu, W, del Castillo, P E J R D and Van Der Zwaag, S, 2009. A combined optimization of alloy composition and aging temperature in designing new UHS precipitation hardenable stainless steels, *Computational Materials Science,* 45:467–473.

Zhang, G H, Chou, K C and Pal, U, 2013. Estimation of Sulfide Capacities of Multicomponent Slags using Optical Basicity, *ISIJ International,* 53:761–767.

Zhao, M G, Li, G and He, S P, 2022. Study of Thermodynamic for Low-Reactive $CaO-BaO-Al_2O_3-SiO_2-CaF_2-Li_2O$ Mold Flux Based on the Model of Ion and Molecular Coexistence Theory, *Metals,* 12:1099.

Zhao, X H, Zhang, Z H and Ju, J T, 2013. Study on Melting Temperature of $CaF_2-SiO_2-Al_2O_3-CaO-MgO$ Slag, *Hot Working Technology,* 42:81–84.

Flux smelting behaviour of pre-reduced Mn ore by hydrogen at elevated temperatures

P Kumar[1] and J Safarian[2]

1. PhD candidate, Department of Materials Science and Engineering, NTNU, N-7034 Trondheim, Norway. Email: Pankaj.kumar@ntnu.no
2. Professor, Department of Materials Science and Engineering, NTNU, N-7034 Trondheim, Norway. Email: jafar.safarian@ntnu.no

ABSTRACT

Understanding how ore interacts with flux particles at elevated temperatures to create molten slag is crucial since it governs the dynamics of a chemical reaction. This study explores the smelting behaviour of pre-reduced Nchwaning manganese ore when combined with lime, with the objective of examining the evolving interaction between pre-reduced ore particles and lime over time. The research sheds light on the interaction between solid and liquid and the phases that emerge during this process. To achieve this, a sessile drop furnace was employed to rapidly heat the materials positioned adjacent to each other on an alumina substrate and to observe the smelting process as it unfolded over time. This method allowed for the direct observation of the melting temperatures and the flux-ore reaction progression rate, and the potential disruptive events that might occur. By comparing the molten interfaces of the fluxed materials at various time intervals, this study provides insights into the relative rate of slag formation from the two materials. The results indicate that the main slag formation initiated at approximately 1400°C and continued to advance with time, with complete mixing occurring around 1500°C. The possible phases formed were identified using Scanning Electron Microscopy and modelled using Fact Sage thermodynamic software. In addition, the iron particles in the pre-reduced Mn ore were separated and settled from a rich MnO-containing slag. It was found that the separation of molten iron droplets from the slag depends on the rate of solid MnO particles dissolution into the adjacent slag phase.

INTRODUCTION

Ferromanganese production consumes anywhere between 2400–2700 kWh energy per ton of metal produced (Tangstad and Olsen, 1995). Ferroalloys such as ferromanganese are mainly produced in submerged arc furnaces (Office of Air Quality Planning and Standards, 1992), where 50–70 per cent of the required thermal energy is given by electrical dissipation and the rest is met by carbon or other carbonaceous material like biomass and charcoal (Monsen *et al*, 2007), which is used as a reductant. The process of manganese production goes through series of reduction steps as shown in Equations 1–3 (Tangstad and Olsen, 1995). Where the higher manganese oxide is reduced to lower oxide by indirect reduction by CO gas followed by final step where the manganese monoxide is reduced to manganese metal by carbon Equation 4 (Tangstad and Olsen, 1995):

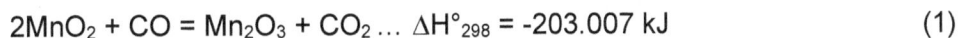

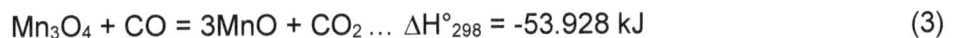

$$2MnO_2 + CO = Mn_2O_3 + CO_2 \ldots \Delta H°_{298} = -203.007 \text{ kJ} \tag{1}$$

$$3Mn_2O_3 + CO = 2Mn_3O_4 + CO_2 \ldots \Delta H°_{298} = -178.371 \text{ kJ} \tag{2}$$

$$Mn_3O_4 + CO = 3MnO + CO_2 \ldots \Delta H°_{298} = -53.928 \text{ kJ} \tag{3}$$

$$MnO + C = Mn + CO \ldots \Delta H°_{298} = 274.206 \text{ kJ} \tag{4}$$

The process emits huge amount of greenhouse gas (GHG) which greatly depends on the source of energy and reductant material (biomass, metallurgical coke, or coal (Haque and Norgate, 2013; Westfall *et al*, 2016). It was found from the previous studies that around 1.04 to 6.0 kg CO_2 is emitted per kg FeMn production (Haque and Norgate, 2013; Olsen, Monsen and Lindstad, 1998; Westfall *et al*, 2016). The CO_2 is responsible for 20 per cent of thermal absorption, which directly causes global warming (Schmidt *et al*, 2010) and climate change, which is recognised as a substantial threat to human health (Costello *et al*, 2009; Mora *et al*, 2017). With more awareness and stricter government rules, industries are forced to move towards greener alternative routes to produce these metal and alloys. One such route is the use of hydrogen, for the reduction of metal oxides like Cr_2O_3 (Davies *et al*, 2022), Fe_2O_3 (Heidari *et al*, 2021), MnO_2 (Barner and Mantell, 1968; Safarian, 2021), to its lower oxides. The use of hydrogen produces water vapor as the main off-gas component which is safe and even can be further used, hence significantly reducing CO_2 emission. Manganese oxide can only be partially reduced with hydrogen following Equations 5–7,(Safarian, 2021) unlike iron oxide that can be completely reduced with hydrogen following Equation 8:

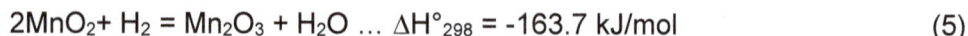

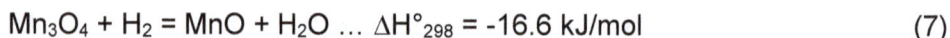

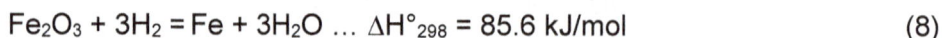

$$2MnO_2 + H_2 = Mn_2O_3 + H_2O \dots \Delta H°_{298} = -163.7 \text{ kJ/mol} \tag{5}$$

$$3Mn_2O_3 + H_2 = 2Mn_3O_4 + H_2O \dots \Delta H°_{298} = -135.1 \text{ kJ/mol} \tag{6}$$

$$Mn_3O_4 + H_2 = MnO + H_2O \dots \Delta H°_{298} = -16.6 \text{ kJ/mol} \tag{7}$$

$$Fe_2O_3 + 3H_2 = Fe + 3H_2O \dots \Delta H°_{298} = 85.6 \text{ kJ/mol} \tag{8}$$

The pre-reduction of manganese ore with hydrogen significantly reduces the net CO_2 emission from the process. In a recent approach, of which this manuscript is a part of, researchers have tried coupling the use of hydrogen for pre-reduction of manganese ore followed by aluminium for complete reduction of the pre-reduced ore following Equation 9. the process is referred to as HAlMan (Safarian, 2021):

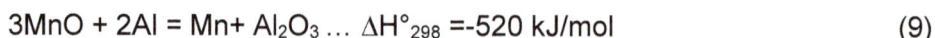

$$3MnO + 2Al = Mn + Al_2O_3 \dots \Delta H°_{298} = -520 \text{ kJ/mol} \tag{9}$$

The reaction being highly exothermic in nature gives enough energy for slag metal formation and separation. The slag formed in the process mostly contains Al_2O_3, CaO and some unreduced MnO along with low amount of SiO_2 and MgO. One of the major aims in the HAlMan process is to generate slag which could be leached to recover alumina and calcia. The recovered alumina and calcia can be re-used in the process, hence further reducing the net energy consumption. But, for this to be achieved, the slag needs to be designed well to be easily leachable. It is found from the literature that calcium aluminate ($CaAl_2O_4$) phase is the most easily leached phase (Azof, Kolbeinsen and Safarian, 2017), while gehlenite phase ($Ca_2Al_2SiO_7$) is tough to be leached.

To achieve the required phase, it is of prime importance to understand the mechanisms of phases formation and transformation with time and temperature. The current manuscript focuses on studying the interaction of flux (CaO) with a pre-reduced Mn ore by H_2 gas on alumina. The study is performed in a sessile drop test furnace, where the lime and pre-reduced ore particles are placed on the alumina substrate and heated at various rates and durations to monitor softening, melting and interaction of material upon heating. After the tests, the microstructure and composition of samples are studied by microscopic examination.

EXPERIMENTAL

Method

The overall experimental procedure of this work is presented in Figure 1. The detail about the experiment is provided in later sections.

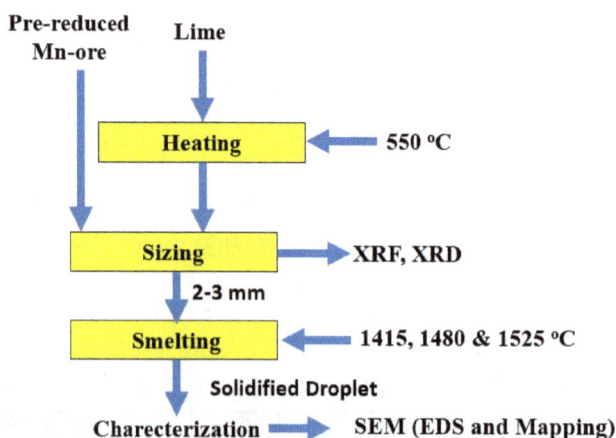

FIG 1 – Schematic of used methodology showing the materials flow.

Materials and preparation

A pre-reduced Nchwaning ore and lime was used for the study. The pre-reduction of the Nchwaning ore was carried out using a vertical tube resistance furnace, which is designed to reach a maximum temperature of 1100°C and can purge Ar, H_2 and CO gas. 50 g of dried Nchwaning ore in the size range of 4–10 mm was reduced with hydrogen at a flow rate of 4 Nl/min, held at 800°C for 1 hr. The reduced sample was then crushed and sieved to obtain particles of size 2–3 mm, which was then sealed in an airtight plastic bag to prevent any reoxidation. Similarly, the lime was pre-heated to 550°C under Ar atmosphere, to decompose any calcium hydroxide formed due to exposure to atmosphere for long time. Following the reaction in Equation 10:

$$Ca(OH)_2 = CaO + H_2O \ldots \Delta H°_{298} = 104.903 \text{ kJ/mol} \tag{10}$$

After pre-heating, the lime samples were collected and sealed in airtight plastic to avoid hydration. The lime particles were carefully sized to 2–3 mm manually, as they are quite soft and may turn fine on pulverising mechanically. These sized particles were then kept for further sessile drop test study.

Sessile drop test

Particles of similar size, from the size range of 2–3 mm were taken from each pre-reduced Nchwaning ore and lime. One particle each of pre-reduced Nchwaning ore and lime was kept on alumina substrate, which was then kept on the graphite holder Figure 2. The sample holder was then pushed inside the furnace sealed, evacuated, and backfilled to atmospheric pressure with argon. The argon flow of 0.5 Nl/min was maintained throughout the test. For all the test similar schedule was used. The schedule comprised of four stages shown in Table 1. Starting of the interaction of lime and pre-reduced ore particle and the completion of melting of the samples were made the two end points of the experiment, and an intermediate stage between them was also considered for study. It was observed that the interaction started at 1415°C, and was completed at 1525°C, these two points were considered as the two end points and designated as T1 and T3, respectively, while the intermediate temperature 1480°C will be marked as T2 for ease of discussion.

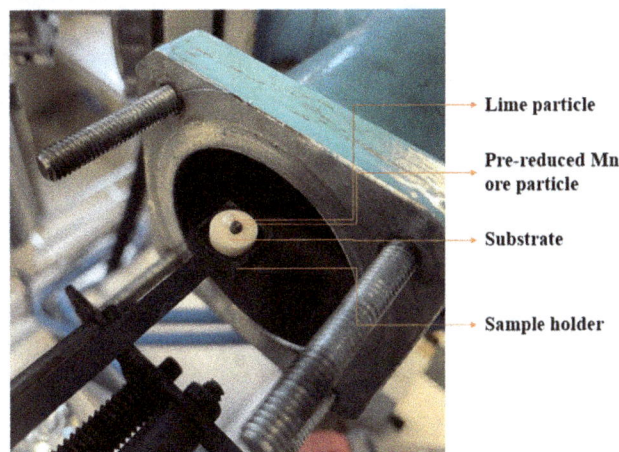

FIG 2 – Samples placed on sample holder.

TABLE 1

Program for sessile drop test furnace.

Total time [HH:MM:SS]	Time interval [HH:MM:SS]	End temp (°C)	Ar flow rate (Nl/min)
00:00:30	00:00:30	25	0.5
00:03:30	00:03:00	900	0.5
00:09:30	00:06:00	1200	0.5
00:49:30	00:40:00	1800	0.5
00:59:30	00:10:00	25	0.5

Figure 3 depicts the schematic of the sessile drop test furnace comprising three key components: the primary heating chamber equipped with heating element and a thermocouple, a firewire digital video camera with a telecentric zoom Lense capable of capturing images at 1280 × 960 pixels resolution situated to the right, and a pyrometer on the left for temperature measurement. The furnace offers a capability to purge CO, hydrogen, and inert gas, facilitating the examination of pre-reduction behaviour or properties such as melting point and surface tension. In this specific instance, the focus was solely on understanding the smelting characteristic of the pre-reduced Nchwaning manganese ore with lime particle and alumina substrate.

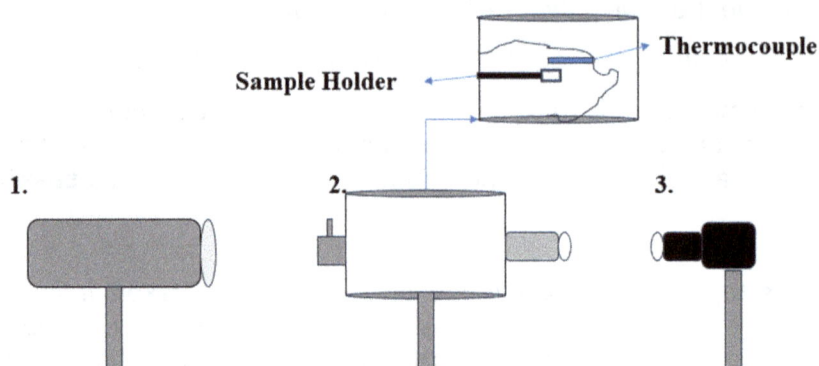

FIG 3 – Schematic of Sessile drop test furnace wherein: (1) represents the pyrometer, (2) main chamber and (3) the camera.

Characterisation technique

The pre-reduced Nchwaning ore and lime were powdered (<75 mm) in a ring mill for 1 min at a speed of 800 revolution per minute (rev/min) and sent for elemental analysis. Elemental analysis was done using the X-ray fluorescence (XRF) technique (Thermo fisher, Degerfors labortorium AB, Sweden). The mineralogical examination through X-ray diffraction (XRD) was done using the Bruker D8 A25 DaVinciTM equipment from Karlsruhe, Germany. For lime samples the XRD analysis was carried out both before and after pre-heating to be sure about complete decomposition of $Ca(OH)_2$ to CaO. The XRD for each sample was done for 2θ ranging from 0–80°, with a step size of 0.2°. The samples after sessile drop test were cold mounted using epoxy and polished using automatic polishing machine Tegrapol 30. The polished sample was observed under scanning electron microscope (SEM) (Zeiss ultra 55LE, Carl Zeiss, Jena, Germany) for microstructural analysis. While the elemental analysis and mapping was done using energy dispersive spectroscopy (EDS) (Bruker, AXS, microanalysis GmbH, Berlin, Germany).

The chemical composition of pre-reduced Nchwaning ore and lime is presented in Table 2. A major fraction of the weight of pre-reduced ore is composed of manganese monoxide and reduced iron. Limestone mostly contains CaO and small amount of MgO and SiO_2.

TABLE 2

XRF elemental analysis of pre-reduced Nchwaning ore and lime sample (wt per cent).

Sample	%MnO	%Fe	%Al₂O₃	%SiO₂	%CaO	%MgO	%LOI
Lime	-	-	0.16	0.19	90.20	0.50	8.45
Pre-reduced Nchwaning ore	72.22**	12.21*	0.43	4.66	8.93	1.48	-

*metallic iron, **in form of MnO.

RESULTS AND DISCUSSION

Melting behaviour

The lime and the pre-reduced manganese ore particle were kept on the alumina substrate and heated. It was observed that the lime particle started melting first, at around 1350°C near the contact point with alumina substrate. Figure 4 shows different stages of melting of lime particle and pre-reduced manganese ore.

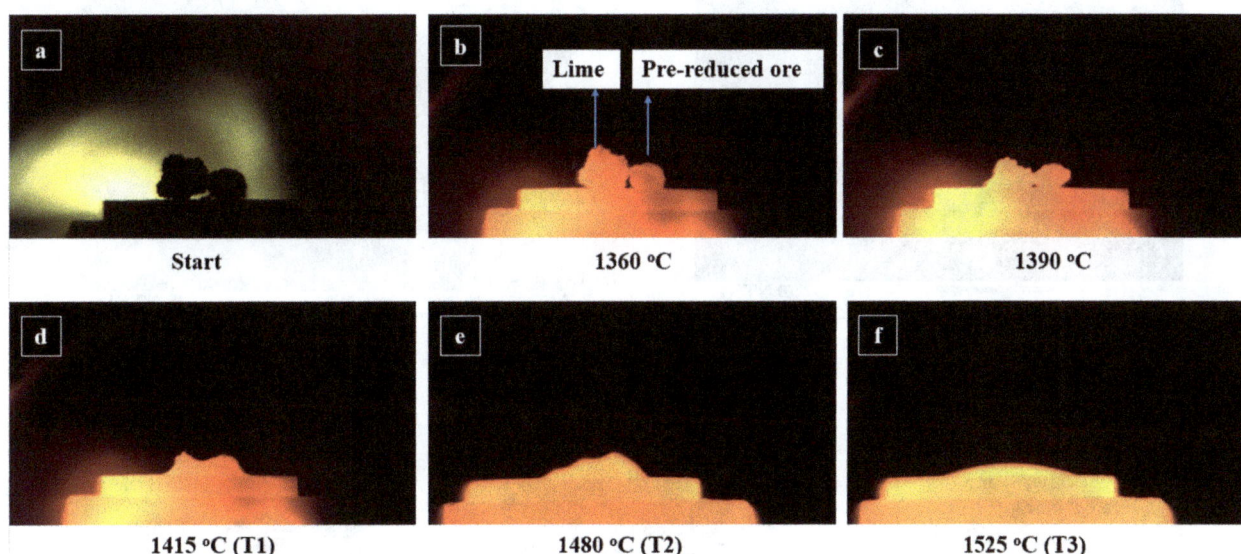

FIG 4 – Different stages of interaction between lime and pre-reduced ore particle.

The interaction of molten lime particle was observed to start at around 1415°C, while the complete melting occurred at around 1525°C Figure 4f, where a single molten pool was observed. To further understand the interaction of lime and ore particle, microscopic examination was done which is discussed in later section.

Microstructural analysis

Figure 5 shows the microstructure of test samples for test T2. From the microstructural images three distinct region were found in samples with lower holding time and temperature (T1 and T2), which was analysed further using EDS and mapping Figure 6. The outer most region (transformed) starting from the pre-reduced manganese ore particle, was mostly observed to be composed of (37–39 wt per cent) calcia and (55–57 wt per cent) alumina with very little (3–5 wt per cent) MnO. The middle region (partially transformed) was found to contain dendritic structure, made up of the ore particle, which shows small amount of undissolved ore particle in this region. While the inner portion (un-transformed) was found to contain larger circular undissolved ore particles. The width of the middle region (partially transformed region) was observed to decrease with increasing holding time and temperature. The width in case of samples heated till 1410°C was found to be around 250 mm which decreased to around 180 mm in case of sample heated till 1450°C and it completely vanished for the samples heated till 1500°C, Figure 6. The observation is obvious as the transformation is time and temperature dependent phenomenon. The reduced iron particles were mostly observed in the inner region, though there were few iron ore particles which was found suspending in the middle region. While no iron particles were observed in the outer region, the slag is already formed Figure 7. It is interesting to note that the iron particles were only observed attached to the manganese particles. As the outer region has no separate manganese particles hence no iron particle was observed as well. The iron particle was supposed to sink towards the bottom, because of the difference in density, tough the presence of small amount of iron particle in middle region was mostly because of higher viscosity due to presence of solid MnO particles in the slag phase. The good contact of metal droplets with solid MnO particles may indicate low interfacial energies between them that causes not detachment of metal droplets. Hence, we may conclude that the separation of tiny iron particles from the slag is dependent on the dissolution of solid MnO particles into the molten slag phase and hence the loss of the MnO/iron interfacial area.

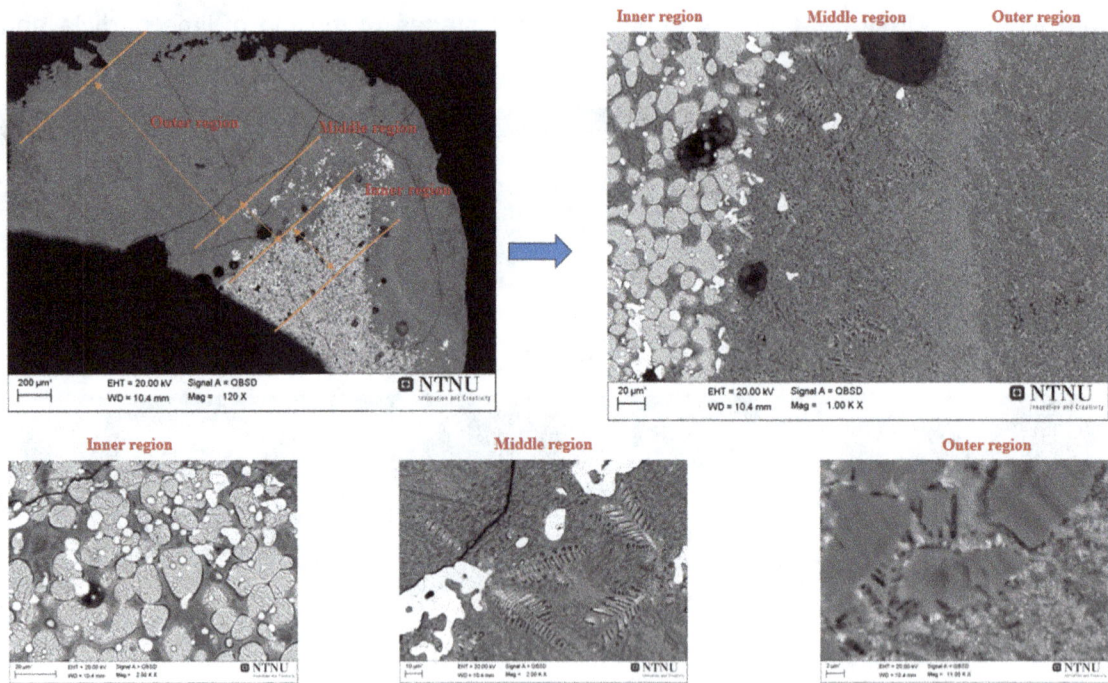

FIG 5 – Microstructure of test sample T2 showing different zones of transformation.

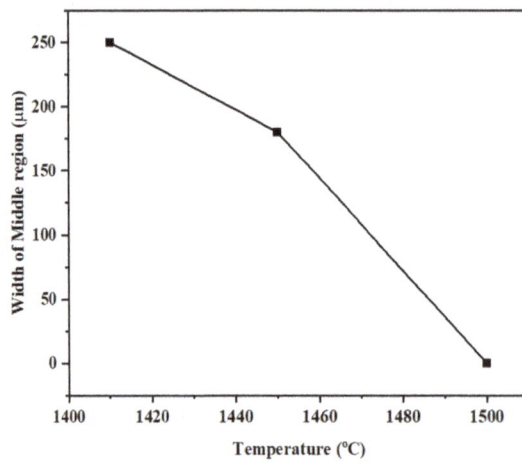

FIG 6 – Change in width of partially transformed zone (middle region), with temperature.

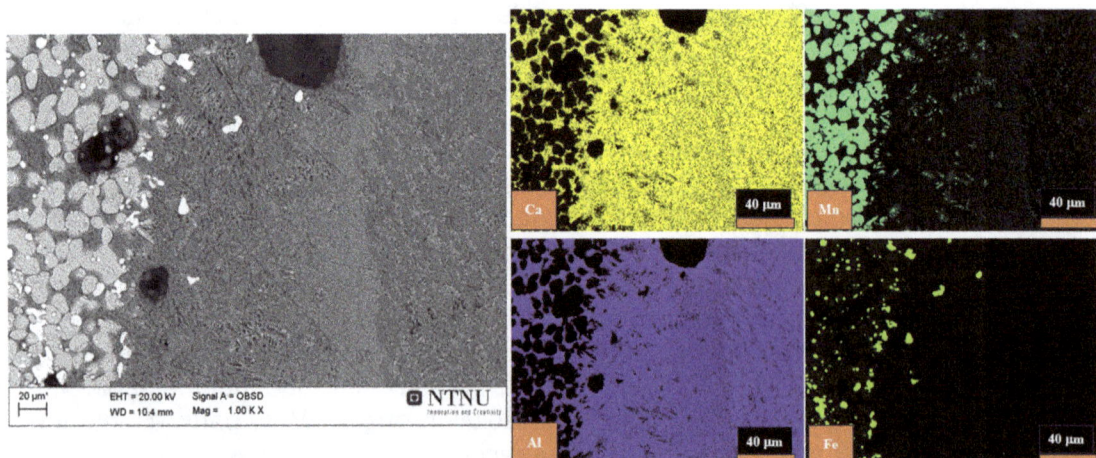

FIG 7 – Elemental X-ray mapping image of sample T2 showing the distribution of the main elements.

For the test sample T3, the phases were completely evolved and there was no demarcation of regions as was observed in other samples (T1 and T2). This shows that the melting and the phase formation was completed at 1525°C. The cross-sectional image of the sample shows a large single metal droplet at the bottom of the melted sample Figure 8, which also signifies complete melting and homogenisation of the molten bath. Two distinct phases were observed in the slag, a darker phase consisting of alumina and calcia, in the matrix of a brighter phase consisting of manganese, alumina and calcia. Figure 9 and Table 3 shows the difference in the chemical compositions of these two phases in higher magnification with small area analysis. The darker phase had a CaO/Al_2O_3 mass ratio close to 0.3, which is like that in $CaO.2Al_2O_3$ phase. While the CaO/Al_2O_3 mass ratio in case of brighter phase was close to 0.5, which is like that in $CaO.Al_2O_3$ phase. Hence, it can be very well said that the darker phases were $CaO.2Al_2O_3$, while the brighter ones were $CaO.Al_2O_3$.

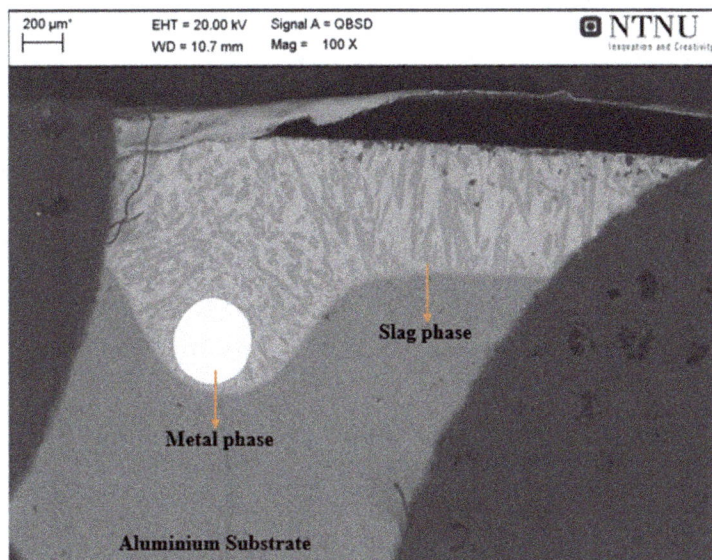

FIG 8 – Cross-section image of test T3 sample showing metal and slag phase.

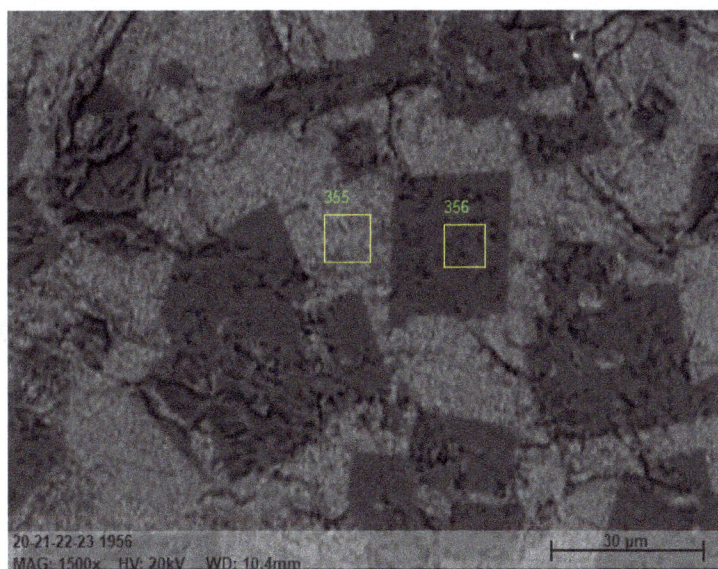

FIG 9 – SEM image showing two distinct phases in test T3.

TABLE 3

EDS result of darker and brighter phases formed in the slag of test T3.

Points/Elements	MnO	CaO	Al$_2$O$_3$	SiO$_2$	MgO
355	19.40	29.7	43.33	3.65	2.314
356	1.53	24.55	73.39	-	-

Probable mechanism

Figure 10 shows the probable mechanism of the melting and dissolution of lime and pre-reduced manganese ore on alumina substrate.

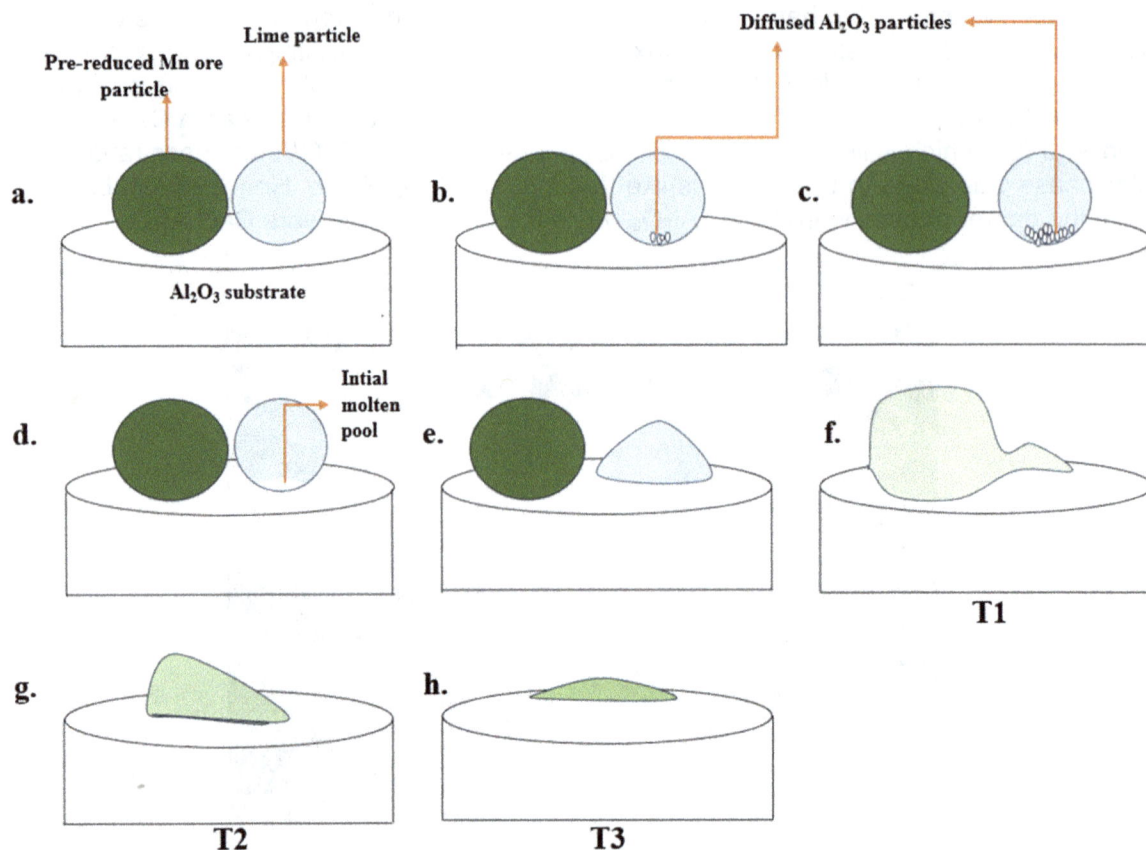

FIG 10 – Mechanism of Lime and pre-reduced manganese ore and Alumina substrate interaction.

As the temperature increases, more and more alumina particle from alumina substrate diffuses into the lime particle, which is obvious as diffusion is highly dependent on temperature Figure 11b and 11c. The diffusion of alumina particle would have been significantly easier, as the lime particles were highly porous. However, we may expect that the CaO and adjacent oxides in the lime such as MgO and Al_2O_3 and SiO_2 (Table 2) may yield some early slaggy phases formation in the lime particle, and the free surfaces of the particle and hence the lime/substrate interface. The slow increase in alumina concentration into the lime particle near the interface causes the CaO/Al_2O_3 ratio to decrease to lower ratios and eventually to 1, which reflects the eutectic point in CaO-Al_2O_3 phase diagram Figure 11, marked by arrow. It is at this point that the molten pool starts to form Figure 11d, which explains why the melting of lime particle was observed at 1360°C, much lower than the theoretical melting point of 2572°C. Once, the molten pool is formed, the further mechanism is a combination of diffusion and dissolution of lime and alumina into the molten pool. The further melting occurs rapidly and within a narrow temperature zone, because of the combined effect. The concentration of alumina in lime increases rapidly with time and temperature, Equations 11–13 show decreasing CaO/Al_2O_3 mass ratio from 1.66 to 0.2:

$$3CaO + Al_2O_3 = 3CaO.Al_2O_3 \ldots CaO/Al_2O_3 = 1.66 \tag{11}$$

$$CaO + Al_2O_3 = CaO.Al_2O_3 \ldots CaO/Al_2O_3 = 0.55 \tag{12}$$

$$CaO + 2Al_2O_3 = CaO.2Al_2O_3 \ldots CaO/Al_2O_3 = 0.28 \tag{13}$$

Once the lime is completely molten and spread over the substrate more it touches the solid pre-reduced manganese ore particle, and it covers the particle very rapidly, in a fraction of a second, so that a single molten pool is seen while there is an increase in temperature as shown in Figure 11f to 11h. The interaction of pre-reduced manganese ore particle with CaO and Al_2O_3 starts the slagging reactions with Mn oxide contribution via Equations 14 and 15 and the mass transport of MnO to the slag occurs simultaneously via Equation 16, which is solid MnO dissolution into the adjacent molten slag:

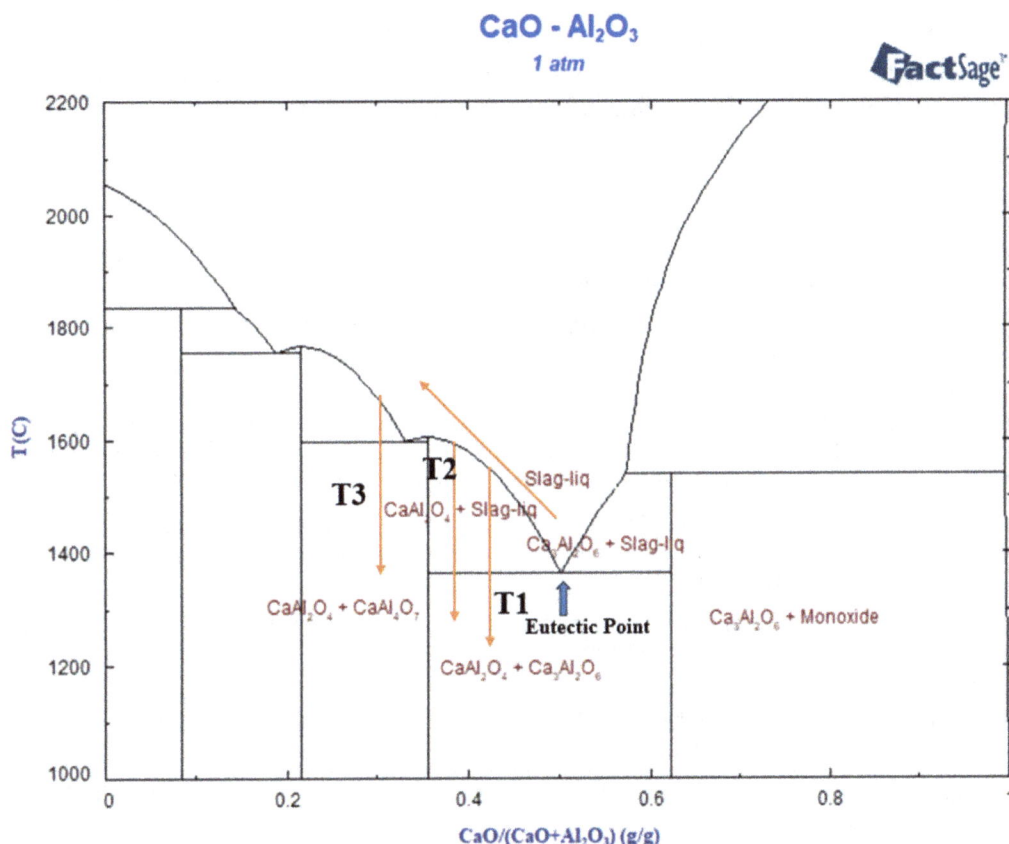

$$SiO_2 + MnO = MnSiO_3 \qquad (14)$$

$$SiO_2 + CaO.Al_2O_3 = CaO.Al_2O_3.SiO_2 \qquad (15)$$

$$MnO\ (s) = MnO \qquad (16)$$

FIG 11 – CaO-Al$_2$O$_3$ phase diagram, showing possible phases formed during T1, T2 and T3.

At the beginning of the interaction two distinct humps were observed, which slowly reduces to single molten pool at around 1525°C, marking the end of the reaction. Throughout the reaction the concentration of alumina increases and follows the path marked by arrow on the CaO-Al$_2$O$_3$ phase diagram Figure 11. Upon cooling the phases formed during T1, T2 and T3 is marked on the phase diagram plot with arrow. The melting point increases with increase in Al$_2$O$_3$ content, and the melting point from the phase diagram seems higher than the original melting point during experiment, this is mainly because of impurity such as silica, which is present in the pre-reduced ore. The marked phases are like that observed using SEM analysis discussed earlier.

CONCLUSION

The interaction of lime particles with pre-reduced manganese ore particles on alumina substrate was studied using sessile drop approach. The main conclusions are summarised as follows:

- It was found that there is a significant drop in the melting point of lime (theoretically 2572°C) due to the presence of impurities and adjacent alumina substrate to temperature of 1380°C.

- It is proposed that CaO and Al$_2$O$_3$ interact at their contact and the two solids yield a molten interfacial phase that facilitates the rapid melting of CaO particle.

- The molten slag and solid Mn ore particle getting united rapidly in the moment the molten slag touches the ore particle. Melting is continued via the dissolution of MnO in the ore into the adjacent molten slag.

- The melting or the dissolution rate of Mn ore into the slag is depending on the applied temperature, not complete melting at around 1410°C, while complete melting at around 1525°C occurred within 2 mins.

- The separation of molten metal (iron) drops from the pre-reduced ore or slag during the smelting process is depending on the rate and extent of the dissolution of MnO solid particles to the bulk slag, when the solid oxide disappears, the metal is completely separated.

ACKNOWLEDGEMENTS

The authors would like to thank NTNU for providing the resources and supporting the research work. The authors would also thank all the supporting staff at NTNU. Special thanks to EU council for HalMan project, of which this is a part of.

REFERENCES

Azof, F I, Kolbeinsen, L and Safarian, J, 2017. The Leachability of Calcium Aluminate Phases in Slags for the Extraction of Alumina, in *Proceedings of the 35th International ICSOBA Conference*, pp 243–253.

Barner, H E and Mantell, C L, 1968. Kinetics of Hydrogen Reduction of Manganese Dioxide, *Ind Eng Chem Process Des Dev*, 7:285–294. https://doi.org/10.1021/i260026a023

Costello, A, Abbas, M, Allen, A, Ball, S, Bell, S, Bellamy, R, Friel, S, Groce, N, Johnson, A, Kett, M, Lee, M, Levy, C, Maslin, M, McCoy, D, McGuire, B, Montgomery, H, Napier, D, Pagel, C, Patel, J, De Oliveira, J A P, Redclift, N, Rees, H, Rogger, D, Scott, J, Stephenson, J, Twigg, J, Wolff, J and Patterson, C, 2009. Managing the health effects of climate change, *The Lancet*, 373:1693–1733. https://doi.org/10.1016/S0140-6736(09)60935-1

Davies, J, Paktunc, D, Ramos-Hernandez, J J, Tangstad, M, Ringdalen, E, Beukes, J P, Bessarabov, D G and Du Preez, S P, 2022. The Use of Hydrogen as a Potential Reductant in the Chromite Smelting Industry, *Minerals*, 12:534. https://doi.org/10.3390/min12050534

Haque, N and Norgate, T, 2013. Estimation of greenhouse gas emissions from ferroalloy production using life cycle assessment with particular reference to Australia, *J Clean Prod*, 39:220–230. https://doi.org/10.1016/j.jclepro.2012.08.010

Heidari, A, Niknahad, N, Iljana, M and Fabritius, T, 2021. A Review on the Kinetics of Iron Ore Reduction by Hydrogen, *Materials*, 14:7540. https://doi.org/10.3390/ma14247540

Monsen, B, Tangstad, M, Solheim, I, Syvertsen, M, Ishak, R and Midtgaard, H, 2007. Charcoal For Manganese Alloy Production, in *INFACON XI Conference*, 11:297–310.

Mora, C, Dousset, B, Caldwell, I R, Powell, F E, Geronimo, R C, Bielecki, C R, Counsell, C W W, Dietrich, B S, Johnston, E T, Louis, L V, Lucas, M P, McKenzie, M M, Shea, A G, Tseng, H, Giambelluca, T W, Leon, L R, Hawkins, E and Trauernicht, C, 2017. Global risk of deadly heat, *Nat Clim Change*, 7:501–506. https://doi.org/10.1038/nclimate3322

Office of Air Quality Planning and Standards, 1992. Background Report: AP-42 Section 12.4: Ferroalloy Production; Prepared for US Environmental Protection Agency. Available from: <https://www.epa.gov/sites/default/files/2020-11/documents/b12s04.pdf>

Olsen, S E, Monsen, B E and Lindstad, T, 1998. CO2 Emissions from the Production of Manganese and Chromium Alloys in Norway, in *Proc. 56th Electr Furn Conf*, pp 363–369.

Safarian, J, 2021. A Sustainable Process to Produce Manganese and Its Alloys through Hydrogen and Aluminothermic Reduction, *Processes*, 10:27. https://doi.org/10.3390/pr10010027

Schmidt, G A, Ruedy, R A, Miller, R L and Lacis, A A, 2010. Attribution of the present-day total greenhouse effect, *J Geophys Res*, 115. https://doi.org/10.1029/2010JD014287

Tangstad, M and Olsen, S E, 1995. The Ferromanganese process - Material and Energy Balance, in *INFACON 7 Conference*, pp 621–630.

Westfall, L A, Davourie, J, Ali, M and McGough, D, 2016. Cradle-to-gate life cycle assessment of global manganese alloy production, *Int J Life Cycle Assess*, 21:1573–1579. https://doi.org/10.1007/s11367-015-0995-3

Machine learning for predicting chemical system behaviour of CaO-MgO-SiO$_2$-Al$_2$O$_3$ steelmaking slags case study

B Laidens[1], W Bielefeldt[2] and D Souza[3]

1. Materials Engineer, RHI Magnesita, Contagem, Minas Gerais 32210-080, Brazil. Email: bruno.laidens@rhimagnesita.com
2. Metallurgical Engineer, Federal University of Rio Grande do Sul (UFRGS), Rio Grande do Sul 91501-970, Brazil. Email: wagner@ct.ufrgs.br
3. Metallurgical Engineer, RHI Magnesita, Contagem, Minas Gerais 32210-080, Brazil. Email: dickson.souza@rhimagnesita.com

ABSTRACT

The CaO-MgO-SiO$_2$-Al$_2$O$_3$ system, characterised by its intricate phases and thermodynamic properties, plays a pivotal role in steel secondary refining processes, encompassing desulfurisation, non-metallic inclusion capture, and refractory protection. Accurate predictions for diverse industrial applications, including metallurgy, ceramics, and materials science, are imperative. To address this challenge, a combination of machine learning techniques will be specifically applied to model the liquid fraction of the slag and the solid fraction of MgO. The development of an artificial intelligence (AI) system, leveraging various machine learning techniques, has gained momentum in this project. The focus of this work is on constructing an AI model, based on machine learning techniques, within the CaO-MgO-SiO$_2$-Al$_2$O$_3$ system, utilising simulation results from FactSage™, version 8.1 (by GTT Technologies). The primary objective is to train the AI model using these simulation outputs to predict the percentage of liquid fraction and MgO saturation based on chemical composition parameters. The AI model will undergo training with a comprehensive data set of simulations within the CaO-MgO-SiO$_2$-Al$_2$O$_3$ system, covering a diverse range of compositional at 1873 K. These simulations, conducted through FactSage™ 8.1 software, provide a robust foundation for AI model training, ensuring generalisability and precise predictions for the liquid fraction of the slag and the solid fraction of MgO, the solid fraction of MgO in this case is determined by the difference between the total MgO and the MgO in the liquid fraction, so it is not the objective of this study to determine which phase of MgO is in the solid state.. The predictive capabilities of this AI model hold significant implications for process optimisation, quality control, and decision-making in CaO-MgO-SiO$_2$-Al$_2$O$_3$-dependent industries. Precise estimations of the liquid fraction and MgO saturation empower researchers and engineers to enhance operational efficiency and quality. This paper explores the methodologies employed for AI model creation and training, achieved results in terms of prediction accuracy, and potential applications in the field. The development of this AI system signifies a notable advancement in utilising machine learning for better comprehension and control of complex chemical systems. Furthermore, to align the study with real-world steel production, we introduce FeO and MnO at concentrations of 2 per cent and 1 per cent at 1873 K, respectively, following the validation of model results using the CaO-MgO-SiO$_2$-Al$_2$O$_3$ system. This adjustment aims to bring the study closer to the observed reality in steel mills globally.

INTRODUCTION

In the contemporary landscape of the steel industry, the relentless pursuit of operational efficiency and sustainability has propelled the adoption of innovative technologies. In this context, artificial intelligence (AI) emerges as a transformative catalyst, facilitating significant advancements in the optimisation of complex processes (Rodriguez *et al*, 2022), With the collection of data from productive sectors and their relationships with artificial intelligence, especially in adoption, there is an expectation that artificial intelligence will be responsible for generating $15.7 trillion of the global gross domestic product by 2030 (Carvalho, 2021). This study aims to underscore the utilisation of AI trained on thermodynamic calculations' results conducted by FactSage™ 8.1 software, especially when applied in the specific context of the CaO-MgO-SiO$_2$-Al$_2$O$_3$ slag system. Precise prediction of MgO is crucial to enhance efficiency during the secondary refining of steel, such as adjustment of the sulfur level, capture of non-metallic inclusions (Bielefeldt, Vilela and Heck, 2014) and prolong the lifespan of ladles in steelmaking (Dahl, Brandberg and Sichen, 2006; Gran, 2011).

The steel industry, inherently linked to steel production, confronts considerable challenges related to the efficient consumption of inputs and waste reduction (Andrade, 2023). The manipulation of slag, composed of $CaO-MgO-SiO_2-Al_2O_3$, plays a critical role in the quality of produced steel (Zhao *et al*, 2016). However, optimising this process requires a profound understanding of the interaction among slag components, with a specific emphasis, for this study, on MgO.

In this context, this research proposes the implementation of AI developed from previously calculated data by FactSage™ 8.1 software. The main objective is to determine, through an AI model, the quantity of MgO in solid phase (indicating supersaturation) in the slag, facilitating the optimisation of the addition of sources of this fundamental oxide. This approach aims to avoid excesses or deficiencies in MgO addition, thus optimising the slag composition and, consequently, enhancing the efficiency of the steelmaking process. FactSage™ 8.1 stands out as a robust software solution with a well-established track record, having undergone a series of studies that validate its ability to interpret the $CaO-MgO-SiO_2-Al_2O_3$ system (Bielefeldt, Vilela and Heck, 2013; Bielefeldt, Vilela and Heck, 2014; Bale *et al*, 2016). The wealth of prior research and validations highlights the reliability of FactSage™ 8.1 in supporting the AI-driven analysis proposed in this study.

By precisely forecasting MgO requirements, the proposed AI contributes not only to input savings but also to the extension of ladle lifespan (Pretorius and Carlisle, 1998). Avoiding unnecessary or insufficient MgO additions not only results in economic benefits but also promotes more sustainable practices by reducing resource waste and minimising environmental impacts.

This article aims to assess the performance of an artificial intelligence (AI) model generated using data produced by FactSage™. The objective is to predict, based on the chemical composition of slag, the percentage of MgO in the solid phase and the percentage of liquid phase in the slag.

MATERIALS AND METHODOLOGY

While the $CaO-MgO-SiO_2-Al_2O_3$ system is crucial to this study, there are also other reducible oxides that may play a significant role in the solubilisation of MgO and the liquid fraction of the slag (Pretorius and Carlisle, 1998). Consequently, FeO and MnO were added into the system, with concentrations of 2 per cent and 1 per cent, respectively. These values are suggested as a starting point, and future research can further explore their impact. Typically observed in secondary metallurgy processes is the FeO content falling within the range of 1 to 3 per cent, and MnO ranging from 0.2 to 1 per cent. However, to better assess the methodology, a database was initially created solely for $CaO-MgO-SiO_2-Al_2O_3$, aiming to compare the results with existing literature. The same training methodology was employed for both systems. The system without the presence of FeO and MnO was trained solely to compare the data with the literature and verify if the calculations performed by FactSage™ align with the established literature. After training, the model's output consists of the MgO content in solid fractions and the percentage of liquid fraction in the slag. Thus, in practice, it would be possible to predict whether the slag used in secondary metallurgy is saturated with MgO, thereby preserving the refractory and with sufficient liquid content to promote reactions at the metal-slag interface.

The primary objective was to attempt to achieve results comparable to those found in the Slag Atlas (Allibert, 1995) using approximately 30 000 slags with different chemical compositions. The objective is to verify the feasibility of training a tool. Because of this, we chose to use a high sample volume. After this study, we will progress to optimise the training by varying the temperature and reducing the number of inputs to compare performance in terms of accuracy and computational resources. These slags were simulated using the Equilib mode of FactSage™. The machine learning method employed in this article is the Random Forest algorithm, which is a technique utilised for classification or regression problems. It relies on an ensemble of decision trees. Given the objective of obtaining numerical values, we are employing Random Forest Regression. The model's operation is elucidated by Lima and Amorin (2020). The step-by-step process is described in Figure 1's flow chart.

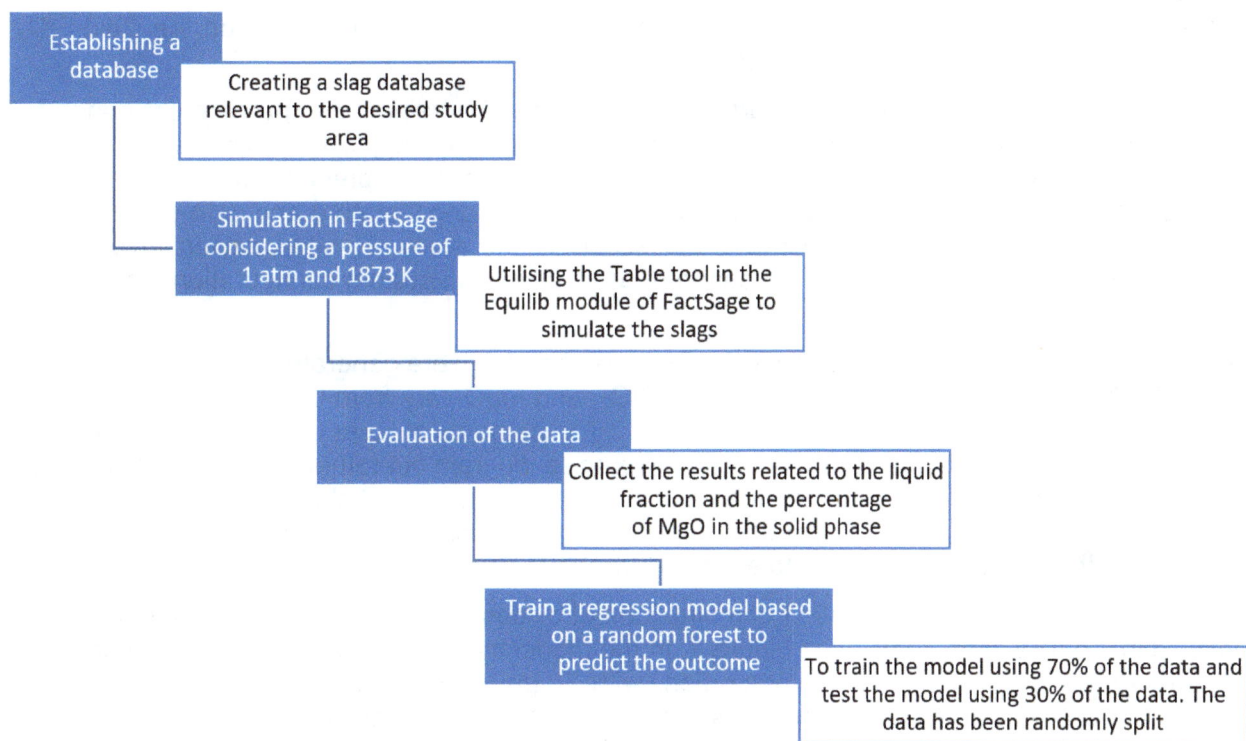

FIG 1 – Simplified flow chart of the methodology for data generation and AI model training.

This approach involved an extensive simulation process to generate a data set that could be compared with the literature data. The Equilib mode of FactSage™ was utilised to simulate the thermodynamic equilibrium of the slags under various conditions, aiming to replicate the range of compositions reported in the literature. The subsequent analysis and comparison of the simulated results with the literature findings constitute a crucial step in validating the accuracy and reliability of the simulation methodology employed in this study.

CaO-SiO$_2$-MgO-Al$_2$O$_3$ system

Understanding the intricacies of the CaO-SiO$_2$-MgO-Al$_2$O$_3$ system is of paramount importance in the context of the steel industry. Researchers such as Xu *et al* (2015), Zhao *et al* (2016), Bielefeldt, Vilela and Heck (2013), and Bielefeldt, Vilela and Heck (2014) have conducted seminal studies on this system, unravelling its critical role in various stages of the steelmaking process. Their works have contributed valuable insights into the thermodynamic and kinetic aspects of the interactions within the CaO-SiO$_2$-MgO-Al$_2$O$_3$ system, offering foundational knowledge for optimising steel production processes.

The Slag Atlas book (Allibert, 1995) provides diagrams of the CaO-MgO-SiO$_2$-Al$_2$O$_3$ system, which are commonly depicted and acknowledged in the literature. In consideration of this system outlined in the Slag Atlas, the primary objective of this study was to assess whether the methodology employed aligns with this significant body of literature. To achieve this, a dedicated AI model was constructed, utilising the same creation flow described in Figure 1, focusing solely on the oxides CaO-MgO-SiO$_2$-Al$_2$O$_3$.

Pseudo ternary diagrams were generated, highlighting the percentages of liquid fraction and MgO supersaturation. The aim was to compare these results with the diagrams presented in the Slag Atlas. Additionally, the data generated for machine learning training was also employed to generate pseudo ternary diagrams, where the information is entirely a result of calculations performed by the FactSage™ 8.1 algorithm.

This approach ensures a comprehensive evaluation of the presented methodology by juxtaposing it with established literature and utilising real-world data from FactSage™. The study not only validates the alignment of the AI model with the Slag Atlas but also provides insights into potential enhancements or divergences in results, contributing to the ongoing discourse on the CaO-SiO_2-MgO-Al_2O_3 system within the steelmaking context.

In the following section, we delve into artificial intelligence (AI) trained using a data set generated from the combination of four different oxides: Al_2O_3, CaO, SiO_2, and MgO. The objective is to attempt the recreation of pseud oternary diagrams using the AI model's predictions. If the diagrams generated by the AI exhibit a strong resemblance to both existing literature and data obtained from FactSage™, we can consider expanding the scope of the study. This expansion involves the inclusion of additional oxides and the creation of alternative slag system configurations that closely mirror the reality of steel production.

For the establishment of the database, compositions of a slag were generated based on the oxides Al_2O_3, CaO, SiO_2, and MgO. The oxides CaO, SiO_2, and MgO vary from 0 to 100 in increments of 5 per cent, while the oxide Al_2O_3 varies from 0 to 30 in increments of 5 per cent. The objective in this phase of the study is not to achieve an extremely precise diagram but rather to demonstrate that the model is capable of handling intermediate values not present in the initial database. This capability allows for the construction of a pseudo ternary diagram akin to those found in the literature, showcasing the model's adaptability to a broader range of compositional inputs.

The AI model was trained using 1211 combinations of oxides, and the process of creating the AI is detailed in the flow chart depicted in Figure 1. In the model training phase, these slag compositions were randomly divided in a 70–30 proportion, with 70 per cent of the data employed for model training and the remaining 30 per cent utilised for testing the model's accuracy. The implementation of the AI model employed the Python library sklearn, utilising the DecisionTreeRegressor algorithm. The diagrams generated from the model are presented using the plotly.figure_factory library in Python.

The results of the simulations and the comparison of the generated diagrams with the literature will be described in Figures 2–9.

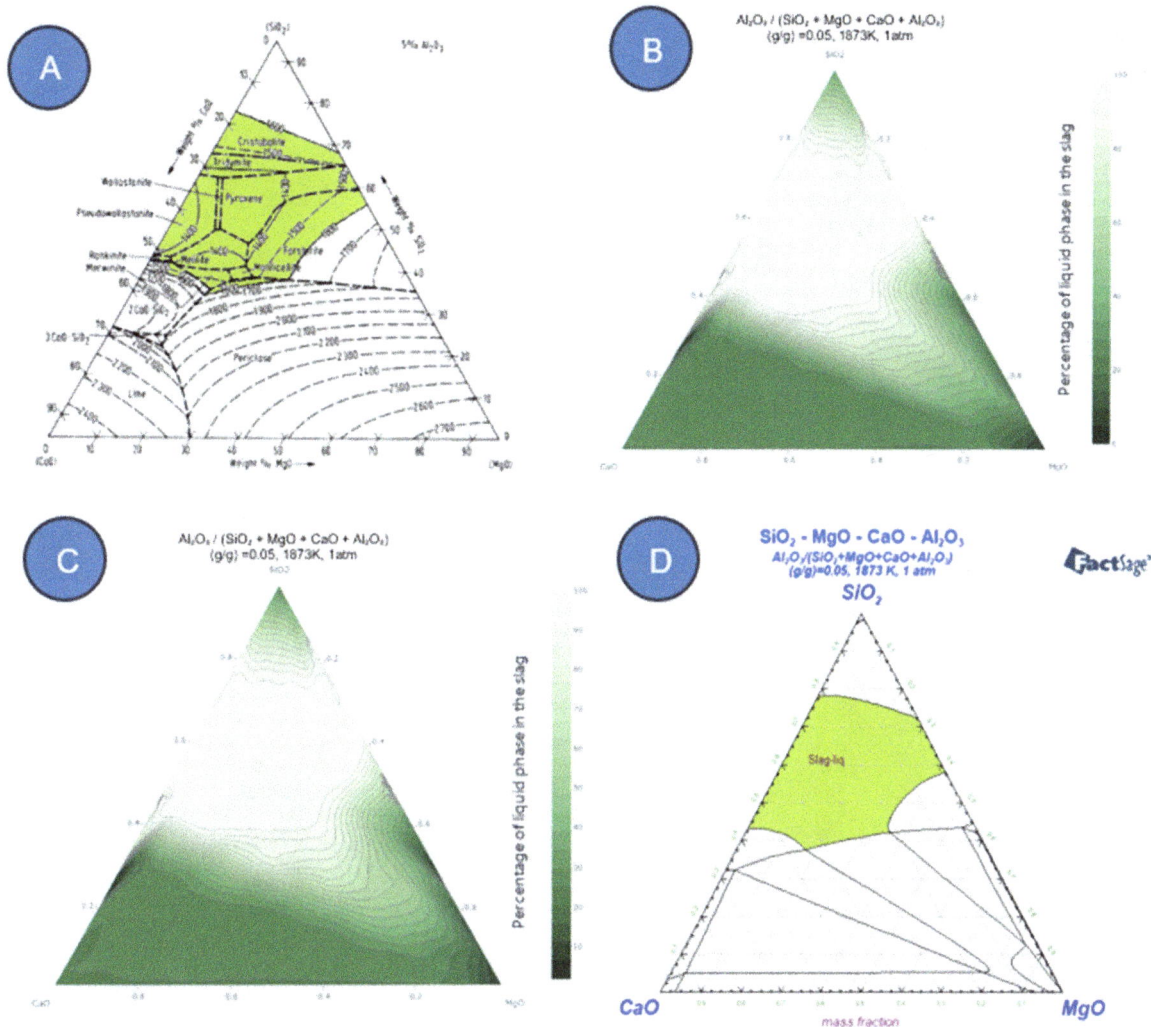

FIG 2 – (a) Liquidus surface: 100 per cent Liquid in green (Osborn *et al*, 1954); (b) Liquidus surface calculated by FactSage™; (c) Liquidus surface calculated by AI; (d) Liquidus surface calculated and built using the Phase Diagram mode in FactSage™. The diagrams are made in the CaO-MgO-SiO$_2$-Al$_2$O$_3$ system with 5 per cent Al$_2$O$_3$ by mass with temperature at 1873 K and pressure at 1 atm.

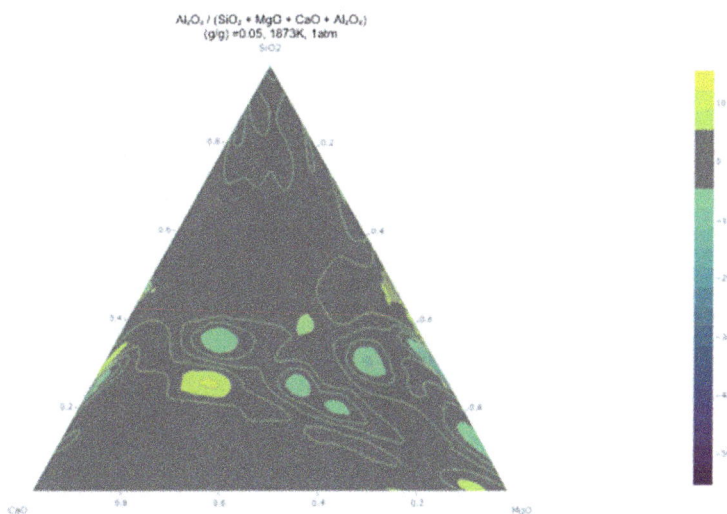

FIG 3 – Difference in the liquid phase (%) values between those calculated by FactSage™ and those predicted by the AI for a slag with 5 per cent Al$_2$O$_3$ at a temperature of 1873 K. The grey-shaded region represents areas where the difference between the values calculated by FactSage™ and those predicted by the AI falls between -5 and +5.

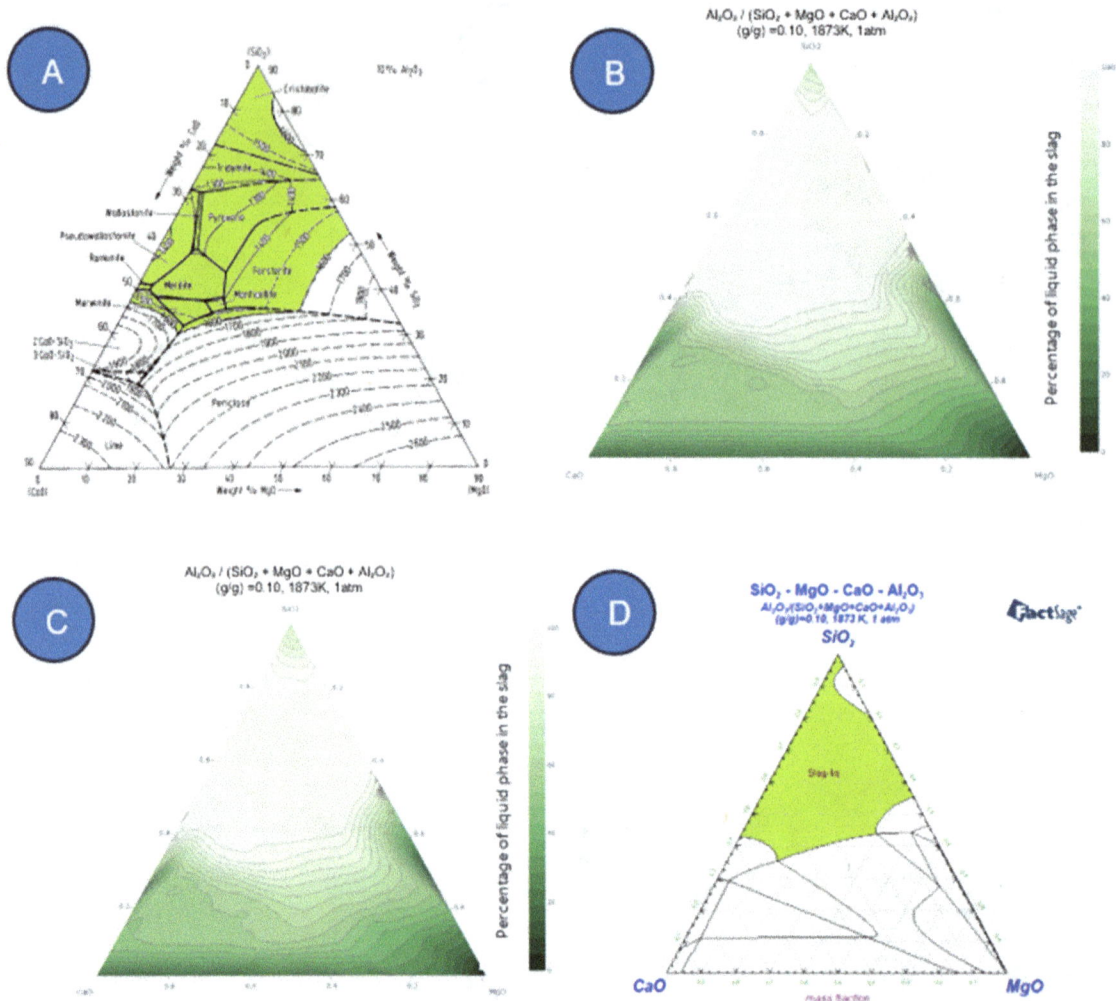

FIG 4 – (a) Liquidus surface: 100 per cent Liquid in green (Osborn *et al*, 1954); (b) Liquidus surface calculated by FactSage™; (c) Liquidus surface calculated by AI; (d) Liquidus surface calculated and built using the Phase Diagram mode in FactSage™. The diagrams are made in the CaO-MgO-SiO₂-Al₂O₃ system with 10 per cent Al₂O₃ by mass with temperature at 1873 K and pressure at 1 atm.

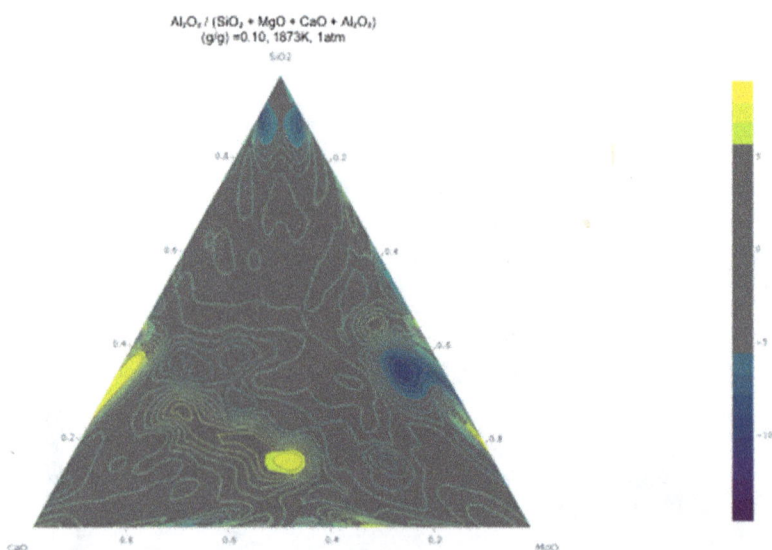

FIG 5 – Difference in the liquid phase (%) values between those calculated by FactSage™ and those predicted by the AI for a slag with 10 per cent Al₂O₃ at a temperature of 1873 K. The grey-shaded region represents areas where the difference between the values calculated by FactSage™ and those predicted by the AI falls between -5 and +5.

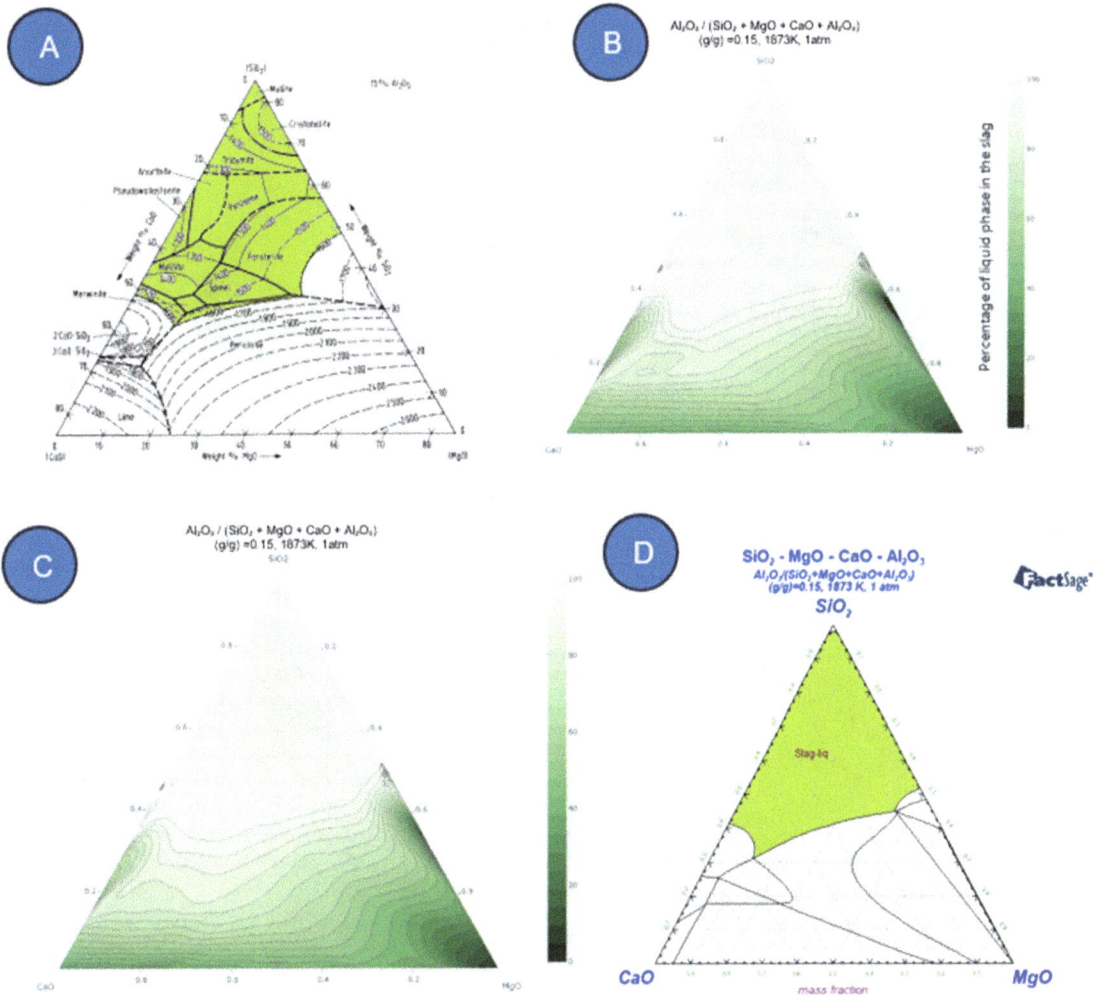

FIG 6 – (a) Liquidus surface: 100 per cent Liquid in green (Osborn *et al*, 1954); (b) Liquidus surface calculated by FactSage™; (c) Liquidus surface calculated by AI; (d) Liquidus surface calculated and built using the Phase Diagram mode in FactSage™. The diagrams are made in the CaO-MgO-SiO₂-Al₂O₃ system with 15 per cent Al₂O₃ by mass with temperature at 1873 K and pressure at 1 atm.

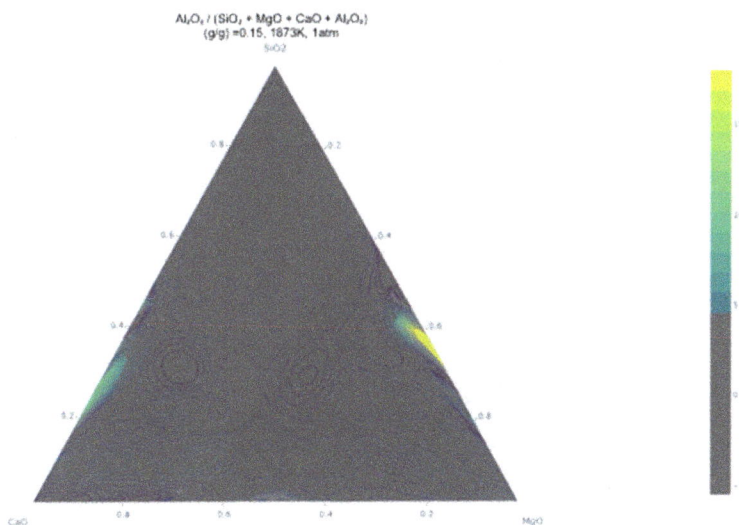

FIG 7 – Difference in the liquid phase (%) values between those calculated by FactSage™ and those predicted by the AI for a slag with 15 per cent Al₂O₃ at a temperature of 1873 K. The grey-shaded region represents areas where the difference between the values calculated by FactSage™ and those predicted by the AI falls between -5 and +5.

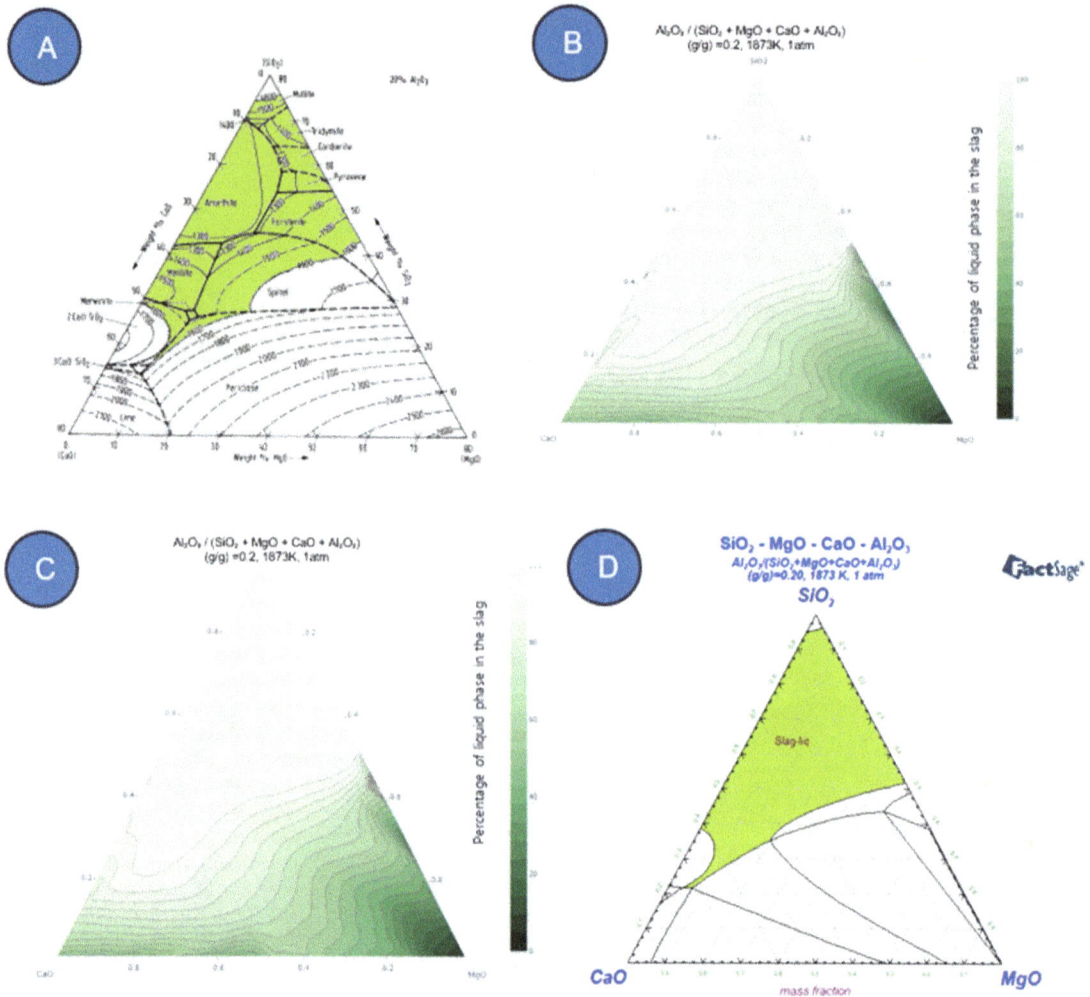

FIG 8 – (a) Liquidus surface: 100 per cent Liquid in green (Osborn *et al*, 1954); (b) Liquidus surface calculated by FactSage™; (c) Liquidus surface calculated by AI; (d) Liquidus surface calculated and built using the Phase Diagram mode in FactSage™. The diagrams are made in the CaO-MgO-SiO₂-Al₂O₃ system with 20 per cent Al₂O₃ by mas with temperature at 1873 K and pressure at 1 atm.

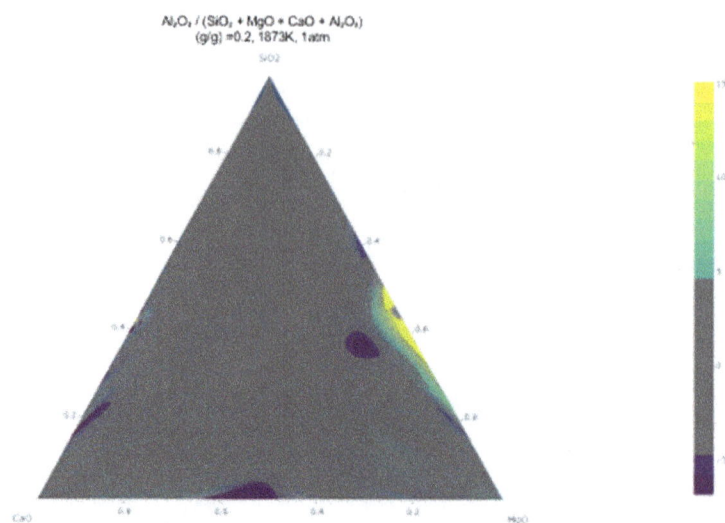

FIG 9 – Difference in the liquid phase (%) values between those calculated by FactSage™ and those predicted by the AI for a slag with 20 per cent Al₂O₃ at a temperature of 1873 K. The grey-shaded region represents areas where the difference between the values calculated by FactSage™ and those predicted by the AI falls between -5 and +5.

Through the analysis of the preceding figures, it can be concluded that, despite lacking precision in the 100 per cent liquid phase contour region, the silhouette of the graph generated by AI closely resembles that calculated by FactSage™ and depicted in the Slag Atlas diagram (Osborn *et al*, 1954). One possible factor contributing to this discrepancy is the database variation in five unit increments during slag generation. In an effort to enhance accuracy, the forthcoming section of this article, delving into the core theme of slag containing 2 per cent FeO and 1 per cent MnO, will adopt a revised approach. The variation will be reduced from five to one, with a specific focus on a chemical composition region based on the prevalent slags found in the steel industry, aiming to refine the model's precision.

CaO-MgO-SiO$_2$-Al$_2$O$_3$ system with 2 per cent FeO and 1 per cent MnO

In the literature, as previously described, there is a common occurrence of studies addressing the CaO-MgO-SiO$_2$-Al$_2$O$_3$ system. However, for the future development of a slag optimisation tool to be employed in real-time refining processes, it is imperative that other oxides present in the secondary refining slag are included in the system. Given that the training methodology of the AI has been validated in the previous section, we will now explore the system with the inclusion of 2 per cent FeO and 1 per cent MnO (Table 1). This system has been chosen for this article to showcase the accuracy of the simulation. However, for future endeavours, it is crucial to train the algorithm using various percentages of FeO and MnO to make the algorithm more comprehensive. It is highly relevant to consider other components such as Na$_2$O, K$_2$O, CaF$_2$, which should be included in future works.

TABLE 1

Variation of oxides in the database used for AI training. The variation steps for each oxide are set at 1. Temperature at 1873 K and pressure at 1 atm.

	Al$_2$O$_3$	CaO	FeO	MgO	MnO	SiO$_2$
Maximum	50	87	2	30	1	50
Minimum	0	20	2	0	1	10

More than 20 000 slag compositions were generated under these conditions. Using this, the artificial intelligence model has two objectives: determining the percentage of liquid fraction and the percentage of MgO in the solid phase of the slag. After performing the same procedure as in Figure 1, we evaluate the model's performance by comparing the predicted value with the calculated value from FactSage™ in Figures 10 and 11. It is important to note that we are using the 30 per cent of data that the model did not use for training.

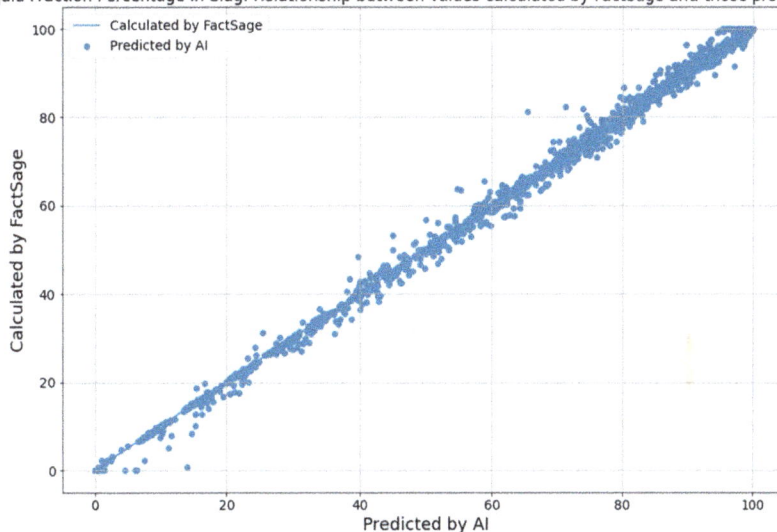

Liquid Fraction Percentage in Slag: Relationship between values calculated by FactSage and those predicted by AI

FIG 10 – Liquid fraction percentage in slag: relationship between values calculated by FactSage™ and those predicted by AI at 1873 K and 1 atm.

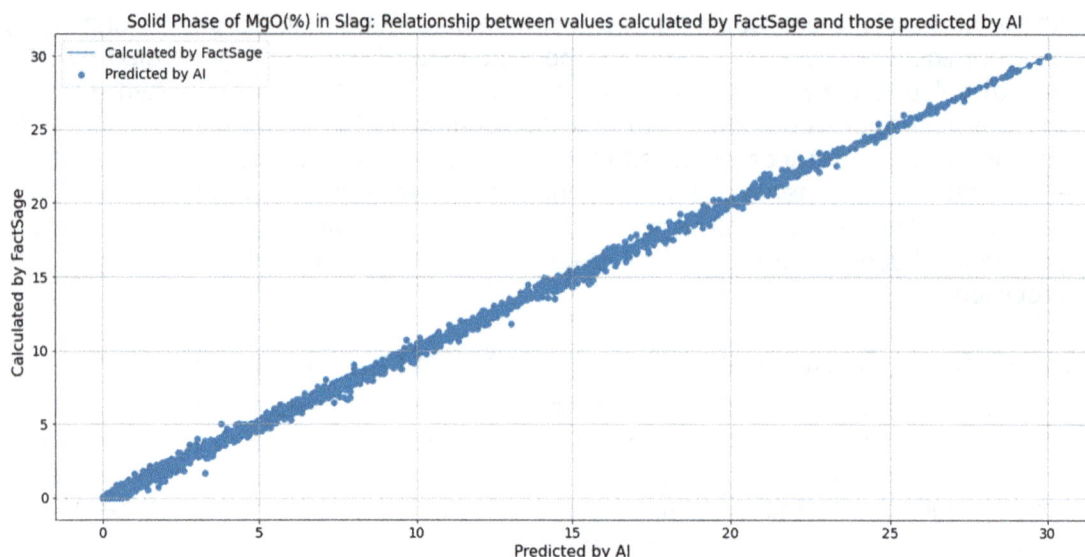

Solid Phase of MgO(%) in Slag: Relationship between values calculated by FactSage and those predicted by AI

FIG 11 – Solid phase of MgO(%) in slag: relationship between values calculated by FactSage™ and those predicted by AI at 1873 K and 1 atm.

In order to compare the results, two histograms were generated, depicting the differences between the predicted values by the model and the values calculated by FactSage™. One histogram pertains to the prediction of the liquid fraction, while the other focuses on the prediction of the percentage of MgO in the solid phase (Figures 12 and 13).

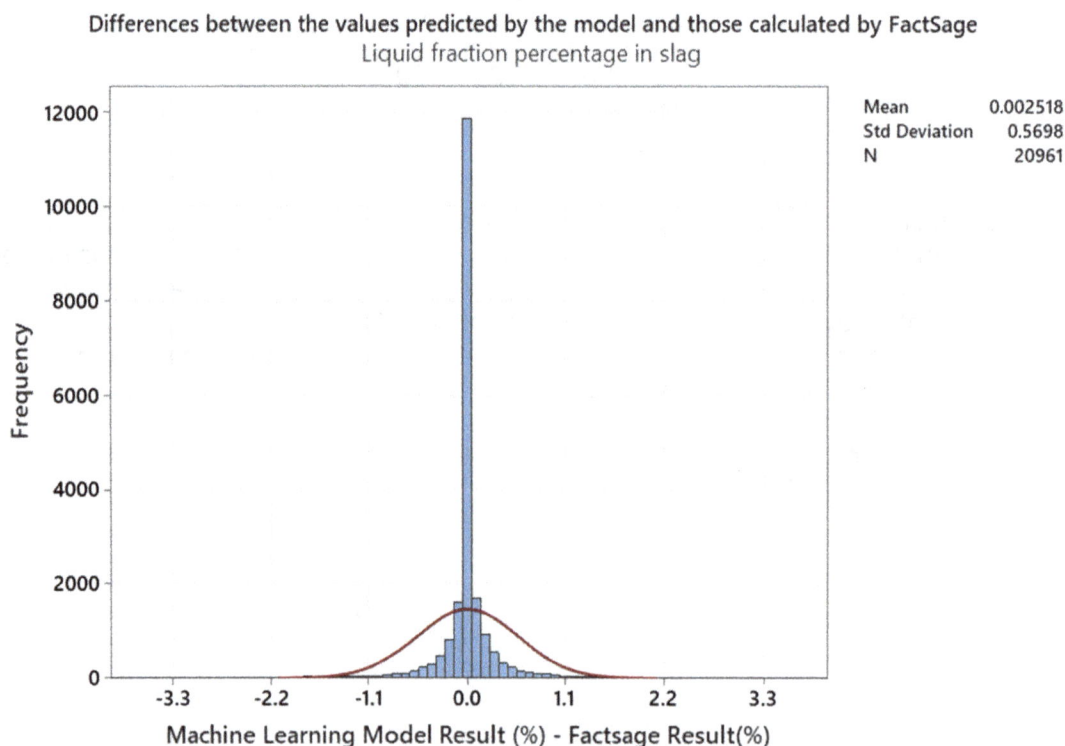

Differences between the values predicted by the model and those calculated by FactSage
Liquid fraction percentage in slag

Mean	0.002518
Std Deviation	0.5698
N	20961

FIG 12 – Liquid Fraction percentage in slag – difference (%) between the value calculated by FactSage™ and the AI model result at 1873 K and 1 atm.

Differences between the values predicted by the model and those calculated by FactSage

Solid phase of MgO in slag

Mean	-0.0006212
Std Deviation	0.1067
N	20961

Machine learning model result (%) - FactSage Result (%)

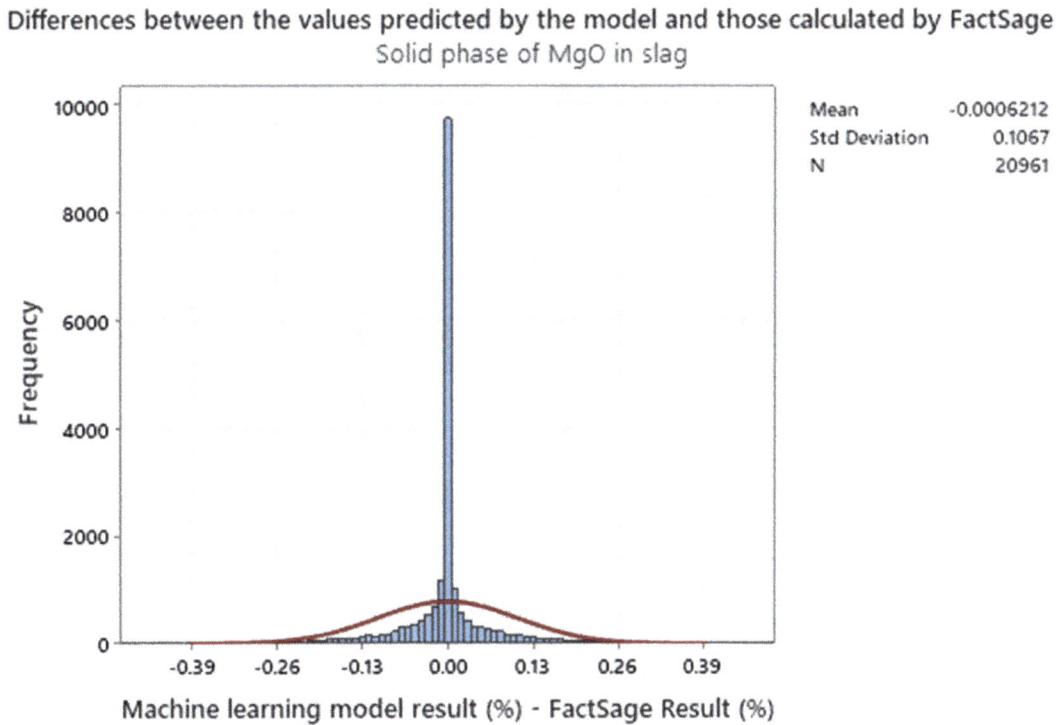

FIG 13 – Solid phase of MgO in slag – difference between the value calculated by FactSage™ and the AI model result at 1873 K and 1 atm.

Looking at the histogram depicting the differences in responses between FactSage™ and the model, it is evident that, for the liquid fraction, we observe an average difference very close to zero, with a standard deviation smaller than 1 per cent. In extreme cases, the error hardly exceeds +2 per cent or -2 per cent. Similarly, the AI model, when predicting MgO in the solid phase, also exhibits low error, with the vast majority of data falling between -0.4 per cent and 0.4 per cent. The implementation of a model with such accuracy seems to be sufficiently reliable to initiate an industrial-scale test.

It's important to note that the red line of the theoretical distribution is not aligned with the data from the histogram. This could indicate an error in the standard deviation generated by a very high number of values close to zero. Therefore, using the data generated by the model, we compare its results with those obtained by FactSage™, which will be displayed in the ternary diagrams described in these final figures.

Subsequently, the generated algorithm was employed to recalculate a diagram illustrating the evolution of the solid fraction of MgO in the slag and the percentage of liquid slag. In this approach, MgO was held constant at values of 5, 10, 15, and 20, while the oxides CaO, MgO, and SiO_2 were varied at the vertices of the ternary diagram. It is important to note that the system now incorporates 2 per cent of FeO and 1 per cent of MnO. The values of CaO, MgO, and SiO_2 needed to be normalised between 0 and 100 to enable the generation of the diagram. The outcomes concerning the liquid percentage in the slag will be elucidated in the diagrams presented in Figures 14–17. The grey regions represent areas outside the training range of the model.

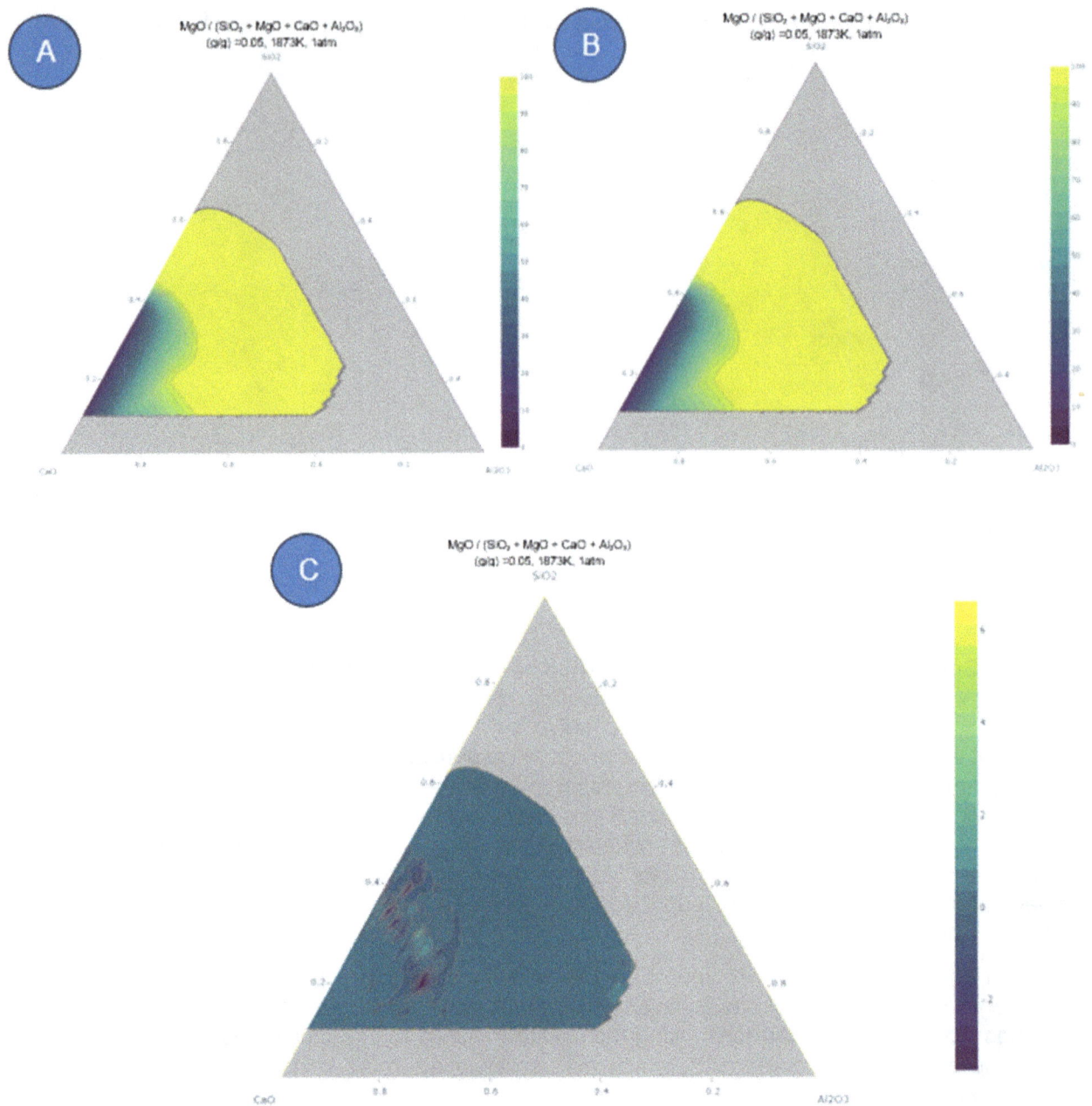

FIG 14 – (a) Diagram depicting the percentage of liquid phase of the slag calculated by FactSage™ at 1873 K and 1 atm with 5 per cent MgO; (b) Diagram illustrating the percentage of liquid phase of the slag predicted by the AI model at 1873 K and 1 atm with 5 per cent MgO; (c) Diagram showcasing the difference between the results calculated by the two methods at 1873 K and 1 atm with 5 per cent MgO.

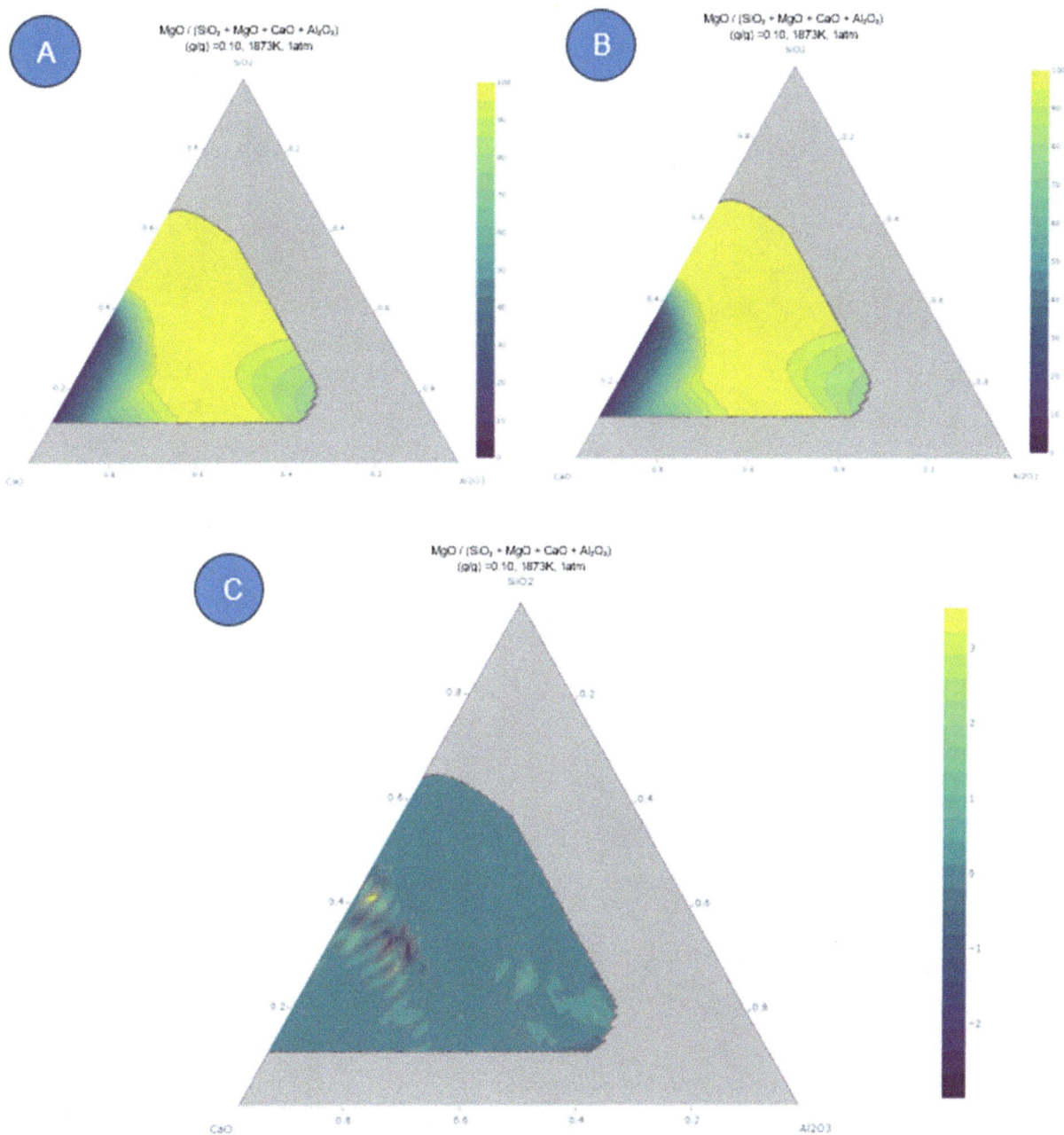

FIG 15 – (a) Diagram depicting the percentage of liquid phase of the slag calculated by FactSage™ at 1873 K and 1 atm with 10 per cent MgO; (b) Diagram illustrating the percentage of liquid phase of the slag predicted by the AI model at 1873 K and 1 atm with 10 per cent MgO; (c) Diagram showcasing the difference between the results calculated by the two methods at 1873 K and 1 atm with 10 per cent MgO.

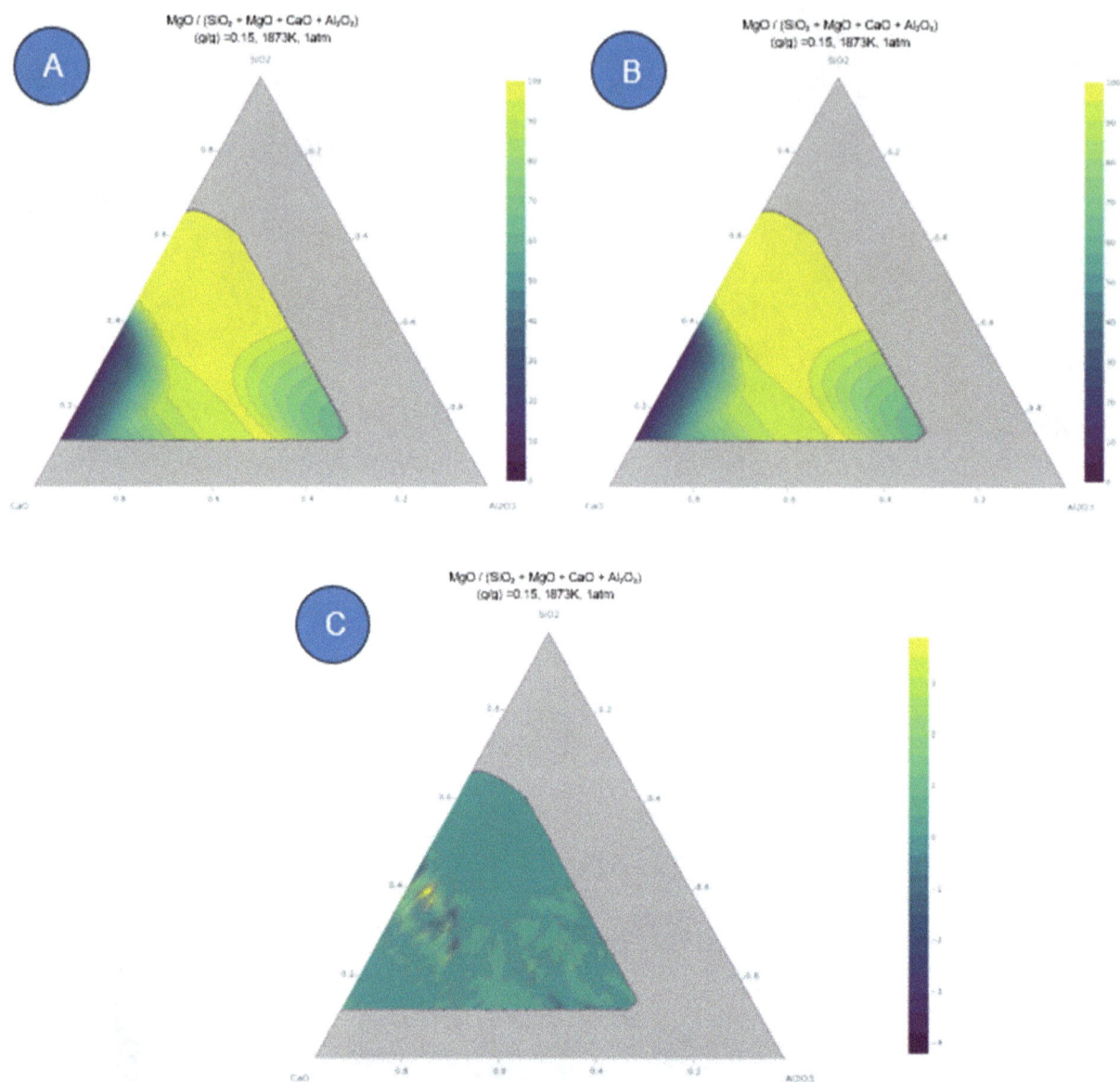

FIG 16 – (a) Diagram depicting the percentage of liquid phase of the slag calculated by FactSage™ at 1873 K and 1 atm with 15 per cent MgO; (b) Diagram illustrating the percentage of liquid phase of the slag predicted by the AI model at 1873 K and 1 atm with 15 per cent MgO; (c) Diagram showcasing the difference between the results calculated by the two methods at 1873 K and 1 atm with 15 per cent MgO.

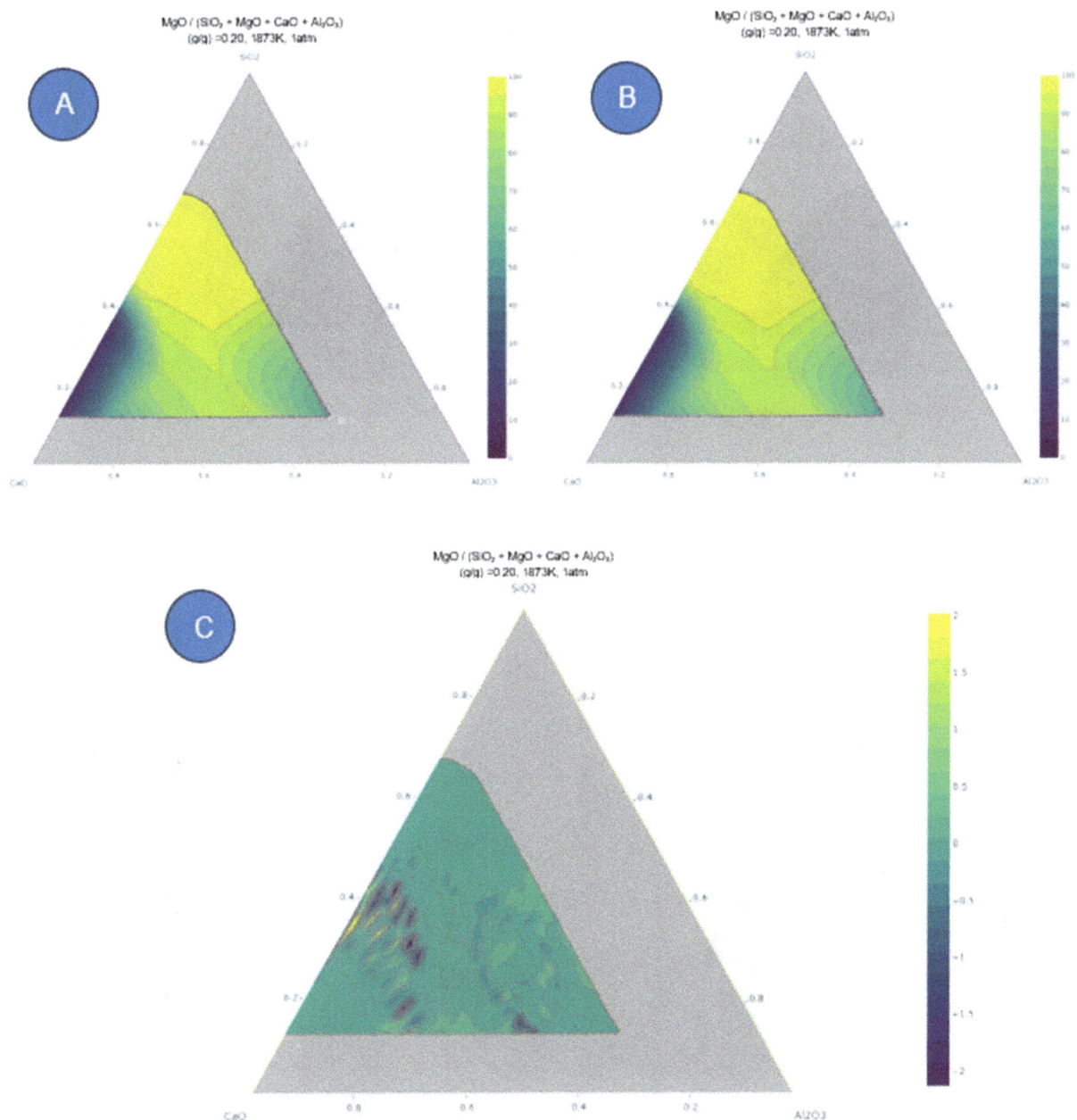

FIG 17 – (a) Diagram depicting the percentage of liquid phase of the slag calculated by FactSage™ at 1873 K and 1 atm with 20 per cent MgO; (b) Diagram illustrating the percentage of liquid phase of the slag predicted by the AI model at 1873 K and 1 atm with 20 per cent MgO; (c) Diagram showcasing the difference between the results calculated by the two methods at 1873 K and 1 atm with 20 per cent MgO.

Analysing the results concerning the liquid fraction in the slag, we can conclude that the AI model produces results very close to those generated by FactSage™. An important factor to note is that for slags with lower CaO content and higher Al_2O_3 levels (> 35 per cent), there are regions where solid precipitates are formed. It is observed that the liquid fraction content begins to decrease, possibly due to the formation of crystalline networks of Al_2O_3-rich oxides. The amount of solid formed in these regions is dependent on the MgO content in the slag; for lower MgO contents, the formation of solid fractions is less than for higher MgO contents.

Next, in Figures 18–21, we will present the data related to the MgO content in the solid fraction of the slag in a similar manner. Once again, we emphasise that the grey regions are not part of the study's sample space.

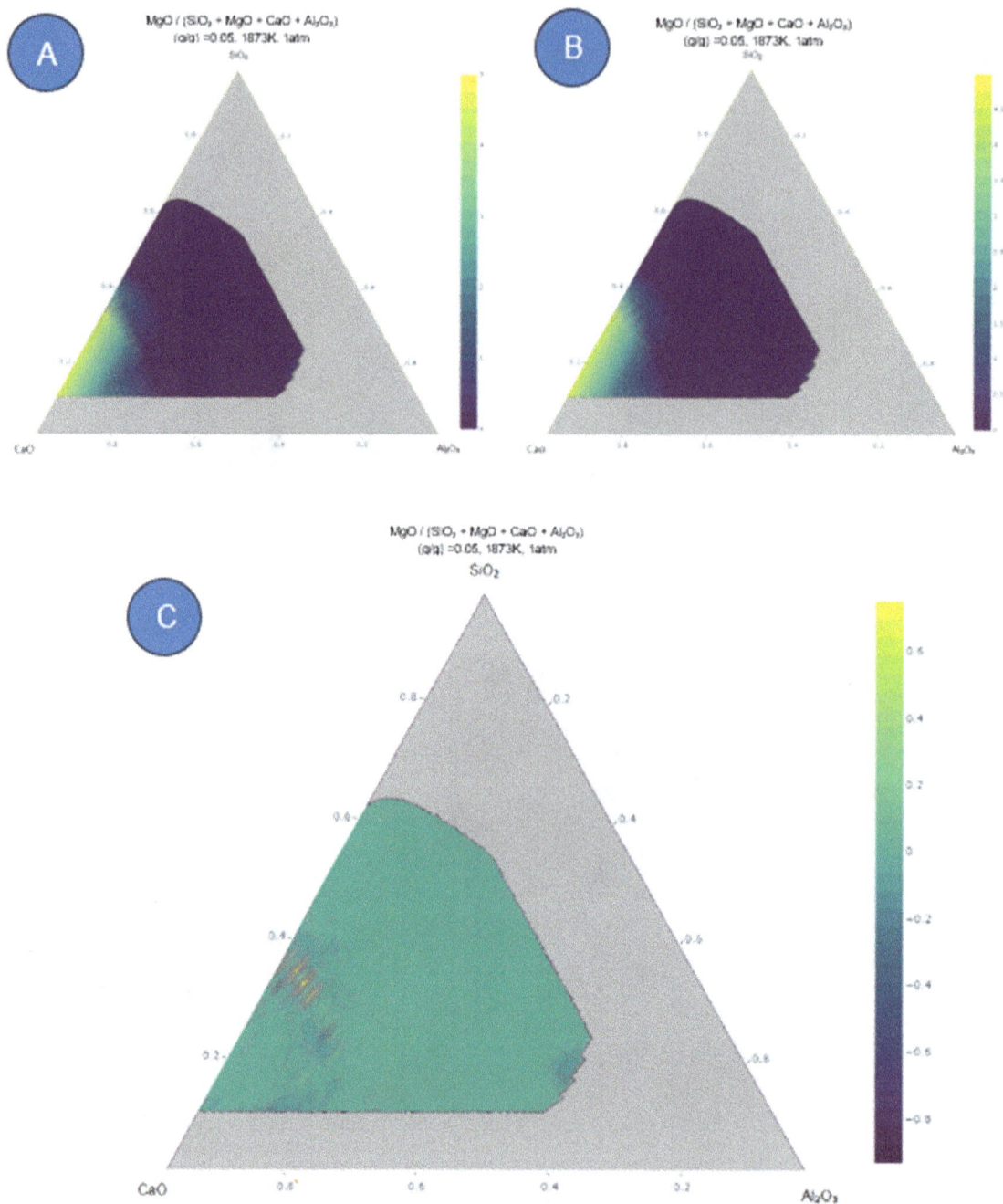

FIG 18 – (a) Diagram depicting the percentage of MgO in the solid fraction of the slag calculated by FactSage™ at 1873 K and 1 atm with 5 per cent MgO; (b) Diagram illustrating the percentage of MgO in the solid fraction of the slag predicted by the AI model at 1873 K and 1 atm with 5 per cent MgO; (c) Diagram showcasing the difference between the results calculated by the two methods at 1873 K and 1 atm with 5 per cent MgO.

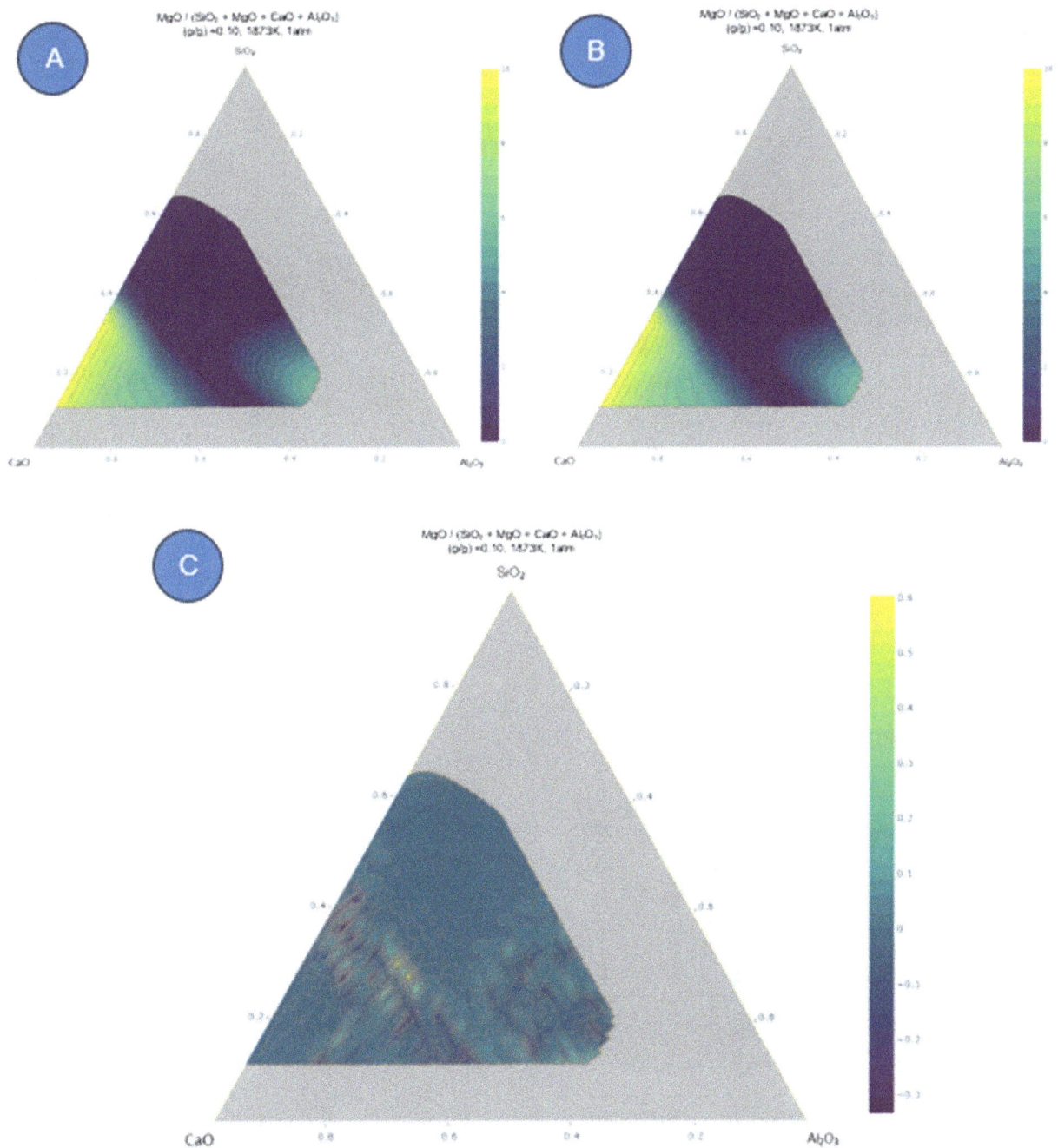

FIG 19 – (a) Diagram depicting the percentage of MgO in the solid fraction of the slag calculated by FactSage™ at 1873 K and 1 atm with 10 per cent MgO; (b) Diagram illustrating the percentage of MgO in the solid fraction of the slag predicted by the AI model at 1873 K and 1 atm with 10 per cent MgO; (c) Diagram showcasing the difference between the results calculated by the two methods at 1873 K and 1 atm with 10 per cent MgO.

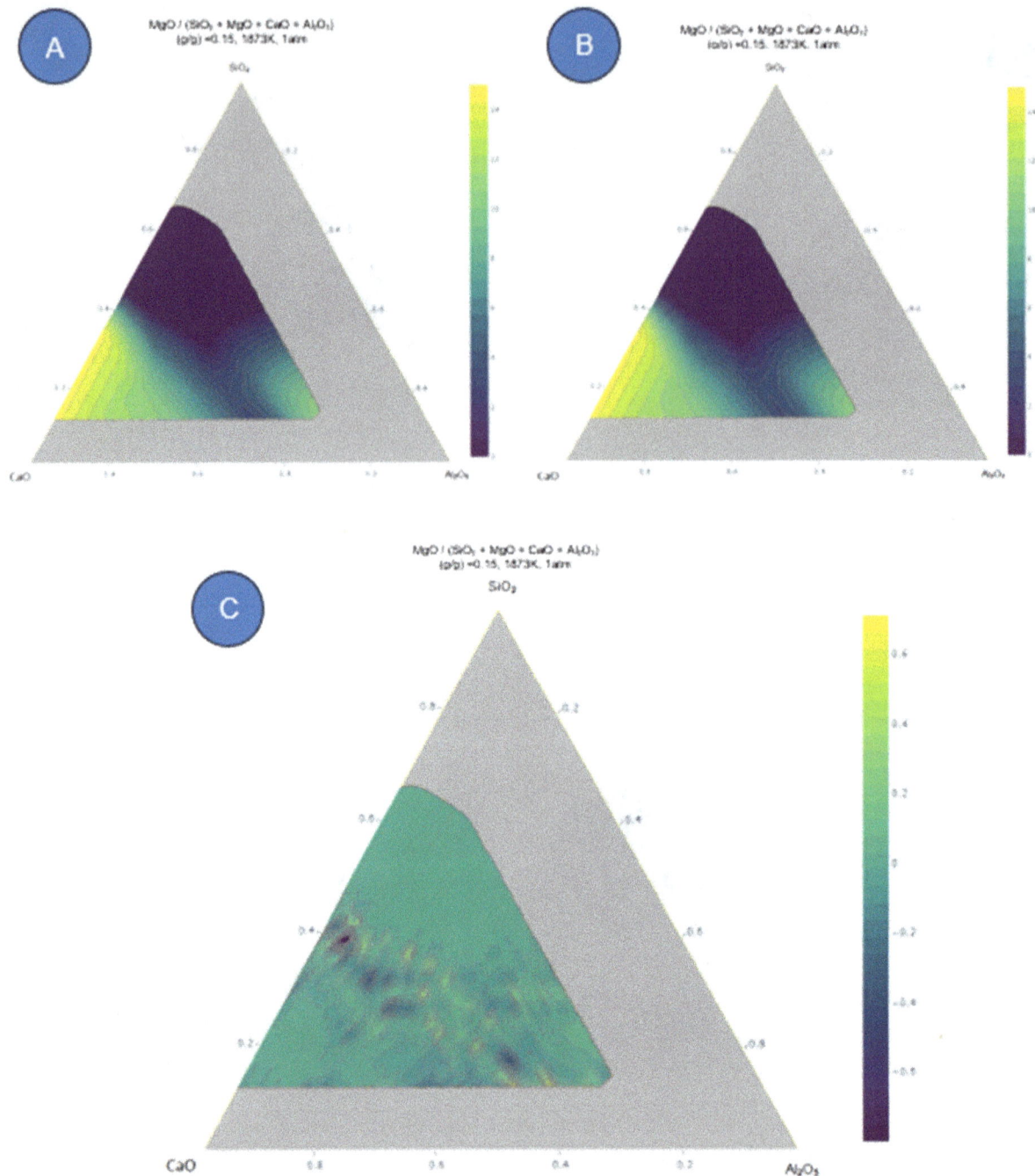

FIG 20 – (a) Diagram depicting the percentage of MgO in the solid fraction of the slag calculated by FactSage™ at 1873 K and 1 atm with 15 per cent MgO; (b) Diagram illustrating the percentage of MgO in the solid fraction of the slag predicted by the AI model at 1873 K and 1 atm with 15 per cent MgO; (c) Diagram showcasing the difference between the results calculated by the two methods at 1873 K and 1 atm with 15 per cent MgO.

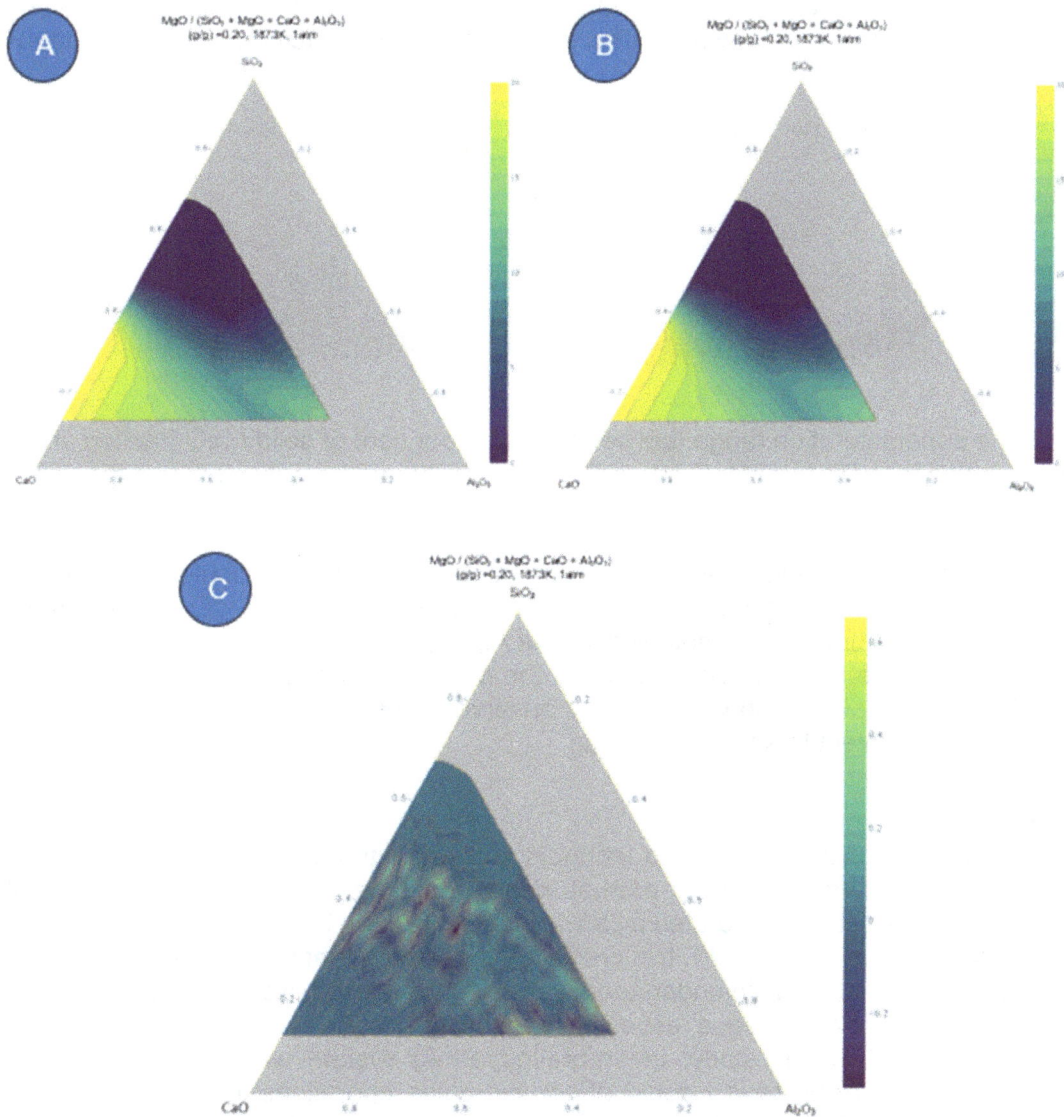

FIG 21 – (a) Diagram depicting the percentage of MgO in the solid fraction of the slag calculated by FactSage™ at 1873 K and 1 atm with 20 per cent MgO; (b) Diagram illustrating the percentage of MgO in the solid fraction of the slag predicted by the AI model at 1873 K and 1 atm with 20 per cent MgO; (c) Diagram showcasing the difference between the results calculated by the two methods at 1873 K and 1 atm with 20 per cent MgO.

Analysing the diagrams depicting the MgO content in the solid fraction as a function of CaO, Al_2O_3, and SiO_2, it is evident that the diagrams align with those in Figures 14–17. It is apparent that MgO plays a crucial role in reducing the liquid fraction, along with the addition of CaO and Al_2O_3. For higher total MgO contents, the amount of MgO increases in the solid phase and the liquid reaches a saturation value. The fraction of solids in the slag increases and that of liquid decreases. For instance, at 5 per cent MgO (Figures 14 and 18), there is no observed reduction in the liquid fraction in slags with elevated Al_2O_3 content. However, as the total MgO content in the slag increases, both the liquid fraction reduces, and the amount of MgO in the solid fraction increases, particularly with the addition of CaO and/or Al_2O_3 to the slag. In conclusion, MgO proves to be a crucial factor in the solid fraction content of the slag, especially when the slag requires higher additions of Al_2O_3 or CaO.

In order to gain a deeper understanding of the behaviour of Al_2O_3, CaO, and SiO_2 in MgO saturation, an alternative perspective was generated, as depicted in Figure 22. Only data calculated by FactSage™, where the MgO content in the solid fraction of the slag ranged between 1 and 2 per cent, was selected for this analysis. The graph considers the total MgO content in the slag (Y-axis) and the binary basicity (B_2) on the X-axis. To elucidate the influence of Al_2O_3, a colour gradient corresponding to the Al_2O_3 content was employed. The MgO supersaturation level ranging from 1 to

2 per cent is designed to have minimal impact on the liquid fraction of the slag (maintaining it close to 100 per cent), while concurrently mitigating the slag's propensity to corrode the refractory lining of the ladle due to its saturation in MgO.

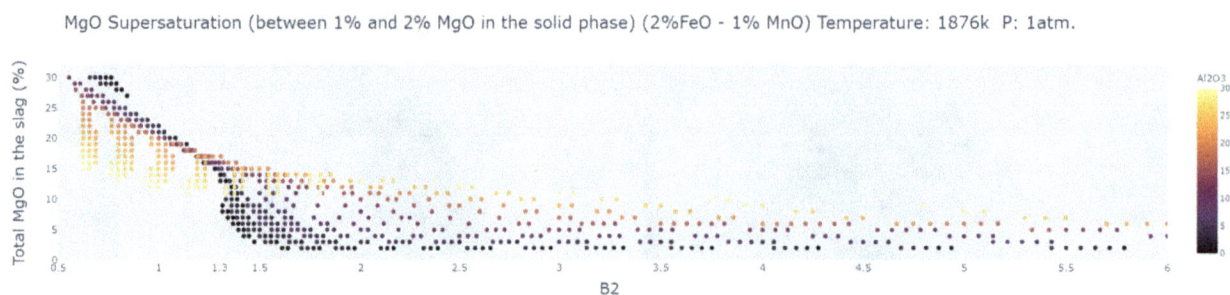

MgO Supersaturation (between 1% and 2% MgO in the solid phase) (2%FeO - 1% MnO) Temperature: 1876k P: 1atm.

FIG 22 – The supersaturation range (between 1 and 2 per cent of solid MgO fraction at 1873 K) is examined in relation to the total MgO content in the slag, binary basicity (B$_2$), and the concentration of Al$_2$O$_3$.

Analysing Figure 22, it becomes evident that both basicity and the Al$_2$O$_3$ content play crucial roles in MgO saturation. It is observed that for lower basicities, a higher total MgO content in the slag is required to achieve saturation. It is noteworthy that Al$_2$O$_3$ plays a significant role for basicities below 1.3, where an increase in its content promotes the precipitation of solid MgO fractions. However, beyond a B$_2$ value of 1.3, an opposing effect is observed, wherein elevated levels of Al$_2$O$_3$ demand a higher total MgO content to saturate the slag.

CONCLUSIONS

The transformative impact of AI-focused technology on industrial processes is substantial, leading to heightened efficiency and providing prompt, nearly real-time responses. This article delves into the comparison of predictions generated by an AI model based on FactSage™ results with the actual outcomes obtained from FactSage™. It is crucial to underscore that FactSage™ remains pivotal for a comprehensive thermodynamic understanding of slag processes. Nonetheless, the integration of AI introduces significant advantages, especially in refining specific process parameters. In practical terms, as exemplified in this study, incorporating an AI system within a level 2 metallurgical framework in a steelmaking facility could effectively guide operations in fine-tuning the chemical composition of the slag. This optimisation ensures the efficient utilisation of inputs, minimising waste and ensuring the availability of slag for anticipated metallurgical processes.

Moreover, the impact of Al$_2$O$_3$, CaO, SiO$_2$, and MgO contents on both the liquid fraction and the solid-phase MgO content is discernible. Comparing these results with widely accepted diagrams from the literature reveals a remarkable similarity, highlighting the AI's potential to accurately predict and replicate the behaviour of slag compositions. This alignment with established theoretical frameworks in the literature further underscores the robustness of AI models.

In comparing the initial model, which solely addressed the CaO-MgO-SiO$_2$-Al$_2$O$_3$ system with limited training data, to the model incorporating FeO and MnO with a more extensive training data set, we observe a reduction in serrations in the contour regions of the graphs. This indicates that as the AI model is supplied with more data, its accuracy tends to improve. Consequently, there arises the possibility of developing specialised models for each region in the ternary phase diagram, continually augmenting data for a more confined space, thereby enhancing accuracy for specific purposes.

The aim of this work was to make a prediction of the liquid fraction and determine whether there is MgO saturation or not. A suggestion for future work in this study would be to understand how the model interprets the formation of solid precipitates by comparing it with laboratory tests and utilising thermodynamic tools.

ACKNOWLEDGEMENTS

I would like to extend my sincere gratitude, firstly to the approval team at AusIMM, for believing in and approving this article for participation in Molten 2024. I also want to express appreciation to RHI Magnesita for their steadfast support in fostering the development of new technologies aimed at assisting their clients' processes. One notable contribution is the Ladle Slag Model, a tool designed to optimise customer slag. Additionally, special thanks are owed to the UFRGS Ironmaking and Steel Laboratory (LaSid), and specifically to Professor Bielefeldt, for his invaluable guidance in the writing of this article. Their collective support has been instrumental in advancing research and innovation in this field.

REFERENCES

Allibert, M, 1995. *Slag Atlas*, 2nd ed (Verlag Stahleisen GmbH: Düsseldorf).

Andrade, A, 2023. Main challenges regarding Steel Industries Operation and Maintenance, Vidya Technology. Available from: <https://vidyatec.com/blog/steel-industries-problems/>

Bale, C W, Bélisle, E, Chartrand, P, Decterov, S A, Eriksson, G, Gheribi, A E, Hack, K, Jung, I-H, Kang, Y-B, Melançon, J, Pelton, A D, Petersen, S, Robelin, C, Sangster, J, Spencer, P and Van Ende, M-A, 2016. FactSage™ thermochemical software and databases, 2010–2016, CALPHAD: Computer Coupling of Phase Diagrams and Thermochemistry, 54:35–53.

Bielefeldt, W V, Vilela, A C and Heck, N C, 2014. Termodynamic evaluation of the slag system $CaO–MgO–SiO_2–Al_2O_3$, in *Proceedings of the Iron Steel Technol Conf,* 2:1433–1445 (AISTech).

Bielefeldt, W V, Vilela, A C F and Heck, N C, 2013. Thermodynamic evaluation of the slag system $CaO-MgO-SiO_2-Al_2O_3$. in *Anais do Congresso Internacional of ABM* [Proceedings of the Annual Congress of ABM], pp 1630–1641.

Carvalho, G, 2021. Artificial Intelligence And The Perspectives Of The World Of Work, Ciência da Computação da Universidade do Sul de Santa Catarina.

Dahl, F, Brandberg, J and Sichen, D, 2006. Characterization of melting of some slags in the $Al_2O_3-CaO-MgO-SiO_2$ quaternary system, *ISIJ International*, 46(4):614–616.

Gran, J, Wang, Y and Sichen, D, 2011. Experimental determination of the liquidus basicity region in the Al_2O_3 (30 mass%)-$CaO-MgO-SiO_2$ system, *CALPHAD*, 35:249–254.

Lima, M and Amorin, F, 2020. Random Forest [in Spanish], Lamfo - UNB. Available from: <https://lamfo-unb.github.io/2020/07/08/Random-Forest/>

Osborn, E F, Devries, R C, Gee, K H and Kraner, H M, 1954. Optimum composition of blast furnace slag as deduced from liquidus data for the quaternary system CaO-MgO-Al2O3-SiO2, *JOM*, 6(1):33–45.

Pretorius, E B and Carlisle, R C, 1998. Foamy slag fundamentals and their practical application to electric furnace steelmaking, in *Proceedings of the Electric Furnace Conference*, pp 275–292.

Rodriguez, J, *et al*, 2022. Strengthening the steel industry with AI, BCG Global. Available from: <https://www.bcg.com/publications/2021/value-of-ai-in-steel-industry>

Xu, J, Su, L, Chen, D, Zhang, J and Chen, Y, 2015. Experimental investigation on the viscosity of $CaO–MgO(–Al_2O_3)–SiO_2$ slags and solid–liquid mixtures, *J Iron Steel Res Int*, 22(12):1091–1097.

Zhao, S, He, S P, Guo, Y T, Chen, G J and Lv, J C, 2016. Effect on cleanliness of molten steel with different refining slag systems for low alloy ship plate, *Ironmak Steelmak,* 43:790. https://doi.org/10.1080/03019233.2016.1223791

Perspectives of chemical metallurgy fundamentals in slag innovation

S Lee[1] and I Sohn[2]

1. Research Professor, Yonsei University, Seoul 03722, Korea. Email: slag@yonsei.ac.kr
2. Professor, Yonsei University, Seoul 03722, Korea. Email: ilsohn@yonsei.ac.kr

ABSTRACT

The ferrous and non-ferrous metallurgical industries have been utilising slags, fluxes and mattes for process optimisation and quality improvements at high temperatures. Understanding the fundamentals of the thermodynamics and kinetics of chemical metallurgy regarding slags, fluxes and mattes has been essential to the technological developments of these optimisation and improvements. This work will discuss how these fundamentals have been studied in the reactions involved and implemented in the ferrous industry from primary refining to continuous casting. In particular, the chemical driving forces that allow these reactions to take place and the limitations of the kinetics that require optimisation to enhance the reactions. In addition, the future research direction for slag technology innovation and the role of these fundamentals will also be described.

INTRODUCTION

Ferrous metals including iron and steel are widely used for structural materials imbedded in nearly all industry sectors including construction, shipbuilding, transportation and many others. Non-ferrous metals are typically added into the basic materials of the ferrous metals to improve the various physical and chemical properties.

According to the World Steel Association (World Steel Association, 2023), approximately 1.9 Bt of crude steel was produced. For non-ferrous metals, the US Geological Survey notes that world aluminium output was 69 Mt (Merrill, 2023) and publications from the Government of Canada indicates global refined copper production to be 25 Mt. Depending on the metals processing method, large amounts of slags and mattes are produced. Global slag generation from these processes have been estimated to be approximately 500 Mt within the ferrous industry (Yang, Firsbach and Sohn, 2022).

The critical role of slags and mattes in chemical metallurgy cannot be overstated in terms of their importance to the final product quality. Although the humble beginnings of the slag and matte may have started out in part as gangue components in the raw materials, the product quality, energy benefits and protection from the environment with optimised design have become essential for process engineering. For slags in ferrous metallurgy, optimised slags have the ability to refine unwanted elements in the molten metal including sulfur, phosphorus, nitrogen, hydrogen and other elements. When a slag covers the molten metal, it acts as a barrier to the outside environment minimising the infiltration of dissolved gaseous elements into the molten metal and can also act as a thermal barrier for increased insulation of the molten metal. For mattes in non-ferrous metallurgy, optimised mattes can increase the recovery of the non-ferrous element and improve the separation efficiency of the oxide slags and also increase the purity of the target element. The typical sulfide mattes used for primary copper smelting operations and the slags that are formed during smelting of the concentrates comprised of oxy-sulfide minerals require process optimisation to ensure greater recovery of copper is possible, while maintaining greater purity with less ferrous impurities in the matte. With higher concentrations of Cu can be achieved in the matte, it can be sent to the converter for sulfur removal and subsequent blister Cu production.

The fundamentals that control the role in slags and mattes are based on thermodynamic and kinetic principles, which have been widely studied in chemical metallurgy. This review looks at some of the fundamentals of ferrous and non-ferrous chemical metallurgy that have been applied to ensure greater process efficiency and product quality.

THERMODYNAMIC FUNDAMENTALS

Slag capacities in steelmaking and refining

Slag capacity in steelmaking and refining is the ability of a slag to absorb and retain various impurities and elements during high temperature metallurgical processes. Several critical factors affect slag capacity, which in turn interacts with the chemical and physical properties of the slag. These factors have been widely studied and includes the slag chemical composition, temperature, and oxygen partial pressure.

Slags can typically be comprised of SiO_2, Al_2O_3, CaO, MgO, and FeO, which can influence the thermophysical properties of the slag including the viscosity and the chemical potential of the slag including the basicity, which have been known to directly impact the absorption and retainment of impurities from molten steel. Temperatures provide the thermal energy to the system for reaction and the kinetic energy of the bondings within the ionic slag melts to form new bonds. Depending on the exothermic or endothermic reactions between the slag and metal, reactions can be hindered or accelerated. Temperatures can also affect the viscosity and fluidity of slags, where higher temperatures lower the viscosity and increase the fluidity.

Oxygen partial pressure or oxygen potential of the system influences the redox reactions occurring at the slag-metal interface, where the impurities of sulfur and phosphorus can be exchanged. Low oxygen potentials or reducing conditions can promote the removal of certain impurities such as sulfur into the slag as sulfides, while high oxygen potentials or oxidising conditions may facilitate the removal of certain impurities such as phosphorus into the slag by forming stable oxide phases as phosphates. While not typical, slag viscosities and phase equilibria can be altered under high pressures, affecting the capacity of a slag to absorb impurities.

Basicity, which refers to the chemical potential of the free oxygen anions, is difficult to directly measure, but is one of the critical parameters determining the capacity of the ionic slag to remove impurities from metals during metallurgical processing. Thus, an indirect measure of the slag basicity was warranted and the Vee ratio defined as the mass ratio of basic oxides (eg CaO, MgO) to acidic oxides (eg SiO_2, Al_2O_3) in the slag composition was utilised. The basicity of an ionic slag influences both the chemical and physical properties, including the viscosity, fluidity, and slag capacity to absorb and retain impurities. Optimal slag basicity through slag composition design is often crucial for effective impurity removal and slag performance. The adage 'Take care of the slag, and the steel will take care of itself' emphasises the importance of managing the slag effectively in metallurgical processes to ensure the production of high-quality steel.

Considering the aforementioned slag capacity, two key impurities are closely controlled in the steelmaking operations, which include sulfur and phosphorus corresponding to the sulfide and phosphate slag capacities respectively.

Desulfurisation of steels typically involves the transfer of dissolved sulfur from the metal into the slag phase. The most common reaction under steelmaking conditions is the formation of calcium sulfide (CaS), when sulfur reacts with calcium oxide (CaO) in the slag, as expressed by Equation 1. The corresponding ionic reaction is express as Equation 2 and the subsequent sulfide capacity under reducing conditions is expressed by Equation 3. Depending on the slag composition, the activities of the reactants and products can be significantly modified. Thus, depending on the impurities existing in the steel, the activity of dissolved S can affect the reaction and the components such as Al_2O_3 or SiO_2 in the slag can also affect the activity of the CaO in the slag, which can inherently affect the S transfer to the slag phase.

$$CaO(slag)+S(metal) \rightarrow CaS(slag)+1/2O_2(g) \tag{1}$$

$$O^{2-}(slag)+S(metal) \rightarrow S^{2-}(slag)+1/2O_2(g) \tag{2}$$

$$C_{S^{2-}} = (\%S) \times \left\{ \frac{P_{O_2}}{a_S} \right\}^{1/2} = K_2 \times \frac{a_{O^{2-}}}{f_{S^{2-}}} \tag{3}$$

According to Figure 1, the sulfide capacities according to the Vee ratio (corresponding to the basicity) and the activity of FeO are shown. It is clearly evident that the sulfide capacities for these systems follow the expected thermodynamic interpretations and a slope of unity corresponding to Equation 3 is shown. Similar results have been identified for dephosphorisation with a slope of 1.5.

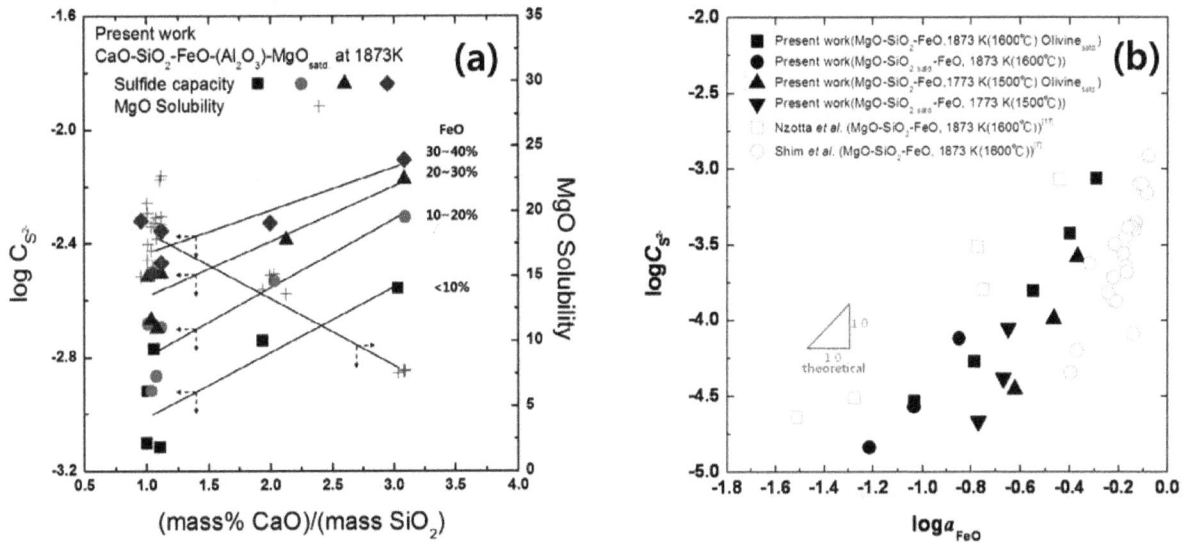

FIG 1 – Sulfide capacities of various slag systems as a function of: (a) Vee ratio (apparent basicity) (Park and Min, 2016); and (b) logarithm of the FeO activity (Kim, Huh and Min, 2014).

In Figure 2, the temperature dependence of the slag capacities are provided. Unlike the basicity, the sulfide and phosphate capacity show opposite trends with the temperature dependence. Higher temperatures are favourable for the sulfide capacity, but lower temperatures are favoured for the phosphate capacity. Dephosphorisation is the process of transferring the dissolved phosphorus in the molten metal to the slag phase resulting in the desired product specifications. Considering phosphorus removal in steelmaking occurs under a high oxygen partial pressure, the typical main reaction corresponds to the formation of a phosphate phase in the slag. Phosphorus reacts with CaO in oxidising conditions to form calcium phosphate ($Ca_3(PO_4)_2$), as described in Equation 4. The reaction in ionic form can be expressed as Equation 5. Accordingly, under oxidising conditions of steelmaking operations, the phosphate capacity can be expressed by Equation 6.

$$3/2CaO(slag)+P(metal)+5/4O_2(g) \rightarrow 1/2Ca_3(PO_4)_2(slag) \qquad (4)$$

$$3/2O^{2-}(slag)+P(metal)+5/4O_2(g) \rightarrow PO_4^{3-}(slag) \qquad (5)$$

$$C_{PO_4^{3-}} = (\%PO_4^{3-}) \times \frac{1}{a_P \times P_{O_2}^{1/2}} = K_5 \times \frac{a_{O^{2-}}^{3/2}}{f_{PO_4^{3-}}} \qquad (6)$$

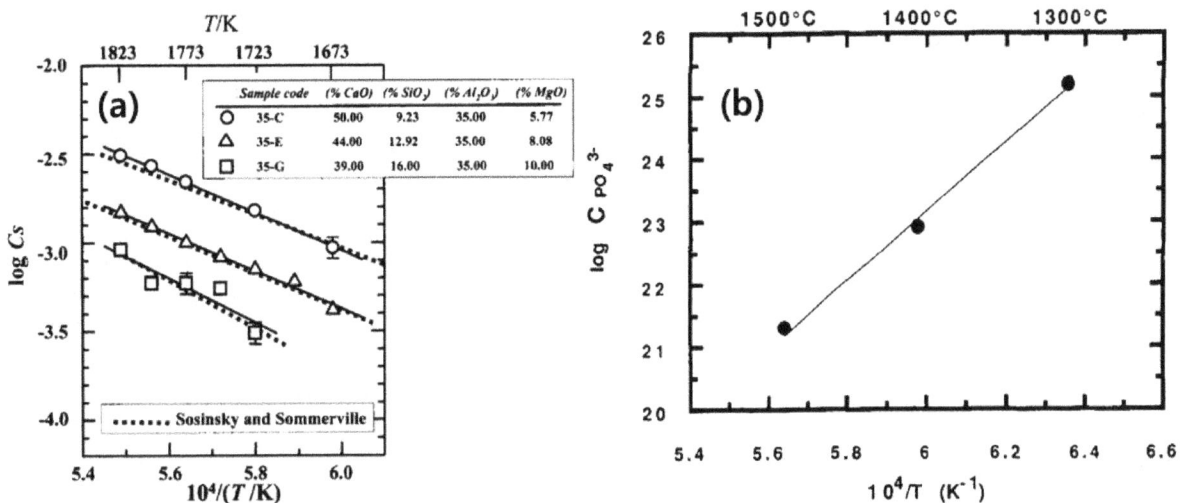

FIG 2 – (a) Sulfide capacity and (b) Phosphate capacity as a function of temperature (Nassaralla and Fruehan, 1992).

Similar to sulfur, the slag capacity for phosphorus removal from the metal to the slag depends on various factors including temperature, slag composition, and activities of the reactants and products. It should be noted that under a fixed temperature and oxygen partial pressure, the slag capacities are determined according to the slag system.

Overall, the thermodynamics of desulfurisation and dephosphorisation in slags involve complex interactions between the metal and slag phases, as well as the equilibrium between different chemical species. Understanding these fundamental thermodynamic principles is crucial to optimise steelmaking processes and achieving the levels of S and P removal into the designed slags for steel quality.

Matte optimisation in non-ferrous metallurgy

Beyond ferrous smelting operations, non-ferrous metallurgy is also a field for chemical metallurgists to make a significant impact on not only the industry, but also the environment considering the significant amount of emissions involved with smelting non-ferrous metals with relatively small concentrations of the product. Unlike ferrous raw materials, non-ferrous raw materials contains typically less than 5 mass per cent. of the target metals. Copper is one of the widely used non-ferrous metals that is critical to the electrification of the society. Copper smelting involves various processing stages, including matte production, converting, and refining. Matte, a sulfide-rich intermediate product before the production of blister copper after converting, plays a critical role in determining the efficiency and quality of copper extraction. Significant challenges are associated with matte optimisation in copper smelting, which can enhance process efficiency, reduce environmental impact, and improve product quality. Key optimisation parameters include flux addition, matte grading, and slag designing. These optimisation factors can maximise copper recovery and minimise energy consumption and emissions, ultimately contributing to sustainable copper processing and production.

Matte optimisation is challenged by the complexity of sulfide ore composition, feed variability, and impurities such as iron, sulfur, and arsenic. Blending different ore types, adjusting the sulfur-to-iron ratio, and flux additions assists in matte composition optimisation and minimise impurities.

Fluxes play a crucial role in copper smelting by facilitating slag formation, promoting metal separation, and controlling impurity levels. Optimising flux addition rates and compositions improves matte quality, reduces slag viscosity, and enhances metal recovery. Adjusting silica-to-alumina ratio, controlling slag basicity, and minimising iron oxide content can improve slag fluidity, reduce matte losses, and enhance overall process performance. Silica is a primary flux used in copper smelting to facilitate a liquid slag phase. It helps lower the melting point of the slag, promotes the dissolution of gangue minerals, and improves slag fluidity. Limestone is commonly added as a flux to neutralise acidic impurities such as S and As to form CaS and $Ca_3(AsO_4)_2$, which are then incorporated into the slag phase. Fayalite (Fe_2SiO_4) is a by-product of copper smelting and often recycled, which contribute to the formation of a stable slag phase, absorbing excess sulfur, and assists in controlling the slag viscosity. As described from Equations 7 and 8, copper smelting from sulfide ores is a stepwise oxidation of iron, where the oxygen activity increases resulting in blister copper production. (Taskinen *et al*, 2019).

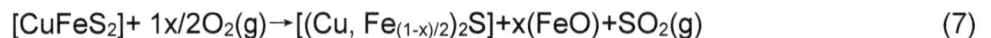

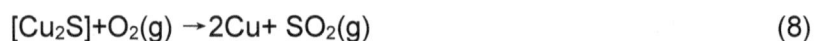

$$[CuFeS_2] + 1x/2O_2(g) \rightarrow [(Cu, Fe_{(1-x)/2})_2S] + x(FeO) + SO_2(g) \qquad (7)$$

$$[Cu_2S] + O_2(g) \rightarrow 2Cu + SO_2(g) \qquad (8)$$

According to the results shown in Figure 3 the Cu content in the matte can be significantly affected by the amount of oxygen activity in the reactor and the subsequent Fe/SiO_2 slag formed during processing. It is generally known that the chemical dissolution affecting Cu loss can be of two possible types: oxidic and sulfidic. However, it is also known that when the grade of matte exceeds 60 per cent, it is influenced by oxidic copper dissolution (Cornejo Mardones, 2020). Therefore, higher flow rates, which correspond to higher oxygen potential and activity can increase the Cu content in the matte. However, excessive oxygen injection can not only oxidise the Fe in the chalcopyrite raw material, but can also oxidise the Cu into the slag phase resulting in lower recovery of Cu. As the Fe oxidises into the slag phase, there are some amounts of SiO_2 fluxing required to optimise the Cu content in the matte. As can be seen in Figure 3b, higher grade Cu mattes can be obtained with

Fe/SiO$_2$ reaching the ratios beyond 1.6. However, this needs to be balanced since higher oxygen potentials can lead to the formation of Fe$_3$O$_4$ in the slag phase, which can increase the melting point of the slag and reduce fluidity and also the separation between matte and slag phase during matte smelting and separation. An optimal Fe/SiO$_2$ ratio seem to be near the low temperature fayalite phase formation region and thus operations typically would target the slag composition to be in this range.

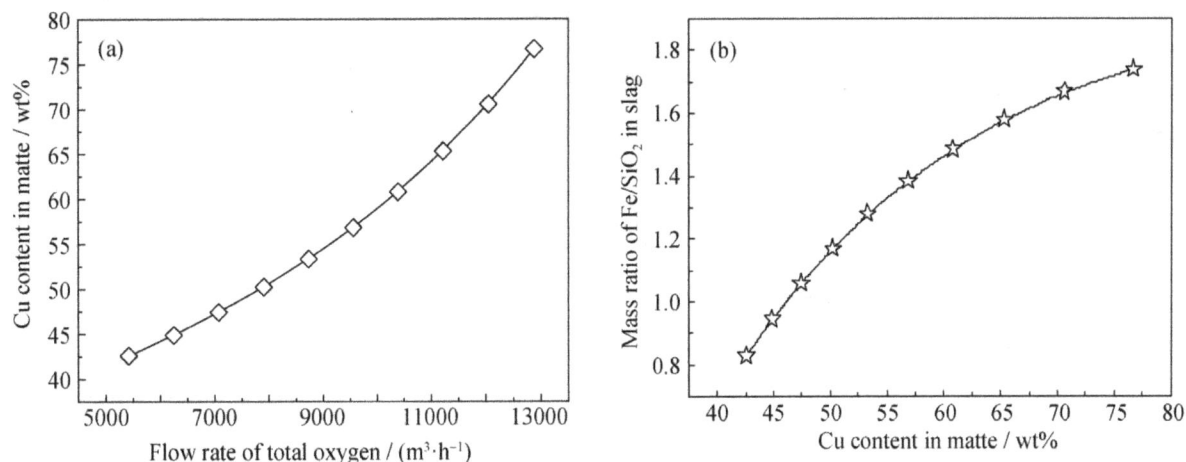

FIG 3 – (a) Cu content in matte as a function of oxygen and (b) relationship between the Fe/SiO$_2$ ratio to the Cu content in matte (Wang *et al*, 2019).

Thermophysical properties of slags and mattes

The thermophysical properties of slags and mattes play a significant role in determining their capacity to interact with molten metal and effectively absorb impurities. These properties influence the behaviour of slags during metallurgical processes, affecting their viscosity, thermal conductivity, heat capacity, and other characteristics.

Viscosity affects fluid flow behaviour and retention of the impurities in slags. Highly viscous slags can impede mass transfer kinetically reducing the slag capacity. This reduced mobility can hinder the absorption of impurities by the slag, limiting its capacity to remove contaminants from the metal. Lower viscosity enhances slag-fluid metal interactions, promoting impurity removal. The viscosity of slag is influenced by temperature, chemical composition, and phase composition. In addition, slags with high viscosity tend to have slower settling rates, prolonging the time required for slag-metal separation. Higher viscosities can also hinder convective heat transfer within the slag, leading to temperature gradients and non-uniform heating. Thus, optimising slag viscosity is essential for enhancing the slag capacity and improving the efficiency of impurity removal and other desired functions in metal production.

Slags and mattes with higher thermal conductivity can efficiently transfer heat from the surrounding environment, influencing temperature distributions within the system. Temperature gradients affect reaction kinetics and phase equilibria, consequently impacting the slag capacity. Like many of the thermophysical properties of slags and mattes, the thermal conductivity of slag is also influenced by its chemical composition, phase fractions, and temperature.

Slags and mattes with higher specific heat capacity can absorb more heat energy before significant temperature changes can be observed, which impact the energy needs of a reacting system in metallurgical processes. The heat capacity also depends on its chemical composition, phase fraction and temperature.

Understanding and controlling these thermophysical properties are essential for optimising metallurgical processes using slags and mattes. Slag and matte compositions and process conditions are often manipulated to achieve desired thermophysical properties and improve process performance.

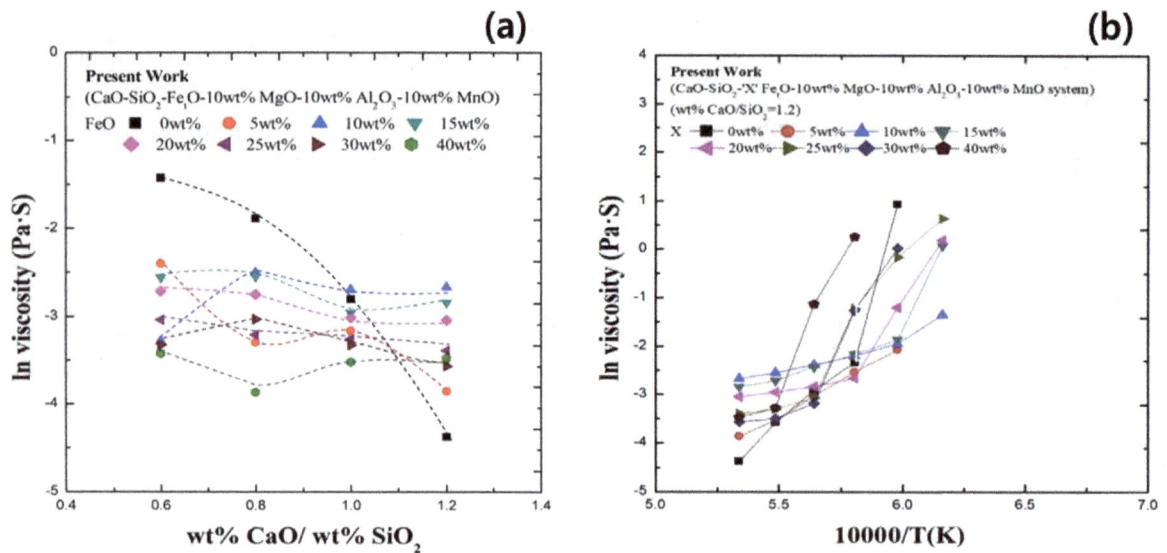

FIG 4 – Natural logarithm of the viscosity (Choi *et al*, 2021) as a function of (a) Vee ratio and (b) temperature.

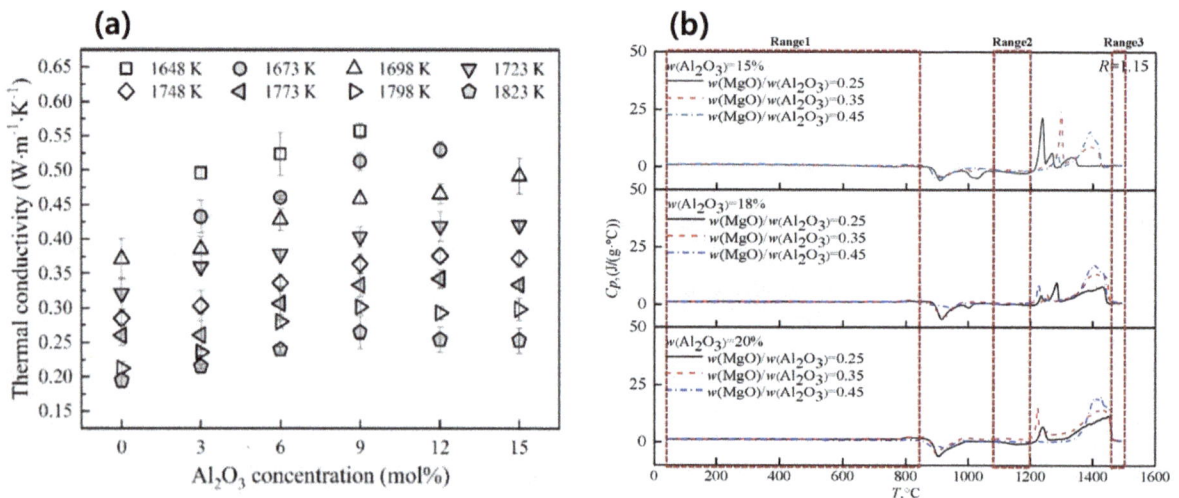

FIG 5 – (a) Thermal conductivity of slags as a function of Al_2O_3 content (Yang, Wang and Sohn, 2022) and (b) Specific heat capacities of slags as a function of temperature and composition (Zheng *et al*, 2021).

SUMMARY

This review looks briefly at the fundamental thermodynamics and thermophysical properties of slags and mattes for chemical metallurgy. Understanding the theory behind the various reactions and controlling the fundamental chemical potential allows the process engineer to achieve metals production at a more efficient and environmentally optimal way.

ACKNOWLEDGEMENTS

This work was supported by the Technology Innovation Program (Commercialization and development of new design on turbulent high temperature melting furnace (2000 tony pilot scale) and separation and (or) recovery of valuable metals from end of the xEV (ESS) battery pack) (20011183) funded by the Ministry of Trade, Industry and Energy (MOTIE, Korea).

REFERENCES

Choi, J S, Park, T J, Min, D J and Sohn, I, 2021. Viscous behavior of high-Fe$_t$O-bearing slag systems in relation to their polymeric structural units, *Journal of Materials Research and Technology*, 15:1382–1394.

Cornejo Mardones, K, 2020. Control of copper loss in flash smelting slags, Master's Thesis, School of Mechanical and Mining Engineering, The University of Queensland.

Kim, K D, Huh, W W and Min, D J, 2014. Effect of FeO and CaO on the Sulfide Capacity of the Ferronickel Smelting Slag, *Metall Mater Trans B,* 45B:889.

Merrill, A M, 2023. US Geological Survey, Aluminum.

Nassaralla, C and Fruehan, R J,1992. Phosphate Capacity of CaO-Al$_2$O$_3$ Slags Containing CaF$_2$, BaO, Li$_2$O, or Na$_2$O, *Metall Trans B,* 23B:117.

Park, Y and Min, D J, 2016. Sulfide Capacity of CaO-SiO$_2$-FeO-Al$_2$O$_3$-MgOsatd Slag, *ISIJ Int*, 56:520–526.

Taskinen, P, Akdogan, G, Kojo, I, Lahtinen, M and Jokilaakso, A, 2019. Matte converting in copper smelting, *Mineral Processing and Extractive Metallurgy*, 128:58–73.

Wang, Q-M, Wang, S-S, Tian, M, Tang, D-X, Tian, Q-H and Guo, X-Y, 2019. Relationship between copper content of slag and matte in the SKS copper smelting process, *International Journal of Minerals, Metallurgy and Materials*, 26:301.

World Steel Association, 2023. World Steel in Figures 2023.

Yang, J, Firsbach, F and Sohn, I, 2022a. Pyrometallurgical processing of ferrous slag 'co-product' zero waste full utilization: A critical review, *Resources, Conservation and Recycling*, 178:106021.

Yang, J, Wang, Z and Sohn, I, 2022b. Topological understanding of thermal conductivity in synthetic slag melts for energy recovery: An experimental and molecular dynamic simulation study, *Acta Materialia*, 234:118014.

Zheng, H, Liang, L, Du, J, Zhou, S, Jiang, X, Gao, Q and Shen, F, 2021. Mineral Transform and Specific Heat Capacity Characterization of Blast Furnace Slag with High Al$_2$O$_3$ in Heating Process, *Steel Research Int*, 92:2000448.

Effect of slag composition on titanium distribution ratio between ferrosilicon melt and CaO-SiO_2-Al_2O_3 slag at 1773 K

M J Lee[1] and J H Park[2]

1. Student, Department of Materials Science and Chemical Engineering, Hanyang University, Ansan 15588, South Korea. Email: dlalswn06@gmail.com
2. Professor, Department of Materials Science and Chemical Engineering, Hanyang University, Ansan 15588, South Korea. Email: basicity@hanyang.ac.kr

ABSTRACT

Because the environmental problems are occurred by greenhouse gases such as CO_2, the industries have tried to increase electric motor efficiency to reduce CO_2 emissions. A core material of electric motor is high-silicon electrical steel. The ferrosilicon (FeSi) alloy, a raw material for the production of electrical steel, affects the electric properties of silicon steel. The main impurities such as Ca, Al and Ti make inclusions or precipitates that causes core loss. The slag treatment has been widely used to remove impurity elements. Thermodynamic principles for the refining of several impurities, specifically Ca and Al, have been investigated, whereas thermodynamic behaviour of Ti has not been fully understood yet. Therefore, in the present study, the titanium distribution ratio (L_{Ti}) between ferrosilicon and CaO-SiO_2-Al_2O_3 slag system at 1773 K was investigated.

The results show that L_{Ti} is influenced by both the basicity and stability of titanium ion in the slag. Specifically, L_{Ti} exhibits a minimum value at about C/S=0.7 indicating a shift in the titanate structure unit from $[TiO_5]$-square pyramid to $[TiO_4]$-tetrahedron because of lack of Ca^{2+} ions in a C/S<0.8 region, contributing to an increase in titanate capacity and L_{Ti}.

INTRODUCTION

The electric vehicle market is growing quickly and study of electric motor efficiency is in progress to reduce CO_2 emission and air pollutants. The raw material for production of high-silicon electrical steel, ie ferrosilicon (FeSi) alloy, affects the electric properties of silicon steel. The common impurities of Ca, Al, Ti, C and B form inclusions or precipitates that result in core loss by decreasing grain size (Nakayama and Tanaka, 1997; Nakayama and Honjou, 2000; Steiner Petrovič et al, 2010). These impurities in electrical steel are generally originated from FeSi alloy which have an adverse effect on the magnetic properties. Unfortunately, Ti equilibrium distribution data between silicon and slag are very limited. Moreover, the thermodynamics of the Ti distribution behaviour between FeSi melt and CaO-based slags has not been investigated. Therefore, in the present study, we measured the Ti distribution ratio between FeSi melt and the CaO-SiO_2-Al_2O_3 slag at 1773 K. Additionally, a structural analysis using Raman spectroscopy was performed to understand the stabilisation mechanism of titanium in aluminosilicate melts.

EXPERIMENTAL PROCEDURE

The slag-metal equilibrium experiments were performed using an electric resistance furnace with a $MoSi_2$ heating element. A schematic diagram of the experimental apparatus is shown in Figure 1. The temperature was controlled within ±2 K. A mixture of 5 g commercial low-carbon FeSi alloy (0.013 wt per cent Ti) and 3 g slag were loaded in a graphite crucible placed in an alumina porous holder at 1773 K. The furnace was filled with purified Ar gas controlled by a mass flow controller at a flow rate 500 mL/min. After the equilibration for 24 hrs, the sample was extracted from the furnace and quenched in water. The compositions of metal and slag samples were determined by X-ray fluorescence (XRF) and inductively coupled plasma – optical emission spectroscopy (ICP-OES).

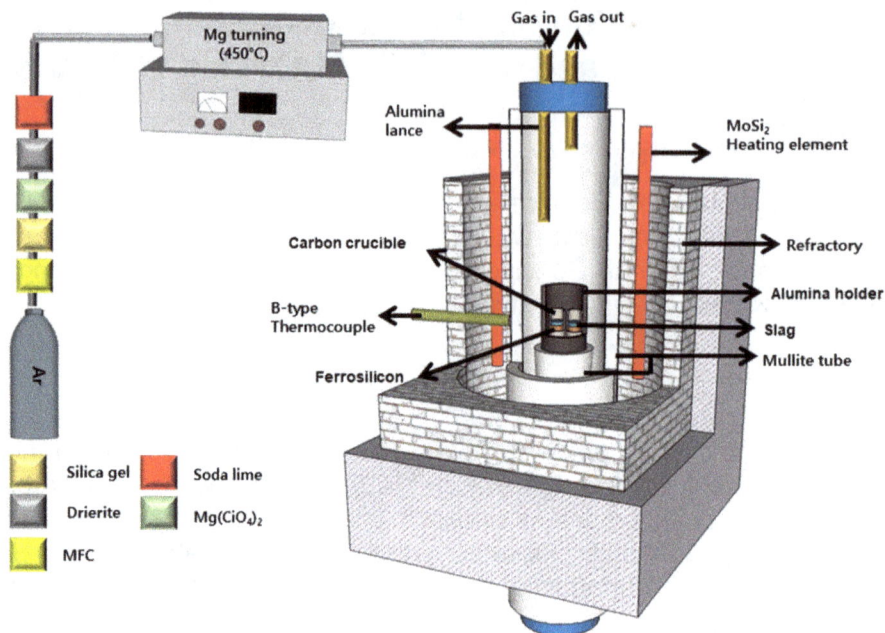

FIG 1 – Schematic diagram of the experimental apparatus.

RESULTS AND DISCUSSION

Influence of basicity on Ti distribution ratio between FeSi melt and CaO-SiO₂-Al₂O₃ slag

The distribution ratio of Ti (L_{Ti}) between FeSi and slags at 1773 K as a function of CaO/SiO₂ (=C/S) ratio is depicted in Figure 2. The distribution ratio of Ti is defined by the following equation.

$$L_{Ti} = \frac{(\text{wt\%TiO}_2)_{\text{slag}}}{[\text{wt\%Ti}]_{\text{metal}}} \tag{1}$$

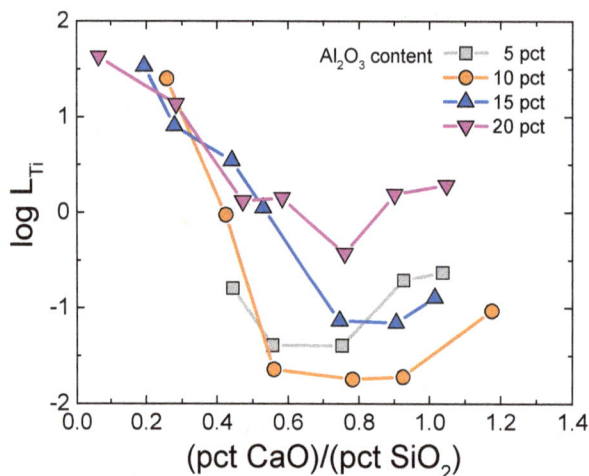

FIG 2 – Distribution ratio of Ti between CaO-SiO₂-Al₂O₃ slags and FeSi melts at 1773 K.

As shown in Figure 2, the V-shaped Ti distribution behaviour is observed in entire slags. In the 20 per cent Al₂O₃ system, the distribution ratio of Ti in a logarithmic scale is consistently greater than zero, ie $L_{Ti} > 1.0$, except for the C/S=0.8 system. This result suggests that the distribution ratio of Ti is significantly affected by the C/S ratio and Al₂O₃ content.

Effect of basicity and Al₂O₃ on titanate capacity and stability of titanium in slag

In the present study, it was assumed that the valence state of titanium is 4, ie Ti⁴⁺ is preferentially stable and the coordination number of titanium with oxygen is 5, ie [TiO₅] square-based pyramid structure mainly exists in the present CaO-SiO₂-Al₂O₃ system (Romano *et al*, 2000; Le Cornec *et al*, 2021). Based on this assumption, the titanium refining reaction and titanate capacity of molten slag are represented as given in Equations 2 and 3, respectively. Here, the titanate capacity is the ability of a slag to absorb titanium as a function of slag basicity, stability of titanate ion in the slag and temperature. To our knowledge, few study have reported titanate capacity, C_{TiO_x}, of molten slag. Hence, in the present study, we propose the application of titanate capacity to understand the thermodynamic behaviour of titanium oxide in the slag.

$$[Ti] + O_2(g) + 3(O^{2-}) = (TiO_5^{6-}) \tag{2}$$

$$C_{TiO_5^{6-}} = \frac{K_{(2)} \cdot a_{O^{2-}}^3}{f_{TiO_5^{6-}}} = \frac{(wt\%TiO_5^{6-})}{f_{Ti} \cdot [wt\%Ti] \cdot p_{O_2}} = \frac{L_{Ti}}{f_{Ti} \cdot p_{O_2}} \tag{3}$$

where $K_{(2)}$ is the equilibrium constant of Equation 2, $a_{O^{2-}}$ and f_{Ti} represent the activity of free oxygen ion in slag and the Henrian activity coefficient of titanium in FeSi melt, respectively. In the present study, f_{Ti} was assumed to be unity because the concentration of Ti in FeSi melt is not more than 0.4 wt per cent. Here, $f_{TiO_5^{6-}}$ is the activity coefficient of titanate ion in slag, and p_{O_2} is the oxygen partial pressure, calculated from the following equilibrium (Turkdogan, 1980).

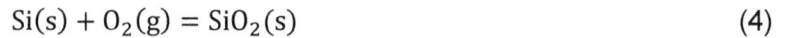

$$Si(s) + O_2(g) = SiO_2(s) \tag{4}$$

$$\Delta G_{[4]}^o = -907,130 + 175.7\,T\,(J/mol) \tag{5}$$

On the other hand, it is suggested that CaO behaves as a basic oxide by contributing free oxygen ion in the slag as shown below (Sano, 1997).

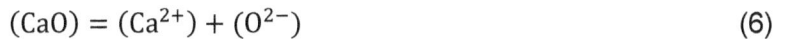

$$(CaO) = (Ca^{2+}) + (O^{2-}) \tag{6}$$

$$a_{O^{2-}} = \frac{K_{(6)} \cdot a_{CaO}}{a_{Ca^{2+}}} \tag{7}$$

Therefore, the titanate capacity is a function of the basicity ($a_{O^{2-}}$) and the stability of titanate ion ($f_{TiO_5^{6-}}$) at fixed temperature and it can be calculated using L_{Ti} and p_{O_2}. The activity of CaO in the slag system at 1773 K was calculated using Factsage software. By combining Equations 3 and 7, the relationship between CaO activity and titanate capacity can be described as follows:

$$log\,C_{TiO_5^{6-}} = 3\,log\,a_{CaO} - 3\,log\,a_{Ca^{2+}} - log\,f_{TiO_5^{6-}} + Const. \tag{8}$$

Based on Equation 8, the titanate capacity is expected to exhibit a linear relationship with the activity of CaO by assuming that $a_{Ca^{2+}}$ and $f_{TiO_5^{6-}}$ terms are not significantly affected by slag compositions. As shown in Figure 3, the titanate capacity linearly increases with increasing activity of CaO in a logarithmic scale at $log\,a_{CaO} > -2.2$, indicating that titanium oxide primarily acts as an acidic component in the relatively basic slag. The fitted lines have slopes of 2.7~3.0, indicating that $a_{Ca^{2+}}$ and $f_{TiO_5^{6-}}$ terms would be constant with increasing CaO activity. Therefore, in the basic slag system (ie $log\,a_{CaO} > -2.2$), it is confirmed that titanate ion forms [TiO₅]-square pyramid based on Equation 2.

However, in the acidic slag system ($log\,a_{CaO} < -2.3$), the titanate capacity increases with decreasing activity of CaO, suggesting that Ti dissolution reaction given in Equation 2 does not simply applicable in the relatively acidic region. Under the same activity of CaO condition, titanate capacity varies with Al₂O₃ content. This also implies that the Al₂O₃ concentration influences the $f_{TiO_5^{6-}}$ term in the slag. Overall, the titanate capacity of the slag is significantly influenced by the activity of CaO and Al₂O₃ under the experimental conditions.

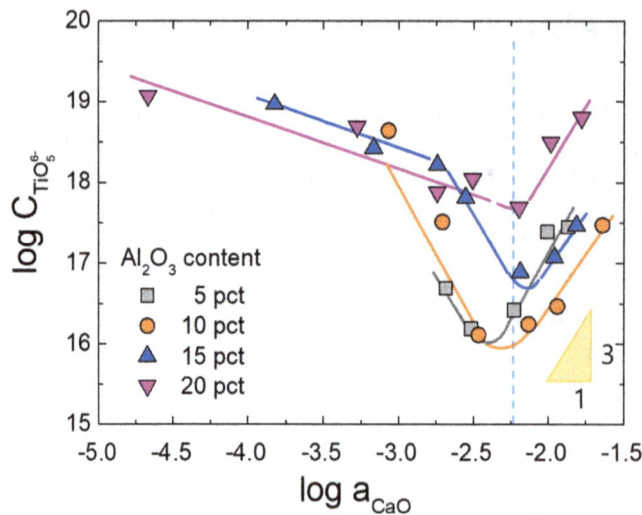

FIG 3 – Titanate capacity as a function of activity of CaO in the CaO-SiO$_2$-Al$_2$O$_3$ slags at 1773 K.

Effect of aluminosilicate structure on titanate stability

From the literature, titanium ion in aluminosilicate glasses have been studied and found that it is mainly Ti^{4+} and has five-fold coordination by Ti K-edge XAFS and thermodynamic analysis (Romano *et al*, 2000; Le Cornec *et al*, 2021). The five-coordinated Ti incorporated into the silicate framework through the Ti–O–Si bonds including the non-bonding oxygen of the short Ti=O titanyl bond requires a local charge compensation by Ca^{2+} ion. A distribution of titanium coordination region in the CaO-SiO$_2$-Al$_2$O$_3$ system is illustrated in Figure 4. The four-fold coordinated Al, ie [AlO$_4$] tetrahedral unit also requires Ca^{2+} ion for charge balancing. When the Al$_2$O$_3$ is relatively low at C/S>0.8 region (Region A in Figure 4), the Ca^{2+} ion interacts with [AlO$_4$] unit to keep the charge balance, and thus the activity coefficient of TiO$_2$ is increased by increasing the Al$_2$O$_3$ concentration due to deficiency of Ca^{2+} ion. Over the 10 per cent Al$_2$O$_3$ at C/S>0.8 (Region B in Figure 4), Ca^{2+} ion is mainly consumed for charge balance of [AlO$_4$] structure, and thus some Ti in five-fold coordination is forced to transform to four-fold coordination (Romano *et al*, 2000). Transformation of titanium coordination from five to four also occurred in C/S<0.8 region (Region C in Figure 4) due to the lack of Ca^{2+} ion. Because [TiO$_4$] group can be incorporated into the aluminosilicate network by replacing [SiO$_4$] and [AlO$_4$] tetrahedrons (Duan *et al*, 1998), the activity coefficient of TiO$_2$ is decreased by decreasing the C/S ratio and increasing Al$_2$O$_3$ concentration.

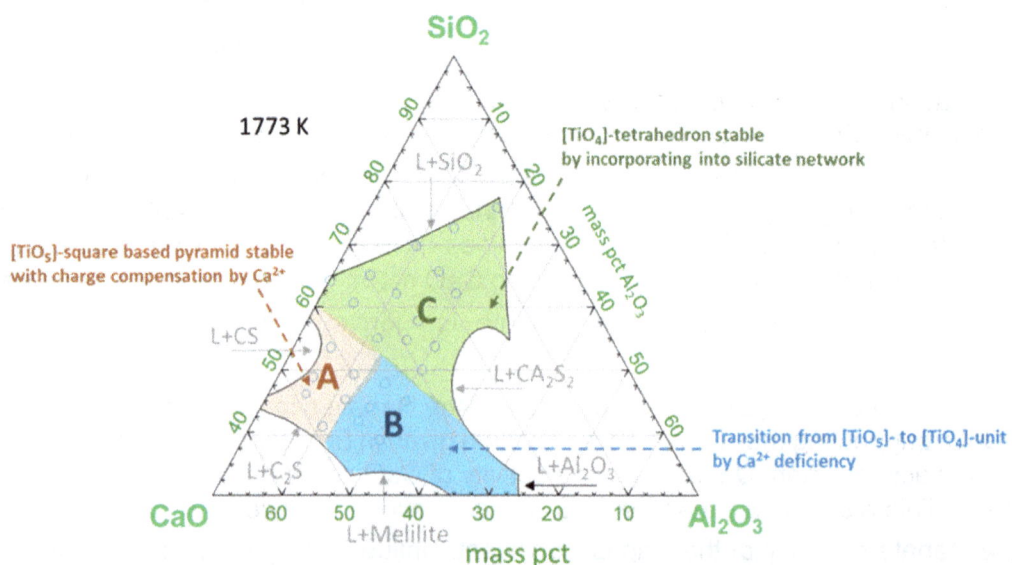

FIG 4 – Titanium coordinations in different regimes in the CaO-SiO$_2$-Al$_2$O$_3$ system at 1773 K.

CONCLUSIONS

The distribution behaviour of titanium between FeSi and CaO-SiO$_2$-Al$_2$O$_3$ slag was investigated at 1773 K. The major results of present study can be summarised as follows:

- The distribution behaviour of Ti is strongly influenced by slag basicity. The activity coefficient of TiO$_2$ decreases with an increase in C/S ratio at C/S>0.8, whereas it significantly decreases by decreasing the C/S ratio at C/S<0.8 system.

- At C/S>0.8 basic system, the addition of CaO stabilises five-fold coordinated Ti, but addition of Al$_2$O$_3$ depletes free Ca^{2+} ion due to charge balancing of [AlO$_4$]. Beyond 10 per cent Al$_2$O$_3$, Ti coordination partly shifts from five to four, resulting in a stabilisation of TiO$_2$ in slag.

- At C/S<0.8 acidic system, Ti in five-fold coordination is forced to transform to four-fold coordination because of Ca^{2+} depletion. The [TiO$_4$] group incorporates into the aluminosilicate network by replacing [SiO$_4$] and [AlO$_4$] structure, which increases the stability of TiO$_2$.

REFERENCES

Duan, R G, *et al*, 1998. The effect of Ti^{4+} on the site of Al^{3+} in the structure of CaO–Al$_2$O$_3$–SiO$_2$–TiO$_2$ system glass, *Materials Science and Engineering: A*, 249:217–222.

Le Cornec, D, *et al*, 2021. Structural role of titanium on slag properties, *Journal of the American Ceramic Society*, 104:105–113.

Nakayama, T and Honjou, N, 2000. Effect of aluminum and nitrogen on the magnetic properties of non-oriented semi-processed electrical steel sheet, *Journal of Magnetism and Magnetic Materials*, 213:87–94.

Nakayama, T and Tanaka, T, 1997. Effects of titanium on magnetic properties of semi-processed non-oriented electrical steel sheets, *Journal of Materials Science*, 32:1055–1059.

Romano, C, *et al*, 2000. Effect of aluminum on Ti-coordination in silicate glasses: A XANES study, *American Mineralogist*, 85:108–117.

Sano, N, 1997. *Advanced Physical Chemistry for Process Metallurgy* (Academic Press: New York).

Steiner Petrovič, D, *et al*, 2010. Correlation of titanium content and core loss in non-oriented electrical steel sheets, *Metalurgija*, 49:37–40.

Turkdogan, E T, 1980. *Physical Chemistry of High Temperature Technology* (Academic Press: New York).

Using novel methods to characterise slag films for continuously casting challenging and innovative steel grades

Z Li[1], T Zhang[2], S Qin[3], X Yang[4], Z Yan[5], P Wilson[6] and M A Williams[7]

1. Professor, University of Warwick, Coventry CV4 7AL, England. Email: z.li.19@warwick.ac.uk
2. Professor, Shanghai University, Shanghai 200444, China.
   Email: tongsheng_zhang@shu.edu.cn
3. Research Fellow, University of Warwick, Coventry CV4 7AL, England.
   Email: shiying.qin@warwick.ac.uk
4. Research Fellow, Brunel University London, Middlesex UB8 3PH, England.
   Email: xinliang.yang@brunel.ac.uk
5. Research Fellow, University of Warwick, Coventry CV4 7AL, England.
   Email: zhiming.yan@warwick.ac.uk
6. Research Fellow, University of Warwick, Coventry CV4 7AL, England.
   Email: paul.wilson@warwick.ac.uk
7. Professor, University of Warwick, Coventry CV4 7AL, England.
   Email: m.a.williams.1@warwick.ac.uk

ABSTRACT

Continuous development of complex new steel grades to meet the ever-increasing demand of high performance causes recurrent issues in steel continuous casting such as surface quality defects (eg cracks, depressions, deep oscillation marks) and productivity challenges (eg faster casting, near-net shape casting). Continuous casting of steel is a highly successful metallurgical process. Much of this success can be attributed to the performance of the casting powder that is added to the top of the mould, creates a liquid slag pool as it is heated, and forms a slag film between the water-cooled copper mould and steel shell during its passage down the mould.

Mould powder selection is a compromise between the conflicting requirements of heat transfer and lubrication. In an EU RFCS (Research Fund for Coal and Steel)-funded project, various techniques are developed to offer the opportunity to adapt the slag film to meet the conflicting needs in different parts of the mould such that mild cooling can be generated in the meniscus, mid-broad face or corner areas whilst maintaining lubrication and offering higher cooling rates in other areas of mould. This has significant potential to address the industrial issues in product quality and productivity of continuous casting.

This paper reports, as part of the EU RFCS project, the determination of crystallinity and porosity in slag films for casting challenging and innovative steel grades using new methods. In comparison with the methods such as X-ray powder diffraction (XRD) analysis and optical microscopy / scanning electron microscopy (OM/SEM) analysis that are currently adopted by the industries, EBSD (electron backscatter diffraction) has been employed to determine the crystallinity of the slag films taken from industry casters. X-ray computed tomography (XCT) plus 3D image reconstruction has been used to determine the slag film % porosity with pore volume, pore size distribution, and pore location, which can be linked with the performance of the slag films during casting different steel grades.

INTRODUCTION

The global crude steel production in 2022 is 1884.2 Mt, 96.8 per cent of which is casted by continuous casting technologies (Worldsteel Association, 2023). The success of the steel continuous casting can be attributed to the performance of the casting powder that is added to the top of the mould, creates a liquid slag pool as it is heated, and forms a slag film between the water-cooled copper mould and steel shell during its passage down the mould. Continuous development of complex new steel grades to meet the ever-increasing demand of high performance causes recurrent issues in steel continuous casting such as surface quality defects (eg cracks, depressions, deep oscillation marks) and productivity challenges (eg faster casting, near-net shape casting). Therefore, continuous endeavours have been made to develop new mould powders for the continuous casting of such challenging and innovative steel grades.

The slag film between the water-cooled copper mould and steel shell in its initial stage consists of a solid slag layer because of its freezing on the water-cooled copper and a liquid layer next to the high temperature steel shell. This structure fulfils the major functions of the mould powder in the continuous casting – the liquid slag layer provides the right level of lubrication to the steel shell and the solid layer determines the heat flux (ie heat transfer from the steel shell to copper mould). Major task for developing new mould powder has always been ensuring its physical and chemical properties (such as viscosity, solidification temperature or break temperature, melting behaviour) to balance the lubrication and horizontal heat transfer and avoid casting problems like longitudinal cracking and sticker breakouts (Sridhar et al, 1998). The slag film evolutes during its travelling down the mould. Depending on the steel grades casted and the designed chemistry of the mould powder, the liquid layer solidifies and may crystallise depending on the chemistry and thermal conditions and the solid layer also crystallises during the thermal processing. The crystals in the slag film scatter infrared radiation and consequently reduce the heat transfer from the steel shell to copper mould. Therefore, the crystallinity of slag film is a major factor controlling the horizontal heat flux in the mould (Susa et al, 2009). Pores are also found during the examination of the slag films taken from the industry mould (Li, Mills and Bezerra, 2004), although the formation mechanisms of pores in slag film have not been well understood. The increased porosity also results in the decrease in heat transfer from the steel shell to the mould (Mills and Dacker, 2017).

In summary crystallinity and porosity are two major characteristics determining horizontal heat transfer from the steel shell to copper mould, which is critically important to cast challenging and new steel grades without defects. This paper investigates the methods for determining the crystallinity and porosity of slag films taken from industrial casting moulds.

CRYSTALLINITY

The initial slag film formed consists of a glassy layer in the water-cooled mould side and a liquid layer in the high-temperature steel shell side. The thin liquid layer solidifies and the slag film crystallises over time until the crystallinity (f_{crys}) in the slag film reaches a steady state. The crystallinity in the slag film is one of the key parameters controlling the horizontal heat transfer from the steel shell to the mould. As summarised by Mills and Dacker (2017), three major parameters controlling the horizontal heat flux from the shell to mould are

1. The thickness of the slag film.

2. The fraction of crystalline phase (ie crystallinity, f_{cryst}) and the size of the crystallites in the slag film.

3. The interfacial resistance between the copper mould and slag film and porosity formed by the shrinkage of slag during crystallisation.

One of the major research topics on crystallinity is to manipulate the crystalline phase and consequently control the crystallinity (f_{cryst}) in the slag film. Cuspidine ($3CaO \cdot 2SiO_2 \cdot CaF_2$) is the primary crystalline phase in the conventional fluorine-containing casting powders but not in the environmentally friendly fluorine-free mould slag or calcium aluminate mould slags. Increasing the slag basicity (CaO/SiO_2) generally increases its crystallinity. Without being involved in too much detail in the mechanisms of crystallinity in slag film, this chapter focuses on the determination of the crystallinity in slag films taken from industry casting moulds. In this chapter, the authors will first determine the crystallinity of slag films using OM (optical microscopy), SEM (scanning electron microscopy) and XRD (X-ray powder diffraction) techniques, and then explore the use of EBSD (electron backscatter diffraction) in quantitatively analysing the crystallinity (f_{cryst}) of the slag film.

OM (Optical Microscopy) image analysis

The procedure to determine the crystallinity of slag film is to observe the well-prepared slag film sample under optical microscope, select the representative range/area(s) of the slag film for analysis, determine the crystalline and glassy phases, take photos of the selected area(s) and determine the crystallinity based on the estimation of the selected area(s). Figure 1 shows how OM is applied to determine the crystallinity of slag film B. Slag film B has two distinguished layers of glass and crystalline, which enables the researcher to estimate the crystallinity to be around 50 per cent.

The OM analysis can allow large areas of slag film to be analysed. This method is simple to provide comparative information for different slag films. However, this purely depends on the researcher's experience to choose the representative area(s) and distinguish the crystalline and glass phases. If there is no clear boundary between the glass and crystalline phases, or the crystallites scatter in the glassy phase or *vice versa*, its accuracy will be influenced.

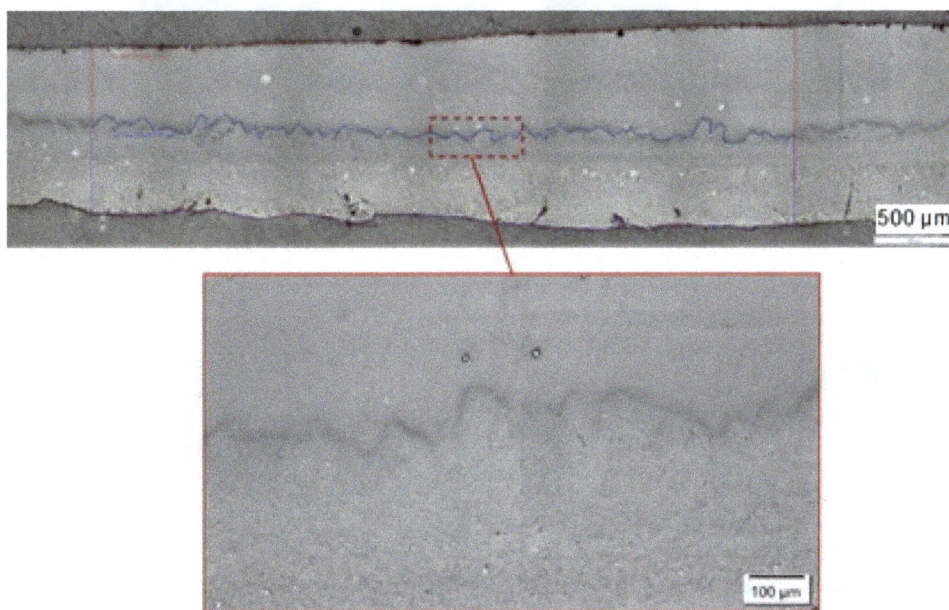

FIG 1 – Optical microscope (OM) image of slag film B for crystallinity estimation.

SEM image analysis

The procedure for SEM image analysis is similar to that for OM image analysis. However, SEM can go higher magnification than OM, which enables clear identification of crystalline phases in a focused area. Figure 2 illustrates the slag films A (top), B (middle) and C (bottom) respectively from industry casting moulds. The crystalline phase, glass phase and pores are clearly shown in the figures.

Slag film A has a thin layer in contact with steel shell consisting of crystalline phase (cuspidine), a thick glassy phase next to mould and pores (Figure 2 top). This thin layer could be the liquid layer in the initial formation of the slag film. Slag film B (Figure 2 middle) composes of two clear, similar thickness layers with the flat, glassy layer in contact with mould. The layer next to the steel shell is mainly crystalline phases (nepheline – $NaAlSiO_4$, cuspidine), pores and iron droplets. Dendrite crystalline is clearly observed to grow at the interface in the direction from the steel shell to the mould. Compared to Figure 1 by OM, SEM image clearly shows the morphology of the slag film. Slag film C (Figure 2 bottom) is mainly crystalline with large pores in the steel shell side.

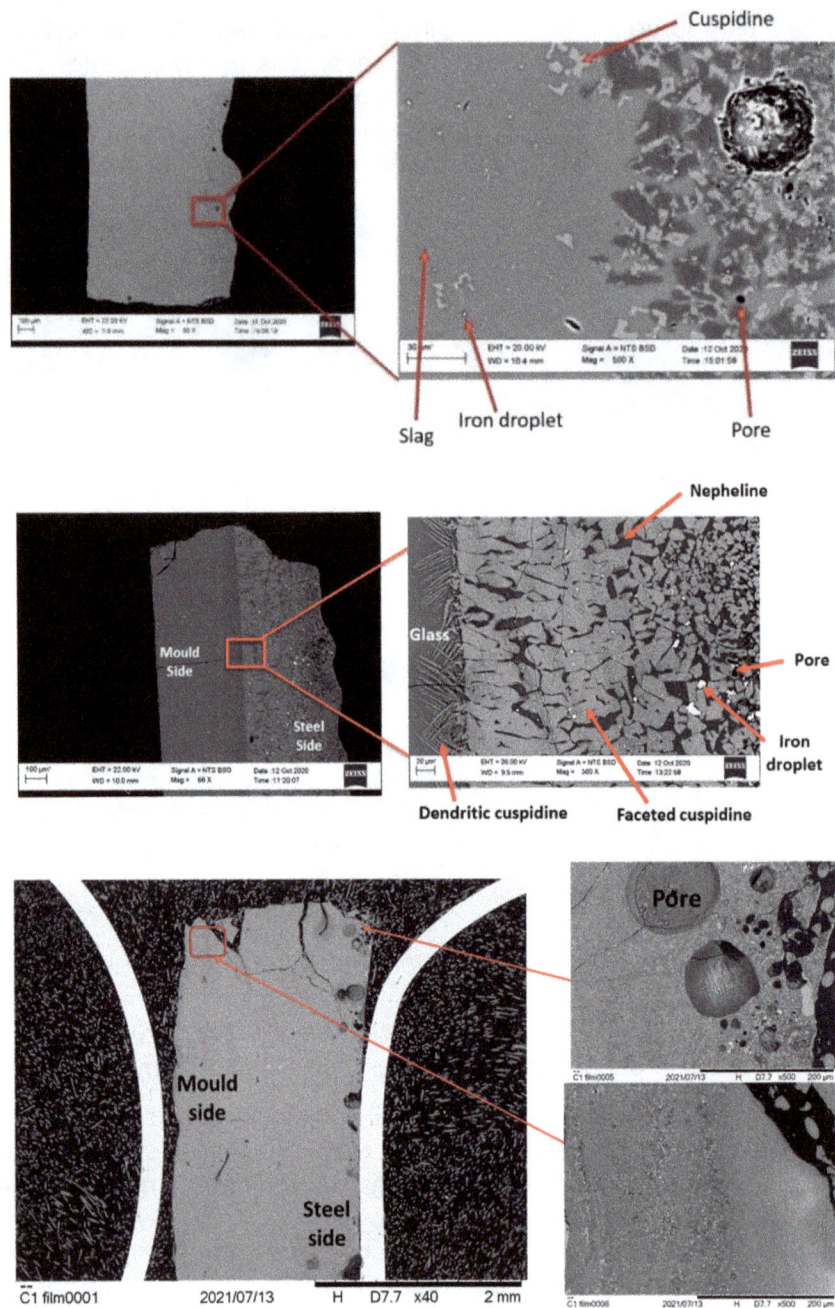

FIG 2 – SEM micrographs of slag films A (top), B (middle) and C (bottom). For each of the slag films, the left is its cross-section image and the right is the focused area.

SEM image can clearly show the crystalline and glassy phases in the slag films. However, the crystallinity determined is limited to the area(s) tested. Reliable results can only be obtained by examining an area which can represent the slag film. Therefore, macro photos of the whole slag film were taken, for example, as shown in Figure 3a. Then a thresholded image was obtained with the help of ImageJ (NIH Image, imagei.net) or other similar software, as shown in Figure 3b. Finally, the crystallinity of the industrial slag film can be calculated according to the colour difference, average f_{cryst} is then calculated using Equation 1:

$$\overline{f}_{cryst} = \frac{\sum f_{cryst} \cdot A_i}{\sum A_i} \tag{1}$$

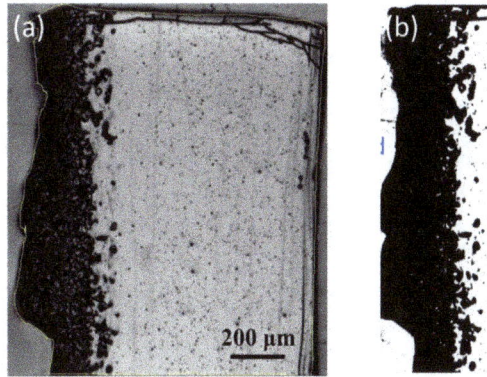

FIG 3 – Analysis process of SEM micrograph of a slag film at higher magnification: (a) SEM micrograph and (b) thresholded image.

Accordingly, the crystallinity determined by SEM image analysis is 21 per cent, 56 per cent and 100 per cent for slag films A, B and C respectively. In summary, OM or SEM can be used to determine the crystallinity of the slag films. The value and accuracy of the crystallinity determined depend on the representative area(s) selected. The difference between crystalline and glass phases may be subject to the researcher's experience. SEM, because of its better identification of crystalline phase, could give better outcomes compared to OM.

XRD analysis

XRD is a proven technique to determine the phases and their quantities in powders. The experimental procedure adopted in this study is to select representative pieces of the slag film and grind them to fine powders for XRD analysis. Panalytical X'Pert Pro MRD™ using a Cu-K$_\alpha$ radiation was employed with an accelerating voltage of 40 kV and a tube current of 40 mA. For XRD scans, the scanning angle was varied from 20° to 80° at a rate of 2°/min.

The type of the crystalline phases in the slag film samples can be obtained by analysing the XRD pattern of the sample, however, this study is more interested in the crystalline phase content (crystallinity f_{cryst}). The diffraction peak of crystal phase presents a high intensity in a narrow degree (ie crystalline area), by contrast, the diffraction peak of the glass phase looks like a steamed bread owing to the low intensity and wide degree (ie amorphous area). The degree of crystallinity (f_{cryst}) can be determined by the total integrated area of the crystal and glassy phase peaks in the XRD pattern of the sample. The f_{cryst} is then calculated by Equation 2:

$$fcryst = \frac{Crystalline\ Area}{Crystalline\ Area + Amorphous\ Area} \qquad (2)$$

Figure 4 is the XRD patterns of three slag film powder samples obtained. For the slag film A, the crystallinity in the mould side and in the steel shell side varies as shown in Figure 4a. So the selected pieces of a slag film were fully crushed for the XRD test. The crystallinity for the three slag films determined is 22 per cent, 65 per cent and 100 per cent for slag films A, B and C respectively. It is concluded that XRD does provide reasonable, consistent results.

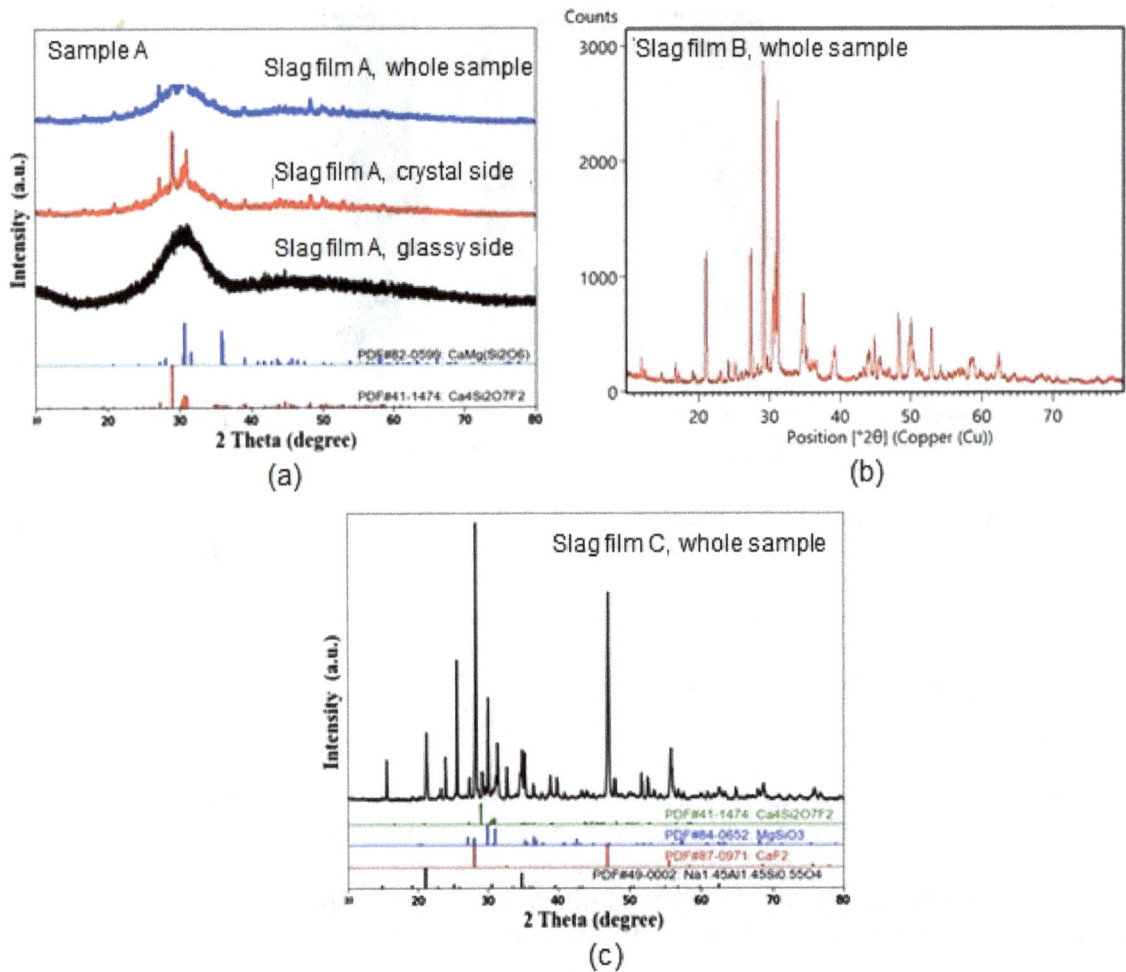

FIG 4 – XRD patterns of three slag film sample powders A, B and C.

EBSD analysis

Electron backscatter diffraction (EBSD), in addition to the conventional techniques of XRD analysis and OM/SEM image analysis, was attempted to obtain the crystallinity of slag film samples. EBSD is a scanning electron microscope-based microstructural-crystallographic characterisation technique commonly used in the study of crystalline or polycrystalline materials, which may be able to provide information about the crystallinity of slag films tested. The industrial slag film C is used as an example for the development. As shown in Figure 5, the presence of many small crystals only 1–2 μm in size (and glass phase if there is) can be detected. The crystalline phases and volumes in the area can be identified by using Kikuchi Patterns in EBSD (Figure 6). The total crystallinity of slag film C is about 71.1 per cent, which is lower than that determined by XRD analysis and SEM image analysis (100 per cent). The main reasons could be:

- To accurately determine the crystallinity of the slag film sample, pre-setting all the crystalline phases in the tested areas is essential, however, some minor phases in the tested areas might not be determined before EBSD analysis.

- The areas tested may not be sufficient in size.

All these factors can be further improved.

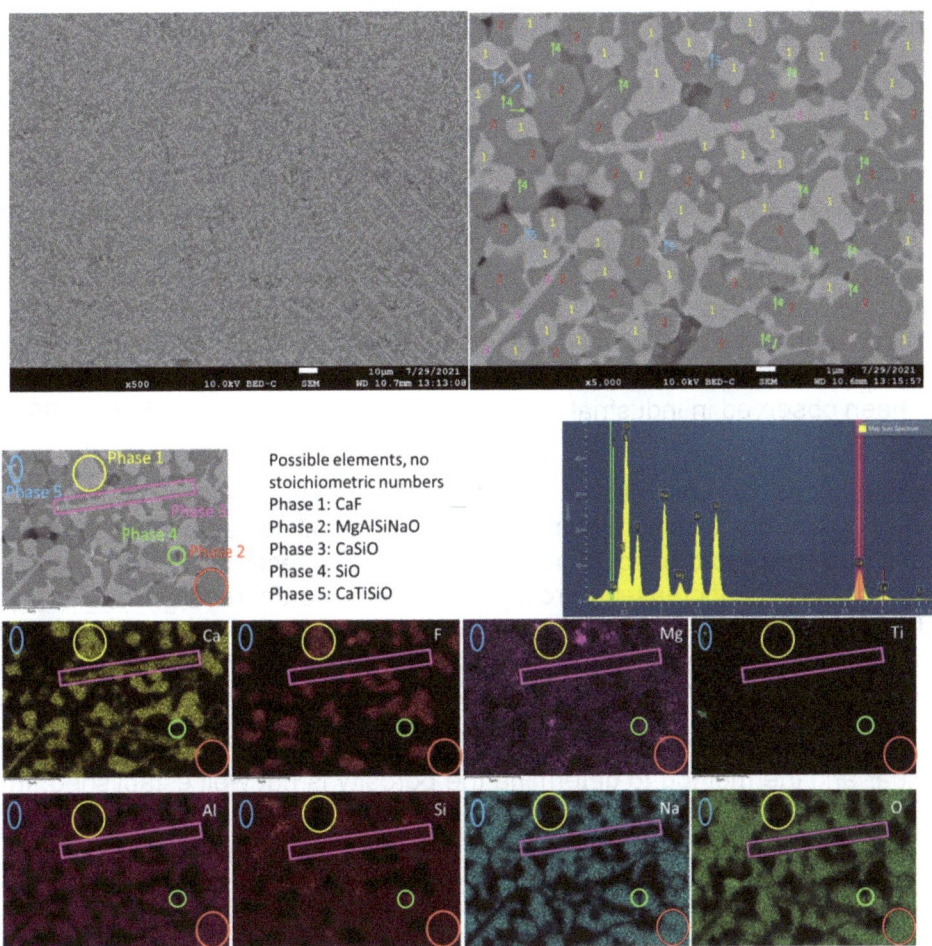

FIG 5 – SEM micrographs and the types of small size crystalline phases in slag film C. (Phase 1: CaF_2, Phase 2: MgAlSiNaO, Phase 3: CaSiO, Phase 4: SiO, Phase 5: CaTiSiO).

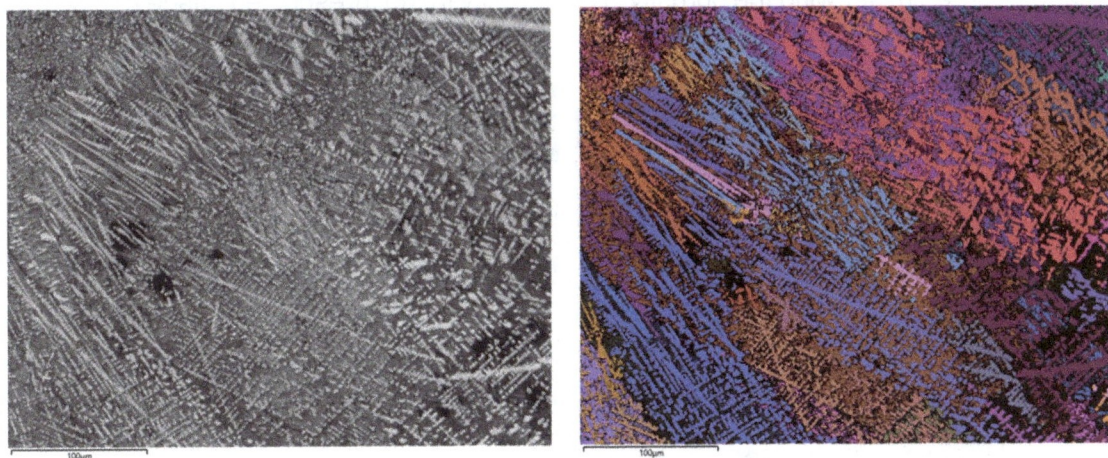

FIG 6 – Identification of crystalline phases using Kikuchi patterns in EBSD. Examination area: 0.4 mm × 0.3 mm, Step size: 0.5 mm, 503930 points measured in total. Acquisition time: 7 mins.

By comparing the crystallinity results from three methods of SEM image analysis, XRD analysis and EBSD analysis, it can be concluded that the values obtained by XRD analysis and SEM image analysis are in reasonably good agreement. The result from EBSD analysis is lower, which could be further improved by presetting all the crystalline phases in the tested area before EBSD scanning and increasing the size of areas. The advantage of SEM image analysis is its clear indication of crystallinity variation with position (tested area), however, due to the limitation of the field of vision, the results of different planes may have a large deviation. To obtain a reliable result, selecting representative area(s) is the key. XRD analysis, partially because of its easiness in selecting representative samples, may be more suitable for the overall crystallinity analysis of slag film samples.

POROSITY

The porosity in the slag film can come from various sources although it is not well understood. Crystal is denser than glass, so crystallisation results in shrinkage and consequently creates micro-pores in the slag film. However, crystallisation is not the only cause of porosity. Casting powder with a high moisture content (or water leaks) can lead to hydrogen pores in the slag film. Blowholes containing $CO_{(g)}$ can be formed from reaction of C and FeO in the slag (Mills and Dacker, 2017). Also the potential reactions within the mould flux at high temperature such as evaporation of CaF_2 and formation of SiF_4 in fluorine-containing mould powder can also form pores. Different morphologies of pores have been observed in industrial slag films, which could be attributed to different formation mechanisms.

Mould flux is distributed between the mould and steel shell, which controls heat transfer and lubrication during continuous casting. The porosity in the slag films has a great influence on the heat transfer from the steel shell to the mould, and consequently affects the casting process and quality of the casting products. The increased porosity could lead to a decrease in heat flux from the shell to mould and to sticker breakouts. Because of the importance of porosity in controlling heat flux, Hunt and Stewart (2016) reported that the use of intumescent coatings on the copper mould could reduce heat transfer by up to 27 per cent in a laboratory cold finger simulation experiment.

Although the importance of porosity in slag film in heat flux is well known, their formation mechanisms are not very clear and the method to determine the porosity is not well established. In this chapter, the authors investigate the methods to determinate the porosity of industrial slag films using techniques from the OM/SEM image analysis to the innovative X-ray computed tomography (XCT) technique for 3D porosity determination. The number, size and distribution of the pores quantified in the three industrial slag films (A, B and C) are summarised in this section.

SEM image analysis

As shown in Figure 7, by applying appropriate threshold, SEM image (Figure 7a) can be processed to 2D image (Figure 7b) showing the porosity. The porosity in the thresholded image can be obtained by using commercial software such as Image J.

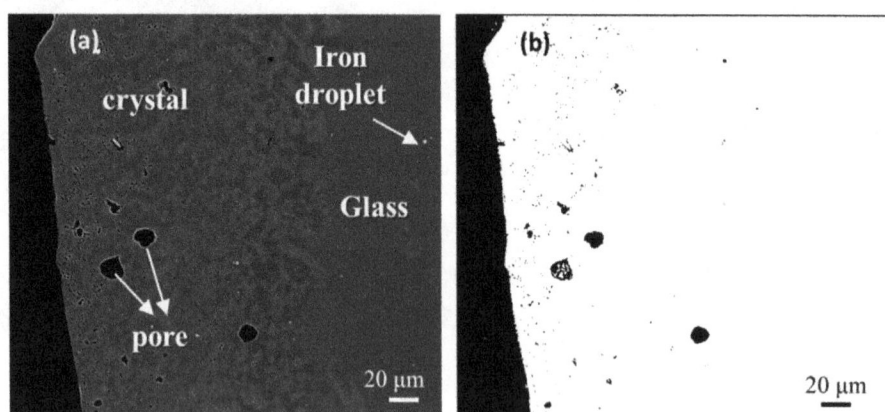

FIG 7 – SEM micrograph of slag film A at higher magnification: (a) SEM micrograph; (b) thresholded image.

As shown in Figure 8, small pores observed in SEM images of slag film C at high magnifications could be formed during crystallisation compared to the large blowholes observed in Figure 7 for slag film A. Using the SEM method, potentially pores in a wide range of size can be measured in a slag film, which may not be possible in the OM image analysis (which is not included in this paper).

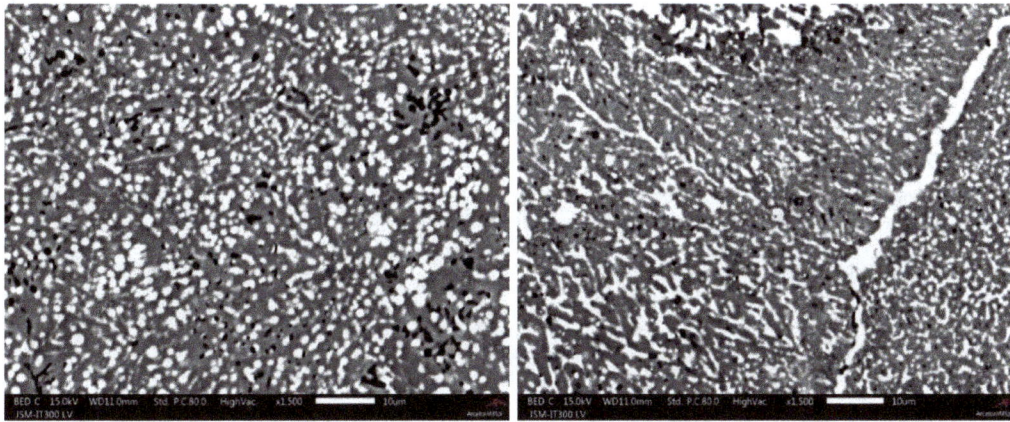

FIG 8 – SEM images at high magnification showing the pores (black spots) in the slag film C probably formed during crystallisation.

The characteristics of porosity in the three slag films A, B and C analysed by the above method are summarised in Figure 9. It should be pointed out that the analyses include all the pores in large areas from 16.18 mm² (slag film A) to 22.70 mm² (slag film C). The porosity obtained is 3.31 per cent (slag film A), 1.83 per cent (Slag film B) and 4.24 per cent (slag film C), respectively.

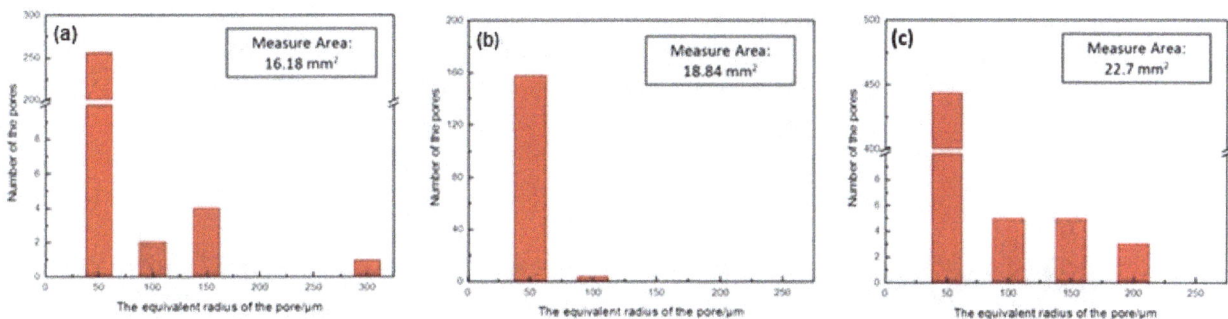

FIG 9 – Size distribution of all internal porosities in industrial slag films: (a) B, (b) A and (c) C. The equivalent radius of the pore is calculated by assuming the pore is circle. Radius 50 μm refers to the pores with radius between 0 and 50 μm and radius 100 μm refers to the pores with radius between 50 and 100 μm and so on. Measured area refers to the size of area measured for the % porosity.

Big difference in the % porosity even for the same slag film determined by different researchers was observed. The possible reasons are:

- The difficulty to balance the measurement area and accuracy due to the characteristics of the thresholded image detection.

- The selected areas – the size of the area and the representative of the area.

XCT analysis

X-ray computed tomography (XCT) is a non-destructive technique for visualising interior features within solid objects and for obtaining digital information on their 3D geometries and properties. An XCT image is typically called a slice, as it corresponds to what the object being scanned would look like if it were sliced along a plane. A complete volumetric representation of an object can then be obtained by acquiring a continuous set of CT slices. In the current study, XCT is used as a novel characterisation technique to visualise the slag films in 3D and identify the pores along with their size distribution. The XCT has the capability of identifying pores of very small dimensions with spatial resolutions of ~1/1000 times the thickness of the slag films. The samples used in the XCT tests are in ~20 mm square or round shape (which could be considered representative) cut from the industrial slag film samples. In this study, a lab-based CT scanner (Zeiss 620 Versa CT system) is used for scanning the industrial slag films prepared. The exposure voltage is 80 kV, exposure power 10 W and exposure time 10 sec. The Voxel size is 9.0895 μm. The number of projections is 3201. Imaging processing including filtering, segmentation and quantification were carried out using the Avizo Software (Thermo Fisher Scientific). As an example, Figure 10 shows the morphologies of the slag film C and porosity in 3D images. From this 3D images, the porosity (%) and pore size distribution can be calculated and the pore distribution across the sample can be observed.

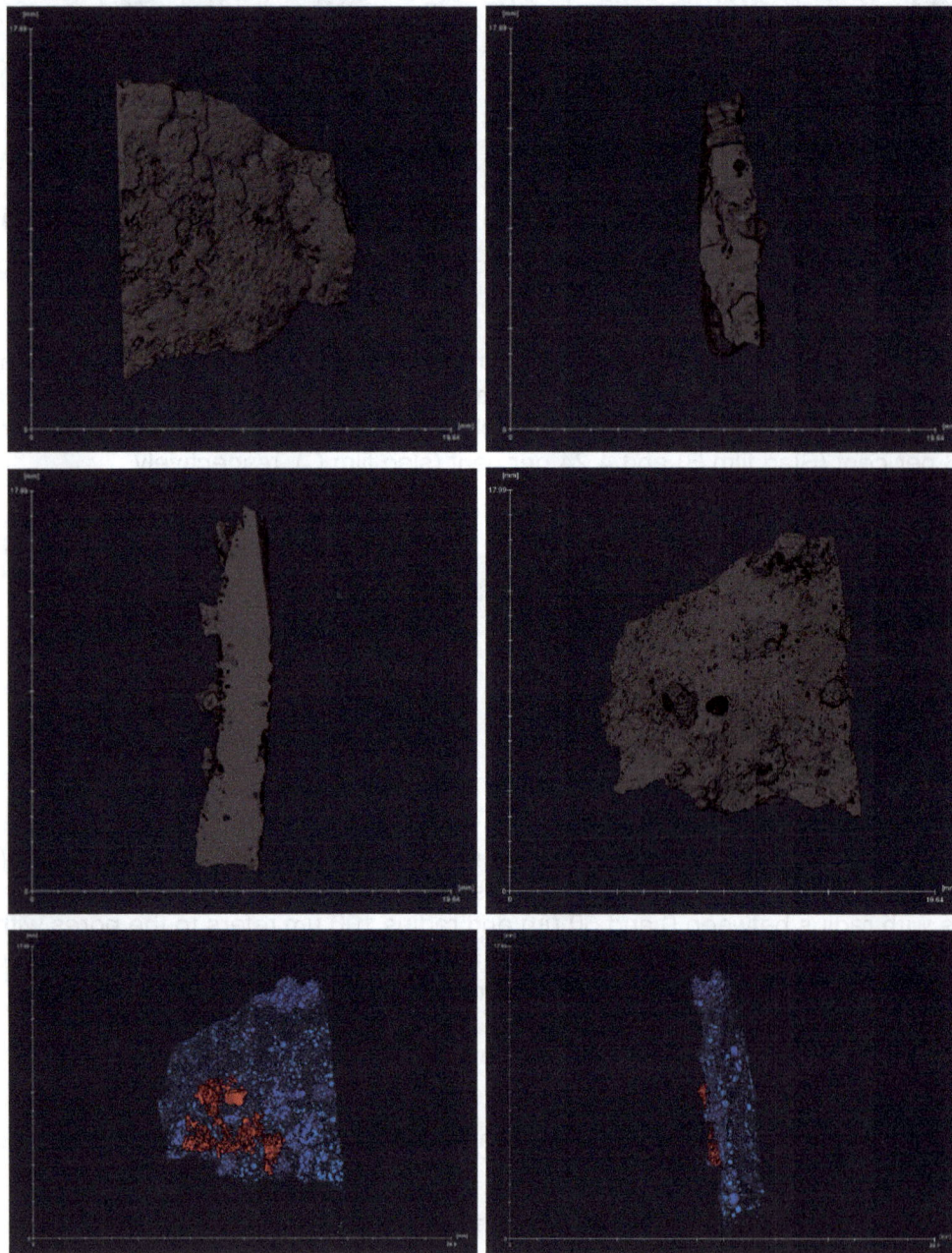

FIG 10 – XCT results for slag film C: morphology (top and middle) and pore distribution in 3D images (bottom).

The number of pores measured by XCT is large (3880 for slag film A sample, 539 for slag film B sample and 9344 for slag film C sample), which makes the data more truly reflect the characteristics of slag films. The measurement accuracy of XCT is 5 μm in this study, so the statistics range from 5 μm to 500 μm with intervals of 50 μm.

It was found that the sizes of all internal pores in slag film B are smaller than 100 μm and the proportion (number density percentage) of porosities with radius smaller than 50 μm is above 99 per cent. For slag film A, although the ratio (ie number density percentage) of the porosities smaller than 50 μm reached 98 per cent, their volume percentage is less than 7.5 per cent. The larger pores between 50 and 400 μm account for 92.5 per cent of the total volume of pores. Similar phenomena are observed for slag film C. In slag film C, the proportion (number density percentage) of pores with radius below 50 µm almost reaches 60 per cent, but its volume percent is only about 3 per cent. The proportion of pores with a radius of 50–100 µm accounts for about 35 per cent and the volume percent accounts for more than 20 per cent. In other words, the larger pores (>50 µm) contributes more to its volume ratio. The proportion of varying pores detected by XCT agrees with that by SEM (Figure 9), at least in trend.

Overall, the pores in industrial slag film B are the smallest in all cases, both in average radius (20 µm) and volume ratio (0.05 per cent). The proportion of pores in industrial slag film A is obviously larger than that in slag film B with the average radius 35 µm and porosity 0.7 per cent. The effect of pores in slag film C is more significant. The proportion of pore volume in slag film C is 3 per cent (average pore radius 85 µm), being one order of magnitude larger than that of slag films A and B.

CONCLUSIONS

The authors have explored various methods to determine the crystallinity (f_{cryst}) and porosity (%) of the industrial slag film samples and the following conclusions can be obtained:

- OM/SEM can be used to determine the crystallinity of the slag film samples and the results does give similar trend to the XRD method. The value and accuracy depend on the representative area(s) selected.

- XRD does provide reasonably consistent results of the crystallinity of slag films.

- EBSD can be a useful tool to determine the crystallinity of the slag film samples, but further development is required for the application.

- The porosity can be measured by using SEM + image analysis, however, variation for the same sample occurs with different researchers. The possible reasons are:
  - the difficulty to balance the measurement area and accuracy due to the characteristics of the thresholded image detection
  - the selected area(s) to test.

- XCT has the advantages of determining % porosity, pore volume, pore size distribution and pore location (steel side or mould side). This could be a useful technique for porosity determination in slag film samples.

ACKNOWLEDGEMENTS

The authors would like to acknowledge the financial assistance received from the Research Fund for Coal and Steel (RFCS) under the Grant Number RFCS-2018–847269. The authors also would like to thank the partners in the project for providing industry slag films and discussions – Sidenor Aceros Especiales S. L., ArcelorMittal Maizières Research, SSAB, Sandvik Materials Technology, Proximion, the Materials Processing Institute, Swerim and the Open University.

REFERENCES

Hunt, A and Stewart, B, 2016. Techniques for controlling heat transfer in the mould-strand gap in order to use fluoride free mould powder for continuous casting of peritectic steel grades, in *Proceedings of the 10th International Conference on Molten Slags, Fluxes and Salts (MOLTEN16)* (eds: R G Reddy, P Chaubal, P C Pistorius and U Pal), pp 349–356 (The Minerals, Metals and Materials Society).

Li, Z, Mills, K C and Bezerra, M C, 2004. Characteristics of mould flux films for casting MC and LC steels, in *Anais do XXXVI Seminário de Fusão, Refino e Solidificação dos Metais (Proceedings of the XXXVI Seminar on Fusion, Refining and Solidification of Metals)*, pp 13–24.

Mills, K C and Dacker, C-A, 2017. *The Casting Powders Book* (Springer: Switzerland AG).

Sridhar, S, Mills, K C, Ludlow, V and Mallaband, S T, 1998. A comparison of the mould powders used to cast slabs, billets and blooms, in *Proceedings of the 3rd European Conference on Continuous Casting*, pp 807–816.

Susa, M, Kushimoto, A, Toyota, H, Hayashi, M, Endo, R and Kobayashi, Y, 2009. Effect of both crystallisation and iron oxides on the radiative heat transfer in mould fluxes, *ISIJ International*, 49(11):1722–1729.

Worldsteel Association, 2022. World Steel in Figures 2023. Available from: <https://unesid.org/descargas_files/World-Steel-in-Figures-2023.pdf>

The recovery of pig iron from the Zimbabwean limonite-coal composite pellet

S Maritsa[1], S M Masuka[2] and E K Chiwandika[3]

1. Master candidate, University of Zimbabwe; Teaching Assistant, Harare Institute of Technology, Harare 263, Zimbabwe. Email: steven.maritsa@gmail.com
2. Lecturer, University of Zimbabwe, Harare 263, Zimbabwe. Email: shebarmasuka@gmail.com
3. Lecturer, Harare Institute of Technology and Module Lecturer, University of Zimbabwe, Harare 263, Zimbabwe. Email: chiwandikae@gmail.com

ABSTRACT

The iron and steel industry is experiencing an increase in the use of iron and steel commodities within societies, decarbonisation pressures, depletion of high iron ore grades and high energy consumption. This necessitates scrap metal recycling, low-grade ore utilisation using the induction or electric arc furnace to achieve the global target of net zero CO_2 emission. The available high-quality scrap metal cannot sustain the steel industry but the world has unused hematite-goethite ores. It is the aim of this research that this resource be utilised using the induction furnace to sustain the steel industry in an environmentally friendly manner. Indurated pellets prepared from the hematite-goethite ore, 5 wt per cent limestone and 5 wt per cent coal addition were smelted in the induction furnace at 1350 ± 50°C with 25 mins holding time. The effects of basicity and carbon content were also investigated to optimise the smelting parameters. The results showed that the careful control of the basicity, and iron-to-carbon ratio maximised total metal recovery. The optimal basicity ratio was determined to be 1.3 and a carbon to iron ratio of 1.4 achieved a total metal recovery of 73 per cent. The metal recovery was closely related to the estimated slag melting temperature. The basicity ratio of 1.3 achieved the lowest slag melting temperature estimated to be around 1380°C based on the $CaO-SiO_2-Al_2O_3$ phase diagram. The recovered metal had a comparable chemical composition and microstructure to that products from an ITmk3 furnace.

INTRODUCTION

The world target of net zero emissions by 2050 has resulted in massive demand for equipment and infrastructure of clean-energy technologies where iron and steel products play a pivotal role (International Energy Agency, 2023). De Carvalho et al (2022) reported the global steel consumption to be around 2 5 billion metric tons and is expected to rebound by 1 per cent in 2023 from a decline of 2.3 per cent in the year 2022. This scenario coupled with the depletion of high-grade iron ores has forced the industry to utilise lower-grade iron ores that have previously been unexploited. The mineralogy of these underutilised ores is often very complex, and requires innovative processing techniques that can yield low-impurity molten metal from these low-grade iron ore resources (Liang et al, 2023).

The recycling of iron and steel has been a form of environmental management and a way of waste control in the industry, with up to 70 per cent recycling of scrap. This scrap is an essential raw material for small plants and foundries where in some instances, producers are almost totally dependent on scrap metal as a raw material (Rehlke, 2001). In such instances, where scrap metal becomes a critical raw material, the availability and prices of scrap metal carry a huge bearing on the productivity of such enterprises.

Zimbabwe is reported to host about 111 million tonnes (Mt) of iron ore at its Ripple Creek deposits in the Midlands Province, and of these deposits, 41 per cent are limonite ores (Magunda, Dube and Simbi, 2004). The Zimbabwean limonite ore was reported to have a total Fe concentrations of approximately 53 wt per cent (Chisahwira et al, 2023). Ores containing an iron (Fe) concentrations below 55 wt per cent are termed low-grade iron ores, where 60 to 70 per cent weight of these lump ores get converted to fines during mining and transportation of run-of-mine (ROM) (Reddy, Sahoo and Kumar, 2023). Limonite ores are also referred to as 'low-grade ores' because of the high amounts of gangue material associated with the ores, and the high α-FeOOH content which causes problems during sintering for the blast furnace iron-making operations (Mochizuki and Naoto, 2019).

The blast furnace has been the major method for the production of iron. However, the arising pressure on environmental resources has necessitated the exploration of other ironmaking technologies. Alternative ironmaking technologies are focusing mainly on reducing CO_2 emissions, lowering of energy consumption and production costs. Some of the new technologies being used include high intensity smelting (HIsmelt) reduction ironmaking, cool Earth 50 (COURSE50), Rotary hearth furnace (RHF) (Zhang *et al*, 2018). One of the technologies that utilises the RHF to produce iron nugget is the Iron-making Technology mark 3 (ITmk3). The ITmk3 technology utilises coal composite pellets that are fed into a RHF as a raw material, heating to temperatures between 1350–1450°C and a residence time of about 8–10 mins (Upakare *et al*, 2018). Product from the ITmk3 technology is of high quality (Fe content > 96 wt per cent). The ability to separate metal and slag in one step via the ITmk3 process allows for effective concentration of the iron ore (Ghosh, Vasudevan and Kumar, 2021).

An investigation by Huang *et al* (2018), on the reduction of low iron ore composite pellets with a high silicon dioxide (SiO_2) of 40.61 wt per cent, sodium chloride (NaCl) was used as an additive to enhance the reduction process at experimental temperatures between 870°C and 990°C. It was concluded that the addition of NaCl increased the reduction of the iron oxide and hence suppressed the formation of the fayalite, a difficult-to-reduce phase, increasing metallisation by enhancing the reduction of wüstite to metallic iron. This was achieved by the enhancement of the gasification of coal, thus improving the reduction of iron oxides (Huang *et al*, 2018).

In smelting experiments at varying basicity conditions ranging from 0.46 to 1.91, where direct reduced iron (DRI) was formed into pellets and smelted at 1600°C for 1–2 hrs by Saleh and Rochani (2015), it was suggested that the reduction of FeO was favoured in alkaline conditions rather than in acidic conditions. At a basicity of 1.18, a recovery of 95.84 per cent was attained from the smelting of the DRI pellets to obtain pig iron with Fe content of 89.95 per cent. The basicity ratio of 1.18 was then considered the optimum basicity ratio. However, the amount of silica in the slag decreased with increasing basicity and the total Fe content in the slag was less than 10 per cent for the entire basicity range (Saleh and Rochani, 2015).

In another study by Sah and Dutta (2010) on the smelting reduction of iron ore-coal composite pellets in a liquid metal bath in an induction furnace, it was determined that the fractional reduction increased with a decrease in the total Fe to fixed C ratio, that is, the extent of reduction increased as the amount of reductant in the pellets increased. The average rate of reduction was calculated by dividing the fractional reduction by the time for which the pellets were in the bath and it was determined that initially the rate of reduction was very fast but decreased with time. This was attributed to the release of volatiles from the coal in the composites, which resulted in rapid solid-gas reaction (Sah and Dutta, 2010).

The aim of this paper was to study the reduction and recovery of cast iron from low-grade iron ore (limonite with 52 wt per cent Fe) by adopting the ITmk3 technology and determining the proper process parameters required to control quantity and quality of the produced iron nuggets. Careful control of basicity and carbon-to-oxygen ratio was studied to determine the best parameters required to produce metallic iron with similar properties to that of pig iron at a laboratory scale.

EXPERIMENTAL

Material preparation

Pellets used in this investigation were prepared using the Zimbabwean limonite ore, carbon, and limestone whose chemical composition is reported by Chisahwira *et al* (2023), according to the methods reported by the same. The iron ore was blended with 2.5 wt per cent coal and 2.5 wt per cent of limestone. The blend was then pelletised using a pelletisation disc and indurated at 1100°C for 25 mins. The indurated compressive strength of 250 kg/pellet was achieved. The chemical composition of the limonite ore and the limestone is shown in Table 1. The chemical composition of the pellets is shown in Table 2. The chemical analysis results were determined using atomic absorption spectrometry (AAS).

TABLE 1

Analysis of Zimbabwean limonite ore and limestone (Chisahwira et al, 2023).

Analyte	T-Fe	SiO$_2$	CaO	Al$_2$O$_3$	MgO	Mn	S	P	LOI
Zimbabwean Limonite Ore	52.51	8.78	1.32	1.30	0.59	2.1	0.005	0.04	11.87
Limestone	1.34	7.10	45.70	1.19	4.22	0.31	0.17	0.01	39.96

TABLE 2

Chemical composition of the indurated pellets made from the limonite ore.

Analyte	Total Fe	SiO$_2$	Al$_2$O$_3$	CaO	MgO	MnO
wt%	66.30	10.96	6.92	0.14	1.29	3.22

The results showed that the total iron concentration had increased from the 52.51 wt per cent in the ore to 66.30 wt per cent in the indurated pellets. Phase analyses in the ore was performed by X-ray diffraction (XRD) (Bruker AXS) using a Cu tube at a scan angle of 20° to 80°, scanning rate of 2°/min, voltage of 40 kV and current of 40 mA.

The results in Figure 1 showed that the major phases in the indurated pellets were Fe$_3$O$_4$, MgO·Fe$_2$O$_3$, Fe$_{2.33}$Si$_{0.67}$O$_4$ and Fe$_2$O$_3$.

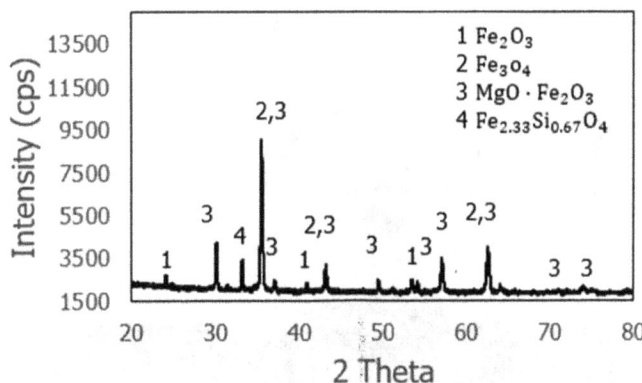

FIG 1 – Phase identification in the pellet by XRD.

Experimental procedure

About 250 g of the pellets were mixed with stoichiometric equivalent carbon to reduce the iron oxides, by placing the pellets in a crucible of 110 mm height and 60 mm diameter and placing the coal and limestone powder in the gaps between the pellets. The equivalent amount of carbon added was evaluated based on the amount of oxygen attached to Fe. The binary basicity of the charge was adjusted by the addition of some limestone with the chemical composition shown in Table 1. The binary basicity (CaO/SiO$_2$) was varied from 0.013 to 1.4, Table 3. The effect of basicity on the metal recovery had to be studied mainly in the alkali region because, smelting of low basicity burdens consumes high energy as compared to the alkali (Saleh et al, 2015). In a study conducted by Zhao et al (2017), an increase in basicity from 1 to 2 resulted in increasing recovery from 75.3 to 78.3 per cent and slightly decreased by adjusting basicity to 3.

TABLE 3

Evaluated chemical composition of the burden material.

Burden	wt%				Basicity ratio (CaO/SiO₂)
	T.Fe	CaO	SiO₂	Al₂O₃	
1	66.300	0.140	10.900	6.920	0.013
2	51.600	10.440	9.400	5.620	1.110
3	50.300	11.300	9.400	5.510	1.200
4	49.110	12.190	9.370	5.400	1.300
5	47.890	13.040	9.320	5.200	1.400

The mixed burden was then charged into a graphite crucible with the following dimensions: an inner diameter of 57 mm and a depth of 120 mm. The charged crucible was placed in an induction furnace. The temperature of the furnace was raised to 1350 ± 50 °C (Ishizaki, Nagata and Hayashi, 2006) and was held at that temperature for 25 mins to allow for the reduction of the pellets. The reduced samples were then removed from the furnace and quenched, the slag was then separated from the metal. The metal could be easily separated from the slag due to the difference in the densities. Figure 2 shows the image of the appearance of the slag and the metal before separation. The chemical composition of the slag and metal after separation is as shown in Tables 4 and 5 respectively to assist in providing a clear difference between the two phases

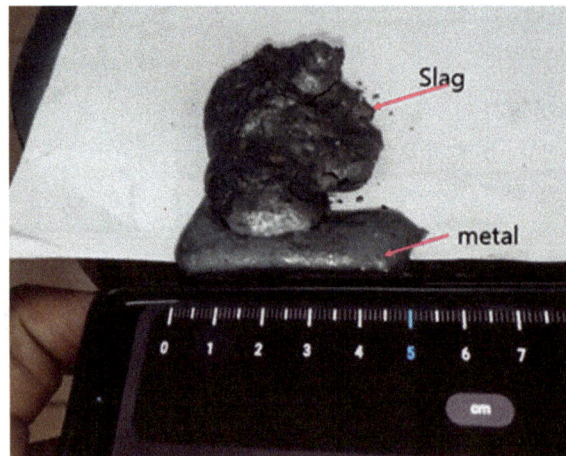

FIG 2 – Image showing metal and slag produced at a basicity of 1.3.

TABLE 4

Chemical composition of the slag phase after separation at a burden binary basicity of 1.3.

Analyte	MnO	TFe	CaO	MgO	Al₂O₃	SiO₂
wt%	4.0	9.55	36.20	2.41	10.30	28.03

TABLE 5

Chemical composition of the metal phase after separation at a burden binary basicity of 1.3.

TFe	C	Si	Al	Mn	P	S
96.37	3.24	0.002	0.001	0.148	<0.001	0.069

Cr	Ni	Mo	V	Co	Ti	CaO/ SiO₂
0.005	<0.004	0.020	0.0011	0.009	<0.001	1.30

The burden material carbon content was also varied in order to investigate the effect of the burden carbon to oxygen ratio on the recovery of iron from the pellets. Since all the experiments were done using the graphite crucible it was assumed that the effect of the carbon from the crucible in all the samples were equal and therefore the burden carbon played a significant role in the reduction and smelting of the pellets. Metal recovery was determined using Equation 1.

$$Total\ metal\ recovery\ (\%) = \frac{W}{W_0 T_{Fe}} \times 100 \qquad (1)$$

where:

W	is the weight of the metal separated from the slag
W_0	is the total weight of the reduced pellets
TF_e	is the iron weight percent in the pellets

The final chemical composition of the cast iron produced was analysed using a spark emission spectrometry.

Microstructural examination was also carried out after some metallographic preparations that begun with sectioning of the samples using a Discotom cutting machine and then mounting the samples using a Struers mounting machine. Grinding and polishing of the samples were completed using the Struers machine. Lastly, etching of the samples was done using a 4 per cent nital reagent. Microstructural examination was carried out at 20× magnification using a metallurgical light microscope.

RESULTS AND DISCUSSION

Effect of basicity on the metal recovery

The effect of basicity on the metal recovery was investigated by varying the basicity using limestone. The results are as shown in Figure 3.

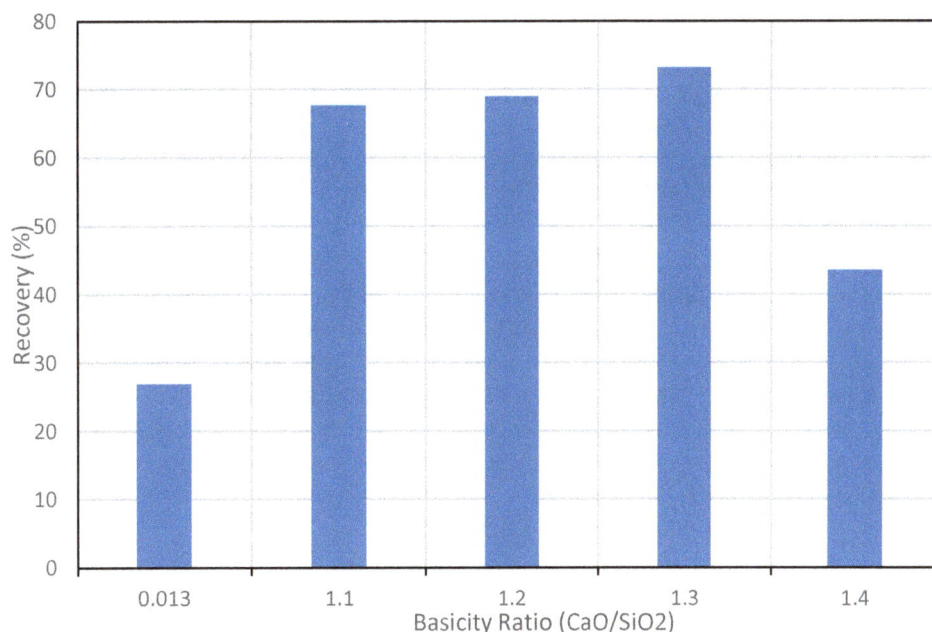

FIG 3 – Percentage metal recovery with increasing basicity.

It is clear from Figure 3 that the percentage metal recovery increased with increasing basicity up to 1.30 and above this value, the recovery decreased from about 73 per cent to about 44 per cent. These results obtained are comparable to those obtained by (Zhao et al, 2017) who studied slag/metal separation from H_2-reduced high phosphorus oolitic hematite. Optimal slag basicity gives the highest recovery of metal iron in smelting because it enhances the fluidity of slags and also suppresses the formation of highly stable compounds such as Fe_2SiO_4 (Saleh and Rochani, 2015).

An increase in basicity up to 1.30 in this current investigation might have facilitated for the easy separation of the metal and slag increasing percentage metal recovery due to the improved fluidity properties. Research has shown that enhanced fluidity in slag may result in higher diffusion coefficient of fine molten metal therefore making them easy to coalesce and form bigger drops that can easily and quickly drip below the slag as seen in Figure 4 where the amount of the total iron was varying inversely with the recovery (Tang *et al*, 2018). However more experimental investigations are required to validate the increased fluidity in this experiment. The binary basicity that gave the highest recovery was therefore determined to be 1.30 that yielded a percentage recovery of about 73 per cent. Large amounts of energy will be required to increase the metal recovery at low basicity, so as to trade-off with fluidity, thus more carbon source have to be used which will have a negative impact on the environment due to large CO_2 emission. A high basicity of 1.30 is favourable both on the energy requirement and on recovery. The total amount of Fe trapped in the slag was investigated using a handheld XRF and the results were as shown in Figure 4.

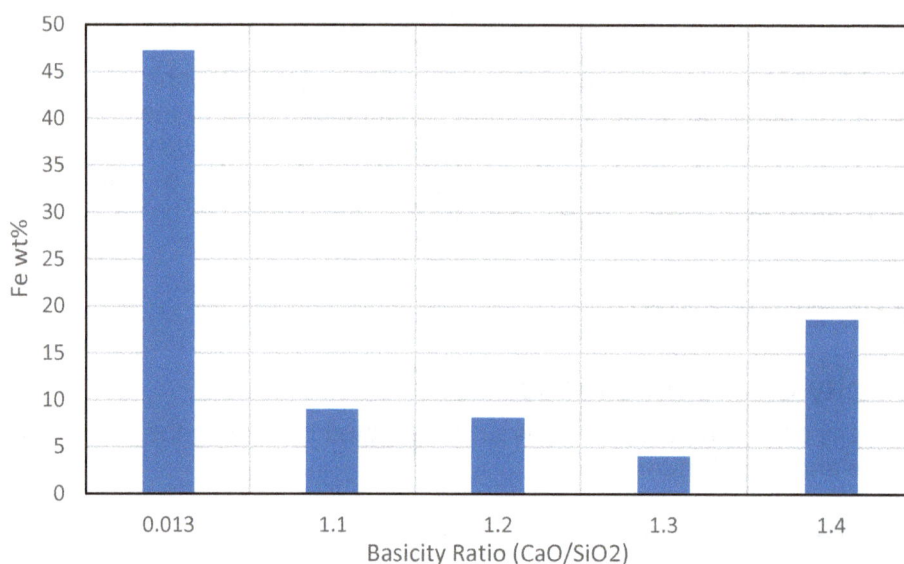

FIG 4 – Percentage of Iron content lost to slags with increasing basicity.

Iron by weight percent lost to slags decreased with increasing basicity from 0.013 to 1.3 following the opposite trend of the total metal recovery shown in Figure 3. Such a trend was observed because suppression of formation of the difficult-to-reduce compounds that might have occurred at suitable basicity increasing the metal recovery (Tang *et al*, 2018). Optimum basicity ratio was selected to be at 1.3 and the amount of Fe trapped in slag was only 4 wt per cent. Beyond the basicity of 1.3 the Fe wt per cent in slags rose to 18.6 wt per cent because of an increase in the slag to metal ratio which might cause the entrapment of more metal within the slags as shown in Figure 4. The recovered metal samples were taken for chemical analysis using spark emission spectrometry to understand the changes in the chemical composition of the cast iron recovered with increasing basicity and the results are shown in Table 6.

TABLE 6

Chemical composition of the cast iron with increasing burden basicity (wt per cent).

T.Fe	C	Si	Al	Mn	P	S	Cr	Ni	Mo	Co	Ti	CaO/SiO₂
97.45	2.270	0.034	0.006	<0.030	<0.02	<0.02	<0.01	0.034	0.052	0.019	<0.002	0.013
96.10	3.430	0.007	0.004	0.199	0.001	0.083	0.032	0.014	0.021	0.009	<0.001	1.110
95.76	3.730	0.009	0.018	0.122	<0.001	0.065	0.002	0.013	0.050	0.010	<0.001	1.200
96.37	3.240	0.002	0.001	0.148	<0.001	0.069	0.005	<0.004	0.020	0.009	<0.001	1.300
96.10	3.480	0.008	0.002	0.132	<0.00005	0.103	0.012	0.012	0.018	0.009	<0.001	1.400

The chemical analysis results showed an Fe concentration above 96 wt per cent in all the cast iron produced in the basicity range of 0.013 to 1.4. Silicon concentration in the metal was found to decrease with increasing basicity of up to 1.2 and increased thereafter. An iron-carbon phase diagram adopted from (Talla Padmavathi College of Engineering, 2017) was then used to predict the microstructure of the samples basing on metal carbon content.

The carbon content in the cast iron produced ranged from 2.27 to 3.73 wt per cent, which is less than the eutectic carbon content of 4.3 wt per cent, therefore the cast iron produced lies within the hypoeutectoid irons. From the iron carbon phase diagram, the samples produced were predicted to consist of a microstructure that include cementite, pearlite and transformed ledeburite. Micrographs in Figure 5, verified the predicted microstructures and confirmed that the castings are white cast irons by showing the ledeburite structure and a network of cementite grains. The ledeburite in Figure 5 is represented by the region with alternating dark and light layers whereas the cementite is the continuous threadlike light region.

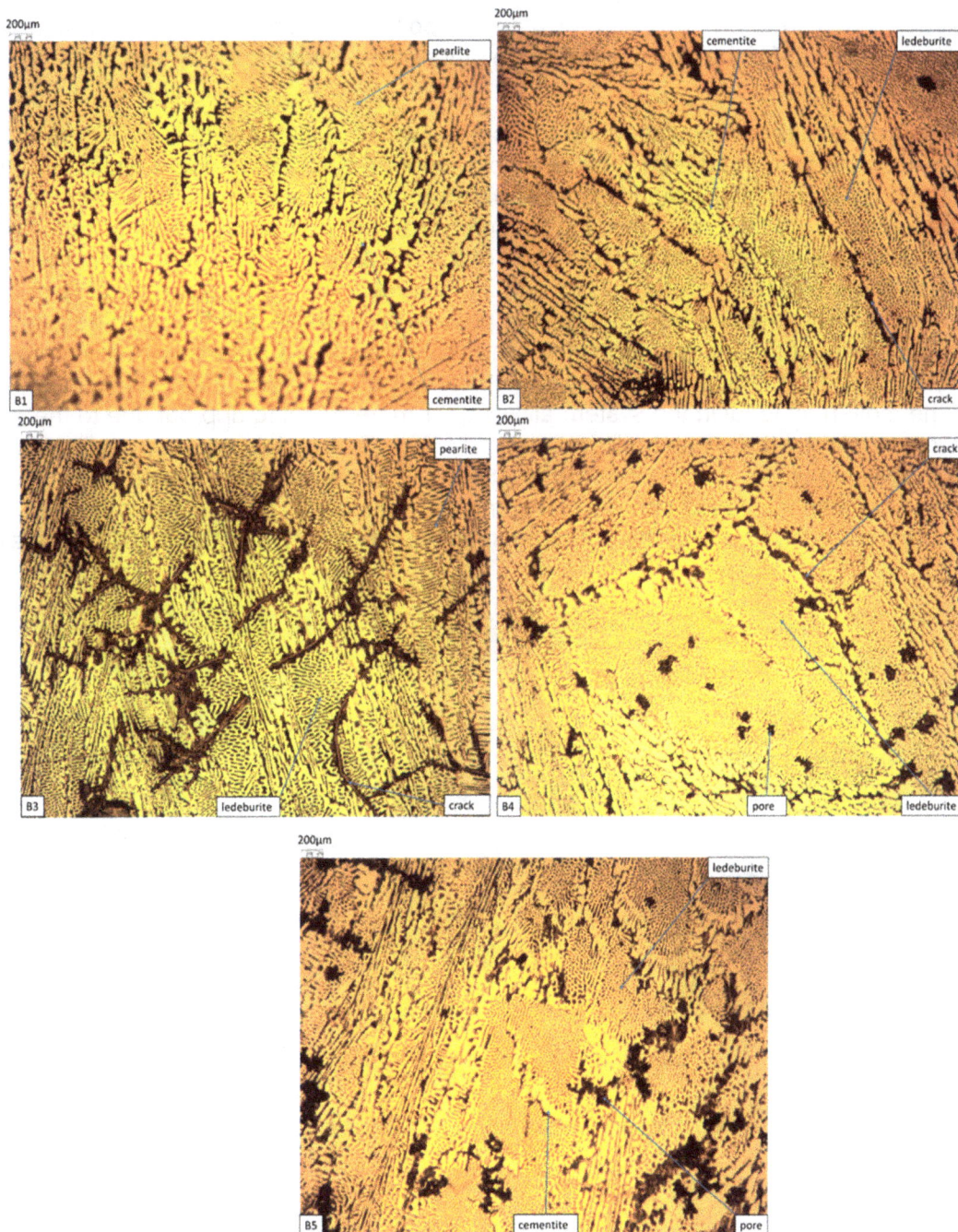

FIG 5 – Metal microstructure at 20× with increasing basicity ratio: B1: 0.034, B2: 1.11, B3: 1.20, B4: 1.30 and B5: 1.40.

The observed pores in Figure 5 in micrograph B4 and B5 may be as a result of the difference in molten metal viscosities as suggested by Cheng *et al* (2014) that viscosity is an important property in casting as it controls the rate of transport of liquid metals and may lead to defects like pores and hot tearing. Cheng *et al* (2014) went on to describe the Kaptay unified equation, which relates melting temperature of a metal to its viscosity. Tsepelev, Starodubtsev and Konashkov (2021) also discussed viscosity as a function related to temperature by associating viscosity to movement of liquid particles relative to each other which depends on diffusion mobility of the particles and in turn, linked to temperature by the Arrhenius equation. The different carbon in the samples shown in Table 6 indicates different melting temperatures of the samples as can be observed in Figure 5, this may lead us to conclude that the samples had different viscosities thus having pores in other castings as cited in literature (Cheng *et al*, 2014).

Effect of burden carbon on the metal recovery

The stoichiometric amount of carbon required for the reduction of iron oxide was evaluated based on the oxygen attached to iron and some excess carbon was added to the burden material in the first experiment to cater for the reduction of some other metal oxides components in the system at a basicity ratio of 1.30.

The percentage metal recovery was found to increase with the increase in the C/O ratio as shown in Figure 6, indicating an increase in the available amount of reducing agent which then became sufficient for the reduction of the metal oxide. The percentage metal recovery was found to decrease with the decrease in the C/O ratio showing a decrease in the available amount of reducing agent, which then became insufficient for the reduction of the metal oxide. The reduction of the iron oxide in this investigation occurred through direct and indirect reduction. The presence of large amounts of carbon might have resulted in an increased possibility of some additional energy through the increased possibility of the indirect reduction. The possibility of the reduction of the metal oxide is also highly expected to be near completion at high carbon content. This might have an effect on the fluidity of the molten material in the system allowing for more metal to drip out due to the increased temperature caused by the exothermic indirect reduction. The metal recovery decreased with the decrease in the amount of carbon due to the effect of the decreased chances of the exothermic reaction as well as an increased chance of incomplete reduction. The cast iron produced was also taken for chemical analysis using the emission spark spectrometry and the results were as shown in Table 7.

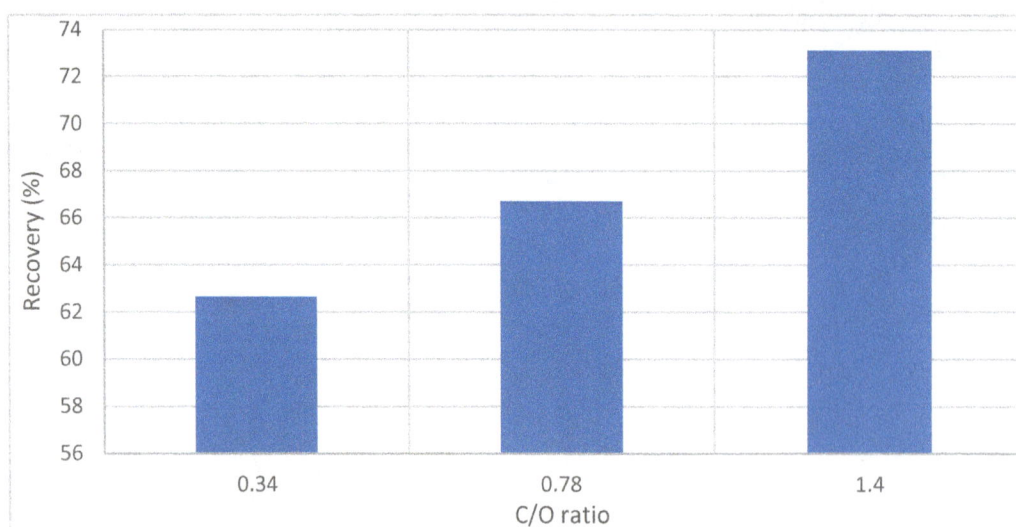

FIG 6 – Percentage metal recovery with change in C/O ratio.

TABLE 7

Chemical composition of the cast iron with decreasing burden carbon content.

TFe	C	Si	Al	Mn	P	S
96.37	3.24	0.002	0.001	0.148	<0.01	0.069
96.13	3.50	<0.001	0.0069	0.148	0.001	<0.001
95.3	4.16	0.003	0.0026	0.222	0.015	0.08

Cr	Ni	Mo	V	Co	Ti	C/O ratio
0.005	<0.004	0.020	0.0011	0.009	<0.001	1.40
<0.002	0.014	0.019	<0.0005	0.009	<0.001	0.78
0.004	0.034	0.019	0.0015	0.01	<0.001	0.34

The results clearly show a decrease in total Fe and an increase in C content in the metal with decreasing carbon content. The decrease in Fe might be an indication of the possibility of the incomplete reduction hence more Fe was trapped in the slag phase with decreasing reducing agent. The increase in C in metal observed might be related to the changes in the fluidity of the metal which might be related to the amount of indirect reduction in the system. A decrease in slag fluidity will offer more resistance to the dripping of the metal allowing more contact time of the metal with C which might result in increasing the amount of C observed in the metal indicated in Table 7. This was confirmed the slag chemical composition shown in Table 8 that showed a little more amount of Fe trapped in the slag with decrease in C/O ratio. The metal trapped in slag was higher when C/O ratio was lower than 1.40.

TABLE 8

Chemical composition of slag with decreasing burden C/O ratio (wt%).

T.Fe%	CaO	SiO$_2$	Al$_2$O$_3$	MgO	MnO	C/O ratio
6.68	36.20	20.03	10.30	2.41	4.55	1.40
9.09	31.00	28.00	13.00	2.50	4.40	0.78
10.80	31.64	29.55	13.67	2.30	4.55	0.32

Figure 7, clearly shows an increased generation of pearlite with a decrease in the burden carbon that resulted in an increase in C content within the metal. The micrographic images provide enough evidence of the increased C in metal that was shown in Table 7.

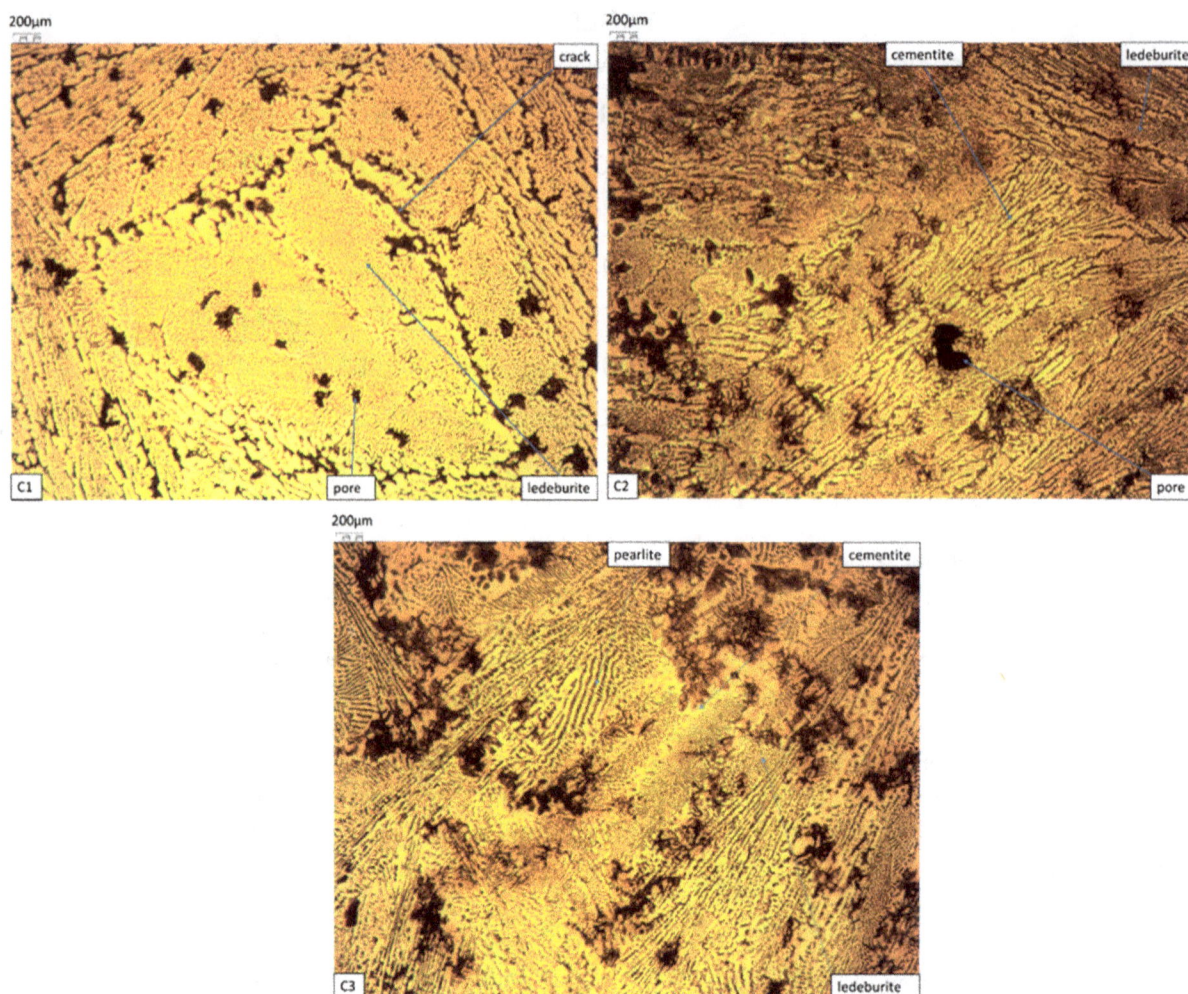

FIG 7 – Metal microstructure at 20× with change in burden C/O ratio: C1: 1.4, C2: 0.78 and C3: 0.34.

Estimation of slag melting temperature

Chemical analysis of slags produced at different basicity ratios was analysed using atomic absorption spectrometry and the results are presented in Table 9. It was important to carry out an analysis of the slags because slags contain unwanted impurities by forming oxides and floating them away from the metal.

TABLE 9

Slags chemical composition with increasing basicity (weight%).

T.Fe%	CaO	SiO_2	Al_2O_3	MgO	MnO	CaO/SiO_2
47.2	2.4	13.7	8.9	0.3	2.4	0.013
9.0	30.9	27.5	11.0	2.3	4.8	1.11
8.1	33.0	25.6	12.4	2.2	4.1	1.20
4.0	36.2	28.0	10.3	2.4	4.6	1.30
18.6	26.4	21.5	12.9	2.1	2.9	1.40

The major constituents of the slag were CaO, SiO_2 and Al_2O_3. The data from the slags chemical analysis was plotted on a CaO-Al_2O_3-SiO_2 phase diagram, Figure 8, to understand the oxide phases that may be in the slags and their melting temperatures at different basicity ratios.

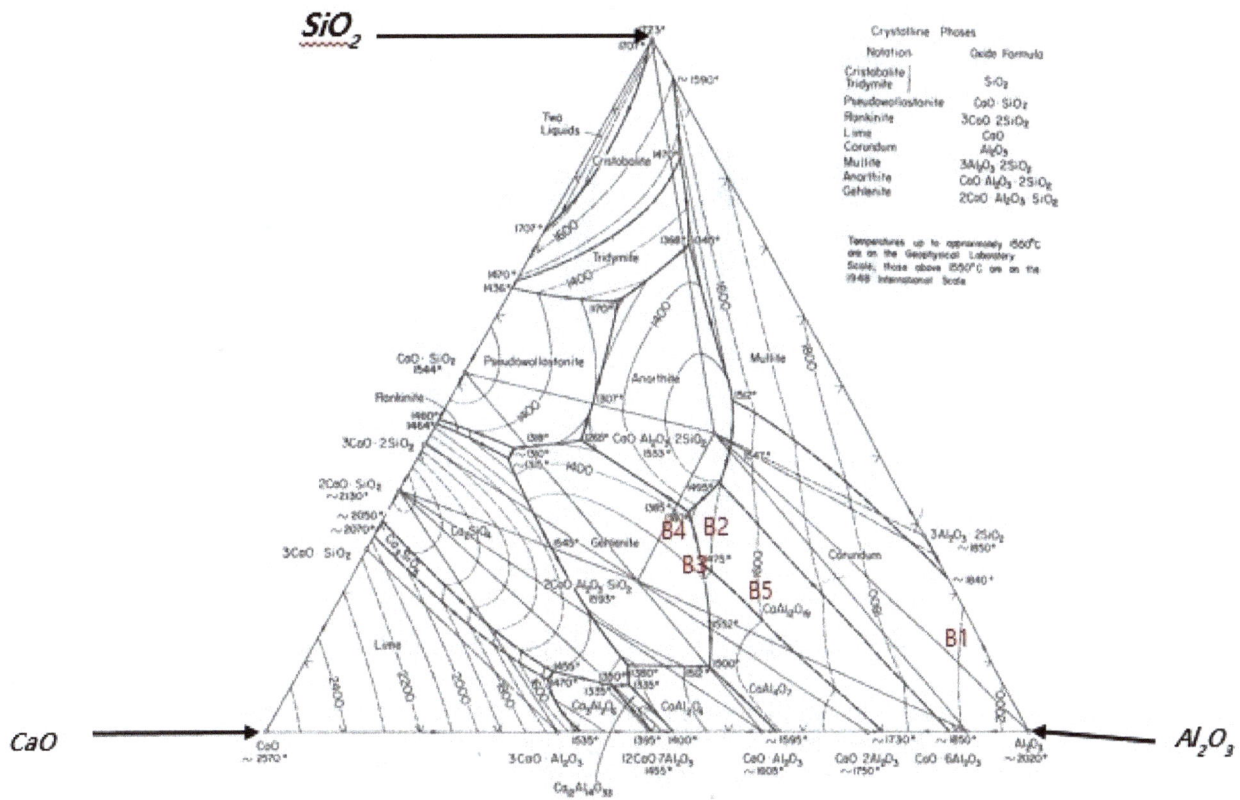

FIG 8 – CaO-Al$_2$O$_3$-SiO$_2$ phase diagram, adopted from (Draper, 1976): Approximate melting temperature with increase in basicity, B1: 0.034, B2: 1.11, B3: 1.20, B4: 1.30 and B5: 1.40.

The expected phases and their melting temperatures presented in Table 10 were estimated from the phase diagram shown in Figure 8.

TABLE 10

Slag's crystalline phases and melting temperature with increasing basicity ratio.

Estimated crystalline phases present	Approximate melting temperature (°C)	Basicity ratio (CaO/SiO$_2$)
CaO·6Al$_2$O$_3$	1850	0.013
CaAl$_{12}$O$_9$ and 2CaO·Al$_2$O$_3$·SiO$_2$	1475	1.11
2CaO·Al$_2$O$_3$·SiO$_2$ and CaAl$_{12}$O$_9$	1400	1.20
2CaO·Al$_2$O$_3$·SiO$_2$	1380	1.30
CaAl$_{12}$O$_9$	1600	1.40

Data from Table 6 was used to plot Figure 9 which shows the effect of the basicity ratio on the melting temperature of slags.

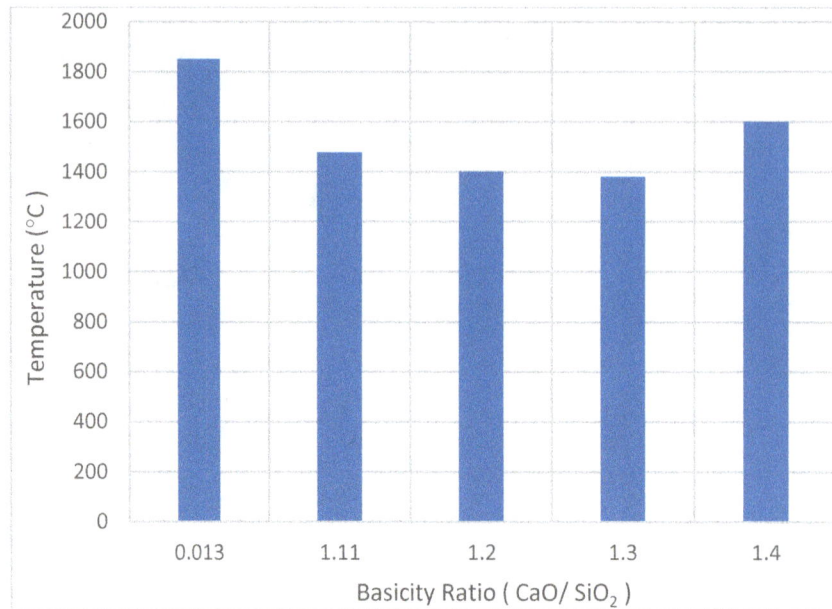

FIG 9 – Effect of basicity on melting temperature of slags.

Figure 9 shows a decrease in the melting temperatures of the slags as the basicity ratio increases. The melting temperature dropped from 1850°C at a basicity ratio of 0.013 to 1 380°C at an optimum basicity of 1.3. This trend is associated with the increase in metal recovery as basicity is increasing as shown in Figure 3 and in the results shown in Figure 4 where a decrease in iron content lost to slags with an increasing basicity ratio is observed. The melting temperature of slag is related to fluidity and viscosity as was discussed by Tang *et al* (2018) who stated that an increase in fluidity increases metal recovery by enhancing the diffusion coefficient of fine molten metal in the slag since there will be less drag due to decreased viscosity. From Figure 9, a clear drop in the melting temperature of the slags towards the induction's operating temperature indicates that the viscosity of the slags will be decreasing similar to the discussion by Cheng *et al* (2014) through the Kaptay unified equation which relates the melting temperature of a metal to its viscosity.

Muller and Erwee (2011) stated that slag viscosity determines metal yield, and impurity removal efficiency through its determination of the slag-metal separation efficiency. Data presented in Table 10 shows that the formed slags at the different basicity ratios do have different melting temperatures. Having furnace temperature fixed at 1350°C with accuracy ±50°C, it is highly likely that slags with melting temperature close to this value may have the best viscosity to allow for better slag-metal separation. Thus having the highest recovery at basicity ratio 1.3 where the slag melting temperature was 1380ºC unlike in other basicity ratios where the slags had melting temperatures above 1400ºC. Highly viscous slags tend to have reduced slag-metal separation efficiencies.

CONCLUSIONS

The reduction and recovery of cast iron from the Zimbabwean limonite carbon composite pellet was carefully studied at various burden basicity and carbon to oxygen ratio and it was found that:

- The careful control of burden basicity can improve the total metal recovery. The basicity with the highest recovery was determined to be 1.3.

- The total iron content in slag first decreased with the increase in burden basicity up to 1.3 and decreased with further increase in basicity.

- The optimal carbon to oxygen ratio was determined to be 1.4. A decrease in the carbon to oxygen ratio resulted in a decreased total metal recovery.

- The Zimbabwean limonite carbon composite pellet can potentially be used as a source of iron to produce cast iron with good quality to supplement the basic oxygen furnace, however more results are required to investigate the correlation between fluidity and the burden basicity in order to improve on metal recovery.

ACKNOWLEDGEMENTS

We would like to thank the Materials Technology and Engineering Department of the Harare Institute of Technology and the Institute of Mining Research of the University of Zimbabwe for providing laboratory equipment to carry out these investigations. We would also like to acknowledge Engineer Aron Mukuya and Mr Quinton Kadenge of Investcast Private limited for providing laboratory reagents.

REFERENCES

Cheng, J, Gribner, J, Hort, N, Kainer, K and Schmid-Fetzer, R, 2014. Measurement and calculation of the viscosity of metals—a review of the current status and developing trends, *Measurement Science and Technology*, (25):1–9.

Chisahwira, T, Masuka, S M, Maritsa, S and Chwandika, E K, 2023. The use of Zimbabwean limonite coal composite pellet as a sustainable feed for pig iron production, in *Proceedings of the Iron Ore 2023 Conference*, p 391 (The Australasian Institute of Mining and Metallurgy: Melbourne).

De Carvalho, A, Mercier, F, Mattera, G and Giua, L, 2022. Steel Market Developments, Q4 2022, Directorate for Science, Technology and Innovation Steel Committee, Organisation for Economic Co-operation and Development, 67 p. Available from: <https://one.oecd.org/document/DSTI/SC(2022)11/FINAL/en/pdf>

Draper, V F, 1976. Mullite Phase Equilibria in the System $CaO-Al_2O_3-SiO_2$, MS thesis, University of California, Berkeley.

Ghosh, A, Vasudevan, N and Kumar, S, 2021. Energy-Efficient Technology Options For Direct Reduction Of Iron Process (Sponge Iron Plants), The Energy and Resources Institute (TERI).

Huang, Z, Zhong, R, Yi, L, Jiang, T, Wen, L and Liang, Z I, 2018. Reduction Enhancement Mechanisms of a Low-Grade Iron Ore–Coal Composite by NaCl, *Metallurgical and Materials Transactions*, 49B:411–422.

International Energy Agency (IEA), 2023. Energy Technology Perspectives 2023, IEA. Available from: <https://www.iea.org/reports/energy-technology-perspectives-2023>

Ishizaki, K, Nagata, K and Hayashi, T, 2006. Production of Pig Iron from Magnetite Ore–Coal Composite Pellets by Microwave Heating, *ISIJ International*, 46(10):1403–1409.

Liang, Z, Peng, X, Huang, Z, Li, J, Yi, L, Huang, B and Chen, C, 2023. Innovative methodology for comprehensive utilisation of refractory low grade iron ores, *Powder Technology*, 418:118238, https://doi.org/10.1016/j.powtec.2023.118283

Magunda, S, Dube, N and Simbi, D J, 2004. Impact of Innovation Science and Technology on National Wealth Creation, paper presented at the HICC, Harare.

Mochizuki, Y and Naoto, T, 2019. Thermal Properties of Carbon-Containing Iron Ore Composite Prepared by Vapor Deposition of Tar for Limonite, *Metallurgical and Materials Transactions B*, 50B:2259–2272.

Muller, J and Erwee, M, 2011. Blast Furnace Control using Slag Viscosities and Liquidus Temperatures with Phase Equilibria Calculations, *Southern African Pyrometallurgy 2011* (eds: R T Jones and P den Hoed), pp 309–326 (Southern African Institute of Mining and Metallurgy: Johannesburg).

Reddy, A L, Sahoo, S K and Kumar, M, 2023. Studies on characterization of properties of low-grade hematite iron ores and their fired pellets, *Ironmaking and Steelmaking*, 50:1215–1223.

Rehlke, R D, 2001. Steel Plants: Size Location and Design, in *Encyclopedia of Materials*: *Science and Technology* (eds: K Buschow, R Cahn and B Ilschner), pp 8824–8832 (Elsevier: Amsterdam).

Sah, R and Dutta, S K, 2010. Smelting reduction of iron ore-coal composite pellets, *Steel Research International*, 81(6):426–433.

Saleh, N and Rochani, S, 2015. Study on basicity in direct reduced iron smelting, *Indonesian Mining Journal*, 18(2):59–70.

Talla Padmavathi College of Engineering (TPCE), 2017. IRON-CARBON Phase Diagram or Equilibrium Diagram, Talla Padmavathi College of Engineering – Mechanical Engineering Reference Books. Available from: <https://tpce-mechbooks.blogspot.com/2017/12/iron-carbon-phase-diagram-or.html> [Accessed: 2 March 2024].

Tang, Z, Ding, X, Yan, X and Dong, Y, 2018. Recovery of Iron, Chromium and Nickel from Pickling Sludge Using Smelting Reduction, *Metals*, 8(936):1–11.

Tsepelev, V S, Starodubtsev, Y N and Konashkov, V V, 2021. The Effect of Nickel on the Viscosity of Iron-Based Multicomponent Melts, *Metals*, 11(11):1724. https://doi.org/10.3390/met11111724

Upakare, R, Tupe, S, Tilak, A and Thosar, A, 2018. An Overview on ITmk3 (Iron-making Technology mark three) Process, *International Journal for Research in Engineering Application and Management*, 4:3.

Zhang, H, Wang, G, Wang, J-S and Xue, Q-G, 2018. Recent Development of Energy-saving Technologies in Ironmaking Industry, *IOP Conference Series: Earth and Environmental Science*, 233(5):052016. https://doi.org/10.1088/1755-1315/233/5/052016

Zhao, D, Li, G, Wang, H and Ma, J, 2017. Slag / Metal Separation from H2-Reduced High Phosphorus Oolitic Hematite, *ISIJ International*, 57(12):2161–2140. http://dx.doi.org/10.2355/isijinternational.ISIJINT-2017-304

Effect of SiO$_2$ on the structure and crystallisation of CaF$_2$-CaO-Al$_2$O$_3$ slag used in electroslag remelting

P M Midhun[1] and S Basu[2]

1. Graduate student, Indian Institute of Technology Bombay, Powai, Mumbai 400076, India.
   Email: 214116001@iitb.ac.in
2. Professor, Indian Institute of Technology Bombay, Powai, Mumbai 400076, India.
   Email: somnathbasu@iitb.ac.in

ABSTRACT

Electro-Slag Remelting (ESR) is a process capable of producing ingots with much lower centre-line segregation and better internal soundness than conventional ingot casting processes, as well as lower inclusion and sulfur concentrations. This process is commonly used for alloyed steels intended for critical applications like components of marine turbines, supercritical powerplants and rocket motor casings. It involves melting a metal electrode using electrical resistive heating in a slag pool and re-solidification of the resulting metal droplets beneath the slag layer, forming an ingot. Conventional ESR fluxes consist primarily of CaF$_2$, Al$_2$O$_3$ and CaO, along with minor concentrations of SiO$_2$, MgO etc. The emission of fluoride vapour from the molten slag causes several environmental and health problems, triggering a need to develop and reduce the fluoride content in the ESR flux.

The present study aims to develop low-CaF$_2$ alternatives to commercial ESR fluxes without compromising on productivity and the surface quality of the refined ingot. Thus, the modification in chemical composition needs to satisfy properties like electrical resistivity, viscosity and liquidus temperature, essential for efficient ESR operation. Since electrical resistivity and viscosity are highly dependent on the ionic structure of the molten slag, it is important to understand the effect of compositional modification on the structure of the flux. The present study employs Fourier-transform infrared spectroscopy (FT-IR) and Raman spectroscopy to investigate the structural variations in the modified fluxes, especially the effect of changing SiO$_2$ concentration, in comparison with conventional CaF$_2$-rich fluxes. Additionally, thermal analysis (Differential Thermal Analysis (DTA) and Thermogravimetric Analysis (TGA)) has been carried out to characterise the melting behaviour of the fluxes. The findings are compared with simulations carried out using the FactSage™ software, ver 8.3 (GTT Technologies, Aachen, Germany). Simultaneous analysis of liquidus temperature and the structure is expected to help in identifying the practical suitability of modified ESR flux compositions for industrial implementation.

INTRODUCTION

Electro-Slag Remelting (ESR) is one of the processes employed for refining of speciality steels. The metal electrode that needs to be refined is subjected to electrical resistive heating within a slag pool, leading to its melting. Owing to density difference, the molten metal droplets descend through the slag pool, held within a water-cooled copper mould and re-solidify at the bottom to form an ingot. The slag plays multiple roles during the ESR process. It acts as the heat source through resistive heating, acts as a medium for refining reactions (eg sulfur removal), dissolves the non-metallic inclusions separating from the molten steel and controls the surface finish of the solidified ingot by forming a solid slag skin at the ingot-mould wall interface. Figure 1 presents a schematic view of the longitudinal section, showing the key components of the process. Commercially used ESR fluxes typically consist of CaF$_2$ (30–70 per cent), Al$_2$O$_3$ (20–40 per cent) and CaO (20–40 per cent); this range of composition is known to provide an optimal combination of liquidus temperature, viscosity and electrical conductivity in the molten slag. However, CaF$_2$-containing slags are prone to evaporation of fluoride vapours during the operation of the ESR, resulting in detrimental effects on the environment as well as the health of the operating personnel (Ju *et al*, 2022; Zheng, Li and Shi, 2020). Due to this concern, reduction in fluoride concentration in ESR fluxes is being attempted across the globe. However, this endeavour is not easy since the fluoride-containing molten slag offers an optimal combination of liquidus temperature, flow behaviour and electrical properties, thus enabling efficient operation of the ESR process. The key challenge is to identify suitable chemical composition(s) for the molten slag such that its electrical conductivity, viscosity and liquidus temperature would remain within an optimal range, in spite of partial or total replacement of the fluoride content (Wroblewski *et al*, 2011, 2016).

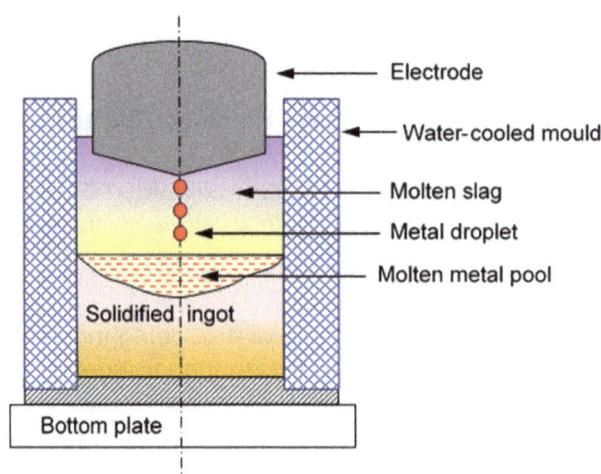

FIG 1 – Schematic view of longitudinal section of the ESR.

In addition to the major constituents (Al_2O_3, CaO and CaF_2), minor quantities of SiO_2, MgO etc are also added to the ESR flux for fine-tuning properties of the molten slag. The presence of SiO_2 in the slag is reported to have a contradicting influence on the ESR process; an increase in SiO_2 concentration tends to improve the surface quality but hinders the overall inclusion content and internal cleanliness of the solidified ingot. Consequently, in most electro-slag remelting practices, the presence of SiO_2 in the flux is considered permissible, but in low concentration (Shi *et al*, 2015).

Numerous studies have been carried out over the past few decades on the influence of varying concentrations of TiO_2, MgO etc on the properties and performance of commercial ESR fluxes (Shi *et al*, 2017; Ju *et al*, 2022; Zheng *et al*, 2020; Zheng, Li and Shi, 2020; Wan *et al*, 2022). However, relatively few reports on the modification of ESR flux through SiO_2 addition (Shi *et al*, 2015; Huang *et al*, 2021; Xu *et al*, 2022; Wan *et al*, 2023) are available in the public domain. Information on the structure as well as crystallisation behaviour of these ESR fluxes is also sporadic in literature. In the present study, SiO_2 concentration in a 20 per cent Al_2O_3 – 30 per cent CaO – 50 per cent CaF_2 commercial ESR flux, commonly employed for refining alloy steels, was modified through small increments, along with a simultaneous reduction in fluoride concentration. Since no studies have been investigated on this commercial flux composition until now, this work will be helpful for researchers and industries involved in ESR. Structural characteristics of the resulting slag were investigated through Raman and FT-IR spectroscopy. The results were crucial for understanding the polymerisation levels in the melt, which would influence the physico-chemical properties such as electrical conductivity and viscosity (Yan *et al*, 2021). The combination of these properties, in turn, would determine the overall suitability of the flux for use in the ESR process. Crystallisation studies were conducted through thermal analysis (Differential Thermal Analysis (DTA); Differential Scanning Calorimeter (DSC); Thermogravimetric Analysis (TGA)), which also allowed estimation of the liquidus and solidus temperatures. Thermodynamic simulations were carried out using the 'Equillib' module of FactSage™ software; the estimated values were compared with those obtained experimentally.

EXPERIMENTAL PROCEDURES

Sample preparation

The slag samples were synthesised using laboratory reagent extra pure grade powders of CaF_2 (≥97 per cent Purity), Al_2O_3 (≥99 per cent Purity), CaO (≥90 per cent Purity) and SiO_2 (≥99 per cent Purity) purchased from Loba Chemie Pvt Ltd. As-received Al_2O_3, CaO and SiO_2 powders were preheated at 1000°C in a muffle furnace to remove adsorbed moisture and decompose any hydroxide/carbonate that might have formed during exposure of the reagents to the ambient atmosphere. Preheating of CaF_2 was carried out at 700°C. A composition of 50 per cent CaF_2,

20 per cent Al_2O_3 and 30 per cent CaO (by mass) was chosen as the base mix for the flux. Requisite quantities of SiO_2 were added to the base mix, along with a corresponding decrease in CaF_2 content, to generate the entire range of compositions having varying concentrations of CaF_2 and SiO_2. The thoroughly mixed samples of the approximate total weight of 100 g were pre-melted in graphite crucibles (having inner diameter = 10 cm and depth = 10 cm) at 1480°C for two hours, followed by rapid quenching in liquid nitrogen. The fluoride content in the slag was measured using the fluoride ion-selective electrode (FISE) method, while calcium, aluminium and silicon concentrations were obtained using Inductively Coupled Plasma – Atomic Emission Spectroscopy (ICP-AES: ARCOS from SPECTRO Analytical Instruments, GmbH, Germany). Solution samples for ICP-AES and FISE analysis were prepared through the digestion of the slag samples via fusion method. Slag samples are well-mixed with lithium tetraborate and lithium metaborate in a ratio of 1:3:3. The platinum crucibles containing these premixed powders are kept inside a muffle furnace at 1050°C for 30 mins and consequently dissolved in 1N HCl. In the measurement of fluorine concentration using FISE method, an electrode has been dipped in the solution to be measured, and the resulting potential difference developed across the membrane has been correlated with the diffusion of fluorine ions from the solution. A few chelating agents (Solutions of Ammonium citrate dibasic, Ammonium tartrate dibasic, Citric acid, Ethylene diamine, EDTA and Sodium chloride) were added to the solution to facilitate the release of fluorine ions from the ionic complexes, ensuring accuracy in measurements (Yeager and Ramanujachary, 2007). Table 1 shows the chemical compositions of the fluxes premelted and the slag generated. After premelting, the CaF_2 concentration has been decreased, and the CaO concentration has been increased, corresponding to the fluoride evaporation (in forms of SiF_4, AlF_3, CaF_2, HF etc) during the chemical reactions at high temperatures (Persson and Seetharaman, 2007). Figure 2 presents the compositional variations within slag sample S2 measured before and after premelting. Even though the CaO/Al_2O_3 ratio in the as-mixed powder was initially 1.5, premelting at high temperature caused compositional variations due to the aforementioned reasons and the ratio was maintained within the range of 1.6–1.7 in every sample.

TABLE 1

Chemical composition of slag (in mass%).

Slag no	Composition before premelting				Composition after premelting			
	CaF_2	CaO	Al_2O_3	SiO_2	CaF_2	CaO	Al_2O_3	SiO_2
S1	50	30	20	0	44.34	35.46	17.52	0
S2	48.54	29.12	19.41	2.91	37.78	38.13	18.73	2.69
S3	47.16	28.31	18.86	5.66	41.49	32.42	18.22	5.21
S4	45.87	27.52	18.35	8.25	39.82	32.51	16.09	8.92

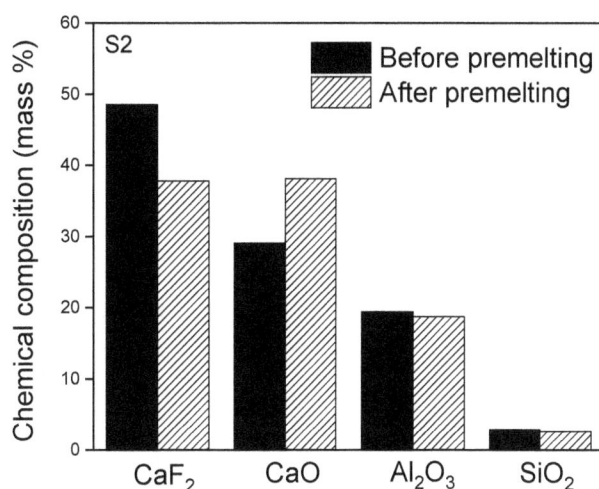

FIG 2 – Chemical composition of slag sample S2 before and after premelting.

Structural analysis using Raman and FT-IR Spectroscopy

Samples of the quenched slag were analysed using Raman Spectroscopy (Labram HR800 UV from Horiba Jobin Yvon, France) as well as Fourier-Transform Infrared (FT-IR) Spectroscopy (FT/IR-300E from JASCO).

Raman Spectroscopy

The Raman spectra for the slag samples were obtained at room temperature within the wavenumber range of 400–2000 cm^{-1}. A laser source with an excitation wavelength of 532 nm was used for all the measurements. The Raman spectra recorded were deconvoluted and fitted using the Fityk software with Gaussian bands.

FT-IR Spectroscopy

IR spectra of the samples were acquired in both absorption and transmission modes, over the wavenumber range of 4000–400 cm^{-1}, employing KBr detector. A spectral resolution of 4 cm^{-1} was chosen for all the measurements. The decision to use transmission and absorption modes arose from the fact that different structural aspects (eg Si-O-Si bond, O-Al-F bond etc) were better revealed under different spectroscopy modes. Each sample was prepared by mixing a particular slag with KBr in a mass ratio of 1:8 in an agate mortar, followed by pressing into pellets. The spectra obtained from the samples, each averaging 50 scans, are subsequently subtracted from the spectrum of 'pure' KBr powder in order to obtain the final spectra. The FT-IR absorbance spectra were deconvoluted using the Fityk and Origin software to better reveal the individual peaks and thus help in understanding the structural aspects. The features better revealed in the transmission spectra did not require any deconvolution.

Melting and solidification behaviour

DTA/TGA analysis

The liquidus temperature of each flux was determined through thermal analysis using a Netzsch™ STA 449 Jupiter unit (from Netzsch Instrument Inc., Germany). The samples were kept in argon atmosphere, with a purge rate maintained at 100 mL/min. For each measurement, close to 30 mg of the pre-melted flux sample was heated at a rate of 20°C/min from room temperature up to 800°C in an alumina crucible. Subsequently, it was further heated at 10°C/min up to 1450°C and held for 1 min for homogenisation. Following this, the melt was cooled at a uniform rate of 10°C/min until it reached 800°C after which it cooled to 100°C at the rate of 20°C/min. Typical temperature profile for the experiment is as shown in Figure 3. The DTA and TGA signals of the sample with time and temperature were captured automatically during each cycle and plotted.

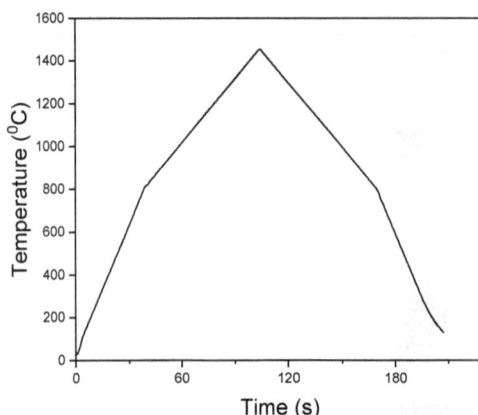

FIG 3 – Temperature profile for DTA/TGA experiment.

Thermodynamic simulations

FactSage™ was utilised to simulate the solidus and liquidus temperatures corresponding to the chemical composition of each sample. The Equilib module within FactSage™ was used, together with the FactPS and FToxid databases.

RESULTS AND DISCUSSION

Analysis of Raman spectra

Figure 4 shows the Raman spectra of the slag samples with varying SiO_2 concentrations. Each sample of molten slag was quenched in liquid nitrogen, with the objective of suppressing phase separations that might otherwise occur during slow solidification. Therefore, the samples were expected to be largely amorphous while retaining any crystalline phase and structural ordering that existed in the melt. In amorphous samples, Raman spectra are often present as envelopes, within which overlapped spectral peaks are responsible for various structural units (Hwa, Hwang and Liu, 1998).

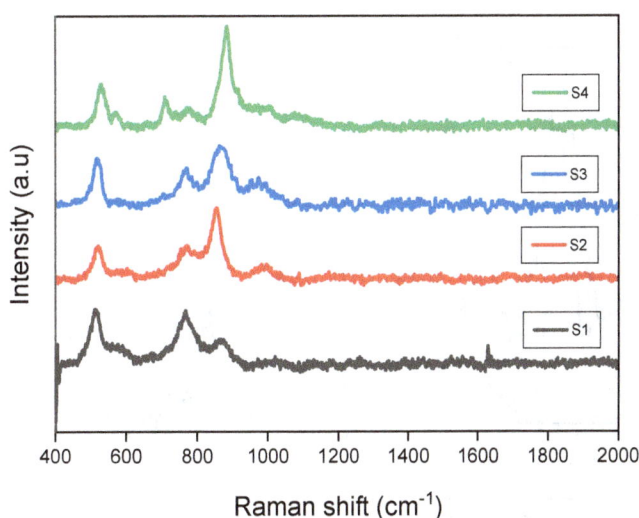

FIG 4 – Raman spectra of samples with various SiO_2 concentrations.

In the Raman spectra obtained from the slag samples, similar envelopes were deconvoluted through the Gaussian function to differentiate the individual structural components. The minimum value of the correlation coefficient (R^2) was maintained at 0.96 for this purpose. Fityk software, utilising Gaussian bands, was used for deconvoluting the Raman spectra and fitting the individual peaks. The resulting deconvoluted spectra for all four slag compositions, shown in Figure 5, provide insights into the relative proportions of the structural units and help to illustrate the influence of chemical composition (*viz* SiO_2 concentration) on the structural feature of each slag.

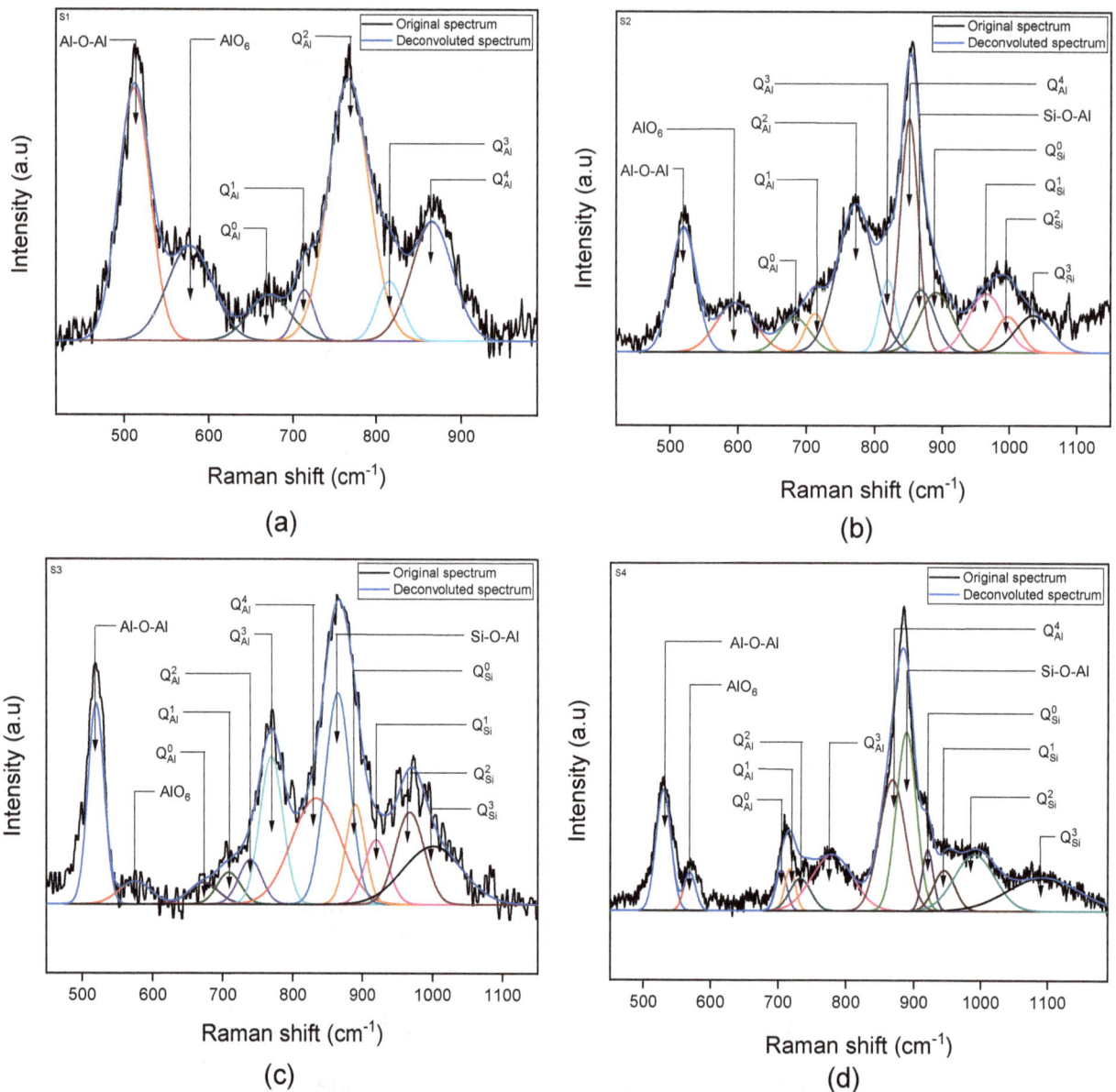

FIG 5 – Deconvolution results of Raman spectra: (a) S1; (b) S2; (c) S3; (d) S4.

In the silicate network, the absorption peak intensity corresponding to the vibrations of the Si-O covalent bonds in 'SiO$_4$' tetrahedral units varies depending on the population of bridging oxygen in the network. The quantity of bridging oxygens in the 'SiO$_4$' unit is denoted by 'n' in 'Q$^n_{Si}$', where n = 0, 1, 2 and 3, correspond to monomer, dimer, chain and sheet structures of silica, respectively. Similarly, 'Q$^n_{Al}$' is defined as the 'AlO$_4$' tetrahedral unit with bridging oxygens expressed as 'n', varying over the range 0–4. As an example, the Raman shift at ~525 cm^{-1} in the spectra corresponds to the transverse vibration of bridging oxygen within 'Al-O-Al' bonds. Absorption peaks at other Raman shift values similarly represent the vibrations associated with other bonds in the structural units. Table 2 lists the ranges of Raman shift identified in the slag samples bands and the corresponding structural information (Huang *et al*, 2021; Zheng *et al*, 2020; Haghdani, Tangstad and Einarsrud, 2022; McMillan and Piriou, 1983; Li, Shu and Chou, 2014; Hwa, Hwang and Liu, 1998; Kim and Park, 2014).

TABLE 2

Assignments of raman bands in spectra of CaF_2-Al_2O_3-CaO-SiO_2 slag (Huang *et al*, 2021; Zheng *et al*, 2020; Haghdani, Tangstad and Einarsrud, 2022).

Raman shift (cm^{-1})	Raman assignment and type of vibrations
500–552	Transverse motion of bridging oxygen in Al-O-Al linkages
578–609	AlO_6 stretching vibrations
665–695	Q^0_{Al}- symmetric stretching vibration of Al-O bonds in AlO_4 tetrahedra (zero bridging oxygen)
700–738	Q^1_{Al}- symmetric stretching vibration of Al-O bonds in AlO_4 tetrahedra (one bridging oxygen)
748–765	Q^2_{Al}- symmetric stretching vibration of Al-O bonds in AlO_4 tetrahedra (two bridging oxygen)
788–805	Q^3_{Al}- symmetric stretching vibration of Al-O bonds in AlO_4 tetrahedra (three bridging oxygen)
820–855	Q^4_{Al}- symmetric stretching vibration of Al-O bonds in AlO_4 tetrahedra (four bridging oxygen)
859–880	Si-O-Al stretching vibrations
890–910	Q^0_{Si}- symmetric stretching vibration of Si-O bonds in SiO_4 tetrahedra (zero bridging oxygen)
915–940	Q^1_{Si}- symmetric stretching vibration of Si-O bonds in SiO_4 tetrahedra (one bridging oxygen)
940–990	Q^2_{Si}- symmetric stretching vibration of Si-O bonds in SiO_4 tetrahedra (two bridging oxygen)
1000–1070	Q^3_{Si}- symmetric stretching vibration of Si-O bonds in SiO_4 tetrahedra (three bridging oxygen)

The integrated area under the curve for each deconvoluted peak may be considered as proportional to the fraction of the corresponding structural unit in the slag sample. Variations in these fractions with respect to the change in SiO_2 concentration in the slag samples are shown in Figure 6. From this figure, it is evident that a rise in SiO_2 concentrations in the slag leads to an enhancement in the intensity of the spectral peak associated with 'Si-O-Al' vibrations. This can be correlated with the more intense vibration of 'Si-O-' bonds with higher SiO_2 content. This vibration is conducive to the 'AlO_4' tetrahedral structures forming complex 'Si-O-Al' bonds in the slag structure. Consequently, there is a rise in complex networks and polymerisation in the slag increases. In the context of alumina-tetrahedral units, an increase in Q^3_{Al} and Q^4_{Al} networks was observed, along with a corresponding decrease in Q^0_{Al}, Q^1_{Al} and Q^2_{Al} networks, as shown in Figure 6. These observations suggest a noticeable increase in the polymerisation of alumina tetrahedra upon the addition of SiO_2 (up to 9 mass per cent) to the selected flux composition. In silicate networks, the simple networks Q^0_{Si} and Q^1_{Si} exhibit a decrease, while the proportions of complex Q^2_{Si} and Q^3_{Si} increase upon the addition of SiO_2. Thus, it can be inferred that the degree of polymerisation of both silicate and alumina networks was enhanced upon increasing SiO_2 concentration.

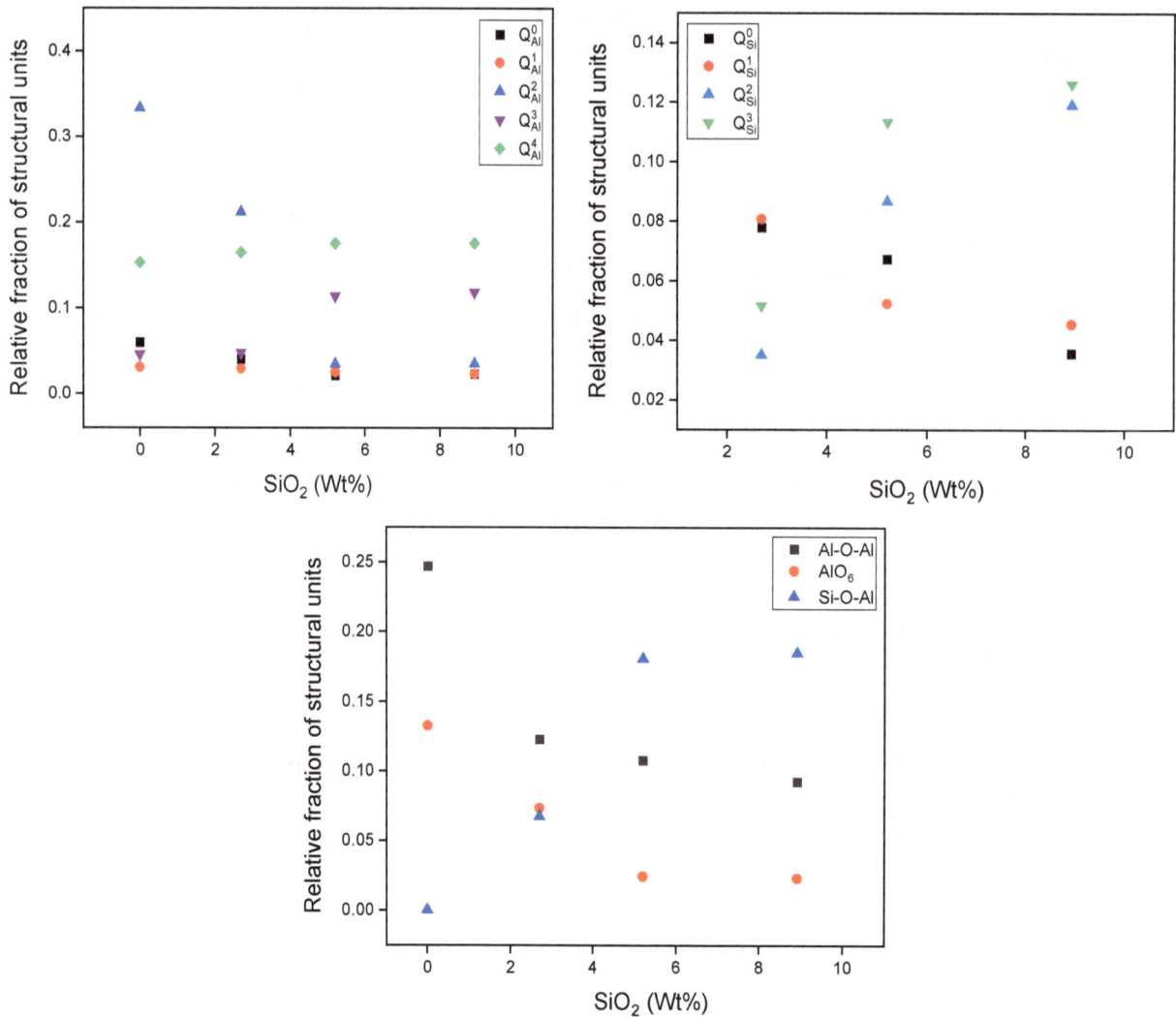

FIG 6 – Relative fraction of structural units in CaF_2-Al_2O_3-CaO-SiO_2 slag from Raman spectra with various SiO_2 concentrations.

Structural analysis of slag using FT-IR spectroscopy

Both the transmittance and absorbance spectra from the slag samples were obtained using FT-IR spectroscopy. Analysis of transmission depths corresponding to various structures present within the slag in the transmittance spectra provided information about changes in the polymerisation behaviour within each slag sample after SiO_2 addition, especially visible variations on the spectra associated with SiO_4 and AlO_nF_{4-n} structures. As well as, the deconvolution of curves from the absorbance spectra of each sample could provide quantitative information about the presence of distinct structural units inside AlO_4 and SiO_4 structures and gave better clarity about variation in absorption associated with AlO_nF_{4-n} structures, which was difficult to differentiate from some smaller structural units of SiO_4 from the analysis of transmission spectra due to overlapping. Combining both of these sets of information provides a more detailed understanding about the variation in the polymerisation behaviour of slag samples after silica addition.

Figure 7 presents the Fourier Transform Infrared (FT-IR) transmittance spectra of the quenched slag samples. By comparison with prior investigations, the different ranges of wave number associated with specific structural units of interest have been identified and are listed in Table 3. For example, the transmission characteristics over the wave number range of 1200–800 cm⁻¹ represents the asymmetric stretching of 'Si-O' bonds within 'SiO_4' tetrahedral units, while that over the 800–600 cm⁻¹ range is associated with the asymmetric stretching of 'Al-O' bonds in 'AlO_4' tetrahedral units (Park, Min and Song, 2002; Hwa, Hwang and Liu, 1998; Ju et al, 2022). Owing to the chemical composition, the slag samples investigated are likely to contain 'SiO_4' and 'AlO_4' tetrahedral units with varying bridging oxygen, represented by the parameters 'Q^n_{Si}' for silicate units and 'Q^n_{Al}' for aluminate units ('n' = 0–4). A similar situation has been indicated in the previous section (on Raman spectra) as well.

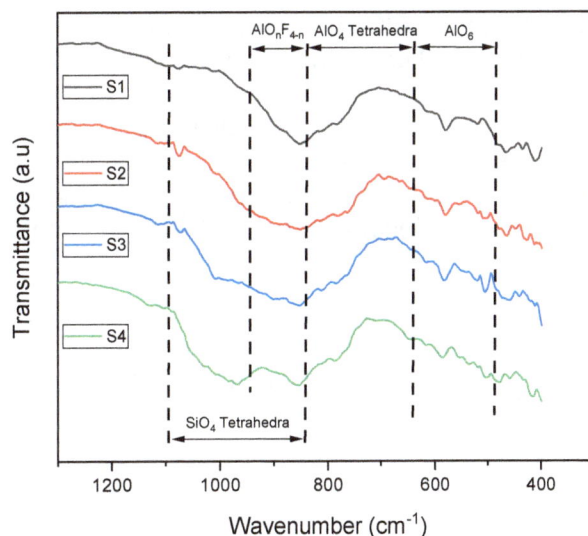

FIG 7 – The FT-IR spectra (transmittance) of the liquid nitrogen quenched CaF_2-Al_2O_3-CaO-SiO_2 flux samples with various SiO_2 contents.

TABLE 3

Assignments of FT-IR bands in spectra of CaF_2-Al_2O_3-CaO-SiO_2 slag (Park, Kim and Sohn, 2011; Lao *et al*, 2019; Ju *et al*, 2021).

Wavenumber range (cm^{-1})	Types of vibrations in FT-IR
850–890	Q^0_{Si}- asymmetric stretching vibration of Si-O bonds in SiO_4 tetrahedra (zero bridging oxygen)
910–930	Q^1_{Si}- asymmetric stretching vibration of Si-O bonds in SiO_4 tetrahedra (one bridging oxygen)
960–990	Q^2_{Si}- asymmetric stretching vibration of Si-O bonds in SiO_4 tetrahedra (two bridging oxygen)
1030–1070	Q^3_{Si}- asymmetric stretching vibration of Si-O bonds in SiO_4 tetrahedra (three bridging oxygen)
800–940	Asymmetric stretching vibrations of AlO_nF_{4-n}
490–600	Asymmetric stretching vibration of Al-O bonds in AlO_6 octahedra
610–630	Q^0_{Al}- asymmetric stretching vibration of Al-O bonds in AlO_4 tetrahedra (zero bridging oxygen)
650–670	Q^1_{Al}- asymmetric stretching vibration of Al-O bonds in AlO_4 tetrahedra (one bridging oxygen)
690–720	Q^2_{Al}- asymmetric stretching vibration of Al-O bonds in AlO_4 tetrahedra (two bridging oxygen)
730–760	Q^3_{Al}- asymmetric stretching vibration of Al-O bonds in AlO_4 tetrahedra (three bridging oxygen)
770–820	Q^4_{Al}- asymmetric stretching vibration of Al-O bonds in AlO_4 tetrahedra (four bridging oxygen)

It can be seen in Figure 7 that the transmission peak corresponding to 'AlO_nF_{4-n}' in sample S1 (0 mass per cent SiO_2) is stronger than in the SiO_2-containing samples over the wave number range

820–940 cm^{-1}. This may be attributed to the possibility of simpler silicate units (Q^0_{Si} and Q^1_{Si}) contributing to the transmission signal over the same wavenumber range in samples S2, S3 and S4, in addition to the contribution from 'AlO$_n$F$_{4-n}$' structural units. Based on this, it can be inferred that sample S1 contains a higher fraction of 'AlO$_n$F$_{4-n}$' units in comparison with the other slag samples. Upon careful observation around 1050 cm^{-1}, the transmission peak broadens and increases in intensity with increasing SiO$_2$ concentration, indicating vibrations associated with more complex Q^2_{Si} and Q^3_{Si} structural units. However, it was difficult to distinguish between the overlapping peaks associated with 'AlO$_4$' and 'AlO$_6$' structural units in the transmittance spectrum, for each of the slag samples. Therefore, FT-IR spectra of all the samples were acquired in the absorption mode as well. The resulting absorbance spectra were deconvoluted to identify the contributions from the individual structural units.

Figure 8 illustrates the deconvoluted FT-IR absorption spectra of all four slag samples, which were fitted using Gaussian functions with the Fityk software. Table 3 lists the FT-IR wave number ranges associated with the individual structural units that were identified after deconvoluting the absorbance spectra (Park, Kim and Sohn, 2011; Kim, Kim and Sohn, 2013; Lao *et al*, 2019; Ju *et al*, 2021).

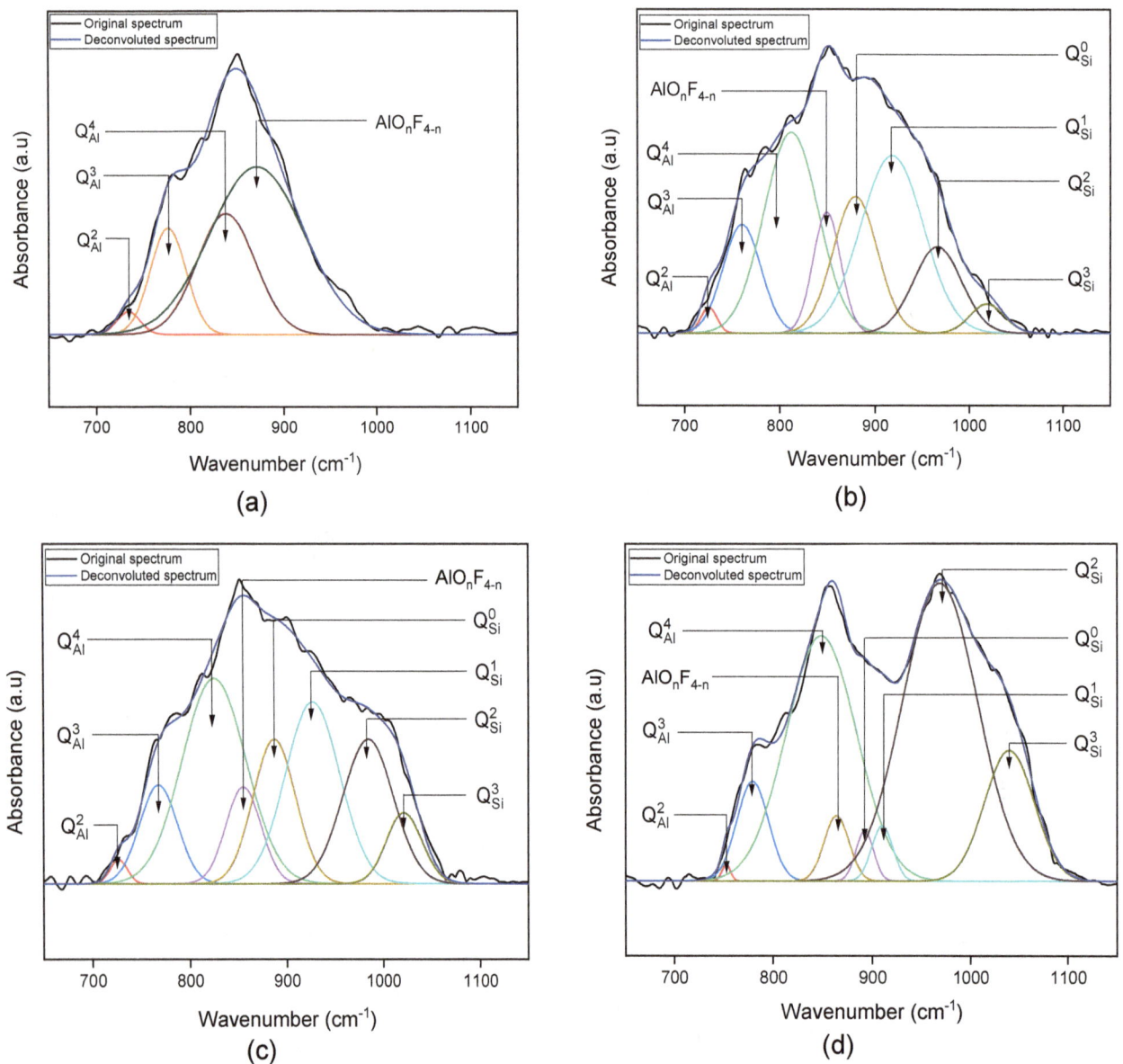

FIG 8 – The deconvolution results of FT-IR absorbance spectra: (a) S1; (b) S2; (c) S3; (d) S4.

In Figure 9, 'relative fraction of structural units' in each sample has been calculated from the deconvoluted peaks of absorption spectra and plotted against the increase in SiO$_2$ concentration in the slag samples. It has been observed from the plot that the addition of SiO$_2$ into the selected base flux composition results in a continuous decrease in the presence of 'AlO_nF_{4-n}' networks. This can be attributed to the relative reduction in the fraction of fluorine and alumina in the system following the addition of SiO_2. From the analysis of the variation of different structural units within the silicate tetrahedral networks, it is observed that complex structural units, ie Q^3_{Si} and Q^2_{Si}, show an increase, while smaller structural units like Q^0_{Si} and Q^1_{Si} decrease. This trend signifies an increase in the polymerisation within silicate network structure. Upon further analysis of the graphs, it is observed that simple alumina tetrahedral units, Q^2_{Al} and Q^3_{Al}, decrease with the addition of SiO_2, while the complex network, Q^4_{Al} increases. These observations align with the inferences obtained from Raman spectroscopy, which also indicate behaviour of increase in the polymerisation of silicate and aluminate networks upon the addition of SiO_2 into the system.

Before introducing silica (SiO_2) into the '50 per cent CaF_2 – 20 per cent Al_2O_3 – 30 per cent CaO' flux system, the system predominantly contained alumina tetrahedral networks with calcium cations serve for charge balance. However, with the addition of SiO_2, silica forms tetrahedral network structures, and basic oxides/fluorides such as CaF_2 and CaO attempt to break these network structures, resulting in the formation of various simple/complex structural units of silica. In the presence of silica networks, aluminium in alumina units will try to enter silicate networks to form 'Si-O-Al' bonds, giving rise to the development of complex networks like Q^3_{Al} and Q^4_{Al}. In the case of silica tetrahedral networks, it is observed that the introduction of SiO_2 into the flux results in an increase in complex tetrahedral units (Q^2_{Si} and Q^3_{Si}), accompanied by a decrease in simple units, such as Q^0_{Si} and Q^1_{Si}. Because the basic oxides/fluorides available in the melt structure are almost constant and when more silica is introduced into the melt, available oxides/fluorides will try to break down more complex networks initially.

The results obtained from FT-IR spectroscopy and Raman spectroscopy show an increase in the level of polymerisation within both silica and Alumina tetrahedral networks due to the addition of SiO_2 into the selected base composition. These structural variations could influence the physicochemical properties such as viscosity and electrical conductivity of the slag. As the polymerisation level within the melt increased gradually upon adding SiO_2, viscosity and electrical resistivity values are expected to rise (Yan *et al*, 2021). However, further experiments are recommended to validate these inferences.

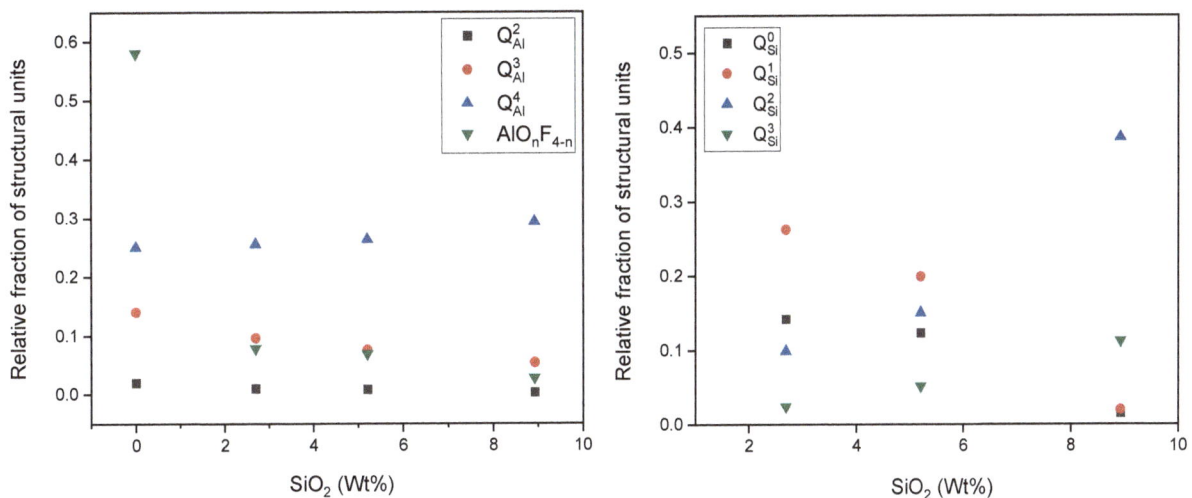

FIG 9 – Relative fraction of structural units in CaF_2-Al_2O_3-CaO-SiO_2 slag from FT-IR absorbance spectra with various SiO_2 contents.

Analysis of melting and solidification behaviour

Analysis of FactSage™ simulations

The Equillib module of FactSage™ software has been used to collect preliminary information on solidus-liquidus temperatures of the selected slag compositions, which can help identify the expected temperature range where the major phase transitions occur for the selected compositions.

Figure 10 shows the phase diagram plotted using FactSage™ simulation, where the X-axis shows the addition of SiO_2 concentration in mass per cent in a slag system containing 50 per cent CaF_2 – 20 per cent Al_2O_3 – 30 per cent CaO. Furthermore, from the Equillib module in the FactSage™, information on the formation of various product phases and their fractions under equilibrium conditions at different temperatures can be collected as output data, and this information has been used to plot the graphs showing the variation of the fraction of phases formed during the solidification of the slag samples, as shown in Figure 11. From these figures, liquidus and solidus temperatures for all four slag samples have been estimated. Liquidus temperature represents the temperature at which solidification initiates from 100 per cent liquid, and solidus temperature represents the temperature at which the whole liquid disappears on cooling.

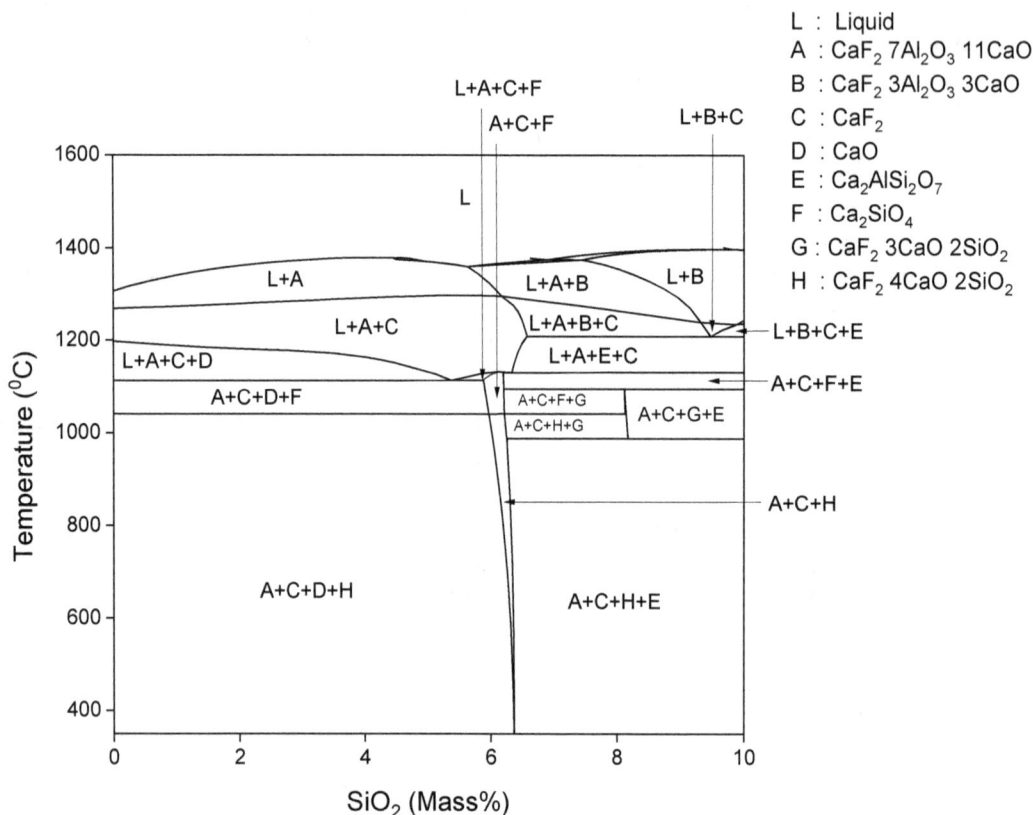

FIG 10 – The phase diagram plotted through FactSage™ simulation (X-axis indicates the concentration of SiO_2 added into 50 per cent CaF_2 – 20 per cent Al_2O_3 – 30 per cent CaO).

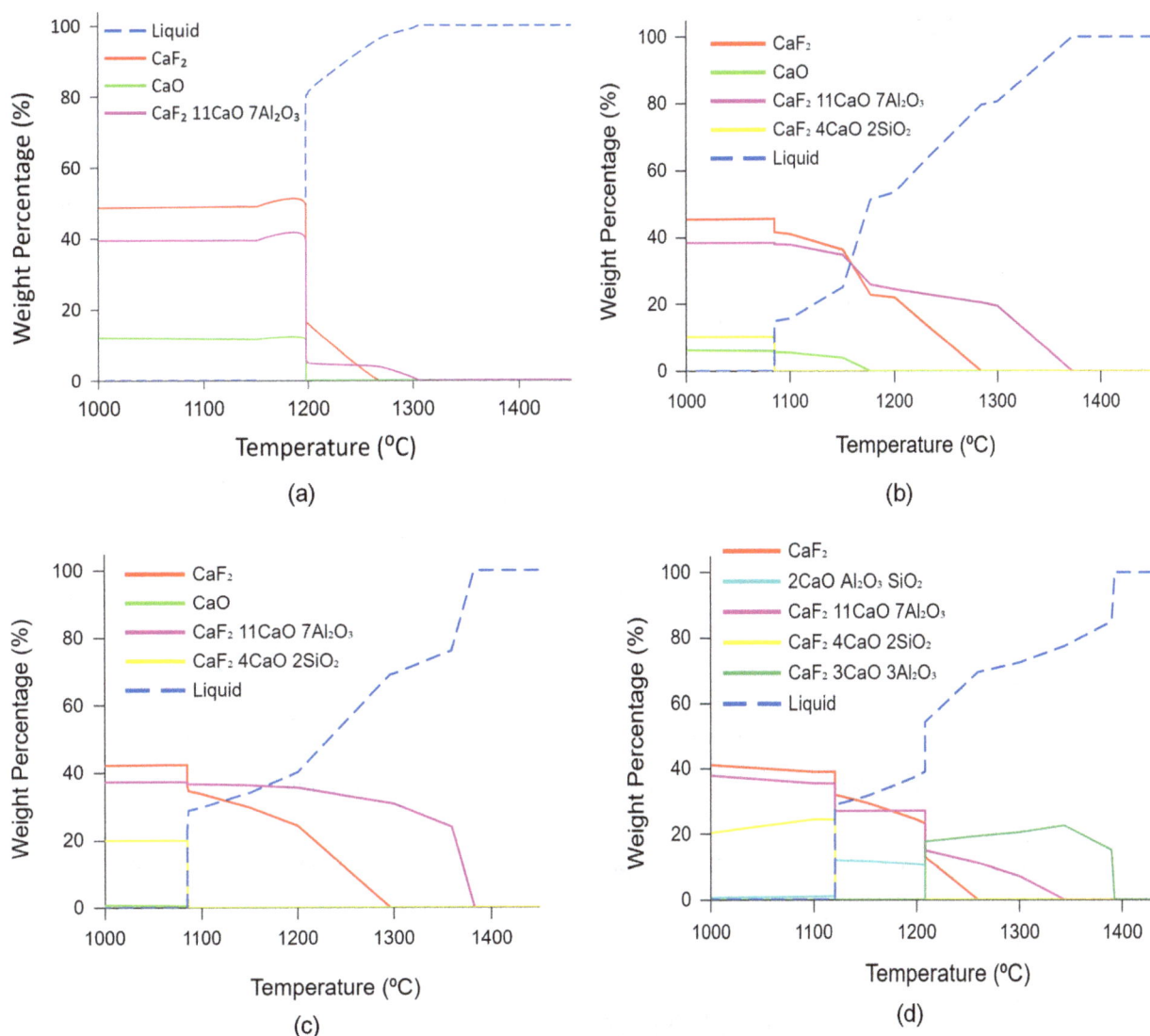

FIG 11 – Crystallisation behaviour of the fluxes calculated using FactSage™: (a) S1; (b) S2; (c) S3; (d) S4.

DTA/TGA analysis

Figure 12 shows the DTA and TGA curves of all four slag samples. The liquidus temperature of a slag is defined as the temperature at which the liquid phase initiates transformation to the solid phase during the cooling cycle or the temperature at which the solid sample is fully transitioned into liquid in the heating cycle. The solidus temperature is the temperature at which the first liquid phase forms in a heating cycle or the temperature at which the whole liquid phase transforms to the solid phase in the cooling cycle. Uniform heating and cooling rates of 10°C/min are performed for all fluxes in the temperature range of 800°C to 1450°C to determine and compare the liquidus and solidus temperatures. In the plots, the downward peaks are considered to be endothermic peaks. For example, from the DTA plots, the endothermic peaks observed during the heating cycle are analysed to estimate the liquidus temperature of the slag samples.

In Figure 12, the sudden variations from the DTA curve while changing the heating/cooling rate, represented by the symbol '#', can be associated with the thermal hysteresis. In the present work, these variations are not considered exo/endothermic peaks during the analysis. The symbols '•' and '×' in the curve denote estimated liquidus and solidus temperature, respectively. Further, the significant peaks observed between the temperature range of 800–1450°C in both heating and cooling cycles for all four slag samples are presented in Figures 13 and 14 respectively.

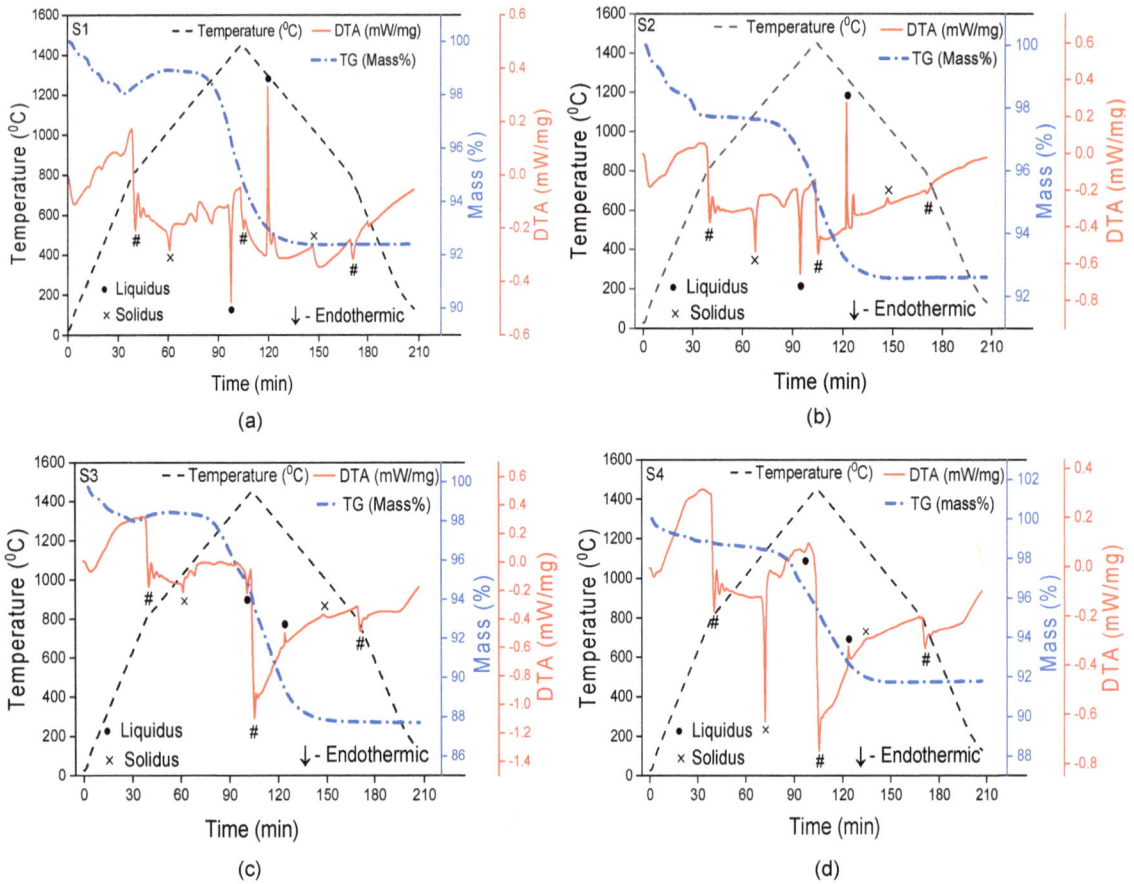

FIG 12 – DTA/TGA curves of the CaF$_2$-Al$_2$O$_3$-CaO-SiO$_2$ slag melts: (a) S1; (b) S2; (c) S3; (d) S4.

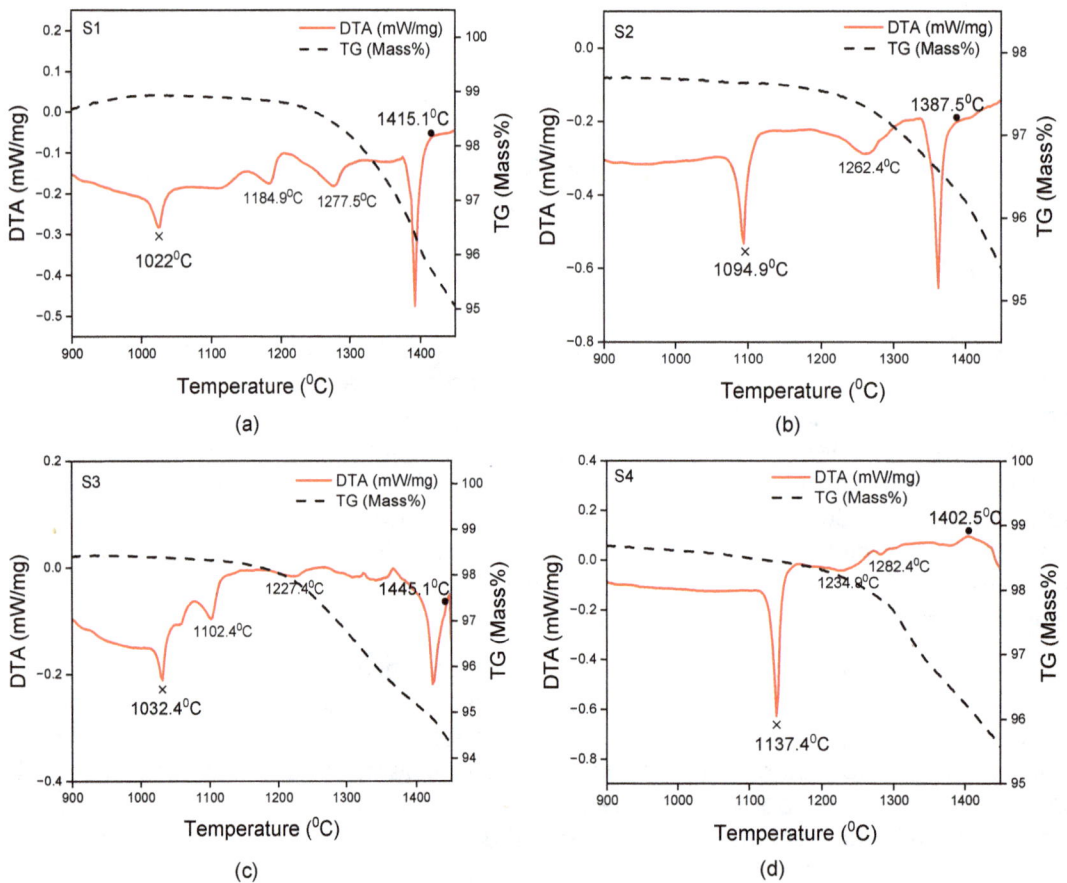

FIG 13 – DTA/TGA curves of the CaF$_2$-Al$_2$O$_3$-CaO-SiO$_2$ slag melts in heating cycle: (a) S1; (b) S2; (c) S3; (d) S4.

12th International Conference on Molten Slags, Fluxes and Salts (MOLTEN 2024) | Brisbane, Australia | 17–19 June 2024

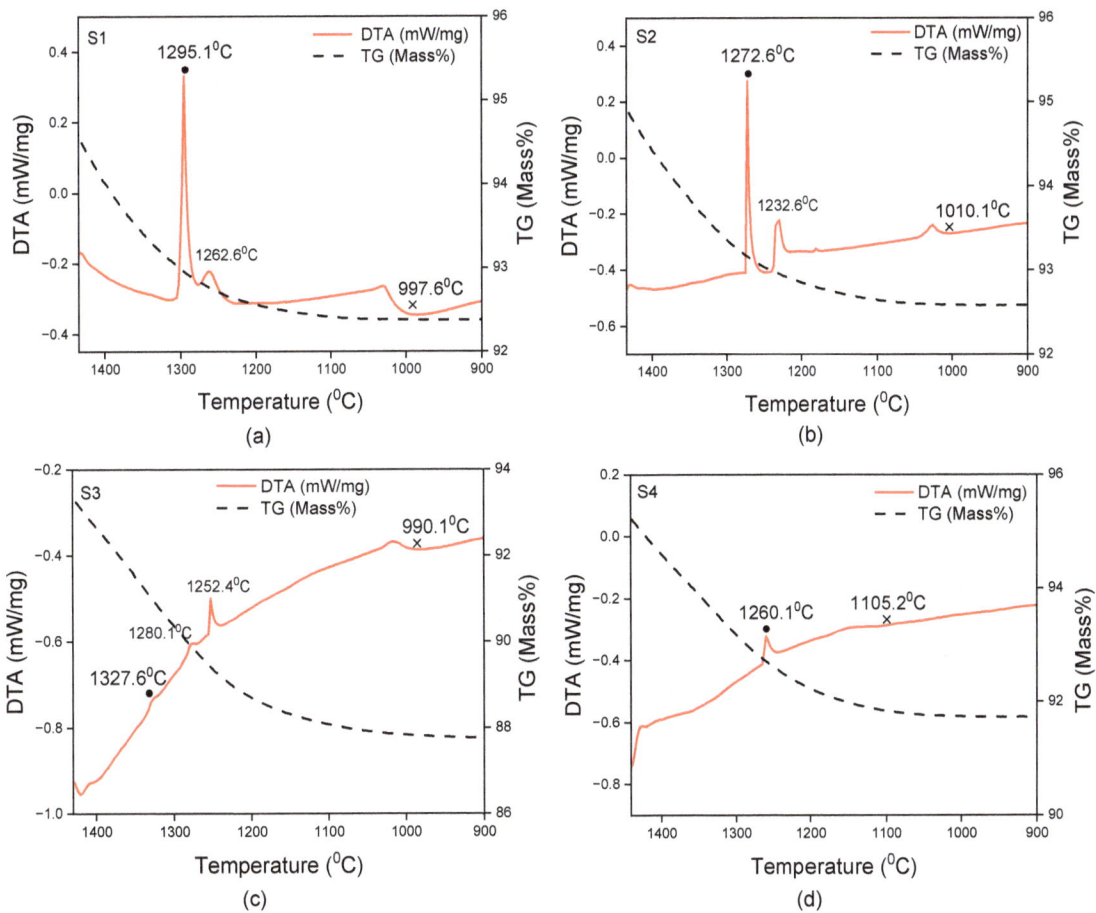

FIG 14 – DTA/TGA curves of the CaF_2-Al_2O_3-CaO-SiO_2 slag melts in cooling cycle: (a) S1; (b) S2; (c) S3; (d) S4.

In the heating cycle, following the energy required to overcome the nucleation barrier, the start of the formation of the liquid phase can be considered as the peak point of the first endothermic peak. Further, following the intermediate phase transformations during melting, the last endothermic peak observed in the cycle indicates the completion of solid-to-liquid phase transformation. The liquidus temperature is estimated from the point at which the curve regains its linearity following the final peak, showing that the slag sample is expected to be fully transitioned to the liquid phase. During the cooling cycle, crystallisation begins at the temperature associated with the first observed peak. After overcoming the energy required for the nucleation barrier, the temperature at which the first crystalline phase formed is considered as the peak point of the initial peak in the cooling cycle, representing the liquidus temperature. Following this, multiple peaks can be observed, corresponding to the possibility of crystallisation temperature of various intermediate phases formed, eventually leading to the last peak associated with the formation of the final phase and the complete solidification of slag. The solidus temperature in the cooling cycle is interpreted as the temperature at which the DTA curve regains its linearity following the final observed peak. Figures 13 and 14 present the estimation of liquidus and solidus temperature according to this convention.

After analysing the DTA curves, it is evident that the studied slag samples exhibit a high degree of undercooling in the cooling cycle, leading to the delayed observation of crystallisation peaks. Figure 15 shows the variation in the presence of exothermic peak associated with liquidus temperature in the cooling cycle due to undercooling from the slag sample S1. In the present work, considering these aspects of undercooling along with the knowledge of the mass loss due to fluoride evaporation at high temperature at the end of the heating cycle causing compositional changes, the selection of solidus-liquidus temperature values is determined from endothermic peaks from the heating cycle.

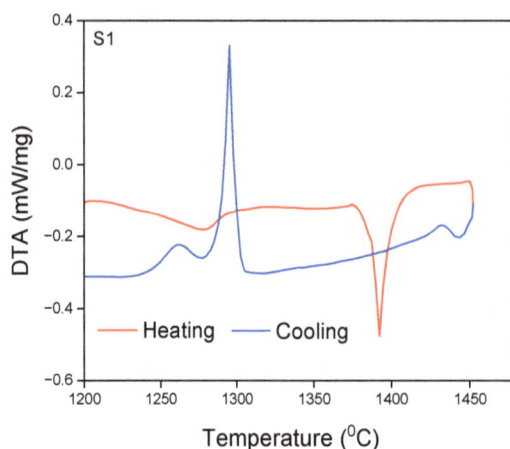

FIG 15 – DTA curve of slag sample S1 showing a high degree of undercooling.

Table 4 presents solidus and liquidus temperature information of all the slag samples determined from Figure 13, ie during the heating cycle. The peaks observed between the solidus and liquidus temperatures in DTA results can be due to the possible potential phase transformations within this temperature range, as reported by some previous works with similar slag samples (Huang *et al*, 2021). The plotted graphs from FactSage™ data shown in Figure 11 can further support this inference, as they show certain thermodynamically possible intermediate phase transitions between solidus and liquidus temperatures. For validating the presence of phases in the slag samples, the X-ray diffraction (XRD) characterisation has been performed on all four samples, quenched from 1000°C. Results are presented in Figure 16. The major phases identified in the analysis are 'CaF_2 $11CaO$ $7Al_2O_3$' and 'CaF_2'. The results closely match with those of FactSage™ simulation as presented in Figure 11, where, in all the slag samples, major phases present below solidus temperature are these identified phases from XRD. However, further studies are recommended to confirm this and identify the reason for the formation of the aforementioned phases and the formation of peaks. TGA results indicate a mass drop of approximately ~10 per cent for all the slag samples following holding at 1450°C, which can possibly be due to the fluoride evaporation from the samples (Ju *et al*, 2023, 2021). Table 4 represents a comparison of the predicted solidus-liquidus temperatures from FactSage™ and results obtained from DTA analysis.

TABLE 4

Estimation of solidus and liquidus temperatures from DTA and FactSage™.

Slag No	Solidus (°C)		Liquidus (°C)	
	FactSage™	DTA	FactSage™	DTA
S1	1197.7	1022	1305.9	1415.1
S2	1084.9	1094.9	1371.8	1387.5
S3	1084.9	1032.4	1382.5	1445.1
S4	1120.7	1137.4	1392.2	1402.5

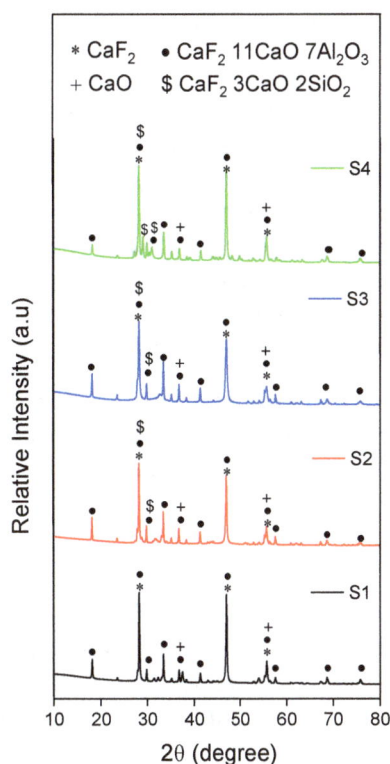

FIG 16 – XRD analysis of slag samples quenched from 1000°C.

After the addition of SiO_2 into the ESR slag selected for the study, which is used for refining special alloy steels, the liquidus temperatures of the fluxes undergo changes with respect to compositional and structural modifications. However, for refining commonly used special alloy steels, the measured values of liquidus and solidus temperatures are within acceptable range, as the liquidus temperatures of the fluxes are below those of commonly used alloy steel grades. Furthermore, it is suggested to conduct the ESR refining using the modified fluxes and analyse the efficiency of the ESR process as well as the quality of refined ingots for a better understanding of the suitability of these fluxes.

CONCLUSIONS

In this study, the influence of SiO_2 concentration on the structural characteristics and melting-solidification behaviour of the 'CaF_2-Al_2O_3-CaO-(SiO_2)' slag was systematically investigated. Structural analyses conducted through Raman spectroscopy and FT-IR spectroscopy techniques revealed significant changes in the polymerisation behaviour of alumina and silica units with varying SiO_2 concentrations. Changes in the relative fraction of structural units have been analysed through Raman and FT-IR spectroscopy techniques, indicating increased level of polymerisation in both alumina and silicate networks. The estimation of liquidus and solidus temperatures, performed using FactSage™ software and DTA/TGA analysis, further supported the suitability of the flux in ESR operation. The conclusions are summarised as follows.

- In raman spectroscopy results, within alumina-tetrahedral units, an increase in Q^3_{Al} and Q^4_{Al} networks was observed, with a corresponding decrease in Q^0_{Al}, Q^1_{Al} and Q^2_{Al} networks, suggesting an increase in alumina tetrahedra polymerisation with SiO_2 addition into the selected commercial flux composition. Silicate networks show a decrease in simple networks (Q^0_{Si} and Q^1_{Si}) and an increase in complex networks (Q^2_{Si} and Q^3_{Si}) after the addition of SiO_2, indicating an increase in the polymerisation of silicate tetrahedral networks.

- In FT-IR analyses, within silicate tetrahedral networks, an increase in complex structural units (Q^3_{Si} and Q^2_{Si}) and a decrease in smaller structural units (Q^0_{Si} and Q^1_{Si}) were observed, indicating increased polymerisation level of the silicate network. Within alumina networks, a decrease in simple alumina tetrahedral units (Q^2_{Al} and Q^3_{Al}) and an increase in the complex network Q^4_{Al} were observed, and results suggest that the addition of SiO_2 increases the polymerisation of alumina tetrahedral networks, consistent with the results from Raman spectroscopy.

- After the addition of SiO_2 in the slag, silica forms tetrahedral networks, and basic oxides/ fluorides (CaF_2 and CaO) try to break these networks, creating various simple and complex structures in silica. Aluminium from alumina units will try to join silicate networks, forming 'Si-O-Al' bonds and leading to complex networks like Q^3_{Al} and Q^4_{Al}. In silica tetrahedral networks, the gradual addition of SiO_2 increases complex units (Q^2_{Si} and Q^3_{Si}) and reduces simpler ones like Q^0_{Si} and Q^1_{Si}. The reason for the rise in complex structural units can be that, after the addition of more silica, the basic oxides/fluorides in the melt, which is at a constant level across all compositions, initially break relatively more complex networks.

- Liquidus temperature measurements obtained from FactSage™ and DTA/TGA analyses show that the fluxes' liquidus temperatures are within permissible ranges, ie they were found to be below the melting point of the ingot to be refined, ensuring the suitability of the flux for the refining of various alloy steel grades.

ACKNOWLEDGEMENTS

The authors are thankful for the support received from the Science and Engineering Research Board (SERB), India, towards fellowship for one of the authors as well as the conduct of the experiments. We express our sincere gratitude to our colleagues in the Ferrous Process Laboratory and Geochemistry lab at IIT Bombay for their invaluable support in conducting this work.

REFERENCES

Haghdani, S, Tangstad, M and Einarsrud, K E, 2022. A Raman-structure model for the viscosity of SiO_2-CaO-Al_2O_3 system, *Metallurgical and Materials Transactions B*, 53(3):1733–1746.

Huang, Y, Shi, C B, Wan, X X, Li, J L, Zheng, D L and Li, J, 2021. Effect of SiO_2 and B_2O_3 on crystallisation and structure of CaF_2-CaO-Al_2O_3-based slag for electroslag remelting of ultra-supercritical rotor steel, *Journal of Iron and Steel Research International*, 28(12):1530–1540.

Hwa, L G, Hwang, S L and Liu, L C, 1998. Infrared and Raman spectra of calcium alumino-silicate glasses, *Journal of non-crystalline solids*, 238(3):193–197.

Ju, J T, Yang, K S, Gu, Y and He, K, 2022. Effect of Na_2O on Viscosity, Structure and Crystallization of CaF_2-CaO-Al_2O_3-MgO-TiO_2 Slag in Electroslag Remelting, *Russian Journal of Non-Ferrous Metals*, 63(6):599–609.

Ju, J, Gu, Y, Zhang, Q and He, K, 2023. Effect of CaF_2 and CaO/Al_2O_3 ratio on evaporation and melting characteristics of low-fluoride CaF_2-CaO-Al_2O_3-MgO-TiO_2 slag for electroslag remelting, *Ironmaking and Steelmaking*, 50(1):13–20.

Ju, J, Ji, G, Tang, C, Yang, K and Zhu, Z, 2021. The effect of Li_2O on the evaporation and structure of low-fluoride slag for vacuum electroslag remelting, *Vacuum*, 183:109920.

Kim, G H, Kim, C S and Sohn, I L, 2013. Viscous behavior of alumina rich calcium-silicate based mold fluxes and its correlation to the melt structure, *ISIJ international*, 53(1):170–176.

Kim, T S and Park, J H, 2014. Structure-viscosity relationship of low-silica calcium aluminosilicate melts, *ISIJ International*, 54(9):2031–2038.

Lao, Y, Gao, Y, Deng, F, Wang, Q and Li, G, 2019. Effects of basicity and CaF_2 on the viscosity of CaF_2-CaO-SiO_2 slag for electroslag remelting process, *Metallurgical Research and Technology*, 116(6):638.

Li, J, Shu, Q and Chou, K, 2014. Structural study of glassy CaO-SiO_2-CaF_2-TiO_2 slags by Raman spectroscopy and MAS-NMR, *ISIJ international*, 54(4):721–727.

McMillan, P and Piriou, B, 1983. Raman spectroscopy of calcium aluminate glasses and crystals, *Journal of Non-Crystalline Solids*, 55(2):221–242.

Park, H S, Kim, H and Sohn, I, 2011. Influence of CaF_2 and Li_2O on the viscous behavior of calcium silicate melts containing 12 wt pct Na_2O, *Metallurgical and Materials Transactions B*, 42:324–330.

Park, J H, Min, D J and Song, H S, 2002. Structural investigation of CaO-Al_2O_3 and CaO-Al_2O_3-CaF_2 slags via Fourier transform infrared spectra, *ISIJ International*, 42(1):38–43.

Persson, M and Seetharaman, S, 2007. Kinetic studies of fluoride evaporation from slags, *ISIJ International*, 47(12):1711–1717.

Shi, C B, Li, J, Cho, J W, Jiang, F and Jung, I H, 2015. Effect of SiO_2 on the crystallisation behaviors and in-mold performance of CaF_2-CaO-Al_2O_3 slags for drawing-ingot-type electroslag remelting, *Metallurgical and Materials Transactions B*, 46:2110–2120.

Shi, C B, Zheng, D L, Shin, S H, Li, J and Cho, J W, 2017. Effect of TiO_2 on the viscosity and structure of low-fluoride slag used for electroslag remelting of Ti-containing steels, *International Journal of Minerals, Metallurgy and Materials*, 24:18–24.

Wan, X, Shi, C, Huang, Y, Shu, Q and Zhao, Y, 2023. Effect of SiO_2 and BaO/CaO Mass Ratio on Structure and Viscosity of B_2O_3-Containing CaF_2-CaO-Al_2O_3-Based Slag for Electroslag Remelting of Rotor Steel, *Metallurgical and Materials Transactions B*, 54(1):465–479.

Wan, X, Shi, C, Zhao, Y and Li, J, 2022. Effect of CaF_2 and Li_2O on structure and viscosity of low-fluoride slag for electroslag remelting of rotor steel, *Journal of Non-Crystalline Solids*, 597:121914.

Wroblewski, K, DiBiaso, B, Fraley, J, Fields, J and Rudoler, S, 2016. Design of ESR Slags According to Requested Physical Properties; Part 2: Density and Viscosity, in *Proceedings of the 2013 International Symposium on Liquid Metal Processing and Casting*, pp 43–46.

Wroblewski, K, Fraley, J, Fields, J, Werner, R and Rudoler, S, 2011. Design of ESR Slags According to Requested Physical Properties; Part 1: Electrical Conductivity, in *Proceedings of the 2011 International Symposium on Liquid Metal Processing and Casting, Nancy, France*, pp 121–126.

Xu, R H, Guo, J, Zhang, M C and Guo, H J, 2022. Effect of SiO_2 and TiO_2 on the Crystal Morphology of CaF_2-Al_2O_3-CaO-Based Electroslag Remelting Slag, *Steel Research International*, 93(3):2100191.

Yan, X, Pan, W, Wang, X, Zhang, X, He, S and Wang, Q, 2021. Electrical conductivity, viscosity and structure of CaO-Al_2O_3-based mold slags for continuous casting of high-Al steels, *Metallurgical and Materials Transactions B*, 52(4):2526–2535.

Yeager, J and Ramanujachary, K, 2007. Method of Measuring Fluoride in Fluxes Using the Fluoride Ion-Selective Electrode, US Patent Application US11/456815.

Zheng, D, Li, J and Shi, C, 2020. Development of low-fluoride slag for electroslag remelting: role of Li_2O on the crystallisation and evaporation of the slag, *ISIJ International*, 60(5):840–847.

Zheng, D, Shi, C, Li, J and Ju, J, 2020. Crystallisation kinetics and structure of CaF_2-CaO-Al_2O_3-MgO-TiO_2 slag for electroslag remelting, *ISIJ International*, 60(3):492–498.

A machine learning model to predict non-metallic inclusion dissolution in the metallurgical slag

W Mu[1], C Shen[2,3], C Xuan[4], D Kumar[5], Q Wang[6] and J H Park[7]

1. Associate Professor, Royal Institute of Technology (KTH), Department of Materials Science and Engineering, Stockholm, SE-10044, Sweden. Email: wmu@kth.se.
2. Visiting scholar, Royal Institute of Technology (KTH), Stockholm SE-10044, Sweden. Email: chshen@kth.se
3. Associate Professor, Hebei University of Technology, Tianjin 300401, China. Email: cgshen@hebut.edu.cn
4. Technology Manager, Metso, Skellefteå SE-93157, Sweden. Email: xuanchangji@msn.com
5. Assistant Professor, IIT Bombay, Department of Metallurgical Engineering and Materials Science, Mumbai, India. Email: deepook@iitb.ac.in
6. Professor, Northeastern University, Shenyang 110819, China. Email: wangq@mail.neu.edu.cn
7. Professor, Hanyang University, Department of Materials Science and Chemical Engineering, Ansan 15588, Korea. Email: basicity@hanyang.ac.kr

ABSTRACT

Dissolution of non-metallic inclusions in the metallurgical slag is of vital importance for cleanliness control of the steel manufacturing. With the development of high temperature confocal laser scanning microscope (HT-CLSM), new insights have been obtained due to its *in situ* observation characteristics, higher resolution and precise control. However, HT-CLSM measurement has the limitation, eg the slag composition cannot include high amount of transition metal oxides. In addition, it is time consuming for the experimental procedure and not so simple to succeed for every measurement. It is known that digitalisation has made a significant progress in recent years. Machine learning (ML), a sub-domain of artificial intelligence (AI), is the key enabling technology for the digitalisation of the material science and industry. The database for ML model is collected using almost all available HT-CLSM experimental data and subsequently the established database is trained by different ML methods. Unseen data is used as the benchmark of the ML model. Al_2O_3 dissolution is the main process to be predicted in the current study. A good agreement between the HT-CLSM data and the ML model prediction results show the possibility to apply ML in process metallurgy.

INTRODUCTION

Inclusion (oxide) dissolution in the metallurgical slag is one of the key issues for the clean steel production. On the one hand, the main components in refractories, eg alumina, magnesia etc have the potency to be dissolved in the slag during contacting. Subsequently, the eroded refractory has a much lower chemical and physical stability at high temperature and tends to be a source for exogenous inclusions formation in the steel. On the other hand, the formed inclusion particles tend to flow up to pass the steel/slag interface and need to be dissolved quickly in the slag (Park and Zhang; 2020; Park and Kang, 2017; Reis, Bielefeldt and Vilela, 2014; Webler and Pistorius, 2020). If the remained inclusions entrapped back into the liquid steel, they may have the risk to make agglomerations resulting in serious industrial problems eg nozzle clogging. In this case, the study of inclusion dissolution attracts the attention of metallurgists during the past decades. Finger rotating test (also named as dip test) is the conventional method to investigate the dissolution of oxide (refractory or synthetic inclusion) in the liquid slag. This method provides a good freedom of varying the experimental conditions (slag composition, temperature, atmosphere etc), however the sample size is much larger than the real inclusion so the obtained knowledge cannot connect to the real steelmaking process directly (Cooper Jr and Kingery, 1964; Aneziris *et al*, 2013).

With the development of the instrumentation in recent 25 years, high temperature confocal laser scanning microscope (HT-CLSM) offers a path to the *in situ* observation of dissolution of inclusion particles in the steelmaking slag in real time. To track the particle dissolution, quasi-three-dimensional images with sharp boundary and good contrast could be taken from the video by the CCD camera equipped with HT-CLSM. Specifically, the size of the used particle which ranges from a few tens to several hundreds of microns fits well with the actual inclusion size. Al_2O_3 dissolution in the slag is the most comprehensively studied by HT-CLSM and the early work can refer to Sridhar and Cramb (2000). They investigated the effect of the temperature on the dissolution of alumina particles in the CaO-Al_2O_3-SiO_2-MgO (CASM) slag. Subsequently, comprehensive research activities focusing on the effect of inclusion type, slag composition, slag viscosity, temperature etc are performed. Specifically, 1630°C is the maximum temperature which could be achieved so far to observe oxide dissolution (Liu *et al*, 2007b). Due to the development of furnace in HT-CLSM, this maximum temperature could be further challenged. CASM slag without transition metal (Fe, Mn etc) oxides is the mainly selected system since the slag needs to be transparent to enable the *in situ* observation, which is one main limitation for the application of HT-CLSM in this field. By utilising the advantage of HT-CLSM, dissolution kinetics of different kinds of non-metallic inclusions besides Al_2O_3, ie MgO, $MgAl_2O_4$, TiO_2, Al_2TiO_5, CaO·$2Al_2O_3$ (CA2); SiO_2, SiC, TiN etc (Fox *et al*, 2004; Liu *et al*, 2007a, 2007b; Miao *et al*, 2018; Michelic *et al*, 2016; Michelic and Bernhard, 2017; Park *et al*, 2010; Ren *et al*, 2022; Sharma, Dabkowska and Dogan, 2019; Sharma, Mu and Dogan, 2018a, 2018b; Yi *et al*, 2003) have been investigated. Except for the chemistry of inclusion, other processing parameters, eg dissolution temperature, particle size, slag composition, have been studied, however, there are still many parameters need to worth further investigated. For instance, the HT-CLSM can only observe the particle dissolution in the transparent slag, which means the influence of some component, eg Fe_tO cannot be studied quantitatively, especially the Fe_tO content in slag is high (Um *et al*, 2022). In this case, the simulation and other technique needs to be performed to understand the inclusion dissolution mechanism with the comprehensive conditions.

Regarding the study of dissolution mechanism, shrinking core (SC) model is believed to be the most frequently applied approach. Using this model, the dissolution mechanism is usually identified through a proper fitting between the SC model prediction and the experimental observation data. If the observed dissolution profile is linear, the dissolution process is considered as the reaction-controlled dissolution (RCD). If the dissolution profile is in a parabolic shape, the boundary layer diffusion-controlled dissolution (BLDD) is the mechanism. When the dissolution profile shows a sigmoidal shape, shell layer diffusion-controlled dissolution (SLDD) may be the controlled mechanism. Alternatively, stationary interface diffusion model (SIM) could also obtain a sigmoidal profile which quite closes to SLDD. For the classical SC models, the experimental profile sometimes could not fit any mechanism curve which is a clear drawback, a few attempts have been made to modify the SC model. For instance, Feichtinger *et al* (2014) introduced a factor 'f' in the SIM model (also named 'invariant interface diffusion control'), which improves the model performance significantly and 'f' is reported to relate to slag viscosity. Very recently, a modified SC-based physical model, ie diffusion-distance-controlled dissolution (DDD) model (Xuan and Mu, 2021) is developed to investigate the shape origin and variation of oxide dissolution in the slag. It is reported that the diffusion-related particle dissolution is actually one mechanism but not a few, the shape of the dissolution profile is controlled by the distance of diffusion region. This model has been verified with HT-CLSM data of different kinds of inclusions dissolution. However, the comprehensive application of this model is relied on the input of physical parameters to present different slag compositions, currently this is the only limitation for the DDD model.

In order to further extend the application of the theoretical model to predict inclusion dissolution in the comprehensive processing conditions, machine learning (ML) has been considered to use to combine with the current DDD physical model. Digitalisation has brought many significant changes to the manufacturing in the recent decade. ML is a sub-domain of artificial intelligence, which is the key enabling technology for industrial-driven research, eg material engineering and metallurgy. The database for ML model is collected using HT-CLSM experimental data, DDD model is used for the augmentation of the current database, to have a continuous distribution of data. Unseen experimental data is used as the benchmark of the ML model. Al_2O_3 is the most common inclusion, so its dissolution in the slag is performed in the current study, to test the accuracy of the current hybrid modelling methodology. The successful implementation of this model can be applied to predict the dissolution of other kinds of inclusions, eg MgO, SiO_2, $MgAl_2O_4$, CaO·$2Al_2O_3$ etc in the further study.

METHODOLOGY

The current ML data set contains 1506 data points collected from the open literatures (Hagemann, Pettsold and Sheller, 2010; Lee *et al*, 2001; Liu *et al*, 2007a, 2007b; Monaghan, Chen and Sorbe, 2005; Michelic *et al*, 2016; Sharma, Mu and Dogan, 2018a, 2018b; Soll-Morris *et al*, 2009; Sridhar and Cramb, 2000; Yi *et al*, 2003), DDD physical model is used for the data augmentation, this combination methodology can be seen in the report by Mu *et al* (2021). The collecting data set is split into 80 per cent for training and 20 per cent for testing. Furthermore, 252 unseen data which are not included in the training and testing data sets (Liu *et al*, 2007a, 2007b; Yi *et al*, 2003; Michelic *et al*, 2016) is used as the validation data set for the benchmark. There are ten parameters including in the data set to predict Al_2O_3 dissolution, ie original inclusion size (radius), temperature, concentration of five components in slag (SiO_2, CaO, Al_2O_3, MgO and Fe_tO), density of inclusion, slag viscosity and radius change (R/R_0, R represents the radius at the certain time and R_0 is the initial radius). The time required for the dissolution of Al_2O_3 particles was selected as the output. For the further work to predict different kinds of inclusion dissolution, one more parameter, inclusion type needs to be added. The detailed information of training and testing data set as well as validation data set are listed in Tables 1 and 2.

TABLE 1

Data distribution in present data set (including training and testing sets).

Parameters	Min	Max	Mean	Std
Radius [μm]	41.3	250.2	176.6	76.9
Temperature [°C]	1673	1873	1772.3	56.7
SiO_2 [wt%]	0	62.4	33.4	19.8
CaO [wt%]	22.4	57.2	35.2	9.1
Al_2O_3 [wt%]	13.3	50.0	29.0	11.6
MgO [wt%]	0	7.5	2.0	3.1
Fe_tO [wt%]	0	9.2	0.4	1.7
Slag Density [kg.m]	3568	3640	3604	20.4
Slag viscosity [Pa.s]	0.1	100.5	4.7	16.4
R/R_0 [-]	0	1	0.7	0.3
Dissolution time [sec]	0	4490	925	1099

TABLE 2

Data distribution in validation set.

Parameters	Min	Max	Mean	Std
Radius [μm]	95.3	249.1	226.6	47.3
Temperature [°C]	1723	1873	1778.2	50.1
SiO_2 [wt%]	0	46.2	39.2	16.5
CaO [wt%]	29.7	49.5	32.7	7.1
Al_2O_3 [wt%]	24.1	50.0	28.0	9.3
MgO [wt%]	0	0.5	0.08	0.18
Fe_tO [wt%]	0	0	0	0
Slag Density [kg.m]	3568	3622	3602	18.0
Slag viscosity [Pa.s]	0.8	3.6	2.5	1.1
R/R_0 [-]	0	1	0.56	0.26
Dissolution time [sec]	0	4490	1527	1290

Standard normalisation was used in this work in order to eliminate the difference in numeric scale of features. Based on the previous modelling results, it is found that the prediction of the trained ML model fluctuates as the different partition of training data set and testing data set, so multiple hold-out method is used in order to reliably evaluate model's performance (Shen *et al*, 2019). In this study, the data set was randomly divided into training and testing sets by 50 times to develop 50 ML models. Moreover, common mean absolute error (MAE) and square correlation coefficient (R^2) were used to evaluate the model's performance (Pal, 2017). Besides, the suitable selection of ML algorithm is also important for good performance. In this work, six common ML algorithms were employed to develop three models, including Support Vector Regression (SVR) (Liu, Wang and Gu, 2021), Multi-Layer Perception (MLP) (Del Campo *et al*, 2021) and three kinds of ensemble learning algorithms, ie AdaBoost (Ada), decision tree regression (DTR), random forest regression (RFR) (Sammut and Webb, 2011) and the modelling results for testing data are plotted in Figure 1. It is seen that ensemble algorithms have the higher prediction accuracy than MLP and SVR algorithms. The R^2 values of these models were greater than 98 per cent, meaning a very good prediction performance. Finally, RFR algorithm was selected to predict the dissolve curve of Al_2O_3 particle owing to the relatively lowest MAE value.

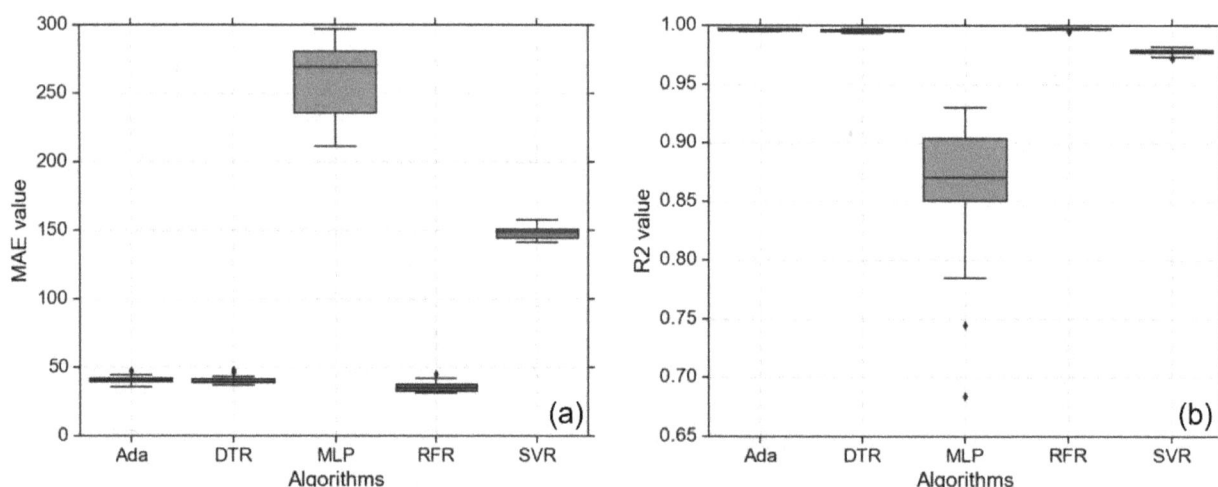

FIG 1 – Prediction performance of different ML algorithms: (a) MAE value; (b) R^2 value.

RESULTS AND DISCISSION

Pearson correlation is one of the simplest methods to determine the linear relation between features and the dependent variable. According to the method, the features that are uncorrelated with the dependent variable are good candidates to exclude from the data set before training the model (Rahaman *et al*, 2019). Here we apply the same methodology to determine the correlation between variables (radius, temperature, slag compositions, inclusion density, slag viscosity and R/R_0) with dissolution time. This method measures the linear correlation between two variables and the resulting value lies between -1 and 1. Negative values mean negative correlation; alternatively, positive values mean the opposite; 0 means that there is no linear correlation between the two variables. The Pearson's correlation coefficient for the inclusion dissolution data set is represented as a heat map in Figure 2. It can be seen that some parameters, eg original inclusion radius and SiO_2 have clear positive correlation with the dissolution time. It is obvious that the increasing inclusion original size as well as the SiO_2 content in slag will increase the dissolution time. The parameter, eg inclusion density has a less influence on the dissolution time. However, since the role of Pearson correlation is to check if some parameters can be removed but not the feature importance, we will discuss deeply on each parameter's linear correlation, also we keep all parameters in this ML model. For the current work, the data amount is not really large, more parameters are always good to use for the accuracy of ML modelling containing small size data set.

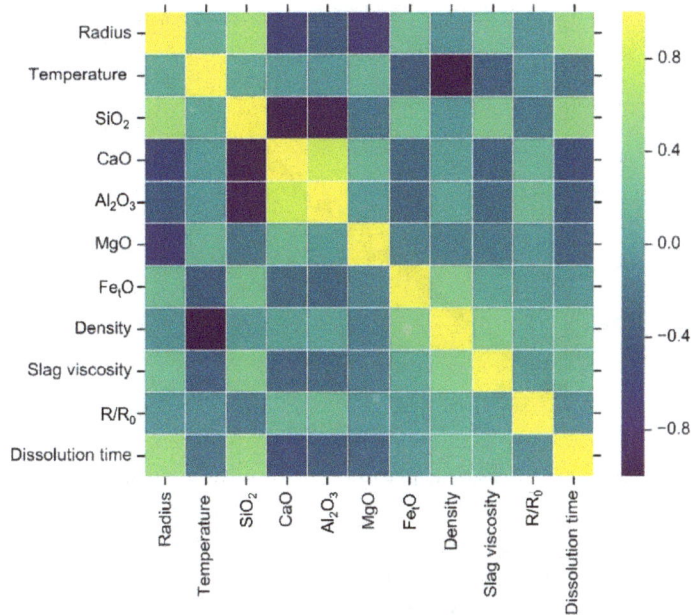

FIG 2 – Pearson correlation.

Figure 3 displays the prediction results of inclusion dissolution. The output is dissolution time of Al_2O_3 inclusion in the slag, and the X and Y axis strand for the actual dissolution time and predicted time by ML model, respectively. Figure 3a shows the actual value versus predicted value for training data. It can be seen that all data points basically concentrated around the diagonal, showing a very high prediction accuracy. The MAE and R^2 values are 17.0 sec and 99.8 per cent, demonstrating that the ML model has learned the reliable relation between dissolution time of inclusions and all input features based on training samples. In order to test the application ability of trained model, 20 per cent of the total data in the testing set were inputted into model and the prediction result is shown in Figure 3b. It is observed that most data points are also located in the line with the slope of 1, showing a good robustness of trained model. The MAE and R^2 values are 30.2 sec and 98.7 per cent. It is also seen that small number of data points deviates from the diagonal, but the largest MAE value is still less than 150 sec, which is within the acceptable scale. In order to further validate the robustness for new data, trained model was applied to the unseen 252 data in the validation data set with different parameters. The prediction result is shown in Figure 4. It can be seen that like training and testing data, present model also has a high prediction accuracy on validation data and most data points locate around the diagonal. The MAE and R^2 values are 70.8 sec and 97.7 per cent, respectively. Compared to training and testing data, the predicted accuracy on validation data slightly reduces, but present prediction level is still high enough to guide the actual experiments.

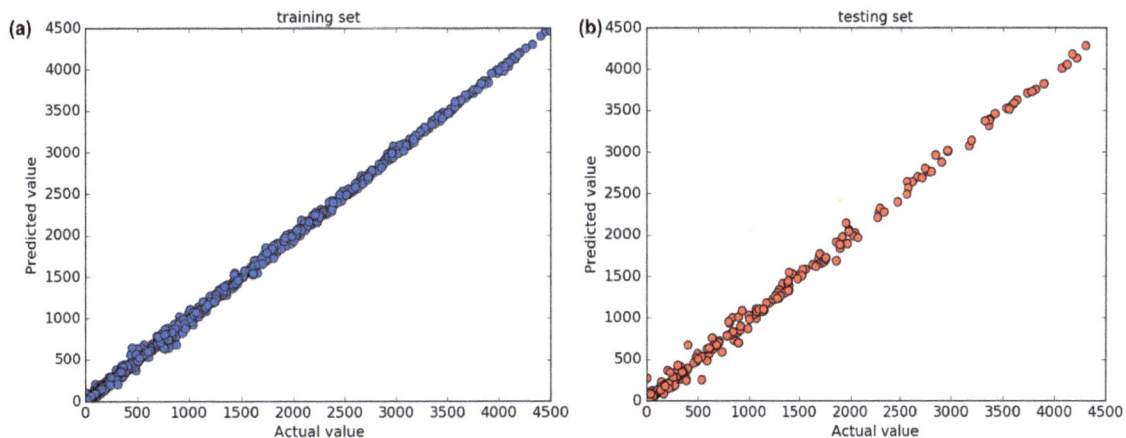

FIG 3 – Predicted results of dissolution time of inclusion using (a) training and (b) testing data set.

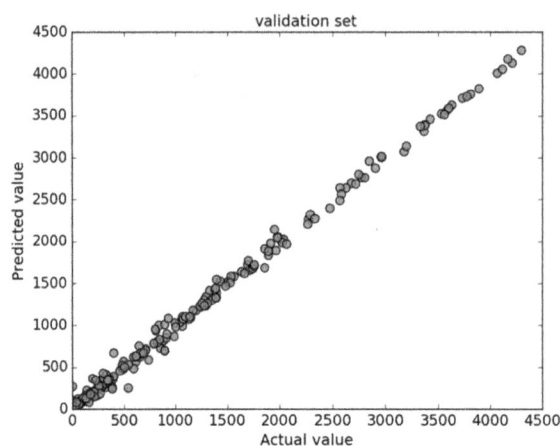

FIG 4 – Predicted results of dissolution time of inclusion using validation data set.

It is known that the parameters, eg slag viscosity is a joint function of temperature and slag composition and the inclusion density is also influenced by the temperature. Since these parameters are already included in the current database, new attempt to remove slag viscosity and inclusion density in the database has been made to see the model prediction performance. It seems the model accuracy decreases slightly but it is still quite acceptable (ie 41.66 of MAE and 98.41 per cent of R^2 for using testing data set in new model). In this case, the new model can be considered to use by inputting slag composition, temperature etc but without using physical parameters, eg slag viscosity, inclusion density etc which will be even easier to operate the model for the application perspective.

In addition, it needs to be mentioned that the correlation importance between dissolution time and each input parameter can be analysed. This type analysis is usually called feature importance (FI). FI is used to evaluate which parameter plays a more important role in within the specific data set (Rahaman *et al*, 2019; Mu *et al*, 2021). The importance of the same parameter can be different if the data set is changed. However, the current data set includes the experimental data are all collected from the HT-CLSM measurement. In this case, some parameters eg contents of different components in the slag are not such comprehensive due to the limitation of HT-CLSM (eg Fe_tO content should be almost zero). In this case, the feature importance may show different tendency compare with the physical understanding of inclusion dissolution, so FI analysis is not included in the current work.

CONCLUSIONS

The current hybrid methodology in combination of machine learning and physical modelling has been established to predict Al_2O_3 dissolution in the steelmaking slag. The successful implementation of applying the method to predict of dissolution time provides the possibility to further application of the model. Different kinds of inclusion dissolution data observed by HT-CLSM and measured by other methodology is needed to be used to establish a robust dissolution model.

ACKNOWLEDGEMENTS

Authors W Mu and C Shen would like to acknowledge the financial support provided by Jernkontoret Gerhard von Hofstens Stiftelse för metallurgisk Forskning foundation to support C Shen visiting at KTH Royal Institute of Technology, Swedish Foundation for International Cooperation in Research and Higher Education (STINT, Project No. IB2020-8781 and IB2022-9228) is acknowledged for the collaboration between KTH (Sweden), Hanyang University (Korea), Northeastern University (China) and IIT Bombay (India).

REFERENCES

Aneziris, C G, Dudczig, S, Emmel, M, Berek, H, Schmidt, G and Hubalkova, J, 2013. Reactive filters for steel melt filtration, *Advanced Engineering Materials*, 15(1–2):46–59.

Cooper Jr, A R and Kingery, W D, 1964. Dissolution in ceramic systems: I, molecular diffusion, natural convection and forced convection studies of sapphire dissolution in calcium aluminum silicate, *Journal of the American Ceramic Society*, 47(1):37–43.

Del Campo, F A, Neri, M C G, Villegas, O O V, Sánchez, V G C, Domínguez, H D J O and Jiménez, V G, 2021. Auto-adaptive multilayer perceptron for univariate time series classification, *Expert Systems with Applications*, 181:115147.

Feichtinger, S, Michelic, S K, Kang, Y B and Bernhard, C, 2014. In situ observation of the dissolution of SiO2 particles in CaO-Al2O3-SiO2 slags and mathematical analysis of its dissolution pattern, *Journal of the American Ceramic Society*, 97(1):316–325.

Fox, A B, Valdez, M E, Gisby, J, Atwood, R C, Lee, P D and Sridhar, S, 2004. Dissolution of ZrO2, Al2O3, MgO and MgAl2O4 particles in a B2O3 containing commercial fluoride-free mould slag, *ISIJ International*, 44(5):836–845.

Hagemann, R, Pettsold, L and Sheller, P R, 2010. The process of oxide non-metallic inclusion dissolution in slag, *Meta Min Indu*, 2:262–266.

Lee, S H, Tse, C, Yi, K W, Misra, P, Chevrier, V, Orrling, C, Sridhar, S and Cramb, A W, 2001. Separation and dissolution of Al2O3 inclusions at slag/metal interfaces, *Journal of Non-crystalline Solids*, 282(1):41–48.

Liu, J, Guo, M, Jones, P T, Verhaeghe, F, Blanpain, B and Wollants, P, 2007a. In situ observation of the direct and indirect dissolution of MgO particles in CaO-Al2O3-SiO2-based slags, *Journal of the European Ceramic Society*, 27(4):1961–1972.

Liu, J, Verhaeghe, F, Guo, M, Blanpain, B and Wollants, P, 2007b. In situ observation of the dissolution of spherical alumina particles in CaO-Al2O3-SiO2 melts, *Journal of the American Ceramic Society*, 90(12):3818–3824.

Liu, Y, Wang, L and Gu, K, 2021. A support vector regression (SVR)-based method for dynamic load identification using heterogeneous responses under interval uncertainties, *Applied Soft Computing*, 110:107599.

Miao, K, Haas, A, Sharma, M, Mu, W and Dogan, N, 2018. In situ observation of calcium aluminate inclusions dissolution into steelmaking slag, *Metallurgical and Materials Transactions B*, 49:1612–1623.

Michelic, S K and Bernhard, C, 2017. Experimental study on the behavior of TiN and Ti2O3 Inclusions in contact with CaO-Al2O3-SiO2-MgO Slags, *Scanning*, 2017.

Michelic, S, Goriupp, J, Feichtinger, S, Kang, Y B, Bernhard, C and Schenk, J, 2016. Study on oxide inclusion dissolution in secondary steelmaking slags using high temperature confocal scanning laser microscopy, *Steel Research International*, 87(1):57–67.

Monaghan, B J, Chen, L and Sorbe, J, 2005. Comparative study of oxide inclusion dissolution in CaO-SiO2-Al2O3 slag, *Ironmaking and Steelmaking*, 32(3):258–264.

Mu, W, Rahaman, M, Rios, F L, Odqvist, J and Hedström, P, 2021. Predicting strain-induced martensite in austenitic steels by combining physical modelling and machine learning, *Materials and Design*, 197:109199.

Pal, R, 2017. Chapter 4 – Validation methodologies, in *Predictive Modeling of Drug Sensitivity* (ed: R Pal), pp 83–107 (Academic Press).

Park, J H and Kang, Y, 2017. Inclusions in stainless steels– a review, *Steel Research International*, 88(12):1700130.

Park, J H and Zhang, L, 2020. Kinetic modeling of nonmetallic inclusions behavior in molten steel: A review, *Metallurgical and Materials Transactions B*, 51:2453–2482.

Park, J H, Park, J G, Min, D J, Lee, Y E and Kang, Y B, 2010. In situ observation of the dissolution phenomena of SiC particle in CaO-SiO2-MnO slag, *Journal of the European Ceramic Society*, 30(15):3181–3186.

Rahaman, M, Mu, W, Odqvist, J and Hedström, P, 2019. Machine learning to predict the martensite start temperature in steels, *Metallurgical and Materials Transactions A*, 50:2081–2091.

Reis, B H, Bielefeldt, W V and Vilela, A C F, 2014. Absorption of non-metallic inclusions by steelmaking slags-a review, *Journal of Materials Research and Technology*, 3(2):179–185.

Ren, Y, Zhu, P, Ren, C, Liu, N and Zhang, L, 2022. Dissolution of SiO2 inclusions in CaO-SiO2-based slags in situ observed using high-temperature confocal scanning laser microscopy, *Metallurgical and Materials Transactions B*, 53(2):682–692.

Sammut, C and Webb, G I (eds), 2011. *Encyclopedia of Machine Learning* (Springer Science and Business Media).

Sharma, M, Dabkowska, H A and Dogan, N, 2019. Application of Optical Floating Zone Method to Dissolution Kinetics of Inclusions in a Steelmaking Slag, *Steel Research International*, 90(1):1800367.

Sharma, M, Mu, W and Dogan, N, 2018a. In situ observation of dissolution of oxide inclusions in steelmaking slags, *JOM*, 70:1220–1224.

Sharma, M, Mu, W and Dogan, N, 2018b. In-Situ dissolution of aluminum titanate inclusion into Al2O3-CaO-SiO2-Type Slag, in *Proceedings of AISTech*, Philadelphia.

Shen, C, Wang, C, Wei, X, Li, Y, van der Zwaag, S and Xu, W, 2019. Physical metallurgy-guided machine learning and artificial intelligent design of ultrahigh-strength stainless steel, *Acta Materialia*, 179:201–214.

Soll-Morris, H, Sawyer, C, Zhang, Z T, Shannon, G N, Nakano, J and Sridhar, S, 2009. The interaction of spherical Al2O3 particles with molten Al2O3-CaO-FeOx-SiO2 slags, *Fuel*, 88(4):670–682.

Sridhar, S and Cramb, A W, 2000. Kinetics of Al2O3 dissolution in CaO-MgO-SiO2-Al2O3 slags: In situ observations and analysis, *Metallurgical and Materials Transactions B*, 31:406–410.

Um, H, Yeo, S, Kang, Y B and Chung, Y, 2022. The effect of FexO content on dissolution behavior of an alumina inclusion in CaO-Al2O3-SiO2-FexO slag by a single hot thermocouple technique, *Ceramics International*, 48(23):35301–35309.

Webler, B A and Pistorius, P C, 2020. A review of steel processing considerations for oxide cleanliness, *Metallurgical and Materials Transactions B*, 51:2437–2452.

Xuan, C and Mu, W, 2021. A mechanism theory of dissolution profile of oxide particles in oxide melt, *Journal of the American Ceramic Society*, 104(1):57–75.

Yi, K W, Tse, C, Park, J H, Valdez, M, Cramb, A W and Sridhar, S, 2003. Determination of dissolution time of Al2O3 and MgO inclusions in synthetic Al2O3-CaO-MgO slags, *Scandinavian Journal of Metallurgy*, 32(4):177–184.

Effect of C/A ratio on the crystallisation behaviour and structure of calcium-aluminate based alternative mold fluxes for casting medium and high Mn/Al steels

A Nigam[1], K Biswas[2] and R Sarkar[3]

1. PhD Student, IIT Kanpur, Kanpur, Uttar Pradesh 208016, India. Email: amann21@iitk.ac.in
2. Professor, IIT Kanpur, Kanpur, Uttar Pradesh 208016, India. Email: kbiswas@iitk.ac.in
3. Assistant Professor, IIT Kanpur, Kanpur, Uttar Pradesh 208016, India. Email: rsarkar@iitk.ac.in

ABSTRACT

Because of the problems with 1st and 2nd generations of Advanced High-Strength Steels (AHSS), a 3rd generation of AHSS steels has become prominent and these steels have properties in between the 1st and 2nd generations of AHSS. However, although the 3rd generation of AHSS is a promising candidate as a replacement for its predecessors, there remain some challenges in processing these steels which are essentially medium Mn (Mn content ~ 5–7 wt per cent) and high Al (Al content ~ 1–3 wt per cent) steels. The use of conventional casting powders based on the $CaO-SiO_2$ system is unsuitable for high and medium Mn/Al steels. This work investigates the development of $CaO-Al_2O_3$-based mold fluxes for casting third-generation AHSS steel. Mold fluxes, with otherwise similar compositions but different C/A ratios, are tested and their crystallisation behaviour is examined using differential scanning calorimetry (DSC). A calcium aluminate phase having a composition $Ca_{12}Al_{14}O_{33}$ was found to be the main crystalline phase in the mold fluxes. A decrease in crystallisation temperature was observed as the CaO/Al_2O_3 ratio increased from 1.00 to 1.33. The effective crystallisation rate constant exhibited an increase with decreasing crystallisation temperature, indicating a potential influence of nucleation rate on the overall crystallisation rate and suggesting an anti-Arrhenius behaviour in the crystallisation process of these mold fluxes.

INTRODUCTION

Mold powders, either in granulated or non-granulated form, are introduced onto the top of liquid steel present within a copper mold. These mold powders generally have low melting point characteristics as their melting point lies usually below the pouring temperature of the liquid steel. Thus the superheat caused by the excessive temperature of the liquid steel helps the mold fluxes to undergo melting which creates a slag pool that infiltrates into the space between the solidifying steel shell and water-cooled copper mold. The primary role of these fluxes is to prevent the adhesion of the solidifying steel to the mold walls, ensuring efficient heat transfer and lubrication for the developing steel shell during solidification. The rapid cooling in the mold region results in the formation of a glassy slag film near the mold wall, which gradually transforms into a crystalline slag layer over time until reaching a stable state and near the shell region there will be liquid slag layer as shown in Figure 1. The solid part of the slag film governs the heat transfer because of its higher thermal conductivity and the liquid part helps in reducing the friction of the solidified steel shell which is moving down and thus will minimise the problems related to the surface quality of steels. Hence maintaining the thickness of the solid and liquid part of the slag film is important according to the type of steel being cast in the case of casting MC (medium carbon), peritectic steels, a thin, even shell is needed to avoid longitudinal and other surface cracking and this can be obtained with a low horizontal heat flux. Similarly in casting HC (high carbon) steels the shell is relatively weak and a thick shell is required to provide mechanical strength so a comparatively high heat flux is required. So for high horizontal heat flux thin slag film is required as heat flux is inversely proportional to thickness and *vice versa* for the low horizontal heat flux.

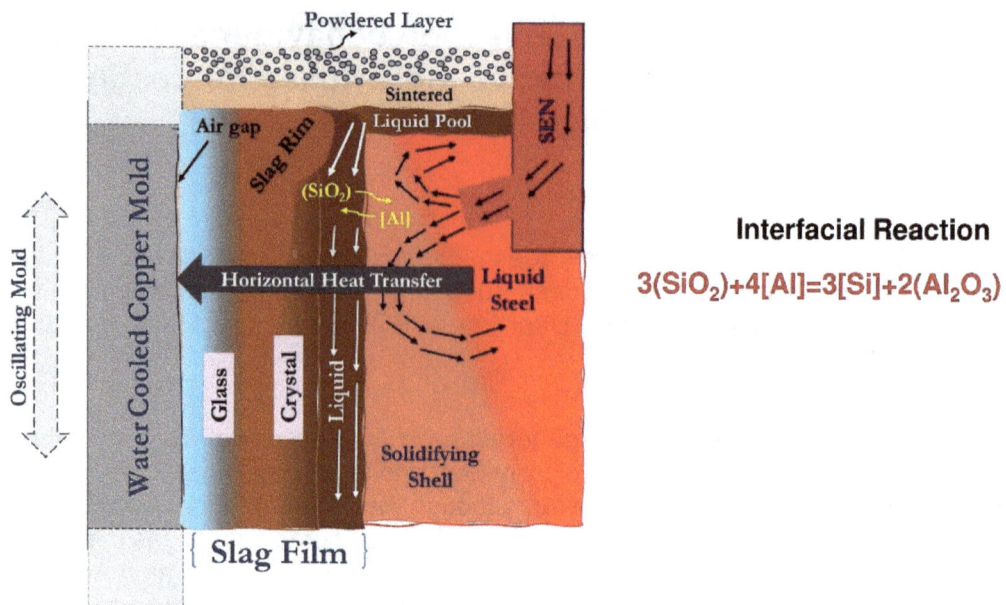

Interfacial Reaction

$$3(SiO_2)+4[Al]=3[Si]+2(Al_2O_3)$$

FIG 1 – Schematic diagram illustrating the function of mold flux in the continuous casting process.

Typically, mold powders are calcium silicate-based, augmented with fluxing agents to tailor their properties for specific steel types. Conventional CaO-SiO_2 mold fluxes are commonly employed for casting low C steels with low Al and Mn content. However, the evolving steel industry, driven by demands for lighter steel in the automotive and aerospace sectors, has led to the development of 3rd Generation AHSS (Advanced High-Strength Steel) with higher Al (0.5–2 wt per cent) and Mn (5–7 wt per cent) (Aydin *et al*, 2013). While using the traditional CaO-SiO_2 based mold fluxes in casting 3rd Generation AHSS there are certain interfacial reactions like the equation in Figure 1 (Kim and Park, 2012) involved between Al and Mn present in steel and SiO_2 present in the slag as Al and Mn has a more tendency to oxidise as they come below Si in the Ellingham Diagram. These interfacial reactions led to an increase in the Al_2O_3 content in the slag which led to an increase in the Al_2O_3/SiO_2 ratio of the slag by increasing the viscosity and melting temperature of the flux. This will create various surface defects in the cast steel like longitudinal cracks and break-out problems.

To address these challenges, there is a growing interest in using calcium aluminate-based mold fluxes, sometimes referred to as 'non-reactive' mold fluxes due to their minimal or absent silica content, eliminating interfacial reactions. Previous studies (Cho *et al*, 2013; Zhang, Wang and Shao, 2019) have found that lime-alumina-based mold fluxes exhibit physical characteristics similar to conventional fluxes, as long as the ratio wt per cent CaO / wt per cent Al_2O_3 remains close to 1.00. Furthermore, maintaining this wt per cent CaO / wt per cent Al_2O_3 ratio near 1.00 is crucial for ensuring that the fluxes have a low melting point because at that region low melting point phases lie, as depicted in Figure 2a. The low melting point area in the ternary phase diagram depicts the formation of a line compound $C_{12}A_7$ phase as can be seen in the binary phase diagram of the CaO-Al_2O_3 system (Figure 2b) indicating that the $C_{12}A_7$ phase has a high melting point depression. So $C_{12}A_7$ phase can be incorporated into the mold flux system while designing the composition, as this phase formation will be quite useful to maintain the low crystallisation temperature. Consequently, it is necessary to formulate these fluxes with a wt per cent CaO / wt per cent Al_2O_3 ratio near 1 and a SiO_2 content ranging from (3–6 per cent) to minimise the interfacial reactions (Liu *et al*, 2014). Also, by increasing the C/A ratio slightly greater than 1, an appropriate fraction of the primary crystalline phase ie may be achieved but this results in increasing the melting point of the flux. However, to reduce the melting point further fluxing agents like Na_2O, B_2O_3, CaF_2 and Li_2O are added to balance other thermophysical properties.

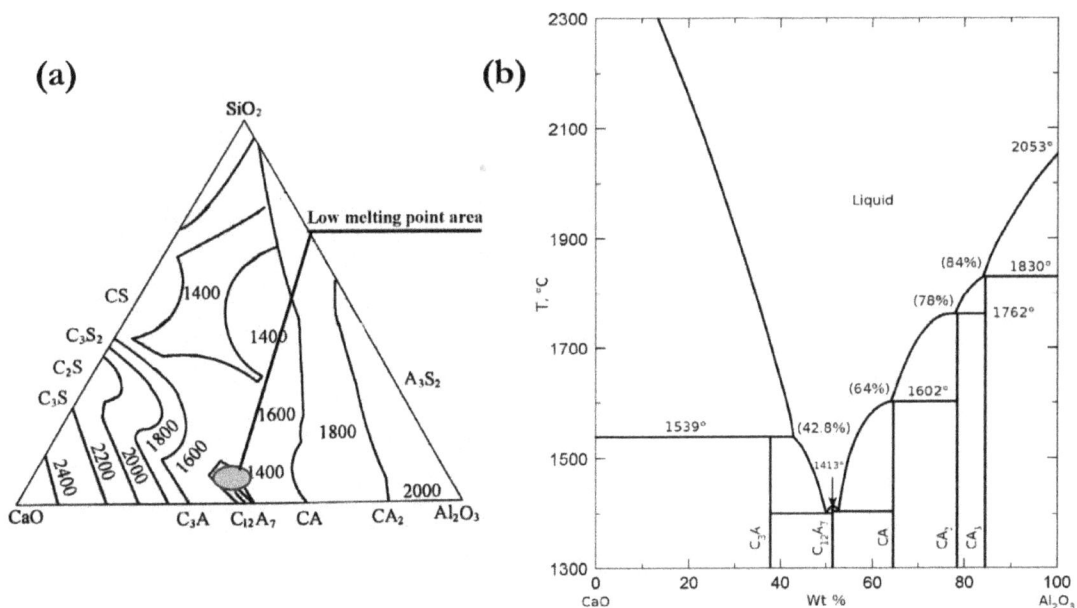

FIG 2 – (a) Liquidus projection of the SiO_2-CaO-Al_2O_3 system showing the low melting region (Liu *et al*, 2014), (b) Binary phase diagram of CaO-Al_2O_3 showing the eutectic composition phase $C_{12}A_7$ (Salasin and Rawn, 2017).

The crucial functions of mold fluxes—heat transfer and lubrication—are governed by the interfacial slag layer's resistance, particularly in horizontal heat transfer. This resistance is influenced by the fraction of crystal transformation (f_{crys}) as a function of temperature and time, which can be determined through methods such as confocal microscopy (CSLM) and DSC/DTA (Differential Thermal Analysis). Therefore, understanding the crystallisation kinetics of the forming phases is essential for comprehending heat transfer.

In the present study, two compositions of calcium aluminate mold fluxes with varying CaO/Al_2O_3 ratios (1.00 and 1.33) were developed. Isothermal crystallisation kinetics were investigated at different temperatures for both compositions and various crystallisation parameters were determined to characterise the type of crystallisation mode and the activation energy involved in that crystallisation phase.

EXPERIMENTAL DETAILS

Sample preparation

Mold flux samples were prepared on a laboratory scale using reagent grade CaO, Al_2O_3, Na_2CO_3, CaF_2, SiO_2 and B_2O_3. Compositional detail of the samples are mentioned in Table 1. Calcination of Na_2CO_3 to Na_2O was done by putting Na_2CO_3 at 800°C for 12 hrs in an air environment in a muffle furnace. The powders were then thoroughly mixed with the help of acetone in a mortar and pestle and then kept in the oven for 2 hrs at 100°C for the removal of moisture. Samples were then heated and melted in a graphite crucible inside a muffle furnace kept at 1400°C for 1 hr. Subsequently, the melted liquid flux was quenched in water and glassy cullets were obtained which was verified through X-ray diffraction (XRD) as seen in Figure 3a. These were then pulverised for further examination of the composition of the samples by X-ray fluorescence (XRF) to determine whether there was any evaporation loss or not. Pulverised samples were then used in the DSC experiments to study the crystallisation kinetics.

TABLE 1

Chemical composition of the mold flux (weight%).

Sample (wt/wt)%	CaO	Al_2O_3	Na_2O	B_2O_3	CaF_2	SiO_2
CaO/Al_2O_3 = 1.00	35	35	15	5	5	5
CaO/Al_2O_3 = 1.33	40	30	15	5	5	5

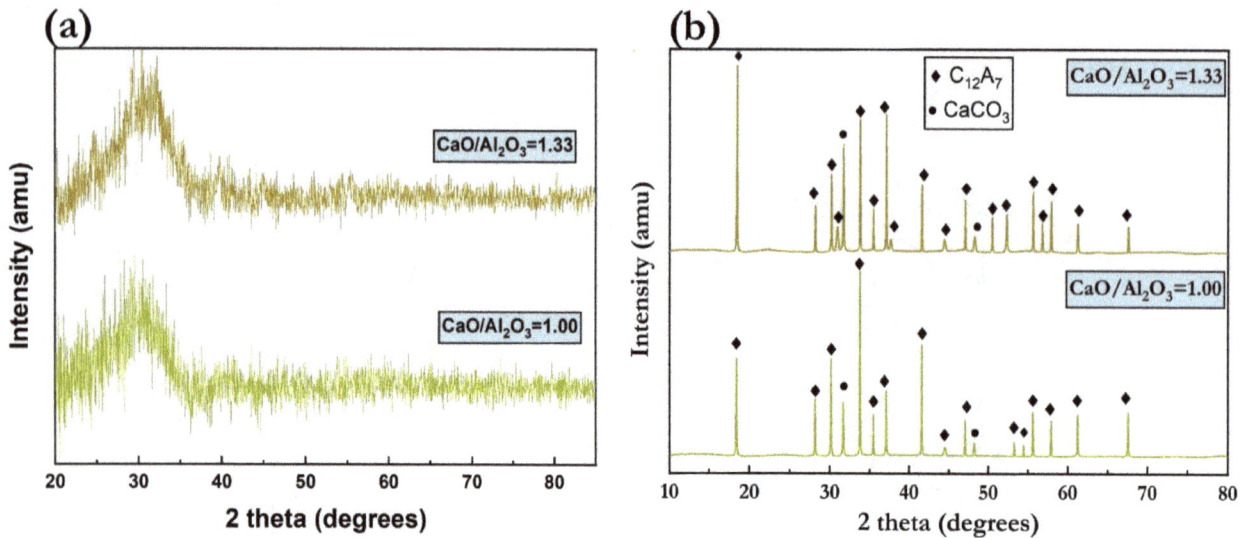

FIG 3 – (a) X-ray diffraction (XRD) plot of the two compositions after quenching, (b) XRD plot of the two compositions for the crystallised sample.

DSC measurements

The crystallisation kinetics of the fluxes were evaluated isothermally at different temperature ranges with DSC (STA 2500 Regulus; NETZSCH Instrument Inc, Germany). Measurement of the samples was done in a platinum crucible with a lid under an N_2 atmosphere at a flow rate of 60 mL/min. For each experiment 50 to 60 mg of the pulverised samples were subjected to a thermal cycle (Seo *et al*, 2015) as shown in Figure 4 under which the samples were heated at a constant rate of 25 K/min up to a target temperature (T_c) (determined by doing a normal DSC scan of the sample upto a temperature of 1400°C, Figure 5a) after which sample was isothermally held at that temperature for 2 hrs followed by subsequent cooling at a faster rate. After each experiment, samples were collected and examined in XRD and Scanning Electron Microscope – Energy Dispersive X-ray Spectroscopy (SEM-EDS) for phase determination.

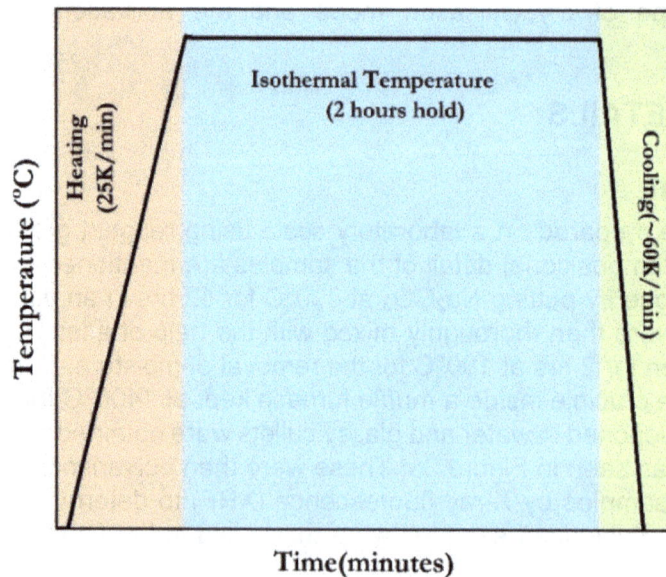

FIG 4 – Thermal cycle for the isothermal experiments.

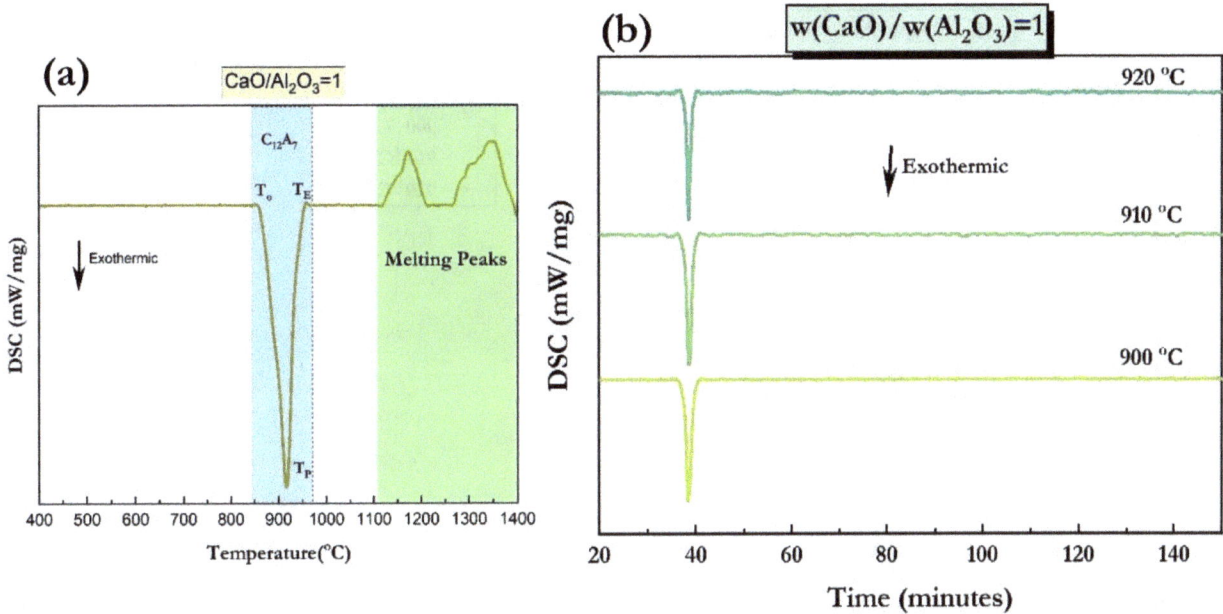

FIG 5 – (a) Differential Scanning Calorimetry (DSC) scan at 10 K/min for the identification of different peaks, (b) DSC scans for isothermal crystallisation of CaO/Al$_2$O$_3$ = 1.00.

RESULTS AND DISCUSSIONS

Isothermal DSC measurement

DSC experiments for two of the fluxes were carried out only after analysing the crystallisation temperature (T_c) through a normal DSC scan as seen in Figure 5. The crystallisation temperature is the peak temperature of the crystallisation event that occurred during the heating of the sample. Figure 5 shows the scans for the isothermal crystallisation of the flux wt per cent CaO / wt per cent Al$_2$O$_3$ = 1.00. There is only one peak of crystallisation which was identified through XRD as C$_{12}$A$_7$ (Ca$_{12}$Al$_{14}$O$_{33}$) (Figure 3b). DSC experiments for wt per cent CaO / wt per cent Al$_2$O$_3$ = 1.33 showed similar profiles.

Isothermal melt crystallisation kinetics

Crystallisation is an exothermic process as can be seen in the DSC scans (Seo *et al*, 2015). As the rate of heat release is proportional to the rate of crystallisation, the relative degree of crystallinity (α) can be obtained as:

$$\alpha(t) = \frac{\Delta Ht}{\Delta Htotal} = \frac{\int_0^t \left(\frac{dH_c}{dt}\right) dt}{\int_0^\infty \left(\frac{dHc}{dt}\right) dt}$$

where:

ΔH$_t$ is the enthalpy as a function of the time from initial to a given crystallisation time

ΔH$_{total}$ is the total enthalpy reached at the end of the isothermal crystallisation process

Figure 6 shows the relative degree of crystallinity as a function of crystallisation time for CaO/Al$_2$O$_3$ = 1.00. So, we can say that the sigmoidal curve shifted towards the right with an increase in temperature indicating the overall crystallisation rate $\left(\frac{d\alpha}{dt}\right)$ for C$_{12}$A$_7$ phase formation decreases with an increase in the crystallisation temperature.

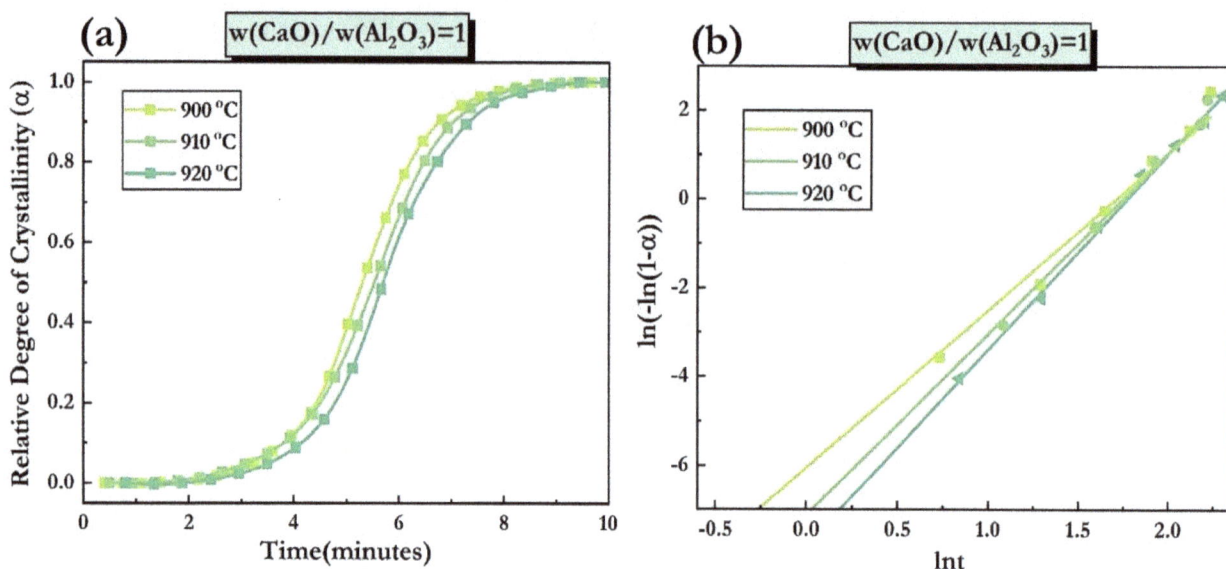

FIG 6 – (a) Relative degree of Crystallinity (α) with time, (b) $\ln\{-\ln[1-\alpha(t)]\}$ versus $\ln t$ plots for isothermal crystallisation of $CaO/Al_2O_3 = 1.00$.

The isothermal crystallisation kinetics of the flux can be understood with the help of the JMAK equation (Avrami, 1939):

$$\alpha(t) = 1 - \exp(-(kt)^n)$$

where:

$\alpha(t)$ is the relative degree of crystallinity at time t (excluding incubation time)

n is the Avrami exponent associated with the crystallisation mode

k is the effective crystallisation rate constant, which is dependent on temperature and rate of nucleation and crystal growth

The values of n and k can be obtained by fitting the double logarithmic form as follows:

$$\ln\{-\ln[1-\alpha(t)]\} = n\ln k + n\ln t$$

The plots of $\ln\{-\ln[1-\alpha(t)]\}$ versus $\ln t$ are also shown in Figure 6. From the slope and intercept the values of n and K can be obtained values of which are summarised in Table 2. The average value of n for $CaO/Al_2O_3 = 1$ lies around 4 and for $CaO/Al_2O_3 = 1.33$ value lies around 3.5. The crystallisation parameters, calculated from isothermal crystallisation experiments for both fluxes are reported in Table 2. It is noted from Table 2 that the values of effective crystallisation rate constant k for both the fluxes increase with the decreasing crystallisation temperature (T_c).

TABLE 2

Results of the Avarami analysis for isothermal crystallisation.

Sample	T_c (°C)	n	k (min^{-1})
$CaO/Al_2O_3 = 1.00$	900	3.548	0.1813
	910	4.082	0.1747
	920	4.144	0.1519
$CaO/Al_2O_3 = 1.33$	790	2.701	0.2119
	800	3.136	0.1919
	810	4.272	0.1832

The effective crystallisation rate constant k can be used to determine the crystallisation activation energy through the Arrhenius equation:

$$(\ln k) = \ln A - \frac{E}{RT_c}$$

where:

A   is the temperature-independent pre-exponential term

n   is the Avrami exponent

E   is the effective activation energy

R   is the universal gas constant

T_c is the crystallisation temperature

Figure 7 shows the plot of $(\ln k)$ versus $(1/T_c)$ for the isothermal crystallisation of both mold fluxes. The effective activation energy determined from the slope of the plot for CaO/Al_2O_3 = 1 and 1.33 is -102.75 KJ/mol and -69.72 KJ/mol, respectively. These values indicate that the mold flux melt crystallisation may be controlled by the free energy change for nucleation related to the degree of undercooling. Activation energy represents the energy barrier that must be overcome for a reaction to occur. In the case of crystallisation, nucleation is the initial step where small clusters of atoms or molecules come together to form a stable nucleus, which then grows into larger crystalline structures. Lower activation energy values, such as those obtained (-102.75 kJ/mol for CaO/Al_2O_3 = 1.00 and -69.72 kJ/mol for CaO/Al_2O_3 = 1.33), imply that the energy barrier for nucleation is lower. This means that it requires less energy for the nucleation process to initiate, making it more favourable and easier for nuclei to form. Undercooling refers to the extent to which a liquid is cooled below its equilibrium melting point before crystallisation occurs. When undercooling is significant, nucleation becomes the rate-limiting step in the crystallisation process because it is the initial barrier that must be overcome for crystals to form. Given the lower activation energy values observed, it suggests that the energy barrier for nucleation is relatively low, indicating that nucleation is more likely to occur even at lower degrees of undercooling. Therefore, nucleation is expected to be the dominant factor governing the overall crystallisation rate for both mold fluxes, as it represents the crucial initial step in the crystallisation process (Vyazovkin and Sbirrazzuoli, 2003; Papageorgiou et al, 2005).

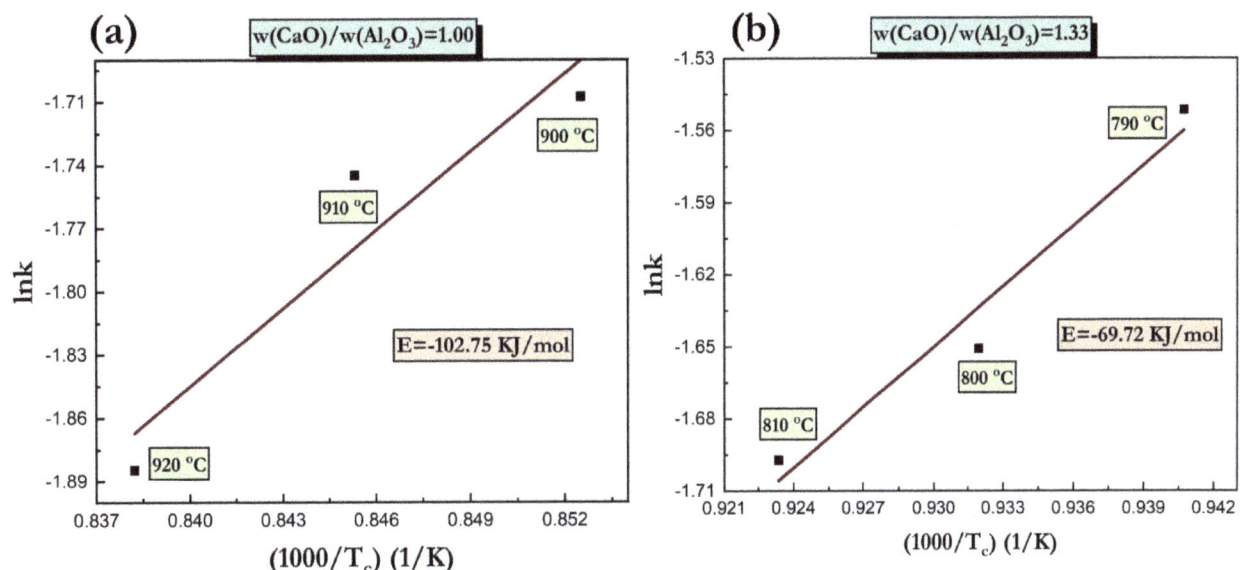

FIG 7 – Plots of $(\ln k)$ versus $1/T_c$ for isothermal crystallisation of (a) CaO/Al_2O_3 = 1.00, (b) CaO/Al_2O_3 = 1.33.

Structural analysis of the CaO-Al$_2$O$_3$ based fluxes

The melt structure of calcium aluminate-based mold fluxes plays a crucial role in influencing both viscosity and crystallisation tendencies. These factors, in turn, impact the lubrication and heat transfer capabilities of the mold fluxes to the steel shell. The structural role of Al_2O_3 in the melt structure is intricate due to its amphoteric nature. The slag system's constituents, which can act as network formers or breakers, depend on the type of steel being cast. Understanding the melt structure of the calcium aluminate slag system and how it evolves with the replacement of SiO_2 by Al_2O_3 has been the subject of numerous studies (Zhou *et al*, 2021).

Figure 8a shows the deconvoluted Raman spectra for the $CaO/Al_2O_3 = 1.00$ glassy mold flux. The spectral pattern at approximately 460 cm^{-1} in the low-frequency range of the Raman spectra corresponds to the Al-O-Al stretching characteristics peak, indicative of the bridging oxygen in the aluminate network containing $[AlO4]^{5-}$ units. Within the mid-frequency range (700–1300 cm^{-1}), the spectra reveal the depolymerisation of the aluminosilicate structure, signifying Al-O⁻ or Si-O⁻ telescopic vibrations. Specifically, the peaks at 818 cm^{-1} and 921 cm^{-1} represent the symmetric Al-O⁻ and Al-O-Si bonds, respectively. The degree of polymerisation or depolymerisation is denoted by Q^n, where 'n' signifies the number of bridging oxygen units. Q^0 corresponds to $[AlO4]^{5-}$ or $[SiO4]^{4-}$ units, with structural units $Q^0(Si)$ and $Q^1(Si)$ identified by peaks at 982 cm^{-1} and 1077 cm^{-1}, respectively (Nigam and Sarkar, 2023):

$$[AlO_4]^{5-} + O^{2-} = [AlO_5]^{7-}$$

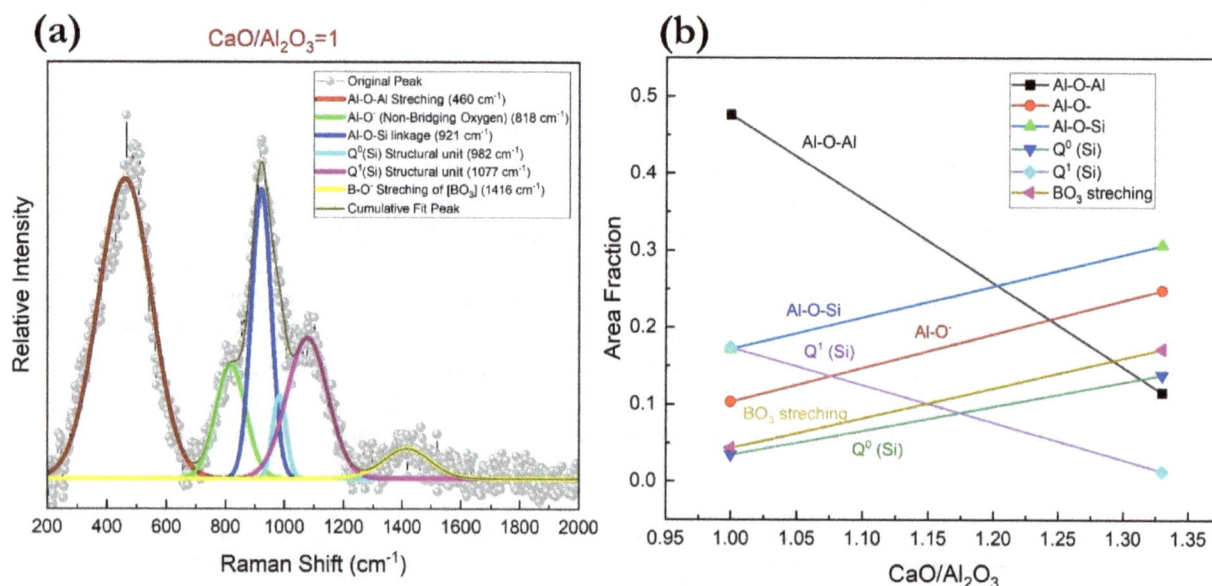

FIG 8 – (a) Deconvoluted Raman Spectra of $CaO/Al_2O_3 = 1.00$, (b) Area fraction of various structural units present in the two compositions.

The peak at 1416 cm^{-1} is assigned to the symmetric stretching vibrations of terminal oxygen atoms in orthoborate units $[BO_3]$. This suggests that the predominant formation of $[BO_3]$ groups in the mold flux is indicative of B^{3+} involvement in the structure.

Figure 8b shows that as wt per cent CaO / wt per cent Al_2O_3 increases from 1.00 to 1.33 the area fraction of Al-O-Al (bridging oxygen) and $Q^1(Si)$ decreases while Al-O⁻, Al-O-Si, $Q^0(Si)$ and $[BO_3]$ increases, this indicates that more complex aluminate and silicate structure is transformed into the simpler one. As the $w(CaO)/w(Al_2O_3)$ increases weight fraction of CaO will increase, as there will be excess CaO the O^{2-} ions in the melt will depolymerise the chain. This indicates that higher CaO/Al_2O_3 ratios will decrease the viscosity of the melt as there will be more mobility of ions due to a higher number of non-bridging oxygen and this might also increase the crystallisation tendency (Shao *et al*, 2019).

CONCLUSIONS

The isothermal crystallisation kinetics of two mold fluxes of different CaO/Al_2O_3 ratios were investigated systematically. Based on the kinetic parameters of the JMAK model the crystallisation mode was determined. The main conclusions are summarised as follows.

- The average value of the Avarami exponent n for the wt per cent CaO / wt per cent Al_2O_3 = 1.00 is ~4 and for wt per cent CaO / wt per cent Al_2O_3 = 1.33 is ~3.5 indicating the $C_{12}A_7$ crystal growth is 3D with constant nucleation rate.

- The effective rate constant (k) for the formation of $C_{12}A_7$ in both mold flux compositions exhibits an increase as the crystallisation temperature decreases. This indicates that elevated temperatures impede the overall crystallisation rate, suggesting that the process is governed by nucleation across a range of crystallisation temperatures.

- The effective activation energy of $C_{12}A_7$ formation for the mold fluxes wt per cent CaO / wt per cent Al_2O_3 = 1.00 and 1.33 is -102.75 KJ/mol and -69.72 KJ/mol respectively. The negative value of activation energy means that it is showing anti-Arrhenius behaviour, indicating that crystallisation is determined by nucleation.

- As the wt per cent CaO / wt per cent Al_2O_3 ratio increased from 1.00 to 1.33 the peak crystallisation temperature of the $C_{12}A_7$ phase decreased 915°C to 800°C.

- The depolymerisation of the various structural units because of the various network breakers led to the formation of simpler structural units by increasing NBO, which in turn will enhance the crystallisation tendency of the melt as we increase the wt per cent CaO / wt per cent Al_2O_3 ratio from 1.00 to 1.33.

ACKNOWLEDGEMENTS

The author wishes to extend thanks to the multiple laboratories within the Department of Materials Science and Engineering at IIT Kanpur for generously providing the necessary facilities for the ongoing research.

REFERENCES

Avrami, M, 1939. Kinetics of phase change, I: General theory, *The Journal of Chemical Physics*, 7(12):1103–1112. https://doi.org/10.1063/1.1750380

Aydin, H, Essadiqi, H, Jung, I H and Yue, H I, 2013. Development of 3rd generation AHSS with medium Mn content alloying compositions, *Materials Science and Engineering A*, 564:501–508. https://doi.org/10.1016/j.msea.2012.11.113

Cho, J W, Blazek, K, Frazee, M, Yin, H, Park, J H and Moon, S W, 2013. Assessment of CaO-Al2O3 based mold flux system for high aluminum TRIP casting, *ISIJ Intl*, 53(1):62–70. https://doi.org/10.2355/isijinternational.53.62

Kim, D J and Park, J H, 2012. Interfacial reaction between CaO-SiO2-MgO-Al2O3 flux and Fe-xMn-yAl (x = 10 and 20 mass pct, y = 1:3 and 6 mass pct) steel at 1873 K (1600 °c), *Metallurgical and Materials Transactions B: Process Metallurgy and Materials Processing Science*, 43(4):875–886. https://doi.org/10.1007/s11663-012-9667-x

Liu, Q, Wen, G, Li, J, Fu, X, Tang, P and Li, W, 2014. Development of mould fluxes based on lime-alumina slag system for casting high aluminium trip steel, *Ironmaking and Steelmaking*, 41(4):292–297. https://doi.org/10.1179/1743281213Y.0000000131

Nigam, A and Sarkar, R, 2023. Development of a Low Silica Calcium Aluminate Based Mould Flux for Casting High Al/Mn Steels, in *Proceedings of the International Conference on Metallurgical Engineering and Centenary Celebration (METCENT 2023)*, pp 3–12 (Springer Nature Singapore). https://doi.org/10.1007/978-981-99-6863-3_1

Papageorgiou, G Z, Achilias, D S, Bikiaris, D N and Karayannidis, G P, 2005. Crystallization kinetics and nucleation activity of filler in polypropylene/surface-treated SiO2 nanocomposites, *Thermochimica Acta*, 427(1–2):117–128. https://doi.org/10.1016/j.tca.2004.09.001

Salasin, J R and Rawn, C, 2017. Structure property relationships and cationic doping in [Ca24Al28O64]4+ framework: A review, *Crystals*, 7(5). https://doi.org/10.3390/cryst7050143

Seo, M D, Shi, C B, Baek, J Y, Cho, J W and Kim, S H, 2015. Kinetics of Isothermal Melt Crystallization in CaO-SiO2-CaF2-Based Mold Fluxes, *Metallurgical and Materials Transactions B: Process Metallurgy and Materials Processing Science*, 46(5):2374–2383. https://doi.org/10.1007/s11663-015-0358-2

Shao, H, Gao, E, Wang, W and Zhang, L, 2019. Effect of fluorine and CaO/Al2O3 mass ratio on the viscosity and structure of CaO–Al2O3-based mold fluxes, *Journal of the American Ceramic Society*, 102(8):4440–4449. https://doi.org/10.1111/jace.16322

Vyazovkin, S and Sbirrazzuoli, N, 2003. Isoconversional Analysis of Calorimetric Data on Nonisothermal Crystallization of a Polymer Melt, *Journal of Physical Chemistry B*, 107(3):882–888. https://doi.org/10.1021/jp026592k

Zhang, L, Wang, W-L and Shao, H-Q, 2019. Review of non-reactive CaO–Al2O3-based mold fluxes for casting high-aluminum steel, *Journal of Iron and Steel Research International*, 26(4):336–344. https://doi.org/10.1007/s42243-018-00226-2

Zhou, L, Wo, H, Wang, W, Luo, H and Li, H, 2021. Crystallization behavior and melt structure of typical CaO–SiO2 and CaO–Al2O3-Based mold fluxes, *Ceramics Intl*, 47(8):10940–10949. https://doi.org/10.1016/j.ceramint.2020.12.213

Improving cobalt extraction through oxidative blowing of copper-nickel matte

R A Pakhomov[1], P V Malakhov[2], L V Krupnov[3] and I M Dymov[4]

1. Senior Researcher, Pyrometallurgy Laboratory, Gipronikel Institute LLC, Saint Petersburg 195220, Russia. Email: pakhomovra@gmail.com
2. Principal Specialist, Pyrometallurgical Technology, MMC Norilsk Nickel's Polar Division, Norilsk 663305, Russia. Email: malahovkm@nornik.ru
3. Principal Specialist, Pyrometallurgical Technology, MMC Norilsk Nickel's Polar Division, Norilsk 663305, Russia. Email: nordleon@mail.ru
4. Engineer 2 category, Gipronikel Institute LLC, Saint Petersburg 195220, Russia. Email: igor99smoke@yandex.ru

ABSTRACT

The article presents results of thermodynamic calculations and laboratory studies, focusing on the cobalt behaviour during oxidative smelting of copper-nickel mattes. The compositions of the studied solutions are related to the products formed during the second stage of the converting process of copper-nickel matte at the Nadezhdinsky Metallurgical Plant (NMP) of MMC Norilsk Nickel company. These sulfide products typically contain from 1.5 to 12 wt per cent of iron. The obtained experimental results and thermodynamic calculations indicate that adjusting the CaO ratio to increase the proportion of CaO or complete replacement of SiO_2 with fluxes containing CaO leads to a significant increase in cobalt recovery rates in the matte phase (with an iron concentration of 2.5 wt per cent) raising them from 30 per cent to 45 per cent. Meanwhile, the recovery rates of copper and nickel remain constant.

INTRODUCTION

Metallic cobalt ranks among the three most treated and valuable non-ferrous metals (Cu, Ni and Co), that are processed at the metallurgical plants of MMC Norilsk Nickel company. These metals are present in sulfide copper-nickel ores, with pentlandite or pyrite serving as the primary minerals containing cobalt. Cobalt concentration in certain particles can reach up to 2.5 wt per cent, averaging between 0.2 and 1.2 wt per cent (Genkin *et al*, 1981; Tsemekhman *et al*, 2010). Sulfide minerals extracted from the mineral deposits are subjected to flotational separation to produce Cu and Cu-Ni concentrates. During this stage approximately 14 per cent of all extracted cobalt is lost with tailings, attributed to technological losses.

The application of flotation separation methods results in the distribution of cobalt among various products: around 9 per cent of Co is transferred into the copper concentrate, while 77 per cent Co goes into copper-nickel concentrate, with the remaining 14 per cent ending up with tailings. Subsequent pyrometallurgical treatment of the copper-nickel concentrate yields a commercially valuable product known as the copper-nickel matte, characterised by a high copper concentration. The recovery rates for Cu and Ni in the final product exceed 95 per cent, whereas cobalt recovery is less than 55 per cent. The low Co recovery rates are primarily attributed to its low concentrations in the feed materials, and losses with slags. Figure 1 shows the schematic flow sheet of the pyrometallurgical treatment of sulfide materials.

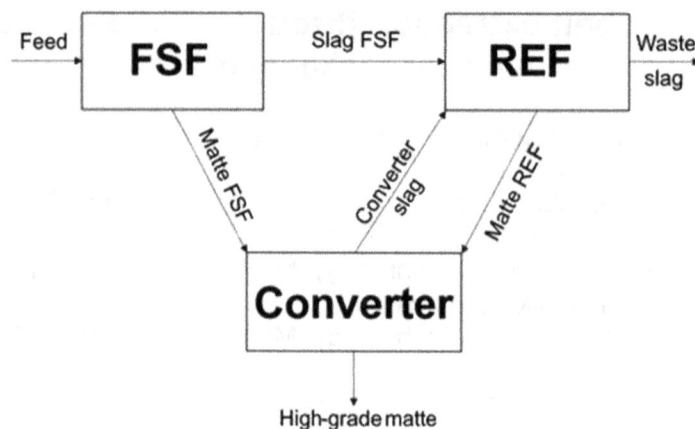

FIG 1 – Scheme of processing nickel-coper sulfide ore at Nadezhdinsky Metallurgical Plant (FSF-flash smelting furnace and REF- slag cleaning furnace).

Sulfide materials are oxidised in a converter to produce high-grade matte. Then the converter slag is conveyed to the slag cleaning furnace, while the matte is returning from the slag cleaning furnace back to converter, leading to circulation of cobalt between the furnaces. Figure 2 illustrates the distribution of cobalt between the materials supplied for converting stage and the products of the slag cleaning furnace. This figure is based on industrial data obtained from a Nadezhdinsky metallurgical plant. The concentration of cobalt in the products is determined by using analytical chemistry methods.

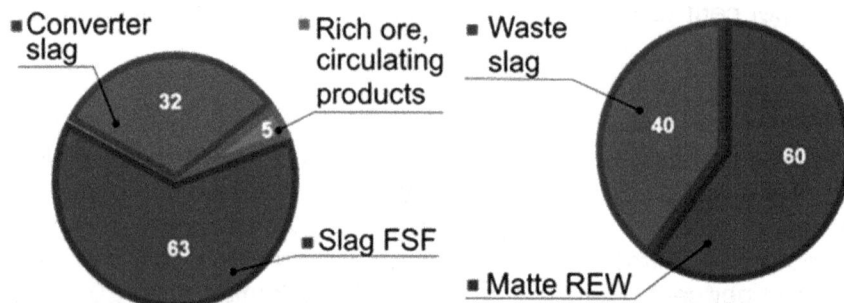

FIG 2 – Distribution of cobalt between processed and obtained products of converter (left image) or slag cleaning furnace (right image) of Nadezhdinsky Metallurgical Plant.

Increasing cobalt recovery at the converter process will reduce the amount of circulated cobalt and enhance overall recovery rates. During converting copper-nickel matte, SiO_2-based fluxes are used as additives to significantly lower the melting point of the formed slag, and to facilitate the bonding of resulting FeO and Fe_3O_4 during matte oxidation. However, the converting of matte into high-grade 'white matte' is accompanied by significant losses of non-ferrous metals into slag, particularly cobalt. Literature and metallurgical practice indicate that dissolved cobalt contributes significantly to total cobalt losses in slags (Krupnov *et al*, 2023; Tsemekhman *et al*, 2010). While iron and cobalt exhibit similar behaviour in these processes, they differ markedly from nickel and copper, where mechanical losses, such as entrained particles of sulfides or metals in slag greatly affect recovery rates. In ferrous metallurgy, fluorite (CaF_2), limestone or lime are the primary fluxing agents used in the processing of iron feed materials. Given the analogous behaviour of cobalt and iron in both non-ferrous and ferrous metallurgy, it is hypothesized that replacing SiO_2 flux with CaO-containing fluxing agents could alter the distribution of cobalt between slag and matte. This hypothesis requires verification through thermodynamic and experimental means.

At the initial stage, thermodynamic calculations were conducted to assess the oxidation of a cobalt-rich sulfide sample using SiO_2 and CaO-based additives in various ratios. Thermodynamic modelling was employed to provide an initial estimation of the distribution of non-ferrous metals. The calculation was focused on determining the required amount of flux and oxygen blow to achieve a matte with the desired composition.

THERMODYNAMIC CALCULATIONS

The calculation of the converting process was performed using the FactSage software version 6.4, utilising the thermodynamic databases FTmisc and FToxid (Bale *et al*, 2016). Throughout the matte oxidation calculation, the SiO_2/CaO ratio was varied within the range of 1 to 0.2 and only SiO_2. Operating conditions included a temperature of 1350°C, pressure at 1 atm, and pure oxygen as the oxidising agent. Table 1 outlines the composition of the matte employed for the calculations. Based on the computational outcomes, a series of graphs depicting the behaviour of non-ferrous metals in both matte and slag were generated and are presented in Figures 3 and 4.

TABLE 1

The concentrations of the main components in the matte used in the calculations.

The mass fraction of the component, wt%				
Cu	Ni	Co	Fe	S
17.19	43.78	2.68	12.68	23.67

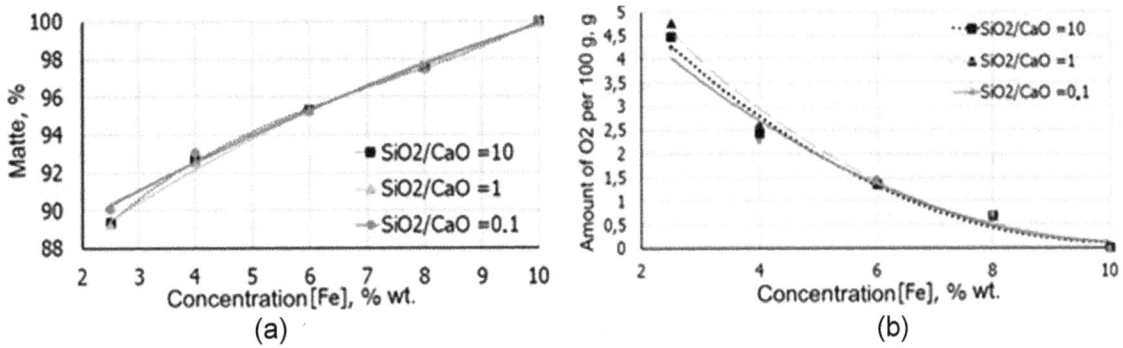

FIG 3 – Output of the matte (a) and oxygen consumption (b) as functions of iron concentration in the matte.

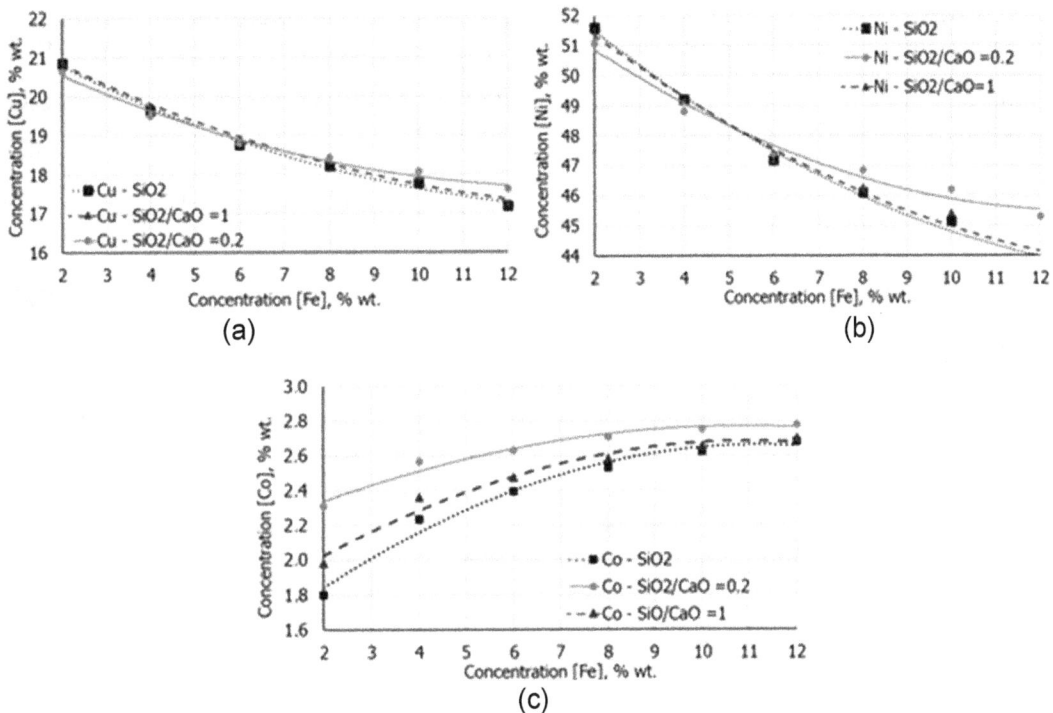

FIG 4 – The concentrations (wt%) of Cu (a), Ni (b) and Co (c) in the matte as functions of iron concentration.

Figure 3 shows the amount of matte produced and oxygen consumption relative to the iron concentration in the matte phase. According to thermodynamic estimation, varying ratio of fluxing agents does not significantly impact matte output and oxygen consumption.

Figure 4 demonstrates that replacement SiO_2 with CaO has negligible effect on the Cu and Ni concentrations in the matte phase. However, there is a notable change in cobalt proportion with the substitution of SiO_2 by CaO. For instance, when obtaining copper-nickel matte with 2.5 wt per cent of Fe, the cobalt concentration increases by 0.5 wt per cent when using CaO. Based on the presented data, the evaluation of cobalt recovery into the matte phase suggests potential savings of cobalt up to 4.5 per cent compared to silicate slags.

Following the assessment of theoretical parameters for cobalt recovery, product compositions, and oxidation process conditions, a series of laboratory experiments were conducted.

RESULTS AND DISCUSSION

In total 11 experiments were performed, including six experiments with SiO_2-based slag (without Ca addition), and five experiments with CaO-based slag with a SiO_2/CaO ratio of 0.2. To estimate the distribution of non-ferrous metals between slag and matte, a synthetic flux composed of FeO, SiO_2 and CaO was used to simulate pyrometallurgical converting slag while maintaining a consistent SiO_2/CaO ratio. The experiments were conducted in an induction furnace at temperatures ranging from 1300 to 1350°C, under an overall pressure of is 1 atm, using pure oxygen as the oxidising agent and employing Al_2O_3 crucible. The compositions of the obtained samples were examined using methods of analytical chemistry and EDX-SEM (energy dispersive X-Ray – scanning electron microscope). Experimental conditions are detailed in Table 2, the results of the chemical analysis are presented in Table 3.

TABLE 2
Experimental conditions.

No	Feed composition, g				O_2		Mass of products, g		SiO_2/CaO
	Matte	SiO_2	CaO	FeO	Vol, L	Gas flow, L/min	Matte	Slag	
1	61.6	12	-	28	2.57	0.7	54.3	46.1	SiO_2
2	61.6	12	-	28	1.23	0.7	60.0	41.1	SiO_2
3	61.6	12	-	28	2.68	0.7	52.5	48.8	SiO_2
4	61.6	12	-	28	4.78	0.7	47.4	51.2	SiO_2
5	61.6	12	-	28	7.23	0.7	47.0	50.0	SiO_2
6	61.6	12	-	28	8.47	0.7	33.1	55.9	SiO_2
7	61.6	2	10.6	28	2.68	0.7	54.7	45.7	~0.2
8	61.6	2	10.6	28	1.46	0.7	57.4	44.6	~0.2
9	61.6	2	10.6	28	4.96	0.7	47.3	53.2	~0.2
10	61.6	2	10.6	28	6.35	0.7	43.8	40.7	~0.2
11	61.6	2	10.6	28	6.40	0.4	45.7	52.8	~0.2

TABLE 3

The concentrations of the main components in the matte and slag phases (wt%).

№	Composition of matte phase, wt%					Composition of slag phase, wt%					
	Cu	Ni	Co	S	Fe	Cu	Ni	Co	Fe	SiO_2	CaO
0	*16.9*	*42.9*	*2.95*	*23.0*	*12.45*						
1	17.4	46.6	2.06	24.0	7.18	0.59	1.47	1.38	47.2	25.2	-
2	16.6	43.9	2.16	24.5	8.97	0.56	1.24	1.18	48.0	25.8	-
3	18.0	47.9	1.97	23.7	6.69	0.88	2.33	1.43	47.5	24.2	-
4	19.0	48.7	2.01	23.0	6.81	1.23	3.25	1.72	44.6	19.8	-
5	20.1	52.4	1.50	22.4	3.25	0.95	3.45	2.03	47.9	20.6	-
6	22.9	55.4	0.81	19.3	1.47	1.62	8.19	2.40	41.9	16.7	-
7	17.8	47.2	2.31	23.6	6.6	0.69	2.02	1.09	46.6	3.8	16.8
8	17.1	45.6	2.46	24.3	8.33	0.67	1.41	0.88	46.3	3.4	18.2
9	20.1	52.0	1.60	24.1	2.2	0.64	2.34	1.12	38.2	2.4	15.6
10	20.8	52.7	1.76	22.0	2.18	3.98	1.05	1.73	42.6	3.6	17.7
11	21.3	53.5	1.57	21.7	1.9	0.83	3.61	1.94	49.3	1.8	16.7

Consider the behaviour of non-ferrous metals in the matte during oxidation and the use of additives based on SiO_2 and CaO, the results are shown in Figure 5.

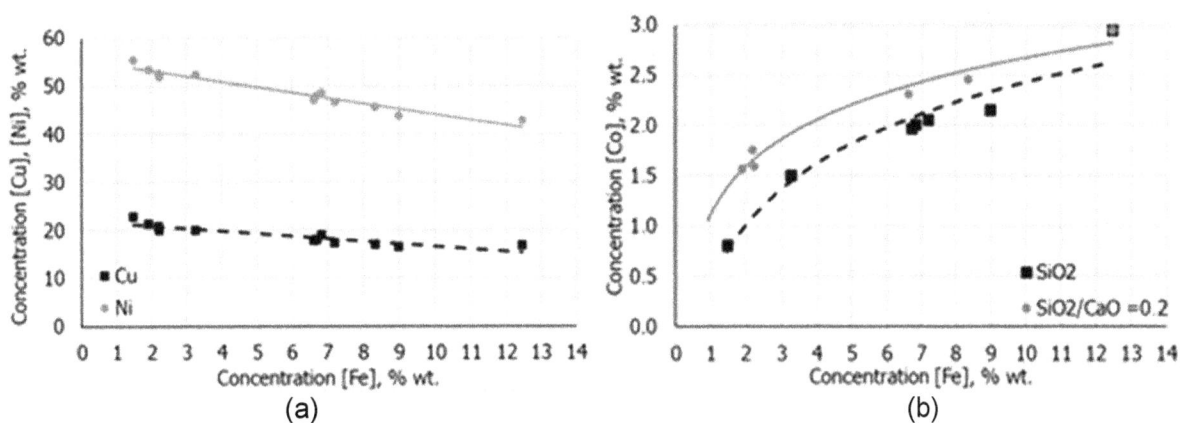

FIG 5 – The concentrations of non-ferrous metals in the matte phase as functions of iron concentration (wt%): a – Cu, Ni; b – Co.

Analysis of the graphs in Figure 5 indicates that the choice of flux does not influence the tonne of copper and nickel concentration in matte, confirming the calculation results. However, the cobalt concentration varies significantly depending on the selected flux. For instance, in the copper-nickel matte (with 2.5 wt per cent Fe) the cobalt concentration increases from 1.24 wt per cent to 1.75 wt per cent. (at 40.5 rel per cent) with the addition of CaO-based slag. Evaluation of cobalt recovery rates using different types of slags is presented in Figure 6.

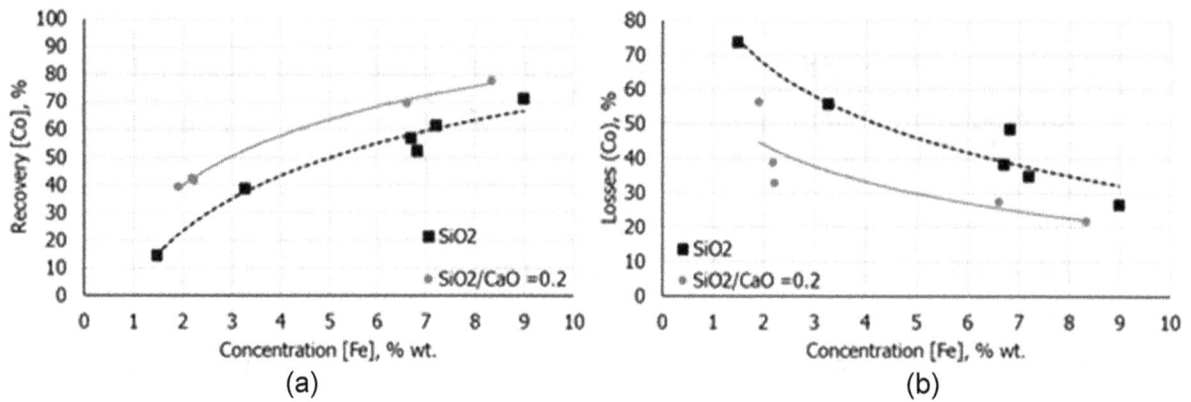

FIG 6 – Recovery of cobalt into the bottom phase (a) and losses with slag (b) from the iron content in the sulfide mass.

The recovery of cobalt in matte or the losses of cobalt with slag were calculated as the ratio of the amount of metal in high-grade matte or slag to the amount of metal in the initial sample, Equations 1 and 2.

$$[Co]_{rec} = \frac{[Co]_{hgm}}{Co_{matte}} \cdot 100\%$$ (1)

$$(Co)_{los} = \frac{(Co)_{slage}}{Co_{matte}} \cdot 100\%$$ (2)

Analysis of cobalt recovery and losses with slags during the second stage of converting reveals the effectiveness of incorporating CaO as fluxing additive. The recovery of Co within the examined range of matte compositions (with 2 to 10 wt per cent Fe) increases from 11 to 20 per cent with the utilising of calcium-ferrite slags. It's worth noting that the concentration of cobalt in the slag is assessed without considering mechanically entrained particles, utilising SEM-EDX methods for analysis. Detailed description of sample preparation methods for SEM-EDX are provided in previous works (Gouldstein *et al*, 1981; Krishtal *et al*, 2009). Figure 7 illustrates the cobalt concentration in slag as function of the iron concentration in the matte phase.

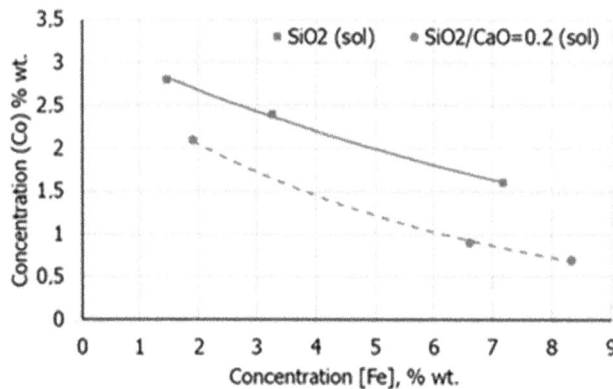

FIG 7 – Cobalt concentration in the matte phase according to SEM-EDX data.

The analysis of the microstructures of the samples reveals that in the first scenario, the matrix base comprises an iron-silicate solution with dispersed sulfide-metal particles of varying composition and structure. In the second scenario, the slag matrix consists primarily of the Fe-Ca-O system with a minor presence of residual silicate forming the slag base. Further analysis of the slags using EDX-SEM methods corroborates the potential benefits of employing CaO-based slags over SiO₂-based ones.

A comparison of the practical and calculated results of the behaviour of cobalt in slag and high-grade matte, the results are presented in Figure 8.

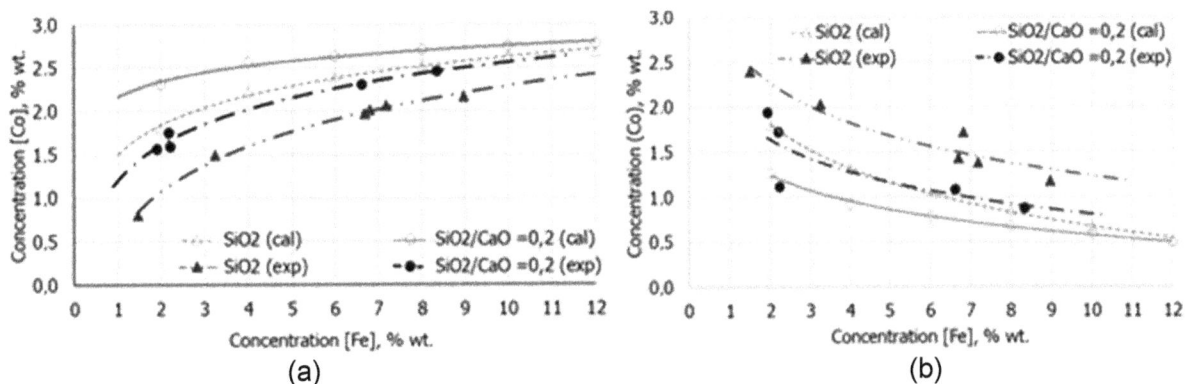

FIG 8 – Cobalt concentration in matte (a) and slag (b) according to calculation and experimental data.

The results of theoretical calculations of cobalt distribution presented in Figure 8 differ significantly from the practical data for two types of slags based on SiO_2 and CaO. The difference in results may be due to incomplete databases that lack data on the distribution of cobalt in sulfide solutions.

CONCLUSIONS

The assessment of cobalt distribution among the products of copper-nickel production of MMS Norilsk Nickel was carried out. The assessment involved thermodynamic calculations and laboratory studies on the converting process of copper-nickel matte to 'white' matte using additives based on SiO_2 and CaO. The utilisation of CaO-based slags resulted in an increase in cobalt concentration in the matte phase, while the concentrations of Cu and Ni remained unchanged. For instance, in high-grade matte with a 2.5 wt per cent Fe concentration, the cobalt concentration increased from 1.24 wt per cent to 1.75 wt per cent, that makes a 40.5 per cent rise, upon transition to calcium-ferrite slag. Furthermore, the recovery of Co within the considered range of Fe compositions from 10 to 2 per cent increases by 11–20 per cent upon switching to calcium-ferrite slag. Thus, the use of calcium oxide as the primary component of fluxing agents at the converting process of copper-nickel matte will have a positive impact on cobalt recovery rates.

REFERENCES

Bale, C W, Bélisle, E, Chartrand, P, Decterov, S A, Eriksson, G, Gheribi, A E, Hack, K, Jung, I-H, Kang, Y-B, Melançon, J, Pelton, A D, Petersen, S, Robelin, C, Sangster, J, Spencer, P and Van Ende, M-A, 2016. FactSage thermochemical software and databases 2010–2016, CALPHAD, 54:35–53.

Genkin, A D, Distler, V V, Gladyshev, G D, Filimonova, A A, Evstigneeva, T L, Kovalenker, V A, Laputina, I P, Smirnov, A V and Grokhovskaya, T L, 1981. *Sulfide copper-nickel ores of the Norilsk deposits* [in Russian], 234 p (Nauka: Moscow).

Gouldstein, J I, Newbury, D E, Etchlin, P, Joy, D C, Fiori, C and Lifshin, E, 1981. *Scanning electron microscopy and X-ray spectral microanalysis* (translated from English), Part 1, 296 p; Part 2, 348 p.

Krishtal, M M, Yasnikov, I S, Polunin, V I, Filatov, A M and Ulyanenkov, A G, 2009. Scanning electron microscopy and X-ray spectral microanalysis in practical application examples, *Technosphere*, 208 p.

Krupnov, L V, Malakhov, P V, Ozerov, S S and Pakhomov, R A, 2023. Analyzing Russian cobalt metallurgy and ways to raise recovery [in Russian], *Tsvetnye Metally*, (7):25–33. https://doi.org/10.17580/tsm.2023.07.03

Tsemekhman, L S, Fomichev, V B, Yertseva, L N, N G, Kaitmazov, S M, Kozyrev, V I, Maksimov, Schneerson Ya, M and Dyachenko, V T, 2010. Atlas of mineralogical raw materials, technological products and commercial products of the Polar Branch of the MMC Norilsk Nickel OJSC [in Russian], 336 p (Ore and Metals Publishing House).

Comparative study of oxide dissolution modelling in secondary steelmaking slags

N Preisser[1] and S K Michelic[2]

1. PhD candidate, Christian Doppler Laboratory for Inclusion Metallurgy in Advanced Steelmaking, Leoben A-8700, Austria. Email: nikolaus.preisser@unileoben.ac.at
2. Full Professor, Christian Doppler Laboratory for Inclusion Metallurgy in Advanced Steelmaking, Leoben A-8700, Austria. Email: susanne.michelic@unileobane.ac.at

ABSTRACT

For certain applications, steels have to feature a very distinct and well-defined level of cleanness in terms of the number and composition of non-metallic inclusions. Therefore, one crucial aspect of steelmaking is removing and controlling said non-metallic inclusions during the process. Particle removal during ladle refining comprises three stages, namely:

1. Flotation

2. Transport through the steel/slag interface

3. Dissolution of the particle in the slag.

Only if all three stages are completed the particle can be considered as fully removed from the steel melt which, in turn, leads to the slag being one of the most influential parameters for inclusion removal and control. Depending on the composition of the particles and the slag, the dissolution can be controlled by different mechanisms. This study focuses on the diffusion-controlled dissolution of solid oxidic particles in slags. For this, a model of particle dissolution has been developed. For modelling diffusion, either the differential equation following Fick's second law can be solved, or an analytical solution to this equation valid for stationary interfaces can be used. Both approaches allow for calculations of concentration profiles of the dissolving species in surrounding slag. From this, the mass flux can be derived, which finally leads to calculating the boundary layer velocity due to considerations of mass balance. In this work, both methods as well as results of dissolution experiments using High Temperature Confocal Scanning Laser Microscopy are compared. This research aims to provide a framework for diffusion-based dissolution modelling in further expanding the understanding of particle dissolution in secondary metallurgy.

INTRODUCTION

The possible applications of steel and steel products are constantly rising. Much of this is due to the fact that enhanced steel cleanness is improving the mechanical, chemical and physical properties of the steel as shown by Garrison and Wojcieszynski (2007) and Chen et al (2019). One of the most essential aspects for steel cleanness is the secondary metallurgy in the steel plant especially the interaction of the steel with the slag phase. And as Sridhar and Cramb (2003) state, slags heavily influence the steel cleanness as the slag's properties are linked to multiple processes essential for steel cleanness. As non-metallic inclusions (NMI) are similar in composition to secondary metallurgical slags, there is a tendency towards chemical equilibrium between particles and slag and the chemical composition of NMIs can be influenced. Moreover, the slag's surface tension and viscosity is important for the transition of NMIs from the steel to the slag-phase as shown by Jimbo, Chung and Cramb (1996). And lastly, slags are responsible for the dissolution and, therefore the ultimate removal of NMIs from the system, as stated by Valdez, Shannon and Sridhar (2006) and Lee et al (2001). only if the NMIs can be transported to the steel-slag interface by mechanisms like flotation, move through this interface due to advantageous wetting conditions and surface tensions, and are then dissolved in the slag promptly to avoid re-entrapment in the steel, the particles can be considered entirely removed. For most applications especially hard and non-deformable NMIs are detrimental. Furuya, Abe and Matsuoka (2003) found that this is because especially with dynamic loads and many load cycles cavities can form around particles, which are not as deformable as the surrounding steel matrix. These cavities can then be the origin of cracks and lead to critical failure of the material. Therefore, it is of interest to gain a better understanding of the mechanisms which are involved in the removal of such particles from the steel melt to ensure the highest possible steel cleanness, avoid fracture of the material, and enable smaller diameters, which in turn will lead to economic and ecological benefits. This work focuses on the last step of particle removal, the dissolution of NMIs in slags. Solid oxide particles like Al_2O_3, which are particularly harmful to the mechanical properties of steel, are studied. This is done by simulations of particle dissolution and experimentally, using high temperature confocal scanning laser microscopy (HT-CSLM), which allows *in situ* investigation of particle dissolution in slags at steelmaking temperatures.

From the work of Levenspiel (1999), it is clear that dissolution can be explained by different mechanisms that lead to very distinct dissolution curves. Suppose the dissolution is for example limited by some kind of chemical reaction between the particle and the matrix in which it dissolves. In that case, the dissolution takes place linearly, where the speed of dissolution can be directly found from the speed of the chemical reaction. In contrast, this work focuses on dissolution purely governed by the diffusion of solid particles into the liquid slag as Huo *et al* (2022) indicate that this shows promising results for the dissolution of solid oxide particles in steelmaking slags.

DESCRIPTION OF THE MODEL

The dissolution behaviour is portrayed in this work, assuming a stationary interface between the particle and the slag phase. According to Guo *et al* (2017) this approach models dissolution behaviour more accurately as the alternative to invariant field or reverse growth approximations. For this model, the interface where mass is transferred from the solid particle to the liquid slag is fixed at a position equal to the initial radius of the dissolving particle. Using this approach, the following calculations are simplified, as a homogeneous distance between spatial steps is used for a finite differences method. In contrast, a moving interface would lead to varying distances between calculation points.

For the diffusion simulation using this method, several parameters have to be provided. Parameters describing the system's spatial layout include the initial radius of the spherical particle, the width of the boundary layer and the number of spatial steps where calculations should be performed. The boundary layer describes a zone surrounding the solid particle where diffusive processes can occur. Outside of this boundary layer, the slag composition is assumed to be unchanged over the whole process of dissolution. All spatial steps where calculations of the concentration profile will be performed lie within this boundary layer between the particles' initial surface at time 0 and the edge of the boundary layer. A visual representation of the spatial layout of this simulation is represented in Figure 1.

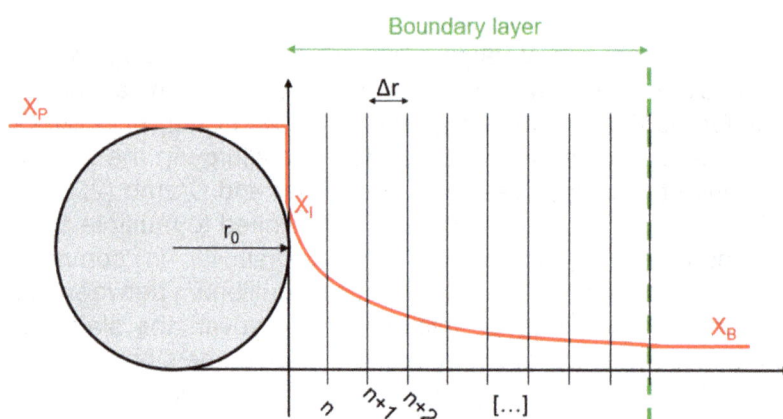

FIG 1 – Spatial layout of the diffusion model.

Additionally, chemical information about the system has to be provided, like the composition of the particle regarding the dissolving species as the dimensionless weight percentage xP (in this case, xP=1 as a homogeneous particle of one species is assumed), the weight percentage xB of the dissolving species in the slag in the bulk zone (initial composition of the slag), as well as the weight percentage of the dissolving species in the slag saturated with the dissolving species (ie this is assumed to be the composition of the slag directly at the interface between particle and slag). The weight percentage of the dissolving species in the saturated slag is calculated via the thermodynamic calculation software FactSage™ ver 8.3 (by Thermfact and GTT-Technologies) using the Equilib module. Lastly, the diffusion coefficient has to be provided to calculate the concentration profile around the dissolving particle at all times. The spatial stepsize can be calculated from the boundary layer thickness and the number of desired spatial steps. From this and the diffusivity the critical temporal stepsize can be calculated following Equation 1 were Δt is the length of the temporal timestep Δr is the size of the spatial step and D is the diffusion coefficient.

$$\Delta t = \frac{\Delta r^2}{2D} \tag{1}$$

With the number of necessary timesteps and the defined number of spatial steps, the corresponding space and time matrices for the simulation of the concentration field around the particle can be set. The initial conditions for the concentration profile are set so that for all timesteps, the mass percentage of dissolving species at the interface is x_I. At the same time, it is x_B at any other spatial step. Additionally, the temporal derivative of the concentration field is set to be 0 at all spatial steps initially. The temporal derivative of the concentration field can be calculated for each timestep following Equation 2, as stated by Ogris and Gamsjäger (2022). Where x_{i-1}, x_i and x_{i+1} are the mass fractions of the dissolving species at a spatial position at different calculation times. The concentration field of the next timestep then follows as the sum of the concentration field of the current timestep and the calculated temporal derivative multiplied by the length of the timestep. Now, the spatial derivative of the concentration profile can be calculated at each spatial step for each timestep as seen in Equation 3. From this, the mass flux is calculated using Fick's second law of diffusion given in Equation 4 where J is the mass flux and $\dot{x}$ is the spatial derivative of the concentration profile. As the dimensionless mass percentage was used for the calculation of the concentration profile instead of actual concentrations, this flux would have to be multiplied with the slags density in order to resemble actual mass flux with unit $kg\,m^{-2}s^{-1}$. Due to this simplification the actual unit of this flux of weight fraction is ms^{-1} Now the reduction in diameter of the particle can be calculated for each timestep. This is performed using a mass balance equation called 'Stefan's interface condition' after Slovenian scientist Jozef Stefan as found in Glicksman (2000). This mass balance is given in Equation 5. On the left hand side, the mass leaving the interface is given as the product of the interface area multiplied by the mass flux at the interface $J(R_0)$, the length of the timestep Δt and the slags density. On the right hand side, the respective difference in the spherical particle volume is multiplied by the particle density. For this simulation, the density of the solid particle and the density of the slag at the interface are assumed to be the same, which allows for simplification and rearrangement of Equation 5 to determine the radius of the particle in the next time step. This is shown in Equation 6.

$$\frac{\Delta x}{\Delta t} = D \left(\frac{x_{i-1} + x_{i+1} - 2x_i}{\Delta r^2} + \frac{x_{i+1} - x_{i-1}}{r_i \Delta r} \right) \tag{2}$$

$$\dot{x} = \frac{\Delta x}{\Delta r} = \frac{x_{n+1} - x_n}{\Delta r} \tag{3}$$

$$J = -D\dot{x} \tag{4}$$

$$4\pi R_0^2 J(R_0) \Delta t\, \rho = \frac{4}{3}\pi \left(r_i^3 - r_{i+1}^3 \right) \rho \tag{5}$$

$$r_{i+1} = \sqrt[3]{\left(r_i^3 - 3 J(R_0) \Delta t\, R_0^2 \right)} \tag{6}$$

PARAMETER ANALYSIS

The impact of changes of initial parameters is presented and discussed. This is necessary to further understand the models' capabilities and restrictions for further application. In Figure 2 the dissolution curves of various simulations are shown. In Figure 2a, the only difference in parameters is the number of spatial steps chosen between the interface and the outer edge of the boundary layer. As absolute values are plotted, it can be seen that with an increasing number of calculation points, the curve is shifted to slightly faster dissolution times. At the same time, the difference between simulations gets smaller as the number of spatial steps increases. Therefore, it can be assumed that the result of the simulation converges, and at some stage, increasing the number of calculation points does not result in significantly more accurate results.

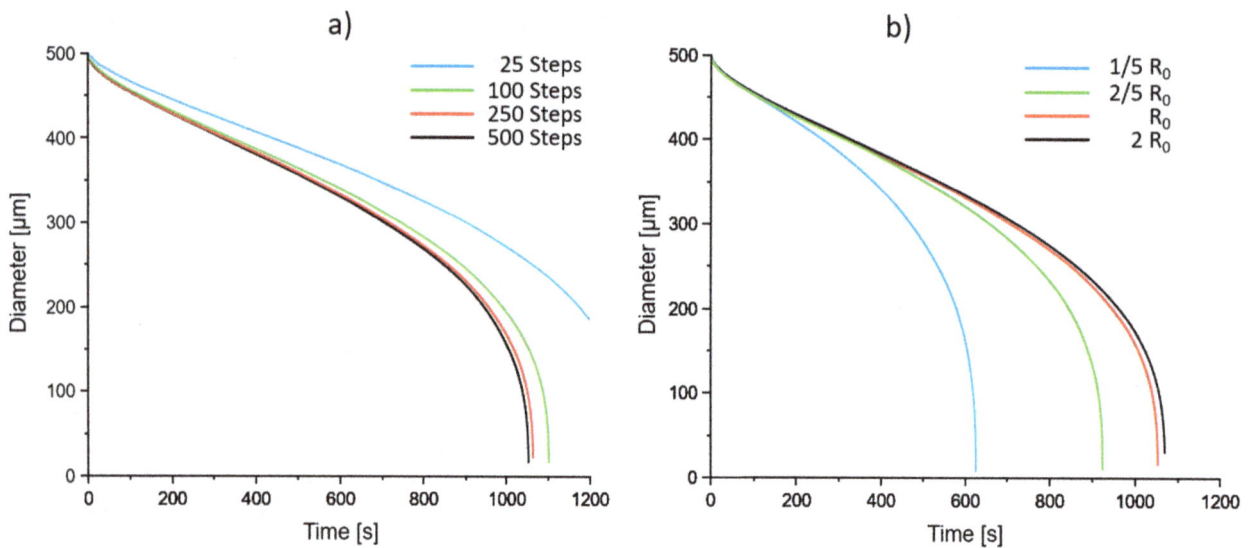

FIG 2 – Variation of number of spatial steps and boundary layer thickness.

In Figure 2b, the impact of the thickness of the boundary layer is shown. This parameter heavily influences the shape of the concentration profile, where smaller values lead to steeper gradients in concentration as the zone where an unchanged composition of the bulk medium is assumed moves closer to the interface between slag and particle. Therefore, smaller values for the boundary layer thickness result in faster particle dissolution. In this example, the slowest dissolution is achieved when the boundary layer thickness is double the initial radius of the particle. With decreasing boundary layer thickness, the dissolution of the particle rapidly speeds up, as the last line representing a boundary layer thickness of one fifth of the initial radius of the particle, shows much faster dissolution. Lastly, the influence of different diffusion coefficients was analysed with fixed amounts of spatial steps, boundary layer thickness and initial radius of the particle. In Figure 3a it can be seen that dissolution is faster with higher values for the diffusion coefficient, as would be expected. The normalised dissolution curves are depicted in Figure 3b. Normalised dissolution curves are derived by dividing particle diameters by their initial values at time zero and time by the total dissolution time. This normalisation technique allows for a direct comparison of dissolution behaviour across different scenarios, independent of specific particle sizes or dissolution durations. Here it is clear that a change in diffusion coefficient does not affect the general shape of the dissolution curve, as the normalised dissolution curves for all values for the diffusion coefficient are congruent. This observation suggests that while the diffusion coefficient influences the rate at which dissolution occurs, it does not alter the overall kinetics of the dissolution process.

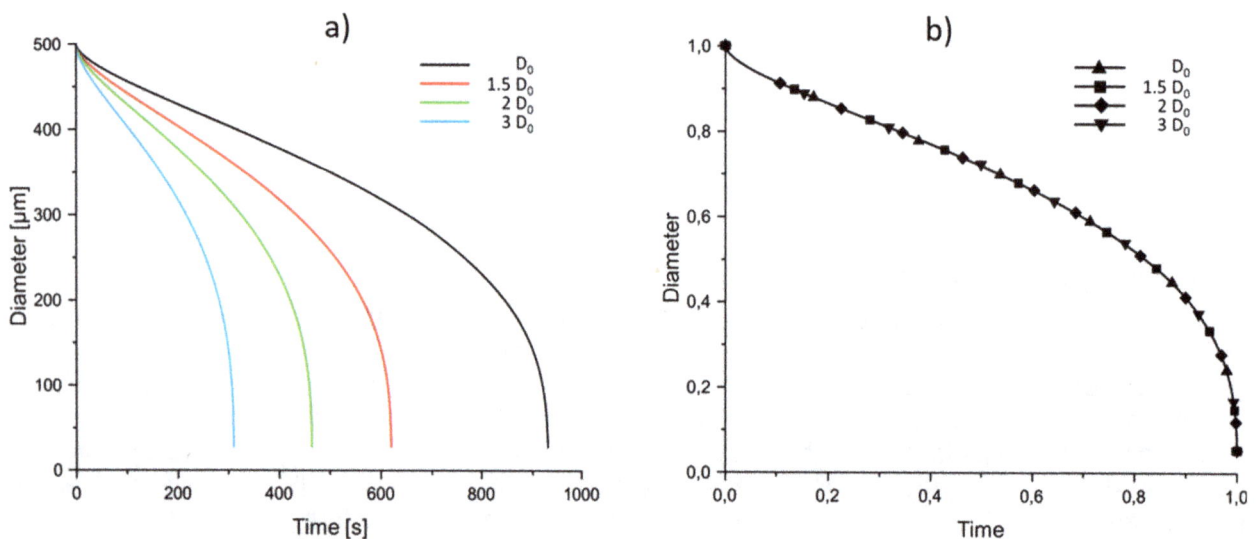

FIG 3 – Influence of changes in diffusion coefficient.

DESCRIPTION OF HT-CSLM AND EXPERIMENTAL SET-UP

For model validation, dissolution experiments have been conducted using high temperature confocal scanning laser microscopy (HT-CSLM). With this, the dissolution of particles in slags can be observed *in situ* and the dissolution curves can be gathered directly. The experimental set-up has already been widely discussed in previous work by Michelic *et al* (2016) as well as Feichtinger *et al* (2014) and will not be presented in detail. For validation of the model, a specific combination of particle, slag and temperature with a known diffusion coefficient has been selected from the literature. The slag composition is given in Table 1. The selection of slag is crucial for the analysis of the dissolution as it is paramount that the slag is transparent for the laser with a wavelength of 405 nm. Therefore, slags without tainting constituents like FeO or MnO should be used. The particles used for the dissolution experiment are synthetic Al_2O_3 particles with nearly perfect spherical geometry and an initial particle diameter of 500 µm. The slag is heated together with the particle to 1500°C. The particle is then manually tracked. The software directly logs the time and individual temperature. The corresponding particle diameter is gathered by extracting single frames from the video data afterwards and measuring the particle's dimensions. From this dissolution curves of diameter over time can be plotted. As time zero of dissolution, the time is chosen at which the experimental temperature of 1500°C is reached. Dissolution, which occurs before this point in time, is disregarded. Third-order polynomial smoothing of the data is performed using the SimpleFit Module of OriginPro (version 2023, OriginLab Corporation, Northampton, MA, USA). The data of both experiments are comparably close, which results in nearly identical dissolution curves after polynomial fitting. The total dissolution is achieved after approximately 900 seconds in both experiments. From both data sets it can be derived that the dissolution rate at the beginning as well as at the end of the dissolution is faster. This behaviour is quite common and must be expected if diffusion is assumed to be the limiting factor of dissolution. This is because of a steeper concentration gradient in the beginning of dissolution and a favourable relation of the particle's surface area to its volume.

TABLE 1

Slag composition.

Name	SiO$_2$ [wt.%]	CaO [wt.%]	Al$_2$O$_3$ [wt.%]	MgO [wt.%]
Slag 1	49.56	32.42	11.16	6.86

COMPARISON OF SIMULATION AND EXPERIMENTAL RESULTS

For validation of the simulation, the experimental results are compared with the dissolution curves calculated by the model using the presented stationary interface approach of dissolution modelling. The boundary layer thickness for the calculation was chosen to be 500 µm, which is double the synthetic particle's initial radius. The diffusion coefficient was set to $4.56*10^{-11}$, as found in the work of Burhanuddin *et al* (2022) for the diffusion of Al_2O_3 in the specific slag used in the HT-CSLM experiments. As pure Al_2O_3 particles are dissolved, the mass fraction of Al_2O_3 of the particle is 1. The mass fraction of Al_2O_3 in the bulk slag is 0.1119 and the mass fraction of Al_2O_3 in the saturated slag is 0.3970, as calculated by Factsage™. For a relatively short computation time, 200 spatial steps were chosen for calculations between the interface and the outer edge of the boundary layer. With these parameters, a total dissolution time of approximately 600 s was calculated. At first glance, this would lead to the conclusion that the simulation does not describe the dissolution well, as this value is about a third lower than the dissolution time gathered from the HT-CSLM experiments. But by closer comparison, it becomes apparent that this discrepancy in dissolution time arises mainly in the second half of the dissolution when the diameter of the particles gets smaller. In contrast, the dissolution rate at the beginning of the dissolution is predicted with high accuracy. While the experimental data only shows a slight pick-up in the dissolution speed, the model leads to a drastic increase in dissolution rate to the point where the particle is expected to dissolve almost instantaneously after a decrease in diameter of about 75 per cent. In Figure 4 the data of both dissolution experiments and the calculated dissolution curve from the model are depicted.

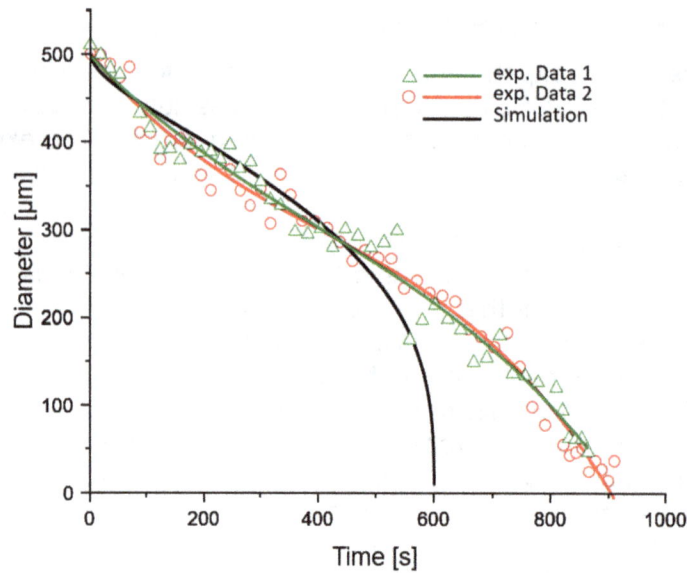

FIG 4 – Comparison of experimental results of dissolution curves and simulated dissolution curve.

DISCUSSION

In this work, a model for the simulation of diffusion-based dissolution of solid particles in liquid slags has been developed. The model follows a stationary interface approach, where for the sake of calculating the concentration profile around the particle and mass balance considerations, the interface is considered as not moving. This aids the calculation as the vectorised space for finite differences methods is homogeneous, where, in contrast, a moving interface would lead to inhomogeneous distances between nodal points of calculation and the concentration profile would need to be recalculated after each timestep as opposed to being calculated once upfront. For comparison, dissolution experiments have been conducted using HT-CSLM, where the dissolution of a synthetic particle in an experimental slag can be observed *in situ* at steelmaking temperatures. Dissolution rates of the experiments and the simulation are in good agreement at the beginning of the dissolution but deviate towards smaller diameters as the simulation predicts a faster pick-up in dissolution speed. This is believed to be an effect of the stationary interface approach, as the surface through which mass flux is assumed stays constant as the surface area of the initial particle, whereas in reality this also decreases in size throughout the dissolution. To clarify this effect, the ratio of surface area for mass flux and particle volume can be calculated for both a moving and a stationary interface approach. In Equation 7 this ratio is expressed for the case of a moving interface as γ_M. Equation 8 states the same ratio γ_S but for the case of a stationary interface.

$$\gamma_M = \frac{A}{V} = \frac{4\pi r^2}{\frac{4}{3}\pi r^3} = \frac{3}{r} \tag{7}$$

$$\gamma_S = \frac{A}{V} = \frac{4\pi R_0^2}{\frac{4}{3}\pi r^3} = \frac{3 R_0^2}{r^3} \tag{8}$$

From these ratios, it is clear that with decreasing values of r the ratio γ_S will increase much faster than the ratio γ_M. This could in turn explain the difference between the experimental results and the simulation of dissolution in later stages of the dissolution process, as the mass flux through the surface area becomes overpronounced with decreasing particle diameters. To better understand the scope of this discrepancy between the different models, a critical radius can be calculated, which marks the radius at which the stationary interface approach theoretically overtakes the moving interface approach, given the concentration field would be the same for both calculations. This can be done by comparing the first derivative of both surface-to-volume ratios, shown in Equations 9 and 10.

$$\dot{\gamma}_M = -\frac{3}{r^2} \tag{9}$$

$$\dot{\gamma}_S = -\frac{9 R_0^2}{r^4} \tag{10}$$

If these derivatives are compared, a critical value r_{crit} can be found. For values of r lower than r_{crit} the stationary interface approach leads to quicker dissolution of the particle. This is shown in Equations 11 and 12.

$$-\frac{3}{r_{crit}^2} = -\frac{9\,R_0^2}{r_{crit}^4} \tag{11}$$

$$r_{crit} = \sqrt{3\,R_0^2} = R_0\,\sqrt{3} \tag{12}$$

As Equation 12 states, this critical value is always higher than the initial particle radius R_0 as the square root of 3 is greater than 1. This shows that from the beginning, the stationary interface approach will already lead to faster dissolution of the particle than a moving interface approach, given the concentration profile would be the same. This assumption is not entirely correct, as a moving interface would lead to slightly different concentration profiles around the particle, but for further comparison of the models in this work, equal concentration fields are assumed. To grasp the magnitude of difference between the two models, $\log(\Delta\gamma)$ is plotted in Figure 5 over the particle radius. At the initial particle radius R_0 the difference is 0 as $\gamma_M = \gamma_S$ for $r = R_0$. As the particle dissolves, $\log(\Delta\gamma)$ grows rapidly at first, but soon the difference between the two modelling approaches somewhat stabilises for the central part of the dissolution. Only towards the end of the dissolution process, a massive increase in difference between γ_M and γ_S can be observed. To locate the radius at which $\Delta\gamma$ starts to increase more rapidly, the second derivative of $\log(\Delta\gamma)$ can be set to 0 for calculating r_{POI}, which represents the point of inflection of the function. This is done in Equations 13 and 14.

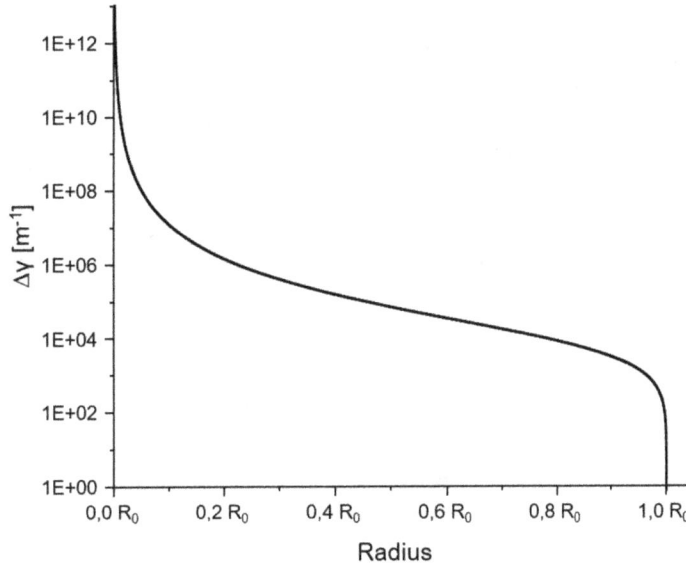

FIG 5 – $\Delta\gamma$ plotted over particle radius.

$$\log_{10}(\Delta\gamma) = \log_{10}\left(\frac{3\,R_0^2}{r_{POI}^3} - \frac{3}{r_{POI}}\right) \tag{13}$$

$$\frac{d^2}{dr^2}(\log_{10}(\Delta\gamma)) = \frac{r_{POI}^4 - 8r_{POI}^2 R_0^2 + 3R_0^4}{\left(r_{POI}^3 - r_{POI}R_0^2\right)^2} = 0 \tag{14}$$

If Equation 14 is solved for r_{POI}, only one solution is positive and smaller than R_0. Only this solution presented in Equation 15 represents the point of inflection for the section of the function of $\log(\Delta\gamma)$ depicted in Figure 5. This shows that $\Delta\gamma$ begins to grow faster as soon as a particle diameter of approximately 63 per cent of R_0 is reached. For an initial radius of R_0 = 250 µm as used for the calculations and HT-CSLM experiments presented in this paper, this leads to r_{POI} = 157 µm or DPO_I = 314 µm, respectively. In Figure 6 this value is marked as a horizontal line. Beyond this point in calculation, the deviation from the experimental results begins to increase rapidly, but before that the simulation is in good agreement to the experimental data.

$$r_{POI} = R_0 \sqrt{(4 - \sqrt{13})} \cong R_0 * 0.628 \tag{15}$$

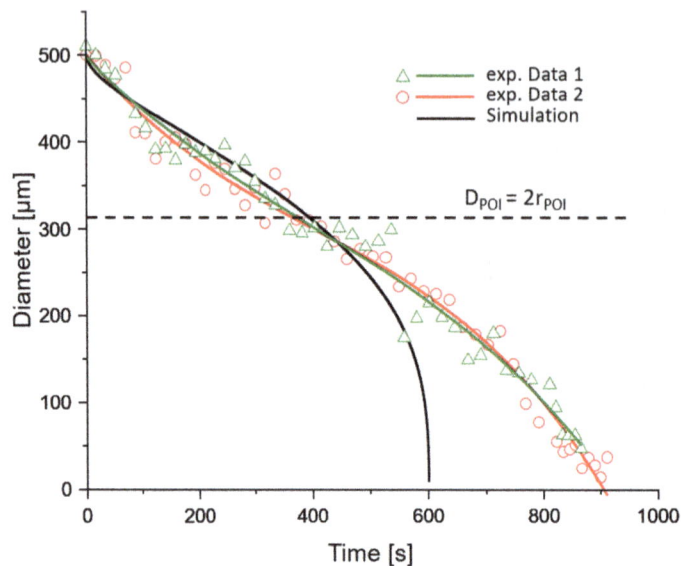

FIG 6 – Comparison of simulation and experimental data. The dashed line represents the diameter of the point of inflection of function $\log(\Delta\gamma)$.

CONCLUSION

Following these observations, it is proposed to expand the presented model to a moving interface approach as to better predict particle dissolution even at later stages of the dissolution. Additionally, the prediction of dissolution time could improve due to the implementation of more accurate density data of the slag and the dissolving particle. Furthermore, experiments for measuring boundary layer thickness around dissolving particles are of great interest as this is one of the most influential parameters apart from the diffusion coefficient. Once the adapted model is validated by further experimental data, the prediction of binary diffusion coefficients based on fitting the simulation data to experimental dissolution curves could be possible.

ACKNOWLEDGEMENTS

The financial support by the Austrian Federal Ministry for Labour and Economy, the National Foundation for Research, Technology and Development and the Christian Doppler Research Association is gratefully acknowledged. The authors declare no conflict of interest.

REFERENCES

Burhanuddin, B, Guarco, J, Harmuth, H and Vollmann, S, 2022. Application of an improved testing device for the study of alumina dissolution in silicate slag, *Journal of the European Ceramic Society*, 8:3652–3659.

Chen, C, Jiang, Z, Li, Y, Zheng, L, Huang, X, Yang, G, Sun, M, Chen, K, Yang, H, Hu, H and Li, H, 2019. State of the Art in the Control of Inclusions in Tire Cord Steels and Saw Wire Steels – A Review, *Steel Research International*, 8:1800547.

Feichtinger, S, Michelic, S K, Kang, Y-B and Bernhard, C, 2014. In-situ observation of the dissolution of SiO_2 particles in $CaO–Al_2O_3–SiO_2$ slags and mathematical analysis of its dissolution pattern, *Journal of the American Ceramic Society*, 1:316–325.

Furuya, Y, Abe, T and Matsuoka, S, 2003. 1010-cycle fatigue properties of 1800 MPa-class JIS-SUP7 spring steel, *Fatigue and Fracture of Engineering Materials and Structures*, 7:641–645.

Garrison, W M and Wojcieszynski, A L, 2007. A discussion of the effect of inclusion volume fraction on the toughness of steel, *Materials Science and Engineering: A*, 1–2:321–329.

Glicksman, M O, 2000. *Diffusion in Solids*, 142 p (John Wiley and Sons: New York).

Guo, X, Sietsma, J, Yang, Y, Sun, Z and Guo, M, 2017. Diffusion-limited dissolution of spherical particles: A critical evaluation and applications of approximate solutions, *AIChE J*, 7:2926–2934.

Huo, Y, Gu, H, Huang, A, Ma, B, Chen, L, Li, G and Li, Y, 2022. Characterization and mechanism of dissolution behavior of Al_2O_3/MgO oxides in molten slags, *J Iron Steel Res Int*, 11:1711–1722.

Jimbo, I, Chung, Y and Cramb, A W, 1996. Interfacial Tensions of Liquid Iron-alloys and Commercial Steels in Contact with Liquid Slags, *ISIJ Int*, Supplement:S42–S45.

Lee, S H, Tse, C, Yi, K W, Misra, P, Chevrier, V, Orrling, C, Sridhar, S and Cramb, A W, 2001. Separation and dissolution of Al_2O_3 inclusions at slag - metal interfaces, *Journal of Non-Crystalline Solids*, pp 41–48.

Levenspiel, O, 1999. *Chemical Reaction Engineering*, pp 582–586 (John Wiley and Sons: New York).

Michelic, S K, Goriupp, J, Feichtinger, S, Kang, Y-B, Bernhard, C and Schenk, J, 2016. Study on Oxide Inclusion Dissolution in Secondary Steelmaking Slags using High Temperature Confocal Scanning Laser Microscopy, *Steel Research International*, 1:57–67.

Ogris, D M and Gamsjäger, E, 2022. Numerical Treatment of Oxide Particle Dissolution in Multicomponent Slags with Local Gibbs Energy Minimization, *Steel Research International*, 2200056.

Sridhar, S and Cramb, A, 2003. Properties of Slags and Their Importance, in Manufacturing Clean Steels, *High Temperature Materials and Processes*, 5–6:275–282.

Valdez, M, Shannon, G S and Sridhar, S, 2006. The ability of slags to absorb solid oxide inclusions, *ISIJ International*, 3:450–457.

Impurity capacities of non-ferrous slags

R G Reddy[1]

1. Professor, The University of Alabama, Tuscaloosa Alabama 35487, USA.
Email: rreddy@eng.ua.edu

ABSTRACT

The complexity in treating environmentally harmful impurities such as As, Sb, Bi from the base metal matte or metal in the smelting stage are responsible for the high cost of the refining process. The impurity capacities (such as arsenic, antimony and bismuth) of slags were calculated a priori using Reddy-Blander (RB) model. The capacity predictions were for a wide range of matte and slag compositions in copper smelting conditions. The calculated impurities capacities and impurity distribution ratios results are in good agreement with the available experimental and industrial slags data. The a priori knowledge of impurities is useful for reduction of energy consumption and enhanced environmental control in the current and future non-ferrous metal processes.

INTRODUCTION

Impurity capacity of a slag is defined as a measure of the ability of slag to hold the impurity. The common minerals that contain As, Sb and Bi in copper are Arsenopyrite (FeAsS), Enargite (Cu_3AsS_4), Lautite (CuAsS), Tennantite [$(Cu,Fe)_{12}As_4S_{13}$], Famatinite (Cu_3SbS_4), Chalcostibite ($CuSbS_2$), Wittichenite (Cu_3BiS_3), Emplectite ($CuBiS_2$) and Aikinite ($PbCuBiS_3$). The lead impurity compounds are Jordanite ($Pb_{14}As_6S_{23}$) and Cosalite (Pb_2Bi_2S) (Larouche, 2001).

In copper, nickel and lead sulfide smelting and refining, removal of sulfur, arsenic, antimony, and bismuth cause metal losses in both entrained and chemically dissolved forms in the slags. As a result, an understanding of impurity capacity of slags is essential for the development of clean metal technology. The Reddy-Blander (RB) model, first proposed in 1987 (Reddy and Blander, 1987), predicted that sulfide capacities can be calculated a prior based on a simple solution model and on the knowledge of chemical and solution properties of sulfides and oxides.

The sulfide capacities of several binary silicate (Reddy and Blander, 1987, 1989; Reddy, Hu and Blander, 1992; Reddy, 2003b), aluminate (Reddy and Zhao, 1995), titanate (Derin, Yucel and Reddy, 2004), multi-component silicate (Yahya and Reddy, 2011; Chen, Reddy and Blander, 1989; Pelton, Eriksson and Romero-Serrano, 1993; Derin, Yucel and Reddy, 2011) and industrial slags (Derin and Reddy, 2003; Derin, Yucel and Reddy, 2005; 2006) were predicted using the RB model. The model was also applied for sulfate (Pelton, 1999, 2000), arsenate (Reddy and Font, 2003), antimonate, Font and Reddy, 2005) capacities in slags and sulfur and oxygen partial pressures in copper slags (Derin and Reddy, 2003). brief description of the RB model is presented on the thermodynamic modelling of impurity capacity section. Impurity capacity expressions for several impurities are presented in Table 1 (Reddy, 2003a).

TABLE 1
Capacities expression for different species.

Species, i	Reaction	Capacity, C_i
Sulfide, [S^{2-}]	$1/2S_2(g) + [O^{2-}] = [S^{2-}] + 1/2O_2(g)$	$C_S = (wt\%S)\left(\dfrac{P_{O_2}}{P_{S_2}}\right)^{\frac{1}{2}}$
Pyrosulfate, [S$_2$O$_7^{2-}$]	$S_2(g) + [O^{2-}] + 3O_2(g) = [S_2O_7^{2-}]$	$C_{S_2O_7^{2-}} = \dfrac{(wt\%S_2O_7^{2-})}{P_{O_2}^3 P_{S_2}}$
Sulfate, [SO$_4^{2-}$]	$1/2S_2(g) + [O^{2-}] + 3/2O_2(g) = [SO_4^{2-}]$	$C_{SO_4^{2-}} = \dfrac{(wt\%SO_4^{2-})}{P_{O_2}^{\frac{3}{2}} P_{S_2}^{\frac{1}{2}}}$
Carbide, [C$_2$]	$3C(gr) + [O^{2-}] = [C_2^{2-}] + CO(g)$	$C_{C_2} = wt\%C_2 \dfrac{P_{CO}}{a_C^3}$
Carbonate, [CO$_3^{2-}$]	$CO_2(g) + [O^{2-}] = [CO_3^{2-}]$	$C_{CO_3^{2-}} = \dfrac{(wt\%CO_3^{2-})}{P_{CO_2}}$
Hydroxyl, [OH$^-$]	$1/2H_2O(g) + 1/2[O^{2-}] = [OH^-]$	$C_{OH^-} = \dfrac{(wt\%OH^-)}{P_{H_2O}^{\frac{1}{2}}}$
Nitride, [N^{3-}]	$1/2N_2(g) + 3/2[O^{2-}] = [N^{3-}] + 3/4O_2(g)$	$C_N = (wt\%N^{3-})\dfrac{P_{O_2}^{\frac{3}{4}}}{P_{N_2}^{\frac{1}{2}}}$
Cyanide, [CN$^-$]	$N_2(g) + 3C + [O^{2-}] = 2[CN^-] + CO(g)$	$C_{CN^-} = (wt\%CN^-)\dfrac{P_{CO}^{\frac{1}{2}}}{a_C^{\frac{3}{2}} P_{N_2}^{\frac{1}{2}}}$
Phosphate, [PO$_4^{3-}$]	$P_2(g) + 3[O^{2-}] + 5/2O_2(g) = 2[PO_4^{3-}]$	$C_{PO_4^{3-}} = \dfrac{(wt\%PO_4^{3-})}{P_{O_2}^{\frac{5}{4}} P_{P_2}^{\frac{1}{2}}}$
Phosphide, [P^{3-}]	$1/2P_2(g) + 3/2[O^{2-}] = [P^{3-}] + 3/4O_2(g)$	$C_{P^{3-}} = (wt\%P^{3-})\dfrac{P_{O_2}^{\frac{3}{4}}}{P_{P_2}^{\frac{1}{2}}}$
Arsenate, [AsO$_4^{3-}$]	$3/2O^{2-} + As(l) + 5/4O_2(g) = AsO_4^{3-}$	$C_{AsO_4^{3-}} = \dfrac{(wt\%AsO_4^{3-})}{a_{As} P_{O_2}^{\frac{5}{4}}}$
Bismuthate, [BiO$_4^{3-}$]	$3/2O^{2-} + Bi(l) + 5/4O_2(g) = BiO_4^{3-}$	$C_{BiO_4^{3-}} = \dfrac{(wt\%BiO_4^{3-})}{a_{Bi} P_{O_2}^{\frac{5}{4}}}$
Antimonate, [SbO$_4^{3-}$]	$3/2O^{2-} + Sb(l) + 5/4O_2(g) = SbO_4^{3-}$	$C_{SbO_4^{3-}} = \dfrac{(wt\%SbO_4^{3-})}{a_{Sb} P_{O_2}^{\frac{5}{4}}}$
Telluride, [Te^{2-}]	$1/2Te_2(g) + [O^{2-}] = [Te^{2-}] + 1/2O_2(g)$	$C_{Te} = (wt\%Te)\left(\dfrac{P_{O_2}}{P_{Te_2}}\right)^{\frac{1}{2}}$
Selenide, [Se^{2-}]	$1/2Se_2(g) + [O^{2-}] = [Se^{2-}] + 1/2O_2(g)$	$C_{Se} = (wt\%Se)\left(\dfrac{P_{O_2}}{P_{Se_2}}\right)^{\frac{1}{2}}$

THERMODYNAMIC MODELLING OF IMPURITY CAPACITY

Reddy-Blander (RB) Model – impurity capacity

The Reddy-Blander model was used to predict the arsenic capacity of slags, metal and mattes. The arsenic capacity model (a measure of the ability of an oxide system or slag to hold arsenic), that can *a priori* predict the arsenic behaviour in copper mattes and slags was derived (Reddy and Font, 2003). For the MO-SiO₂ system, the arsenic equilibrium reaction can be written as:

$$\frac{3}{2}\text{MO}(l) + \text{As}(l) + \frac{5}{4}O_2(g) = M_{3/2}\text{AsO}_4(l) \tag{1}$$

where, M is arsenate compound forming element (such as Fe, Ca, Mg,.). The most stable arsenic compound in copper smelting slags is $M_{3/2}\text{AsO}_4$. At high oxygen partial pressures, the arsenic dissolve into the slag as As_2O_5 (Kojo, Taskinen and Lilius, 1984). The equilibrium constant, K_M, for the above reaction is:

$$K_M = \frac{a_{M_{3/2}\text{AsO}_4}}{a_{MO}^{3/2}\, a_{As}\, p_{O_2}^{5/4}} \tag{2}$$

The arsenic capacity, C_{As}, in terms of measurable quantities was defined by Reddy (2003a) as:

$$C_{\text{AsO}_4^{3-}} = \frac{(\text{wt pct AsO}_4^{3-})}{a_{As}\, p_{O_2}^{5/4}} \tag{3}$$

Combining Equations 2 and 3, then Equation 4 can be obtained.

$$C_{\text{AsO}_4^{3-}} = (\text{wt pct AsO}_4^-)\, \frac{K_M\, a_{MO}^{3/2}}{a_{M_{3/2}\text{AsO}_4}} \tag{4}$$

Development of this expression was made considering two compositions ranges.

A. Arsenic capacity for basic melts (0 $\leq X_{SiO_2} \leq$ 0.33)

The arsenic capacity for the basic melt in the MO-SiO₂ binary system becomes:

$$C_{\text{AsO}_4^{3-}} = \frac{100\, K_M\, a_{MO}^{3/2}\, W_{\text{AsO}_4}\, (1-2\, X_{SiO_2})}{\gamma_{M_{3/2}\text{AsO}_4}\, [W_{MO} + X_{SiO_2}(W_{SiO_2}-W_{MO})]} \tag{5}$$

Using RB model, similar expressions for the sulfide capacity of ferrous and non-ferrous slags in the silicate and aluminates systems were derived (Reddy and Blander, 1987, 1989; Reddy, Hu and Blander, 1992; Reddy, 2003b; Reddy and Zhao, 1995; Derin, Yucel and Reddy, 2004; Yahya and Reddy, 2011; Chen, Reddy and Blander, 1989; Pelton, Eriksson and Romero-Serrano, 1993; Derin, Yucel and Reddy, 2011; Derin and Reddy, 2003; Derin, Yucel and Reddy, 2005, 2006). By using the thermodynamic data for the equilibrium constant K_M, $\gamma_{M_{3/2}\text{AsO}_4}$, a_{MO} in the MO-SiO₂ binary system, and W is molecular weight of compounds arsenic capacities were calculated for several binary arsenic systems and are discussed in the section B.

B. Arsenic capacity for acidic melts (0.33 $< X_{SiO_2} <$ 1)

In this composition range, the arsenic is dissolved in the MO-SiO₂ binary acidic melt that contains polymeric species. It is also assumed that the AsO_4^{3-} ion and the SiO_4 units in the polymer are similar in size and forms a chain with no free O^{2-} ions. For dilute solutions, the volume fraction of As ions sites in solution can be expressed as:

$$\varphi_{As} = \frac{n_{As}}{n_{Si}} \tag{6}$$

The $C_{\text{AsO}_4^{3-}}$ for acidic melts is expressed as:

$$C_{\text{AsO}_4^{3-}} = \frac{100\, K_M\, a_{MO}^{3/2}\, X_{SiO_2}\, W_{\text{AsO}_4}}{[W_{MO} + X_{SiO_2}(W_{SiO_2}-W_{MO})]}\, e^{\left[\frac{1}{m}-1-\mu\right]} \tag{7}$$

where m is the average polymer chain length and μ is the interaction energy between the ions. The arsenic capacity of binary MO-SiO$_2$ system can be predicted using Equations 5 and 7 for the entire composition range ($0 < X_{SiO_2} < 1$) and at a fixed temperature. As the composition crosses between the basic and acidic melts (at X_{SiO_2} of 0.33), the transition in the arsenic capacity is predicted by Equations 5 and 7 to a smooth and continuous.

For the FeO-SiO$_2$ binary system, the arsenate formation reaction can be written as:

$$\frac{3}{2}FeO(l) + As(l) + \frac{5}{4}O_2(g) = Fe_{3/2}AsO_4(l), \Delta G° = -279.7 \text{ K J/mol at 1573 K} \tag{8}$$

It is important to note that in calculating the $\Delta G°$ for Equation 8, the values of $\Delta H°$ and $\Delta S°$ for As (l) and Fe$_{3/2}$AsO$_4$(s) (Roine, 2022) were extrapolated from 1200 K and 811 K, respectively. The $\Delta G°$ for the liquid Fe$_{3/2}$AsO$_4$ (l) was estimated from the experimental arsenic solubility data at 1573 K. The aF$_e$O in FeO-SiO$_2$ binary system was calculated at 1573 K using FactSage software (Bale, Pelton and Thompson, 2002). The arsenic capacity, as predicted using Equation 5, depends on temperature and activity coefficient of Fe$_{3/2}$AsO$_4$ (l). At a constant $\gamma_{Fe_{3/2}AsO_4}$, the arsenic capacity of FeO-SiO$_2$ melts decreases with an increase in temperature. Also, at a constant temperature, the arsenic capacity deceases with increase in $\gamma_{Fe_{3/2}AsO_4}$. The calculated arsenic capacities of melts using Equations 5 and 7 at 1573 K and $\gamma_{Fe_{3/2}AsO_4}$ equal to 1. The arsenic capacity increases with increase in FeO content and shows a strong dependence on the activity of FeO in the melt. The K$_M$ is the equilibrium constant for the R (arsenate, antimonate and bismuthate) forming reaction. The values of K$_M$ arsenates, antimonates and bismuthate at 1573 K are given in Table 2 (Font and Reddy, 2003).

$$\frac{3}{2}MO\ (l) + R\ (l) + \frac{5}{4}O_2\ (g) = M_{3/2}RO_4\ (s, l) \tag{9}$$

TABLE 2

Equilibrium constants for impurity forming reactions at 1573 K.

	Log K_M		
M*	Arsenate	Antimonate	Bismuthate
Fe**	11.0	10.4	9.6
Ca	10.3	NA	NA
Mg	10.2	9.6	NA
Cu	-4.7	-6.0	NA
Ni	3.4	3.1	NA

NA: Not Available. * Reference state: Solid. ** Reference state: Liquid.

The arsenic capacities of CaO-SiO$_2$ and MgO-SiO$_2$ binary melts were calculated using Equations 5 and 7. The arsenic reactions for CaO-SiO$_2$ and MgO-SiO$_2$ melts are as follows:

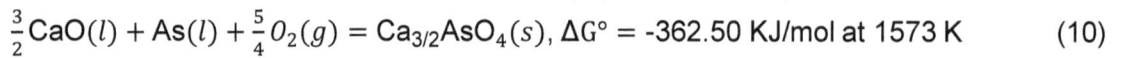

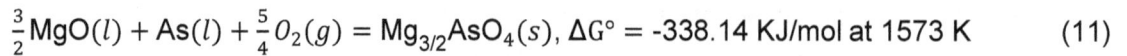

$$\frac{3}{2}CaO(l) + As(l) + \frac{5}{4}O_2(g) = Ca_{3/2}AsO_4(s), \Delta G° = -362.50 \text{ KJ/mol at 1573 K} \tag{10}$$

$$\frac{3}{2}MgO(l) + As(l) + \frac{5}{4}O_2(g) = Mg_{3/2}AsO_4(s), \Delta G° = -338.14 \text{ KJ/mol at 1573 K} \tag{11}$$

The $\Delta G°$ for the reaction of Ca$_{3/2}$AsO$_4(s)$ in Equation 10 is taken from the reported data (Bale, Pelton and Thompson, 2002), and the $\Delta G°$ for Equation 11, the values of $\Delta H°$ and $\Delta S°$ for As(l) and Mg$_{3/2}$AsO$_4$(s) (Roine, 2022) were extrapolated from 1200 K and 1225 K, respectively. The activities of CaO and MgO in melts at 1573 K were calculated (Bale, Pelton and Thompson, 2002). The arsenic capacity in the hypothetical melts of CaO-SiO$_2$ and MgO-SiO$_2$ binary systems were calculated. In the CaO-SiO$_2$ system, a sharp increase in arsenic capacity above the X_{CaO} value of 0.6 was observed (Reddy and Font, 2003). At higher compositions X_{CaO} equal to 0.8 and greater, no significant changes in arsenic capacities were observed. This is mainly due to the variation in the activity of CaO, a$_{CaO}$ in the CaO-SiO$_2$ binary melts, which shows a strong negative deviation. Similar observations were made for MgO-SiO$_2$ melts. But the decrease in arsenic capacity in MgO-SiO$_2$ melts with an increase in the concentration of SiO$_2$ is much smaller than in the CaO-SiO$_2$ melts.

C. Arsenic capacity in the multi-component systems

For multi-component system which contains only one acidic component such as SiO_2 (eg FeO-CaO-SiO_2 ternary system or FeO-CaO-MgO-SiO_2 quaternary system), the arsenate formation reactions for CaO- and FeO- can be expressed as:

$$\frac{3}{2}(\text{FeO, CaO})(l) + \text{As}(l) + \frac{5}{4}O_2(g) = (\text{Fe, Ca})_{3/2}\text{AsO}_4(l) \tag{12}$$

Using the Flood-Grjotheim approximation (Chen, Reddy and Blander, 1989; Flood and Grjotheim, 1952), the partial Gibbs energies of mixing for different oxidative species are comparably similar. Thus, for a ternary system, the standard Gibbs energy change of mixing is expressed as:

$$\Delta G_{(Fe,Ca)O} = N_{FeO}\Delta G_{FeO} + N_{CaO}\Delta G_{CaO} \tag{13}$$

where $\Delta G_{(Fe, Ca)O}$, ΔG_{FeO} and ΔG_{CaO} are the Gibbs energy changes for Equations 12, 8 and 10, respectively. The N_{FeO} and N_{CaO} are the electrical equivalent cationic fractions ($N_{FeO} = \frac{X_{FeO}}{X_{FeO}+X_{CaO}}$ and $N_{CaO} = \frac{X_{CaO}}{X_{FeO}+X_{CaO}}$), and by considering the definition of Gibbs energy ($\Delta G = -RT \ln K$), Equation 13 is further simplified as:

$$\log K_8 = N_{FeO} \log K_5 + N_{CaO} \log K_6 \tag{14}$$

where K_8, K_5, and K_6 are the equilibrium constant for Equations 12, 8 and 10 respectively. For a constant X_{SiO_2} in the FeO-SiO_2 and CaO-SiO_2 binary systems, the arsenic capacity is expressed as $C'_{AsO_4^{3-}(FeO)} = K_{FeO}\, a_{FeO}^{3/2}$ and $C'_{AsO_4^{3-}(CaO)} = K_{CaO}\, a_{CaO}^{3/2}$. Thus, after substituting in Equation 14 and rearranging, Equation 15 is obtained.

$$\text{Log } C'_{AsO_4^{3-}, (Fe, Ca)O} - \frac{3}{2}\log a_{(Fe, Ca)O} = N_{FeO}\left(\log C'_{AsO_4^{3-}, FeO} - \frac{3}{2}\log a_{FeO}\right) + N_{CaO}\left(\log C'_{AsO_4^{3-}, CaO} - \frac{3}{2}\log a_{CaO}\right) \tag{15}$$

Furthermore, taking into consideration only (Fe, Ca)O as a solution, then the integral Gibbs energy of solution for Equation 13 becomes $\log a_{(Fe,Ca)O} = N_{FeO} \log a_{FeO} + N_{CaO} \log a_{CaO}$. After substituting in Equation 15 and rearranging it, the arsenic capacity for multi-component system is expressed as:

$$\log C_{AsO_4^{3-}, (Fe, Ca)O} = N_{FeO} \log C_{AsO_4^{3-}, FeO} + N_{CaO} \log C_{AsO_4^{3-}, CaO} \tag{16}$$

The arsenic capacity of multi-component system using Equation 16 can be calculated at a constant composition of acidic component (ie $X_{SiO_2} + X_{FeO_{1.5}}$) and is further discussed in the following section.

Evaluation of the arsenic capacity model

The phase equilibrium studies for arsenic between the FeO-MgO-SiO_2, FeO-CaO-MgO-SiO_2 slags and copper mattes at 1573 K were reported (Roghani, Takeda and Itagaki, 2000). The arsenic experimental data were reported in the form of distribution coefficients and solubility of arsenic in the slag. In the present study, an expression was derived between the distribution coefficient and arsenic capacity of slags in equilibrium with copper mattes.

Expression between distribution coefficient (LAs) and arsenic capacity (CAsO43-)

The arsenic distribution coefficient between slag and matte phases is defined as:

$$L_{As} = \frac{(\text{wt pct As in slag-oxidic phase})}{\{\text{wt pct As in matte} - \text{metal phase}\}} \tag{17}$$

The weight pct of As in matte is expressed as:

$$\{\text{Weight pct of As in matte}\} = \frac{a_{As(l)}\, W_{As}\{n_T\}}{\gamma_{As}} \tag{18}$$

where $a_{As(l)}$ is activity of arsenic in matte, γ_{As} is activity coefficient of As in matte and W_{As} is the molecular weight of As, and $\{n_T\}$ is the total number of moles of matte phase. Combining

Equations 17, 18 and 3, after making the conversion of weight pct of As in slag, and rearranging, the relationship between the arsenic capacity and the L_{As} is obtained as:

$$C_{AsO_4^{3-}} = \frac{L_{As}\, W_{AsO_4}\, \{n_T\}}{\gamma_{As}\, p_{O_2}^{5/4}}$$ (19)

The experimental arsenic capacities were derived using Equation 19 for each of the experimental L_{As} at 1573 K. The reported data of γ_{As} and p_{O_2} for Cu matte, were also used in the calculations.

Arsenic capacity and distribution ratios between Cu mattes and slags

The RB model calculated *a priori* and experimental data for arsenic capacity and distribution ratios for FeO-FeO$_{1.5}$-CuO$_{0.5}$-MgO-SiO$_2$ slags and Cu mattes at 1573 K are shown in Figures 1 and 2 respectively. The equilibrium constants for impurities reactions are presented in Table 2. The *a priori* predictions were calculated considering the reported data of γ_{As} for Cu matte, the $a_{M}O$ for the MO-FeO$_{1.5}$-SiO$_2$ system and the experimental pO_2 and $\{n_T\}$. A good agreement between the experimental data and RB model *a priori* calculated arsenic capacity and distribution ratio.

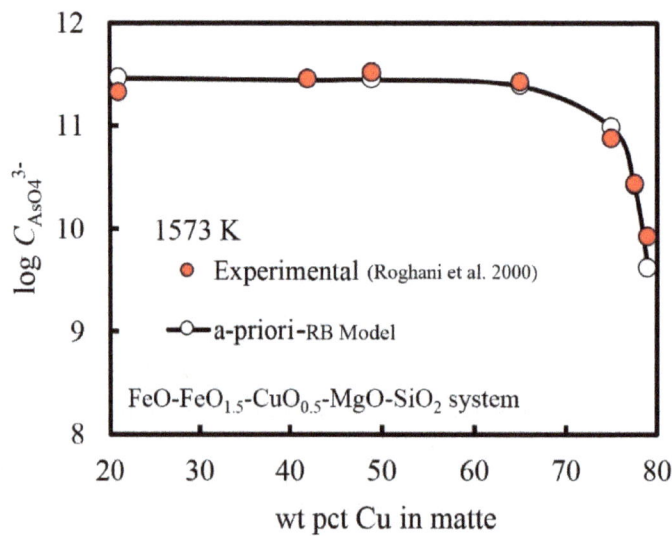

FIG 1 – Arsenic capacity of FeO-FeO$_{1.5}$-CuO$_{0.5}$-MgO-SiO$_2$ slag versus wt pct Cu in matte at 1573 K.

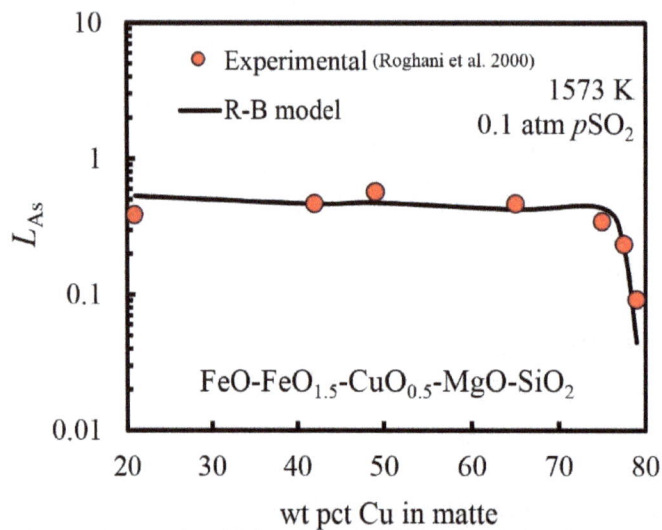

FIG 2 – Distribution coefficient of arsenic of FeO-FeO$_{1.5}$-CuO$_{0.5}$-MgO-SiO$_2$ slag versus wt pct Cu in matte at 1573 K.

The slight deviations between the experimental and the calculated arsenic capacity may be due to the corresponding uncertainties of the experimental L_{As} and the γ_{As} and p_{O_2} values (Roghani, Takeda and Itagaki, 2000; Rosenqvist, 1978; Nikolov, Jalkanen and Kyto, 1992; Roghani, Hino and Itagaki, 1997) used in Equation 19 for deriving the experimental arsenic capacity. For the RB model arsenic capacity calculations using Equation 16, the MgO content in the multi-component system was estimated by using the data of the FeO-$FeO_{1.5}$-$CuO_{0.5}$-MgO-SiO_2 system (Font, Hino and Itagaki, 1998a, 1998b, 1999, 2000). Due to a lack of availability of thermodynamic data on liquid arsenates, the solid $Ca_{3/2}AsO_4$ and $Mg_{3/2}AsO_4$ data were used in calculating the Gibbs energy of the Equations 10 and 11. Use of liquids data for these compounds will lower the arsenic capacity of these systems. The availability of reliable thermodynamic data for impurities in slags and mattes is essential. Further studies are in progress for extending this model in *a priori* prediction of other impurities capacities such as Bi, Sb in copper matte and other non-ferrous metal smelting slags.

Antimonate capacity and distribution ratio between Cu mattes and slags

The antimonate capacities were calculated using RB model for each of the slag composition and corresponding distribution ratios of Cu matte in equilibrium with the FeO-$FeO_{1.5}$-$CuO_{0.5}$-MgO-SiO_2 slag at 1573 K and pSO_2 of 0.1 atm, using expressions similar to Equations 5, 7 and 16. The antimony dissolved into the slag as Sb_2O_5 (Kojo, Taskinen and Lilius, 1984). The equilibrium constants for impurities reactions are presented in Table 2. The calculated using RB model for antimonate capacities and distribution ratios are shown in Figures 3 and 4 respectively. The data of γ_{Sb} for the Cu matte, the a_MO in the MO-SiO_2 binary system and the experimental pO_2 and $\{n_T\}$ were used in the calculations. The *a priori* predictions and experimental data for antimony capacities and distribution ratios in slags and Cu mattes are in good agreement. The observed good agreement may be due to the including copper oxide data in slag system. This is particularly important at higher matte grades because higher solubility of copper into slags is reported (Roghani, Takeda and Itagaki, 2000).

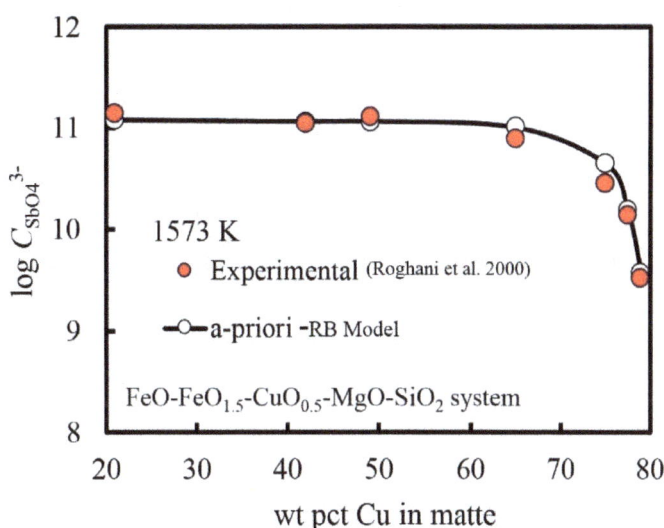

FIG 3 – Antimony capacity of FeO-$FeO_{1.5}$-$CuO_{0.5}$-MgO-SiO_2 slag versus wt pct Cu in matte at 1573 K.

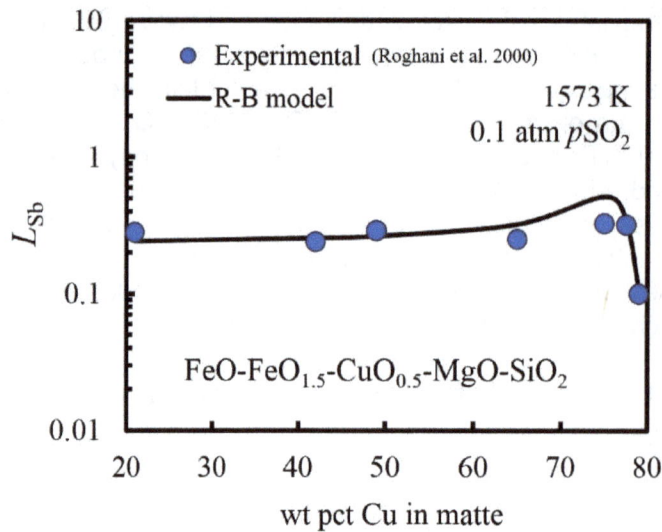

FIG 4 – Distribution coefficient of antimony of FeO-FeO$_{1.5}$-CuO$_{0.5}$-MgO-SiO$_2$ slag versus wt pct Cu in matte at 1573 K.

Bismuth capacity and distribution ratios between Cu mattes and slags

The bismuthate capacities and distribution ratios for Cu matte in equilibrium with the FeO-FeO$_{1.5}$-SiO$_2$ slag at 1573 K and pSO_2 of 0.1 atm were evaluated using the RB model, using an approach similar to arsenic expressions developed using expressions similar to Equations 5, 7 and 16. The calculated data for bismuthate capacities and distribution ratios are shown in Figures 5 and 6 respectively. The equilibrium constants for impurities reactions are presented in Table 2. The Gibbs Energy for the M$_{3/2}$BiO$_4$ (M = Mg, Cu) compounds are not available in the literature. Hence, the bismuthate capacity was calculated for a hypothetical FeO-FeO$_{1.5}$-SiO$_2$ slag. The equilibrium constant (KF_e) for the Fe$_{3/2}$BiO$_4$ formation reaction was used from the Table 2. Because an absence of data the quantities of MgO and CuO$_{0.5}$ in Equations 5, 7 and 16 were excluded from the slag composition at 1573 K and pSO_2 of 0.1 atm. The available experimental pO_2 and $\{n_T\}$, data of γ_{Bi} in the Cu matte and the aF_eO in the FeO-FeO$_{1.5}$-SiO$_2$ system were used in these calculations.

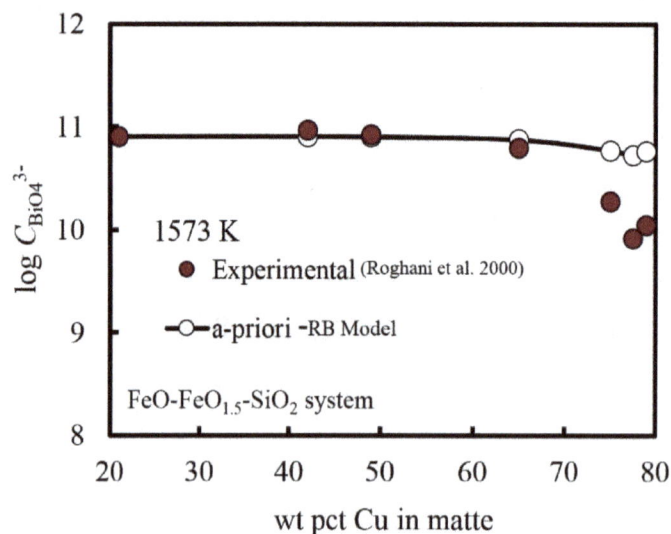

FIG 5 – Bismuth capacity of FeO-FeO$_{1.5}$-CuO$_{0.5}$-MgO-SiO$_2$ slag versus wt pct Cu in matte at 1573 K.

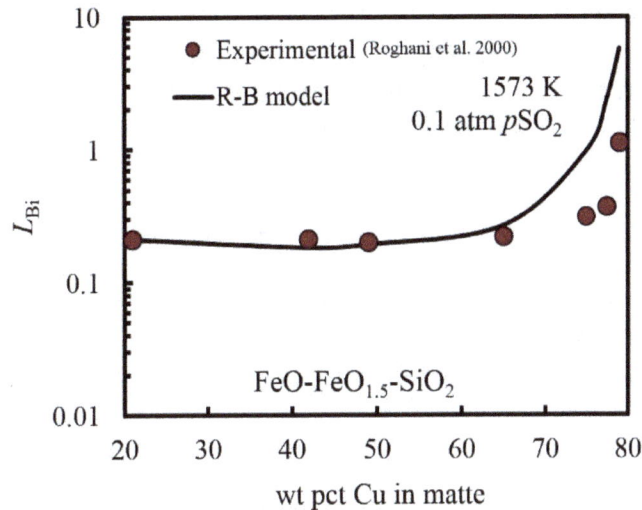

FIG 6 – Distribution coefficient of Bismuth of $FeO\text{-}FeO_{1.5}\text{-}CuO_{0.5}\text{-}MgO\text{-}SiO_2$ slag versus wt pct Cu in matte at 1573 K.

For the $FeO\text{-}FeO_{1.5}\text{-}SiO_2$ slags at 1573 K, the calculated RB model $\log C_{BiO_4^{3-}}$ and L_{Bi} shows that the model data agrees well with the experimental data below the Cu matte grade of 70 wt pct of Cu. But for the RB model data of $\log C_{BiO_4^{3-}}$ and L_{Bi} above the Cu matte grade of 70 wt pct of Cu are higher than the experimental data. As mentioned above, the data for $M_{3/2}BiO_4$ for M = Cu or Mg was not included in the calculations. The addition of basic oxides such as CaO to iron oxide slags increases the Fe_2O_3 content up to about 20 wt per cent (Rosenqvist, 1978), which decreases the capacity of the slag to retain the impurity. The Fe_2O_3, which is known as an acidic component due to its tendency to consume rather than supply oxygen, lowers the $a_{Fe}O$ value in slag, resulting in a decrease in the impurity capacity value. Thus, addition of MgO in the calculation of Bismuthate capacities using Equations 5, 7 and 16, expected to decrease the bismuthate capacity and also decrease the distribution ratios, by which their calculated data will be closer to the experimental data. Hence, the availability of reliable thermodynamic data for slag components and impurity compounds in slags and in mattes or liquid metals are essential for the accurate prediction of impurity capacities and their distribution ratios. Further studies are in progress in extending the RB model for the prediction of capacities and distribution ratios of other impurities, such as Se and Te in copper and other non-ferrous metal slags.

CONCLUSIONS

The impurity (As, Sb and Bi) capacities of iron silicate slags in equilibrium with copper mattes at 1573 K were calculated *a priori* using the RB model. The predicted impurity capacities are in very good agreement with the experimental data. An expression for the relationship between the impurity capacity and the impurity distribution ratio for copper slags and the copper mattes was derived. The derived impurity distribution ratios between the slags and the copper mattes found to be in particularly good agreement with the experimental data for multi-component slags. Such predictions are useful in understanding the behaviour of impurities in the current and eventually future non-ferrous metal process. The impurity capacities of slags are directly proportional: (i) to the equilibrium constant K_M, and (ii) to the values of a_MO, which are related to the solution properties. The availability of reliable thermodynamic data for slag components and impurity compounds in slags and in mattes or liquid metals are essential for the accurate prediction of impurity capacities and their distribution ratios.

The RB model is an invaluable tool for the optimisation of impurity removal in the existing processes and for the development of new processes. The *a priori* prediction of other impurity capacities such as Se and Te in non-ferrous metal smelting slags and mattes using RB model is possible and such predictions are very useful in understanding the behaviour of impurities in the current and eventually future non-ferrous metals technologies.

ACKNOWLEDGEMENTS

The author acknowledges the financial support from the National Science Foundation (NSF) and ACIPCO for this research project. Author also thanks the Department of Metallurgical and Materials Engineering, The University of Alabama for providing the experimental and analytical facilities.

REFERENCES

Bale, C W, Pelton, A D and Thompson, W T, 2002. FactSage 5.1: Thermochemical Software for Windows. TM Montreal, Quebec: Thermfact Ltd.

Chen, B, Reddy, R G and Blander, M, 1989. Sulfide capacities of CaO-FeO-SiO$_2$ Slags, *3rd International Conference on Molten Slags and Fluxes*, pp 270–272.

Derin, B, Yucel, O and Reddy, R G, 2006. Predicting of Sulfide Capacities of Industrial Lead Smelting Slags, *Advanced Processing of Metals and Materials*, 1:237–244.

Derin, B and Reddy, R G, 2003. Sulfur and Oxygen Partial Pressure Ratios Prediction in Copper Flash Smelting Plants Using Reddy-Blander Model, in *Yazawa International Symposium: Metallurgical and Materials Processing: Principles and Technologies; Materials Processing Fundamentals and New Technologies* (eds: F Kongoli, K Itagaki, C Yamauchi, H Y Sohn, F Kongoli, K Itagaki, C Yamauchi and H Y Sohn), 1:625–632.

Derin, B, Yucel, O and Reddy, R G, 2004. Modelling of Sulfide Capacities of Binary Titanate Slags, *EPD Congress 2004*. pp 155–160.

Derin, B, Yucel, O and Reddy, R G, 2005. Sulfide Capacities of PbO-SiO$_2$ and PbO-SiO$_2$-AlO$_{1.5}$ (sat.) Slags, *Mining and Materials Processing Institute of Japan (MMIJ)*, Kyoto, Japan, pp 1279–1287.

Derin, B, Yucel, O and Reddy, R G, 2011. Sulfide Capacity Modelling of FeOx-MO-SiO$_2$ (MO=CaO, MnO, MgO) Melts, *Minerals and Metallurgical Processing*, 28(1):33–36.

Flood, H and Grjotheim, K, 1952. Thermodynamic calculation of slag equilibria, *Journal of the Iron and Steel Institute*, 171:64–70.

Font, J M and Reddy, R G, 2003. Modeling of Impurity Distribution between Mattes and Slags, *Copper 2003*, IV:301–313.

Font, J M and Reddy, R G, 2005. Modelling of Antimonate Capacity in copper and Nickel Smelting Slags, *Trans Inst Min Metall C*, 114:C160–C164.

Font, J M, Hino, M and Itagaki, K, 1998a. Thermodynamic Evaluation of Distribution Behaviour of VA Elements in Nickel Matte Smelting, *Met Rev of MMIJ*, 15(2):202–220.

Font, J M, Hino, M and Itagaki, K, 1998b. Minor Elements Distribution between Iron-Silicate Base Slag and Ni$_3$S$_2$-FeS Matte under High Partial Pressures of SO$_2$, *Mat Trans JIM*, 39(1998):834–840.

Font, J M, Hino, M and Itagaki, K, 1999. Phase Equilibrium Minor and Elements Distribution between Iron-Silicate Base Slag and Nickel-Copper-Iron Matte at 1573 K under High Partial Pressures of SO$_2$, *Mat Trans JIM*, 40:20–26.

Font, J M, Hino, M and Itagaki, K, 2000. Phase Equilibrium and Minor-Element Distribution between Ni$_3$S$_2$-FeS Matte and Calcium Ferrite slag under High Partial Pressures of SO$_2$, *Metall Trans B*, 31B:1231–1239.

Kojo, I V, Taskinen, P A and Lilius, K R, 1984. The Thermodynamics of copper fire refining by Sodium Carbonate, *Second international Symposium on Metallurgical Slags and Fluxes*, pp 723–737.

Larouche, P, 2001. Minor Elements in Copper Smelting and Electrorefining, MS thesis, Mining and Metallurgical Engineering, McGill University, 165 p.

Nikolov, S, Jalkanen, H and Kyto, M, 1992. Distribution of some impurity elements between high grade copper matte and calcium ferrite slag, *4th Inter Conf on Molten Slags and Fluxes*, ISIJ, pp 560–565.

Pelton, A D, 1999. Thermodynamic Calculation of gas solubilities in oxide melts and glasses, *Glastechnische berichte*, 72:40–62.

Pelton, A D, 2000. Thermodynamic Modelling of Complex Solutions, *The Brimacombe Memorial Symposium*, pp 763–780 (MetSoc: Canada).

Pelton, A D, Eriksson, G and Romero-Serrano, A, 1993. Calculation of sulfide capacities of multi-component slags, *Metall Trans B*, 24B:817–825.

Reddy, R G and Blander, M, 1987. Modeling of sulfide capacities of silicate melts, *Metallurgical Transactions*, 18B:591–596.

Reddy, R G and Blander, M, 1989. Sulfide capacities of MnO-SiO$_2$ slags, *Metallurgical Transactions*, 20B:137–140.

Reddy, R G and Font, J M, 2003. Arsenic Capacities of Copper Smelting Slags, *Metallurgical and Materials Transactions B*, 34B:565–571.

Reddy, R G and Zhao, W, 1995. Sulfide capacities of Na2O-SiO$_2$ melts, *Metallurgical and Materials Transactions*, 26B:925–928.

Reddy, R G, 2003b. Emerging Technologies in Extraction and Processing of Metals, *Metallurgical and Materials Transactions B*, 34B:137–152.

Reddy, R G, 2003a. Impurity Capacities in Metallurgical Slags, in *Yazawa International Symposium: Metallurgical and Materials Processing: Principles and Technologies; Materials Processing Fundamentals and New Technologies* (eds: F Kongoli, K Itagaki, C Yamauchi, H Y Sohn, F Kongoli, K Itagaki, C Yamauchi and H Y Sohn), 1:25–48.

Reddy, R G, Hu, H and Blander, M, 1992. Sulfide capacities of silicate slags, in *Proceedings of the Fourth International Conference on Molten Slags and Fluxes*, pp 144–148.

Roghani, G, Hino, M and Itagaki, K, 1997. Phase Equilibrium and Minor Elements Distribution between SiO_2-CaO-FeO_x-MgO Slag and Copper Matte at 1573 K under High Partial Pressures of SO_2, *Mater Trans Jpn Inst Met JIM*, 38:707–713.

Roghani, G, Takeda, Y and Itagaki, K, 2000. Phase Equilibrium and Minor Element Distribution between FeO_x-SiO_2-MgO-Based Slag and Cu_2S-FeS Matte at 1573 K under High Partial Pressures of SO_2, *Metall Trans B*, 31B:705–712.

Roine, A, 2022. HSC Chemistry software, ver. 7.1, Outokumpu Research Oy, Pori, Finland

Rosenqvist, T, 1978. Phase Equilibria in the Pyrometallurgy of Sulfide Ores, *Metallurgical Transaction B*, 9B:337–351.

Yahya, A and Reddy, R G, 2011. Sulfide Capacities of CaO-MgO-$AlO_{1.5}$, MgO-MnO-$AlO_{1.5}$ and CaO-MgO-MnO-$AlO_{1.5}$ Slags, *Trans Inst Min Metall C*, 120(1):45–48.

Flow investigation of multiphase manganese slags

V Rimal[1] and M Tangstad[2]

1. PhD candidate, Department of Materials Science and Engineering, Norwegian University of Science and Technology (NTNU), Trondheim, Norway. Email: vishal.rimal@ntnu.no
2. Professor, Department of Materials Science and Engineering, Norwegian University of Science and Technology (NTNU), Trondheim, Norway. Email: merete.tangstad@ntnu.no

ABSTRACT

Multiphase slags form the basis of the reduction pathway for manganese ferroalloy production. The present work aims to understand the flow of slag through the coke bed based on experimentation in controlled conditions. Synthetic slags of different basicities were evaluated for their phase composition, viscosity, and flow. The two-phase slag will have two zones: a liquid slag zone and a multiphase zone. The multiphase zone will consist mostly of a solid phase with some liquid slag between the solid particles. The multiphase part will hence have a high viscosity. The liquid zone will flow into the coke bed while the multiphase area will stay. With increasing reduction, the multiphase area will decrease in size until the solid phase is gone and the whole slag will flow into the coke bed. The effective viscosity of slag will decrease with the lowering of the solid oxide phase. The results from flow experiments confirmed that the segregation of phases occurs when slag flows through the coke bed. It is observed that the flow is dependent both on the size of the void and the viscosity of slag.

INTRODUCTION

High-carbon ferromanganese are alloys containing Fe and C with a major proportion of Mn. They are produced by the carbothermic reduction of oxidic raw materials using coke in a submerged arc furnace (SAF). The SAF has two main reaction zones: (i) pre-reduction zone and (ii) reduction zone. In the pre-reduction zone, CO reduces iron oxides to metallic iron and higher Mn oxides to solid MnO. Carbonates will also decompose to basic oxides. Gas reduction of MnO to Mn is not possible due to the high stability of MnO. In the reduction zone, the pre-reduced components meet the coke bed. A slag mixture of liquid and solid phases is formed at this stage. The liquid phase consists of CaO, MnO, MgO, Al_2O_3, and SiO_2. The solid phase occurs due to the saturation of MnO in the liquid phase. The MnO present in the liquid phase reacts with the C in the coke bed through the following reaction to yield metallic Mn:

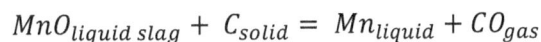

$$MnO_{liquid\ slag} + C_{solid} = Mn_{liquid} + CO_{gas}$$

As the reaction proceeds, the solid phase decreases and the reaction will rapidly slow down when a single-phase slag is achieved (Kumar, Ranganathan and Sinha, 2007; Tangstad *et al*, 2021; Hockaday, Dinter and Reynolds, 2023; Larssen and Tangstad, 2023).

In a SAF the coke bed does not act as a source of fuel rather it provides the carbon reductant, but more importantly the electrical resistance for energy development (Davies *et al*, 2023). The flow behaviour of the slag into the coke bed significantly affects the efficiency and the environmental emissions in the SAF (Oh and Lee, 2016). A consistent flow with characteristic fluid behaviour is desired for the smooth operation of SAF (Natsui *et al*, 2020; Dong *et al*, 2021). The coke bed can be described as a packed bed with the coke particles acting as the packing material. The packing is random and generally results in the formation of interconnected voids of variable sizes through which the liquids trickle down and gases flow upwards. The coke bed should be permeable to allow the flow of the different phases. The structure of the packed bed and packing density affects the size of voids, their interconnectivity as well as the permeability and liquid holdup (Geleta, Siddiqui and Lee, 2020; Natsui *et al*, 2021).

The flow of fluids through the coke bed is often interpreted in terms of void, micro-, and macro-scales. The majority of the studies on relations affecting the flow of slag have been based on cold models and non-moving packed beds, so inferences in high-temperature real-world conditions can vary significantly. Manuscripts have generally considered the coke bed as stationary since the descent of coke bed is very slow (Fukutake and Rajakumar, 1982). A number of studies at high temperatures have been reported, however, the inconsistent physical and chemical properties of coke make it difficult to generalise the characteristics like reactivity, wetting, and strength (Safarian and Tangstad, 2009; George *et al*, 2014). The liquid slag/metal often flows as droplets or channels in the coke bed (Jeong, Kim and Sasaki, 2014). The size of voids has been found to influence the rate of flow of liquids as it progresses through the coke bed (George *et al*, 2014). Prior studies have shown that the overall resistance to the liquid flow increases with the decrease in particle size in the packed bed (Bando *et al*, 2005; Kawabata, Liu *et al*, 2006a; Kawabata, Shinmyou *et al*, 2006b).

For manganese ferroalloys, a significant portion of reduction generally takes place at the top section of the coke bed. The charge (oxides and coke) will settle on top of the permanent coke bed. The proportion of coke fed into the furnace will determine if the permanent coke bed is consumed in the process or not. The charge provides the necessary carbon source. According to theory, negligible reduction takes place when the slag has entered the coke bed as a single phase since the activity of MnO is very low when a single-phase slag is achieved (Olsen, Tangstad and Lindstad, 2007). The interaction of slag with the coke bed is thus dependent on the reduction of Mn in the upper sections of the coke bed.

The flow of slag through the coke bed will depend on the size of voids and the viscosity of Mn slag. The viscosity of Mn slag is high when the slag exists as a solid-liquid mixture and decreases with a decrease in the proportion of solid phase. When the melt initially encounters the coke bed, the viscosity of slag is high. As the reduction proceeds the viscosity decreases enabling the flow of slag through the coke bed (Tang and Tangstad, 2007; Muller, Zietsman and Pistorius, 2015).

This manuscript aims to understand the flow of slag in controlled conditions under the influence of gravity. Limited literature exists elucidating the Mn slag behaviour in the coke bed. The specific interest here is thus to comprehend the segregation of phases as the reduction proceeds and to find the parameters that affect the flow of slag.

EXPERIMENTAL DETAILS

Materials

MnO, SiO$_2$, Fe$_2$O$_3$, CaO, MgO, Al$_2$O$_3$, and CaS of analytical grade were procured from Sigma Aldrich and used as received. Double distilled water was used for the preparation of briquettes. Two types of graphite substrates were used for the experiments: (i) flat, and (ii) funnel (Figure 1). The flat substrate was used to evaluate the phase composition and thus calculate the viscosity. The funnel substrate was used to comprehend the flow of slag under the influence of gravity.

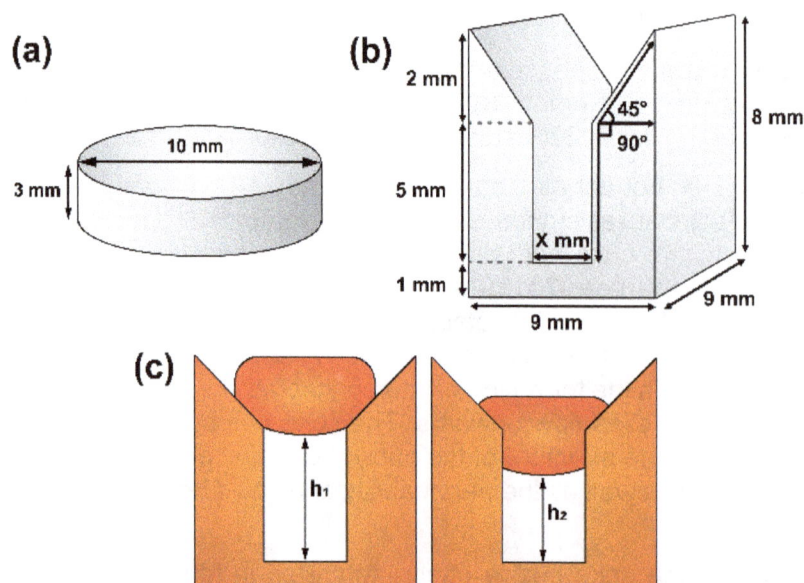

FIG 1 – Illustration of (a) Flat graphite substrate, (b) Graphite funnel diameter 'X = 2, 3, 4, 6 mm', (c) Displacement measurement protocol used in this study. Drawing not to scale.

Methodology

Synthetic slags of different basicity were prepared based on the ratio of chemical compounds as shown in Table 1. The basicity B, in this manuscript is defined using a wt per cent basis as:

$$B = \frac{CaO + MgO}{SiO_2 + Al_2O_3}$$

TABLE 1

Chemical composition of reactants used for experiments.

Basicity	Compound (%)						
	MnO	SiO₂	Fe₂O₃	CaO	MgO	Al₂O₃	CaS
0.8	72.86	10.28	7.85	7.26	1.27	0.30	0.17
1	70.31	9.71	9.31	8.60	1.51	0.36	0.20
1.22	76.45	6.69	7.85	7.26	1.27	0.30	0.17

The compounds 'as received' were mixed, briquetted using a press, and allowed to dry overnight. The samples were then placed on the graphite substrate and heated in a sessile drop furnace in an inert atmosphere with Ar airflow at 0.1 L/min. The heating rate was set at 300°C/min till 900°C followed by 50°C/min till 1500°C. The experiments were independently done for different holding durations (0 min, 20 min, 60 min, and 120 min) at 1500°C as shown in Figure 2. Additionally, slag was also prepared by (i) grinding the mixture before heating as well as by (ii) using a master slag powder of SiO₂, CaO, MgO, Al₂O₃, and CaS prepared at 1700°C. It was observed that identical slags were produced in all three routes. The master slag route was used for all experiments in this manuscript. The basicity was adjusted by adding additional SiO₂ to the mixture before heating.

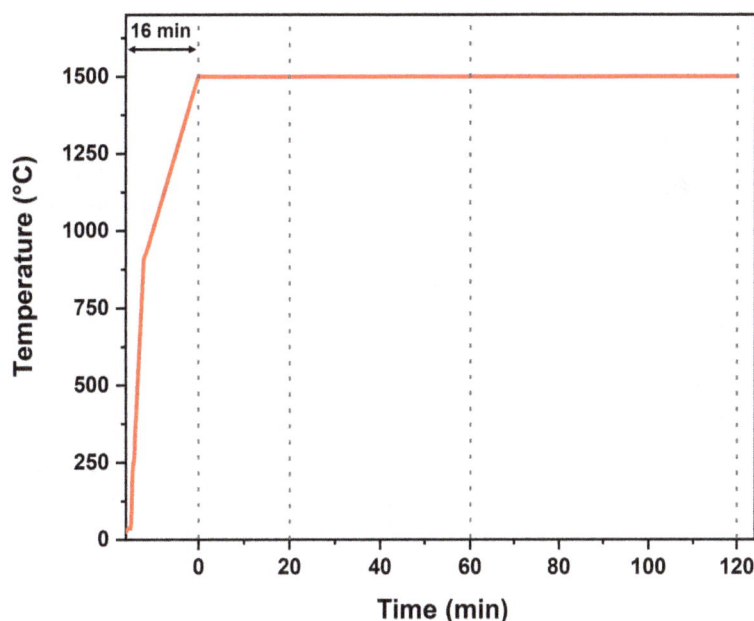

FIG 2 – Heat profile for experiments.

The proportions of the different phases were found from picture analyses from SEM micrographs. The area covered by the different phases was calculated using the measurement feature in Adobe Photoshop 2023. The flow was estimated from the wetting images by evaluating the y-axis displacement of the slag mixture with respect to the centre of the graphite funnel as shown in Figure 1c. The flow measurements were made till the completion of flow or 2 hrs of holding (whichever preceded earlier).

Apparatus

A sessile drop furnace equipped with a pyrometer and a C-type thermocouple was used for heating the samples. The wetting images were taken using a digital video camera (Allied Vision Prosilica GT2000, Edmund Optics, Inc, Barrington) with a telecentric lens (Navitar 1-50993D) at a resolution of 2048×1088 pixels. The distribution of different phases was obtained from SEM (Zeiss Ultra 55LE) images. The images were recorded at 200x magnification with an accelerating voltage of 10 kV. Due to the dimensional limitations, multiple images were captured to cover the periphery of the sample and subsequently compiled into a single image. EPMA (JEOL JXA-8500) was used for determining the composition.

RESULTS AND DISCUSSION

Phase composition and viscosity

The phase composition and viscosity were studied for a basicity of 0.8 and 1.22 on a flat substrate for different durations of holding ie 0 min, 20 min, 60 min, and 120 min at 1500°C. For all the holding durations, three phases *viz* liquid slag phase containing $MnO-CaO-Al_2O_3-MgO$ and SiO_2, solid slag phase containing mainly MnO, and reduced Mn-Fe-metal could be observed in the SEM images (Figure 3). SEM image showing the solid particles surrounded by the liquid phase as well as the dendrites in the liquid slag, which are assumed to be the MnO phase precipitated during cooling is shown in Figure 4. The amount of solid phase was found to decrease with time as the reduction of MnO progressed. The composition analysis obtained using WDS/EPMA shows that the per cent MnO in the liquid phase is higher for 0.8 basicity (Figure 5). For the solid phase, the per cent MnO is the same for both basicities and close to 90 per cent. Here the solid phase does not include dendrites. The composition remains constant throughout holding for both phases except values at 0 min being found to deviate from others. The region has been highlighted in grey in the figure and represents the uncertainty in values in the studied range. Subsequent plots also have similar deviations at 0 min holding and are highlighted similarly. The per cent MnO in the liquid phase was also plotted against basicity (Figure 6). It is seen that the per cent MnO in liquid slag increases with a decrease in basicity. The experimental values were lower than theoretical predictions, however, the trend is comparable.

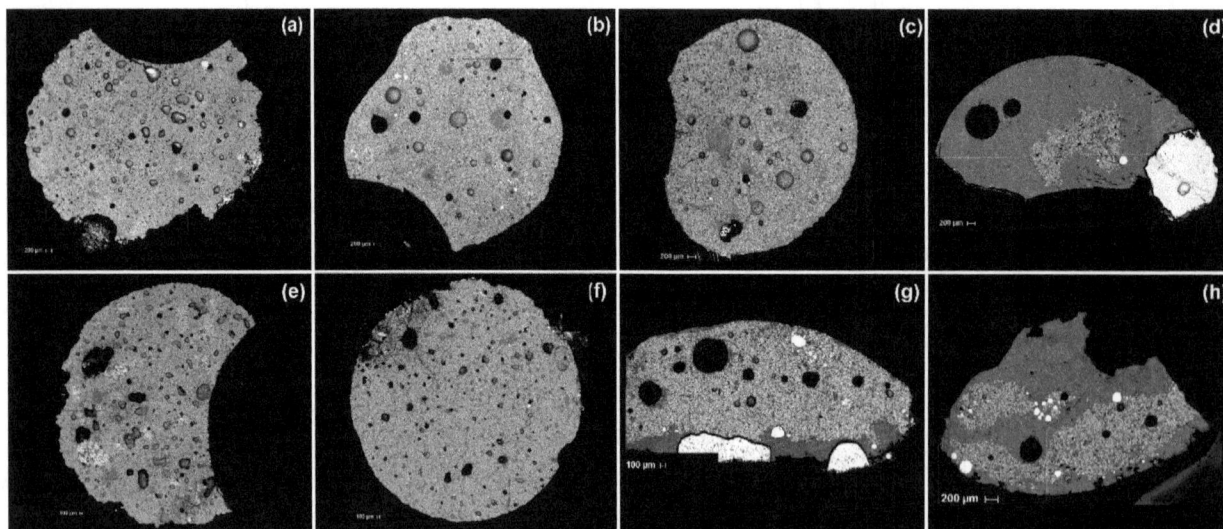

FIG 3 – SEM imaging of (a) 0.8 basicity at 0 min, (b) 0.8 basicity at 20 min, (c) 0.8 basicity at 60 min, (d) 0.8 basicity at 120 min, (e) 1.22 basicity at 0 min, (f) 1.22 basicity at 20 min, (g) 1.22 basicity at 60 min, (h) 1.22 basicity at 120 min.

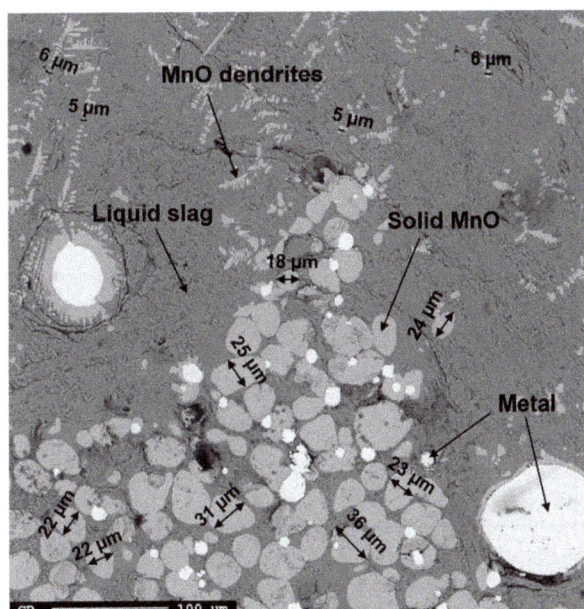

FIG 4 – SEM image showing the different phases. In the lower part one can see the multiphase, the mixture of solid MnO particles in a liquid, and in the top part the liquid.

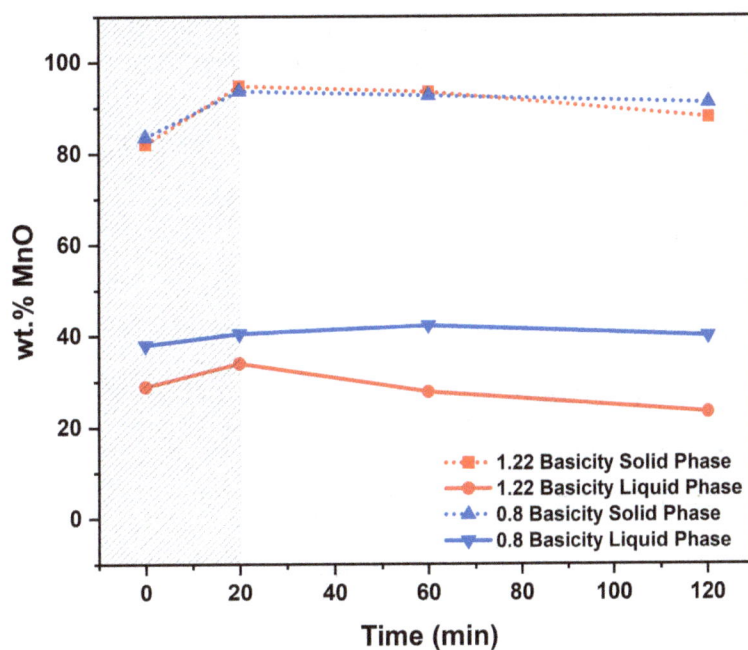

FIG 5 – wt per cent MnO versus Time for both solid MnO phase and liquid phase.

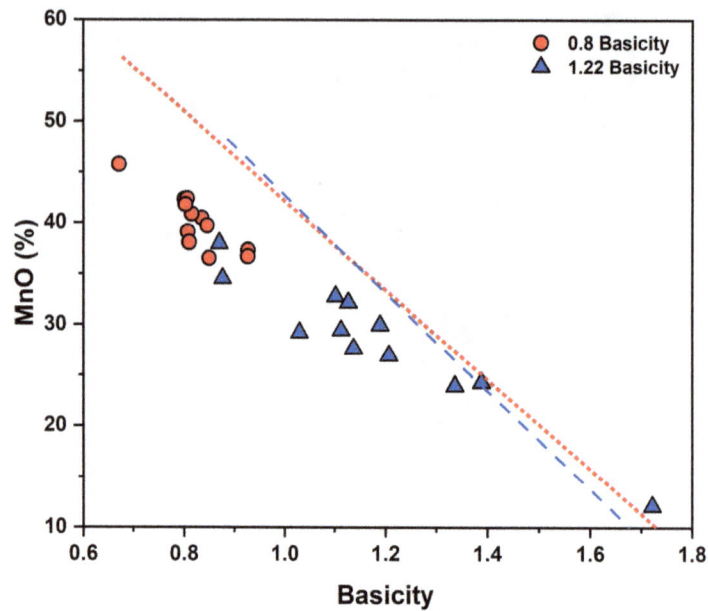

FIG 6 – per cent MnO versus Basicity (red and blue line indicate theoretical trend for 0.8 basicity and 1.22 basicity respectively).

The phase distribution is presented in Figure 7. As explained in the preceding paragraph, the solid phase decreases with reduction, that is the holding time. Using the phase distribution data, the overall MnO content was calculated using the formula:

$$\%MnO \ = \ \theta_{solid} \cdot \%MnO_{solid} + \ \theta_{liquid} \cdot \%MnO_{liquid}$$

where:

%MnO $\qquad$ = total MnO content

θ_{solid} $\qquad$ = fraction of solid phase

$\%MnO_{solid}$ $\qquad$ = %MnO in the solid phase

θ_{liquid} $\qquad$ = fraction of liquid phase

$\%MnO_{liquid}$ $\qquad$ = %MnO in the liquid phase

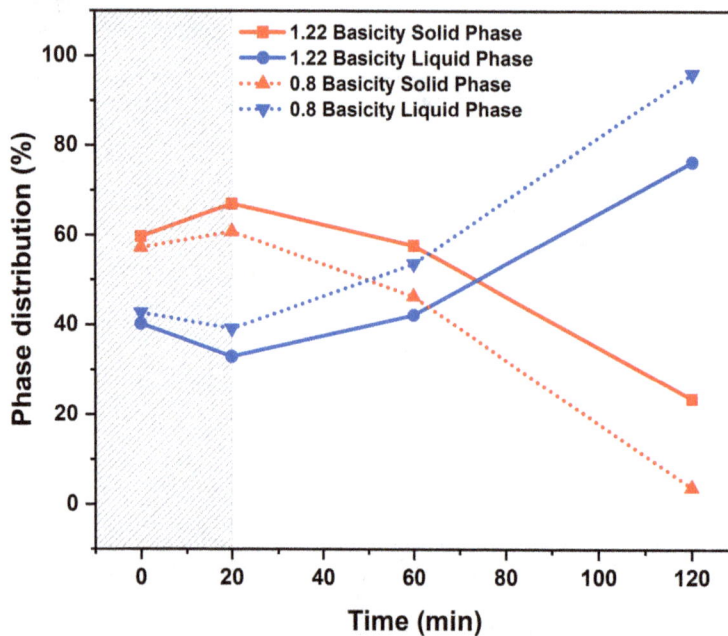

FIG 7 – Calculated amount of solid MnO phase and liquid phase.

Figure 8 shows the rate of MnO reduction with respect to time for 0.8 and 1.22 basicity. The rate of MnO reduction is constant for both basicities. The reduction of MnO is highly endothermic and literature suggests that the rate of reduction is chemically controlled and not transport-controlled (Ostrovski et al, 2002; Olsen, Tangstad and Lindstad, 2007). The present result counters earlier reports wherein the rate of reduction was found to be lower for high basicity solid-liquid mixture. Reports have also concluded that higher basicity was found to give a faster reduction in the case of a homogenous single-phase slag (Olsø, Tangstad and Olsen, 1998).

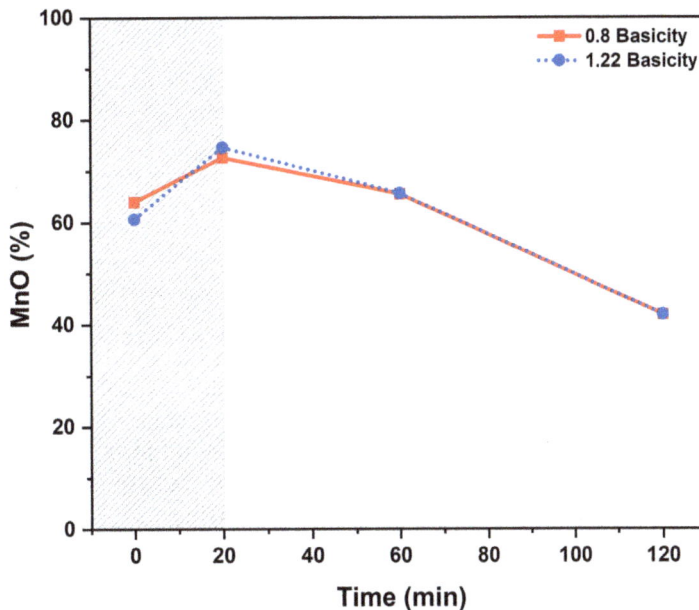

FIG 8 – Total per cent MnO versus Time showing the reduction of slag with time.

The viscosity of the liquid slag phase was calculated using the Urbain model developed by Mills, Yuan and Jones (2011). Figure 9 shows the viscosity of the liquid phase for 0.8 and 1.22 basicity at 0 min and 120 min of holding. The experimental composition at 1500°C was also extended for viscosity calculation at 1400°C and 1600°C. The effect of temperature on viscosity was well-defined in the model used. In the absence of a solid phase, the viscosity is higher for low-basicity systems and *vice versa*. However, for systems with a solid phase in existence the effective viscosity will have to be described using the Einstein-Roscoe equation (Roscoe, 1952):

$$\mu e = \mu(1 - 1.35\theta)^{-\left(\frac{5}{2}\right)}$$

where:

μ_e = effective viscosity of the slag

μ = viscosity of liquid phase

θ = the fraction of precipitated solid phases

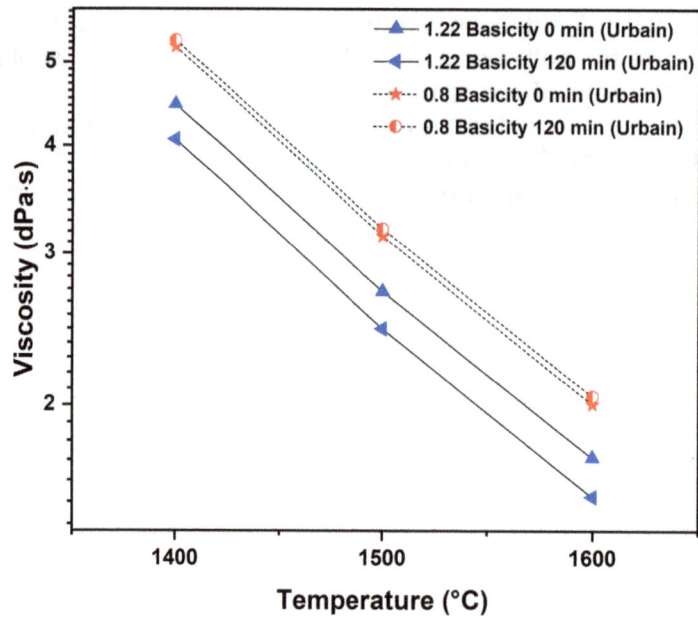

FIG 9 – Viscosity of liquid slag at different temperatures.

It is found that for a solid-liquid mixture, the effective viscosity is higher for high basicity systems as it contains a higher amount of solid MnO and the viscosity gradually lowers with lowering the amount of solid phase (Figure 10). For this calculation, the entire sample has been considered as a mixture of liquid and solid phases. Upon attainment of homogenous slag, the viscosity trend will reverse as discussed earlier.

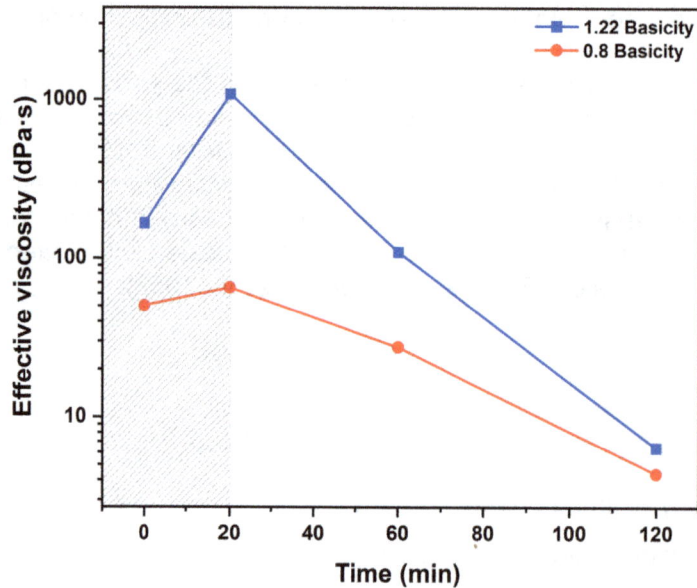

FIG 10 – Effective viscosity versus time.

The effective density of the slag was also determined using the equation:

$$\rho = \frac{(m_1 + m_2)}{(V_1 + V_2)}$$

where:

ρ = effective density

m_1 = mass of solid phase

m_2 = mass of liquid phase

V_1	= volume of solid phase
V_2	= volume of liquid phase

The effective density of the slag mixture was found to be roughly constant for 0.8 and 1.22 basicity at different holding durations (Figure 11). The effective density of the slag mixture lowers with the decrease in the solid phase.

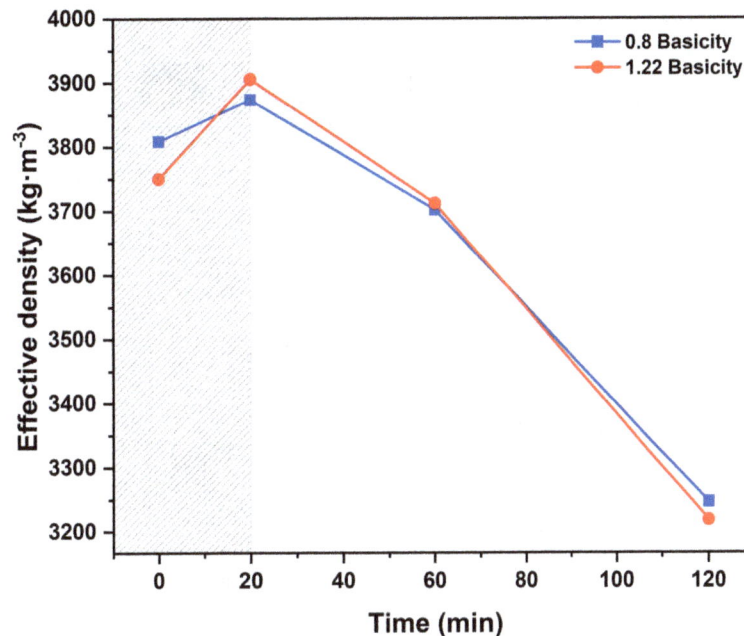

FIG 11 – Effective density versus time.

Flow measurement

Slags of three different basicities (0.8, 1, and 1.22) were evaluated using the funnel substrate to replicate the flow of slag through the coke bed in controlled conditions under the influence of gravity. The funnels of varying dimensions represent the different void sizes in the coke bed. The upper portion of the funnel represents the neck of the voids in the coke bed. The samples were heated using the same temperature profile as the flat substrate. Figure 12 shows the displacement of slag with respect to time at a constant temperature of 1500°C. The time ranges from 0 to 120 min of holding. The duration to complete the flow was typically quicker for funnels with larger dimensions like 3 mm and 4 mm. The slowest flow was observed for samples studied in 2 mm funnels for all basicities. This indicates that the flow of slag is influenced by the size of voids in the coke bed.

The live images obtained from the sessile drop furnace have been superimposed in the displacement plots (Figure 12). The reduction proceeds from the slag surface in contact with carbon and the produced metal phase typically appears away from the slag-carbon interface, here in our case the top or bottom portion of the funnel. The low-viscosity and higher-density metal phase is immiscible in the slag phase and separates as droplets that accumulate together resulting in a larger metal phase. Correlating the images and displacement, the metal phase is thus assumed to bring the sudden rapid displacement in the studied plots.

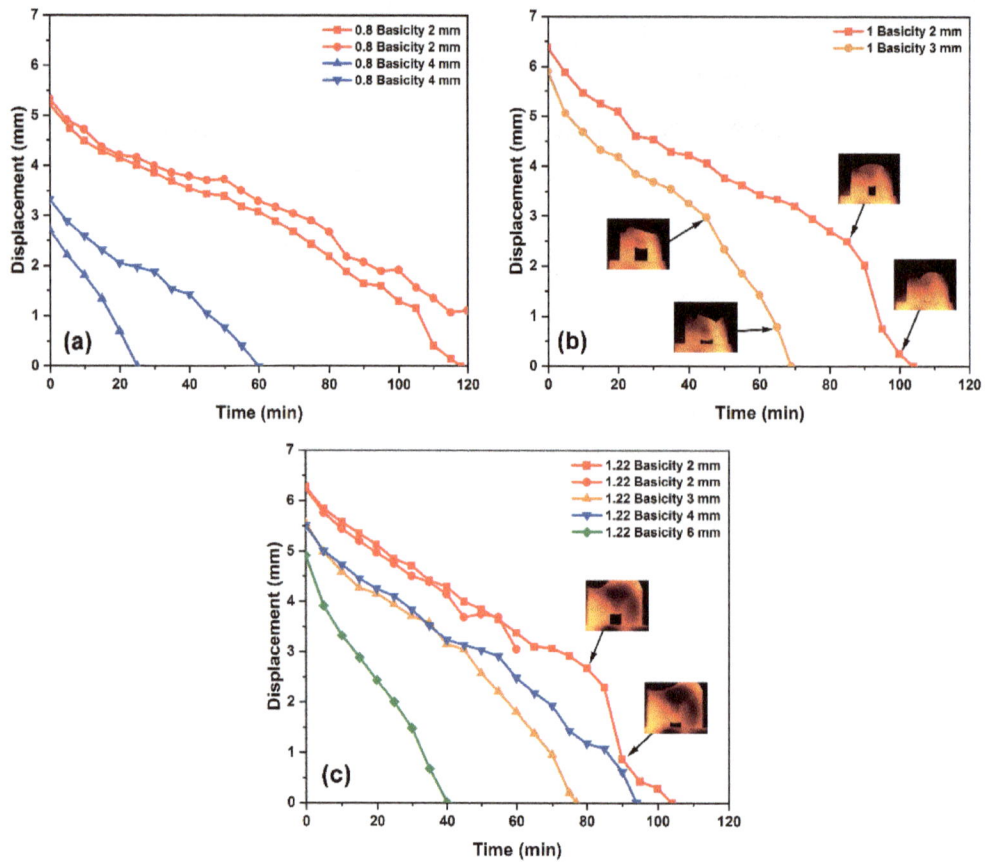

FIG 12 – Displacement versus time plots for different basicity and funnel dimensions.

SEM images of the funnels after completion of flow show the presence of solid, liquid, and metal phases (Figure 13). It is however observed that the phases segregate as the reduction proceeds. The slag is seen to comprise of two regions: a homogenous liquid slag of low viscosity and a multiphase solid-liquid mixture possessing high viscosity. The segregation pattern thus follows the trend of the multiphase slag at the top, homogenous slag in the middle, and metal phase at the bottom of the funnel. The region covered by the multiphase slag has high viscosity, which limits it from deforming. The low viscosity of the homogenous liquid slag allows it to seep under the multiphase region despite possessing a lower density. The immiscibility between the metal and liquid slag results in the clear demarcation between the two phases. A major portion of the metal phase is found at the bottom section of the funnel. However, it could be seen that in situations where the metal phase is produced in the upper sections of the funnel, the metal phase is unable to pass through the multiphase region. The viscosity of the multiphase region can be said to have a considerable impact on determining the flow of slag as it proceeds through the coke bed.

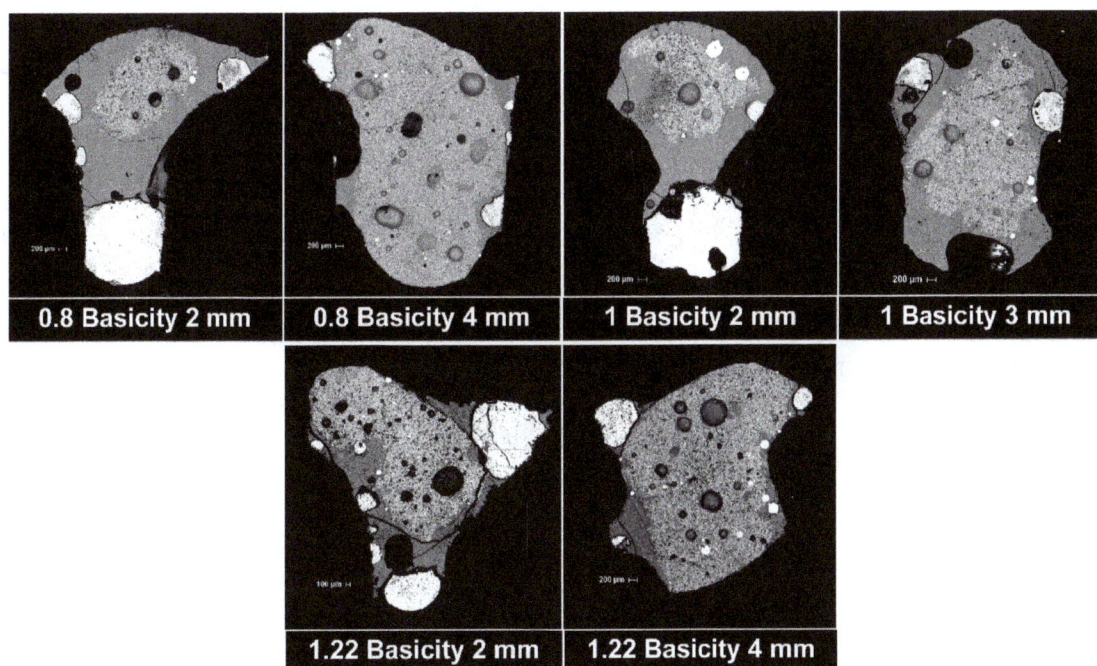

FIG 13 – SEM images of the different funnels after completion of flow.

As previously discussed, the proportion of solid phase decreases with increasing Mn reduction over time. It is seen in SEM images that for the samples investigated in funnel experiments, the quicker flow is generally accompanied with a greater fraction of multiphase. For the 2 mm funnels, the amount of multiphase is considerably less compared to 3 mm and 4 mm funnels. One presumption for this inference is that the dimension of the viscous multiphase in 2 mm is larger than the lower portion of the funnel, which prevents the slag from flowing and enables the reduction to proceed for longer durations. For 4 mm funnels, the multiphase region slides down enabling quick flow, accompanied with low reduction. Figure 14 shows the total per cent MnO after completion of flow for the studied funnels. The per cent MnO is greater for 4 mm funnel compared to 2 mm funnel for all basicities, which confirms the lowering in reduction with an increase in the size of funnels. It can be mentioned that 3 mm with 1.22 basicity is an outliner from this theory.

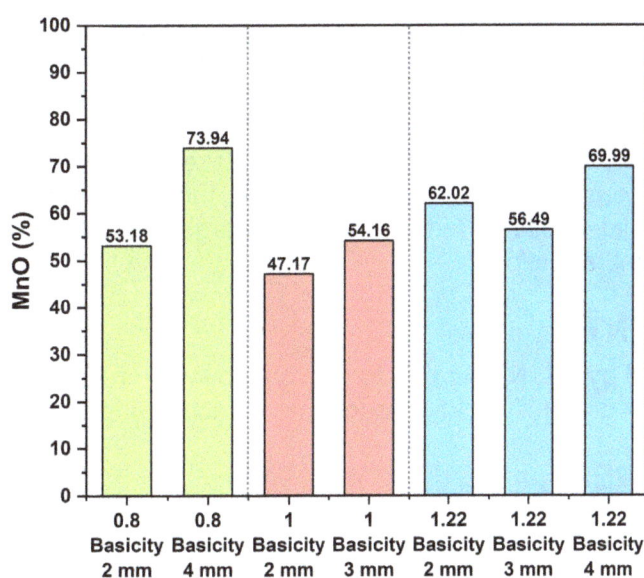

FIG 14 – Total per cent MnO for different basicities and funnel sizes.

The flow of Mn slag through the coke bed thus has two prime factors affecting its flow: (i) dimension of void neck, and (ii) the viscosity of slag. As discussed in the previous section, the flow rate increases with the increase in size of voids. For Mn slag, the high viscosity slag at first encounters the top or outline of the coke bed. Initially, the flow will be extremely difficult even for big void neck dimensions primarily because the viscous forces will dominate and prevent the flow. As the reduction proceeds the metal phase would trickle down as droplets and eventually result in a metal phase with a clear demarcation with the slag phase. The multiphase slag on the other hand is unable to flow unless the void neck is larger than the multiphase volume at the neck. However, the homogeneous slag (with low viscosity) that is produced during the process drains into the voids. The situation is depicted in Figure 15. If the void neck is large enough to allow the multiphase to flow, this will result in a highly viscous slag being taken into the coke bed making it even more detrimental for smooth flow. Previous excavations have put explanations that the reduction generally completes at the upper sections and the coke bed is generally devoid of significant multiphase slag presence making the large void neck situation unlikely to be encountered in practical scenarios. The type of coke and ash percent are also influential during the flow of slags inside the coke bed however for Mn slag, the viscosity, wetting, and size of voids will most likely be the prime criterion that need to be optimised before other factors are taken into consideration.

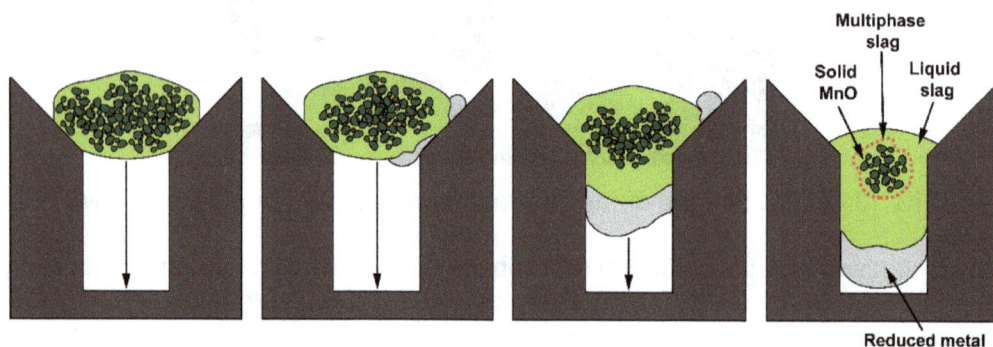

FIG 15 – Flow of slag in FeMn production.

Correlating with the discussion in the introduction, it can be safely concluded that the reduction generally proceeds at the top section of the coke bed in the presence of small voids. A single-phase slag and metal phase is anticipated inside the coke bed. The presence of small voids results in minimum reduction taking place inside the coke bed, but the majority on top of the coke bed.

CONCLUSION

The reduction rate was found to be constant for the studied basicity. The effective viscosity and density decrease with the lowering of the proportion of solid phase. The flow of slag was found to be influenced by the size of funnels, indicating that void size in the coke bed is an important parameter influencing the flow. It can be concluded that a slower flow is expected for coke bed with smaller voids. As the slags flow, the segregation of phases takes place. The solid-liquid multiphase possesses a distinct viscous character. For small voids, the multiphase is unable to flow until a substantial amount of solid phase is reduced whilst a larger void will result in quick flow at the expense of any substantial reduction.

ACKNOWLEDGEMENTS

This publication is funded by the Norwegian Research Council through the KPN project Recursive (326581), and the authors gratefully acknowledge the financial support received.

CONFLICT OF INTEREST

On behalf of all authors, the corresponding author states that there is no conflict of interest.

REFERENCES

Bando, Y, Hayashi, S, Matsubara, A and Nakamura, M, 2005. Effects of packed structure and liquid properties on liquid flow behavior in lower part of blast furnace, *ISIJ International*, 45(10):1461–1465. https://doi.org/10.2355/isijinternational.45.1461

Davies, J, Tangstad, M, Schanche, T L and du Preez, S P, 2023. Pre-reduction of United Manganese of Kalahari Ore in CO/CO2, H2/H2O and H2 Atmospheres, *Metallurgical and Materials Transactions B*, 54(2):515–535.

Dong, X F, Jayasekara, A, Sert, D, Ferreira, R, Gardin, P, Monaghan, B J, Chew, S J, Pinson, D and Zulli, P, 2021. Investigation of Molten Liquids Flow in the Blast Furnace Lower Zone: Numerical Modelling of Molten Slag Through Channels in a Packed Bed, *Metallurgical and Materials Transactions B: Process Metallurgy and Materials Processing Science*, 52(1):255–266. https://doi.org/10.1007/s11663-020-02009-1

Fukutake, T and Rajakumar, V, 1982. Liquid Packed in the Holdup Beds Dropping and under Abnormal Conditions Flow Phenomena the in Flow Simulating Furnace Zone of a Blast, *Transactions ISIJ*, 22:355–364.

Geleta, D D, Siddiqui, M I H and Lee, J, 2020. Characterization of Slag Flow in Fixed Packed Bed of Coke Particles, *Metallurgical and Materials Transactions B: Process Metallurgy and Materials Processing Science*, 51(1):102–113. https://doi.org/10.1007/s11663-019-01750-6

George, H L, Longbottom, R J, Chew, S J and Monaghan, B J, 2014. Flow of molten slag through a coke packed bed, *ISIJ International*, 54(4):820–826. https://doi.org/10.2355/isijinternational.54.820

Hockaday, S A C, Dinter, F and Reynolds, Q G, 2023. The thermal decomposition kinetics of carbonaceous and ferruginous manganese ores in atmospheric conditions, *Journal of the Southern African Institute of Mining and Metallurgy*, 123(8):391–398. https://doi.org/10.17159/2411-9717/2527/2023

Jeong, I H, Kim, H S and Sasaki, Y, 2014. Trickle flow behaviors of liquid iron and molten slag in the lower part of blast furnace, *Tetsu-To-Hagane/Journal of the Iron and Steel Institute of Japan*, 100(8):925–934. https://doi.org/10.2355/tetsutohagane.100.925

Kawabata, H, Liu, Z, Fujita, F and Usui, T, 2006a. Characteristics of liquid hold-ups in a soaked and unsoaked fixed bed, *Tetsu-To-Hagane / Journal of the Iron and Steel Institute of Japan*, 92(12):885–892. https://doi.org/10.2355/tetsutohagane1955.92.12_885

Kawabata, H, Niina, K, Harada, T and Usui, T, 2006b. Influence of channeling factor on liquid hold-ups in an initially unsoaked bed, *Tetsu-To-Hagane / Journal of the Iron and Steel Institute of Japan*, 92(12):893–900. https://doi.org/10.2355/tetsutohagane1955.92.12_893

Kumar, M, Ranganathan, S and Sinha, S N, 2007. Kinetics of reduction of different manganese ores, in *Innovations In The Ferro Alloy Industry - Proceedings of the XI International Conference on Innovations in the Ferro Alloy Industry, Infacon XI*, pp 241–246.

Larssen, T A and Tangstad, M, 2023. Effect of Raw Materials on Temperature Development during Prereduction of Comilog and Nchwaning Manganese Ores, *Minerals*, 13(7):920.

Mills, K C, Yuan, L and Jones, R T, 2011. Estimating the physical properties of slags, *Journal of the Southern African Institute of Mining and Metallurgy*, 111(10):649–658.

Muller, J, Zietsman, J H and Pistorius, P C, 2015. Modeling of Manganese Ferroalloy Slag Properties and Flow During Tapping, *Metallurgical and Materials Transactions B: Process Metallurgy and Materials Processing Science*, 46(6):2639–2651. https://doi.org/10.1007/S11663-015-0426-7

Natsui, S, Tonya, K, Nogami, H, Kikuchi, T, Suzuki, R O, Ohno, K-I, Sukenaga, S, Kon, T, Ishihara, S and Ueda, S, 2020. Numerical study of binary trickle flow of liquid iron and molten slag in coke bed by smoothed particle hydrodynamics, *Processes*, 8(2). https://doi.org/10.3390/pr8020221

Natsui, S, Tonya, K, Hirai, A and Nogami, H, 2021. Comprehensive numerical assessment of molten iron–slag trickle flow and gas counter-current in complex coke bed by Eulerian–Lagrangian approach, *Chemical Engineering Journal*, 414(January):128606. https://doi.org/10.1016/j.cej.2021.128606

Oh, J S and Lee, J, 2016. Composition-dependent reactive wetting of molten slag on coke substrates, *Journal of Materials Science*, 51(4):1813–1819. https://doi.org/10.1007/s10853-015-9588-6

Olsen, S E, Tangstad, M and Lindstad, T, 2007. *Production of Manganese Ferroalloys*, 247 p (Tapir Academic Press).

Olsø, V, Tangstad, M and Olsen, S E, 1998. Reduction kinetics of MnO-saturated slags, in *8th International Ferroalloys Congress*, 8:279–283.

Ostrovski, O, Olsen, S E, Tangstad, M and Yastreboff, M, 2002. Kinetic modelling of MnO reduction from manganese ore, *Canadian Metallurgical Quarterly*, 41(3):309–318. https://doi.org/10.1179/cmq.2002.41.3.309

Roscoe, R, 1952. The viscosity of suspensions of rigid spheres, *British Journal of Applied Physics*, 3(8):267–269. https://doi.org/10.1088/0508-3443/3/8/306

Safarian, J and Tangstad, M, 2009. Wettability of silicon carbide by CaO-SiO$_2$ slags, *Metallurgical and Materials Transactions B: Process Metallurgy and Materials Processing Science*, 40(6):920–928. https://doi.org/10.1007/s11663-009-9292-5

Tang, K and Tangstad, M, 2007. Modeling viscosities of ferromanganese slags, *Innovations In The Ferro Alloy Industry - Proceedings of the XI International Conference on Innovations in the Ferro Alloy Industry, Infacon XI*, pp 344–357.

Tangstad, M, Bublik, S, Haghdani, S, Einarsrud, K E and Tang, K, 2021. Slag Properties in the Primary Production Process of Mn-Ferroalloys, *Metallurgical and Materials Transactions B: Process Metallurgy and Materials Processing Science*, 52(6):3688–3707. https://doi.org/10.1007/s11663-021-02347-8

Improvement of the copper flash smelting furnace (FSF) and the slag cleaning furnace (SCF) process by advice-based control of silica and coke addition

A Schmidt[1], S Winkler[2], E Klaffenbach[3], A Müller[4], V Montenegro[5] and A Specht[6]

1. Process Engineer, Aurubis AG, 20539 Hamburg, Germany. Email: am.schmidt@aurubis.com
2. Senior Process Engineer, Aurubis AG, 20539 Hamburg, Germany.
   Email: s.winkler@aurubis.com
3. Senior Manager R&D, Aurubis AG, 20539 Hamburg, Germany.
   Email: e.klaffenbach@aurubis.com
4. FSF/SCF Manager, Aurubis AG, 20539 Hamburg, Germany. Email: ar.mueller@aurubis.com
5. Senior Manager R&D, Aurubis AG, 20539 Hamburg, Germany.
   Email: v.montenegro@aurubis.com
6. Executive Director Pyrometallurgy R&D, Aurubis AG, 20539 Hamburg, Germany.
   Email: a.specht@aurubis.com

ABSTRACT

In the process of primary copper smelting in flash smelting furnace (FSF), the ratio of iron to silica (Fe/SiO_2 mass ratio) primarily determines the slag viscosity and magnetite formation. Viscous slag does not only cause difficulties in daily operation, such as in tapping, it also increases the copper losses to slag. Thus, keeping the Fe/SiO_2 mass ratio at the desired level is an important aspect of FSF and slag cleaning furnace (SCF) operation.

In this study, the improvement of the Fe/SiO_2 mass ratio was conducted in two steps: First, the optimal Fe/SiO_2 mass ratio was derived based on thermodynamic data. Based on this analysis, appropriate control limits were determined. Second, this range of favourable Fe/SiO_2 mass ratios was implemented into operation by an advice-based control system, which suggests the addition of silica based on the analysis of SCF slag. As this approach was successful, the same advice-based control system principle was adapted for the coke addition to the SCF.

After testing the new advice-based control for several months, the production data confirm the predictions by thermodynamics: since the prediction was implemented as an advisor, the copper and magnetite content in SCF slag is significantly reduced and viscous slag is reported less often by operations.

INTRODUCTION

In flash smelting furnace (FSF) operation, the Fe/SiO_2 mass ratio is an important target for controlling the slag quality, the balance is crucial as both too high and too low ratios can lead to undesirable effects. Too high Fe/SiO_2 mass ratios promote magnetite formation and increased slag viscosity, resulting in higher copper losses (Wang *et al*, 2021). This effect occurs also at too low Fe/SiO_2 mass ratios, as the excessive addition of silica increases the slag viscosity as well (Shen *et al*, 2022). An additional drawback of low Fe/SiO_2 mass ratios is that the overall slag amount increases with higher silica addition, leading also to additional copper losses. Moreover, the presence of magnetite and silica in the slag has been linked to the formation of mechanically entrapped matte particles, which contribute significantly to copper losses in the slag (Imris, Sánchez and Achurra, 2005). Research indicates that the concentration of copper and minor elements in slags can be optimised by controlling the slag composition (Klaffenbach *et al*, 2021). In this study a two-step approach is outlined: First, the optimal Fe/SiO_2 mass ratio is derived from thermodynamic calculations (Step 1). Secondly, this optimal Fe/SiO_2 mass ratio is implemented as a target for slag control into operations by an advice-based control system (Step 2). As an additional step, the same advice-based control system was adapted for the coke charging into the slag cleaning furnace (SCF) for improving the slag quality further and ensuring comprehensive control over the slag quality of both the FSF and the SCF. This study describes a case of optimising the industrial operation of Aurubis Hamburg's primary smelting process with the objective to reduce magnetite formation and hence copper losses to the slag.

Step 1 – Thermodynamic derivation of the optimal Fe/SiO$_2$ ratio

Equilibrium calculations are crucial procedure in determining the conditions, at which a fully liquid slag without solid particle formation is achieved within the FSF. The calculations were conducted using FactSage™ 8.0 using FactPS and UQPY private database. Various solution phases including matte/metal (Liq(Matte/Metal)), slag (UQPY-SLAG), spinel (UQPY-SPIN) were considered along with the stochiometric compounds tridymite (SiO$_2$) and fayalite (Fe$_2$SiO$_4$). Within the calculation, the operating conditions (such as feed composition, slag composition and oxygen enrichment) were set to model the behaviour of the metallurgical system in the FSF of Aurubis Hamburg. In Figure 1 the result of the calculation is illustrated, in which the impact of temperature and Fe/SiO$_2$ mass ratio on the calculated liquidus is shown.

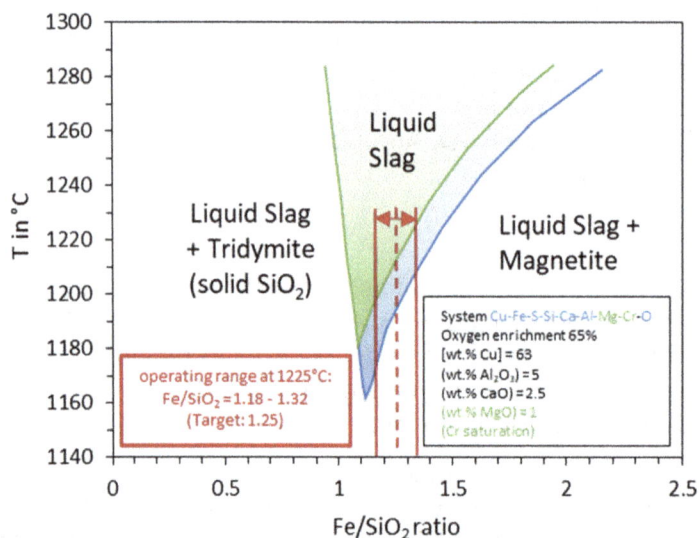

FIG 1 – Impact of temperature and Fe/SiO$_2$ mass ratio on the calculated liquidus of slag under conditions of Aurubis Hamburg flash smelter.

The calculation was conducted with and without presence of MgO and Cr$_2$O$_3$ in slag, in order to also consider the effect of interaction with magnesia chromite refractory on the slag liquidus. It can be seen that the presence of these components decreases the liquid range of the slag (green area in the diagram) as opposed to a slag without MgO and Cr$_2$O$_3$ components (blue and green area in the diagram combined). As MgO and Cr$_2$O$_3$ are likely to be present in an industrial slag, the derivation of the optimal Fe/SiO$_2$ mass ratios is derived from the respective liquid range.

At low Fe/SiO$_2$ mass ratios, tridymite (solid SiO$_2$) would be formed, meaning that in practice incomplete melting of silica would occur. At high Fe/SiO$_2$ mass ratios magnetite would be formed. Both should be avoided to achieve a slag with low viscosity, ensuring undisturbed settling of the copper containing particles within the given time. At the same time, only a minimum amount of slag should be produced, so that a high Fe/SiO$_2$ mass ratio is favoured. It was found that the Fe/SiO$_2$ mass ratio in operation has a variability of ±0.07 around the set point. For this reason, the targeted Fe/SiO$_2$ mass ratio at a process temperature of 1225°C is Fe/SiO$_2$ = 1.25 with an upper limit of 1.32 and a lower limit of 1.18. By the upper limit it is ensured that magnetite formation is prevented, the lower limit prevents excessive production of slag and also occurrence of non-molten silica.

Step 2 – Implementation into operation by an advice-based control system

The Fe/SiO$_2$ mass ratio in the FSF slag is controlled by sampling and analysing the SCF slag. Depending on the iron and silica content in slag, the ratio is then adjusted by manual changes to the silica addition to the feed. In practice, this leads to the following stages:

1. Sampling and analysis of SCF slag. Thereby, the chemical components are analysed by X-ray diffraction (XRD) and standard Satmagan.

2. Review of the SCF slag composition by the operators.

3. Manual calculation of the Fe/SiO$_2$ mass ratio and decision on the addition or reduction of silica.

4. Adjustment of the silica flow.

This process contains an inevitable time delay from sampling until the SCF slag analysis is available. Hence, the in-house developed advice-based control system for SiO_2 aims to shorten the reaction time from this point on until the ratio is adjusted by the operators by giving advice on the necessary addition or reduction of silica, thus simplifying step 2 and 3.

The slag analysis is directly retrieved from laboratory data into a separate calculation tool, which then calculates the instantaneous Fe-SiO_2 ratio. Afterwards, the Fe/SiO_2 mass ratio is compared automatically to the target Fe/SiO_2 mass ratio, giving direct advice to the operators in the control room, on whether to increase or reduce the silica flow.

As shown in Figure 2, this approach has been well accepted by the operators and was very successful in reducing the amplitude of fluctuations in the Fe/SiO_2 mass ratio, causing a reduction of outliers outside of the target Fe/SiO_2 mass ratio range by 70 per cent compared to before the implementation of the new advice-based control system.

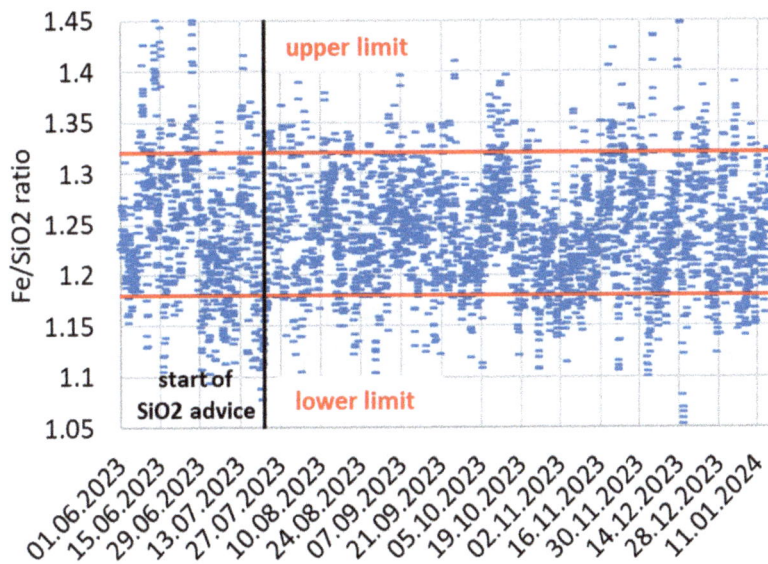

FIG 2 – Fe/SiO_2 mass ratio in SCF slag before and after implementation of the advice-based control system for SiO_2.

Since the Fe-SiO_2 was well implemented in the FSF, this control system was adapted to also give advice on the coke charging into the SCF in order to improve the slag quality further. Thereby, the coke already fed into the SCF is calculated based on the bunker weights, giving real-time advice on when and how much coke to charge until the daily target value is reached. The development of the coke charging is shown in Figure 3, proving the significant stabilisation of the process after implementation of the advice-based control system.

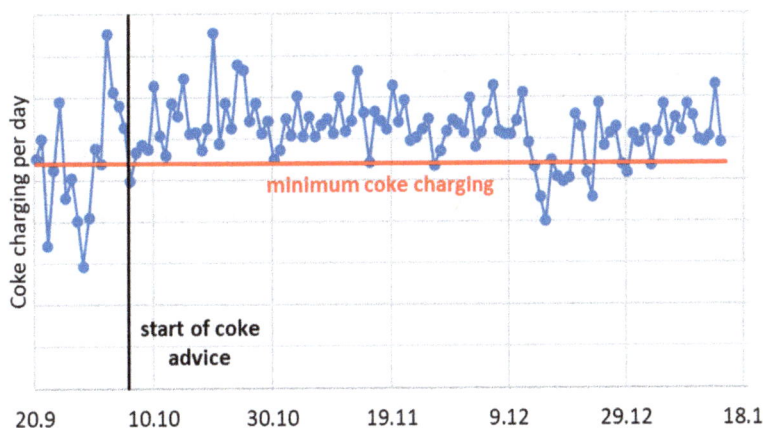

FIG 3 – Daily coke charging before and after implementation of the advice-based control system for coke.

In accordance with the thermodynamic predictions described in Step 1, the magnetite content of SCF stabilised and dropped significantly after the SiO$_2$ advice and the coke advice were implemented, as shown in Figure 4. Finally, a magnetite reduction by 41.9 per cent was observed, this improvement is stable for several months by now. Accordingly, the reports of viscous slag by operation have dropped by 56 per cent.

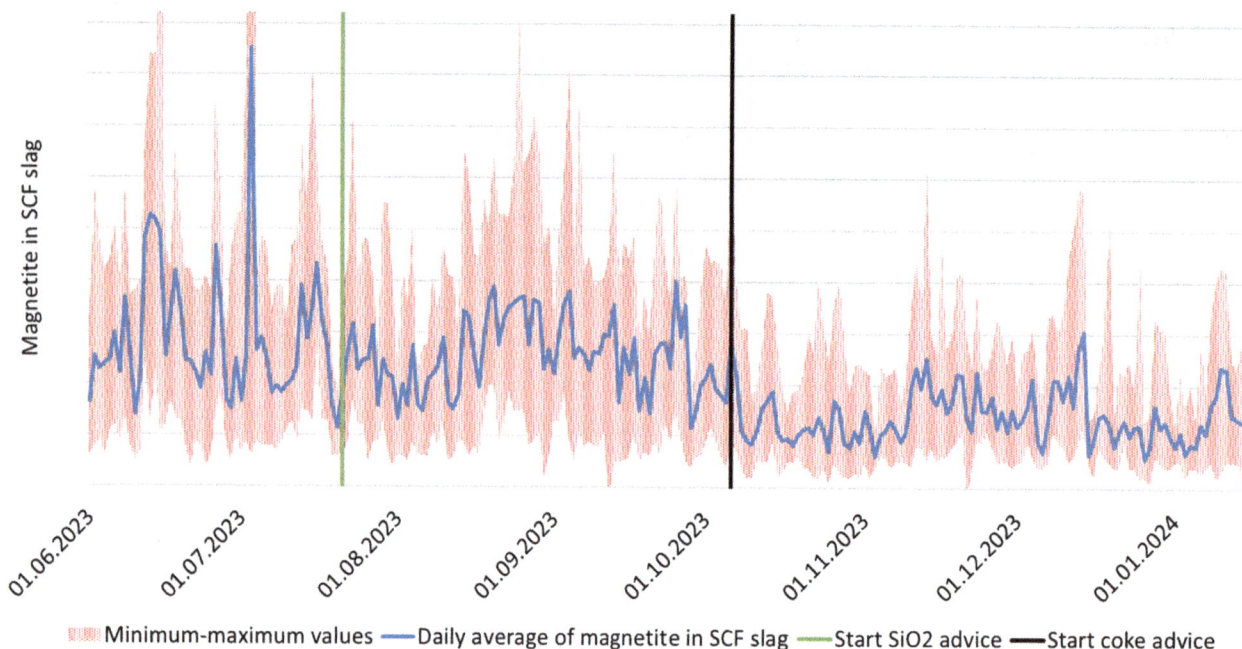

FIG 4 – Development of magnetite content in SCF slag after implementation of the SiO$_2$ advice and the coke advice.

The implementation of the advisors in the FSF and the SCF also caused a drop in copper losses, which currently accounts to a reduction of 6 per cent compared to before implementation of the advice-based control system.

CONCLUSIONS

The new advice-based control system was tested for several months and caused a visible stabilisation of the Fe/SiO$_2$ mass ratio and of coke addition to the SCF. In total, both measures caused a reduction of magnetite in SCF slag by 41.9 per cent. This reduction in magnetite led to a significant reduction in reports of viscous slag by operations and caused a reduction in copper losses by 6 per cent, proving that the thermodynamic predictions are correct and that they are an effective measure for improving slag quality. Both measures are still in operation and further data will be evaluated.

REFERENCES

Imris, I, Sánchez, M and Achurra, G, 2005. Copper losses to slags obtained from the El Teniente process, *Mineral Processing and Extractive Metallurgy*, 114(3):135–140. https://doi.org/10.1179/037195505X49769

Klaffenbach, E, Mostaghel, S, Guo, M and Blanpain, B, 2021. Thermodynamic Analysis of Copper Smelting, considering the Impact of Minor Elements Behavior on Slag Application Options and Cu-recovery, *Journal of Sustainable Metallurgy*, 7(2):664–683. https://doi.org/10.1007/s40831-021-00354-2

Shen, Y, Ma, Y, Peng, Y, Lu, B, Dou, Y, Tian, J, Guo, W and Zhang, Z, 2022. Effect of FeO content on melting characteristics and structure of nickel slag, *Journal of Mining and Metallurgy, Section B: Metallurgy*, 58(3):427–438.

Wang, L, Wei, Y, Zhou, S, Li, B and Wang, H, 2021. Matte separated behavior from slag during the cleaning process by using waste cooking oil as carbon neutral reductant, *Journal of Mining and Metallurgy Section B Metallurgy*, 57(3):379–388.

Characterisation and assessment of B_2O_3 added LF slag

S S Varanasi[1], S Kakara[2], B V R Mandalika[3], M B Shaik[4] and A Kamaraj[5]

1. Senior Manager (R&D), Rashtriya Ispat Nigam Limited, Visakhapatnam Steel Plant, Visakhapatnam 530031, India; PhD student, Department of Materials Science and Metallurgical Engineering, IIT Hyderabad 502285, India. Email: ms22resch14005@iith.ac.in
2. PhD student, Department of Materials Science and Metallurgical Engineering, IIT Hyderabad 502285, India. Email: ms23resch02001@iith.ac.in
3. General Manager, Corporate office, Rashtriya Ispat Nigam Limited, Visakhapatnam Steel Plant, Visakhapatnam 530031, India. Email: mbvenkatarao@vizagsteel.com
4. PhD student, Department of Materials Science and Metallurgical Engineering, IIT Hyderabad 502285, India. Email: ms22resch14003@iith.ac.in
5. Assistant Professor, Department of Materials Science and Metallurgical Engineering, IIT Hyderabad 502285, India. Email: ashokk@msme.iith.ac.in

ABSTRACT

The properties of top slags play a very important role in secondary steelmaking. Fluxes should be chosen based on the ladle furnace (LF) slag requirements like viscosity, melting temperature, sulfide capacity, etc. CaF_2 and $CaO-Al_2O_3$-based synthetic slags are the most used fluxes in secondary steelmaking to improve slag fluidity and interfacial chemical reactions. CaF_2 is obsolete due to fluorine pollution and $CaO-Al_2O_3$ based slags have 40–50 wt per cent Al_2O_3 and 30–40 per cent CaO which is not beneficial for desulfurisation. Yet another drawback is that LF slag disintegrates into fine powder during cooling due to phase transformation C_2S ($2CaO.SiO_2$). Hence, there is a need for the development of low melting fluxes with high CaO content and the capability to prevent slag disintegration. The current work studied the effect of 0–9 wt per cent B_2O_3 on physico-chemical properties and characterisation of LF slags. X-ray diffraction (XRD) analyses were carried out to identify mineralogical phases present in slag and Fourier-transform infrared spectroscopy (FT-IR) analysis was done to understand the changes in bonds of B_2O_3 added LF slag. The characteristic temperature of slag was determined by a hemispherical method using high-temperature microscopy. Empirical models were used to calculate viscosity, break temperature, and sulfide capacity. FactSage™, ver 8.2 (by GTT-Technologies) simulations were carried out to find the equilibrium phase fractions and percentage of liquid slag at different temperatures. New phases such as $Ca_{11}B_2Si_4O_{22}$ and $Ca_2B_3O_5$ were formed in LF slag with B_2O_3 addition. Based on the results, 0.5–1.0 wt per cent B_2O_3 was found to be sufficient to prevent C_2S-driven disintegration of LF slag. Also, the melting temperature and viscosity of LF slag were reduced by adding 3–8 wt per cent B_2O_3 due to the formation of BO_3 planar triangular structure and low melting eutectics in slag. Liquid slag started forming at a temperature as low as 800°C when B_2O_3 is >6 wt per cent. It was found that the sulfide capacity, viscosity, and break temperature of the slag decreased with increasing wt per cent B_2O_3. Based on the analysis carried out, B_2O_3 can be a promising fluxing material in secondary steelmaking, for improving the desulfurisation kinetics and valourisation potential of LF slag.

INTRODUCTION

Fluxes are important in the secondary steelmaking process, particularly in ladle furnace (LF) refining (Varanasi *et al*, 2019a). Conventional ladle fluxes are either CaF_2 or $CaO-Al_2O_3$ based which ensure optimum fluidity, viscosity, sulfide capacity, Liquidus temperature, thermal and and electrical conductivity etc (Varanasi *et al*, 2019b). However, the emission of fluorine causes corrosion to the equipment, environmental pollution, and health hazards (Yan *et al*, 2016). Even though $CaO-Al_2O_3$ based synthetic slags are good alternatives to CaF_2, the optical basicity of $CaO-Al_2O_3$ based synthetic slags is low and increased Al_2O_3 content in LF slag is not beneficial for desulfurisation, deoxidation, and absorption of impurities. So, the content of Al_2O_3 in refining flux should be controlled (Wang *et al*, 2011a). Another problem with LF slags is, that they crumble into fine powder during cooling, due to the volume expansion associated with phase transformation of $2CaO.SiO_2$ phase (Gollapalli *et al*, 2020). Even though LF slag has 50–60 wt per cent CaO and 15–30 wt per cent Al_2O_3 as major constituents, its valourisation potential is low due to the disintegrating phenomena

(Varanasi *et al*, 2022). There is a requirement for alternative fluxes, which help achieve optimum slag properties and prevent the disintegration of LF slag.

B_2O_3 is considered a promising fluxing material as an alternative to CaF_2 and CaO-Al_2O_3 based synthetic slags. It was reported by Babenko *et al* (2017) and Wang *et al* (2011c) that B_2O_3 reduces the melting point and viscosity of CaO-Al_2O_3-SiO_2-MgO based slags and for CaO-based refining flux, the fluxing action of B_2O_3 is better than that of Al_2O_3 and CaF_2. The viscosity of (composition per cent) 53–62 CaO, 7.5–12 SiO_2, 15–28 Al_2O_3, 8 MgO, 4 B_2O_3 slags did not exceed 0.8 Pa.s at 1500–1600°C range and are very much suitable for steel desulfurisation. Even though B_2O_3 is an acidic oxide and network former, it simplifies the complex silicate structure by forming BO_3^{3-} planar triangular structure (Bi *et al*, 2021). It was reported by Pontikes *et al* (2010) that the addition of B_2O_3 prevents the β-Ca_2SiO_4 to γ-Ca_2SiO_4 transformation in slags due to the replacement of SiO_4^{4-} units by BO_3^{3-} units. However, B_2O_3 can also form BO_4^{5-} tetrahedral structure and lead to complex structures in slag. Wang *et al* (2010a) reported that B_2O_3 may react with CaO present in slag to form low melting point eutectics such as $CaO.2B_2O_3$, $CaO.B_2O_3$, $2CaO.B_2O_3$ with melting points 986°C, 1154°C, 1298°C respectively. Bi *et al* (2021) reported that $[SiO_4]^{4-}$-$[BO_3]^{3-}$ structures are formed due to addition of B_2O_3 and polymerisation degree of the system will decrease. In addition, $[BO_3]^{3-}$ is a planar triangular structure that is unstable as the bond energy of B-O (787 kJ/mol) is much larger than that of Si-O (600 kJ/mol) and Al-O (485 kJ/mol). Wang *et al* (2012) reported that B_2O_3 has significant fluxing effect on CaO-($2CaO·SiO_2$)-B_2O_3 slag system. When the content of B_2O_3 is more than 5 per cent, the melting temperature of $2CaO·SiO_2$-B_2O_3 system is lower than 1300°C. CaO-B_2O_3 slags were used by Wang *et al* (2007) to prevent sticking of slag to the snorkel refractory. They reported that the viscosity of CaO (26–42 wt per cent) – Al_2O_3(40–65 Wt per cent) – SiO_2 (9–15 wt per cent) based slag decreased from 6.5 Pa.s to < 2 Pa.s at 1500°C. Also, the melting temperature of the slag decreased by 100°C with the addition of 10 wt per cent CaO-B_2O_3 (mass ratio 1:1) to the slag. The addition of B_2O_3 helps in improving slag metal reaction kinetics and hence desulfurisation efficiency. A highly basic slag with a low melting temperature can be achieved by the addition of B_2O_3 to the LF slag system. It was reported elsewhere (Wang *et al*, 2011d, 2011b, 2010b) that, just by 4 per cent B_2O_3 addition in the high basicity (5.75~7.75) CaO-based refining flux, melting temperature as low as 1250°C can be achieved. From studies conducted by Zhuchkov, Salina and Sychev (2019) there is a possibility of boron microalloying using B_2O_3 based slags. Boron can be reduced from B_2O_3 present in slag by Al or Si present in steel. Zhang *et al* (2019) reported that the addition of B_2O_3 to CaO-SiO_2 based slags increased the electrical conductivity due to the formation of BO_3^{3-} and BO_4^{5-} units whose effect is stronger than the increase in degree of polymerisation caused by B_2O_3. Babenko *et al* (2019) reported that for slags with basicity (CaO/SiO_2) 4–5, the MgO saturation concentration was <7 per cent for 4 wt per cent B_2O_3. Hence, by using B_2O_3 as a fluxing agent, there is a possibility to develop a slag system with ultra-high basicity (>5), ultra-low alumina content as well as low melting temperature and low viscosity.

However, the structure-property correlation of B_2O_3 modified LF slags was not reported clearly. In the present study, an attempt was made to study the effect of B_2O_3 additions on CaO-MgO-SiO_2-Al_2O_3 based LF slag for slag stabilisation and properties of slag. High-end characterisation techniques such as FT-IR, scanning electron microscope (SEM), X-ray diffraction (XRD), and high-temperature microscopy were used to study the effect of B_2O_3 addition on CaO-MgO-SiO_2-Al_2O_3 based LF slag collected from an integrated steel plant. Empirical models were used for calculating sulfide capacity, viscosity, and break temperature. Thermodynamic calculations were performed using FactSage™ ver 8.2 to estimate equilibrium phases of slag.

MATERIALS AND METHODS

Melting experiments

LF slag was collected from an integrated steel plant and laboratory grade boric acid (H_3BO_3) with min purity of 95 per cent is used as the source of B_2O_3. The composition of raw materials is given in Table 1.

TABLE 1

Chemical composition of raw materials.

Material/wt%	CaO	MgO	SiO$_2$	Al$_2$O$_3$	FeO	MnO	B$_2$O$_3$
LF slag	55	10.3	17.12	9.97	1.55	1	-
H$_3$BO$_3$	-	-	-	-	-	-	56

B$_2$O$_3$ was added in 0.25, 0.5, 1, 2, 3, 4, 6, 8 wt per cent in LF slag and mixed thoroughly. The samples each weighing 100 g were put into a graphite crucible and loaded in a muffle furnace with a capacity of 1700°C. The slags were melted at 1600°C with a heating rate of 5°C/min for 30 mins. After heating, the crucible was removed, and air-cooled at room temperature as shown in Figure 1.

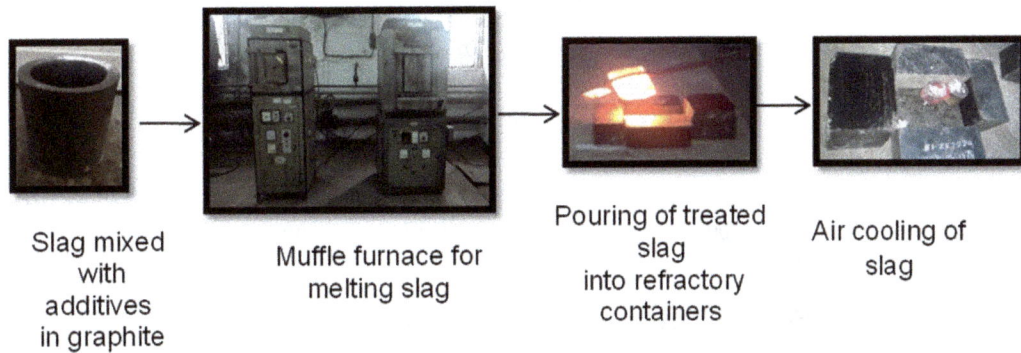

Slag mixed with additives in graphite → Muffle furnace for melting slag → Pouring of treated slag into refractory containers → Air cooling of slag

FIG 1 – Schematic of slag melting experiments carried out in a muffle furnace.

SPECTROSCOPY ANALYSIS

Elemental chemical analysis was done for slag before and after stabilisation using Panalytical axios wavelength dispersive X-ray fluorescence (WD-XRF). Particle size analysis of LF slag was carried out using ROTAP sieve shaker. XRD analysis was carried out in Rigaku ultima IV with 2θ between 10–90°, Cu Kα radiation with a wavelength of 1.54A and step size of 0.02. High-temperature microscopy was used to measure the melting temperature of slag. The hemisphere method was followed to estimate slag melting temperature. Ash fusion equipment with a heating rate of 8°C/min was used to analyse slag melting properties and the hemispherical temperature was considered as a base for comparison. FT-IR was carried out to study molecular vibrations and rotation related to covalent bonds. FT-IR spectra were recorded in the wavenumber range of 600 to 1600 cm^1.

THERMODYNAMIC CALCULATION

The effect of B$_2$O$_3$ addition on crystallisation phase fractions formed in the slag and melting temperature was studied by thermodynamic software FactSage™ 8.2. The composition of the slag considered is given in Table 2. At constant basicity ie 2.9, equal amounts of Al$_2$O$_3$ is replaced by B$_2$O$_3$ for analysis.

TABLE 2

Composition of slags considered for thermodynamic calculation.

CaO	MgO	SiO$_2$	Al$_2$O$_3$	B$_2$O$_3$	Basicity
56	12	19	13	0	2.9
56	12	19	12.5	0.5	2.9
56	12	19	12	1	2.9
56	12	19	10	3	2.9
56	12	19	7	6	2.9
56	12	19	4	9	2.9

RESULTS

Particle size analysis

Some 200 g of LF slag is taken, and sieving is done for 20 min using ROTAP sieve shaker. Weight retained on each sieve is measured, and graphs are plotted for mesh No versus weight retained on each sieve and Mesh No versus Cumulative weight as shown in Figure 2. It was found that 80 per cent of slag fell in the 53–106 μm size range. This fine slag as shown in Figure 3 is very difficult to recycle, store and handle. It also increases suspended particulate matter in the shop floor.

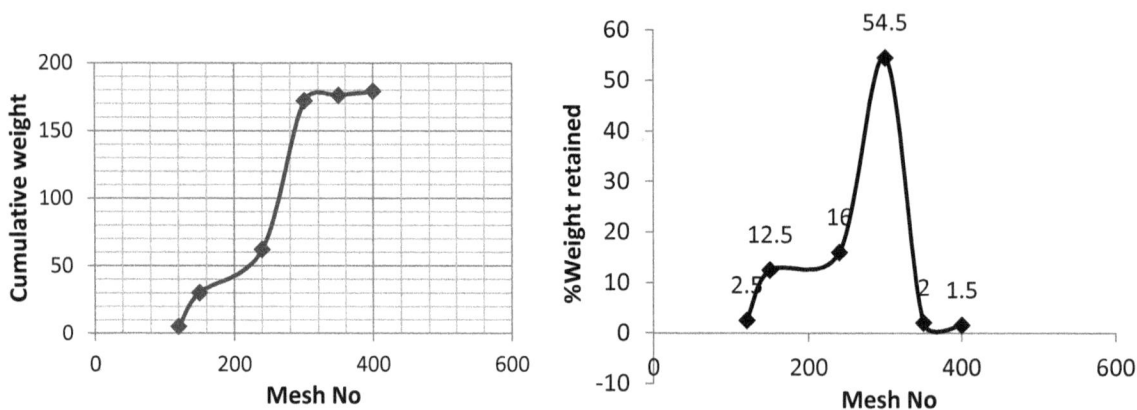

FIG 2 – Sieve analysis of as received LF slag.

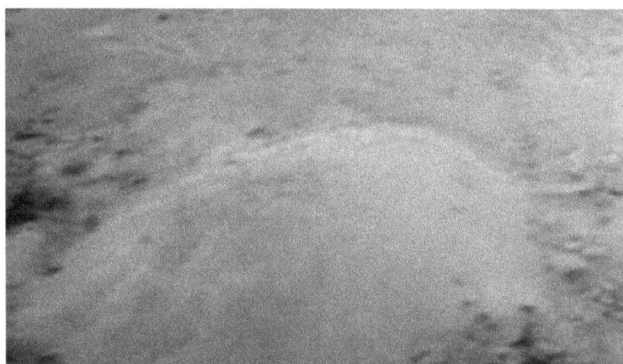

FIG 3 – As received disintegrated LF slag.

EFFECT OF B$_2$O$_3$ ON C$_2$S STABILISATION

The amount of B$_2$O$_3$ in LF slag for C$_2$S stabilisation was varied from 0.25 to 2 wt per cent. The treated slags are presented in Figure 4. After melting the slag, the slag was allowed to cool in the air to study the effect of B$_2$O$_3$ on the disintegration of LF slag. The figure reveals that with 0.25 wt per cent B$_2$O$_3$, some part of the slag disintegrated and for 100 per cent slag stabilisation minimum of 0.5 wt per cent B$_2$O$_3$ addition is required. The treated slag is hard like a rock and can be used as a replacement for natural rock for road construction purposes. This process of slag stabilisation is very economical as a very small quantity of borate is required and there is no need for an additional heat source to melt borates in slag. The addition of B$_2$O$_3$ can be done during steel refining or in slag pots before dumping the slag. The addition of borate sources needs to be optimised depending on slag composition and logistics in the steel industry. The compact slag generated is easier to handle and more environmentally friendly as it can be used as a partial replacement to lime and CaO-Al$_2$O$_3$ based synthetic slags, or else sold to the construction industry for use as a filler. However, optimisation of B$_2$O$_3$ addition should be done as there is a chance of B pickup in steel from slag.

FIG 4 – Visual appearance of B_2O_3 modified LF slag: (a) 0.25 wt per cent B_2O_3; (b) 0.5 wt per cent B_2O_3; (c) 1 wt per cent B_2O_3; (d) 2 wt per cent B_2O_3.

EFFECT OF B_2O_3 ON LF SLAG MELTING TEMPERATURE

A slag sample of 14 mm in size was prepared in the form of a cone. It is then heated in a reducing atmosphere to analyse the melting behaviour. Temperature is recorded at four points, when the rounding off of the sample tip is observed it is called deformation temperature (DT), when the height of the sample equals to width it is called softening temperature (ST), when the height of the sample is half the with it is called hemisphere temperature (HT) considered as melting temperature. And lastly, when the slag fuses and flows, it is called flow temperature (FT) as shown in Figure 5.

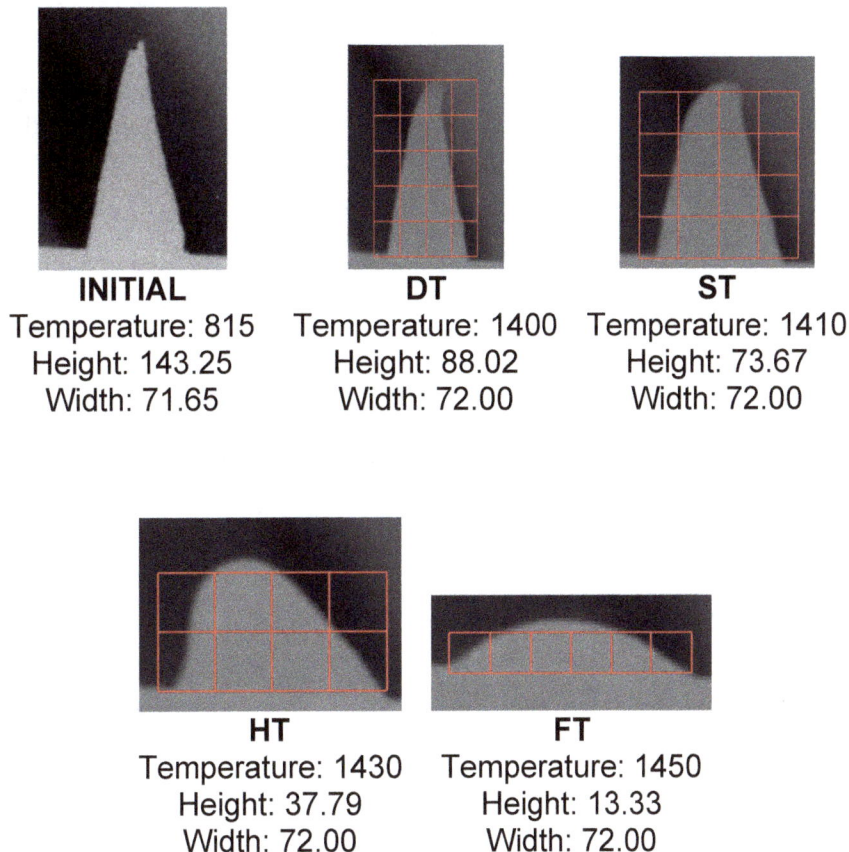

INITIAL
Temperature: 815
Height: 143.25
Width: 71.65

DT
Temperature: 1400
Height: 88.02
Width: 72.00

ST
Temperature: 1410
Height: 73.67
Width: 72.00

HT
Temperature: 1430
Height: 37.79
Width: 72.00

FT
Temperature: 1450
Height: 13.33
Width: 72.00

FIG 5 – High temperature microscopy for estimating slag melting temperature. (DT) deformation temperature; (ST) softening temperature; (HT) hemisphere temperature; (FT) flow temperature.

From lab scale experiments, we found that with 3–8 wt per cent B_2O_3 addition, the melting temperature of slag reduced from 1400 to 1110°C as shown in Figure 6. This may be due to the formation of low melting point eutectics such as $CaO.B_2O_3$, $MgO.B_2O_3$, $CaO.2B_2O_3$, etc. When the content of B_2O_3 is 3 wt per cent the melting temperature is reduced from 1380 to 1300°C. When the content of B_2O_3 is from 3–6 wt per cent there is a gradual decrease in melting temperature from 1270 to 1225°C. When the content of B_2O_3 is 6 wt per cent the slag melting temperature is 1225°C. When the content of B_2O_3 exceeds 6 wt per cent, the temperature is reduced drastically to 1110°C. Comprehensively considering the fluxing effect and cost economics, 3–4 wt per cent B_2O_3 is sufficient to induce the slag fluxing effect. As a result, the melting speed of lime can be improved, which is beneficial for steel desulfurisation and refining efficiency.

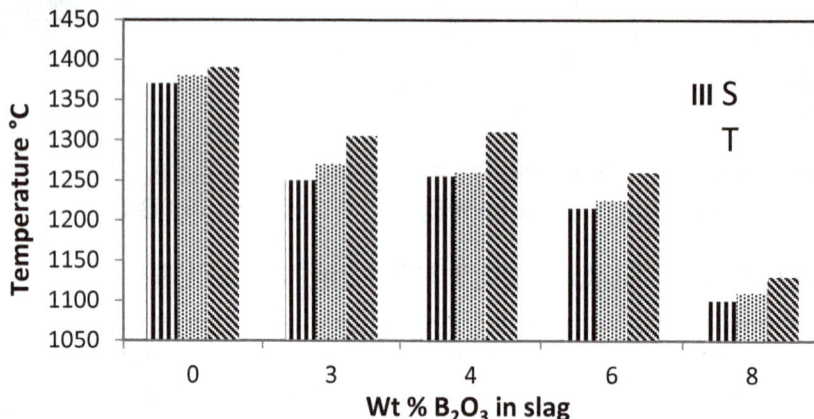

FIG 6 – Effect of B_2O_3 on LF slag melting temperature from lab scale experiments.

Effect of B_2O_3 on structural changes in LF slag

According to extensive studies, three basic units existed in the B_2O_3 bearing slag melts: boroxol ring, non-ring BO_3 triangular, and non-ring BO_4 tetrahedral, as shown in Figure 7. To investigate the influence of B_2O_3 addition on the structure and viscosity variation, FT-IR analysis was carried out. The results are shown in Figure 8. The FT-IR spectra curves were divided into three domains: the 600–800, 800–1200, and 1200–1600 cm^{-1} bands (Lai, Yao and Li, 2020) as shown in Table 3.

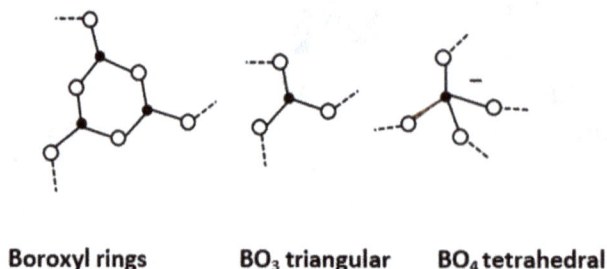

Boroxyl rings   **BO_3 triangular**   **BO_4 tetrahedral**

FIG 7 – Basic structural units of B_2O_3 in slag melts.

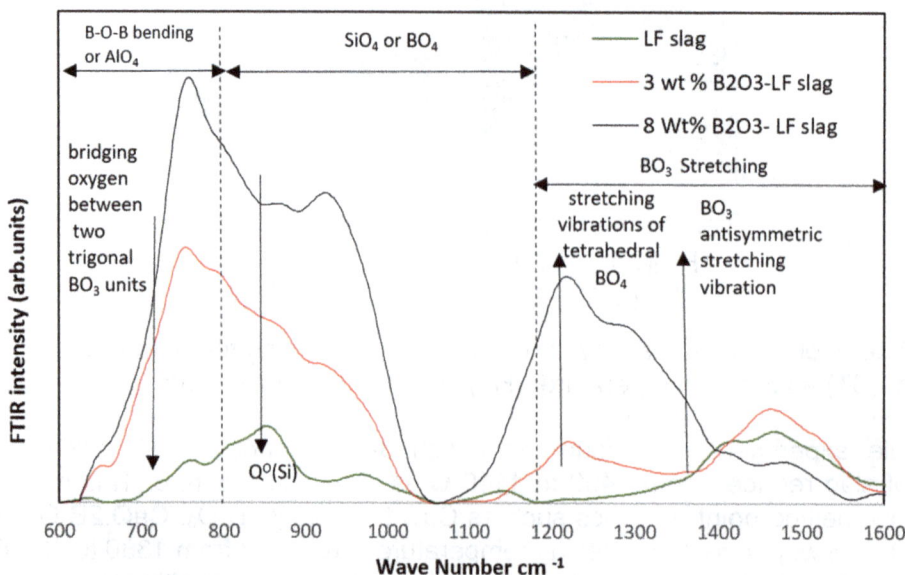

FIG 8 – FT-IR spectra of B_2O_3 added CaO-MgO-SiO$_2$-Al$_2$O$_3$ based LF slag between 600–1600 cm^{-1}.

TABLE 3

Assignments of FT-IR bands associated with B-O, Al-O, Si-O bonds (Sun *et al*, 2014; Lai, Yao and Li, 2020; Sadaf *et al*, 2020; Huang *et al*, 2014).

Wave number (cm^{-1})	FT-IR assignment
600–800	bending vibrations of $[AlO_4]^{5-}$ and B–O–B
800–1200	SiO_4, AlO_4, BO_4
1200–1600	BO_3 Stretching vibrations
~710	bending vibrations of bridging oxygen formed by two trigonal BO_3 units
~845	Q^0 (Si)
~1210	stretching vibrations of tetrahedral BO_4
~1350	BO_3 antisymmetric stretching vibration

The signal of the absorption region at 710 cm^{-1} became more pronounced as the B_2O_3 content increased, which indicated that BO_3 may be the main structural unit. It was observed that the intensity of the band centred at about 845 cm^{-1} gradually weakened with the increase of B_2O_3, which was assigned to Q^0(Si). The signal of trigonal BO_3 and BO_4 gradually became stronger with the increase of B_2O_3. It can be concluded that trigonal BO_3 and BO_4 were the main types of boron-related structural groups in the investigated LF slag.

Effect of B_2O_3 on LF slag mineralogy

The XRD of slags with different B_2O_3 from 0.5 to 8 wt per cent in LF slag is shown in Figure 9. The addition of B_2O_3 significantly influences the slag mineralogy. It depresses the formation of the crystalline phase and induces the formation of an amorphous glassy phase beyond 6 wt per cent. Similar results were reported by (Yan *et al*, 2014; Priven, 2001) and consequently improves the valourisation of LF slag where glass content is important. The major phases present in LF slag are mayenite, merwinite, gehlenite, and γ-dicalcium silicate. By addition of 0.5–2 wt per cent B_2O_3 in LF slag disintegration was minimised and slag β-dicalcium silicate was formed due to partial replacement of SiO_4^{4-} units by BO_3^{3-} units. This replacement suppresses the Ca^{2+} migrations and SiO_4^{4-} rotations required for the β to γ transformation (Seki *et al*, 1986). When B_2O_3 wt per cent was increased from 3–8 in LF slag, phases such as $CaO.B_2O_3$, $CaO.MgO.B_2O_3$ were formed. These are low melting phases and help in decreasing the melting point of slag. Analysis of mineralogical phases is given in Table 4.

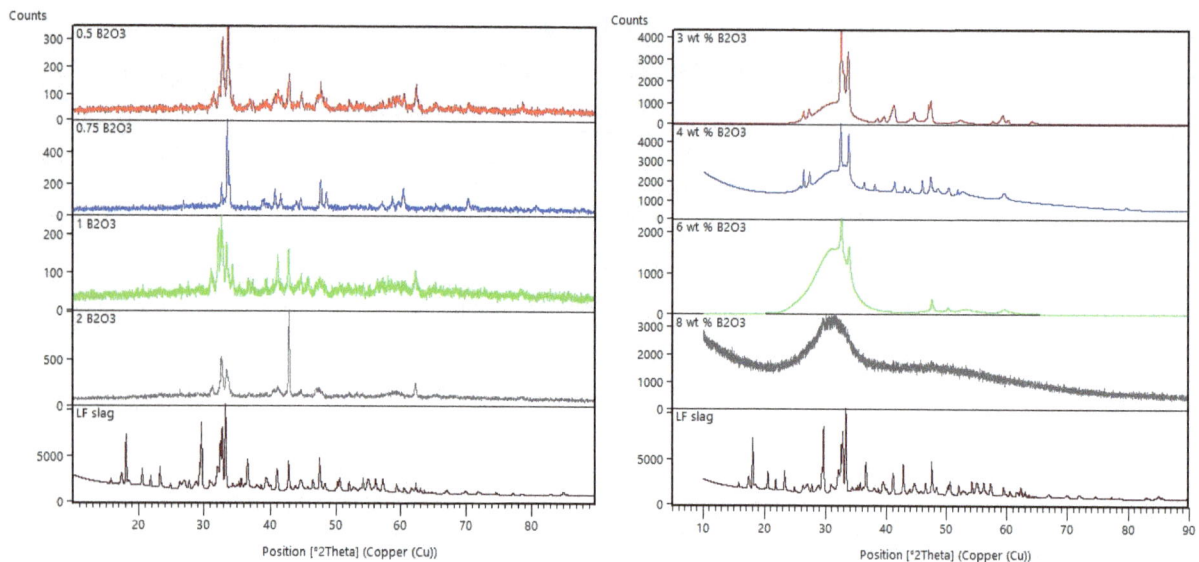

FIG 9 – XRD pattern of 0.5–8 wt per cent B_2O_3 added CaO-MgO-SiO$_2$-Al$_2$O$_3$ based LF slag.

TABLE 4

Mineralogical phases present in as received LF slag and B_2O_3 added LF slag.

Mineral phase	LF slag	LF slag stabilised with 0.5–2 wt%B_2O_3	LF slag stabilised with 3–8 wt%B_2O_3
Mayenite ($Ca_{12}Al_{14}O_{33}$)	✓	✓	✓
Merwinite ($Ca_3MgSi_2O_8$)	✓	✓	-
γ-dicalcium silicate (Ca_2SiO_4)	✓	-	-
β-dicalcium silicate (Ca_2SiO_4)	✓	✓	✓
Hatrurute (Ca_3SiO_5)	✓	✓	-
Akermanite ($Ca_2MgSi_2O_7$)	✓	✓	-
Gehlenite ($Al_2Ca_2SiO_7$)	✓	✓	-
Spinel (Mg_2SiO_4)	✓	✓	-
Periclase (MgO)	✓	-	-
Lime (CaO)	✓	-	-
calcium bisborate ($CaO.B_2O_3$)	-	-	✓
Kurchatovite ($CaO.MgO.B_2O_3$)	-	-	✓
$Ca_{11}B_2Si_4O_{22}$	-	✓	✓

FactSage™ analysis

From the analysis it was found that the main phases in LF slag when B_2O_3 is 0 wt per cent are Ca_2SiO_4, Ca_3SiO_5, and Ca_3MgAlO_4. The effect of B_2O_3 on LF slag at 500°C was studied to understand the phase changes during slag cooling as shown in Figure 10. With the addition of B_2O_3 a new phase, $Ca_{11}B_2Si_4O_{22}$ was found at 500°C. The per cent of this new phase increased with B_2O_3 content from 0.5 to 6 wt per cent and remained constant thereafter. When B_2O_3 is >6 wt per cent a new phase $Ca_3B_2O_6$ is formed. With the increase in B_2O_3 from 0.5 to 3 wt per cent, the per cent of Ca_2SiO_4 decreased from 55 to 25, and with further increase of B_2O_3, Ca_2SiO_4 phase disappeared completely at 500°C.

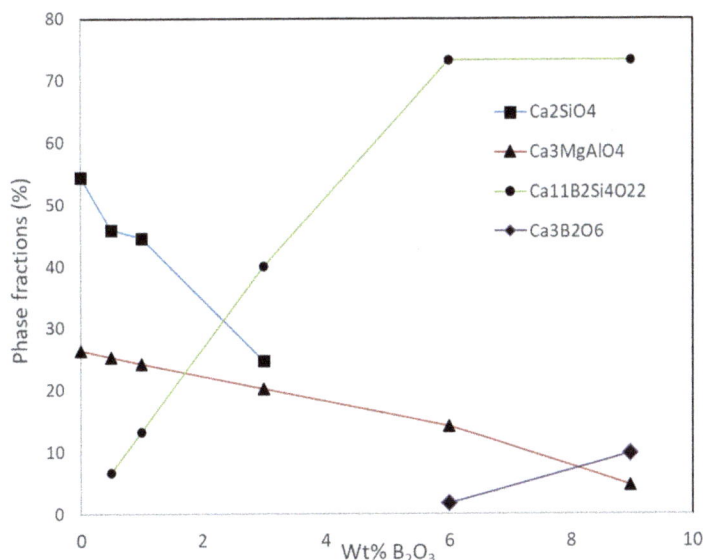

FIG 10 – Effect of B_2O_3 on phase equilibria of LF slag at 500°C.

The percentage of liquid phase slag formed with B_2O_3 addition was studied as shown in Figure 11. Analysis was carried out in slag with basicity of 2.9 by replacing equal amounts of Al_2O_3 with B_2O_3. When B_2O_3 is 0 wt per cent, the liquid phase slag started forming at 1350°C. With the addition of B_2O_3, liquid phase slag started forming at 1100°C. When 6 wt per cent Al_2O_3 is replaced with B_2O_3, liquid phase slag started forming at temperatures as low as 800°C. The solidus temperature of the slag decreased with increase in wt per cent B_2O_3 in slag. At 1550°C, around 80 per cent of slag is liquid when B_2O_3 is 9 wt per cent whereas with 0 wt per cent B_2O_3 only 68 per cent of slag is liquid. Whereas with 6 wt per cent B_2O_3 only 45 per cent of slag is liquid. Even though there is decrease in solidus temperature the per cent of liquid slag at steelmaking temperatures ie at 1550°C decreased till 6 wt per cent B_2O_3 and then increased at 9 wt per cent B_2O_3 as shown in Figure 12. This shows that different structural units are formed in slag when Al_2O_3 is replaced with B_2O_3. However, to understand the effect of B_2O_3 on slag melting temperature, further analysis needs to be carried out by varying slag basicity.

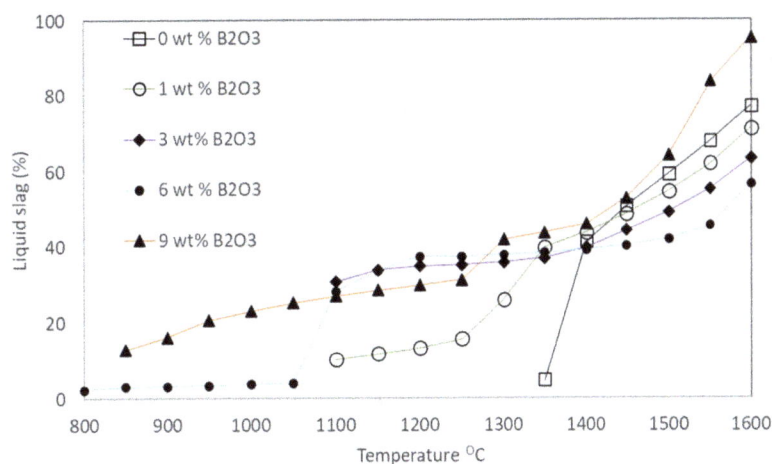

FIG 11 – Effect of replacing Al_2O_3 with B_2O_3 on percentage of liquid phase slag formed at different temperatures in 56 per centCaO, 12 per cent MgO, 19 per cent SiO_2, 13 per cent (Al_2O_3+B_2O_3)-slag wt per cent.

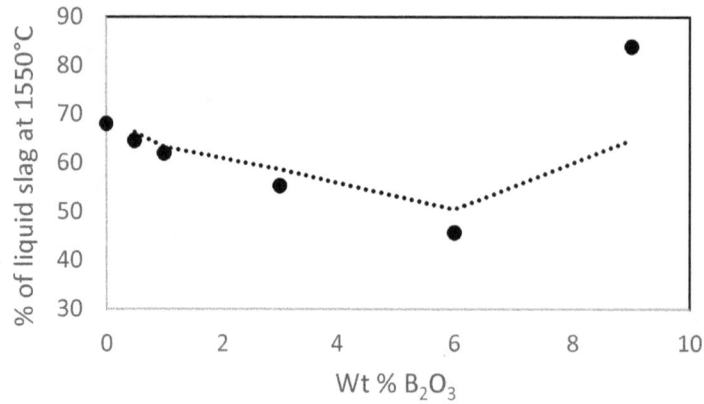

FIG 12 – Effect of replacing Al_2O_3 with B_2O_3 on percentage of liquid phase slag formed at 1550°C in 56 per centCaO, 12 per cent MgO, 19 per cent SiO_2, 13 per cent (Al_2O_3+B_2O_3) slag- wt per cent.

Effect Of B_2O_3 on LF slag viscosity, sulfide capacity, and break temperature

Several models have been developed by various researchers to estimate the sulfide capacity of slags. Earlier models were empirical and depended mainly on the optical basicity of slag and temperature. The optical basicity of various glassy and slag systems (denoted Λ) can be determined from their Pauling electronegativities. The optical basicity of a slag can be calculated from the optical basicity of individual oxides present in the slag using could be derived from Equation 1 where:

$$\Lambda = \frac{\sum(X_1 n_1 \Lambda_1 + X_2 n_2 \Lambda_2 + \cdots)}{\sum(X_1 n_1 + X_2 n_2 + \cdots)} \tag{1}$$

Where Λ is the optical basicity of the slag; Λ^{th} is the optical basicity of individual oxides as shown in Table 5; as calculated from Pauling electronegativities. X is the mole fraction of individual oxides and n is the number of oxygen atoms associated with acidic and basic oxides, respectively. The sulfide capacity of B_2O_3 based LF slags was calculated from a model developed by (Sosinsky and Sommerville, 1986) as shown in Equation 2 where C_S is the sulfide capacity of the slag:

$$\log C_S = \frac{(22690 - 54640\Lambda)}{T} + 43.6\Lambda - 25.2 \tag{2}$$

TABLE 5

Optical basicity values of individual oxides (Sosinsky and Sommerville, 1986).

Oxide	CaO	MgO	SiO_2	Al_2O_3	B_2O_3
Optical basicity	1	0.78	0.48	0.61	0.42

The sulfide capacity of CaO, MgO, SiO_2, Al_2O_3, FeO, MnO (55, 11, 19, 5–14, 1, 1 wt per cent respectively) was calculated at 1773 K by varying B_2O_3 from 0–9 wt per cent where an equal amount of B_2O_3 replaced with Al_2O_3. Sulfide capacity was also calculated at constant basicity by varying wt per cent CaO and SiO_2 (CaO/SiO_2 = 2.5) with MgO, Al_2O_3, FeO, MnO, B_2O_3 (10, 11, 1, 1, 0–9 wt per cent respectively). For both cases, it was found that the sulfide capacity of slag decreased with B_2O_3 wt per cent as shown in Figure 13 indicating that B_2O_3 is an acid oxide and decreases the sulfide capacity. Similar results were reported by (Yan *et al*, 2014).

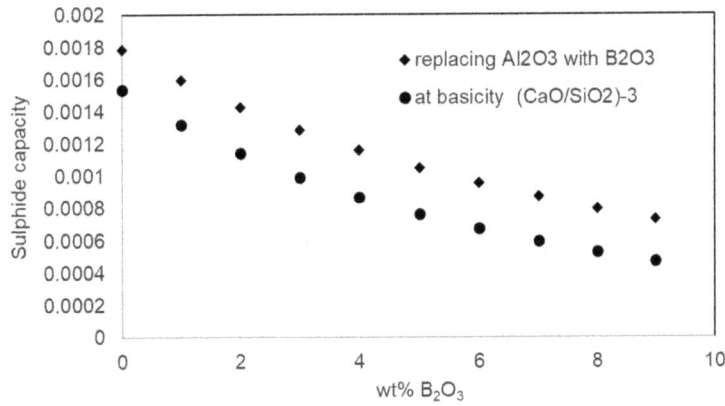

FIG 13 – Effect of B_2O_3 on sulfide capacity of CaO-MgO-SiO$_2$-Al$_2$O$_3$-FeO-MnO based LF slags.

The viscosity of slag plays an important role in kinetic conditions of steel refining particularly in desulfurisation as it is slag-metal interfacial reaction.

The viscosity of molten slag depends on the internal structure of oxide melt and is affected by changes in temperature, slag composition, and oxygen partial pressure. Many models have been developed for estimating viscosities for molten oxide slag systems. For the B_2O_3 based LF slag system, the Riboud model modified by (Wang *et al*, 2013) was used.

$$\eta = ATexp\left(\frac{B}{T}\right)$$

where:

$$A = exp[-22.47 - 2.46(X_{CaO} + X_{MgO}) + 43.07\,X_{TiO_2} + 72.61X_{B_2O_3} + 7.02(X_{Na_2O} + X_{K_2O}) - 35.76X_{Al_2O_3}]$$

$$B = 34428 - 7342(X_{CaO} + X_{MgO}) - 84121X_{TiO_2} - 130586X_{B_2O_3} - 39159(X_{Na_2O} + X_{K_2O}) + 68833X_{Al_2O_3}$$

T   is temperature in K

η   is viscosity in Pa.S

The viscosity of CaO, MgO, SiO$_2$, Al$_2$O$_3$ (60, 10, 17, 4–13 wt per cent respectively) based slag was calculated by varying B_2O_3 from 0–9 wt per cent, where an equal amount of B_2O_3 is replaced with Al$_2$O$_3$. Viscosity was also calculated at constant basicity varying wt per cent CaO and SiO$_2$ (CaO/SiO$_2$ = 2.6) with MgO, Al$_2$O$_3$, B_2O_3 (10, 10, 0–9 wt per cent respectively). For both cases, the viscosity of slags decreased with increasing wt per cent B_2O_3 as shown in Figure 14. This confirms that B_2O_3 can be used as a slag fluxing agent despite being an acidic oxide. This may be due to the formation of the BO$_3$ planar triangular structure. Similar results are reported by (Wang *et al*, 2006, 2015).

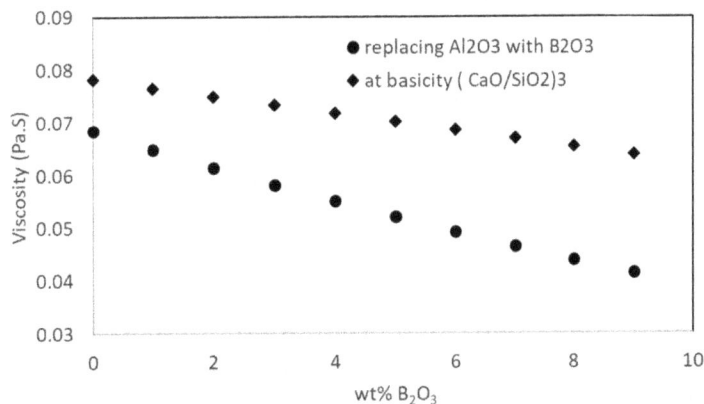

FIG 14 – Effect of B_2O_3 on viscosity of CaO-MgO-SiO$_2$-Al$_2$O$_3$ based LF slags.

The break temperature is the temperature below which there is a dramatic increase in viscosity and the slag becomes non-Newtonian in behaviour. The addition of B_2O_3 plays an important role in affecting the break temperature. Break temperature (T_{br}) of the LF type slags with CaO/SiO_2 = 3, Al_2O_3, B_2O_3 (15, 0–9 mol per cent respectively) were calculated using a model developed by (Huang et al, 2014) as shown in Equation 3 where T_{br} is in K. The break temperature reduced with an increase in B_2O_3 per cent as shown in Figure 15.

$$T_{br} = 1502.73 + 0.56\% \text{ CaO} + 0.54\% \text{ SiO}_2 - 1.41\% \text{ Al}_2\text{O}_3 - 4.49\% \text{ Na}_2\text{O} - 15.3\% \text{ B}_2\text{O}_3 \qquad (3)$$

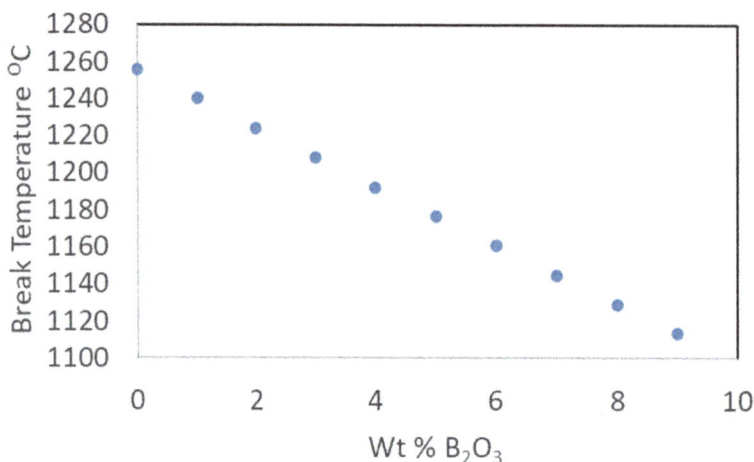

FIG 15 – Effect of B_2O_3 on break temperature of CaO, SiO_2, Al_2O_3 LF slag system.

CONCLUSIONS

- From the study, it was found that B_2O_3 can act as a slag stabiliser and slag fluxing agent.

- Around 0.5–1 wt per cent B_2O_3 in the studied slag system is sufficient to prevent di-calcium silicate-driven disintegration as only β-C_2S was present in B_2O_3 added LF slag. By increasing B_2O_3 wt per cent in LF slag, the melting point reduced from 1380°C to 1100°C with 8 wt per cent B_2O_3.

- From XRD analysis it was found that the slag slowly transformed from crystalline to amorphous/glassy phase after 6 wt per cent B_2O_3 addition and melting temperature reduced drastically beyond 6 wt per cent B_2O_3. Low melting phases such as CaO.B_2O_3 and CaO.2B_2O_3 were also formed in investigated LF slag. Considering cost economics, and requirements in steel desulfurisation 3–5 wt per cent B_2O_3 may be sufficient to induce a fluxing effect in LF slag.

- FT-IR analysis also confirmed the fluxing effect of B_2O_3 as trigonal BO_3 was the main type of boron-related structural group present in the investigated LF slag. However, the effect of B_2O_3 on the properties of slag depend on CaO/SiO_2 and CaO/Al_2O_3 ratios.

- FactSage™ analysis also confirmed the formation of liquid slag phases at 800°C when B_2O_3 is >6 wt per cent. New phases such as $Ca_{11}B_2Si_4O_{22}$ and $Ca_2B_3O_5$ were formed in slag. At 500°C when B_2O_3 is >3 wt per cent, Ca_2SiO_4 phase disappears proving that B_2O_3 helps in preventing slag disintegration. The solidus temperature of slag decreased by replacing equal amounts of Al_2O_3 with B_2O_3.

- From empirical models, it was found that with B_2O_3 addition the sulfide capacity, viscosity, and break temperature of LF slag decreased. As desulfurisation is a kinetic phenomenon compromise should be made between sulfide capacity and slag fluidity for effective interfacial chemical reactions.

- Further studies need to be carried out to optimise B_2O_3 content for different LF slag systems considering the slag fluxing effect, sulfide capacity, and effect on basic refractory. However, B_2O_3 additions should be optimised based on slag composition and process parameters by carrying out industrial trials.

- There is also a need for the development of a reliable model for predicting viscosity, melting temperature, sulfide capacity, and break temperature of CaO, MgO, SiO$_2$, Al$_2$O$_3$, B$_2$O$_3$ based LF slag systems.

REFERENCES

Babenko, A A, Istomin, S A, Zhuchkov, V I, Sychev, A V, Ryabov, V V and Upolovnikova, A G, 2017. Construction of viscosity diagrams for CaO-SiO2-Al2O3-8% MgO-4% B2O3 slags by the simplex lattice method, *Russian Metallurgy (Metally)*, 2017:372–375. https://doi.org/10.1134/S0036029517050020

Babenko, A A, Smetannikov, A N, Zhuchkov, V I and Upolovnikova, A G, 2019. Influence of B2O3 and Basicity of CaO-SiO2-B2O3-Al2O3 Slag on the Saturation Concentration of Magnesium Oxide, *Steel in Translation*, 49:87–90. https://doi.org/10.3103/S0967091219020037

Bi, Z, Li, K, Jiang, C, Zhang, J, Ma, S, Sun, M, Wang, Z and Li, H, 2021. Effects of B2O3 on the structure and properties of blast furnace slag by molecular dynamics simulation, *J Non Cryst Solids*, 551. https://doi.org/10.1016/j.jnoncrysol.2020.120412

Gollapalli, V, Tadivaka, S R, Borra, C R, Varanasi, S S, Karamched, P S and Venkata Rao, M B, 2020. Investigation on Stabilization of Ladle Furnace Slag with Different Additives, *Journal of Sustainable Metallurgy*, 6:121–131. https://doi.org/10.1007/s40831-020-00263-w

Huang, X H, Liao, J L, Zheng, K, Hu, H H, Wang, F M and Zhang, Z T, 2014. Effect of B2O3 addition on viscosity of mould slag containing low silica content, *Ironmaking and Steelmaking*, 41:67–74. https://doi.org/10.1179/1743281213Y.0000000107

Lai, F, Yao, W and Li, J, 2020. Effect of B2O3 on Structure of CaO-Al2O3-SiO2-TiO2-B2O3Glassy Systems, *ISIJ International*, 60:1596–1601. https://doi.org/10.2355/isijinternational.ISIJINT-2019-679

Pontikes, Y, Jones, P T, Geysen, D and Blanpain, B, 2010. Options to prevent dicalcium silicate-driven disintegration of stainless steel slags, *Archives of Metallurgy and Materials*, 55:1167–1172. https://doi.org/10.2478/v10172-010-0020-6

Priven, A I, 2001. Calculation of the viscosity of glass-forming melts: VI, The R2O-B2O3-SiO2 and RO-B2O3-SiO2 ternary borosilicate systems, *Glass Physics and Chemistry*, 27:360–370. https://doi.org/10.1023/A:1011372328059

Sadaf, S, Wu, T, Zhong, L, Liao, Z and Wang, H C, 2020. Effect of basicity on the structure, viscosity and crystallization of CaO-SiO2-B2O3 based mold fluxes, *Metals (Basel)*, 10:1–12. https://doi.org/10.3390/met10091240

Seki, A, Aso, Y, Okubo, M, Sudo, F and Ishizaka, K, 1986. Development of Dusting Prevention Stabilizer for Stainless Steel Slag, Kawasaki Steel Technical Report.

Sosinsky, D J and Sommerville, I D, 1986. The Composition and Temperature Dependence of the Sulfide Capacity of Metallurgical Slags, *Metallurgical and Materials Transactions B: Process Metallurgy and Materials Processing Science*, 17:331–337.

Sun, Y, Liao, J, Zheng, K, Wang, X and Zhang, Z, 2014. Effect of B2O3 on the Structure and Viscous Behavior of Ti-Bearing Blast Furnace Slags, *JOM*, 66:2168–2175. https://doi.org/10.1007/s11837-014-1087-8

Varanasi, S S, More, V M R, Rao, M B V, Alli, S R, Tangudu, A K and Santanu, D, 2019a. Recycling Ladle Furnace Slag as Flux in Steelmaking: A Review, *Journal of Sustainable Metallurgy*. https://doi.org/10.1007/s40831-019-00243-9

Varanasi, S S, Pathak, R K, Sahoo, K K, More, V M R, Santanu, D and Alli, S R, 2019b. Effect of CaO-Al2O3-Based Synthetic Slag Additions on Desulphurisation Kinetics of Ladle Furnace Refining, *Transactions of the Indian Institute of Metals*, 72:1447–1452. https://doi.org/10.1007/s12666-019-01616-0

Varanasi, S S, Rao, M V M, Santanu, D, Alli, R, Kumar, D S, Tangudu, A K, Gollapalli, V, Pathak, R K and Santhamma, C S, 2022. Effect of recycling ladle furnace slag as flux on steel desulphurization during secondary steel making secondary steel making, *Ironmaking and Steelmaking*, 8 p. https://doi.org/10.1080/03019233.2022.2060459

Wang, H M, Li, G R, Dai, Q X, Li, B, Zhang, X J and Shi, G M, 2007. CAS-OB refining: Slag modification with B2O3-CaO and CaF2-CaO, *Ironmaking and Steelmaking*, 34:350–353. https://doi.org/10.1179/174328107X155277

Wang, H M, Li, G R, Li, B, Zhang, X J and Yan, Y Q, 2010a. Effect of B2O3 on Melting Temperature of CaO-Based Ladle Refining Slag, *Journal of Iron and Steel Research International*, 17:18–22. https://doi.org/10.1016/S1006-706X(10)60177-X

Wang, H M, Li, G R, Li, B, Zhang, X J and Yan, Y Q, 2010b. Effect of B2O3 on Melting Temperature of CaO-Based Ladle Refining Slag, *Journal of Iron and Steel Research International*, 17:18–22. https://doi.org/10.1016/S1006-706X(10)60177-X

Wang, H M, Li, G, Dai, Q, Lei, Y, Zhao, Y, Li, B, Shi, G and Ren, Z, 2006. Effect of Additives on Viscosity of LATS Refining Ladle Slag, *ISIJ International*, 46(5):637–640. https://doi.org/10.2355/isijinternational.46.637

Wang, H M, Li, P S, Li, G R, Zhang, M, Zhao, Z and Zhao, Y, 2012. The melting temperature of CaO-(2CaO·SiO2)-B2O3-SiO2-(Al2O3) slag system, *App Mech Mat*, pp 35–38. https://doi.org/10.4028/www.scientific.net/AMM.217-219.35

Wang, H, Yang, L, Zhu, H and Yan, Y, 2011a. Comparison of effects of B2O3 and CaF2 on metallurgical properties of high basicity CaO-based flux, *Adv Mat Res*, pp 966–969. https://doi.org/10.4028/www.scientific.net/AMR.311-313.966

Wang, H, Yang, L, Zhu, H and Yan, Y, 2011b. Comparison of effects of B2O3 and CaF2 on metallurgical properties of high basicity CaO-based flux, *Advanced Materials Research*, 311–313:966–969. https://doi.org/10.4028/www.scientific.net/AMR.311-313.966

Wang, H, Zhang, T, Zhu, H, Li, G, Yan, Y and Wang, J, 2011c. Effect of B2O3 on Melting Temperature, Viscosity and Desulfurization Capacity of CaO-based Refining Flux, *ISIJ International*.

Wang, H, Zhang, T, Zhu, H, Li, G, Yan, Y and Wang, J, 2011d. Effect of B2O3 on melting temperature, viscosity and desulfurization capacity of CaO-based refining flux, *ISIJ International*, 51:702–706. https://doi.org/10.2355/isijinternational.51.702

Wang, L, Cui, Y, Yang, J, Zhang, C, Cai, D, Zhang, J, Sasaki, Y and Ostrovski, O, 2015. Melting Properties and Viscosity of SiO2-CaO-Al2O3-B2O3 System, *Steel Res Int*, 86:670–677. https://doi.org/10.1002/srin.201400353

Wang, Z, Shu, Q and Chou, K, 2013. Viscosity of Fluoride-Free Mold Fluxes Containing B2O3 and TiO2, *Steel Res Int*, 84:766–776. https://doi.org/10.1002/srin.201200256

Yan, P, Nie, P, Huang, S, Blanpain, B and Guo, M, 2014. Sulphide capacity and mineralogy of BaO and B2O3 modified CaO-Al2O3 top slag, *ISIJ International*, 54:1570–1577. https://doi.org/10.2355/isijinternational.54.1570

Yan, W, Chen, W, Yang, Y, Lippold, C and McLean, A, 2016. Evaluation of B_2O_3 as replacement for CaF_2 in CaO-Al_2O_3 based mould flux, *Ironmaking and Steelmaking*, 43:316–323. https://doi.org/10.1179/1743281215Y.0000000062

Zhang, P, Liu, J, Wang, Z, Qian, G and Ma, W, 2019. Effect of B2O3 Addition on Electrical Conductivity and Structural Roles of CaO-SiO2-B2O3 Slag, *Metallurgical and Materials Transactions B: Process Metallurgy and Materials Processing Science*, 50:304–311. https://doi.org/10.1007/s11663-018-1472-8

Zhuchkov, V I, Salina, V A and Sychev, A V, 2019. The Study of the Process of Metal-Thermal Reduction of Boron from the Slag of the System CaO-SiO2-MgO-Al2O3-B2O3, *Materials Science Forum* (Trans Tech Publications Ltd), pp 423–429. https://doi.org/10.4028/www.scientific.net/MSF.946.423

Slag-steel reactions in the refining of advanced high-strength steel

P Su[1], P C Pistorius[2] and B A Webler[3]

1. PhD candidate, Carnegie Mellon University, Pittsburgh PA 15213, USA.
   Email: panwens@andrew.cmu.edu
2. POSCO Professor, Carnegie Mellon University, Pittsburgh PA 15213, USA.
   Email: pistorius@cmu.edu
3. Professor, Carnegie Mellon University, Pittsburgh PA 15213, USA. Email: webler@cmu.edu

ABSTRACT

Advanced high-strength steels (AHSS) typically have much higher aluminium concentrations (by an order of magnitude or more) than conventional low-carbon aluminium-killed steels. The resulting lower oxygen activity at the steel-slag interface changes the kinetics and thermodynamics of steel-slag reactions. Previous work showed that the rapid transformation of alumina inclusions to spinel inclusions, and spinel to periclase, occurs because of the relatively high concentration of dissolved magnesium. In this paper, experimental results on nitrogen removal by ladle slag are compared with predictions based on the available thermodynamic databases. As in previous work, the kinetics was modelled by assuming mass transfer control, with steel or slag mass transfer limiting. The results show that significant removal of nitrogen by steel-slag reaction is possible.

INTRODUCTION

The third-generation advanced high-strength steels (3rd GEN AHSS), with ≥ 1 GPa tensile strength and ~30 per cent elongation (World Steel Association, 2021), contain intermediate levels of alloying elements (0.05 to 0.5 per cent C, 0–4 per cent Al, 0–12 per cent Mn and 0–4 per cent Si) (Tang and Pistorius, 2021). Nitrogen control is important for steel quality: Al, Si and Mn are strong nitride formers (Paek $et\ al$, 2016). The resulting high solubility of nitrogen in these steels precludes nitrogen removal by vacuum degassing (Tang and Pistorius, 2022). A method was proposed to remove nitrogen by intentionally forming AlN precipitates that can be removed in the slag (Tada and Matsumura, 2011), but the AlN precipitates may redissolve before leaving the liquid steel. Instead, in the work presented here, the focus is on slag-based nitrogen removal from liquid 3rd GEN AHSS, using conventional calcium aluminate ladle slag. To study nitrogen removal, thermodynamic and kinetics calculations with FactSage 8.1 (Bale $et\ al$, 2016) and laboratory experiments with liquid steel and slag were employed.

For the conditions in this work (steel, not carbon saturated, temperature around 1600°C, CaO-rich slag), nitrogen can dissolve in the slag by the following equations (Jung, 2006):

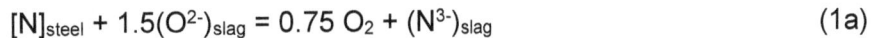

$$[N]_{steel} + 1.5(O^{2-})_{slag} = 0.75\ O_2 + (N^{3-})_{slag} \tag{1a}$$

$$[Al]_{steel} + 0.75\ O_2 = 0.5(Al_2O_3)_{slag} \tag{1b}$$

Combination of Equations (1a) and (1b) gives the following net reaction:

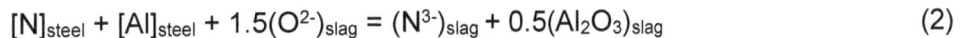

$$[N]_{steel} + [Al]_{steel} + 1.5(O^{2-})_{slag} = (N^{3-})_{slag} + 0.5(Al_2O_3)_{slag} \tag{2}$$

Equation (2) illustrates why nitrogen removal by slag from AHSS steels might be feasible: the high aluminium concentration in these steels (one to two orders of magnitude higher than in conventional low-carbon aluminium-killed steel) would drive the nitrogen removal reaction to the right.

The high aluminium concentration in such steels leads to rapid transformation of alumina (initial deoxidation production), to spinel (approximately $MgAl_2O_4$) and periclase (MgO). The rapid transformation results from the relatively high concentration of dissolved magnesium (several tens of parts per million) at the steel-slag interface (Tang and Pistorius, 2021).

CALCULATIONS AND EXPERIMENTAL PROCEDURE

Kinetics calculations

The expected rate of steel-slag reactions – for the conditions of the laboratory trials – was calculated based on the assumption that mass transfer in the steel or slag (to the steel-slag interface) was rate-controlling, with local equilibrium at the steel-slag interface. The calculations were performed with FactSage macros, based on the effective equilibrium reaction zone (EERZ) approach (Van Ende *et al*, 2011): within a chosen time interval, portions of slag and steel are transported to the slag-steel interface to equilibrate; the amount of the phase reacting is equal to the product of the time interval, the effective mass transfer coefficient, the projected slag-steel area and the density of the phase (Pistorius and Vermaak, 1999). After equilibration, the reaction products are mixed back into the slag and steel and each phase is homogenised. Gaseous reactions were not considered, given the low rate of gas-based removal of nitrogen from AHSS (Tang and Pistorius, 2022). The databases and parameters employed for the FactSage simulations are listed in Table 1. The initial slag and steel compositions for the simulations and for the laboratory slag-steel reactions are given in Tables 2 and 3. The slag compositions were chosen to be close to double saturation with both periclase and lime (based on FactSage calculations), with different initial SiO_2 concentrations. The FTOxCN slag database was found to agree with previous experimental results for slag-based nitrogen removal (Jung, 2006) (albeit not for higher-Al steels as considered in this work).

TABLE 1

Solution models and simulation parameters for modelling slag-steel reactions.

Phase		Solution model
Liquid steel		FTmisc FeLQ
Slag		FTOxCN-slag
Solid oxides		FToxid-A-monoxide; FToxid-B-spinel; FToxid-corundum; FToxid-a-$(Ca,Sr)_2SiO_4$
Simulation parameters	Slag density, ρ_{slag}	2500 kg/m^3
	Steel density, ρ_{steel}	7000 kg/m^3
	Slag-to-steel mass ratio	1:5 to 1:12
	Steel mass transfer coefficient, m_{steel}	3.1×10^{-5} m/s (Piva and Pistorius, 2021)
	Slag mass transfer coefficient, m_{slag}	$0.1 m_{steel}$

TABLE 2

Initial steel compositions for simulations and steel-slag reactions (mass percentages).

Version	Fe	Al	C	Mn	N	O	S	Si
High Al	Balance	0.83	0.2	2.1	0.002-0.02	0.01	~0.001	0.45
Low Al	Balance	0.087	0.2	2.1	0.002-0.02	0.01/0.15	~0.001	0.45

TABLE 3

Initial slag compositions and temperatures for experiments and simulations, with the experimental steel-to-slag mass ratio and the estimated increase in slag area by emulsification (A/A_0).

Experiment	T (°C)	%Al$_2$O$_3$	%CaO	%MgO	%SiO$_2$	W_{steel}/W_{slag}	A/A_0
E1	1550	35	54	6	5	12.8	5
E2	1600	42.3	51.2	6.5	0	6.3	2
E3	1600	36.9	53.5	5.6	4	4.6	2
E4	1600	35	54	6	5	5.0	1.5
E5	1600	42.3	51.2	6.5	0	5.3	2
E6	1600	42.3	51.2	6.5	0	5.0	5

Note: Experiments E1 to E5 used the higher-Al steel (0.83 per cent Al before reaction), while E6 used the lower-Al steel (0.087 per cent Al before reaction).

For reactions involving substantial transfer of oxygen between slag and steel (as in this work, due to reduction of SiO$_2$ from the slag, by reaction with Al in the metal), emulsification can be expected at the slag-steel interface (Riboud and Lucas, 1981; Assis *et al*, 2015; Song *et al*, 2021). To approximate the observed time constants of the steel-slag reactions, it was necessary to increase the interfacial area in the simulations (to be larger than the cross-sectional area of the crucible); the extent of this increase is given by the column 'A/A_0' in Table 3.

To match the final Al and Si concentrations in the steel, it was needed to add additional oxygen to the initial steel composition, for the kinetic simulations. This reflects entry of oxygen into the furnace chamber when the slag is added. The total oxygen load in the steel at the start of steel-slag reaction was taken to be 2000 ppm for the results shown here.

Slag-steel reactions

Experiments were performed in an induction furnace, as described previously (Mu *et al*, 2018; Piva, Kumar and Pistorius, 2017; Roy, Pistorius and Fruehan, 2013; Song *et al*, 2021). These experiments were performed in slip-cast MgO crucibles (OD = 64 mm, ID = 56 mm, height = 138 mm), using approximately 600 g of metal and 120 g of slag, in a high-purity argon atmosphere (flow rate ~0.6 dm^3/min at room temperature and ambient pressure). Raw materials for the metal were placed into the crucible before the experiment. Electrolytic iron (with ~400 ppm O, based on previous work (Piva, Kumar and Pistorius, 2017)), Al shot, graphite powder, electrolytic Mn pieces, pure Si pieces and crushed pieces of nitrided electrolytic manganese were used to make up the metal composition. Although the nominal nitrogen concentration in the nitrided manganese was approximately 6 per cent, the actual nitrogen at melt-in varied (likely because of inhomogeneity of the manganese briquettes), as the results show.

Slag for the experiments was prepared from pure oxides that were mixed, pressed into pellets and then premelted in a graphite crucible under argon (heating to and holding at approximately 1600°C for 10 mins). After cooling to room temperature, the graphite crucible was broken to remove the premelted slag. The slag was broken up and ground using a tungsten carbide puck mill. Following grinding, the slag was decarburised in air at 1000°C for 24 hrs. The decarburised slag powder was then pelletised and sintered at 1000°C for 12 hrs in an alumina crucible (62×56×60 mm) before the slag-steel experiments.

The slag-steel reactions were conducted in an induction furnace (maximum power = 10 kW). The MgO crucible containing the steel was placed in a graphite susceptor (OD = 71.5 mm, ID = 65.5 mm, height = 160 mm, bottom thickness = 15 mm) on an insulating alumina pedestal. Alumina felt was used as thermal insulation around the outside of the susceptor. A disc cut from porous alumina brick was used as a radiation shield on top of the crucible (with holes through which the sheath of the upper thermocouple and the feeding tube passed). The atmosphere was controlled by enclosing the susceptor in a fused-quartz tube (80 mm ID, 85 mm OD, 430 mm long), with a water-cooled stainless-steel end cap sealing onto Viton gaskets at each end of the fused-quartz tube. Temperature was monitored with both upper and lower B-type thermocouples, sheathed with 6.35 mm OD alumina tubes. The sheathed tip of the bottom thermocouple (B-type) ended in a shallow hole (7.5 mm diameter, 9 mm deep) in the bottom centre of the graphite susceptor. The offset between the reading of the bottom thermocouple and the interior of the MgO crucible was measured with an empty crucible and found to be approximately 150°C. The tip of the upper thermocouple was placed approximately 5 mm above the estimated top of the slag layer. An alumina feeding tube passed through the upper end cap and was sealed with a silicone rubber stopper when slag was not being fed into the crucible.

After charging the crucible and sealing the working tube with the end caps, the crucible with the steel mixture was heated to the experimental temperature (1600°C for all but the first case; see Table 3) at a rate of approximately 30 K/min, by manually adjusting the power of the induction supply. The steel was held for 10 mins for melting and homogenisation, before taking the first steel sample.

Steel rod samples were taken by inserting the tip of a fused-quartz tube (4 mm ID, 6.35 mm OD, length 600 mm) through the feeding tube into the crucible and using a manual pipette pump to draw liquid steel into the fused-quartz tube.

After taking the first steel sample, slag pieces (diameter ≤ 3 mm) were added to the crucible through the unplugged feeding tube. The addition took several minutes, during which time the Ar flow was turned off to avoid slag powder blowing out of the feeding tube. During slag addition, oxygen would have entered the working tube, affecting the aluminium balance (as mentioned earlier). After slag addition, the feeding tube was sealed with the silicone stopper and Ar flow restarted at 0.6 dm^3/min. The system was subsequently held for ~10 mins for complete melting of the slag, before taking the first sample after slag addition. Several samples were taken subsequently while maintaining a constant temperature, for total times up to 3 hrs.

The nitrogen concentrations in the steel rod samples were analysed by IGA (instrumental gas analysis) at an external laboratory. The steel remaining in the crucible was analysed at another laboratory, using spark optical emission spectroscopy (OES) for all elements other than nitrogen and IGA for nitrogen. Slag was manually separated from the crucible, crushed and analysed by X-ray fluorescence (for all except experiment E6). Given uncertainties around the analysis of nitrogen in slag, the analysed nitrogen concentration in the steel was used to calculate the nitrogen distribution coefficient between the steel and slag, as follows:

The nitrogen distribution coefficient is given by:

$$L_N = (\%N)_{slag}/[\%N]_{steel,} \tag{3}$$

where $(\%N)_{slag}$ is the mass percentage of nitrogen in the slag, and $[\%N]_{steel}$ is the mass percentage of nitrogen in the steel. From a simple mass balance, based on the assumptions that the steel and slag are homogeneous, and that the slag contained no nitrogen before reaction, the nitrogen distribution coefficient after the steel-slag reaction can be calculated as follows from the measured steel compositions:

$$L_N = ([\%N]_i/[\%N]_f - 1) \, (W_{steel}/W_{slag}), \tag{4}$$

where $[\%N]_i$ is the nitrogen concentration in the steel before reaction, $[\%N]_f$ the nitrogen concentration after reaction, and W_{steel} and W_{slag} are the total masses of steel and slag.

Microscopy

Selected steel and slag samples were examined by scanning electron microscopy after the experiments. To test for possible AlN formation, a steel rod sample (N content = 190 ppm as given by IGA analysis) was mounted in conductive bakelite and polished to a 1 μm finish with diamond. Possible slag-crucible reactions were examined by mounting a piece of crucible (with attached slag) in cold-mounting epoxy resin, followed by grinding and polishing (with diamond) while avoiding any contact with water. Samples were examined by back-scattered electron imaging, using an accelerating voltage of 10 kV. The slag sample was carbon-coated before microscopy.

RESULTS AND DISCUSSION

Extent of steel-slag reactions

The time variation of %N in steel, for both FactSage simulations and experiments, are given in Figure 1; the measured steel and slag compositions after reaction are given in Tables 4 and 5. The measured MgO concentration in the samples of runs E2-E4 was much higher than expected for MgO-saturated slags; a likely reason is incomplete separation of slag from the MgO crucible before the slag was crushed for X-ray fluorescence (XRF) analysis.

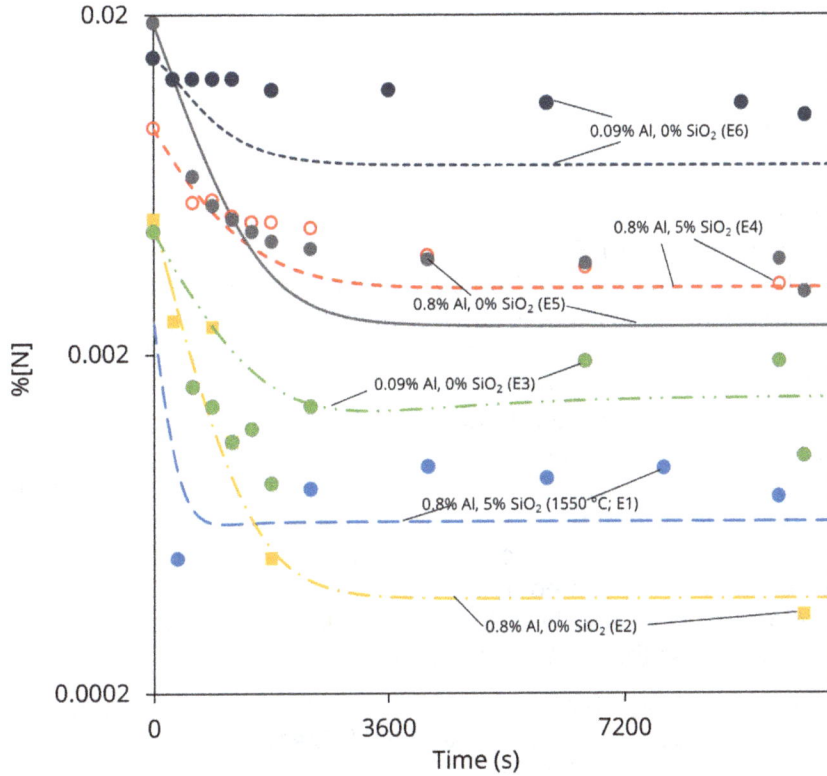

FIG 1 – Change in [%N] over time, for slag-steel experiments. Markers show the experimental results and the lines the results of FactSage simulations.

TABLE 4

The final steel composition after slag-steel experiments (nitrogen measured by IGA; other elements by spark OES; mass concentrations).

Experiment	Al%	C%	Mn%	Mg (ppm)	Si%	N (ppm)
E1	0.33	0.19	2.0	43	0.63	4.1
E2	0.41	0.15	1.9	49	0.53	3.4
E3	0.11	0.14	2.0	15	0.78	10
E4	0.36	0.20	2.4	71	0.58	47
E5	0.48	0.19	2.6	44	0.54	31
E6	0.06	0.20	2.4	12	0.37	101

TABLE 5

Initial slag compositions (double-saturated with CaO and MgO, based on FactSage) and the actual final slag compositions as analysed by XRF.

Slag	Al_2O_3%		CaO%		MgO%		SiO_2%	
	Initial	Final	Initial	Final	Initial	Final	Initial	Final
E1	35.6	42.53	53.95	48.66	5.45	7.72	5	1.09
E2	41.81	42.9	51.04	45.72	7.15	10.41	0	0.97
E3	36.45	33.6	53.26	40.61	6.29	24.97	4	0.81
E4	35.15	29.04	53.71	39.29	6.14	30.4	5	1.27
E5	41.81	43.98	51.04	49.42	7.15	6.61	0	0

Note that a large decrease in aluminium concentration occurred (from around 0.9 per cent to 0.4 per cent, for experiments E1 to E5), with a slight increase in silicon concentration in the steel (from 0.45 per cent to around 0.6 per cent on average) and greatly reduced silica concentration in the slag. However, the decrease in [%Al] is larger than can be accounted for just by the reduction of SiO_2: From stoichiometry, the reaction of 4 moles of Al with 3 moles of SiO_2 would yield 3 moles of Si (and 2 moles of Al_2O_3), with result that the ratio of the change in [%Al] to the change in [%Si] would be -1.3. The larger observed ratio of the aluminium to silicon changes (around -3) indicates that additional oxygen entered the experiment, causing additional loss of aluminium from the steel; this is also shown by the decrease of both [%Al and [%Si] in the low-Al experiment (E6). As mentioned earlier, oxygen likely entered during slag addition. The resulting loss of aluminium would have decreased the extent of nitrogen removal (as shown by Equation 2). (As Equation 2 indicates, some Al would also have been consumed by the nitrogen removal reaction, but this amount is small: given the 1:1 stoichiometry of Al and N in the reaction, removal of up to 200 ppm N would have resulted in consumption of less than 400 ppm of Al.)

Despite the loss of aluminium, substantial nitrogen removal did occur, as illustrated by Figure 1 and by the summary of nitrogen distribution coefficients in Table 6. The extent of nitrogen removal was insignificant only in the case of the low-Al control experiment (E6; distribution coefficient approximately 2). In the cases with higher final [%Al], the L_N values were much larger.

TABLE 6

Nitrogen distribution coefficients between slag and steel, as predicted from FactSage and observed in experiments.

Experiment	[N]$_{initial}$, ppm	[N]$_{final}$, ppm		W_{steel}/ W_{slag}	L_N	
		FactSage	Expt.		FactSage	Expt.
E1	25	3.5	4.6	12.8	79	56.8
E2	50	3.8	3.4	6.3	77	86.3
E3	46	5.0	10.3	4.6	38	15.9
E4	93	12.9	32.0	5.0	31	9.5
E5	190	12.9	30.5	5.3	73	27.7
E6	150	53/80 (for initial [O] 100/1500 ppm)	101	5.0	9.2/4.3 (for initial [O] 100/1500 ppm)	2.4

The strong effect of [%Al] on nitrogen removal is emphasised by Figure 2, which shows the experimental L_N values for the different final [%Al]. The relationship is scattered, because not only [%Al] varied between experiments, but also slag composition – which would have changed the activities of Al_2O_3 and O^{2-} and the N^{3-} activity coefficients in the slag. Despite the scatter, fitting a power-law expression to the data shows that L_N is approximately proportional to [%Al] (the fitted exponent is 1.17, similar to the expected value of 1). The conclusion is that nitrogen can be removed from high-Al AHSS by reaction with ladle slag.

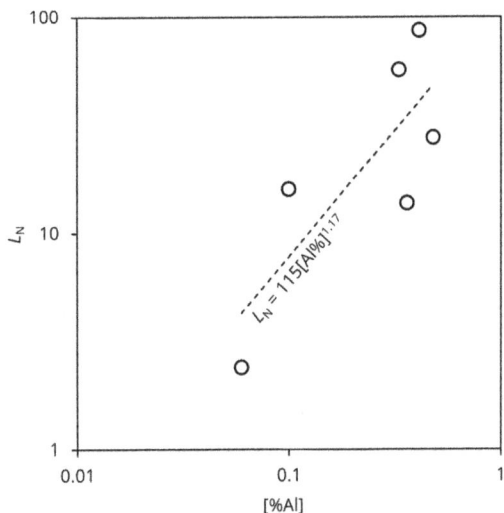

FIG 2 – Observed relationship between the nitrogen distribution coefficient and the final Al concentration in the steel, for all the experiments.

The extent of nitrogen removal in the experiments was large also because of the high mass ratio of slag to steel. This high mass ratio was essential to ensure that the steel was fully covered by slag. However, significant nitrogen removal would also be possible under industrial conditions, with much lower slag-to-steel mass ratios. Taking a typical ladle slag to steel mass ratio of $W_{slag}/W_{steel} = 0.015$ and $L_N = 60$ (feasible for [%Al]=1 per cent, as indicated by the experiments), the fraction of nitrogen removed from the steel (if the slag contained zero nitrogen before reaction) would be $1 - 1/(1+L_N W_{slag}/W_{steel}) = 0.47$. That is, approximately half of the nitrogen would be removed from the steel.

Figure 1 indicates that the observed time constants for nitrogen removal could be matched approximately by increasing the modelled steel-slag interfacial area (by the factors listed in Table 3). This increased area is expected for a system such as this, where high-Al steel reduces SiO_2 from the slag (Riboud and Lucas, 1981).

However, Figure 1 indicates that the observed extent of nitrogen removal was smaller than that predicted with FactSage. Part of this difference is due to the larger-than-predicted loss of aluminium due to the presumed ingress of oxygen. However, even when compensating for the loss of Al, the observed nitrogen distribution coefficients are generally smaller than those predicted by FactSage, as illustrated by Table 6. The conclusion is that, while the FactSage databases used here give useful indications of the reaction trends, the actual nitrogen removal – while substantial – would be less than predicted.

Slag microstructure

The solidified slag (after reaction with the Al-bearing steel at 1600°C) (Figure 3) contained the expected phases. As noted in Table 1, matching the observed nitrogen removal rate required increasing the effective steel-slag contact area by a factor of 2 (for the slag example in Figure 3). However, no emulsification was observed in the steel and slag after the experiments, likely because of the long duration of the experiments: Once the steel-slag reactions cease, no driving force for emulsification remains.

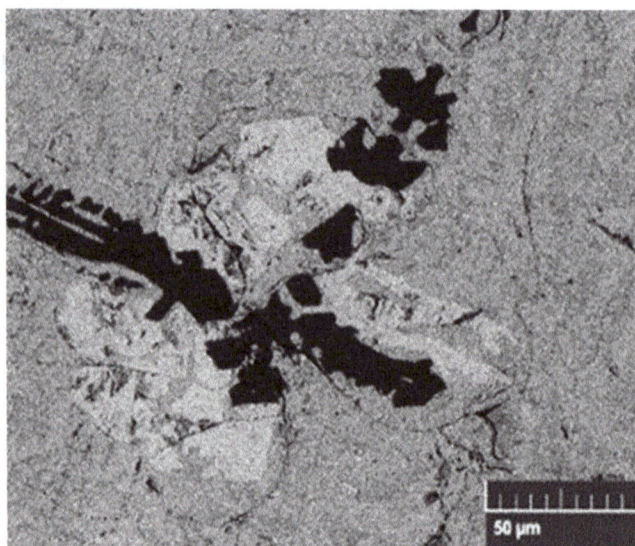

FIG 3 – SEM-BSE image of the 0 per cent SiO_2 slag after reaction at 1600°C with higher-Al steel (E5). The major phases are $Ca_3Al_2O_6$ (brightest phase), $Ca_3MgAl_4O_{10}$ (mid-grey region) and MgO (darkest dendrites).

Aluminium nitride in solidified steel

Given the high Al and N concentrations in the experimental steels (before nitrogen removal by the slag), the solid AlN may form in the steel (Paek *et al*, 2013). Scanning electron microscopy (10 kV accelerating voltage) of the first sample taken from Run E5 (containing 190 ppm) did show the presence of small AlN precipitates; examples are given in Figure 4. In some cases, the AlN appeared to have precipitated on an MgO-containing oxide core, as indicated by the energy dispersive X-ray (EDX) spectra in the figure. The predicted phase equilibria, calculated with Thermo-Calc 2024a (TCFE13 database) (Andersson *et al*, 2002) support the formation AlN during solidification (Figure 5). While some AlN forms during solidification, it is not stable in the liquid steel during ladle treatment (for the compositions considered here) and does not contribute to nitrogen removal during steel refining.

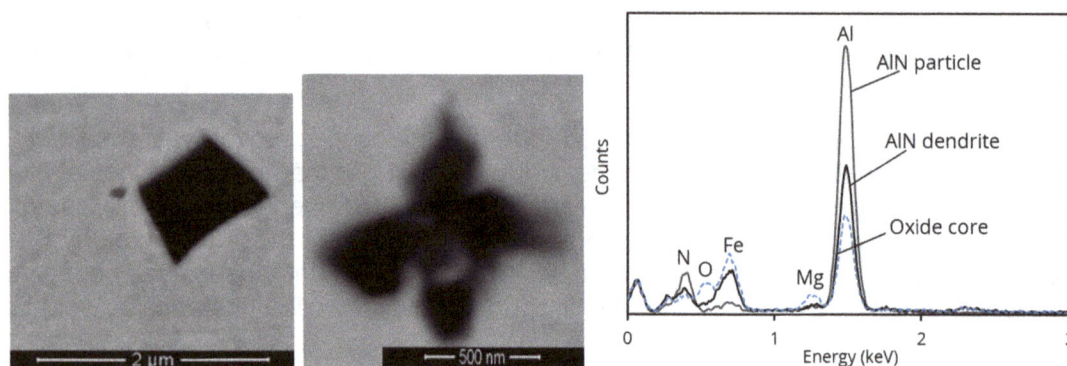

FIG 4 – Backscattered electron micrographs of AlN precipitates in the first steel sample from run E5, with EDX spectra at right.

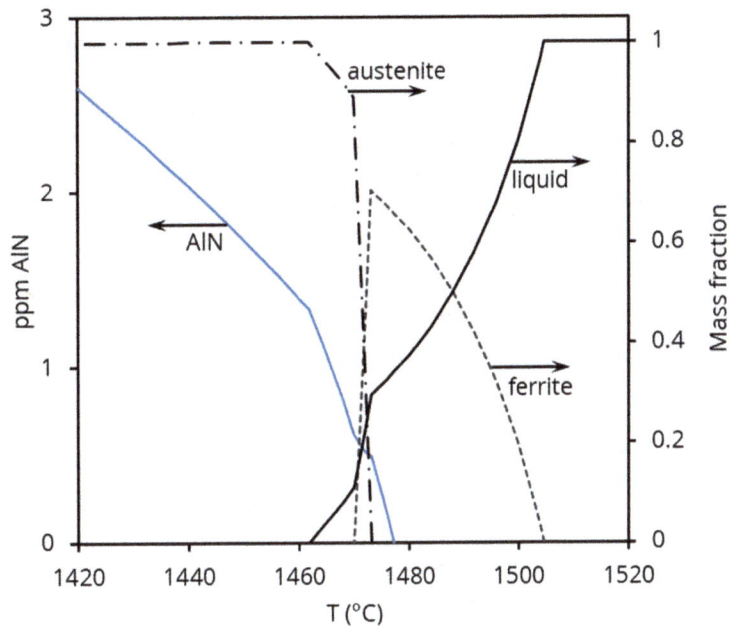

FIG 5 – Calculated equilibrium phases in the steel of run E5, with 190 ppm nitrogen.

Comparison of IGA and OES analysis of nitrogen in steel

Optical emission spectroscopy (OES) provides rapid analyses, but is less accurate for low concentrations of nitrogen. However, comparison of the reported compositions of the final steel samples (Table 7) shows close agreement between the IGA and OES results, for samples containing more than approximately 30 ppm nitrogen. This indicates that suitably calibrated OES can be used to rapidly assess nitrogen removal (or pick-up) during industrial ladle processing of liquid steel.

TABLE 7

Comparison of nitrogen analyses of the final steel samples, obtained by IGA and Spark-OES.

Experiment	N ppm, IGA	N ppm, Spark OES
E1	4.6	<5
E2	3.4	<5
E3	10.3	13.6
E4	32	32
E5	32	37
E6	110	101

CONCLUSIONS

Considerable nitrogen removal with calcium aluminate slag is possible during the ladle processing of AHSS containing >0.1 per cent dissolved aluminium. The available FactSage database provides a useful indication of the expected reactions, though it slightly overpredicts the extent of nitrogen removal.

ACKNOWLEDGEMENTS

The authors acknowledge use of the Materials Characterization Facility at Carnegie Mellon University supported by grant MCF-677785, and support of this work by the industrial members of the Center for Iron and Steelmaking Research.

REFERENCES

Andersson, J-O, Helander, T, Höglund, L, Shi, P and Sundman, B, 2002. Thermo-Calc and DICTRA, computational tools for materials science, *CALPHAD*, 26:273–312. https://doi.org/10.1016/S0364-5916(02)00037-8

Assis, A N, Warnett, J, Spooner, S, Fruehan, R J, Williams, M A and Sridhar, S, 2015. Spontaneous Emulsification of a Metal Drop Immersed in Slag Due to Dephosphorization: Surface Area Quantification, *Metall Mater Trans B*, 46:568–576. https://doi.org/10.1007/s11663-014-0248-z

Bale, C W, Bélisle, E, Chartrand, P, Decterov, S A, Eriksson, G, Gheribi, A E, Hack, K, Jung, I-H, Kang, Y-B, Melançon, J, Pelton, A D, Petersen, S, Robelin, C, Sangster, J, Spencer, P and Van Ende, M-A, 2016. FactSage thermochemical software and databases, 2010–2016, *CALPHAD*, 54:35–53. https://doi.org/10.1016/j.calphad.2016.05.002

Jung, I-H, 2006. Thermodynamic Modeling of Gas Solubility In Molten Slags (I)- Carbon and Nitrogen, *ISIJ International*, 46:1577–1586. https://doi.org/10.2355/isijinternational.46.1577

Mu, H, Zhang, T, Fruehan, R J and Webler, B A, 2018. Reduction of CaO and MgO Slag Components by Al in Liquid Fe, *Metall Mater Trans B*, 49:1665–1674. https://doi.org/10.1007/s11663-018-1294-8

Paek, M-K, Chatterjee, S, Pak, J-J and Jung, I-H, 2016. Thermodynamics of Nitrogen in Fe-Mn-Al-Si-C Alloy Melts, *Metall Mater Trans B*, 47:1243–1262. https://doi.org/10.1007/s11663-016-0588-y

Paek, M-K, Jang, J-M, Jiang, M and Pak, J-J, 2013. Thermodynamics of AlN Formation in High Manganese-Aluminium Alloyed Liquid Steels, *ISIJ Int*, 53:973–978. https://doi.org/10.2355/isijinternational.53.973

Pistorius, P and Vermaak, M, 1999. Modelling pyro metallurgical kinetics: ladle desulphurization, *South African Journal of Science*, 95:377–380.

Piva, S P T and Pistorius, P C, 2021. Ferrosilicon-Based Calcium Treatment of Aluminium-Killed and Silicomanganese-Killed Steels, *Metall Mater Trans B*, 52:6–16. https://doi.org/10.1007/s11663-020-02017-1

Piva, S P T, Kumar, D and Pistorius, P C, 2017. Modeling Manganese Silicate Inclusion Composition Changes during Ladle Treatment Using FactSage Macros, *Metall and Materi Trans B*, 48:37–45. https://doi.org/10.1007/s11663-016-0764-0

Riboud, P V and Lucas, L D, 1981. Influence of Mass Transfer Upon Surface Phenomena in Iron and Steelmaking, *Canadian Metallurgical Quarterly*, 20:199–208. https://doi.org/10.1179/cmq.1981.20.2.199

Roy, D, Pistorius, P C and Fruehan, R J, 2013. Effect of Silicon on the Desulfurization of Al-Killed Steels: Part II, Experimental Results and Plant Trials, *Metall Mater Trans B*, 44:1095–1104. https://doi.org/10.1007/s11663-013-9888-7

Song, S, Tang, D, Kumar, D and Pistorius, P C, 2021. Recycling of Chromium-Containing Waste Oxide as Alloying Addition in Ladle Metallurgy, *Metall Mater Trans B*, 52:2612–2618. https://doi.org/10.1007/s11663-021-02212-8

Tada, C and Matsumura, C, 2011. Removal method of nitrogen in molten steel, US Patents, US7901482B2.

Tang, D and Pistorius, P C, 2021. Non-metallic Inclusion Evolution in a Liquid Third-Generation Advanced High-Strength Steel in Contact with Double-Saturated Slag, *Metall Mater Trans B*, 52:580–585. https://doi.org/10.1007/s11663-021-02084-y

Tang, D and Pistorius, P C, 2022. Kinetics of Nitrogen Removal from Liquid Third Generation Advanced High-Strength Steel by Tank Degassing, *Metall Mater Trans B*, https://doi.org/10.1007/s11663-021-02417-x

Van Ende, M-A, Kim, Y-M, Cho, M-K, Choi, J and Jung, I-H, 2011. A Kinetic Model for the Ruhrstahl Heraeus (RH) Degassing Process, *Metall Mater Trans B*, 42:477–489. https://doi.org/10.1007/s11663-011-9495-4

World Steel Association, 2021. Advanced High-Strength Steel (AHSS) Definitions [online], *WorldAutoSteel News*. Available from: <https://www.worldautosteel.org/steel-basics/automotive-advanced-high-strength-steel-ahss-definitions/> [Accessed]:13 April 2024.

Crystallisation control of CaO-SiO$_2$-Al$_2$O$_3$-MgO system inclusion

Y Wang[1,3,4], S Sukenaga[3], W Z Mu[4], H Zhang[2], H W Ni[2] and H Shibata[3]

1. The State Key Laboratory of Refractories and Metallurgy, Wuhan University of Science and Technology, Wuhan 430081, China. Email: wangyong6@wust.edu.cn
2. The State Key Laboratory of Refractories and Metallurgy, Wuhan University of Science and Technology, Wuhan 430081, China.
3. Institute of Multidisciplinary Research for Advanced Materials, Tohoku University, Sendai, Miyagi 980-8577, Japan.
4. Department of Materials Science and Engineering, KTH Royal Institute of Technology, Stockholm 10044, Sweden.

ABSTRACT

The crystallisation behaviour of low melting point CaO-SiO$_2$-Al$_2$O$_3$-MgO inclusions greatly affect the deformability of inclusions during the rolling process. The crystallisation characteristics of CaO-SiO$_2$-Al$_2$O$_3$-MgO melts representing the oxide inclusions in Si-Mn killed steel were systematically investigated. The effect of Al$_2$O$_3$ and MgO contents on the crystallisation change were analysed under a fixed CaO/SiO$_2$ in the complex inclusions. The continuous-cooling-transformation (CCT) and time-temperature-transformation (TTT) experiments were conducted. The results showed that the increase of MgO content increases the crystallisation ability of CaO-SiO$_2$-Al$_2$O$_3$-MgO inclusions, while the increase of Al$_2$O$_3$ content has the opposite effect. The viscosity, liquidus temperature of oxide melt and also the initial crystallisation potential might be the reasons for the change in crystallisation behaviour. Also, the crystallisation activation energy could semi-quantitatively characterise the crystal growth during the crystallisation of inclusions. To obtain low melting point plasticized CaO-SiO$_2$-Al$_2$O$_3$-MgO (CaO/SiO$_2$=1) system inclusions, the Al$_2$O$_3$ content of the system needs to be controlled larger than 15 wt per cent and the MgO content should be kept at a small amount.

INTRODUCTION

Inclusion control is one of the key areas in the clean steel production research since it directly affect the quality of final steel products. Till now, the inclusion evolution during refining and solidification processes has been widely studied in various steel grades (Zhang and Thomas, 2003). Therefore, there is quite enough information on the inclusion control in the molten and solidification state of steel. However, the inclusion characteristics in the final products are usually not the same as them in the as cast steel. This is due to the fact that inclusions can also evolve changes in the heating and rolling process because the combined effect of thermal and mechanical treatments. Kitamura (2011) pointed out the new directions of inclusion control should be focused on the heating, rolling deformation and cooling processes instead of refining and casting processes. Recent years, the inclusion characteristics during heat treatment as well as hot and cold rolling process have been investigated (Zhang *et al*, 2018; Yang *et al*, 2022). The deformation ability of inclusions has a direct impact on the fatigue resistance of steel. Controlling the deformation performance of inclusions is one of the keys to improving the quality of high-end special steel. In order to improve the deformability of inclusions, the high melting point and non-deformable inclusions (such as Al$_2$O$_3$, MgO-Al$_2$O$_3$ spinel) should be removed as much as possible in steel (Kirihara, 2011). This is especially important in some specific steel types, such as spring steel and tire cord steel (Chen et al, 2019; 2020). Therefore, Si-Mn deoxidation is used instead of Al deoxidation in these steels (Kim, Kim and Kim, 2021; Lyu *et al*, 2019). CaO-SiO$_2$-Al$_2$O$_3$-MgO system oxide is a typical inclusion type in Si-Mn deoxidised steel (Bertrand *et al*, 2003). A larger number of studies aimed to control these inclusions in low melting point region to obtain better deformability. However, low melting point inclusions with high crystallisation ability tend to crystalise during the soaking process before hot rolling, which can greatly decrease the deformation ability of inclusions during the rolling process. The crystallisation behaviour of oxide melts is commonly studied in glass ceramics, mold fluxes, and molten slags. However, there are few studies concerning the crystallisation of oxide non-metallic inclusions in steel (Meng *et al*, 2022; Rocabois *et al*, 2001; Li *et al*, 2022; Liang *et al*, 2022).

In this study, the effects of MgO and Al_2O_3 contents on the crystallisation behaviour of CaO-SiO_2-Al_2O_3-MgO system inclusions were investigated using the sessile drop technique and thermodynamic calculations. This study provides a theoretical basis for understanding the crystallisation behaviour of CaO-SiO_2-Al_2O_3-MgO system inclusions and optimisation of its compositions with the lowest crystallisation ability.

EXPERIMENTAL METHODS

Sample preparation

The compositions of oxide melt were firstly designed based on the typical compositions of the inclusions in Si-Mn killed steel, the range of composition is presented in Table 1. In total, there are seven groups of sample, where the basicity (CaO/SiO_2) keeps as one. In groups S1, S2, S3, S4, the MgO content fixes as 8 per cent, and Al_2O_3 content varies from 5 per cent to 20 per cent. While in groups S2, S5, S6, S7, the Al_2O_3 content fixes as 10 per cent, and MgO content varies from 4 per cent to 16 per cent. The oxide samples were prepared using reagent-grade powders of $CaCO_3$ (≥99.0 mass per cent), Al_2O_3 (≥99.99 mass per cent), MgO (≥98.50 mass per cent), SiO_2 (≥99.99 mass per cent). About 20 g oxide mixtures were placed in a platinum crucible and melted in $MoSi_2$ furnace at 1600°C for 1 hr to ensure the homogenisation of chemical composition. Thereafter, the molten melt was quickly quenched on copper plate to obtain glass samples. The chemical compositions of the quenched samples were analysed using X-Ray fluoroscopy (XRF) and listed in Table 1.

TABLE 1

Chemical compositions of oxide mixtures before and after pre-melting (mass per cent).

Sample No.	Designed composition			
	CaO	SiO_2	Al_2O_3	MgO
S1	43.5	43.5	5	8
S2	41	41	10	8
S3	38.5	38.5	15	8
S4	36	36	20	8
S5	43	43	10	4
S6	39	39	10	12
S7	37	37	10	16

The crystallisation behaviour of the oxide melt was determined using a sessile drop measurement apparatus, which widely used for the wettability measurements (Xuan et al, 2015a, 2015b) as well as crystallisation control (Tashiro, Sukenaga and Shibata, 2017). The gas-tight furnace with customised apparatus is shown in Figure 1, which can allow us to observe the morphology change of melt and crystallisation behaviour in both vertical and horizontal directions. The whole heating and cooling process of samples were recorded by a CCD camera at the top, which was connected to a computer. About 50 mg glass sample was placed on a platinum substrate, which in turn, was placed on a platinum plate (diameter 13.0 mm; thickness 1.4 mm) on an Al_2O_3 tube pedestal. A sapphire plate was placed between the two-platinum holder. The temperature was measured using a thermocouple that was welded to the back of the platinum plate. Before the start of the experiment, the chamber was cleaned thoroughly by a cycle of vacuum and purging with a high purity Ar (purity > 99.9999 per cent). Then, the samples were heated up through quartz plates by two halogen lamps under an Ar protected atmosphere. The temperature profile for continuous-cooling-transformation (CCT) and time-temperature-transformation (TTT) experiments are shown in Figure 2. For both CCT and TTT experiments, the sample was heated to 1500°C and kept for 3 mins to eliminate the bubbles and uniform the chemical composition. Thereafter, the molten samples were continuously cooled at different cooling rates ranging from 5°C/min to 900°C/min in the CCT tests. In TTT experiments, the molten melt was cooling down rapidly to different temperatures ranging from 950°C to 1250°C at a cooling rate of 900°C/min. The iso-thermal treatment time for TTT tests are about 2 hrs. In TTT experiments, the temperature and time at which 5 vol per cent. of melt crystallised was determined as the start crystallisation temperature and time, respectively.

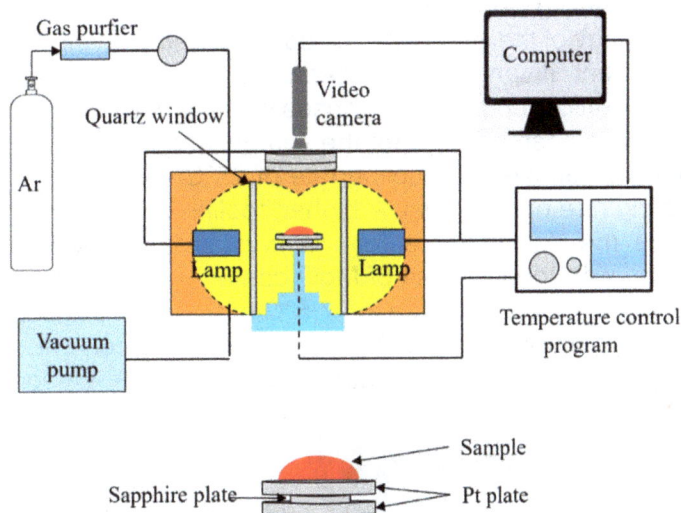

FIG 1 – Schematic diagram of the apparatus used to observe crystallisation behaviour.

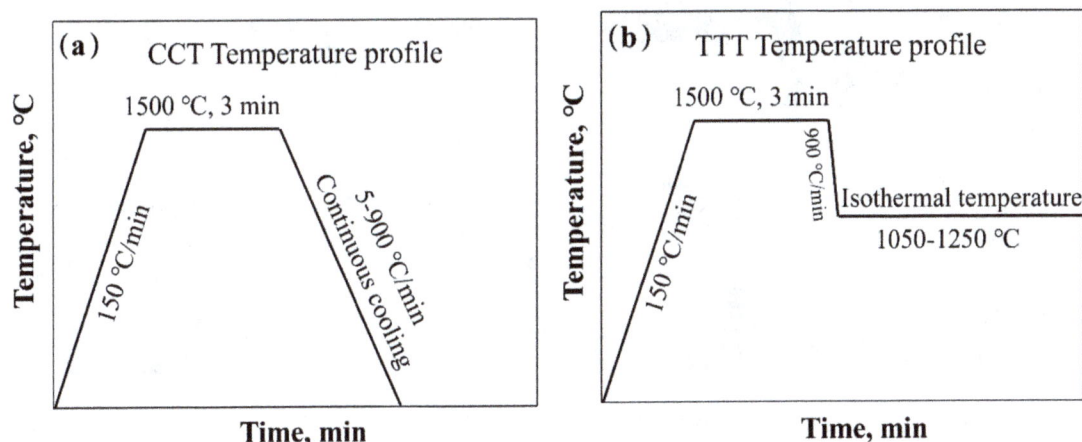

FIG 2 – Temperature profile of: (a) CCT and (b) TTT tests.

Characterisation methods

The crystallisation characteristics of each sample were measured by ImageJ software, according to the difference of brightness and contrast between the crystals and the melt. In this study, the structures of the quenched oxide melt were analysed using a laser confocal Raman spectrometer in order to reveal the relationship between the composition change and structure. The Raman spectra of the samples were recorded in the range of 200^{-1} to 2000 cm^{-1} at room temperature with the light source of a 1 mW semiconductor and an excitation wavelength of 500 nm. The solidified samples after the isothermal heat treatment experiments were divided into two parts, one part was crushed and grounded for X-ray diffraction (XRD) analysis in order to obtain the information of crystalline phases. The XRD data were collected using Cu Kα radiation in a range of $2\theta = 10–90°$. The other part was cold mounted in epoxy resin, and grinded, polished, then the crystallised phases were determined by a field-emission scanning electron microscope (FESEM) equipped with energy dispersive X-ray spectroscopy (EDS; Bruker).

RESULTS AND DISCUSSION

Continuous cooling crystallisation behaviour

During the continuous cooling experiment, the crystallisation happens under a certain supercooling degree. Besides, the supercooling degree varies with the cooling rates and chemical compositions. Figure 3 shows the morphology of the crystals in the samples with varying MgO contents and cooling rates after the CCT experiment. It can be clearly seen that crystallisation easily happens at a lower cooling rate and the crystallisation tendency decreases with the cooling rate. When the MgO content is 4 per cent, fully crystallisation can be seen for the sample at a cooling rate of 7.5°C/min. Two bulk crystals can be found for the sample at a cooling rate of 25°C/min, and the samples is transparent with no crystals at a cooling rate of 45°C/min. A higher cooling rate results in a higher supercooling degree, which decreases the driving force for the crystallisation. Moreover, the crystallisation tendency greatly increases with the increase of MgO contents. This is due to the fact that the fully crystallisation happens under a cooling rate of 7.5°C/min with 4 per cent MgO content, and it happens under a cooling rate of 240°C/min with 16 per cent MgO content. In terms of Al_2O_3 effect, the crystallisation tendency greatly decreases with the increase of Al_2O_3 contents.

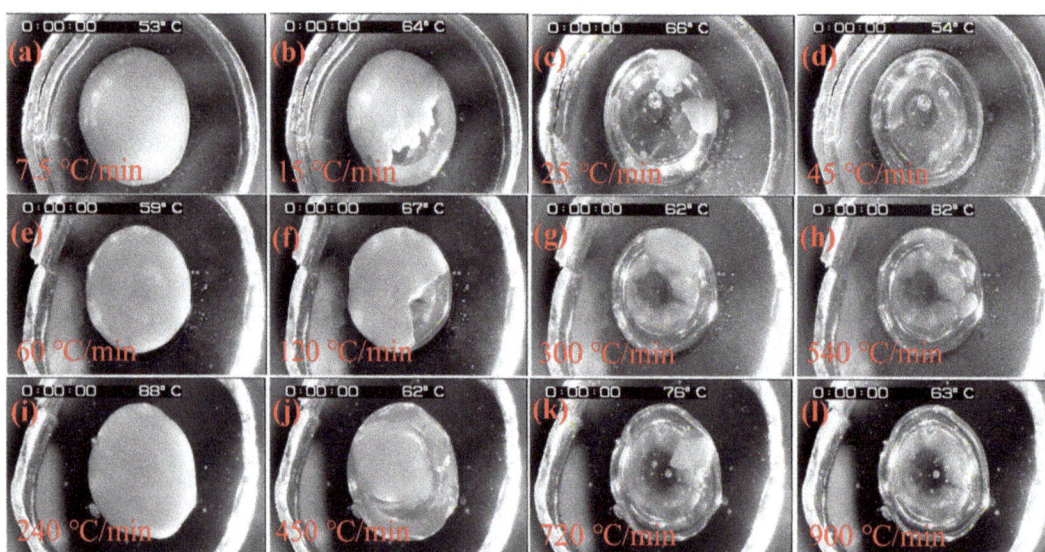

FIG 3 – The morphology of crystals after the CCT experiments with varying MgO contents.

Systematic analysis of the crystallisation characteristics based on the CCT experiments were performed. Figure 4 presents the CCT diagrams of the samples with varying Al_2O_3 and MgO contents. There are three obvious areas, a liquid area in a higher temperature, a crystallisation area in the middle and a glass area in a lower temperature. The critical cooling rate and the start crystallisation temperature are important parameters for evaluating the crystallisation ability of the samples. Taken the sample with 5 per cent Al_2O_3 as an example (Figure 4a), crystals can be formed when the cooling rate smaller than 180°C/min, and the sample will be glass when the cooling rate larger than 180°C/min. Thus, the critical cooling rate can be determined to be 180°C/min. It should be noticed that the critical cooling rate greatly increases from 30°C/min to 600°C/min when the MgO content increases from 8 per cent to 12 per cent. Further increase of MgO to 16 per cent slightly increases the critical cooling rate. Figure 5 shows the start crystallisation temperature of the samples with varying Al_2O_3 and MgO contents. The start crystallisation temperature under a lower cooling rate decreases approximately 100°C when the Al_2O_3 content increases from 10 per cent to 20 per cent, while that increases approximately 120°C when the MgO content increases from 8 per cent to 16 per cent. In addition, the start crystallisation temperature decreases with the increase of cooling rate under each composition due to the nucleation and growth of crystals are closely related to the change of viscosity and undercooling degree of the samples, which is discussed in detail later. It should be noticed that the start crystallisation temperature only slightly decreases with the increase of the cooling rate under higher MgO contents (eg 12 per cent and 16 per cent), which is still above 1200°C at a condition close to the critical cooling rate. It means that cooling rate has a small effect on the crystallisation temperature when the sample has a larger crystallisation tendency.

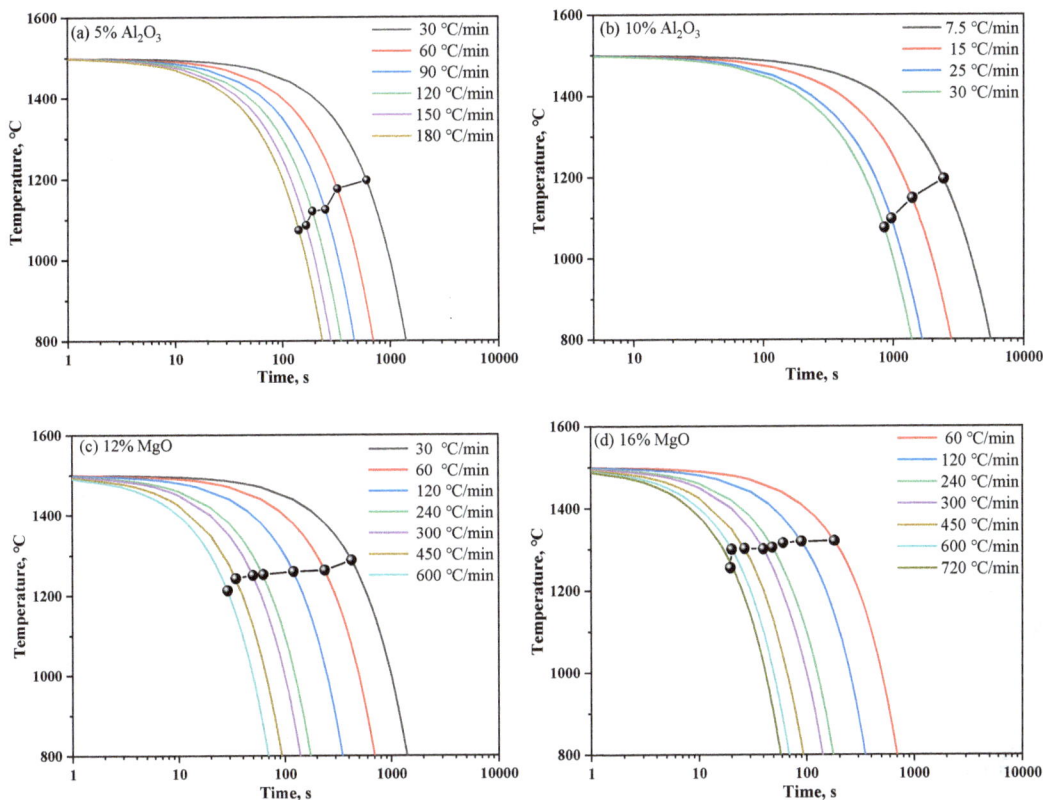

FIG 4 – CCT diagrams of CaO-SiO$_2$-Al$_2$O$_3$-MgO system inclusions with different compositions of (a) 5 per cent Al$_2$O$_3$, (b) 10 per cent Al$_2$O$_3$, (c) 12 per cent MgO, and (d) 16 per cent MgO.

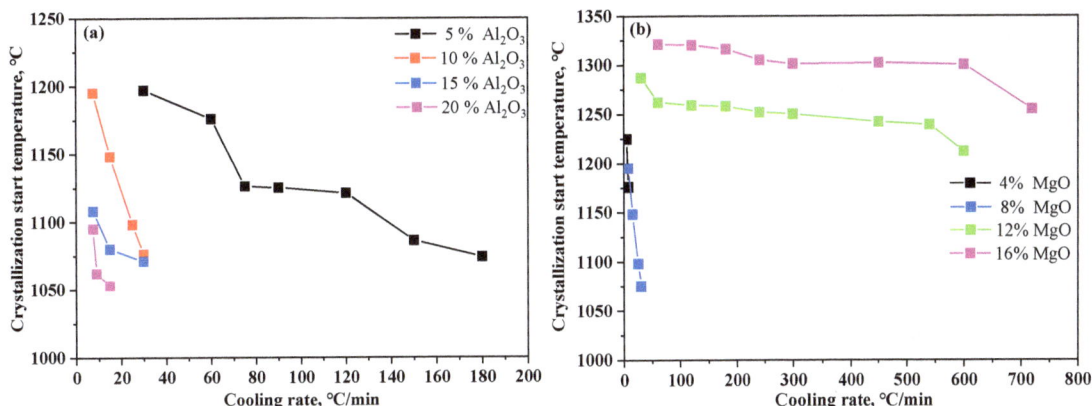

FIG 5 – The start crystallisation temperature in sample with varied (a) Al$_2$O$_3$ and (b) MgO contents.

Specifically, the increase of MgO content has a smaller effect on the start crystallisation temperature than the change of Al$_2$O$_3$ content. This indicates that the crystallisation ability greatly increases with the increase of MgO content, especially the MgO content is larger than 8 per cent. Under different conditions of cooling rates and compositions, the final area fractions of crystallisation phases can be obtained, as shown in Figure 6. The larger cooling rate, which result in lower crystallisation ability and therefore smaller area fractions of crystallisation phases can be obtained. This can partially reflect the nucleation and growth ability of crystals. The sample with more than 15 per cent Al$_2$O$_3$ and less than 4 per cent MgO under the CaO/SiO$_2$ ratio of 1 has the lowest crystallisation tendency. The change of cooling rate and composition can result in the change of viscosity, melting temperature and structure, which can directly affect the crystallisation behaviour. The detailed discussion of different factors on the crystallisation behaviour will be explained together with the TTT results in the following section.

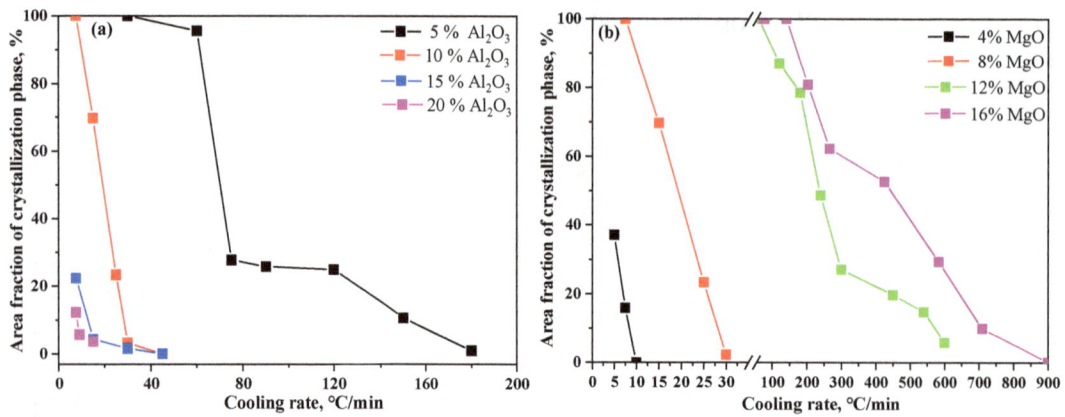

FIG 6 – The area fractions of crystallisation phases in sample with varied (a) Al$_2$O$_3$ and (b) MgO contents.

Isothermal crystallisation behaviour

The TTT diagrams of the samples with varying Al$_2$O$_3$ and MgO contents are shown in Figure 7 and Figure 8. It can be seen that the crystallisation period (time between the start crystallisation and the end of crystallisation) first decrease and then increase with the increase of temperature when the Al$_2$O$_3$ content is less than 10 per cent. Higher temperature greatly delays the crystallisation process due to the smaller supercooling degree and lower driving force for nucleation of crystals. This can be obviously seen in Figure 7c and 7d, since no crystallisation can be observed after two hrs holding time when the temperature is larger than 1150°C and 1100°C, respectively. With the increase of Al$_2$O$_3$ content, the crystallisation process becomes much later as the start crystallisation time becomes larger. Moreover, the crystallisation period (time between the start of crystallisation and the end of crystallisation) expands with the increase of Al$_2$O$_3$ content. Specially, fully crystallisation cannot be observed when the Al$_2$O$_3$ content is 20 per cent.

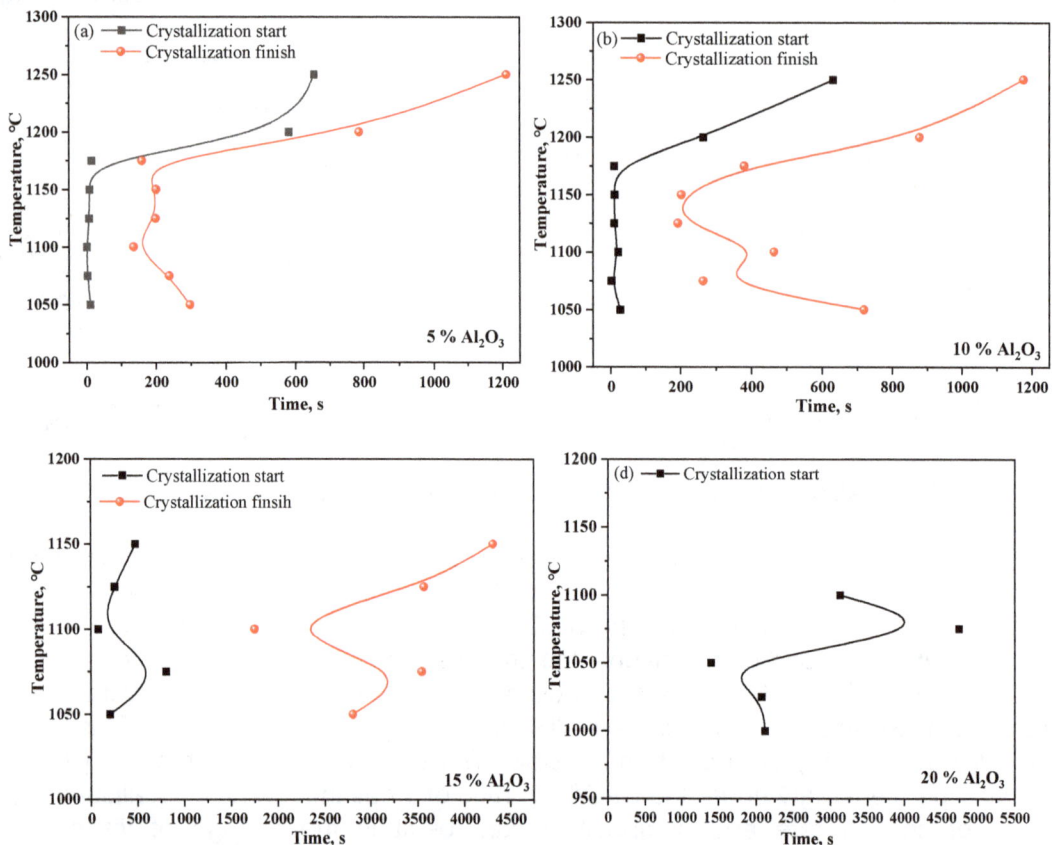

FIG 7 – TTT diagram for samples with varied Al$_2$O$_3$ contents (a) 5 per cent, (b) 10 per cent, (c) 15 per cent, (d) 20 per cent.

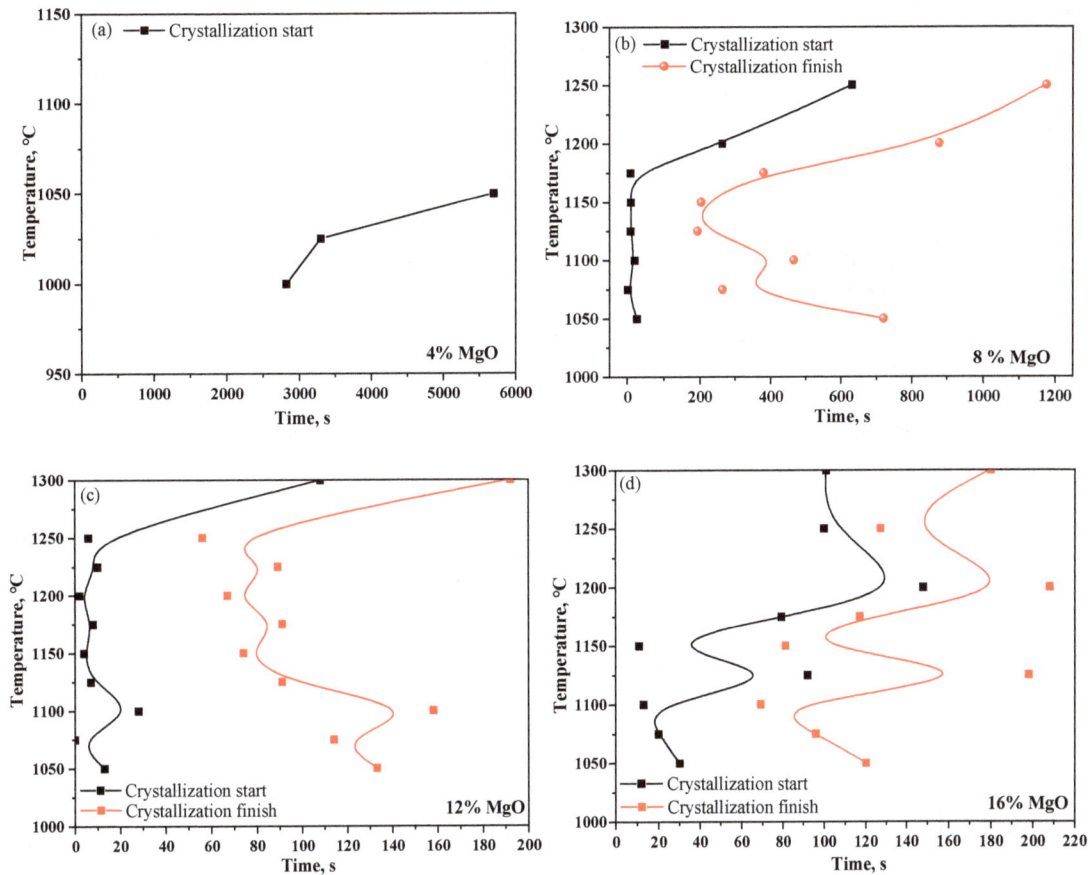

FIG 8 – TTT diagram for samples with varied MgO contents (a) 4 per cent, (b) 8 per cent, (c) 12 per cent, (d) 16 per cent.

In terms of MgO effect, an opposite tendency can be obtained. When the MgO content is about 4 per cent, crystallisation can only be observed at lower temperatures, and the start crystallisation time is much larger compared to those when the Al_2O_3 content is 20 per cent. The start crystallisation time becomes much smaller when the MgO content increases from 4 per cent to 8 per cent. With further increase of MgO content, the crystallisation period getting shorter, indicating that the crystallisation of inclusions occurred relatively easily. Moreover, the crystallisation period of inclusions under different MgO contents is much shorter compared to those with different Al_2O_3 contents in the similar temperature range. This indicates that the change of MgO content has a larger effect on the crystallisation rate compared to that of Al_2O_3. Thus, the increase of Al_2O_3 content and decrease of MgO content can greatly suppress the crystallisation of the CaO-SiO_2-Al_2O_3-MgO system inclusions (CaO/SiO_2 = 1).

Besides, the TTT curve of inclusions shows multiple 'C' shape, especially for the case of different MgO content, indicating that different crystallisation processes occurred in different temperature zones. However, it is difficult to identify the different phases since single crystallisation cannot be easily obtained. Based on the TTT results, the proper holding temperature, time and inclusion composition can be controlled to obtain the optimised soaking process.

CONCLUSIONS

- The increase of MgO content increases the crystallisation ability of CaO-SiO_2-Al_2O_3-MgO inclusions, while the increase of Al_2O_3 content has an opposite effect.

- The Al_2O_3 content of the system needs to be controlled larger than 15 wt per cent and the MgO content should be kept at a small amount (less than 5 wt per cent) to obtain low melting point plasticized CaO-SiO_2-Al_2O_3-MgO (CaO/SiO_2=1) system inclusions.

ACKNOWLEDGEMENTS

The authors are grateful for support from the National Natural Science Foundation of China (52304355 and U21A20113) and Japan Society for the Promotion of Science (JSPS).

REFERENCES

Bertrand, C, Molinero, J, Landa, S, Elvira, R, Wild, M, Barthold, G, Valentin, P and Schifferl, H, 2003. Metallurgy of plastic inclusions to improve fatigue life of engineering steels, *Ironmaking and Steelmaking*, 30:165–169.

Chen, C, Jiang, Z, Li, Y, Sun, M, Wang, Q, Chen, K and Li, H, 2020. State of the Art in the Control of Inclusions in Spring Steel for Automobile-a Review, *ISIJ International*, 60:617–627.

Chen, C, Jiang, Z, Li, Y, Zheng, L, Huang, X, Yang, G, Sun, M, Chen, K, Yang, H and Hu, H, 2019. State of the art in the control of inclusions in tire cord steels and saw wire steels–A Review, *Steel Research International*, 90:1800547.

Kim, W Y, Kim, K S and Kim, S Y, 2021. Evolution of Non-metallic Inclusions in Si-Killed Stainless Steelmaking, *Metallurgical and Materials Transactions B*, 52:652–664.

Kirihara, K, 2011. Production technology of wire rod for high tensile strength steel cord, *Kobelco Technol Rev*, 30:62–65.

Kitamura, S Y, 2011. Preface to the special issue on fundamentals and applications of non-metallic inclusions in solid steel, *ISIJ international*, 51:1943–1943.

Li, Z, Yang, W, Yao, H and Zhang, L, 2022. Effect of Na2O addition on crystallization behavior and properties of 25 wt% Al_2O_3–SiO_2–CaO non-metallic inclusion-type oxides, *Ceramics International*, 48:23849–23861.

Liang, Y, Shi, C, Huang, Y, Ju, J and Li, J, 2022. Effect of CaO/SiO2 mass ratio and Li2O on structure and phase precipitation behaviors of CaO–SiO2–MgO–Al_2O_3 oxide inclusions, *Journal of Non-Crystalline Solids*, 597:121911.

Lyu, S, Ma, X, Huang, Z, Yao, Z, Lee, H-G, Jiang, Z, Wang, G, Zou, J and Zhao, B, 2019. Understanding the formation and evolution of oxide inclusions in Si-deoxidized spring steel, *Metallurgical and materials Transactions B*, 50:1862–1877.

Meng, Y, Li, J, Wang, K and Zhu, H, 2022. Effect of the Bloom-Heating Process on the Inclusion Size of Si-Killed Spring Steel Wire Rod, *Metallurgical and Materials Transactions B*, 53:2647–2656.

Rocabois, P, Pontoire, J, Lehmann, J and Gaye, H, 2001. Crystallization kinetics of Al_2O_3–CaO–SiO_2 based oxide inclusions, *Journal of Non-Crystalline Solids*, 282:98–109.

Tashiro, M, Sukenaga, S and Shibata, H, 2017. Control of crystallization behaviour of supercooled liquid composed of lithium disilicate on platinum substrate, *Scientific Reports*, 7:6078.

Xuan, C, Shibata, H, Sukenaga, S, Jönsson, P G and Nakajima, K, 2015a. Wettability of Al_2O_3, MgO and Ti_2O_3 by liquid iron and steel, *ISIJ International*, 55:1882–1890.

Xuan, C, Shibata, H, Zhao, Z, Jönsson, P G and Nakajima, K, 2015b. Wettability of TiN by liquid iron and steel, *ISIJ International*, 55:1642–1651.

Yang, Y, Zhan, D, Qiu, G, Li, X, Jiang, Z and Zhang, H, 2022. Inclusion evolution in solid steel during rolling deformation: A review, *Journal of Materials Research and Technology*, 18:5103–5115.

Zhang, L and Thomas, B G, 2003. State of the art in evaluation and control of steel cleanliness, *ISIJ international*, 43:271–291.

Zhang, L, Guo, C, Yang, W, Ren, Y and Ling, H, 2018. Deformability of oxide inclusions in tire cord steels, *Metallurgical and Materials Transactions B*, 49:803–811.

Effect of Al$_2$O$_3$/SiO$_2$ ratio on structure and properties of mould flux for high-Al steel continuous casting

Q Wang[1], J Zhang[2], O Ostrovski[3], C Zhang[4] and D Cai[5]

1. The University of New South Wales, Sydney NSW 2052; Midea Group Laundry Appliance Division, Wuxi, 214028 Jiangsu, China.
2. The University of New South Wales, Sydney NSW 2052. Email: j.q.zhang@unsw.edu.au
3. The University of New South Wales, Sydney NSW 2052.
4. Baosteel Group Corporation Research Institute, Baoshan, 201900 Shanghai, China.
5. Baosteel Group Corporation Research Institute, Baoshan, 201900 Shanghai, China.

ABSTRACT

The conventional CaO-SiO$_2$-based mould fluxes are not suitable for high-Al steel casting because of the strong reaction between silica in the flux and aluminium in the steel strand. In the process of casting of high-Al steel, flux composition changes; with the decrease of the silica concentration and increase of alumina. Knowledge and understanding of the effect of the Al$_2$O$_3$/SiO$_2$ ratio on flux structure and properties are useful for flux design for the high Al-steel continuous casting.

This paper investigated the effect of the Al$_2$O$_3$/SiO$_2$ ratio on structure, viscosity, phase composition of fluxes quenched at different temperatures and heat transfer of CaO-Al$_2$O$_3$-SiO$_2$-B$_2$O$_3$-Na$_2$O-Li$_2$O-MgO-F fluxes. It was found that flux melting temperature increased with the increase in Al$_2$O$_3$/SiO$_2$ ratio. Viscosity of the flux melts increased significantly with the increase of the Al$_2$O$_3$/SiO$_2$ ratio from 0.7 to 1.2, reaching the maximum value, and then decreased with further increase of the Al$_2$O$_3$/SiO$_2$ ratio. Raman spectroscopy analysis revealed that the change of the Al$_2$O$_3$/SiO$_2$ ratio led to the change of aluminate and silicate structural units. The turning point for viscosity was attributed to the change in the degree of flux polymerisation. X-ray diffraction (XRD) analysis showed that increasing Al$_2$O$_3$/SiO$_2$ ratio increased crystallisation tendency of the fluxes. Heat transfer measurement by infrared emitter technique (IET) revealed that increasing Al$_2$O$_3$/SiO$_2$ ratio led to the decrease in heat flux which is correlated well with the increased crystallinity of the flux. The results suggested that the flux with Al$_2$O$_3$/SiO$_2$ ratio 4.3 is the best candidate among the studied CaO-Al$_2$O$_3$-based mould fluxes for casting of high-Al steel.

INTRODUCTION

Conventional CaO-SiO$_2$-based mould fluxes contain high concentrations of SiO$_2$ (up to 56 mass per cent (Brandaleze *et al*, 2012)) which is essential to ensure the required flux properties for steel casting. However, for casting of high aluminium steel such as advanced high strength steel (0.5–2.0 mass per cent Al), [Al] in the steel can react with (SiO$_2$) in the flux, increasing the Al$_2$O$_3$ content and decreasing the SiO$_2$ content in the mould flux, ie increasing the ratio of Al$_2$O$_3$/SiO$_2$. As a result, it leads to an inevitable variation of flux properties, and consequently, affecting the casting of high-Al steel (Kim *et al*, 2013; Cho *et al*, 2013; Kang *et al*, 2013; Chung and Cramb, 2000; Zhou *et al*, 2017; Zhou, Wang and Zhou, 2015).

This work investigated the effect of Al$_2$O$_3$/SiO$_2$ ratio on the physicochemical properties, structure, and heat transfer of mould fluxes for continuous casting of high-Al steel. The ratio of Al$_2$O$_3$/SiO$_2$ changed from a relatively low value to a high one to reflect the flux composition change during high-Al steel casting process. The information of the effect of this ratio on flux structure and properties is useful for the flux design for high Al-steel continuous casting.

MATERIALS AND EXPERIMENTAL PROCEDURE

Materials

Flux samples were prepared by using reagent grade CaCO$_3$, Na$_2$CO$_3$, Li$_2$CO$_3$, SiO$_2$, Al$_2$O$_3$, B$_2$O$_3$ and CaF$_2$ which were fully mixed, and then melted in a high-purity graphite crucible at 1400°C for 20 mins. After that, the melted flux was quenched into water, then dried at 120°C for 2 hrs, and ground into fine powders using a ring mill. The contents of B$_2$O$_3$ and Li$_2$O were analysed using inductively coupled plasma (ICP, Thermo Scientific IRIS Intrepid II, MA), while the other flux components were determined by X-ray fluoroscopy (XRF, PANalytical AXIOS-Advanced WDXRF spectrometer, Netherland). The measured flux compositions are shown in Table 1. The Al_2O_3/SiO_2 ratio of fluxes varied from 0.7 to 10.8 while concentrations of all other components were set as constants. All five flux samples are based on the $CaO-Al_2O_3-SiO_2-B_2O_3$ quaternary system with the addition of Na_2O, Li_2O, MgO, and F. It should be mentioned that variation of the Al_2O_3/SiO_2 mass ratio changed the CaO/Al_2O_3 mass ratio within the range 2.2–1.1.

TABLE 1

Measured chemical composition of mould fluxes, mass per cent and Al_2O_3/SiO_2 ratio.

Flux	CaO	Al_2O_3	SiO_2	Na_2O	B_2O_3	Li_2O	MgO	F	Al_2O_3/SiO_2
1	32.7	15.1	21.5	6.0	12.4	3.9	2.1	6.3	0.7
2	32.6	19.8	16.5	6.1	12.7	3.8	2.0	6.5	1.2
3	33.2	24.6	11.8	6.1	11.7	3.7	2.1	6.8	2.1
4	31.6	30.0	7.0	6.2	12.7	4.1	2.0	6.3	4.3
5	33.6	32.0	2.9	6.4	12.4	4.0	2.1	6.5	10.8

Experimental procedure

Melting properties of mould fluxes were investigated using a hot stage microscopy. As-quenched flux powders were pressed to pellets (Φ 3×3 mm). The pellets were continuously heated in a horizontal tube furnace at a rate of 15°C/min; their appearance change was monitored by a video camera. To characterise the melting properties of mould fluxes, the softening, hemispherical, and fluidity temperatures were defined as the temperatures at which the height of the flux pellet dropped to 75, 50 and 25 per cent of its original height, respectively. The details of melting property measurement were described elsewhere (Yang et al, 2017, 2018).

Viscosity of mould fluxes was investigated using a rotation viscometer (model ZC-1600, China). Approximately 140 g pre-melted flux was heated up to 1400°C in a graphite crucible in nitrogen atmosphere. After holding the molten flux at 1400°C for 1200 secs, a Mo bob was slowly submerged into the homogenised melt and rotated at a rate of 12 rev/min for the measurement of viscosity. After measurement of the flux viscosity at 1400°C, the viscosity was measured in the process of flux continuous cooling with a rate of 5°C/min until the Mo bob stopped rotation.

The structure of as-quenched mould fluxes from 1400°C was studied using Raman spectroscopy (Renishaw inVia Raman Microscope, UK). The pulverised mould flux was illuminated by Ar-ion laser beam. The excitation wavelength of Ar-ion laser was 514 nm with a beam spot size of 1.5 μm. The measurement was conducted in the Raman shift range from 400 to 1700 cm^{-1}. The obtained Raman spectra were deconvoluted using WiRE 4.4 software.

The heat transfer rate across the mould fluxes was measured using infrared emitter technique (IET, Central South University, China). The details of IET system were described elsewhere (Wang and Cramb, 2005). The flux disk was prepared using 13 g as-quenched flux powders which were melted at 1400°C and held for 1200 secs before pouring into a copper cylinder (Φ 40 mm) and pressing to a disk. The pressed mould flux disk was immediately placed in a muffle furnace in which the temperature gradually decreased from 800°C to 25°C with a slow cooling rate of 1°C/min to minimise the internal stress within the disk. The fabricated flux disk was ground to the thickness of 4 ± 0.01 mm using a diamond grinding wheel, and then carefully polished using sandpapers from 300 to 1200 grits. In the heat flux measurement, the mould flux disk was placed on the copper base of the IET system. The incident thermal radiation was increased up to 1.6 MW/m^2, which is close to the radiation released from the steel strand in the continuous casting. The temperatures recorded by the embedded thermocouples were used for the heat flux calculation according to the Fourier's law (Park et al, 2016):

$$q = \frac{-1}{n}\sum_i k \left(\frac{dT}{dx}\right)_i \tag{1}$$

where:

q	is the heat flux
n	presents the total number of thermocouples
k	is the thermal conductivity of copper

The heat fluxes measured at an incident radiation of 1.6 MW/m^2 were compared to reflect the heat transfer ability of fluxes during continuous casting.

The crystalline phases in the mould fluxes heated to different temperatures and quenched were determined using X-ray diffraction (XRD) with Cu-Kα radiation in the scanning range of 2θ from 10 to 80 degree with a scanning speed of 0.021 deg/s. Eight-gram samples for the XRD analyses were prepared separately by holding the fluxes at 950°C (low temperature zone), 1050°C (medium temperature zone), and 1150°C (high temperature zone) for 30 mins. Then the heat-treated samples were quenched into water and subjected to the XRD analysis. The selection of these three temperatures was based on the TTT measurement to reflect the phase changes at low, medium and high temperatures. The XRD spectra were analysed using HighScore Plus 4.2.

RESULTS

Melting properties

Figure 1 illustrates the influence of Al$_2$O$_3$/SiO$_2$ ratio on softening temperature (T_s), hemispherical temperature (T_h) and fluidity temperature (T_f) of the CaO-Al$_2$O$_3$-based mould fluxes.

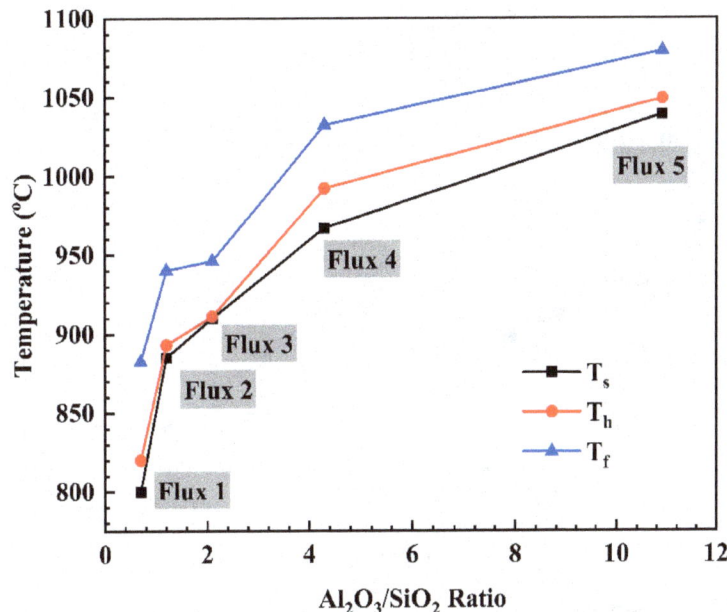

FIG 1 – Softening temperature (T_s), hemispherical temperature (T_h) and fluidity temperature (T_f) as functions of the Al$_2$O$_3$/SiO$_2$ ratio.

With the increase in the mass ratio of Al$_2$O$_3$/SiO$_2$ from 0.7 to 10.8, all three characteristic temperatures T_S, T_h and T_f continuously increased (Figure 1). The value of T_f is much higher than those of T_s and T_h in all cases.

Viscosity

Figure 2 shows the measured viscosity as a function of temperature. It was observed that for all mould fluxes, viscosity increased with the decrease in temperature. The enlarged scale of viscosity in the high-temperature zone from 1300 to 1400°C is shown in Figure 3. It was observed that when increasing the mass ratio of Al$_2$O$_3$/SiO$_2$ from 0.7 to 1.2, the viscosity increased significantly. With further increasing this ratio, the viscosity decreased. As a result, Flux 8-2 had the highest viscosity in all the fluxes. The viscosities at 1400°C (η_{1400}) of all examined mould fluxes are presented in

Table 2, further confirming that the viscosity reaches the highest point when $Al_2O_3/SiO_2 = 1.2$, and reduces gradually with further increase in Al_2O_3/SiO_2 ratio.

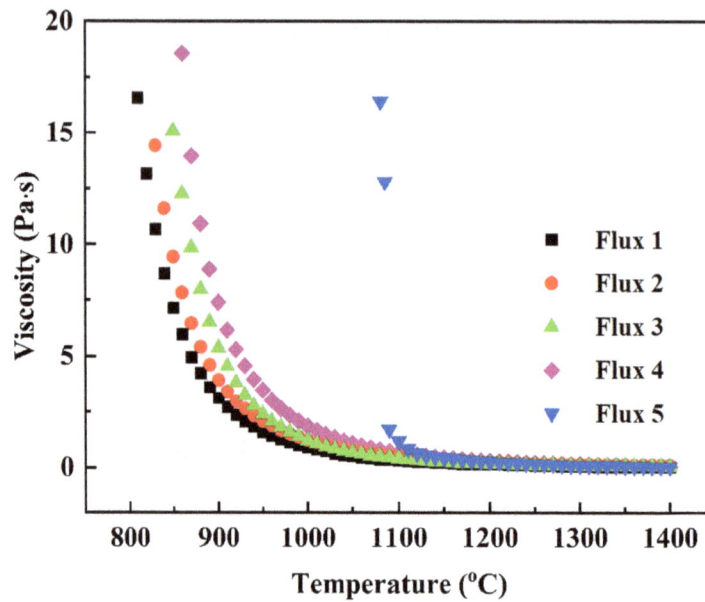

FIG 2 – Viscosity of mould flux with different Al_2O_3/SiO_2 ratios at different temperatures.

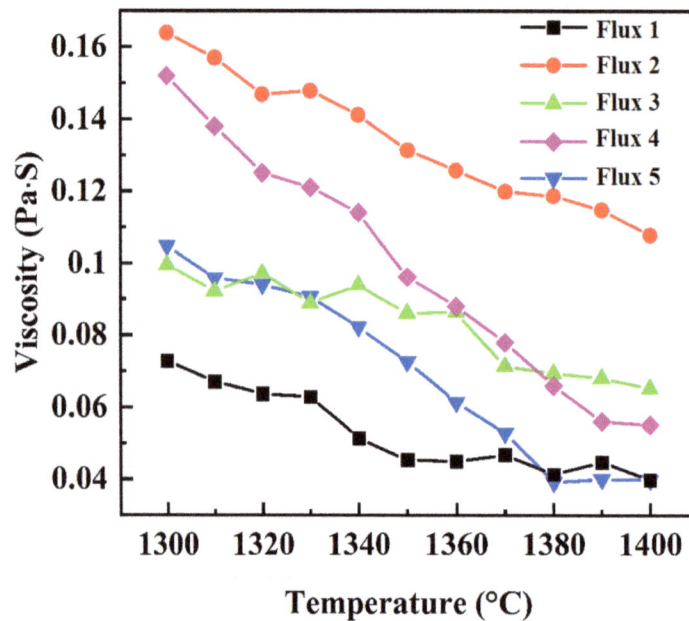

FIG 3 – Highlight the viscosity at temperatures 1300–1400°C.

TABLE 2

Calculated apparent activation energy (Ea), break temperature (T_{br}), and viscosity at 1400°C (η_{1400}) for the fluxes with different Al_2O_3/SiO_2 ratios.

Flux	Ea (kJ·mol^{-1})	R^2	T_{br}, °C	η_{1400}, Pa·s
1	137	0.991	-	0.040
2	98	0.996	-	0.108
3	117	0.989	-	0.065
4	156	0.978	-	0.055
5	188	0.985	1105	0.039

Liquid flux can be assumed to be a Newtonian fluid and its viscosity can be fitted by the Arrhenius equation.

$$\ln \eta = \ln A + \frac{Ea}{RT} \tag{2}$$

where:

η represents viscosity

A is viscosity constant

T is the temperature

Ea activation energy

Figure 4 shows the Arrhenius plot of $\ln \eta$ as a function of $1/T$. It appears that only Flux 5 shows a clear break temperature at 1105°C, while all other fluxes do not have apparent break temperatures. The activation energies determined in the temperature range of 1100–1330°C are 137 kJ·mol^{-1} for Flux 1, 98 kJ·mol^{-1} for Flux 2, 117 kJ·mol^{-1} for Flux 3, 156 kJ·mol^{-1} for Flux 4, and 188 kJ·mol^{-1} for Flux 5 in temperature range of 1100–1330°C. All these results are shown in Table 2.

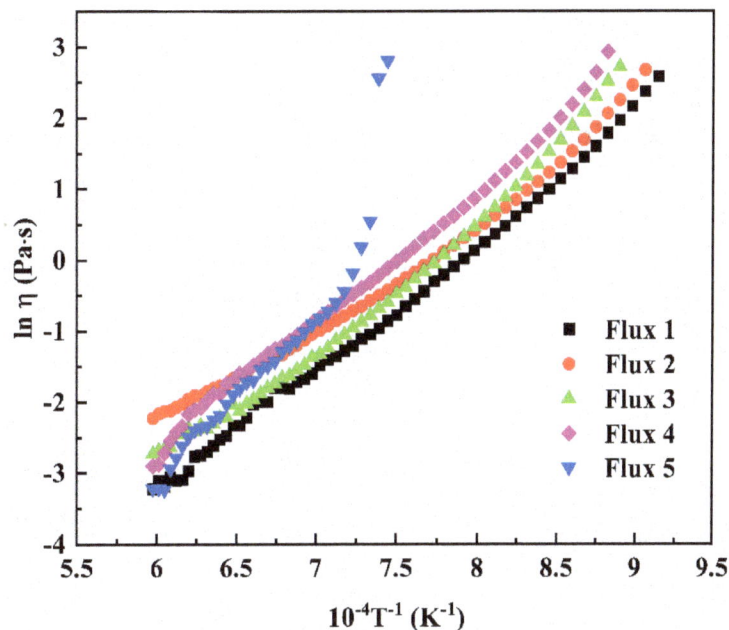

FIG 4 – Plots of $\ln \eta$ vs $\frac{1}{T}$ for fluxes with Al$_2$O$_3$/SiO$_2$ ratio increased from 0.7 to 10.8.

Raman analysis

The structure of mould flux with varying Al$_2$O$_3$/SiO$_2$ mass ratios was studied using the Raman spectroscopy. Figure 5 illustrates the Raman spectra of as-quenching mould fluxes and Table 3 lists the assignments of deconvoluted Raman bands. In the low frequency region (between 400 and 800 cm^{-1}) the deconvoluted bands were assigned to aluminate group in which peaks around 500 cm^{-1} were assigned to Al-F stretching vibration in AlF$_6$; peaks centred around 550 cm^{-1} correspond to Al-O-Al linkage which is the major bond in the 3D aluminate network; peaks located around 590 cm^{-1} are related to Al-O$^-$ stretching vibration in AlO$_6$ units; peaks located around 770 cm^{-1} are related to Al-O$^-$ stretching vibration in Al-NBO units with 1 or 2 NBOs (non-bridging oxygen); peaks centred around 865 cm^{-1} correspond to Al-O-Si linkage. Among these characteristic peaks, structural units with Al-O-Si and Al-O-Al linkages had the most prominent peaks; with the increase of Al$_2$O$_3$/SiO$_2$ mass ratio, peaks assigned to AlF$_6$ and Al-O-Al linkages became more and more prominent, while peaks assigned to Al-O-Si linkage became weaker.

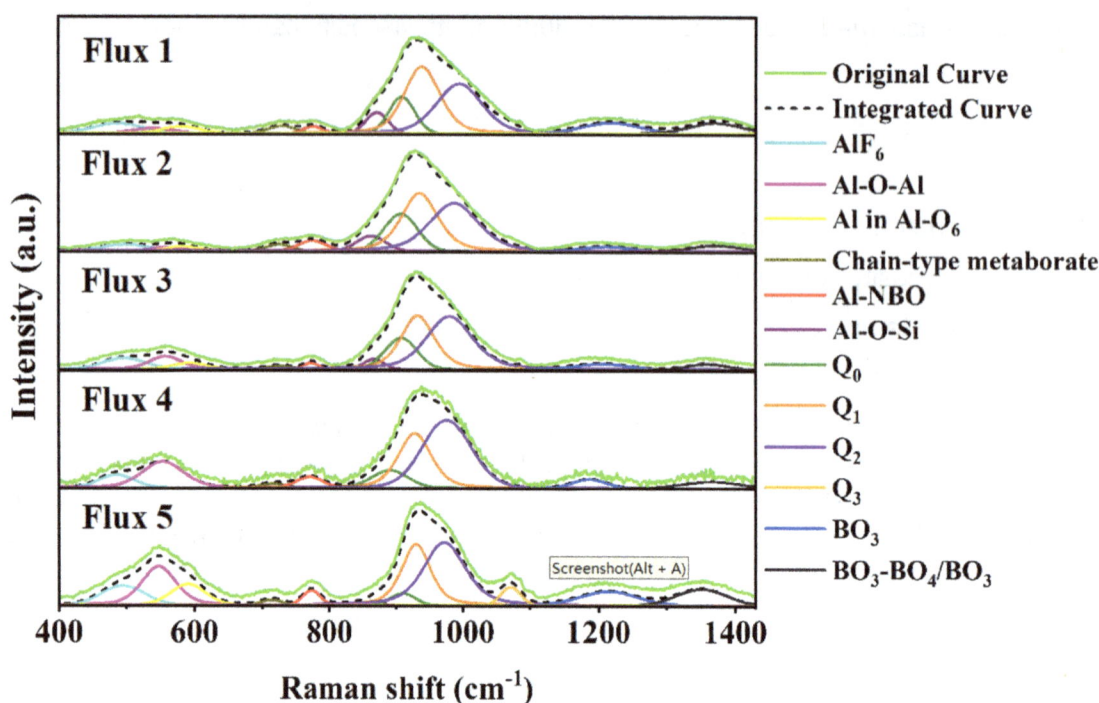

FIG 5 – Deconvoluted Raman spectra for mould flux with various Al_2O_3/SiO_2 ratios.

TABLE 3
Assignments of deconvoluted Raman bands.

Flux 1	Flux 2	Flux 3	Flux 4	Flux 5	Structural unit	Reference
488.1	498.5	497.0	485.2	494.3	AlF_6	Park, Min and Song (2002); Ma *et al* (2018)
539.3	558.0	557.8	553.5	547.5	Al-O-Al	Park, Min and Song (2002)
582.2	584.1	594.2	-	590.9	AlO_6	Park, Min and Song (2002); McMillan and Piriou (1983)
729.2	724.4	724.7	715.6	712.9	Chain-type metaborate	Kim and Sohn (2014); Kamitsos, Karakassides and Chryssikos (1987)
777.0	775.4	772.9	770.2	772.4	Al-NBO	Gao, Wang and Zhang (2017); Kim and Park (2014)
872.0	863.1	866.8	851.0	870.4	Al-O-Si	Gao (2016a, 2016b)
909.5	907.0	908.0	890.9	910.0	Q^0	
940.0	936.2	933.3	928.5	930.0	Q^1	Kim and Sohn (2012); Zheng *et al* (2014)
995.1	987.4	980.7	976.3	972.6	Q^2	
-	-	-	-	1071.5	Q^3	
1217.8	1210.9	1205.0	1186.3	1214.8	BO_3	Kim and Sohn (2014); Kamitsos, Karakassides and Chryssikos (1987)
1370.7	1371.8	1358.1	1364.2	1350.3	BO_3-BO_4/BO_3	

In the medium frequency range (900 to 1100 cm^{-1}), the deconvoluted Raman peaks were assigned to silicate structures, including Si-O stretching vibration in SiO_4^{4-} (Q^0), $Si_2O_7^{6-}$ (Q^1), SiO_3^{2-}(Q^2) and $Si_2O_5^{2-}$ (Q^3). In all fluxes, Q^1 and Q^2 were the dominant units, Q^0 and Q^3 were less noticeable in the

Raman spectra. No Q^4 unit was detected for all fluxes. Based on the spectra profile between 900 and 1100 cm^{-1} (Figure 5), with the increase of Al_2O_3/SiO_2 ratio, the area fraction of Q^0 decreased while the fraction of Q^2 increased, which means the simple silicate structure became less significant with the increase of Al_2O_3/SiO_2 mass ratio; the fraction of Q^1 did not change.

In the high frequency range, the Raman bands were assigned to different borate structures, where the peaks around 1210 cm^{-1} correspond to the B-O stretching vibration in BO_3 units which was a necessary component to form 3D borate groups, while the peak around 1360 cm^{-1} was assigned to B-O stretching vibration in BO_3^- units attached to other 3D borate groups. There is also a borate band located around 720 cm^{-1}, corresponding to chain-type metaborate groups. With the increase of Al_2O_3/SiO_2 ratio from 0.7 to 1.2, the fractions of these two borate units became depressed; with further increase the ratio to 10.8, the fractions of these units gradually increased.

The percentage of some main peaks from deconvoluted Raman spectra is shown in Figure 6. Increasing Al_2O_3/SiO_2 ratio increased the fraction of Al-O-Al, AlF_6 and Q^2, but decreased that of Q^0, Al-O-Si. The fraction of Q^1 remained no change.

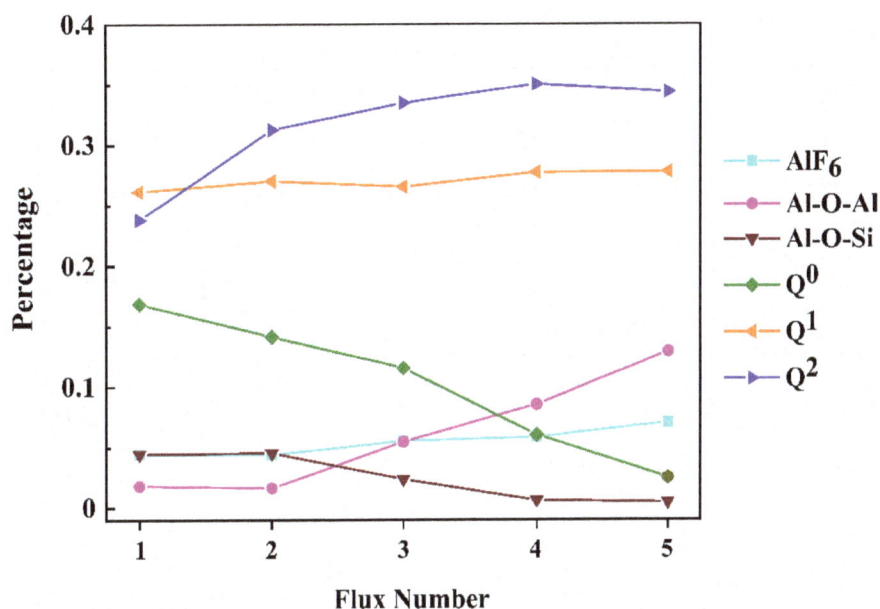

FIG 6 – Percentage of main peaks based on the deconvoluted Raman spectra.

Flux phase analysis by XRD

Figures 7 and 8 show the XRD patterns of mould fluxes 1 to 5 heat treated at 900°C and 1000°C; the XRD patterns of Flux 4 heat-treated at 950°C and 1050°C were also shown in these two figures, respectively. Table 4 lists the crystal phases identified by XRD at different temperatures. For Fluxes 2 and 3 (Al_2O_3/SiO_2 ratio of 1.2 and 2.1, respectively), $Ca_2Al_2SiO_7$ was the only phase at both 900°C and 1000°C. However, for other fluxes, different phases formed. At 900°C, $Ca_4Si_2O_7F_2$ (cuspidine) was detected to be the main phase in Flux 1 with Al_2O_3/SiO_2 ratio of 0.7, while the crystalline phases of Flux 5 with Al_2O_3/SiO_2 ratio of 10.8 were $Ca_5B_3O_9F$, $CaSiO_3$ and $LiAlO_2$ ($LiAlO_2$ was the main phase). For Flux 4, $Ca_2Al_2SiO_7$, $MgAl_6O_{10}$, $LiAlO_2$ were detected at 950°C. At 1000°C, the crystal phases detected in Flux 4 were the same as those at 900°C, while there was only trace amount of $Ca_4Si_2O_7F_2$ (cuspidine) detected in amorphous matrix for Flux 1 after heat treatment at 1000°C. For Flux 4 at 1050°C, $MgAl_6O_{10}$, $LiAlO_2$, and $Ca_2Al_2SiO_7$ were identified.

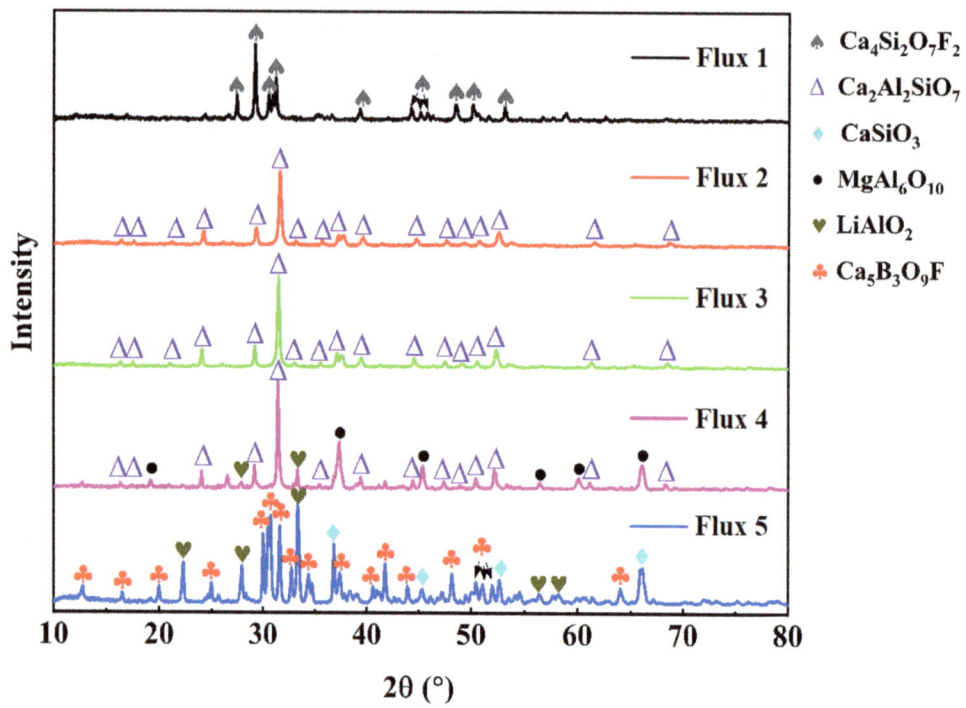

FIG 7 – XRD patterns of Fluxes 1, 2, 3 and 5 heated at 900°C and Flux 4 at 950°C.

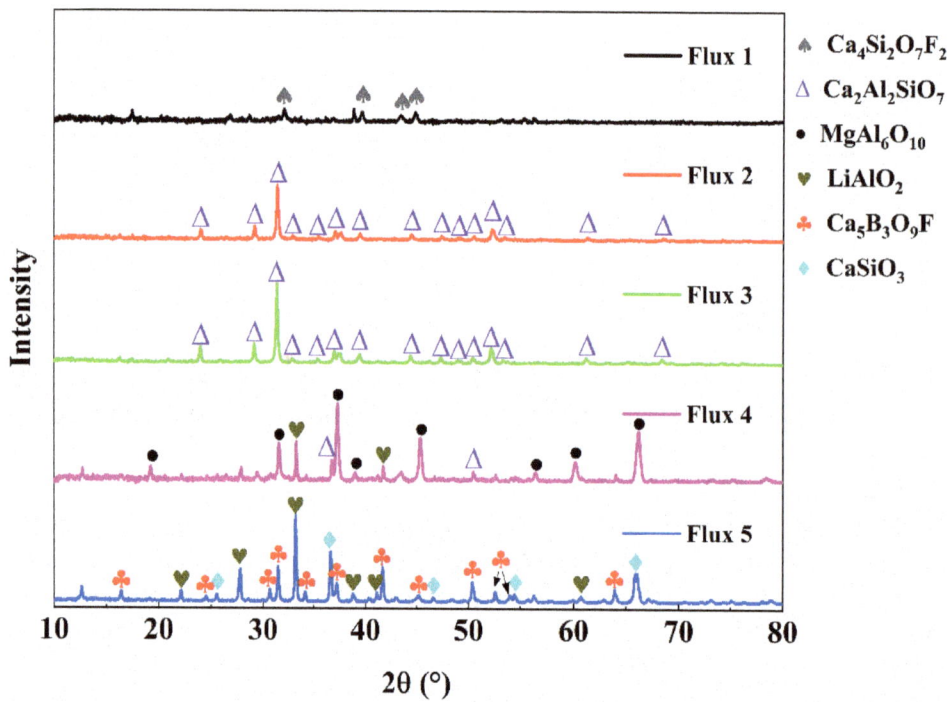

FIG 8 – XRD patterns of Fluxes 1, 2, 3 and 5 heated at 1000°C and Flux 4 at 1050°C.

TABLE 4

Summary of the XRD results at different temperatures.

T/°C	900	1000
Flux 1	$Ca_4Si_2O_7F_2$	Amorphous + trace amount of $Ca_4Si_2O_7F_2$
Flux 2	$Ca_2Al_2SiO_7$	$Ca_2Al_2SiO_7$
Flux 3	$Ca_2Al_2SiO_7$	$Ca_2Al_2SiO_7$
Flux 4	(950°C) $Ca_2Al_2SiO_7$, $MgAl_6O_{10}$, $LiAlO_2$	(1050°C) $MgAl_6O_{10}$, $LiAlO_2$, $Ca_2Al_2SiO_7$
Flux 5	$LiAlO_2$, $Ca_5B_3O_9F$, $CaSiO_3$	$LiAlO_2$, $CaSiO_3$, $Ca_5B_3O_9F$

Heat transfer of mould flux

Figure 9 shows results of IEF experiments for five mould fluxes. The heat flux tended to be stable after 2100 secs. The average heat flux values of the five fluxes at the final stage were 760.6, 720.0, 700.1, 686.0, 440.3 kW/m^2, respectively. It means that with the increase of Al_2O_3/SiO_2 ratio from 0.7 to 10.8 the heat flux decreases. Among these values, the heat flux for Flux 5 had a much lower value than those for other fluxes.

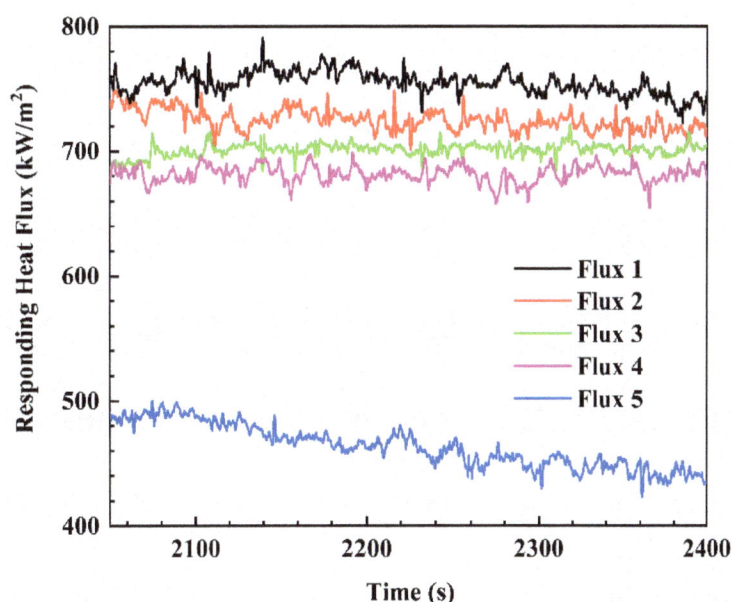

FIG 9 – Heat fluxes of five mould fluxes determined in IET experiment.

Flux discs after IET experiments were subjected to XRD analysis; the results are shown in Figure 10. Flux 1 was fully amorphous, and Flux 2 was basically amorphous with a couple of very small peaks. With the increase of Al_2O_3/SiO_2 ratio, more crystal peaks appeared with the increased intensity, indicating the crystallisation tendency of the mould flux was enhanced. The main phases identified were $Ca_2Al_2SiO_7$ for Flux 3; $LiAlO_2$, $MgAl_6O_{10}$, $Ca_2Al_2SiO_7$ and $Ca_5B_3O_9F$ for Flux 4; and $LiAlO_2$, $Ca_5B_3O_9F$, $CaAl_4O_7$ for Flux 5.

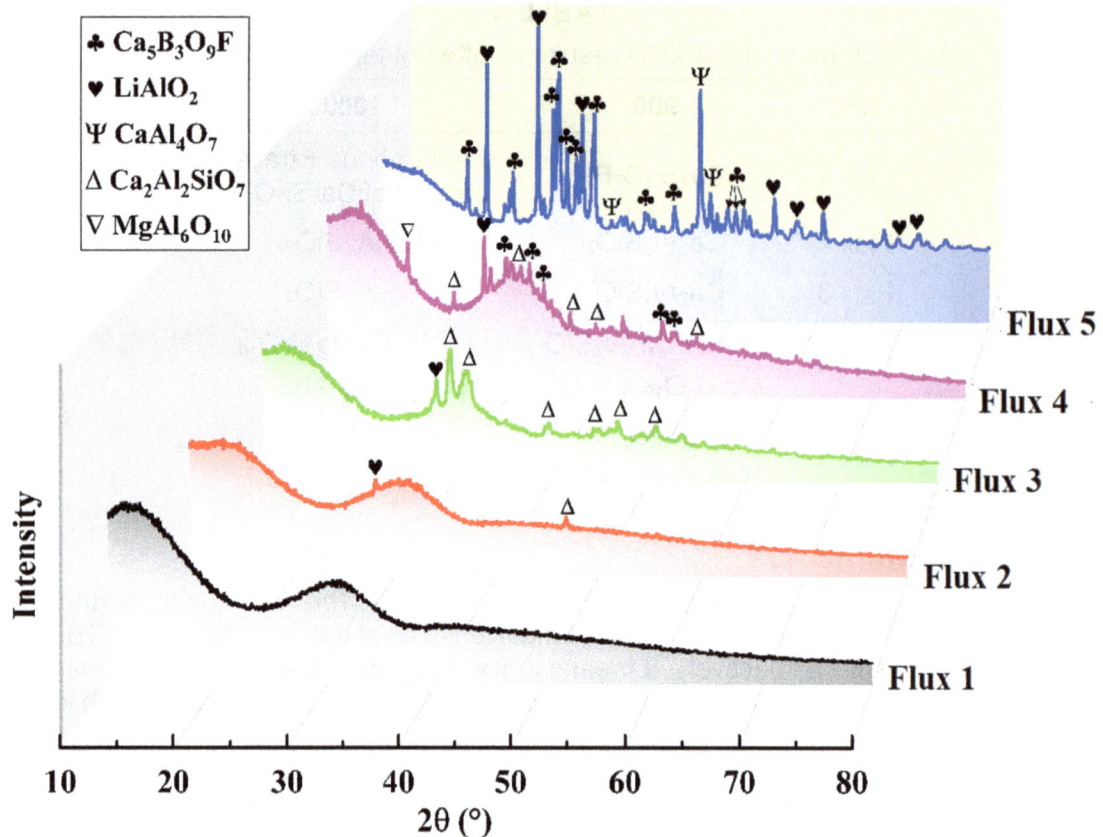

FIG 10 – XRD patterns of the flux discs after IET experiments.

DISCUSSION

Experimental results showed that the change of Al_2O_3/SiO_2 ratio led to the variation of flux properties. With the increase of Al_2O_3/SiO_2 ratio, the flux melting temperature and viscosity increased, while the heat transfer decreased. The dominate silicate structure shifted to the aluminate structure, affecting the viscosity. The increasing Al_2O_3/SiO_2 ratio raised the crystallisation tendency of fluxes, which explains a decreasing trend of heat transfer and an increasing flux melting temperatures.

Effect of Al_2O_3/SiO_2 ratio on the structure and the viscosity of mould fluxes

According to the results of Raman spectroscopy shown in Figures 5 and 6, the increase of Al_2O_3/SiO_2 ratio, ie the increase of Al_2O_3 content and the decrease of SiO_2 content, promoted the formation of aluminate structural units, eg AlF_6, AlO_6, and Al-O-Al, but decreased the Al-O-Si linkage. Both Al-O-Al and Al-O-Si were 3D structure components, while AlF_6, and AlO_6 were 2D structure components in the melt with octahedral coordination. The increased Al-O-Al 3D structural units lead to an increased degree of polymerisation, while the decreased Al-O-Si (3D structure) and the enhanced AlF_6 and AlO_6 2D structural units result in a decreased degree of polymerisation. Overall, the increasing Al_2O_3 content in the flux facilitated the accumulation of Al-related network.

In addition to Al-O related structures, with the increase of Al_2O_3/SiO_2 ratio, the silicate structures were also changed (Figures 5 and 6). The area fraction of Q^2 increased, and the area fraction of Q^0 decreased. There was even Q^3 presented in Flux 5. The degree of polymerisation referring to the silicate structure can be characterised by non-bridging oxygen per silicon NBO/Si, which is determined as:

$$NBO/Si = 4X_{Q^0} + 3X_{Q^1} + 2X_{Q^2} + X_{Q^3} \qquad (3)$$

The values of NBO/Si are shown in Figure 11 where NBO/Si decreased slightly with the increase in the Al_2O_3/SiO_2 ratio from 0.7 to 10.8, indicating an increased polymerisation of silicate structures. (Zhang *et al*, 2008) also found that the increase in Al_2O_3/SiO_2 ratio increased the [SiO_4]-tetrahedral structure and therefore increased polymerisation.

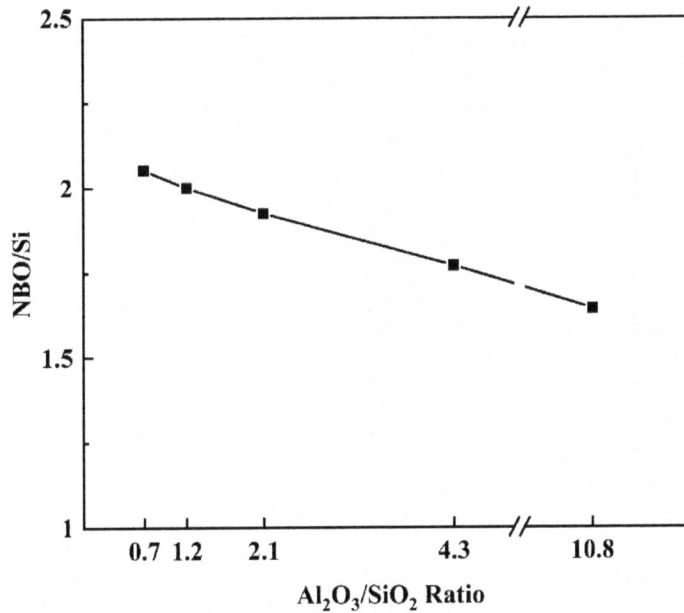

FIG 11 – Values of NBO/Si as a function of Al$_2$O$_3$/SiO$_2$ ratio.

For borate region, within the range of Al$_2$O$_3$/SiO$_2$ ratio from 0.7 to 2.1, changes of the area fractions of borate-related structural units were marginal compared to those of aluminate-related and silicate-related structural units. With further increasing Al$_2$O$_3$/SiO$_2$ ratio from 2.1 to 10.8, the area fraction of borate-related structures increased, correlating with the decrease of SiO$_2$ content. Silicate and aluminate structure can be treated as the backbone of the whole melt (Mills, 1993). The borate structure change is much less significant than those of aluminate and silicate structures. The effect of Al$_2$O$_3$/SiO$_2$ ratio on polymerisation of the whole flux depends mainly on the combined contribution of aluminate and silicate structures.

The viscosity of mould fluxes is generally correlated well with the flux structures, or more specifically, with the degree of polymerisation. As shown in Figure 3, viscosity sharply increased first when Al$_2$O$_3$/SiO$_2$ ratio increased from 0.7 to 1.2, but then rapidly decreased when Al$_2$O$_3$/SiO$_2$ ratio further increased from 1.2 to 10.8. This observation indicates that the flux polymerisation experiences a turning point when Al$_2$O$_3$/SiO$_2$ reaches 1.2 (Al$_2$O$_3$ 20.5 mass per cent, Flux 2). Similar results were also reported by Kim *et al* (2012) in their work on CaO-SiO$_2$-Al$_2$O$_3$-Na$_2$O-Li$_2$O flux and Chen *et al* (2019) on CaO-SiO$_2$-MgO-Al$_2$O$_3$ slag where they found that the viscosity initially increased to a maximum at about 20 mass per cent Al$_2$O$_3$ and then decreased with further addition of Al$_2$O$_3$. This phenomenon cannot be explained by the decreased NBO/Si shown in Figure 11 which indicates an increased degree of polymerisation. Similarly, NBO/T ratio (non-bridging O/tetragonal O) calculated using the following equation:

$$\frac{NBO}{T} = \frac{2\left(X_{CaO}+X_{MgO}+X_{Na2O}+X_{Li2O}-X_{Al2O3}-X_{B2O3}\right)}{X_{SiO2}+2X_{Al2O3}+2X_{B2O3}} \tag{4}$$

also showed the decreased NBO/T with the Al$_2$O$_3$/SiO$_2$ ratio (Figure 12). Therefore, there are other factors to cause this change.

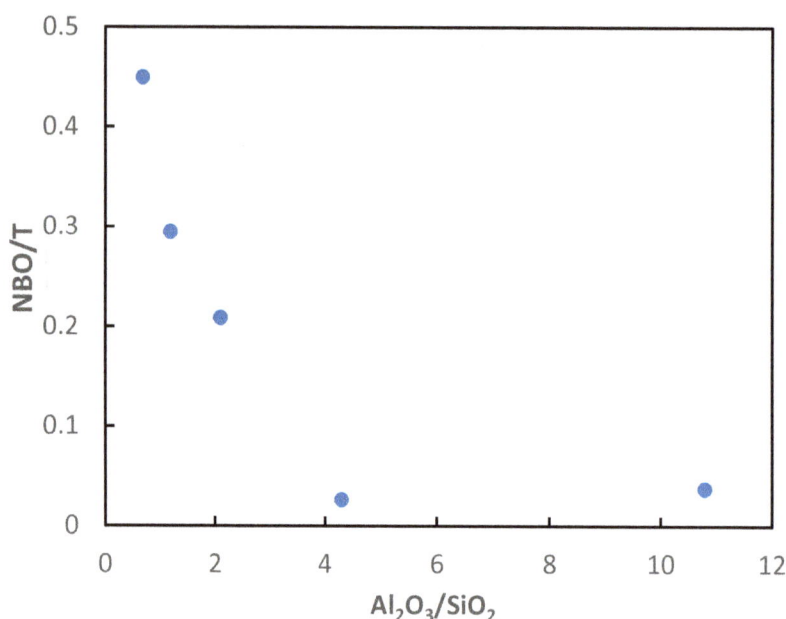

FIG 12 – NBO/T versus Al_2O_3/SiO_2.

When Al_2O_3 is introduced into the silicate network, the Al^{3+} ions can be absorbed into the silicate structure, exhibiting fourfold coordination like Si^{4+}. However, there is a charge difference between Al^{3+} and Si^{4+} which needs to be compensated by M^{2+} and/or M^+ ions, eg Na^+, Li^+, Ca^{2+} and Mg^{2+}. Thus, Al_2O_3 additions act principally as network formers, eg in the case of Fluxes 1 and 2. However, when large amounts of Al_2O_3 are added to the slag (eg Fluxes 4 and 5), the Al^{3+} ions can also act as network breakers forming five- or sixfold coordination (Stebbins Farnan and Xue, 1992). Therefore this phenomenon can be attributed to the amphoteric effect of Al_2O_3, serving as a network former when Al_2O_3 content is low but a network modifier when its concentration is high. The turning point is when the Al/M ratio ($=2X_{Al2O3}/(2X_{Na2O}+2X_{Li2O}+X_{CaO}+X_{MgO})$) approaches 1 which is the case of Flux 4 (Al/M = 0.9).

Chen *et al* (2019) found a transition point (minimum) in the molar Gibbs energy of the mixing of the CaO-SiO_2-MgO-Al_2O_3 slag system at 17 mass per cent Al_2O_3. This calculated transition point is close to the viscosity turning point observed in this work and other reports (Zhang *et al*, 2008; Kim *et al*, 2012).

Effect of Al_2O_3/SiO_2 ratio on heat transfer of mould flux

XRD analysis of isothermally treated fluxes in Figures 7 and 8 reveals that changing Al_2O_3/SiO_2 ratio changes the phase composition of the fluxes. For the highest SiO_2 content (Flux 1, lowest Al_2O_3/SiO_2 ratio), caspidine appeared which is a key component of CaO-SiO_2-Al_2O_3-based commercial mould fluxes. Because of high SiO_2 concentration, this flux and Flux 2 were converted to predominantly amorphous phases after a fast cooling after the heat transfer measurements (Figure 10). In the Flux 3 with Al_2O_3/SiO_2 ratio 2.1, no cuspidine was observed but only $Ca_2Al_2SiO_7$ was detected at both 900 and 1000°C and in the samples after heat transfer measurements (Figure 10). Further increasing Al_2O_3/SiO_2 ratio to 4.3 and 10.8 led to a more complex multi-crystal phase formation, including $LiAlO_2$, $MgAl_6O_{10}$, $CaSiO_3$ or $Ca_5B_3O_9F$ (Figures 7, 8 and 10). Clearly, increasing Al_2O_3/SiO_2 ratio increases significantly the crystallisation tendency of the fluxes.

Heat transfer of mould flux depends strongly on flux crystallinity and types of crystals. With the increase in Al_2O_3/SiO_2 ratio, the heat flux through the slag film decreased gradually first (Flux 1 to Flux 4) and then dropped tremendously (Figure 9, Flux 5). As discussed above, increasing Al_2O_3/SiO_2 ratio increased flux crystallinity which therefore led to the decrease in heat transfer because of the heat scattering effect by grain boundaries and defects. The higher the crystallinity, the lower the heat flux. The significant decrease in the heat flux when Al_2O_3/SiO_2 is 10.8 could be also related to the crystals morphology although it is not the crucial factor in the heat transfer.

Flux selection in terms of Al$_2$O$_3$/SiO$_2$ ratio

As discussed above, CaO-SiO$_2$-based mould fluxes cannot be used for casting of high-Al steel because of the strong flux-steel reaction, leading to the significant variation of the flux properties demonstrated in this work. Above discussion reveals that the turning point of the viscosity reflects the difference in the flux properties and structure between CaO-SiO$_2$ based fluxes and CaO-Al$_2$O$_3$ based mould fluxes. This work also indicates that when SiO$_2$ is very low, eg Flux 5, flux crystallisation is too strong to provide an appropriate lubrication. Therefore, Flux 3 (Al$_2$O$_3$/SiO$_2$ = 2.1 mass per cent ratio) and Flux 4 (Al$_2$O$_3$/SiO$_2$ = 4.3 mass per cent ratio) could be considered to provide suitable structure and crystallinity to achieve optimal flux properties. By minimising the flux-steel reaction, the lower SiO$_2$ flux is preferred, ie Flux 4.

CONCLUSIONS

The effect of Al$_2$O$_3$/SiO$_2$ ratio on structure, viscosity, crystallisation behaviour and heat transfer of CaO-Al$_2$O$_3$-SiO$_2$-B$_2$O$_3$-Na$_2$O-Li$_2$O-MgO-F fluxes was investigated. The major findings are summarised as follows:

- Flux melting temperatures increase with the increase in the Al$_2$O$_3$/SiO$_2$ ratio.

- Viscosity of the flux melts increase significantly with the increasing Al$_2$O$_3$/SiO$_2$ ratio from 0.7 to 1.2, reaching the maximum value, and then decrease with the further increase of Al$_2$O$_3$/SiO$_2$ ratio. Raman spectroscopy analysis revealed that the change of Al$_2$O$_3$/SiO$_2$ ratio leads to the change of flux structures. The turning point for viscosity can be attributed to the amphoteric effect of Al$_2$O$_3$, serving as a network former when Al$_2$O$_3$ content is low but a network modifier when its concentration is high.

- XRD analysis showed that increasing Al$_2$O$_3$/SiO$_2$ ratio increased crystallisation tendency of the fluxes. Heat transfer measurement by IET revealed that increasing Al$_2$O$_3$/SiO$_2$ ratio led to the decrease in the heat flux which is correlated well with the increased crystallinity of the flux.

- It can be concluded from this work that Al$_2$O$_3$/SiO$_2$ = 4.3 (SiO$_2$ 7 mass per cent, Flux 4) is the best candidate among studied CaO-Al$_2$O$_3$-based mould fluxes to provide optimal flux properties but minimise the flux-steel reaction.

ACKNOWLEDGEMENTS

Financial supports from Baosteel-Australia Joint Research and Development Centre (BAJC) and Australian Research Council (ARC) Industrial Transformation Hub are greatly acknowledged.

REFERENCES

Brandaleze, E, Di Gresia, G, Santini, L, Martín, A and Benavidez, E, 2012. Mould fluxes in the steel continuous casting process, in *Science and Technology of Casting Processes*, 1st ed, InTech, pp 205–233.

Chen, Z, Wang, H, Sun, Y, Liu, L and Wang, X, 2019. Insight into the Relationship Between Viscosity and Structure of CaO-SiO$_2$-MgO-Al$_2$O$_3$ Molten Slags, *Metallurgical and Materials Transactions B*, 50(6):2930–2941.

Cho, J W, Blazek, K, Frazee, M, Yin, H, Park, J H and Moon, S W, 2013. Assessment of CaO–Al$_2$O$_3$ Based Mold Flux System for High Aluminum TRIP Casting, *ISIJ Int*, 53:62.

Chung, Y and Cramb, A W, 2000. Dynamic and equilibrium interfacial phenomena in liquid steel-slag systems, *Metall Mater Trans B*, 31B:957.

Gao, E, Wang, W and Zhang, L, 2017. Effect of alkaline earth metal oxides on the viscosity and structure of the CaO-Al$_2$O$_3$ based mold flux for casting high-al steels, *Journal of Non-Crystalline Solids*, 473:79–86.

Gao, J, Wen, G, Huang, T, Bai, B, Tang, P and Liu, Q, 2016a. Effect of slag-steel reaction on the structure and viscosity of CaO-SiO$_2$-based mold flux during high-Al steel casting, *Journal of Non-Crystalline Solids*, 452:119–124.

Gao, J, Wen, G, Huang, T, Bai, B, Tang, P and Liu, Q, 2016b. Effect of Al Speciation on the Structure of High-Al Steels Mold Fluxes Containing Fluoride, *Journal of the American Ceramic Society*, 99(12):3941–3947.

Kamitsos, E I, Karakassides, M A and Chryssikos, G D, 1987. Vibrational spectra of magnesium-sodium-borate glasses, 2, Raman and mid-infrared investigation of the network structure, *The Journal of Physical Chemistry*, 91(5):1073–1079.

Kang, Y B, Kim, M S, Lee, S W, Cho, J W, Park, M S and Lee, H G, 2013. A Reaction Between High Mn-High Al Steel and CaO-SiO$_2$-Type Molten Mold Flux: Part II. Reaction Mechanism, Interface Morphology, and Al$_2$O$_3$ Accumulation in Molten Mold Flux, *Metall Mater Trans B*, 44B:309.

Kim, G H and Sohn, I, 2012. Effect of Al_2O_3 on the viscosity and structure of calcium silicate-based melts containing Na_2O and CaF_2, *Journal of Non-Crystalline Solids,* 358(12–13):1530–1537.

Kim, G H and Sohn, I, 2014. Role of B_2O_3 on the Viscosity and Structure in the CaO-Al_2O_3-Na_2O-Based System, *Metallurgical and Materials Transactions B,* 45(1):86–95.

Kim, H, Matsuura, H, Tsukihashi, F, Wang, W, Min, D J and Sohn, I, 2012. Effect of Al_2O_3 and CaO/SiO_2 on the Viscosity of Calcium-Silicate–Based Slags Containing 10 Mass Pct MgO, *Metallurgical and Materials Transactions B,* 44(1):5–12.

Kim, M S, Lee, S W, Cho, J W, Park, M S, Lee, H G and Kang, Y B, 2013. A Reaction Between High Mn-High Al Steel and CaO-SiO_2-Type Molten Mold Flux: Part I, Composition Evolution in Molten Mold Flux, *Metall Mater Trans B,* 44B:299.

Kim, T S and Park, J H, 2014. Structure-Viscosity Relationship of Low-silica Calcium Aluminosilicate Melts, *ISIJ International,* 54(9):2031–2038.

Ma, N, You, J, Lu, L, Wang, J, Wang, M and Wan, S, 2018. Micro-structure studies of the molten binary K_3AlF_6–Al_2O_3 system by in situ high temperature Raman spectroscopy and theoretical simulation, *Inorganic Chemistry Frontiers,* 5(8):1861–1868.

McMillan, P and Piriou, B, 1983. Raman spectroscopy of calcium aluminate glasses and crystals, *Journal of Non-Crystalline Solids,* 55(2):221–242.

Mills, K C, 1993. The Influence of Structure on the Physico-chemical Properties of Slags, *ISIJ International,* 33(1):148–155.

Park, J H, Min, D J and Song, H S, 2002. Structural Investigation of CaO-Al_2O_3 and CaO-Al_2O_3-CaF_2 Slags via Fourier Transform Infrared Spectra, *ISIJ International,* 42(1):38–43.

Park, J Y, Kim, G H, Kim, J B, Park, S and Sohn, I, 2016. Thermo-Physical Properties of B_2O_3-Containing Mold Flux for High Carbon Steels in Thin Slab Continuous Casters: Structure, Viscosity, Crystallization, and Wettability, *Metallurgical and Materials Transactions B,* 47:2582.

Stebbins, J F, Farnan, I and Xue, X, 1992. The structure and dynamics of alkali silicate liquids: A view from NMR spectroscopy, *Chem, Geol.* 96:371–385.

Wang, W and Cramb, A, 2005. The Observation of Mold Flux Crystallization on Radiative Heat Transfer, *ISIJ International,* 45:1864.

Yang, J, Zhang, J, Ostrovski, O, Zhang, C and Cai, D, 2018. Effects of B_2O_3 on Crystallization, Structure, and Heat Transfer of CaO-Al_2O_3-Based Mold Fluxes, *Metallurgical and Materials Transactions B,* 50:291.

Yang, J, Zhang, J, Sasaki, Y, Ostrovski, O, Zhang, C, Cai, D and Kashiwaya, Y, 2017. Effect of B_2O_3 on Crystallization Behavior, Structure, and Heat Transfer of CaO-SiO_2-B_2O_3-Na_2O-TiO_2-Al_2O_3-MgO-Li_2O Mold Fluxes, *Metall Mater Trans B,* 48B:2077.

Zhang, Z, Wen, G, Tang, P and Sridhar, S, 2008. The Influence of Al_2O_3/SiO_2 Ratio on the Viscosity of Mold Fluxes, *ISIJ International,* 48(6):739–746.

Zheng, K, Zhang, Z, Liu, L and Wang, X, 2014. Investigation of the Viscosity and Structural Properties of CaO-SiO_2-TiO_2 Slags, *Metallurgical and Materials Transactions B,* 45(4):1389–1397.

Zhou, L, Li, J, Wang, W and Sohn, I, 2017. Wetting Behavior of Mold Flux Droplet on Steel Substrate With or Without Interfacial Reaction, *Metall Mater Trans B,* 48B:1943.

Zhou, L, Wang, W and Zhou, K, 2015. Effect of Al_2O_3 on the Crystallization of Mold Flux for Casting High Al Steel, *Metallurgical and Materials Transactions E,* 2:99.

Pushing the boundaries of slag operability – processing of high-MgO nickel concentrates with the Ausmelt TSL process

J Wood[1], J Coveney[2], S Creedy[3], D Grimsey[4] and A Rich[5]

1. Manager Technology, Metso Australia, Melbourne Vic 3175. Email: jacob.wood@metso.com
2. Principal Process Engineer, Metso Australia, Melbourne Vic 3175.
   Email: james.coveney@metso.com
3. Customer Account Manager, Metso Australia, Melbourne Vic 3175.
   Email: stefanie.creedy@metso.com
4. Principal Smelter Development Metallurgist, BHP, Kalgoorlie WA 6430.
   Email: david.grimsey@bhp.com
5. Smelter Development Manager, BHP, Kalgoorlie WA 6430. Email: anthony.rich2@bhp.com

ABSTRACT

In 2021, BHP and Metso examined processing of high-MgO nickel sulfide concentrates using Metso's Ausmelt Top-Submerged-Lance (TSL) technology. Test work was conducted in Metso's pilot test work facility in Dandenong, Australia to explore operability of the SiO_2-FeO-MgO-CaO-NiO slag system across a wide range of compositions, temperatures and bath oxygen potentials. Pilot-scale testing aimed to define slag 'operability limits', representing the lowest bath temperature at which stable process and equipment operation could be maintained. This work was supported by FactSage™, version 8.2 (by Thermfact/CRCT and GTT-Technologies) thermodynamic modelling, slag viscosity measurements, physical characterisation of quenched slag samples performed by the University of Queensland and benchmarking of commercial-scale TSL nickel smelting operations.

A wide range of slag compositions were examined, with Fe/SiO_2 ratios varying from 0.4–1.1, CaO content from 0.8–7.0 wt per cent and MgO content from 6–19 wt per cent. Slag SiO_2/MgO ratio, wt per cent CaO and matte grade were found to have the greatest impacts on identified operability limits.

Operability limits were found to be influenced by both the solids content in slag and viscosity of the remaining liquid slag phase, with the relative contribution of these parameters heavily influenced by the slag composition. In the majority of trials, limits were defined by a theoretical solids content of 40–50 per cent, however in trials with a low Fe/SiO_2 ratio and/or low wt per cent CaO, limits were characterised by a much lower solids content due to the increased effect of the liquid slag viscosity in determining the behaviour of these slags.

The test work highlighted inherent flexibility of the Ausmelt TSL process to operate across a wide slag range of slag compositions and recover from process disturbances without an interruption to feeding. The trials also demonstrated the possibility for Ausmelt TSL technology to process concentrates with an Fe/MgO ratio as low as 1.4, which has important implications to the commercial-scale processing of high MgO feeds. Arsenic rejection across the trials was very good, with only 30 per cent of arsenic in the feed inputs reporting to the matte phase. Such high levels of arsenic removal provide the Ausmelt TSL process with a notable advantage over alternative smelting technologies.

INTRODUCTION

Metso's Ausmelt TSL process has achieved widespread acceptance as a leading smelting technology for the treatment of copper, nickel, lead, tin and zinc bearing feeds (Wood, Hoang and Hughes, 2017). Core to the technology is a vertically suspended lance operated with its tip submerged in the slag layer of the molten bath (Figure 1). Oxygen enriched air and fuel are injected via the lance resulting in significant bath mixing and agitation with consequently high rates of energy and mass transfer. The Ausmelt TSL process is used at more than 50 non-ferrous metals production facilities globally.

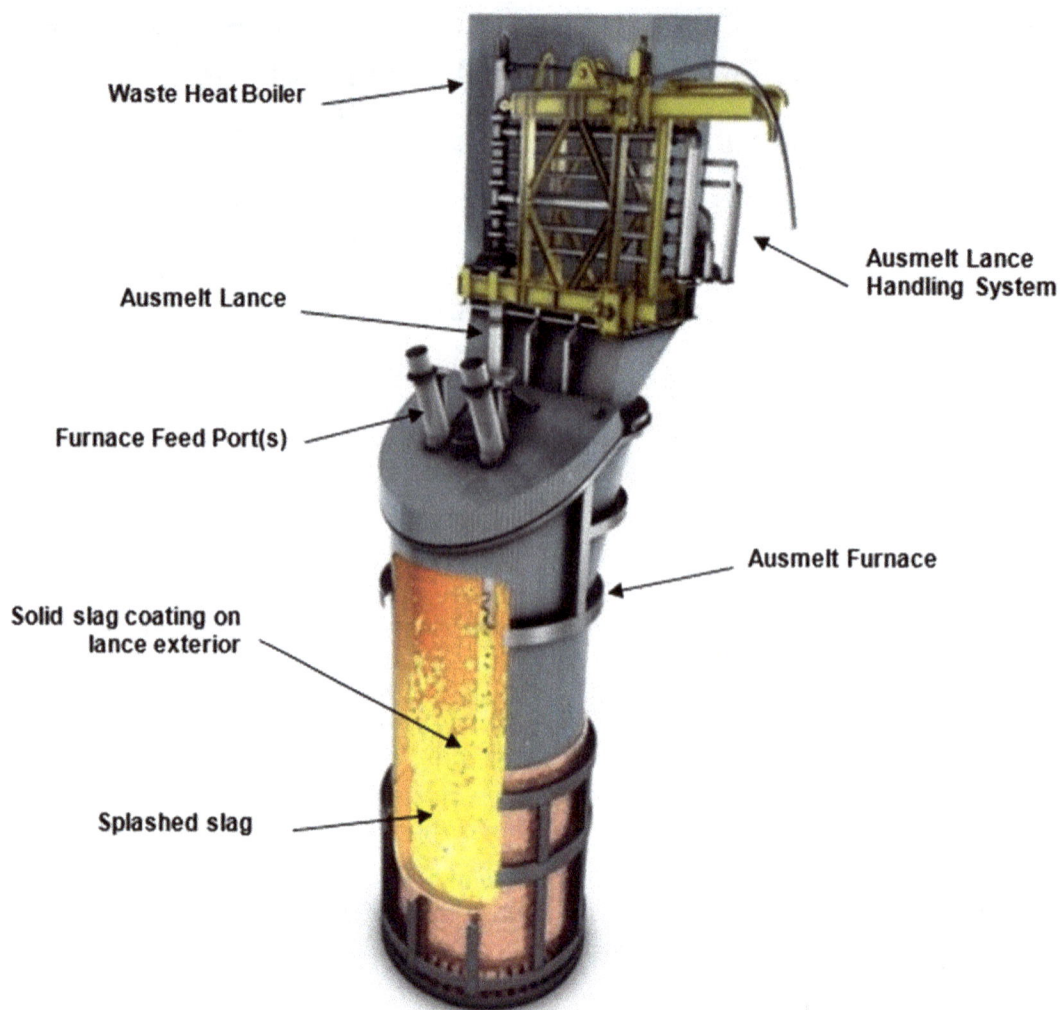

FIG 1 – Ausmelt TSL process schematic.

The Ausmelt TSL process has been applied at commercial scale for the treatment of nickel-bearing concentrates, matte and residues (Table 1). Jinchuan Nickel Mining Company (JNMC) utilises Ausmelt TSL technology to process more than 1 100 000 t/a of nickel sulfide concentrates to produce a low-grade matte, which is further upgraded in downstream Peirce Smith (PS) converters (Zhou *et al*, 2010). Jilin Ji'en Nickel Industry Co. Ltd. (JJNI) employ a similar process flow sheet for the processing of 275 000 t/a of nickel concentrates (Aspola *et al*, 2012).

TABLE 1

Ausmelt TSL nickel references.

Customer	Location	Start-up	Feed	Capacity (t/a)
Jilin Ji'en Nickel (JJNI)	China	2009	Concentrates	275 000
Jinchuan Nickel (JNMC)	China	2008	Concentrates	1 100 000
Anglo American Platinum	South Africa	2002	Matte	213 000
Bindura Nickel	Zimbabwe	1995	Residues	10 000
Rio Tinto Zimbabwe	Zimbabwe	1992	Residues	7700

Anglo American Platinum have installed two Ausmelt TSL converting furnaces to upgrade approximately 210 000 t/a of low-grade electric furnace matte via a continuous converting process (Hundermark *et al*, 2011). The resultant composition of the high-grade matte from the TSL process is critical to recovery of contained platinum and palladium.

As with other bath smelting processes, the Ausmelt process relies on regulation of the slag fluidity to ensure effective mass and energy transfer (Wood and Hughes, 2016), straightforward transfer of molten products to downstream handling/refining operations and to minimise the propensity for bath foaming. The process typically aims to achieve operation at the lowest possible temperature through a combination of slag compositional and bath temperature control. Silica, limestone and in some cases hematite flux addition as part of the overall furnace charge is used to regulate the slag composition whilst bath temperatures are controlled by adjusting lance fuel and/or oxygen flow rates.

Unlike some other smelting technologies, the Ausmelt TSL process is capable of operating at temperatures well below the slag liquidus, with elevated concentrations of solids in slag. The highly turbulent nature of the Ausmelt TSL process means solid particles remain suspended, without settling out to form a layer of build-up on the furnace hearth.

TEST WORK

In 2021, BHP and Metso conducted 15 individual trials in Metso's pilot scale test work facility in Dandenong, Australia to examine the processing of high-MgO nickel concentrates using Ausmelt TSL technology. The test work examined operability of the SiO_2-FeO-MgO-CaO-NiO slag system across a wide range of compositions, temperatures and bath oxygen potentials and aimed to define slag 'operability limits', representing the lowest bath temperature at which stable process and equipment operation could be maintained.

Materials and equipment

The following materials were used in the test work:

- Leinster nickel sulfide concentrate
- Mt Keith nickel sulfide concentrate
- magnesia powder
- silica-revert flux mixture
- limestone flux.

The feed rates of Leinster concentrate, Mt Keith concentrate and magnesia powder were varied during the test work to achieve a specific Fe/MgO ratio in feed blend for each trial. Magnesia powder was used to simulate processing of high-MgO concentrates, to achieve a target MgO concentration in slag. A mixture of silica and crushed revert from BHP's Kalgoorlie Nickel Smelter was introduced to achieve the target Fe/SiO_2 ratio, with limestone flux also added in the majority of trials to achieve a target wt per cent CaO in slag. Concentrates and fluxes were dispensed individually and combined with a small quantity of water in a screw mixer to minimise material carryover with the process offgas. The combined feed charge was then directed to the furnace roof with an inclined conveyor. Figure 2 depicts the arrangement of equipment used for concentrate and flux handling in the test work.

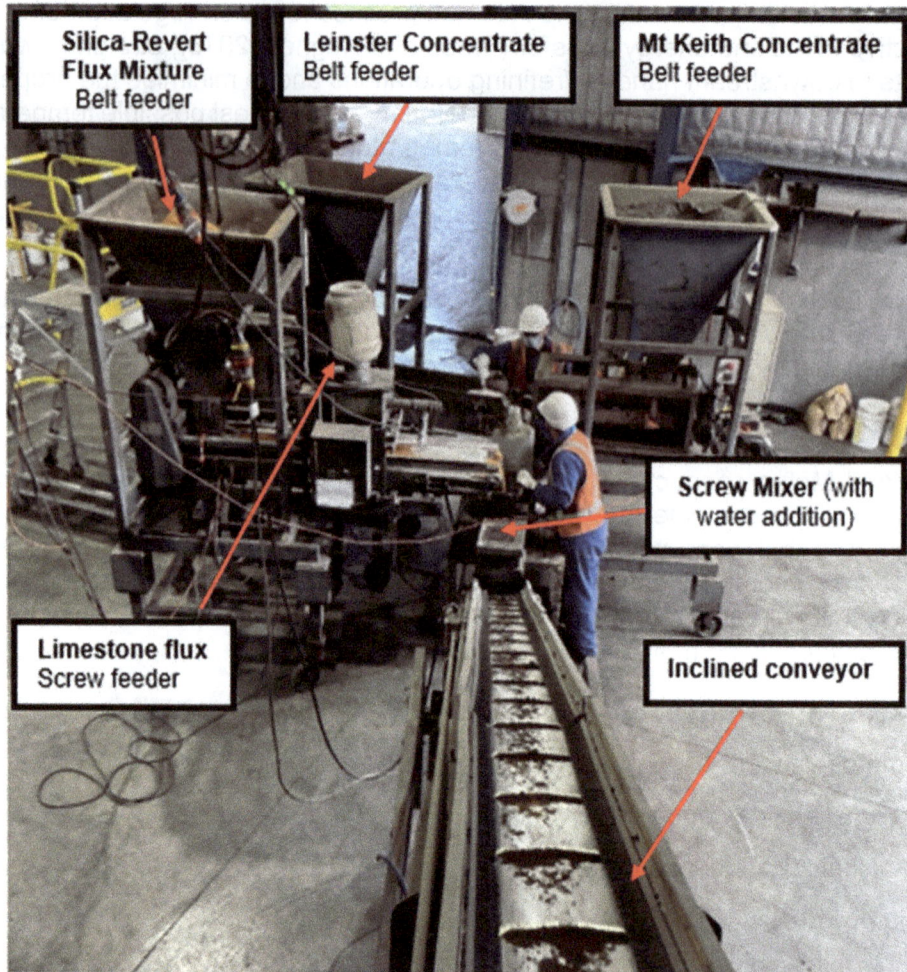

FIG 2 – Feed system arrangement.

The test work employed a refractory-lined furnace with internal dimensions of 1900 mm height and 500 mm diameter. The furnace roof contained four openings; the gas offtake, lance port, feed port and sampling/inspection port. The furnace was lined with magnesia-chrome refractories. A single water-cooled, copper taphole block with graphite inserts was employed in the trials. The taphole had two openings, one at the hearth level and another at a height of 200 mm. Figure 3 presents furnace equipment used in the test work.

FIG 3 – Furnace arrangement.

Natural gas was used as a fuel in all trials. Flow rates of natural gas and oxygen, delivered via the Ausmelt lance were adjusted to achieve the desired bath temperature and to maintain a relatively consistent lance injection volume between the trials.

Matte and slag were tapped into sand-lined moulds (Figure 4) from both the top and bottom tapholes. The tapholes were opened using oxygen tapping rods and closed (when required) using fireclay. In instances where a mixture of slag and matte were tapped, these phases were subsequently separated once the mould contents had solidified and cooled sufficiently.

FIG 4 – Molten product handling.

Process offgas was directed to downstream cooling, de-dusting and cleaning operations, prior to discharge to the atmosphere. Two induced draught fans provided suction to maintain the furnace under constant negative pressure.

Objectives

The primary objective of the test work was to define 'operability limits' across a wide range of slag compositions and matte grades. The 'operability limit' represented the lowest bath temperature for which stable process and furnace operation could be maintained. In particular, the test work examined links between these identified operability limits and the following process parameters:

- feed Fe/MgO ratio
- slag Fe/SiO_2 ratio
- slag MgO/SiO_2 ratio
- slag wt per cent CaO
- slag wt per cent MgO
- matte wt per cent Fe.

Four main indicators were used to determine when an operability limit had been reached, namely:

1. Difficulties tapping slag and/or matte.
2. Formation of build-up ('mush') on the furnace hearth.
3. Rapid and excessive accretion formation around the furnace roof ports and/or offtake.
4. 'Foamy' bath condition due to injected gases being unable to disengage from the slag bath.

Methodology

Prior to commencing each trial, FactSage™ software was used to calculate the slag liquidus temperature, based on target slag and matte compositions for the trial, with this information used to enable exploration of process operability limits in a controlled manner as follows:

- Commence operation with a bath temperature at or slightly above the slag liquidus.

- Confirm slag composition and matte grade are stable and within the target range.

- Progressively reduce the bath temperature (measured via immersion thermocouple and confirmed with optical pyrometer during tapping) until the operability limit for a particular set of conditions has been identified.

- Collect quenched (water-granulated) slag samples during tapping and/or via the furnace roof inspection port at the identified operability limit.

- Increase the bath temperature slightly and maintain stable operation at or slightly above the identified operability limit for as long as possible.

Quenched slag samples taken during the test work were subsequently analysed by the University of Queensland (UQ) using an Electron Probe Micro-Analyser (EPMA) to visually determine the slag solids content for operability limits identified in each trial (Figure 5). This work was supported by results from FactSage™ thermodynamic modelling, which was used to calculate the expected solids content in slag at a particular set of operating conditions (slag composition, matte grade and bath temperature).

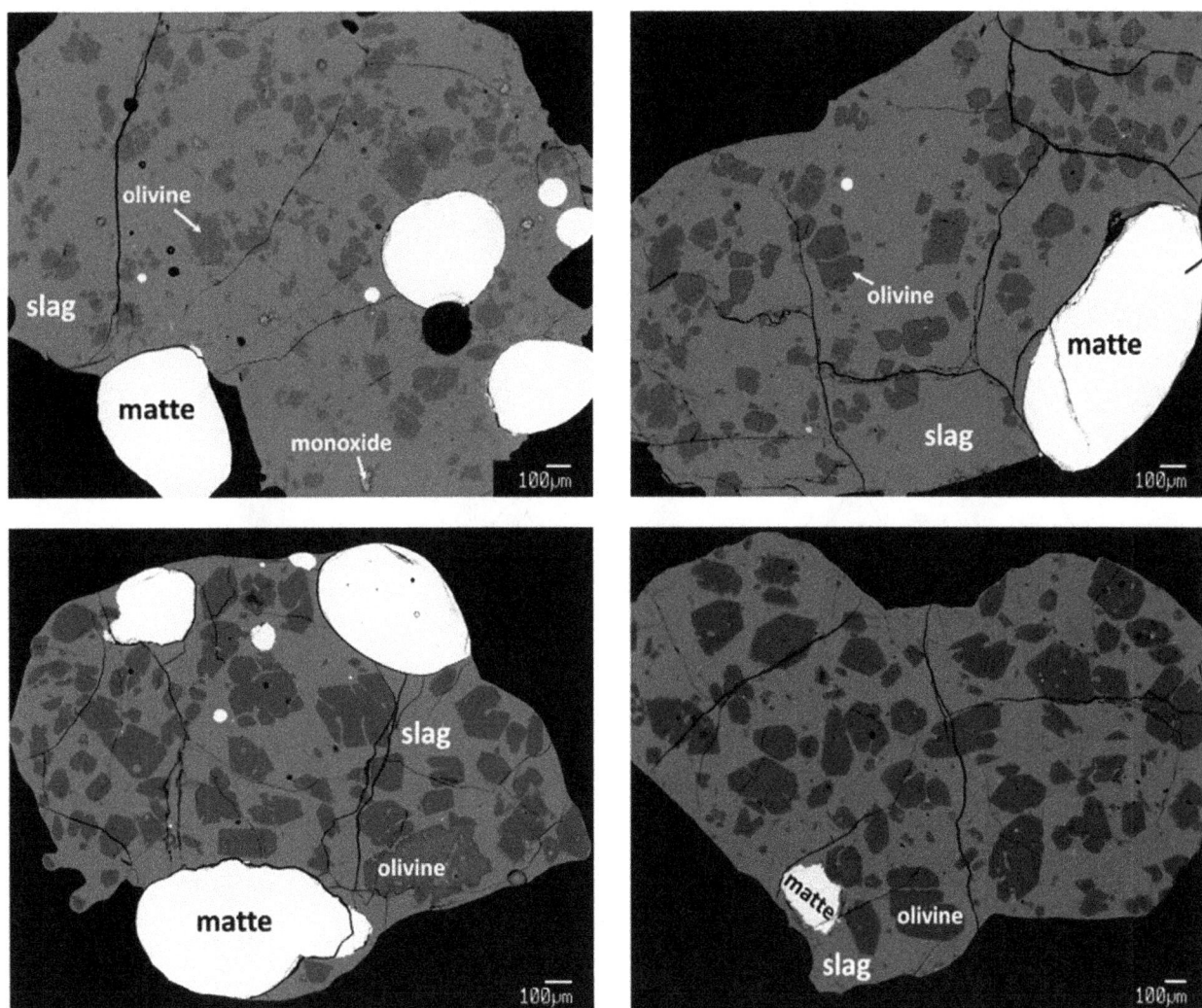

FIG 5 – Quenched slag analysis with EPMA.

Operating conditions

A wide range of slag compositions and matte grades were examined in the test work (Table 2). This presented challenges in directly comparing operability limits for each trial, particularly given the combined effects of multiple parameters in determining the limit for a particular set of conditions. An isothermal phase diagram generated with FactSage™ is also shown in Figure 6, with the slag compositional range investigated in the test work presented by the shaded area of the figure. The $p(O_2)$ chosen for this diagram is typical for slag in equilibrium with matte containing 15 wt per cent Fe at a temperature of 1300°C.

TABLE 2

Test work operating conditions.

Parameter	Minimum	Maximum
Feed Fe/MgO	1.4	3.0
Matte wt% Fe	2.2	32.1
Slag SiO_2/MgO	1.6	5.6
Slag Fe/SiO_2	0.4	1.3
Slag wt% CaO	0.6	8.4

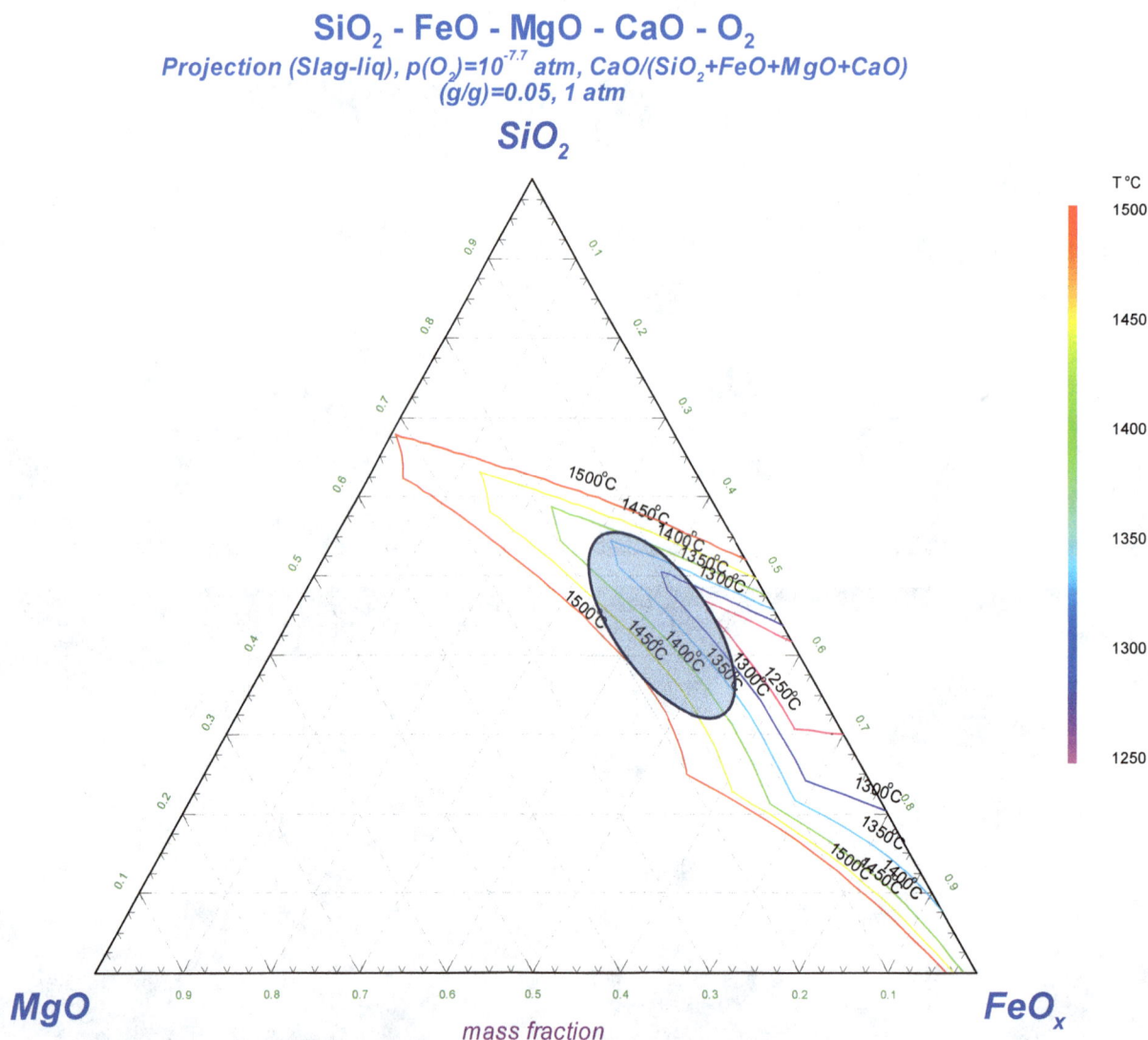

FIG 6 – Isothermal phase diagram.

Results and discussion

Operability limits were successfully identified in ten of the trials. Prior to the test work, it was hypothesised that the slag solids content would be the dominant parameter influencing the operability limit for a particular set of conditions. Results from the test work indicated however, that limits were defined by both the slag solids fraction and viscosity of the remaining liquid slag phase, with the relative contribution of these parameters heavily influenced by the slag composition. Operability limits were found to be largely dictated by the slag SiO_2/MgO ratio, bath oxygen potential (matte wt per cent Fe) and slag wt per cent CaO (Figure 7), with minimal impact from the slag Fe/SiO_2 ratio. Based on this finding, operability limits were grouped into three distinct data subsets (Table 3). The effects of these parameters are discussed in more detail in the following sections.

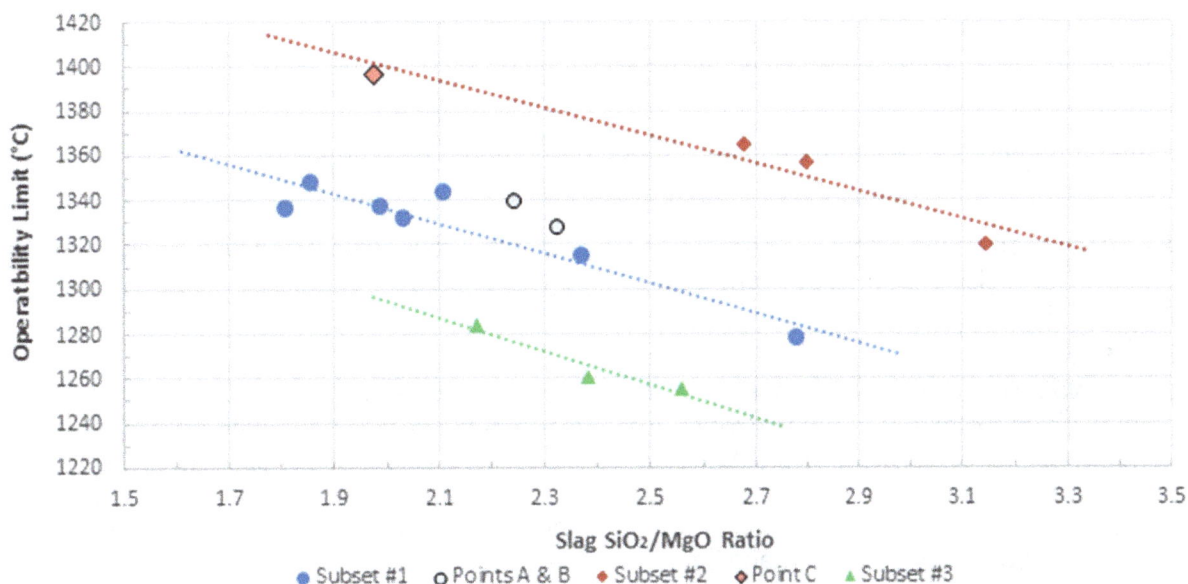

FIG 7 – Operability limit correlation with Slag SiO_2/MgO ratio, %CaO and matte %Fe.

TABLE 3

Data subset operating conditions.

	Data subset	Slag wt% CaO	Slag Fe/SiO$_2$	Matte wt%Fe
●	Subset #1	Mid (5–8 wt%)	Mid (0.8–1.1)	Mid (8–13 wt%)
○	Points A & B	Mid (6–7 wt%)	Low (0.5–0.6)	Low (5 wt%)
◆	Subset #2	Low (1 wt%)	Low (0.5)	Mid (9–13 wt%)
◇	Point C	Low (2 wt%)	Mid (0.9)	Mid (8 wt%)
▲	Subset #3	Mid (5–7 wt%)	Mid (0.8–1.0)	High (25–28 wt%)

Slag SiO₂/MgO ratio

Operability limits were strongly influenced by the slag SiO_2/MgO ratio, evidenced by the consistent slope of data sets in Figure 7. Modelling in FactSage™ confirmed the strong effect of SiO_2/MgO ratio on the slag liquidus temperature and hence, solids fraction in slag but also suggested minimal impact of slag SiO_2/MgO ratio on the liquid slag viscosity (Figure 8).

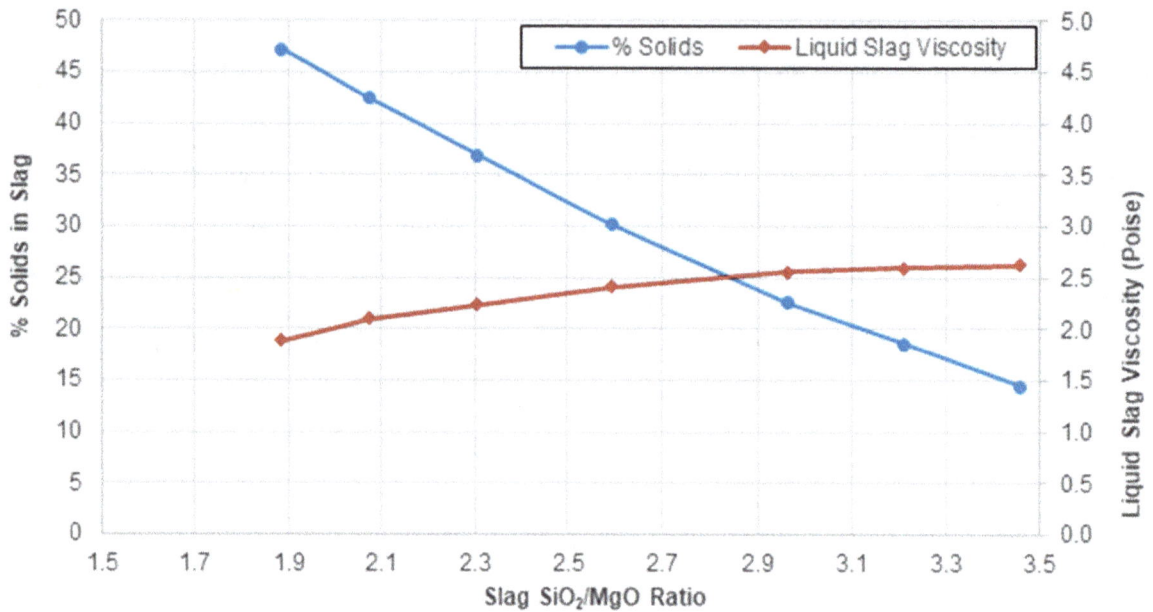

FIG 8 – Slag% solids and viscosity correlation with SiO$_2$/MgO ratio (calculated with FactSage™ at 1320°C, 5 wt per cent CaO and 15 wt per cent Fe in matte).

Slag% CaO

Decreasing slag wt per cent CaO necessitated operation at higher temperatures across the range of slag compositions examined in the test work. This finding is supported by results from Factsage™ modelling, which indicated not only an increased solids content at low wt per cent CaO, particularly at low SiO$_2$/MgO ratio (Figure 9) but also a significant increase in liquid slag viscosity (Figure 10). Consequently, it was reasoned that operability limits for data Subset #2 and Point C, were dictated by both the slag solids concentration and liquid slag viscosity.

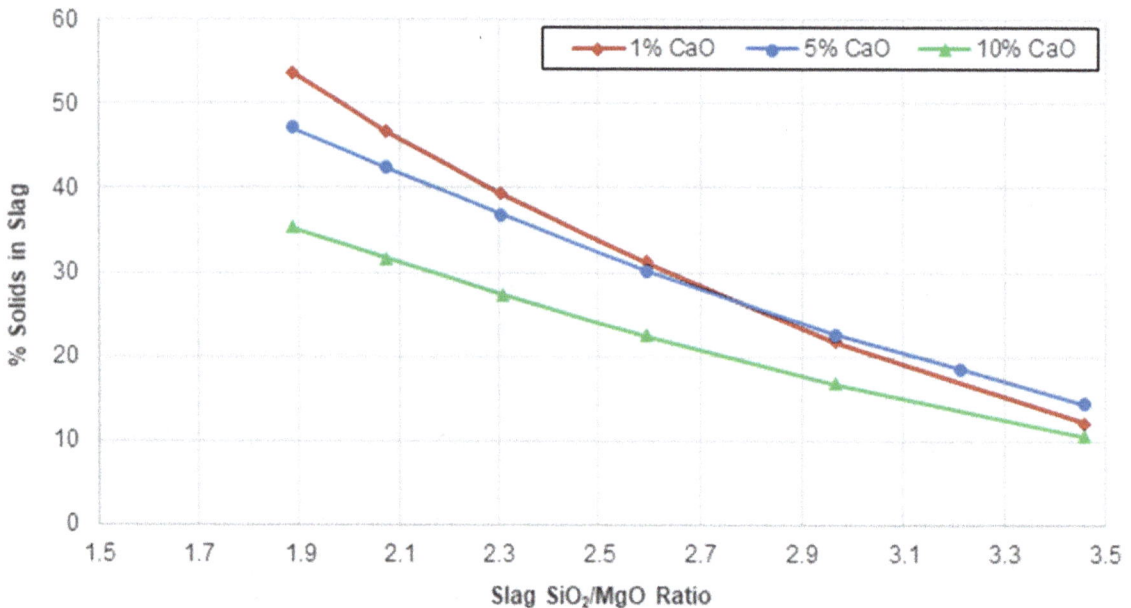

FIG 9 – Slag% solids correlation with SiO$_2$/MgO ratio at variable wt per cent CaO (calculated with FactSage™ at 1320°C and 15 wt per cent Fe in matte).

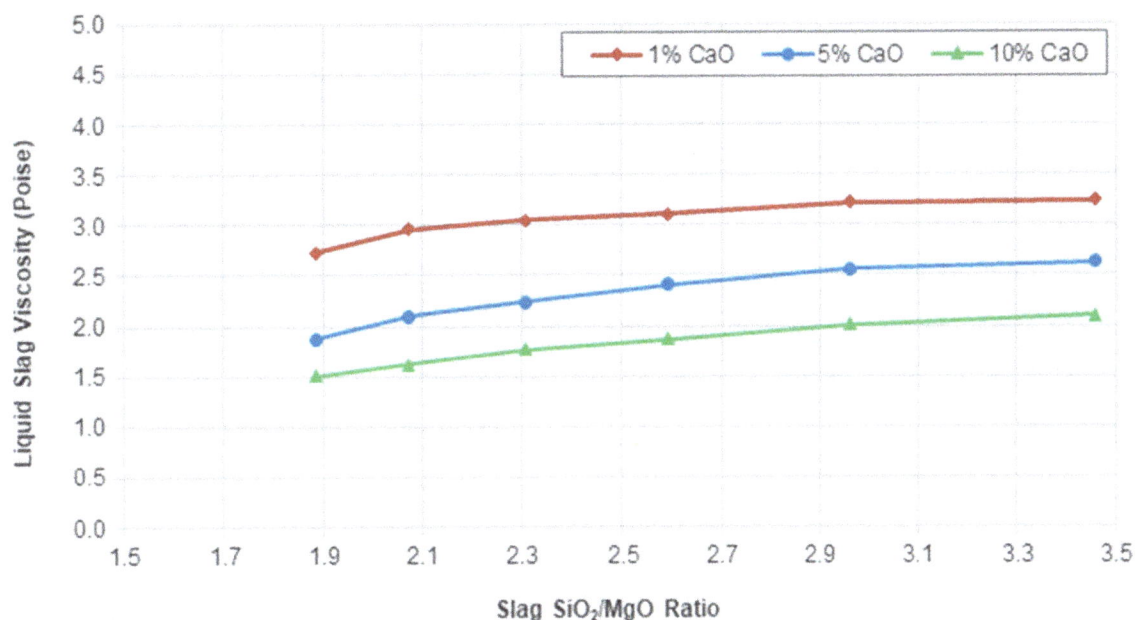

FIG 10 – Slag viscosity correlation with SiO_2/MgO ratio at variable wt per cent CaO (calculated with FactSage™ at 1320°C and 15 wt per cent Fe in matte).

Matte%Fe

Increasing wt per cent Fe in matte (ie decreasing matte grade) enabled operation at lower temperatures across the range of matte grades examined in the test work. Modelling in FactSage™ indicated a significantly higher slag solids content with decreasing wt per cent Fe in matte, particularly at high SiO_2/MgO ratio (Figure 11), due to displacement of MgO by NiO within the olivine structure and consequently, an increasing quantity of MgO reporting to the liquid slag phase. FactSage™ modelling also suggested minimal effect of matte grade on the liquid slag viscosity (Figure 12), operability limits for data Subset #3, were almost entirely dictated by the slag solids content.

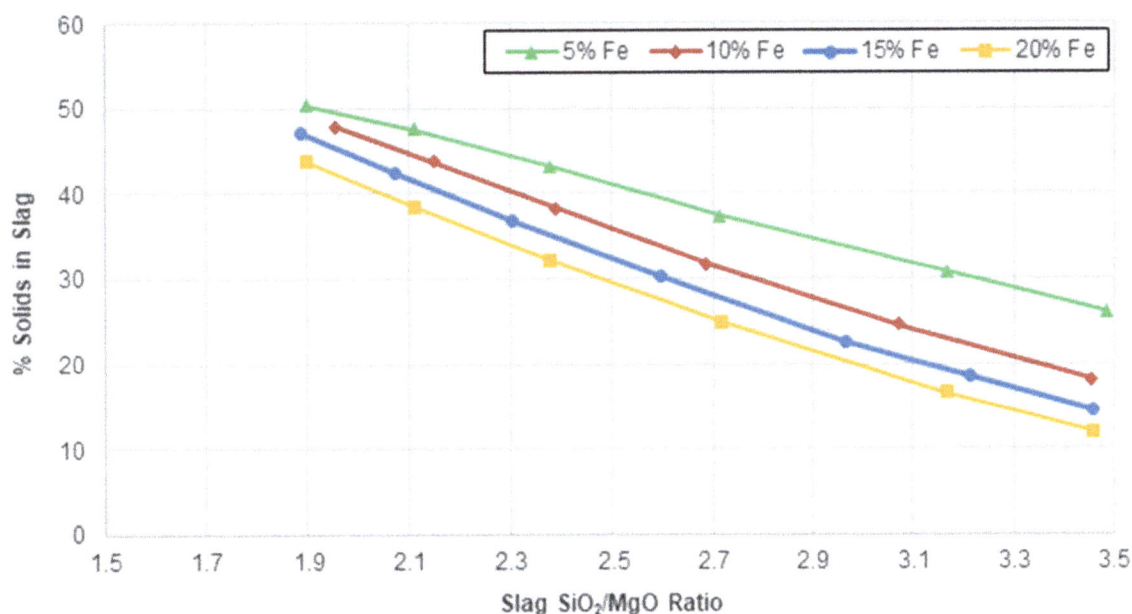

FIG 11 – Slag% solids correlation with SiO_2/MgO ratio at variable wt per cent Fe in matte (calculated with FactSage™ at 1320°C, $Fe/SiO_2 = 0.8$ and 5 wt per cent CaO in slag).

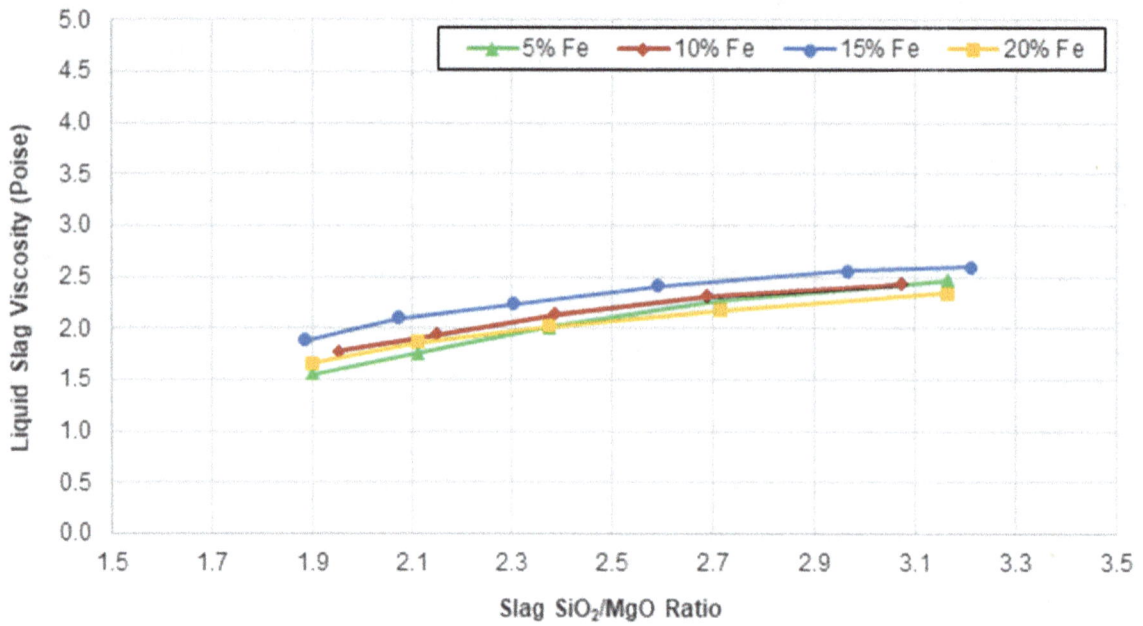

FIG 12 – Slag viscosity correlation with SiO_2/MgO ratio at variable wt per cent Fe in matte (calculated with FactSage™ at 1320°C, $Fe/SiO_2 = 0.8$ and 5 wt per cent CaO in slag).

Slag Fe/SiO₂ ratio

Variability in slag Fe/SiO_2 ratio was found to have minimal impact on identified operability limits across the range of slag compositions examined in the test work. Whilst limits at low Fe/SiO_2 ratio were typically higher, this was rationalised by variability in matte grade (Points A and B) and wt per cent CaO (Point C). Modelling in FactSage™ indicated limited influence of Fe/SiO_2 ratio on the slag solids concentration, except at elevated Fe/SiO_2 ratios (Figure 13). Of greater significance, however, is the significant increase in liquid slag viscosity predicted at low Fe/SiO_2 ratio (Figure 14), which is likely to have impacted findings from the test work.

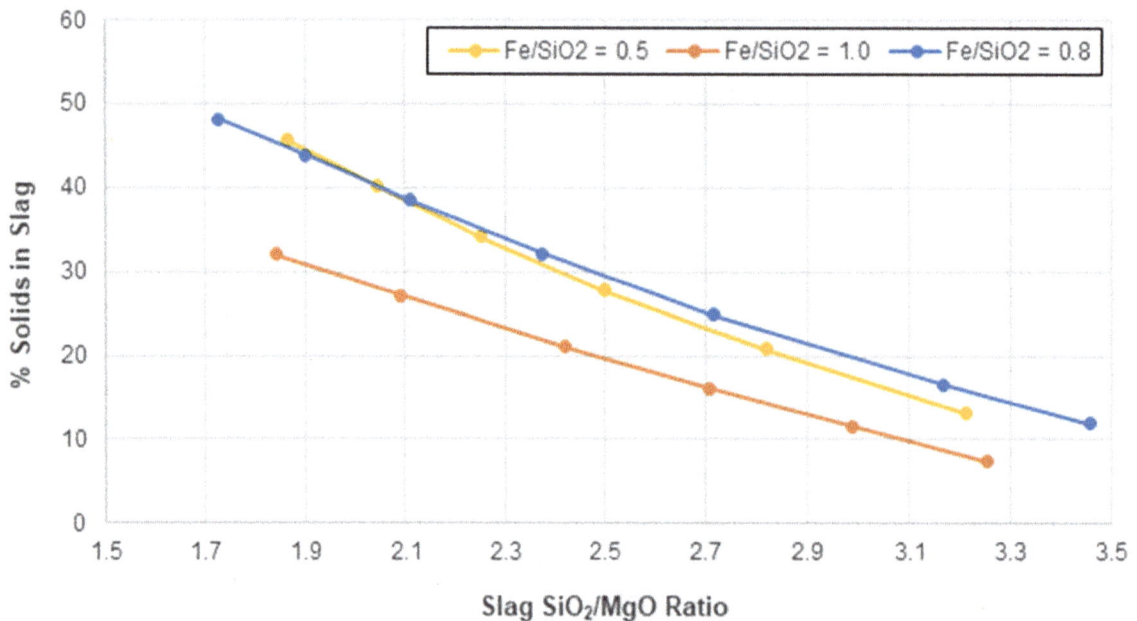

FIG 1 – Slag% solids correlation with SiO_2/MgO ratio at variable Fe/SiO_2 ratio (calculated with FactSage™ at 1320°C, 20 wt per cent Fe in matte and 5 wt per cent CaO in slag).

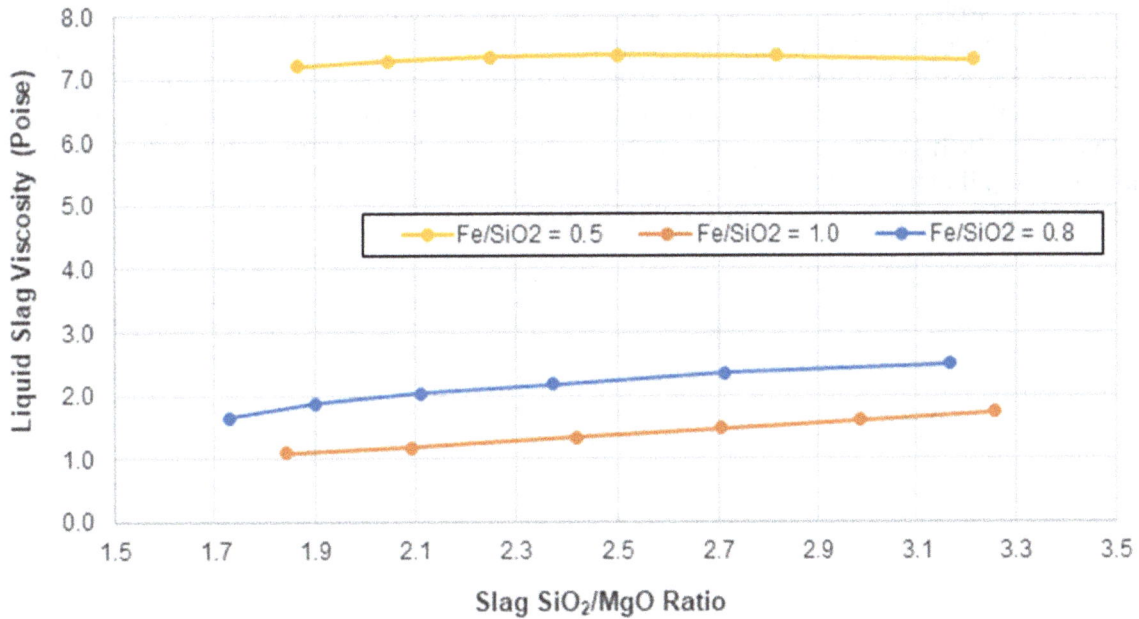

FIG 14 – Slag viscosity correlation with SiO_2/MgO ratio at variable Fe/SiO_2 ratio (calculated with FactSage™ at 1320°C, 20 wt per cent Fe in matte and 5 wt per cent CaO in slag).

Slag EPMA analysis

Quenched slag samples taken throughout the test work were analysed by the UQ to visually determine the solids concentration in slag at the operability limit(s) identified in each trial. Slag samples were for the most part, characterised by a visual solids content in slag of 15–30 per cent. FactSage™ was also used to determine the theoretical slag solids content for each operability limit, which suggested a much higher solids fraction than observed via EPMA (Figure 15). The reasons for discrepancies between these two data sets were not clear and as this was not a primary objective of the test work, this was not investigated further.

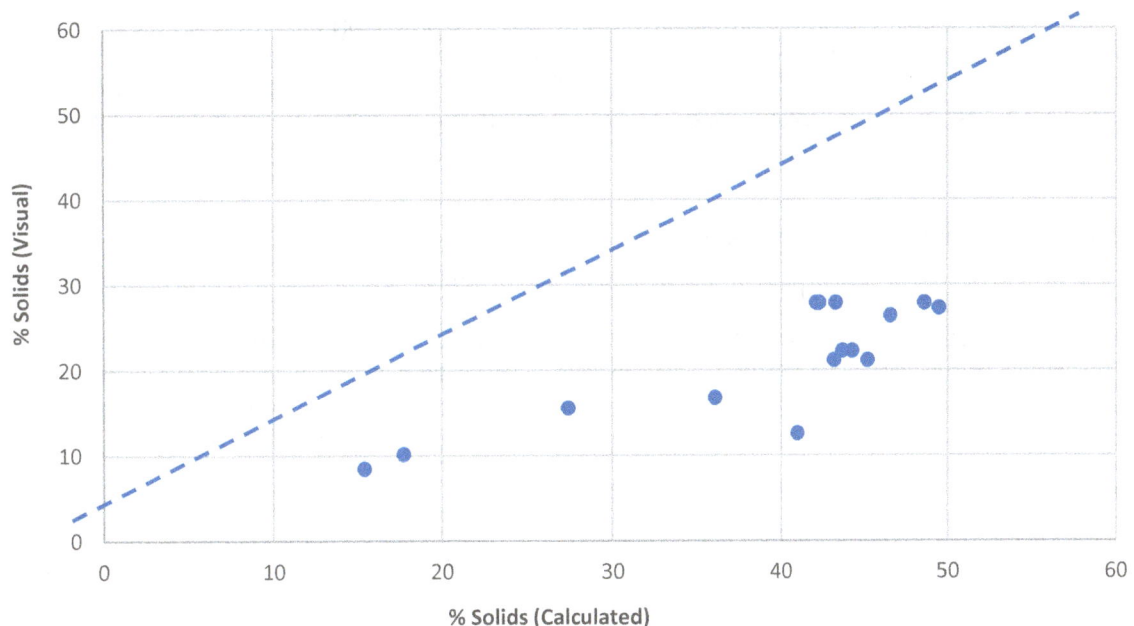

FIG 15 – Correlation in slag solids content.

Operability limit definition

Operability limits were for the most part, characterised by a slag solids content of 40–50 per cent and liquid slag viscosity of 1–3 Poise (as determined with FactSage™), with the exception of data Subset #2 (low per cent CaO and Fe/SiO_2 ratio) which was had only 15–30 per cent solids and viscosity of 6–7 Poise. These results indicated the relative contribution of solids content and liquid slag viscosity in defining the operability limit for a particular set of conditions, varied with slag composition. Figure 16 presents the relationship between per cent solids and liquid slag viscosity (determined with FactSage™) for operability limits identified in the test work. A suggested range of conditions, for which stable operation could be achieved is also illustrated by the shaded region in the figure, noting that further test work would be required to expand the understanding of permissible operating conditions, particularly at low Fe/SiO$_2$ ratio.

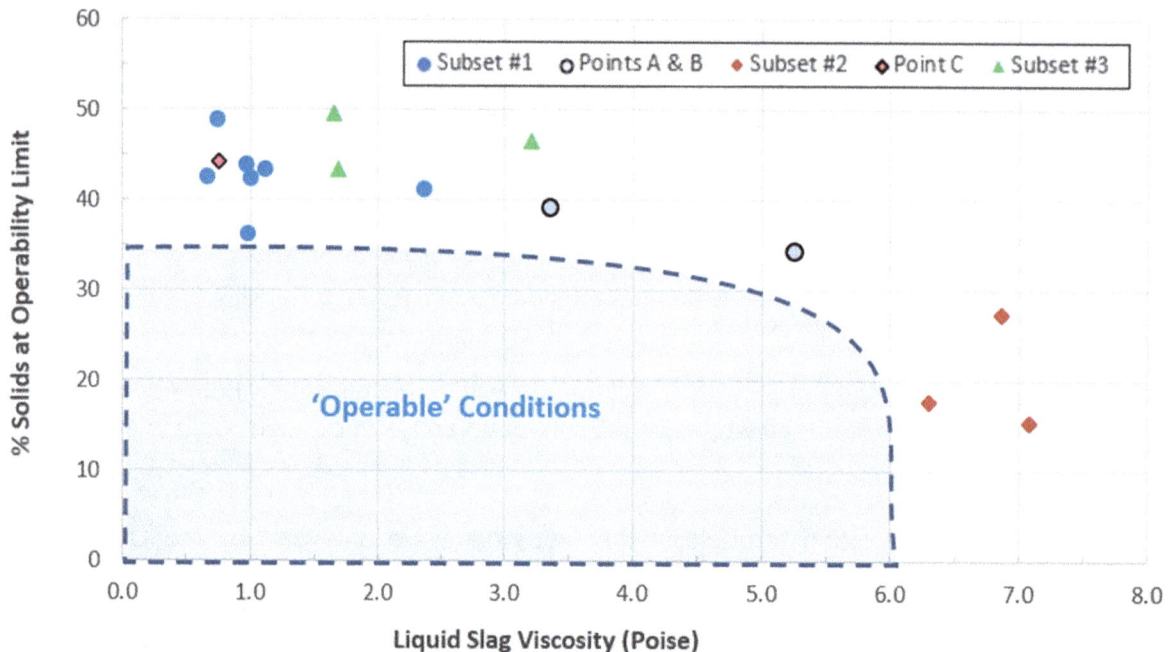

FIG 16 – Operability limit variation with per cent solids and liquid slag viscosity.

COMMERCIAL SCALE OPERATIONS

Results from the test work indicated the Ausmelt TSL nickel smelting process can operate with 30–40 per cent solids in slag (determined with FactSage™) across a wide range of conditions and particularly at low matte grades (> 20 wt per cent Fe). This is supported by data from commercial-scale, Ausmelt TSL nickel smelters operated by JJNI (Si *et al*, 2017) and JNMC (January 2021, personal communication), for which FactSage™ predicts a slag solids content in the range from 25–35 per cent.

The ability to operate with a significant solids content in slag, provides enormous operational and economic benefits. In addition to a reduction in process oxygen and fuel consumptions by roughly 10 per cent and 20 per cent respectively, refractory wear rates are greatly reduced at lower temperatures, ensuring an extended furnace campaign life and thus reduced expenditure on refractory lining demolition and installation activities. The possibility of operating well below the slag liquidus also provides enhanced flexibility to handle short and long-term variability in feed material compositions and easily recover from periods of bath temperature instability.

Operability limits in the test work manifested primarily as difficulties tapping molten slag and to a lesser extent, accretion build-up around the furnace roof ports and offtake. Owing to the design and sizing of the pilot-scale furnace employed in the trials, these issues were somewhat exaggerated, meaning operation at lower temperatures may be possible on a commercial scale. The combination of an intensely-cooled copper taphole block and small taphole openings, resulted in material freezing without regular 'working' of the taphole to keep slag flowing. These issues are readily mitigated on a commercial scale with appropriate furnace design (large taphole openings, short taphole length and multiple tapholes). For reference, JNMC utilise three slag tapholes with openings of 80 mm (January 2021, personal communication), with one or more tapholes in use more or less continuously (Figure 17). Similarly, the Ausmelt TSL converter operated by Anglo Platinum in South Africa, utilises slag taphole inserts with an opening size of 75 mm (Hoosen, Schione and Ramblyana, 2018).

FIG 17 – Commercial scale taphole configuration.

To a lesser extent, accretion build-up around the furnace roof ports and offtake also dictated operability limits observed in the trials. This is a persistent issue in the pilot-scale furnace due to both the relatively small freeboard height and significant air-ingress. On a commercial scale these issues are addressed with appropriate furnace design (sufficient freeboard height, tight sealing of ports and provision for freeboard burners to deliver localised heat input for melting of accretions).

CONCLUSIONS

Test work conducted in Metso's pilot test work facility in Dandenong, Australia confirmed the possibility of treating high high-MgO, nickel sulfide concentrates using Ausmelt TSL technology. The test work examined behaviour of the SiO_2-FeO-MgO-CaO-NiO slag system across a wide range of conditions with the goal of determining the lowest bath temperature at which stable process and equipment operation could be maintained. Slag SiO_2/MgO ratio, wt per cent CaO and matte grade found to have the greatest impacts on identified operability limits.

Contrary to initial thinking, results from the test work indicated operability limits were not solely determined by the slag solids content, with the liquid slag viscosity also playing a role, particularly at low wt per cent CaO and Fe/SiO_2 ratio.

The test work highlighted inherent flexibility of the Ausmelt TSL process to operate across a large slag compositional range and recover from process disturbances an interruption to feed introduction. Results from the trials indicated the Ausmelt process is capable of operating with a high concentration of suspended solids in slag, without negative impacts to slag fluidity. The test work also demonstrated the possibility for Ausmelt TSL technology to process concentrates with an Fe/MgO ratio as low as 1.4, which has important implications to the commercial-scale processing of high-MgO content feeds.

REFERENCES

Aspola, L, Matusewicz, R, Haavanlammi, K and Hughes, S, 2012. Outotec Smelting Solutions for the PGM Industry, in *Proceedings Platinum 2012*, pp 235–250 (The Southern African Institute of Mining and Metallurgy).

Hoosen, A, Schione, M and Ramblyana, I, 2018. Improvements to the Anglo Converting Process (ACP) Tap-Block Management, in *Proceedings from Furnace Tapping 2018*, pp 49–56 (The Southern African Institute of Mining and Metallurgy).

Hundermark, R, Mncwango, S, de Villiers, L and Nelson, L, 2011. The smelting operations of Anglo American's Platinum Business: An Update, in *Proceedings Southern African Pyrometallurgy 2011*, pp 295–307 (The Southern African Institute of Mining and Metallurgy).

Si, J, Zhao, Y, Wang, C and Geng, J, 2017. A Brief Analysis of the Method of Reducing the Nickel Content of Slag in Nickel Smelting Production, *China Nonferrous Metallurgy*, August 2017, (4):36–39.

Wood, J, Hoang, J and Hughes, S, 2017. Energy Efficiency of the Outotec® Ausmelt Process for Primary Copper Smelting, *JOM*, 69(6):1013–1020.

Wood, J and Hughes, S, 2016. Future Development Opportunities for the Outotec® Ausmelt Process, in *Proceedings 9th International Copper Conference 2016*, pp 361–372 (The Mining and Materials Processing Institute of Japan).

Zhou, M, Wan, A, Li, G, Baldock, R and Li, H, 2010. Industrial Operation of JAE Nickel Smelting Technology at Jinchuan Nickel Smelter, in *Supplemental Proceedings: Volume 1: Materials Processing and Properties 2010*, pp 525–534 (The Minerals, Metals and Materials Society).

A fundamental investigation on welding flux tunability geared towards high heat input submerged arc welding for shipbuilding applications

H Yuan[1], H Tian[2], Y Zhang[3], Z Wang[4] and C Wang[5]

1. PhD Candidate, School of Metallurgy, Northeastern University, Shenyang 110819, China. Email: yuanhangneu@foxmail.com
2. PhD Candidate, School of Metallurgy, Northeastern University, Shenyang 110819, China. Email: tianhuiyu@omgmail.cn
3. Postdoctoral Researcher, School of Metallurgy, Northeastern University, Shenyang 110819, China. Email: yanyun_zhang@foxmail.com
4. Associate Professor, School of Metallurgy, Northeastern University, Shenyang 110819, China. Email: wangzhanjun@smm.neu.edu.cn
5. Professor, FASM, School of Metallurgy, Northeastern University, Shenyang 110819, China. Email: wangc@smm.neu.edu.cn

ABSTRACT

Submerged arc welding (SAW) is one of the significant metal-joining processes for manufacturing marine vessels, steel pipes, and offshore structures with high deposition rate and engineering reliability. Welding flux serves several essential functions, including atmospheric shielding, arc stabilisation, bead morphology control and weld metal (WM) refinement. Therefore, from a thermodynamic point of view, the focus for flux design and WM compositional/microstructural modification has been placed to elucidating the transfer pathways and mechanisms of major alloying elements, such as Si, Mn, Ti and O during welding. To this end, a thermodynamic model has been established to predict alloying element contents in the WM. Such functions are enabled by the physicochemical properties of the fluxes, which are inherently rooted in the nature of the fluxes. A unique yet systematical investigation, including physicochemical property changes and structural evolution behaviours, has been conducted over the wide range of fluxes applied to actual welding of EH36 shipbuilding steels. Combined with spectroscopic methods, structural behaviours of network formers such as SiO_2, Al_2O_3 and TiO_2 and network modifiers such as MgO, MnO and CaO have been illustrated. Viscosity and ionic conductivity have been found to be positively associated with the degree of polymerisation.

INTRODUCTION

High heat input submerged arc welding (SAW) technology, known for high deposition efficiency and excellent weld metal formability, has found widespread applications in various engineering fields (Yuan *et al*, 2023). However, high heat input could produce coarse yet brittle microstructural constituents in the weld metal (WM), thus deteriorating mechanical properties of the entire weldment (Wu *et al*, 2023). During SAW, various redox reactions and heat transfer events could occur in the arc cavity, in which fluxes could play significant roles in forming atmospheric shielding, refining WM and preventing heat loss (Wang and Zhang, 2021). Such needed functions are enabled by appropriate flux thermophysical properties, such as viscosity and thermal conductivity (Zhang *et al*, 2022a). It is well known that thermophysical properties are structure dependent and largely dictated by the extent of chemical and topological disorder in the molten state (Wang *et al*, 2023a). Therefore, a clear yet comprehensive understanding of flux structures is crucial to optimised design and enhanced welding performance. However, the current flux design mainly relies on empirical methods, lacking a comprehensive and systematic exploration of the flux structure and physicochemical properties. This deficiency leads to an ambiguous evolution of the flux structure-physicochemical properties.

Furthermore, fluxes should improve their metallurgical properties aside from satisfying the welding process. The WM mechanical properties under high heat input welding can be improved by fine-tuning the flux composition to precisely optimise the metallurgical properties. Specifically, the flux composition contributes significantly to the WM chemistry through chemical reactions occurring in the welding process (Zhang *et al*, 2022b). For example, Zhang, Leng and Wang (2019) designed TiO_2-containing basic-fluoride-type fluxes, enabling the transfer of Ti, Mn and Si to the WM. They emphasised that optimal mechanical properties were achieved with the addition of 6 wt per cent TiO_2 in the flux. Therefore, addressing the roles of fluxes in regulating WM composition becomes urgent, as they are directly linked to the element transfer behaviours between the flux and WM.

This study aims to concentrate on the microstructure and physicochemical properties of flux modification. This work will gain an in-depth understanding of the flux characteristics and help to determine the guidelines for flux design in high heat input welding.

EXPERIMENTAL

All the experimental samples for the present welding fluxes were prepared from the reagent-grade powders of SiO_2, Al_2O_3, TiO_2, MnO, MgO and CaO. The samples were thoroughly mixed and placed inside a pure molybdenum crucible, and were then pre-melted in the high-temperature region of the resistance furnace at 1550°C under a pure argon atmosphere (>99.999 per cent, 0.3 L/min). After being held at that temperature for 1 hr, the molten slags were quenched in cold water rapidly and then dried at 200°C for 4 hrs to remove any moisture. The as-quenched samples were crushed and ground into powders of less than 200 mesh. Viscosity measurements were carried out using the rotating cylinder method with Brookfield viscometer, where one molybdenum crucible filled with 150 g flux sample was placed in an electric resistance furnace.

Structural information was identified through spectroscopic analyses using Raman spectroscopy and X-ray photoelectron spectroscopy (XPS). Raman was carried out by a laser confocal Raman spectrometer (LabRam HR800, Horiba, United States) with a 1-mW semiconductor laser using a 532 nm excitation wavelength. The spectra were recorded with a 1 cm^{-1} resolution ranging from 400 to 1600 cm^{-1}. At least ten spectra were acquired for each sample to increase the reliability of the measures. XPS analysis was performed using an imaging photoelectron spectrometer (Thermo Scientific K-Alpha, United States) at room temperature with a monochromatic Al-Kα X-ray source with a pass energy of 50 eV. The spectra were calibrated by taking the C 1s peak as the reference binding energy at 284.8 eV.

Double-wire SAW was performed on the base metal of EH36 shipbuilding steel. Welding parameters (electrode forward: DC-850 A/32 V, electrode backward: AC-625 A/36 V) were kept constant with the travel speed at 500 mm/min.

RESULTS AND DISCUSSIONS

Transfer behaviours of Si, Mn, Ti and O

The element transfer between slag and WM is quantified by a Δ value and refers to the difference between analytical composition and nominal composition of the WM (Zhang, Wang and Coetsee, 2021). The Δ value indicates the contribution of flux to the WM composition. A positive Δ value implies that a specific element has been transferred from flux (slag) to WM. A negative Δ value means that a given element has been lost from WM to slag. For a specific element, a zero Δ value implies that an apparent equilibrium condition has been reached between the slag and WM (Zhang, J et al, 2020).

In the droplet zone of the SAW, where droplet forms at the electrode tip and travels through the arc, most oxides are susceptible to decompose and thus increase the O level of the droplet by increasing the local pO_2 via Equations 1 and 2 under the presence of the arc. In this period, O level of the droplet increases, and negligible amount of alloy transfer, such as Si and Mn, may occur.

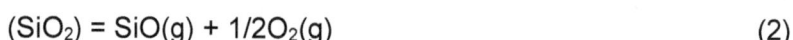

$$(MnO) = Mn(g) + 1/2O_2(g) \qquad (1)$$

$$(SiO_2) = SiO(g) + 1/2O_2(g) \qquad (2)$$

Subsequently, the droplet is diluted by the weld pool and the transfer of Mn and O at slag-metal interface would be governed by Equation 3.

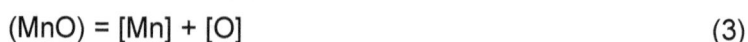

$$(MnO) = [Mn] + [O] \qquad (3)$$

Table 1 summarises the Δ values of the WMs. As can be seen from Table 1, the 'neutral point' of Mn for WMs, where equilibrium is built between slag and WM apparently, lies at between

10~20 wt per cent MnO. For Flux with MnO addition of 0 and 10 wt per cent, the WMs lose 0.318 and 0.162 wt per cent Mn to the slags, respectively; for the MnO-rich fluxes, the WMs gain up to 0.588 wt per cent Mn from the slags. All WMs gain Si from the slags with an average value of about 0.187 wt per cent. The transfer of Si appears to be independent of MnO content in the fluxes.

TABLE 1
Changes of chemical compositions of WMs (wt%).

CaF$_2$-40%SiO$_2$-MnO				CaO-40%SiO$_2$-MnO				25%CaF$_2$-SiO$_2$-CaO-TiO$_2$		
MnO	ΔSi	ΔMn	ΔO	MnO	ΔSi	ΔMn	ΔO	TiO$_2$	ΔTi	ΔO
0	0.212	-0.318	0.015	10	0.081	0.218	0.058	0	-0.002	0.018
10	0.193	-0.162	0.033	20	0.071	0.244	0.061	5	0.002	0.021
20	0.164	0.125	0.038	30	0.082	0.278	0.069	10	0.005	0.023
30	0.16	0.092	0.045	40	0.118	0.344	0.084	15	0.007	0.025
40	0.163	0.318	0.058	50	0.167	0.327	0.092	20	0.01	0.027
50	0.216	0.437	0.085	60	0.16	0.377	0.085			
60	0.201	0.588	0.08							

Flux plays a major role in O uptake of the WM in SAW (Zhang *et al*, 2023). MnO and SiO$_2$ are reported as primary sources of O in SAW. It is observed from Table 1 that there is a significant increase in ΔO values from 150 to 850 ppm as MnO content increases from 0 to 60 wt per cent. In the droplet zone, O is transferred to the droplet via the increase of pO$_2$ in Equations 1 and 2 as mentioned previously. It is deduced from Equation 1 that pO$_2$ would increase with higher MnO content in the fluxes for a given amount of SiO$_2$. In the weld pool zone, the transfer of O to the WM increases with MnO addition as Equation 3 is promoted to the right side.

Although the transfer mechanism of O from CaO to WM remains ambiguous, it is accepted that CaO is one of the most stable oxides with a lower O potential than MnO. The amount of O contributed from the flux, *viz* ΔO value, increases from 576 to 920 ppm at higher MnO content. It is seen that the ΔSi value generally increases with higher MnO additions in the flux. The addition of MnO tends to increase the activity of SiO$_2$, which, in turn, promotes the transfer of Si from flux to the weld metal. Only a slight improvement of measured ΔMn value from 0.218 to 0.377 wt per cent is observed, although the MnO addition level dramatically increases from 10 to 60 wt per cent. This can be attributed to the fact that the evaporation of Mn from the weld pool tends to occur at the plasma-metal interface, reducing the magnitude of ΔMn value.

Gas–slag–metal model

Results above have quantified the element transfer behaviours. Therefore, to precisely predict the essential element contents, a gas–slag–metal equilibrium model is established, based on the assumption that the the O level is controlled by pO$_2$ (derived from the decomposition of the oxide in flux) in the arc plasma. Figure 1 shows the predicted contents by gas–slag–metal equilibrium calculations for Si, Mn and O. It can be seen that this model offers excellent prediction accuracy for O content, and can differentiate the O content of the weld metals produced by fluxes with varying formulas but same basicity index. Furthermore, when the gas–slag–metal equilibrium model is applied, the prediction error for Si and Mn contents is significantly reduced as compared to the slag–metal equilibrium model. Thermodynamic calculation data indicates that the consideration of gas formation, which essentially controls the predicted flux O potential and oxide activity, is necessary to improve the overall prediction accuracy.

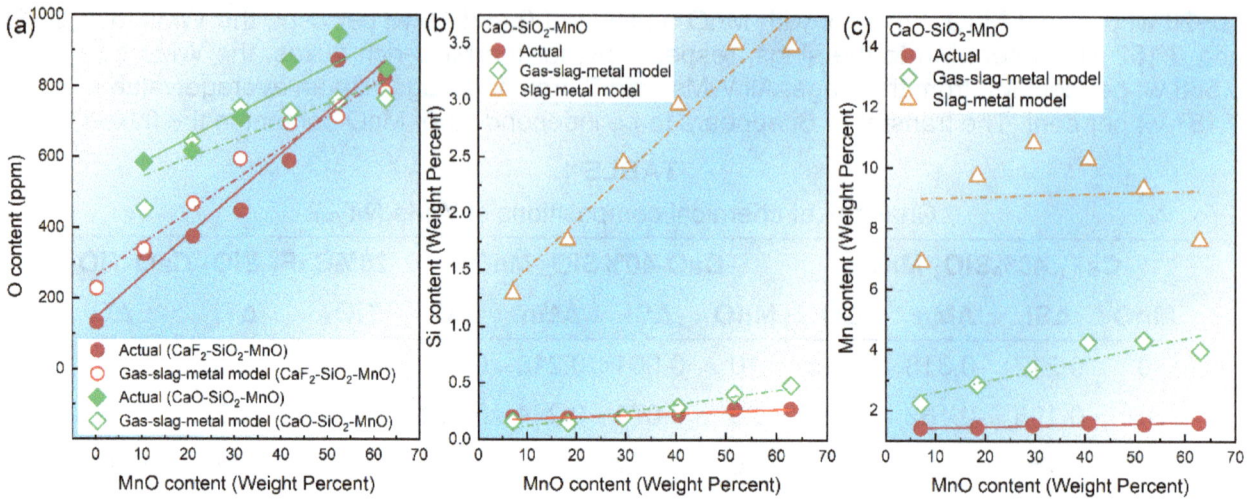

FIG 1 – Predicted contents by gas–slag–metal equilibrium calculations for: (a) O, (b) Si and (c) Mn content as a function of MnO level in fluxes.

Transfer pathway of oxygen

It is revealed that O transfer behaviour could likely be a dynamic process involving O gain and loss. A practical challenge is that the dynamic process is covered under molten flux, which is difficult to observe and quantify directly. O transfer processes are related to O transfer pathways that can be deciphered by WM O gain and loss. However, controversies remain regarding the transfer pathway for O, which dims the understanding of the O transfer mechanism. Hence, further investigations are called upon to account for the transfer pathway of O in pertinent welding fluxes.

At the slag–metal reaction interface, the transfer of O is largely enabled simultaneously with the transfer of other alloying elements via Equations 4 and 5 in CaF_2-SiO_2-CaO-(0~20 wt per cent)TiO_2 system. Table 2 shows the WM O gain by slag–metal reactions. O contributions from SiO_2 and TiO_2 are denoted as ΔO_{SiO_2} and ΔO_{TiO_2}, respectively. It is seen that positive ΔO_{SiO_2} and ΔO_{TiO_2} values increase with higher TiO_2 content in the flux, indicating O transfer from the flux to the WM. WM O gain is favoured with enhanced TiO_2 content, which will increase the activities of SiO_2 and TiO_2. What stands out is that, for any given TiO_2 content, ΔO_{SiO_2} value is over ten times that of ΔO_{TiO_2}, which could be attributed to the fact that the O potential of SiO_2 is reportedly higher than that of TiO_2 during the welding process, as manifested by the high WM O levels for multi-component acidic fluxes with high SiO_2 contents. Moreover, Table 2 also shows actual changes of the O content in the WM. Total O gain of the WM through slag–metal reactions increases from 0.267 to 0.389 wt per cent. However, actual WM ΔO ranges from 0.018 to 0.027 wt per cent, suggesting that there is O loss in the WM after WM gains O by slag–metal reactions.

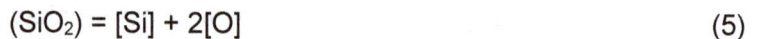

$$(TiO_2) = [Ti] + 2[O] \tag{4}$$

$$(SiO_2) = [Si] + 2[O] \tag{5}$$

TABLE 2

Weld metal ΔO content as a function of TiO_2 content in the flux (wt%).

TiO$_2$(%)	ΔO	Gain O			Loss O		
		ΔSiO_2	ΔTiO_2	gas	C	Mn	Fe
0	0.018	0.269	0	0.003	-0.008	-0.029	-0.217
5	0.021	0.272	0.004	0.018	-0.01	-0.033	-0.23
10	0.023	0.288	0.007	0.036	-0.014	-0.04	-0.254
15	0.025	0.313	0.012	0.045	-0.014	-0.054	-0.277
20	0.027	0.338	0.017	0.053	-0.018	-0.057	-0.306

O loss incurred by Fe, Mn and C is denoted as ΔOF_e, ΔO_{Mn} and ΔO_C. It is demonstrated that ΔOF_e, ΔO_{Mn} and ΔO_C are always negative, and their absolute values increase with higher TiO_2 content, indicating O loss from the WM to the slag. O loss by oxidation of Fe from the WM increases from 0.23 to 0.31 wt per cent, which is significantly larger than those of ΔO_{Mn} and ΔO_C, suggesting that deoxidation of the WM is mainly attributed to oxidation of Fe. In addition, the WM O content reduced by deoxidation is 0.268 to 0.389 wt per cent with TiO_2 increase from 0 to 20 wt per cent, which is greater than that of the O gain (0.267 to 0.355 wt per cent) by slag–metal reactions, indicating another pathway for the WM to gain O is through gas–metal reactions. It is concluded that gas–metal reactions mainly occur in the 'droplet zone', the region where droplets form at the electrode tip and travel through the arc cavity. Wt conservation calculations of WM can be used to calculate O gain (ΔO_{gas}) through gas–metal reactions, which can be calculated by Equation 6. ΔO_{gas} content picks up from 0.017 to 0.062 wt per cent as TiO_2 content in the flux increases from 0 to 20 wt per cent.

$$\Delta O_{WM} = (\Delta O_{SiO_2} + \Delta O_{TiO_2} + \Delta O_{gas}) - (\Delta OF_eO + \Delta O_{Mn}O + \Delta OCO) \tag{6}$$

WM O gain is primarily driven by gas–slag–metal reactions, while O loss is largely enabled by slag–metal reactions. The total O gain and loss increase from 0.286 to 0.416 and 0.268 to 0.389 wt per cent, respectively and WM O gain is greater than its loss. This result indicates a net O gain (Total O gain minus total O loss, 0.018 to 0.027 wt per cent) in the WM, resulting in increased WM O content compared to the BM or electrode. Moreover, ΔO_{SiO_2} contributes above 80 per cent to the total O gain, indicating that SiO_2 in the flux is the main O source for the WM. Fe deoxidation is the largest contributor to the WM O loss, demonstrated by ΔOF_e, which accounts for more than 80 per cent of the total O loss.

It can be concluded that oxides dissociate under the high temperature of the arc plasma, and O generated by TiO_2 and SiO_2 dissolves into the electrode droplets and WM. O generated by TiO_2 and SiO_2 dissolves into the electrode droplets and WM by gas–metal reactions, which accounts for 5.91 to 14.68 per cent of the total O gain. In addition, O also transfers from the molten flux into the WM through Equations 4 and 5, and SiO_2 in the flux is the most significant contributor to the O gain for the WM, accounting for more than 80 per cent of the total O gain. The solubility of O in WM decreases after the molten welding pool cools down from higher temperature to the solidification temperature. Therefore, Fe, Mn and C will react with dissolved O, and then the deoxidation products will transfer into the slags. Oxidation of Fe is the main way for O loss from the WM, which has been demonstrated by ΔOF_e accounts for more than 80 per cent of the total O loss.

Structure and physicochemical properties

The optical basicity (Λ) is a measure of the electron donor properties of different ions (Mills, Yuan and Jones, 2011; Mills $et\ al$, 2012). It was also used as a theoretical measure of the depolymerisation of the melt. In terms of experiment methods, XPS analysis possesses a unique function by showcasing features of various oxygen species, including O^- (non-bridging oxygen), O^0 (bridging oxygen) and O^{2-} (free oxygen), which are closely related to the degree of polymerisation of the flux structure (Wang $et\ al$, 2023b; Zhang, Y $et\ al$, 2020). Therefore, the non-bridging oxygen distributions as a function of optical basicity for varied flux systems are shown in Figure 2a. It can be seen that the fraction of non-bridging oxygen increases with an increase in optical basicity for all the fluxes, which is consistent with earlier study (Yang, Wang and Sohn, 2022).

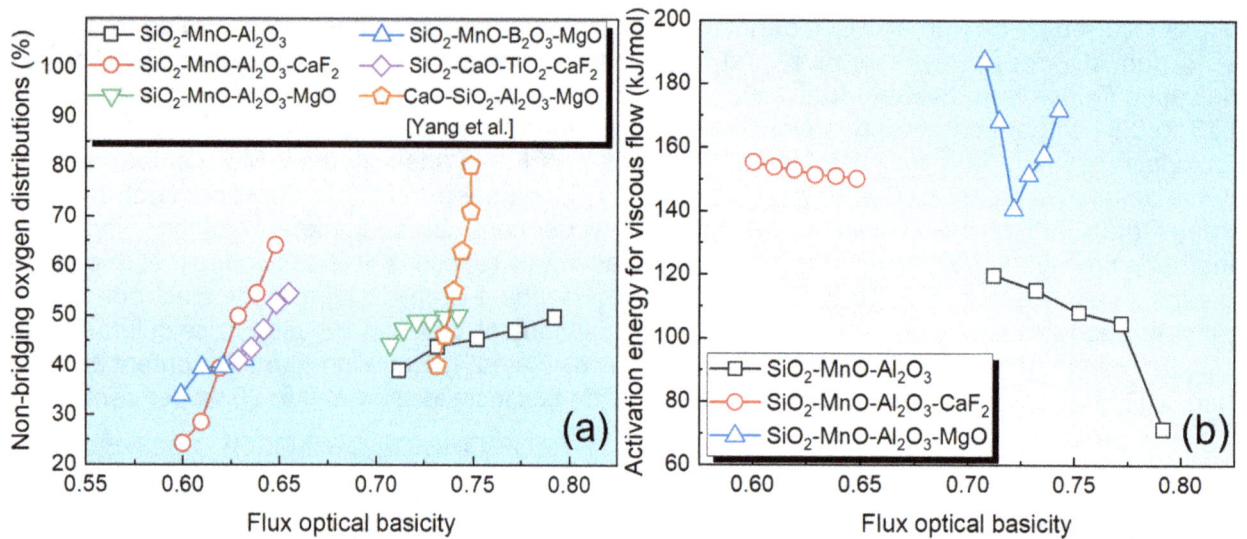

FIG 2 – Relationship between flux optical basicity and: (a) non-bridging oxygen, and (b) activation energy for viscous flow.

The activation energy (E_a) represents the frictional resistance within the structural units of the liquid flux that needs to be overcome during shearing (Wang *et al*, 2022). For Newtonian fluids, E_a can be obtained from the Arrhenius equation, as shown in Equation 7:

$$Ln\eta = ln\eta_0 + (E_a)/R \cdot 1/T \qquad (7)$$

where:

η	is the viscosity (Pa s)
η_0	is a pre-exponential factor (Pa s)
E_a	is the activation energy for viscous flow (J/mol)
T	is the absolute temperature (K)
R	is the ideal gas constant (8.314 J mol^{-1} K^{-1})

The calculated E_a values with different Al_2O_3 contents are shown in Figure 2b. As can be seen, E_a gradually decreases as the optical basicity increases. However, for the SiO_2-MnO-Al_2O_3-CaF_2 system, E_a decreases from 171.68 to 140.53 kJ mol^{-1} as the Al_2O_3 content increases from 0 to 15 wt per cent and then E_a shows a significant increase from 168.05 to 187.26 kJ mol^{-1} with further addition of Al_2O_3, which correlates well with the 'V'-shaped variation trend of the viscosity.

CONCLUSIONS

In summary, the transfer pathways and mechanisms of major alloying elements, such as Si, Mn, Ti and O during welding have been elucidated. A thermodynamic model has been established to predict alloying element contents in the WM. Combined with spectroscopic methods, the flux optical basicity has been found to be positively associated with the degree of polymerisation. Moreover, the activation energy for viscous flow gradually decreases as the optical basicity increases.

ACKNOWLEDGEMENTS

The authors sincerely thank the National Natural Science Foundation of China (Grant Nos. U20A20277 and 52350610266) and National Key Research and Development Program of China (Grant No. 2022YFE0123300).

REFERENCES

Mills, K C, Yuan, L and Jones, R T, 2011. Estimating the physical properties of slags, *Journal of the Southern African Institute of Mining and Metallurgy,* 111:649–658.

Mills, K C, Yuan, L, Li, Z, Zhang, G H and Chou, K C, 2012. A review of the factors affecting the thermophysical properties of silicate slags, *High Temperature Materials and Processes,* 31(4–5):301–321.

Wang, C and Zhang, J, 2021. Fine-tuning weld metal compositions via flux optimization in submerged arc welding: An overview, *Acta Metallurgica Sinica*, 57(9):1126–1140.

Wang, Z, Li, Z, Zhong, M, Li, Z and Wang, C, 2023a. Elucidating the effect of Al_2O_3/SiO_2 mass ratio upon SiO_2-MnO-CaF_2-Al_2O_3-based welding fluxes: Structural analysis and thermodynamic evaluation, *Journal of Non-Crystalline Solids*, 601:122071.

Wang, Z, Shen, B, Zhong, M, Li, Z and Wang, C, 2023b. A structure-oriented elucidation for viscous flow of SiO_2-MnO-Al_2O_3 fused submerged arc welding fluxes, *Journal of Non-Crystalline Solids*, 612:122360.

Wang, Z, Zhang, J, Zhong, M and Wang, C, 2022. Insight into the viscosity–structure relationship of MnO–SiO_2–MgO–Al_2O_3 fused submerged arc welding flux, *Metallurgical and Materials Transactions B*, 53:1364–1370.

Wu, Y, Yuan, X, Kaldre, I, Zhong, M, Wang, Z and Wang, C, 2023. TiO_2-Assisted microstructural variations in the weld metal of EH36 shipbuilding steel subject to high heat input submerged arc welding, *Metallurgical and Materials Transactions B*, 54(1):50–55.

Yang, J, Wang, Z and Sohn, I, 2022. Topological understanding of thermal conductivity in synthetic slag melts for energy recovery: An experimental and molecular dynamic simulation study, *Acta Materialia*, 234:118014.

Yuan, H, Wang, Z, Zhang, Y and Wang, C, 2023. Roles of MnO and MgO on structural and thermophysical properties of SiO_2-MnO-MgO-B_2O_3 welding fluxes: A molecular dynamics study, *Journal of Molecular Liquids*, 386:122501.

Zhang, J, Coetsee, T, Dong, H and Wang, C, 2020. Element transfer behaviors of fused CaF_2-SiO_2-MnO fluxes under high heat input submerged arc welding, *Metallurgical and Materials Transactions B*, 51(3):885–890.

Zhang, J, Leng, J and Wang, C, 2019. Tuning weld metal mechanical responses via welding flux optimization of TiO_2 content: application into EH36 shipbuilding steel, *Metallurgical and Materials Transactions B*, 50(5):2083–2087.

Zhang, J, Wang, C and Coetsee, T, 2021. Thermodynamic evaluation of element transfer behaviors for fused CaO-SiO_2-MnO fluxes subjected to high heat input submerged arc welding, *Metallurgical and Materials Transactions B*, 52:1937–1944.

Zhang, Y, Coetsee, T, Yang, H, Zhao, T and Wang, C, 2020. Structural roles of TiO_2 in CaF_2-SiO_2-CaO-TiO_2 submerged arc welding fluxes, *Metallurgical and Materials Transactions B*, 51(5):1947–1952.

Zhang, Y, Liu, H, Coetsee, T, Wang, Z and Wang, C, 2023. Identifying oxygen transfer pathways during high heat input submerged arc welding: a case study into CaF_2-SiO_2-CaO-TiO_2 fluxes, *Metallurgical and Materials Transactions B*, 54(6):2875–2880.

Zhang, Y, Wang, Z, Zhang, J, Li, Z, Basu, S and Wang, C, 2022a. Probing viscosity and structural variations in CaF_2–SiO_2–MnO welding fluxes, *Metallurgical and Materials Transactions B*, 53(5):2814–2823.

Zhang, Y, Zhang, J, Liu, H, Wang, Z and Wang, C, 2022b. Addressing weld metal compositional variations in EH36 shipbuilding steel processed by CaF_2-SiO_2-CaO-TiO_2 fluxes, *Metallurgical and Materials Transactions B*, 53:1329–1334.

Electrodeposition and electrochemical behaviour of molybdenum ions in ZnCl$_2$-NaCl-KCl molten salt

H Zhang[1], S Li[1], Z Lv[1] and J Song[2]

1. Zhongyuan Critical Metals Laboratory, Zhengzhou University, Zhengzhou, Henan 450001, China; School of Material Science and Engineering, Zhengzhou University, Zhengzhou, Henan 450001, China.
2. Zhongyuan Critical Metals Laboratory, Zhengzhou University, Zhengzhou, Henan 450001, China; School of Material Science and Engineering, Zhengzhou University, Zhengzhou, Henan 450001, China. Email: jianxun.song@zzu.edu.cn

ABSTRACT

In this paper, a method for extracting and refining metallic molybdenum using low-temperature ZnCl$_2$-NaCl-KCl molten salt was proposed. The electrochemical behaviour of molybdenum ions in ZnCl$_2$-NaCl-KCl molten salt was investigated in detail, and pure metallic molybdenum was collected on the cathode. The reduction and diffusion processes of Mo(V) ions in ZnCl$_2$-NaCl-KCl molten salt were determined through a series of electrochemical methods. The diffusion coefficients and nucleation modes of Mo(V) ions were also studied, and electrolysis was conducted for a prolonged period at a constant current density. The results indicated that the electrode reduction of Mo(V) ions in ZnCl$_2$-NaCl-KCl molten salt proceeds through a three-step reaction: Mo(V) → Mo(IV) → Mo(III) → Mo at 250°C.

The Mo(V) → Mo(IV) reaction was reversible and diffusion-controlled, with a Mo(V) diffusion coefficient of 6.02×10^{-6} cm^2 s^{-1}. Additionally, the nucleation mode of Mo(V) ions was instantaneous nucleation. The products electro-deposited were confirmed of metallic molybdenum with a particle size of less than 1 μm.

INTRODUCTION

Molybdenum is a metal with the high melting point, high hardness, good thermal and electrical conductivity and strong corrosion resistance (Huang *et al*, 2016; Lv *et al*, 2021a, 2021b; Zhang *et al*, 2023), and is widely used in the fields of iron and steel, metallurgy, medical treatment, semiconductors and other fields (Luo *et al*, 2019; Kuroda *et al*, 2020). Among the various extraction and refining technologies for molybdenum metal, the electrolysis method has shown great potential for development and application with its high efficiency and environmentally friendly characteristics. In the process of electrolytic extraction and refining, the electrolyte composition has an important influence on the electrolysis process due to its different physical and chemical properties. Therefore, the choice of electrolyte composition is extremely important issue for the extraction and purification of molybdenum metal. Depending on the type of electrolyte used in the electrolysis extraction and refining of molybdenum metal, it can be divided into ionic liquid electrolytes and molten salt electrolytes.

Ionic liquid electrolyte is a special liquid substance mainly composed of a large number of cations and anions. Researchers attempted to successfully deposit high-purity and dense molybdenum metal using ionic liquids with high concentrations of acetic acid (Morley *et al*, 2012), ammonium acetate solution (Kuznetsov *et al*, 2018), and 1-butyl-3-methylimidazolium tetrafluoroborate (BMIMBF$_4$) (Tian *et al*, 2023) as the electrolyte. However, although the ionic liquid electrolyte has the advantages of fine-tuning and low-temperature electrolysis in the preparation of metallic molybdenum, its efficiency is low, with most of the current being used to electrolyse water and produce hydrogen, resulting in a high oxygen content in the product and a slow electrolysis rate.

Compared with ionic liquid electrolytes, molten salt electrolyte electrolysis has the advantages of high current efficiency, fast electrolysis rate and high product purity. High purity molybdenum metal was successfully prepared by adding K$_3$MoCl$_6$ to the molten salts of KCl-NaCl and LiCl-KCl, respectively, and electrolysing them at 900°C in an inert atmosphere (Senderoff and Brenner, 1954). After that, molybdenum metal with higher purity was successfully prepared from the molten salt system of NaF-KF-MoF$_6$ (Senderoff and Mellors, 1967) at 600°C. In addition to molybdenum halide as solute in molten salt for molten salt electrolysis, researchers have also tried to use molybdenum oxide (Kou *et al*, 2023; Malyshev *et al*, 2018; Kushkhov and Adamokova, 2007) and metal molybdates (Kōyama, Hashimoto and Terawaki, 1987) as solutes, which can also successfully prepare molybdenum metal, but the oxygen content in the product will inevitably be higher. Therefore, the extraction and refining of molybdenum metal using molten salt electrolyte, although the current efficiency is high and the electrolysis speed is fast, the temperature required for electrolysis is usually above 450°C, and the energy consumption is high and the requirement of equipment is high. To address this problem, Nakajima, Nohira and Hagiwara (2006) investigated a new low-temperature molten salt system, ie $ZnCl_2$-NaCl-KCl system. $MoCl_3$ was added to the $ZnCl_2$-NaCl-KCl system at 250°C, and molybdenum metal was successfully obtained by electrodepositing on the cathode. The advantage of this molten salt system is that it can realise the electrodeposition of metallic molybdenum at a lower temperature, which will reduce the energy consumption and the volatilisation loss of molybdenum chloride. Carrying on electrodepositing in such melt, the current efficiency is much higher than that of ionic liquids, which is a more ideal electrolyte.

However, the results of the present study only demonstrate the feasibility of extracting metal molybdenum from the $ZnCl_2$-NaCl-KCl molten salt electrolyte. There are few studies on the electrochemical behaviour of Mo(V) in low-temperature molten salts, and no uniform conclusions have been drawn. In addition, the molybdenum metal has a complex dissolved valence state during the refining and extraction process. Therefore, it is of great significance to study the electrochemical behaviour of high-valent molybdenum ions in the $ZnCl_2$-NaCl-KCl (Moon *et al*, 2022). In summary, $ZnCl_2$-NaCl-KCl (3:1:1) was chosen as the electrolyte in this study, and thermodynamic calculations and various electrochemical tests were performed to probe deeply into the reduction and diffusion processes of Mo(V) ions. The diffusion coefficient and nucleation mode of molybdenum ions were investigated, and electrodeposition experiments were carried out at a constant current, and molybdenum metal was successfully obtained at the cathode.

EXPERIMENTAL

In this study, a total of 100 g of $ZnCl_2$-NaCl-KCl salt mixture was used as the electrolyte with a molar ratio of 3:1:1. The specific compositions were $ZnCl_2$ (99 per cent purity), NaCl (99.8 per cent purity), and KCl (99.8 per cent purity), which were purchased from McLean Biochemicals Ltd. The salt mixtures were first kept at a constant temperature of 200°C for 4 hrs to remove residual moisture, followed by a temperature increase to 250°C and maintained for 4 hrs to ensure that the molten salt mixtures were fully melted and well mixed. This was followed by a 12 hr pre-electrolysis treatment, followed by electrochemical tests and deposition experiments. Molybdenum ions were added to the molten salt as $MoCl_5$ (99.9 per cent purity, purchased from Macklin Biochemicals Ltd).

The whole experiment was carried out under argon atmosphere. A three-electrode system was used during the electrochemical tests. In this system, a tungsten wire (1 mm diameter, 99.99 per cent purity) was used as the working electrode (WE), while another tungsten wire was used as the reference electrode (RE) and a graphite rod (10 mm diameter, 99.99 per cent purity) was used as the counter electrode (CE). When constant current electrolysis was performed, a two-electrode system was used, in which molybdenum sheets (50 mm in length, 10 mm in width, and 1 mm in thickness) were used as cathode and anode, respectively, in different electrolysis steps. The electrochemical experiments were recorded by Nova 2.1 software (by Metrohm Autolab) controlled by AutoLab (PGSTAT 302N) and transient electrochemical techniques were used to study the electrochemical behaviour of Mo(V) in $ZnCl_2$-NaCl-KCl molten salt. The experiments were carried out at a current density of 0.06 A cm^{-2} in a constant current electrolysis test and the cathodic products were collected. In order to investigate the microstructure and morphology of the cathode products, scanning electron microscopy (SEM, including TESCAN MIRA LMS and ZEISS Sigma 300), as well as energy spectrometry (EDS), were used for the analysis.

RESULTS AND DISCUSSION

Redox behaviour of molybdenum ions in molten salt

The decomposition reactions corresponding to NaCl, KCl, $ZnCl_2$, and $MoCl_x$ ($x \leq 5$) were shown in Equations 1–4. Through thermodynamic calculations and the study of Nakajima, Nohira and

Hagiwara (2006), it was found that the decomposition potential of $ZnCl_2$ is the smallest when compared to NaCl and KCl under 250°C. Therefore, for extraction and refining of molybdenum metal by using this molten salt, the $MoCl_x$ ($x \leq 5$) decomposition potential must be lower than the $ZnCl_2$ decomposition potential. The Gibbs free energy changes for the Reactions 3 and 4 at 250°C were calculated using HSC Chemistry 6.0, from which the decomposition potentials of molybdenum chloride at different valence states as well as the theoretical electrochemical window of the molten salt electrolyte were determined by Equation 5.

$$2NaCl(s) = 2Na(l) + Cl_2(g) \quad (1)$$

$$2KCl(s) = 2K(l) + Cl_2(g) \quad (2)$$

$$ZnCl_2 = Zn + Cl_2(g) \quad (3)$$

$$2MoCl_x = 2MoCl_y + (x-y)Cl_2(g) \quad (4)$$

$$\Delta G = -nEF \quad (5)$$

Where ΔG represents the change in Gibbs free energy (kJ mol^{-1}); the relationship between x and y is $0 \leq y < x \leq 5$; n is the number of electrons transferred by the reaction (mol); E refers to the reduction potential (V) of the above chloride and sulfide, and F is the Faraday constant (C mol^{-1}).

The common valence states of molybdenum chloride were more complex (including 5, 4, 3, 2 and 0 valence states) and the calculated decomposition voltages were shown in Figure 1a. The electrolyte selection criteria for the electrolysis of molten salts was satisfied because the decomposition voltage of $MoCl_x$ (maximum -1.13 V) was lower than that at molten NaCl-KCl (-1.74 V) at 750°C. In addition, the reduction order of each valence state of each molybdenum ion in Figure 1a provided theoretical support for the subsequent molybdenum ion reduction process.

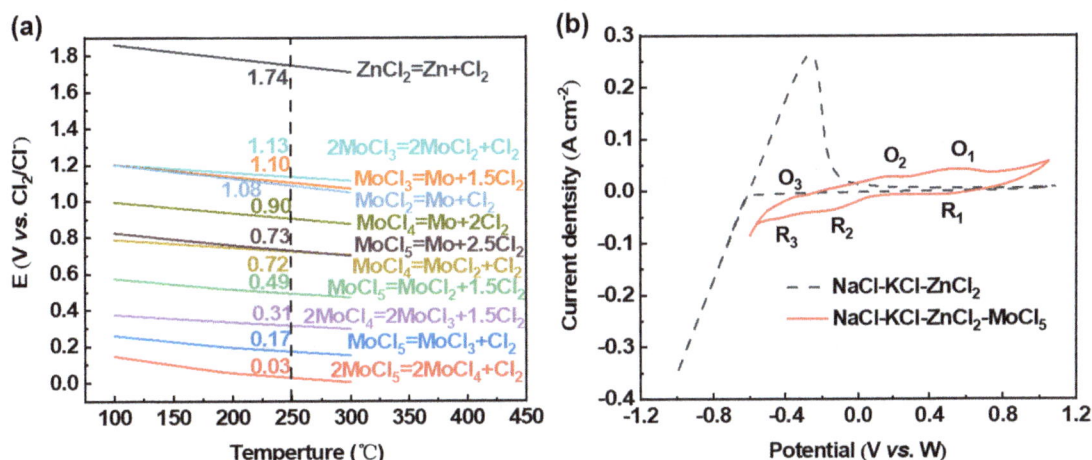

FIG 1 – (a) Molybdenum chloride and zinc chloride theoretical decomposition voltage as a function of temperature (all thermodynamic data from HSC 6.0 in activity (α = 1)); (b) Cyclic voltammograms before and after the addition of $MoCl_5$ in the molten $ZnCl_2$-NaCl-KCl salt; scan rate: 0.25 V s^{-1}

The results of the cyclic voltammetry (CV) tests, conducted before the addition of $MoCl_5$, were shown as the grey line in Figure 1b. The reduction of Zinc ions began near -0.6 V versus W, and there were no significant current density peaks observed between 1.1 V and -0.6 V, suggesting that no other reduction reactions occurred in this voltage range. Therefore, it was determined that the electrolyte system was suitable for performing electrochemical tests. After the addition of $MoCl_5$, the voltammetric curve appeared as the red line in Figure 1b. Three pairs of distinct current density peaks (R_1/O_1, R_2/O_2, and R_3/O_3) were observed, indicating that molybdenum ions were involved in redox reactions. Furthermore, all these reduction reactions of molybdenum ions occurred before the precipitation of Zinc ions, demonstrating that the selection of the $ZnCl_2$-NaCl-KCl-$MoCl_5$ system was feasible for the extraction and refining of molybdenum metal.

The cyclic voltammograms obtained after the addition of $MoCl_5$ were further analysed through Figure 2a, where it can be seen that the reduction peaks R_1, R_2, and R_3 occurred at reduction potentials of 0.52 V, -0.08 V, and -0.42 V (versus W), respectively. Based on these observations, it can be inferred that the reduction of molybdenum ions occurred in three distinct steps. Subsequently, the electrochemical behaviour of Mo(V) was further investigated using the more precise square wave voltammetry (SWV) technique. The results obtained from this analysis are presented in Figure 2b, where three distinct electrode reduction reactions R_1, R_2, and R_3 can be observed at potentials of 0.53 V, -0.05 V, and -0.45 V (versus W), respectively. These findings suggest that the conversion of Mo(V) to metal in the $ZnCl_2$-NaCl-KCl molten salt involves three reduction processes, which is consistent with the results obtained from the previously conducted cyclic voltammetry tests.

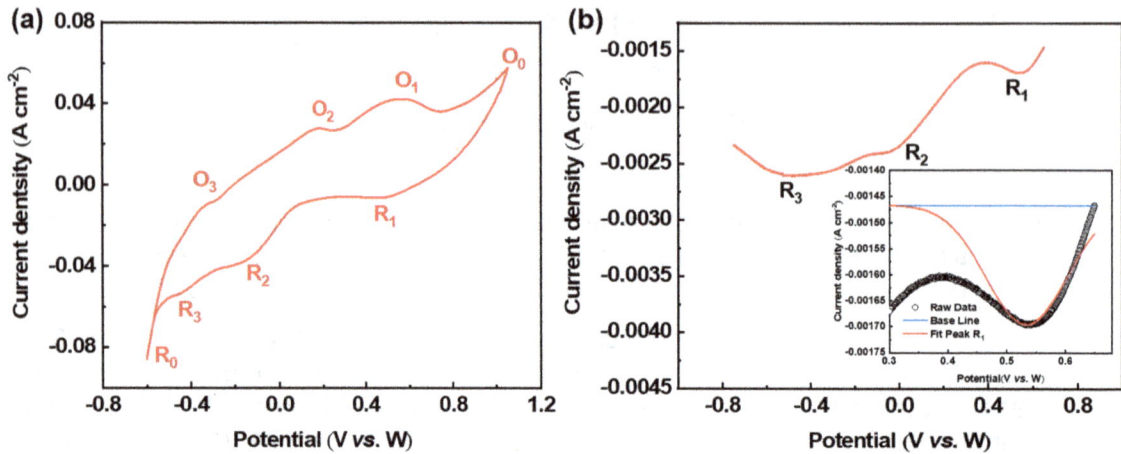

FIG 2 – Electrochemical test plots after addition of $MoCl_5$ to $ZnCl_2$-NaCl-KCl molten salt: (a) cyclic voltammetry; scan rate: 0.25 V s^{-1}; (b) square-wave voltammetry; scan frequency: 25 Hz; inset: Gaussian fit to R_1 peaks.

In addition, the peak potentials of R_1, R_2, and R_3 did not change with increasing scan rate or scan frequency in Figure 3a and Figure 4a, which indicates that R_1, R_2, and R_3 are reversible reactions. Where the number of transferred electrons corresponding to the reduction peak R_1 can be calculated by Equation 6 (Krause and Ramaley, 1969; O'Dea, Osteryoung and Osteryoung, 1981; Aoki $et\ al$, 1986):

$$W_{1/2} = 3.52 \frac{RT}{nF} \tag{6}$$

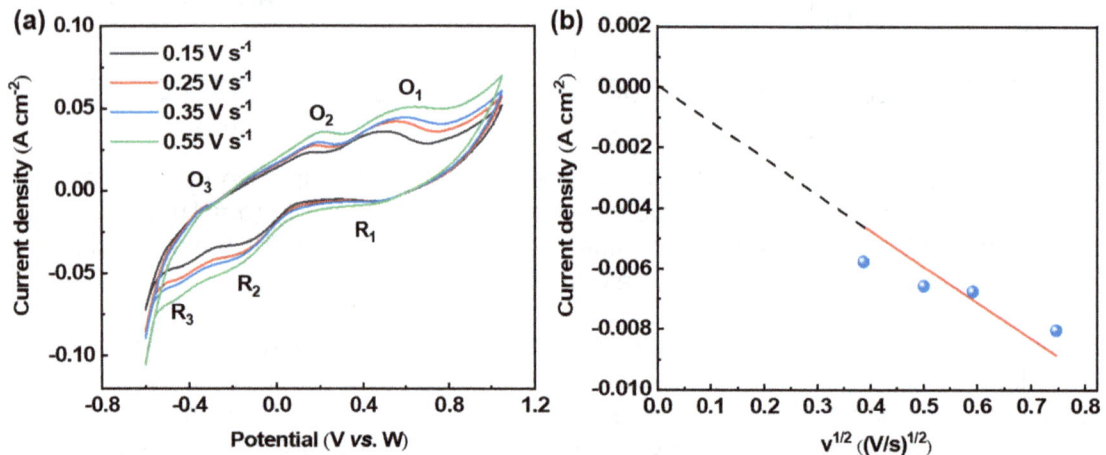

FIG 3 – (a) Cyclic voltammograms of $MoCl_5$ in molten $ZnCl_2$-NaCl-KCl on a tungsten wire electrode at different scan rates; (b) The relationship between the current density and the square root of scan rate.

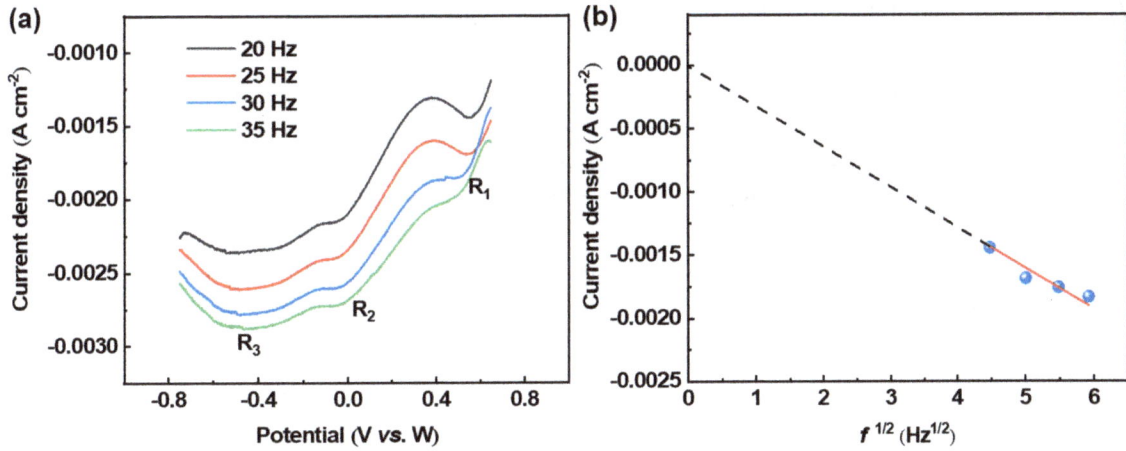

FIG 4 – (a) Square wave voltammetry of MoCl$_5$ in molten ZnCl$_2$-NaCl-KCl on a tungsten wire electrode at different frequencies; (b) The relationship between the current density and the square root of frequency.

A Gaussian fit was applied to R$_1$ in the square-wave voltammogram, and the fitted curve is presented in the inset of Figure 2b. The half-wave width of R$_1$ was determined to be 0.16. The number of electrons transferred corresponding to R$_1$ was calculated to be approximately 0.99, which was close to 1, indicating that the reduction step represented by the R$_1$ peak was Mo(V) → Mo(IV). Subsequently, Mo(IV) was reduced to metallic molybdenum through two further steps, R$_2$ and R$_3$. The reduction reaction of the Mo(IV) ion can be referenced to our previous determination in NaCl-KCl (Zhang *et al*, 2024): Mo(IV) → Mo(III) → Mo. Therefore, it is inferred that R$_2$ and R$_3$ correspond to the reduction steps of Mo(IV) → Mo(III) and Mo(III) → Mo, respectively. In conclusion, the reduction process of Mo(V) in the ZnCl$_2$-NaCl-KCl molten salt was Mo(V) → Mo(IV) → Mo(III) → Mo. This was consistent with the order of the Mo(V) reduction potential calculated thermodynamically in Figure 1a.

Diffusion coefficient of molybdenum ions in molten salt

To further investigate the kinetics of the electrode process of molybdenum ions in ZnCl$_2$-NaCl-KCl molten salt, cyclic voltammetry experiments were conducted at various scan rates within the scan interval of 0.9 V to -0.7 V (versus W). The resulting cyclic voltammetry curves were presented in Figure 3a, and the correlation between the potential of the reduction peak R$_1$ and the logarithm of the scan rate was depicted in Figure 3b. The peak potential of R$_1$ hardly varied with increasing scan rate and tended to level off with a value of 0.53 V (versus W), which indicated that the R$_1$ reduction was reversible (Yuan *et al*, 2021). The relationship between the maximal current density of reaction R$_1$ and the square root of the scan rate was subsequently investigated. As demonstrated in Figure 3b, the peak current density of the reduction reaction R$_1$ was linear with the square root of the scan rate, indicating that the electrode reduction reaction R$_1$ was controlled by diffusion.

The diffusion coefficient of Mo(V) ions in molten ZnCl$_2$-NaCl-KCl was determined by Equation 7 (Liu *et al*, 2019).

When the reactants and products were both soluble:

$$I_p = 0.4463 \frac{(nF)^{3/2} A D^{1/2} C_0 v^{1/2}}{(RT)^{1/2}} \tag{7}$$

When the product was insoluble, Equation 8 could be used to determine the diffusion coefficient:

$$I_p = 0.6102 \frac{(nF)^{3/2} A D^{1/2} C_0 v^{1/2}}{(RT)^{1/2}} \tag{8}$$

Where I_p is the peak current (A), v is the scan rate (V s^{-1}), and C_0 is the bulk concentration of the reducible ion (mol cm^{-3}). The concentration C_0 was determined to be 2.24 × 10^{-5} mol cm^{-3}. A is the surface area of the W working electrode (cm^2), and D is the diffusion coefficient of Mo(V) (cm^2 s^{-1}).

Mo(V) ions were calculated to have a diffusion coefficient of 6.02×10^{-6} cm^2 s^{-1} in molten ZnCl$_2$-NaCl-KCl at 250°C.

To further confirm the findings above, we conducted square wave voltammetry at various frequencies. The test results are displayed in Figure 4a. The responsive R$_1$, R$_2$ and R$_3$ potentials barely changed as the applied frequency increased. The linear fitting results were given in Figure 4b, which related the peak current density linearly to the square root of frequency. This was consistent with the finding made by cyclic voltammetry and leads to the conclusion that the electrode reduction reaction R$_1$ was a reversible diffusion reaction. The diffusion coefficient was calculated by Equations 9 and 10 (Song *et al*, 2016).

$$I_p = nFAC_0 \frac{1-\Gamma}{1+\Gamma}\left(\frac{Df}{\pi}\right)^{1/2} \tag{9}$$

$$\Gamma = \exp\left(\frac{nF\Delta E}{2RT}\right) \tag{10}$$

Where ΔE is the amplitude of SWV (0.02 V), and f indicates the frequency. The diffusion coefficient of Mo(V) in molten ZnCl$_2$-NaCl-KCl at 250°C was calculated to be 5.61×10^{-6} cm^2 s^{-1}. This was about on the same scale as what was discovered during cyclic voltammetry testing.

Nucleation mode of molybdenum ions

In this paper, the MoCl$_5$ nucleation mode was further investigated by using the time-current method, as shown in Figure 5a. Its time-current curve is shown in Figure 5a. The applied potential was set to -0.55 V versus W. The current decreased rapidly and stabilised with time, which was attributed to the fact that molybdenum would be deposited on the electrode at this potential, resulting in a decrease in the molybdenum ion concentration in the vicinity of the electrode, which led to a rapid decrease in the current because molybdenum ions in the molten salt were unable to diffuse to the surface of the electrode in time to replenish it. However, as the reaction proceeds, the diffusion rate of molybdenum ions on the cathode surface and the deposition rate of molybdenum gradually reached a dynamic equilibrium, which led to the gradual stabilisation of the current. This also showed that the reduction of Mo(V) ions was controlled by diffusion.

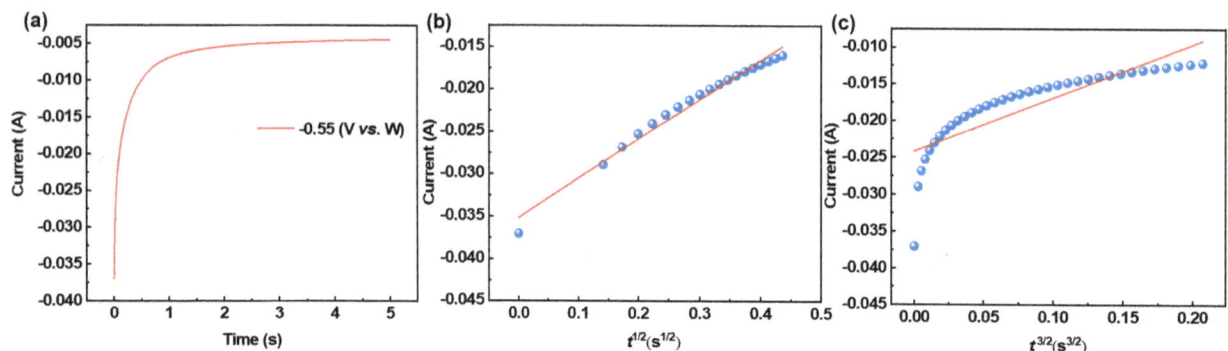

FIG 5 – (a) Timing current curve of MoCl$_5$ in ZnCl$_2$-NaCl-KCl molten salt; applied potential: -0.55 V versus W; (b) The relationship between the current and the square root of time; (c) The relationship between the current and the square root to the third power of time.

The nucleation mode of molybdenum ions can also be determined by the relationship between time and current in the chronoamperometry method. The nucleation modes are generally categorised into two: progressive nucleation and transient nucleation (Khelladi *et al*, 2009; Tylka, Willit and Williamson, 2017):

For instantaneous nucleation:

$$I(t) = \frac{zFN_0\pi(2DC)^{3/2}M^{1/2}}{\rho^{1/2}} t^{1/2} \tag{11}$$

For progressive nucleation:

$$I(t) = \frac{2zFK_nN_0\pi(2DC)^{3/2}M^{1/2}}{3\rho^{1/2}} t^{3/2} \tag{12}$$

where I (t) is the polarisation current at time t; t is the polarisation time (s); Z is the ion valence state; N_0 is the initial nucleation number; M is the atomic weight of the deposit (g mol^{-1}); K_n is the nucleation constant; and ρ is the density of the deposit (g cm^{-3}).

The current versus square root of time was shown in Figure 5b and 5c. The relationship between the current and the square root of time in Figure 5b remained linear after fitting, whereas the relationship between the current and the third power of the square root of time in Figure 5c had a poor correlation after fitting. Thus, molybdenum ions nucleated in a transient mode.

The diffusion coefficients and nucleation modes of Mo ions in different molten salt systems are shown in Table 1, and there are differences in the results in the molten salt system. This phenomenon may be due to different ionic species in the molten salt electrolyte, different experimental temperatures and the use of electrodes.

TABLE 1

Diffusion coefficients and nucleation modes of Mo ions obtained in different molten salt systems.

Molten salt	T (°C)	Diffusion coefficient (cm^2 s^{-1})		n	Nucleation mode	Reference
		CV	SWV			
NaCl-KCl-MoCl$_5$	750	1.22×10^{-4}	1.67×10^{-4}	2	progressive	Zhang et al (2024)
ZnCl$_2$-NaCl-KCl-MoCl$_5$	250	6.02×10^{-6}	5.61×10^{-6}	1	instantaneous	This work

Electrodeposition of molybdenum

After a series of electrochemical tests, in this paper, constant current electrolysis was carried out in molten ZnCl$_2$-NaCl-KCl-MoCl$_5$ (1.0 wt per cent) at 250°C in a two-electrode system. The current density was set at 0.06 A cm^{-2} and the electrolysis time was 20 hrs. Molybdenum sheets were used as cathode and anode, respectively. During the electrolysing of molybdenum metal, the anode molybdenum metal was oxidised and dissolved as molybdenum ions, which diffused to the cathode and were reduced to metal. The voltage change during the constant electrolysis process was recorded as shown in Figure 6a, and it can be found that the voltage was maintained near 1.65 V with little change during the electrolysis process, indicating that the electro-refining process was stable. After 20 hrs of electrolysis, a black deposit was collected at the cathode as shown in the inset in Figure 6a. The product was ultrasonically cleaned and vacuum-dried. The electrolysis product was analysed and detected using SEM and EDS, as shown in Figure 6b, the cathode product after electrolysis was metallic molybdenum and showed an irregular granular shape with particle size less than 1 μm, and this shape may be due to the transient nucleation of molybdenum ions in the electrochemical deposition process.

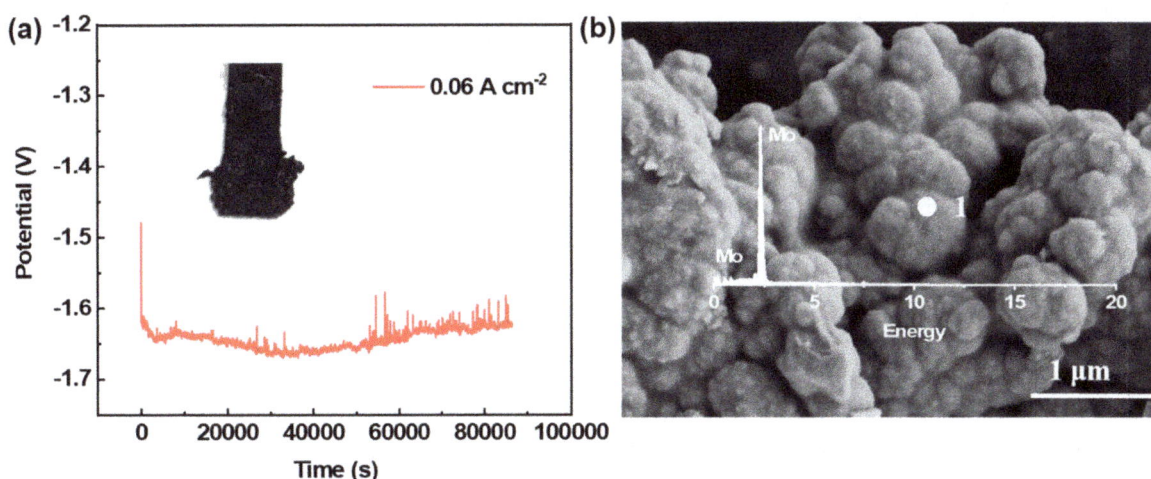

FIG 6 – (a) Time voltage curve at constant current density; inset: Cathode after 20 hr electrolysis; (b) SEM image of cathode product and the results of EDS analysis at point 1.

CONCLUSIONS

In this paper, the electrochemical behaviour of Mo(V) ions in molten $ZnCl_2$-NaCl-KCl at 250°C was investigated in detail. In the three-electrode system, thermodynamic calculations and a series of electrochemical tests were conducted to determine that Mo(V) in $ZnCl_2$-NaCl-KCl molten salt was reduced to molybdenum metal in three steps, namely, Mo(V) → Mo(IV) → Mo(III) → Mo, and the electrode reduction process was a diffusion-controlled reversible reaction.

The diffusion coefficients of Mo(V) were calculated to be 6.02×10^{-6} cm^2 s^{-1} and 5.61×10^{-6} cm^2 s^{-1}, respectively. Subsequently, the nucleation of Mo(V) ions in $ZnCl_2$-NaCl-KCl molten salts was determined to be a transient nucleation process through the use of the constant current method. Finally, electrolytic refining was performed for a prolonged duration at a current density of 0.06 A cm^{-2}, and the electrolytic process remained stable, resulting in the collection of metallic molybdenum at the cathode with a particle size of less than 1 μm. It indicates that $ZnCl_2$-NaCl-KCl holds potential as an electrolyte molten salt for the electrolytic extraction and refining of molybdenum metal.

ACKNOWLEDGEMENT

The authors thank the National Key Research and Development Program of China (Grant No. 2021YFC2901600). The work was also supported by project of Zhongyuan Critical Metals Laboratory (No. GJJSGFJQ202302), Youth Science and technology innovation of Henan Province (No. 23HASTIT009), Henan Province Youth Talent Support Program (2022).

REFERENCES

Aoki, K, Tokuda, K, Matsuda, H and Osteryoung, J G, 1986. Reversible square-wave voltammograms independence of electrode geometry, *Journal of Electroanalytical Chemistry and Interfacial Electrochemistry*, 207(1–2):25–39. https://doi.org/10.1016/0022-0728(86)87060-7

Huang, Z, Liu, J H, Deng, X G, Zhang, H J, Lu, L L, Hou, Z and Zhang, S W, 2016. Low temperature molten salt preparation of molybdenum nanoparticles, *International Journal of Refractory Metals and Hard Materials*, 54:315–321. https://doi.org/10.1016/j.ijrmhm.2015.08.01

Khelladi, M R, Mentar, L, Azizi, A, Sahari, A and Kahoul, A, 2009. Electrochemical nucleation and growth of copper deposition onto FTO and n-Si(100) electrodes, *Materials Chemistry and Physics*, 115(1):385–390. https://doi.org/10.1016/j.matchemphys.2008.12.017

Kou, Q, Jin, W L, Ge, C T, Pang, J, Zhang J, Haarberg, G M, Xiao, S J and Wang, P, 2023. Preparation of molybdenum coatings by molten salt electrodeposition in Na_3AlF_6-NaF-Al_2O_3-MoO_3 system, *Coatings*, 13(7):1266. https://doi.org/10.3390/coatings13071266

Kōyama, K, Hashimoto, Y and Terawaki, K, 1987. Smooth electrodeposits of molybdenum from KF-$Li_2B_4O_7$-Li_2MoO_4 fused salt melts, *Journal of the Less Common Metals*, 132(1):57–67. https://doi.org/10.1016/0022-5088(87)90174-3

Krause, M S and Ramaley, L, 1969. Analytical application of square wave voltammetry, *Analytical Chemistry*, 41(11):1365–1369. https://doi.org/10.1021/ac60280a008

Kuroda, P A B, Lourenço, M L, Correa, D R N and Grandini, C R, 2020. Thermomechanical treatments influence on the phase composition, microstructure and selected mechanical properties of Ti–20Zr–Mo alloys system for biomedical applications, *Journal of Alloys and Compounds*, 812:152108. https://doi.org/10.1016/j.jallcom.2019.152108

Kushkhov, K B and Adamokova, M N, 2007. Electrodeposition of tungsten and molybdenum metals and their carbides from low-temperature halide-oxide melts, *Russian Journal of Electrochemistry (Translation of Elektrokhimiya)*, 43(9):997–1006. https://doi.org/10.1134/s1023193507090030

Kuznetsov, V V, Volkov, M A, Zhirukhin, D A and Filatova, E A, 2018. Electroreduction of Mo(VI) compounds in ammonium-acetate solutions, *Russian Journal of Electrochemistry (Translation of Elektrokhimiya)*, 54(11):1006–1011. https://doi.org/10.1134/s1023193518130256

Liu, S Y, Wang, L J, Chou, K C and Kumar, R V, 2019. Electrolytic preparation and characterization of VCr alloys in molten salt from vanadium slag, *Journal of Alloys and Compounds*, 803:875–881. https://doi.org/10.1016/j.jallcom.2019.06.366

Luo, L M, Zhou, Y F, Zhang, Y X, Zan, X, Liu, J Q, Zhu, X Y and Wu, Y C, 2019. Current status and development trend of toughening technology of molybdenum-based materials, *Chinese Journal of Nonferrous Metals*, 29(3):525–537. https://doi.org/10.19476/j.ysxb.1004.0609.2019.03.12

Lv, C, Jiao, H D, Li, S L, Che, Y S, Li, S X, Liu, C, He, J L and Song, J X, 2021a. Liquid zinc assisted electro-extraction of molybdenum, *Separation and Purification Technology*, 279:119651. https://doi.org/10.1016/j.seppur.2021.119651

Lv, C, Li, S L, Che, Y S, Chen, H T, Shu, Y C, He, J L and Song, J X, 2021b. Study on the molybdenum electro-extraction from MoS$_2$ in the molten salt, *Separation and Purification Technology*, 258:118048. https://doi.org/10.1016/j.seppur.2020.118048

Malyshev, V, Gab, A, Shakhnin, D, Donath, C, Neacsu, E I, Popescu, A M and Constantin, V, 2018. Influence of electrolysis parameters on mo and w coatings electrodeposited from tungstate, molybdate and tungstate-molybdate melts, *Revista de Chimie*, 69(9):2411–2415. https://doi.org/10.37358/rc.18.9.6544

Moon, J, Myhre, K, Andrews, H and McFarlane, J, 2022. Potential of electrolytic processes for recovery of molybdenum from molten salts for [99]Mo production, *Progress in Nuclear Energy*, 152:104369. https://doi.org/10.1016/j.pnucene.2022.104369

Morley, T J, Penner, L, Schaffer, P, Ruth, T J, Bénard, F and Asselin, E, 2012. The deposition of smooth metallic molybdenum from aqueous electrolytes containing molybdate ions, *Electrochemistry Communications*, 15(1):78–80. https://doi.org/10.1016/j.elecom.2011.11.026

Nakajima, H, Nohira, T and Hagiwara, R, 2006. Electrodeposition of metallic molybdenum films in ZnCl$_2$–NaCl–KCl–MoCl$_3$ systems at 250°C, *Electrochimica Acta*, 51(18):3776–3780. https://doi.org/10.1016/j.electacta.2005.10.041

O'Dea, J J, Osteryoung, J and Osteryoung, R A, 1981. Theory of square wave voltammetry for kinetic systems, *Analytical Chemistry*, 53(4):695–701. https://doi.org/10.1021/ac00227a028

Senderoff, S and Brenner, A, 1954. The electrolytic preparation of molybdenum from fused salts: I, Electrolytic studies, *Journal of the Electrochemical Society*, 101(1):16–17. https://doi.org/10.1149/1.2781198

Senderoff, S and Mellors, G, 1967. Electrodeposition of coherent deposits of refractory metals: V, Mechanism for the deposition of molybdenum from a chloride melt, *Journal of the Electrochemical Society*, 114(6):556. https://doi.org/https://10.1149/1.2426648

Song, Y, Jiao, S Q, Hu, L W and Guo, Z C, 2016. The cathodic behavior of Ti(III) ion in a NaCl-2CsCl melt, *Metallurgical and Materials Transactions B: Process Metallurgy and Materials Processing Science*, 47(1):804–810. https://doi.org/10.1007/s11663-015-0521-9

Tian, H X, Gang, Q J, Ying, Y, Xin, L, Jun, L T and Li J, 2023. Approaches to electrodeposit molybdenum from ionic liquid, *Rare Metals*, 42(7):2439–2446. https://doi.org/10.1007/s12598-018-1040-z

Tylka, M M, Willit, J L and Williamson, M A, 2017. Electrochemical nucleation and growth of uranium and plutonium from molten salts, *Journal of the Electrochemical Society*, 164(8):H5327–H5335. https://doi.org/10.1149/2.0471708jes

Yuan, R, Lv, C, Wan, H L, Li, S L, Che, Y S, Shu, Y C, He, J L and Song, J X, 2021. Electrochemical behavior of vanadium ions in molten LiCl-KCl, *Journal of Electroanalytical Chemistry*, 891:115259. https://doi.org/10.1016/j.jelechem.2021.115259

Zhang, H K, Li, S L, Lv, Z P, Fan, Y and Song, J X, 2023. The role of zinc sulfide in the electrochemical extraction of molybdenum, *Separation and Purification Technology*, 311:123290. https://doi.org/10.1016/j.seppur.2023.123290

Zhang, H K, Lv, Z P, Li, S L, He, J L, Fan, Y and Song, J X, 2024. Electrochemical behavior and cathodic nucleation mechanism of molybdenum ions in NaCl-KCl, *Separation and Purification Technology*, 329:125121. https://doi.org/10.1016/j.seppur.2023.125121

Development of solid waste-based flux and its application in out-of-furnace dephosphorisation of low-silicon hot metal

Z Zhao[1] and Y Zhang[2]

1. Phd Student, University of Science and Technology Beijing, Haidian District, Beijing 100083, China. Email: suc_zheng@163.com
2. Professor, University of Science and Technology Beijing, Haidian District, Beijing 100083, China. Email: ustbzly1108@163.com

ABSTRACT

The green production of the steelmaking process has urgent requirements for fluorine-free fluxes. In this article, a new solid waste-based slagging flux was applied in the external dephosphorisation test of low silicon hot metal to replace the traditional fluorite-lime dephosphorisation slag. The results showed that under laboratory conditions, the new flux was utilised for smelting low-silicon molten iron with a phosphorus content of 0.10~0.13 wt per cent, the temperature of the molten iron was 1350°C and the slag amount was controlled at 2 per cent. Within ten minutes of smelting, the dephosphorisation rate of the molten iron reached more than 50 per cent. When making slag with new flux, the slag melting time was about 40 seconds and the P_2O_5 content in the final slag was higher than 8 wt per cent. In the industrial test of a 150 t molten iron ladle, under the same molten iron conditions, an out-of-furnace dephosphorisation test was conducted in the Kambara Reactor (KR) process. The flux material consumption per ton of steel was 8~10 kg. Within 15 minutes of smelting, the dephosphorisation rate of the molten iron was about 30 per cent. During flux slag making, the slag melting time was about three minutes and the P_2O_5 content in the final slag was higher than 6 wt per cent. This new solid waste-based flux has significant cost advantages and out-of-furnace dephosphorisation effects and is expected to become the benchmark for a new generation of fluorine-free steelmaking fluxes.

INTRODUCTION

The pre-dephosphorisation process in molten iron significantly reduces converter slag emissions, offering notable advantages in smelting high-quality low-phosphorus steel. Additionally, it can lower the production costs of converter steelmaking, enhance production efficiency and allow for a modest relaxation of phosphorus content in blast furnace ores, demonstrating important practical significance (Gao et al, 2013; Li et al, 2014; Wu et al, 2020; Bai et al, 2020).

The lower temperature and higher phosphorus activity coefficient in the pre-treatment phase of molten iron are thermodynamically favourable for phosphorus removal, particularly in low-silicon conditions. However, insufficient heat and challenges in slag formation constrain the kinetic conditions for phosphorus removal (Li, Zhang and Guo, 2017; Gupta et al, 2023). Low-silicon molten iron can achieve desulfurisation through the KR process while also facilitating off-furnace phosphorus removal. However, off-furnace phosphorus removal processes typically employ fluorine or sodium-containing fluxes, resulting in multiple temperature drops, challenging operational procedures and causing environmental pollution. Few enterprises in China adopt this process (Kitamura et al, 2002; Zhou et al, 2017), highlighting the crucial importance of developing fluorine-free, low-melting-point fluxes.

Calcium ferrite, an excellent fluorine-free flux, possesses a low melting point and has the ability to quickly form a primary dephosphorisation slag with high oxidation and high alkalinity. Wright et al (2001) and Sukenaga et al, (2010) found the viscosity of the 19 wt per cent CaO–81 wt per cent FeO_X slag system to be 0.065 Pa·s, significantly lower than the 2.4 Pa·s of the $CaO\text{-}SiO_2\text{-}MgO\text{-}Al_2O_3$ slag system. A slag system with good fluidity can promote the transfer of phosphorus into the molten slag. Jeon, Jung and Sasaki (2010) and Paananen et al (2010) had investigated the formation mechanism of calcium ferrite. They had discovered that the reaction rate of Fe_3O_4 and CaO to form $Ca_2Fe_2O_5$ is faster than the reaction of Fe_2O_3 and CaO and that the higher the oxygen partial pressure, the slower the formation rate. This research has provided valuable insights into the formation of calcium ferrite and its influencing factors. Sayama et al (2002) and Lee and Barr (2002) examined the mechanism and effect of calcium ferrite flux on lime melting. When the surface layer of lime had been coated with iron oxide, compared with the uncoated sample, more low-melting point calcium ferrite phases had been formed, which had reduced the mechanical properties of the lime and had been more easily dissolved and absorbed in steelmaking slag. Sato, Nakashima and Mori (2001) conducted high-carbon hot metal dephosphorisation experiments using three calcium ferrite fluxes: $CaO \cdot Fe_2O_3$(CF), $2CaO \cdot Fe_2O_3$(C2F), and $3CaO \cdot Fe_2O_3$(C3F). Results had indicated that C2F and C3F, with a higher CaO concentration in the flux, had higher dephosphorisation rates, while the decarburisation reaction had been inhibited. However, due to the higher melting points of C2F and C3F, their dephosphorisation rates had been slower than that of CF. Mi *et al* (1999) investigated the dephosphorisation behaviour of mechanically synthesised calcium ferrite under molten iron conditions of 1300°C. When 20 per cent of the $CaFe_2O_4$ content had replaced part of the CaF_2 and Fe_2O_3, it had resulted in an increase in the alkalinity of the slag and an improvement in its oxidation property. This had led to a dephosphorisation rate of 88.6 per cent, which had been higher than the 79.3 per cent dephosphorisation rate achieved when only adding CaF_2 flux slag. Wu *et al* (2022) conducted industrial experimental research on the dephosphorisation of hot metal using calcium ferrite synthesised from iron oxide scale and lime. Results had demonstrated that calcium ferrite had reduced the slag-making melting temperature in the early stage of the converter and had increased the slag discharge amount in the double-slag process. When slagging had been performed with calcium ferrite, the average dephosphorisation rate of hot metal had been 88.06 per cent and the average melting temperature had been 1137°C. These values had been superior to the 77.7 per cent dephosphorisation rate and 1296°C slag melting temperature that had been observed when slagging with fluorite. The aforementioned research findings suggest that calcium ferrite-based slag may emerge as the optimal flux to replace fluorite, yielding favourable outcomes in dephosphorisation and slagging. Indeed, the majority of calcium ferrite slag products are manufactured using iron concentrate powder as a primary ingredient. The high temperatures required for processing contribute to substantial costs. These factors can pose significant challenges to the large-scale application of these products.

Based on a new low-melting-point solid waste-based slagging flux, this study proposed a process concept for implementing outside-furnace dephosphorisation of low-silicon hot metal using the Kambara Reactor (KR) process. Dephosphorisation tests were carried out under laboratory conditions and 150 t-level industrial test conditions, and the key factors affecting the dephosphorisation effect were clarified, such as flux melting time, flux addition amount, phosphorus distribution and phosphorus content in the final slag. This further verified the huge advantages of the new flux in reducing production costs and improving the dephosphorisation effect of molten iron, thereby providing an important basis for industrial application.

DEVELOPMENT OF SOLID WASTE-BASED FLUXES – MINERAL COMPOSITION AND MELTING PROPERTIES

Among many pre-melted fluxes, calcium ferrite was an important slagging material due to its low melting point and its content of calcium oxide and iron oxide. In foreign countries, it was widely used in the hot metal dephosphorisation process, but there were few application research reports in China. The then-current standard YBT-4266-2011 showed the physical and chemical indicators of pre-melted calcium ferrite CF-55 for metallurgical use in Table 1. This study used an industrial solid waste (bayer red mud) produced by the aluminium industry to prepare a new type of flux. The specific preparation method referred to the authorised international invention patent (Zhang *et al*, 2023) that our team had obtained. Its composition was shown in Table 1. According to the $CaO-Fe_2O_3$ binary phase diagram (Figure 1a), the melting point of calcium ferrite ($CaO \cdot Fe_2O_3$) was about 1216°C. According to the $CaO-Fe_2O_3-Al_2O_3$ ternary phase diagram (Figure 1b), since Al^{3+} and Fe^{3+} had similar electrical properties, they were easily replaced by Al^{3+}. The melting point of composite calcium ferrite, $Ca(Al, Fe)_2O_4$, was about 1142°C.

TABLE 1

Physical and chemical indicators of pre-melted calcium ferrite and composition of solid waste-based flux (wt%).

Item	Fe$_2$O$_3$	CaO	SiO$_2$	MgO	Al$_2$O$_3$	TiO$_2$	Na$_2$O	P	S	density
CF-55	55~60	30~35	0~4	0~6	-	-	-	0.10	0.10	3.2 g/cm^3
SW-Flux	50.74	29.46	2.45	0.51	10.29	3.46	1.38	0.12	0.08	3.5 g/cm^3

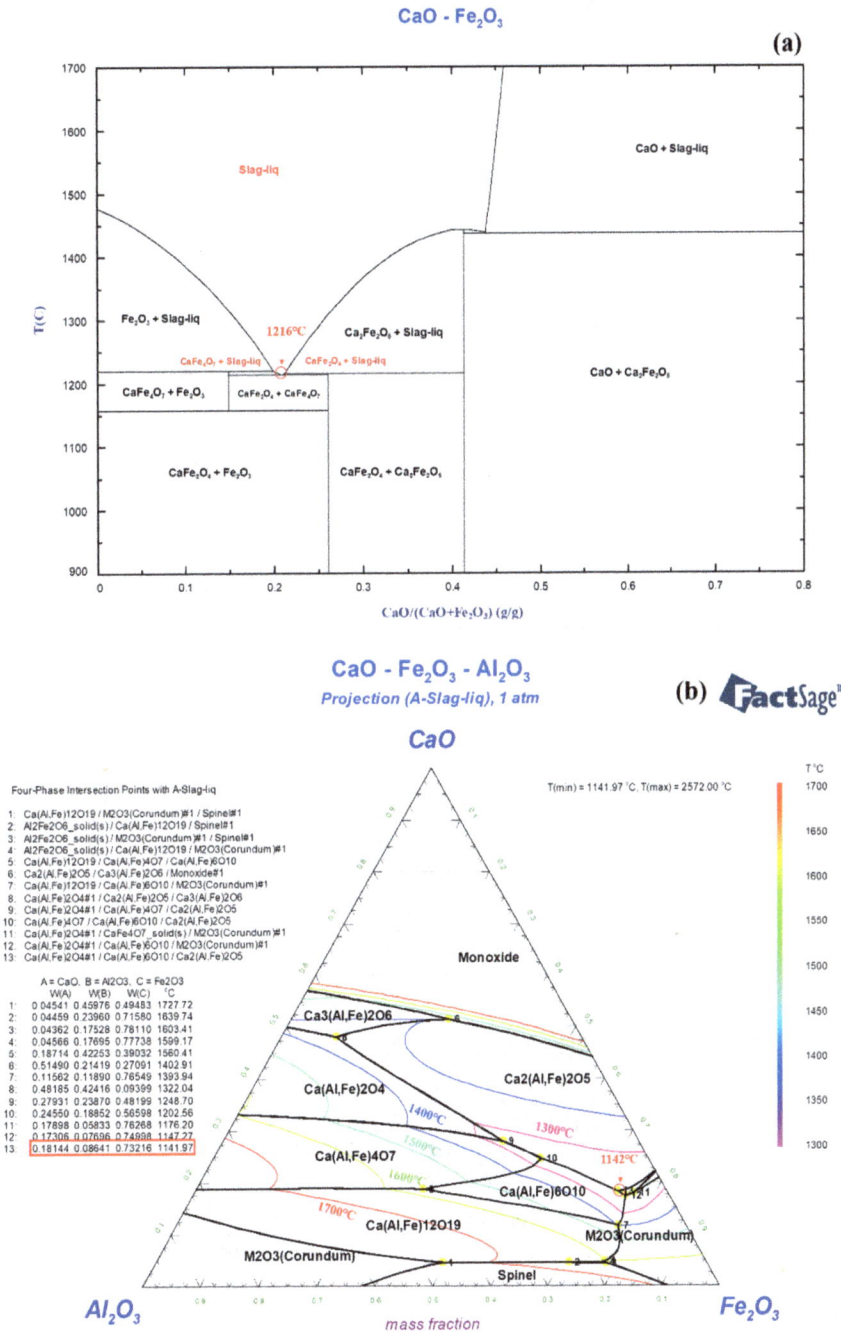

FIG 1 – (a) Preparation principle of flux: CaO-Fe$_2$O$_3$; and (b) CaO-Fe$_2$O$_3$-Al$_2$O$_3$ phase diagrams.

The two fluxes CF-55 and SW-Flux were prepared under laboratory conditions, as shown in Figure 2a. The main phases of CF-55 were CaFe$_2$O$_4$ and Ca$_2$Fe$_2$O$_5$, which were low melting point phases. The main phases in SW-Flux flux were CaFe$_2$O$_4$, Ca$_2$Fe$_2$O$_5$ and Ca$_2$(Fe,Al)$_2$O$_5$. It could be seen from the above components and phase characteristics that CF-55 flux had a lower melting point, was oxidising and could increase alkalinity. Industrially, calcium ferrite was usually obtained by pre-melting and cooling iron ore powder with lime and the preparation cost was high. In addition to the calcium ferrite mineral phase, SW-Flux flux also had a calcium aluminoferrite mineral phase, which had a lower melting point, high oxidation and high alkalinity. This made SW-Flux flux possess the characteristics of calcium ferrite and could significantly reduce the preparation cost.

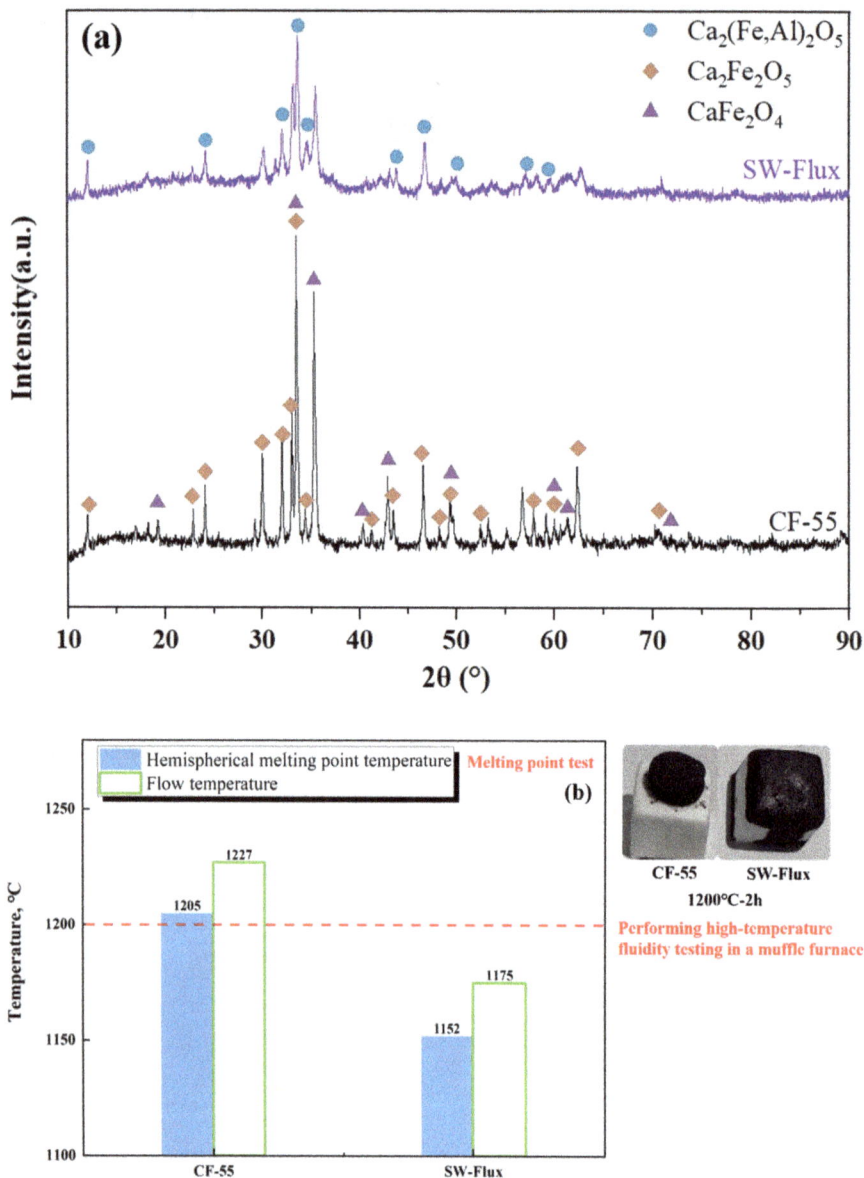

FIG 2 – (a) The XRD pattern; (b) and melting properties of flux.

As depicted in Figure 2b, the melting point test result of the flux indicates that the SW-Flux flux possesses a lower hemispheric softening point temperature and flow temperature, which are 1152°C and 1175°C respectively. These temperatures are much lower than 1300°C, sufficient to meet the temperature requirements of the desiliconisation and dephosphorisation process of molten iron (1300~1400°C), ensuring quick slag formation. The flux was placed on the refractory material and its high-temperature fluidity was tested in a muffle furnace. After being maintained at the same temperature of 1200°C for two hours, CF-55 exhibited significant shrinkage, a slight slump, and a sintered appearance. SW-Flux melted into a fluid state.

APPLICATION OF SOLID WASTE-BASED FLUX IN DEPHOSPHORISATION OF LOW SILICON MOLTEN IRON

According to the above research, the newly developed flux derived from solid waste exhibits a low melting point, high alkalinity and elevated oxidation levels. In Japanese steel companies, industrial calcium ferrite flux has been utilised as a slagging material in the production of low-phosphorus or ultra-low-phosphorus steel. The new flux studied in this study was used as a dephosphorisation agent in the laboratory stage and in 150 t-level KR industrial tests, providing a basis for its application.

Equipment and methods

The laboratory dephosphorisation test was conducted in a vertical high-temperature $MoSi_2$ furnace, as shown in Figure 3. During the entire experiment, argon gas (99.9 per cent purity) was introduced from the top of the shaft furnace to prevent oxidation of the molten iron and maintain a relatively stable oxygen partial pressure and atmosphere. The pig iron used in the test was prepared from industrial pure iron (Fe 99.9 wt per cent), phosphorus iron (P 21.5 wt per cent) and sulfur iron (S 32.5 wt per cent). The composition is shown in Table 2.

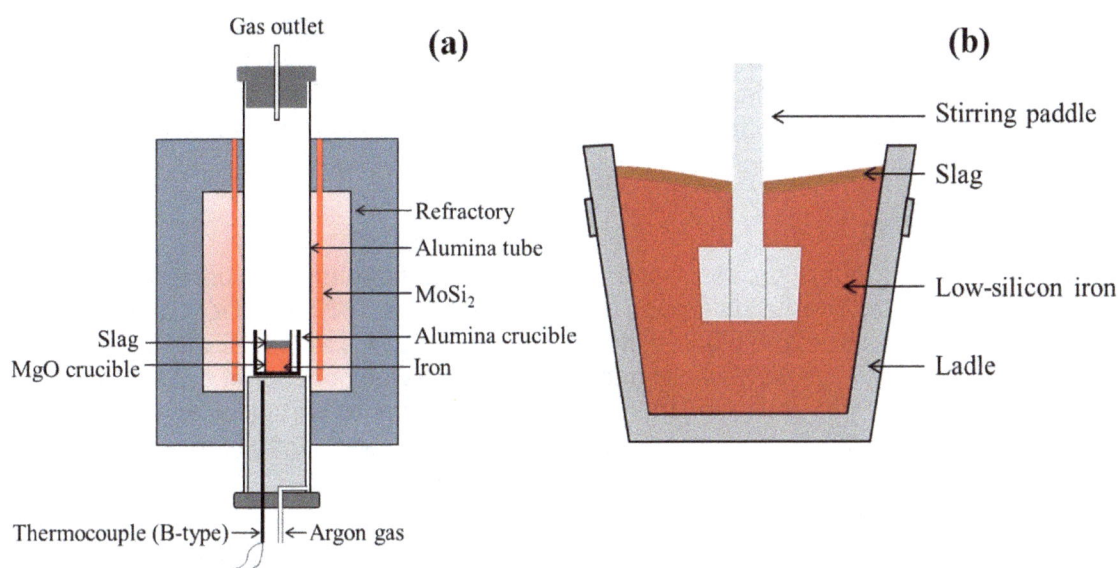

FIG 3 – (a) Laboratory; and (b) 150 t KR industrial test equipment.

TABLE 2

Semi-steel composition (wt%).

C	P	S	Si	Mn
3.8~4.0	0.10~0.13	0.04~0.06	0.01~0.10	0.01~0.05

First, the raw materials were put into a MgO crucible (D45 × H110 mm) and heated from room temperature to 300°C at a rate of 25°C/min, and then heated to 1350°C at a rate of 8°C/min. After the sample was kept at 1350°C for 20 minutes, dephosphorisation slag with a weight of 2 per cent of the molten iron was added to the surface of about 600 g of molten iron. In the test, refractory Al_2O_3 rods were used for stirring. After the slag gold reacted for a certain period of time, a high-purity quartz tube was used to take a molten iron sample. After the test, the crucible containing the quenched sample was put into the oven to completely dry the water and the slag and pig iron blocks were physically separated for further analysis. The experimental plan is shown in Table 3.

TABLE 3

Experimental programme (wt%).

No. 1	(Si)	(P)	Slag ratio	Stirring
1	0.01	0.11	2%	Without
2	0.01	0.13	2%	Yes
3	0.10	0.13	2%	Yes

The industrial test of dephosphorisation outside the furnace was conducted in the KR process. The capacity of the molten iron ladle is 150 t, the loading amount of low silicon molten iron is 142 t, the size of the molten iron ladle is H5200 × Φ3300 mm, the stirring paddle size of the KR process is H820 mm × L1100 mm, the rotation speed can be adjusted in the range of 10~90 rev/min and the stirring depth is fixed at 0.5 m. The molten iron composition was similar to the above and the temperature was 1260–1360°C. The specific experimental method involved opening the molten iron ladle to the desulfurisation position to perform desulfurisation operations first, and then removing the slag and taking samples after desulfurisation was completed. At the beginning of the dephosphorisation process outside the furnace, 15 per cent to 20 per cent dephosphorisation flux was added. 30 per cent to 40 per cent dephosphorisation flux was added while lowering the stirring paddle. After stirring started, the remaining dephosphorisation flux was added in batches for a total of two minutes. A low speed of 30~50 rev/min was used initially and then switched to a high speed of 70~90 rev/min. The duration of the mixing process was about 15 minutes. After the dephosphorisation process outside the furnace was completed, slag samples and metal samples were taken respectively, and the dephosphorisation slag was then removed and entered into the converter smelting process.

Analysis and testing

The sample analysis of laboratory tests and industrial tests is conducted using the same testing methods and instruments. The slag is ground to a size of less than 200 mesh for composition testing. Weigh out 0.1 g of scraps from the pig iron block and use 3 mL HCl and 1 mL HNO_3 in a Teflon test tube to completely digest it in a 45°C water bath. Examination was conducted by inductively coupled plasma mass spectrometry to determine the content of (P). The slag sample was ground to a size of less than 80 μm and the slag composition was analysed using X-ray fluorescence spectroscopy.

Experimental results and analysis

As depicted in Figures 4a and 4c, under laboratory conditions, a new flux was used to smelt low-silicon molten iron with a (P) content of 0.10~0.13 wt per cent. No. 1–No. 3 respectively simulated the pretreatment dephosphorisation of the flux in the molten iron ladle, the dephosphorisation of the flux outside the furnace during the KR stirring process and the KR stirring dephosphorisation under low silicon conditions. The (Si) content of No. 1 was extremely low. Without stirring, the slagging time was 40 seconds and the dephosphorisation rate was 57.3 per cent. No. 2 was under the same conditions as No. 1 molten iron (Si), but the stirring process was added. The slagging time of No. 2 was about 32 seconds and the dephosphorisation rate was 52.3 per cent. Stirring could shorten the slag removal time, but the dephosphorisation rate decreased slightly. The main reason is that there was a certain slag layer at the contact end of the Al_2O_3 refractory material during stirring, which reduced the amount of slag participating in the reaction and reduced the dephosphorisation effect. The (Si) of No. 3 was 0.10 wt per cent, the slagging time was about 29 seconds and the dephosphorisation rate was 28.5 per cent. When there was a small amount of (Si) in the molten iron, there was competition between (Si) removal and (P) removal. (Si) removal consumed the oxidant in the flux. At the same time, the amount of slag was too small, which had a more significant impact on dephosphorisation. It was worth noting that the C-O reaction was inhibited at this temperature and the main source of oxidant consumption was desiliconisation and dephosphorisation.

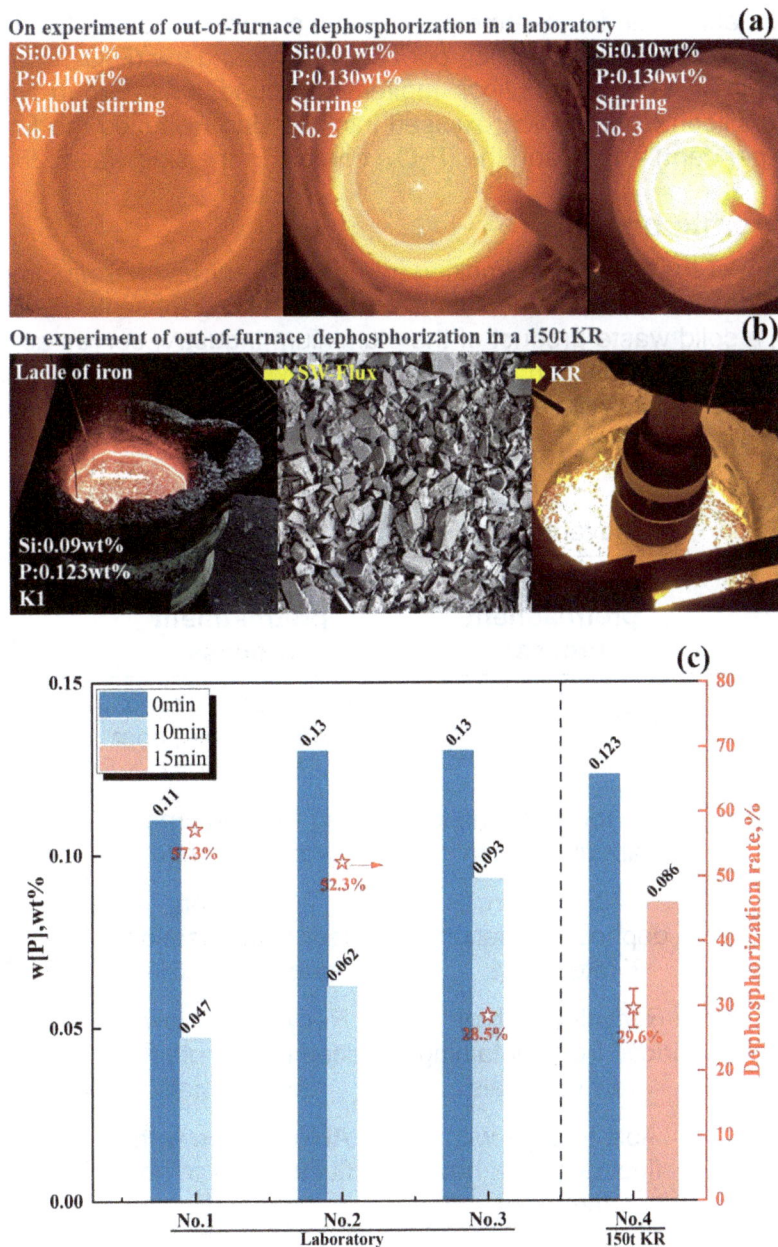

FIG 4 – (a) Simulation of out-of-furnace dephosphorisation in the laboratory; (b) on experiment of out-of-furnace dephosphorisation in a 150t KR; and (c) w(P) content and dephosphorisation rate in the experiments.

As depicted in Figures 4b and 4c, an out-of-furnace dephosphorisation experiment was conducted in a 150 t KR using solid waste-based flux as the slagging material. The test was performed for a total of five heats. When smelting low-silicon molten iron with (P) content of 0.10~0.13 wt per cent and (Si) content of 0.08~0.10 wt per cent, the average flux slagging material consumption was 10 kg/t iron and the average dephosphorisation rate of the molten iron was 29.6 per cent. The flux was added to KR and stirred. The material would completely melt in about 3 minutes and 40 seconds. The chemical composition of the slag at the end of the laboratory and 150 t KR out-of-furnace dephosphorisation industrial trials is depicted in Table 4.

TABLE 4

Final slag composition (wt%).

No.	SiO₂	CaO	MgO	Al₂O₃	Fe₂O₃	TiO₂	MnO	P₂O₅
1	2.92	32.60	2.92	12.40	25.10	3.28	-	8.26
4	16.71	36.22	1.54	8.69	27.67	4.03	4.52	6.73

As can be seen from Table 4, under extremely low silicon conditions, Group No. 1 could still maintain a high final slag alkalinity (CaO per cent/SiO$_2$ per cent) = 11.2 and the (P$_2$O$_5$) content in the slag was 8.26 per cent. In contrast, under low silicon molten iron conditions, Group No. 4 had a smaller amount of slag. The silicon oxide, which oxidised into the slag, reduced the final slag alkalinity to (CaO per cent/SiO$_2$ per cent) = 2.17 and the (P$_2$O$_5$) content in the slag was 6.73 per cent.

Analysis of economic benefits

Firstly, the application of the flux in steelmaking introduces very few harmful impurity elements (P, S), eliminating the need to consider the additional cost of removing these newly introduced impurities. Secondly, red mud, a solid waste product of the aluminium industry, serves as a highly useful and low-cost feedstock for steel production. Therefore, the new pre-melting fluxes can offer significant advantages over the use of commercial calcium ferrate and CaF$_2$, as demonstrated in Table 5 (Ban-Ya et al, 1989; Wu et al, 2022).

TABLE 5

Advantages of new pre-melted fluxes.

BOF smelting process	Hot metal pretreatment process	Hot metal pretreatment process	Effect of process improvement
CaF$_2$, cost USD205/t	Commercial calcium ferrate, cost USD310/t	Solid waste-based calcium ferrate, cost USD211/t	Reduction of raw material cost
Slag forming time is about 5~6 min	Slag forming time is about 5~6 min	Slag forming time is about 3~4 min	Reduction of smelting time
2~5 kg/t iron	40 kg/t iron, dephosphorisation rate >80%	20 kg/t iron, dephosphorisation rate is 50~60%	Better dephosphorisation
Iron-free raw materials	Recycling of high-priced iron-containing raw materials	Recycling of low-priced iron-containing raw materials	Increased revenue
Need to control hot metal (P) stabilisation	Able to cope with fluctuations in hot metal (P)	Able to cope with fluctuations in hot metal (P)	Appropriately relax the requirements for hot metal (P)

In terms of process, by adopting a KR-like dephosphorisation method outside the furnace, SW-Flux can replace CF-55. Currently, the price of CF-55 is USD310/t, while the cost of SW-Flux is estimated to be USD211/t. Given the addition of 20 kg/t iron, the dephosphorisation rate of iron can be between 50 per cent and 60 per cent and the cost can be reduced by USD1.98/t. Assuming that the dephosphorisation process is adopted in the BOF furnace, the SW-Flux can be used as a substitute for CaF$_2$, serving as a non-fluorine steelmaking flux. At present, the price of fluorite is USD205/t and the cost is about USD0.62/t for the addition of 3 kg/t iron. For the same dephosphorisation effect, the cost of pre-melting flux is about USD1.27/t for the addition of 6 kg/t iron and the cost of dephosphorisation of molten iron only increases by USD0.65/t. Therefore, this study suggests that the utilisation of this melt is more beneficial in terms of dephosphorisation efficiency and cost and it is entirely viable to substitute fluorite and costly commercial calcium ferrate.

In practical application, using the full amount of calcium ferrate as a flux for out-of-furnace dephosphorisation is suitable for the smelting of high-quality steel grades with ultra-low phosphorus steel. It can also replace fluorite in BOF smelting to reduce the environmental threat of fluorine-containing melts.

CONCLUSIONS

- The flux was prepared from industrial solid waste and primarily consisted of $CaFe_2O_4$, $Ca_2Fe_2O_5$ and $Ca_2(Fe,Al)_2O_5$. These mineral phases were referred to as complex calcium ferrite phases. These phases exhibited low melting points, high oxidising properties and high alkalinity.

- This flux was utilised as a dephosphorisation slagging material. Under laboratory conditions, the phosphorus content of the smelted low-silicon molten iron was 0.10~0.13 wt per cent. The temperature of the molten iron reached 1350°C and the slag amount was controlled at 2 per cent. Within ten minutes of smelting, the dephosphorisation rate of the molten iron exceeded 50 per cent. When making slag with new flux, the slag melting time approximated 40 seconds and the P_2O_5 content in the final slag exceeded 8 wt per cent.

- An industrial test of 150 t hot metal ladle was conducted. Under the same hot metal conditions, an out-of-furnace dephosphorisation test was performed in the KR process. The flux material consumption for per ton of steel was 8~10 kg. Within 15 minutes of smelting, the dephosphorisation rate of the hot metal approximated 30 per cent. During flux slag making, the slag melting time approximates 3 minutes and 40 seconds and the P_2O_5 content in the final slag exceeds 6 wt per cent.

- In the development of steelmaking technology, this new solid waste-based flux possesses significant cost advantages and out-of-furnace dephosphorisation effects. It is anticipated to represent a new generation of fluorine-free steelmaking flux and possesses extremely high promotion and application value.

ACKNOWLEDGEMENTS

This work was supported by the National Key R&D Program of China (2019YFC1905701), the National Natural Science Foundation of China (U1960201).

REFERENCES

Bai, X, Sun, Y, Luo, L and Zhao C, 2020. Effect of direct charging of hot recycled slag on hot metal pretreatment dephosphorization in a dephosphorization furnace, *Journal of Iron and Steel Research International*, 27:148–159. https://doi.org/10.1007/s42243-019-00296-w.

Ban-Ya, S, Hino, M, Nagabayashi, R and Terayama, O, 1989. Dephosphorization and Desulphurization of Hot Metal with $CaO-Al_2O_3-FexOy$ Flux, *Tetsu-to-Hagané*, 75(1):66–73. https://doi.org/10.2355/tetsutohagane1955.75.1_66.

Gao, X, Matsuura, H, Miyata, M and Tsukihashi, F, 2013. Phase Equilibrium for the $CaO-SiO_2-FeO-5mass\%P_2O_5-5mass\%Al_2O_3$ System for Dephosphorization of Hot Metal Pretreatment, *ISIJ International*, 53(8):1381–1385. https://doi.org/10.2355/isijinternational.53.1381.

Gupta, P, Kumar, S, Babu, J H, Mali, G S and Roy, T K, 2023. Production of low phosphorous steel from low silicon high phosphorous hot metal in BOF, *Ironmaking and Steelmaking*, 50:1014–1021. https://doi.org/10.1080/03019233.2023.2173877.

Jeon, J-W, Jung, S-M and Sasaki, Y, 2010. Formation of Calcium Ferrites under Controlled Oxygen Potentials at 1273 K, *ISIJ International*, 50:1064–1070. https://doi.org/10.2355/isijinternational.50.1064.

Kitamura, S-Y, Yonezawa, K, Ogawa, Y and Sasaki, N, 2002. Improvement of reaction efficiency in hot metal dephosphorization, *Ironmaking and Steelmaking*, 29:121–124. https://doi.org/10.1179/030192302225004070.

Lee, M S and Barr, P V, 2002. Production and properties of burnt lime coated with dicalcium ferrite, *Ironmaking and Steelmaking*, 29(2):96–100. https://doi.org/10.1179/030192302225001983.

Li, F, Zhang, Y and Guo, Z, 2017. Pilot-Scale Test of Dephosphorization in Steelmaking Using Red Mud-Based Flux, *JOM*, 69:1624–1631. https://doi.org/10.1007/s11837-017-2449-9.

Li, J, Wang, S-J, Xia, Y-J and Kong, H, 2014. Study on dephosphorization of hot metal pretreatment with Al_2O_3 to replace CaF2 in slag, *Ironmaking and Steelmaking*, 42(1):70–73. https://doi.org/10.1179/1743281214y.0000000213.

Mi, G, Murakami, Y, Shindo, D, Saito, F, Shimme, K and Masuda, S, 1999. Mechanochemical Synthesis of Calcium Ferrite by a Planetary Mill and Its Dephosphorization Characteristic from Molten Iron, *Shigen-to-Sozai: Journal of the Mining and Materials Processing Institute of Japan*, 115(9):683–687. https://doi.org/10.2473/shigentosozai.115.683.

Paananen, T, Heikkinen, E, Kokkonen, T and Kinnunen, K, 2010. Preparation of mono-, di- and hemicalcium ferrite phases via melt for reduction kinetics investigations, *Steel Research International*, 80(6):402–407. https://doi.org/10.2374/sri09sp016.

Sato, T, Nakashima, K and Mori, K, 2001. Dephosphorization rate of high carbon iron melts by CaO-based slags, *Tetsu-to-Hagané*, 87:643–649. https://doi.org/10.2355/tetsutohagane1955.87.10_643.

Sayama, H, Tanaka, S, Ito, M and Machinaga, O, 2002. Calcination process of limestone coated with iron oxide and aluminum oxide, *J Soc Inorg Mat*, 9:218–223.

Sukenaga, S, Gonda, Y, Yoshimura, S, Saito, N and Nakashima, K, 2010. Viscosity Measurement of Calcium Ferrite Based Slags during Structural Relaxation Process, *ISIJ International*, 50(2):195–199. https://doi.org/10.2355/isijinternational.50.195.

Wright, S, Zhang, L, Sun, S and Jahanshahi, S, 2001. Viscosities of calcium ferrite slags and calcium alumino-silicate slags containing spinel particles, *Journal of Non-Crystalline Solids*, 282(1):15–23. https://doi.org/10.1016/S0022-3093(01)00324-6.

Wu, W, Yang, Q, Gao, Q and Zeng, J, 2020. Effects of calcium ferrite slag on dephosphorization of hot metal during pretreatment in the BOF converter, *Journal of Materials Research and Technology*, 9(3):2754–2761. https://doi.org/10.1016/j.jmrt.2020.01.009.

Wu, W, Zhao, B, Zhao, B and Meng, H-D, 2022. Hot metal dephosphorization process using calcium ferrite slag without fluorite, *Ironmaking and Steelmaking*, 49(7):661–668. https://doi.org/10.1080/03019233.2022.2037040.

Zhang, Y, Zhao, Z, Zhang, W, Yu, K and Zhang, Y, 2023. Red Mud-Based Composite Calcium Ferrite and Preparation Method and use Thereof, United States Patent US11773025B2.

Zhou, J, Bi, X, Yue, R and Yang, F, 2017. Phosphorus Distribution Ratio between Multi Phase $CaO–FeOt–SiO_2–P_2O_5$ (6%–13%) Slags with MP Near Hot Metal Temperature and C-saturated Molten Iron at 1573 K, *ISIJ International*, 57(4):706–712. https://doi.org/10.2355/isijinternational.isijint-2016-592.

Analysis on composition and physico-chemical property change of PbO-bearing slag in Pb smelting

J X Zhao[1], Y R Cui[2], G Cao[2], G H Wang[2], B Li[2], M M Ren[2], Z M Wang[3] and H P Gui[3]

1. School of Metallurgical Engineering, Xian University of Architecture and Technology, Xian, Shaanxi 710055, China. Email: zhaojunxue1962@126.com
2. School of Metallurgical Engineering, Xian University of Architecture and Technology, Xian, Shaanxi 710055, China.
3. Hanzhong Zinc Co. LTD, Hanzhong, Shaanxi 724200, China.

ABSTRACT

Pyro-metallurgy is still the main process for lead production and a lot of lead direct smelting processes have been developed in recent years. All the processes will experience two steps in common. The first is the oxidisation of PbS to PbO and then reduction from PbO to Pb, furthermore, from ZnO to Zn. So, the composition of melts, especially the slag composition at reduction step change continuously. Up to now, there are few systematic results on the slag physical-chemical properties with its composition change. In this work, the typical lead direct smelting processes were analysed from theoretical and practical standpoints. The full-period slag composition and slag melting point change were compared. Especially, to the PbO and ZnO reduction steps, based on the full-period slag composition change, the slag properties were predicted. Based on dust composition and test results, the mechanism of Pb compound evaporation and its effect on practice and measurement of high PbO bearing slag physical-chemical property is discussed. It is suggested that to restrain the evaporation in practice and to obtain more accurate measurement results of physicochemical properties of smelting slag are the key problem not only to set high efficiency, energy saving process, but also to environmental pollution control.

INTRODUCTION

Up to now, pyro-metallurgy is still the main route for lead production. At earlier stage, sintering and blast furnace was used with galena concentrate as raw material. Since the 1970s a lot of direct reduction processes have been developed, such as QSL (Queneau and Siegmund, 1996), SKS (Zhang, 2013), Kivcet process (Slobodkin, Sannikov and Grinin, 2013), Ausmelt/ISAsmelt (Wang, Zhou and Feng, 2004), Kaldo, bottom blowing/side submerged blowing and so on (Li, Yang and Cheng, 2011; Cui, Li and He, 2013; Chen, Yang and Bin, 2014a; Chen, Hao and Yang, 2015). At the beginning, blast furnace is widely used to reduce the higher PbO-content slag from upstream smelter, such as SKS process, Ausmelt/ISA smelter (Queneau and Siegmund, 1996; Zhang, 2013; Wang, Zhou and Feng, 2004; Wang and Chen, 2016), and then it is being gradually replaced by other oxygen-enriched blowing smelters.

In all these processes, two steps are experienced, that is oxidation first and then reduction. In the oxidising smelting step, Pb sulfide concentrate is oxidised and melted. Some molten lead is formed and slag with high-PbO concentration produced concurrently. In the reduction step, slag with high-PbO concentration is reduced with coal, semicoke or natural gas as reductant (Chen, Yang and Liu, 2014b; Li, Zhan and Fan, 2017; Zhang, Li and Zhan, 2018; Yang, Cui and Hao, 2020), such as Kivcet/flash smelting process (Jiang, Li and Guo, 2018). To recover the Zn in the reduction slag, further reduction is necessary sometimes. To keep the process working at high temperature, oxygen enriched blowing is widely adopted.

In recent years, the triple-furnace system of molten high-lead slag direct reduction smelting applied in the metallurgical industry in China (Slobodkin, Sannikov and Grinin, 2013; Wang and Chen, 2016). Bottom blowing furnace or side submerged blowing furnace is used as oxidising smelter or reduction furnace respectively (Li, Yang and Cheng, 2011; Chen, Yang and Liu, 2014b; Li, Zhan and Fan, 2017; Zhang, Li and Zhan, 2018). In addition, fuming furnace is used for further reduction and ie ZnO recovery (Li, Zhan and Fan, 2017; Zhang, Li and Zhan, 2018). In these processes, the slag with high PbO, ZnO bearing slag or melt is involved.

Generally, the oxidising step is a continuous process, PbS concentrate is added with oxygen blowing, and PbO (with some molten Pb) formed until it accumulated up to tapped. The reduction step is a batch process, reducing reaction with C or natural gas as reductant to the slag formed until the high-PbO bearing slag changes to the low-PbO bearing slag (contains about 2.5 per cent Pb) and molten Pb forms at the same time, and then high-ZnO bearing slag changes to the low-ZnO bearing slag. So, the composition of melts, especially the slag composition at reduction step will changes continuously. Up to now, there are few reports on the slag physical-chemical properties with its composition change systematically (Huang, Zhang and Cheng, 2021; Cui, Guo and Chen, 2016; Jak and Hayes, 2002). In this paper, we will try to draw up the relationship between slag composition and physico-chemical property change in whole process.

BASIC REACTION AND TYPICAL SLAG FORMING AND COMPOSITION CHANGE

Basic reaction

To take the typical direct reduction process as example (as shown in Figure 1), the process can be divided into three steps, ie oxidisation smelting, PbO reduction and further reduction for Zn recovery.

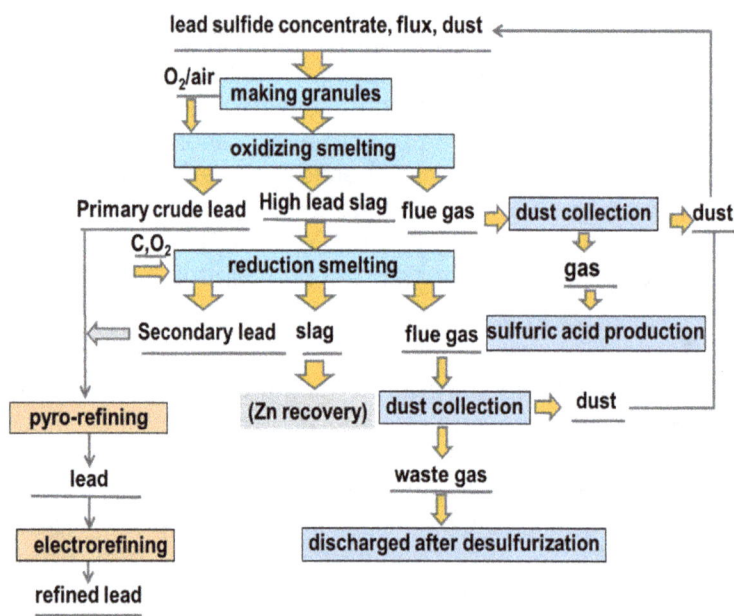

FIG 1 – Typical direct reduction process.

Oxidising smelting

The sulfide concentrates (oxides or lead sulfate can be used as regulator for heat balance), flux (limestone and quartz), coal or semicoke, and recycled dust are prepared in a suitable proportion, and mixed. Then, the mixture is added into oxidising smelter, ie side submerged or bottom blowing smelter. Oxygen enriched air is blown into the smelter to oxidise the sulfide concentrate and fuel, if needed, to generate heat to keep the molten bath at high temperature and quick reaction.

The main reaction in this step is showing as following Equations 1–3, high-PbO slag and some crude lead is formed as a result in this step. The FeS and ZnS is changed into oxides, and turn into slag with other gangue. So the main slag system is $PbO-ZnO-FeO-CaO-SiO_2$. There is less fuel added because the reactions are exothermal. To prevent PbS evaporation, the temperature should be kept at a relatively lower level. The predominance diagram of Pb-Zn-S-O system is displayed in Figure 2 (Yang, Cui and Hao, 2020) district I.

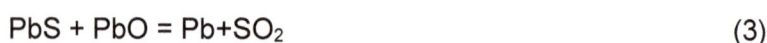

$$PbS + O_2 = Pb + SO_2 \tag{1}$$

$$PbS + O_2 = PbO + SO_2 \tag{2}$$

$$PbS + PbO = Pb + SO_2 \tag{3}$$

Zn-Pb-S-O, 1050 C

0 < Pb/(Zn+Pb) < 1

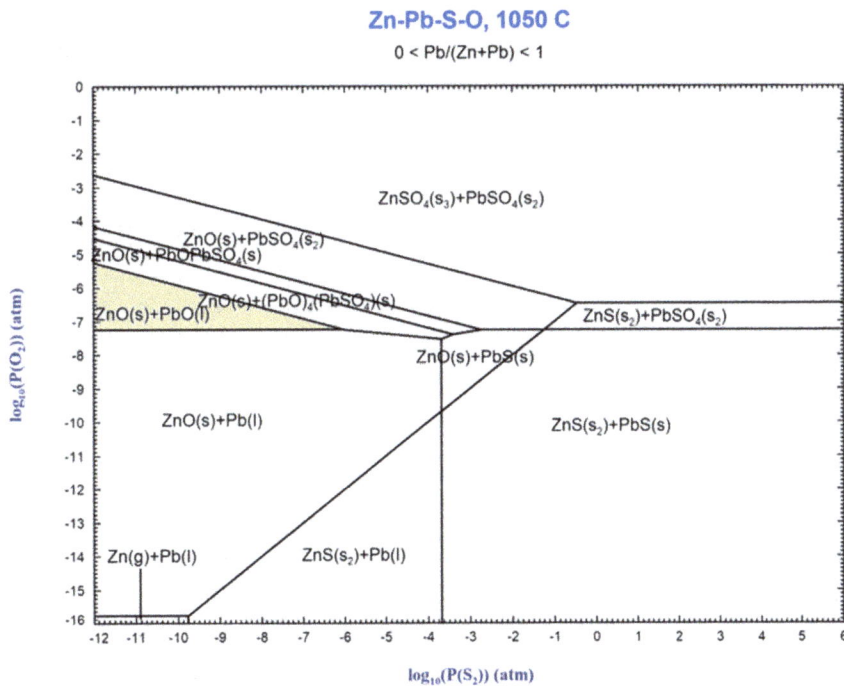

FIG 2 – Predominance diagram of Pb-Zn-S-O system, 1050°C.

The atmosphere is controlled between the Pb(l) and +PbO(l) regions to ensure that ZnO in slag is not reduced, some crude lead and high lead bearing slag is formed. In practice, a relatively high oxidation smelting temperature (1000~1050°C) is selected, the PO_2 is controlled between 10^{-5} ~ 10^{-6} atm, and $P_SO_2 \leq 10^{-4}$ atm.

PbO reduction

High-PbO slag is added into the reduction smelter, ie side-submerged or bottom blowing furnace with proportional flux and fuel. Oxygen enriched air is blown in as well. The main reaction in this step is as Equations 4–5.

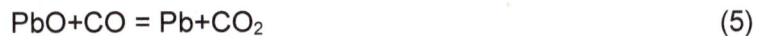

$$PbO+C = Pb+CO \tag{4}$$

$$PbO+CO = Pb+CO_2 \tag{5}$$

The temperature and atmosphere are controlled to prevent ZnO reduction. So crude lead and slag with higher ZnO content is formed in this step, as shown in Figure 2 district II. The reactions are endothermic. Coal or semicoke or natural gas as fuel and reductant have to be added.

Because FeO and CaO can replace PbO from silicate and promote activity of PbO, increasing FeO/SiO_2 and CaO/SiO_2 is in favour of PbO reduction.

Further reduction for Zn recovery

ZnO bearing slag is reduced in fuming furnace to recover Zn in general. The reaction is as Equation 6.

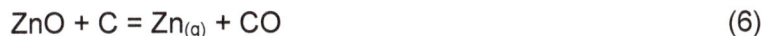

$$ZnO + C = Zn_{(g)} + CO \tag{6}$$

Zinc is reduced and evaporates, as shown in Figure 2 district III. Then Zn is reoxidised to ZnO and collected for further use. increasing FeO/SiO_2 and CaO/SiO_2 is in favour of ZnO reduction.

Above reactions are common to all direct reduction processes.

Parameters in practice

The smelting temperature is usually selected at 1000–1350°C according to the slag properties. The melting temperature should be 80–150°C higher (superheat) than the liquidus temperature (complete fusion temperature).

According to relevant results that the slag liquidus temperature decreases with the Pb content or $w(FeO)/w(SiO_2)$ increase, and ZnO content decrease. It is determined that $w(FeO)/w(SiO_2)$ = 1.0–2.0, $w(CaO)/w(SiO_2)$ = 0.4–0.8. In practical application, the relevant parameters can be adjusted properly.

Oxidising stage

Generally, the composition of mixed ore prepared for smelting is at about Pb 45–60 per cent, S 15–17 per cent, Zn 6–7 per cent, Cu < 0.7 per cent. At the same time, appropriate amount of the flux limestone (CaO > 40 per cent), silica sand (SiO_2 > 85 per cent), coal (C > 73 per cent) is added.

The slag composition in this stage is with a high PbO content and relatively lower ZnO content. When Pb content is above 40 per cent, the slag physico-chemical properties is mainly determined by Pb content.

Some slag composition and smelting temperature in practice is given in Table 1 (Li and Jia, 2010).

TABLE 1

Typical slag composition and smelting temperature.

No.	Pb,%	Zn,%	$w(FeO)/w(SiO_2)$	$w(CaO)/w(SiO_2)$	Smelting temp °C
1	42	7–8	1.0	0.33–0.49	1040–1090
2	39.57	9.95	0.93	0.64	1100–1150
3	53.47	6.40	1.46	0.65	980
4	45.02	4.81	1.19	0.48	866
5	49.59	4.42	1.62	0.44	807

Different plants may have different operation parameters. In an oxygen-enriched bottom blowing smelting plant, $w(FeO)/w(SiO_2)$ = 1.3–1.6 (target: 1.5), $w(CaO)/w(SiO_2)$ = 0.4–0.7 (target: 0.5) in oxidising step. The bath temperature is 1000–1100°C. The crude lead output is 40–60 per cent. Fume rate is 11–16 per cent. Pb content in high-PbO slag is 40–48 per cent (Li and Jia, 2010).

The typical raw materials, slag composition and dust composition to a triple-furnace system, ie side oxygen-enriched submerged blowing + side oxygen-enriched submerged blowing+ fume furnace, is listed in Table 2 (Chen, 2021).

TABLE 2

The raw materials, slag composition and dust composition in side oxygen-enriched blowing.

Stage	Materials	Composition, %							Fluxes/fuel
		Pb	Zn	Cu	Fe	CaO	SiO_2	S	
Oxidising step	Mixed ore	55.0	4.7	0.80	8.25	0.35	3.8	17.0	Lime stone and silica sand Coal
	Dust	49.0	4.7	0.12	0.83	0.04	0.38	7.17	
	High-PbO slag	42.11	6.54	0.45	11.66	7.42	10.60	0.19	
Reduction step	Dust	50.78	5.28	0.05	1.88	1.20	1.71	0.10	Small amount of limestone coal
	slag with higher Zn	2.38	12.35	0.41	22.15	18.99	21.10	0.07	
Further reduction	Final slag	0.42	1.46	0.47	26.40	24.63	24.40	0.06	Coal
	Dust	10.94	60.0	0.06	2.99	25.71	2.86	0.12	

It can be seen that the sulfide in mixed ore changes to oxides and part of Pb compound is reduced. As a result, slag with the remaining PbO is formed. The Pb content in slag is about 42 per cent.

The dust rate is about 15 per cent (to take mixed ore as reference) in this step. Dust with higher PbO content and much lower FeO, SiO_2 and CaO content than mixed ore and slag is obtained. That means there is selective evaporation of Pb compound is the main reason, rather than mechanical carrying of exit gas.

With higher PbO content, the slag can be kept in melt state well in this stage.

PbO and ZnO reduction stage

As shown in Table 2, In the PbO reduction step, the Pb content in slag decreases gradually from about 42 per cent to < 2 per cent, and Zn content in the slag increases accordingly. The smelting temperature is at about 1200–1250°C.

The dust rate is about 13 per cent (to take high-Pb slag as reference) in PbO reduction step. As in oxidising step. The dust with higher PbO content and much lower FeO, SiO_2 and CaO content than mixed ore and slag is obtained. It can also be concluded that selective evaporation of Pb and its compounds takes place rather than mechanical carrying of exit gas.

In the further reduction step, the ZnO content in the slag gradually decreased from 12.35 per cent to 1.5 per cent, and Pb content decreased further to 0.42 per cent. The smelting temperature is at about 1300–1400°C.

Higher temperature can promote the reduction anyway.

Apart from the high lead slag, there is no detailed results on slag composition and property change for reference up to now.

ANALYSIS ON PHYSICO-CHEMICAL PROPERTIES OF SLAG IN WHOLE PROCESS

Basic research results

As listed in Table 2, the high-PbO slag is formed at oxidising step. And then the PbO is reduced gradually in reduction step. Finally, the ZnO and rest PbO is further reduced.

Slag physico-chemical properties is the key reference for smelting process. Up to now, there is fewer results on slag physico-chemical properties of lead direct smelting.

Some works are focused on final slag, ie FeO-CaO-SiO_2 system when PbO and ZnO is completely reduced (Zhang and Dai, 2020). Jak and Hayes (2002; 2003a; 2003b) researched the phase equilibrium of PbO-CaO, PbO-CaO-SiO_2, and PbO-ZnO-SiO_2 system. Yang, Dou and Zhang (2019) studied the high-lead slag with Zn content of 15–32 per cent. It is found that the decrease of ZnO/PbO and FeO/SiO_2 leads to the drop of viscosity (Yang, Dou and Zhang, 2019). Cui, Li and He (2013) and Cui, Guo and Chen (2016) investigated phase diagram and melting characteristic of PbO-ZnO-FeO-CaO-SiO_2 system using FactSage™ ver 7.1 (by CRCT-ThermFact Inc.& GTT-Technologies). It is pointed out that the melting temperature reduces with FeO/SiO_2 increase (1.2–2.0) when the PbO content in slag is more than 10 per cent, in accordance with Perez-Labra, Romero-Serrano and Hernandez-Ramirez (2012) results.

Huang, Zhang and Cheng (2021) measured the melting temperature and viscosity of PbO-ZnO-FeO-CaO-SiO_2 slag with high-Pb content (about Pb 40 per cent), Fe/SiO_2 (0.9–1.3), CaO/SiO_2 (0.4–0.8) and ZnO content (5–11 per cent), and obtained a similar measurement result of physico-chemical properties as Cui, Li and He (2013). It is suggested that the appropriate Fe/SiO_2 is 1.0–1.1 and CaO/SiO_2 0.6 and ZnO content <7 per cent in oxygen-enriched smelting process. The main results are as Figure 3. This is relatively comprehensive and can reflect the research state of high Pb-bearing slag. Anyway, the slag property in whole reduction process is in lacking up to now.

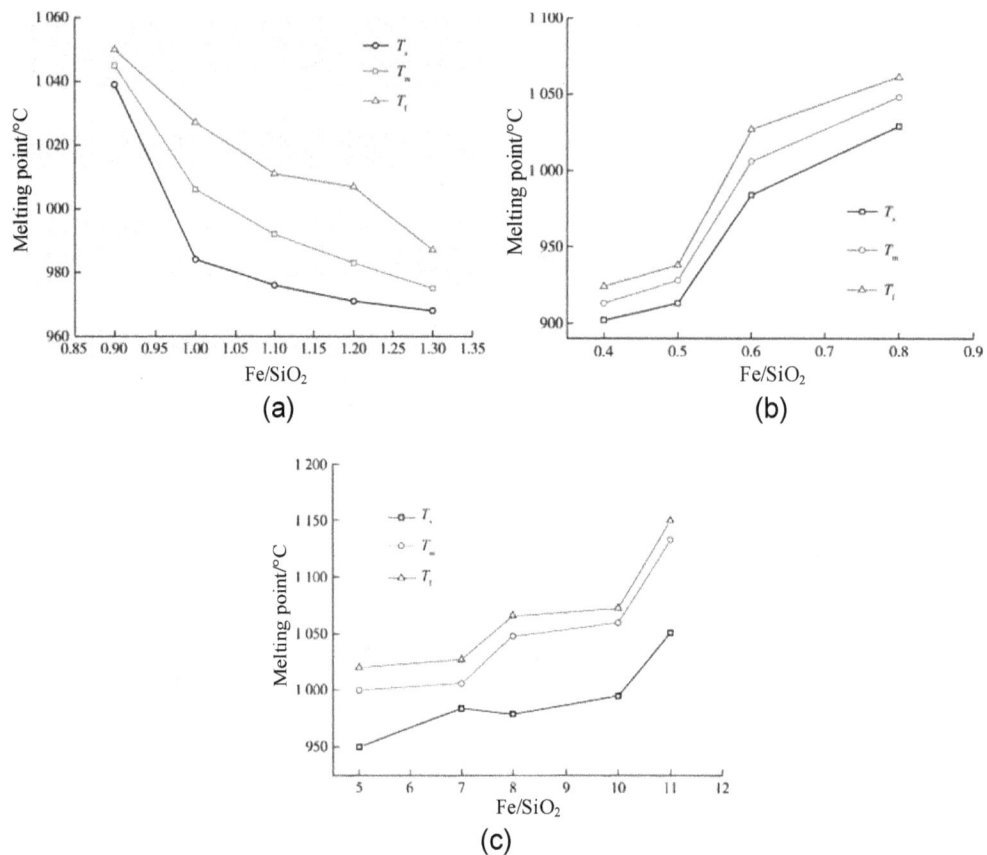

FIG 3 – Effects of Fe/SiO$_2$, CaO/SiO$_2$ and ZnO on melting point respectively. (a) (CaO/SiO$_2$ = 0.6, Pb per cent = 38, ZnO = 7 per cent); (b) (Fe/SiO$_2$ = 1.0, Pb per cent = 38, ZnO = 7 per cent); (c) (Fe/SiO$_2$ = 1.0, Pb per cent = 38, CaO/SiO$_2$ = 0.6).

Theoretical analysis on slag properties in whole reduction process

To demonstrate the slag property with the composition change, an iso-liquidus temperature curve based on PbO-FeO-CaO-SiO$_2$-ZnO system and relevant composition is drawn as Figure 4. The point A and B is the beginning and end of PbO reduction step respectively. The point B and C is the beginning and end of ZnO reduction step respectively. In PbO reduction process (as shown in line A to B), FeO/SiO$_2$ = 1.4, CaO/SiO$_2$ = 0.7, ZnO/FeO = 0.54, only PbO content decreases. As the result, FeO+CaO+SiO$_2$+ZnO content increase proportionally. As a result, the ZnO content can reach about 20 per cent in mass. In ZnO reduction process (as shown in line BC), FeO/SiO$_2$ = 1.4, CaO/SiO$_2$ = 0.7, PbO = 0.54, only ZnO content decreases and FeO+CaO+SiO$_2$ increase proportionally. It can be seen that the whole reduction process is in a composition range with lower melting point (1100–1250°C).

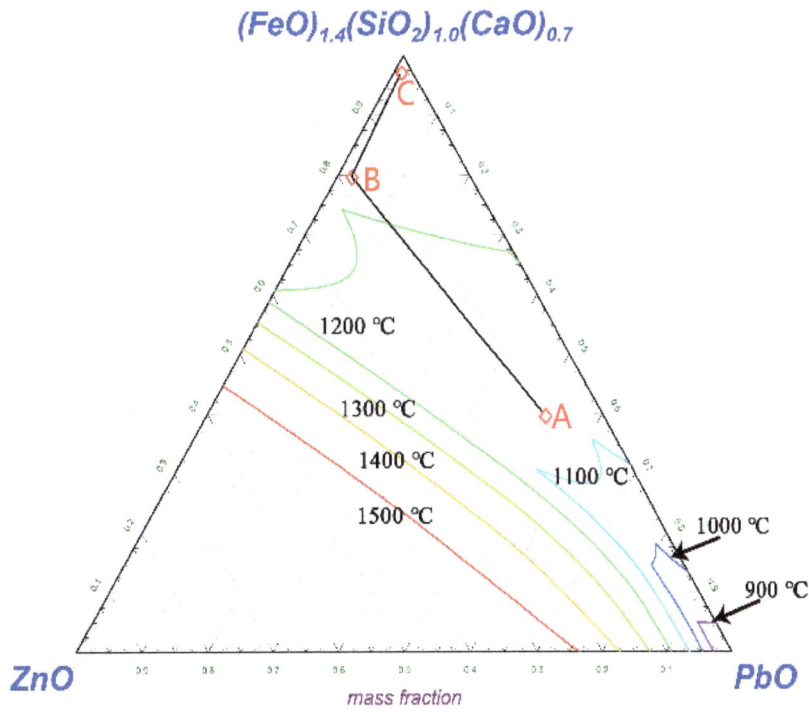

FIG 4 – Calculated PbO-FeO-CaO-SiO$_2$-ZnO liquid phase diagram with liquidus isotherms.

For further analysis, the liquidus melting point was calculated with FactSage™ and shown in Figure 5.

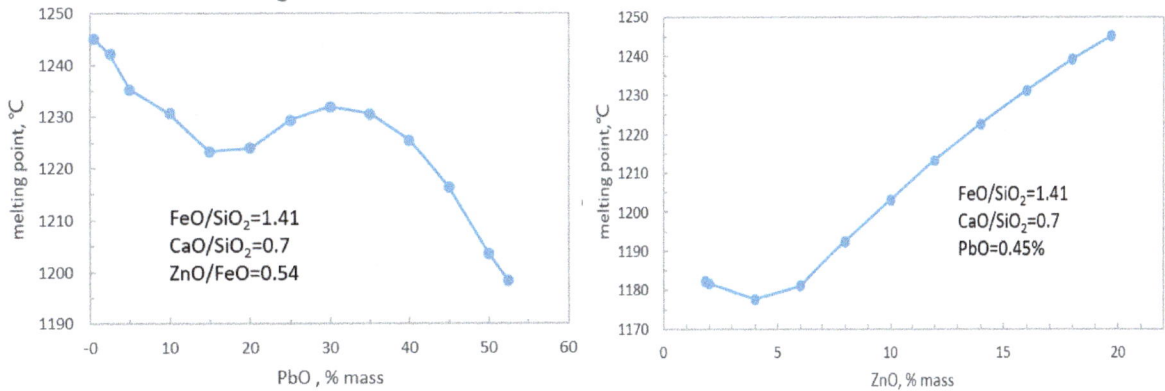

FIG 5 – Relation among melting point with PbO and ZnO content in slag.

It can be seen that with the PbO reduction, the slag melting point will increase gradually from 1198°C to 1245°C. And then to further reduction of ZnO, the slag melting point will decrease gradually from 1245°C to 1182°C with the ZnO content decrease. The smelting temperature should be at 1300–1350°C. The slag melting point will experience increase until the end of PbO reduction and then decrease with the PbO reduction. This is in accordance with relevant practical results. The smelting temperature and slag composition should be adjusted properly according to these changes.

The results of some relevant slag property measurement

Based on high PbO slag reduction process, the properties of some slag samples with different composition were measured. The results is listed in Table 3. It can be seen that with the PbO reduction, the ZnO and other components in the slag increased. The melting point increased accordingly. When PbO is reduced from 53.86 per cent to 2.69, the melting point increased from 841°C to 1296°C. Higher than that of theoretical calculation with FactSage™.

TABLE 3

Some properties measurement results of slag samples with different composition.

Slag No.	Composition, %					Liquidus, °C	Viscosity, Pa.s		
	PbO	ZnO	FeO	CaO	SiO$_2$		1180°C	1230°C	1280°C
1	53.86	6.0	21.04	5.46	13.64	841	0.05		
2	43.09	6.0	26.69	6.92	17.3	1158	0.31	0.1	0.05
3	32.32	8.0	31.29	8.11	20.28				
4	21.54	10.0	35.89	9.3	23.26	1285	0.62	0.12	0.11
5	10.77	13.0	39.96	10.36	25.9				
6	2.69	13.0	44.2	11.46	28.65	1296	0.38	0.11	0.098

THE EVAPORATION OF SLAG COMPONENTS AND ITS EFFECT

Evaporation and its effect in practice

In the whole process, Pb, PbO, PbS, Zn, ZnO and even relevant sulfate appear. It is well known that PbS, PbO, Pb and Zn are volatile material at high temperature (Osada, Osada and Kokado, 2004; Nakada, Mihara and Kawaguchi, 2008; Guo, Cui and Zhao, 2016).

In practice, the evaporation information can be found in collected dust as listed in Table 2.

In oxidising step, the dust composition is obviously different from either raw material concentrate or high-Pb slag. Low Fe and SiO$_2$ concentrations reflect that the dust is neither concentrate mechanical carrying nor high-Pb slag splashing. To take the Fe or SiO$_2$ content as a reference, the concentrate mechanical carrying is about one tenth, a small part. A higher sulfur content reflects that PbS evaporates in main. It is imagined that the PbS evaporates first and then some oxidised to PbO.

In reduction step, there is more Pb and low Fe, Si, and Ca in dust than in slag. This reflects that the dust is not high-Pb slag splashing, but evaporation in main. The main volatiles should be PbO and Pb.

The dust and fume rate are about 13–15 per cent of the charge in relevant steps. This exposed that the importance of evaporation to slag composition and properties, as well as the process control in mass and heat balance.

In another point, dust is circulated in the smelting processes. The Pb in dust per unit of bullion lead is about 0.31 t/t Pb to oxygen-enriched bottom blowing process (Zhong, 2014). This can not only lead to the low efficiency of smelting process, but also potential risk of Pb emission and pollution. How to reduce the amount of recycling dust should be studied.

Zn is more easily to evaporate in further reduction step. That is good for Zn recovery. But the Pb compound evaporation is not good for Pb recovery and should be limited.

Evaporation and its effect to property measurement

In the physicochemical property measurement, these volatiles will evaporate continuously in heating or high temperature keeping process, so to the production process. The higher, the temperature is, or the longer the keeping time is, the great the loss of volatile from the slag is. That means the slag composition will be changed continuously. As a result, the measured slag physicochemical properties will change (Guo, Cui and Zhao, 2016; Cui, Wang and Zhao, 2018; Wang, Cui and Yang, 2020). It is referred as uncertainty of slag property for brief. Some weight loss results with TG-DSC on lead bearing slag is listed in Table 4. To consider the weight loss at 410°C–736°C is the crystal water and decomposition of compound, it can be seen that the weight loss is very evident at temperature 736°C–1450°C. This part of weight loss represents the evaporation effect. From the results in Table 4, almost all of the PbO in the slag can exit from the slag at 1450°C. With the increase of FeO/SiO$_2$ and CaO/SiO$_2$, the weight loss increases.

TABLE 4

The weight loss of slag sample with PbO at high temperature (Guo, Cui and Zhao, 2016).

Slag type	Composition			Weight loss 410°C–736°C	Weight loss 736°C–1450°C
	FeO/SiO$_2$, CaO/SiO$_2$	PbO			
Slag with low PbO content	1.54, 0.4	2.69%		12.88%	3.95%
	1.8, 0.6	2.69%		13.3%	4.66%
Slag with medium PbO content	1.54, 0.4	21.54%		9.34%	21.62%
	1.8, 0.6	20%		9.69%	24.4%
Slag with high PbO content	1.54, 0.4	43.09%		7.92%	38.69%
	1.8, 0.6	40%		8.09%	43.3%

So either to physico-chemical property measurement or production process, the evaporation of PbS, PbO, Pb and Zn should be given more attention. Relevant work should be push forward further. The evaporation can lead the results of melting point higher than it should be.

It is suggested to use premelted Pb-bearing slag in physicochemical property measurement to prevent evaporation effect. But anyway, the way should be evaluated seriously. No credible results are reported in detail. Zhao *et al* (2023) suggested a method to measure the physicochemical properties in traditional way, and modify the in-time composition based on evaporation to the measured property data. This is a promising method worthy of detected.

CONCLUSION

- Sulfide concentrates direct reduction processes are the main trend for crude Pb production, bottom blowing/side submerged blowing based processes is well developed. Two steps (oxidising and reduction) or three steps (oxidising, reduction and further reduction for Zn recovery) is common in these processes. That is reasonable and practical, but there are less basic research results for reference on direct reduction processes. The practical processes are controlled mainly by experience.

- The slag composition at reduction step will changes continuously. The slag melting point will increase first until the end of PbO reduction and then decrease with the ZnO reduction. The variation range is about 50–60. Proper slag composition selection of (FeO)/(SiO$_2$) at 1.0–2.0, (CaO)/(SiO$_2$) at 0.4–0.8 and ZnO < 7 per cent–8 per cent can guarantee smelting process go on wheels.

- Evaporation in practical oxidising step and PbO reduction step is mainly PbS and PbO respectively. These consist of the main composition in the dust. Mechanical carrying of exit gas is only a small part unless the dust recycled is carried out in a mass.

- PbO in slag evaporates evidently and can causes physical-chemical property measurement uncertainty to high PbO-containing slag. Relevant fundamental research should be pushed forward to give references for Pb-smelting parameters control and process optimisation.

- How to restrain the evaporation and to obtain more accurate measurement results of physicochemical properties of smelting slag are the key not only to set high efficiency, energy saving process, but also to environmental pollution control.

ACKNOWLEDGEMENTS

The work is supported by NSFC (No.51674185) and relevant enterprises.

REFERENCE

Chen, L, 2021. Molecular dynamics simulation on composition regulation of PbO-FeOx-CaO-SiO$_2$-ZnO system for the high-lead slag, Master's thesis, Xi'an University of architecture and technology, Xi'an.

Chen, L, Hao, Z and Yang, T, 2015. A Comparison Study of the Oxygen-Rich Side Blow Furnace and the Oxygen-Rich Bottom Blow Furnace for Liquid High Lead Slag Reduction, *JOM,* 67(5):1123–1129.

Chen, L, Wang, Z H and Chen, W, 2018. Thermodynamic simulation on elements distribution of lead concentrate oxidative smelting in oxygen-rich bottom-blow smelting process, *Nonferrous Metals (Extractive Metallurgy),* (09):1–6.

Chen, L, Yang, T Z and Bin, S, 2014a. An Efficient Reactor for High-Lead Slag Reduction Process: Oxygen-Rich Side Blow Furnace, *JOM,* 66(9):1664–1669.

Chen, L, Yang, T Z and Liu, W F, 2014b. Distribution of valuable metals in liquid high lead slag during reduction process, *The Chinese Journal of Nonferrous Metals,* 24(04):1056–1062.

Cui, Y R, Guo, Z L and Chen, A L, 2016. Phase equilibria of PbO-FeO$_x$-CaO-SiO$_2$-ZnO high-lead slag system for direct reduction process, *Chinese Journal of Rare Metals,* 40(09):928–933.

Cui, Y R, Li, K M and He, J S, 2013. Melting point of molten high-lead slag in direct reduction process, *Chinese Journal of Rare Metals,* 37(03):473–478.

Cui, Y R, Wang, G H and Zhao, J X, 2018. Volatilization Kinetics of PbO-FeO$_x$-CaO-SiO$_2$-ZnO Lead-bearing Slag, *The Chinese Journal of Process Engineering,* 18(02):393–398.

Guo, Z L, Cui, Y R and Zhao, J X, 2016. Deviation prediction on physicochemical properties of high-lead slag, *Journal of Liaoning University of Science and Technology,* 39(1):47–51.

Huang, H, Zhang, J L and Cheng, Y Q, 2021. Study on Melting Temperature and Viscosity of PbO-FeO-CaO-SiO$_2$-ZnO System, *Nonferrous Metals (Extractive Metallurgy),* (5):11–16.

Jak, E and Hayes, P C, 2002. Experimental liquidus in the PbO-ZnO-Fe$_2$O$_3$-(CaO+SiO$_2$) system in air, with CaO/SiO$_2$ = 0.35 and PbO/(CaO+SiO$_2$) = 3.2, *Metallurgical and Materials Transactions B,* 33B(5):851–863.

Jak, E and Hayes, P C, 2003a. The effect of the CaO/SiO$_2$ ratio on the phase equilibria in the ZnO-Fe$_2$O$_3$-(PbO+CaO+SiO$_2$) system in air: CaO/SiO$_2$ = 0.1, PbO/(CaO+SiO$_2$) = 6.2 and CaO/SiO$_2$ = 0.6, PbO/(CaO+SiO$_2$) = 4.3, *Metallurgical and Materials Transactions B,* 34B(4):369–382.

Jak, E and Hayes, P C, 2003b. Experimental Study of Phase Equilibria in the PbO-ZnO-Fe$_2$O$_3$ -(CaO-SiO$_2$) System in Air for the Lead and Zinc Blast Furnace Sinters (CaO/SiO$_2$)Weight Ratio of 0.933 and PbO/(CaO+SiO$_2$) Ratios of 2.0 and 3.2), *Metallurgical and Materials Transactions B,* 34B(8):383–397.

Jiang, J X, Li, X R and Guo, H J, 2018. Kivcet lead smelting process practice, *World Nonferrous Metal,* (18):7–9.

Li, W F and Jia, Z H, 2010. Lead sulphide concentrate smelting with bottom or top oxygen-enriched blowing, 12 p (Central South-University Press: Changsha).

Li, W F, Yang, A G and Cheng, H C, 2011. Technology Study on Direct Reduction of Lead-Rich Slag, *Nonferrous Metals (Extractive Metallurgy),* (4):10–13.

Li, W F, Zhan, J and Fan, Y Q, 2017. Research and Industrial Application of a Process for Direct Reduction of Molten High-Lead Smelting Slag, *JOM,* 69(4):784–789.

Nakada, H, Mihara, N and Kawaguchi, Y, 2008. Volatilization behavior of lead from molten slag under conditions simulating municipal solid waste melting, *Journal of Material Cycles and Waste Management,* 10(1):19–23.

Osada, S, Osada, M and Kokado, M, 2004. Thermodynamic Study of the Behavior of Elements with Low Boiling Points in the Melting Process, *Journal of the Japan Society of Waste Management Experts,* 15(5):353–362.

Perez-Labra, M, Romero-Serrano, A and Hernandez-Ramirez, A, 2012. Effect of CaO/SiO$_2$ and Fe/sio$_2$ ratios on phase equilibria in PbO-ZnO-CaO-SiO$_2$-Fe$_2$O$_3$ system in air, *Transactions of Nonferrous Metals Society of China,* 22(3):665–674.

Queneau, P E and Siegmund, A, 1996. Industrial-scale lead making with the QSL continuous oxygen converter, *JOM,* 48(48):38–44.

Slobodkin, L, Sannikov, Y and Grinin, Y, 2013. The Kivcet Treatment of Polymetallic Feeds In book: *Lead-Zinc 2000,* pp 687–692 (John Wiley and Sons, Inc: New York).

Wang, C Y and Chen, Y Q, 2016. The lead and zinc metallurgy technology situation and development trend of China: lead metallurgy, *Nonferrous Metals Science and Engineering,* 7(6):1–7.

Wang, G H, Cui, Y R and Yang, Z, 2020. Volatilization characteristics of high-lead slag and its influence on measurement of physicochemical properties at high temperature, *Journal of Mining and Metallurgy – Section B – Metallurgy,* 56(1):59–68.

Wang, J K, Zhou, T X and Feng, G L, 2004. ISA-YMG Smelting Process for Lead Bullion, *Chinese Engineering Science,* (4):61–66.

Yang, K O, Dou, Z H and Zhang, T A, 2019. Viscosities in PbO-ZnO-Fe$_x$O-SiO$_2$–CaO system for lead and zinc smelting slags, *Metallurgical Research and Technology,* 116(6):606.

Yang, Z, Cui, Y R and Hao, Y, 2020. Optimization on bottom blowing-side blowing lead smelting for utilizing of lead-silver leaching residues, *The Chinese Journal of Process Engineering,* 20(11):1304–1312.

Zhang, L R, 2013. *Modern Lead Metallurgy*, pp 180–186 (Central South University Press: Changsha).

Zhang, Z T and Dai, X, 2020. Effect of Fe/SiO_2 and CaO/SiO_2 mass ratios on metal recovery rate and metal content in slag in oxygen-enriched direct smelting of jamesonite concentrate, *Transactions of Nonferrous Metals Society of China*, 30(2):501–508.

Zhang, Z T, Li, W F and Zhan, J, 2018. The Effect of Coal Ratio on the High-lead Slag Reduction Process, *Journal of Mining and Metallurgy Section B – Metallurgy*, 54B(2):179–184.

Zhao, J X, Wang, Z and Wang, G H, 2023. A method on evaluating the effect of evaporation of component in slag, Chinese Patent 2021114176146.

Zhong, Q D, 2014. Pb element flow in typical Pb smelting process, Master thesis, Chinese Academy of Environmental Sciences, Beijing.

Slag volume effects on direct-reduced iron (DRI)-based electric furnace steelmaking

Q Zhuo[1], P C Pistorius[2] and M N Al-Harbi[3]

1. PhD candidate, Carnegie Mellon University, Pittsburgh PA 15213, USA.
   Email: qzhuo@andrew.cmu.edu
2. POSCO Professor, Carnegie Mellon University, Pittsburgh PA 15213, USA.
   Email: pistorius@cmu.edu
3. Research Fellow, Hadeed, Jubail 31961, Saudi Arabia. Email: harbimn@hadeed.sabic.com

ABSTRACT

A likely increasingly important pathway for future low-carbon ironmaking and steelmaking is the combination of gas-based direct reduction with electric furnace steelmaking (DRI-EAF route), or with electric smelting (DRI-ESF route). In the DRI-EAF route, the gangue that is present in the iron ore is fluxed and removed as slag in the melting (steelmaking) step. This is in contrast with integrated steelmaking, in which the gangue is removed as part of blast furnace ironmaking. For DRI-EAF steelmaking, the amount and composition of slag depend on the iron ore composition, flux additions and iron yield. Slag volume and composition affect phosphorus removal, foaming behaviour and the process energy requirements. In this work, heat data from a year of DRI-EAF production was analysed to test whether the expected relationship between slag volume and electricity consumption for EAF steelmaking was observed. The data include slag analyses (one sample per shift), and heat-level information on all inputs, final steel temperature and dissolved oxygen concentration, electricity consumption, tapped steel mass and tap-to-tap time. In analysing the data, the slag volume was estimated from the assayed slag CaO concentration and CaO inputs to the EAF. The expected effect of slag volume on steelmaking energy consumption was calculated with a mass and energy balance, considering the heat of mixing of the slag. The calculated energy requirement of slag formation is in line with previous estimates reported in the literature.

INTRODUCTION

Gas-based direct reduction is expected to become an increasingly important ironmaking route, since it could utilise green hydrogen to produce iron from ore with low or zero carbon dioxide emissions. The direct-reduced iron (DRI) product retains the impurities (mainly gangue oxides and phosphorus) that are present in the ore feed; these impurities need to be removed as slag during subsequent melting or electric arc furnace (EAF) steelmaking. Currently, relatively high-grade pellets (total iron concentration > 67 per cent) are mainly used to produce DRI to serve as EAF feed (Kim and Sohn, 2022); in comparison, blast furnace pellets typically contain 63–65 per cent Fe (Poveromo, 1999). As the globally consumed tonnage of pellets for DRI increases, the availability of high-grade pellets is expected to become a constraint for DRI-EAF production (Barrington, 2022).

DRI that is produced from lower-grade pellets can be used in EAF steelmaking, but with disadvantages due to the increased slag volume (mass of slag relative to steel mass): the disadvantages include the increased energy consumption (to heat and melt the slag), lower iron yield (more iron lost to the slag), the production of more by-product and lower production rate (because of the higher electricity requirement). Already in 1980 effects of increased DRI used in EAF steelmaking were noted – in particular, the increased electricity consumption related to slag formation (Kishida et al, 1980). In addition, the release of acidic gangue from DRI during the initial part of the heat, before lime and doloma fluxes have dissolved, tends to cause the formation of acidic slag, leading to attack on the refractory lining of the furnace (Song, Zhao and Pistorius, 2020). If lime additions are insufficient to maintain a high slag basicity when melting DRI with acidic gangue, poorer dephosphorisation results (Heo and Park, 2018).

In this work, data from an operating plant that utilises a high proportion of DRI in EAF steelmaking was used to test whether the expected effects of slag volume on electricity consumption, iron yield and power-on time are observed. Details of the plant operation and data are summarised in the next section, followed by analysis of the data and comparison with the theoretically expected trends.

PLANT DETAILS

The plant uses a mixture of scrap and cold DRI as metallic feed, with lime and doloma as fluxes. Natural gas is used in oxyfuel burners and additional oxygen is injected. The DRI contains around 2 per cent carbon (by mass); additional carbon is fed through the top of the furnace and also injected. Data from one year of production was analysed. Slag analyses were recorded approximately three times per day, resulting in data for around 800 heats (after removing missing analyses). Out of these, 'pour back' – return of steel from secondary metallurgy to the EAF – occurred during 16 heats.

DRI compositions were available for the year of production, but were not related to specific heats in the analysis presented here. The variability in the gangue content of the DRI was taken to be the main cause of changes in slag volume. The DRI analysis reported the mass percentages of carbon, metallic iron and oxidic iron (taken to be FeO); the balance of the DRI was assumed to be gangue. The distributions of DRI compositions for the year of production are given in Figure 1. The median DRI composition was 2.2 per cent carbon, 90.3 per cent total iron, 83.2 per cent metallic iron and 5.3 per cent gangue (calculated value).

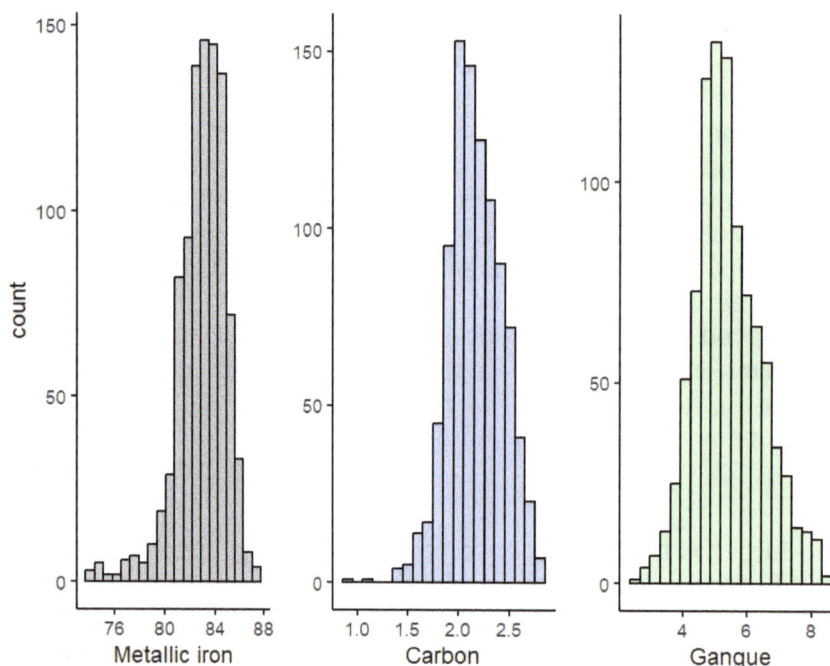

FIG 1 – Distribution of direct-reduced iron (DRI) compositions (mass percentages).

The main metallic input is DRI (Figure 2), with the scrap input accounting for about 40 per cent of the tap mass on average.

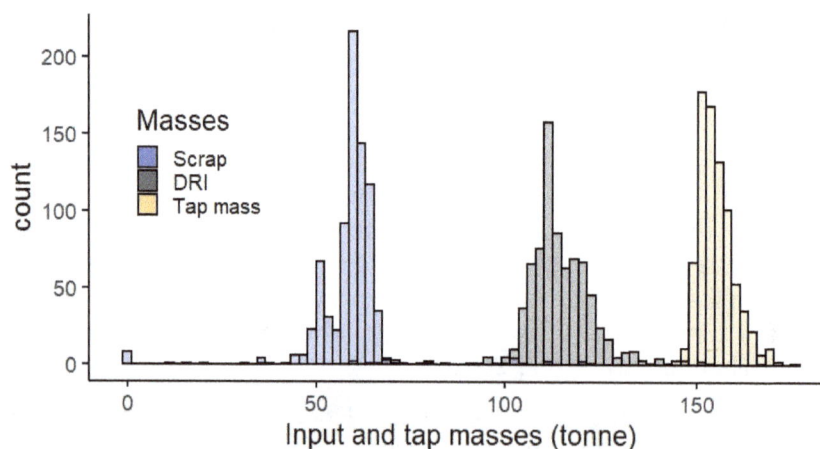

FIG 2 – Steel input and tap masses.

The power-on time was around 50 minutes (Figure 3), with a wider range of tap-to-tap times. The power during arcing was near the median value of 95 MW for all the heats (Figure 4). The slag volume was estimated from the (%CaO) in the slag and additions of lime and doloma (the average slag composition is given in Table 1, and the distribution FeO and MgO concentrations, and slag basicity, in Figure 5). In this calculation, it was assumed that the only sources of CaO were the fluxing additions and that the lime was pure CaO and the doloma pure stoichiometric CaO.MgO. The calculated slag volume varied significantly between heats (Figure 4), allowing evaluation of the effects of slag volume from the recorded data. The Fe yield (also shown in Figure 4) was estimated from the tap mass of steel (approximated as 100 per cent Fe), the scrap mass (also taken to be 100 per cent Fe) and the DRI input (using the median total Fe in DRI of 90.3 per cent). In some cases the calculated iron yield was greater than 100 per cent, likely reflecting variations in the amount of steel retained in the furnace between taps, as a hot heel. Similar values for the Fe yield were found using the analysed iron concentration in the slag, the calculated slag volume and the tap mass of steel: both approaches gave median Fe yields of 95.3 per cent.

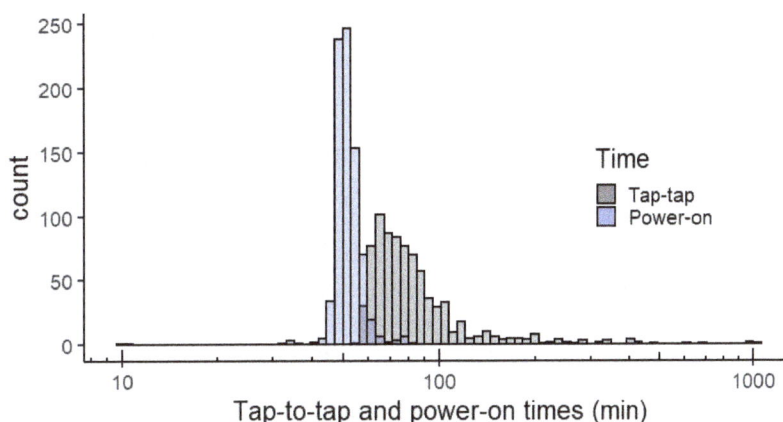

FIG 3 – Distribution of tap-to-tap and power-on times.

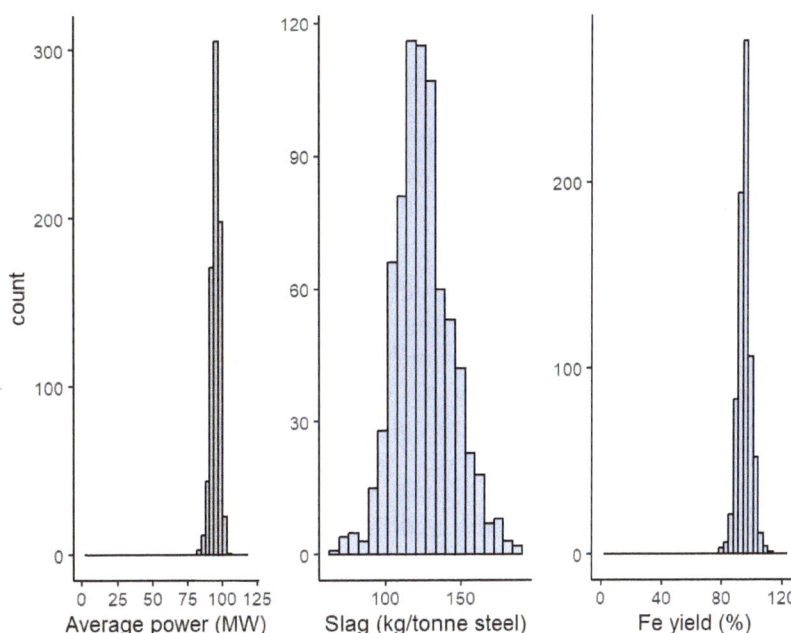

FIG 4 – Furnace power during arc heating, with the calculated slag volume and Fe yield.

TABLE 1

Average slag composition (mass percentages).

%CaO	%SiO$_2$	%MgO	%Al$_2$O$_3$	%MnO	%'FeO'	B3*
24.0	14.2	5.99	4.93	2.39	42.9	1.33

*B3 = (%CaO) / (%SiO$_2$ + %Al$_2$O$_3$).

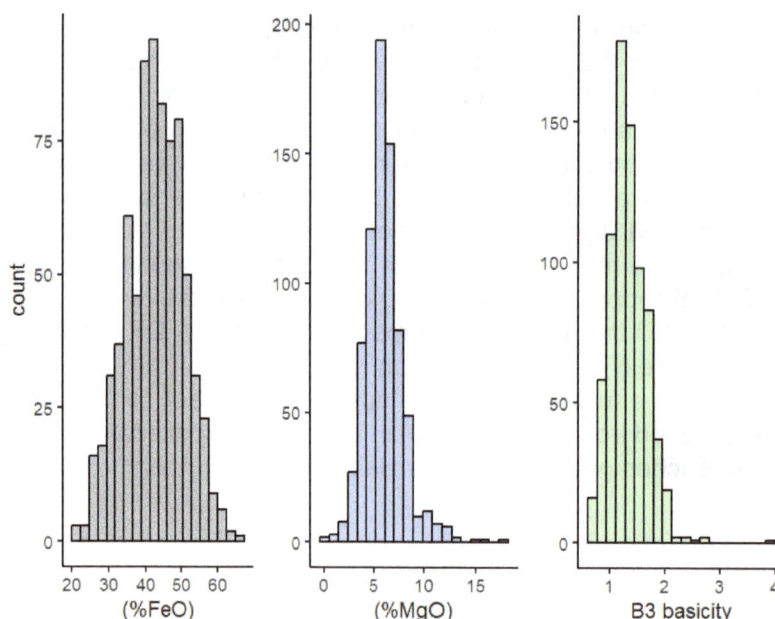

FIG 5 – Slag composition parameters: iron content, percentage MgO and
B3 = (%CaO) / (%SiO$_2$ + %Al$_2$O$_3$).

The reported concentration of iron in the slag likely includes a contribution of entrained metallic iron. One mechanism for iron entrainment is incomplete settling: the DRI is top-fed into the furnace, requiring the iron to travel through the slag layer before reaching the metal bath. However, because information on the metallic iron concentration in the slag was not available, in these calculations all the iron in the slag was taken to be oxidic.

FITTED AND EXPECTED TRENDS

Correlations of electricity consumption, power-on time and iron yield are shown in Figure 6. Before fitting these trends (and plotting the data) outliers were removed using the approach of Iglewicz and Hoaglin (National Institute of Standards and Technology, 2020) to account for measurement and reporting errors; the total number of heats removed as outliers was 51 out of a total of 808. In Figure 6, the lines are separate linear fits to each of these correlations, with the slopes of the lines reported in Table 2. While the R^2 values of the correlations are low, the small P values do indicate significance and the fitted slopes are close to the theoretically expected values, as discussed below.

The expected slopes were calculated as follows: The effect of slag volume on energy consumption was estimated from the estimated average oxide composition of DRI. In the absence of detailed DRI analyses, it was assumed that the relative masses of SiO$_2$, Al$_2$O$_3$ and MnO in the DRI were the same as in the slag (that is, it was assumed that the main source of these species in the slag was the DRI). The ratio of FeO to gangue (SiO$_2$, Al$_2$O$_3$ and MnO) in the DRI was found from the DRI analyses. The required additions of CaO and CaO.MgO were calculated from the analysed MgO/CaO ratio in the slag and the B3 basicity value. The resulting mass balance (for 1 kg additional slag mass) is summarised in Figure 7. As this figure indicates, the composition of the additional slag that is estimated using this approach is similar to the average slag composition of Table 1.

For the energy balance, the input species were taken to be the simple compounds listed in Figure 7. To calculate the enthalpy of the slag, the heat of mixing of the liquid slag (relative to pure oxides) was based on a slag thermodynamics model (Björkvall, Sichen and Seetharaman, 2001). Based on this energy balance, the heat transfer required to melt and heat 1 kg of slag to 1600°C was estimated as 0.55 kWh/kg. This is close to the value of 0.53 kWh/kg mentioned in the classic study of the effects of DRI on EAF steelmaking (Rigaud, Marquis and Dancy, 1976). In comparison, the fitted slope was 0.78 kWh/kg.

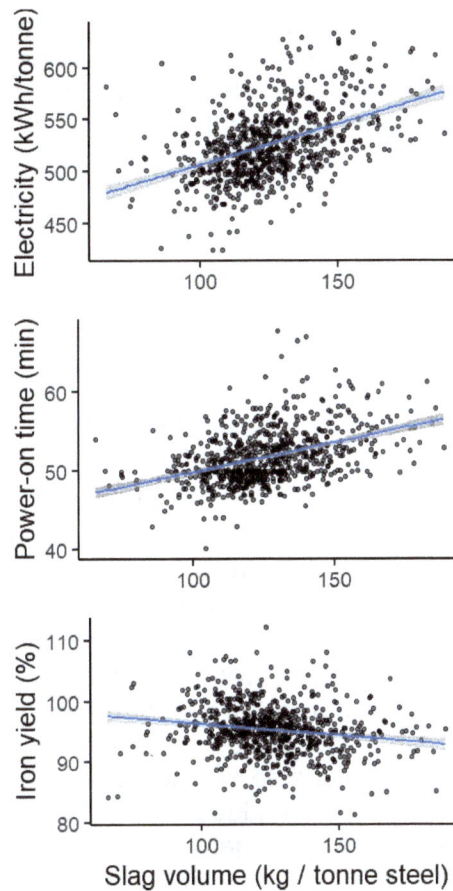

FIG 6 – Observed correlations between the slag volume, electricity consumption, power-on time and Fe yield. The lines show fitted linear relationships, with 95 per cent confidence intervals around these lines.

TABLE 2

Fitted and expected effects of changes in slag volume on electricity consumption, power-on time and Fe yield.

Output variable	Units	Theor coeff	Fitted coeff	Std error	Units of coefficient	R^2	P value
Electricity	kWh/tonne	0.55	0.78	0.06	kWh / kg slag	0.19	$<2 \times 10^{-16}$
Power-on time	min	0.076	0.075	0.006	min / (kg slag / tonne steel)	0.17	$<2 \times 10^{-16}$
Fe yield	%	-0.031	-0.038	0.008	% / (kg slag / tonne steel)	0.03	1.2×10^{-6}

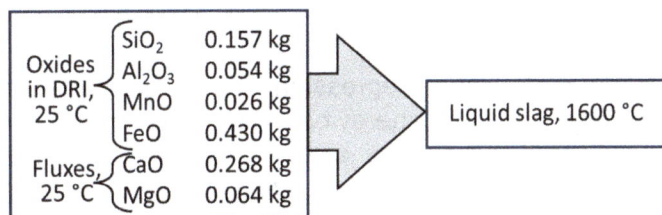

FIG 7 – Schematic of the inputs to the mass and energy balance used to estimate the effect of slag volume on electric arc furnace (EAF) energy consumption.

The expected effect of slag volume on the power-on time was calculated by using the observed relationship between the slag volume and electricity consumption (0.78 kWh/kg slag), together with the median power during arcing (95 MW); this gave an expected increase of 0.076 mins for every 1 kg/tonne increase in slag volume. The observed slope was nearly the same, at 0.075 min/(kg/tonne) (Table 2).

The effect of slag volume on iron yield was estimated by assuming that the slag composition would remain unchanged with increases in slag volume (Figure 4 does show that the relative variation in slag volume is much greater than the variation in the iron concentration in the slag). The yield is then given by the steel tap mass, divided by the sum of the tap mass and the mass of iron in the slag:

$$\text{Yield} = (100\%) / [1 + (W_{slag}/WF_e^{tap})(\%Fe)_{slag}/100]$$

where:

W_{slag} is the slag mass

WF_e^{tap} is the tap mass of steel

$(\%Fe)_{slag}$ is the iron concentration in the slag

This dependence was well approximated by a linear relationship between yield and slag volume, for values of the slag volume from 90 kg/tonne to 160 kg/tonne. The slope of this expected relationship is -0.031 per cent/(kg/tonne), which is similar to the fitted slope of -0.038 per cent/(kg/tonne) listed in Table 2.

The low R^2 value of the fitted correlation between slag volume and electricity consumption reflects the reality that several other process conditions strongly influence energy consumption. A previously fitted correlation for the electricity consumption includes tap-to-tap time, tapping temperature and injection of oxygen and natural gas, together with variables related slag volume (DRI and flux additions) (Kleimt et al, 2005). A similar approach was tested here, using the process variables listed in Table 3. These variables differed from those used by Kleimt et al in that the effects of DRI and flux additions were captured with the single variable of slag volume and carbon input was added as a variable. The carbon input was the sum of top-added carbon, injected carbon and the carbon in the DRI (calculated from the DRI mass and its average carbon concentration of 2.16 per cent).

TABLE 3

Process variables considered in multiple linear correlation for electricity consumption, with the observed ranges of the variables in the plant data.

Variable	Units	Min	Lower quartile	Upper quartile	Max
Slag	kg/tonne	65.6	113	137	189
Natural gas	Nm³/tonne	0	0.94	1.36	2.65
Carbon	kg/tonne	12.2	23.8	28.7	39.8
Oxygen	Nm³/tonne	13.7	19.7	22.5	27.3
Tap temp.	°C	1578	1620	1648	1714
Tap-to-tap	minutes	54.5	66.4	89.6	430
Pour-back*	tonne/tonne	0	0	0	0.32

*Only 16 heats out of 757 had non-zero pour-back amounts.

The coefficients from the multiple linear regression are reported in Table 4. The theoretical coefficients were calculated from mass and energy balances, with inputs and outputs as summarised in Table 5. The results of these calculations are summarised in Table 4 as the 'Theoretical coefficients'. The expected effect of pour-back is equal to the enthalpy of liquid iron at 1600°C, which is 377 kWh/tonne, similar to the previously used value of 385 kWh/tonne (Rigaud, Marquis and Dancy, 1976).

Table 4 shows that the fitted effects of slag volume, natural gas injection and carbon addition are similar to the theoretically calculated effects. Burning natural gas with oxygen decreases the electricity requirement: the natural gas burners are effective at melting scrap early in the heat. Carbon addition **increases** the electricity requirement, because the added carbon tends to reduce FeO from the slag; the reduction reaction (C + FeO → CO + Fe) is endothermic. The fitted effect of tap temperature is weaker than expected, but with a relatively large P value. The fitted benefit of pour-back is not as strong as expected, but this is based on a small number of heats.

Oxygen injection would be expected to decrease the electricity consumption because oxidation of iron is strongly exothermic; this is the basis for the theoretical coefficient of -5.4 kWh per Nm^3 of oxygen, as listed in Table 4. However, the fitted trend is that increased oxygen injection correlates with **higher** electricity consumption. A likely mechanism is that increased oxygen injection would increase the iron oxide concentration in the slag. An increased iron oxide concentration lowers the slag foaming index (Jung and Fruehan, 2000); poorer slag foaming can lead to increased heat loss to the furnace side walls, resulting in increased electricity consumption despite the exothermicity of iron oxidation.

TABLE 4

Theoretically calculated effects of process variables on electricity consumption (in kWh/tonne), with the coefficients from multiple linear correlation.

Variable	Theor coeff	Fitted coeff	Std error	P value
Slag	0.55	0.31	0.06	1.7×10^{-8}
Natural gas	-8.1	-7.9	2.7	0.0038
Carbon	3.3	3.0	0.26	$<2 \times 10^{-16}$
Oxygen	-5.4	4.6	0.49	$<2 \times 10^{-16}$
Tap temp.	0.23	0.08	0.04	0.056
Tap-to-tap	–	0.24	0.02	$<2 \times 10^{-16}$
Pour-back	-377	-169	44	0.00012

$R^2 = 0.47$; Residuals: 1st quartile -16.7 kWh/tonne; 3rd quartile 16.1 kWh/tonne.

TABLE 5

Summary of inputs and outputs used to estimate the effects of process variations on energy consumption.

Case	Inputs, with temperature	Outputs, with temperature
Natural gas combustion	CH_4 and stoichiometric O_2 (25°C)	CO_2, H_2O (1200°C)
Oxygen lancing	Fe (1600°C), stoich. O_2 (25°C)	FeO (1600°C)
Carbon injection	FeO (1600°C), stoich. C (25°C)	Fe (1600°C), CO (1200°C)
Melting DRI gangue	See Figure 7 for inputs (25°C)	Molten slag (1600°C)

The fitted and actual electricity consumption for all the heats is compared in Figure 8. The background shading in the figure shows the distribution of the data points. Most of the values cluster close to the median electricity consumption of 522 kWh/tonne. As noted in Table 4, half the fitted values of electricity consumption lie within 17 kWh/tonne of the actual values.

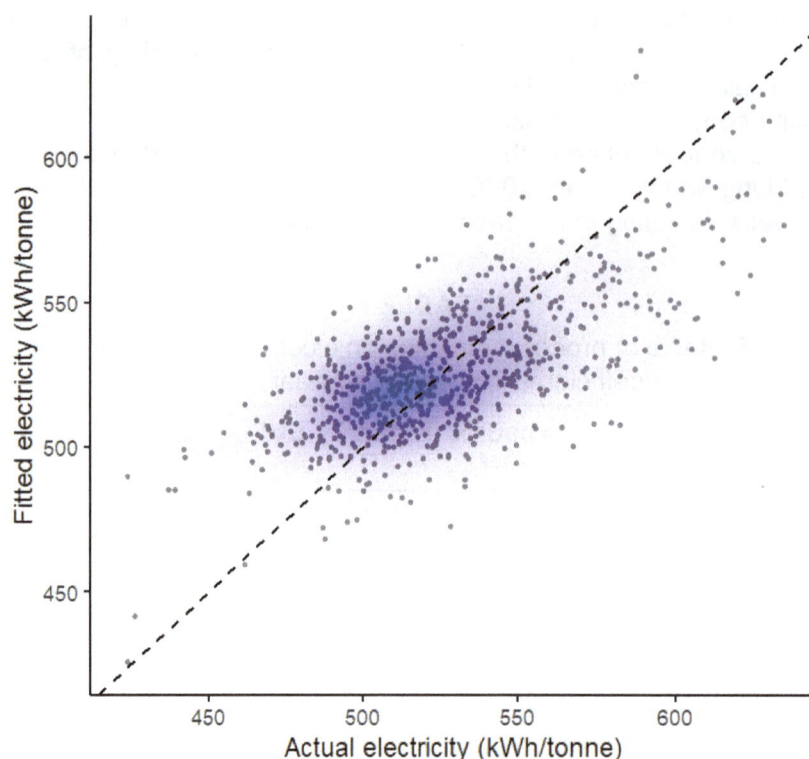

FIG 8 – Comparison of actual electricity consumption per heat and the consumption calculated with the coefficients from the multiple linear regression. The broken line shows the 1:1 relationship and the shading indicates the density of the distribution of points.

CONCLUSIONS

Analysis of a year's production data of an EAF plant that uses a large proportion of DRI confirmed the expected effects of the increased slag volume that would result from a higher input of gangue in DRI: Higher electricity consumption per tonne of steel, longer power-on times and decreased iron yield. The effects of process variables on process heating requirements – from multiple linear regression – generally agree with the quantitative relationships derived from simple mass and energy balances. A notable exception is oxygen usage: For the conditions considered here, increased oxygen injection was associated with higher electricity consumption, possibly because of a negative effect on slag foaming.

ACKNOWLEDGEMENTS

Support of this project by the industrial members of the Center for Iron and Steelmaking Research is gratefully acknowledged.

REFERENCES

Barrington, C, 2022. OBMs and Carbon Neutral Steelmaking, Paper 3 Future DRI Production and Iron Ore Supply [online], International Iron Metallics Association. Available from: <https://www.metallics.org/assets/files/Public-Area/Decarbonisation/Paper3_DRIProduction.pdf>

Björkvall, J, Sichen, D and Seetharaman, S, 2001. Thermodynamic model calculations in multicomponent liquid silicate systems, *Ironmaking and Steelmaking*, 28(3):250–257. https://doi.org/10.1179/030192301678118

Heo, J H and Park, J H, 2018. Effect of Direct Reduced Iron (DRI) on Dephosphorization of Molten Steel by Electric Arc Furnace Slag, *Metallurgical and Materials Transactions B*, 49(6):3381–3389. https://doi.org/10.1007/s11663-018-1406-5

Jung, S-M and Fruehan, R J, 2000. Foaming Characteristics of BOF Slags, *ISIJ International*, 40(4):348–355. https://doi.org/10.2355/isijinternational.40.348

Kim, W and Sohn, I, 2022. Critical challenges facing low carbon steelmaking technology using hydrogen direct reduced iron, *Joule*, 6(10):2228–2232. https://doi.org/10.1016/j.joule.2022.08.010

Kishida, T, Fukumoto, Y, Mogi, K and Kitazawa, T, 1980. The Result of Melting Direct Reduced Iron by UHP Arc Furnace, *Steel Times*, 208(2):137–143.

Kleimt, B, Köhle, S, Kühn, R and Zisser, S, 2005. Application of models for electrical energy consumption to improve EAF operation and dynamic control, 8th European Electric Steelmaking Conference, Birmingham.

National Institute of Standards and Technology, 2020. *NIST/SEMATECH e-Handbook of Statistical Methods (NIST Handbook 151)*. https://doi.org/10.18434/M32189

Poveromo, J J, 1999. Iron ores, *The Making, Shaping and Treating of Steel, Ironmaking Volume* (ed: D H Wakelin), AISI Steel Foundation.

Rigaud, M, Marquis, H A and Dancy, T E, 1976. Electric arc furnace steelmaking with prereduced pellets, *Ironmaking and Steelmaking*, 3(6):366–372.

Song, S, Zhao, J and Pistorius, P C, 2020. MgO Refractory Attack by Transient Non-saturated EAF Slag, *Metallurgical and Materials Transactions B*, 51(3):891–897. https://doi.org/10.1007/s11663-020-01788-x

Research on –
New energy and metal
production technologies

An attempt towards ferrochromium production using molten oxide electrolysis

J Biswas[1], L Klemettinen[2], D Sukhomlinov[3], W Malan[4] and D Lindberg[5]

1. Assistant Professor, IIT Bombay, Powai, Mumbai, Maharashtra 400076, India.
   Email: biswasj@iitb.ac.in
2. Staff Scientist, Aalto University, Espoo, 00076 Aalto, Finland. Email: lassi.klemettinen@aalto.fi
3. Post-doctoral researcher, Aalto University, Espoo, 00076 Aalto, Finland.
   Email: dmitry.sukhomlinov@aalto.fi
4. Post-doctoral researcher, Aalto University, Espoo, 00076 Aalto, Finland.
   Email: willem.malan@aalto.fi
5. Associate Professor, Aalto University, Espoo, 00076 Aalto, Finland.
   Email: daniel.k.lindberg@aalto.fi

ABSTRACT

In order to mitigate the effects of climate change, our society must be able to drastically reduce its CO_2 emissions. As the metallurgical industry is a considerable contributor to these emissions, new greener technologies must be invented and developed. One option could be the direct production of metals from oxides using renewable electricity.

The aim of this study was to investigate the possibility of using high-temperature molten oxide electrolysis (MOE) for iron and ferrochrome production. FactSage™, ver 8.2 (by Thermfact Ltd and GTT-Technologies) was utilised for estimating a feasible equilibrium electrolyte composition, and the experiments were conducted at 1550–1580°C in conical alumina crucibles with iridium wire as the cathode and platinum wire as the anode. The sources of iron and chromium were either pure FeO powder or industrial chromite pellets. Voltage was applied to the electrolysis cell for 6 hrs and the resulting current was measured, along with oxygen concentration in the off-gas line.

The electrolyte comprised of SiO_2, Al_2O_3, MgO and CaO. First, the effect of FeO concentration on iron reduction efficiency was investigated using only pure oxide powders as starting materials. The results indicated that 10 wt per cent FeO mixed with the electrolyte resulted in more efficient iron reduction compared to 20 wt per cent. In further experiments, industrial chromite pellets were ground and mixed with the CaO-free electrolyte. Iron and chromium reduction efficiencies were higher when approximately 19 wt per cent of the total sample mass consisted of pellets compared to increasing the pellet amount to 37 wt per cent. Generally, the chromium solubility in the liquid electrolyte was relatively low, and most of the chromium was confined to the spinel solid solution, from where its dissolution to the liquid electrolyte was very slow, resulting in slow reduction kinetics. For larger scale applications, economically more viable electrode materials should be investigated.

INTRODUCTION

The global challenge to mitigate the greenhouse gas emission has impacted steel industry, which accounts for 7 to 9 per cent of global greenhouse gas emissions (Kim *et al*, 2022). The traditional iron production process involves coke for iron ore reduction, resulting in a carbon footprint of 1.6–2 tons of CO_2 per ton of crude steel (Somers, 2021). Currently, the EU aims to decrease the greenhouse gas emissions by 80–95 per cent by 2050 compared to the level in 1990 (European Commission, 2011), which will also require a technological breakthrough in the steel industry. In order to address this challenge, extensive research is ongoing globally for example regarding hydrogen steelmaking (Patisson and Mirgaux, 2020), carbon capture and storage (Raza *et al*, 2019) and molten oxide electrolysis (Allanore, 2015). Molten oxide electrolysis (MOE) is a process where metals can be produced directly by electrolysis of oxidic ores or concentrates (Allanore, 2015). In recent years, this technology has drawn the attention of several steelmakers as a possible clean metals production route. The main challenges in iron ore electrolysis can be summarised as: 1) the operating temperature should be above the melting temperature of iron (1538°C); 2) most metals that could be used as anodes do not survive under the corrosive and oxidising conditions at the anode; 3) the multivalent state of iron causes loss of current due to electronic conduction in the melt (Allanore, Yin and Sadoway, 2013).

In recent years, there has been some progress about anode materials and more information has been obtained regarding the electrochemical nature of iron ions in oxide melts (Zhou *et al*, 2017a). Initially, Kim *et al* (2011) proposed iridium as an anode material, but only for acidic melts. Later, Allanore, Yin and Sadoway (2013) proposed a chromium-based alloy for anode, which exhibited limited consumption during iron extraction and oxygen evolution due to formation of an electronically conductive solid solution of chromium and aluminium oxide. Wiencke *et al* (2018) studied production of iron from acidic melt by electrolysis, utilising platinum as anode and an alloy of platinum and 30 per cent rhodium as cathode. In another work, Zhou *et al* (2017a) produced Fe and Fe-Ni alloys from $CaO-MgO-SiO_2-Al_2O_3-Fe_2O_3$ and $CaO-MgO-SiO_2-Al_2O_3-Fe_2O_3-NiO$ melts using a graphite anode and molybdenum cathode. Jiao *et al* (2018) produced Ti-Fe alloys via MOE from $CaO-Al_2O_3-MgO-TiO_2$ melt using liquid iron as cathode and graphite as anode. Liu, Zhang and Chou (2015) have also demonstrated iron production from $CaO-Al_2O_3-SiO_2-Fe_2O_3$ melt via electrolysis using molybdenum cathode and graphite anode. Although some progress has been made, the technology is still facing several other challenges, such as low current efficiency and lack of inexpensive, inert anode materials. The use of graphite as an anode is feasible, but as the oxygen formed at the anode reacts with carbon, some direct CO_2 emissions will be produced. More research is required to mature this technology for industrial iron production without any direct CO_2 emissions.

In this study, a set-up for molten oxide electrolysis experiments was established and experiments were conducted with synthetic $MgO-SiO_2-Al_2O_3-CaO-FeO$ mixtures to produce molten iron by electrolysis. Some experiments were also conducted with industrial chromite pellets, with an objective to test the feasibility of producing ferrochrome alloy via molten oxide electrolysis in laboratory scale.

EXPERIMENTAL

Materials

The raw materials for the study were synthetic oxide powders and industrial chromite pellets supplied by Outokumpu (Finland), as presented in Tables 1 and 2. For each experiment, a total of approximately 16 g of the oxide mixture was prepared by grinding the oxides at different mass ratios using mortar and pestle, followed by pressing the mixtures into pellets. The starting compositions for all four experiments presented in this work are shown in Table 3.

TABLE 1

Materials used in the experiments.

Material	Supplier	Purity	Notes
Al_2O_3	Sigma Aldrich	> 99%	
SiO_2	Sigma Aldrich	99%	
MgO	Sigma Aldrich	> 99%	
CaO	Sigma Aldrich	99.9%	
FeO	Sigma Aldrich	99.7%	
Chromite pellets	Outokumpu, Finland	See Table 2	From Kemi mine, diameter approximately 10 mm
Pt (anode and wiring)	Johnson-Matthey Noble Metals	> 99.9%	Diameter 0.5 mm (anode) and 0.25 mm (wires)
Ir (cathode)	Johnson-Matthey Noble Metals	> 99.9%	Diameter 0.5 mm
Al_2O_3 (crucible)	Kyocera	> 99.5%	Conical, max diameter 29 mm, height 38 mm
Al_2O_3 (tubes)	Kyocera	> 99.5%	OD 8 mm (supporting rod), OD 1 mm (electrode and wiring protection)
Al_2O_3 cement (two-component)	Morgan Advanced Materials Haldenwanger GmbH		Used for building the supporting rod and attaching electrodes

TABLE 2
Composition of chromite pellets.

	SiO$_2$	Al$_2$O$_3$	MgO	Fe$_2$O$_3$	FeO	Cr$_2$O$_3$	MnO	TiO$_2$	CaO
wt%	3.3	13.8	11.2	23.8	2.9	43.8	0.20	0.50	0.6

TABLE 3
Starting compositions for the experiments, in wt%.

Oxide melt code	Al$_2$O$_3$	SiO$_2$	MgO	FeO	Fe$_2$O$_3$	Cr$_2$O$_3$	CaO	MnO	TiO$_2$
A	20.0	48.0	12.0	10.0			10.0		
B	17.5	42.0	10.5	20.0			10.0		
C	18.5	31.8	23.8	1.0	8.5	15.7	0.2	0.07	0.18
D	19.9	39.7	27.3	0.5	4.3	7.9	0.1	0.04	0.09

Apparatus

A vertical laboratory furnace (Nabertherm RHTV 120-150/18, equipped with molybdenum disilicide heating elements) with 38 mm inner diameter alumina work tube was used for the experiments. The furnace gas outlet was connected to an oxygen sensor (Rapidox 2100ZF) for the measurement of the oxygen concentration using Rapidox software, ver 7.0.85 (by Cambridge Sensotec). The furnace was also connected to a potentiostat-galvanostat (VersaSTAT 4, Princeton Applied Research) for the application of potential and VersaStudio software, version 2.62.2 (by Ametek Scientific Instruments) was used for recording the data. The electrolysis cell consisted of a conical alumina crucible of 29 mm max diameter and 38 mm outer height, and the final design used an iridium wire cathode of 0.5 mm diameter and a platinum wire anode of 0.5 mm diameter. Trials were conducted using thinner electrode wires, which resulted in the wires breaking during the experiments. Similar problems were also encountered when iridium was tested as the anode.

The electrodes were immersed from the top of the crucible and secured with alumina cement at approximately diametric opposite locations (Figure 1). The cathode wire was placed at the bottom of the electrolysis cell, and the vertical part was covered by a thin alumina tube. The anode was placed on the other side of the cell, approximately 12 mm above the bottom of the cell. At the beginning of each experiment, the distance from the tip of the cathode to the tip of the anode was approximately 12 mm. To hold the electrolysis cell, a bigger cylindrical crucible (35 mm outer diameter and 50 mm height) was connected to a hollow supporting rod. The anode and cathode wires from the cell were welded to the electrical wires (0.25 mm diameter platinum) and passed through the supporting rod to the outside of the furnace and connected to the potentiostat. One of the electrical wires within the supporting rod was covered by a thin alumina rod, to avoid any possibility of a short circuit.

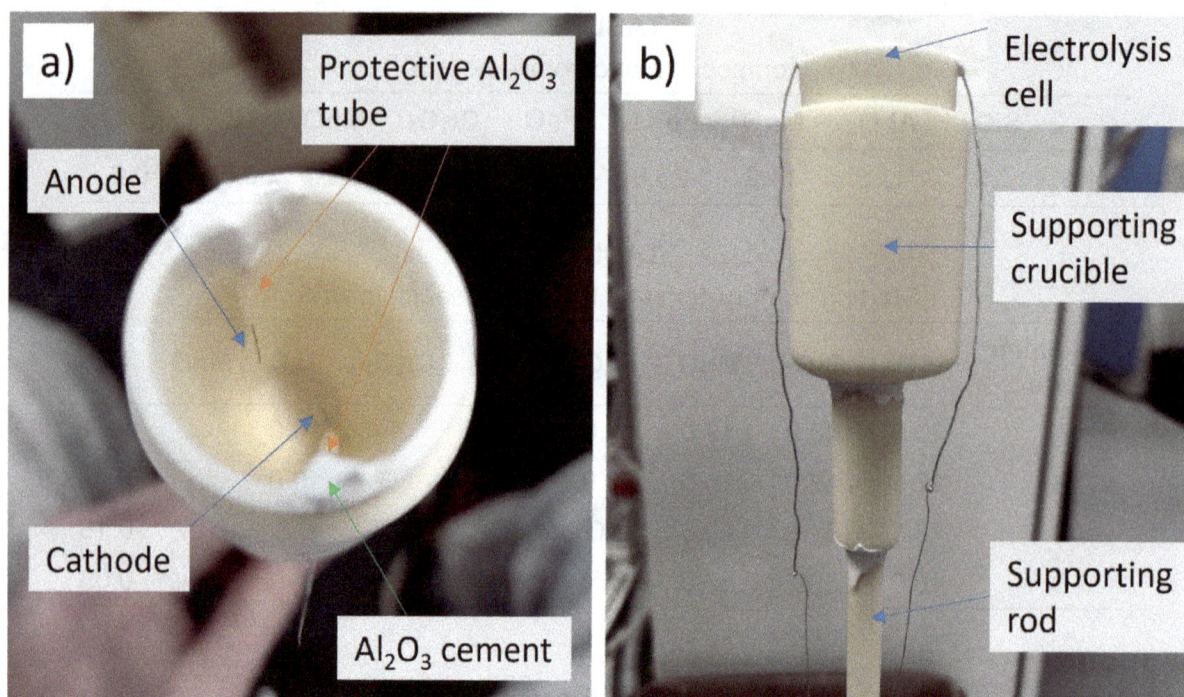

FIG 1 – (a) Electrolysis cell from the top, showing the cathode and anode, (b) Electrolysis cell inside the supporting crucible, attached to the supporting rod.

Procedure

For the experiments, the electrolysis cell with the oxide mixture was first lifted to the middle of the furnace at room temperature using the support rod, after which the furnace was sealed. The furnace was heated to the target temperature at 4°C/min under continuous flow of argon with 120 mL/min flow rate. The oxygen measurement was started approximately 30 mins after starting to heat the furnace, and the measurement interval was 10 secs. At the target temperature, the experiments were conducted in potentiostatic mode of the potentiostat-galvanostat, ie a constant potential was applied to the cell for a certain period of time and the resulting electrical current data was recorded every 5 secs. After the completion of the experiment, the furnace was cooled to room temperature under argon atmosphere, after which the cell was removed from the furnace, the sample was collected and broken using a hammer, and the weight of the formed metal alloy droplet at the cathode was recorded.

Characterisation

The preliminary elemental compositions of the metal alloys, slags and spinels were analysed using a scanning electron microscope (SEM, Mira3, Tescan, Czech Republic) equipped with an energy dispersive spectrometer (UltraDry Silicon Drift EDS, Thermo Fisher Scientific, USA). The final chemical compositions of all three phases were characterised using EPMA (SX100, Cameca SAS, France) with 20 kV acceleration voltage and 40 nA beam current. Depending on the phase, either focused, 5 µm or 20 µm defocused beam diameters were employed. The used standards as well as X-ray lines analysed and peak and background dwell times have been presented in Table 4. The elements Mn and Ti were present only in trace quantities and originated from the industrial chromite pellets.

TABLE 4

TABLE 4

EPMA parameters – analysed elements, X-ray lines, peak and background dwell times and standard materials utilised.

Element	O	Si	Al	Cr	Mg	Ir	Pt	Ca	Mn	Fe	Ti
Line	Kα	Kα	Kα	Kα	Kα	Lα	Lα	Kα	Kα	Kα	Kα
Peak dwell time (s)	20	20	20	30	20	20	30	20	20	20	20
Total background dwell time	20	20	20	30	20	20	30	20	20	20	20
Standard material	Al_2O_3	Quartz	Al_2O_3	Chromite	Diopside	Ir metal	Pt metal	Diopside	Rhodonite	Hematite	Rutile

RESULTS AND DISCUSSION

Electrolysis for iron production

The first electrolysis experiment for iron oxide reduction was performed in potentiostatic mode with 10 per cent CaO, 12 per cent MgO, 48 per cent SiO_2, 20 per cent Al_2O_3, 10 per cent FeO (melt A) at 1540°C. The phase diagrams for the melt, in two oxygen partial pressures at 1550°C, are presented in Figure 2. In a molten oxide electrolysis cell, oxygen pressure at the anode increases due to oxygen evolution reaction. The slag composition for this experiment was selected in the fully liquid region as marked in Figure 2. When the oxygen partial pressure increases during electrolysis, the slag should still remain fully liquid throughout the electrolysis cell, according to Figure 2b.

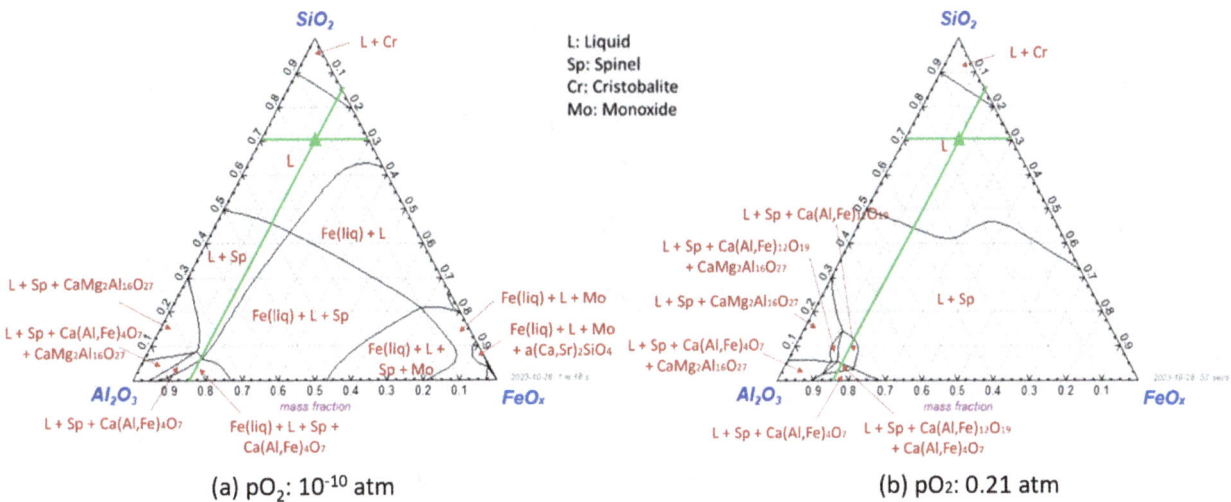

(a) pO_2: 10^{-10} atm   (b) pO_2: 0.21 atm

FIG 2 – SiO_2-Al_2O_3-FeO_x ternary phase diagrams at 1550°C with fixed CaO and MgO concentrations (10 wt per cent each) at oxygen partial pressure of 10^{-10} and 0.21 atm in (a) and (b), respectively. The green triangle represents the starting composition in the first experiment (melt A). The lower oxygen partial pressure represents the area around the cathode and the higher represents the area around the anode.

A potential of 3V was applied to the oxide melt for 6 hrs. The resulting electrical current and the oxygen concentration in the off-gas line as a function of time have been presented in Figure 3. For the experiments where only iron oxide reduction occurred, 3V potential was chosen in order to imitate the experimental conditions chosen by Wiencke *et al* (2018).

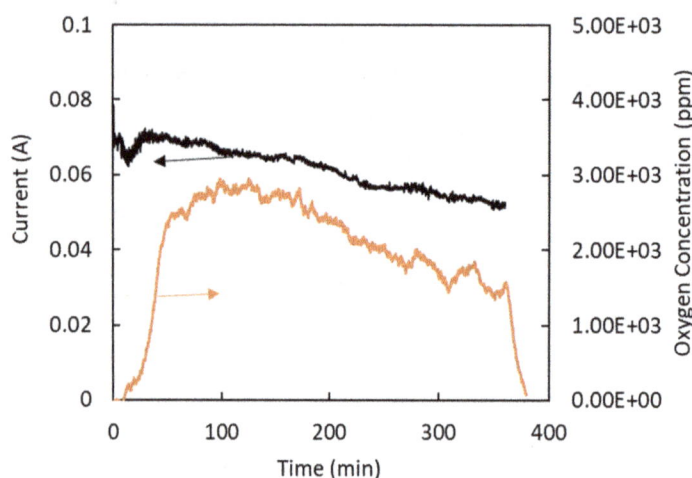

FIG 3 – Current in the electrolysis cell (black line) and oxygen concentration (orange line) as a function of time for electrolysis of 10 per cent CaO, 12 per cent MgO, 48 per cent SiO$_2$, 10 per cent Al$_2$O$_3$, 10 per cent FeO melt at 1540°C with 3V potential.

Initially, the current through the cell was approximately 80 mA and it decreased with the progression of reduction, reaching a value of slightly over 50 mA at the end of the 6 hr experiment. Previous studies on electrical properties of FeO-CaO, FeO-SiO$_2$, CaO-SiO$_2$ and CaO-FeO-SiO$_2$ slags/melts have revealed that cations (such as Ca^{2+} or Fe^{2+} ions) are the exclusive ionic charge carriers in these melts and contribute significantly to the ionic conductivity (Barati and Coley, 2006a, 2006b). Other ions, such as (SiO$_4$)$^{4-}$ and (FeO$_4$)$^{5-}$ are relatively immobile due to their large tetrahedral structures, hence their contribution to ionic conductivity can be ignored. Therefore, for CaO-MgO-SiO$_2$-Al$_2$O$_3$-FeO melt, it can be assumed that ionic conduction happens through only Ca^{2+}, Fe^{2+} and Mg^{2+} ions. As the electrolysis time increases, the concentration of Fe^{2+} ions decreases in the melt due to reduction into metallic iron, which in turn decreases the ionic conductivity of the melt. This is in agreement with observations from electrolysis in aqueous medium (Yuan and Haarberg, 2009; King and Botte, 2011), and is the reason for the slow decrease in the measured current value with time in Figure 3. The corresponding oxygen concentration could be seen to increase to as high as approximately 2750 ppm during the electrolysis. It could also be noted that there exist few small peaks, indicating fluctuations and sudden rises in the oxygen concentration. Generally, these oxide melts are viscous, which may cause difficulties in releasing the generated oxygen bubbles from the anode. There is a possibility of a fraction of the anode surface area being blocked by oxygen bubbles, which would decrease the available surface area for further anodic reactions. The sudden release of these bubbles from anode possibly results in sudden rises in oxygen concentrations (small peaks at different times). This phenomenon at the anode has also been reported by Zhou *et al* (2017b).

After 6 hrs of electrolysis for this melt, a metal droplet weighing 0.14 g was recovered at the cathode and the composition of this metal alloy droplet was 79.1 wt per cent Fe and 20.4 wt per cent Ir according to EPMA analysis. An SEM image of the metal droplet is shown in Figure 4b. Even though a significant amount of Ir dissolved in the metal alloy droplet, the cathode retained roughly its original shape after several hours of electrolysis. An image of the electrolyte, showing spinel formation at the crucible-electrolyte interface and within the electrolyte, is shown in Figure 4a.

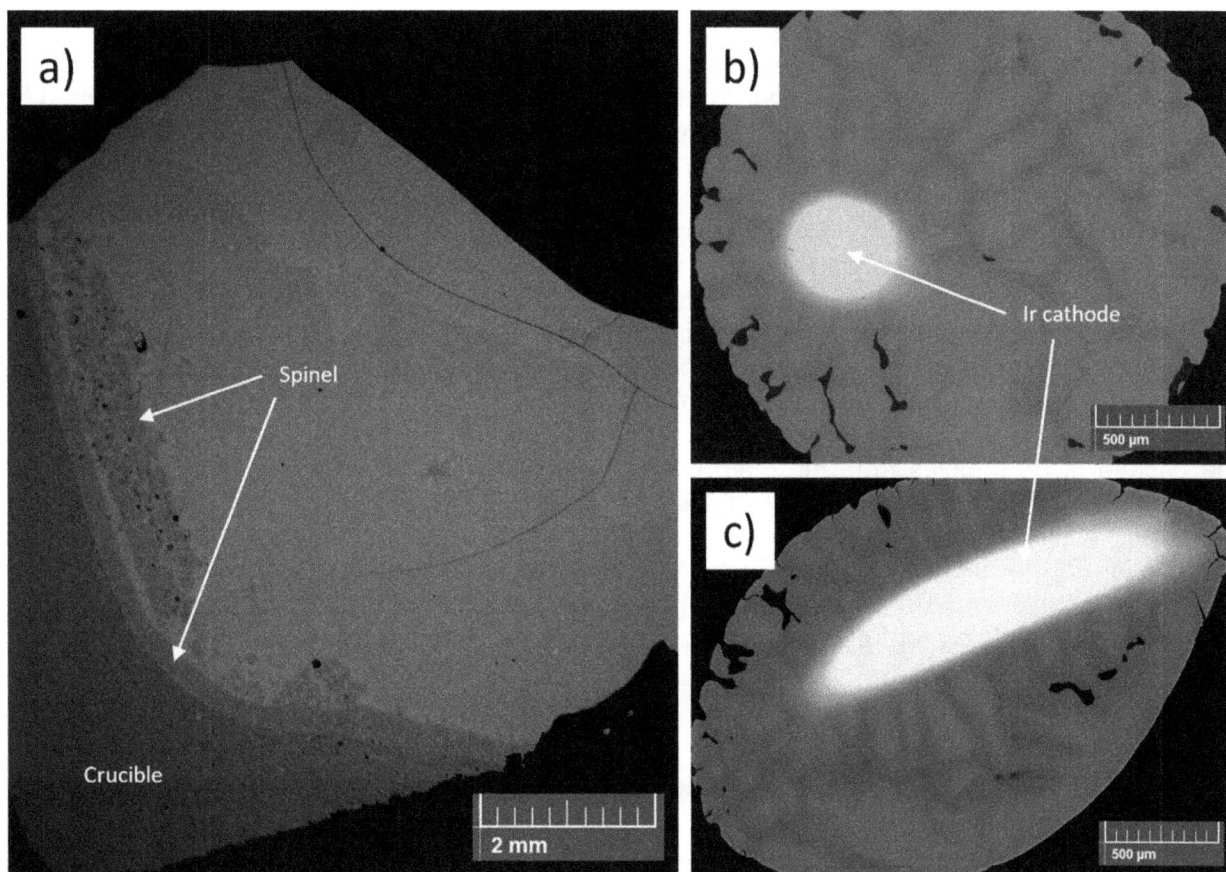

FIG 4 – (a) Microstructure of the electrolyte (slag) after FeO reduction with 10 wt per cent FeO and 10 wt per cent CaO (melt A); (b) metal alloy droplet formed during the experiment with 10 wt per cent FeO and 10 wt per cent CaO; (c) metal alloy formed during the experiment with 20 wt per cent FeO and 10 wt per cent CaO (melt B). 3V potential was applied in all cases.

The cathodic efficiency was calculated according to Equation (1) from the iron recovered at the cathode and the total transferred charge from the cathode, assuming all charge transfer was due to the cathodic reaction of iron oxide reduction.

$$Cathodic\ efficiency = \frac{Fe_{recovered}}{Fe_{current}} * 100\% \qquad (1)$$

Where $Fe_{recovered}$ is the total moles of iron recovered from the cell and $Fe_{current}$ is the moles of Fe generation expected based on the cell current, assuming all charge is involved in iron reduction.

The results from the current study were compared with the results obtained by Wiencke *et al* (2018). They performed electrolysis by applying 3V potential for 6 hrs on 15 per cent FeO, 17 per cent Al$_2$O$_3$, 56.1 per cent SiO$_2$, 11.9 per cent MgO melt at 1550°C. Interestingly, the current in their electrolysis cell seemed to increase from 70 mA to around 80 mA, whereas in our case, the current decreased with time. Wiencke *et al* obtained a cathodic efficiency of 36 per cent, which is quite comparable with the value of 27.6 per cent obtained in the current study.

To gain a better understanding of the iron reduction process with MOE, another experiment was conducted using a higher FeO starting concentration (20 wt per cent, melt B). The iron oxide concentrations of 10 and 20 wt per cent in this work were chosen based on the work of Sadoway (1995). A potential of 3 V was applied for 6 hrs, and the formed metal alloy droplet is shown in Figure 4c. Interestingly, the cell current almost doubles (from 70 mA to 118 mA) by doubling the Fe-ion concentration in the melt, as presented in Figure 5. The cathodic efficiencies for the experiments with 10 and 20 wt per cent FeO were calculated according to Equation (1). Table 5 presents the viscosities from Factsage™, cathodic efficiencies, as well as ionic and electronic conductivities for both melts. The ionic conductivities were calculated from Nernst Einstein Equation (Barati and Coley, 2006b), where cations Ca^{2+}, Fe^{2+} and Mg^{2+} were assumed to be the exclusive ionic charge carriers. For this calculation, the diffusivities of Ca^{2+} and Fe^{2+} were adopted from the values reported by Barati and Coley (2006b) and modified according to the respective slag viscosities. As the data for Mg^{2+} diffusivity was not available, it was assumed to be the same as Ca^{2+}. The electronic conductivities of the melts were calculated based on the diffusion assisted charge transfer model developed by Barati and Coley (2006b). Both ionic and electronic conductivities in the melt increase with total Fe-ion concentration increase. This illustrates the reason for higher cell current with higher FeO concentration. Although the cell current rises with increasing FeO concentration, the cathodic efficiency decreases and the Fe recovery from melt also decreases, indicating very poor performance with the melt containing more iron oxide. According to Table 5, the ionic conductivity increases by 33 per cent with increasing total FeO concentration from 10 per cent to 20 per cent, while at the same time the electronic conductivity increases by 470 per cent. The higher electronic conductivity increases cell current losses, and this possibly results in lower cathodic efficiency and decreased Fe recovery from electrolysis of higher iron oxide melt. On the other hand, FeO being a network breaker, viscosity decreases with increasing iron oxide concentration, which should ease the escape of O_2 bubbles from the melt. It should also be noted that the electrolysis cell broke towards the end of the experiment with 20 wt per cent FeO (current decrease after 300 min in Figure 5), resulting in some of the electrolyte leaking to the support crucible. This also has some contribution to the poor reduction performance. The breakage was most likely due to significant alumina dissolution from the crucible (electrolysis cell) because the starting composition of the electrolyte was relatively far from the liquid-spinel saturation boundary shown in the phase diagrams of Figure 2.

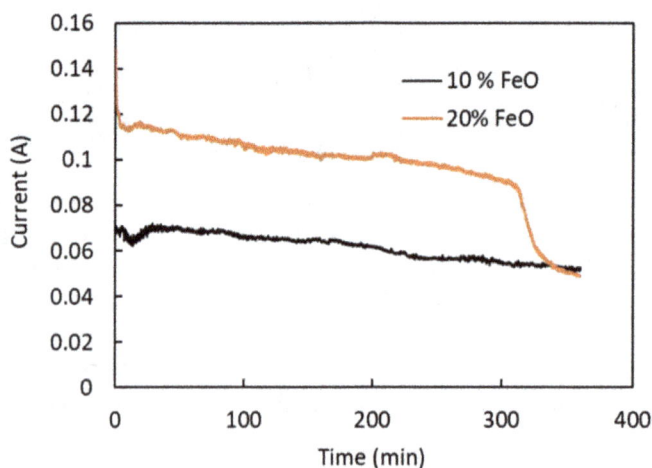

FIG 5 – Cell current profile for electrolysis of $CaO-SiO_2-Al_2O_3-MgO-FeO$ melts with two different concentrations of FeO at 1540°C.

TABLE 5

Comparison of $CaO-SiO_2-Al_2O_3-MgO-FeO$ melt electrolysis performance at 1540°C with 3V potential by varying the FeO concentration.

Oxide melt code	Electrolyte FeO (wt%)	Viscosity (poise)	Cathodic efficiency (%)	Electronic conductivity (S/m)	Ionic conductivity (S/m)	Fe recovery (%)
A	10	9.30	27.6	0.03	18.62	9.15
B	20	3.36	13.1	0.17	24.86	3.43

The overall Fe recovery percentage was calculated based on Equation (2):

$$Fe\ Recovery\ (\%) = \frac{Fe_{recovered}}{Fe_{total}} * 100\% \qquad (2)$$

where Fe_{total} is the total moles of Fe in the oxide melt at the beginning of the experiment. The iron recovery was approximately 9.2 per cent with 10 wt per cent FeO in the starting mixture and decreased to 3.4 per cent when FeO concentration was increased to 20 wt per cent. These values are very low compared to the traditional ironmaking processes.

Electrolysis for ferrochrome production

The feasibility of MOE for ferrochrome production was also investigated with two experiments. The electrolyte consisted of Al_2O_3-SiO_2-MgO mixed with chromite pellets (Table 2) at different mass ratios. Two phase diagrams for the pellet-containing system are presented in Figure 6 at 1550°C for $pO_2 = 10^{-10}$ and 0.21 atm. The green triangle represents the starting composition for the experiment with melt C (Table 3). The single-phase liquid region almost disappears when the oxygen pressure increases from 10^{-10} atm to 0.21 atm, indicating that with any melt composition, there will always be spinels present near the anode, ie high oxygen partial pressure region. In this case, when using chromite pellets in the starting mixture, the composition will always be in the liquid-spinel region. This means that the melt viscosity will increase, but the alumina crucible dissolution will decrease, increasing the lifetime of the electrolysis cell.

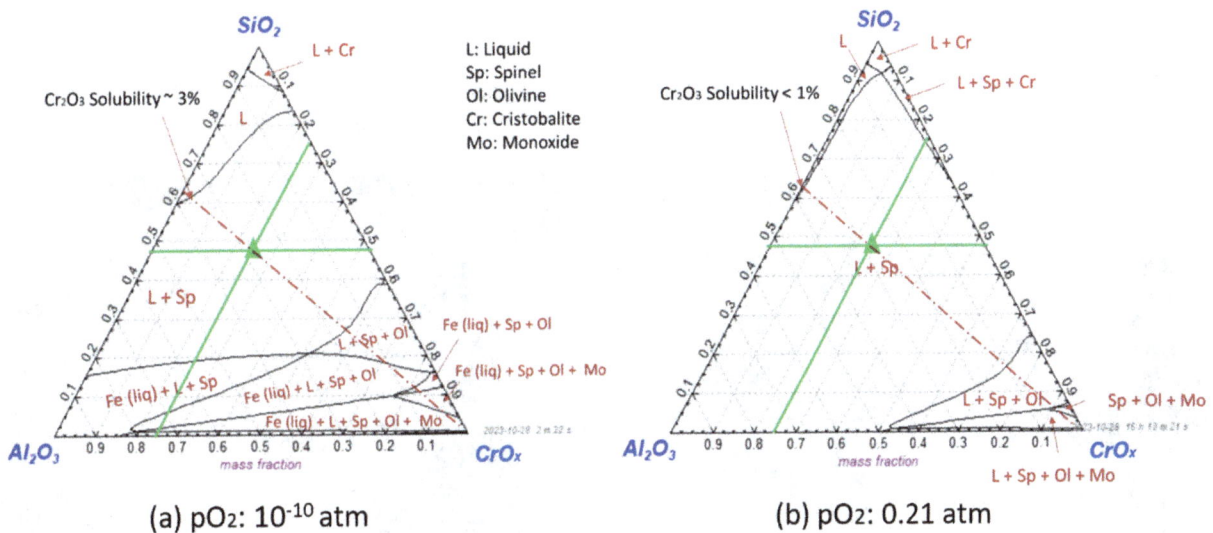

(a) pO_2: 10^{-10} atm (b) pO_2: 0.21 atm

FIG 6 – SiO_2-Al_2O_3-CrO_x ternary phase diagrams at 1550°C with MgO, Fe_2O_3 and CaO concentrations fixed at 23.8 wt per cent, 9.71 wt per cent and 0.21 wt per cent, respectively, for oxygen partial pressures of 10^{-10} and 0.21 atm.

The first experiment with chromite pellets was performed at approximately 1580°C with 5.8 g of pellets, resulting in 8.5 wt per cent Fe_2O_3, 1.0 wt per cent FeO and 15.7 wt per cent Cr_2O_3 in the melt (oxide melt C in Table 3). As the reduction potential of chromium is higher than that of iron, a higher cell potential was selected for this case. A potential of 6V was applied for 6 hrs, and a metal droplet of approximately 0.45 g (64.5 wt per cent Fe, 2.4 wt per cent Cr and 32.6 wt per cent Ir according to EPMA analyses) was reduced at the cathode. The cell current profile as well as the oxygen concentration profile are presented in Figure 7, and the formed metal alloy droplet is shown in Figure 8b. The cell current slowly increases during the first 100 mins, after which it remains stable for about 150 mins and then it decreases. The oxygen evolution profile shows almost similar trend, but with few additional peaks. Interestingly, the cell current for electrolysis of synthetic iron oxide, as presented in Figures 3 and 5, was observed to decrease continuously as a function of time. In the synthetic iron oxide case, all the conducting ions (Ca^{2+}, Mg^{2+}, Fe^{2+}) were more or less uniformly distributed in the liquid oxide, where the concentration of Fe^{2+} ions decreased due to reduction, and this resulted in decreasing ionic conductivity with increasing electrolysis time. In contrast to that, for the oxide system with pellets, a large amount of the conducting ions are expected to be bound inside the solid spinel phase (Figure 6 and Figure 8a). These ions (Cr^{3+}, $Fe^{3+,2+}$, Mg^{2+}) slowly dissolve to the liquid electrolyte, maintaining its composition almost the same during the electrolysis process. Therefore, the conductivity variation over time should be lower in this case, resulting in a relatively stable cell current. The peaks in oxygen profile could be due to sudden escape of O_2 bubble clusters from the anode. However, these peaks are more prominent during chromite pellet reduction compared to synthetic FeO reduction. Oxygen evolution in molten oxides is likely controlled by the oxygen-containing reactant transport (Allanore, 2013), and some explanations may be found from spatiotemporal differences in transport properties of the melts during the electrolysis process.

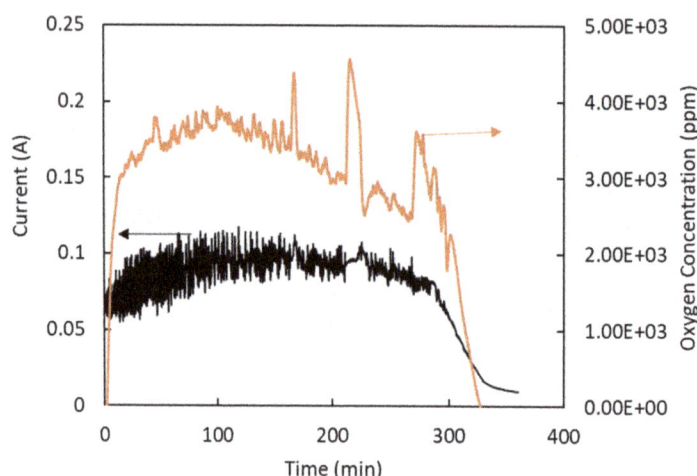

FIG 7 – Cell current (black line) and oxygen concentration (orange line) as a function of time for electrolysis of oxide melt C at 1580°C with 6V cell potential.

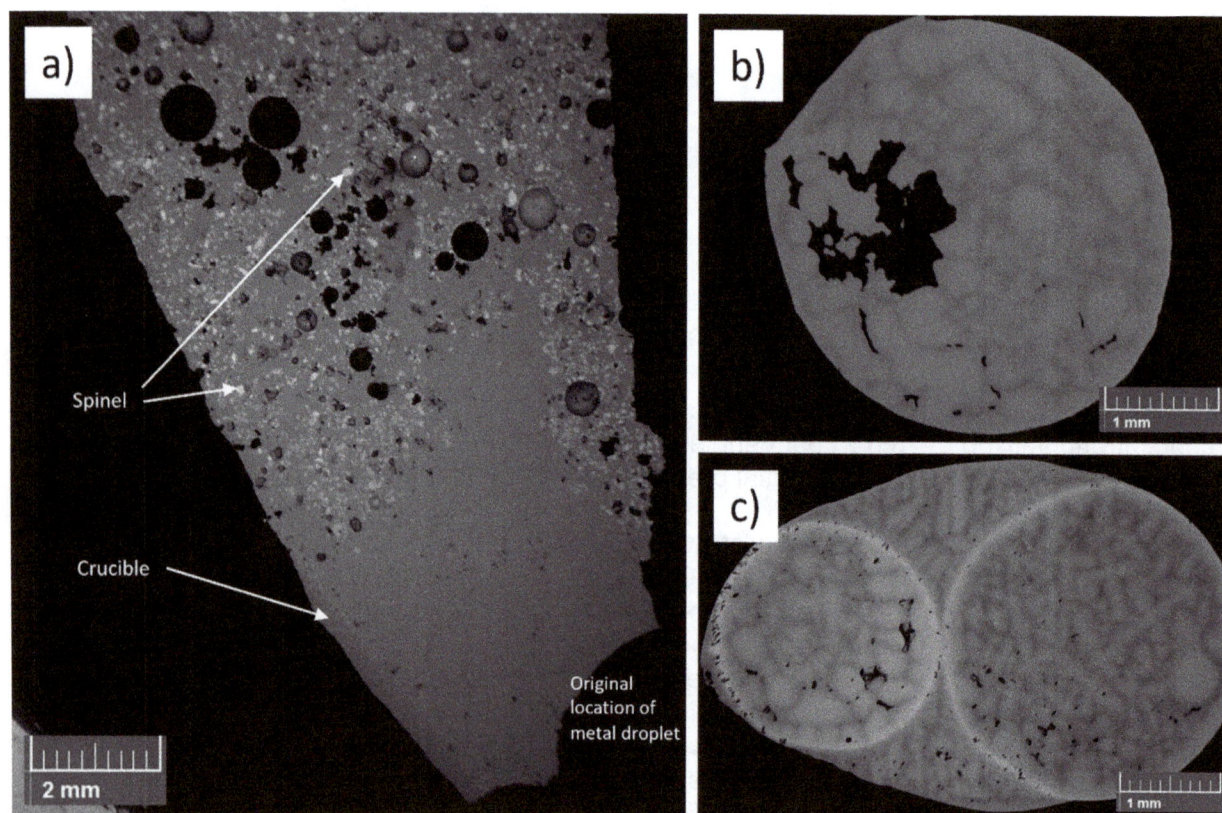

FIG 8 – a) Microstructure of the electrolyte (slag) after iron and chromium reduction from chromite pellets with melt D; b) metal alloy droplet formed during reduction with melt C; c) alloy formed during reduction with melt D. 6V potential was applied in all cases.

The recoveries of Fe and Cr, according to Equation (2), were calculated to be 26.4 per cent and 0.6 per cent, respectively. Although the iron recovery is higher with chromite pellets compared to synthetic FeO reduction, the Cr recovery seems to be very low. The main challenge for chromium reduction is the low solubility of chromium oxide in the liquid electrolyte at the experimental temperature. The chromium reduction seems to be a three-step process: 1) chromium oxide dissolution from the spinel to the liquid phase; 2) transport of chromium ions in the liquid phase to the cathode; and 3) reduction of chromium ions at the cathode. Furthermore, chromium is a multivalent ion, therefore its reduction occurs firstly from 3+ to 2+, followed by reduction into metallic

Cr at the cathode. It is unclear which stage is the rate limiting in this case, but the complexity of chromium reduction process is evident.

The second electrolysis experiment containing chromite pellets was conducted by decreasing the pellet mass (melt D). SEM images of the electrolyte and formed metal alloy are shown in Figure 8a and 8c, respectively. Cell current profile comparison for melts C and D is presented in Figure 9 and the starting composition comparison is shown in Table 3. Interestingly, the cell current is much lower for the higher iron and chromium oxide containing melt compared to the melt with lower concentration of these oxides. The recovery of iron and chromium could also be observed to increase with decreasing pellet mass, as shown in Table 6. When decreasing the pellet mass, the starting composition is closer to the liquid phase boundary shown in Figure 6, indicating a lower fraction of spinel and lower total viscosity, which increases the mobility of the conducting ions and could explain the higher cell current as well as higher Fe and Cr recoveries.

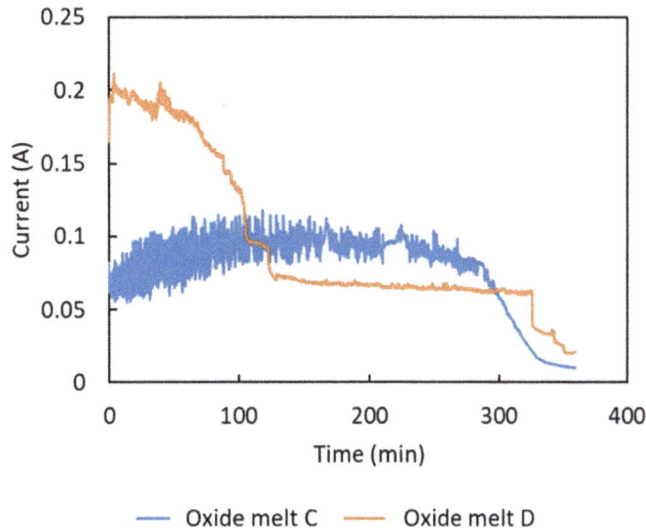

FIG 9 – Cell current profiles for electrolysis of CaO-SiO$_2$-Al$_2$O$_3$-MgO-Fe$_2$O$_3$-Cr$_2$O$_3$ melts with different masses of ferrochrome pellets at 1580°C.

TABLE 6

Electrolysis performance comparisons of ferrochrome containing oxide systems C and D at 1580°C with 6V cell potential.

Oxide system code	Mass of pellets / total sample mass (g)	Viscosity (Poise)	Fe recovery (%)	Cr recovery (%)
C	5.8 / 16	Higher	26.4	0.62
D	3.0 / 16	Lower	30.9	1.77

According to Sadoway (1995), the MOE technology could be capable of producing ferrochromium (80 per cent Fe – 20 per cent Cr) with less energy than the existing processes, and naturally with considerably less process carbon. However, based on the results of the current study, a huge amount of work is required before ferrochromium production using MOE can be considered feasible and attractive.

CONCLUSIONS

In this study, the performance of molten oxide electrolysis for the reduction of iron oxide and chromite pellets was investigated. For CaO-SiO$_2$-Al$_2$O$_3$-MgO-FeO melt, the iron recovery was less than 10 per cent and it decreased with increasing concentration of FeO from 10 to 20 wt per cent in the starting mixture. When adding industrial chromite pellets to the melt without CaO, the iron recovery improved significantly compared to synthetic iron oxide reduction. Decreasing the pellet mass from

5.8 g to 3 g (total mass approximately 16 g) resulted in a slight increase in iron recovery while chromium recovery almost tripled, however, it still remained below 2 per cent.

Provided that green electricity is available, molten oxide electrolysis is a clean metal production technology and could be a potential route for larger iron production in the future. However, extensive research is required to make the process efficient enough for production of chromium-containing alloys, and in general to make the process industrially attractive.

ACKNOWLEDGEMENTS

This study is part of the Business Finland-funded TOCANEM Project (Grant number 2118452). During the work, the Academy of Finland's RawMatTERS Finland Infrastructure (RAMI) based at Aalto University, GTK, and VTT in Espoo was utilised.

REFERENCES

Allanore, A, 2013. Electrochemical engineering of anodic oxygen evolution in molten oxides, *Electrochimica Acta*, 110:587–592.

Allanore, A, 2015. Features and Challenges of Molten Oxide Electrolytes for Metal Extraction, *Journal of The Electrochemical Society*, 162(1):E13–E22.

Allanore, A, Yin, L and Sadoway, D, 2013. A new anode material for oxygen evolution in molten oxide electrolysis, *Nature*, 497:353–356.

Barati, M and Coley, K S, 2006a. Electrical and Electronic Conductivity of CaO-SiO$_2$-FeO$_x$ Slags at Various Oxygen Potentials: Part I, Experimental Results, *Metallurgical & Materials Transactions B*, 37B:41–49.

Barati, M and Coley, K S, 2006b. Electrical and Electronic Conductivity of CaO-SiO$_2$-FeO$_x$ Slags at Various Oxygen Potentials: Part II, Mechanism and a Model of Electronic Conduction, *Metallurgical & Materials Transactions B*, 37B:51–60.

European Commission, 2011. A Roadmap for Moving to a Competitive Low Carbon Economy in 2050. Available from: <https://eur-lex.europa.eu/LexUriServ/LexUriServ.do?uri=COM:2011:0112:FIN:en:PDF> [Accessed: 11 January 2024].

Jiao, H, Tian, D, Tu, J and Jiao, S, 2018. Production of Ti-Fe alloys via molten oxide electrolysis at a liquid iron cathode, *RSC Advances*, 8:17575–17581.

Kim, H, Paramore, J, Allanore, A and Sadoway, D R, 2011. Electrolysis of Molten Iron Oxide with an Iridium Anode: The Role of Electrolyte Basicity, *Journal of The Electrochemical Society*, 158(10):E101–E105.

Kim, J, Sovacool, B K, Bazilian, M, Griffiths, S, Lee, J, Yang, M and Lee, J, 2022. Decarbonizing the iron and steel industry: A systematic review of sociotechnical systems, technological innovations, and policy options, *Energy Research & Social Science*, 89:102565.

King, R L and Botte, G G, 2011. Hydrogen production via urea electrolysis using a gel electrolyte, *Journal of Power Sources*, 196(5):2773–2778.

Liu, J-H, Zhang, G-H and Chou, K-C, 2015. Electrolysis of Molten FeO$_x$-Containing CaO-Al$_2$O$_3$-SiO$_2$ Slags under Constant Current Field, *Journal of the Electrochemical Society*, 162(12):E314–E318.

Patisson, F and Mirgaux, O, 2020. Hydrogen Ironmaking: How It Works, *Metals*, 10:922.

Raza, A, Gholami, R, Rezaee, R, Rasouli, V and Rabiei, M, 2019. Significant aspects of carbon capture and storage – A review, *Petroleum*, 5(4):335–340.

Sadoway, D R, 1995. New opportunities for metals extraction and waste treatment by electrochemical processing in molten salts, *Journal of Materials Research*, 10:487–492.

Somers, J, 2021. Technologies to decarbonise the EU steel industry, EUR 30982 EN, Publications Office of the European Union, Luxembourg, JRC127468. https://doi.org/10.2760/069150

Wiencke, J, Lavelaine, H, Panteix, P J, Petitjean, C and Rapin, C, 2018. Electrolysis of iron in a molten oxide electrolyte, *Journal of Applied Electrochemistry*, 48:115–126.

Yuan, B and Haarberg, G M, 2009. Electrowinning of Iron in Aqueous Alkaline Solution Using Rotating Disk Electrode, *Rev Met Paris*, 106(10):455–459.

Zhou, Z, Jiao, H, Tu, J, Zhu, J and Jiao, S, 2017a. Direct Production of Fe and Fe-Ni Alloy via Molten Oxides Electrolysis, *Journal of The Electrochemical Society*, 164(6):E113–E116.

Zhou, Z, Wang, S, Jiao, H and Jiao, S, 2017b. The Feasibility of Electrolytic Preparation of Fe-Ni-Cr Alloy in Molten Oxides System, *Journal of The Electrochemical Society*, 164(14):D964–D968.

Ferroalloy extraction from a Zimbabwean chrome ore using a closed DC furnace

S Dandi[1], M J Masamvu[2], S M Masuka[3], S Bright[4] and E K Chiwandika[5]

1. BSc student, University of Zimbabwe, Harare 263, Zimbabwe.
   Email: dandisimbarashe1@gmail.com
2. Lecturer, University of Zimbabwe, Harare 263, Zimbabwe.
   Email: malbeniajomasamvu@gmail.com
3. Lecturer, University of Zimbabwe, Harare 263, Zimbabwe. Email: shebarmasuka@gmail.com
4. Lecturer, University of Zimbabwe, Harare 263, Zimbabwe. Email: sharrydonbright@gmail.com
5. Lecturer, Harare Institute of Technology and module Lecturer, University of Zimbabwe, Harare 263, Zimbabwe. Email: chiwandikae@gmail.com

ABSTRACT

The depletion of high-grade hard and lumpy chrome ores has forced Zimbabwean ferrochrome producers to resort to low-grade friable ores that have an average of 28 per cent fines against a set limit of 12 per cent. Submerged open arc furnaces currently in use are associated with eruptions within the furnace because of the high amounts of fines, high energy consumption, as well as high SO_2 and CO_2 emissions. Traditionally employed agglomeration methods such as pelletising and the use of coke in the furnace have been discarded because of the pressure to lower operational costs and to achieve higher profit margins. This research aims to develop a method to incorporate low-grade friable ore fines and unprocessed coal into production while lowering CO_2 and SO_2 emissions through a closed direct current (DC) arc furnace. The closed DC arc furnace could incorporate the friable ores and coal while maintaining above 90 per cent reduction rates. Preliminary findings also show that energy consumption could be reduced by up to 35 per cent by incorporating a pre-heating and pre-reduction system using the flue gases from the furnace. In the proposed circuit, the cost of production per ton of ferrochrome might be potentially lowered by an average of 30 per cent.

INTRODUCTION

Global ferrochrome production has been steadily increasing over the past few decades, primarily due to the growing demand for ferrochrome by-products. This trend is expected to continue as industrial and usage demands for ferrochrome continue to rise (Fortune Business Insights, 2023) due to the rise in demand for stainless steel (du Preez et al, 2023). In Zimbabwe, ferrochrome production is mainly from four smelters (Chitambira, Miso-Mbele and Gumbie, 2011) where the most common method used for ferrochrome production is the submerged arc furnace (SAF) that uses an electric current to increase temperature enhancing the reduction reactions of chromite (Yu et al, 2023).

The depletion of high-grade lumpy ores has resulted in the utilisation of fine ores in open alternative current (AC) furnaces causing frequent eruptions and high energy consumption, approximately 3.8 to 4.0 MWh/MT FeCr (Chitambira, Miso-Mbele and Gumbie, 2011). Agglomeration techniques have proved to be expensive for Zimbabwean ferrochrome producers and as such, there is a need to find new ways to smelt chromite fines to produce ferrochrome using methods that are cost-effective and environmentally friendly. A possible alternative to an open AC furnace is the closed direct current (DC) furnace (Sager et al, 2010), which will incorporate pre-treatment techniques and utilise low-grade fine ores.

The conventional smelting of ferrochrome is done in electric reduction furnaces with electrodes submerged in the burden materials, ore-bearing material, a carbonaceous reductant, and fluxes. For high efficiencies, the gas flow should be uniform, and channelling of the CO produced during the reactions should not be avoided. This is achieved by the use of hard lumpy ores that are not friable to avoid excessive degradation and generation of fines in the furnace. The depletion of hard lumpy ores and the increased mechanisation in mining operations has resulted in considerable utilisation of fine ore material of much smaller grain sizes that require agglomeration (Alison, 2019), and the use of friable ores that generate fines (Goel, 1997).

Agglomeration of chromite fines and friable material currently utilises a combination of sintering and briquetting methods (Xiaohai *et al*, 2019; Alison, 2019). This improves the smelting efficiency while simultaneously reducing the power consumption of the furnace. Electrical energy consumption typically ranges from 3.0 to 3.5 MWh/MT alloy for sintered or briquette feed, compared to 3.8 to 4.0 MWh/MT alloy for raw fines (Chitambira, Miso-Mbele and Gumbie, 2011).

The DC arc furnace was developed as a way of allowing for the direct smelting of fine chromite that emanated from highly friable ores without the need for costly agglomeration techniques (Jone and Erwee, 2016). Non-coking coal can also be used as the carbonaceous reductant and theoretically, the smelter power input is independent of the burden composition since the resistance is manipulated by the arc length, making it a low-cost technology for ferrochrome production (Grant *et al*, 2010). The DC arc uses a single solid carbon electrode as the cathode and is normally open or semi-submerged. Burden materials can be charged directly into the furnace or through a hollow electrode.

Innovations in smelting technology, such as the Premus process have been developed that can decrease electrical energy consumption during smelting by partly reducing pelletised chromite ores in a rotary kiln using energy obtained from combustion of pulverised coal and hot gases generated from the closed submerged arc furnace (Naiker, 2007; Chima, 2012). Pre-reduction and pre-heating techniques that have been employed enhanced the quality and the grade of the ferrochrome product, by increasing the chromium recovery and the chromium-to-iron ratio and decreasing the carbon and silicon contents. Pre-reduction of pelletised chromite ore was the best option with the lowest specific energy consumption using the submerged arc furnace (Neizel *et al*, 2013). Generally, higher temperatures (+1000°C) longer times (+60 mins), finer particles (-10 mm), and higher reductant ratios are favourable for achieving higher degrees of pre-reduction (Jian, Yang and Deqing, 2014).

In this study, a novel structure of the DC arc furnace is being investigated for ferrochrome production in an attempt to utilise fine chromite ore while maintaining low power consumption. In addition, this new structure is expected to process low-grade fine ore material and friable ores while avoiding agglomeration of the fine ore through sintering and briquetting techniques which are expensive and economically non-viable for local smelters. It is also the aim of this study to investigate the effect of pre-heating and pre-reduction techniques on the recovery of metal using off-gas.

EXPERIMENTAL

Materials

The chromite ore fines used in this study were low-grade (A), high-grade (B) and middle-grade (C). which were obtained from Shurugwi, Mutorashanga, and Ngezi chrome fields. The grades of the ores are classified according to the percentage of Cr_2O_3 as shown in Table 1. The coal was obtained from Zambezi Gas Company in Hwange, and the quartz from Kwekwe.

TABLE 1

Chemical composition of different chromite ore grades.

Sample ID	% Cr_2O_3	% FeO	Cr/Fe	% SiO_2	% MgO	% Al_2O_3	% CaO	% TiO_2	% P	% Others
Low-grade fines (A)	25.13	18.71	1.19	17.29	24.45	5.23	1.38	0.17	0.0011	7.64
High-grade Fines(B)	40.23	17.7	2.18	7.46	15.03	10.41	0.43	0.33	0.001	8.41
Middle grade fines (C)	32.71	16.63	1.89	11.73	20.42	7.40	1.24	0.2	0.0013	9.67

Sample preparations and experimental procedures

Ten samples of 1 kg each were randomly taken from stockpiles of fine ores using an ore sample cutter of 3 m height to obtain a good representation of ore. To determine the particle size distribution of the ores, the samples were first dried using drying ovens at 110°C for 20 hrs and then sieved with different sieve sizes to obtain a representative particle size distribution shown in Figure 1.

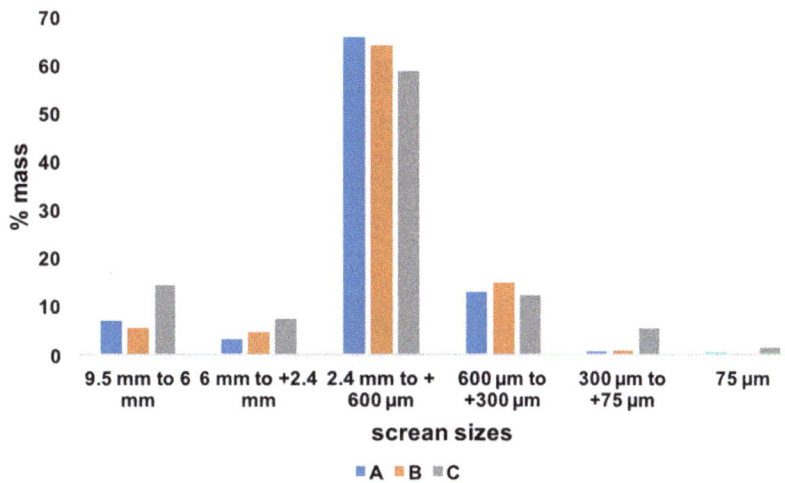

FIG 1 – Particle size distribution of the chromite ore sample.

An average of 65.8 per cent of particles fall within the size range of 2.4 mm to +600 µm. This indicates that the chromite ore samples are primarily composed of fine-sized particles. A noticeable 4 per cent of those with sizes of 75 µm and below, indicate that the sample contains a relatively small proportion of ultra-fine particles. The sample also includes a notable proportion of large particles above 6 mm, constituting 18 per cent of the samples.

To obtain a homogeneous feed of less than 2 mm, the chromite ores were first crushed using a laboratory jaw crusher, then pulverised then assayed for the average chrome head grade using a wet chemical analysis and an Inductively Coupled Plasma - Optical Emission Spectroscopy (ICP-OES). The chemical composition of the chromite ore is shown in Table 1. Quartz, used for basicity adjustment, was crushed using a laboratory jaw crusher from 40–50 mm to below 2 mm and then dried using an oven, followed by pulverisation and chemical analysis. The coal fines were sieved and screened to 90 per cent fines in the range of 1 mm to 2 mm. Results from the characterisations of the coal and quartz are shown in Tables 2 and 3 respectively.

TABLE 2

Coal proximate analysis.

Material (wt%)	Volatile combusted matter	Ash	Fixed carbon	S	P	Moisture (%)
ZG Coal	22.05	9.92	66.8086	1.01	0.0014	0.21

TABLE 3

Quartz characterisation.

Material (wt%)	SiO$_2$	FeO
Quartz	99.37	0.63

The proximate analysis for the coal fines was done according to the Indian Standard IS:1350 (Part-I)-1984. The samples were blended according to the desired mass proportions of different chromite ores as shown in Table 4 and the chemical composition of the blends is shown in Table 5.

TABLE 4
The mass blend proportions of the chromite ores.

Sample ID	Blend 1 (%)	Blend 2 (%)	Blend 3 (%)	Blend 4 (%)	Blend 5 (%)	Blend 6 (%)
A	100	80	70	60	40	0
B	0	10	15	30	50	100
C	0	10	10	10	10	0

TABLE 5
Chemical composition of the different blends.

Sample ID	% Cr_2O_3	% FeO	Cr/Fe	% SiO_2	% MgO	% Al_2O_3	% CaO	% TiO_2
Blend 1	25.13	17.91	1.79	11.70	19.32	8.13	0.85	0.25
Blend 2	27.03	18.71	1.79	12.01	19.85	7.92	0.90	0.24
Blend 3	28.33	19.46	1.84	11.83	19.90	7.97	0.89	0.22
Blend 4	30.73	20.96	1.95	11.46	20.01	8.07	0.88	0.23
Blend 5	33.73	23.21	1.89	11.65	19.95	8.02	0.89	0.23
Blend 6	40.23	17.7	2.18	7.46	15.03	10.41	0.43	8.41

A mixture of 100 g of chromite and quartz (flux material for separation between gangue and ferrochrome), in proportions shown in Table 6, firstly underwent pre-treatment in a muffle furnace using carbon monoxide from gas cylinders. The temperature within the muffle furnace was deliberately altered in order to observe and analyse the production of metallic chromium and iron (pre-reduction). Then subsequently for smelting in a laboratory-scale closed DC furnace made from the magnisite refacrory paste, coal as reducing agent was added into the mixture. The amount of coal used in every reaction was based on stoichiometry. The operational voltage was established at 500 volts, while a current of 60 amperes was utilised. The power supply employed for this operation was a robust DC welding machine designed for heavy-duty applications. The power-conducting media within the furnace consisted of gauging rods, with 80 per cent graphite content. The furnace was sealed with a lid which was made from the same material, which has two holes where the electrodes are introduced (Shotanov et al, 2023).

TABLE 6
The amount of raw material used in each blend.

Sample ID	Chromite (g)	Coal (g)	Quartz (g)
Blend 1	100	18.608	54.831
Blend 2	100	19.8	53.67
Blend 3	100	20.695	54.45
Blend 4	100	22.39	53.23
Blend 5	100	24.655	54.17
Blend 6	100	25.547	53.987

Each sample was smelted for 30 secs after carrying out preliminary experiments on reaction time, taking into consideration the size of the sample and the crucible used as shown in Figure 2, as elucidated by Asish, Somnath and Sarada (2020) and Pankaj *et al* (2024). The products were discharged from the top of the furnace and allowed to cool down with natural air. Once cooled, the alloy and slag were analysed using an ICP-OES technique and wet chemical analysis. Figure 3 illustrates the experimental stages.

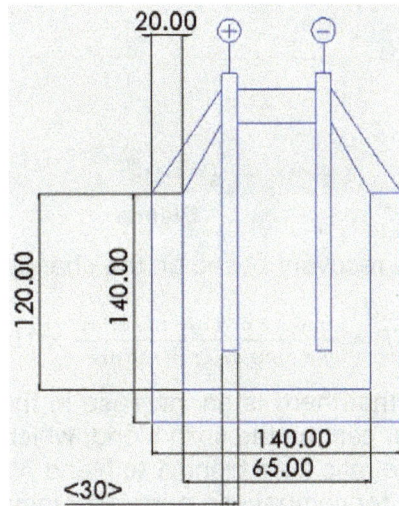

FIG 2 – Technical drawing of the furnace used for experiments (dimensions in mm).

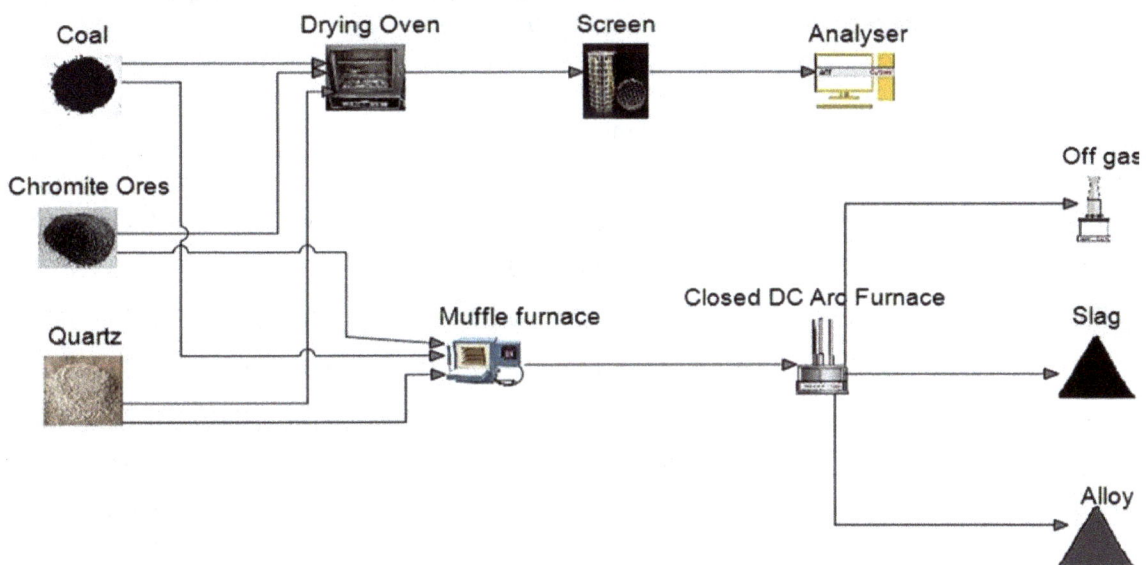

FIG 3 – Schematic presentation of ferrochrome production using DC arc furnace.

RESULTS AND DISCUSSION

Effect of chrome grade on metal recovery

This was done to investigate the effect of chromite grade on the metal recovery. The chromite grade was achieved by varying the different proportions of the chromite ores, as shown in Table 4. The results from the six blends are shown in Figure 4 showing the tapping stage recovery (TSR) based on the changes in blends. Equation 1 shows how the TSR is determined.

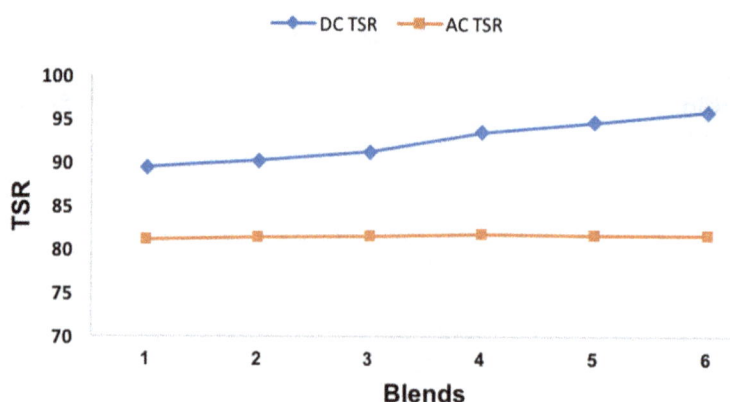

FIG 4 – Trends in tapping stage recovery based on the changes in Cr_2O_3 and the Cr/Fe ratio.

$$TSR = \frac{Amount\ of\ Cr\ in\ the\ alloy}{Amount\ of\ Cr\ in\ the\ ore} \times 100 \qquad (1)$$

It can be observed from Figure 4 that there is an increase in the TSR for the DC arc from about 89 per cent to approximately 96 per cent on the sixth blend, which can be attributed to the increase in both Cr_2O_3 content and the Cr/Fe ratio from blend 1 to blend 6. The results obtained from the AC furnace serve as a reference point for comparison purposes, indicating the high energy conversion efficiency of the closed DC arc furnace. This implies that a significant portion of the input energy is effectively utilised in the DC furnace, resulting in improved recovery rates during the tapping stage. The chemical composition of the tapped alloy is shown in Table 7.

TABLE 7

Alloy chemical composition.

Sample ID	Cr (%)	Fe (%)	Cr/Fe	Si (%)	Mn (%)	P (%)	Ni (%)	Co (%)	Ti (%)	V (%)	C (%)
Blend 1	59.75	28.97	2.06	1.00	0.28	0.03	0.33	0.06	0.30	0.20	7.79
Blend 2	61.45	32.27	1.90	1.76	0.20	0.02	0.37	0.06	0.12	0.21	7.52
Blend 3	62.54	33.04	1.89	1.33	0.18	0.02	0.39	0.06	0.08	0.21	7.46
Blend 4	62.86	32.96	1.91	1.29	0.18	0.02	0.39	0.06	0.08	0.21	7.35
Blend 5	64.98	33.70	1.92	1.23	0.17	0.02	0.41	0.07	0.08	0.21	7.09
Blend 6	67.56	33.90	1.99	1.30	0.17	0.02	0.41	0.07	0.08	0.20	7.15

Generally, the percentage of Cr and Fe in the alloy increases as the Cr_2O_3 and FeO grades in the blends increase. There is a noticeable increase in the percentage of Cr and Fe in the alloy from 59.75 per cent to 62.54 per cent and 28 per cent to 33 per cent for blend 1 to blend 3 respectively, but between blend 3 and 4, there is a slight increase due to smaller difference between Cr_2O_3, FeO and the Cr/Fe ratio in these blends. From blend 4 to blend 6, there is a significant increase in Cr and Fe in the alloy because of an increase in Cr/Fe. The amount of Si and C was almost constant due to a constant reductant mass used across all blends. The amount of Cr that reported to slag is shown in Figures 4 and 5.

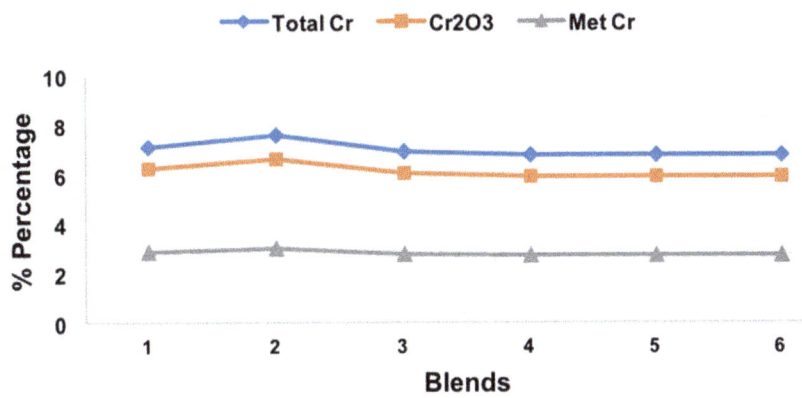

FIG 5 – Slag analysis.

Figure 5 shows the slag analysis from the closed DC arc furnaces, indicating the total Cr, unreduced Cr_2O_3, and metallic Cr in the slag. The total Cr content in the slag is uniformly low at around 8 per cent across all the blends, with Cr_2O_3 content less than 7 per cent. This indicates the high reduction efficiency of the closed DC arc furnace. A minimum metallic Cr content remains in the slag at around 3 per cent across all blends, which is a positive indication of effective separation by the flux.

Effect of basicity on separation efficiency

The MgO/SiO_2 and MgO/Al_2O_3 ratios of the blends are shown in Figure 6 and these ratios might affect slag fluidity.

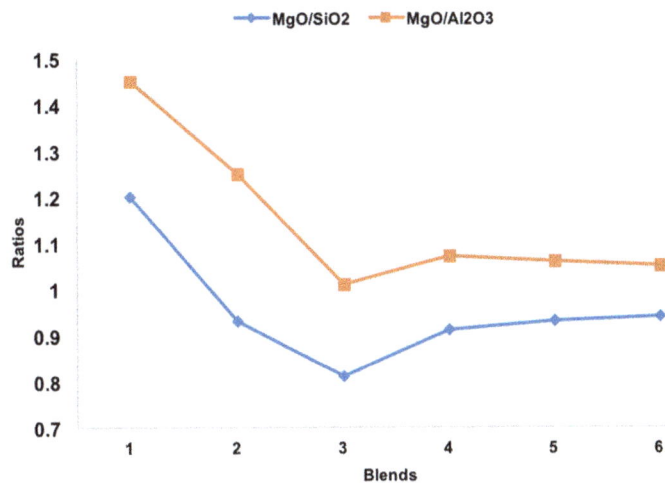

FIG 6 – Slag analysis on separation efficiency.

The MgO/SiO_2 ratio shows a decreasing trend from blend 1 to 3, suggesting a decrease in separation efficiency due to an increase in SiO_2 relative to MgO since there was a corresponding increase of the Cr reporting to slag in blend 2, as shown in Figure 5. This was also supported by the slow increase in tapping stage recovery between Blend 1 and 3 shown in Figure 6 followed by a rapid increase in recovery afterwards. This shift in ratio from blend 1 to 3 might have increased the melting point of slag which results in a specific energy consumption (SEC) increase and an increase in viscosity of the slag decreasing the separation efficiency between the slag and alloy. On the other hand, the MgO/Al_2O_3 ratio remains relatively constant, indicating stable separation efficiency and consistent proportions of these compounds across the samples. This implies that the difference in melting points during slag chemistry due to the difference in MgO, Al_2O_3 and SiO_2 affects the separation efficiency.

Specific energy required

The different trends of SEC with and without pre-treatment techniques is shown in Figure 7. These were calculated from Equation 2 (How to determine the specific energy required).

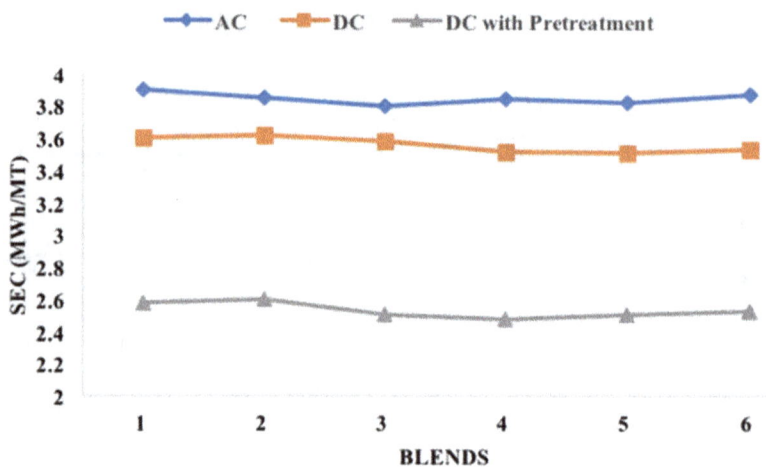

FIG 7 – SEC comparison of with and without pre-treatment methods.

$$SEC = \frac{\text{Amount of Electrical energy used}}{\text{amount of alloy produced}} \qquad (2)$$

Figure 7 compares the SEC in MWh/MT of alloy between AC furnaces, DC furnaces, and DC with pre-treatment. AC furnaces consistently show low SEC, averaging 3.9 MWh/MT due to the inherent characteristics of AC furnaces. DC furnaces exhibit better SEC than AC furnaces due to higher operating temperatures, minimum heat losses and more efficient energy transfer. DC furnaces with pre-heating and pre-reduction have much better SEC, with an average SEC of 2.5 MWh/MT. This suggests that pre-heating and pre-reduction significantly improve the performance of DC arc furnaces, by reducing the energy required for smelting. However, the SEC remains constant throughout the samples, indicating that the efficiency of DC furnaces is not significantly affected by the grade of chromite fines. There was a general increase in SEC from blend 1 to 2 in the DC with a pre-heating curve that corresponded to an increase in the MgO/SiO_2 that resulted in a slight increase in Cr reporting to slag, shown in Figure 6. Blend 4 was then considered the optimal blend based on both recovery and SEC. This blend was used in the subsequent experiments.

SEC and tapping stage recovery

This smelting experiment was performed to check on the tapping stage recovery using blend 4 and pre-calculated SEC. Figure 8 shows the effects of SEC on the taping stage recovery.

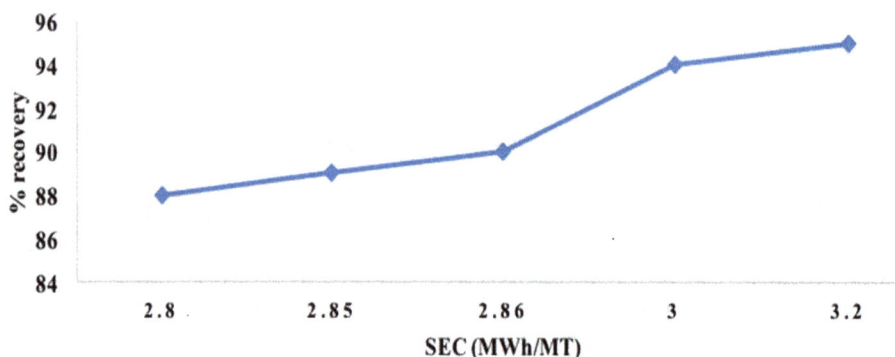

FIG 8 – The effects of Specific energy consumption on Recovery.

There is a clear correlation between SEC and percentage recovery in the given data. As the SEC increases from 2.8 MWh/MT to 3.2 MWh/MT, there is a noticeable increase in the percentage recovery from 88 per cent to 95 per cent. The most significant increase in recovery occurs between the SEC levels of 2.86 MWh/MT and 3 MWh/MT, where the recovery surges from 88 per cent to 94 per cent. Although there is only a slight increase in recovery from 3 MWh/MT to 3.2 MWh/MT, it is still evident that better SEC leads to higher recovery rates. However, for optimum operation, the range of 3 MWh/MT to 3.2 MWh/MT appears to be more viable. Within this range, the cost of SEC remains relatively lower, while achieving a high recovery rate of 94 per cent.

Effect of arc length on TSR

Figure 9 shows the recovery of Cr in ferrochrome at different arc lengths during the tapping stage in smelting using a closed DC arc furnace.

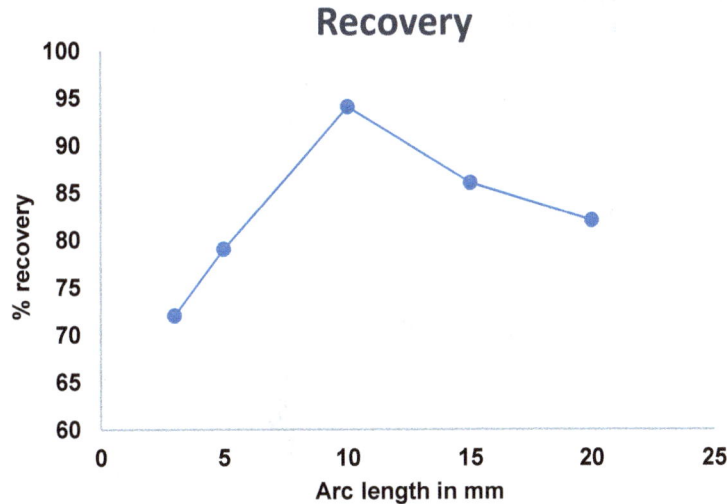

FIG 9 – Effects of arc length on the tapping stage recovery.

The recovery increases with an increase in arc length up to 10 mm, after which it starts to decline as the arc length continues to increase. This suggests that 10 mm is an optimal arc length for maximum recovery of Cr in ferrochrome of 94 per cent, this could be due to the optimal balance between heat generation and heat loss at this arc length, which facilitates the reduction of Cr_2O_3 to Cr. A decline in recovery for arc length beyond 10 mm, is due to excessive heat loss at larger arc lengths, which reduces the efficiency of the smelting process.

Effect of pre-heating and pre-reduction

Figure 10 shows the % Cr and Fe reduced after pre-treatment of the sample at different temperatures using CO as a reducing agent in blend 4. During the pre-heating and pre-reduction process, the temperature of the chromite ore fines raised to approximately 450°C, when pretreated with CO at a temperature of 700°C which is the estimated temperature of the off-gas leaving the furnace. The difference in temperature between the off-gas and the ore is due to the partial energy lost during the transfer of CO to ore and the energy used to raise the temperature of the ores. This therefore limits the overall reduction efficiency. At 400°C, only 10 per cent of FeO was reduced of Fe and an insignificant amount of Cr_2O_3 was reduced.

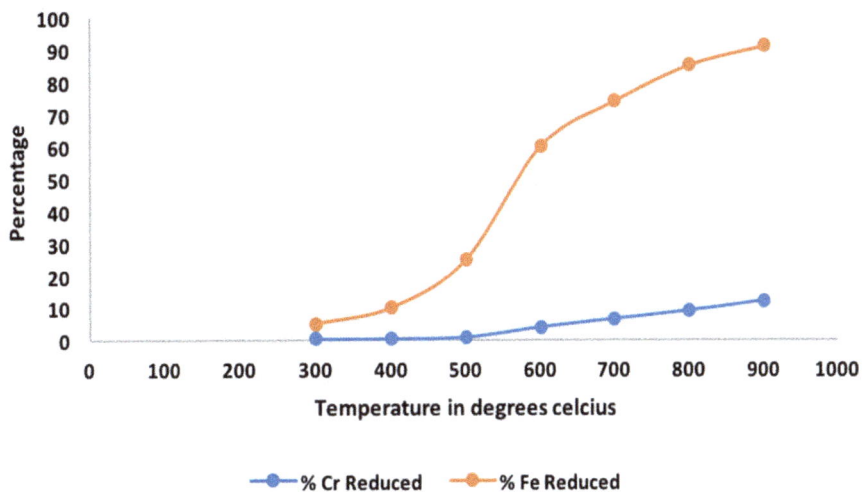

FIG 10 – Pre-reduction percentages of Cr and Fe using CO.

There is minimal reduction in both Cr and Fe at low temperatures less than 300°C as indicated in Figure 10. Above 400°C, Fe shows a significant increase in metallic percentage of Cr and Fe as temperature increases, reaching close to 100 per cent reduction at 1000°C. The reduction of FeO to Fe is highly temperature-dependent and becomes more efficient at higher temperatures. In contrast, Cr reduction is relatively steady and minimal across the entire temperature range. The reduction of Cr_2O_3 to Cr is less efficient than that of FeO to Fe under the same conditions as can easily be seen by the thermodynamic data for Equations 3 and 5 (Lee, 1999).

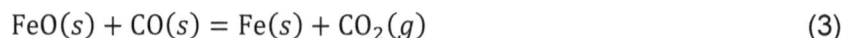

$$FeO(s) + CO(s) = Fe(s) + CO_2(g) \tag{3}$$

$$\Delta G° = 11000 - 156.9T \text{ J/mol (T = 298–973 K)} \tag{4}$$

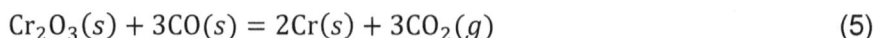

$$Cr_2O_3(s) + 3CO(s) = 2Cr(s) + 3CO_2(g) \tag{5}$$

$$\Delta G° = 279400 - 452.7T \text{ J/mol (T = 298–973 K)} \tag{6}$$

ENERGY REQUIREMENTS FOR FERROCHROME SMELTING

The probable reactions that may occur during the smelting of chromite fines with pre-heating and pre-reduction effects were predicted using basic thermodynamics, taking into account the chemical composition of the blends. It should be emphasised that the thermodynamic analysis was conducted to provide theoretical insight on the energy requirement despite the use of natural chromite ore with varying chemical interactions.

In order to determine the energy required in Figure 10 to elevate the temperature of each compound from the reference temperature to the pre-heating temperature using the off-gas from the furnace, the specific heat capacities of each compound within the chromite ore were employed. Shomate constant data obtained from the thermodynamics database (Chase, 1998) as depicted in Table 8, which shows the specific heat capacity (Cp) and total enthaplies (H-H$_{298}$). The masses of the compounds were ascertained by computing the proportions of each compound based on the percentage composition of the 0.1 kg sample utilised.

The Cp (J/mol.K) data from Table 8, was converted to Cp (Kwh/Kg.K). The specific energy in Table 9 was calculated using Equation 7 and calculated Cp (Kwh/Kg.K). The total energy consumption for pre-heating the chromite ore to 700 K was found to be 0.012386 kWh. This value represents the cumulative energy needed to heat all the components in the ore to the desired pre-heating temperature.

TABLE 8

Specific heat capacity and total enthalpy from shamote data (Chase, 1998).

Compound	Temp (K)	Cp (J/mol.K)	H-H$_{298}$
Preheating			
FeO		56.11	-
Cr$_2$O$_3$		122.7	-
SiO$_2$	700	68.77	-
CaO		51.54	-
Al$_2$O$_3$		116.5	-
MgO		148.69	-
Pre-reduction			
FeO		56.11	21.42
CO	700	7.45	2.87
CO$_2$		49.57	17.75
Fe		34.48	12.08
Smelting zone			
FeO		-	104.5
Cr$_2$O$_3$		-	216.9
SiO$_2$		-	111.5
Cr		-	61.06
Fe	2000	-	68.85
Si		-	46.28
CO		-	13.56
CaO		58.48	-
Al$_2$O$_3$		136.6	-
MgO		55.70	-

TABLE 9

Energy required for pre-heating to 700 K.

Compounds	Mass (Kg)	Cp*10^{-5} (kWh/Kg.K)	Energy (kWh)
FeO	0.011558	21.61	0.0010166
Cr$_2$O$_3$	0.033812	22.42	0.0030853
SiO$_2$	0.0115	31.83	0.0014898
MgO	0.01737	48.321	0.003416
Al$_2$O$_3$	0.01186	31.73	0.0015316
CaO	0.0139	32.64	0.001847
		Total energy	**0.012386**

$$Q = mc_p\Delta T \tag{7}$$

To determine the energy required for pre-reduction using the off-gas from the furnace and the smelting process inside the furnace, the total enthalpies of each component within the chromite ore were employed. These enthalpies were obtained from Shomate constant data as presented in Table 8 (Chase, 1998).

Equations 8, 9 and 10 were used in the evaluation of the values obtained in Tables 10 and 11 (Toulouevski and Zinurov, 2009). The energy/mol was calculated using enthalpies data from Table 8 and Equation 8. Reaction for the pre-reduction stage shown in Equation 11 and its associated energy consumption are shown in Table 10. The mass of Fe is based on the percentage of metallic iron produced in the pre-reduction graph. By combining the specific energy requirement and the mass of Fe reduced, an estimation of the energy consumption for Fe reduction during pre-reduction can be obtained. This energy consumption value is typically expressed in kWh and quantifies the amount of electrical energy needed to facilitate the reduction reaction.

$$\text{Energy per mol} = \text{total enthalpies of the final products} - \text{total enthalpies of the reactants} \tag{8}$$

$$\text{Specific energy required(SER)} = \text{Energy per mol} \times \text{molecular mass of the species} \tag{9}$$

$$\text{Energy used} = \text{SER} \times \text{mass of the species} \tag{10}$$

TABLE 10
Energy used in the pre-reduction.

Product species	Mass of product species (Kg)	Energy/mol (kWh/mol)	SER (kWh/kg)	Energy (kWh)	
				Alloy	Off-gas
Fe	0.001631	0.001539	0.0275	0.000044825	
CO_2	0.00128	0.001539	0.03498		0.000044825

TABLE 11
Reactions and Energyrequired in the smelting phase.

Reduction reactions	Product species	Product species Mass (kg)	Energy/mol (kWh/mol)	SER (kWh/kg)	Energy (KWh)	
					Alloy	Off-gas
$Cr_2O_3 + 3C \Rightarrow$ 2Cr + 3CO	Cr	0.01997	-0.0446	0.7964	0.0159	
	CO	0.015	-0.0446	1.014		0.0159
FeO + C $\Rightarrow$ Fe + CO	Fe	0.013885	-0.01598	0.2854	0.00396	
	CO	0.0069	-0.01598	0.5707		0.00396
$SiO_2 + 2C \Rightarrow$ Si + 2CO	Si	0.00114	-0.0303	1.0809	0.00123	
	CO	0.00113	-0.0303	1.0802		0.00123
				Total energy	0.02109	

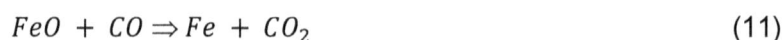

$$FeO + CO \Rightarrow Fe + CO_2 \tag{11}$$

The energy/mol required for smelting are shown in Table 11, were calculated using enthalpies data from Table 8 and Equation 8.

From Tables 11 and 12, the amount of energy required for ferrochrome production is 0.05226 kWh. However, after subtracting the pre-heating and pre-reduction energies, the overall energy required for smelting becomes 0.03983 KWh. Based on the nature of the experiments, which were performed on a small scale, considering this factor, the approximate amount of energy converted to smelting is 52 per cent of the total energy supplied. Thus, the total energy required for smelting is:

$$\frac{0.03983}{0.52} = 0.07659 \ kWh$$

TABLE 12

Amount of Energy used in raising temperature of materials which reported to slag.

Compounds	Mass (Kg)	$Cp*10^{-5}$ (kWh/Kg.K)	Energy (kWh)
CaO	0.00089	40.61	0.0006059
Al_2O_3	0.00807	37.2	0.005295
MgO	0.0201	38.7	0.01317
SiO_2	0.054	36.11	0.03319
		Total	0.05226

The amount of ferrochrome alloy produced in the reaction amounted to 0.03051 kg. The SEC was therefore evaluated to be 2.51 MWh/MT using Equation 1.

CONCLUSIONS

The results shown in this study provide a baseline for the viability of the transition from the traditionally used AC furnace to the closed DC arc furnace with pre-treatment stages. This confirms that fine chromite ores can be processed using a closed DC arc furnace without going through agglomeration techniques.

The DC arc furnace proved to be efficient regardless of the quantity of fines in the blend but rather recovery was determined by the grade of the ore. Comparison between the DC and AC furnace TSR indicated high energy conversion efficiency of the closed DC arc furnace because the input energy is effectively utilised in the DC furnace, resulting in improved recovery rates during the tapping stage.

SEC was significantly lowered by 35.8 per cent from 3.9 MWh/MT to 2.5 MWh/MT when pre-treatment techniques were employed. Also, the implementation of coal as the reductant and use of low-grade fines chromite ore aid in lowering the overall cost of production by an average of 30 per cent. This significant difference is important because it shows the potential of the closed DC arc furnace to achieve the goal of lowering energy consumption in ferrochrome production.

ACKNOWLEDGEMENTS

This research was funded by the Zimbabwe Mining and Smelting Company (ZIMASCO).

REFERENCES

Alison, C, 2019. Sintering: A Step Between Mining Iron Ore and Steelmaking [online]. Available from: <https://www.thermofisher.com/blog/mining/sintering-a-step-between-mining-iron-ore-and-steelmaking/> [Accessed: 5 January 2024].

Asish, K D, Somnath, K and Sarada, P, 2020. Processing of Low-Grade Chromite Ore for Ferroalloy Production: A Case Study from Ghutrigaon, Odisha, India, Transactions of the Indian Institute of Metals, 73:2309–2320.

Chase, M, 1998. NIST-JANAF Thermochemical Tables [online]. Available from: <https://webbook.nist.gov/cgi/cbook.cgi?ID=C1345251&Mask=2> [Accessed: 16 April 2024].

Chima, U, 2012. Innovative Systems Design and Engineering, Technology Innovations in the Smelting of Chromite Ore, 3(12):48–54.

du Preez, S P, van Kaam, T P M, Ringdalen, E, Tangstad, M, Morita, K, Bessarabov, D and van Zyl, P, 2023. An overview of currently applied ferrochrome production processed and their waste management practices, Minerals, 13(06):809.

Fortune Business Insights, 2023. Ferrochrome Market Size, Share and COVID-19 Impact Analysis, By Product Type (High Carbon, Low Carbon and Others), By Application (Stainless Steel, Specialty Steel and Others) and Regional Forecast, 2023–2030. Specialty and Fine Chemicals.

Figure 5 shows the cross-section micrographs and EDS analysis results of the siliconised molybdenum samples after hot dipping in Bi-Si bath at 1000°C for 15 and 60 mins. Similar with the results when using a Sn-Si bath, SEM-EDS results suggest the formation of about 12 μm thick layer, which contains both molybdenum and silicon, on the surface of the Mo-substrate. Point analysis of the layer reveals a Mo:Si atomic ratio of 1:2, confirming the formation of $MoSi_2$. However, the presence of other silicide compounds of molybdenum, such as Mo_5Si_3 which were determined via XRD, was not detected using SEM-EDS. In contrast, the presence of 0.39 per cent and 3.52 per cent Bi in the formed $MoSi_2$ layer was measured using EDS point analysis, but peaks of bismuth were not detected via XRD analysis.

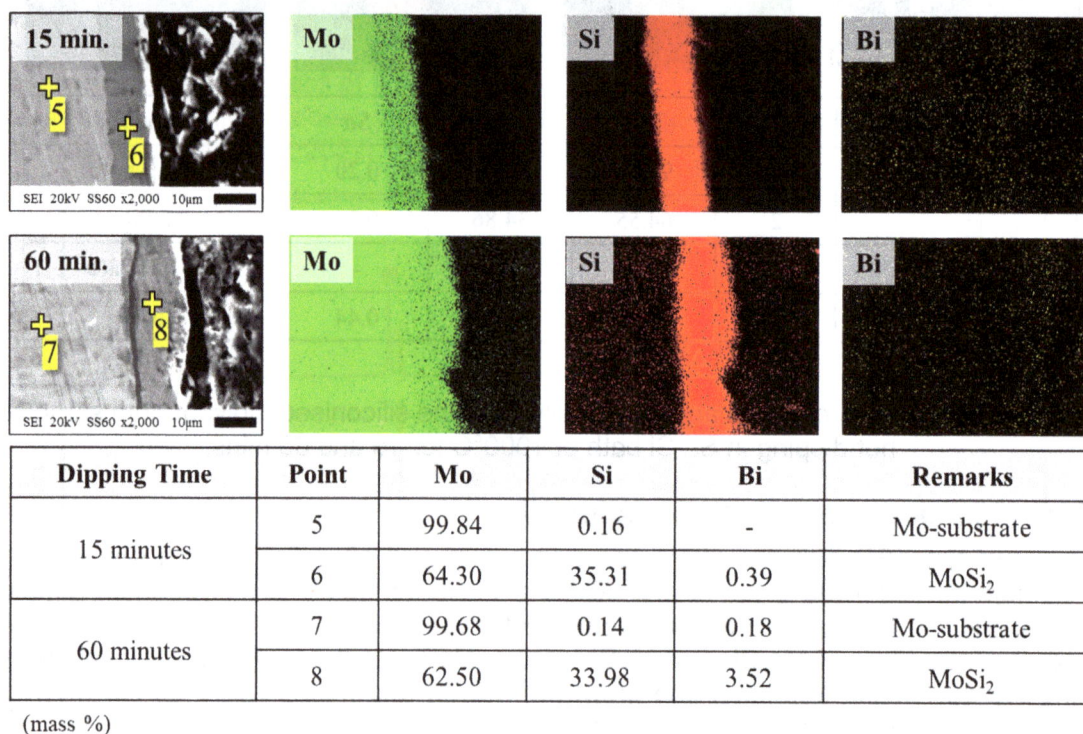

Dipping Time	Point	Mo	Si	Bi	Remarks
15 minutes	5	99.84	0.16	-	Mo-substrate
	6	64.30	35.31	0.39	$MoSi_2$
60 minutes	7	99.68	0.14	0.18	Mo-substrate
	8	62.50	33.98	3.52	$MoSi_2$

(mass %)

FIG 5 – Cross-section micrographs and EDS analysis of the siliconised molybdenum samples after hot dipping in Bi-Si bath at 1000°C for 15 and 60 mins.

Cu-Si bath (73.5 per cent Cu – 26.5 per cent Si)

Due to the irregularity in shape and the difficulty in obtaining good measurement results, only the XRD patterns of the sample immersed in molten Cu-Si bath for 15 mins at 1000°C were obtained in this work, as shown in Figure 6. After dipping for 15 mins, formation of the characteristic peaks of $MoSi_2$ were found, while the formation of new Cu_3Si peaks were also observed. However, continuous dipping for 60 mins resulted in the formation of thick, dendrite-like protrusions along the sample edges, impeding true and representative sampling for XRD analysis without breaking off or damaging the surface.

Figure 7 shows the cross-section micrographs and EDS analysis results of the samples after dipping for 15 and 60 mins. These results confirm the formation of $MoSi_2$ on the surface of the Mo-substrate and a new layer composed primarily of copper and silicon. Point analysis of this layer revealed a Cu:Si atomic ratio of approximately 3, confirming the formation of Cu_3Si, which was also identified via XRD analysis. The Cu_3Si layer was found primarily located on the outermost surface of the sample, but it also appeared in between the Mo-substrate and formed $MoSi_2$ layers, particularly in the sample immersed for 60 mins.

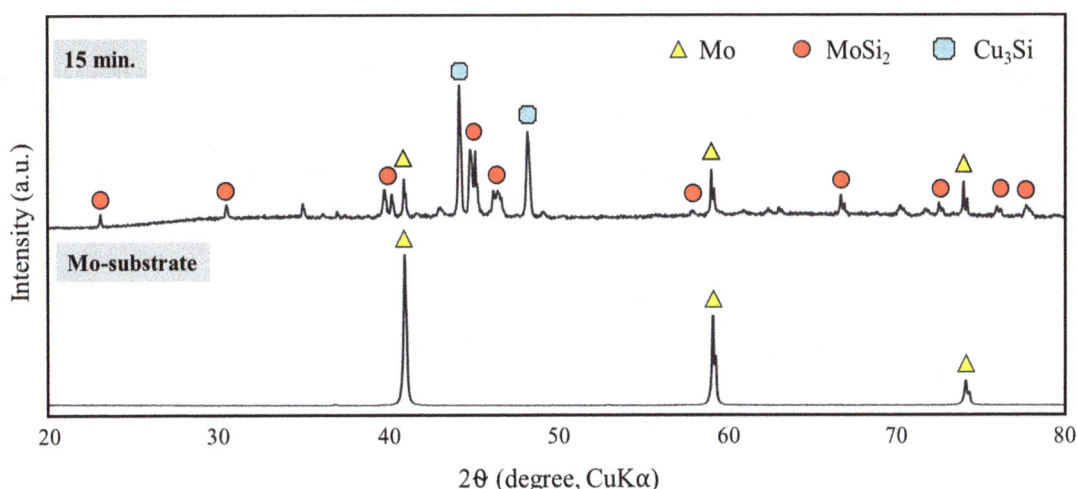

FIG 6 – XRD patterns of the bulk Mo-substrate before after hot dipping in a Cu-Si bath for 15 mins at 1000°C.

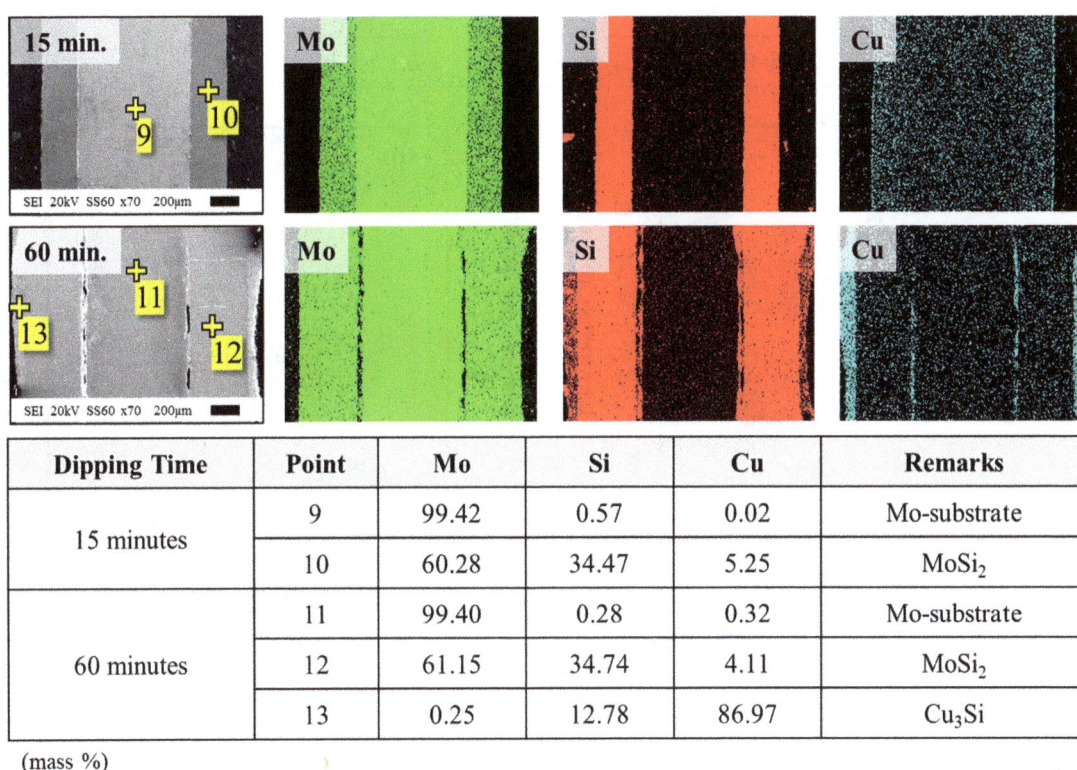

Dipping Time	Point	Mo	Si	Cu	Remarks
15 minutes	9	99.42	0.57	0.02	Mo-substrate
	10	60.28	34.47	5.25	$MoSi_2$
60 minutes	11	99.40	0.28	0.32	Mo-substrate
	12	61.15	34.74	4.11	$MoSi_2$
	13	0.25	12.78	86.97	Cu_3Si

(mass %)

FIG 7 – Cross-section micrographs and EDS analysis of the siliconised molybdenum samples after hot dipping in Cu-Si bath at 1000°C for 15 and 60 mins.

The formation of other molybdenum silicide compounds was not detected using either XRD or SEM-EDS analysis, but the presence of about 4–5 per cent Cu in the formed $MoSi_2$ layer was identified using EDS point analysis. While we cannot draw definitive conclusions without XRD data for longer dipping times, the extremely rapid growth of the $MoSi_2$ layer when using molten Cu-Si bath and the substantial concentration of copper within the layer suggest that penetration of copper caused significant expansion of the $MoSi_2$ lattice, facilitating the faster diffusion of silicon towards the unreacted Mo-substrate (Gamutan and Miki, 2022).

The thickness of the formed $MoSi_2$ layers after dipping for 15 and 60 mins, measured at 268 μm and 441 μm, respectively, showed a tenfold increase in growth rate when using molten Cu-Si bath compared to samples immersed in molten Sn-Si and Bi-Si baths.

Mechanism and kinetics of MoSi$_2$ formation

The above-mentioned findings indicate that the formation of MoSi$_2$ at the surface of the Mo-substrate was possible by dipping molybdenum in various silicon-containing tin (98.8 per cent Sn – 1.2 per cent Si), bismuth (98.0 per cent Bi – 2.0 per cent Si) and copper (73.5 per cent Cu – 26.5 per cent Si) baths. The concentration of silicon in these baths were designed such that the thermodynamic activity of silicon in this melt was close to unity and molybdenum reacts with silicon in the bath according to the following equation:

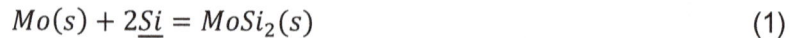

$$Mo(s) + 2\underline{Si} = MoSi_2(s) \tag{1}$$

Meanwhile, the formation of intermediate compounds of molybdenum and silicon, such as Mo$_5$Si$_3$, also suggest the occurrence of following equation:

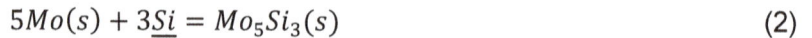

$$5Mo(s) + 3\underline{Si} = Mo_5Si_3(s) \tag{2}$$

The measured thickness of the MoSi$_2$ layer formed by hot dipping in various silicon-containing molten baths is shown in Figure 8a. The thickness was an average of six measurements made at different positions in the sample. It was found that the MoSi$_2$ layers grew with dipping time and growth was fastest when using molten Cu-Si bath, followed by molten Sn-Si and Bi-Si baths.

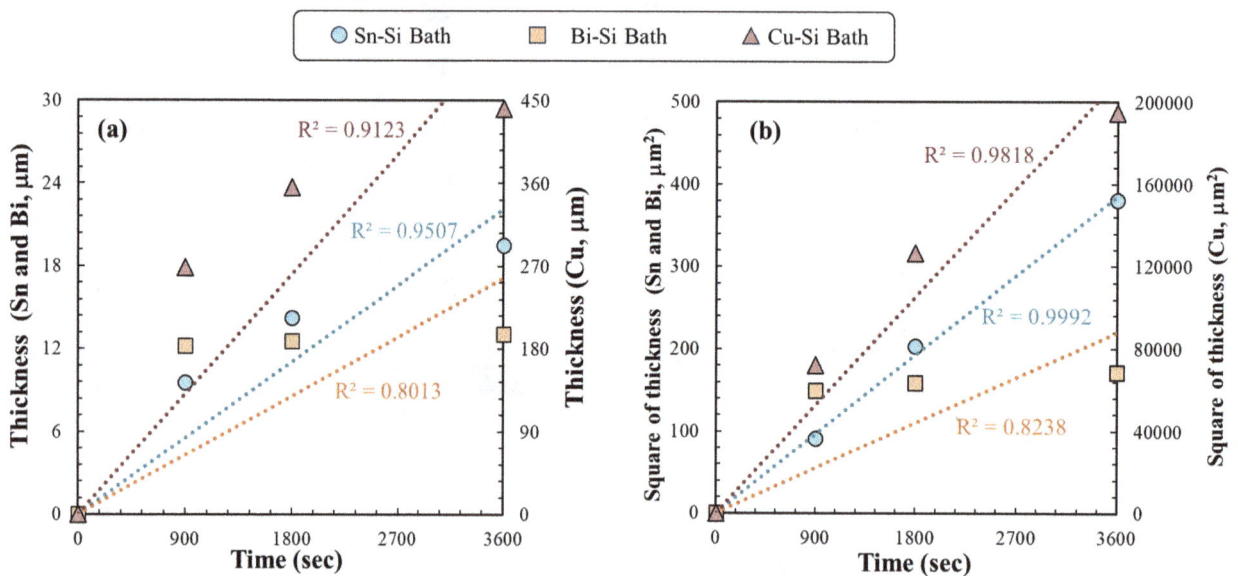

FIG 8 – Growth of MoSi$_2$ layer thickness as a function of time according to: (a) reaction-controlled growth mechanism, and (b) diffusion-controlled growth mechanism.

To determine whether the growth of the MoSi$_2$ layer was controlled either by the reaction at the Mo-substrate interface or by diffusion through the formed MoSi$_2$ product layer, variation in thickness versus time was plotted as follows. For a reaction-controlled growth, it is known that MoSi$_2$ layer thickness should increase linearly with time according to the following equation:

$$y = kt \tag{3}$$

where y is the MoSi$_2$ layer thickness (µm) at a certain dipping time t (sec). For a diffusion-controlled growth, the thickness should increase parabolically with time according to the following equation:

$$y^2 = kt \tag{4}$$

Plotting the measured MoSi$_2$ layer thickness from this study revealed through linear regression that R^2 values were closer to unity for the parabolic growth rate model, as shown in Figure 8b. This suggests that siliconisation of the Mo-substrate via the hot dipping method was primarily a diffusion-controlled process. This is particularly true for samples immersed in molten Sn-Si and Cu-Si bath, with very high R^2 values of 0.9992 and 0.9818, respectively. Consequently, promoting diffusion of silicon through the formed MoSi$_2$ layer could potentially improve growth, which aligns with the findings above, particularly in the case of silicon-containing copper bath.

However, attempts to apply both reaction and diffusion-controlled growth models to samples dipped in a molten Bi-Si bath did not achieve very strong linear relationships. In fact, the thickness of the formed $MoSi_2$ layer barely increased. These results imply that another factor influenced the growth kinetics of the $MoSi_2$ layer. Based on our observations, the formation and presence of the intermediate phase, Mo_5Si_3, might be the rate-limiting step. Further investigation is required to solidify this conclusion.

To summarise, the mechanism of growth of the $MoSi_2$ layer on the surface of Mo-substrate can be schematically described as shown in Figure 9. When pure molybdenum is dipped into the silicon-containing bath, silicon initially diffuses through the molten bath, as shown in Figure 9a, until it comes into contact with the molybdenum surface. Due to the elevated temperature conditions employed in the hot dipping experiment, diffusion of silicon through the molten bath is assumed to be fast enough such that its influence on the kinetics of growth could be neglected.

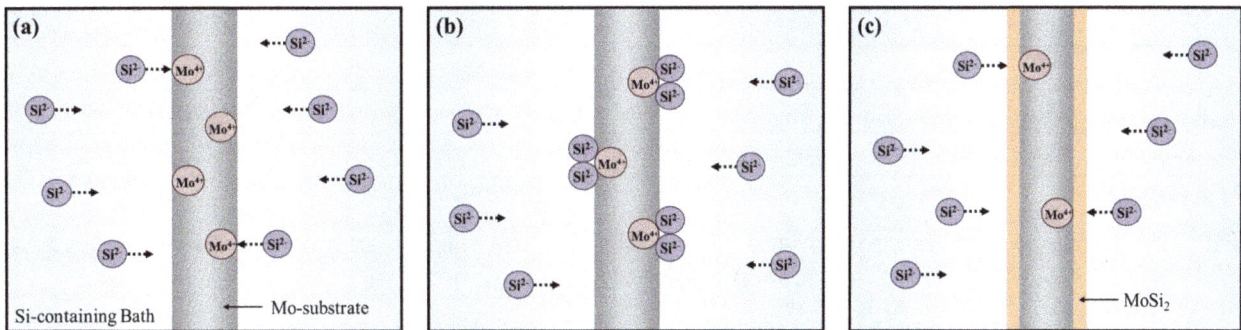

FIG 9 – Proposed mechanism of $MoSi_2$ formation via the hot dipping method.

Next, once the silicon, with thermodynamic activity close to unity, comes into contact with molybdenum, both elements react together according to Equation 1 to produce $MoSi_2$ as shown in Figure 9b. Potentially, Equation 2 also occurs to form Mo_5Si_3, particularly when using molten Bi-Si bath, due to low silicon concentrations within the interface. Eventually, the $MoSi_2$ layer grows and thickens, forming a product layer, which substantially slows down the rate at which silicon reaches the unreacted molybdenum surface, as shown in Figure 9c.

Oxidation resistance behaviour

Fully coated samples produced via hot dipping in molten Sn-Si bath at 1000°C for 60 mins were subjected to oxidation test carried out at 1150°C for 2 hrs in air. While the bare uncoated molybdenum sample was completely oxidised and lost in the form of gaseous molybdenum oxides, the fully coated samples suffered a mass loss of about 4 per cent only. These results suggest that formation of a $MoSi_2$ layer on the surface of molybdenum imparts oxidation resistance at elevated temperatures, a promising solution to address the challenges to both $MoSi_2$ and molybdenum alloys.

Figure 10 shows the cross-section micrograph and EDS analysis results of the siliconised molybdenum samples after the oxidation test. As evident from the micrograph, the density of the $MoSi_2$ layer did not change significantly and the oxidised surface consisted of three layers – an outer SiO_2 layer followed by the $MoSi_2$ layer and a thin intermediate layer on top of the Mo-substrate. Area analysis of this intermediate layer showed a Mo:Si atomic ratio of about 5:3, indicating that the main phase composition of this layer is Mo_5Si_3. Line scan also supports the formation of these three layers, revealing a clear change in molybdenum and silicon concentrations. To determine the exact oxidation resistance mechanism of the $MoSi_2$ layer, further tests with higher temperatures, longer holding times and enhanced sensitivity is necessary.

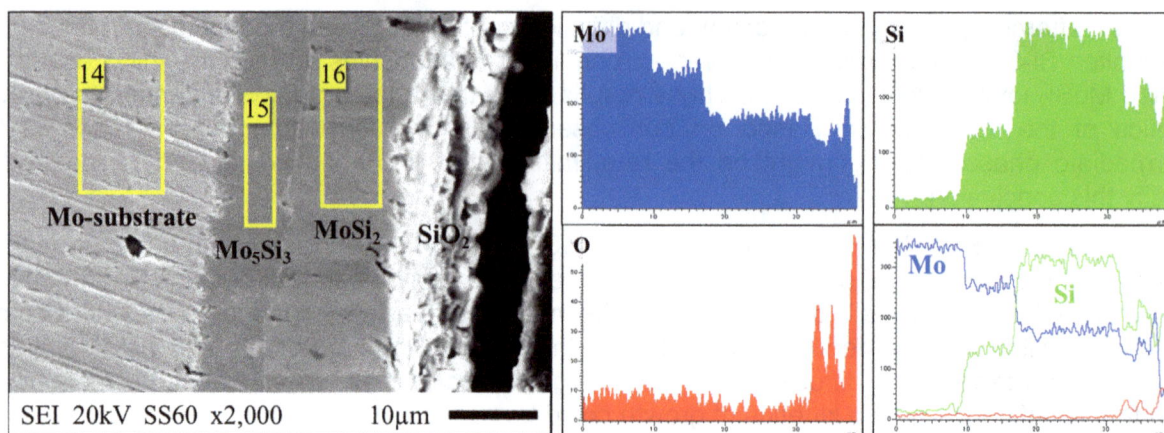

Area Analysis	Si	Mo	Sn	Remarks
14	-	99.78	0.22	Mo-substrate
15	14.30	85.69	0.01	Mo_5Si_3
16	35.41	64.59	-	$MoSi_2$

(mass %)

FIG 10 – Cross-section micrograph and EDS analysis of the siliconised molybdenum samples after the oxidation test at 1150°C for 2 hrs in air.

CONCLUSIONS

The present work demonstrates the feasibility of a novel hot dipping method to produce a protective $MoSi_2$ layers on pure Mo-substrate using silicon-containing tin (98.8 per cent Sn – 1.2 per cent Si), bismuth (98.0 per cent Bi – 2.0 per cent Si) and copper (73.5 per cent Cu – 26.5 per cent Si) baths. This simple and straightforward method enabled the formation of a dense $MoSi_2$ layer after hot dipping at 1000°C for 15–60 mins with thickness of about 10–20 µm, 12 µm and 268–441 µm, respectively. The silicon-containing molten copper bath allowed $MoSi_2$ to be synthesised most rapidly, followed by tin and bismuth.

Siliconisation of the Mo-substrate using Sn-Si and Cu-Si baths was found to be mainly a diffusion-controlled process. Meanwhile, other factors, such as the formation of an intermediate Mo_5Si_3 layer, were found to be potentially rate-limiting when using a Bi-Si bath. Finally, oxidation test of the fully coated samples confirms the potential of $MoSi_2$ to protect molybdenum and its alloys from rapid oxidation loss at elevated temperatures.

REFERENCES

Brewer, L and Lamoreaux, R H, 1980. The Mo-O system, *Bulletin of Alloy Phase Diagrams*, 1:85–89.

Cai, Z, Liu, S, Xiao, L, Fang, Z, Li, W and Zhang, B, 2017. Oxidation behavior and microstructural evolution of a slurry sintered Si-Mo coating on Mo alloy at 1650°C, *Surface and Coatings Technology*, 324:182–189.

Christian, F and Narita, T, 1998. Siliconizing of molybdenum metal in indium-silicon melts, *Materials Transaction JIM*, 39:658–662.

Cox, A R and Brown, R, 1964. Protection of molybdenum from oxidation by molybdenum dilicide based coatings, *Journal of the Less-Common Metals*, 6:51–69.

Fu, T, Zhang, Y, Chen, L, Shen, F and Zhu, J, 2024. Micromorphology evolution, growth mechanism and oxidation behaviour of the silicon-rich MoSi2 coating at 1200°C in air, *Journal of Materials Research and Technology*, 29:491–503.

Gamutan, J and Miki, T, 2022. Surface modification of a Mo-substrate to form an oxidation-resistant MoSi2 layer using a Si-saturated tin bath, *Surface and Coatings Technology*, 448:128938.

Govindarajan, S, Mishra, B, Olson, D L, Moore, J J and Disam, J, 1995. Synthesis of molybdenum disilicide on molybdenum substrates, *Surface and Coatings Technology*, 76–77:7–13.

Ito, K, Murakami, T, Adachi, K and Yamaguchi, M, 2003. Oxidation behavior of Mo-9Si-18B alloy pack-cemented in a Si-base pack mixture, *Intermetallics*, 11:763–772.

Jeng, Y L and Lavernia, E J, 1994. Review: Processing of molybdenum disilicide, *Journal of Materials Science*, 29:2557–2571.

Kamata, S H, Kanekon, D, Lu, Y, Sekido, N, Maruyama, K, Eggeler, G and Yoshimi, K, 2018. Ultrahigh-temperature tensile creep of TiC-reinforced Mo-Si-B based alloy, *Scientific Reports*, 8:10487.

Li, W, Fan, J, Fan, Y, Xiao, L and Cheng, H, 2018. MoSi2/(Mo,Ti)Si2 dual-phase composite coating for oxidation protection of molybdenum alloy, *Journal of Alloys and Compounds*, 740:711–718.

Majumdar, S, 2012. Formation of MoSi2 and Al doped MoSi2 coatings on molybdenum base TZM (Mo-0.5Ti-0.1Zr-0.02C) alloy, *Surface and Coatings Technology*, 206:3393–3398.

Nanko, M, Kitahara, A, Ogura, T, Kamata, H and Maruyama, T, 2001. Formation of Mo(Si,Al)2 layer on Mo dipped in Al melt saturated with Si and the effects of transition metals added in the melt, *Intermetallics*, 9:637–646.

Nomura, N, Suzuki, T, Yoshimi, K and Hanada, S, 2003. Microstructure and oxidation resistance of a plasma sprayed Mo-Si-B multiphase alloy coating, *Intermetallics*, 11:735–742.

Okamoto, H, 2012. Cu-Si (Copper-Silicon), *J Phase Equilibria and Diffusion*, 33:415–416.

Olesinski, R W and Abbaschian, G J, 1984. The Si-Sn (silicon-tin) system, *Bulletin of Alloy Phase Diagrams*, 5:273–276.

Olesinski, R W and Abbaschian, G J, 1985. The Si-Bi (silicon-bismuth) system, *Bulletin of Alloy Phase Diagrams*, 6:359–361.

Reisel, G, Wielafe, B, Steinhäuser, S, Morgenthal, I and Scholl, R, 2001. High temperature oxidation behavior of HVOF-sprayed unreinforced and reinforced molybdenum disilicide powders, *Surface and Coatings Technology*, 146–147:19–26.

Rice, M J and Sarma, K R, 1981. Interaction of CVD silicon with molybdenum substrates, *Journal of the Electrochemical Society*, 128:1368–1373.

Sakidja, R, Park, J S, Hamann, J and Perepezko, J H, 2005. Synthesis of oxidation resistant silicide coatings on Mo-Si-B alloys, *Scripta Materialia*, 53:723–728.

Shah, D M, 1992. MoSi2 and other silicides as high temperature structural materials, in *Proceedings of the 7th International Symposium on Superalloys* (eds: S D Antolovich, R W Stusrud, R A MacKay, D L Anton, T Khan, R D Kissinger and D L Klarstrom), pp 409–422 (The Minerals, Metals and Materials Society).

Sharifitabar, M, Oukati Sadeq, F and Shafiee Afarani, M, 2021. Synthesis and kinetic study of Mo(Si,Al)2 coatings on the surface of molybdenum through hot dipping into a commercial Al-12wt.%Si alloy melt, *Surface and Interfaces*, 24:101044.

Subramanian, P R and Laughlin, D E, 1990. The Cu-Mo (copper-molybdenum) system, *Bulletin of Alloy Phase Diagrams*, 11:169–172.

Sun, J, Fu, Q G, Guo, L P, Liu, Y, Huo, C X and Li, H J, 2016. Effect of filler on the oxidation protective ability of MoSi2 coating for Mo substrate by halide activated pack cementation, *Materials and Design*, 92:602–609.

Sun, J, Li, T and Zhang, G P, 2019. Effect of thermodynamically metastable components on mechanical and oxidation properties of the thermal-sprayed MoSi2 based composite coating, *Corrosion Science*, 155:146–154.

Suzuki, R, Ishikawa, M and Ono, K, 2000. MoSi2 coating on molybdenum using molten salt, *Journal of Alloys and Compounds*, 306:285–291.

Vasudévan, A K and Petrovic, J J, 1992. A comparative overview of molybdenum disilicide composites, *Mat, Sci and Eng*, A155:1–17.

Yanagihara, K, Maruyama, T and Nagata, K, 1994. Dip-coating of Mo(Si,Al)2 on Mo with an Al-Si melt, *Tetsu-to-Hagane*, 80:178–182.

Yoon, J K, Byun, J Y, Kim, H G, Kim, J S and Choi, C L, 2002. Growth kinetics of three Mo-silicide layers formed by chemical vapor deposition of Si on Mo substrate, *Surface and Coatings Technology*, 155:85–95.

Zhang, Y, Cui, K, Fu, T, Wang, J, Shen, F, Zhang, X and Yu, L, 2021. Formation of MoSi2 and Si/MoSi2 coatings on TZM (Mo-0.5Ti-0.1Zr-0.02C) alloy by hot dip silicon-plating method, *Ceramics International*, 47:23053–23065.

Zhang, Y, Fu, T, Yu, L, Shen, F, Wang, J and Cui, K, 2022. Improving oxidation resistance of TZM alloy by deposited Si-MoSi2 composite coating with high silicon concentration, *Ceramics International*, 48:20895–20904.

Zhu, L, Zhu, Y, Ren, Z, Zhang, P, Qiao, J and Feng, P, 2019. Microstructure, properties and oxidation behavior of MoSi2-MoB-ZrO2 coating for Mo substrate using spark plasma sintering, *Surface and Coatings Technology*, 375:773–781.

Reduction and melting behaviours of carbon – iron oxide composite using iron carbides and free carbon obtained by vapour deposition

R Higashi[1], D Maruoka[2], Y Iwami[3] and T Murakami[4]

1. Graduate Student, Graduate School of Environmental Studies, Tohoku University, Sendai 980-8579, Japan. Email: ryota.higashi.r3@dc.tohoku.ac.jp
2. Assistant Professor, Graduate School of Environmental Studies, Tohoku University, Sendai 980-8579, Japan. Email: daisuke.maruoka.e6@tohoku.ac.jp
3. Senior Researcher, JFE Steel Corporation, Fukuyama 721-8510, Japan. Email: yu-iwami@jfe-steel.co.jp
4. Professor, Graduate School of Environmental Studies, Tohoku University, Sendai 980-8579, Japan. Email: taichi@material.tohoku.ac.jp

ABSTRACT

The ironmaking industry consumes a large amount of fossil fuel derived carbon as heat source, reducing agent of iron ores and carburising agent of reduced iron. Although the demand for drastic decrease of carbon dioxide emission, carbon is an essential element for smelting process of molten iron. The carbon recycling ironmaking process by circulating CO has been already proposed to achieve carbon neutrality. However, the production of molten hot metal is not considered in this process because sufficient amount of carbon does not dissolve in reduced iron by CO. Therefore, our group has suggested a new carbon recycling ironmaking process which can produce hot metal. In this process, free carbon and iron carbides produced by carbon deposition reaction using metallic iron as a catalyst are used. It is known that only Fe_3C is obtained as iron carbide by using CO gas, however, Fe_5C_2 is also produced by adding H_2 gas. The composite agglomerated with these carbonaceous materials and fine iron ore (Deposited Carbon-Iron oxide Composite: DCIC) is reduced and melted in a furnace. It is reported that Fe_3C in DCIC accelerates the reduction reaction and melting of the composite. In this study, the effects of iron carbides and free carbon on the melting behaviour of DCIC are investigated.

Fe_3C, Fe_5C_2 and free carbon were produced by vapour deposition using porous iron whiskers and CO-CO_2-H_2 gas. These were agglomerated with hematite reagent at a certain ratio to prepare DCIC samples with and without Fe_5C_2. The samples were heated up to 1300°C in inert atmosphere. The DCIC containing Fe_5C_2 completely melted and iron nuggets were obtained after the experiment. This behaviour was not observed in the composite without Fe_5C_2. This indicates that using Fe_5C_2 is more preferable than Fe_3C for molten iron production using DCIC.

INTRODUCTION

The blast furnace ironmaking process has been adopted as the main process at integrated steel mills because of its high efficiency. However, it needs a large amount of fossil fuel derived carbon for its heat source, reducing agent of iron ores and carburising agent of reduced iron. Thus, it is essential to develop a new process to achieve carbon-neutral steelmaking. Currently, research and development for the commercialisation of hydrogen reduction ironmaking with shaft furnace have been conducted around the world. However, ironmaking processes based on Direct Reduced Iron (DRI) production such as MIDREX® process, are largely limited by the grade of the iron ore. This is because the gangue in the DRI obstructs the conduction of heat to the metallic iron. Another matter is the increase in the amount of molten slag in the melting process of DRI with Electric Arc Furnace (EAF). To improve this problem, it is effective to produce hot metal separated from molten slag before the refining process. Nonetheless, carbon must be added to reduced iron for the production of the hot metal. Therefore, attempts using biomass char as a carburising agent have been made to produce hot metal carbon-neutrally (Robinson *et al*, 2022; Norgate *et al*, 2012). Practical application, however, is difficult in countries with large crude steel production such as China and Japan. This is because of difficulty to supply the required amount of carbon with biomass. As long as the amount of available biomass char is limited, the application of Carbon Capture and Utilization (CCU) is inevitable to achieve carbon neutrality in ironmaking process.

Examples of CCU application in steel mills include iACRES (Kato, 2010) and carbon recycling blast furnace (Kawashiri, Nouchi and Kashihara, 2022). iACRES is designed to recycle CO reformed by high-temperature electrolysis of CO_2 and the reverse water gas shift reaction as shown in Equation 1 and Equation 2, respectively. CH_4 produced by the methanation reaction described as Equation 3, on the other hand, is reused in the carbon recycling blast furnace. In both cases, the carbon cycle is achieved by reforming and reusing the CO_2 recovered from the exhaust gas, but its scope is limited to the reduction process. To extend the scope to oxidation refining, a carburising agent for hot metal production should be reused.

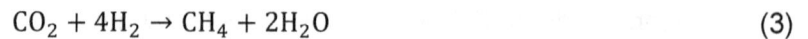

$$2CO_2 = 2CO + O_2 \tag{1}$$

$$CO_2 + H = CO + H_2O \tag{2}$$

$$CO_2 + 4H_2 \rightarrow CH_4 + 2H_2O \tag{3}$$

There are mainly two routes for carburisation of iron. One is direct carburisation by solid carbon and another is indirect one by CO gas. The reaction equations are described as Equation 4 and Equations 5–6, respectively.

$$Fe(s) + C(s) = Fe(s) + \underline{C} \tag{4}$$

$$CO(g) = C^* + O^* \tag{5}$$

$$CO(g) + O^* = CO_2(g) \tag{6}$$

$$C^* = \underline{C} \tag{7}$$

(*) and (_) mean the atom absorbing on and dissolving in Fe, respectively. Murakami and Nagata (2003) compared the melting rate of metallic iron plate by direct carburisation with a graphite piece and gas carburisation with CO at 1250°C and reported that the melting proceeded more than 1000 times faster with direct contact carburisation than with gas carburisation. This is because the mass transfer of carbon to molten iron at the boundary between the carburising agent and solid iron is the rate-limiting factor for iron melting. Indirect carburisation, which requires the desorption process of gas molecules, supplies carbon to molten iron slower than direct one. Therefore, considering hot metal production in carbon recycling ironmaking, it is desirable to reform the CO_2 in the exhaust gas to solid carbon. However, the solid carbon is recovered in a fine powder state, which causes handling problems.

Therefore, the authors propose the following new ironmaking process named Carbon Recycling Ironmaking Process (CRIP) using Deposited Carbon-Iron Ore composite (DCIC) (CRIP-D), which extends the scope of carbon recycling to steel production without fossil fuel derived carbon. Hydrogen and energy are added to the gas generated from iron ore reduction process and hot metal refining process. The gas is reformed to CO-rich gas through reverse water gas shift reaction catalysed by porous iron. Regarding the production of the porous iron, it has been reported that metallic iron whiskers were produced by reducing carbon-iron ore composite prepared using charcoal (Murakami et al, 2017b). Due to intertwining the fibrous structure, the porous iron showed about 95 per cent of porosity. This porous iron is also used as a substrate for carbon deposition reaction using reformed CO-rich gas. The carbon in the gas, therefore, can be recovered as not only free carbon but also iron carbide. Furthermore, the reaction rate can be improved by increasing the catalytically active area, and the handling of carbon powder can be more easily by incorporating the deposited carbon into the pores. Nishihiro et al (2019) conducted carbon deposition reaction catalysed by metallic iron particles. They reported that a larger amount of carbon was deposited by using $CO-H_2$ mixed gas than pure CO gas. Sawai, Iguchi and Hayashi (1999a; 1999b) also reported that cementite (Fe_3C) was produced by using $CO-CO_2$ mixed gas for carbonising metallic iron particles, while not only cementite but also Hägg iron carbide (Fe_5C_2) were produced by using $CO-CO_2-H_2$ mixed gas. Regarding the utilisation of iron carbides, Sato et al (2011) conducted melting experiment under elevated temperature using tablets made of metallic iron powder with cementite or graphite powder. The melting rate of the tablet with cementite was higher than that with graphite. This indicates iron carbides are suitable for a carburising agent. Murakami et al (2017a) also conducted elevated temperature test on the carbon-iron ore composite using coal and graphite. The reduction and carburisation reactions were reported to be accelerated by dividing the function of carbonaceous materials as reducing and carburising agents. Thus, DCIC with high reactivity may be prepared by using the free carbon and iron carbides recovered in the proposed process as reducing and carburising agents, respectively. Hot metal can be produced with DCIC in rotary hearth furnaces and shaft furnaces (Hasanbeigi, Arens and Price, 2014). By replacing some of the raw materials in the blast furnace with DCIC, it is expected to reduce CO_2 emissions while using the facilities in the proven integrated steelmaking technology (Yokoyama *et al*, 2014). The gas after the carbon deposition reaction is introduced in to hot metal production process as a reducing agent and heat source since the gas contains certain amount of CO and H_2. Carbon recycling is achieved by reforming the gases emitted in the reducing, hot metal production and refining process.

In this study, the porous iron whiskers with high porosity are focused on, as a substrate for carbon deposition reaction. Furthermore, the reduction and melting behaviours of carbon-iron oxide composite prepared using recovered deposited carbon are investigated to examine the fundamental possibility of CRIP-D.

EXPERIMENTAL

Sample preparation

Hematite reagent (1 µm size, purity 99.9 per cent) and biomass char (53–150 µm size) were well mixed to have the molar ratio of fixed carbon in the char to oxygen in the hematite, C/O ratio, was 0.75. The biomass char contained 95.2 mass per cent of the fixed carbon. The porous iron was prepared by carbothermal reduction of the mixture in the crucible with 26 mm diameter at 1000°C in an inert atmosphere. The reduced iron showed high porosity of 94.6 per cent due to its fibrous structure. The sample for carbon deposition experiment was prepared by cutting the porous iron with 5 mm in thick and was set in an experiment apparatus shown in Figure 1. The sample holder has meshed bottom woven by Pt wire with 0.5 mm in diameter. The wire was coated with ceramic to avoid the carbon deposition reaction on the wire. The sample holder was prepared by sandwiching the porous iron sample between the ring with preheater zone and another mullite ring. The gap between the sample side and the ring was filled with inorganic adhesive. In addition, the surface of the adhesive was coated with ceramic wool to protect the reaction tube. Curing of the inorganic adhesive and pre-reduction were conducted in Ar-10 per centH_2 atmosphere at 250°C and 800°C, respectively. The carbon deposition reaction proceeded for 120 mins in two conditions. One is used pure CO at 800°C. The other is used CO-CO_2-H_2 gas mixture at 600°C. The gas flow rate was set as 500 mL/min. After the carbon deposition reaction, the samples were pulverised. The Sample labelled 'a' is obtained from the carbon deposition experiment using CO gas. Whereas, the sample obtained using the gas mixture had grey and black layers, which were separated into two samples.

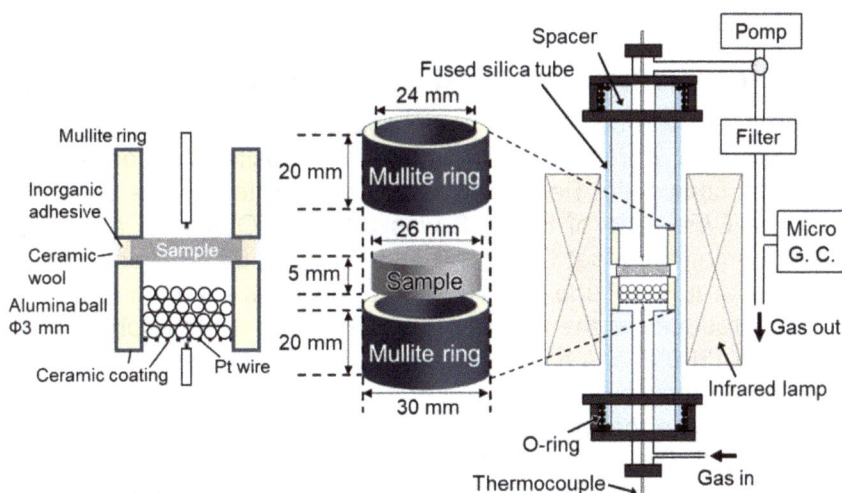

FIG 1 – Schematic diagram of an apparatus for carbon deposition using an image furnace.

Each layer was also pulverised separately into Samples b and c. Figure 2 shows XRD profiles obtained for each deposited carbon sample using a sealed X-ray tube with a Fe anode (tube voltage 40 kV, tube current 30 mA) as an X-ray source. Some noises appeared in Sample c. Fe_3C is the only iron carbide phase in Sample a. Fe_5C_2, on the other hand, exists in Samples b and c. A broad graphite peak is observed in Sample c, suggesting free carbon deposition. The carbon contents analysed by infrared absorption method after combustion are 14.3 per cent, 6.62 per cent and 48.5 per cent in Samples a, b and c, respectively.

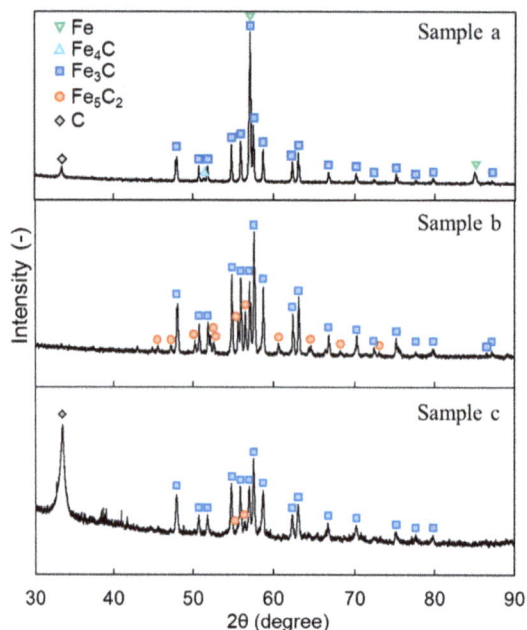

FIG 2 – XRD profiles of samples with deposited carbon obtained by using CO gas and $CO-CO_2-H_2$ gas mixture.

In situ XRD Measurement

Decomposition behaviour of different types of iron carbides was investigated *in situ* using XRD at elevated temperatures. X-ray enters the chamber through a window made of beryllium. The chamber can be sealed, allowing control of the internal atmosphere. Sample b was filled in a black quartz glass sample holder and heated by an infrared heating equipment attached to the chamber. The sample temperature was monitored using a thermocouple with a sheath inserted into the sample holder. The sample temperature elevated by 10°C/min to 700°C under N_2 atmosphere. The gas flow rate was set as 200 mL/min. XRD measurement were repeatedly conducted in the range of $2\theta = 55°\sim58°$.

Reduction experiment using deposited carbon-iron oxide composite

The deposited carbon-iron oxide composite was prepared by press-shaping of the mixture of the hematite reagent (1 μm size, purity 99.9 per cent) and deposited carbon samples sieved under 106 μm into a cylindrical composite. $Fe_2O_3-\theta$ was the composite sample using Sample 'a' so that C/O ratio was 1.0. The C/Fe ratio of $Fe_2O_3-\theta$ was 0.52. $Fe_2O_3-\chi$ was also prepared by mixing Samples b and c with hematite so that C/O = 1.0 and C/Fe = 0.52. The composite sample was set in the experimental apparatus as shown in Figure 3. After evacuating air in the chamber, Ar-5 per centN_2 gas was introduced at the rate of 500 mL/min under atmospheric pressure. Then, the sample was heated up to 1300°C at a heating rate of 10°C/min using an infrared image furnace and cooled down by turning off the power. The temperature at 1 mm upper the surface of the sample was monitored using an R-type thermocouple. The concentrations of CO, CO_2 and N_2 of the outlet gas were measured during the experiment at 90 sec intervals by a gas chromatography. N_2 gas was used as a tracer to estimate the amount of gas generated from the sample. Reduction degree (RD) of the sample was calculated by Equation 8 using the amounts of generated gas.

$$RD = \frac{M_{CO} + 2M_{CO_2}}{M_{\text{total O}}} \tag{8}$$

M_{CO} and M_{CO_2} are the molar amounts of CO and CO_2 gases detected by the gas chromatography, respectively. $M_{\text{total O}}$ is the molar amount of monatomic oxygen in the Fe_2O_3 reagent.

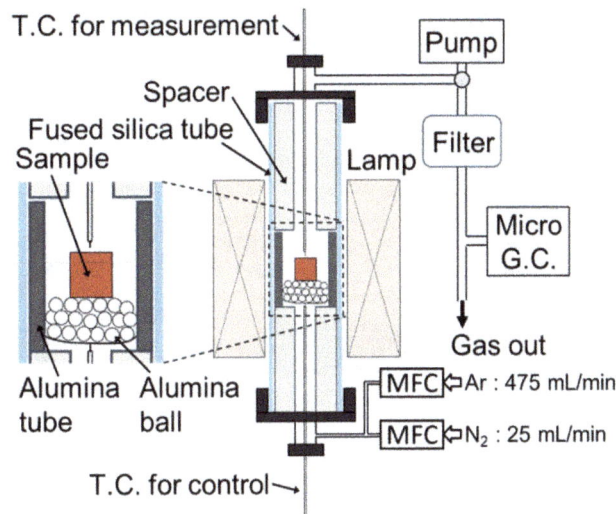

FIG 3 – Schematic diagram of an experimental apparatus for reduction experiment.

RESULTS AND DISCUSSION

In situ XRD measurement

Figure 4 shows changes in XRD profiles obtained for Sample b with temperature. The peaks of Fe_3C, Fe_5C_2 and Fe are detected at room temperature before heating up. However, the peaks of Fe_5C_2 are decreasing from 620°C and disappeared at 660°C. The peaks of Fe_3C, on the other hand, remains after heating up to 700°C. This indicates that the two different iron carbides obtained by carbon deposition reaction are expected to show different reactivity as a reducing and carburising agent. Fe_5C_2 is more easily decomposed than Fe_3C.

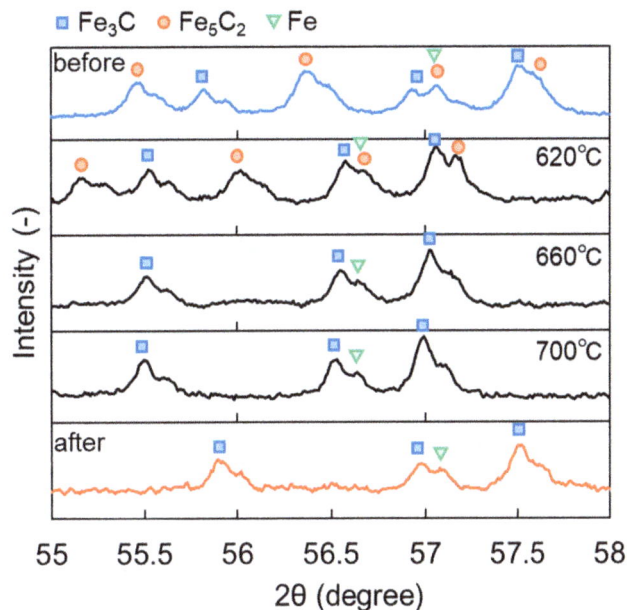

FIG 4 – XRD profiles of sample b heated up to 700°C and at room temperature before and after heating.

Reduction experiment using deposited carbon-iron oxide composite

The appearance of Fe_2O_3-θ and Fe_2O_3-χ after the reduction experiment are shown in Figure 5. Each sample shows metallic luster. Fe_2O_3-θ keeps cylindrical shape, while Fe_2O_3-χ becomes an iron nugget. This indicates the reactivity of iron carbide affects the melting behaviour of carbon-iron oxide composite.

FIG 5 – Appearances of Fe_2O_3-θ and Fe_2O_3-χ after reduction heated up to 1300°C.

Figure 6 shows the changes in the partial pressure ratio of CO to $CO+CO_2$ gas (CO gas ratio) generated from the composite, and the reduction degree of iron oxide with temperature, drawn on the Fe-O phase diagram. The equilibrium line of the Boudouard reaction ($C + CO_2 = 2CO$) is represented as dashed line in the diagram. The CO gas ratio obtained for Fe_2O_3-θ starts to be increase along with the equilibrium line of the Boudouard reaction. Then, the ratio reaches the Fe_3O_4/FeO equilibrium at 650°C and FeO/Fe equilibrium at 800°C. The ratio obtained for Fe_2O_3-χ, on the other hand, starts to increase above the equilibrium line of Boudouard reaction at 620°C, which corresponds to the decomposition of Fe_5C_2. This suggests the possibility of improving the reducibility of the carbon-iron oxide composite since the carbon generated by the decomposition of Fe_5C_2 is easily gasified. The ratio then directly reaches the FeO/Fe equilibrium at 720°C. The reduction degrees are drastically increased soon after the gas ratio moves to Fe-single phase region in both composites. The metallic iron has a catalytic effect for gasification of carbon and accelerates the reduction reaction of iron oxides (Higashi *et al*, 2023). Therefore, almost all of the iron oxides in the composites are reduced to metallic iron by heating up to 1300°C.

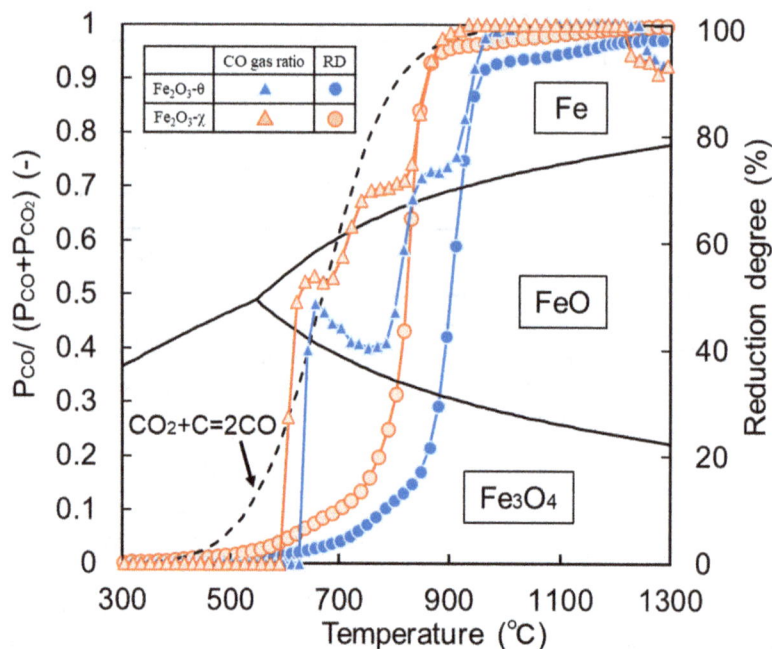

FIG 6 – Changes with temperature in the ratio of CO to $CO+CO_2$ for the Fe_2O_3-θ and Fe_2O_3-χ and in the reduction degree, plotted on the phase diagram of the Fe-O system.

CONCLUSIONS

In this study, the porous iron whiskers with high porosity were focused on, as a substrate for carbon deposition reaction. Furthermore, the reduction and melting behaviours of carbon-iron oxide composite prepared using recovered deposited carbon were investigated. The following results were obtained.

- Fe_3C is obtained as iron carbides by using CO gas for carburising the porous iron. Fe_5C_2 is also produced by using CO-CO_2-H_2 gas mixture.

- Fe_5C_2 is more easily decomposed than Fe_3C.

- Carbon-Iron oxide composite containing not only Fe_3C but also Fe_5C_2 shows rapid reduction reaction at lower temperature and complete melting of metallic iron at 1300°C.

ACKNOWLEDGEMENT

This work was supported by JSPS Grant-in-Aid for JSPS Fellows Grant Number 22KJ0282, Steel Foundation for Environmental Protection Technology and JFE 21st Century Foundation.

REFERENCES

Hasanbeigi, A, Arens, M and Price, L, 2014. Alternative emerging ironmaking technologies for energy-efficiency and carbon dioxide emissions reduction: A technical review, *Renew Sust Energ Rev*, 33:645–658.

Higashi, R, Maruoka, D, Kasai, E and Murakami, T, 2023. Low Temperature Reduction Mechanism of Carbon-Iron Ore Composite Using Woody Biomass, *ISIJ Int*, 63(12):1972–1978.

Kato, Y, 2010. Carbon Recycling for Reduction of Carbon Dioxide Emission from Iron-making Process, *ISIJ Int*, 50(1):181–185.

Kawashiri, Y, Nouchi, T and Kashihara, Y, 2022. Reduction of CO_2 Emissions from Blast Furnace with Carbon Recycling Methane, *JFE TECHNICAL REPORT*, 28:9–15.

Murakami, T and Nagata, K, 2003. New ironmaking process from the viewpoint of carburisation and iron melting at low temperature, *Extr Metall Rev*, 24:253–267.

Murakami, T, Ohno, M, Suzuki, K, Owaki, K and Kasai, E, 2017a. Acceleration of Carburisation and Melting of Reduced Iron in Iron Ore–Carbon Composite Using Different Types of Carbonaceous Materials, *ISIJ Int*, 57(11):1928–1936.

Murakami, T, Takahashi, T, Fuji, S, Maruoka, D and Kasai, E, 2017b. Development of Manufacturing Principle of Porous Iron by Carbothermic Reduction of Composite of Hematite and Biomass Char, *Material Transactions*, 58(12):1742–1748.

Nishihiro, K, Maeda, T, Ohno, K and Kunitomo, K, 2019. Effect of Concentration on Carbon Deposition Reaction by CO-H_2 Gas Mixture at 773 K to 973 K, *ISIJ Int*, 59(4):634–642.

Norgate, T, Haque, N, Somerville, M and Jahanshahi, S, 2012. Biomass as a Source of Renewable Carbon for Iron and Steelmaking, *ISIJ Int*, 52(8):1472–1481.

Robinson, R, Brabie, L, Pettersson, M, Amovic, M and Ljunggren, R, 2022. An Empirical Comparative Study of Renewable Biochar and Fossil Carbon as Carburiser in Steelmaking, *ISIJ Int*, 62(12):2522–2528.

Sato, K, Noguchi, T, Miki, T, Sasaki, Y and Hino, M, 2011. Effect of Fe_3C on Carburisation and Smelting Behavior of Reduced Iron in Blast Furnace, *ISIJ Int*, 51(8):1269–1273.

Sawai, S, Iguchi, Y and Hayashi, S, 1999a. Carbidization Rate of Reduced Iron with CO-CO_2 Gas Mixture (in Japanese), *Tetsu-to-Hagané*, 85(1):6–13.

Sawai, S, Iguchi, Y and Hayashi, S, 1999b. Carbidization Rate of Reduced Iron Ore in H_2-CO-CO_2 Gas Mixture (in Japanese), *Tetsu-to-Hagané*, 85(1):20–26.

Yokoyama, H, Higuchi, K, Ito, T and Oshio, A, 2014. Decrease in the Carbon Consumption of a Commercial Blast Furnace by Using Carbon Composite Iron Ore (in Japanese), *Tetsu-to-Hagané*, 100(5):601–609.

Electrolytic reduction of metal sulfides/oxides in molten salts for sustainable metal production

X Hu[1], L Sundqvist Ökvist[2] and J Björkvall[3]

1. Senior Researcher, SWERIM AB, Luleå 974 37, Sweden. Email: xianfeng.hu@swerim.se
2. Associate Professor, Luleå University of Technology, Luleå 971 87, Sweden.
   Email: lena.sundqvist-oqvist@ltu.se
3. Senior Researcher, SWERIM AB, Luleå 974 37. Email: johan.bjorkvall@swerim.se

ABSTRACT

The metal production industry is a significant contributor to global CO_2 emissions due to the use of fossil fuels such as coal and coke. To mitigate these emissions and meet climate goals, innovative and sustainable technologies are required. Molten salt electrolysis is a promising technology that directly produces metals from their precursor sulfides or oxides using electricity. When combined with renewable electricity and an inert anode, the electrolysis process can be carbon neutral.

This paper presents the results of two pilot-scale studies on the electrolytic reduction of metal oxides and sulfides in molten salts. The first study focuses on the electrolytic reduction of chalcopyrite in molten NaCl-KCl salt. The results demonstrate that *in situ* separation of copper, iron, and sulfur is possible, enabling the extraction of all valuable elements without CO_2 emissions. Furthermore, the findings underscore the capability to eliminate impurities like zinc, antimony, arsenic, and mercury from the electrolysis product. The second study investigates the electrolytic reduction of pure/synthetic chemicals of wüstite, hematite, and magnetite, as well as a magnetite-type iron ore in molten NaOH salt. The findings reveal a stepwise reduction of iron oxides from high valence to low valence, ultimately leading to the production of metallic iron electrolytically. Notably, this study underscores the challenges associated with the selection of an economically viable and durable inert anode material for efficient oxygen generation.

These results indicate that molten salt electrolysis provides a sustainable and green route for base metal production. The use of this technology has the potential to significantly reduce CO_2 emissions in the metal production industry, contributing to achieving climate goals.

INTRODUCTION

The base metal production industry is a major emitter of CO_2 due to its heavy reliance on fossil fuels for reduction and heating processes. For instance, steelmaking in blast furnaces and basic oxygen furnaces emits 1800 kg of CO_2 per metric tonne of steel produced (Suopajärvi *et al*, 2018). Achieving carbon neutrality in this industry requires transitioning to renewable energy sources or agents for metal extraction from compounds, typically sulfides and oxides. Promising candidates for renewable energy sources or agents include bio-based materials like biocarbon, green hydrogen, and renewable electricity. While bio-based materials show potential to replace fossil fuels in metallurgical processes, their supply is limited (Pei *et al*, 2020). Hydrogen, produced via water electrolysis using renewable electricity, serves as a green energy and reducing agent for metal oxide reduction. However, its application is limited by thermodynamic constraints, particularly in reducing metal sulfides and oxides like chromium oxide (Davies *et al*, 2022), manganese oxide (Safarian, 2022), and silicon oxide (Itaka *et al*, 2015). Moreover, the two-step process of hydrogen production and subsequent metal oxide reduction lowers process efficiency. Electrons derived from renewable electricity are potent reducing agents suitable for electrolysis processes to convert metal compounds to metals. Two notable examples of electrolysis approaches are molten salt electrolysis (FFC process (Chen and Fray, 2020; Mohandas and Fray, 2004)) and molten oxide electrolysis (MOE process (Allanore, Ortiz and Sadoway, 2011; Allanore, Yin and Sadoway, 2013; Sadoway, 2017)). The FFC process involves controlled liberation of oxygen or sulfur ions from metal oxides or sulfides within a molten salt electrolyte, primarily chloride salt, positioned at the cathode. This process yields sponge-like metal at the cathode while generating oxygen or sulfur at an inert anode (Cox and Fray, 2008; Tan *et al*, 2016). Operating at lower temperatures compared to conventional methods, the FFC process reduces energy consumption. The MOE process dissolves iron ore in a molten oxide electrolyte, leading to electrolytic reduction of iron ions and continuous liquid metal production with an inert anode and renewable electricity, thereby avoiding CO_2 emissions. Despite the environmental benefits, challenges remain in both processes, particularly in finding durable and economically viable anode materials (Allanore, Yin and Sadoway, 2013).

This study investigates the electrolytic reduction of chalcopyrite and iron oxides in molten NaCl-KCl and NaOH salts in a pilot-scale electrolysis reactor to further explore the potential of molten salt electrolysis for sustainable metal production.

METHODOLOGY

Materials and their preparation

Materials and preparation for electrolytic reduction of chalcopyrite/pyrite

Two samples, designated as Concentrate A and Concentrate B, were chosen for the electrolysis trials:

- Concentrate A, a chalcopyrite material, is composed of 33.9 per cent iron (Fe), 31.2 per cent copper (Cu), and 18.5 per cent sulfur (S).

- In contrast, Concentrate B, a pyrite material, contains 50.6 per cent Fe, 0.75 per cent Cu, and 22.7 per cent S.

These materials were subjected to thorough analysis utilising X-ray Fluorescence (XRF) for elemental composition, LECO combustion for carbon and sulfur content, and X-ray Diffraction (XRD) for mineralogical characterisation. The analyses confirmed that Concentrate A is primarily chalcopyrite ($CuFeS_2$), with a Fe to Cu ratio of approximately 1.1:1. Meanwhile, Concentrate B is predominantly composed of pyrite (FeS_2), with a substantially higher Fe to Cu ratio of 67.5:1 and a marked presence of impurities such as antimony (Sb) and arsenic (As).

To prepare these concentrates for the electrolysis experiments, they were processed into briquettes. This involved first sieving the concentrates to ensure 90 per cent of the particles were finer than 250 µm (D_{90} = 250 µm). Following sieving, the concentrates were mixed with 2 per cent bentonite and 2 per cent water to form a mixture. This mixture was then compacted under a pressure of 200 kN for 5 mins, resulting in briquettes measuring 40 mm in diameter and 6 mm in thickness (Φ40×6 mm). The final step involved drying and sintering the briquettes in an oven set at 150°C for about 40 hrs. Once completed, the briquettes were stored in plastic bags, ready for subsequent use in electrolysis processes.

Materials and preparation for electrolytic reduction of iron oxides

Wüstite, hematite, magnetite, and iron ore concentrate served as the primary materials in this study. Wüstite was synthesised through the thorough blending of chemical-grade iron powder and magnetite powder in a precise stoichiometric ratio. This mixture was then subjected to a heat treatment process in a vertical furnace at 900°C for 24 hrs, under a continuous flow of argon gas. The successful formation of wüstite was subsequently verified through XRD analysis. Both hematite and magnetite used in the experiments were of high purity, classified as pure chemicals. The iron ore concentrate (total iron content 64.6 per cent), characterised as a magnetite-type ore with a particle size where 80 per cent of the material is below 82 µm (D_{80} = 82 µm), was procured from a local iron ore producer. For the electrolysis experiments, the iron oxides underwent a process to form briquettes, employing a methodology akin to that previously described. This process ensures the creation of briquettes from the iron oxide materials to facilitate their use in the electrolysis experiments.

Experimental set-up and procedure

Figure 1 shows the experimental set-up for the electrolytic reduction of chalcopyrite and pyrite. A steel crucible (inner diameter 230 mm and height 508 mm) with lid was placed in a pit furnace and then the crucible was filled with ca. 25 kg mixture of chemical pure NaCl (purity > 98 per cent) and KCl (purity > 99.8 per cent) of equal mole. The salt in the crucible was heated to a desired temperature (around 680–870°C) into a molten salt at a heating rate of 10°C/min. Two electrodes (namely, anode and cathode) were placed in the molten salt. The anode is a graphite block connected on copper conductor; the cathode is a molybdenum net, which is used to hold the as-prepared briquettes. The net was made by folding molybdenum mesh into a bag hold either two or four briquettes. The net was connected to a stainless steel plate and then to a copper conductor. This is to avoid the damage of the molybdenum net and copper conductor, which could otherwise easily get oxidised at the molten slag/gas interface by oxygen and the generated sulfurous gas.

During the electrolysis the chamber above the molten salt was purged by N_2 or Ar gas in order to protect the electrodes from oxidation and expel the generated sulfurous gas. For electrolytic reduction of iron oxides, a similar experimental set-up, as it is shown in Figure 1, was applied. In this case, the steel crucible was filled with ca. 25 kg NaOH (purity > 99.8 per cent) and the operation temperature was set to be ca. 500°C.

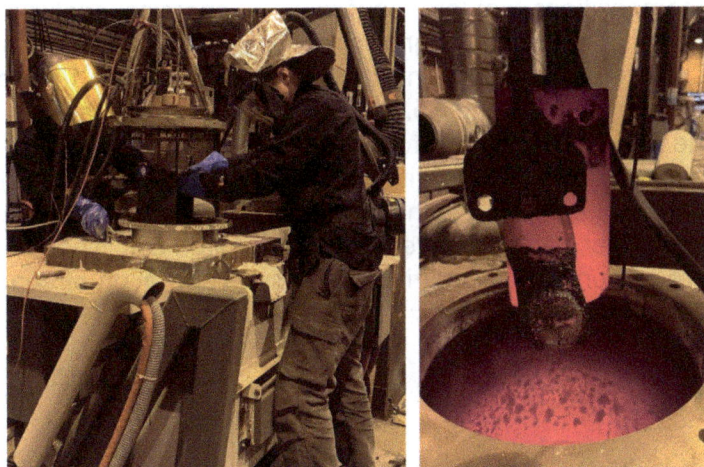

FIG 1 – Experimental set-up for molten salt electrolysis at SWERIM.

In all the electrolysis trials, constant voltage electrolysis was conducted. Each electrolysis trial began with a short series, during which the voltage was increased from 0 to 2.5 V for chalcopyrite electrolysis in molten NaCl-KCl salt or 1.7 V for iron oxides in molten NaOH salt. The electrolysis process was terminated after several hours. After each experiment, the molybdenum net with sample was detached from the cathode. Figure 2 shows the appearances of the obtained sample. The sample with the net was placed in distilled water. Then the glass was put in an ultrasonic water bath to make the salt in the sample dissolved in the water. Thereafter, the sample for the chalcopyrite/pyrite trials was able to be separated into two sub-samples. These samples were carefully washed with distilled water and filtrated several times to get rid of the salt. The sample from the iron oxide trials only has one single sample for each experiment. The resulting samples were then filtered out and dried in an oven at a temperature of 105°C for over 12 hrs to remove any remaining water. Selected samples were analysed by XRD, XRF and scanning electron microscope (SEM) methods.

FIG 2 – Appearance of the cathode sample after electrolysis.

RESULTS AND DISCUSSION

Electrolytic reduction of chalcopyrite/pyrite

Sulfur content in the sample or subsample is an important index to evaluate the extent of electrolytic reduction. A lower sulfur content in the sample indicates a higher degree of electrolytic reduction. After electrolysis, the sulfur contents in the samples were analysis by LECO analysis. It is seen that the samples had quite low S contents (as low as 0.01 per cent) compared to that in the original samples (18.5 per cent S in the chalcopyrite sample and 22.7 per cent S in the pyrite sample). This indicates an almost complete reduction of chalcopyrite/pyrite in the electrolysis trials.

Figure 3 shows an example of the obtained two sample fractions from one electrolysis trial using chalcopyrite. It is seen that the one sample fraction had reddish colour, while another sample fraction had dark green colour. The difference in colours in the two sample fractions is attributed to the difference in chemical compositions that later confirmed by chemical analysis. The reddish fraction is referred to as Cu-rich fraction and the dark green fraction as Fe-rich fraction. The chemical analysis results show that the Cu-rich fractions contain 43.1 per cent Cu and 40.1 per cent Fe, while the Fe-rich fractions contain 4.9 per cent Cu and 55.3 per cent Fe. The high amount of Fe in the Cu-rich fraction is attribute to the fact that during scratching the coagulated materials from the Mo net, some iron particles were also taken down. Further, it should be noted that the collected Cu-rich fraction is far below the balanced amount according to the amount of the charged materials during electrolysis. This is due to that some copper particles were lost in the molten salt. These results clearly indicate that during electrolytic reduction of chalcopyrite, the *in situ* separation of copper, iron and sulfur is possible.

FIG 3 – Appearances of the obtained samples of iron-rich and copper-rich fractions from one electrolysis trials using chalcopyrite.

After electrolytic reduction of pyrite, it was found that the materials coagulated on the Mo were found quite less and not be able to be collected. Therefore, this electrolysis sample was taken a whole and analysed by XRD and XRF methods. By combining the LECO analysis result for the same sample, it can conclude that iron and copper were completely reduced after the electrolysis of pyrite, the leading phase in the obtained product is iron. The SEM analysis of the reduced pyrite sample shows that the reduced iron particles have a cubic shape in the scale of several to several tens of microns.

It is known that in the used pyrite sample it contains several trace elements, the concentrations of which are in noticeable level. To evaluate how these trace elements behave during the electrolysis, the removal efficiency of these trace elements was calculated, by combining the chemical compositions of the original pyrite and the reduced products. It is seen that elements Zn, Sb, As and Hg can be removed; elements Pd and Cd can be partially removed after electrolysis.

Electrolytic reduction of iron oxides

During electrolytic reduction of wüstite, it is found that wüstite was reduced to iron in a single step. However, for the electrolytic reduction of hematite, magnetite, and magnetite ore, co-existence of FeO, Fe_2O_3, and Fe_3O_4 were found in the incompletely reduced samples. This indicates a stepwise reduction of iron oxides from high valence to low valence, ultimately leading to the production of metallic iron electrolytically. It should be noted, during electrolytic reduction of iron oxides, graphite was used as the anode. This has led to the generation of CO/CO_2. The future research focus will be investigating the use of inert anode to reach the carbon neutrality of the process. Scaling up the process to a scale comparable to the conventional ironmaking process can be quite challenging; however, the process may still find its use to produce high-quality iron product or product other metal products that cannot be produced in a conventional or sustainable way.

CONCLUSIONS

A pilot-scale molten salt electrolysis reactor was established at SWERIM to study the molten salt electrolysis of sulfur ore, chalcopyrite and pyrite, and various iron oxides, including wüstite, hematite, magnetite, and magnetite-type iron ore. The results from electrolytic reduction of chalcopyrite/pyrite demonstrate that *in situ* separation of copper, iron, and sulfur is possible, enabling the extraction of all valuable elements without CO_2 emissions. Furthermore, the findings underscore the capability to eliminate impurities like zinc, antimony, arsenic, and mercury from the electrolysis product. The results from electrolytic reduction of iron oxides/ore demonstrate the stepwise reduction of iron oxides from high valence to low valence, ultimately leading to the production of metallic iron.

ACKNOWLEDGEMENTS

The work on electrolytic reduction of chalcopyrite/pyrite was funded by Hugo Carlsson foundation via the Swedish Iron producers' Association (Jernkontoret). The work on electrolytic reduction iron oxides/ore is funded by the Swedish Energy Agency (Energimyndigheten). Centre for Advanced Mining and Metallurgy, is acknowledged for supporting the work for preparing this manuscript.

REFERENCES

Allanore, A, Ortiz, L A and Sadoway, D R, 2011. Molten Oxide Electrolysis for Iron Production: Identification of Key Process Parameters for Large scale Development, *TMS Annual Meeting*, pp 121–129.

Allanore, A, Yin, L and Sadoway, D R, 2013. A new anode material for oxygen evolution in molten oxide electrolysis, *Nature*, 4977449:353–356. https://doi.org/10.1038/nature12134

Chen, G Z and Fray, D J, 2020. Invention and fundamentals of the FFC Cambridge Process, *Extractive Metallurgy of Titanium: Conventional and Recent Advances in Extraction and Production of Titanium Metal*, chpt 11, pp 227–286. https://doi.org/10.1016/B978-0-12-817200-1.00011-9

Cox, A and Fray, D J, 2008. Electrolytic formation of iron from haematite in molten sodium hydroxide, *Ironmaking and Steelmaking*, 35(8):561–566. https://doi.org/10.1179/174328108X293444

Davies, J, Paktunc, D, Ramos-Hernandez, J J, Tangstad, M, Ringdalen, E, Beukes, J P, Bessarabov, D G and Du Preez, S P, 2022. The Use of Hydrogen as a Potential Reductant in the Chromite Smelting Industry, *Minerals*, 12(5). https://doi.org/10.3390/min12050534

Itaka, K, Ogasawara, T, Boucetta, A, Benioub, R, Sumiya, M, Hashimoto, T, Koinuma, H and Furuya, Y, 2015. Direct carbothermic silica reduction from purified silica to solar-grade silicon, *Journal of Physics: Conference Series*, 596(1). https://doi.org/10.1088/1742-6596/596/1/012015

Mohandas, K S and Fray, D J, 2004. FFC Cambridge process and removal of oxygen from metal-oxygen systems by molten salt electrolysis: An overview, *Transactions of the Indian Institute of Metals*, 57(6):579–592.

Pei, M, Petäjäniemi, M, Regnell, A and Wijk, O, 2020. Toward a Fossil Free Future with HYBRIT: Development of Iron and Steelmaking Technology in Sweden and Finland, *Metals*, 10(7):972. https://doi.org/10.3390/met10070972

Sadoway, D R, 2017. Towards Carbon-free steelmaking by molten oxide electrolysis, Presentation. http://web.mit.edu/course/3/3.a30/www/refs/present_env_brf_sadoway.pdf

Safarian, J, 2022. A sustainable process to produce manganese and its alloys through hydrogen and aluminothermic reduction, *Processes*, 10(1). https://doi.org/10.3390/pr10010027

Suopajärvi, H, Umeki, K, Mousa, E, Hedayati, A, Romar, H, Kemppainen, A, Wang, C, Phounglamcheik, A, Tuomikoski, S, Norberg, N, Andefors, A, Öhman, M, Lassi, U and Fabritius, T, 2018. Use of biomass in integrated steelmaking – Status quo, future needs and comparison to other low-CO_2 steel production technologies, *Applied Energy*, 213(January):384–407. https://doi.org/10.1016/j.apenergy.2018.01.060

Tan, M, He, R, Yuan, Y, Wang, Z and Jin, X, 2016. Electrochemical sulfur removal from chalcopyrite in molten NaCl-KCl, *Electrochimica Acta*, 213:148–154. https://doi.org/10.1016/j.electacta.2016.07.088

Aluminothermic production of silicon using different raw materials

K Jakovljevic[1], N Simkhada[2], M Zhu[3], M Wallin[4] and G Tranell[5]

1. Researcher, Norwegian University of Science and Technology, Trondheim 7034, Norway. Email: katarina.jakovljevic@ntnu.no
2. Research Engineer, Norwegian University of Science and Technology, Trondheim 7034, Norway. Email: nishan.simkhada@ntnu.no
3. Postdoctoral Researcher, Norwegian University of Science and Technology, Trondheim 7034, Norway. Email: mengyi.zhu@ntnu.no
4. Researcher, Norwegian University of Science and Technology, Trondheim 7034, Norway. Email: maria.wallin@ntnu.no
5. Professor, Norwegian University of Science and Technology, Trondheim 7034, Norway. Email: gabriella.tranell@ntnu.no

ABSTRACT

Silicon is a vital element in many products today, such as electronic components, solar devices, high-quality alloys, and many others. The growing global demand highlights the need for the development of sustainable production methods to meet this demand as an alternative to the current carbothermic reduction, submerged arc furnace (SAF) based process. An alternative to this is the aluminothermic reduction of quartz in a CaO-SiO_2 slag, which not only reduces direct carbon dioxide emissions but also promotes the utilisation of secondary raw materials such as quartz fines, aluminium dross and scrap as well as secondary alumina (SA) from dross recycling.

In the current study, the effects of SA and CaF_2 additions to slag on the resulting metal composition and metal yield were explored. Results were compared with thermodynamic simulations using FactSage™ 8.1.

Experimental results show that, in agreement with thermodynamic simulations, the silicon content of the alloy is increased, while the Ca is decreased for starting slags where CaO-SiO_2 is partly replaced by CaF_2. Similarly, the addition of SA to the initial slag results in an alloy with a higher silicon content.

INTRODUCTION

Silicon is the second most abundant element in the Earth's crust and one of the most important elements in high-tech applications. Traditionally, metallurgical-grade silicon (MG-Si), typically containing 96–99 per cent Si, is produced by reducing silicon dioxide (SiO_2) with carbon in a submerged arc furnace (SAF) (Schei, Tuset and Tveit, 1998). In this process, a mixture of quartz (crystalline SiO_2) and carbon (coke, coal, charcoal, and/or wood chips) is heated to produce silicon. The energy is supplied through electrodes, and it takes 10–13 MWh to produce one tonne of silicon. Using carbon materials as reductants leads to emissions primarily of carbon dioxide (approximately 5 t CO_2 per tonne of silicon produced), NO_x, SO_x, methane, polycyclic aromatic hydrocarbons (PAH) etc (Kero, Grådahl and Tranell, 2017). In addition to the high greenhouse gas (GHG) emissions and high energy demands, the requirements for the quartz used in the process need to be met. Aside from purity, there are strict requirements for size since particles that are too small reduce gas permeability in the furnace, increasing the likelihood of blow-outs due to gas build-up, which means that sand of high purity cannot be used in the traditional production process (Schei, Tuset and Tveit, 1998).

Aluminium, having a stronger affinity for oxygen than silicon, may hence be utilised as a reducing agent for the production of silicon, and this process can be considered a viable alternative to the carbothermic process. The SisAl process, based on the aluminothermic reduction of silica in slag, aims to produce silicon in a more environmentally beneficial way (Tranell, Safarian and Wallin, 2020). This is achieved by eliminating the need for primary carbon reductants and lumpy quartz raw materials and using residues from the silicon and aluminium industries instead. No carbon use means no direct emission of CO_2 and other gaseous pollutants.

In the SisAl process, SiO_2 can be used in the form of fines to produce slag, which means that less costly quartz sand can also be used as a SiO_2 source. By introducing an aluminium source (end-of-life scrap, dross) for the reduction of SiO_2 in $CaO\text{-}SiO_2$ slag, silicon alloy and $Al_2O_3\text{-}CaO$ slag are produced according to the following equation:

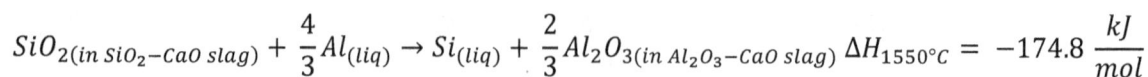

$$SiO_{2(in\ SiO_2-CaO\ slag)} + \frac{4}{3}Al_{(liq)} \rightarrow Si_{(liq)} + \frac{2}{3}Al_2O_{3(in\ Al_2O_3-CaO\ slag)}\ \Delta H_{1550°C} = -174.8\ \frac{kJ}{mol}$$

The generated slag is isolated from the metal, and the CaO and Al_2O_3 may be separated through a hydrometallurgical technique. CaO can be reintroduced back into the SisAl process, while the Al_2O_3 can be sent to primary aluminium production. Due to lower operating temperatures and an exothermic reduction reaction, the SisAl process consumes less energy compared to the carbothermic reduction in SAF.

A diverse range of waste by-products is generated in the aluminium sector. Dross is a byproduct created during the handling of liquid alloy and is a mixture of Al alloy and oxides as well as some amount of carbides, nitrides etc) and is formed as the melt oxidises. The quantity of waste material produced during the process of aluminium production is influenced by factors such as the quality and type of raw materials, operation conditions, the operator's skill, the kind of furnace utilised, and the dimensions, types, and techniques of alloying (Lazzaro, Eltrudis and Pranovi, 1994; Yoshimura *et al*, 2008). Approximately 15–25 kg of dross is produced for every tonne of molten aluminium (Liu and Chou, 2013).

The primary objective of the present study was to investigate the impact of various input materials on the overall process, with a particular emphasis on the final products. This study includes an examination of the suitability of aluminium by-products, such as dross and secondary alumina from treated dross obtained from Befesa Alumino, Spain, as reductants and additions in the process of aluminothermic production of silicon in terms of alloy composition. Recovery of materials in dross at Befesa is made through two cycles; in the first, the aluminium contained in the dross is recovered in a salt-based rotary furnace process, producing aluminium alloys, and in the second, recycling of salt slags, the waste that was traditionally discarded in landfills is recovered, transforming them into secondary raw materials, applicable in different industrial sectors. The recovered high-alumina fraction, in the following called secondary alumina (SA), contains some amount of residual fluorine, as discussed below.

The addition of CaF_2 to silicate slags typically decreases the viscosity and lowers the melting temperature (Park, Min and Song, 2002). As such, it was of interest to investigate the effect of limited fluorine additions on the SisAl process and its product compositions.

EXPERIMENTAL PROCEDURE

Overview of experimental procedure

The study on the impact of different input materials and process parameters on the overall process of aluminothermic production of silicon is organised into two subsections, as shown in the flow chart (Figure 1).

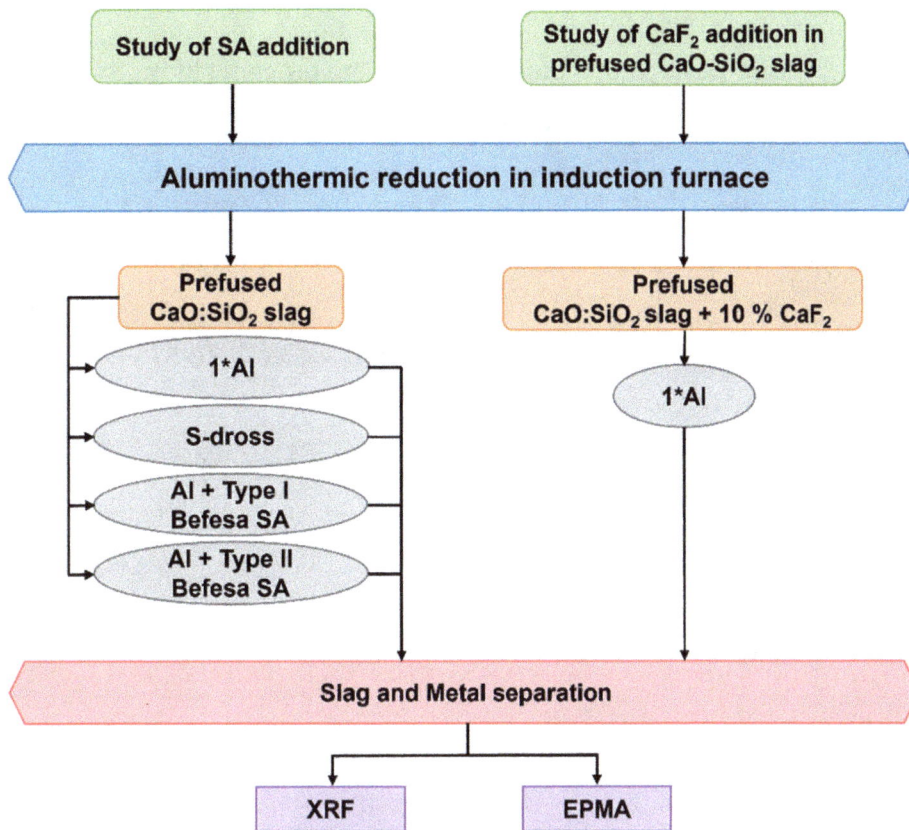

FIG 1 – The flow chart representing an overview of the experimental procedure.

Materials

The pre-fused slag utilised in this study as a SiO_2 source had a $CaO:SiO_2$ ratio of 1.018. A fine fraction of dross (denoted S-dross), with particle size ≤5 mm, which contains 30 per cent Al and 70 per cent Al_2O_3, and pure aluminium (~99 per cent pure) was acquired. Two different SA samples were used along with pure aluminium as a reductant in the study of SA addition. The chemical composition of the materials used is presented in Table 1.

TABLE 1

Composition of materials used in the present study.

Species	Material composition (%)		
	Prefused CaO-SiO$_2$ slag	Befesa Type I SA	Befesa Type II SA
CaO	49.9	2.76	3.56
MgO	0.42	7.69	6.14
SiO$_2$	49.00	4.97	10.57
Al$_2$O$_3$*	0.43	75.20	67.07
Fe$_2$O$_3$	0.15	2.14	2.39
K$_2$O	0.03	0.90	1.14
SO$_3$	0.03	~	~
Cl	~	0.82	0.59
CuO	~	0.65	0.82
F	~	0.85	2.40
MnO	~	0.34	0.35
Na$_2$O	~	1.18	2.48
NiO	~	0.04	0.05
P$_2$O$_5$	~	0.09	0.10
TiO$_2$	~	0.95	0.76
ZnO	~	0.25	0.35
S	~	0.07	0.11
Cr$_2$O$_3$	~	~	~
ZrO$_2$	~	~	~
SrO	~	~	~
BaO	~	~	~
Co$_3$O$_4$	~	~	~

* The values for Befesa Type I SA and Befesa Type II SA include both Al$_2$O$_3$ and metallic aluminium converted to oxide. The metallic aluminium content is estimated to be between 2.6 and 5.3 per cent, based on Befesa analysis.

Apparatus and procedure

The aluminothermic reduction tests in this work were carried out in an Induction Furnace (75 kW, 3000 Hz). Each set of experiments utilised three small-sized resin crucibles, which were placed within a single large crucible and then inserted into the furnace. The inner sides and the base of the crucible were wrapped with graphite paper to prevent input from interacting with the graphite and to facilitate the effortless removal of the final product from the crucible. A C-type thermocouple was used to measure the temperature. The pre-fused slag (with CaF$_2$ in the second set of experiments) and reductant mixture were kept in crucibles, heated to a selected temperature, and held for a specific time. Throughout the experiment, argon gas was continuously purged to maintain an inert environment inside the system.

The experimental matrix of the present study is shown in Table 2. Prior to the experiment, dross, SA, and pre-fused CaO-SiO$_2$ slag were dried to remove all the moisture content. Pure aluminium was used along with Befesa SA in relative weights to mimic the same Al:Al$_2$O$_3$ ratio as that in the S-dross.

TABLE 2

Experimental matrix of the present study.

Study	Type	Expt no	SiO₂/slag source		Reductant/addition			Process parameters	
			Prefused CaO:SiO₂ slag (g)	CaF₂ (g)	Pure Al (g)	S-dross (g)	SA (g)	Temp (°C)	Holding time (min)
Study of SA addition	Pure Al	1	141.9	-	41.58	-	-	1650	30
		2	141.6	-	41.49	-	-	1650	30
	S-dross	1	100.03	-	-	87.7	-	1700	60
		2	100.03	-	-	87.7	-	1700	60
		3	100.02	-	-	87.7	-	1700	60
	Al + Befesa Type I SA	1	100.03	-	26.46	-	82.08	1650	60
		2	100.05	-	26.66	-	82.00	1650	60
		3	100.03	-	26.40	-	82.05	1650	60
	Al + Befesa Type II SA	1	100.02	-	26.60	-	92.03	1650	60
		2	100.17	-	26.65	-	92.15	1650	60
		3	100.02	-	26.58	-	92.05	1650	60
Study of CaF₂ addition in prefused CaO-SiO₂ slag	Effect of CaF₂	1	90	10	25.98	-	-	1650	60

Characterisation

Electron probe microanalyzer (EPMA)

Metal samples were investigated using a JXA-8500F Field Emission Electron Probe Microanalyzer (EPMA). Backscattered electrons were used for sample imaging to visually represent different phases in the metal sample. EPMA's focus was on determining the elemental distribution and composition of silicon metal produced. Wavelength dispersive X-ray spectrometer (WDS) helped in performing elemental analysis of the different phases, while energy-dispersive X-ray spectrometer (EDS) was used over a particular area to determine an estimate of the element distribution of the different main elements.

X-ray fluorescence (XRF)

Samples from the Study of CaF₂ addition in prefused slag were also analysed using the X-ray fluorescence (XRF) technique at Degerfors Laboratorium AB in Bruksparken, Sweden. The slag and alloy samples were analysed using Thermo Fischer Scientific's ARL 9900 Series XRF device. The samples obtained from tests using pure Al and S-dross were analysed at RWTH, IME-Process Metallurgy and Metal Recycling, Aachen, Germany, using the RFA Omnian 37/S via combustion method. Similarly, the Department of Geoscience and Petroleum (IGP) at NTNU in Trondheim, Norway, conducted an analysis of the samples obtained from trials using Befesa Type I and Type II SA.

Theoretical calculations- FactSage™

Thermodynamic simulations were performed using the software FactSage™, version 8.1 (by Thermfact Ltd. and GTT-Technologies) with databases FToxid and FTlite (see https://www.crct.polymtl.ca/fact/documentation/FSData.htm).

RESULTS AND DISCUSSION

Results from SA addition experiments

This section gives details regarding the observations derived from experiments conducted using different Al sources. These observations encompass the resulting product's physical characteristics, the alloy's yield, and its composition.

Physical structure

This section presents the variations in the physical structure and appearance of aluminothermic reduction products obtained using different reductants.

It was observed that the structure of almost all the samples upon dissection exhibited a consistent pattern: a hemispherical Si alloy located at the uppermost part, with a slag layer positioned beneath the silicon metal (Figure 2). However, in some products, the silicon alloy was not observed at the top. In those products, the silicon metal was slightly below the top portion, entirely encircled by slag in every direction. When pure aluminium was employed as a reducing agent, the resulting product consisted entirely of a silicon alloy at the uppermost section and a slag at the bottom (Figure 2a). In the experiment where S-dross was used as a reductant, tiny drops of silicon alloy were seen above the topmost silicon part, as depicted in Figure 2b. The slag obtained was greenish brown in colour, which deviated slightly from the slags produced in experiments involving alternative reductants. XRF analysis of the slag revealed that the amount of Si in the slag from the trial with S-dross is relatively lower than that using pure Al and Befesa SA, which can be the reason for such appearance of slag. When SA, both Type I and Type II were utilised as a reductant, the topmost part was not entirely covered with silicon alloy. Observations revealed the presence of circular silicon alloys encased by slag. In one sample from the trial involving Al+Type II SA as a reductant, the silicon alloy was found to be positioned slightly below the top layer (Figure 2d). XRF analysis of SA identified several other impurities like Cl, CuO, F, Fe_2O_3, K_2O, and MgO. These impurities might have hindered the rapid reduction reaction at 1650°C and the non-uniform deposition of the reduced Si alloy.

FIG 2 – Different structures of products obtained from aluminothermic reduction with the use of: (a) pure Al, (b) S-dross, (c) Al +Type I SA, and (d) Al +Type II SA.

Metal yield and mass loss

The loss in mass was calculated from the difference between the total input feed and output. The amount of silicon alloy yielded/added Al unit from each crucible, where different reductants were used, was measured. The percentage mass loss and the average amount (from three parallels) of Si per unit Al reductant produced using different reductants are shown in Figure 3.

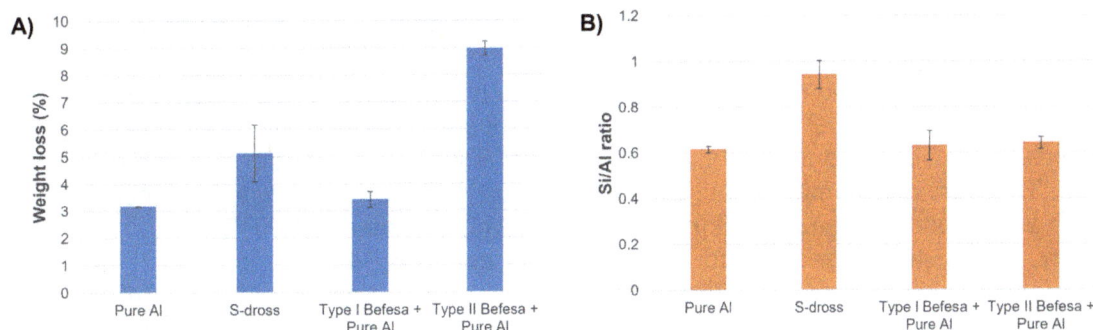

FIG 3 – (a) Mass loss, and (b) Si/Al ratio in experiments using different reductants. The values shown for reductant are the average of three trials conducted, except for pure Al, where values represent the average of two trials.

The Si/Al ratio, when the reductant used was S-dross, was higher than that yield obtained in all other experiments, while the trials using SA gave a comparatively similar but slightly higher metal output than the experiments using pure Al. Many small silicon droplets were present in different locations of the slag, making it difficult to collect the entire silicon alloy produced and consequently leading to comparatively high metal loss.

The weight loss could be attributed to a combination of SiO production/losses and the presence of halides and as such, since SA samples contain more volatiles (alkali, F etc), evaporation is expected to be higher for these materials, leading to a considerably higher loss in weight when SA was utilised.

Chemical composition

Figure 4 displays the predicted chemical composition of metal and slag by FactSage™ and the EDS results for metal and XRF results for slag as mean values of three areas measured in each of the three crucibles, in which different reductants were used.

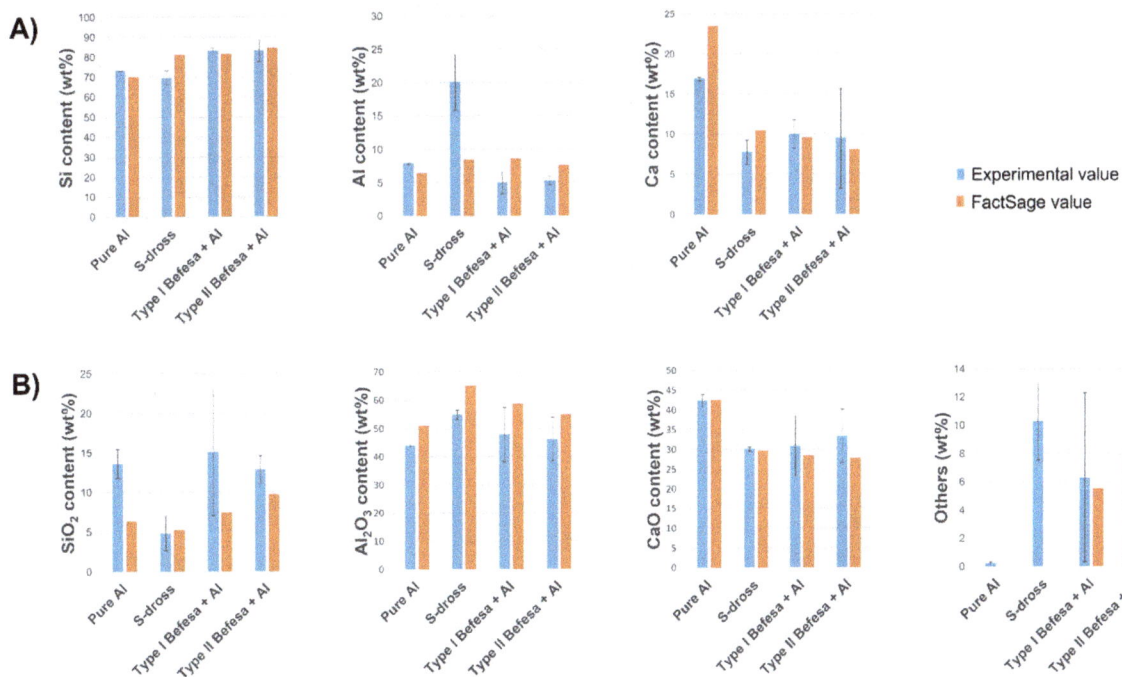

FIG 4 – The chemical composition of (a) metal and (b) slag obtained from experiments utilising different reductants, compared with the chemical composition based on FactSage™ simulations. The given experimental values represent the average of three trials for each experiment, except for the one with pure aluminium as a reductant, where values represent the average of two trials.

The composition of metal samples from each crucible was analysed with EDS. The samples produced in experiments employing pure aluminium had a higher Ca content and lower or similar Al content than other samples, which mostly aligned with the observed pattern in the FactSage™ simulation. The reverse result was obtained from experiments where S-dross was used, ie the content of Al was higher, and the content of Ca was lower. Since the metallic Al accounts for 2.6–5.3 per cent of SA, it has no significant impact on the graph. While the concentration trend for the produced metal in each experiment aligns with the simulation pattern, the most significant deviation in concentration is observed for S-dross, ie Si content is lower while Al content is much higher than predicted. This occurrence may be attributed to the composition of S-dross. The metallic Al content in S-dross was found by dissolving it in a NaCl-KCl-CaF$_2$ salt melt at 800°C, a method with certain inaccuracy. Those values are taken for Al content for FactSage™ calculation, which may, due to the heterogenous nature of dross, have affected the variation between experimental and predicted values. By using the two types of SA as additions to metallic Al, a metal with a higher Si content and a lower Al and Ca content was obtained compared to the metal obtained using pure Al.

An XRF analysis of SiO$_2$, Al$_2$O$_3$, and CaO content in the slag was conducted to examine the composition of the slag formed in experiments utilising different reductants. The remaining slag contained various impurities such as Cl, Mg, P, K, Fe, etc. No significant differences were seen while comparing the SiO$_2$, Al$_2$O$_3$, and CaO content in these samples from trials with different reductants. However, the composition of SiO$_2$ in slag samples from experiments using S-dross as a reductant was somewhat lower than the SiO$_2$ content present in other samples. As in the case of metal samples, the concentration trend aligns well with the FactSage™ simulation, with a slight deviation in concentration values compared to the predicted ones.

BSE images

EPMA analysis of metal was conducted to provide backscatter electron imaging of the phases present. A significant difference in phase composition was not observed, as most of the EPMA results for metal showed similar phases (Figure 5a–5d). Regardless of the type of reductant used, three distinct phases were consistently seen in all metal samples: Si, Si$_2$Al$_2$Ca, and Si$_2$Ca. Agreeing with the higher Si content measured by XRF, the presence of the Si phase was more prominently observed in BSE images of metal samples from trials with SA Type I and Type II. In the metal samples obtained from experiments using S-dross as a reductant, traces of Si$_{13}$AlMnFe$_4$ were observed, resulting from impurities in the dross.

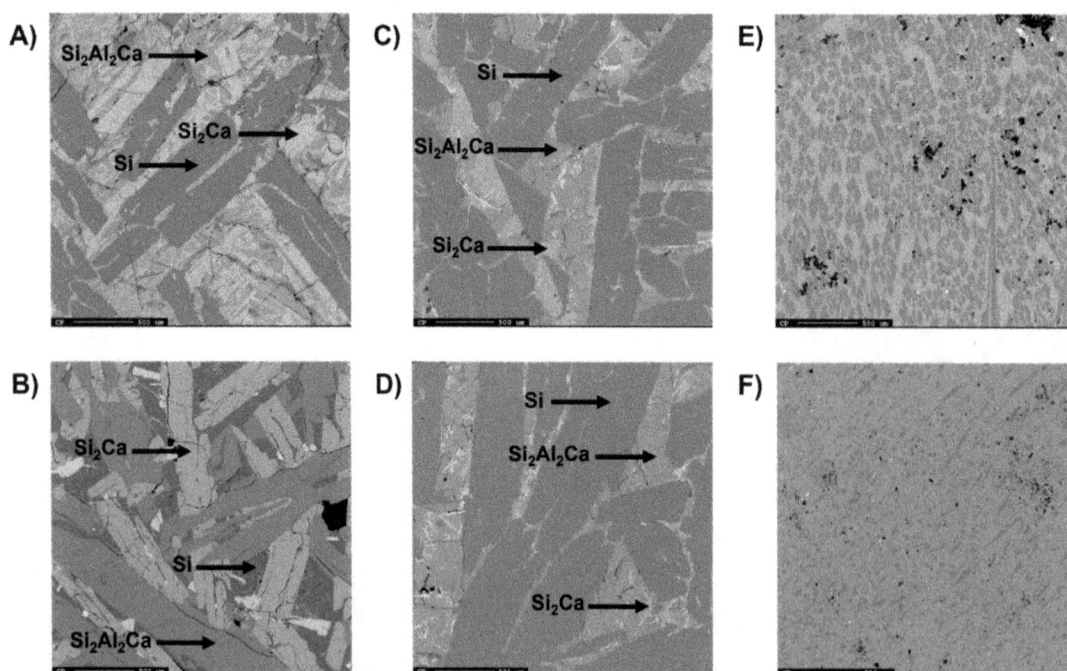

FIG 5 – BSE-image of: (a) metal from Pure Al experiment, (b) metal from S-dross experiment, (c) metal from Type I SA + Pure Al experiment, (d) metal from Type II SA + Pure Al experiment, (e) slag from Type I SA + Pure Al experiment, and (f) slag from Type II SA + Pure Al experiment.

The BSE images of the slags obtained from experiments using two different reductants, Type I SA and Type II SA, revealed the presence of two distinct phases (Figure 5e and 5f). One phase, characterised by a darker appearance, mainly consisted of Al_2O_3-CaO, with Al_2O_3 being the predominant component, while the second phase, appearing lighter in colour, was primarily composed of CaO-Al_2O_3, with CaO being the dominant constituent. Small concentrations of SiO_2 (lighter) were rarely seen on BSE images of slag samples from trials with Type II SA, as EDS analysis revealed the composition of SiO_2 to be much lower compared to slag from experiments with Type I SA.

Results from experiments using CaF2

The results of an aluminothermic reduction experiment in which a set concentration of CaF_2 was added to the slag included observations of its physical structure, the metal yield determination, and the chemical composition of the alloy and slag produced. A comparison was also made between an aluminothermic reduction experiment using CaF_2 in slag and an experiment without using CaF_2. Furthermore, this study's findings encompass comparing the composition of metal produced and simulated composition using FactSage™ software.

The product obtained from the experiment using CaF_2 in pre-fused slag with stoichiometric aluminium as a reductant exhibits a similar structure to a product from the experiment without CaF_2 (Figure 2a). A rigidly structured product contains hemispherical-shaped silicon alloy at the top and heavily dense bluish-grey slag at the bottom, as shown in Figure 6.

FIG 6 – Physical structure of the product obtained from the experiment using CaF_2 in prefused slag and pure aluminium as a reductant.

A loss in mass and amount of Si per unit Al reductant, compared with the experiment without using CaF_2, are shown in Figure 7a. Compared with experiments with stoichiometric Al and rec slag without CaF_2, weight loss has been significantly lower in experiments using CaF_2. Weight loss might be caused by some formation/losses of SiO. However, calculating total SiO loss accurately by FactSage™ is not trivial and will be the focus of future studies. Si yield from the product is similar for both cases, as a stoichiometric reductant in the same amount was used in both.

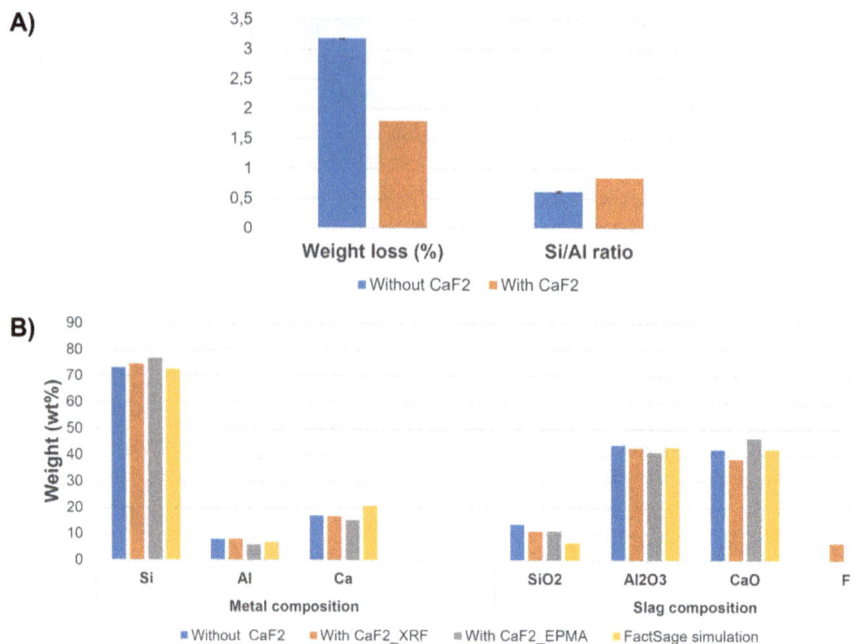

FIG 7 – (a) Mass loss and metal yield, and (b) chemical composition of metal and slag obtained from the experiment using CaF_2 in prefused slag compared to FactSage™ simulation values, and the metal and slag composition obtained from the experiment without CaF_2.

The slag and metal obtained from this experiment were subjected to both XRF and EPMA analysis, and results were compared with the experiment without using CaF_2 and also with FactSage™ simulation values, as shown in Figure 7b. When compared to the experiments without CaF_2, minor differences in metal and slag composition were seen, ie a slight increase in Si content, while the Al and Ca content in metal decreased in the experiment where CaF_2 was used. In the case of slag, the results are similar to experiments without CaF_2; however, SiO_2 and Al_2O_3 contents were slightly lower. When comparing XRF and EPMA results, the only difference was a slightly lower concentration of CaO in slag in the XRF analysis results. The XRF analysis of the slag obtained from this experiment indicates the presence of 6.63 wt per cent of F.

The experimental results closely resemble the predictions made by the FactSage™ simulation but with a slightly elevated Si concentration and reduced Al and Ca content in the metal. The same is the case with the content of SiO_2 in the slag, but the opposite is true in the case of CaO, where the FactSage™ simulation predicted a lower value than the one obtained by the EPMA analysis.

Backscattered image obtained from EPMA for metal sample obtained in this experiment is shown in Figure 8.

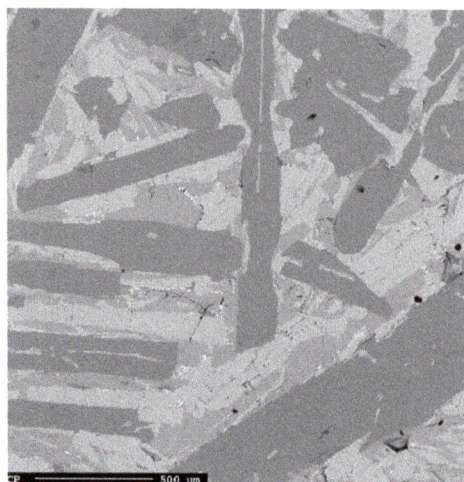

FIG 8 – BSE image of metal obtained from an aluminothermic reduction experiment in which CaF_2 was added to the silicate slags.

Three dominant phases have been identified in metal, similar to what we observed in the metal produced in experiments without CaF_2. The portion of dark-coloured Si is relatively equal in both cases, which is expected considering the similar Si content observed in XRF analysis results. Si_2Ca and Si_2Al_2Ca phases have been abundant, whereas only a few minor phases of impurities such as Fe, Mn, and Mg have been detected.

CONCLUSIONS

The primary aim of this study was to assess the feasibility of producing Si via an aluminothermic reduction process, focusing on achieving a high yield and minimising the calcium content. This study focused on investigating the impact of several factors on the aluminothermic reduction process, including the usage of different aluminium by-products as reductants/additions and the influence of calcium fluoride.

Using SA combined with pure aluminium as a reductant resulted in a slightly higher purity of Si alloy than using other reductants. When pure aluminium was used as a reductant, the alloy produced had a higher concentration of Ca. When S-dross was added as a reductant, the situation was reversed, but the Al concentration was elevated. The dross contains a significant amount of Al_2O_3, so its use resulted in an increase in the concentration of Al_2O_3 in the slag.

Results from the study of CaF_2 addition also showed constructive results in obtaining pure Si. A small mass loss and high metal yield were observed from the experiment using pre-fused slag with 10 per cent CaF_2. Si content in metal has been increased with the use of CaF_2. At the same time, Al and Ca content has been reduced, which is a favourable outcome.

ACKNOWLEDGEMENTS

This work was funded by the EU H2020 through the research project 'SisAl Pilot' (Grant No 869268).

REFERENCES

Kero, I, Grådahl, S and Tranell, G, 2017. Airborne Emissions from Si/FeSi Production, *JOM*, 69(2):365–380.

Lazzaro, G, Eltrudis, M and Pranovi, F, 1994. Recycling of aluminium dross in electrolytic pots, *Resources, Conservation and Recycling*, 10(1):153–159.

Liu, N W and Chou, M S, 2013. Reduction of secondary aluminum dross by a waste pickling liquor containing ferrous chloride, *Sustainable Environment Research*, 23(1):61–67.

Park, J H, Min, D J and Song, H S, 2002. The effect of CaF_2 on the viscosities and structures of $CaO-SiO_2(-MgO)-CaF_2$ slags, *Metallurgical and Materials Transactions B*, 33(5):723–729.

Schei, A, Tuset, J K and Tveit, H, 1998. *Production of High Silicon Alloys* (Tapir: Trondheim).

Tranell, G, Safarian, J and Wallin, M, 2020. SisAl - A New Process for Production of Silicon, in *Silicon for the Chemical and Solar Industry XV*, pp 129–139 (The Norwegian University of Science and Technology: Trondheim).

Yoshimura, H N, Abreu, A P, Molisani, A L, de Camargo, A C, Portela, J C S and Narita, N E, 2008. Evaluation of aluminum dross waste as raw material for refractories, *Ceramics International*, 34(3):581–591.

Manufacturing of FeSiB high-temperature phase change material by silicothermic reduction

J Jiao[1], M Wallin[2], W Polkowski[3] and M Tangstad[4]

1. Postdoc Researcher, Norwegian University of Science and Technology, Department of Materials Science and Engineering, 7491 Trondheim, Norway. Email: jian.m.jiao@ntnu.no
2. Researcher, Norwegian University of Science and Technology, Department of Materials Science and Engineering, 7491 Trondheim, Norway. Email: maria.wallin@ntnu.no
3. Researcher, Norwegian University of Science and Technology, Department of Materials Science and Engineering, 7491 Trondheim, Norway. Email: polkowski.wojciech@ntnu.no
4. Professor, Norwegian University of Science and Technology, Department of Materials Science and Engineering, 7491 Trondheim, Norway. Email: merete.tangstad@ntnu.no

ABSTRACT

Fe-26Si-9B (wt per cent) alloy has been identified as a potential high-temperature phase change material (PCM) due to its attractive properties, such as its high latent heat of fusion and low volumetric change during solid/liquid transition. For the successful utilisation of this alloy into thermal energy storage (TES) systems, the development of a cost-effective production method is essential. Presently, the Fe-26Si-9B alloy is produced by mixing FeSi alloys with either pure boron element or FeB alloys. However, the use of pure boron is financially prohibitive, and the carbothermic reduction results in high greenhouse gas emissions and high energy consumption in the production of FeB alloys. In this regard, our study proposes a silicothermic reduction method to produce Fe-26Si-9B PCM by using FeSi alloys and B_2O_3-based oxides. Accordingly, the influence of various parameters on the production process was investigated, including operating temperature, holding time, B_2O_3 content in the added oxides, and initial Slag/Metal (S/M) ratio. Based on the experimental results, the optimal parameters for producing FeSiB alloys with over 9 wt per cent boron were determined. Consequently, it was documented that the FeSi alloys and the added oxides enriched with 50–65 wt per cent B_2O_3 should be subjected to temperatures ranging from 1550–1650°C, maintain an initial S/M ratio exceeding 1, and ensure a holding duration beyond 1 hr. Moreover, the energy consumption of this process was estimated to be ~1.86 MWh/t metal and the mass loss was lower than 7 per cent. Therefore, silicothermic reduction offers a sustainable approach for producing FeSiB alloys with a boron content above 9 wt per cent.

INTRODUCTION

Renewable energy sources, such as solar and wind, play an important role in achieving climate neutrality (EASE, 2023). However, a mismatch between energy supply and energy demand is the main challenge in their applications. By addressing this problem, thermal energy storage (TES) techniques have been developed, in which using of phase change materials (PCMs) has emerged as a particularly promising solution (IEA, 2019). Recently, Fe-26Si-9B (wt per cent) eutectic alloy stood out due to its moderate melting temperature, high energy storage capacity, low volumetric change, and high thermal conductivity (Grorud, 2018; Sellevoll, 2018, 2019; Sindland, 2018; Jiao *et al*, 2019a, 2019b; Jiao, 2020; Jiao, Safarian and Tangstad, 2022). However, the traditional methods of producing this alloy are based either on mixing pure iron (Fe), silicon (Si), and boron (B) elements, or on mixing ferroboron (FeB) and ferrosilicon (FeSi) alloys (Jiao *et al*, 2023). The former approach is financially prohibited, particularly due to the cost of pure boron. The latter suffers from its high greenhouse gas emissions and a high energy consumption.

To tackle these challenges, a new route is proposed for manufacturing Fe-26Si-9B PCM from raw and waste materials in the Thermobat project funded by the European Commission. The strategy is to produce this alloy by using a metallothermic reduction, for which silicon is selected as the reducing agent. The primary reaction between silicon and B_2O_3 is exothermic, and thus, this method has a low energy consumption. The process starts with mixing FeSi and B_2O_3-based oxides at high temperatures followed by a production of molten mixture of FeSiB alloys and SiO_2-based slags. Additionally, the chemical composition of FeSiB alloys can be adjusted at its molten state before tapping. Besides, the raw materials are abundant and cheap and can be obtained from iron and FeSi scraps, silicon kerf-loss waste and discarded FeSi and silicon fines. Thus, this process is expected to produce FeSiB alloys with a low cost.

The present study focused on optimising the production process of FeSiB alloys. Firstly, the experimental process is established theoretically through thermodynamic analyses. Then, we explore the effect of process parameters on the boron content in the resulting FeSiB alloys. The investigated process variables include holding time, B_2O_3 content in the added oxides, initial Slag/Metal (S/M) ratio, and operating temperature.

MATERIALS AND METHODS

In the production of FeSiB alloys, Fe-75Si and iron alloys, CaO, B_2O_3, and colemanite powders ($CaO \cdot 2B_2O_3 \cdot 5H_2O$) were used in the experiments. Fe-75Si alloy was sourced from Finnfjord AS, Norway, and the chemical composition of iron and silicon was analysed to be 25.78 wt per cent and 74.22 wt per cent using Inductively Coupled Plasma Sector Field Mass Spectrometry (ICP-SFMS), respectively. Iron metal (99.99 wt per cent) and CaO powder (99.95 wt per cent) came from Alfa Aesar. B_2O_3 powder (98 wt per cent) obtained from Thermo Scientific, and commercial colemanite powder (particle size <75 μm) was provided by Eti Maden company (ETi Maden, 2015). It is important to note that prior to use colemanite, it must undergo calcination to eliminate its crystal water. This involved maintaining the colemanite powders in a muffle furnace at 600°C for 180 min. The content of B_2O_3 was analysed to be ~55 wt per cent in the calcinated colemanite using the ICP-SFMS technique.

The experiments were conducted in a resistance furnace under an argon (99.999 per cent Ar) atmosphere. 7 to 24 g of B_2O_3-based oxides and 5 to 15 g of FeSi alloys were placed in the BN-coated SiC crucibles. The thin BN spray coating was applied to prevent the oxides creeping from the crucible due to its good wetting behaviour with SiC. Notably, the B_2O_3-based oxides were either pre-produced by mixing pure B_2O_3 and CaO powders or taken directly from the calcinated colemanite. The added oxides were placed at the bottom of crucible, and the Fe-41Si master alloy pieces were layered at the top of these oxides, where the Fe-41Si master alloy was produced by mixing Fe-75Si and iron metals in an induction furnace. This arrangement aimed to enhance mass transfer during the process, as after melting, the molten alloys would settle at the crucible's bottom, and the molten slags would ascend due to their density differentials. Subsequently, the holding temperature was set in the range of 1500–1650°C, maintaining this for a duration of 5 to 180 min. The purpose was to find the optimal parameters to produce FeSiB alloy having the boron content above 9 wt per cent. After experiments, the chemical composition of the produced alloys was analysed using ICP-SFMS. Furthermore, thermodynamic analyses were conducted by FactSage™ 8.1, using databases such as FactPS, FTlite, FToxide, and Melts database within the viscosity module. The energy consumption associated with the process was estimated using the Heat and Material Balance module of HSC Chemistry 9 software.

RESULTS AND DISCUSSION

Silicothermic reduction process design

The Fe-26Si-9B eutectic alloy was initially designed by FactSage 7.2 using the FTlite database (Jiao et al, 2019a). However, the experimental results have revealed that the actual eutectic alloy's composition was near but not precisely at the eutectic point (Jiao, 2020). The Fe-29Si-10B alloy was further regarded as the eutectic alloy based on the phase distribution in the observed eutectic structures (Jiao, 2020). Then, when a new version of FactSage 8.1 was applied in the thermodynamic analyses, the composition of this eutectic FeSiB alloy was changed to Fe-24Si-11B. This led to the assumption that the eutectic point for this system was in a range of 61–65 wt per cent iron, 24–29 wt per cent silicon, and 9–11 wt per cent boron. For this research, the silicothermic reduction process was designed based on the FactSage 8.1 using the FTlite and FToxide databases. The chemical reactions were expressed as:

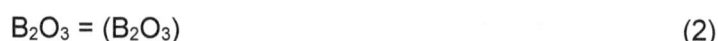

$$CaO = (CaO) \qquad (1)$$

$$B_2O_3 = (B_2O_3) \qquad (2)$$

$$Fe = \underline{Fe} \tag{3}$$

$$Si = \underline{Si} \tag{4}$$

$$3\underline{Si} + 2(B_2O_3) = 3(SiO_2) + 4\underline{B} \tag{5}$$

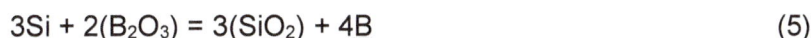

Here, the parentheses indicate that the oxides are in a molten state, and the underline denotes that the metal elements are in their molten state.

Figure 1 shows the calculated iso-boron lines of FeSiB alloys within the CaO-SiO_2-B_2O_3 liquidus region at 1500°C, fixing a constant Fe content at 65 wt per cent. It is observed that an increase in the boron content in FeSiB alloys is directly proportional to the increased content of B_2O_3 in the equilibrium slags. Here, the 11 per cent boron line shows the composition of the equilibrium CaO-SiO_2-B_2O_3 slags corresponding to the production of Fe-24Si-11B alloy. So, to produce an FeSiB alloy with a boron content over 11 per cent, the added CaO-SiO_2-B_2O_3 oxides should ideally be situated within the light green area in the diagram, where the B_2O_3 content exceeds 25 wt per cent in the equilibrium CaO-SiO_2-B_2O_3 slags.

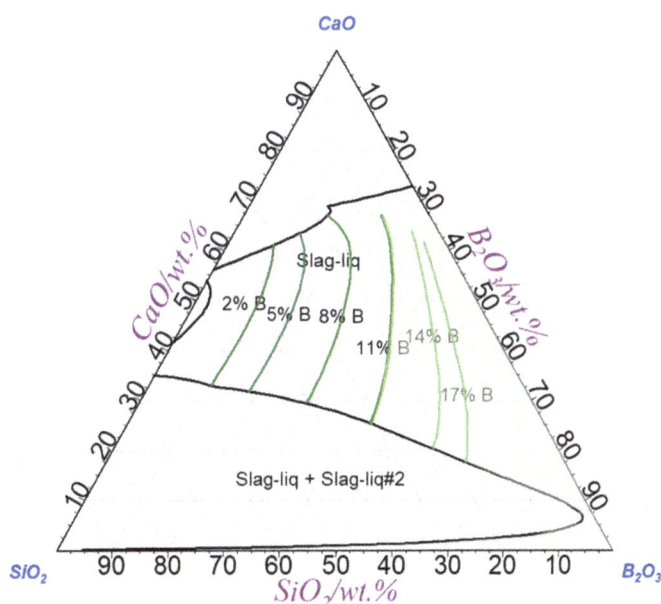

FIG 1 – CaO-SiO_2-B_2O_3 phase diagram at 1500°C, showing iso-boron lines (in green) for Fe-Si-B alloys (65 wt per cent iron) with their equilibrium slags. The light green region indicates the compositional range of the added oxides to produce the Fe-24Si-11B alloys (FTlite + FToxide databases).

Figure 2 presents the predicted effects of B_2O_3 content in the added oxides on both the mass of CaO-B_2O_3 oxides added and the viscosity of the equilibrium CaO-SiO_2-B_2O_3 slags. In the production of 1 ton Fe-24Si-11B alloy, it is found that an increase of B_2O_3 content and consequently a decrease in the CaO content, less slag will be produced and hence it leads to a decrease of the requested mass of CaO-B_2O_3 oxides. Conversely, the viscosity of the equilibrium slags increases with higher B_2O_3 content in the added CaO-B_2O_3 oxides. In the production process, the goal is to use as little of the added CaO-B_2O_3 oxides as possible while keeping the slags fluid enough. However, it's hard to achieve both low mass of the CaO-B_2O_3 oxides added and its low viscosity at the same time. According to Figure 2, a critical intersection point of these parameters is ~55 wt per cent B_2O_3. Hence, the added CaO-B_2O_3 oxides containing the B_2O_3 in the range of 50–65 wt per cent are investigated in the experiments. Ideally, this B_2O_3 composition range corresponds to an input mass for the added oxides between 1.6–2.7 t and a viscosity range for the equilibrium slags of 1–1.8 poise in the production of 1 ton Fe-24Si-11B alloy.

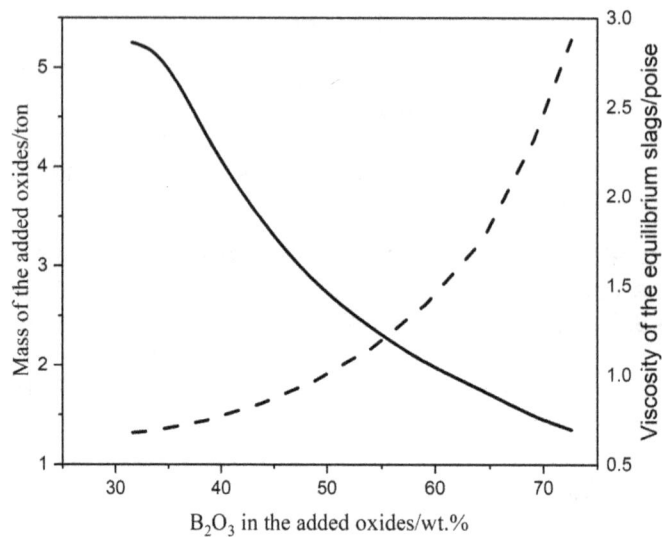

FIG 2 – The solid line shows the relationship between the content of B_2O_3 in CaO-B_2O_3 added oxides and the mass for producing 1 ton Fe-24Si-11B alloy at 1500°C, and the dashed line shows its corresponding viscosity of the equilibrium CaO-SiO_2-B_2O_3 slags (Melts, FTlite, and FToxide databases).

Effect of holding time

Figure 3 shows the influence of holding time on the boron content in the produced FeSiB alloys, where the Fe-41Si master alloy and CaO-65B_2O_3 oxides were subjected to 1650°C for the holding time ranging from 5 to 120 min with the fixed initial S/M ratio of 1.6. It is observed that the boron content increased with the increase in holding time, tending the towards to the equilibrium concentration of 10.3 wt per cent, as shown in the dashed line. It reveals a rapid initiation of the S/M reaction within the initial 5 min, achieving a boron content ranging from 2.6 to 5 wt per cent. Then, a more gradual increase was observed over time. The system was close to its equilibrium state after 60 min holding time, where the boron content was increased to ~9 wt per cent. Interestingly, the boron content was higher than the predicted equilibrium content after 120 min holding time, which was impossible. This deviation might be caused by the inaccuracy in the FactSage database. These results show that a holding time of 60 min is necessary to achieve a boron content over 9 wt per cent in the FeSiB alloys.

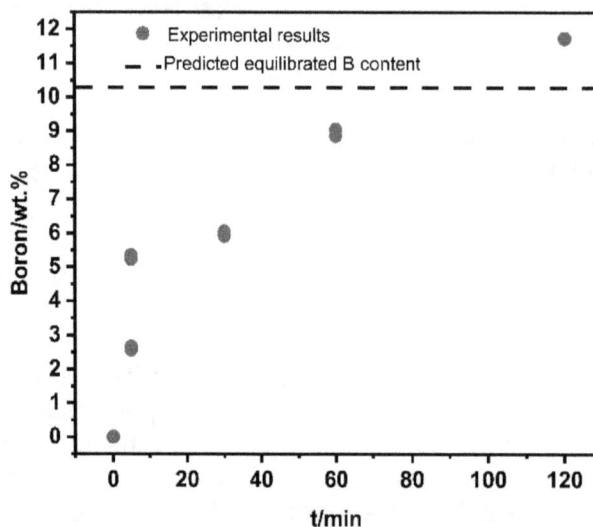

FIG 3 – The effect of holding time on boron content in the produced FeSiB alloys. Fe-41Si and CaO-65B_2O_3 oxides is reacted at 1650 °C with the initial S/M ratio of 1.6. The dashed line represents the predicted equilibrium B content (FTlite, and FToxide databases).

Effect of B₂O₃ content in the added oxides

Figure 4 presents a comparison of the experimental results and thermodynamic predictions regarding the impact of B_2O_3 content in the added oxides on the boron content in the produced FeSiB alloys. In the experiments, Fe-41Si master alloys reacted with a range of B_2O_3-based oxides at an initial S/M ratio of 1.6 during 60 min holding time at 1650°C. It is noted that 55 wt per cent B_2O_3 based oxides were used directly from the calcinated colemanite, whereas 50 wt per cent and 65 wt per cent B_2O_3 based oxides were synthesized by blending pure CaO and B_2O_3 powders. It is seen from the modelling results that the boron content increased slightly in the produced FeSiB alloys with a higher B_2O_3 content in the added oxides. Experimental results showed that the boron content in the alloys had a good agreement with the thermodynamic predictions by using 50 wt per cent and 55 wt per cent B_2O_3 based oxides. Conversely, when employing 65 wt per cent B_2O_3 based oxides, the analysed boron content was lower than the predicted equilibrium boron content. indicating that the reaction did not reach equilibrium within a 60 min holding time. It implicates that the mass transfer was likely the limiting step in this process. Therefore, an increase in the B_2O_3 content in the added oxides from 50 wt per cent to 65 wt per cent, a longer holding time is expected to achieve equilibrium.

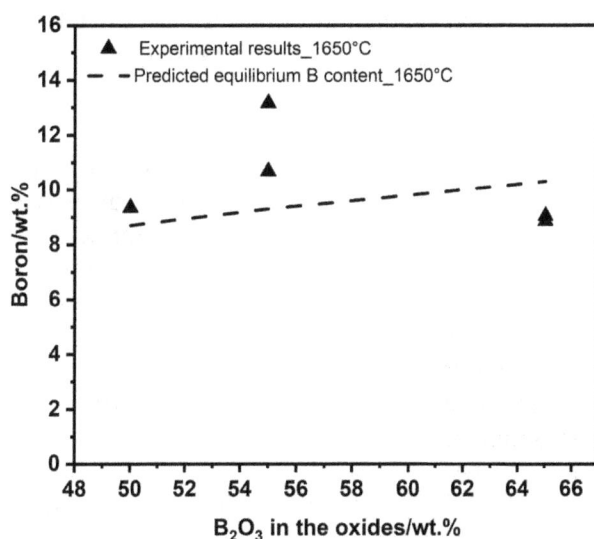

FIG 4 – The effect of B_2O_3 content in the added oxides on boron content in the produced FeSiB alloys. The points represent the experimental results, and the dotted line represents the predicted results based on FactSage 8.1 using FToxide and FTlite databases. The calcinated colemanite was used as the 55 wt per cent B_2O_3 based oxides.

Effect of the initial S/M ratio

Figure 5 shows the relationship between the initial S/M ratio and boron content in the produced FeSiB alloys. These experiments involved a reaction of Fe-41Si master alloys with the calcined colemanite (~55 wt per cent) at different initial S/M ratios, maintaining the reaction for 60 min at 1650°C. The analysed boron content in the produced alloys was summarised in the figure, accompanied by the predicted equilibrium boron content represented by the dashed line. There was a clear trend indicating that the boron content increased with an increasing initial S/M ratio, as a higher initial S/M ratio would give a higher B_2O_3 for a given boron content, and hence the driving force would be higher. Significantly, the boron content was higher than 9 wt per cent in the FeSiB alloys when the initial S/M ratio was over 1. In the comparison of the experimental results and the predicted equilibrium boron content, it is found that the experimental results were close to or higher than that from the thermodynamic modelling, as mentioned before, this deviation might be caused by an inaccurate FactSage database. According to the experimental results, to achieve a boron content higher than 9 wt per cent in our target FeSiB alloys, the initial S/M ratio should be greater than 1.

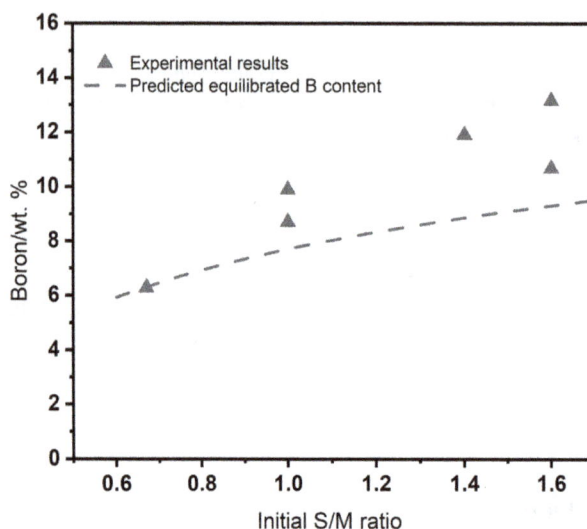

FIG 5 – The effect of initial Slag/Metal (S/M) ratio on boron content in the produced FeSiB alloys. The points represent the experimental results, and the dash line represents the predicted result by FactSage 8.1 using FToxide and FTlite databases.

Effect of operating temperature

Figure 6 shows the impact of operating temperature on the boron content in the produced FeSiB alloys. These alloys were produced by reacting Fe-41Si master alloys with the calcined colemanite at the initial S/M ratio of 1 and 1.6, within a 60 min holding time at temperatures range of 1500–1650°C. The figure summarised the experimental results, with the dashed lines indicating the predicted equilibrium boron content for these initial S/M ratios. The experimental data reveals that an increase in temperature at a constant initial S/M ratio led to a higher boron content. In contrast, the predicted equilibrium boron content decreases with increasing temperature, indicating that equilibrium was not achieved at lower temperatures (1550°C and 1500°C) within 60 min holding time. It is noted that a lower temperature was preferable for the silicothermic reduction process. However, a lower temperature also resulted in a higher viscosity of the molten slag, which impeded mass transfer during the process. Thus, at a holding time of 60 min, mass transfer became the limiting factor.

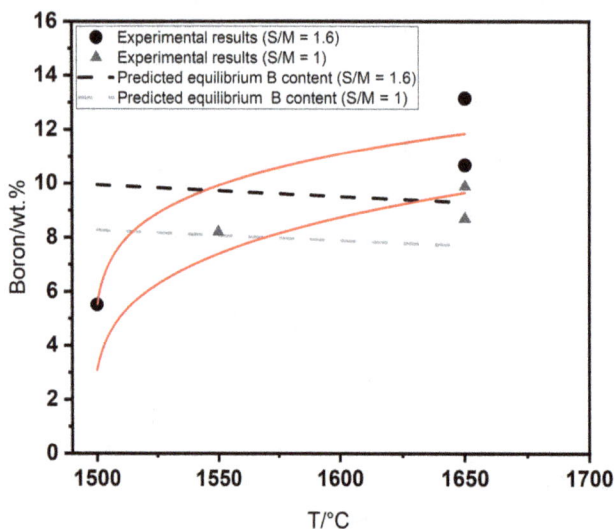

FIG 6 – The effect of temperature on boron content in the produced FeSiB alloys. The points represent the experimental results, and the dash lines represent the predicted results based on FactSage 8.1 using FToxide and FTlite databases.

The two red lines in the figure shows the trend of B content with increasing temperature at initial S/M ratios of 1 and 1.6, respectively. These two lines were drawn based on the experimental data. It was observed that a higher initial S/M ratio resulted in a higher boron content at a given temperature and holding time, due to a stronger driving force in the slag/metal systems with a higher initial S/M ratio. Consequently, the system with a higher S/M ratio was likely to reach equilibrium in a shorter holding time at a constant temperature. Therefore, for a 60 min holding time and an initial S/M ratio greater than 1, the optimal operating temperature is approximately 1650°C for the calcinated colemanite. To achieve equilibrium at temperatures below 1650°C, a longer holding time would be necessary.

Mass loss

Figure 7 presents a plot of mass loss as a function of holding time under various temperatures, in the production of FeSiB alloys using B_2O_3-based oxides. The measurement of mass loss was conducted by comparing the total weight of raw materials before and after conducting the experiments. It reveals that the mass loss was below 7 per cent. In the experiments using CaO-$65B_2O_3$ at 1500°C, the mass loss increased proportionally with holding time, from 0.4 per cent after 5 min to 3.5 per cent after 180 min. Additionally, when examining the impact of temperature on mass loss with a constant holding time of 60 min using the CaO-$65B_2O_3$, an increase in temperature led to an increase in mass loss, from 0.1 per cent at 1400°C to 4.1 per cent at 1650°C. In contrast, when it came to the calcinated colemanite, the behaviour was less predictable. It varied in the range of 2–7 per cent after 60 min holding time at 1650°C. A review of the data indicates that the use of calcinated colemanite led to a higher mass loss compared to CaO-$65B_2O_3$ master oxides. Figure 8 shows the partial pressure of potential species in the CaO-SiO_2-B_2O_3 system across a temperature range of 1200–1700°C. This suggests that the mass loss was primarily due to the evaporation of B_2O_3 in the system. Given these findings, it is suggested to compensate for this loss by adding an additional B_2O_3 to the oxides charged in the FeSiB alloy production process.

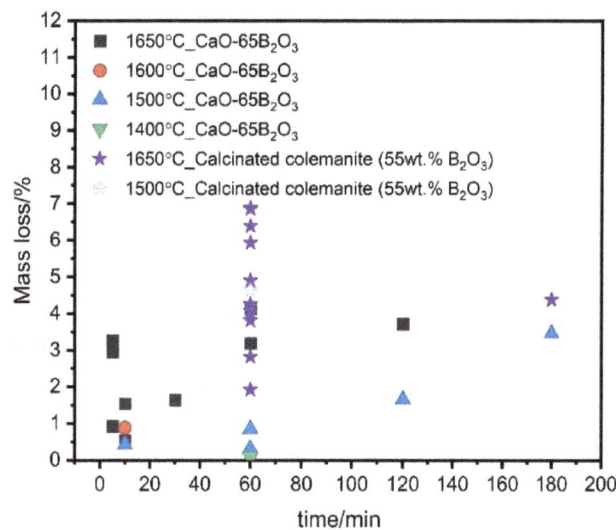

FIG 7 – Mass loss versus holding time in the silicothermic reduction process.

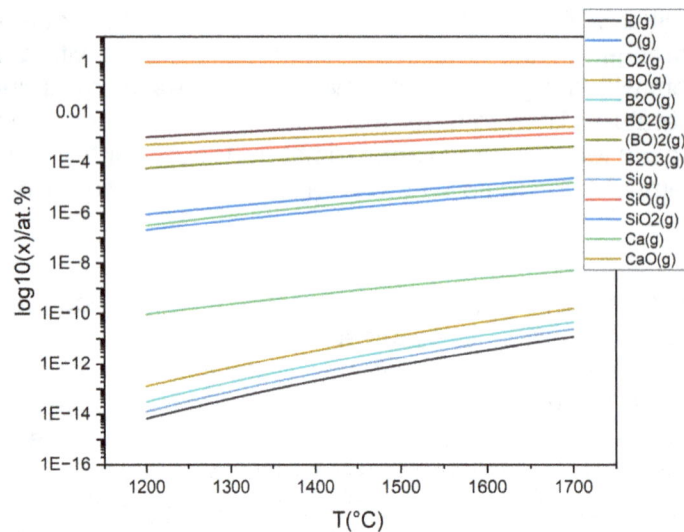

FIG 8 – The partial pressure of possible species in the CaO-SiO$_2$-B$_2$O$_3$ system at a standard atmosphere. It was calculated by FactSage 8.1 using FactPS database.

Energy balance

In the production of 1 ton Fe-24Si-11B alloy, the estimated theoretical energy consumption was 1860 kWh/t metal. This estimation was calculated based on a mass balance achieved at 1650°C under thermodynamic equilibrium. The relevant stoichiometry equation is as follows:

$$1.7t(35CaO - 65B_2O_3) + 1.1t(59Fe - 41Si) = 1.8t(33CaO - 24SiO_2 - 43B_2O_3) + 1t(65Fe - 24Si - 11B) \quad (6)$$

Further analysis of the electrical energy consumption was carried out by using HSC Chemistry software (version 9). This analysis was based on the principle of enthalpy conservation, as shown in the equation:

enthalpy in input materials + electrical energy = enthalpy in output products   (7)

Subsequently, the energy flow involved in the production of 1 ton Fe-24Si-11B alloy was illustrated in the Sankey diagram, as it is presented in Figure 9. It is observed that the energy from the input metal was slightly higher than the energy from the output metal, attributed to the exothermic reaction between silicon an B$_2$O$_3$. The electrical energy (1860 kWh/t of metal) was directly transformed to the heat in the output metal and slag. Given these observations, the challenge was to develop an efficient method to reuse the heat in the products. This is crucial for improving energy efficiency and sustainability of the proposed production process.

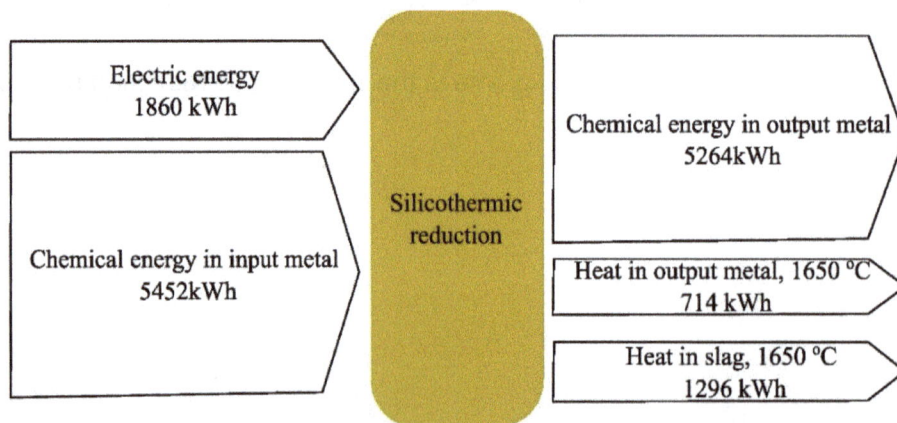

FIG 9 – Sankey diagram, illustrating the energy flow for the production of 1 ton Fe-24Si-11B alloy. Fe-41Si and CaO-65B$_2$O$_3$ oxides reacted at 1650°C with an initial S/M ratio of 1.6. Thermal loss was not included.

CONCLUSIONS

This research aimed to develop a method to produce FeSiB alloys, targeting a composition of 61–65 wt per cent iron, 24–29 wt per cent silicon, and 9–11 wt per cent boron. This method was based on the silicothermic reduction between silicon and B_2O_3. In this regard, the experiments were conducted in a resistance furnace under an argon atmosphere. The parameters in the influence of the boron production in the alloys were investigated. According to the experimental results, the optimal procedure involved reacting an Fe-41Si master alloy with oxides containing over 50 wt per cent B_2O_3 by maintaining an initial Slag/Metal (S/M) ratio of 1–1.6. The mixture was then subjected to the temperatures above 1550°C for a duration over 60 min. The mass loss was observed to be below 7 per cent, and the energy consumption was estimated to be ~1860 kWh/t metal. These findings provided a basis to develop reliable and efficient method for producing the targeted FeSiB alloys with the boron content above 9 wt per cent.

ACKNOWLEDGEMENTS

The THERMOBAT project received funds from European Commission under grant agreement 10105754. The sole responsibility for the content of this publication lies with the authors. It does not necessarily reflect the opinion of the European Union. Neither the REA nor the European Commission is responsible for any use that may be made of the information contained therein.

REFERENCES

ETi Maden, 2015. Product Specification – Ground Colemanite. Available from: <https://www.etimaden.gov.tr/storage/uploads/2018/01/16-2017-Ground-Colemanite-75-Micron.pdf> [Accessed: 18 April 2023].

European Association for Storage of Energy (EASE), 2023. Thermal Energy Storage. Available from: <https://ease-storage.eu/publication/thermal-energy-storage/>

Grorud, B, 2018. Interaction of Eutectic Fe-Si-B Alloy with Graphite Crucibles, Master thesis, Norwegian University of Science and Technology.

International Energy Agency (IEA), 2019. World Energy Outlook 2019. Available from: <https://www.iea.org/reports/world-energy-outlook-2019>

Jiao, J M, 2020. Si-based phase change materials in thermal energy storage systems, PhD thesis, Norwegian University of Science and Technology (NTNU).

Jiao, J Safarian, J and Tangstad, M, 2022. The Use of Fe-26Si-9B Alloy as Phase Change Material in Si3N4 Container, *Crystals*, 12(3):376. https://doi.org/10.3390/cryst12030376

Jiao, J, Grorud, B, Safarian, J and Tangstad, M, 2019b. Wettability of molten Fe-Si-B alloy on graphite, Al2O3, and h-BN substrates, in *Proceedings of the Liquid Metal Processing and Casting Conference 2019,* pp 425–433.

Jiao, J, Grorud, B, Sindland, C, Safarian, J, Tang, K, Sellevoll, K and Tangstad, M, 2019a. The use of eutectic Fe-Si-B alloy as a phase change material in thermal energy storage systems, *Materials*, 12(14). https://doi.org/10.3390/ma12142312

Jiao, J, Jayakumari, S, Wallin, M and Tangstad, M, 2023. Graphite crucible interaction with Fe–Si–B phase change material in pilot-scale experiments, *High Temperature Materials and Processes*, 42(1):20220288. https://doi.org/10.1515/htmp-2022-0288

Sellevoll, K, 2018. Interactions of Eutectic Fe-Si-B Alloy with Graphite Crucibles, Specialization project report, Norwegian University of Science and Technology (NTNU).

Sellevoll, K, 2019. Interactions of FeSi Alloys with Graphite Crucibles, Master thesis, Norwegian University of Science and Technology (NTNU).

Sindland, C, 2018. Production of Eutectic Si-Fe-B and Si-Cr-B alloy and their Interaction with Graphite Crucibles, Summer job report, Trondheim: Norwegian University of Science and Technology (NTNU). https://doi.org/10.1093/jicru/os9.1.report16

Physical properties optimisation of the Zimbabwean limonite ore-carbon composite pellets as a sustainable feed for pig iron production

S M Masuka[1], D Simbi[2], S Maritsa[3] and E K Chiwandika[4]

1. Lecturer, University of Zimbabwe, Harare 263, Zimbabwe. Email: shebarmasuka@gmail.com
2. BTech student, Harare Institute of Technology, Harare 263, Zimbabwe.
   Email: h200741h@hit.ac.zw
3. Lecture, Harare Institute of Technology, Harare 263, Zimbabwe.
   Email: steven.maritsa@gmail.com
4. Lecturer, Harare Institute of Technology, Harare 263, Zimbabwe.
   Email: chiwandikae@gmail.com

ABSTRACT

The steel industry is facing challenges on a global scale that include depletion of resources, huge energy consumption, and the emission of CO_2. The demand for iron products is increasing due to the increased infrastructural development. Zimbabwe produced cast iron from scrap metals. However, there is currently a shortage of scrap, and yet Zimbabwe is rich in limonite ores that are currently being underutilised. The possibility of using limonite ore as a sustainable feed for pig iron production was investigated by preparing some limonite ore carbon composite pellets. The results showed that the addition of coal to the limonite ore and calcium carbonate mix to form the composite pellets resulted in a decrease in the drop number as well as the dry compressive strength of the composite pellets. This research aims to improve the physical properties of the green pellets by the careful addition of hydroxyethyl cellulose as a binder that was found to improve the physical properties of the green pellet. This is important for materials handling during the production process of the composite pellets. Results showed that the drop number was substantially improved by the addition of 0.4 wt per cent hydroxyethyl cellulose while the dry compression strength improved from 2.5 kg/pellet to around 23 kg/pellet irrespective of the amount of binder added. The binder improved the physical strength of the iron-carbon composite pellets enough to allow for large-scale production of the pellets that can be an alternative and sustainable feed for cast iron production. However more results on the indurated compression strength and other properties such as the reduction degradation index, swelling index, and the reducibility test are required.

INTRODUCTION

The huge demand for equipment and infrastructure for clean energy technologies as a result of the need to achieve net zero emissions by 2050 has resulted in massive demand for iron and steel products since these play a pivotal role in the development of such infrastructure (International Energy Agency, 2023). Global steel consumption is expected to continue rising to meet this demand and yet the high-grade iron ores have depleted, forcing players in the industry to utilise lower-grade iron ores. The mineralogy of these underutilised ores is often very complex and requires innovative processing techniques that can yield low-impurity molten metal from these low-grade iron ores resources (Liang et al, 2023).

Zimbabwe has low-grade limonite ores that are not being utilised in the Ripple Creek Deposits, located in the Midlands Province of central Zimbabwe which are estimated to be 111 million tonnes (Mt). Of these iron ores, 59 per cent meet the blast furnace requirements and the remainder are soft limonite ores that generate fines during mining and processing. Sintering of these limonite ores results in a product of poor quality because of more pores in the sinter due to the high water of crystallisation content and the high loss on ignition of the same affects the sinter strength, making sinter made from these ores unsuitable for blast furnace feed. The production of high-quality sinter from this resource would require a high coke rate and subsequently high carbon dioxide emissions (Clout and Manuel, 2015). Hence investigations on the utilisation of this resource through the production of some limonite ore-carbon composite pellets are being done by Chisahwira et al (2023).

The preparation of limonite–carbon composite pellets will increase reduction efficiency by improving reactivity and lower carbon dioxide emissions thus contributing towards Zimbabwe's aim to reduce emissions by 40 per cent per capita by 2030. However, the physical strength of the green pellets was found to decrease with increasing coal addition (Chisahwira *et al*, 2023) which is detrimental to the productivity during mass production of the pellets hence the need to investigate ways to improve the strength by the addition of an organic binder, hydroxyethyl cellulose.

The use of binders in the agglomeration of fine iron ores through pelletisation has always been a major step in the utilisation of lower-grade iron ores. Bentonite has in the past been the most commonly used binder because it controls moisture and adds to the physical strength of the pellets but the major disadvantage is the high content of acid constituents, namely, silica and alumina which are approximately 65 per cent and 20 per cent respectively (Devashayam, 2018). In addition, bentonite has been associated with availability challenges and high costs. Binders that can produce the same quality of iron ore pellets in terms of pellet strength, at sustainable prices while also avoiding the addition of more impurities would be ideal for use in the limonite ore-carbon composite pellets.

The most extensively researched group of binders are organic binders since they are free of silica and have exhibited good binding properties and wet strength (Eisele and Kawatra, 2003). However, they have lacked in terms of fired pellet strength because of their low burning temperature. At high firing temperatures, organic binders that burn with no residue cannot provide the necessary bonding strength to iron oxide grains (Sivrikaya and Arol, 2014). On the other hand, this elimination during firing can be viewed as an advantage because this ensures that there are no impurities in the iron ore pellets, which has not been the case with inorganic binder (Guanzhou *et al*, 2002). As such, research and development on the use of organic binders in iron ore pelletisation has continued through the years.

A review of organic binders in a report by Halt and Kawatra (2014) stated that a range of organic materials from synthetic chemicals such as acrylamides and naturally occurring materials such as starch and cellulose and their derivatives have been explored as binders in the agglomeration of iron ore. Some binders that were successful at laboratory scale production failed at pilot scale production and underlying reasons are often difficult to determine. However, the general overview is that purely organic binders enhance the green-ball formation, and the green-ball quality increases with the binder's ability to thicken and the ability to form an adhesive film after drying.

In an investigation by Ngara *et al* (2023), where organic binders were used as alternatives to bentonite in the pelletisation of low-grade iron ore, and known industrial standards used for references, it was determined that organic binders could be used as effective alternatives in iron ore pelletisation. The organic binders, corn starch, sodium lignosulfonate, and carboxymethyl cellulose produced pellets whose drop indices were comparable to bentonite and could be used in synthesizing green pellets. However, indurated pellets produced from lignosulfonate and carboxymethyl were weaker than those with bentonite even though they surpassed the minimum industrial standards (Ngara *et al*, 2023). As such, these organic binders were deemed viable alternatives to bentonite in low-grade iron ore pelletisation.

Research by Guanzhou *et al* (2002) on the functions and molecular structure of organic binders based on molecular design and interface science indicated that structurally good binders should have sufficient polar groups and hydrophilic groups. Functionally, a good binder should improve the wettability of the iron ore particles, have great adhesive force to the iron ore, as well as good cohesive strength and thermal stability. According to this study, the structure of an organic binder for iron ore should have a polar group, a hydrophilic group, an organic chain skeleton, and a degree of polymerisation. The binder used in this study, (hydroxyethyl cellulose) based on the findings by Guanzhou *et al* (2002) could potentially be used as a binder for limonite-carbon composite pellets because of its relatively higher molecular weight, which is reported to increase mechanical strength. The presence of the hydroxyl and ethyl groups, which act as the hydrophilic and polar parts of the polymer, respectively, is also expected to aid in the effective bonding of the iron ore and carbon particles and offer significant mechanical strength to the composite pellets. It is expected that the binder will increase the physical strength of the iron-carbon composite pellets enough to allow for large-scale production of the pellets as an alternative and sustainable feed for cast iron production. It should be noted that limonite-carbon composite pellets have been previously produced with the indurated compression strength greater than 250 kg/pellet, but the green physical properties were not sufficient enough for mass production in our previous research (Chisahwira *et al*, 2023). Hence this research only focused on the improvement of the green pellet physical properties.

EXPERIMENTAL

Materials

Limonite ore obtained from Kwekwe, in the Midlands province of central Zimbabwe, and coal fines obtained from waste dumps at Hwange Colliery Company (HCC) were used in this study. The chemical composition of the limonite and limestone is shown in Table 1 while proximate analysis for the coal fines was done according to the Indian Standard IS:1350 (Part I) (Bureau of Indian Standards, 1984) (Table 2).

TABLE 1

Chemical composition of the Zimbabwean Limonite ores and limestone.

Analyte	T-Fe	SiO$_2$	CaO	Al$_2$O$_3$	MgO	Mn	S	P	LOI
Zimbabwean limonite ore	52.51	8.78	1.32	1.30	0.59	2.1	0.005	0.04	11.87
Limestone	1.32	7.10	45.70	1.19	4.22	0.31	0.17	0.01	39.96

TABLE 2

Coal proximate analysis.

Fixed carbon	Volatile matter	Ash content	Moisture content
64.28	24.12	10.40	1.20

The major phases in the iron ore were determined using X-ray diffraction at a scanning speed of 0.5 with increments of 0.02 at a 2θ range of 10–80° and were determined to be hematite, goethite and silica as shown in Figure 1.

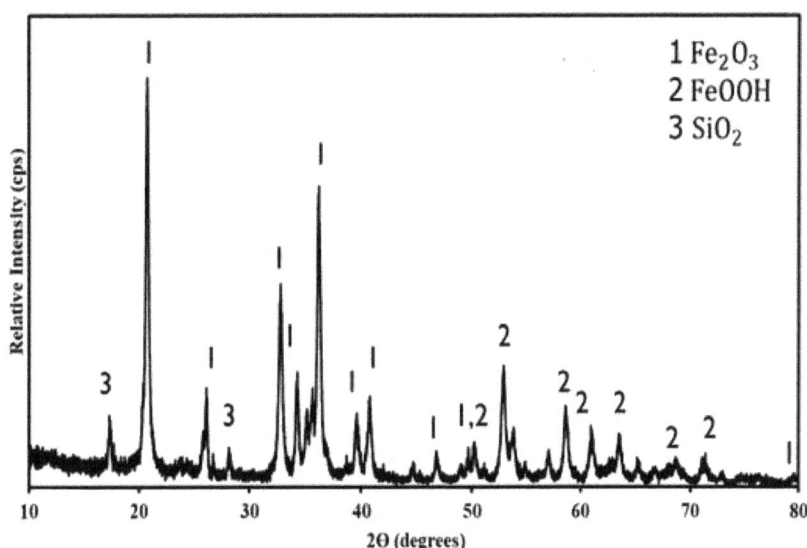

FIG 1 – Phase identification of limonite ore.

Preparation of raw materials

Moisture was removed from limestone and limonite by oven drying at 383 K for 24 hrs. Size reduction for the ore, limestone, and coal was done to achieve 80 per cent passing 75 µm using a pulveriser. The samples were sieved on a 75 micron sieve and the oversize material was re-pulverised until the target size of less than 75 µm was achieved to ensure uniform mixing of the raw materials for efficient pelletising. The limestone and coal optimised sample was then mixed with 0.1–1 wt per cent hydroxyethyl cellulose as the binder with the representative molecular structure shown in Figure 2.

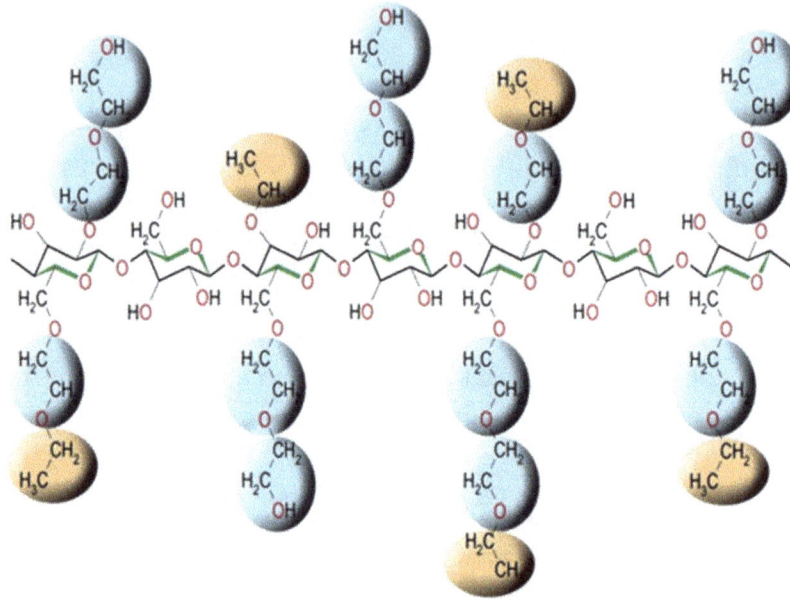

FIG 2 – Representative molecular structure of the hydroxyethyl cellulose used as the binder.

After homogenisation, a pelletising disc of 610 mm diameter, rotating at 28 rev/min and inclined at 40° was then used to make green pellets. The necessary amount of water was added to the rolling pellet feed for green pellet formation which were removed when an average target size of between 12–16 mm was attained. Drop tests were performed on the pellets by dropping randomly selected ten pellets from each batch on a steel plate at a height of 45 cm. The number of drops that were required to break an individual pellet was noted as the drop number. The average of the ten pellets was then considered as the drop number of that batch.

$$Drop\ number_{batch} = \frac{\sum drop\ number\ of\ each\ pellet}{10}$$

Dry compressive strength was done using the Electronic Universal Testing Machine (ADW-50S) on 20 pellets. The average maximum load a pellet was able to crack was recorded as compressive strength in units of kg/pellet.

RESULTS AND DISCUSSION

Effect of hydroxyethyl cellulose on the drop index and the dry compressive strength

Hydroxyethyl cellulose was added as a binder and the effects on the drop index of the composite pellets are shown in Figure 3. Drop tests are essential to ensure that the pellets have enough strength to minimise fractures and breakages during green pellet handling (Geerdes, 2009).

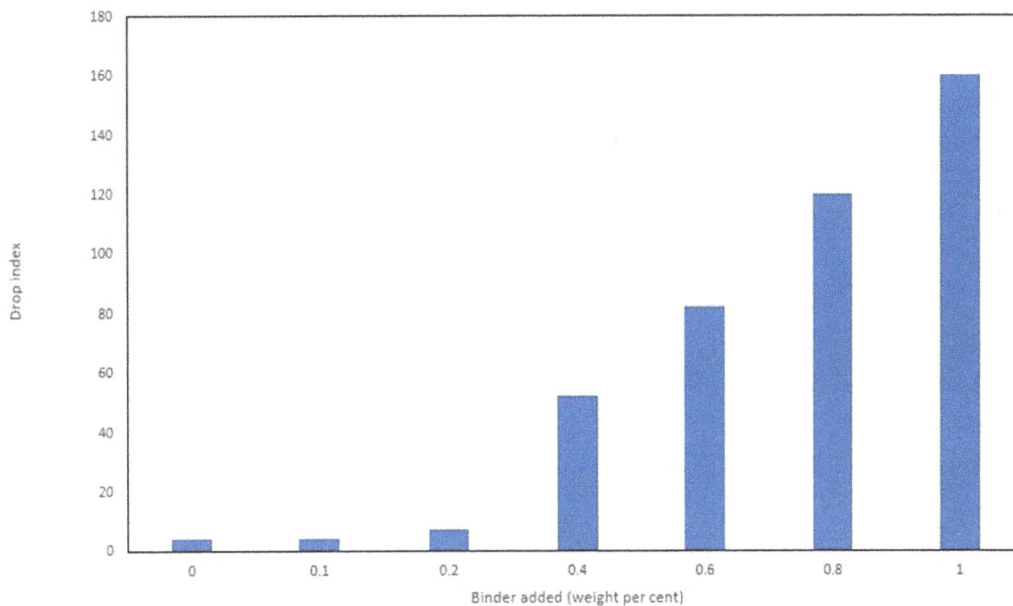

FIG 3 – Drop index with increasing binder.

The addition of a small amount of hydroxyethyl cellulose binder of up to 0.2 wt per cent showed little improvement in the drop index. The drop index was largely improved from 0.4 wt per cent addition onwards. This was largely due to the capillary forces that are imparted on the material through the action of the binder (Halt and Kawatra, 2014) the hydrophilic groups on the organic groups allow for the binder to be effectively dispersed and form strong capillary forces that result in an operative binding effect (Kawatra and Claremboux, 2022). Since the industrial required drop index is above 4 (Sivrikaya and Ali, 2012), the proposed recommended addition level of this binder was 0.4 wt per cent otherwise any addition above this will result in an unnecessary cost of production.

A minimum dry pellet strength is necessary so that pellets can withstand a load of pellet layers or pressure from gases flowing through the charge on traveling grates (Sivrikaya and Arol, 2014). Figure 4 shows the effect of binder addition on the dry compressive strength of the composite pellets.

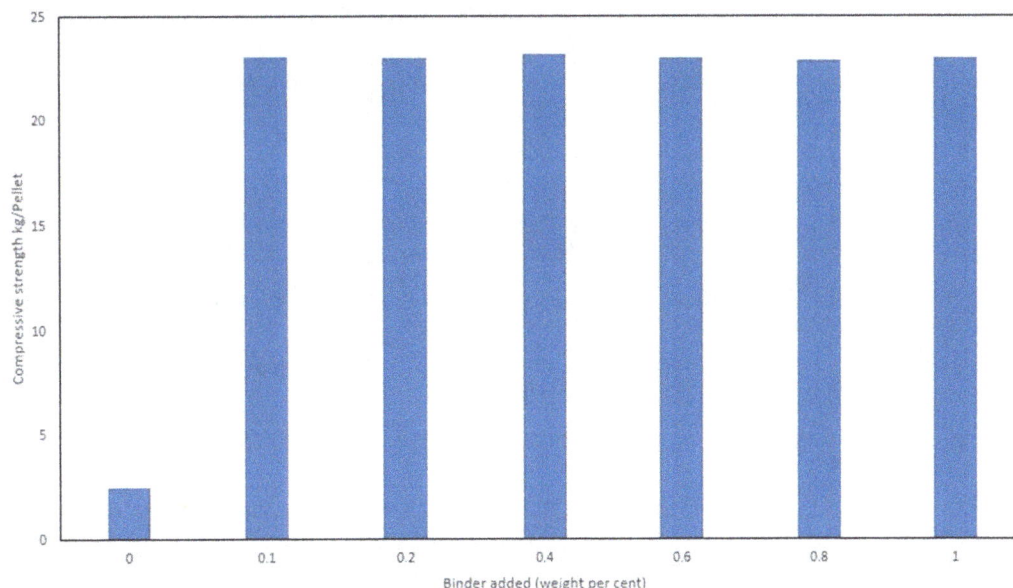

FIG 4 – Dry compressive strength with an increasing binder.

The addition of the binder improved the dry compressive strength from 2.5 kg/pellet to around 23 kg/pellet irrespective of the amount of binder added. This can again be attributed to the chemical structure of the organic binder where the hydrophilic groups aid in the dispersion of the binder into the binding medium resulting in a good binding effect (Ngara *et al*, 2023). Investigations show that hydroxyl groups are good hydrophilic groups and their presence in hydroxyethyl cellulose may be attributed to the increase in the physical strength of the pellets (Guanzhou *et al*, 2002).

This improvement in the strength of the green pellet is a desirable characteristic during material handling in mass production of the pellets. The dry compressive strength of the pellets was well above the industrial requirement of at least 2.2 kg/pellet (Sivrikaya and Ali, 2012). The recommended hydroxyethyl cellulose addition level was determined to be 0.4 wt per cent basing on both the drop number and the dry compression strength. However more investigation into the indurated compression strength is needed to fully understand the behaviour of the produced pellets.

CONCLUSIONS

The organic binder used in this study, hydroxyethyl cellulose, increased the physical strength of the green pellets as indicated by the results of the drop index and the dry compressive strength. The drop index increases significantly from 0.4 wt per cent to 52 which was then recommended as the optimum amount of binder addition. Any further increase in the amount of binder was deemed to be an unnecessary cost on the production of the pellets since the drop index value at 0.4 wt per cent was above the industrial minimum requirement. The dry compressive strength of the composite pellets upon the addition of the organic binder increased, ranging above 20 kg/pellet for all values of binder from 0.1–1 wt per cent. It can be concluded that the optimum amount of hydroxyethyl cellulose binder for the palletisation of Zimbabwean limonite-ore carbon composite pellets is 0.4 wt per cent. However more results on the indurated compression strength and other properties like the reduction degradation index, swelling index, and the reducibility test are required to fully understand the behaviour of the produced pellets.

REFERENCES

Bureau of Indian Standards, 1984. IS:1350-1 (1984): Methods of Test for Coal and Coke, Part I: Proximate Analysis [PCD 7: Solid Mineral Fuels] [online]. Available from: <https://law.resource.org/pub/in/bis/S11/is.1350.1.1984.pdf>

Chisahwira, T, Masuka, S M, Maritsa, S and Chwandika, E K, 2023. The use of Zimbabwean limonite-coal composite pellet as a sustainable feed for pig iron production, in *Proceedings of Iron Ore 2023 Conference*, p 391 (The Australasian Institute of Mining and Metallurgy: Melbourne).

Clout, J M F and Manuel, J R, 2015. Mineralogical, Chemical and physical characteristics of iron ore, in *Iron Ore: Mineralogy, Processing and Environmental Sustainability* (ed: L Liming), pp 45–84 (Woodhead Publishing).

Devashayam, S, 2018. A novel iron ore pelletization for increased strength under ambient conditions, *Sustainable Materials and Technologies*, 17:1–27.

Eisele, T C and Kawatra, S K, 2003. A review of binders in iron ore pelletization, *Mineral Processing and Extractive Metallurgy Review*, 24:1–90.

Geerdes, M, 2009. The Ore Burden: Sinter, Pellets, Lump Ore, in *Morden Blast Furnace Iron Making: An Introduction*, pp 31–35 (IOS Press: Amsterdam).

Guanzhou, Q, Tao, J, Hongxu, L and Dianzou, W, 2002. Functions and molecular structure of organic binders for iron ore pelletization, *Colloids and Surfaces A: Physicochemical Engineering Aspects*, 224:11–22.

Halt, J A and Kawatra, S, 2014. Review of organic binders for iron ore concentrate agglomeration, *Minerals and Metallurgical Properties*, 24(1):73–94.

International Energy Agency (IEA), 2023. Energy Technology Perspectives 2023, IEA. Available from: <https://www.iea.org/reports/energy-technology-perspectives-2023>

Kawatra, S K and Claremboux, V, 2022. Iron ore pelletisation: part, I, fundamentals, *Mineral Processing and Extractive Metallurgy Review*, 43(4):1–16.

Liang, Z, Peng, X, Haung, Z, Li, J, Yi, L, Houng, B and Chen, C, 2023. The innovative methodology for comprehensive utilization of refractory low-grade iron ores, *Powder Technology*, 418(15):118283.

Ngara, T, Mavengere, S, Bright, S and Mapamba, L, 2023. Investigating the effectiveness of organic binders as an alternative to bentonite in the pelletization of low-grade iron ore, *Physicochemical Problems of Mineral Processing*, 59(6):1–12.

Sivrikaya, O and Ali, I, 2012. The bonding/strengthening mechanism of colemanite added organic binders in iron ore pelletization, *International Journal of Minerals Processing*, pp 90–100.

Sivrikaya, O and Arol, A, 2014. Alternative binders to bentonite for iron ore pelletizing, Part I: Effects on physical and mechanical properties, *Holos*, 3:94–103. https://doi.org/10.15628/holos.2014.1758

State-of-the-art of electroslag refining and challenges in the control of ingot cleanness

L Medovar[1], G Stovpchenko[2] and G Jianjun[3]

1. Foreign Chief Scientist, Tianjin Heavy Industries Research and Development Co, Ltd, Tianjin 300457, China; EO Paton Electric Welding Institute of NASU, Kyiv 03150, Ukraine. Email: medovar@ukr.net; medovar@cfhi.com
2. Foreign Chief Scientist, Tianjin Heavy Industries Research and Development Co, Ltd, Tianjin 300457, China; EO Paton Electric Welding Institute of NASU, Kyiv 03150, Ukraine. Email: anna_stovpchenko@ukr.net; stovpchenko@cfhi.com
3. Head of Steelmaking Department, Tianjin Heavy Industries Research and Development Co, Ltd, Tianjin 300457, China. Email: guojj@cfhi.com

ABSTRACT

Electroslag remelting (ESR) today and into the future, will still be the main process for producing high-quality ingots by suppressing segregation, not just refining chemical composition from impurities. This paper analyses state-of-the-art electroslag refining with a focus on ESR with consumable electrodes, and recent technology with liquid metal supply (ESR LM) in view of their refining abilities and development trends, taking into consideration that the size of the ingots required by industry continues to grow, and the composition of new materials sophisticates to withstand increasingly higher mechanical and temperature loads . The yield of suitable metal in large ingots produced by conventional casting is low due to problems of element segregation by height and cross-section. For example, conventional ingots for nuclear powerplant rotors today reach nearly 700 t in weight. Attempts to directly replace such giant ingots with big-diameter ESR ingots weighing approximately 400 t were also unsuccessful because of growing segregation. Modern ESR equipment and technologies produce 200–250 t of ingots of satisfactory quality, reaching 2.5 m in diameter, implementing the change of consumable electrodes during remelting to increase ingot length. Another way to mitigate segregation is to enlarge the smaller diameter ESR ingot by a 500–800 mm coaxial layer to the desirable diameter by ESR in the current-supplying mould (CSM). Modelling and experiments prove that refining from impurities and non-metallic inclusions at ESR most effectively occurs at the slag bath and liquid metal pool interface because the ESR LM has provided the same desulfurisation. ESR's special refining capabilities in CSM come from the much longer time of LM residence shown at titanium purification from hard-melting nitride inclusions that are unremovable in vacuum arc remelting (VAR) and electron beam melting (EBM). Implementation of ESR technologies in CSM, especially with LM supply, is prospective to produce heavy enlarged and hollow forging ingots with prevented development of segregation due to dividing cross-section and reducing overheat of solidifying metal.

INTRODUCTION

Electroslag remelting (ESR) is currently the leading process for producing high-alloyed steels and alloy ingots used in critical industries. This position is due to its unique ability to refine both the chemical composition through slag treatment and the structure of the ingot by suppressing segregation, achieved through the constant renewal of the liquid metal bath of small depth.

Since Medovar and Paton made the first ingot in 1952 (Paton, Medovar and Latash, 1958; Medovar and Boyko, 1991), areas of application of electroslag remelting were widened and partially changed due to rethinking its role and importance. The development of ladle refining, and later the widespread use of additional degassing at vacuum treatment of steel, have led to a widespread reduction in the application of ESR furnaces for nearly two decades at the end of the last century.

The primary objectives now are to improve ESR metal's homogeneity and cleanness for producing sophisticated steel and alloy grades, reduce production costs and increase the competitiveness of the ESR. Today's trends include increasing the ESR ingot weight, improving automation and efficiency of equipment and recent interest in new specified slag systems intended for certain steel and alloy grades. Improvements in ESR technology and new innovative technical solutions for ESR equipment are crucial for product competitiveness, given that the demand for high-quality steels and alloys is expected to continue to rise in the coming years.

Heavy mechanical engineering has significantly progressed in recent years. Specifically, there has been an impressive intensification in the power of certain machines, industrial vessels, and aggregates. This expansion in the power of a single unit has been accompanied by an equally impressive rise in the size of their components and parts and the temperature and pressure in production processes where these machines are in use.

One of the biggest consumers of heavy metal parts is the power sector. Renewables (wind and photovoltaic power) become major electric power supply sources worldwide, but their output fluctuates daylong, destabilising the power system. That means that fossil fuel energy now and nuclear energy in the future are necessary to compensate power systems to deal with fluctuations in renewable output and customer demand. As a result, the manufacturing of big-size steel ingots and forgings from them will grow. The same tendency is in nuclear energy, the petrochemical industry, metallurgy, etc. For example, advanced ultra-supercritical steam turbine rotors operating at temperatures exceeding 700°C and pressure surpassing 30 MPa require clean steels and alloys with dense, stable, homogenous structures.

What makes ESR unique is its ability to combine remelting and gradual solidification of a continuously renewed liquid metal pool in a single process. This results in the formation of ingots that are much more chemically homogeneous and have a surface ready to be deformed due to a thin slag skin. Naturally, the constant complication requirements to products evoke new ESR technologies and equipment, which have undergone several advancements to keep up with the industry's ever-increasing demands. Therefore, the article aims to analyse the state-of-the-art electroslag remelting and identify trends for further development and improvement of technologies and equipment.

A SHORT LOOK AT THE HISTORY OF ELECTROSLAG REMELTING TECHNOLOGIES AND EQUIPMENT DEVELOPMENT

The priority in the ESR invention in its modern state (Figure 1a) with the use of alternating current of industrial frequency belongs to Borys Medovar (1916–2000) and Borys Paton (1918–2021). To eliminate misunderstandings that sometimes arise, we emphasise that in the 'Kellogg process', invented a decade earlier in the USA by H Hopkins, direct current was used as it was mistakenly assumed that the current flow through the slag was in the form of a soft arc (Hopkins, 1940) and the purpose of direct current usage was to minimise arc instability.

The first pilot furnace for melting 500 kg round ingots (type R-909) was built in 1956–1957 under Borys Medovar's leadership (Figure 1b).

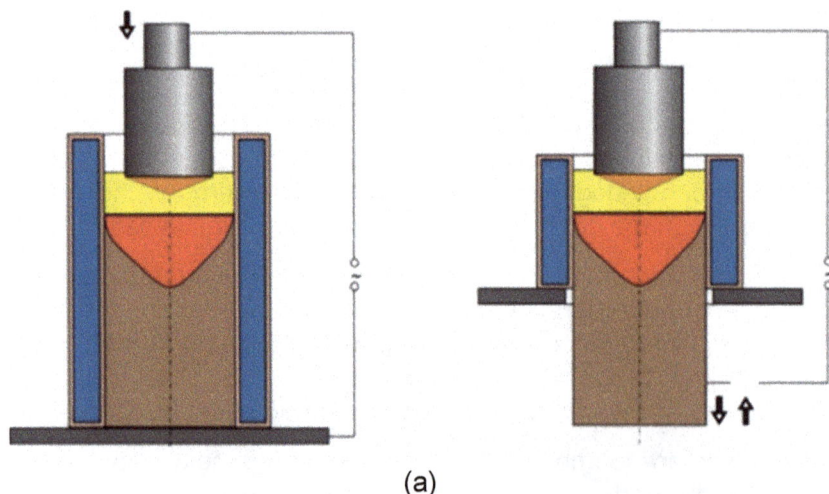

(a)

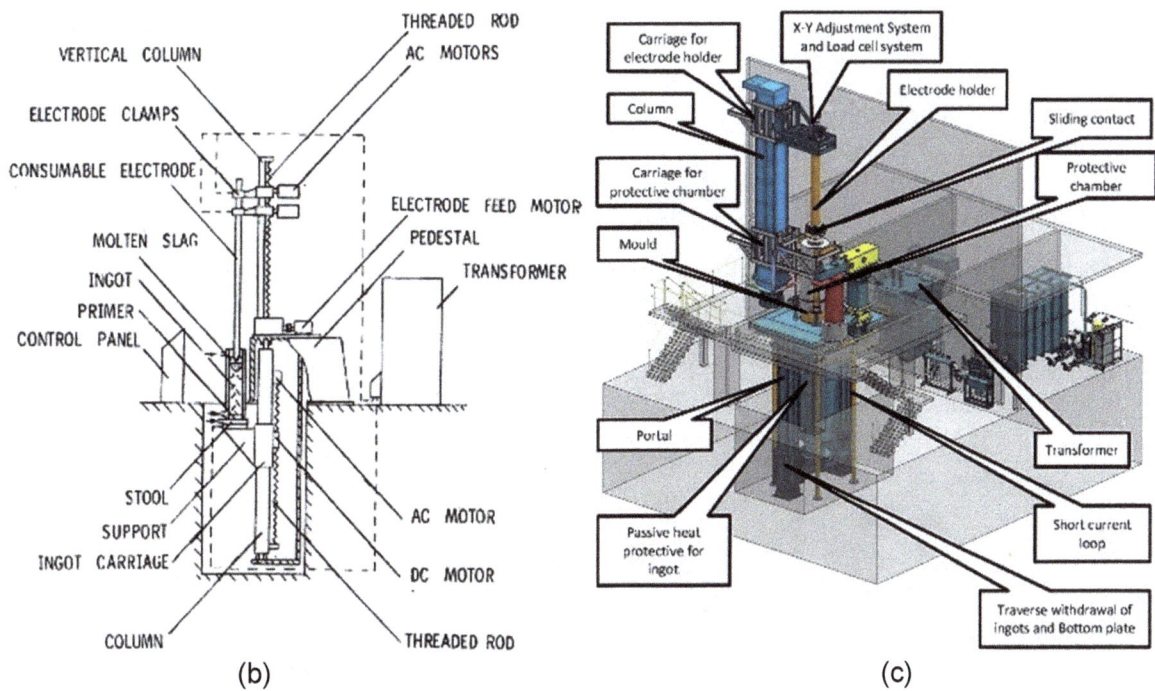

FIG 1 – The most common diagrams (a) of ESR remelting in a stationary mould (left) and short collar mould (right), the layout of the first furnace R-909 (b) of modern ESR process, and today's ESR plant structure (c).

The world's first electroslag furnace was put into industrial operation in 1958 at the Dniprospetsstal electrometallurgical plant in Zaporizhzhia, Ukraine. During the 70 years of electroslag remelting, engineers from many countries have invented, worked out, tested, or commercialised in the industry several drastic improvements of the original ESR of the consumable electrode and new types of related technologies.

Among them, the most widespread ESR technologies have become:

- Inert Gas Electroslag Remelting (IESR) – single phase one consumable electrode remelting at atmospheric pressure in a closed chamber filled with protective inert gas (argon) is today's benchmark process. Change of the consumable electrodes is used to reduce the height of the furnace (Holzgruber and Holzgruber, 2001; Jarczyk and Franz, 2012; Arh, Podgornik and Burja, 2016).

- Pressure Electroslag Remelting (PESR) – single phase one consumable electrode remelting in a closed chamber filled with nitrogen for melting steels with an excess nitrogen content (Stein and Menzel, 2014; Ritzenhoff et al, 2013).

- Three-phase ESR, using three parallel electrodes connected to a three-phase line or six electrodes arranged as three bifilar pairs, is still in use in China (Liu et al, 2021) for stainless and other steel remelting. Despite difficulties in providing reliable protection atmosphere and uneven melting of electrodes in bifilar pairs, it is a very good decision due to even electrical load.

The following options of ESR with a consumable electrode in the copper water-cooled mould are less general but have a certain potential in areas of their use:

- Electroslag Remelting Under Vacuum (VAC-ESR or VSR-Vacuum Electroslag Remelting) is designed for deep degassing during remelting of superalloys and titanium alloys, the efficiency of which was not high enough due to the barrier effect of the slag layer or gas evacuation and because of evaporation of slag components (Radwitz, Scholz and Friedrich, 2013).

- Arc Slag Remelting (ASR) is an energy-saving process of obtaining high-nitrogen steels and titanium and its alloys by remelting a consumable electrode due to an electric arc between it and the surface of the liquid slag bath (Medovar, 1997; Paton et al, 2004).

The recent technologies using the short collar current-supplying (synonym – current conductive) mould (Figure 2a) are still of interest:

- Electroslag Rapid Remelting Process (ESRR) for high-productive manufacturing of high-alloyed steels and superalloys by remelting a single large-diameter electrode (3–10 times larger than the ingot) in a T-shaped mould connected with a power source to increase process productivity (Alghisi, Milano and Pazienza, 2005; Karimi-Sibaki *et al*, 2018).

- Electroslag Remelting with Two Circuits (ESR TC) for gradual formation of the homogenous ingot from segregation-prone steels and superalloys (Medovar *et al*, 2005) due to breaking the rigid dependence between the speed of electrode melting and ingot formation to control of heat input and temperature of slag and metal baths to form shallow liquid metal bath, which is difficult to achieve in a standard ESR (Figure 2b).

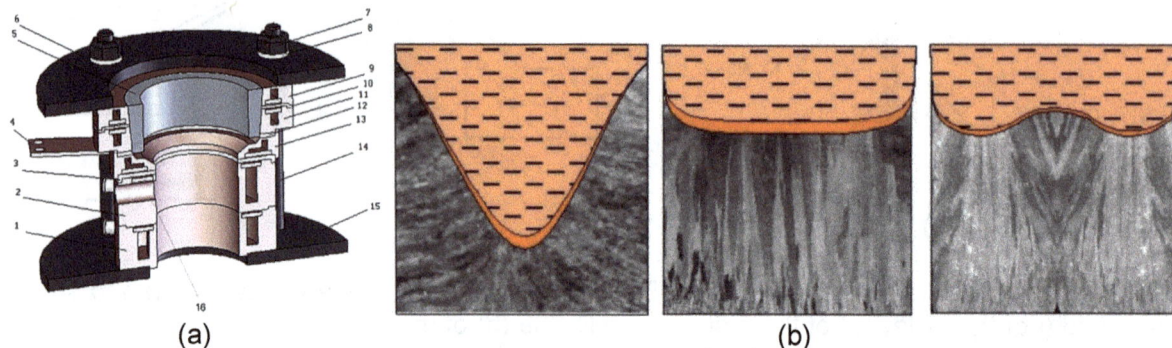

(a)   (b)

FIG 2 – Current supplying mold structure: (a) 1, 2 – forming section; 3, 8, 10, 12 – electric insulation; 7 – nut; 4 – current supplying bus; 5 – protective graphite ring; 6 – upper flange; 9, 11 – current supplying section; 13 – dividing section; 14 – stud; 15 – lower flange; 16 – metal level sensor); and liquid metal pool shape; (b) at different ESR methods: from left to right – standard ESR, ESR TC, ESR LM (Medovar, *et al*, 2005).

Extra opportunities are provided by electroslag processes using a current-supplying mould (Medovar, Stovpchenko and Petrenko, 2016; Paton *et al*, 2007; Medovar *et al*, 2018) without consumable electrodes:

- Electroslag Surfacing by and recycling of discrete materials (fines, powder, shavings etc) in a current supplying mould for the production or renewal of a working layer of rolling mill rolls, rotors of electric motors, stamps of high-speed and tool steels, tungsten carbides etc (Kuskov *et al*, 2018).

- Electroslag Refining using Liquid Metal (ESR LM) and Electroslag Surfacing by Liquid Metal – ESS LM (Figure 2b) for the production of highly homogeneous ingots of solid cross-section, hollow ingots and composites with coaxial or horizontal arrangement of layers from steels and alloys prone to segregation ((Medovar *et al*, 2018; Medovar *et al*, 2020).

This article compares the two most different technologies—classical ESR with consumable electrodes and ESR LM—to better understand and highlight their abilities to refine a metal's chemical composition and structure.

ELECTROSLAG TECHNOLOGIES' ABILITY TO REFINE THE CHEMICAL COMPOSITION OF METALS

Along with the term 'electroslag remelting', another name for this technology in English-language sources is 'electroslag refining'. The latter retained the context associated with one of the first purposes of the process – remelting ingots or rolled products rejected due to the high content of impurities, primarily sulfur and non-metallic inclusions (NMI). ESR technologies are limited in refining agent addition during the process, first, because of the lack of ability to control resulting changes in both slag and metal phases.

Refining from impurities and NMI during ESR is ensured by the use of fluoride and fluoride-oxide slags (system of CaF₂–Al₂O₃–CaO–(SiO₂, MgO)), which were initially selected from the experience of electro-slag welding. Pure calcium fluoride was the first slag for the ESR to maximise the refining effect from oxide, sulfide and nitrides. CaF₂ is still used today for remelting the most critical alloys (including titanium alloys) despite its high price, low conductivity (which makes an inversely proportional effect on energy consumption at ESR) and some technological limitations.

From the point of view of refining, the main impurity and alloying elements involved in the electroslag remelting of steels and nickel-based alloys can be divided into four main groups (Table 1).

TABLE 1

Elements classification on their behaviour in the ESR slag-metal system concerning Fe- and Ni-based alloys.

Process components	Elements present in metal, absent in initial slag and able to react with slag	Elements present in both phases, and able to transfer from one to another via exchange and redox reactions	Elements present in slag, absent in metal due to extra low solubility and acting in slag-metal reactions	Trace elements present in steel, absent in slag and do not react with slag
Consumable electrode	Fe, Mn, C, Cr, Ti, Mo, V, W, B	O, N, S, H, Al, Si		P, Ni, Cu, Sn, Pb, Zn, Sb
Slag: CaF₂–Al₂O₃–CaO–(SiO₂, MgO)		S^{2-}, O^{2-}, N^{4-}, $3Al^{3+}$, Si^{4+}	F^-, Ca^{2+}, Mg^{2+}, Na^+, K^+	
Reactions (compounds are coming into slag or forms NMI)	$X[Me]+Y\{O^{2-}\} = \{Me_xO_y\}$ $[Me] + \{S^{2-}\} = \{3MeS\}$ $\{Me_xO_y\}+\{O^{2-}\}=Me_xO_{y+1}$ $[Me] + \{N^{2-}\} = \{MeN\}$			No
Ingot	Fe↓, Mn↓, C↓, B↓, Cr↓, Ti↓, V↓, Mo↔, W↔,	S↓↓, N↔ O↑↓, Al↑↓, Si↑↓- Depending on concentrations	NMI of CaS, CaO, MgO, MgS and their complexes with other oxides and nitrides	P, Ni, Cu, Sn, Pb, Zn, Sb etc: all ↔
In process slag	FeO↑↑, MnO↑↑, MnS↑↑, B₂O₃↑, BN↑, Cr₂O₃↑, TiO₂↑, TiN, V₂O₅↑, MoO₃↔, WO₃↔	CaS↑↑ Al₂O₃↑↑↓, SiO₂↑↑↓↓, MgS↑, MgO↑ Depending on concentrations		No

1. Elements that are contained in the metal and are absent in the initial slag but can react with its components, which can pass from the metal to the slag and *vice versa* as a result of displacement or redox reactions, mainly forming oxides, sulfides, nitrides, or carbides. These are manganese, iron, carbon, chromium, vanadium, molybdenum, and other alloying elements, the oxidation of which violates the specified chemical composition, structure and steel properties. The same situation applies to titanium and boron, in which oxide addition in slag is made to prevent their enhanced oxidation, trying to shift the equilibrium in favour of metal.

2. Elements are contained in the slag and participate in slag-metal reactions but are not in the metal due to their vanishingly low solubility. These are fluorine, calcium, magnesium, alkali and alkaline earth metals. The oxides of the last group of elements ensure the removal of sulfur and change the composition of the initial non-metallic oxide-type inclusions from a consumable electrode.

3. Elements are present in both phases, which can transfer from one phase to another via an exchange in redox reactions. These are, first of all, sulfur, oxygen and nitrogen, the removal

of which from the metal composition is one of the tasks of the process, the realisation degree of which depends on their initial concentration (at high initial content in the metal the refining occurs and *vice versa*). In addition to impurities, aluminium and silicon play a substantial role in the refining processes, competing in redox reactions involving oxygen being an active steel deoxidiser and, simultaneously, are contained as oxides in the initial ESR slag in significant quantities.

4. Trace elements are present in the steel in low concentrations and are not existent in and interact with an initial slag. These are phosphorus, nickel, copper, lead, tin and others. Accordingly, metal refining from these elements in the ESR process using slag is practically impossible.

The following refining actions occur at the remelting with slags of CaF_2–Al_2O_3–CaO–(SiO_2, MgO) system that is typical for steels and nickel-base alloys:

- chemical interactions between the slag and metal (redox reactions, ie sulfides, oxides, nitrides formations/removal), which happen at all interfaces between slag and metal (Figure 3) and NMI removal to slag accompanied by change in their chemical composition and new inclusions formation occurring by:

  o NMI interaction with slag on all interfaces resulting in their composition change or complex particle formation.

  o Elements from NMI dissolve in a metal (thermal dissolution with no contact with slag) and re-precipitate at a temperature reducing while cooling and solidification form new particles in the metal, ie saturated by liquates near the growing solidification front (Burja *et al*, 2018; Persson *et al*, 2020).

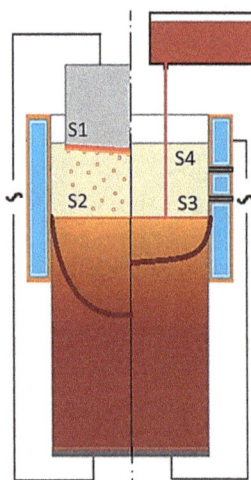

FIG 3 – The slag-metal interaction surfaces S1–S4 (description in text) at the ESR with a consumable electrode, left part, and ESR with liquid metal, right part (Stovpchenko *et al*, 2020).

The efficiency of NMI removal from metal depends on their properties (size, density, solid or liquid states and their surface tension with metal and slag) and the shapes and intensity of macro- and microflows in a metal pool. Depending on the combination of these factors, the non-metallic inclusions can be either assimilated by the slag surface, captured by the solidification front or still be circulating until the metal's last portions solidify.

Many years of ESR refining practice have proven a drastic reduction in the size and number of NMI and their type and composition changes. Endogenous particles of oxides, nitrides, sulfides and oxysulfides mostly represent NMI in ESR metals. However, exogenous inclusions from the lining of the steel-teeming ladle or tundish still occur (often, they differ by a high magnesium or silicon content depending on the lining type and casting powder). The size of inclusions found in laboratory-scale trials is usually between 1–5 µm but more commonly ≤2 µm. However, larger inclusions are often present in industrial-size ingots, especially when larger sample areas are analysed (Persson *et al*, 2021). The high thermal inertia of ESR and steady remelting conditions make it possible to organise microalloying the metal from slag. Still, control of results is post-melted, which is impossible to correct. The most used recently is boron-bearing additions to slag to keep stable boron content in the ESR ingot.

THE CONTRIBUTION OF SLAG-METAL INTERACTION SURFACES IN REFINING AT ELECTROSLAG TECHNOLOGIES

The issue about the main place of metal refining at electroslag remelting was disputable for years. To answer the question of the contribution of different interaction surfaces, we made two ingots using two different technologies in comparable conditions. In our comparative experiments, we use the same T-shape current supplying mould (Figure 1a) and produce two ingots from the same chemistry metal by standard ESR and by electroslag refining by liquid metal pouring (ESR LM). The same current supplying mould (180 mm in diameter), steel grade St45 and ANF29 slag (layer height of 200 mm) were used at the ESR and ESR LM to ensure comparable conditions. Process productivity was 120–160 kg/h (speed of ingot withdrawal – 15–17 mm/min). A comparison of the interaction surfaces was made for a 160 kg/h feed rate, which is typical for the stationary stage. The size and number of drops were measured and counted on a 160 mm electrode after its tip was rapidly frozen. The specific power consumption at ESR averaged 1350 kWh/ton of ingot (due to the necessity to melt a solid electrode), while at ESR LM – 900 kWh/ton at the same productivity level (Stovpchenko *et al*, 2020).

We compared cross-sections of contact surfaces and measured sulfur content before and after both technologies. In the standard ESR process, slag-metal interaction occurs on three contact surfaces (Figure 3).

The first surface (S1) belongs to the liquid metal film at the tip of the melting consumable electrode contacting with slag. The second area (S2) is the surface of the drops of liquid metal rain inside a slag layer. The third area (S3) is the interface between the liquid slag bath and metal pool – the same for both ESR processes with and without electrodes. The S4 is the analogue of S2, which refers to the square area of a liquid metal stream in the ESR LM process that works without consumable electrodes. A formal comparison of the values of slag-metal contact surfaces of a standard ESR with a consumable electrode and an ESR LM without an electrode supply was performed.

According to the characteristic filling ratio, the molten metal film surface (S1) on the electrode tip is 0.5 to 0.7 of the ingot surface (Figure 4a).

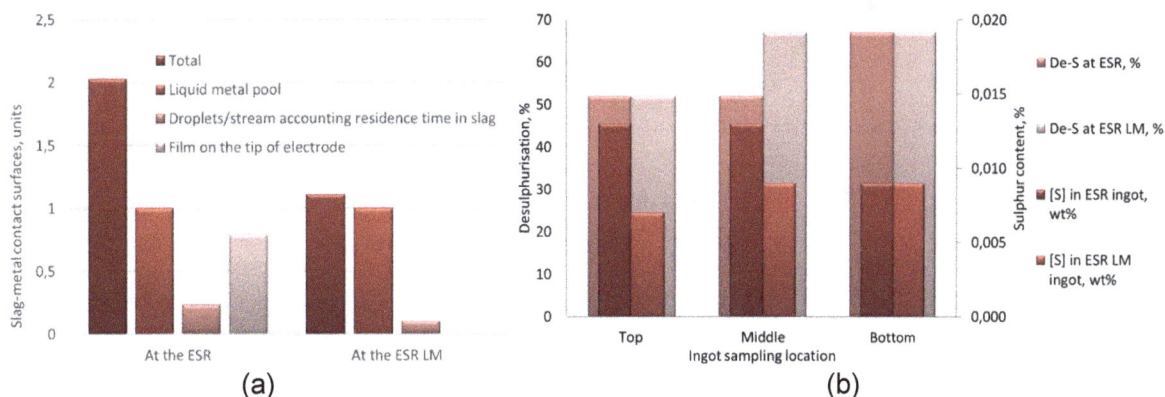

(a)   (b)

FIG 4 – Geometrically defined values of slag-metal contact surfaces (a) and the sulfur content and degree of desulfurisation (b) in ESR and ESR LM ingots at comparable conditions (Stovpchenko *et al*, 2020).

In our direct experiment, the total interface of the metal and slag phases in ESR was twice as large as that in ESR LM. Specifically, it was 2.03 in ESR and 1.12 in ESR LM. The main difference was due to the electrode surface, which was 0.79 in our experiment (due to the T-shape mould used).

The size of drops is directly related to the diameter of an electrode and can range from 1 to 10 mm found by direct experiments (Campbell, 1970) and numerical simulations (Kharicha, Ludwig and Wu, 2011; Wang *et al*, 2016; Dong *et al*, 2016; Liu *et al*, 2021). The largest drops have an equivalent diameter of 10 to 15 mm and the smallest drops measure 1 to 4 mm. Due to the small size and short residence time in the slag of the drops, the part of their surface constitutes 0.23 of the liquid metal pool. During the ESR LM process, the liquid metal stream (S4) has permanent contact with the slag bath and can be broken into smaller drops. For the sake of simplicity, we can ignore the drop formation and express the interaction surface as a cylinder having a stream diameter (up to 20 mm – 0.11 of the surface of the metal bath) and a length that matches the mass of the metal at the same productivity as a standard ESR process.

Direct comparison of sulfur content before and after ESR and ESR LM (Figure 4b) shows the same value of desulfurisation (Stovpchenko *et al*, 2020), proving that the contact surface between a slag bath and metal pool is the most effective interface in refining (first of all from sulfide and oxysulfide non-metallic inclusions).

It is worth highlighting that despite the absence of film at the tip of the electrode and another type of metal transfer through the slag (a stream instead of drops), the refining ability of the ESR-LM is the same. The reason is that, due to the lower density of non-metallic inclusions, they move towards the slag under the influence of the Archimedes force, and, in addition, the value of surface tension at a flat interface between slag and metal bath is less than at near-spherical shape surface for metal droplets in slag. In the film at the tip of the electrodes, the conditions for NMI removal are even worse because the Archimedes force counteracts their movement to the slag bath (Figure 5).

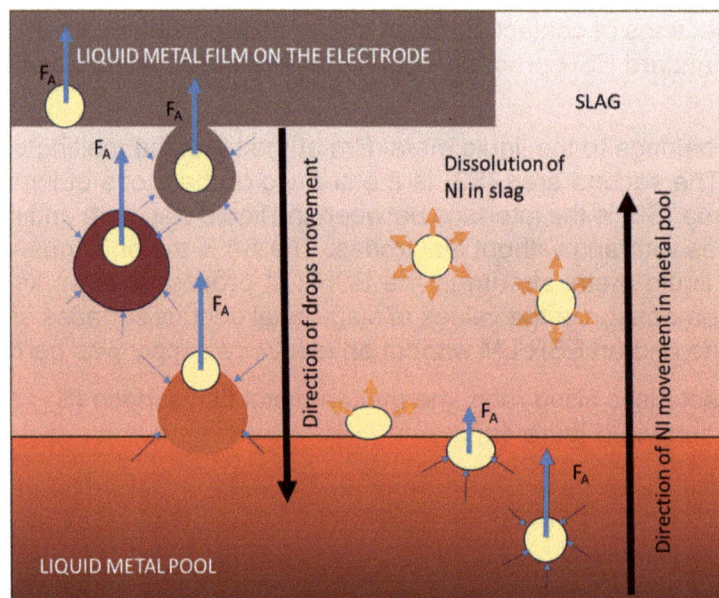

FIG 5 – Simplified scheme illustrating NMI (light spheres) movement on the different slag-metal contact at standard ESR: F_A is Archimede force, thin arrows show elements exchange between metal drops and slags.

The physical reason is that non-metallic inclusions move towards the slag under the influence of hydrostatic lift force (Archimedes' force – F_A) due to their lower density in comparison with liquid metal (three times and more difference). In addition, the value of surface tension at a flat interface between the slag bath and metal pool is less than at droplets having a near-spherical surface in slag. In a film at the tip of electrodes, the action of Archimedes' force on non-metallic inclusions is directed opposite to the location of the slag bath. In the liquid drops, the inner friction forces make them turbulent inside, make more difficult for NMI to be assimilated by slag. For these reasons, the main refining action of NMI removal much more easily occurs on the contact surface of the slag and metal baths and not in a film on the tip of the electrode.

Performed simulation of liquid metal drops' movement in the slag also shows that already during the passage of metal drops having size 2–10 mm through a 200 mm layer of slag, they heat up to the temperature of the slag (modelling parameters and more detailed results are available in (Stovpchenko *et al*, 2020). Large droplets move faster than small ones due to their internal flows and turbulence, making Stokes' law-based predictions inaccurate. Classical ESR has smaller droplets that move more slowly and heat up more than the ESR LM being compared. Turbulent flows (inside bigger liquid drops (from 6 to 10 mm) deformed their shape from sphere to flatter spheroid and also can cause their splitting into smaller parts, increasing the contact surface between metal and slag (Krivtsun et al, 2021).

The extrapolation of the established features of the movement of solid and liquid drops explains the better removal of liquid-state non-metallic inclusions. At smaller sizes, the difference in movement speeds between solid and liquid particles tends to be zero due to smaller inner friction in liquids, but the separation of liquid inclusions from steel proceeds similarly. In contrast, the assimilation of liquid inclusions in slag goes much easier without diffusion restrictions. Moreover, at a residence of liquid metal drops in oxygen-bearing ionic slag with free O^{2-} anion, oxidation of active elements from a liquid metal and other reactions between it and slag components are unavoidable and forming non-metallic inclusions can be removed from the liquid metal pool mainly.

Slag temperature also affects refining ability at the ESR because of its increasing enhanced dissolution of previously formed NMI and their assimilation due to lower viscosity and rate of chemical reactions between slag and metal, resulting in new NMI formation in the liquid metal drops and pool. The temperature of the slag bath at ESR with a consumable electrode is typically estimated to be 100–250 K higher than the liquidus temperature of the steel or alloy and of metal droplets overheat – 90–100°C (Klyuev and Volkov, 1984). The conventional ESR process uses the resistive type of heating that requires high overheating of slag to ensure contact melting of an electrode immersed in a slag, but ESR LM is devoid of this shortcoming. Due to an induction heated furnace, the liquid metal overheating is minimised to keep metal liquid during the short period (seconds) to be poured into the current supplying mould.

A decrease in the temperature of the slag bath is also favourable due to reducing the evaporation of volatile fluorides, which, along with exchange reactions, causes changes in the chemical composition and properties of the slag. No less important is the slag bath temperature for ESR ingot formation. It is the most important factor defining ESR process productivity, flow conditions and depth of the liquid metal pool and creating a high gradient of temperature is crucial for the solidification pattern and structure of the formed ingot.

Comparing the standard ESR and ESR LM allows asserting that the physicochemical conditions of refining in the ESR LM are more favourable due to the less slag-metal contact at the droplet stage and the consumable electrode absence with liquid metal film on it. At these contact surfaces in the standard ESR, conditions for non-metallic inclusion removal are less convenient and high temperatures can cause oxidation of elements from the metal composition. There is no need for electrode melting, resulting in reduced slag heating and no overheat of metal that comes with preset temperature from induction heated furnace (temperature 70–95° less than for drops at ESR), allowing the forming of a shallower liquid metal bath that is favourable for the solidification control of ingots from segregation-prone grades of steels and alloys, where ESR LM technology can be prospective. Moreover, no need to melt the electrodes reduces electricity consumption on the value of heat required for the solid-liquid phase transition.

Despite the attractiveness of ESR with the liquid metal, it has a specific niche. The standard ESR with consumable electrodes in protective gas is often the perfect decision, especially when strict gas protection is critical and stationary mould is preferable. The current supplying mould works with ingot withdrawing, which requires a more precise operation; thus, it is worth using the ESR with consumable electrodes when a very low melting rate and flat liquid metal pool are needed to reduce segregation or for hollow ingot manufacturing.

Summarising, all ESR technologies stand apart from the steelmaking primary and secondary refining methods: it is refining by high-temperature slag only; other possible refining actions are not effective because of no direct contact of gas phase or alloying additives with liquid metal, causing difficulties with vacuum treatment and alloying (microalloying) organisation and control. Despite the much higher temperature of the ESR process than any treatment in traditional steelmaking, single melts go hours and decent hours to receive a very dense dendrite structure with minimal segregation due to good conditions in water-cooled copper mould with the gradual renewal of liquid metal pool for metal solidification control and ingot structure refining.

Specific case – ESR of titanium as a way to dissolve titanium nitrides

It is often overlooked that the first ESR plant was constructed to manufacture titanium ingots. Being the first to attempt this was difficult and led to dangerous accidents, which prevented commercialisation for many years. High reactivity of titanium at liquidus temperatures, severe limitations on flux compositions and only alkaline-earth and lanthanide-series fluorides are suitable for titanium ESR (Paton *et al*, 1999).

Nevertheless, the research made worldwide on lab and pilot scales. The produced commercial batches of titanium and titanium alloys prove that it can compete with other remelting technologies (such as vacuum arc remelting (VAR) and electron beam cold hearth melting (EBCHM)) for titanium manufacturing in terms of ingot quality and product efficiency and even have an advantage – titanium refining from nitride inclusions, which was shown experimentally at titanium sponge remelting using ESR TC technology in the current supplying mould. The problem with hard-alpha phase defects is that the melting temperatures of these nitrogen-rich inclusions are much higher than the melting temperature of titanium itself. As a result, the time for dissolution of large inclusions coming from titanium sponge is longer than the time of their hovering in melting bath at VAR or EBCHM – standard now processes for titanium remelting.

ESR with active slags containing metallic calcium helps to accelerate the dissolution process (Figure 6), which value observed at level $2 \cdot 10^{-4}$ m/s (Paton *et al*, 1999) that is much faster than at VAR – $2.2 \cdot 10^{-6}$ m/s (Bewlay and Gigliotti, 1997) and EBM – 4.2×10^{-5} to 4.9×10^{-5} m/s (Xu *et al*, 2022).

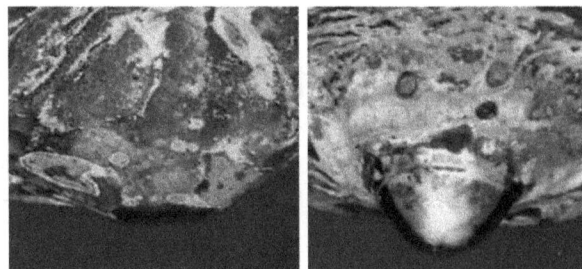

FIG 6 – The appearance of frozen after ESR tips of the consumable electrode with the films of nitrogen-rich phase-light spots (Paton *et al,* 1999).

The successful experiments proved a perspective to include electroslag remelting of sponge as a consumable electrode or lump pieces for refining titanium from large-size nitride inclusions to achieve its uniform distribution in ingot volume before the final VAR. The ability to refine titanium from nitride inclusions gives opportunities for cost-effective commercial ingot production using ESR.

The results allow us to conclude that the slag layer at ESR works like refining media and heat buffer, helping to provide enough time for inclusions to remove being assimilated by the slag bath or to be dissolved fully or partially to be reborn in smaller sizes and even distribution. That is why ESR refines titanium from hard-to-melt titanium nitrides and white spot defects are not characteristic of the ESR ingots.

ESR TECHNOLOGIES' ABILITY IN SOLIDIFICATION CONTROL FOR INGOT STRUCTURE REFINING

The industry has three primary technologies for receiving huge forge ingots – conventional casting, casting with electroslag hot topping and electroslag remelting. It is well known that the bigger an ingot diameter is, the longer its solidification takes, resulting in an elevated degree of macro-segregation. The final portions of solidifying metal in conventionally cast ingot hold the highest concentration of elements displaced by the solidification front, shrinkage and porosity. These parts are concentrated near the centre and top of the ingot. Therefore, huge forging ingots are designed with extra-hot top parts meant to be removed. Sometimes, the whole length of the central part is removed through trepanning. Naturally, this results in a low yield from ingot to ready forging, making sense for hollow ingot manufacturing. For ordinary people, it is difficult to imagine that for manufacturing backup roll or turbine rotors for nuclear powerplants weighing 250–260 t, the cast forging ingots of 700 t are in use (Figure 7). At heavy hollow vessels and reactors (for nuclear and petrochemical industry) manufacturing by forging from huge cast solid ingots, the metal yield to ready part also does not exceed 25–35 per cent (Medovar *et al*, 2023).

FIG 7 – Low-pressure rotor forging 408 t in weight maximal diameter 3650 mm forged from 715 t conventional casted ingot of 30Cr2Ni4MoV (Courtesy of China First Heavy Industries).

Electroslag hot topping eliminates shrinkage, and increases yield but only partially weakens segregation as it does not change the residual melt composition saturated by impurities. Chemical composition purifying helps but does not fully resolve this problem. Even at very low concentrations of segregation-prone impurities, inner defects can arise like well visible A-segregation in 20 t slab ingot (1680 × 780 mm) of 09Mn2SiV grade with sulfur content 0.004 per cent (Figure 8).

FIG 8 – Macrostructure of 20 t conventional cast slab ingot 780 mm side from 09Mn2SiV steel with sulfur content 0.004 per cent and heavy A segregation in the central part (hot etching in 50 per cent vol HCl).

The lines of A segregation form when solidification front stop – the point of columnar-to-equiaxed growth transition (CET) – a place where the heat sink is reduced, melt mixing flows low and impurities accumulate – so the non-metallic inclusions and excessive phases have enough time to grow and agglomerate. The mitigation of segregation is crucial in producing sound huge forging ingots, especially for sophisticated grades containing many alloying elements of varying physical properties and chemical activities in their composition.

The ESR rate of ingot formation is much lower than that of traditional metallurgy methods of pouring liquid metal into ingots or continuously cast billets. ESR deliberately significantly reduces the melting process productivity for the sake of solidification control for the high quality of an ingot, minimising the volume of the liquid metal pool, which helps prevent the segregation-caused redistribution of elements and formation of shrinkage defect.

The charge for the ESR is ingots or even forged or rolled billets after ladle treatment and vacuum degassing, which are pre-final products for most metallurgical applications. Naturally, the ESR with consumable electrodes is quite costly. However, the quality and yield of ESR ingot are significantly higher (85–95 per cent) because the volume of metal that solidifies in one time is smaller, its composition is constantly renewed and heat removal in a copper water-cooled mould is better than in an iron mould. Therefore, the efficiency rule is easy: the bigger the ingot size and alloy grade price, the more effective the use of ESR. The sulfur print of the ESR ingot of the same cross-section we planned to put aside, but it occurs white because for such a cross-section, due to progressive renewal of liquid metal, the pool provides a 100 per cent dendrite structure with a density close to the theoretical value.

Today, the ESR heavy forging ingots reach 250 t in weight and 2.6 m in diameter (Kubin *et al*, 2013; Bettoni *et al*, 2014). Nevertheless, the attempts to directly replace giant cast ingots with ESR-made ingots weighing approximately 400 t were unsuccessful because of growing segregation at increasing the ingot diameter to 3.5 m and, accordingly, the volume of simultaneously solidifying metal and depth of liquid metal pool.

The bigger the diameter of the ESR ingot, the deeper the liquid metal bath became and the solidification conditions tend to approach those in a conventional ingot. The zone of equiaxed crystal appears and as a result, segregation defects worsen.

The practice has shown that the ultimate diameter (mass) of the electroslag ingot, which retains quality advantages over the usual one (cast in iron mould), is different for steels and alloys of various degrees of alloying. Thus, for carbon and low-alloy steels, this ultimate diameter of ESR ingot is 2000–2600 mm (up to 250 t), for high-alloy steels (stamping, corrosion-resistant, and high-speed steels) and alloys (including some superalloys) it is significantly less – near 1000 mm (usually it is to 20 t), and for grade Inconel 718, the critical diameter is just 500 mm (2.5–3 t). The last limitation is well known due to Professor Alec Mitchel's calculation of Local Solidification Time (LST) showing that the critical point could be reached at an ESR ingot size of 1050 mm in diameter (Figure 9).

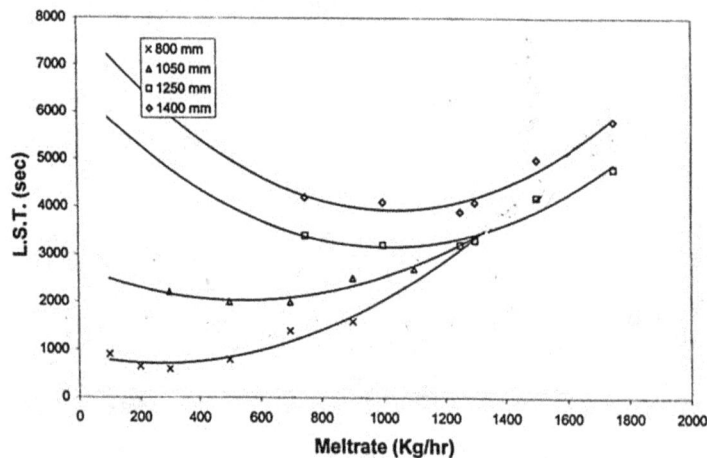

FIG 9 – LST versus ESR melt rate for different diameter ingots (Persson *et al*, 2021).

Detailed investigation (Persson *et al*, 2021) of NMI in ESR and PESR Remelted Martensitic Stainless Steel commercial ingots shows that as the ingot diameter increases and dendrite growth approaches the centre, the growth of inclusion number and size is near linear. Still, when the centre of the ingot solidifies in equiaxial mode, the increase in inclusion number and size is much higher. For the martensitic stainless steel grade 0.4Cr13MnSiV, the transition from dendritic to equiaxial solidification mode happens in the 800 to 1050 mm diameter ingot. It is also well-known that consumable electrode quality is critical for ESR ingot cleanness and casting is the critical procedure causing NMI entraining, mostly reoxidation products (Campbell, 2023).

The reason is that at traditional ESR with one consumable electrode, the ingot is formed under direct dependence of the melting speed on the electric power (which cannot be less than a certain value to provide a smooth surface). The central supply and peripheral removal of heat lead to the formation of a deep metal bath of a conical shape, which solidification can be accompanied by segregation whose manifestation is the bigger, the larger the cross-section. Therefore, the problem of producing high-quality, huge-size ingots must be resolved using approaches other than increasing diameter.

There are three ways to resolve this task with increasing ingot weight at keeping or increasing ingot diameter:

1. Increase the ESR ingot length.

2. Maximally reduce ESR technology productivity as was shown at ESR TC technology (Medovar *et al*, 2005; 2018) to keep shallow liquid metal bath and at ESR LM with an additional reduction of an overheat of slag and metal (Medovar *et al*, 2018; 2020).

3. Reduce the cross-section of solidifying metal by using an inner cooler – hollow ingot manufacturing or enlargement of smaller diameter and weight ingot into larger one. It is obvious that the reduction of a cross-section of an ingot diminishes the volume of simultaneously solidifying metal and weakens segregation (Medovar *et al*, 2011, 2023; Medovar, Stovpchenko and Petrenko, 2016).

Colleagues from INTECO (personal communication, 2023) decided to use the first approach: to keep the already reached diameter of the ingot with satisfying quality and to increase its length and weight to 300 t for a confidential customer in China. This equipment and technology implement the change of consumable electrodes during remelting, ie in fact, there is a kind of liquid phase welding of individual parts of a large ingot into a single whole and an increase in the mass of the ingot is achieved through increased length. Such giant production takes weeks and its handling is not easy.

We suppose the realisation of the last listed approach using technology with the current mould supply and producing huge weight ingots with central coolers as it is realised in hollow or enlarged ingots. Much smaller consumable electrodes (ESR TC technique) with less inherited segregation or liquid metal supply can be used for such ingot manufacturing. In the last case, we avoid the stage of consumable electrode casting/forging (rolling), where large-size NMI can be captured from the casting mixture or formed in the central part of the ingot when equiaxed dendrite growth occurs.

In traditional ESR, a depth of liquid metal bath approximates an ingot radius or more. However, at hollow or enlarged (by surfacing) ingots formation by ESR LM in current supplying mould, the liquid metal bath can be drastically reduced and reshaped, as shown in Figure 1 and proved by macrostructures in Figure 10a, 10b. The electroslag enlargement technology divides the cross-section of an electroslag ingot into coaxial layers, which are formed sequentially. Even better conditions to form the sound low-segregation ingot of heavy weight with fewer cross-sections of simultaneously solidifying metal are in ESR hollow ingot for large pipes and shell manufacturing.

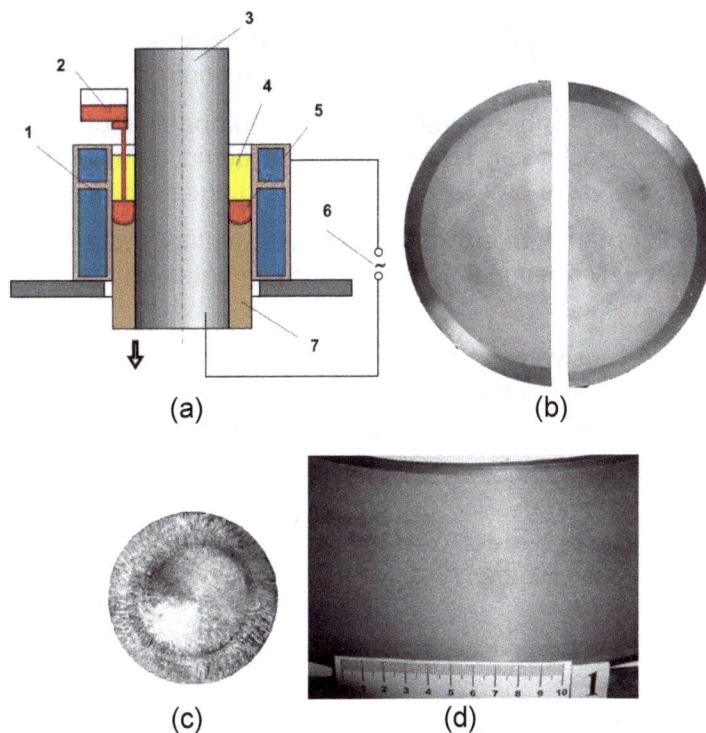

(a) (b)

(c) (d)

FIG 10 – ESR enlargement/surfacing/hollow ingot manufacturing principle: (a) 1 – current supplying mould; 2 – pouring furnace; 3 – ingot to enlarge/billet for surfacing/water-cooled inner mould; 4 – slag bath; 5 – liquid metal pool; 6 – power source; 7 – ready product; (b) cross-section of 740 mm HSS surfaced rolled mill roll; (c) lab-scale enlarged ingot 70/110 mm; (d) hollow ingot wall; were produced by ESR/ESS LM (Medovar *et al*, 2018).

The enlargement, in principle, is the same method as electroslag surfacing by liquid metal (ESS LM- Figure 10a), successfully developed for surfacing a working layer of rolling mill rolls at NKMZ (Ukraine). The fusion zone is of high quality (Figure 10b), even when dissimilar steels in the central ingot and the deposited layer are connected.

Central ingot can be done by ESR or other methods providing the required quality. When the central ESR ingot is settled upside down, the parquet-like arrangement of its dendrites to dendrites of deposited layers increases metal properties.

Having an inner cooler inside the enlarged ingot solidification occurs at much better conditions than a solid one with the same diameter: sizes of carbides and the distance between axes of secondary dendrites in each layer, including the central ingot, are sufficiently less according to both modelling and experimental results (Medovar *et al*, 2009). Due to the two-stage process, the enlargement of ingots by surfacing is *a priori* more expensive than conventional ESR or ESR LM, so it is advisable to use this method for expensive alloys prone to segregation. We believe the enlargement method will find its place primarily for large-diameter forging ingots from superalloys. Since today, the maximum diameter of such ingots has been less than 1050 mm because of freckle formation. The deposition of a 500–800 mm thick layer around the central ingot with a critical diameter of 1050 mm makes it possible to obtain an ESR ingot of a size desirable for forging rotors or large rolling rolls. The appropriate equipment and slags were designed and tested to ensure the reliable connection of similar or different composition layers in composite ingots into a single whole and also hollow ingot manufacturing (Medovar *et al*, 2018; Stovpchenko *et al*, 2018), making possible start wide implementation of the ESR/ESS LM technologies.

Employing ESR technology with current supplying mould and, especially, using liquid metal, including for electro slag enlargement and hollow ingots, to produce large-diameter forging ingots with prevented development of segregation due to dividing cross-section and reducing metal overheat is prospective.

CONCLUSION

The concept of metal cleanness in ESR ingots is considered in terms of refining the metal from sulfur and non-metallic inclusions in combination with solidification control.

Analysis of trends in engineering further development and comparison of modern ESR technological varieties show prospects of electroslag technologies, including techniques with a liquid metal to achieve high-quality ingots and final products, including forgings for power engineering.

It has been demonstrated through modelling and experiments that metal melt refining from sulfides and oxysulfides most effectively goes in the contact surface between the slag bath and liquid metal pool.

The experimental studies conclusively prove that ESR has the potential to purify titanium from nitride inclusions (hard alpha phase) having high melting temperature, which are otherwise impossible to remove using other remelting methods for producing ingots from titanium and titanium alloys.

The potential of employing ESR technology with liquid metal, including electroslag enlargement, to produce large-diameter forging ingots with prevented development of segregation due to dividing cross-section and reducing slag and metal overheating is substantiated.

REFERENCES

Alghisi, D, Milano, M and Pazienza, L, 2005. From ESR to continuous CC-ESRR process: development in remelting technology towards better products and productivity, *La Metallurgia Italiana*, 1:21–32.

Arh, B, Podgornik, B and Burja, J, 2016. Electroslag remelting: A process overview, *Materiali in Tehnologije*, 50(6):971–979. https://doi.org/10.17222/mit.2016.108

Bettoni, P, Biebricher, U, Franz, H, Lissignoli, A, Paderni, A and Scholz, H, 2014. Large ESR forging ingots and their quality in production, *La Metallurgia Italiana*, 10:13–21.

Bewlay, B P and Gigliotti, M F X, 1997. Dissolution rate measurements of TiN in Ti-6242, *Acta Materialia*, 45(1):357–370. https://doi.org/10.1016/s1359-6454(96)00098-5

Burja, J, Tehovnik, F, Godec, M, Medved, J, Podgornik, B and Barbič, R, 2018. Effect of electroslag remelting on the non-metallic inclusions in H11 tool steel, *Journal of Mining and Metallurgy, Section B, Metallurgy*, 54(1):51–57. https://doi.org/10.2298/jmmb160623053b

Campbell, J, 1970. Fluid flow and droplet formation in the electroslag remelting process, *JOM*, 22(7):23–35. https://doi.org/10.1007/bf03355649

Campbell, J, 2023. A Future for Vacuum Arc Remelting and Electroslag Remelting: A Critical Perspective, *Metals*, 13(10):1634. https://doi.org/10.3390/met13101634

Dong, Y-W, Jiang, Z-H, Fan, J-X, Cao, Y-L, Hou, D and Cao, H-B, 2016. Comprehensive mathematical model for simulating electroslag remelting, *Metall Mater Trans B*, 47:1475–1488. https://doi.org/10.1007/s11663-015-0546-0

Holzgruber, W and Holzgruber, H, 2001. Development trends in electroslag remelting, *Medovar Memorial Symposium 2001*, pp 71–77.

Hopkins, R K, 1940. Production of High quality metal in a water cooled mould, US Patent No. US2191479.

Jarczyk, G and Franz, K, 2012. Vacuum melting equipment and technologies for advanced materials, *Archives of Materials Science and Engineering*, 56(2):82–88.

Karimi-Sibaki, E, Kharicha, A, Wu, M, Ludwig, A, Bohacek, J, Holzgruber, H, Ofner, B, Scheriau, A and Kubin, M, 2018. A multiphysics model of the electroslag rapid remelting (ESRR) process, *Applied Thermal Engineering*, 130:1062–1069. https://doi.org/10.1016/j.applthermaleng.2017.11.100

Kharicha, A, Ludwig, A and Wu, M, 2011. 3D Simulation of the Melting during an Electro-Slag Remelting Process, in *EPD Congress 2011* (eds: S N Monteiro, D E Verhulst, P N Anyalebechi and J A Pomykala), pp 770–778 (John Wiley & Sons, Inc.: Hoboken). https://doi.org/10.1002/9781118495285.ch84

Klyuev, M, and Volkov, S, 1984. Electroslag remelting, Moscow, *Metallurgiya* [in Russian], 208 p.

Krivtsun, I V, Sidorets, V M, Sybir, A V, Stovpchenko, G P, Polyshko, G O and Medovar, L B, 2021. Effect of deformation of molten metal drops on their movement and heating in a slag layer at ESR, *Sovremennaâ Èlektrometallurgiâ*, 2021(1):9–16. https://doi.org/10.37434/sem2021.01.01

Kubin, M, Scheriau, A, Knabl, M, Holzgruber, H and Kawakami, H, 2013. Operational Experience of Large Sized ESR Plants and Attainable Quality of ESR Ingots with a Diameter of up to 2600mm, in *Proceedings of the 2013 International Symposium on Liquid Metal Processing & Casting*, pp 57–64. https://doi.org/10.1007/978-3-319-48102-9_8

Kuskov, Y M, Soloviov, V G, Osechkov, P P and Osin, V V, 2018. Electroslag surfacing of billet end faces with application of consumable and nonconsumable electrodes, *The Paton Welding Journal*, 2018(2):38–41. https://doi.org/10.15407/tpwj2018.02.08

Liu, Z-L, Medovar, L, Stovpchenko, G, Petrenko, V, Sybir, A and Volchenkov, Y, 2021. Phenomena at three-phase electroslag remelting, *China Foundry*, 18(6):557–564. https://doi.org/10.1007/s41230-021-0125-8

Maknenko, V, Medovar, L, Saenko, V and Korolyova, T, 2009. Modeling of the ESR ingot enlargement, *International Symposium on Liquid Metal Processing and Casting (LMPC-2009)*, pp 287–294.

Medovar, B I and Boyko, G A (eds), 1991. *Electroslag Technology* (Springer eBooks). https://doi.org/10.1007/978-1-4612-3018-2

Medovar, B I, 1997. Arc-Slag remelting of steel and alloys (Cambridge International Science Publication). https://ci.nii.ac.jp/ncid/BA27851308

Medovar, L B, Petrenko, V L, Tsykoulenko, A K, Saenko, V Y and Fedorovsky, B B, 2005. ESR with two power sources and process control, *International Symposium on Liquid Metal Processing and Casting (LMPC-2005)*, pp 131–135.

Medovar, L B, Polishko, G, Petrenko, V and Stovpchenko, G, 2020. Modern electroslag technologies of electrode remelting and processing of liquid metal (review), *Theory and Practice of Metallurgy*, 125:17–25. https://doi.org/10.34185/tpm.2.2020.03

Medovar, L B, Stovpchenko, A P, Saenko, V Ya, Noshchenko, G V, Fedorovskii, B B, Petrenko, V L, Lantsman, I A and Zhuravel, V M, 2011. Concept of a universal ESR furnace for the production of large ingots, *Russian Metallurgy (Metally)*, 2011(12):1118–1123. https://doi.org/10.1134/s003602951112010x

Medovar, L, Polishko, G, Stovpchenko, G, Kostin, V, Tunik, A and Sybir, A, 2018. Electroslag refining with liquid metal for composite rotor manufacturing, *Archives of Materials Science and Engineering*, 2(91):49–55. https://doi.org/10.5604/01.3001.0012.5489

Medovar, L, Stovpchenko, G and Petrenko, V, 2016. Current Supplying Mould (CSM®) — Possibilities and Restrictions, *Medovar Memorial Symposium, MMS-100*, pp 180–182.

Medovar, L, Stovpchenko, G, Sybir, A, Gao, J, Ren, L and Kolomiets, D, 2023. Electroslag hollow ingots for nuclear and petrochemical pressure vessels and pipes, *Metals*, 13(7):1290. https://doi.org/10.3390/met13071290

Paton, B E, Medovar, B I and Latash, Y V, 1958. The Electroslag Melting of Steels and Alloys in Water Cooled Copper Moulds, *Automaticheskaya Svarka*, 11:5–15.

Paton, B E, Medovar, B I, Benz, M G, Nafziger, R H and Medovar, L B, 1999. ESR for titanium: yesterday, today, tomorrow, *Titanium-9, Science and Technology 1999*, 3:1385–1398.

Paton, B E, Medovar, L B and Saenko, V Y, 2007. ESS LM as a Way for Heavy Ingots Manufacturing, *International Symposium on Liquid Metal Processing and Casting (LMPC-2007)* pp 23–28.

Paton, B E, Saenko, V Ya, Pomarin, Yu M, Medovar, L B, Grigorenko, G M, Fedorovskii, B B, Petrenko, V L and Chernets, A V, 2004. Arc slag remelting for high strength steel and various alloys, *Journal of Materials Science*, 39(24):7269–7274. https://doi.org/10.1023/b:jmsc.0000048741.47509.b3

Persson, E S, Karasev, A, Mitchell, A and Jönsson, P G, 2020. Origin of the inclusions in Production-Scale electrodes, ESR ingots and PESR ingots in a martensitic stainless steel, *Metals*, 10(12):1620. https://doi.org/10.3390/met10121620

Persson, E S, Brorson, S, Mitchell, A and Jönsson, P G, 2021. Impact of solidification on inclusion morphology in ESR and PESR remelted martensitic stainless steel ingots, *Metals*, 11(3):408. https://doi.org/10.3390/met11030408

Radwitz, S, Scholz, H and Friedrich, B, 2013. Investigation of Slag Compositions and Pressure Ranges Suitable for Electroslag Remelting under Vacuum Conditions, in *Proceedings of the 2013 International Symposium on Liquid Metal Processing & Casting*, pp 87–93. https://doi.org/10.1007/978-3-319-48102-9_12

Ritzenhoff, R, Medovar, L, Petrenko, V and Stovpchenko, G, 2013. Comparison of ARC Slag Remelting vs, P-ESR Melting for High Nitrogen Steels, in *Proceedings of the 2013 International Symposium on Liquid Metal Processing & Casting*, pp 159–162. https://link.springer.com/chapter/10.1007/978-3-319-48102-9_24

Stein, G and Menzel, J, 2014. High pressure electroslag remelting – a new technology of steel refining, *International Journal of Materials and Product Technology*, 10(3–6):478–488. https://doi.org/10.1504/IJMPT.1995.036470

Stovpchenko, G P, Sybir, A V, Polishko, G O, Medovar, L B and Gusiev, Ya V, 2020. Mass Transfer in Electroslag Processes with Consumable Electrode and Liquid Metal, *Uspehi Fiziki Metallov*, 21(4):481–498. https://doi.org/10.15407/ufm.21.04.481

Stovpchenko, G, Medovar, L, Polishko, G, Goncharov, I and Lisova, L, 2018. Self-disintegrating slag for electroslag remelting of hollow ingot, *Ironmaking and Steelmaking*, 46(8):782–788. https://doi.org/10.1080/03019233.2018.1428418

Wang, H, Zhong, Y, Li, Q, Fang, Y, Ren, W, Lei, Z and Ren, Z, 2016. Visualisation study on the droplet evolution behaviours in electroslag remelting process by superimposing a transverse static magnetic field, *ISIJ International*, 56(2):255–263. https://doi.org/10.2355/isijinternational.isijint-2015-581

Xu, J, Ou, J, Lane, C D, Cockcroft, S L, Maijer, D M, Akhtar, A and Marciano, Y, 2022. Dissolution of Ti–N inclusions in liquid titanium during electron beam melting, *Journal of Materials Research and Technology*, 17:1522–1539. https://doi.org/10.1016/j.jmrt.2022.01.096

Hydrogen plasma in extractive metallurgy application

B Satritama[1], D Fellicia[2], M I Pownceby[3], S Palanisamy[4], A Ang[5], G A Brooks[6] and M A Rhamdhani[7]

1. PhD Student, Swinburne University of Technology, Melbourne Vic 3122.
   Email: bsatritama@swin.edu.au
2. PhD Student, Swinburne University of Technology, Melbourne Vic 3122.
   Email: dfellicia@swin.edu.au
3. Senior Principal Research Scientist, CSIRO Mineral Resources, Clayton Vic 3168.
   Email: mark.pownceby@csiro.au
4. Director of Manufacturing Futures Research Platform, Swinburne University of Technology, Melbourne Vic 3122. Email: spalanisamy@swin.edu.au
5. Co-Director of Space Technology and Industry Institute, Swinburne University of Technology, Melbourne Vic 3122. Email: aang@swin.edu.au
6. Professor, Swinburne University of Technology, Melbourne Vic 3122.
   Email: gbrooks@swin.edu.au
7. Professor, Swinburne University of Technology, Melbourne Vic 3122.
   Email: arhamdhani@swin.edu.au

ABSTRACT

Metal production have long been using carbon sources as both reducing agents and energy sources. Consequently, the global extractive metal sector contributes significantly to greenhouse gas emissions, accounting for approximately 9.5 per cent. Hydrogen gas offers as a promising eco-friendly alternative to carbon in metallurgical processes, serving as both a reductant and energy supplier with a by-product being only water vapour. However, the implementation of molecular hydrogen faces certain challenges related to the thermodynamics and kinetics of metal oxide reduction. In addressing these challenges, researchers have explored the application of hydrogen plasma, generated by subjecting molecular hydrogen to high energy to produce atomic, ionic and excited hydrogen species. Hydrogen plasma offers thermodynamic and kinetic advantages over molecular hydrogen and carbon-based reductants, exhibiting lower standard Gibbs free energy of reaction and activation energy. Therefore, hydrogen plasma can produce metal in fewer steps, process any oxide feed and feed size and even be used to refine metals. Despite these advantages, challenges exist in utilising hydrogen plasma in extractive metallurgy, including electricity costs, potential reverse reactions and industrial-scale implementation. This study provides a mini review of prior research on hydrogen plasma for metal oxides reduction, particularly iron oxide, as well as state-of-the-art techniques for its use in extractive metallurgy applications by mentioning several reactor types. Future prospects and scale-up possibilities of the hydrogen plasma in extractive metallurgy will also be presented.

INTRODUCTION

The field of extractive metallurgy continues to grow, driven by the growing demand for metals in human needs. Traditionally, carbon-based reductants and energy sources have been heavily relied upon due to their affordability and abundance. However, this reliance contributes to climate change, with carbon dioxide (CO_2) emissions accounting for a significant 9.5 per cent of global greenhouse gas emissions (Carvalho, 2023). Recently, there is a rising interest in hydrogen as a sustainable alternative for various sectors, including extractive metallurgy, as it holds the potential to eliminate scope 1 CO_2 emissions. Unlike traditional methods, using hydrogen in metal production results in the production of water vapour as a byproduct. Despite its promise, hydrogen faces challenges as a weaker reductant compared to carbon at higher temperatures based on thermodynamics. One alternative to address this challenge is the utilisation of hydrogen plasma, a higher state of hydrogen gas. Hydrogen plasma is achieved by subjecting molecular hydrogen to high energy until it transforms into excited (H_2^*), atomic (H), and/or ionised (H^+) hydrogen. The main advantages of hydrogen plasma include its ability to enhance the feasibility of metal oxide reduction and its reaction rates. Its high thermal conductivity facilitates efficient heat transfer, while low viscosity enables rapid mass transfer (Sahu, 2014). This mini review paper systematically compiles and reviews recent and current studies on the application of hydrogen plasma in the context of extractive metallurgy. The aim is to provide information that supports future scale-up efforts. A more comprehensive and complete review on the topic is presented by the authors elsewhere (Satritama *et al*, 2024).

OVERVIEW OF HYDROGEN PLASMA

To generate a plasma state in a specified reactor, energy in the form of DC or AC electricity and/or electrodeless microwave electromagnetic must be applied to the gas. When an electric arc forms between two electrodes or an electromagnetic field is present in the gas, the gas becomes electrically conductive, leading to excitation, dissociation and eventual ionisation. Hydrogen fully dissociating into H atoms at 5000°C and complete ionisation occurring at 25 000°C under atmospheric pressure (Satritama *et al*, 2024). Lowering the pressure significantly decreases the required temperatures for both dissociation and ionisation. Plasma is broadly categorised as thermal or non-thermal, also known as non-equilibrium or cold plasma. Thermal plasma is characterised by uniform temperature among all plasma particles (electrons, ions and neutral atoms), while non-thermal plasma exhibits a temperature differences, with heavy particles (T_i and T_n denoting ions and neutral species temperature) significantly lower than free electrons (T_e). This temperature difference arises from low particle collisions at low pressure conditions, typically lower than 1 atm (Inui *et al*, 2010). There are a few plasma reactors used for metallurgical application which include batch, continuous feeding and in-flight reactors (Figure 1). The difference between continuous feeding and in-flight reactors is that in-flight reactors combine continuous feed and plasma gas into a single stream.

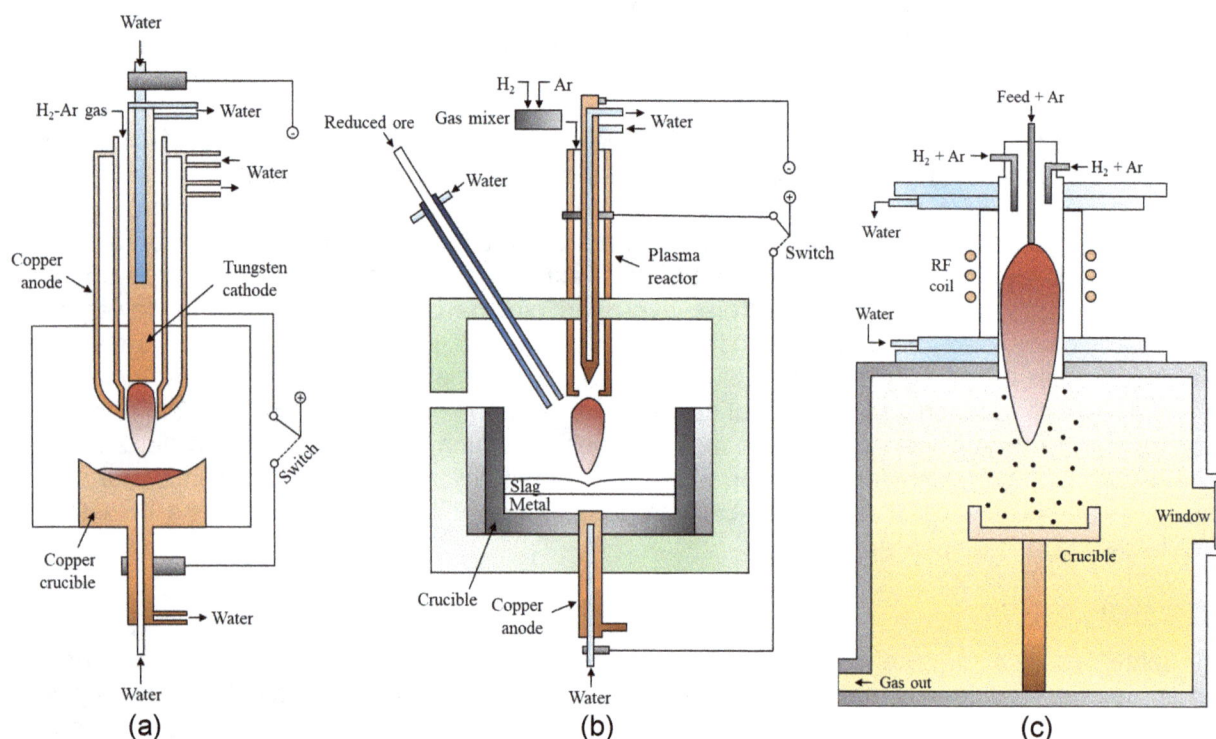

FIG 1 – Type of plasma reactors, including: (a) batch DC non-transferred and transferred plasma, (b) continuous DC non-transferred and transferred plasma, (c) in-flight microwave plasma. Adapted from (Yvon *et al*, 2003; De Sousa *et al*, 2016; Rains and Kadlec, 1970) by (Satritama *et al*, 2024).

HYDROGEN PLASMA FOR EXTRACTIVE METALLURGY

Thermodynamics and kinetics of hydrogen plasma in extractive metallurgy

Hydrogen plasma demonstrates the ability to reduce a broad spectrum of metal oxides, overcoming limitations observed with molecular hydrogen. Thermodynamic calculations reveal that the use of molecular hydrogen for most metal oxide reductions tends to yield positive Gibbs free energy, hinders spontaneous reaction. In contrast, hydrogen plasma addresses this challenge due to its enhanced energy profile. An Ellingham diagram (Figure 2) was generated using FactSage 8.2 database, illustrating the Gibbs free energy of metal oxide formation (from its elemental state with 1 mol of O_2) alongside reductants like carbon, hydrogen gas and hydrogen plasma at a total pressure of 1 atm. It is evident that carbon can serve as a reductant for most metal oxides at high temperatures, while hydrogen gas can reduce specific metal oxides at all temperatures and others at higher temperatures. In contrast, the application of hydrogen plasma in the form of atomic H and/or H^+ significantly broadens the range of reducible oxides, offering the additional advantage of lower temperatures favouring the reduction process.

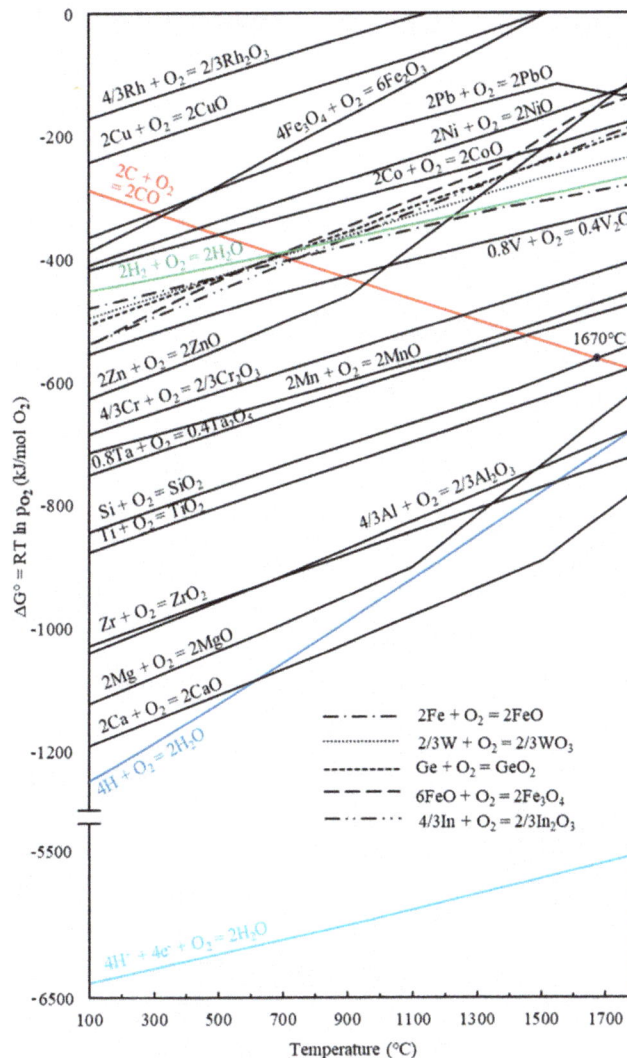

FIG 2 – Ellingham diagram of metal oxides stability in a function of temperature (Satritama *et al*, 2024).

In addition to meeting the Gibbs free energy criteria, the reactants must surpass the activation energy for a chemical reaction to take place. Activation energy (E_a) represents the minimum energy required for a reaction to proceed. The reduction of metal oxide to metal using molecular hydrogen may face challenges due to a high activation energy ($E_{a,1}$). Hydrogen plasma, in atomic ($E_{a,2}$) and excited atomic hydrogen forms ($E_{a,3}$), can significantly decrease the activation energy of the process, with the order $E_{a,1} > E_{a,2} > E_{a,3}$ (Sabat *et al*, 2013). There is limited research on the kinetics of metal oxide reduction using hydrogen plasma. Cobalt oxide reduction involves multiple steps, with the activation energy from Co_3O_4 to CoO (26.2 kJ/mol) higher than from CoO to Co (13.3 kJ/mol) metal (Sabat *et al*, 2015). Two kinetic studies on Cu_2O reduction to Cu metal indicate that the activation energy is nearly four times lower at lower temperatures (300–400°C) (Ramos *et al*, 2021), that is 19.06 kJ/mol, compared to higher temperatures (470–900°C) (Sabat, Paramguru and Mishra, 2016), that is 75.64 kJ/mol. These results align with hydrogen plasma's preference for lower temperatures as a reductant. A similar trend is observed in the study of Fe_2O_3 reduction to Fe metal by Rajput *et al*

(2013, 2014) with the activation energy four times lower at lower temperatures (300–800°C), that is 5.069 kJ/mol, compared to higher temperature (690–1150°C), that is 20.54 kJ/mol. Gonoring *et al* (2022) reported a significantly higher activation energy for Fe_2O_3 reduction (68.5 kJ/mol), attributed to the lower frequency parameter of microwave plasma (3.4 kHz) compared to Rajput *et al* (2013) (2.45 GHz), resulting in lower energy during the induction process by the microwave plasma and incomplete dissociation and/or ionisation of hydrogen gas.

This mini review briefly discusses the utilisation of hydrogen plasma for metal oxides reduction, where a more comprehensive discussion of its broader application in extractive metallurgy, which include for reduction, metals refining and wastes processing, is presented in other work (Satritama *et al*, 2024).

Reduction of metal oxides using hydrogen plasma

The reduction of metal oxides using hydrogen plasma is categorised into heterogeneous reactions, involving the interaction between hydrogen plasma and solid/liquid metal oxides and homogeneous reactions. Certain metals, such as aluminium and chromium, are efficiently produced through homogeneous reactions. This process involves vapourising the metal oxide, breaking the metal-oxygen bond, utilising hydrogen plasma to bind with the oxygen and depositing the pure metal vapour in a cooler zone. Referred to as dissociative reduction, this method requires a water-cooled reactor wall and a quench probe that can optionally inject quench gas (Rains, 1968). General reaction mechanism is given in the Figure 3 by distilling from several experimentation results (Sabat *et al*, 2015; Sabat, Paramguru and Mishra, 2018; Sabat, 2019; Zarl *et al*, 2022).

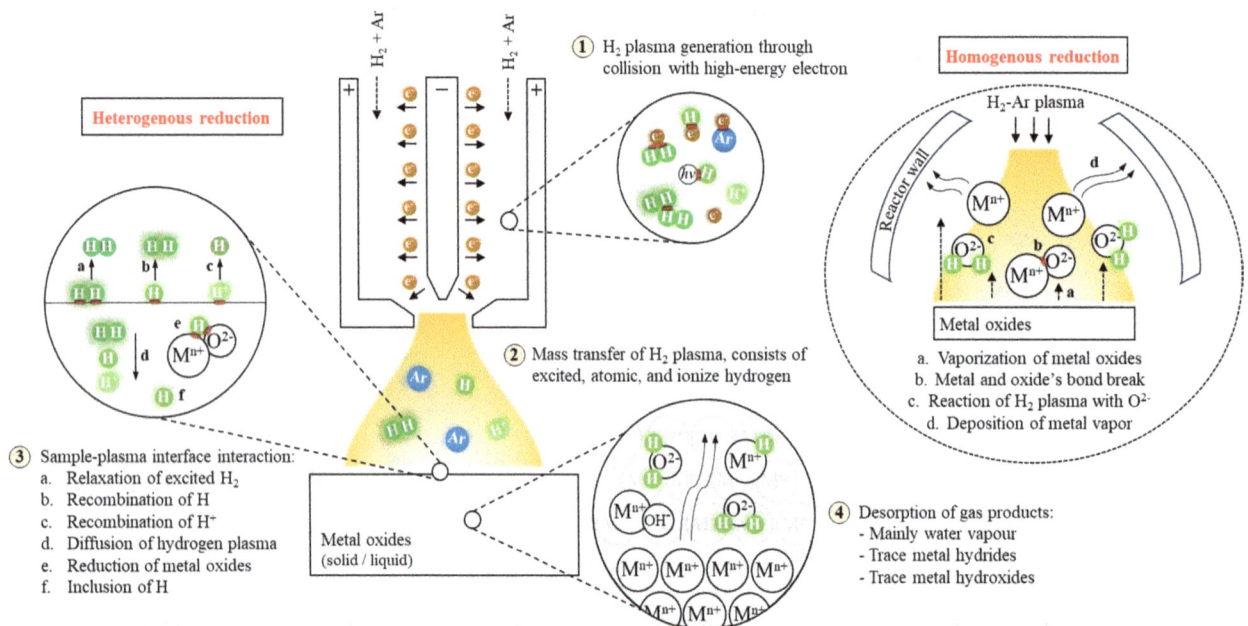

FIG 3 – General reaction mechanism of metal oxides reduction with hydrogen plasma (Satritama *et al*, 2024).

In the heterogeneous process, the initial step involves generating plasma by injecting gas between electrodes with different potentials, illustrated in the schematic mechanism using DC plasma. Microwave energy can also serve as a source to produce hydrogen plasma, as discussed earlier. H_2 plasma is formed through collisions with high-energy electrons sputtering from the cathode to the anode. These collisions elevate hydrogen to an excited state, leading to dissociation and ionisation. Additionally, collisions with photons (*hv*) can occur, further enhancing the hydrogen state. Photons are generated when high-energy electrons collide with hydrogen gas, reducing their state and releasing photons. Heat transfer between gas molecules allows higher-state hydrogen to transfer energy to lower-state hydrogen, improving the equilibrium of hydrogen plasma. The hydrogen plasma then flows through a constricted anode into the sample. In many instances, argon is injected alongside hydrogen to broaden the radial distribution of plasma within the sample, as hydrogen's high specific heat results in a more constricted plasma arc.

The reduction process initiates from the upper interface of the solid/liquid-gas layer in the sample and gradually proceeds downward as more interfaces of metal oxides undergo reduction. The interaction between hydrogen plasma and the metal oxides interface is complex. Some hydrogen plasmas undergo relaxation and recombination at the interface, experiencing sudden temperature changes upon contact with the sample surface. These processes induce local heating on the sample's surface, enabling the energy to surpass the activation energy required for oxide reduction. Other types of hydrogen plasma can bind oxygen at the interface and diffuse further into the sample to reduce additional oxide. This process requires intensive control, as several studies have noted that reduction with hydrogen plasma may result in the inclusion of trace amounts of hydrogen inside the sample. A higher entrapment of hydrogen may lead to hydrogen embrittlement (Sabat, Paramguru and Mishra, 2018).

Certain metals are efficiently produced through homogeneous reactions with hydrogen plasma in the vapour state (Huczko and Meubus, 1988; Kitamura, Shibata and Takeda, 1993). A study by Kitamura, Shibata and Takeda (1993) revealed that hydrogen plasma can evaporate metal oxides much faster than argon plasma. Using hydrogen plasma, Fe_2O_3, Cr_2O_3, TiO_2 and Al_2O_3 can be evaporated in 1.4, 1.8, 2 and 5.9 ms, respectively. Once metal oxides evaporate, the bonds between metal and oxide breakdown, allowing hydrogen plasma to bind free oxygen. The resulting metal vapour can then be deposited in the cooler part of the reactor. Rains (1968) employed an additional quench probe inserted into the bottom of the vertical reactor, alongside its water-cooled reactor casing. This water-cooled quench probe injects cooled gas to enhance metal vapour deposition.

Iron oxide reduction

Several studies have examined the effectiveness of different hydrogen plasma methods in reducing iron ore (Fe_2O_3) to metallic iron, positioning hydrogen plasma as a leading technology for oxide reduction compared to other metals. Microwave non-thermal plasma with H_2 has been particularly investigated, demonstrating rapid reduction of bulk Fe_2O_3. Bergh (1965) observed colour changes indicative of iron oxide reduction with atomic H at a relatively low temperature of 40°C, significantly lower than the 310°C required for molecular hydrogen. Rajput et al (2014, 2013) achieved over 95 per cent reduction within two hours using low power (500–1500 W), attributing this efficiency to the lower activation energy of hydrogen plasma compared to molecular hydrogen. Microwave power density (MWPD) emerges as a critical factor in microwave plasma reduction, influenced by reaction temperature and pressure, reflecting the efficiency of microwave energy absorption by plasma. Study from Sabat, Paramguru and Mishra (2017) estimated an MWPD of 4–9 W/cm³ in their process, translating to a mere 1–2 per cent conversion of molecular hydrogen into atomic H based on Hassouni, Grotjohn and Gicquel (1999) findings.

Extensive research on DC thermal plasma reduction with hydrogen plasma, commonly known as Hydrogen Plasma Smelting Reduction (HSPR), has been conducted at Montanuniversität Leoben as part of the SuSteel project, a collaboration between Montanuniversität Leoben and Voestalpine aimed at scaling up the process to a pilot scale (Seftejani et al, 2020). Zarl et al (2022) developed a promising six-step batch and continuous plasma reduction process using H_2 for potential industrial-scale applications. The process includes argon purging, pre-melting (electrode contact with a steel pin), pre-reduction to complete melting with Ar+H_2, reduction and sample charging, post-reduction until H_2 content in the off-gas reached 40 per cent and Ar/N_2 flushing. Filho et al (2022) examined a hybrid approach, combining molecular hydrogen direct reduction with hydrogen plasma. This method resulted in complete conversion in just 15 minutes for the plasma step, as opposed to 40–70 minutes for direct reduction alone.

Increased microwave power and temperature generally result in higher rates and degrees of reduction (Gilles and Clump, 1970; Rajput et al, 2013; Filho et al, 2022; Gonoring et al, 2022). However, accurately measuring temperature during hydrogen plasma reduction poses a challenge due to uneven heat distribution within the reactor. Kumar et al (2023) proposed an innovative method involving the placement of reflective aluminium foil beneath the sample and analysing the emitted light to estimate temperature based on black-body radiation principles. Beyond assessing reaction rate and degree, optimising hydrogen utilisation is critical for efficient reduction. Although higher hydrogen concentrations and flow rates enhance reduction rates, they typically lead to decreased utilisation (Behera et al, 2018; Plaul, Krieger and Bäck, 2005). Seftejani et al (2020) observed an initial peak in hydrogen utilisation (60 per cent), followed by a gradual decline to near zero by the conclusion of the reduction process.

Other oxides reduction

The initial endeavour to reduce alumina using hydrogen plasma dates to 1970, conducted by Rains and Kadlec (1970). They achieved a dissociative reduction, separating aluminium from oxygen through evaporation and the reaction of oxygen with hydrogen plasma. Subsequent studies have expanded on this work. Lyubochko et al (2000) achieved a higher conversion of 50–60 per cent in a batch system with higher power input, surpassing Rains and Kadlec, who achieved a 30 per cent alumina conversion with an in-flight system. Rains and Kadlec successfully produced aluminium with a quench system and lower power input, while Kitamura, Shibata and Takeda (1993) encountered challenges in producing Al due to reoxidation into alumina. This underscores the significance of an effective quenching system in homogeneous reduction.

Reduction of chromium oxide can occur through both heterogeneous and homogeneous reactions. Although prior studies reported incomplete reduction or the need for optimised parameters, homogeneous reactions generally exhibit higher efficiency. Huczko and Meubus (1988) found that vapour-phase reduction is more efficient than solid-vapor reduction, showing increased efficiency (up to 47 per cent) with higher H_2 flow rates. Their calculations indicated that the diffusivity of Cr in H_2-Ar plasma is higher than that of Cr_2O_3, with the value increasing with temperature. This suggests that Cr can more easily diffuse in the gas system, facilitating its deposition. Kitamura, Shibata and Takeda (1993) stressed the importance of exceeding 2000 K for Cr vapourisation and subsequent quenching to form solid Cr. Dalaker and Hovig (2023) achieved a heterogeneous reaction, obtaining a metallic phase with 90 per cent Cr, primarily due to the low reaction time (only six mins, with 12 cycles of 30 s each), vacuum conditions and relatively low current (30–100 A), preventing Cr oxide from vapourising. They proposed a potential mechanism of Cr oxide reduction as Cr_2O_3 to CrO to Cr.

Although copper oxides can be reduced using molecular hydrogen, research has been conducted on their reduction with hydrogen plasma. The typical reduction steps for Cu oxides involve CuO → Cu_4O_3 → Cu_2O → Cu (Sabat, Paramguru and Mishra, 2016). Ramos et al (2021) observed that the reduction step from Cu_2O is directly to Cu, without the presence of metastable phases. However, Sabat, Paramguru and Mishra (2017) suggested a different reaction sequence, stating that the steps are CuO → Cu_2O → Cu when iron oxide is added to the process (Sabat, Paramguru and Mishra, 2017). The reported activation energy for Cu_2O reduction is 75.64 kJ/mol (Sabat, Paramguru and Mishra, 2016), while Ramos et al reported an activation energy of 19.06 kJ/mol for Cu_2O reduction. Furthermore, Ramos et al found that the activation energy for H_2 plasma reduction was almost four times lower than that for molecular H_2 reduction (73.79 kJ/mol compared to 19.06 kJ/mol). Atomic hydrogen seems to be more efficient for reduction compared to molecular hydrogen, as observed by Bergh (1965). They noted that the reduction temperature for Cu_2O and CuO ranges from 225–450°C and 100–225°C, with a colour change occurring at 265°C and 140°C, respectively. In contrast, the colour changes with H occurred at 25°C, consistent with findings by Fleisch and Mains (1982).

Same case applies to nickel oxide as it can be reduced using molecular hydrogen, but some researchers also tried to reduce it with hydrogen plasma. Bergh (1965) observed a reduction in NiO occurring at a notably lower temperature (62°C) when exposed to atomic hydrogen, in contrast to molecular hydrogen (250°C). Sabat (2021) achieved complete reduction of pure NiO within five mins, employing a power input of 750 W. The reduction exceeded 100 per cent due to excess lattice oxygen in non-stoichiometric NiO. They calculated the specific enthalpy for NiO reduction as 32.8 kWh/kg, with the potential for reduction to 7.37 kWh/kg Ni when considering a 60 per cent reactor power efficiency and simultaneous multi-pellet reduction. Despite these intriguing results, justifying the additional energy consumption of converting hydrogen to plasma for CuO and NiO reduction requires stronger arguments as these oxides are easily reduced with molecular hydrogen.

Most research on SiO_2 reduction with hydrogen plasma primarily results in the formation of SiC (Salinger, 1972; Hollabaugh et al, 1983; Meyer et al, 1987; Asakami, Hokazono and Kato, 1988; Pirzada, 1990; Kong, Huang and Pfender, 1986), SiO (Kong, Huang and Pfender, 1986), or Si_3N_4 (Lee, Eguchi and Yoshida, 1990; Allaire and Dallaire, 1991) with the production of silicon metal proving challenging due to the complex behaviour of SiO_2 reduction. Only two studies, conducted by Watanabe *et al* (1999) and De Sousa *et al* (2016), have successfully produced silicon metal from silicon oxide samples. Watanabe *et al* utilised a mixture of silica and alumina with argon and hydrogen plasma in a 4 kW DC non-transferred thermal plasma, discovering that argon plasma could generate silicon, while hydrogen was crucial for alumina reduction. Silicon with 98.3–99 per cent purity was formed within a 15-minute treatment. De Sousa *et al* investigated the deoxidation of silicon kerf, a byproduct generated during wafer slicing in solar panel manufacturing, using in-flight DC non-transferred plasma. They found that the reaction time between hydrogen plasma and the silicon kerf sample was crucial. The deoxidation rate increased with an increasing nozzle diameter at low hydrogen content due to the resulting lower gas velocity. The oxygen wt per cent in silicon kerf decreased from 17.3–23.4 per cent to 5.4–8.7 per cent and the deoxidation rate increased with an increasing nozzle diameter with low hydrogen content (<5 L/m).

FUTURE PROSPECTS AND SCALE UP POSSIBILITY

While there is considerable research on metal oxide reduction with hydrogen plasma, its application to reduce complex ores, is still in the early stages of exploration. However, building on successful studies of pure metal oxides reduction with hydrogen plasma, there is potential to produce various metals from complex oxide ores by carefully controlling the process parameters. Future investigations should also consider exploring the use of hydrogen plasma for rare earth metal production to support advanced applications, as these metals often necessitate high temperatures or strong reductants for extraction. Implementation of in-flight plasma process holds promise for future applications. This method offers higher reaction times, leading to increased reduction degrees and enhanced hydrogen utilisation, facilitated by the extensive surface area of the sample. Additionally, this approach appears economically viable for industrial-scale implementation, given its higher productivity compared to batch processing. When considering a larger-scale application, a thorough techno-economic assessment is crucial by evaluating cost-effectiveness, energy efficiency and practical challenges associated with implementing hydrogen plasma technology. Furthermore, interdisciplinary collaboration between metallurgists and plasma physicists is essential to address existing knowledge gaps and facilitate the successful scale-up of hydrogen plasma technology in extractive metallurgy.

CONCLUSIONS

Hydrogen plasma is emerging as a viable option for low-carbon extractive metallurgy, addressing the limitations associated with molecular hydrogen as a weaker reductant, particularly at elevated temperatures. Typically, hydrogen plasma is generated by subjecting molecular hydrogen to high energy, resulting in vibrationally and rotationally excited, atomic hydrogen and/or ionised hydrogen. This combination of species finds diverse applications in extractive metallurgy, particularly in the reduction of various metal oxides. While existing research predominantly focuses on the reduction of iron ore or iron oxides, numerous studies have explored the efficacy of hydrogen plasma in reducing oxides such as Al_2O_3, Co_3O_4, Cr_2O_3, CuO, NiO, SiO_2 and others, revealing promising results and showcasing its superiority over molecular hydrogen. Hydrogen plasma demonstrates effective reduction even for more stable oxides like Al_2O_3, TiO_2 and SiO_2. Less stable oxides such as NiO and CuO are more efficiently and sustainably reduced using molecular hydrogen, as it proves sufficient for less stable oxides. The added cost of electricity required to generate plasma does not justify the benefits in terms of reaction time, based on current research findings. Despite the manifold advantages, hydrogen plasma utilisation in extractive metallurgy poses several challenges, which ongoing studies are actively addressing and must be further optimised in the future.

ACKNOWLEDGEMENTS

This work was conducted as part of the PhD study of Bima Satritama and Dian Mughni Fellicia. Bima Satritama is co-funded by Swinburne University of Technology (SUT), Victorian Hydrogen Hub (VH2) and Commonwealth Scientific and Industrial Research Organisation (CSIRO). Dian Mughni Fellicia is co-funded by SUT and CSIRO.

REFERENCES

Allaire, F and Dallaire, S, 1991. Synthesis and characterization of silicon nitride powders produced in a dc thermal plasma reactor, *Journal of Materials Science,* 26:6736–6740.

Asakami, O, Hokazono, S and Kato, A, 1988. Preparation of SiC Powders from SiO_2 Powders by RF-Plasma Technique, *Journal of the Ceramic Society of Japan,* 96:1203–1205.

Behera, P R, Bhoi, B, Paramguru, R K, Mukherjee, P S and Mishra, B K, 2018. Hydrogen Plasma Smelting Reduction of Fe_2O_3. *Metallurgical and Materials Transactions, B,* 50:262–270.

Bergh, A A, 1965. Atomic hydrogen as a reducing agent, *The Bell System Technical Journal,* 44:261–271.

Carvalho, D, 2023, 25 August. Beyond electricity: is hydrogen the key to greener smelting and refining? [online], Wood Mackenzie. Available from: <https://www.woodmac.com/news/opinion/beyond-electricity-is-hydrogen-the-key-to-greener-smelting-and-refining/> [Accessed 10 December 2023].

Dalaker, H and Hovig, E W, 2023. *Hydrogen Plasma-Based Reduction of Metal Oxides, Advances in Pyrometallurgy,* pp 85–94 (Springer).

De Sousa, M, Vardelle, A, Mariaux, G, Vardelle, M, Michon, U and Beudin, V, 2016. Use of a thermal plasma process to recycle silicon kerf loss to solar-grade silicon feedstock, *Separation and Purification Technology,* 161:187–192.

Filho, I R, Springer, H, Ma, Y, Mahajan, A, Da Silva, C C, Kulse, M and Raabe, D, 2022. Green steel at its crossroads: Hybrid hydrogen-based reduction of iron ores, *Journal of Cleaner Production,* 340:130805.

Fleisch, T H and Mains, G J, 1982. Reduction of copper oxides by UV radiation and atomic hydrogen studied by XPS, *Applications of Surface Science,* 10:51–62.

Gilles, H L and Clump, C W, 1970. Reduction of iron ore with hydrogen in a direct current plasma jet, *Industrial and Engineering Chemistry Process Design and Development,* 9(2):194–207. https://doi.org/10.1021/i260034a007.

Gonoring, T B, Franco, A R, Vieira, E A and Nascimento, R C, 2022. Kinetic analysis of the reduction of hematite fines by cold hydrogen plasma, *Journal of Materials Research and Technology,* 20:2173–2187.

Hassouni, K, Grotjohn, T A and Gicquel, A, 1999. Self-consistent microwave field and plasma discharge simulations for a moderate pressure hydrogen discharge reactor, *Journal of Applied Physics,* 86:134–151.

Hollabaugh, C M, Hull, D E, Newkirk, L R and Petrovic, J J, 1983. RF-plasma system for the production of ultrafine, ultrapure silicon carbide powder, *Journal of Materials Science,* 18:3190–3194.

Huczko, A and Meubus, P, 1988. Vapor phase reduction of chromic oxide in an Ar-H2 Rf Plasma, *Metallurgical and Materials Transactions, B,* 19:927–933.

Inui, H, Takeda, K, Kondo, H, Ishikawa, K, Sekine, M, Kano, H, Yoshida, N and Hori, M, 2010. Measurement of hydrogen radical density and its impact on reduction of copper oxide in atmospheric-pressure remote plasma using H2 and Ar mixture gases, *Applied Physics Express,* 3:126101. https://doi.org/10.1143/apex.3.126101.

Kitamura, T, Shibata, K and Takeda, K, 1993. In-flight reduction of Fe_2O_3, Cr_2O_3, TiO_2 and Al_2O_3 by Ar-H2 and Ar-CH4 plasma, *ISIJ International,* 33:1150–1158.

Kong, P, Huang, T T and Pfender, E, 1986. Synthesis of ultrafine silicon carbide powders in thermal arc plasmas, *IEEE Transactions on Plasma Science,* 14:357–369.

Kumar, S, Xiong, Z, Held, J, Bruggeman, P and Kortshagen, U R, 2023. Rapid carbon-free iron ore reduction using an atmospheric pressure hydrogen microwave plasma, *Chemical Engineering Journal,* 472.

Lee, H J, Eguchi, K and Yoshida, T, 1990. Preparation of ultrafine silicon nitride and silicon nitride and silicon carbide mixed powders in a hybrid plasma, *Journal of the American Ceramic Society,* 73:3356–3362.

Lyubochko, V A, Malikov, V V, Parfenov, O G and Belousova, N V, 2000. Reduction of aluminum oxide in a nonequilibrium hydrogen plasma, *Journal of Engineering Physics and Thermophysics,* 73:568–572. https://doi.org/10.1007/BF02681800.

Meyer, T N, Becker, A J, Edd, J F, Smith, F N and Liu, J, 1987. Plasma synthesis of ceramic powders, *ISPC-8 Tokyo,* pp 2006–2011. Available from: <https://plas.ep2.rub.de/ispcdocs/ispc8/content/8/08-2006.pdf>

Pirzada, S A, 1990. Silicon carbide synthesis and modeling in a non-transferred arc thermal plasma reactor, PhD thesis, University of Idaho.

Plaul, J F, Krieger, W and Bäck, E, 2005. Reduction of Fine Ores in Argon-Hydrogen Plasma, *Steel Research International,* 76:548–554.

Rains, R K and Kadlec, R H, 1970. The reduction of Al_2O_3 to aluminum in a plasma, *Metallurgical Transactions,* 1:1501–1506. https://doi.org/10.1007/BF02641992.

Rains, R K, 1968. Reduction of aluminum-oxide to aluminum in radio frequency generated plasmas, thesis, Chemical Engineering, University of Michigan. https://dx.doi.org/10.7302/15672.

Rajput, P, Bhoi, B, Sahoo, S, Paramguru, R K and Mishra, B K, 2013. Preliminary investigation into direct reduction of iron in low temperature hydrogen plasma, *Ironmaking and Steelmaking*, 40:61–68.

Rajput, P, Sabat, K C, Paramguru, R K, Bhoi, B and Mishra, B K, 2014. Direct reduction of iron in low temperature hydrogen plasma, *Ironmaking and Steelmaking*, 41:721–731.

Ramos, S V, Cisquini, P, Nascimento Jr, R C, Franco Jr, A R and Vieira, E A, 2021. Morphological changes and kinetic assessment of Cu_2O powder reduction by non-thermal hydrogen plasma, *Journal of Materials Research and Technology*, 11:328–341.

Sabat, K C, 2019. Formation of CuCo alloy from their oxide mixtures through reduction by low-temperature hydrogen plasma, *Plasma Chemistry and Plasma Processing*, 39:1071–1086.

Sabat, K C, 2021. Production of nickel by cold hydrogen plasma, *Plasma Chemistry and Plasma Processing*, 41:1329–1345.

Sabat, K C, Paramguru, R K and Mishra, B K, 2016. Reduction of copper oxide by low-temperature hydrogen plasma, *Plasma Chemistry and Plasma Processing*, 36:1111–1124.

Sabat, K C, Paramguru, R K and Mishra, B K, 2017. Reduction of oxide mixtures of (Fe_2O_3 + CuO) and (Fe_2O_3 + Co3O4) by low-temperature hydrogen plasma, *Plasma Chemistry and Plasma Processing*, 37:979–995.

Sabat, K C, Paramguru, R K and Mishra, B K, 2018. Formation of copper–nickel alloy from their oxide mixtures through reduction by low-temperature hydrogen plasma, *Plasma Chemistry and Plasma Processing*, 38:621–635.

Sabat, K C, Paramguru, R K, Pradhan, S and Mishra, B K, 2015. Reduction of cobalt oxide (Co3O4) by low temperature hydrogen plasma, *Plasma Chemistry and Plasma Processing*, 35:387–399.

Sabat, K C, Rajput, P, Paramguru, R K, Bhoi, B and Mishra, B K, 2013. Reduction of oxide minerals by hydrogen plasma: an overview, *Plasma Chemistry and Plasma Processing*, 34:1–23.

Sahu, M, 2014. An Investigation on Treatment of Bauxite through Hydrogen Plasma, thesis, Masters of Technology, National Institute of Technology.

Salinger, R M, 1972. Preparation of silicon carbide from methylchlorosilanes in a plasma torch, *Industrial and Engineering Chemistry Product Research and Development*, 11:230–231.

Satritama, B, Cooper, C, Fellicia, D, Pownceby, M I, Palanisamy, S, Ang, A, Mukhlis, R Z, Pye, J, Rahbari, A, Brooks, G A and Rhamdhani, M A, 2024. Hydrogen plasma for low-carbon extractive metallurgy: oxides reduction, metals refining and wastes processing, submitted to *Journal of Sustainable Metallurgy*.

Seftejani, M, Schenk, J, Spreitzer, D and Zarl, M, 2020. Slag formation during reduction of iron oxide using hydrogen plasma smelting reduction, *Materials (Basel)*, 13.

Watanabe, T, Soyama, M, Kanzawa, A, Takeuchi, A and Koike, M, 1999. Reduction and separation of silica-alumina mixture with argon–hydrogen thermal plasmas, *Thin Solid Films*, 345(1):161–166.

Yvon, A, Fourmond, E, Ndzogha, C, Delannoy, Y and Trassy, C, 2003. Inductive plasma process for refining of solar grade silicon silicon, in *Proceedings of EPM 2003, Fourth International Conference on Electromagnetic Processing of Materials*, HAL Open Science, pp 125–130. https://hal.science/hal-00551473.

Zarl, M A, Ernst, D, Cejka, J and Schenk, J, 2022. A new methodological approach to the characterization of optimal charging rates at the hydrogen plasma smelting reduction process part 1: method, *Materials*, 15:4767.

Challenges facing non-ferrous metal production

P Taskinen[1] and D Lindberg[2]

1. Professor Emeritus, Aalto University, Espoo FI-00076, Finland. Email: pekka.taskinen@aalto.fi
2. Professor, Aalto University, Espoo FI-00076, Finland. Email: daniel.k.lindberg@aalto.fi

ABSTRACT

The increase in metals demand in the electrifying globe means significant growth in the smelting of copper, nickel, zinc, and lead, produced from primary sulfide sources or using sulfide mattes as the intermediates of the process chains. This means that leaner and complex mineral deposits will be evaluated as ores and are in the future traded in the commodity market for smelting and refining to pure metals.

An important issue in the trend is the technology metals, like antimony, tellurium, and gallium, which exist as trace elements in sulfide ores and form no ores of their own. Their recovery becomes important in the coming decades as will be the case with growing slag amounts without use in other industries. The key is to produce environmentally acceptable slags in the smelting operations. It sets new boundary conditions to the treatment of flue dusts. This means that all material streams of smelting and refining must be re-evaluated for the deportments of the main and minority metals.

In copper smelting, the recoveries of precious metals are today important for the feasibility of the custom smelters but due to low prices of many minority metals they are discarded in slag landfills. It is one of the emerging issues also in the secondary copper smelting today and once the demand grows, the same question will be faced also in the mining-beneficiation-smelting-refining chain of the primary production of nickel, zinc, and lead. The distributions of many technology metals in the copper and nickel smelting have been recently studied using methods where the chemically bound trace elements in the slag and its phases at the smelting conditions have been studied. Thus, the key data about options for process modifications and additional processing steps are piling up.

Short processing routes in the metals smelting and refining are attractive due to their simplicity. At the same time, complexity of many raw materials challenges the fluxing at high oxygen partial pressures in low silica slags with high metal concentrations. The compromise between high primary recovery and safe operation is a demanding task in conditions where slag foaming outside the processing window is evident.

The increase in the demand of pure nickel is challenging the raw material basis where low-grade sulfide ores are soon smelted along with nickel laterites to matte. It is a demanding task at high MgO concentrations. The low solubility of MgO in iron silicate slags requires new fluxing strategies and new smelting end points for the operation at reasonable temperatures; the direct nickel matte smelting in one-matte mode may be an option.

INTRODUCTION

The increase in metals' demand anticipated in the electrifying world means growth in the smelting of, for example copper, nickel, and lead which primarily are produced from primary sulfide minerals or using sulfide mattes as the intermediate product. This means that leaner and more complex mineral deposits will be evaluated and classified as ores and are traded in the commodity market for smelting and refining. Whether this is possible, the key is in the development of agile smelting and refining technologies. The question also is, who will have 'patient money' to carry out and commit to the long-term basic and applied research, aiming at CO_2-lean and environmentally lean technologies. Another side of the coin is the common need of less special knowledge and skills to operate the plants and to do that in a safe way. In that, for example novel online measurements, like direct oxygen partial pressure from the smelting vessels, robotics, and AI, will play vital roles.

A simple calculation of global demand of refined copper indicates that 3 per cent annual growth ends up from the present about 25 Mt pa (International Copper Study Group, 2023) to 50 Mt pa in 2050. This means a significant increase in the mine production, due to the long end-of-lifetime for copper products. In terms of smelter capacity, it means one to two new large scale greenfield smelters each year in 2025 and two to three in 2050. This requirement of 'new metal' is stricter for copper than in, say alloy steels, where substituting alloys can be developed, whereas most copper use and essentially all future growth is in high-purity metal to be used for heat and in particular for electricity transfer purposes. This means that 99.99 per cent pure A-grade cathode must be able to be produced also using the low-grade and complex raw materials. At the same time, the discard slag production is at least doubled which lowers the primary yield and increases the production costs, when the slag recycling will be the preferred option instead of landfilling. The same is valid with nickel and zinc as well as to some degree with lead.

Running large scale smelting and refining operations today includes reliable life cycle management of the equipment and forecasting the maintenance needs several months, even more than half a year beforehand, due to the availability and long delivery times of many critical spares. Such vulnerable areas in the non-ferrous smelting and refining are the refractory bricks, cooling jackets, and many electric and electronic components.

The recovery of all trace elements in copper, nickel, zinc, and lead smelting is a future challenge which requires re-thinking of the process chains for optimising the recoveries and allowing the circular economy also in the small so-called technology metals. They today deport and deplete in slags without hope of recovery. On the other hand, the global distributions of many toxic elements in copper, nickel and lead smelting are very uneven (Risopatron, 2018) which emphasises the need of using available technological advancements for limiting mercury, arsenic and lead emissions also in the current facilities when increasing the production.

The regulation in industrialised countries today is hostile to metal making. That discourages the activities to develop the present technologies and makes long-term commitment of the industry less attractive. This again has a major impact in the technology development and innovations which lead to new and more green metal making processes in a field where major technology improvements need decades to mature. The development is evident when looking at the distribution of fundamental and engineering publications by country over the last two to three decades.

THE IMPURITY PROBLEM

In the non-ferrous metallurgy, most trace elements in the primary and secondary raw materials are either technology metals or harmful elements. The former must be recovered, and latter eliminated for environmental reasons. The fundamental knowledge in finding the optimal end points, eg for the processing of matte, converting, and refining steps, includes the elements' equilibrium distributions between matte, slag and gas, and the thermodynamic properties of their species in the phases. Thus, the individual conditions for the maximum deportments of the impurity elements in the various side streams can be evaluated. Such accurate and reliable fundamental data also allows digitalisation of the processing operations.

In the non-ferrous industry, with exception of aluminium production, the use of hydrogen and fossil-free metals in terms of Scope 1, 2 and 3 greenhouse gas emissions (US EPA, 2023) will not be as demanding as in the ferrous and especially in ferroalloys industries. There is no fundamental restriction from thermodynamic point of view even if the degrees of utilisation of hydrogen in some processing steps, like slag cleaning, may be realised on a relatively low level.

The smelter flow sheets of today include closed flue dust circulation which forces all volatile elements into the two existing outlets of the smelter, namely the crude metal/matte and the slag, see Figure 1. Independently from the slag cleaning technology used, the closed operation mode leads to high trace element concentrations in the metal product and the discard slag when/if they are not removed from the material streams.

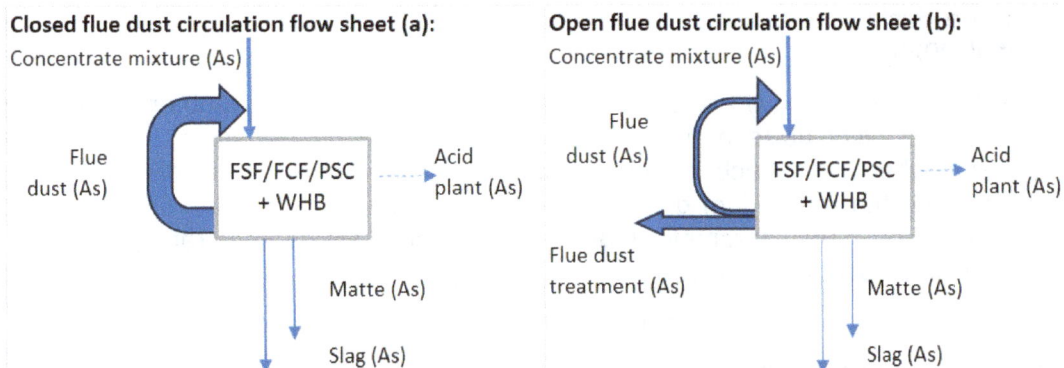

Closed flue dust circulation flow sheet (a):

Open flue dust circulation flow sheet (b):

FIG 1 – Schematic flow sheets of the present closed-type flue dust treatments and an open circuit mode with the suspension smelting technology as an example (FSF – flash smelting furnace; FCF – flash converting; PSC – Peirce Smith converting; WHB – Waste heat boiler).

The utilisation of the opportunity of high particle temperature in the gas-concentrate suspension in the copper and nickel in the flash smelting processes and also the vigorous agitation in the bath smelting processes allow us to split several impurity elements from the main metal stream and redirect them into safe treatment instead of mixing them in the discard slag and crude metal. The key trace elements in the copper production by International Copper Study Group (ICSG) today and in near future are arsenic, lead and bismuth (Risopatron, 2018).

RECOVERY OF THE TRACE ELEMENTS

The sulfide ores are important sources of several minority metals some of which do not have ores of their own. Typical examples are selenium, tellurium, antimony, and bismuth. Their recoveries are part of the smelting and refining steps in cases there is economy behind the extraction. Today, a lot of these minor elements (often called technology metals) are lost in the earlier steps prior to smelting due to commercial reasons (Moats, Alagha and Awuah-Offei, 2021; Nassar *et al*, 2022), ie lack of demand and low price.

The deportment of minority components will be a more vital issue in the case where the trace elements from secondary resources will be recovered. The use of standard primary copper smelting circuits is out of question because many elements deport in the smelting and converting slags and cannot be recovered from there in the slag cleaning operations due to their thermodynamic properties and low concentrations (Lennartsson *et al*, 2018; Faraij *et al*, 2022). Thus, the focus of Waste Electrical or Electronic Equipment (WEEE) recycling has been so far on the recovery of the platinum metals (PM) and platinum group metals (PGM) values only. Therefore, the optimal routes for the primary and secondary raw feeds must be evaluated in future from this perspective (Alvear Flores, Risopatron and Pease, 2020).

Losses of trace elements in beneficiation-smelting-refining chain are evident and often intentional due to penalties involved in concentrate trading. Recent detailed analyses on some trace elements in the primary copper ores and their recovery in the current beneficiation-smelting-refining chains and technologies (Moats, Alagha and Awuah-Offei, 2021; Nassar *et al*, 2022) indicate that significant fractions are lost or actively suppressed in the gangue or slag streams. Thus, they are led to the tailings at mine site and slag landfills at smelters which seem to contain large quantities of them accumulated as reserves over the past decades.

An important issue in this context is the technology metals in copper, zinc, and nickel smelting, like, eg antimony, tellurium, molybdenum in copper extraction, and gallium in zinc and alumina processing, which exist as trace elements in many sulfide ores and form no ores of their own. Thus, their increased demand is essentially linked with the use and availability the primary sources of those metals. Many of them are on the critical metals/materials lists of the geographical areas, eg in EU and USA (European Commission, 2023; USGS, 2022). Their effective recovery may become important in the coming decades as will be the case with growing slag amounts without sustainable use in the other industries. The key is to produce environmentally acceptable slags and other by-products in the smelting operations which sets new boundary conditions to such things as the treatment of flue dusts and its volatile elements as well as the hydrometallurgical iron residues (Salminen *et al*, 2020).

The recovery of many technology metals in copper and nickel smelting is problematic because they do not volatilise, and they oxidise in the matte smelting step or converter stage. In many cases they deport in the slag when the extraction process advances from reducing to oxidising. This is clearly seen when studying the equilibrium distributions as the function of the prevailing oxygen partial pressure. The thermodynamic parameter independent of the scale demonstrating the issue can be defined (Park, Takeda and Yazawa, 1984) as:

$$L^{s/m}(Me) = [wt\%Me]_{matte/metal}/(wt\% \ Me)_{slag}. \tag{1}$$

And it can be derived from the thermodynamic properties of the element in the phases in contact or from the assays of the phases in the industrial or laboratory conditions. Typically, they are measured in equilibrium conditions using chemical analyses made from the homogeneous part of the phase at process conditions. That was often neglected in the past, and bulk analyses were used instead of *in situ* phase analyses which leads to too small (apparent) distribution coefficient values if they deviate strongly (>1000) from unity (Klemettinen, Avarmaa and Taskinen, 2017). For this reason, the experiments where the phases were sampled from a large pool of slag and quenched the presence of inclusions is difficult to trace and separate from the 'clean' phase. Then also the chemical composition is only average and contains average values of the phase including the uncertainties of composition gradients (from volatile species) and inclusions (not belonging to the phase) (Jak *et al*, 2022). An important issue is that in such cases, volatilisation and inhomogeneities in the phases remain undetected.

A separate recovery step is needed for preventing the loss of trace elements to the discard slag or other residues. This may include re-thinking of the flue dust treatment, as discussed earlier, due to volatile nature of some trace elements. The situations for indium in black copper and copper matte smelting can be seen in Figure 2 and those of tin in Figure 3. The matte/metal-slag distribution coefficients are unbiased to the gas-melt equilibria and thus, the volatility of each element and its deportment in the flue dust must be studied as a separate case. For that purpose, typically mass distributions of elements obtained from the mass balance are used (Nakajima *et al*, 2009). Particularly problematic is the situation in cases like indium and tin in black copper smelting where the distribution coefficient in reducing conditions is strongly on the metal side and in oxidising conditions in a similar way on the slag side.

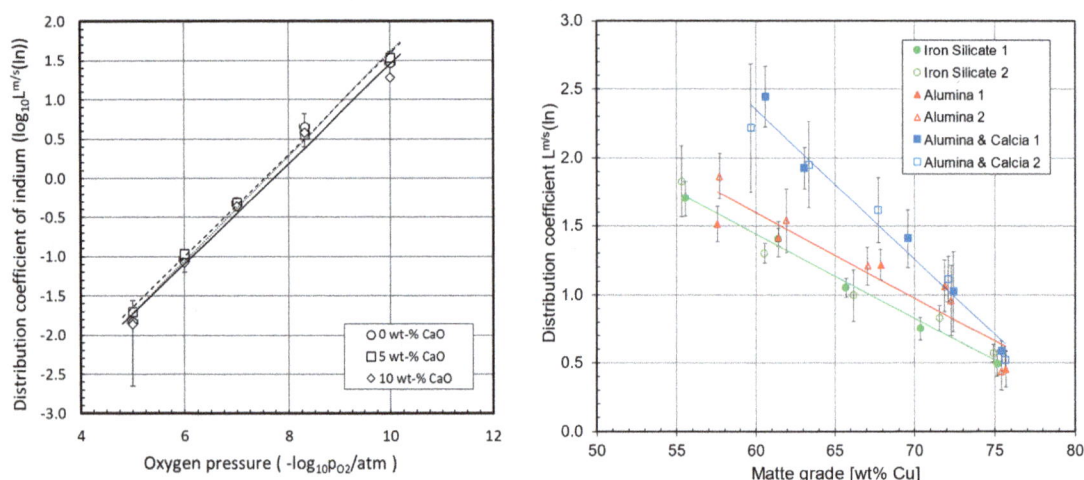

FIG 2 – The metal-slag equilibrium distribution coefficients of indium between copper/copper matte and slag as a function of the prevailing oxygen partial pressure at 1300°C (Sukhomlinov *et al*, 2020a; 2020b).

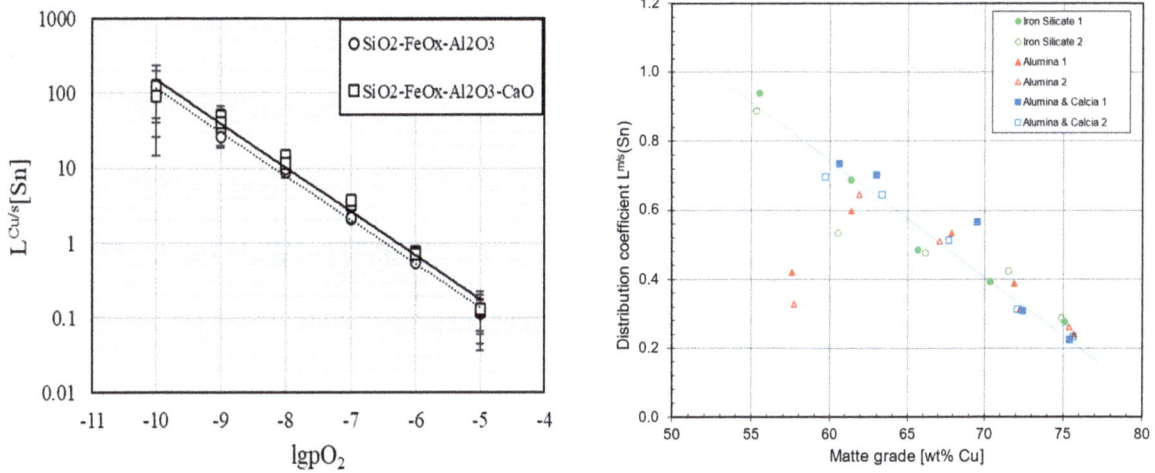

FIG 3 – The metal-slag distribution coefficients of tin between (a) copper (b) copper matte and slag as a function of the prevailing oxygen partial pressure at 1300°C (Avarmaa, Yliaho and Taskinen, 2018; Sukhomlinov *et al*, 2020b).

Figure 4 shows an example of copper matte smelting which is one important source of gold and palladium. All elements with a high distribution coefficient between matte/metal and slag suffer from the systematic technical problem of matte/metal entrainment in the slag. A simple mathematical exercise shows that when the distribution coefficient is $Lm^{/s}[M] > 1000$ the apparent values of the experimental results remain at around 10–100 if the separation of matte/metal droplets in the slag is ignored, ie assumed negligible without proof, or made unsuccessfully.

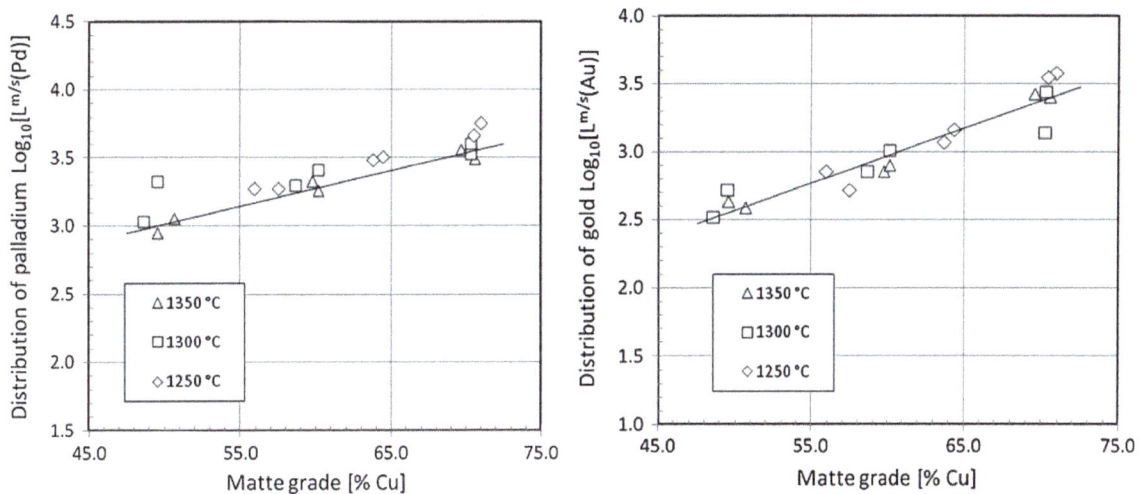

FIG 4 – The experimental distribution coefficients of (a) gold and (b) palladium between copper matte and iron silicate slag saturated with silica as a function of the matte grade (Avarmaa *et al*, 2015).

The high accuracy and significantly larger experimental equilibrium values of the distribution coefficients compared with the previous data (Baba and Yamaguchi, 2013) was justified by the spatial accuracy of the used LA-ICP-MS (laser ablation – inductively coupled plasma – mass spectrometry) technique which allowed selecting inclusion free phase areas of the slag for the analysis (Klemettinen, 2021). Also, the isotope fluxes as a function of time in the MS analyses (ie time resolved analysis (TRA) signals) provide information about the presence of micro/nano nuggets in the samples (Richter *et al*, 2004).

The behaviours of many technology metals in the primary and secondary copper and nickel smelting have been recently studied to facilitate their extraction. The used methods have enabled to locate where the chemically bound trace elements in the slag and its phases at the smelting conditions deport in the various processing steps. Thus, the key data for process modifications and additional processing steps for separating them from the slag, matte and flue dust streams, is gradually piling up and soon allows the assessments of their thermodynamic properties.

IMPORTANCE OF THE SLAG CHEMISTRY AND FLUXING

The short processing routes in the metals smelting and refining are attractive due to their better recoveries and potentially, smaller capital expenditures. The key is to eliminate internal circulations within the smelter. A fundamental issue is limiting wider use of such processes, namely the shortage of low-iron or high-copper ores. Thus, only less than ten such smelters exist today. An option for increasing the use of such processing routes is a pre-concentration of standard sulfide concentrates and cleaning of complex bulk concentrates (Tuominen and Kojo, 2005; Tuominen, Anjala and Björklund, 2007; Awe, 2010; Fuentes, Vinals and Herreros, 2009). For such purposes, straightforward processing concepts are needed which eliminate most minority and trace metals and gangue from the copper minerals and at the same time, allow increasing the copper-to-iron ratio.

Another option may be the production of a bulk concentrate and to split that into two streams at mine site, one with a high Cu/Fe ratio and the other with most impurities and low copper for hydro-metallurgical processing which can handle the impurities in a comprehensive way (Risopatron, 2018). This strategy would allow deposition of the nasties in a concentrated form and avoid their spreading at low concentrations, eg in the slag landfills.

In the direct-to-blister copper smelting, the shorter process chain means the absence of the entire converter isle which allows more compact building and the movement of all molten materials in covered launders which eliminates scattered fugitive emissions at the smelter. An example of the streamlined lay-out is in Figure 5 which also indicates the lack of internal material circulations in the process and the option of selecting the slag cleaning technology based on the local needs.

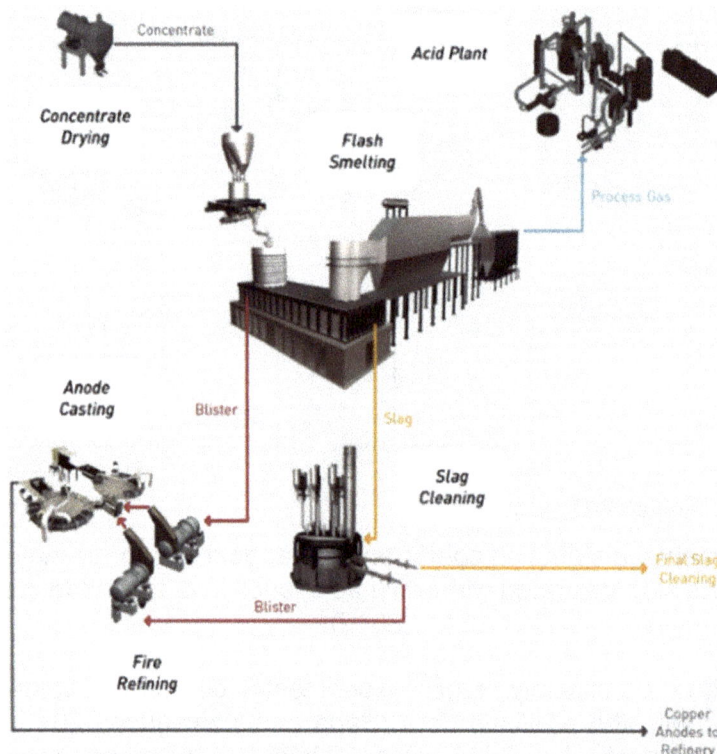

FIG 5 – A schematic of the direct-to-blister flash smelter: feed preparation, the furnace-waste heat boiler- electrostatic precipitator (ESP), and the auxiliary crude copper, off-gas, and slag treatments steps prior to the electrolytic refining to 4N copper cathodes; the material flows are also demonstrated (Outotec).

In the high-grade matte or direct blister production, major increase in the oxidation degree exposed to the charge brings challenges to the fluxing at the prevailing high oxygen partial pressures in low silica slags with high copper concentrations. The compromise between high primary recovery and safe operation is a demanding task in the conditions where slag foaming outside the processing window is evident.

The complexity of sulfide concentrates in nickel smelting does not limit the use of the direct high-grade nickel matte smelting which allows in a flexible way the use of all possible nickel sulfide concentrates either as such or together with the electric furnace matte which fulfil the heat balance at reasonable oxygen enrichments.

Case studies – direct-to-blister copper smelting

The shortest way and smelter chain in sulfide smelting is the direct production of metal or its intermediate high-grade matte for refining in one smelting step. Such industrial examples are the direct-to-blister smelting of copper (Kojo and Huppe, 2011; Taskinen, 2011), direct-to-high-grade matte smelting of nickel (Mäkinen et al, 2005) and direct lead smelting by suspension technologies (Bryk, Malmström and Nyholm, 1966). In those technologies, the feed mixture with its gangue minerals goes through the heating and combustion steps in the reaction shaft of the smelting vessel. There in copper and lead smelting, the oxidation conditions have been adjusted for oxidising almost all sulfur and iron from the matte allowing production of low-sulfur metal. In the direct nickel matte smelting, the product is essentially iron-free sulfide matte. Thus, the fluxing in metal making conditions must take into consideration also the gangue components.

The forming slag is determined in such conditions, at much higher oxygen partial pressures than conventionally, by the end point of the crude metal in copper smelting and in nickel that of the high-grade matte smelting. To achieve this condition, the metal droplets generate a large volume of SO_2 in the reaction shaft and the forming slag must dissolve the iron oxides together with gangue and maintain exactly the correct degree of oxidation for the iron oxides present. An example of the challenges is given in Figure 6 where the high-silica sulfide concentrate will form silica-saturated slags without any fluxing ie it locates within the primary field of silica (the red circle). Note the steep saturation boundaries of silica (tridymite on the right) and pseudo wollastonite (on the left side) which indicate that the correct fluxing is the key for avoiding foaming.

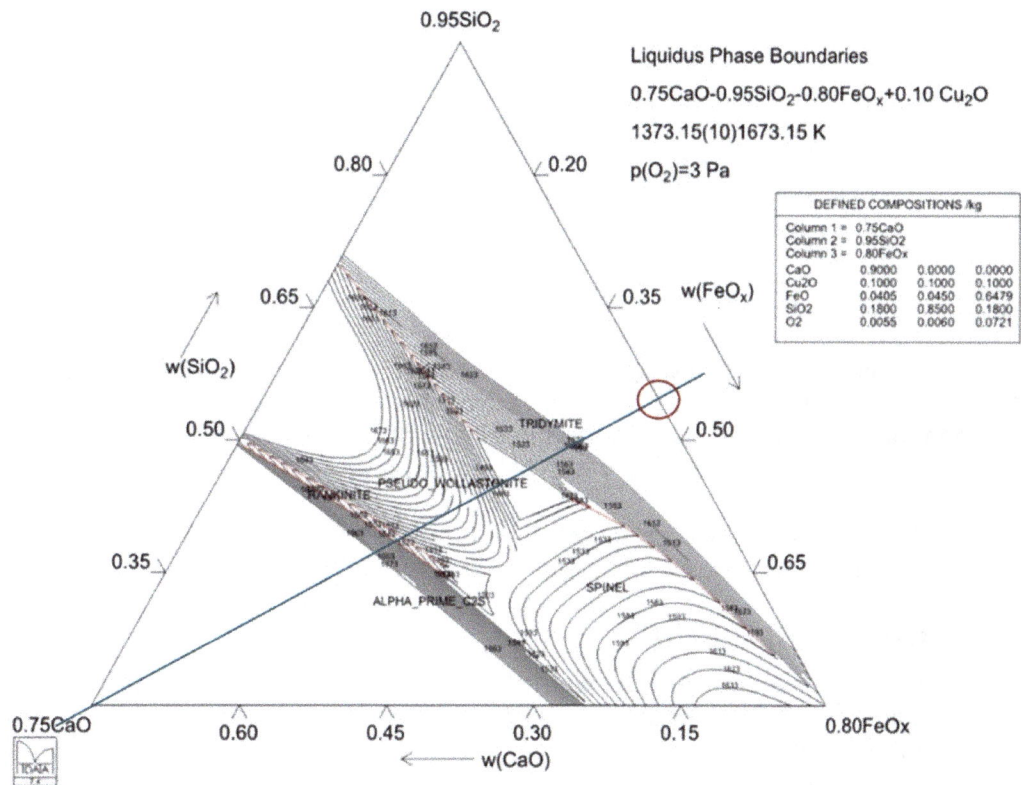

FIG 6 – Operation line of slag in a DtoB smelter when it is fluxed with lime depicted on an isothermal, quasiternary Gibbs triangle at constant oxygen partial pressure of 3×10^{-5} atm and 10 wt per cent Cu_2O; blue line shows fluxing as a function of CaO and the red circle is the unfluxed slag of the feed mixture (MTDATA).

The correct operation of the smelting may need some sort of preprocessing for the complex and low calorific value concentrates for removing selected harmful sulfide minerals and gangue components. This will stabilise the smelting step and remove slag fluidity problems when the composition of the feed mixture can be kept within narrow bounds in terms of its iron and sulfur concentrations.

Auxiliary fuel in sulfide smelting

Another issue is the smelting of low-calorific value sulfide concentrates which need additional fuel for the smelting for maintaining the process temperature. For effective recovery of minority components, the primary smelting step or matte making may end up at matte grades lower than the present 60–65 per cent Cu which additionally limits the heat generation in the smelting step. A fossil-free low-CO_2 alternative is the use of pyrite as fuel, but the downside is increased S/M ratio and the growing slag amount which requires improved copper recovery and lower residual copper in the discard slag.

Thermodynamic equilibrium calculations at fixed matte grade of 65 per cent Cu indicate that the slag amount is doubled at a pyrite addition of slightly less than 50 wt per cent of chalcopyrite content of the feed mixture (in case $CuFeS_2$ is the only copper carrier). The extra heat provided by pyrite is, however, significant as shown in Figure 7. An addition of about 42 wt per cent FeS_2 doubles the heat input in copper smelting in a case where high oxygen enrichment is used in the Flash Smelting Furnace (see the used constraints of Figure 7). No process gas preheating was assumed (Pesonen *et al*, 2022), all raw materials were at 25°C, and the products at 1300°C.

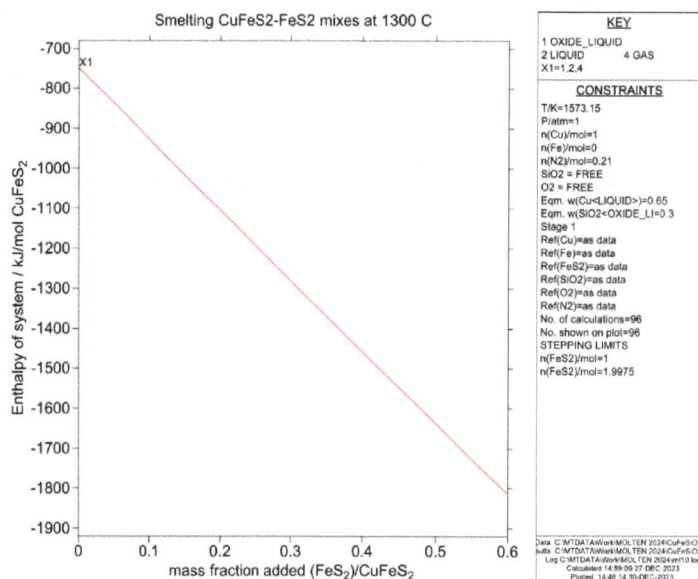

FIG 7 – Extra enthalpy generated by FeS_2 when added in a lean chalcopyrite concentrate, producing a fixed matte grade of 65 per cent Cu and fluxing the slag to 30 wt per cent (SiO_2); a simplified heat balance without considering the effects of gauge minerals on the slag chemistry (MTDATA 7.4 with MTOX database 8.2).

Thus, from the point of view of the CO_2 footprint, it is more sustainable to use pyrite as additional fuel than the present hydrocarbons or coal/coke.

SUMMARY AND CONCLUSIONS

If countries and geographic areas are truly worried about the criticality of metals and minerals, that should be not the interest of commercial actors only. In the copper making the environmental and fugitive emissions must be eliminated and the recovery of trace elements need significant technological advancements. The deficiency of nickel and cobalt output will be a clear challenge which needs totally new technologies for the extraction.

Radical improvements in technology have so far not emerged without thorough fundamental information about the physics, chemistry, and engineering behind the inventions. This fact requires significant Research and Technology development (RTD) input from the mining and smelting community for decarbonising their operations and maximising the metal value yields to the thermodynamic limits.

The environmental awareness of the consumers is an important issue of today which supports transparency to the entire production chain of metals. Carbon free metals and sulfuric acid over the production chain are today's slogan but when that option is reality in many mining operations in remote areas is a question mark. It will be seen whether the consumers are willing to pay premium for environmentally friendly metals' options in the end use. That also means, to cover the cost of recycling the metals when the demand will be 2–5 per cent larger next year to the present and some investment goods have a life span on several decades instead of six months of the cell phones of today. Or is such an option viable that you're allowed to get a new cell phone, car battery, or PC only when you after payment return the old one to an authorised dealer for reuse or recycling?

In copper and nickel sulfide smelting, the recoveries of precious metals are today important for the feasibility of the custom smelters, but due to low prices of many minority metals they are discarded in slag landfills. It is one of emerging issues in the secondary copper smelting today. Once the demand of the trace elements 'technology elements' grows, the same question will be faced also in the mining-beneficiation-smelting-refining chains of the primary production of nickel, zinc, and lead. The distributions of many technology metals in the copper and nickel smelting have been recently studied using methods where the chemically bound trace elements in the slag and its phases at the smelting conditions have been studied. Thus, the key data about options for process optimisation, modifications, and additional processing steps are piling up.

An important issue in the process development is the needs of mode accurate fluxing control in the intensive processing technologies compared with the old processes. Then, the practical operation must more and more rely on on-line or off-line computational tools for maintaining the heat balance and the intended fluxing. For that purpose, digital tools and AI may bring added value.

ACKNOWLEDGEMENTS

The authors are indebted to several R&D programs funded by Tekes and Business Finland over the years 2013–2023 (eg ARVI, SIMP, SYMMET, TOCANEM) as well as Metso-Outotec and Boliden for their encouragement, industrial contribution, and financial support.

REFERENCES

Alvear Flores, G, Risopatron, C and Pease, J, 2020. Processing of complex materials in the copper industry: Challenges and opportunities ahead, *JOM*, 72(10):3447–3461.

Avarmaa, K, O'Brien, H, Johto, H and Taskinen, P, 2015. Equilibrium distribution of precious metals between slag and copper matte at 1250–1350°C, *J Sustain Metall*, 1(3):216–228.

Avarmaa, K, Yliaho, S and Taskinen, P, 2018. Recoveries of rare elements Ga, Ge, In and Sn from waste electric and electronic equipment through secondary copper smelting, *Waste Mngmnt*, 71:400–410.

Awe, S, 2010. Hydrometallurgical upgrading of a tetrahedrite rich copper concentrate, Lic Thesis, Tech Univ Luleå, Sweden, 106 p.

Baba, K and Yamaguchi, K, 2013. The solubility of platinum in the FeO_x-SiO_2 slag at 1573 K, *JMMIJ*, 129:208–212.

Bryk, P, Malmström, R and Nyholm, E, 1966. Flash smelting of lead concentrates, *JOM*, 18(12):1298–1302.

European Commission, 2023. European Critical Raw Materials Act 2023, Directorate-General for Internal Market, Industry, Entrepreneurship and SMEs. Available from: <https://single-market-economy.ec.europa.eu/publications/european-critical-raw-materials-act_en>

Faraij, F, Golmohammadzadeh, R and Pickles, C, 2022. Potential and current practices of recycling waste printed circuit boards: A review of the recent progress in pyrometallurgy, *J Environm Mngmnt*, 316:115242.

Fuentes, G, Vinals, J and Herreros, O, 2009. Hydrothermal purification and enrichment of Chilean copper concentrates, Part 2: The behavior of the bulk concentrates, *Hydrometall*, 95(1–2):113–120.

International Copper Study Group (ICSG), 2023. *World Copper Factbook 2023*. Available from: <https://icsg.org/copper-factbook/>

Jak, E, Shishin, D, Shevchenko, M and Hayes, P, 2022. Integrated experimental phase equilibria and thermodynamic modelling research and implementation in support of pyrometallurgical copper processing, in *Copper-Cobre 2022 The Igor Wilkomirsky Symposium of Pyrometallurgy*, pp 609–634.

Klemettinen, L, 2021. Behaviour of trace elements in copper smelting processes- LA-ICP-MS as a tool for sample characterisation, PhD thesis, Aalto University, 64 p.

Klemettinen, L, Avarmaa, K and Taskinen, P, 2017. Trace element distributions in black copper smelting, *World of Metallurgy-Erzmetall*, 70(5):257–264.

Kojo, I and Huppe, H, 2011. Sustainable copper production processes from mine to cathode copper: some process selection considerations, in *9th Int Conf on Clean Technologies for the Mining Industry*, 10 p.

Lennartsson, A, Engström, F, Samuelsson, C, Björkman, B and Pettersson, J, 2018. Large-scale WEEE recycling integrated in an ore-based Cu-extraction system, *J Sust Metall*, 4(2):222–234.

Mäkinen, T, Fagerlund, K, Anjala, Y and Rosenback, L, 2005. *Outokumpu's Technologies for Efficient and Environmentally Sound Nickel Production* in COM 2005, pp 71–80 (CIM-MetSoc: Montreal).

Moats, M, Alagha, L and Awuah-Offei, K, 2021. Towards resilient and sustainable supply of critical elements from the copper supply chain: A review, *J Clean Prod*, 307:127207.

Nakajima, K, Takeda, O, Miki, T and Nagasaka, T, 2009. Evaluation of methods of metal resource recyclability based on thermodynamic analysis, *Mater Trans*, 50(4):453–460.

Nassar, N, Kim, H, Frenzel, M, Moats, M and Hayes, S, 2022. Global tellurium supply potential from electrolytic copper refining, *Res Conserv Recycl,* 184:106434.

Park, M, Takeda, Y and Yazawa, A, 1984. Equilibrium relations between liquid copper, matte and calcium ferrite slag at 1523 K, *Trans JIM,* 25(10):710–715.

Pesonen, L, Johto, H, Jyrkönen, S and Lindgren, M, 2022. Expanded comparison of modern copper smelting technologies and related downstream processes, in *Proceedings of Copper-Cobre 2022 Symposium The Igor Wilkomirsky Symposium of Pyrometallurgy*, pp 165–181.

Richter, K, Campbell, A, Humayun, M and Hervig, R, 2004. Partitioning of Ru, Rh, Pd, Re, Ir and Au between Cr-bearing spinel, olivine, pyroxene and silicate minerals, *Geochim Cosmochim Acta*, 68(4):867–880.

Risopatron, C, 2018. Impurities in copper raw materials and regulatory advances in 2017, A global overview, in *Int Semin Impurities in Copper Raw Materials*, p 85 (International Copper Study Group: Portugal). Available from: <http://www.jogmec.go.jp/content/300358430.pdf>

Salminen, J, Nyberg, J, Imris, M and Heegard, B, 2020. Smelting jarosite and sulphur residue in a plasma reactor, in *9th Int Symp PbZn* (eds: A Siegmund, S Alam, J Grogan, U Kerney and E Shibata), pp 391–403.

Sukhomlinov, D, Avarmaa, K, Virtanen, O, Taskinen, P and Jokilaakso, A, 2020a. Slag-copper equilibria of selected trace elements in black-copper smelting, Part II, Trace element distributions, *Miner Process Extr Metall*, 41(3):171–177.

Sukhomlinov, D, Klemettinen, L, O'Brien, H, Taskinen, P and Jokilaakso, A, 2020b. Behavior of Ga, In, Sn and Te in copper matte smelting, *Metall Mater Trans B*, 50(6):2723–2732.

Taskinen, P, 2011. Direct-to-blister smelting of copper concentrates: the slag fluxing chemistry, *Miner Process Extr Metall*, 120(4):240–246.

Tuominen, J and Kojo, I, 2005. Blister flash smelting-Efficient a flexible low-cost continuous copper process, in *Converter and Fire Refining Practices* (eds: A Ross, T Warner and K Scholey), pp 271–282.

Tuominen, J, Anjala, Y and Björklund, P, 2007. Slag cleaning of Outokumpu direct-to-blister flash smelting slags, in *Copper-Cobre 2007 International Conference*, pp 339–350 (MetSoc: Montreal).

United States Geological Survey (USGS), 2022. US Geological Survey releases 2022 list of critical minerals (media release). Available from: <https://www.usgs.gov/news/national-news-release/us-geological-survey-releases-2022-list-critical-minerals>

US EPA, 2023. Scope 3 Inventory Guidance. Available from: <https://www.epa.gov/climateleadership/scope-3-inventory-guidance>

The extraction of white phosphorus in molten salt

X Yang[1]

1. Principle Investigator, Westlake University, Hangzhou, China.
   Email: yangxiao@westlake.edu.cn

ABSTRACT

White phosphorus (P_4) is a critical element in industry, but its current production from phosphate rock using the carbothermal reduction method raises significant environmental and health concerns. This pressing issue necessitates the development of new technologies that facilitate the sustainable production of P_4. In response, my team is actively developing an innovative, coke-free approach for white phosphorus production. This method involves the dissolution of phosphate rock in molten $CaCl_2$, followed by the extraction of P_4 through electrochemistry at relatively low temperatures. Moreover, this approach has the potential to utilise phosphate-bearing solid wastes as raw materials, further enhancing its environmental benefits. By employing this cutting-edge technique, we aim to revolutionise the phosphorus chemical industry towards greener and more sustainable practices. This paper provides evidence of the feasibility and effectiveness of our proposed approach.

INTRODUCTION

White phosphorus (P_4) is the starting material for various critical chemicals, such as phosphoric acids, chlorides, sulfides and phosphides. The global demand for P_4 has been steadily increasing in recent years. However, the P_4-manufacturing industry is facing significant challenges as the production process is highly energy-intensive and environmentally harmful. The supply of P_4 is susceptible to national trade policies, further complicating its availability in the market. Consequently, securing a stable supply of P_4 has become an urgent concern for many countries.

The industrial production of P_4 involves heating a mixture of phosphate rock [$Ca_5(PO_4)_3F$], sand (SiO_2) and coke (C) in an electric furnace at around 1500°C. This energy-intensive process, depicted in Figure 1a, consumes a substantial amount of electricity (13.5~15 MWh per ton of P_4) due to the highly endothermic nature of the carbon-driven reduction of phosphate. Ensuring product quality is challenging as the high temperatures make the process susceptible to impurity contamination, necessitating additional purification steps for high-purity product production. Furthermore, the emission of significant quantities of hazardous by-products poses severe environmental concerns.

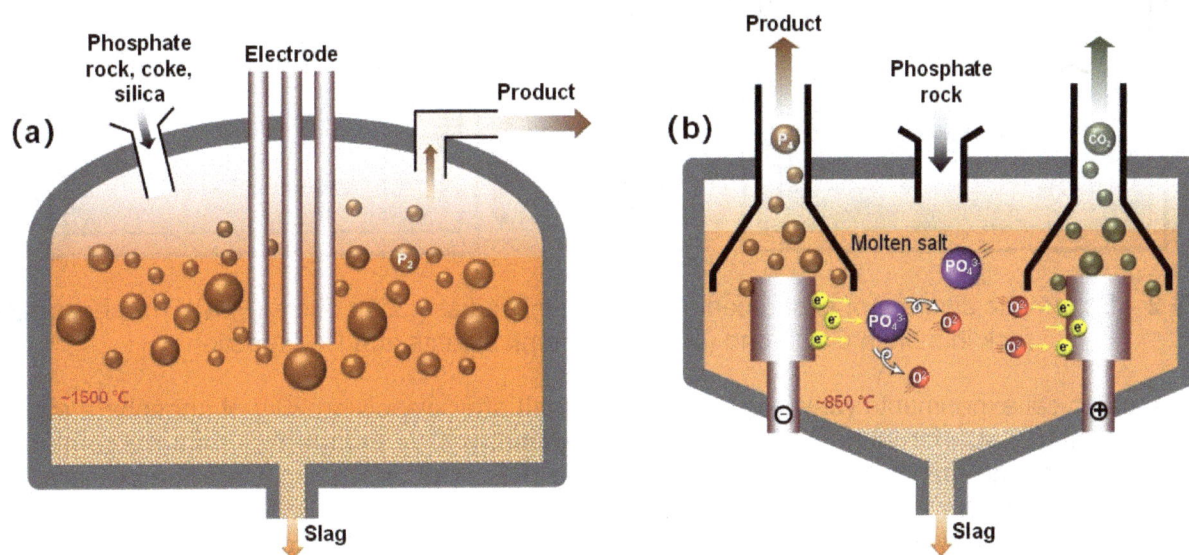

FIG 1 – Illustration of white phosphorus production processes: (a) the carbothermal reduction process; and (b) the electrochemical reduction process.

The advancement of P_4-manufacturing technology is crucial for sustainable development. In this study, we present a novel process for producing P_4 (Yang and Nohira, 2020; Liu, Ma and Yang, 2024), as illustrated in Figure 1b. This innovative process involves dissolving phosphate rock in molten $CaCl_2$, followed by the electrochemical extraction of P_4. Notably, this approach eliminates the need for coke and operates at a lower temperature of approximately 850°C. As a result, the electrochemical method is expected to offer substantial advantages over the traditional carbothermal process, particularly in terms of simplicity and cleanness. This paper demonstrates the feasibility and effectiveness of our proposed approach.

EXPERIMENT

Molten $CaCl_2$ was employed as the solvent and electrolyte. Pure $Ca_5(PO_4)_3F$ and natural phosphate rock (mined in Hubei, China) were added into molten $CaCl_2$ at 850°C, respectively. Figure 2a shows the set-up used to study the dissolution behaviour of $Ca_5(PO_4)_3F$ or natural phosphate rock in molten $CaCl_2$. The experiment was performed in a crucible furnace. The melts sampled at regular time intervals by dipping an Al_2O_3 rod into the melt bath were analysed by inductively coupled plasma – mass spectroscopy (ICP-MS) to reveal the concentration of soluble phosphorus.

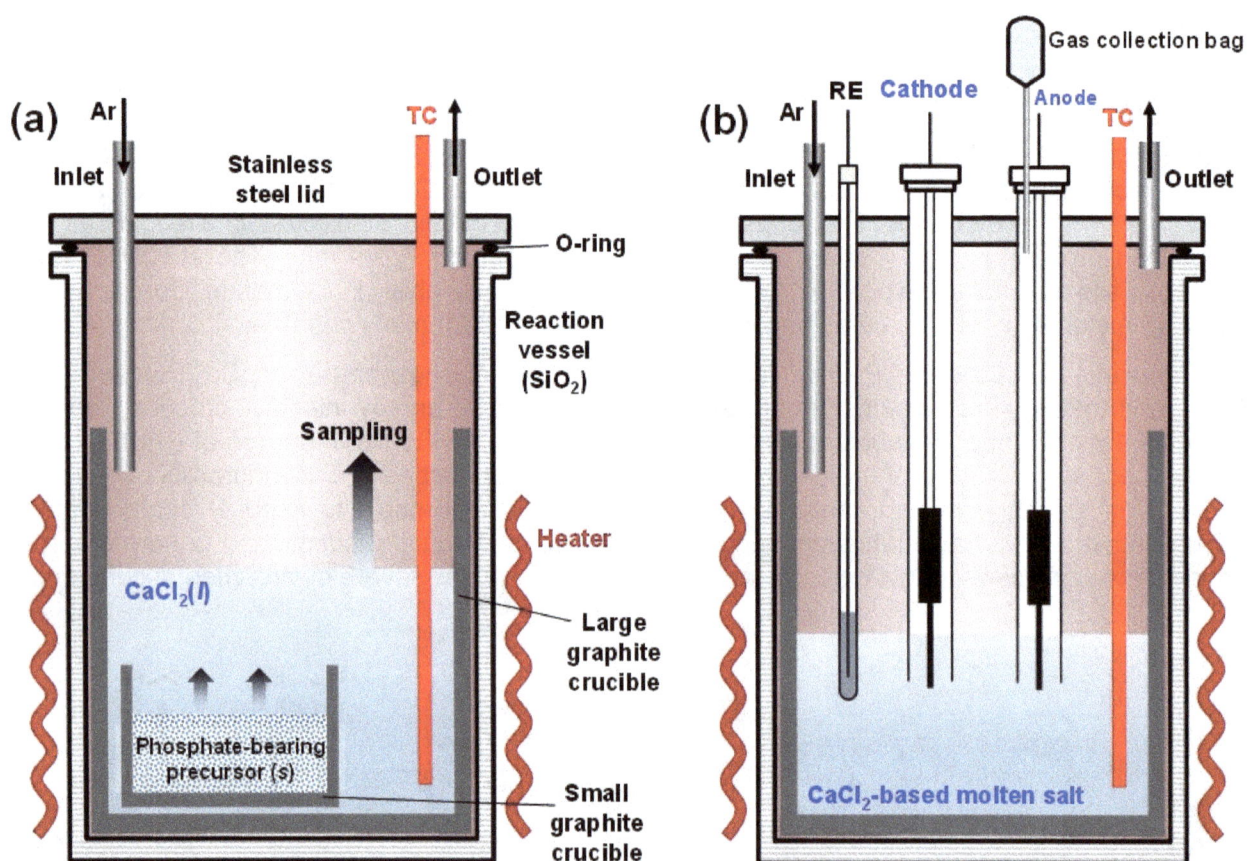

FIG 2 – Schematic diagram of experimental set-ups for (a) dissolution of $Ca_5(PO_4)_3F$ or phosphate rock in molten $CaCl_2$ and (b) electrochemical tests and P_4 extraction.

Electrochemical experiments were conducted in the melt after dissolving 20 g of phosphate rock in 300 g of molten $CaCl_2$ for 48 hr at 850°C. Cyclic voltammetry (CV) and electrolysis were conducted in a three-electrode manner. The reference electrode (RE) was an Ag^+/Ag electrode prepared by immersing an Ag wire in $CaCl_2$ containing 0.5 mol per cent AgCl in a SiO_2 tube. A graphite rod (diameter 5 mm) enclosed by a quartz tube was used as the cathode and anode, which were immersed in the melt to a depth of 25 mm. The cathodic product after dissolving in CS_2 was tested by 600 MHz Solution NMR Spectrometer. The anodic gas was analysed by Gas Chromatography.

RESULTS AND DISCUSSION

In Figure 3a, the dissolution profiles of P from excessive phosphate rock in 300 g of molten $CaCl_2$ at different temperatures are presented. The results indicate that the concentrations of P in the melt increase rapidly and reach a plateau in less than 10 mins, suggesting a fast and efficient dissolution process. The plateau values represent the solubility of P at each temperature, as they are lower than the theoretical concentration of P assuming complete dissolution of P in phosphate rock. The solubility of P is observed to increase with temperature. It should be noted that, in comparison to the dissolution of $Ca_5(PO_4)_3F$, the dissolution rate of P from phosphate rock is evidently higher, while the solubility is slightly lower. This difference can be attributed to the presence of impurities in the phosphate rock.

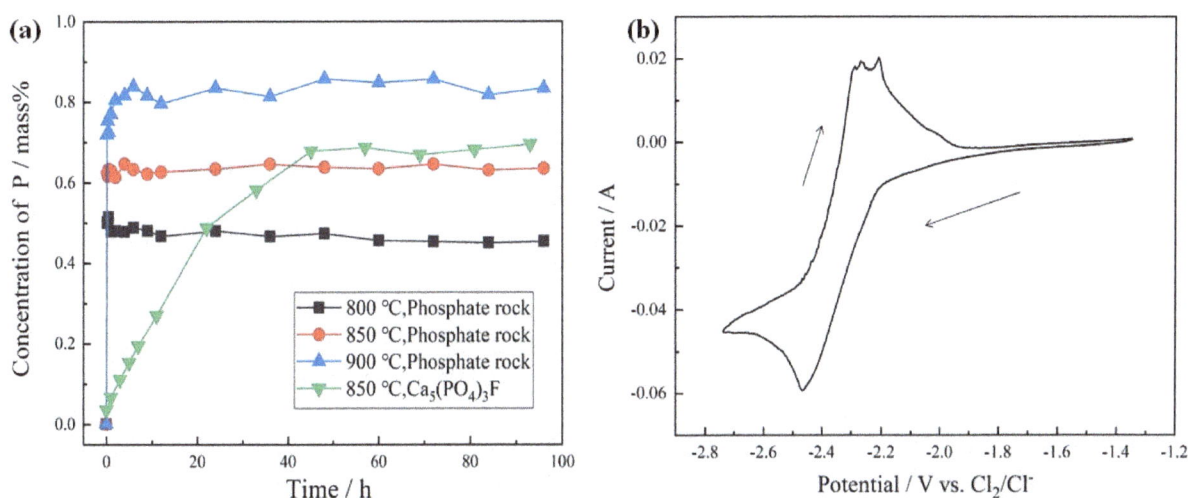

FIG 3 – (a) Dissolution profiles of P from phosphate rock and $Ca_5(PO_4)_3F$ in molten $CaCl_2$ at various temperatures. (b) Cyclic voltammograms at a graphite rod electrode (diameter 3 mm, immersion depth 1 mm) in 300 g of molten $CaCl_2$ after dissolving 20 g of phosphate rock for 48 hr at 850°C.

Figure 3b shows the cyclic voltammograms in the melt. Considering the electrical resistance of the melt, 70 per cent IR compensation was applied. Cathodic currents increase at potentials negative than –2.0 V (vs Cl_2/Cl^-), which is attributed to the electrochemical reduction of PO_4^{3-}. Subsequent investigations confirm that the product is P_4. A single cathodic peak can be clearly observed in most cases, suggesting that the electrochemical reduction of PO_4^{3-} can be approximated as a one-step (five-electron transfer) reduction process. The anodic peaks observed in the CV curves correspond to the oxidation of newly formed P_4 adhered on the cathode surface. The presence of multiple anodic peaks suggests that the oxidation process is more complex than the reduction process and may involve several intermediate products and electron transfer steps.

Following the cyclic voltammetry test, constant current electrolysis was conducted. Figure 4a shows the photographs of the upper section of the cathode after electrolysis at -0.1 A cm^{-2} for 10 hrs. Yellow substance condensed on the inner wall of the quartz tube was confirmed to be P_4 by ^{31}P NMR analysis. Chemical analysis indicates that the obtained P_4 product through molten salt electrolysis is of significantly higher purity compared to the industrial-grade commercial product produced by the carbothermal reduction process.

Figure 4b shows the chromatograms of the anodic gas collected at different moments during the electrolysis. Both CO and CO_2 were identified in the anodic gas and the quantitative analysis showed that CO_2 was the dominant one. These results suggest that the following electrode reactions occur during electrolysis:

$$\text{Cathode: } 4PO_4^{3-} + 20e^- \rightarrow 16O^{2-} + P_4(g) \tag{1}$$

$$\text{Anode: } C(s) + 2O^{2-} \rightarrow CO_2(g) + 4e^- \tag{2}$$

$$\text{Overall: } 4PO_4^{3-} + 5C(s) \rightarrow 6O^{2-} + 5CO_2(g) + P_4(g) \tag{3}$$

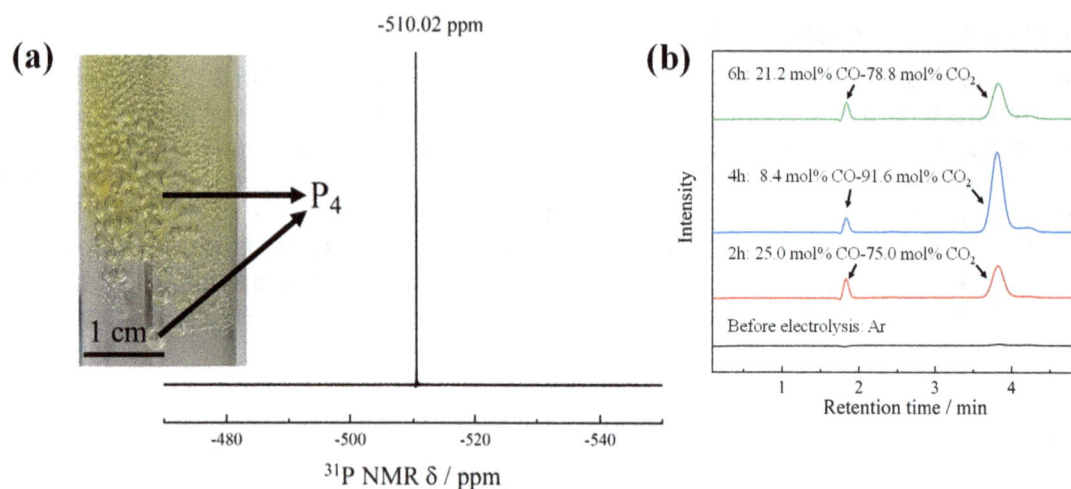

FIG 4 – (a) Photograph of the upper section of the quartz tube enclosing the cathode and ^{31}P NMR spectrum of the product after electrolysis at -0.1 A cm^{-2} for 10 hrs in the melt after dissolving 20 g of phosphate rock in 300 g of molten CaCl$_2$ for 24 hrs at 850°C. (b) Chromatograms of three anodic gas samples collected at different moments during the electrolysis and pure Ar.

CONCLUSIONS

This work demonstrates a more environmentally friendly and efficient approach for extracting white phosphorus from phosphate rock through molten salt electrolysis. In this process, phosphate rock dissolves in molten CaCl$_2$, allowing the soluble PO$_4^{3-}$ to be electrochemically reduced. Notably, the operating temperature required for producing white phosphorus is only 850°C, which is significantly lower than that required in the conventional carbothermal reduction process. Additionally, white phosphorus produced through molten salt electrolysis exhibits a higher purity level compared to the industrial-grade commercial product obtained through carbothermal reduction. The electrochemical process also shows decreased electricity and carbon consumption. These findings underscore the potential of the electrochemical reduction process as a future method for extracting high-purity white phosphorus.

REFERENCES

Liu, G, Ma, Y and Yang, X, 2024. Cleaner extraction of white phosphorus from phosphate rock through molten salt electrolysis, *J Cleaner Prod,* 434(1):1403.

Yang, X and Nohira, T, 2020. A new concept for producing white phosphorus: electrolysis of dissolved phosphate in molten chloride, *ACS Sustain Chem Eng,* 8(36):13784–13792.

Research on –
Recycling and environment sustainability

The effect of FeO/SiO$_2$ ratio on the feasibility of utilising iron silicate slags as supplementary cementitious materials

A Andersson[1], J Isaksson[2], L Brander[3], A Lennartsson[4], Å Roos[5] and F Engström[6]

1. Associate Senior Lecturer, Luleå University of Technology, Luleå 971 87, Sweden.
   Email: anton.andersson@ltu.se
2. Associate Senior Lecturer, Luleå University of Technology, Luleå 971 87, Sweden.
   Email: jenny.isaksson@ltu.se
3. Senior Scientist, Research Institutes of Sweden, Mölndal 431 53, Sweden.
   Email: linus.brander@ri.se
4. Senior Lecturer, Luleå University of Technology, Luleå 971 87, Sweden.
   Email: andreas.lennartsson@ltu.se
5. Product Manager, Boliden Smelters, Stockholm 101 20, Sweden. Email: ake.roos@boliden.com
6. Associate Professor, Luleå University of Technology, Luleå 971 87, Sweden.
   Email: fredrik.i.engstrom@ltu.se

ABSTRACT

Slags are the most voluminous solid by-products generated in pyrometallurgical operations, and, as such, finding an application for these oxidic materials is pertinent to maintaining resource efficiency. In particular, pyrometallurgical copper production is associated with high slag rates, typically ranging from 2.2 to 3.0 tons of slag per ton produced copper, which necessitates slag valourisation. A possible application for these iron silicate slags is as supplementary cementitious materials (SCMs), which effectively lowers the CO$_2$ emissions per ton of cementitious material. Although utilising iron silicate slags as SCMs has been studied in previous work, the scientific literature has limited data on the effect of composition on the inherent reactivity in cementitious systems. More specifically, no reports on the impact of FeO/SiO$_2$ ratio have been presented in previous publications. Therefore, the present study aimed to isolate this parameter in a synthetic FeO-SiO$_2$-Al$_2$O$_3$-CaO-MgO-Cr$_2$O$_3$ system. Since the amorphous content is a pertinent parameter for SCMs, a high-temperature confocal laser scanning microscope (HT-CLSM) was utilised to assess the crystallisation behaviour at continuous cooling conditions. Furthermore, high-temperature rheological experiments were conducted to measure the viscosities of the slags in relation to the crystallisation behaviour. The experiments showed that depolymerising the slag by increasing the FeO/SiO$_2$ ratio generated a less viscous slag, which showed higher tendencies for crystallisation. Furthermore, experiments assessing the inherent reactivity as an SCM showed that depolymerising the slag by increasing the FeO/SiO$_2$ ratio as the sole independent parameter generated an initially more reactive slag. However, replacing SiO$_2$ with FeO was found to negatively affect the reactivity as measured over seven days within the tested compositional range.

INTRODUCTION

The pyrometallurgical extraction of copper is associated with slag rates in the range of 2.2 to 3.0 tons of slag per ton produced copper (Shi, Meyer and Behnood, 2008; Gabasiane *et al*, 2021), and this necessitates slag valourisation to ensure a resource-efficient process. Multiple external applications are reported for the recycling of copper slags, eg sand blasting media, auxiliary to sand in cement, and various construction applications excluding cement (Al-Jabri *et al*, 2006; Shi, Meyer and Behnood, 2008; Murari, Siddique and Jain, 2015; Gabasiane *et al*, 2021; Kero Andertun *et al*, 2022). An attractive valourisation route for the slag is the utilisation as an supplementary cementitious material (SCM), which lowers the carbon dioxide emissions per ton cementitious material since the SCM requires neither calcination nor clinkering (Lothenbach, Scrivener and Hooton, 2011).

As summarised in a previous publication (Andersson *et al*, 2023a), several studies have been published addressing iron silicate copper slags in SCM applications, and the assessed compositions span wide ranges in the major components. For a specific slag composition, several parameters affect the reactivity of the slag as an SCM. These parameters include, eg the amorphous content, specific surface area, and thermal history (Skibsted and Snellings, 2019). Furthermore, in addition to generating specific surface area, the milling can be operated to cause amorphisation (Feng *et al*,

2019a) or activation by introducing surface defects (Romero Sarcos *et al*, 2021; Andersson *et al*, 2023a), both of which improve reactivity. The review of previously studied slag compositions showed that the slags were generated using different cooling processes, amounting to different amorphous contents, and, in addition, the slags were milled to different specific surface areas and evaluated for their performance as SCMs using different methodologies (Andersson *et al*, 2023a). Consequently, the previous data cannot be used across studies to evaluate the compositional aspect since essential parameters are not fixed between studies. Nevertheless, research on the isolated effects of CaO content (Feng *et al*, 2019b) and Al_2O_3 content (Zhang *et al*, 2023) on the performance of iron silicate slag in SCM applications have been performed. However, no study on the implications of changing the FeO/SiO_2 ratio has been published.

Assuming that the slag composition is fixed, two main parameters affect the appropriateness of using the slag as an SCM. Firstly, the cooling process, which correlates to the degree of crystallinity of the slag and possibly the polymerisation (Stebbins, 1988). Crystalline phases are generally considered inert in SCM applications (Snellings, 2013; Kucharczyk *et al*, 2018; Glosser *et al*, 2021), which means that the cooling rate must supersede crystallisation. Secondly, the subsequent milling of the solidified slag, which determines the specific surface area, as well as the possible amorphisation (Feng *et al*, 2019a), or activation from introducing surface defects (Romero Sarcos *et al*, 2021; Andersson *et al*, 2023a). Ramanathan, Tuen and Suraneni (2022) argued that the reactivity of SCMs is controlled by the surface area, extent of amorphous phases, and compositions of the amorphous phases. Consequently, to compare the effect of chemical composition, the slags require similar specific surface areas and amorphous contents.

Based on the above, the present study aimed to offer insight into the effect of FeO/SiO_2 ratio on the suitability of forming a reactive SCM. The aim was addressed using a synthetic $FeO-SiO_2-CaO-Al_2O_3-MgO-Cr_2O_3$ system and evaluating how the FeO/SiO_2 ratio influences the crystallisation behaviour under continuous cooling conditions and how this relates to the viscosity of the slag. Furthermore, the effect of the FeO/SiO_2 ratio on the inherent reactivity in the SCM application for slags cooled under repeatable conditions and milled to similar specific surface areas was assessed.

METHOD

Synthesis of samples

Two iron silicate slag samples in the $FeO-SiO_2-CaO-Al_2O_3-MgO-Cr_2O_3$ system were synthesised, and their intended compositions given in atomic percent (at%) are presented in Table 1. The methodology of synthesising and granulating the samples has been accounted for in detail in a previous publication (Andersson *et al*, 2023b), and, therefore, the present description offers a summary.

TABLE 1

Composition of the samples included in the study [at%].

Sample	FeO/SiO_2	FeO	SiO_2	CaO	Al_2O_3	MgO	Cr_2O_3
A	1.25	51.2	41.0	1.2	5.0	1.5	0.08
B	2.00	61.5	30.7	1.2	5.0	1.5	0.08

The compositions in Table 1 were mixed based on the reagent grade chemicals Fe, Fe_2O_3, SiO_2, $CaCO_3$, Al_2O_3, MgO, and Cr_2O_3. Each synthetic slag was melted in an iron crucible (>99.82 per cent Fe), which ensures that a partial pressure of oxygen relevant to iron silicate slags undergoing the zinc fuming process is attained (Andersson *et al*, 2023b). Figure 1 illustrates the graphite resistance-heated furnace employed in the experiments, and this set-up was operated under inert atmosphere achieved by introducing nitrogen (99.996 per cent N_2) and argon (99.999 per cent Ar) at flow rates of 12 L/min and 3 L/min, respectively. The temperature was ramped at a rate of 10 K/min to 100 K above the calculated liquidus temperature of each respective slag composition (disregarding the spinel phase in case of primary crystallisation, ie 100 K above the first crystal to co-precipitate). These calculations were performed using the Equilib module of FactSage™ ver 8.2 (by ThermFact

Inc. and GTT-Technologies) employing the GTOx database (Bale *et al*, 2016; Yazhenskikh *et al*, 2019). The idea of including Cr_2O_3 and subsequently not considering the primary crystallisation of the spinel was accounted for in the previous publication (Andersson *et al*, 2023b). Based on the calculations, slag A and B were synthesised at 1509 and 1517 K, respectively. After an isothermal section of two hrs at the respective temperature, the crucible was removed manually from the furnace, and the slag was granulated in water jets operating with cold tap water at a flow rate of 1.1 L/s. The methodology has been shown to produce synthetic iron silicate slags with relevant and repeatable results (Andersson *et al*, 2023b).

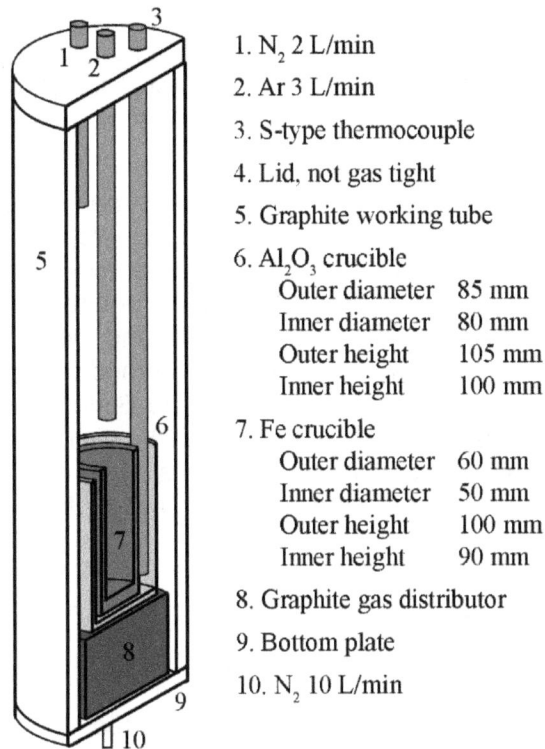

1. N_2 2 L/min

2. Ar 3 L/min

3. S-type thermocouple

4. Lid, not gas tight

5. Graphite working tube

6. Al_2O_3 crucible
   Outer diameter   85 mm
   Inner diameter   80 mm
   Outer height   105 mm
   Inner height   100 mm

7. Fe crucible
   Outer diameter   60 mm
   Inner diameter   50 mm
   Outer height   100 mm
   Inner height   90 mm

8. Graphite gas distributor

9. Bottom plate

10. N_2 10 L/min

FIG 1 – Schematic overview of the experimental set-up.

The synthesised slags were analysed for their chemical composition by an accredited laboratory using inductively coupled plasma mass spectrometry. The digestion of the samples prior to the analyses was achieved via microwave-assisted dissolution in a mixture of hydrochloric acid, nitric acid, and hydrofluoric acid after fusion with lithium metaborate.

Rietveld powder X-ray diffraction (XRD) with an internal standard was used to determine the amorphous content of the two slags. The internal standard calcite (99.5 per cent $CaCO_3$) was mixed to obtain 10 wt per cent in each sample. Mixing was made using a ring mill, and the subsequent scans were performed between 10 and 90 °2θ with copper K_α generated at 45 kV and 40 mA using a Malvern Panalytical Empyrean X-ray diffractometer (Malvern Panalytical, Malvern, UK). The refinement was performed using HighScore+ and the COD database (Gražulis *et al*, 2009).

Crystallisation and viscosity

A high-temperature confocal laser scanning microscope (HT-CLSM) (Yonekura VL2000DX-SVF17SP) was utilised to study the crystallisation behaviour of the two slags under continuous cooling experiments. The measurements were performed on sample sizes of 115 mg in molybdenum crucibles. The crucible choice was based on the alloying of platinum with iron experienced for iron silicate slags. Prior to the cooling cycle, the slags were allowed to homogenise at 1673 K for 300 secs. The primary crystallisation was determined for the cooling rates of 10, 25, 50, 100, 250, 500, and 1000 K/min.

In accordance with the evaluation of using different crucible and spindle materials for measuring the viscosity of iron silicate slags (Isaksson *et al*, 2023), the viscosity measurements in the present study were made using the rotating cylinder technique with a spindle and crucible made of molybdenum (tzm molybdenum 364). Details related to the experimental set-up, eg manufacturer, placement of the thermocouple, gas atmosphere, calibration, and control of performance, have been accounted for in a previous publication (Isaksson *et al*, 2023).

The viscosity was measured at increments of 50 K during a cooling cycle from 1723 to 1423 K. Prior to each measurement at a new temperature, the slag was allowed to homogenise at a constant shear rate until stable values of the viscosity were achieved. The actual measurements were performed at shear rates of 1, 2, 4, 8, and 16 1/sec.

Inherent reactivity

The synthetic slags were milled iteratively to a similar specific surface area using a planetary ball mill operated at 600 revolutions per minute with 18 tungsten carbide balls of 10 mm in diameter in 45 mL tungsten carbide grinding bowls. The Brunauer, Emmet, and Teller (BET) specific surface area was measured after degassing at 573 K for 60 min.

The milled samples were subjected to the isothermal calorimetry-based rapid screening test for SCMs outlined by Snellings and Scrivener (2016), further developed and referred to as the rapid, relevant, and reliable (R^3) test by Avet *et al* (2016). This method was chosen based on the study presented by Li *et al* (2018), which demonstrated that the R^3 test results correlated strongly with the strength of mortars, independent of the type of SCM and with great repeatability between laboratories. Furthermore, this calorimetry-based testing method tests the inherent reactivity of the SCM without the influence of cement hydration (Hallet, De Belie and Pontikes, 2020; Sivakumar *et al*, 2021). Mixtures were prepared in accordance with Li *et al* (2018), and the heat flow was measured for seven days using a TAM Air isothermal calorimeter. A detailed description of the procedure employed at the current laboratory has been accounted for in a previous publication (Andersson *et al*, 2023c).

RESULTS AND DISCUSSION

Chemical and mineralogical composition

The analysed chemical compositions of the slags are presented in Figure 2a. For slag A, the FeO and SiO_2 contents were analysed to 53.3 and 38.2 at per cent, respectively. Analogously, for slag B, the FeO and SiO_2 contents were analysed to 61.7 and 30.4 at per cent, respectively. As no volatile components were included in the slags, deviations from the mixed composition should be solely attributed to the possible interaction with the crucible or to analytical errors. Stable oxides such as CaO, Al_2O_3, MgO, and SiO_2 are expected to interact less with the crucible, which is consistent with Figure 2a. More specifically, in comparison to the theoretical or mixed compositions, CaO, Al_2O_3, and MgO were within the experimental error reported by the accredited laboratory. The FeO/SiO_2 ratio of slag B was satisfactory, while slag A deviated from the mixed composition. Additional FeO might enter the slag from oxidation of the ferrous iron to ferric and the subsequent reduction of ferric iron to ferrous iron by the crucible, which could constitute the higher FeO/SiO_2 ratio. However, the experimental set-up was operated in an inert atmosphere, which should mitigate the proposed mechanism. Accounting for the reported error in the analysis, the quotient of slag A, determined to be 1.39, is within the experimental error of the theoretical ratio of 1.25. In conclusion, although the FeO/SiO_2 ratio of slag A was analysed to a higher quotient than aimed for, the attained chemical compositions successfully isolated the FeO/SiO_2 ratio as a parameter.

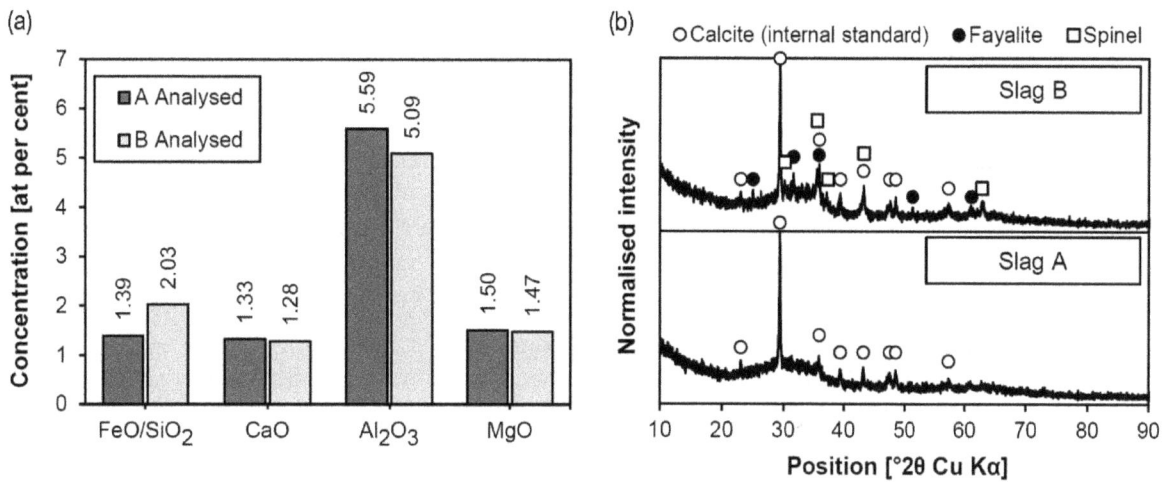

FIG 2 – (a) Chemical compositions in at% and quotient of at% for the FeO/SiO$_2$ ratio. (b) X-ray diffractograms of both slags including calcite as internal standard.

The mineralogical compositions of the samples are presented in Figure 2b. Slag A, having a lower FeO/SiO$_2$ ratio, was not analysed for any crystalline phases except the mixed internal standard. Therefore, an entirely amorphous slag was generated. On the other hand, slag B, with an FeO/SiO$_2$ ratio representing stoichiometric fayalite, was analysed for both fayalite and spinel. Based on the Rietveld refinement accounting for the internal standard, slag B was calculated to contain 90.1 wt per cent amorphous slag. Furthermore, spinel and fayalite were determined to constitute 3.3 and 6.6 wt per cent of the sample, respectively.

Crystallisation behaviour and measured viscosities

Based on the equilibrium-type cooling calculations, the liquidus temperatures of slag A and B were estimated to 1553 and 1610 K, respectively. Both slags experienced primary crystallisation of spinel with olivine, ie fayalite-forsterite solid solution, as the first crystal to co-precipitate. The co-precipitation was calculated to start at 1409 and 1417 K for slag A and B, respectively. Figure 3a presents the results of the HT-CLSM experiments. The dashed lines in the figure represent the continuous cooling temperature curves, and the plotted circles represent the crystallisation, ie the continuous cooling transformation curves. In comparison to the calculated temperature of primary crystallisation, the slowest cooling rate suggests lower temperatures than those predicted by the GTOx database.

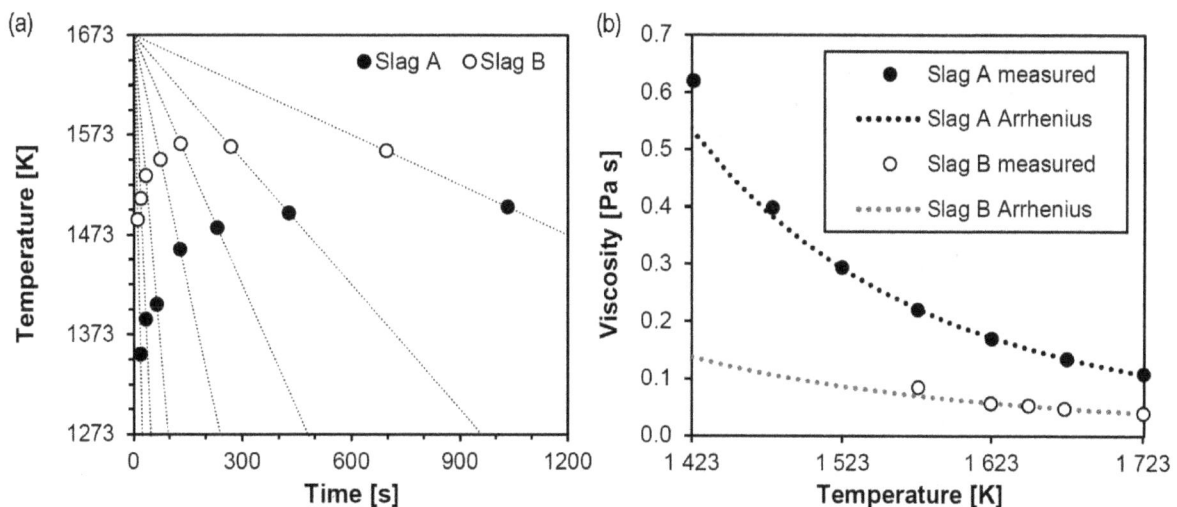

FIG 3 – (a) continuous cooling transformation diagram of the slags. (b) Measured viscosities of the slags.

The juxtaposition of the continuous cooling transformation curves of slag A and B in Figure 3a shows that the former slag is less prone to crystallisation. Both the absolute and relative temperature difference between the crystallisation at 10 K/min and 1000 K/min is more considerable for slag A in comparison to slag B, and, therefore, the granulation of slag A should more easily generate an entirely amorphous material. During the synthesis of the samples, the temperature of onset granulation for slag B (1517 K) was below the temperature of observed primary crystallisation, Figure 3a, which explains the presence of spinel in slag B, Figure 2b. However, in addition to spinel, fayalite was found in slag B, and this observation complies with the data related to the higher tendencies towards crystallisation.

Since low ion migration is related to energy barriers for crystallisation (Deng *et al*, 2022), the viscosities of slag A and B were measured to indirectly assess the differences in the degree of polymerisation and ion mobility. The average viscosity of 20 measurement points at a shear rate of 8 1/sec, at each temperature, is presented as circles in Figure 3b. For slag B, temperatures below 1573 K did not generate consistent values, which indicated that significant crystallisation had occurred. Therefore, the measurements in the cooling cycle were discontinued at 1523 K, and a supplementary measurement at 1648 K was included to incorporate an additional data point in the liquidus region. The viscosity data was used to calculate the activation energy of flow using an Arrhenius-type equation (Equation 1).

$$\eta = A exp\left(\frac{E_a}{RT}\right) \tag{1}$$

where:

η	is the viscosity [Pa s]
A	is the pre-exponential factor [Pa s]
E_a	is the activation energy of flow [J/mol]
R	is the gas constant [J/mol K]
T	is the temperature [K]

Since Equation 1 is valid for a fully liquid slag, the calculations of the activation energy of flow were performed for the data points above the liquidus temperature. For slag A, the lowest temperature included in the calculations was 1523 K, which is consistent with the HT-CLSM data presented in Figure 3a. Analogously, for slag B, the calculations were performed with the four data points at 1623 K and above. This approach resulted in activation energies of 109.1 and 84.9 kJ/mol for slag A and B, respectively. Furthermore, the pre-exponential factors were calculated to 53 and 106 µPa s for slag A and B, respectively. The values represent what can be seen in Figure 3b, ie slag A has higher internal friction of flow. In addition, the activation energies highlight the structural changes with temperature, suggesting that lowering the temperature has higher implications on the structural changes of slag A compared to slag B. Finally, the plotted calculated viscosity for fully liquid slags using the parameters from Equation 1, labelled Arrhenius in Figure 3b, deviates from the measured viscosity in accordance with the crystallisation shown in Figure 3a.

The measured viscosities and the calculated activation energies of flow agree with the degree of polymerisation of the two slags, which can be compared using the ratio between non-bridging oxygen and tetragonally coordinated oxygen (NBO/T) calculated according to the method accounted for in the literature (Mills, 1995). More specifically, the NBO/T was calculated to be 1.92 and 2.91 for slag A and B, respectively, highlighting the network-breaking character of the ferrous oxide. Thus, the data support the observations from the crystallisation experiments and the mineralogical composition after the water granulation. A caveat can be introduced in accordance with the review on the viscosity of glass-forming systems (Zheng and Mauro, 2017), ie the viscosity-temperature relationship in the fully liquidus state does not necessarily represent the behaviour for a fully liquid supercooled melt. However, as SiO_2 behaves as a strong liquid with Arrhenius behaviour under supercooling while more depolymerised silicate melts are classified as fragile, the viscosity difference of slag A and B during supercooling might be enhanced. In conclusion, generating an entirely amorphous slag during granulation is comparatively more feasible for slag A than for slag B, which is positive for an intended application as an SCM.

Reactivity as a supplementary cementitious material

A previous study on SCMs have shown that the reactivity measured using the R^3 isothermal calorimetry-based experiment correlates to dissolved aluminium and silicon in cement pore solution conditions (Ramanathan *et al*, 2021). This is consistent with the formation of calcium-silicate-hydrate gel and calcium-aluminium-silicate-hydrate gel, as well as the alumina ferric oxide tri-substituted and mono-substituted phases upon the SCMs consumption of portlandite (Hallet, De Belie and Pontikes, 2020; Sivakumar *et al*, 2021). In the present study, the iterative milling generated samples with BET specific surface areas of 1.04 and 1.11 m^2/g for slag A and B, respectively. Therefore, the differences in the reactivities measured in the R^3 test, presented in Figure 4, can be attributed to the chemical and mineralogical characteristics of the slags, translating to the dissolution of silicon and aluminium from the slags to the cementitious system.

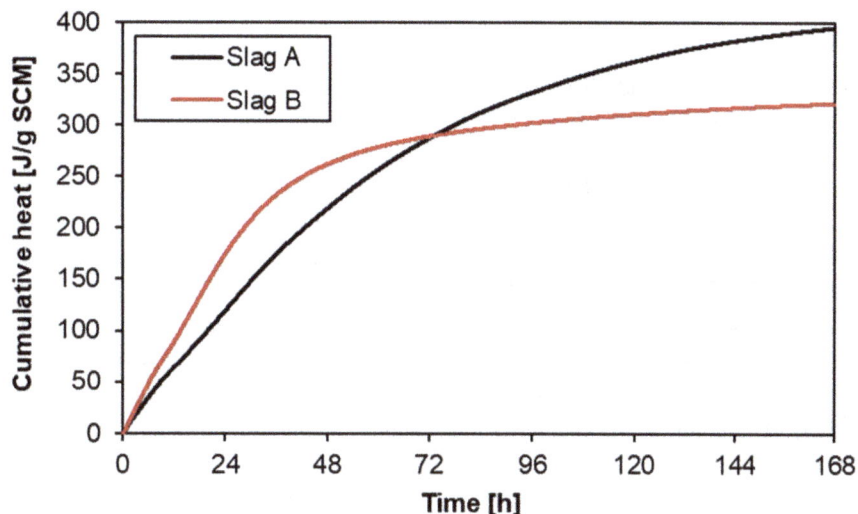

FIG 4 – Cumulative heat from the R^3 isothermal calorimetry experiment.

By depolymerising the slag via the replacement of silica with ferrous oxide, the contribution of silicon per mass unit of dissolved slag is less. Therefore, according to the data presented in Figure 4, slag B has an initially higher dissolution rate, which is consistent with previous studies showing that more depolymerised SCMs have higher dissolution rates in cement pore solution conditions (Snellings, 2013; Schöler *et al*, 2017).

Despite having an initially higher reaction rate, slag B is less reactive, as seen over the whole experimental period. This observation is still valid if the crystalline content of 9.9 wt per cent is considered inert, ie if the reactivity in Figure 4 is normalised for the amorphous part of slag B. Dissolving iron into the solution at alkaline pH and redox potentials relevant for cement pore solution conditions has been indicated to be less favourable (Andersson *et al*, 2023c), which might facilitate the generation of diffusion barriers upon dissolution of the slag. Such a diffusion barrier would be more pronounced for a higher FeO/SiO_2 ratio. Furthermore, depolymerising the slag by replacing silica with ferrous iron oxide lowers the total potential silicon available for the cementitious system. Consequently, the progressively lower heat flow of slag B, generating the intersection observed at 74 hrs in Figure 4, can be attributed to either mechanism or a combination of both.

CONCLUSIONS

In the present study, two synthetic slags within the $FeO-SiO_2-CaO-Al_2O_3-MgO-Cr_2O_3$ system were studied to isolate the effect of the changed FeO/SiO_2 ratio on the feasibility of utilising the slags as SCMs. Based on the experimental work, the following was concluded:

- By increasing the FeO/SiO_2 ratio from 1.39 to 2.03 (at%/at%), the viscosity was lowered at comparable temperatures above the liquidus of both slags owing to the network-breaking characteristics of FeO.

- Continuous cooling transformation experiments showed that the slag with higher FeO/SiO_2 ratio had higher inclination towards crystallisation, which corresponds to the mobility within the melt.

- Although the more depolymerised slag had higher initial reactivity as an SCM, the seven-day reactivity was lower, which in combination with the tendency to crystallise suggest that the lower FeO/SiO_2 ratio of the present study is more attractive for slag valourisation as an SCM.

ACKNOWLEDGEMENTS

The present study was financed by Boliden AB and conducted within the Centre for Advanced Mining and Metallurgy (CAMM) at Luleå University of Technology. The assistance of Britt-Louise Holmqvist and Jakob Kero Andertun is gratefully acknowledged.

REFERENCES

Al-Jabri, K S, Taha, R A, Al-Hashmi, A and Al-Harthy, A S, 2006. Effect of copper slag and cement by-pass dust addition on mechanical properties of concrete, *Construction and Building Materials*, 20:322–331.

Andersson, A, Brander, L, Lennartsson, A, Roos, Å and Engström, F, 2023a. Ground Granulated Iron Silicate Slag as Supplementary Cementitious Material: Effect of Prolonged Grinding and Granulation Temperature, *Journal of Cleaner Materials*, 10:100209.

Andersson, A, Brander, L, Lennartsson, A, Roos, Å and Engström, F, 2023b. A method for synthesizing iron silicate slags to evaluate their performance as supplementary cementitious materials, *Applied Sciences*, 13(14):8357.

Andersson, A, Isaksson, J, Lennartsson, A and Engström, F, 2023c. Insights into the valorization of electric arc furnace slags as supplementary cementitious materials, *Journal of Sustainable Metallurgy*, pp 1–14.

Avet, F, Snellings, R, Diaz, A A, Haha, M B and Scrivener, K, 2016. Development of a New Rapid, Relevant and Reliable (R^3) Test Method to Evaluate the Pozzolanic Reactivity of Calcined Kaolinitic Clays, *Cement and Concrete Research*, 85:1–11.

Bale, C W, Bélisle, E, Chartrand, P, Decterov, S A, Eriksson, G, Gheribi, A E, Hack, K, Jung, I-H, Kang, Y-B, Melançon, J, Pelton, A D, Petersen, S, Robelin, C, Sangster, J, Spencer, P and Van Ende, M-A, 2016. Reprint of: FactSage Thermochemical Software and Databases, 2010–2016, *CALPHAD*, 55:1–19.

Deng, L, Yao, B, Lu, W, Zhang, M, Li, H, Chen, H, Zhao, M, Du, Y, Zhang, M, Ma, Y and Wang, W, 2022. Effect of SiO_2/Al_2O_3 ratio on the crystallization and heavy metal immobilization of glass ceramics derived from stainless steel slag, *Journal of Non-Crystalline Solids*, 593:121770.

Feng, Y, Kero, J, Yang, Q, Chen, Q, Engström, F, Samuelsson, C and Qi, C, 2019a. Mechanical Activation of Granulated Copper Slag and Its Influence on Hydration Heat and Compressive Strength of Blended Cement, *Materials*, 12:772.

Feng, Y, Yang, Q, Chen, Q, Kero, J, Andersson, A, Ahmed, H, Engström, F and Samuelsson, C, 2019b. Characterization and evaluation of the pozzolanic activity of granulated copper slag modified with CaO, *Journal of Cleaner Production*, 232:1112–1120.

Gabasiane, T S, Danha, G, Mamvura, T A, Mashifana, T and Dzinomwa, G, 2021. Environmental and Socioeconomic Impact of Copper Slag—A Review, *Crystals*, 11:1–16.

Glosser, D, Suraneni, P, Burkan Isgor, O and Jason Weiss, W, 2021. Using glass content to determine the reactivity of fly ash for thermodynamic calculations, *Cement and Concrete Composites*, 115:103849.

Gražulis, S, Chateigner, D, Downs, R T, Yokochi, A F T, Quirós, M, Lutterotti, L, Manakova, E, Butkus, J, Moeck, P and Le Bail, A, 2009. Crystallography Open Database - an Open-Access Collection of Crystal Structures, *Journal of Applied Crystallography*, 42:726–729.

Hallet, V, De Belie, N and Pontikes, Y, 2020. The impact of slag fineness on the reactivity of blended cements with high-volume non-ferrous metallurgy slag, *Construction and Building Materials*, 257:119400.

Isaksson, J, Andersson, A, Lennartsson, A and Samuelsson, C, 2023. Interactions of Crucible Materials With an FeO_x–SiO_2–Al_2O_3 Melt and Their Influence on Viscosity Measurements, *Metallurgical and Materials Transactions B*, 54B:3526–3541.

Kero Andertun, J, Samuelsson, C, Peltola, P and Engström, F, 2022. Characterisation and leaching behaviour of granulated iron silicate slag constituents, *Canadian Metallurgical Quarterly*, 61:14–23.

Kucharczyk, S, Sitarz, M, Zajac, M and Deja, J, 2018. The effect of CaO/SiO_2 molar ratio of $CaO-Al_2O_3-SiO_2$ glasses on their structure and reactivity in alkali activated system, *Spectrochimica Acta Part A: Molecular and Biomolecular Spectroscopy*, 194:163–171.

Li, X, Snellings, R, Antoni, M, Alderete, N M, Haha, M B, Bishnoi, S, Cizer, Ö, Cyr, M, De Weerdt, K, Dhandapani, Y, Duchesne, J, Haufe, J, Hooton, D, Juenger, M, Kamali-Bernard, S, Kramar, S, Marroccoli, M, Joseph, A M,

Parashar, A, Patapy, C, Provis, J L, Sabio, S, Santhanam, M, Steger, L, Sui, T, Relesca, A, Vollpracht, A, Vargas, F, Walkley, B, Winnefeld, F, Ye, G, Zajac, M, Zhang, S and Scrivener, K L, 2018. Reactivity Tests for Supplementary Cementitious Materials: RILEM TC, 267-TRM Phase 1, *Materials and Structures*, 51:151.

Lothenbach, B, Scrivener, K and Hooton, R D, 2011. Supplementary cementitious materials, *Cement and Concrete Research*, 41:1244–1256.

Mills, K C, 1995. Structure of liquid slags, in *Slag Atlas* (ed: D Springorum), pp 1–8 (Verlag Stahleisen GmbH: Düsseldorf).

Murari, K, Siddique, R and Jain, K K, 2015. Use of waste copper slag, a sustainable material, *Journal of Material Cycles and Waste Management*, 17:13–26.

Ramanathan, S, Perumal, P, Illikainen, M and Suraneni, P, 2021. Mechanically activated mine tailings for use as supplementary cementitious materials, *RILEM Technical Letters*, 6:61–69.

Ramanathan, S, Tuen, M and Suraneni, P, 2022. Influence of supplementary cementitious material and filler fineness on their reactivity in model systems and cementitious pastes, *Materials and Structures*, 55(5):136.

Romero Sarcos, N, Hart, D, Bornhöft, H, Ehrenberg, A and Deubener, J, 2021. Rejuvenation of granulated blast furnace slag (GBS) glass by ball milling, *Journal of Non-Crystalline Solids*, 556:120557.

Schöler, A, Winnefeld, F, Haha, M B and Lothenbach, B, 2017. The effect of glass composition on the reactivity of synthetic glasses, *Journal of the American Ceramic Society*, 100(6):2553–2567.

Shi, C, Meyer, C and Behnood, A, 2008. Utilization of copper slag in cement and concrete, *Resources, Conservation and Recycling,* 52:1115–1120.

Sivakumar, P P, Matthys, S, De Belie, N and Gruyaert, E, 2021. Reactivity assessment of modified ferro silicate slag by R^3 method, *Applied Sciences*, 11(1):366.

Skibsted, J and Snellings, R, 2019. Reactivity of supplementary cementitious materials (SCMs) in cement blends, *Cement and Concrete Research*, 124:105799.

Snellings, R and Scrivener, K L, 2016. Rapid Screening Tests for Supplementary Cementitious Materials: Past and Future, *Materials and Structures*, 49:3265–3279.

Snellings, R, 2013. Solution-controlled dissolution of supplementary cementitious material glasses at pH 13: the effect of solution composition on glass dissolution rates, *Journal of the American Ceramic Society*, 96(8):2467–2475.

Stebbins, J F, 1988. Effects of temperature and composition on silicate glass structure and dynamics: Si-29 NMR results, *Journal of Non-Crystalline Solids*, 106:359–369.

Yazhenskikh, E, Jantzen, T, Hack, K and Müller, M, 2019. A New Multipurpose Thermodynamic Database for Oxide Systems, *Rasplawy,* 2:116–124.

Zhang, Q, Deng, D, Feng, Y, Wang, D, Liu, B and Chen, Q, 2023. Effect of Al_2O_3 on the Structural Properties of Water-Quenched Copper Slag Related to Pozzolanic Activity, *Minerals*, 13(2):174.

Zheng, Q and Mauro, J C, 2017. Viscosity of glass-forming systems, *Journal of the American Ceramic Society*, 100(1):6–25.

Pyrometallurgical treatment of nickel smelting slag with biochar

D Attah-Kyei[1], D Sukhomlinov[2], M Tiljander[3], L Klemettinen[2], P Taskinen[2], A Jokilaakso[2] and D Lindberg[2]

1. Department of Chemical and Metallurgical Engineering, School of Chemical Engineering, Aalto University, Kemistintie 1, FI-00076 Aalto, Finland. Email: desmond.attah-kyei@aalto.fi
2. Department of Chemical and Metallurgical Engineering, School of Chemical Engineering, Aalto University, Kemistintie 1, FI-00076 Aalto, Finland.
3. Geological Survey of Finland, Vuorimiehentie 2, 02150 Espoo, Finland.

ABSTRACT

In this study, the use of carbon neutral biochars as reducing agents in the pyrometallurgical treatment of nickel slag was investigated at 1400°C in order to recover valuable elements such as copper and nickel from the slag. A large amount of slag is generated during the nickel matte smelting. Nickel slag contains valuable elements such as copper, nickel and cobalt, which can be recovered. Disposal of this slag results in loss of resources and may cause pollution of the environment. It is important to retrieve these metals for environmental and economic reasons.

In this study, the slag was reacted with non-fossil reducing agents (biochar) which were produced from hydrolysis lignin and black pellet biomass pyrolyzed at 600 and 1200°C and with metallurgical coke for comparison. The reduction experiments were done at 1400°C for 15, 30 and 60 mins under inert gas atmosphere. The samples were quickly quenched and analysed with Electron Probe X-ray Microanalysis (EPMA). The results showed that the use of biochar resulted in faster reaction kinetics in the reduction process compared to coke. The fast reaction kinetics is attributed to the relatively high content of volatiles in this biochar, leading to gas formation and thus mixing of the sample material. Moreover, thermodynamic modelling was also performed using FactSage™ to simulate equilibria with different amounts of biochar. The metal to slag distribution coefficient calculated from the results of thermodynamic modelling were consistent with experimental results. In addition, thermodynamic calculations confirmed that nickel is reduced rapidly and it deports to the metal alloy phase. As the reduction progresses or extra reductant is available for reactions, more iron is deported to the metal phase. The calculations also revealed that Zn vaporises into the gas phase.

INTRODUCTION

Nickel is an important non-ferrous metal which is widely used in machinery, architecture, steelmaking, military and other fields due to its physical and chemical properties. During the pyrometallurgical production of nickel, about 6–16 tons of nickel smelting slag is generated per ton of nickel produced (Sun *et al*, 2021; Zhang *et al*, 2020). The slag usually contains some valuable elements such as copper, cobalt and nickel and these valuable metals can exist in the form of sulfides, oxides, or dissolved elements. It is important to effectively extract valuable elements from nickel slag to ensure environmental sustainability and efficient resource utilisation since negligent disposal of this waste stream results in environmental pollution and loss of valuable elements (Zhang *et al*, 2020).

Dańczak *et al* (2021) employed anodic graphite of spent batteries as reductant in nickel slag reduction at 1350°C. They reported that metal alloy, matte, and slag phases were formed after reduction, and concluded that the metal/matte to slag distribution coefficient increases with an increase in the amount of spent battery. Avarmaa *et al* (2020), on the other hand, studied the reduction of nickel slag at 1400°C with biochar and battery scrap mixture to recover valuable metals and compared the results to the reduction with coke. They reported that biochar enhanced reaction kinetics compared to coke and the presence of battery scrap greatly increased the distribution coefficients of valuable metals.

While nickel slag cleaning aims to recover valuable metals such as nickel, cobalt, and copper as much as possible in the metal alloy phase, it is important to minimise the iron content as it is very difficult to separate during downstream processes. Since some of the metals have similar reduction properties as iron, some losses of valuable metals are unavoidable in the slag cleaning (Jones *et al*, 2002).

The metallurgical industry consumes great amounts of fossil fuels such as coke, which is mainly used as fuel and reducing agent resulting in large emissions of carbon dioxide and other greenhouse gases. In recent times, the use of non-fossil reductants in metal production have gained more attention (Cholico-González *et al*, 2021; Guo *et al*, 2017). One such option of non-fossil reductant is biocoke or biochar obtained from pyrolyzed biomass. Since biomass absorbs CO_2 from the atmosphere during their growth, they release the absorbed CO_2 in the reduction process, thus creating a closed loop carbon cycle (Adrados *et al*, 2016; Suopajärvi, Pongrácz and Fabritius, 2013). Several research groups have investigated the use of biomass or biochar as reductant in metal production (Avarmaa *et al*, 2020; Demey *et al*, 2021; Wiklund *et al*, 2017).

Adrados *et al* (2016) compared biocokes pyrolyzed from olive and eucalyptus trees with commercial reducing agents (metallurgical coke, petroleum coke and anthracite). They reported that based on proximate and ultimate analysis, olive, and eucalyptus biocokes (bioreducers) have better quality than the usual reducing agents, since the bioreducers have lower ash and sulfur contents. Moreover, the bioreductants used have much higher specific surface area and porosity than that of the commercial reductants, making them much more reactive. In their study on the effectiveness of using biomass-based reducing agent in blast furnace, Koskela *et al* (2019) concluded that the utilisation of biomass as raw material in blast furnace ironmaking can have an enormous contribution to the mitigation of CO_2 emissions in steelmaking. Moreover, Mousa *et al* (2016) reported that the use of biomass as a source of energy and reducing agents provides a promising alternative solution for green steel production.

The aim of this study was to ascertain the potential and effectiveness of using biochar as reducing agent to recover valuable metals like Co, Ni and Cu from nickel smelting slag as well as to understand the effects of reaction time.

EXPERIMENTAL

Materials

Industrial (ground) nickel smelting slag provided by Boliden Harjavalta Oy (Harjavalta, Finland) was used experiments. The slag used was XRF (X-ray fluorescence) as well as *magnetite content analyses (Satmagan) for the slag were conducted at Boliden Harjavalta Oy, see Table 1.

TABLE 1
Composition of Ni slag by XRF and it's *magnetite content.

Fe	Ni	Cu	Co	Cr	Zn	S	SiO$_2$	MgO	Al$_2$O$_3$	CaO	*Fe$_3$O$_4$
37	3.62	0.77	0.42	0.08	0.08	0.11	33.7	7.6	2.1	1.74	22

Four different kinds of reductants were tested in the experiments, ie black pellets pyrolyzed at 600°C (B600) and 1200°C (B1200) and hydrolysis lignin pyrolyzed at 600°C (L600) and 1200°C (L1200). The ultimate and proximate analysis (shown in Table 2) of these reductants were done with Elemental Analyzer Flash 2000. Ash analysis of lignin and black pellet was conducted using inductively coupled plasma-optical emission spectroscopy (ICP-OES) and is shown in Table 3.

Table 2 shows that a biomass pyrolyzed at a higher temperature has higher fixed carbon and lower volatile matter contents. In addition, the ash content of black pellet is much higher than that of lignin. The ash analysis revealed that black pellets contain relatively high concentrations of Ca, K and Si. Most of the elements in the ash deport to the slag phase during reduction reactions. It can be seen in Table 2 that lignin has higher fixed carbon content compared to black pellet. The analysis reveals that biochar pyrolyzed at higher temperature contain higher carbon and lower hydrogen and nitrogen composition (Table 2). When biochar is employed as fuel or reducing agent in some metallurgical industries, low ash contents and high volatile matter may be advantageous (Wiklund *et al*, 2017).

TABLE 2
Ultimate and Proximate analyses of biochar.

	Properties	B600	B1200	L600	L1200
Ultimate analysis	C	76.62	81.21	87.90	96.73
	H	1.56	0.00	1.73	0.07
	O	7.33	5.73	8.50	2.15
	N	0.47	0.12	0.97	0.13
	S	0.00	0.00	0.00	0.00
Proximate analysis	Volatile matter	10.82	3.57	6.39	1.31
	Fixed Carbon	76.16	83.19	91.62	97.54
	Ash content	12.32	12.83	1.29	0.92

TABLE 3
Ash analysis of biochar used for the reduction study.

Element		Ca	Mg	Na	K	P	S	Fe	Al	Si	Mn	Ba
Lignin	mg/kg	290	51.0	<10	130	74	1100	810	14	<10	41	4.8
Black Pellet	mg/kg	13900	960	210	2600	560	-	680	680	7400	400	200

The stoichiometric amount of carbon required for reduction was calculated based on the composition of nickel slag after converting the metal concentrations to oxides, as shown in Equations (1) to (5). The stoichiometric mass of carbon to be used was calculated using Equations (6) to (11). The amount of reductant needed was calculated from the amount of carbon required.

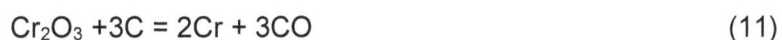

$$2Cu + 0.5O_2 = Cu_2O \tag{1}$$

$$Zn + 0.5O_2 = ZnO \tag{2}$$

$$Ni + 0.5O_2 = NiO \tag{3}$$

$$Co + 0.5O_2 = CoO \tag{4}$$

$$2Cr + 1.5O_2 = Cr_2O_3 \tag{5}$$

$$Fe_3O_4 + C = 3FeO + CO \tag{6}$$

$$Cu_2O + C = 2Cu + CO \tag{7}$$

$$ZnO + C = Zn + CO \tag{8}$$

$$NiO + C = Ni + CO \tag{9}$$

$$CoO + C = Co + CO \tag{10}$$

$$Cr_2O_3 + 3C = 2Cr + 3CO \tag{11}$$

Method

The vertical-tube furnace (Figure 1) used was Lenton LTF 16/450 (Lenton Furnaces and Ovens, Hope Valley, UK) The furnace temperature was set to 1400°C. The furnace was purged with argon gas (99.999 vol per cent purity, Wiokoski Oy, Finland) with a flow rate of 300 mL/min, controlled by a rotameter (Kytölä Instruments, Jyväskylä, Finland ±5 per cent full scale) during the experiments. The cone-shaped silica glass Heraeus HSQ®300 crucibles (fused quartz with purity of >99.998 wt per cent by Finnish Special Glass Oy, Espoo, Finland), with a diameter of 25 mm and height of 15 mm were employed for nickel slag containment.

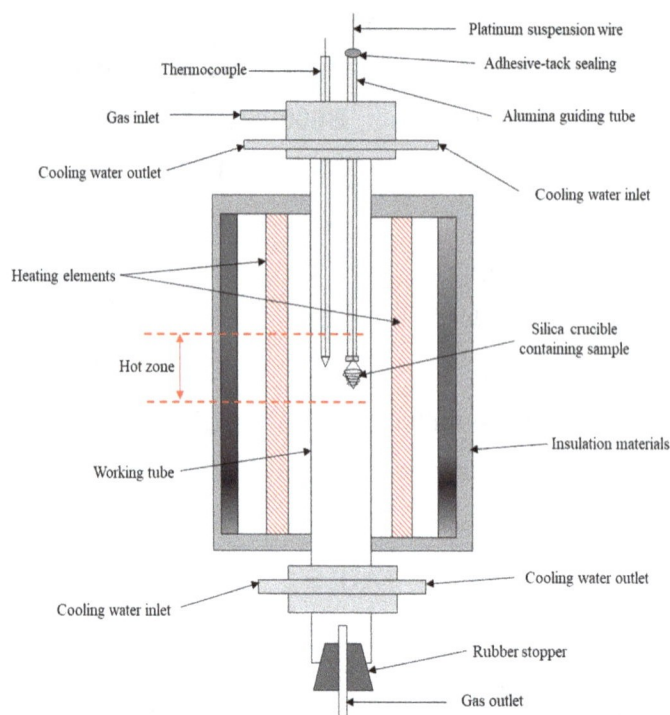

FIG 1 – Schematic of the vertical furnace used for nickel slag reduction experiments.

The drop-quench technique which have been employed by Avarmaa *et al* (2020) was used in this study. One gram of the slag was mixed with the calculated amount of biochar, pelletised and placed in a crucible. The crucible was placed in a basket and attached to platinum wire hanging from inside of the furnace and subsequently lifted to the lower end of the furnace of the furnace before sealing the work tube with a rubber plug. 300 mL/min argon gas was used to purge the furnace to create an inert atmosphere. After about 15 min, the sample was lifted to the hot zone where the reduction reactions took place. After reaching the desired reaction time, the sample was dropped into ice water by pulling the platinum wire, which caused the sample-basket assembly to detach and fall. Reduction experiments were carried out at 1400°C for 15, 30 and 60 mins for each of the reductants.

Characterisation of Sample

The samples were dried and cast in epoxy resin and after curing, they were cut in half and again mounted in epoxy resin and cured. The samples were then polished and carbon coated for scanning electron microscopy (SEM) and electron probe microanalysis (EPMA). Mira 3 Scanning Electron Microscope (Tescan, Brno, Czech Republic) equipped with an UltraDry Silicon Drift Energy Dispersive X-ray Spectrometer (EDS) was used to perform preliminary elemental analyses and microstructural imaging. The compositions of the metal and slag phases were measured using EPMA with an SX100 (Cameca SAS, Gennevilliers, France) microprobe equipped with five wavelength dispersive spectrometers (WDS).

RESULTS AND DISCUSSION

Reduction experiments

At the hot zone of the furnace (1400°C), the sample melts. Metal oxides (MeO) present in the slag react with the carbon in biochar producing metal (Me) as shown in Equation (12). Carbon monoxide produced may also take part in the reduction of metal oxides (see Equation (13)). As time progresses, more carbon and carbon monoxide react leading to a decrease in concentration of metals in the slag. The metal species produced from reduction of metal oxides as well as the mechanically droplets entrapped grow and form a metal alloy:

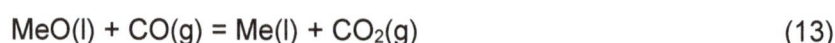

$$MeO(l) + C(s) = Me(l) + CO(g) \tag{12}$$

$$MeO(l) + CO(g) = Me(l) + CO_2(g) \tag{13}$$

Figure 2 shows the SEM micrographs of sample polished sections prepared from nickel slag reduction with biochar at 15, 30 and 60 mins. It can be observed that a metal alloy droplet was obtained with every contacting time, which suggests that the contact between the slags and reductants was good and the reactions proceeded rapidly. The round shaped bright regions seen in the images are the metal alloy droplets and the grey regions are the slag areas (see Figure 2). The samples from reduction were rapidly quenched (<3 sec) and the slag phase was generally homogenous.

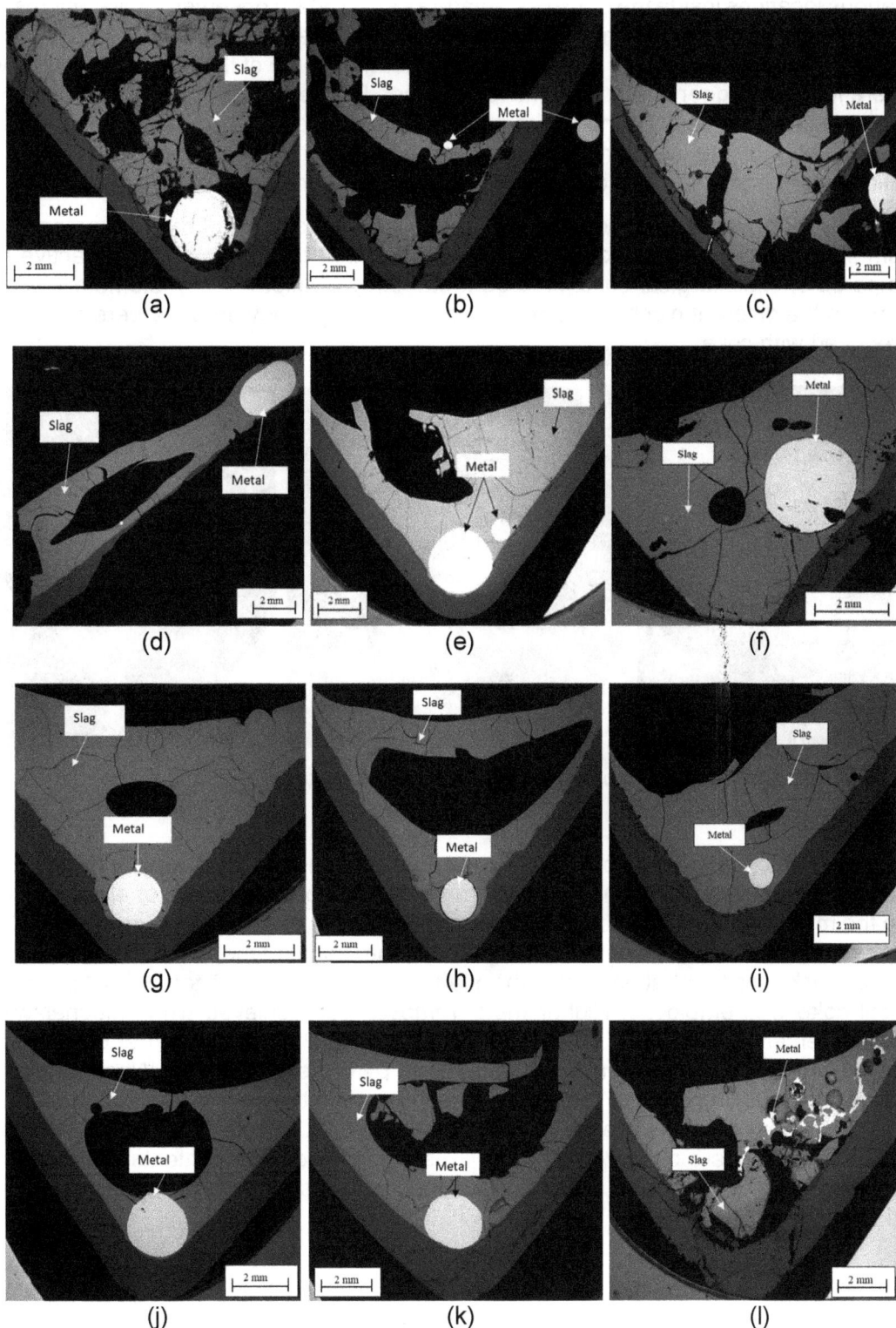

FIG 2 – SEM micrographs of sample polished sections of nickel slag reduction with: (a) L1200 for 15 mins, (b) L1200 for 30 mins, (c) L1200 for 60 mins (d) L600 for 15 mins, (e) L600 for 30 mins,

(f) L600 for 60 mins, (g) B1200 for 15 mins, (h) B1200 for 30 mins, (i) B1200 for 60 mins, (j) B600 for 15 mins, (k) B600 for 30 mins, and (l) B600 for 60 mins.

Since some of the slags were brittle because of rapid quenching, the metal droplets as well as slag pieces were sometimes detached from the rest of the sample while cutting. The dark regions (epoxy) found within the slag are attributed either to this or to gas formation within the melt.

The formed metal generally settled at the bottom of the crucible although some smaller droplets were found in various locations inside the crucibles. Avarmaa *et al* (2020) observed similar microstructures and attributed the presence of the smaller droplets found at the edges of the crucibles to the surface and interfacial tensions between the phases of the system. In some of the reduction tests, two metal droplets were found (a bigger one and a smaller one: see Figure 2b, 30 mins). It is believed that the bigger droplet (>1 mm diameter) is the first metal that was formed and the smaller one(s) is(are) later generated. Analysis of the smaller metal droplets revealed that the nickel, cobalt and copper concentrations were higher but iron concentration was lower compared to the bigger droplet.

Nickel slag reduction was also conducted with metallurgical coke (Figure 3) to compare with the results from slag reduction with biochar. It was observed that biochar is a more reactive reducing agent compared to metallurgical coke. A large metal alloy droplet (~2 mm diameter) was formed (Figure 2) when the different biochars were used, but such large alloy droplets were not found in the samples reduced with coke.

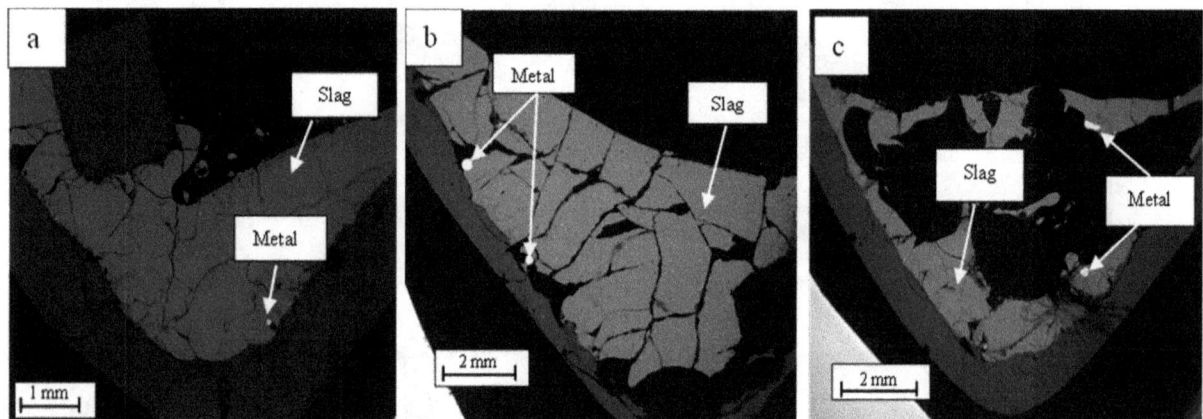

FIG 3 – SEM micrographs of sample polished sections of nickel slag with coke for (a) 15, (b) 30, and (c) 60 mins.

Figure 4 shows the concentrations of nickel, cobalt, copper and iron in the slag before (0 mins) and after reduction (15, 30 and 60 mins). This gives an indication of the amount of elements that have reduced to the metal phase or have been volatilised. The concentrations of nickel, cobalt and copper in the slag decrease as reduction time increases for all the biochar types. Within the first 15 mins of the reduction, nickel concentration within the slag reduced from 3.62 to 1.06 wt per cent when metallurgical coke was employed. Higher metal removal was achieved when biochar was used, yielding nickel percentages in slag below 0.5 wt per cent. Biochars pyrolyzed at 600°C (L600 and B600) showed higher reduction potentials than biochars pyrolyzed at 1200°C. When B600 and L600 were employed as reducing agents, the nickel concentration in the slag after 15 mins was 0.06 and 0.07 wt per cent, respectively, while those for B1200 and L1200 were 0.48 and 0.24 wt per cent, respectively. While biochars pyrolyzed at 1200°C have higher carbon contents, the higher volatile matter concentrations in L600 and B600 may be indirectly responsible for their higher reactivity, ie showed faster reduction kinetics.

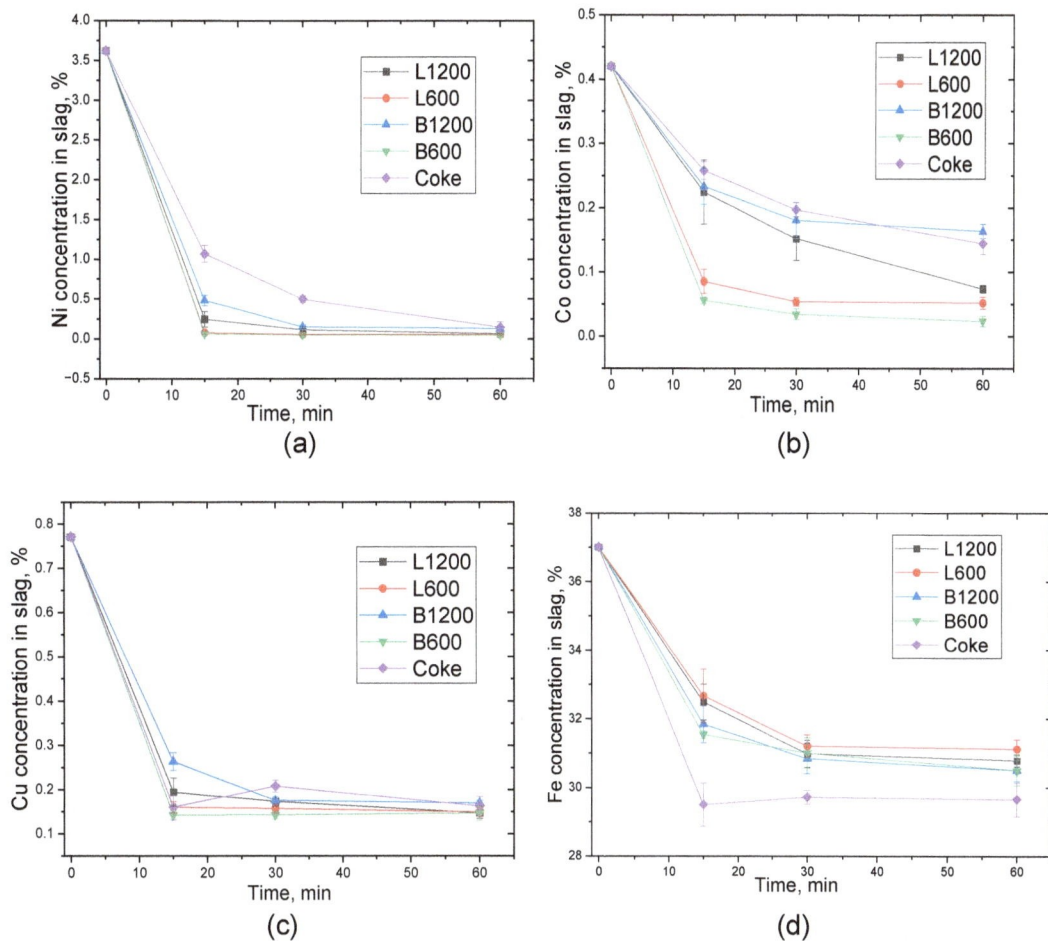

FIG 4 – Concentrations of (a) nickel, (b) cobalt, (c) copper, and (d) iron in slag as a function of time during nickel slag reduction with L1200, L600, B1200 and B600 biochar and metallurgical coke.

It is commonly stated that in slag cleaning processes (at 1300–1400°C) and eg in titania smelting, volatiles are not contributing the reduction processes. This means that they are not considered when calculating the stoichiometric amounts of reductant needed in the process. However, the volatile matter upon moving to the gas phase increases the surface area for sufficient slag and reductant reaction (Lotfian *et al*, 2017). Moreover, compared to the coke and biochar pyrolyzed at 1200°C, more CO_2 is produced by biochar pyrolyzed at 600°C which quickly reacts with solid carbon as shown in Equation 14 (Boudouard reaction). Also, the generated CO take part in the reduction. Since reduction with solid carbon is slower compared to CO, the higher CO present during biochar reduction increases its effectiveness as reductant compared to coke (Lahijani *et al*, 2015):

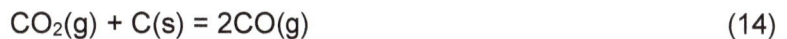

$$CO_2(g) + C(s) = 2CO(g) \qquad (14)$$

After 60 mins reduction, the final concentrations of nickel in the slags were very close to each other with all reductants studied. Similar patterns regarding reduction kinetics of different reductants were observed for cobalt and copper as well, with the difference that the final concentration of cobalt in slag after 60 mins was more than double with coke and B1200 compared to L600 and B600 reductants. Regarding iron concentration in slag, all the biochars behaved similarly, however, the use of coke resulted in the lowest iron concentrations.

Biochar reactivity has been compared to that of coke. The reactivity of biochar was calculated from the ratio of volatile matter to fixed carbon (VM/FC) (Cui *et al*, 2023). Biochar with a high VM/FC ratio is more active and more advantageous in biofuel and chemical applications. It can be observed in Table 4 that the biochars are more reactive than metallurgical coke with B600 have reactivity close to 15 time that of coke.

TABLE 4
Reactivity of reductant calculated from volatile matter and fixed carbon.

Reductant	B600	B1200	L600	L1200	Coke
Reactivity (VM/FC)	0.1420	0.0428	0.0697	0.01343	0.0097

Distribution coefficients of the elements

The distribution coefficient is a parameter that is used to estimate the progress of the reduction reaction. It gives an indication of the behaviour of the various elements in the reduction study. In this work, the distribution coefficient between metal alloy and slag is used, and thus, a higher distribution coefficient indicates higher deportment to the alloy. The distribution coefficient of an element, Me, was calculated using Equation (15) (Avarmaa *et al*, 2020):

$$L^m/_s = \frac{wt\% \, Me \, in \, metal}{wt\% \, Me \, in \, slag} \tag{15}$$

where *m* represents the metal phase and *s* the slag phase.

The uncertainty (*ΔL*) of the distribution coefficient was calculated using the relation in Equation (16):

$$\Delta L^m/_s = \left\{ \left[\frac{\Delta Me}{wt\% \, Me \, in \, metal} \right] + \left(\frac{\Delta Me}{wt\% \, Me \, in \, slag} \right) \right\} \times L^m/_s \tag{16}$$

where *ΔMe* is the standard deviation of element Me calculated from the EPMA results.

Figure 6 shows the distribution coefficient of nickel, cobalt, copper and iron between metal alloy and slag at 1400°C as a function of time. The results from this work have been compared to other studies where battery scrap (Dańczak *et al*, 2021) and mixture of battery scrap and biochar (Avarmaa *et al*, 2020) were used as reductants for nickel slag reduction.

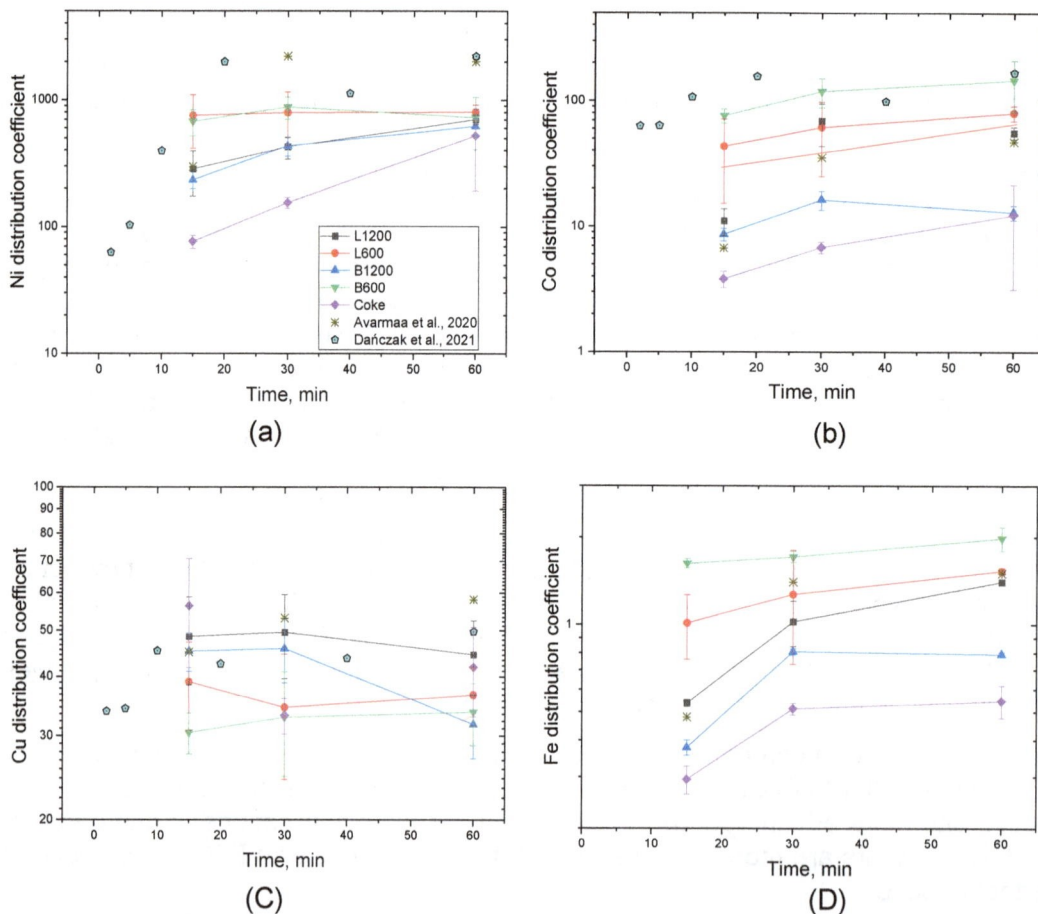

FIG 6 – Metal alloy to slag distribution coefficient of (a) Ni, (b) Co, (c) Cu and (d) Fe as time progresses at 1400°C. Note the difference in the y-axis scale.

Generally, the distribution coefficients of all the metals using the various reductants increase with time. It can be inferred that the reduction proceeds quickly, and the metal oxides are efficiently reduced to the alloy already after 15 mins reduction time since distribution coefficient values increase only slightly afterward. The use of B600 resulted in the highest distribution coefficients for nickel, cobalt and iron, closely followed by L600 biochar. The distribution coefficients of nickel, cobalt, and iron had the lowest values with coke reduction, although these values have the highest increase as a function of time. This is attributed to the slower reaction kinetics of coke compared to the different biochars. The distribution coefficient values for Ni and Cu are similar for coke and the biochars after 60 mins of reduction. It can be inferred from Figure 6 that in terms of the distribution coefficient values, the reductants can be ranked according to effectiveness or reactivity as B600 > L600 > L1200 > B1200 > Coke. This corresponds to reactivity of biochar that has been calculated.

Thermodynamic modelling

FactSage™ thermodynamic software package, version 8.0 was used in this study to investigate the effect of addition of biochar on the composition of metal and slag from nickel slag reduction. The composition of nickel slag, based on XRF analysis, as well as the ultimate and ash analyses of biochar were used as inputs in the software. The databases used for the calculations were custom collected based on the databases FactPS (pure substances), FToxid (optimised for oxide systems) and FSCopp (optimised for copper-containing solid and liquid alloys). The phases selected for the calculations were the spinel (solid solution phase with stoichiometry AB_2O_4, A, B = divalent and trivalent metals), slag (liquid oxide silicate phase), monoxide phase (solid solution), FCC, BCC, HCP-A3 (three solid multicomponent alloys; Face-Centred Cubic, Body-Centred Cubic, Hexagonal Close-Packed) and liquid metal. Ideal gas and pure solids were also selected before the calculations were done at 1400°C.

Effect of addition of biochar

One gram of nickel slag was reacted with different masses of biochar. The phases formed during reduction, as predicted by FactSage™, plotted as a function of addition of biochar are shown in Figure 7. It is observed that as the reduction progresses, the mass of slag decreases while the masses of metal and gas increase until about 0.15 g of biochar addition. After this point the masses of metal as well as the gas and slag remain constant. About 0.45 g of metal is formed with >0.15 g of biochar addition. Moreover, the thermodynamic simulation reveals that relatively more gas is produced when L600 and B600 are employed as reductants compared to their respective biochars pyrolyzed at 1200°C.

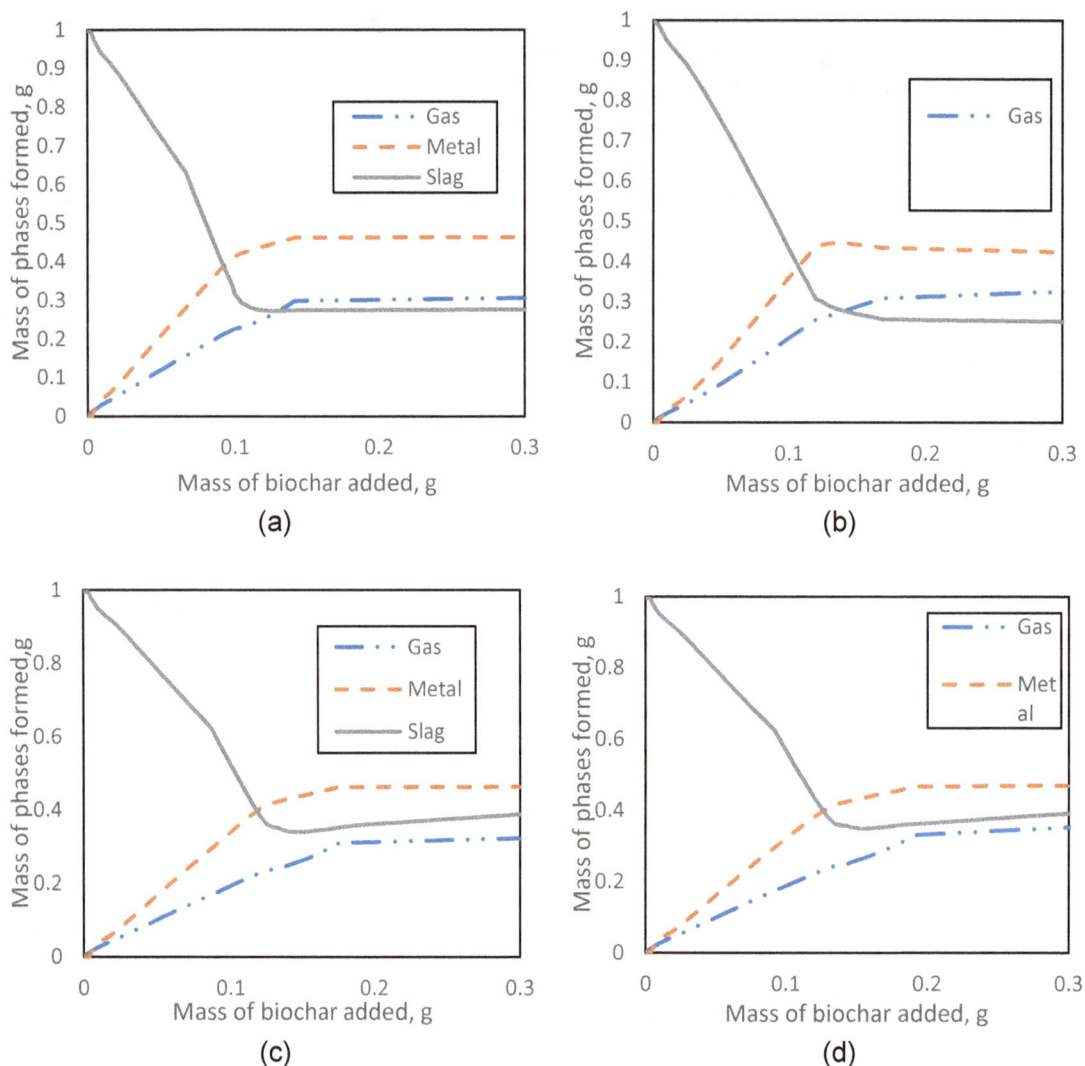

FIG 7 – Formation of gas, metal and slag phases during nickel slag reduction with varying addition of (a) L1200 (b) L600, (c) B1200, and (d) B600 biochar to 1 g of nickel slag at 1400°C as predicted by FactSage™.

The mass percentages of selected metals (Zn, Cu, Ni, Co and Fe) that were present in the gas, metal and slag phases have been shown in Figure 8. The results from reduction of Ni slag with B600 have been selected for plotting since the calculated results were remarkably similar with all investigated biochars. When 0.01 g of biochar is added to 1.0 g of nickel slag, at equilibrium, 80 wt per cent of the metal formed is nickel. This percentage continually decreases when higher amounts of biochars are added. Although the mass of nickel formed does not decrease with biochar addition, more iron is reduced, which consequently decreases the relative nickel concentration in the metal phase. In addition, the relative concentrations of copper and cobalt in the metal alloy increase to their highest values, 11 per cent and 4 per cent respectively, with less than 0.03 g biochar addition, after which they begin to decrease. The mass content and mass balance were not determined in the experimental section of this work, as it was not possible to separately collect all the phases from the quenched samples.

Nickel and copper have the highest concentrations of 80 per cent and 10 per cent, respectively, when 0.01 g of biochar is reacted with 1 g slag. The concentration of iron in the metal phase is the lowest at this addition. Since the formation of solid iron is undesired in industrial operation, it is important to recover other valuable metals like nickel and copper with low iron concentration.

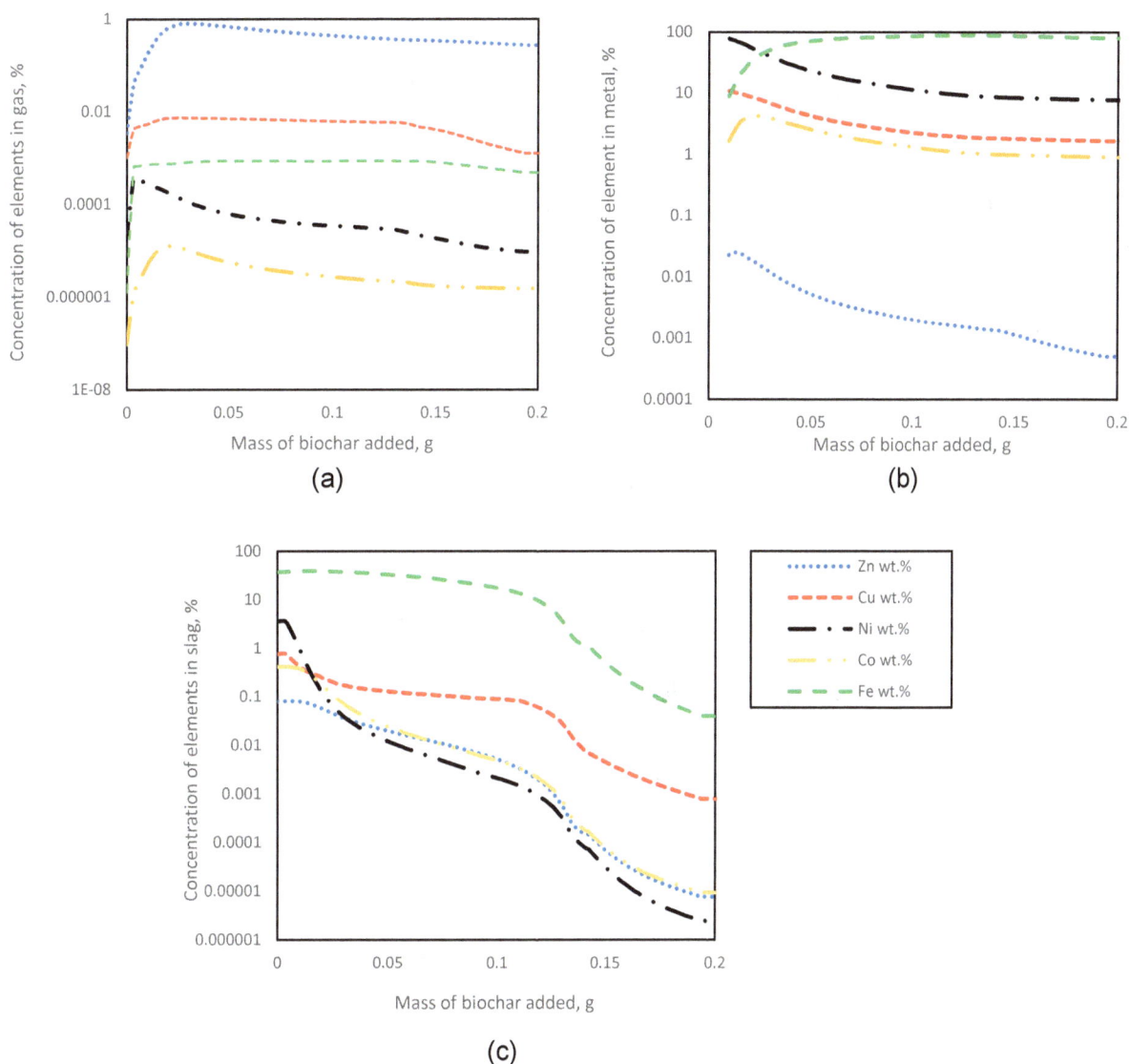

FIG 8 – Concentrations of copper, nickel, cobalt, zinc, and iron in (a) gas, (b) metal and (c) slag phase during nickel slag reduction with varying addition of B600 biochar to 1 g of nickel slag as predicted by FactSage™.

In the slag, nickel concentration plummets to about 0.1 wt per cent when 0.02 g of biochar is added to a gram of slag, and it continuously decreases until the nickel oxide in slag is almost completely reduced and Ni deported to the metal phase. Cobalt followed a similar trend as nickel in the slag. However, efficient copper oxide reduction from the slag requires a higher amount of reductant compared to nickel and cobalt. According to the calculations, approximately 0.15 g of biochar (to 1 g of slag) is enough to reduce all the iron oxides from the slag to metallic iron. As can be seen from Figure 8 a, zinc evaporates into the gas phase.

SUMMARY AND CONCLUSIONS

In this study, the use of carbon neutral biochars as reducing agents in the pyrometallurgical treatment of nickel slag was investigated at 1400°C to recover valuable elements such as copper and nickel from the slag. Black pellet and lignin, each pyrolyzed at 600 and 1200°C, were employed as reductants and the recoveries of nickel, copper and cobalt were investigated and compared to the results when using fossil coke as the reductant. The biochars pyrolyzed at 1200°C had higher fixed carbon and lower volatile matter contents compared to those pyrolyzed at 600°C. While this may be preferred in industrial applications, the cost of energy required for pyrolyzing at higher temperatures might deter the use of biomass pyrolyzed at higher temperatures. B600 appeared to be the most efficient biochar, although it has lower carbon content compared to the other chars. The fast reaction kinetics is thought to be attributed to the relatively high content of volatiles in this biochar, leading to gas formation and thus mixing of the sample material. During the slag cleaning, biochars showed higher reactivity and faster reduction kinetics compared to coke. This, again, may be attributed to high concentration of volatile matter present. Although it is stated that at high temperatures (1300–1400°C) volatile matter do not contribute to the reduction process, it is possible that the volatiles create pores upon moving into the gas phase, increasing thus the surface area for reaction. In addition, CO_2 generated reacts with carbon (Boudouard reaction) resulting in more CO produced which takes part in the reduction reaction. Higher distribution coefficients between metal alloy and slag were obtained for Ni, Co and Fe using any biochar, compared to using coke.

Thermodynamic modelling was performed with FactSage™ and the distribution coefficients calculated were compared to the experimental results, which were consistent with the thermodynamic modelling. In addition, thermodynamic calculations confirmed that nickel is reduced rapidly, and it deports to the metal alloy phase when enough ferric oxide has been removed from the slag. As the reduction progresses or extra reductant is available for reactions, more iron is deported to the metal phase. The calculations also revealed that Zn vaporises into the gas phase.

ACKNOWLEDGEMENTS

This study received financial support from Aalto University, School of Chemical Engineering TOCANEM Project (Business Finland grant 2118452), BOLIDEN Harjavalta Oy for providing the nickel slag; University of Oulu for preparing and providing the biochars. The utilisation of the Academy of Finland's RawMatTERS Finland Infrastructure (RAMI) based at Aalto University, GTK and VTT in Espoo is appreciated.

REFERENCES

Adrados, A, De Marco, I, López-Urionabarrenechea, A, Solar, J, Caballero, B M and Gastelu, N, 2016. Biomass pyrolysis solids as reducing agents: Comparison with commercial reducing agents, *Materials*, 9(1). https://doi.org/10.3390/ma9010003

Avarmaa, K, Järvenpää, M, Klemettinen, L, Marjakoski, M, Taskinen, P, Lindberg, D and Jokilaakso, A, 2020. Battery scrap and biochar utilization for improved metal recoveries in nickel slag cleaning conditions, *Batteries*, 6(4):1–21. https://doi.org/10.3390/batteries6040058

Cholico-González, D, Lara, N O, Miranda, M A S, Estrella, R M, García, R E and Patiño, C A L, 2021. Efficient metallization of magnetite concentrate by reduction with agave bagasse as a source of reducing agents, *International Journal of Minerals, Metallurgy and Materials*, 28(4):603–611. https://doi.org/10.1007/s12613-020-2079-z

Cui, D, Zheng, X, Zou, J, Yu, S, Kong, X, Cai, J and Zhang, X, 2023. Physicochemical Characterization and Chemical Reactivity of Biochar from Pyrolysis of Dried Distiller's Grains with Solubles (DDGs), *Combustion Science and Technology*. https://doi.org/10.1080/00102202.2023.2204521

Dańczak, A, Ruismäki, R, Rinne, T, Klemettinen, L, O'Brien, H, Taskinen, P, Jokilaakso, A and Serna-guerrero, R, 2021. Worth from waste: Utilizing a graphite-rich fraction from spent lithium-ion batteries as alternative reductant in nickel slag cleaning, *Minerals*, 11(7). https://doi.org/10.3390/min11070784

Demey, H, Rodriguez-Alonso, E, Lacombe, E, Grateau, M, Jaricot, N, Chatroux, A, Thiery, S, Marchand, M and Melkior, T, 2021. Upscaling severe torrefaction of agricultural residues to produce sustainable reducing agents for non-ferrous metallurgy, *Metals*, 11(12). https://doi.org/10.3390/met11121905

Guo, D, Li, Y, Cui, B, Chen, Z, Luo, S, Xiao, B, Zhu, H and Hu, M, 2017. Direct reduction of iron ore/biomass composite pellets using simulated biomass-derived syngas: Experimental analysis and kinetic modelling, *Chemical Engineering Journal*, 327:822–830. https://doi.org/10.1016/j.cej.2017.06.118

Jones, R T, Denton, G M, Reynolds, Q G, Parker, J A L and Van Tonder, G J J, 2002. Recovery of cobalt from slag in a DC arc furnace at Chambishi, Zambia, *Journal The South African Institute of Mining and Metallurgy*, 102(1):5–9.

Koskela, A, Suopajärvi, H, Mattila, O, Uusitalo, J and Fabritius, T, 2019. Lignin from bioethanol production as a part of a raw material blend of a metallurgical coke, *Energies*, 12(8). https://doi.org/10.3390/en12081533

Lahijani, P, Zainal, Z A, Mohammadi, M and Mohamed, A R, 2015. Conversion of the greenhouse gas CO_2 to the fuel gas CO via the Boudouard reaction: A review, *Renewable and Sustainable Energy Reviews*, 41:615–632. https://doi.org/10.1016/j.rser.2014.08.034

Lotfian, S, Ahmed, H, El-Geassy, A H A and Samuelsson, C, 2017. Alternative Reducing Agents in Metallurgical Processes: Gasification of Shredder Residue Material, *Journal of Sustainable Metallurgy*, 3(2):336–349. https://doi.org/10.1007/s40831-016-0096-y

Mousa, E, Wang, C, Riesbeck, J and Larsson, M, 2016. Biomass applications in iron and steel industry: An overview of challenges and opportunities, *Renewable and Sustainable Energy Reviews*, 65:1247–1266. https://doi.org/10.1016/j.rser.2016.07.061

Sun, W, Li, X, Liu, R, Zhai, Q and Li, J, 2021. Recovery of valuable metals from nickel smelting slag based on reduction and sulfurization modification, *Minerals*, 11(9):1–14. https://doi.org/10.3390/min11091022

Suopajärvi, H, Pongrácz, E and Fabritius, T, 2013. The potential of using biomass-based reducing agents in the blast furnace: A review of thermochemical conversion technologies and assessments related to sustainability, *Renewable and Sustainable Energy Reviews*, 25:511–528. https://doi.org/10.1016/j.rser.2013.05.005

Wiklund, C M, Helle, M, Kohl, T, Järvinen, M and Saxén, H, 2017. Feasibility study of woody-biomass use in a steel plant through process integration, *Journal of Cleaner Production*, (142):4127–4141. https://doi.org/10.1016/j.jclepro.2016.09.210

Zhang, G, Wang, N, Chen, M and Cheng, Y, 2020. Recycling nickel slag by aluminum dross: Iron-extraction and secondary slag stabilization, *ISIJ International*, 60(3):602–609. https://doi.org/10.2355/isijinternational.ISIJINT-2019-173

An investigative study on the interfacial behaviour of waste graphite resource with liquid iron

S Biswal[1], S Udayakumar[2], F Pahlevani[3], N Sarmadi[4] and V Sahajwalla[5]

1. Researcher, Centre for Sustainable Materials Research and Technology (SMaRT@UNSW), UNSW Sydney, Sydney NSW 2052. Email: s.biswal@unsw.edu.au
2. Research Assistant, Centre for Sustainable Materials Research and Technology (SMaRT@UNSW), UNSW Sydney, Sydney NSW 2052. Email: s.udayakumar@unsw.edu.au
3. Associate Professor, Centre for Sustainable Materials Research and Technology (SMaRT@UNSW), UNSW Sydney, Sydney NSW 2052. Email: f.pahlevani@unsw.edu.au
4. Research Assistant, Centre for Sustainable Materials Research and Technology (SMaRT@UNSW), UNSW Sydney, Sydney NSW 2052. Email: n.sarmadi@unsw.edu.au
5. Professor, Centre for Sustainable Materials Research and Technology (SMaRT@UNSW), UNSW Sydney, Sydney NSW 2052. Email: veena@unsw.edu.au

ABSTRACT

Recent times have witnessed a notable momentum in the interest surrounding technology dedicated to harnessing waste as valuable resources, prompted by the mounting challenges posed by waste accumulation and the swift depletion of natural resources. The carbon dissolution process within a molten iron bath stands as a pivotal reaction within the metal processing sector. The current study explores the utilisation of waste graphite electrodes (WGE) (obtained after being utilised in an electric arc furnace) as a secondary resource of carbon for dissolution into liquid iron. Experimental investigations were carried out at 1550°C using the sessile drop approach. High purity iron powder was used and the carbonaceous substrates were prepared from the ground WGE. The carbon pickup was determined using a suitable carbon analyser. Furthermore, an examination of the interfacial area between the molten iron and the WGE substrate was conducted beneath the metallic droplet, which effectively represents the iron/carbon interface. This interfacial analysis was carried out on two different samples obtained after a varying period of heat treatments at 1550°C. The interface layer was investigated using scanning electron microscopy (SEM) in conjunction with energy-dispersive X-ray spectroscopy (EDS).

INTRODUCTION

Steel production is vital for global infrastructure, industry and job creation, with applications ranging from construction to transportation. The industry is responding to sustainability concerns by embracing cleaner technologies, such as furnaces powered by renewable energy and adopting circular economy principles through increased recycling. Governments are implementing regulations, certifications like ResponsibleSteel™ (2018) are promoting responsible practices, and there is a growing emphasis on supply chain transparency. While the economic importance of steel is undeniable, efforts to balance it with environmental and social considerations are underway to ensure a more sustainable future (Muslemani *et al*, 2021; Ben Ellis, 2020). Global steel production in the year 2022 stands at approximately 1885 Mt (World Steel Association, 2023).

The transition to steel production through electric arc furnace (EAF) is gradually gaining momentum. Sponge iron (also known as direct reduced iron) can be used in conjunction with scrap steel in EAFs. This dual-feed approach allows for a flexible raw material mix, optimising the use of both recycled and virgin materials based on availability and market conditions. Moreover, the steel industry has been witnessing investments in EAF technology, with some companies expanding their EAF capacities. This suggests a positive outlook for the role of EAFs in future steel production. It's been predicted that the EAF steel production will increase to around 50 per cent by the year 2050 (Wood Mackenzie, 2023).

Carbon is a crucial element in steel production, influencing its strength, hardness and overall performance. To enhance sustainability, the steel industry is exploring innovative ways to utilise waste resources. By incorporating recycled steel, not only does it reduce the demand for virgin raw materials, but it also minimises the environmental impact associated with traditional production processes. Similarly, efforts to optimise the carbon content in steel while embracing circular economy principles play a vital role in mitigating the environmental footprint of the industry. Carbon in the form of coke/coal is used as a fuel source, reducing agent and carburiser at various steps in steelmaking production (Ahmed, 2018). Incorporating secondary carbon resources into the steelmaking process helps reduce the overall carbon footprint. The use of recycled materials, such as carbon-containing waste, requires less energy compared to the extraction and processing of virgin raw materials. Incorporating secondary carbon resources, especially recycled materials, can lead to cost savings for steel producers. This flexibility also enhances resilience in the face of market fluctuations and supply chain disruptions.

Waste graphite electrodes from EAFs in steelmaking is one such notable byproduct with potential for recycling and environmental sustainability. Graphite electrodes are essential components in EAFs, conducting the electrical charge needed for melting scrap steel. The usage of graphite electrodes in electric furnaces is commonly regarded as the third most significant cost factor in the production of molten steel, following scrap steel and electrical energy (Xuran New Materials Limited, 2024). Consumption typically falls within the range of 1.8–9.9 kg per ton of crude steel (Babich and Senk, 2013). Once these electrodes reach the end of their usable life, they are considered a waste. However, with appropriate treatment and processing, the carbon content in these electrodes can be recovered and reused, contributing to the circular economy within the steel industry.

The incorporation of carbon into a molten iron bath is a crucial reaction. Numerous investigations have concentrated on examining the principles of carbon dissolution by utilising graphite, coke and coal as carburising materials. The current research work investigates the dissolution of carbon present in the waste graphite electrode (WGE) through the sessile drop technique.

MATERIALS AND EXPERIMENTAL METHODS

The WGE was obtained from a steel manufacturing facility in New South Wales, Australia. The as-received WGE was first pulverised to a powder (100–150 µm) using a ring mill. The powdered sample was then subjected to CHNS analysis (Carbon (C), Hydrogen (H), Nitrogen (N), and Sulphur (S); Vario MACRO CUBE) as shown in Table 1.

TABLE 1

CHNS analysis (Carbon (C), Hydrogen (H), Nitrogen (N), and Sulphur (S)) of waste graphite electrodes (WGE) sample.

Sample	C(%)	H(%)	N(%)	S(%)
WGE	97.2	0.83	0.01	0.34

Sessile drop approach was used to investigate the interaction between WGE and molten iron. The WGE sample weighing 1 gram (each substrate) was transferred into cylindrical steel die set. The set of WGE sample and the die were subjected to a pressure of 200 bar for 5 mins at room temperature using a hydraulic press equipment. The prepared tabular substrate of WGE sample was placed on a bed of alumina powder. Approximately 0.25 g of high purity iron powder (Sigma Aldrich, ≥ 99.8 per cent) was placed on the top of the substrate. The assembly was then put inside a horizontal tube furnace which was continuously purged with an argon flow of one litre/min. Initially, it was held in the cold zone of the furnace for 10 mins to avoid any kind of thermal shock, after which it was pushed to the hot zone at a temperature of 1550°C for different durations of time. Carburised metallic droplets were obtained and further cleaned with ethanol to remove any superficially attached carbonaceous particles. The percentage of carbon pickup in the metallic droplets was measured using LECO carbon analyser. Some metallic droplets were selected for cross-section study using a Scanning Electron Microscope (Hitachi S3400N) coupled with an Energy Dispersive Spectroscopy (SEM-EDS). Metallic droplets obtained at the end of 1 and 15 mins along with the WGE substrate were cold mounted using epoxy resin in a cylindrical mold and then subjected to a vacuum environment for removal of air bubbles. After that, it was cured for 15 hrs followed by cutting with a diamond blade on a Minitom equipment to obtain the cross-sections. The cross-sectioned samples were polished to 1 micron (using diamond suspension) and then placed on suitable SEM stubs. The operating conditions used for the imaging are 15 kV accelerating voltage and a probe current of 50.

RESULTS AND DISCUSSIONS

Figure 1 represents the percentages of carbon pick-up from WGE into molten iron as a function of time.

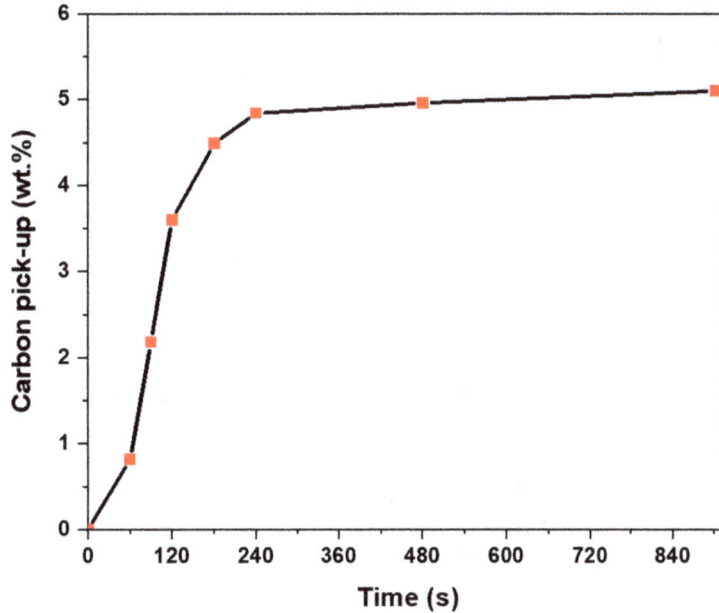

FIG 1 – Carbon pick-up from waste graphite electrodes (WGE) as a function of time at 1550°C.

As observed from Figure 1, there was a steep increase in carbon pick-up within the first 3 mins with a slight increase at a time period of 4 mins followed by a very slow pick-up phase till 15 mins.

The general rate constant for carbon dissolution (K) was determined through the application of the following equations (Mansuri *et al*, 2013, 2018):

$$\frac{dCt}{dt} = \frac{Ak}{V} \; x \; (C_s - C_t)$$

$$\ln\frac{C_s - C_t}{C_s - C_o} = -K \; x \; t$$

In these equations, C_s and C_t denote the saturation solubility and carbon concentration (wt per cent) in liquid iron at time t respectively. The first-order rate constant is represented by k (m·s⁻¹) and A and V correspond to the interfacial area of contact and the volume of the liquid iron bath, respectively. C_o is the initial carbon concentration in the liquid metal (wt per cent) and since 99.98 per cent pure iron powder was utilised, the value of C_o is assumed as zero. The overall carbon dissolution rate constant, $K = (Ak/V)$, was determined from the negative slope of the natural logarithm of $(C_s-C_t)/(C_s-C_o)$ versus time plot. It was assumed that there was negligible change in the contact area during this brief initial contact period.

The saturation carbon limit for a particular melt composition can be calculated using the below equation (Wu, Wiblen and Sahajwalla, 2000):

$$C_s \; (wt\%) = 1.3 + 0.00257T - 0.31Si - 0.33P - 0.45S + 0.28Mn$$

As high purity iron powder was used, the saturation carbon limit was calculated to be 5.28 wt per cent. Figure 2 represents the plot of natural logarithm of $(C_s-C_t)/(C_s-C_o)$ versus time for the duration of first 3 mins.

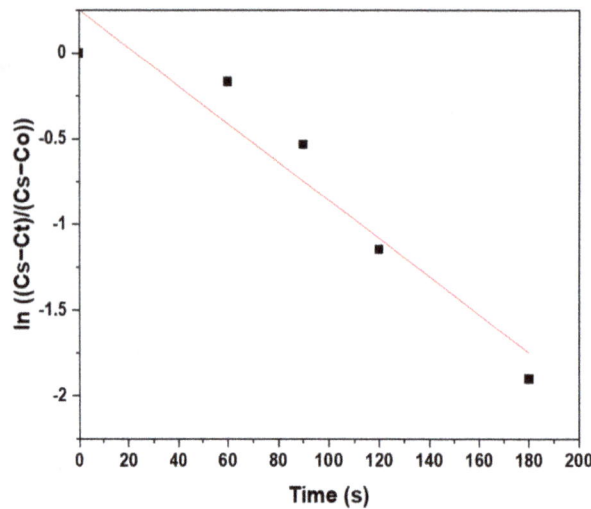

FIG 2 – The plot of ln $(C_s-C_t)/(C_s-C_o)$ versus time for waste graphite electrodes (WGE) for the duration of initial 3 mins.

The apparent rate constant K was obtained from the slope of Figure 2 and was found to be 11.1×10^{-3} s^{-1}. The following rate constants mentioned in Table 2 have been reported from literature using the sessile drop approach and various other carbonaceous materials.

TABLE 2

Apparent rate constants calculated using sessile drop approach with various carbonaceous materials.

Sample material	Apparent rate constant, $K \times 10^{-3}$ s^{-1}	References
Coke	0.003	Kongkarat *et al* (2012)
Coal char-1	0.1	McCarthy (2005)
Coal char-4	0.3	
Natural graphite	10.68	Wu, Wiblen and Sahajwalla (2000)

It could be observed from Table 2 that the apparent rate constant of WGE is similar to that of natural graphite. Cham *et al* (2004) observed an augmented dissolution of carbon, corresponding to an elevated crystalline order of coke. Studies (Cham *et al*, 2004; Wu and Sahajwalla, 2000) have also underscored an interconnected relationship between structural ordering and the rate of carbon dissolution, indicating that carbonaceous materials with greater Lc (crystallite size) values generally manifest a heightened carbon dissolution rate. In the case of graphitic carbonaceous materials, the dissociation of carbon atoms occurs at a swifter pace owing to the arrangement of carbon atoms in two-dimensional arrays, as opposed to coke and char (Mansuri *et al*, 2013). Consequently, a higher Lc value is conducive to facilitating the dissociation of carbon atoms. This specific occurrence can be characterised as the separation of carbon and the subsequent breakdown of the carbonaceous structure. The extent to which C (carbon) dissolves in molten iron is not determined by the mass transfer of C but rather by C's ability to dissociate from the carbonaceous matrix. The carbonaceous matrix structure influences the dissociation of carbon atoms, leading to the notable difference in the carbon dissolution limits, with graphite exhibiting a considerably higher limit than others. Reports indicate that the crucial step governing the rate of carbon dissolution from graphite is the diffusion of C from the liquid boundary layer at the interface into the bulk melt (Sun *et al*, 2021; Jang *et al*, 2012).

The following reactions highlight the phenomenon of carbon dissolution into molten iron mass (Wu and Sahajwalla, 2000):

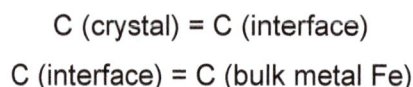

$$C \text{ (crystal)} = C \text{ (interface)}$$

$$C \text{ (interface)} = C \text{ (bulk metal Fe)}$$

The examination of the interface of the iron-carbon alloy situated between molten iron and the char substrate on the underside of the iron droplet was conducted through SEM/EDS analysis. Figure 3 presents SEM images integrated with EDS for the interfacial layer on the iron side. The pivotal role of ash in the carbon dissolution reaction has been thoroughly elucidated by numerous researchers. Various studies have shown the impact of ash impurities on carbon dissolution, highlighting that the ash present in diverse carbonaceous materials can distinctly shape the composition of the liquid iron during the reaction (Mansuri *et al*, 2018). The results of the SEM/EDS analysis unequivocally establish the total absence of an ash layer at the interface connecting molten iron and the char substrate. The EDS spectra exhibited solely carbon and iron peaks, with no detection of any other elements or phases at the interfacial layer. This comprehensive analysis provides robust evidence supporting the exclusive presence of carbon and iron within the examined interface, underscoring the elemental composition and purity of the interface. It is also interesting to observe that 15 min-SEM image show the random distribution of carbon-based specks and laths along the interface compared to that of 1 min-SEM image. It is evident that with longer duration of exposure, the carbon particles penetrated into the iron mass and this mass transfer across the interface would have occurred till the maximum saturation. This investigation conclusively validates the potential utilisation of discarded graphite electrodes as a viable supplementary carbon source in the production of Iron-Carbon alloys as a carburiser.

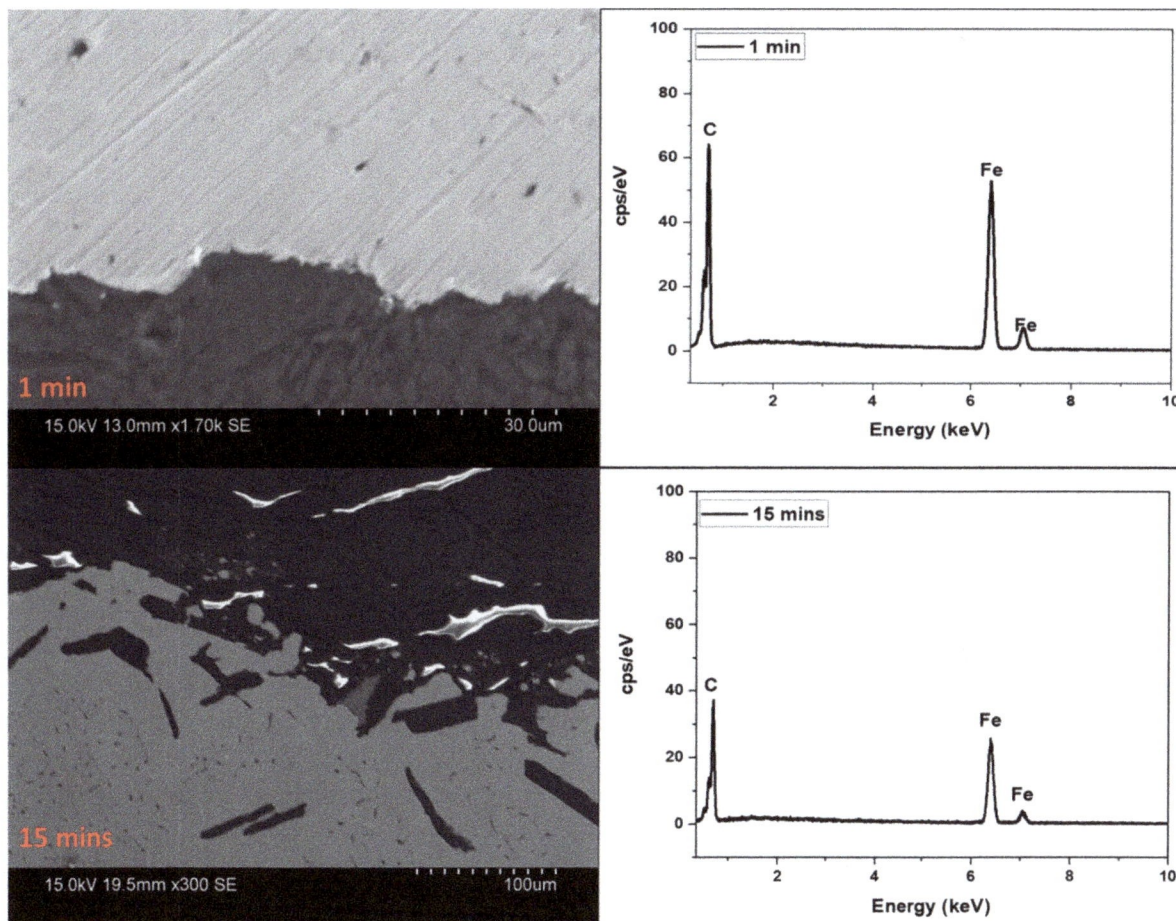

FIG 3 – Scanning electron microscopy (SEM) images coupled with energy-dispersive X-ray spectroscopy (EDS) of metal/carbon interface of waste graphite electrodes (WGE) substrate and iron.

CONCLUSIONS

The elevated rate of carbon dissolution into molten iron was evidently observed when utilising waste graphite electrode as a carbon source, resulting in a significant 5.1 wt per cent of carbon pick-up. Notably, the carbon dissolution rate achieved with graphite electrode surpassed that of various other carbonaceous materials, including metallurgical coke and coal chars and was similar to that of natural graphite. This heightened efficiency can be attributed to the substantial carbon content and low ash content characteristic of waste graphite electrode. This indicates the potential reuse of waste graphite electrodes, even after going through numerous cycles of operation in the EAF. The study underscores the sustainable and effective utilisation of waste graphite electrodes, emphasising their transformation into a higher carbon iron-carbon alloy as a carburiser. This approach not only establishes waste graphite electrodes as a valuable resource but also aligns with environmentally conscious practices, contributing to the advancement of sustainable metallurgical processes.

ACKNOWLEDGEMENTS

This research was cooperated under the Australian Research Council's Research Hub for microrecycling of battery and consumer wastes (grant no. IH190100009). The authors acknowledge the staff at Mark Wainwright Analytical Centre at the University of New South Wales for their help and technical support.

REFERENCES

Ahmed, H, 2018. New Trends in the Application of Carbon-Bearing Materials in Blast Furnace Iron-Making, *Minerals*, 8:561. https://doi.org/10.3390/min8120561

Babich, A and Senk, D, 2013. Chapter 12 – Coal use in iron and steel metallurgy, in *The Coal Handbook: Towards Cleaner Production* (ed: D Osborne), (Woodhead Publishing).

Ellis, B and Bao, W, 2020. Pathways to decarbonisation episode two: steelmaking technology [online], BHP Insights. Available from: <https://www.bhp.com/news/bhp-insights/2020/11/pathways-to-decarbonisation-episode-two-steelmaking-technology>

Cham, S, Sahajwalla, V, Sakurovs, R, Sun, H and Dubikova, M, 2004. Factors Influencing Carbon Dissolution from Cokes into Liquid Iron, *ISIJ International*, 44:1835–1841.

Jang, D, Kim, Y, Shin, M and Lee, J, 2012. Kinetics of Carbon Dissolution of Coke in Molten Iron, *Metallurgical and Materials Transactions B*, 43:1308–1314.

Kongkarat, S, Khanna, R, Koshy, P, Paul, Kane, O and Sahajwalla, V, 2012. Recycling Waste Polymers in EAF Steelmaking: Influence of Polymer Composition on Carbon/Slag Interactions, *ISIJ International*, 52:385–393.

Mansuri, I A, Khanna, R, Rajarao, R and Sahajwalla, V, 2013. Recycling Waste CDs as a Carbon Resource: Dissolution of Carbon into Molten Iron at 1550°C, *ISIJ International*, 53:2259–2265.

Mansuri, I, Farzana, R, Rajarao, R and Sahajwalla, V, 2018. Carbon Dissolution Using Waste Biomass–A Sustainable Approach for Iron-Carbon Alloy Production, *Metals*, 8:290.

McCarthy, F, 2005. Interfacial Phenomena and Dissolution of Carbon from Chars into Liquid Iron During Pulverised Coal Injection in a Blast Furnace, PhD thesis, University of New South Wales.

Muslemani, H, Liang, X, Kaesehage, K, Ascui, F and Wilson, J, 2021. Opportunities and challenges for decarbonizing steel production by creating markets for 'green steel' products, *Journal of Cleaner Production*, 315:128127.

ResponsibleSteel™, 2018. What we do [online]. Available from: <https://www.responsiblesteel.org/what-we-do> [Accessed: 17 Jan 2024].

Sun, M, Pang, K, Zhang, J, Li, K and Li, H, 2021. In Situ Monitoring and Dissolution Limit of Carbon Dissolution in Hot Metal, *Steel Research International*, 92:2100111. https://doi.org/10.1002/srin.202100111

Wood Mackenzie, 2023. Steel decarbonisation to redefine supply chains by 2050 [online]. Available from: <https://www.woodmac.com/press-releases/steel-decarbonisation-to-redefine-supply-chains-by-2050> [Accessed: 19 Jan 2024].

World Steel Association, 2023. World Steel in Figures 2023 [online]. Available from: <https://worldsteel.org/steel-topics/statistics/world-steel-in-figures-2023/> [Accessed: 17 Jan 2024].

Wu, C and Sahajwalla, V, 2000. Dissolution rates of coals and graphite in Fe-C-S melts in direct ironmaking: Dependence of carbon dissolution rate on carbon structure, *Metallurgical and Materials Transactions B*, 31:215–216.

Wu, C, Wiblen, R and Sahajwalla, V, 2000. Influence of ash on mass transfer and interfacial reaction between natural graphite and liquid iron, *Metallurgical and Materials Transactions B*, 31:1099–1104.

Xuran New Materials Limited, 2024. Graphite Electrode Overview [online]. Available from: <https://www.graptek.com/products/graphite-electrode.html> [Accessed: 19 Jan 2024].

Distribution of impurity elements in oxygen-enriched top-blowing nickel smelting with Fe extraction-oriented slag adjustment

G Cao[1], J Zhao[2], J Wang[1], S Yue[1], B Li[1], J Zheng[3], Y Cui[1] and H Zong[3]

1. School of Metallurgical Engineering, Xi'an University of Architecture and Technology, Xi'an, Shaanxi 710055, China.
2. School of Metallurgical Engineering, Xi'an University of Architecture and Technology, Xi'an, Shaanxi 710055, China. Email: zhaojunxue1962@126.com
3. Jinchuan Nonferrous Metal Group Corporation, Jinchang, Gansu 737100, China.

ABSTRACT

The iron (Fe) concentration in nickel slag is high, making its recovery and utilisation a hot and challenging topic. In the nickel smelting process, using CaO instead of SiO_2 to adjust slag's composition, and then directly reducing Fe from the output of molten slag is a more feasible method in practice for Fe recovery in nickel slag. The adjustment of slag's composition in the nickel smelting process will have an impact on the distribution of impurity elements such as Pb, Zn, and As. In this paper, the slag adjustment of nickel smelting process was carried out with the oxygen-enriched top-blowing system. The distribution of impurity elements in the smelting products with each composition slag was conducted by using thermodynamic software. Experimental verification was performed with rich oxygen top-blowing, and the distribution characteristics of impurity elements in the smelting products were characterised to explore the differences in mineral phases of impurity elements with each smelting conditions, elucidating the distribution patterns of Pb, Zn, and As in each phase. The results indicate that after slag adjustment, impurity elements are more easily to volatilisation and removal into the gas, reducing their total concentration in slag and matte. In the slag, the existence form of Pb, Zn is changed from silicate to ferrate, and As is changed from arsenate to low-valent arsenic oxide. It is easier to decompose the impurity-containing mineral phase and remove the impurity elements in the subsequent iron extraction process.

INTRODUCTION

In the nickel pyrometallurgical process, approximately 6~16 tons of slag are generated per ton of nickel produced (Marenych and Kostryzhev, 2020; Wu *et al*, 2018; Xia *et al*, 2018). To the smelting with sulfide concentrate, the slag contains a small amount of valuable metal elements such as Ni, Cu, Co, as well as a significant amount of Fe, with Fe typically constituting around 40 per cent of the nickel slag (Guo *et al*, 2018b; Li *et al*, 2020a; Park *et al*, 2011; Sun *et al*, 2021; Wu *et al*, 2020; Zhang *et al*, 2020a). The output of the same kind of slag can be estimated at 5 000 000 tons per annum (tpa) in China (Zhang *et al*, 2020b). The iron in the nickel slag is taken as potential resource, but up to now remains unextracted and unused. Currently, nickel slag is primarily disposed of through stacking and landfill. With the long-term effects of chemical weathering and physical erosion, impurity elements in unprotected nickel slag may can be released into the surrounding environment, posing a potential threat to human health (Dimitrijevic *et al*, 2009; Dosmukhamedov and Kaplan, 2016; Guo *et al*, 2018a; Kobayashi and Hirano, 2016; Yang *et al*, 2017; Zhong, Li and Tan, 2017).

The recovery of Fe from nickel slag has consistently been a focus of research. Methods such as high-temperature oxidation and reduction smelting have been attempted. In the process of high-temperature oxidation of nickel slag, CaO is introduced as a flux and the slag oxidised. Fe in the slag ultimately transform into iron oxides, and then selected with–magnetic separation. This process is relatively long and difficult to put into use for extracting Fe from nickel smelting slag (Gyurov *et al*, 2014; Shen *et al*, 2018). Most of research introduced the method of molten reduction for Fe extraction from nickel slag (Li *et al*, 2013, 2020b; Long *et al*, 2016; Zhang *et al*, 2015). During the reduction process, a significant amount of CaO has to be added to displace FeO in olivine, which leads to a substantial consumption of CaO and reducing agent (Wang *et al*, 2014; 2015; Yu *et al*, 2021). as a results, a large consumption of energy is necessary and a lot of slag in Fe extraction is produced. This makes the Fe extraction process unfeasible. Currently, most research on Fe extraction from nickel slag remains in the experimental stage, and due to cost constraints, it cannot be implemented in industrial production (Li *et al*, 2021; Wang *et al*, 2020, 2021). Some industrial practice failed.

In the traditional nickel concentrates smelting process, SiO_2 is utilised as a slag-making agent for the separation of Fe and other gangue from the matte formed. The reaction formula is depicted in Equation 1. The iron silicate is generated during the nickel smelting stage, and it is stable and very difficult to be reduced, leading to difficulties in Fe extraction from the slag. In all of the works mentioned above, this kind of slag was used as raw material for Fe extraction and deduced the process problems. Our team suggested that adjustments were to use CaO as flux instead of SiO_2 in traditional process. The slag forming reaction is changed, as indicated in Equation 2. Without affecting the concentration of Fe and other valuable elements in the slag of nickel smelting, the new slag composition suitable for Fe extraction can be obtained through slag adjustment (Zhao *et al*, 2018; Cao *et al*, 2023; Wang *et al*, 2023).

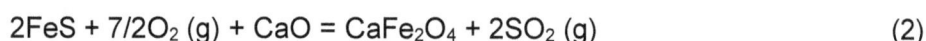

$$2FeS + SiO_2 + 2O_2 \text{ (g)} = Fe_2SiO_4 + SO_2 \text{ (g)} \tag{1}$$

$$2FeS + 7/2O_2 \text{ (g)} + CaO = CaFe_2O_4 + 2SO_2 \text{ (g)} \tag{2}$$

Because of the presence of impurity elements such as Pb, Zn, and variety of nickel concentrates sources, the control of impurity elements in products is a key problem for high attention. Impurity elements in the nickel smelting process can distribute in nickel matte, slag, dust and gas. The distribution of impurity elements depends on factors such as smelting process, temperature, reaction intensity, oxygen potential, slag and matte composition and so on. During the smelting stage of nickel matte, increasing carbon concentration and Fe/SiO_2 ratio can reduce the As concentration in the charge, enhancing the stability of As in the matte and slag. As the grade of nickel matte increases, Pb and Zn are more likely to transfer from the matte to the slag (Wang *et al*, 2017; Yang *et al*, 2017). In this paper, the slag adjustment is carried out with CaO instead of SiO_2 as flux in oxygen-enriched top-blowing nickel smelting process. There must be some changes in the transforming of impurity elements in the slag during smelting. The distribution of impurity elements in nickel smelting was studied to provide a theoretical basis for subsequent parameter control and pollution control in the smelting process.

EXPERIMENTAL METHODOLOGIES

Based on the oxygen-enriched top-blowing smelting and the idea of slag adjustment for Fe extraction, theoretical calculations on the distribution of Pb, Zn, and As after reactions were conducted using FactSage™ ver 7.1 (by CRCT-ThermFact Inc.& GTT-Technologies). Smelting experiments were conducted with the slag samples which chemical composition for matte production was chosen in previous work as adjustment for Fe extraction. The concentration of impurity elements in the produced slag and matte after the reactions were analysed through chemical analysis. The forms of impurity elements in the slag and matte samples were investigated by using X-ray photoelectron spectroscopy (XPS) and scanning electron microscopy with energy-dispersive X-ray spectroscopy (SEM-EDS).

Experimental materials

The materials used in the experiment were based on the traditional nickel smelting furnace slag with top-blowing (as shown in Table 1). Analytically pure reagents are used for adjustment, including CaO, SiO_2. The relevant materials were crushed to a size below 0.045 mm and subsequently dried at 473 K for 3 hrs before use.

TABLE 1

Composition of nickel concentrates and slag (per cent).

Element	Ni	Cu	Fe	Co	S	CaO	MgO	SiO₂	Pb	Zn	As
Nickel concentrates	3.44	3.54	22.69	0.16	19.48	1.12	6.25	16.54	0.053	0.169	0.039
Top blowing slag	1.26	1.62	28.9	0.14	3.03	3.33	10.99	29.11	0.019	0.181	0.011
Nickel matte	8.60	14.53	25.07	0.46	26.33	0.14	0.10	0.91	-	-	-

Theoretical calculation

Using the Reaction module in Factsage 7.1, the simulations were conducted under oxygen-enriched top-blowing smelting conditions, introducing different flux agent additions for matte smelting. The analysis on the distribution of Pb, Zn, and As among products after the reactions was carried out. The theoretical calculation set the reaction temperature at 1400°C (the actual temperature of Nickel smelting). The slag composition from traditional smelting was taken as the reference for comparison. The selected range for variations in slag composition for slag adjustment was defined as follows: CaO/SiO_2 ratio from 0.4 to 1.2, Fe/SiO_2 ratio from 0.6 to 1.4, MgO concentration from 7 per cent to 15 per cent. In investigating the effect of CaO/SiO_2 ratio, only the concentration of CaO and SiO_2 in the slag is altered, while keeping their total mass constant. For examining the impact of the Fe/SiO_2 ratio, the total mass of Fe and SiO_2 is maintained unchanged. When studying the effect of MgO concentration on the distribution of impurity elements, as the MgO concentration increases, the mass of the remaining substances is proportionally reduced, ensuring the total mass of the slag remains constant. In the calculations, the mass ratio of slag to nickel matte is set to 2:1. This approach ensures that the mass of nickel matte remains constant under different conditions.

Test apparatus and procedures

Slag and balance system

Smelting experiments were carried out with the slag samples which chemical composition for matte production was chosen in previous as shown in Table 2. In the traditional oxygen-enriched top-blowing smelting process of nickel sulfide ore, 8 per cent silicon dioxide is added as a flux. The traditional nickel smelting slag is shown in 1#. During the slag adjustment process, the silicon dioxide concentration in the slag is reduced by adding other oxides, followed by additional calcium to adjust slag's composition. 3# is a scheme that no SiO_2 added but 10 per cent CaO added as flux in smelting process, as a result, SiO_2 changed from 36.11 per cent to 27.56 per cent and CaO from 4.13 per cent to 13.85 per cent. 2# represents an intermediate state between 1# and 3#, where the SiO_2 concentration is adjusted from 36.11 per cent to 30.5 per cent, and CaO from 4.13 per cent to 10.91 per cent. To replace SiO_2 with CaO in the reaction process, SiO_2 concentration in the slag is diluted by adding CaO, MgO and FeO. 4# involves adding more CaO than 3# to alter the form of Fe in the produced nickel slag and minimise CaO addition in next step of Fe extraction, resulting a composition change of SiO_2 from 36.11 per cent to 25.56 per cent, and CaO from 4.13 per cent to 20.11 per cent. With the increase in CaO concentration, the hemispherical temperature gradually decreases and can meet the requirement of nickel smelting well.

TABLE 2

The initial slag composition used in the experiments.

Slag type	FeO/%	CaO/%	MgO/%	SiO_2/%	CaO/SiO_2	Fe/SiO_2
1#	46.12	4.13	13.63	36.11	0.11	0.99
2#	45.22	10.91	13.36	30.50	0.36	1.15
3#	45.21	13.85	13.36	27.56	0.50	1.27
4#	41.92	20.11	12.39	25.56	0.79	1.27

Test procedure and equipment

based on the conditions outlined in Table 2. After mixing nickel slag with nickel matte in 2:1 ratio, the materials were introduced into a well-type of electric furnace (as illustrated in Figure 1). The heating rate during the matte smelting process was controlled at 5°C/min to 1400°C. Ar is introduced at a flow rate of 100 mL/min. After maintaining the temperature at 1400° for 6 hrs, the crucible is removed from the furnace and quickly cooled. Subsequently to the smelting reaction, samples were separately collected from the slag and matte for analysis.

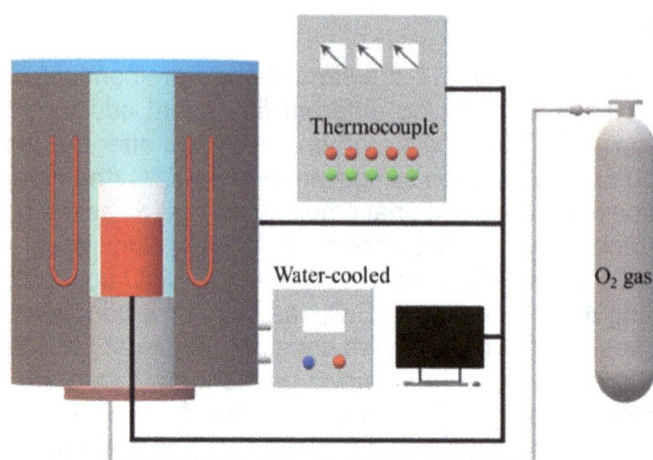

FIG 1 – Schematic diagram of the smelting and matte-making process.

Sample analysis and characterisation

The concentration of Pb, Zn, and As in the samples before and after the reaction were determined using inductively coupled plasma atomic emission spectroscopy (ICP-AES). The form of impurity elements in the resulting slag was analysed using XPS. The samples were crushed to a particle size of 200 mesh. XPS (Thermo Scientific K-Alpha) was employed to investigate changes in the valence states of each element, with orbital and binding energy ranges specified in Table 3. After embedding the post-reaction bulk slag and matte samples in resin molds, surface polishing and gold sputtering treatment were applied. Subsequently, SEM-EDS (JSM-7001F, GeminiSEM500) was carried out to analyse the microstructure and mineral phase composition of the post-reaction slag and matte.

TABLE 3

XPS detection orbits and binding energy ranges.

Elements	Orbitals	Binding energy range (eV)
Pb	Pb4f	130~150
Zn	Zn2p	1010~1060
As	As3d	40~50

RESULTS AND DISCUSSION

Theoretical analysis of the distribution of impurity elements with the composition of slag changes

Thermodynamic calculations of the nickel smelting process were performed using Factsage 7.1 to explore the influence of each factor on the distribution of impurity elements at 1400°C.

Distribution of Pb

The primary reactions involving Pb are illustrated in Equations 3 and 4. The Main form of Pb in the nickel concentrates is sulfide. In the process of matte smelting from nickel concentrates, Part of lead sulfide in the nickel slag is oxidised into lead oxide and forms slag. In the molten smelting process, with a low oxygen partial pressure, typically around 10^{-7} atm, part of lead oxide reacts with iron sulfide to generate lead sulfide into matte and ferrous oxide into slag. Because of the relatively low boiling point of PbS (1114°C), part of Pb directly vaporises into the flue gas during the smelting process and is finally collected as dust.

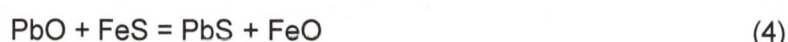

$$2PbS + 3O_2 \text{ (g)} = 2PbO + 2SO_2 \text{ (g)} \tag{3}$$

$$PbO + FeS = PbS + FeO \tag{4}$$

The impact of changes in the CaO/SiO_2 ratio on the distribution of Pb in each phase is depicted in Figure 2a. With an increase in the CaO/SiO_2 ratio, the distribution of Pb shows an upward trend in the gas and slag and a downward trend in the matte. During the process of increasing the CaO/SiO_2 ratio from 0.4 to 1.2, the distribution of Pb in the matte decreases from 26.3 per cent to 22.7 per cent. With an increase in the CaO concentration in the slag, the activity coefficient of PbO in slag increases. As a result, most of Pb exists in the form of oxides, making it more prone to entering the slag. The effect of changes in the Fe/SiO_2 ratio on the distribution of Pb in each phase is illustrated in Figure 2b. With an increase in the Fe/SiO_2 ratio, the distribution of Pb shows an initial increase followed by a decrease in the gas, increase in the matte, and decrease in the slag. During the process of increasing the Fe/SiO_2 ratio from 0.6 to 1.4, the distribution of Pb in the slag decreases from 15.6 per cent to 7.4 per cent. The distribution of Pb in the matte increases from 20.6 per cent to 25.6 per cent, and the distribution of Pb in the gas increases from 63.8 per cent to 67.1 per cent. The impact of changes in the MgO concentration on the distribution of Pb in each phase is shown in Figure 2c. With an increase in the MgO concentration, the distribution of Pb increases in the gas, decreases in the matte, and increases in the slag.

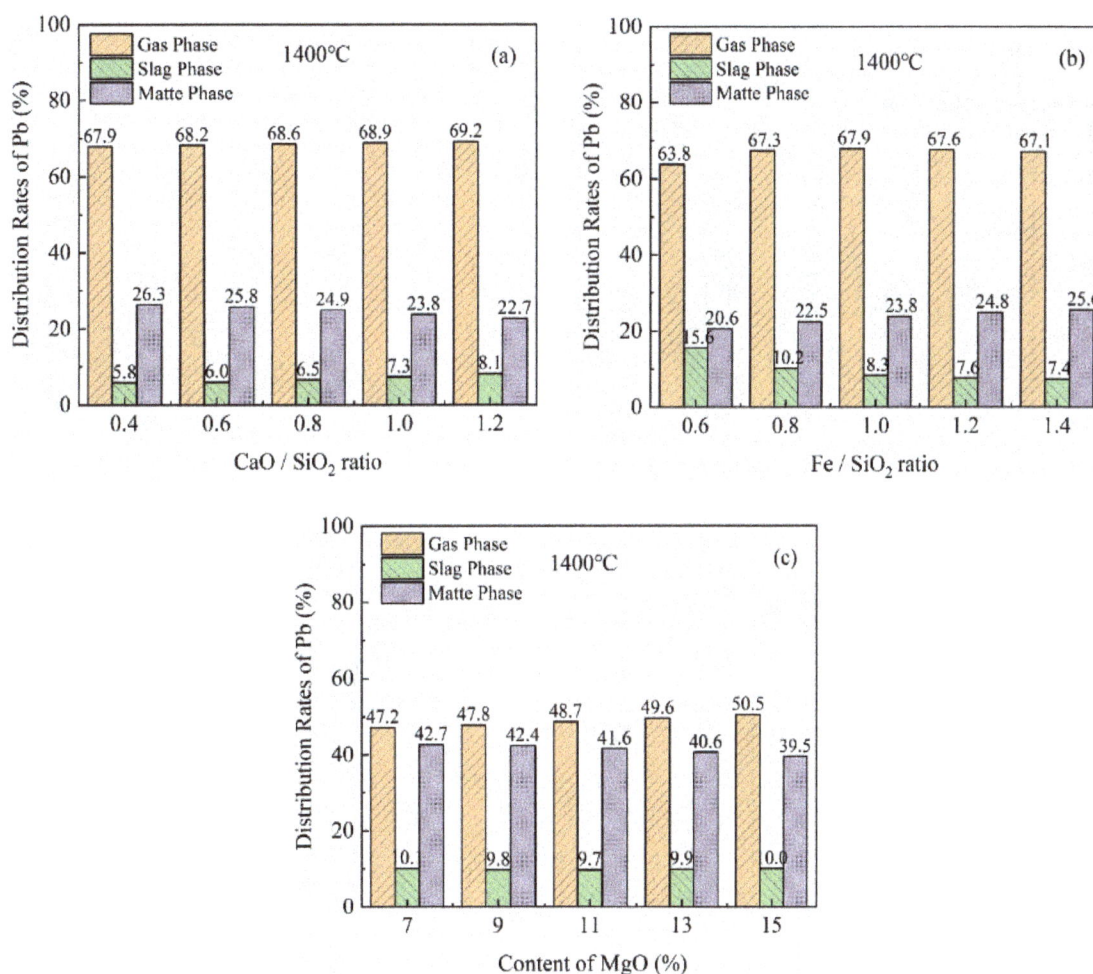

FIG 2 – The influence of slag composition changes on the distribution ratio of Pb in each phase: (a) CaO/SiO_2 ratio; (b) Fe/SiO_2 ratio; (c) MgO).

Distribution of Zn

The main form of Zn in the nickel concentrates is ZnS. The main reactions involving Zn in the smelting process are represented by Equations 5–8. A substantial portion of ZnS is oxidised by iron oxides to ZnO and then combines with silica to form zinc silicate. Part of ZnS undergoes carbon reduction, and this portion of Zn volatilises into gas. The remaining portion of Zn, in sulfide form, turns into the matte along with nickel and copper.

$$ZnS + 9Fe_2O_3 = ZnO + 6Fe_3O_4 + SO_2 \text{ (g)} \tag{5}$$

$$ZnS + 3Fe_3O_4 = ZnO + 9FeO + SO_2 \text{ (g)} \tag{6}$$

$$2ZnO + SiO_2 = Zn_2SiO_4 \tag{7}$$

$$ZnS + 2C = 2Zn \text{ (g)} + 2CO \text{ (g)} + S \text{ (g)} \tag{8}$$

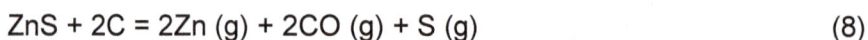

At 1400°C, the impact of changes in the CaO/SiO_2 ratio on the distribution of Zn in each phase is illustrated in Figure 3a. With an increase in the CaO/SiO_2 ratio, the distribution of Zn exhibits an upward trend in the gas and slag and a downward trend in the matte. During the process of increasing the CaO/SiO_2 ratio from 0.4 to 1.2, the distribution of Zn in the matte decreases from 29.2 per cent to 24.0 per cent. The increase in the CaO/SiO_2 ratio is conducive to more FeO being displaced from silicate by CaO, the activity coefficient of ZnO in slag decreases. The influence of changes in the Fe/SiO_2 ratio on the distribution of Zn in each phase is shown in Figure 3b. With an increase in the Fe/SiO_2 ratio, the distribution of Zn shows an increasing trend in the gas and matte and a decreasing trend in the slag. Changes in the MgO concentration and their impact on the distribution of Zn in each phase are depicted in Figure 3c. With an increase in the MgO concentration, the distribution of Zn increases in the gas and decreases in the slag and matte.

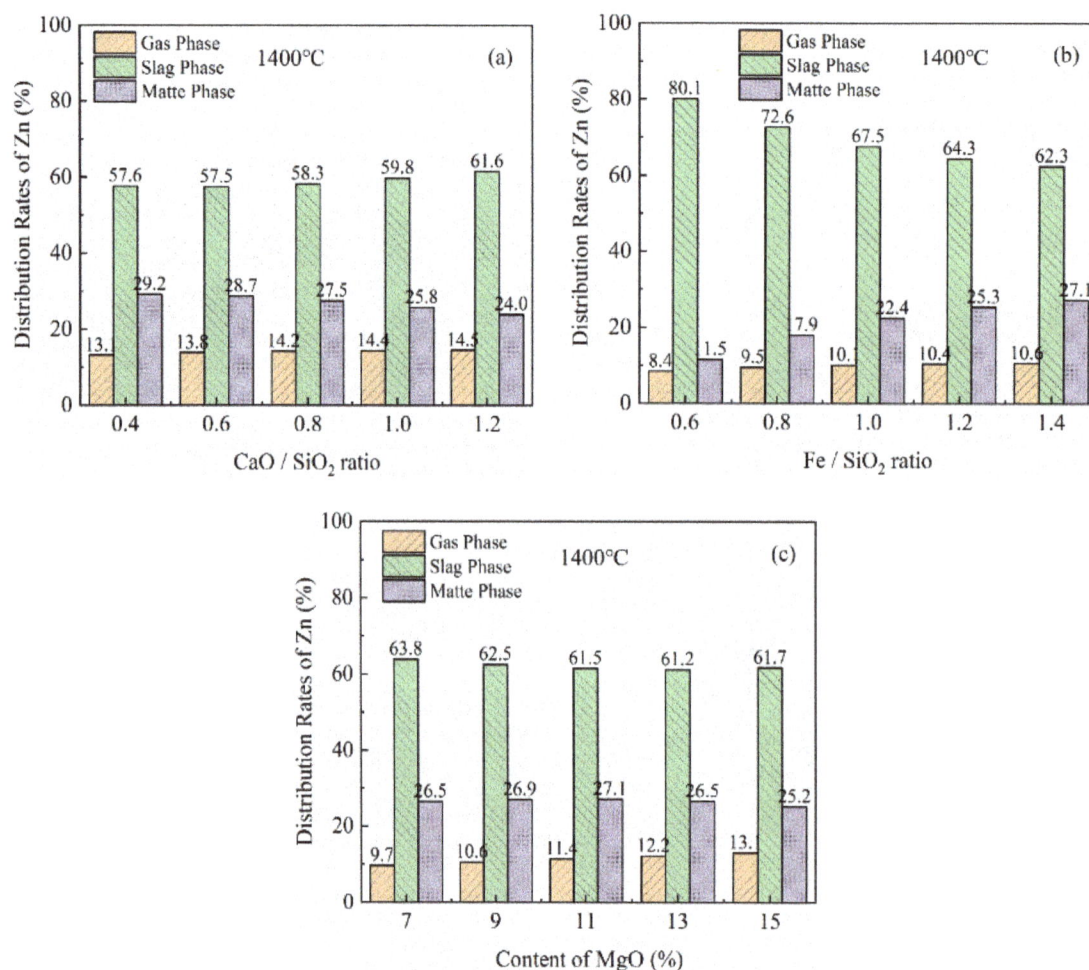

FIG 3 – The influence of changes slag composition on the distribution ratio of Zn in each phase (a) CaO/SiO_2 ratio; (b) Fe/SiO_2 ratio; (c): MgO.

Distribution of As

In the nickel concentrates, As primarily exists in the form of As_2S_3 or FeAsS. During the smelting process, arsenic sulfides undergo thermal decomposition to elemental As and then is oxidised. Most of the As oxides evaporates into the gas (Chen and Jahanshahi, 2010; Chen, Zhang and Jahanshahi, 2010; Swinbourne and Kho, 2012), while some arsenic oxides are embedded in the silicate matrix. The main reactions are illustrated in Equations 9–11.

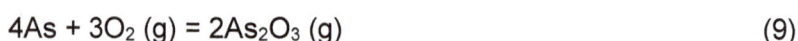

$$4As + 3O_2 \text{ (g)} = 2As_2O_3 \text{ (g)} \tag{9}$$

$$2As_2S_3 + 9O_2 \text{ (g)} = 2As_2O_3 + 6SO_2 \text{ (g)} \tag{10}$$

$$As_2O_3 + 2Fe_2O_3 = As_2O_5 + 4FeO \tag{11}$$

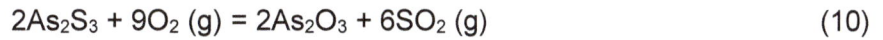

At 1400°C, the influence of the CaO/SiO_2 ratio changes on the distribution of As in each phase is illustrated in Figure 4a. With an increase in the CaO/SiO_2 ratio, the distribution of As shows an upward trend in the slag and a downward trend in the gas and matte. During the process of increasing the CaO/SiO_2 ratio from 0.4 to 1.2, the distribution of As in the matte decreases from 1.0 per cent to 0.9 per cent. The distribution of As in the gas decreases from 60.2 per cent to 53.7 per cent, and the distribution of As in the slag decreases from 38.7 per cent to 45.4 per cent. Increase of CaO in slag will reduce the activity coefficient of As_2O_3 and As_2O_5, hence increase As capacity in slag. The volatilisation of arsenic oxides has been reduced. The impact of changes in the Fe/SiO_2 ratio on the distribution of As in each phase is shown in Figure 4b. With an increase in the Fe/SiO_2 ratio, the distribution of As exhibits an initial increase followed by a decrease in the gas, a decrease in the matte, and an initial decrease followed by an increase in the slag. Changes in the MgO concentration and their impact on the distribution of As in each phase are depicted in Figure 4c. With each temperature condition, with an increase in the MgO concentration, the distribution of As shows an upward trend in the matte and slag and a downward trend in the gas.

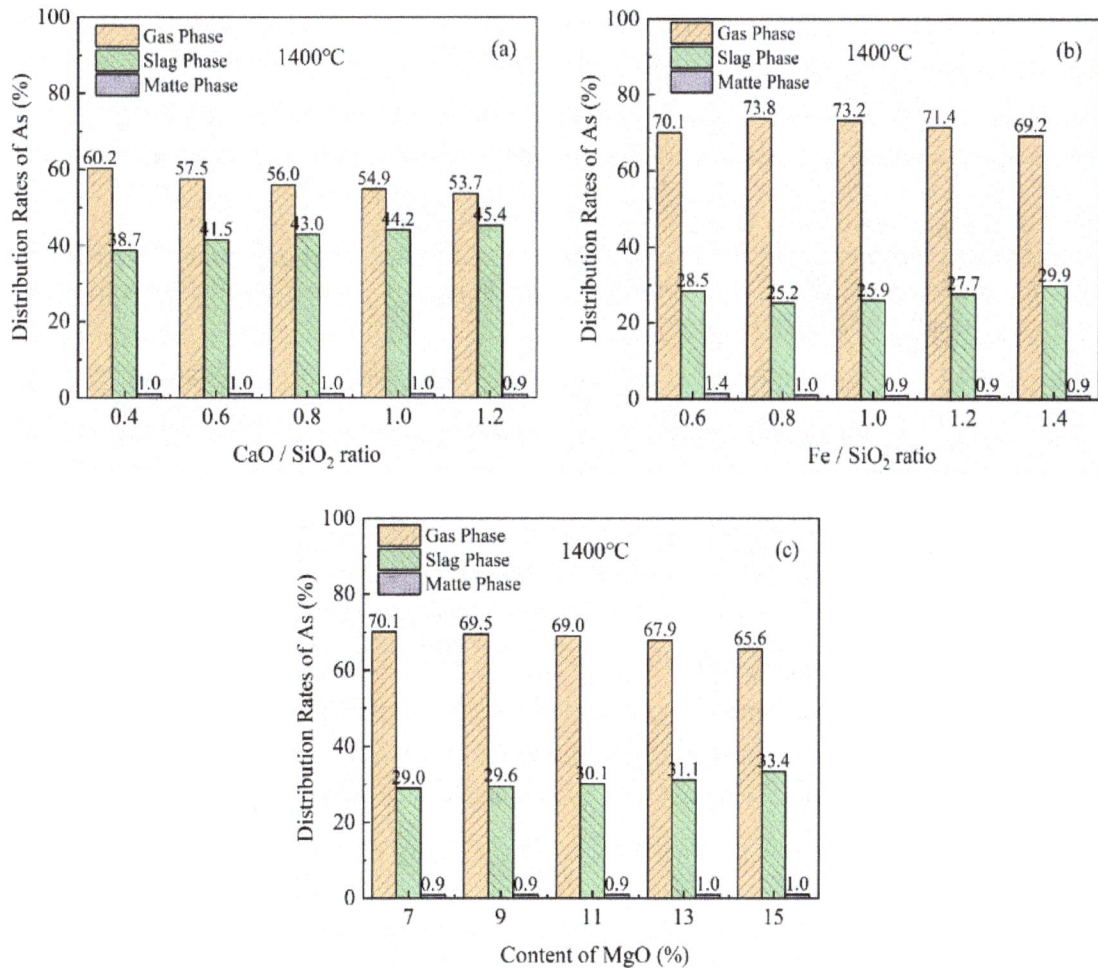

FIG 4 – The influence of slag composition changes on the distribution ratio of As in each phase (a) CaO/SiO_2 ratio; (b) Fe/SiO_2 ratio; (c) MgO.

Test results on the migration trends of impurity elements

The concentration of Pb, Zn, and As in each phase from tests is shown in Figure 5. The impurity element concentration in nickel slag and nickel matte was determined by chemical analysis, while the impurity element concentration in the gas was calculated using the formula:

$$g_x = (a \times S_{x1} + b \times M_{x1} - c \times S_{x2} - d \times M_{x2})/(S_{x1} + M_{x1} - S_{x2} - M_{x2}), \% \tag{12}$$

where:

g_x	the proportion of impurity elements in the gas, per cent
a	the proportion of impurity elements in the nickel slag before reaction, per cent
b	the proportion of impurity elements in the nickel matte before reaction, per cent
c	the proportion of impurity elements in the nickel slag after reaction, per cent
d	the proportion of impurity elements in the nickel matte after reaction, per cent
S_{x1}	the mass of nickel slag before reaction, g
M_{x1}	the mass of nickel matte before reaction, g
S_{x2}	the mass of nickel slag after reaction, g
M_{x2}	the mass of nickel matte after reaction

From Figure 5a, it can be observed that with an increase in alkalinity, the concentration of Pb in nickel slag and nickel matte shows a decreasing trend, while in the flue gas, it exhibits an increasing trend. In comparison to the traditional process (scheme 1#), the Pb concentration in nickel matte in the scheme 4# decreases from 0.105 per cent to 0.093 per cent, the Pb concentration in the slag decreases from 0.017 per cent to 0.009 per cent, and the Pb concentration in the flue gas increases from 0.113 per cent to 0.197 per cent. Figure 5b presents the distribution rate of Pb in each phase. It can be observed that with the traditional process, 19 per cent Pb, 59 per cent Pb and 21 per cent Pb distribute in slag, matte and gas, respectively. In contrast to the Pb distribution with the traditional process (scheme 1#), in the scheme 4#, 10 per cent Pb, 53 per cent Pb, 37 per cent Pb distribute in slag, matte and gas, respectively. The change in the Pb distribution may be attributed to the decrease in melting temperature resulting from the addition of CaO during slag adjustment, providing higher superheat for matte smelting, thereby enhancing the fluidity of reactant and facilitating the easier release of volatile substances in gas form.

Figure 5c displays the proportion of Zn in each phase. With an increase in alkalinity, the concentration of Zn in nickel slag shows a decreasing trend, while in nickel matte and flue gas, it exhibits an increasing trend. In comparison to the traditional process (scheme 1#), the Zn concentration in the slag in the scheme 4# decreases from 0.097 per cent to 0.062 per cent, the Zn concentration in nickel matte increases from 0.172 per cent to 0.177 per cent, and the Zn concentration in the flue gas increases from 0.592 per cent to 0.787 per cent. Figure 5d illustrates the distribution rate of Zn in each phase. With the traditional process, 18 per cent Zn, 52 per cent Zn and 30 per cent Zn distribute in slag, matte and gas, respectively. In contrast to the Zn distribution with the traditional process (scheme 1#), in the scheme 4#, 22 per cent Zn, 31 per cent Zn, 47 per cent Zn distribute in slag, matte and gas respectively. with CaO addition for slag adjustment, more Zn volatilises into the gas, and this phenomenon is similar with the distribution trend of Pb.

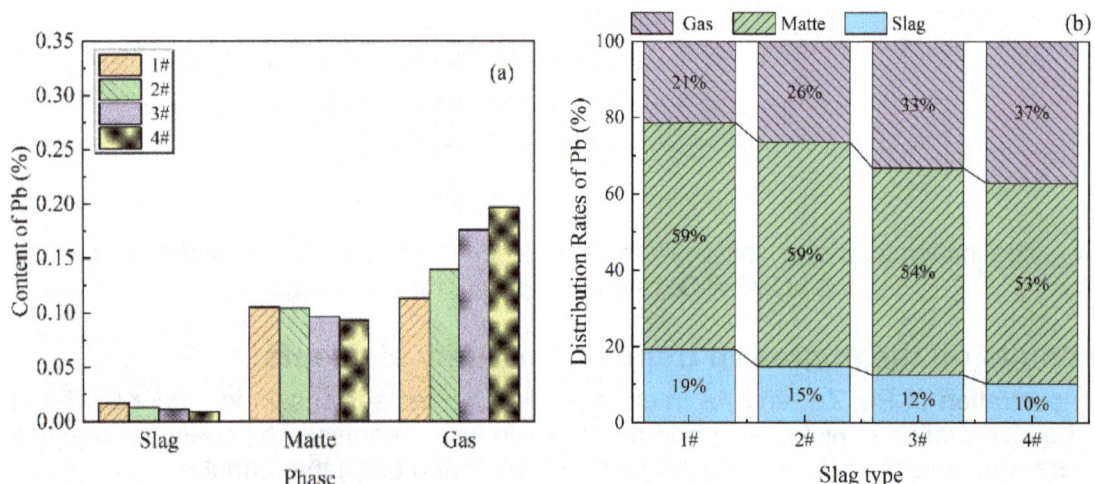

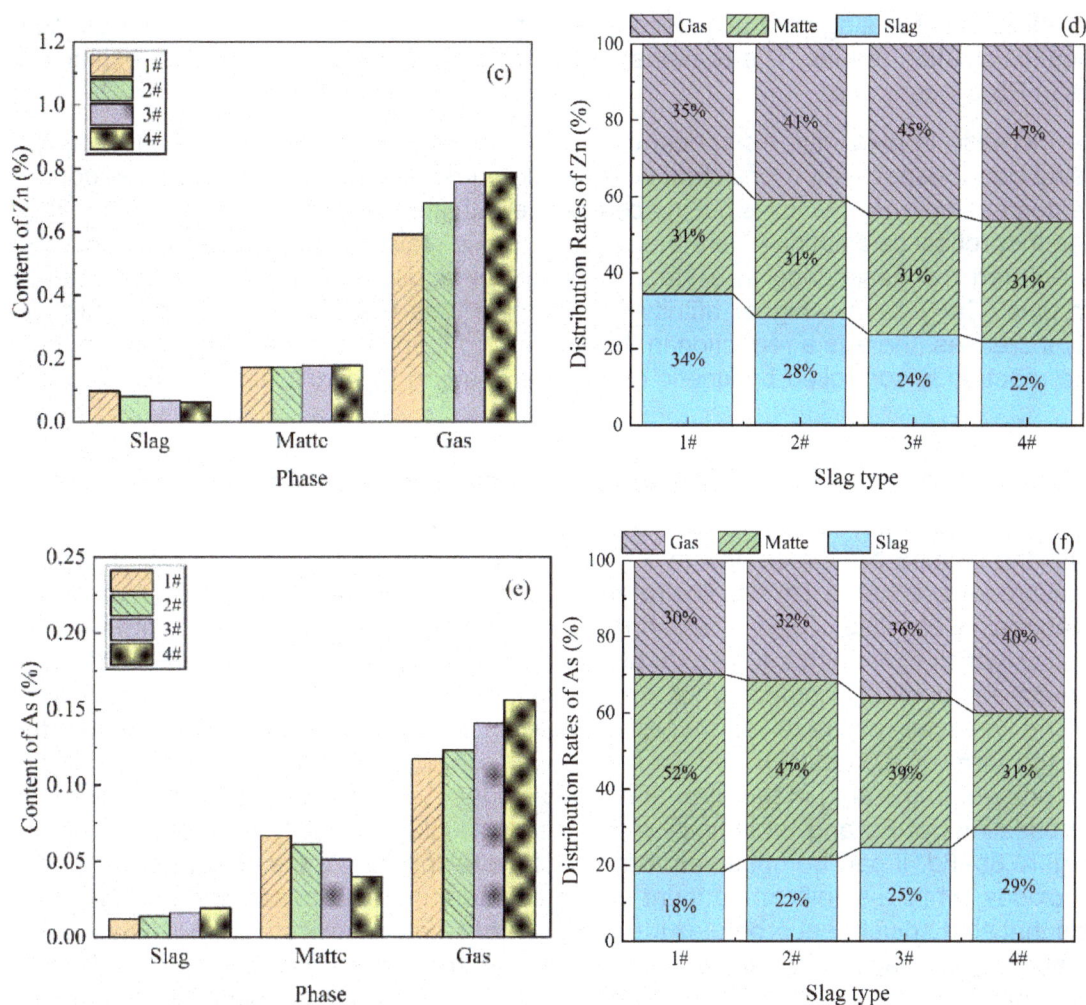

FIG 5 – The distribution of impurity elements in each phase after smelting and refining:
(a) Proportion of Pb in each phase; (b): Distribution rate of Pb; (c) Proportion of Zn in each phase;
(d) Distribution rate of Zn; (e) Proportion of As in each phase; (f) Distribution rate of As.

Figure 5e depicts the proportion of As in each phase. With an increase in alkalinity, the concentration of As in nickel slag shows a decreasing trend, while in nickel matte and flue gas, it exhibits an increasing trend. In comparison to the traditional process (scheme 1#), the As concentration in nickel matte in the scheme 4# decreases from 0.067 per cent to 0.04 per cent, the As concentration in the slag increases from 0.012 per cent to 0.019 per cent, and the As concentration in the flue gas increases from 0.117 per cent to 0.156 per cent. Figure 5f illustrates the distribution rate of As in each phase. It can be seen that to the traditional process, 72 per cent As, 6 per cent As and 21 per cent As distribute in slag, matte and gas respectively. In contrast to the As distribution in the traditional process (scheme 1#), in the scheme 4#, 29 per cent As, 31 per cent As and 40 per cent As distribute in the slag, matt and gas respectively. with CaO addition for slag adjustment, As volatilises more from the slag into the gas.

The theoretical calculation results of impurity element distribution indicate that with an increase in alkalinity, the Pb concentration decreases in the matte and an increase in gas. This is consistent with the experimental results. Contrary to the experimental results, the Pb concentration in the slag increases, attributed to the elevated alkalinity enhancing the oxidative atmosphere in the reaction. This enhanced the migration of lead sulfides from the matte to the slag. In the theoretical calculation, process reaches to equilibrium state, the amount of Pb transformed from the matte to the slag exceeds the volatile concentration in the slag, resulting in an increase in Pb concentration in the slag with the increasing alkalinity.

The Zn concentration in the slag and gas gradually increases with the rise in alkalinity. While in the matte, the Zn concentration gradually decreases. This trend aligns with the variations observed in Pb concentration and the results of theoretical calculations. In the traditional smelting process, the

Zn concentration in the slag gradually decreases with the increase in alkalinity, contrasting with the theoretical calculation results. The reasons for the difference are similar to those explained for the changes in Pb concentration.

With the increase in alkalinity, As concentration in the slag and matte exhibits an increasing and decreasing trend respectively, consistent with the experimental results. In the gas, As concentration shows a decreasing trend, which is not in accordance with the experimental results. The discrepancy lies in the theoretical calculation where the reaction is in a state of complete equilibrium; most of the As has entered the gas during in a state of complete equilibrium. In the experimental process, Calcium ions (Ca^{2+}) have a higher affinity for arsenide ions (As^{3-}) than oxide ions (O^{2-}). With the alkalinity increases, there is a reduction in the emitted concentration of arsenic oxide in the furnace gas. The reaction is controlled by kinetic conditions, leading to an increase in As concentration in the gas.

The morphology analysis of impurity elements in nickel slag

Utilising SEM-EDS analysis to examine the mineralogical embedding characteristics of impurity elements, the results are presented in Figure 6. The major phases in the nickel slag with the traditional process (scheme 1#) are illustrated in Figure 6a. The matrix composition of the produced nickel slag mainly consists of olivine, with brighter regions representing magnetite phases composed of Fe and O, which constitute the primary phases in the slag. The bright portions precipitated in the nickel matte are primarily composed of copper sulfide, while the matrix consists mainly of nickel sulfide. The line scan results reveal that, after smelting, Pb and Zn are mainly present in the nickel matte in the form of sulfides. Pb and Zn in the slag exhibit a relatively uniform distribution. As is primarily present in the slag in the form of arsenic oxides. The baseline of As in the slag fits better with the baselines of Fe and Si, indicating that As in the slag is mainly enveloped by iron silicate. From Figure 6b–6d, it can be observed that crystals precipitated in the slag after slag adjustment are iron oxides, while a significant amount of iron sulfate mineral phases appear in the matrix. This suggests that slag adjustment promotes the transformation of iron silicate to iron sulfate. Comparing the morphology of slag and matte produced with different slag condition, it is evident that with increasing alkalinity during the reaction process, there are partially entrapped nickel mattes in the slag. This phenomenon might be related to differences in density, viscosity, and interfacial tension between sulfides and slag. As primarily exists in the slag in the form of arsenic oxides, entangled and embedded with olivine, rather than combined with iron sulfate.

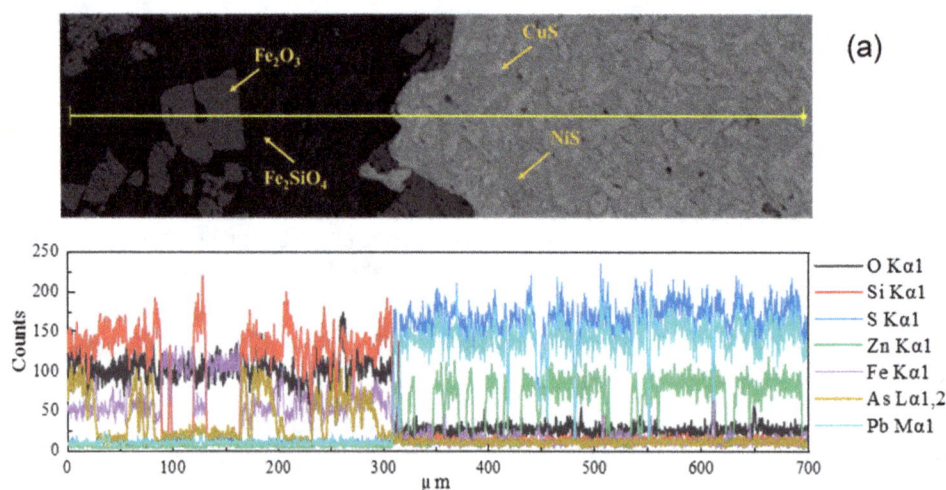

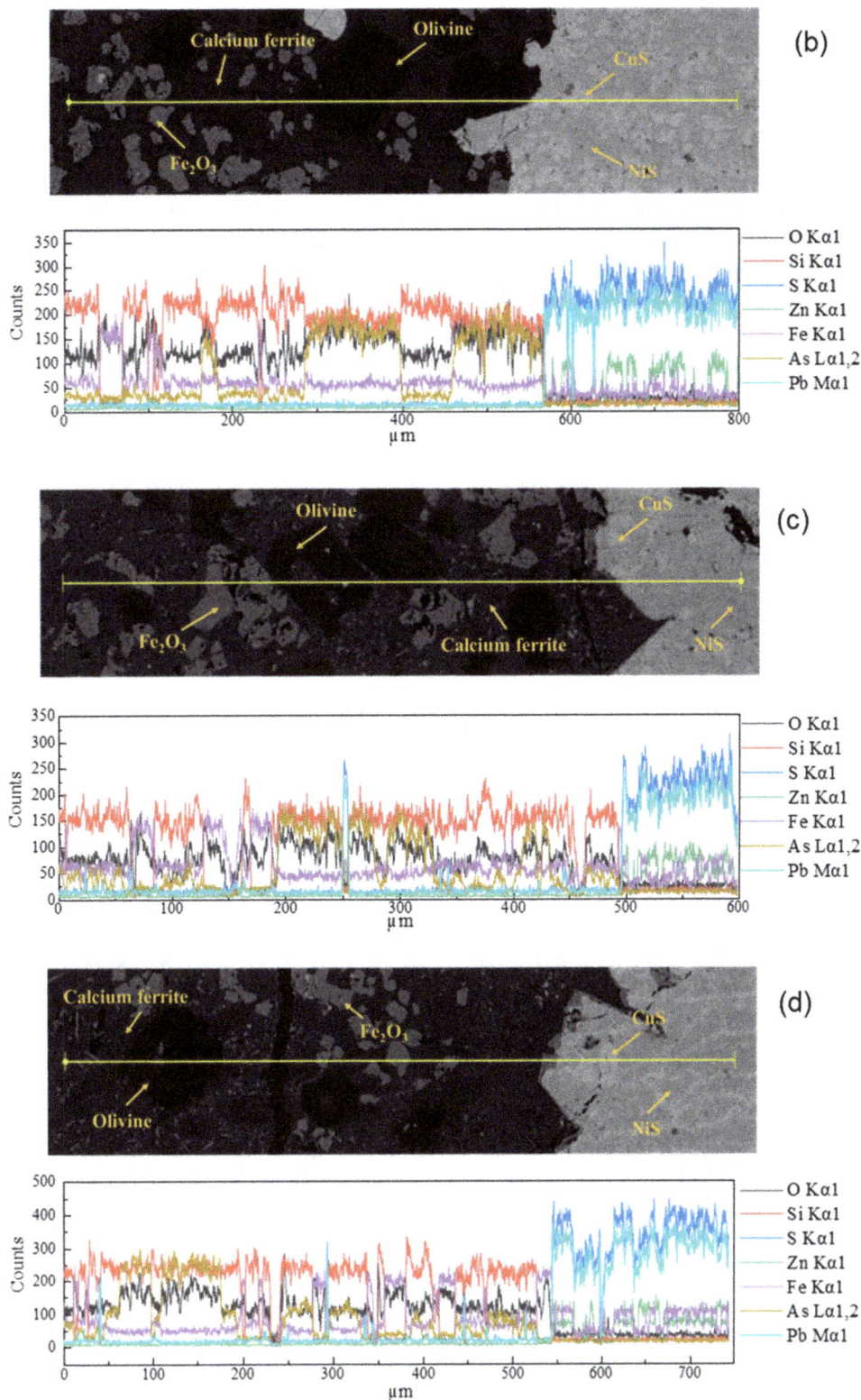

FIG 6 – The microstructure and surface scanning spectrum results near the slag-matte interface: (a) scheme 1#; (b) scheme 2#; (c) scheme 3#; (d) scheme 4#.

The chemical valence analysis of impurity elements in nickel slag

The chemical valence of Pb in nickel slag with different slag condition are depicted in Figure 7, typically represented as Pb, PbO, PbO_2, $PbSO_4$. The binding energy difference between Pb4f 7/2 and Pb4f 5/2 is 4.87 eV. The peak position indicating metallic Pb in the main peak of Pb4f 7/2 is approximately between 136~137 eV. The binding forms of Pb oxide in nickel slag may exhibit two peak positions, one in the range of 137~138 eV and the other around 138.5 eV. There is a gap in the binding energy of Pb oxide in each mineral phase, with the binding energy peaks of Pb, Pb-O

(Pb/Fe/O), and Pb-O (Pb/Si/O) increasing sequentially. The primary forms of Pb in nickel slag include metallic lead, lead silicate, and lead sulfate. With the traditional process, a semi-quantitative analysis of Pb in the nickel slag produced indicates that the metallic Pb concentration is 45 per cent, lead sulfate concentration is 20 per cent, and lead silicate concentration is 35 per cent. Through adjusting slag's composition, introducing CaO as a flux during the nickel matte smelting stage, the overall form of metallic Pb in the slag tends to decrease. During the smelting process, adjusting slag's composition enhance the stability of Pb compounds in the slag, and reducing the risk of Pb release, while the concentration of lead iron sulfate gradually increases. When the slag has higher alkalinity, CaO more easily combines with silicate ions, displacing Fe and resulting in lower lead silicate concentration and higher lead iron sulfate concentration in the slag.

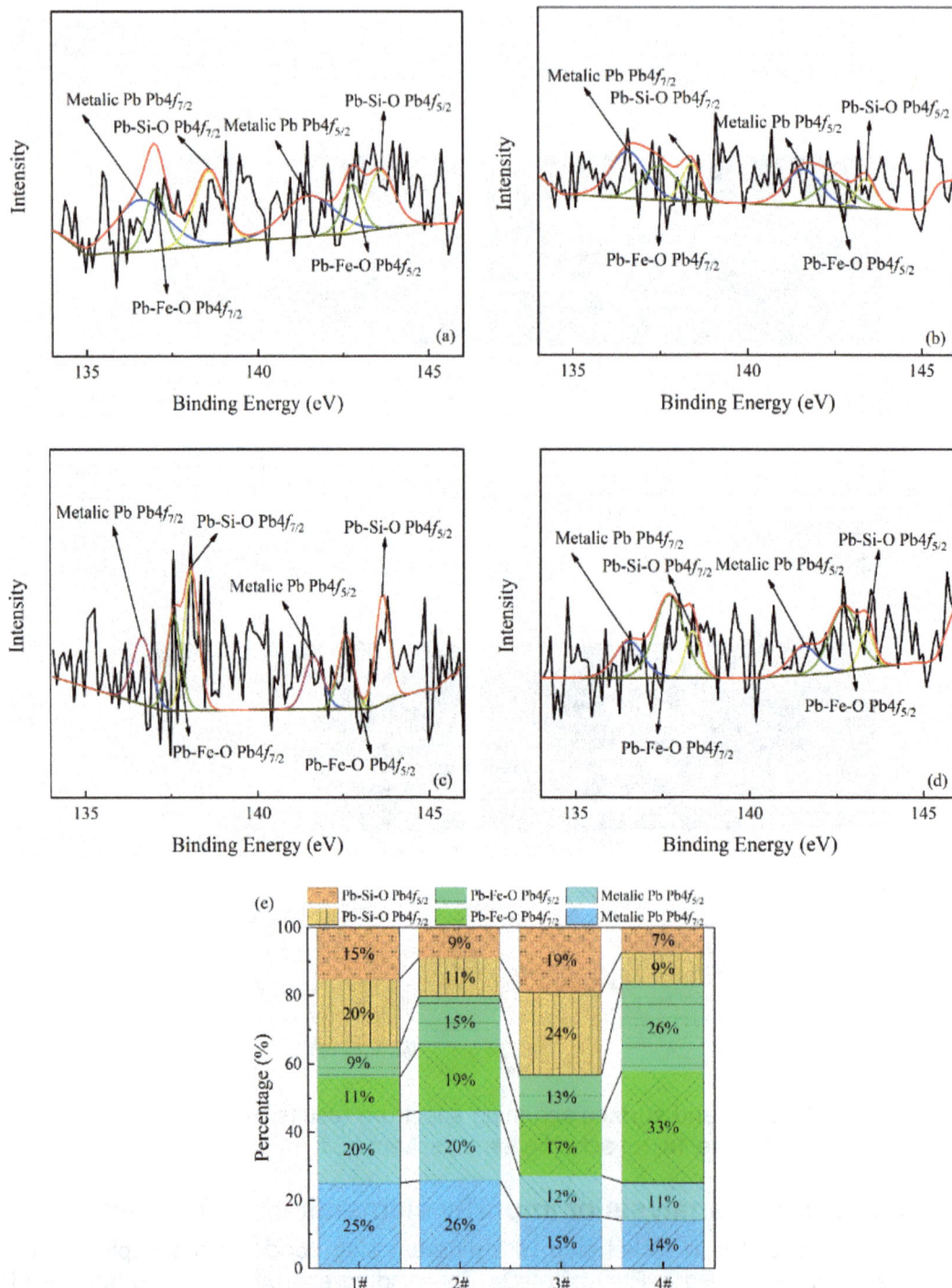

FIG 7 – The chemical forms of Pb in nickel slag with each composition: (a) scheme 1#; (b) scheme 2#; (c) scheme 3#; (d) scheme 4#; (e) Intensity ratio of each valence state of Pb.

The chemical forms of Zn in nickel slag with each condition are illustrated in Figure 8. The XPS analysis for Zn is similar to Pb, and zinc oxide exhibits two peak positions in each mineral phase. The binding energy of the Zn2p 3/2 fitting peak for zinc oxide (Zn/Fe/O) is roughly between 1021 and 1021.9 eV. Peaks above 1022 eV in the Zn2p 3/2 fitting are considered indicative of zinc oxide (Zn/Si/O). The fitting peak position for Zn sulfide is close to that of metallic Zn, with a binding energy difference of 23 eV between Zn2p 3/2 and Zn2p 1/2. XPS analysis of the main forms of Zn in nickel slag produced during nickel smelting with each scenario reveals that Zn in nickel slag is mainly present in the forms of zinc silicate and zinc iron sulfate. Zinc iron sulfate is typically found or associated with Fe-containing silicate mineral phases, and its peak position is lower compared to zinc silicate. Using CaO instead of SiO$_2$, there is an increasing trend in the proportion of zinc iron sulfate in the slag, while the proportion of zinc silicate decreases. With high alkalinity, the increased combination of sulfate ions with calcium ions may be influenced by the dissolution and loss of metal sulfides in the slag.

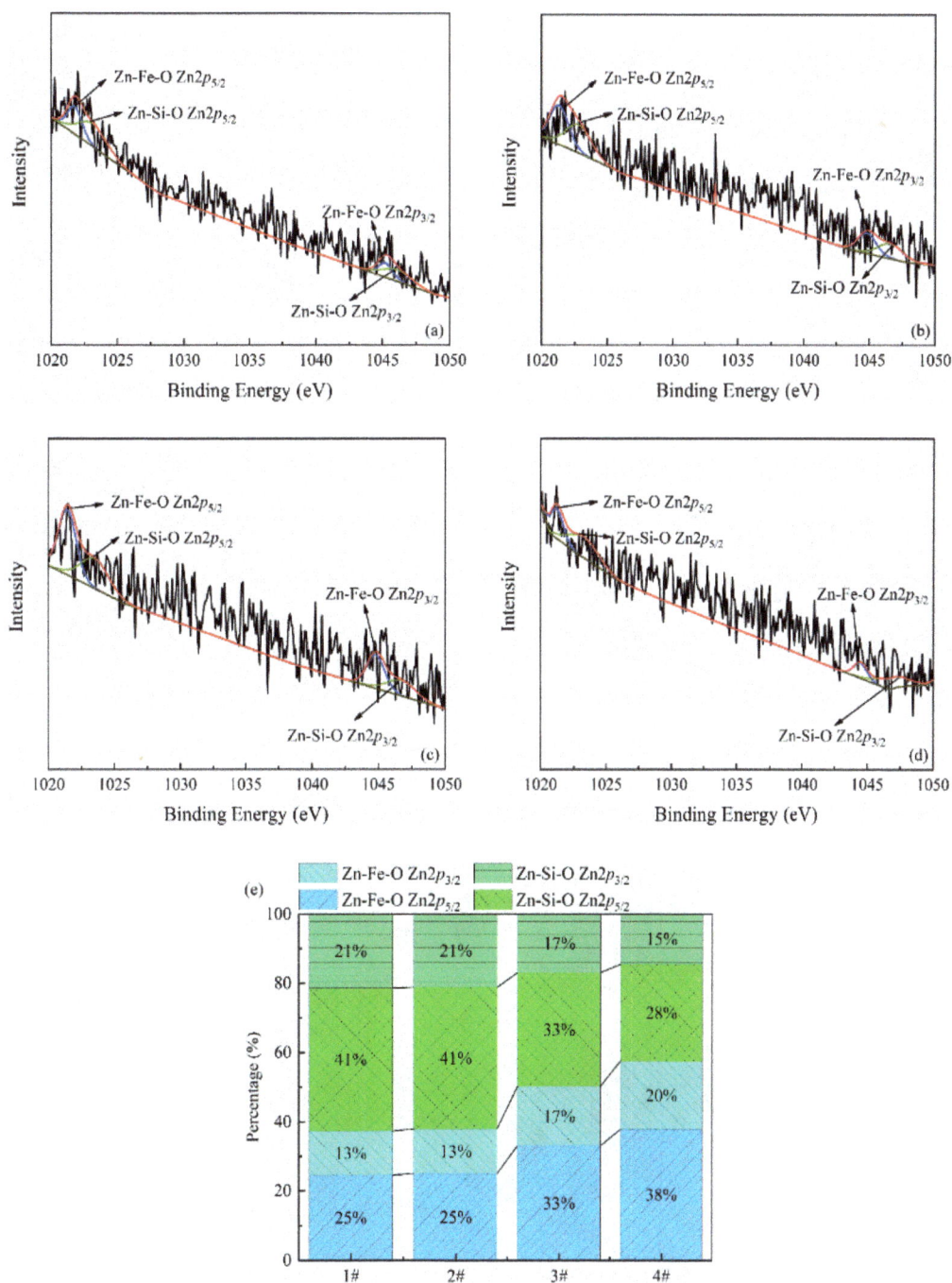

FIG 8 – The chemical forms of Zn in nickel slag with each composition: (a) scheme 1#; (b) scheme 2#; (c) scheme 3#; (d) scheme 4#; (e) Intensity ratio of each valence state of Zn.

The chemical forms of As typically include metallic As, arsenic sulfides, As^{3+} (As_2O_3), and As^{5+} (As_2O_5). In this research, As3d was employed investigate the chemical states of As in nickel slag. The As3d orbital comprises As3d 5/2 and As3d 3/2 double peaks (with a set binding energy difference of 0.7±0.1 eV), where the As (III) peak in As3d is generally 1 eV lower than As (V). Analysis of the chemical forms of As in nickel slag with each scenario, as shown in Figure 9, reveals that blue represents metallic As or arsenic sulfides, green represents As(III) oxide, and yellow represents As(V)-O oxide. In the traditional process, As in nickel slag mainly exists in the form of As(V)-O arsenate, with small amounts of trivalent arsenic oxide and metallic or sulfide arsenic. In scheme 2#, scheme 3#, and scheme 4#, As exists mainly in the form of trivalent arsenic oxide. When using CaO instead of SiO_2 as a flux in the nickel smelting process, it is observed that the overall presence of sulfide arsenic or metallic As in the slag tends to decrease, while the proportion of As_2O_3 in nickel slag shows an increasing trend. Compared to arsenates, As_2O_3 is more easily removed in the subsequent Fe reduction process.

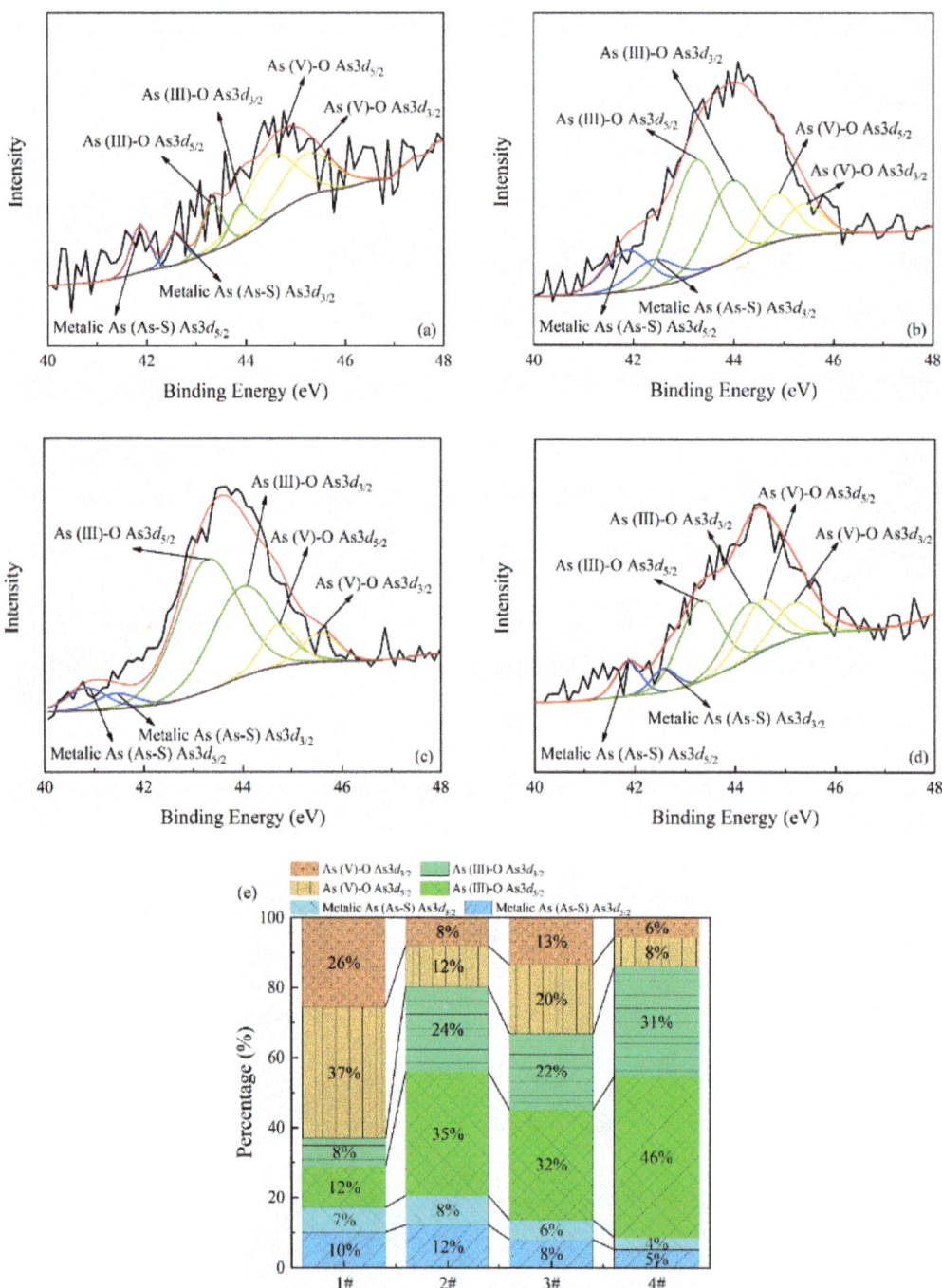

FIG 9 – The chemical forms of As in nickel slag with each composition: (a) scheme 1#; (b) scheme 2#; (c) scheme 3#; (d) scheme 4#; (e) Intensity ratio of each valence state of As.

CONCLUSIONS

Based on the oxygen-enriched top-blowing process, the distribution patterns of Pb, Zn, and As were explored in nickel sulfide concentrates smelting with different slag. The main conclusions are as follows:

- In theoretical equilibrium state, Pb has the highest concentration in the gas, followed by the matte, with the lowest concentration in the slag. As also has the highest concentration in the gas, followed by the slag, with the lowest concentration in the matte. Zn has the highest concentration in the slag, followed by the matte, with the lowest concentration in the gas. The increase in the calcium-to-silicon ratio results in an increase of impurity element concentration in the slag.

- To traditional operating conditions, 79 per cent of Pb, 34 per cent of Zn, and 72 per cent of As enter the slag. 6 per cent of Pb, 31 per cent of Zn, and 6 per cent of As enter the matte. 14 per cent of Pb, 35 per cent of Zn, and 21 per cent of As enter the gas, respectively. The deviation between test results and theoretical calculations is attributed to the difference between equilibrium state and the nonequilibrium state in experimental process.

- With increasing CaO, the proportion of Pb entering gas increases, and the proportion entering slag, matte decreases, respectively. The proportion of Zn entering the slag decreases, the proportion entering the matte remains unchanged, and the proportion entering gas increases. The proportion of As entering matte decreases, the proportion entering slag, gas increases, respectively. Adjusting slag's composition is advantageous for the volatilisation and enrichment of impurity elements in the gas and slag.

- With slag composition adjustment with CaO addition, the chemical forms of Pb in the slag produced during the nickel smelting process transform from metallic lead and lead silicate to lead iron sulfate; the chemical form of Zn changes from zinc silicate to iron zinc sulfate; the form of As transforms from arsenic sulfide and arsenate to As_2O_3. Iron sulfate and As_2O_3 are more easily reduced compared to silicate and arsenate, making it easier to remove impurity elements in the next iron extraction process.

- With slag composition adjustment with CaO addition, Impurity elements tend to be entrained into the gas, reduces the impact of impurity elements on product performance during subsequent matte refining and iron extraction from nickel slag process.

ACKNOWLEDGEMENTS

This research was financially supported by the National Natural Science Foundation of China (No. 51674185).

REFERENCES

Cao, G, Zhao, J X, Wang, G J, Zheng, J H, Zong, H X, Li, B and Cui, Y R, 2023. The research status and development analysis of iron resource recovery from nickel smelting slag, *Nonferrous Metals Science and Engineering*, 9 p. Available from: <http://kns.cnki.net/kcms/detail/36.1311.TF.20230920.1642.002.html> [Accessed: 11 September 2023].

Chen, C and Jahanshahi, S, 2010. Thermodynamics of Arsenic in FeOx-CaO-SiO2 Slags, *Metallurgical and Materials Transactions B*, 41:1166–1174.

Chen, C, Zhang, L and Jahanshahi, S, 2010. Thermodynamic Modeling of Arsenic in Copper Smelting Processes, *Metallurgical and Materials Transactions B*, 41:1175–1185.

Dimitrijevic, M, Kostov, A, Tasic, V and Milosevic, N, 2009. Influence of pyrometallurgical copper production on the environment, *Journal of Hazardous Materials*, 164(2–3):892–899.

Dosmukhamedov, N K and Kaplan, V, 2016. Efficient Removal of Arsenic and Antimony During Blast Furnace Smelting of Lead-Containing Materials, *JOM*, 69:381–387.

Guo, Z, Pan, J, Zhu, D and Zhang, F, 2018a. Green and efficient utilization of waste ferric-oxide desulfurizer to clean waste copper slag by the smelting reduction-sulfurizing process, *Journal of Cleaner Production*, 199:891–899.

Guo, Z, Zhu, D, Pan, J and Zhang, F, 2018b. Innovative methodology for comprehensive and harmless utilization of waste copper slag via selective reduction-magnetic separation process, *Journal of Cleaner Production*, 187:910–922.

Gyurov, S, Rabadjieva, D, Kovacheva, D and Kostova, Y, 2014. Kinetics of copper slag oxidation under nonisothermal conditions, *Journal of Thermal Analysis and Calorimetry*, 116(2):945–953.

Kobayashi, Y and Hirano, S, 2016. Distribution and Excretion of Arsenic Metabolites after Oral Administration of Seafood-Related Organoarsenicals in Rats, *Metals*, 6(10).

Li, K Q, Ping, S, Wang, H Y and Ni, W, 2013. Recovery of iron from copper slag by deep reduction and magnetic beneficiation, *Int J Min Met Mater*, 20(11):1035–1041.

Li, X M, Yang, H B, Ruan, J B, Li, Y, Wen, Z Y and Xing, X D, 2020a. Effect of mechanical activation on enhancement of carbothermal reduction of nickel slag, *Journal of Iron and Steel Research International*, 27(11):1311–1321.

Li, X, Li, Y, Xing, X, Wang, Y, Wen, Z and Yang, H, 2021. Effect Of Particle Sizes Of Slag On Reduction Characteristics Of Nickel Slag-Coal Composite Briquette, *Archives of Metallurgy and Materials*, 66(1):127–134.

Li, X, Zhang, X, Zang, X and Xing, X, 2020b. Structure and Phase Changes of Nickel Slag in Oxidation Treatment, *Minerals*, 10(4).

Long, H, Chun, T, Di, Z, Wang, P, Meng, Q and Li, J, 2016. Preparation of Metallic Iron Powder from Pyrite Cinder by Carbothermic Reduction and Magnetic Separation, *Metals-Basel*, 6(4).

Marenych, O and Kostryzhev, A, 2020. Strengthening Mechanisms in Nickel-Copper Alloys: A Review, *Metals-Basel*, 10(10).

Park, J G, Eom, H S, Huh, W W, Lee, Y S, Min, D J and Sohn, I, 2011. A Study in the Thermodynamic Behavior of Nickel in the MgO-SiO2-FeO Slag System, *Steel Research International*, 82(4):415–421.

Shen, Y, Huang, Z, Zhang, Y, Zhong, J, Zhang, W, Yang, Y, Chen, M and Du, X, 2018. Transfer Behavior of Fe Element in Nickel Slag during Molten Oxidation and Magnetic Separation Processes, *Materials Transactions*, 59(10):1659–1664.

Sun, W, Li, X, Liu, R, Zhai, Q and Li, J, 2021. Recovery of Valuable Metals from Nickel Smelting Slag Based on Reduction and Sulfurization Modification, *Minerals*, 11(9).

Swinbourne, D and Kho, T S, 2012. Computational Thermodynamics Modeling of Minor Element Distributions During Copper Flash Converting, *Metallurgical and Materials Transactions B*, 43:823–829.

Wang, G J, Zhao, J X, Cao, G, Zheng, J H, Li, B, Mang, Y F, Cui, Y R and Zong, H X, 2023. Study of nickel smelting slag melting temperature regulation based on slag transfer and iron lifting, *The Chinese Journal of Nonferrous Metals*, 17 p. Available from: <http://kns.cnki.net/kcms/detail/43.1238.TG.20230906.1517.002.html> [Accessed: 7 September 2023].

Wang, G, Cui, Y, Li, X, Yang, S, Zhao, J, Tang, H and Li, X, 2020. Molecular Dynamics Simulation on Microstructure and Physicochemical Properties of FexO-SiO2-CaO-MgO-'NiO' Slag in Nickel Matte Smelting under Modulating CaO Concentration, *Minerals*, 10(2).

Wang, G, Cui, Y, Yang, J, Li, X, Yang, S, Zhao, J and Tang, H, 2021. Fe/SiO2 Ratio on the Properties, Microstructure and Fe-Containing Phases of Nickel Matte Smelting Slag, *Metallurgical and Materials Transactions B-Process Metallurgy and Materials Processing Science*, 52(3):1463–1471.

Wang, Q, Guo, X, Tian, Q, Chen, M and Zhao, B, 2017. Reaction Mechanism and Distribution Behavior of Arsenic in the Bottom Blowing Copper Smelting Process, *Metals-Basel*, 7(8):302.

Wang, S, Ni, W, Li, K, Wang, C and Wang, J, 2014. Effect of basicity on recovering iron, nickel and copper by deep reduction process of nickel slag pellets, *Transactions of Materials and Heat Treatment*, 35(9):23–28.

Wang, S, Ni, W, Li, K, Wang, C and Wang, J, 2015. Effect of carbide slag as additive on recovering iron by deep reduction process of nickel slag, *Transactions of Materials and Heat Treatment*, 36(12):7–12.

Wu, Q, Chen, Q, Huang, Z, Gu, B, Zhu, H and Tian, L, 2020. Preparation and characterization of porous ceramics from nickel smelting slag and metakaolin, *Ceramics International*, 46(4):4581–4586.

Wu, Q, Wu, Y, Tong, W and Ma, H, 2018. Utilization of nickel slag as raw material in the production of Portland cement for road construction, *Construction and Building Materials*, 193:426–434.

Xia, B D, Li, R F, Zhao, X Y, Dang, Q L, Zhang, D P and Tan, W B, 2018. Constraints and opportunities for the recycling of growing ferronickel slag in China, *Resources Conservation and Recycling*, 139:15–16.

Yang, W, Tian, S, Wu, J, Chai, L and Liao, Q, 2017. Distribution and Behavior of Arsenic During the Reducing-Matting Smelting Process, *JOM*, 69:1077–1083.

Yu, J, Qin, Y, Gao, P, Sun, Y and Ma, S, 2021. The Growth Characteristics and Kinetics of Metallic Iron in Coal-Based Reduction of Jinchuan Ferronickel Slag, *Minerals*, 11(8).

Zhang, G, Wang, N, Chen, M and Cheng, Y, 2020a. Recycling Nickel Slag by Aluminum Dross: Iron-extraction and Secondary Slag Stabilization, *ISIJ International*, 60(3):602–609.

Zhang, J, Qi, Y H, Yan, D L and Xu, H C, 2015. A New Technology for Copper Slag Reduction to Get Molten Iron and Copper Matte, *Journal of Iron and Steel Research International*, 22(5):396–401.

Zhang, T T, Zhi, S W, Guo, L J, Wu, Z L and Han, J N, 2020b. Research Progress of Resource Utilization of Copper-Nickel Smelting Slag, *Gold Science and Technology*, 28(05):637–645.

Zhao, J, Zhao, Z, Cui, Y, Shi, R, Tang, W, Li, X and Shang, N, 2018. New Slag for Nickel Matte Smelting Process and Subsequent Fe Extraction, *Metallurgical and Materials Transactions B-Process Metallurgy and Materials Processing Science*, 49(1):304–310.

Zhong, D P, Li, L and Tan, C, 2017. Separation of Arsenic from the Antimony-Bearing Dust through Selective Oxidation Using CuO, *Metallurgical and Materials Transactions B*, 48:1308–1314.

Lessons learned from attempts at minimising CO_2 emissions in process metallurgy – pyrolysed secondary raw materials, bio-coke, and hydrogen as alternative reducing agents

F Diaz[1], M Sommerfeld[2], G Hovestadt[2], D Latacz[2] and B Friedrich[2]

1. Group Leader Automation and Digital Transformation for Process Metallurgy; Institute of Process Metallurgy and Metal Recycling IME, RWTH Aachen University, 52056 Aachen, Germany. Email: fdiaz@ime-aachen.de
2. Institute of Process Metallurgy and Metal Recycling IME, RWTH Aachen University, 52056 Aachen, Germany.

ABSTRACT

The metallurgical sector significantly contributes to the global carbon footprint and encounters the challenge of developing more sustainable production methods. This study introduces novel approaches aimed at reducing CO_2 emissions within the industrial sector, some of which have been developed up to the demonstration scale.

In two case studies, urban waste from the agricultural sector (corn, olives, coconut etc) and waste electrical or electronic equipment (WEEE) recycling residues (shredder light fractions (SLF)) were subjected to thermal treatment. The resulting materials were then used as substitutes for conventional fossil-based reducing agents in both the copper and ferroalloy industries. Furthermore, hydrogen was employed at different scales to assess its efficacy as a reducing agent for recovering metal oxides from fayalitic copper slags.

Our findings reveal promising prospects and defined specific challenges for the integration of these alternative reducing agents in the industry. The synergistic use of pyrolysed SLF, bio-coke, and hydrogen presents a viable pathway to significantly diminish CO_2 emissions while simultaneously improving the sustainability of the metallurgical sector.

INTRODUCTION

The metallurgical industry produced approximately 1.9 billion tonnes (Bt) of steel, 69 Mio tonnes of aluminium and 22 Mio tonnes of copper in 2022 (US Geological Survey, 2023), contributing to roughly 5.2 per cent of the total anthropogenic CO_2 emissions together with other industries like cement and chemicals. In comparison to other industrial sectors, the metallurgical industry accounts for approximately 30 per cent of these emissions, closely competing with the cement industry (Malischek, Baylin-Stern and McCulloch, 2019).

Greenhouse gas emissions are a significant concern due to the climate crisis, especially within the metallurgical industry. The primary source of CO_2 emissions is iron production, specifically in blast furnaces, which employ carbothermal reduction, resulting in substantial direct CO_2 emissions. Furthermore, processes like pelletising, coking, sintering, and steelmaking also contribute to direct CO_2 emissions (Cavaliere, 2016). In addition to the direct CO_2 emissions generated during carbothermal reduction processes for metal production, there are other, less-discussed sources of CO_2 emissions that are also relevant. These include the production of metallurgical slags and the subsequent handling of such metallurgical waste materials, often involving high temperatures that require fuels or energy and carbothermal reduction, which requires reducing agents. This is especially significant, given that for instance in the copper sector, flash smelters can yield up to 3 t of slag per tonne of copper (Gorai, Jana and Premchand, 2003; Zhang, Zhang and Zheng, 2022), containing 1 per cent to 5 per cent copper by weight, equivalent to up to 10 per cent of the copper in the input material (Gorai, Jana and Premchand, 2003; Gonzalez et al, 2005; Roy, Datta and Rehani, 2015), thus post-handling of such a slag is ultimately required.

The European Union has already established significant targets and signed the 'Green Deal', aiming to achieve a net CO_2-free economy by 2050 (European Commission, 2021). This initiative creates a compelling demand for innovative process solutions within the ferrous and non-ferrous industry for the metal production and the handling of metallurgical slags.

In the context of circular economy and sustainable practices, there is extensive research underway to identify viable alternatives to fossil reducing agents. These alternatives can be broadly classified into two categories: alternative solid reducing agents sourced from 'waste' materials and alternative reducing gases. Figure 1 illustrates these alternative reducing materials in relation to their role in metallurgical applications. This, in the context of decarbonisation of the metallurgical sector.

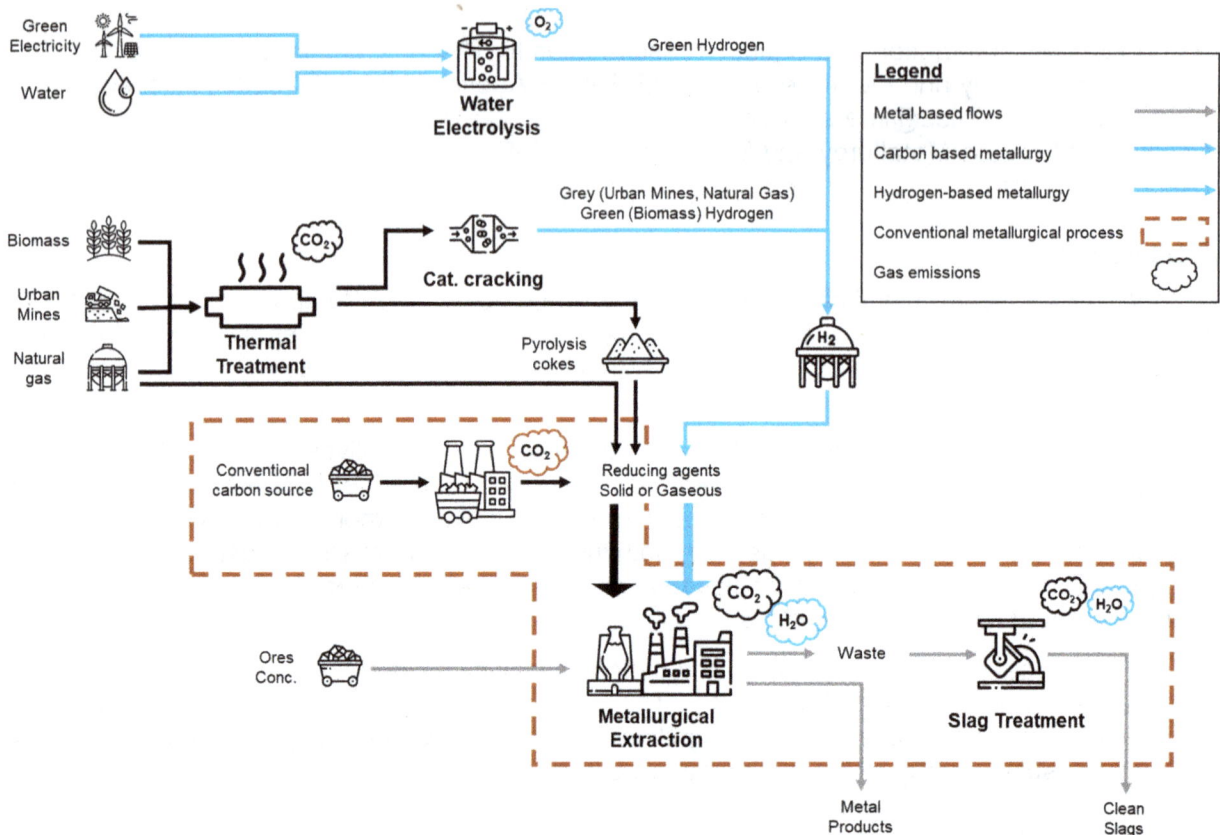

FIG 1 – Decarbonisation alternatives for the metallurgical industry.

Alternative solid reducing agents are produced through the thermal treatment or cracking of organic materials using methods such as pyrolysis, torrefaction, or hydrothermal carbonisation (Alamgir Ahmad *et al*, 2023). These processes entail elevated temperatures and in the case of pyrolysis, the absence of oxygen. During such a process, polymers or organic structures are broken down into smaller intermediate substances, including oil, gas, and solid (Alamgir Ahmad *et al*, 2023; Diaz, Latacz and Friedrich, 2023). The solid product of the cracking process can be referred to as 'Bio-coke' when derived from agricultural waste or 'Pyrolysed secondary raw material' if sourced in the context of urban mines, from complex organic anthropogenic materials such as plastics, textiles, or electronic scrap.

In the case of alternative reducing gases, hydrogen has emerged as a primary focus in recent years. However, to be sustainable, the method of hydrogen production also plays a crucial role. For example, hydrogen can be generated through energy-intensive electrolysis, wherein water molecules are split into hydrogen and oxygen. To produce 1 t of hydrogen direct reduced iron, approximately 50–54 kg (545 Nm^3) of hydrogen is needed, requiring between 2.5 to 3.5 MWh of energy for the water electrolysis (Vogl, Åhman and Nilsson, 2018; Patisson and Mirgaux, 2020). To put this in perspective, it is equivalent to the energy consumption of an individual in Germany over a span of two years. Therefore, it is imperative that hydrogen used for metallurgical applications is produced from renewable sources. In this case, this reducing gas can refer as 'green' hydrogen.

In addition to electrolysis, hydrogen can also be produced through the thermal treatment or cracking of organic material-rich substances such as biomass and materials from urban mines. These processes are used to generate bio-coke or to pyrolyse secondary raw materials. During such treatments, hydrogen is often produced alongside with carbon dioxide and other reducing gases like carbon monoxide, methane, and various hydrocarbons (Diaz, Latacz and Friedrich, 2023). This mixture, depending on its composition, is commonly referred to as 'syngas'. While syngas and natural gas share some similar characteristics, they are distinct in composition. Syngas, often derived from sources like coal or biomass, and natural gas can undergo different processing methods to produce hydrogen. When natural gas is catalytically cracked down to produce hydrogen, and the CO_2 byproduct is not captured, the hydrogen produced is typically referred to as 'grey hydrogen'. In contrast, hydrogen produced from biomass can be considered 'green hydrogen', especially when the biomass is sourced sustainably and the energy used in the process is from renewable sources (AlHumaidan *et al*, 2023).

This article aims to offer insights into the lessons learned, key characteristics, advantages, and the identification of essential comparative criteria in the utilisation of bio-coke, pyrolysed secondary raw materials, and hydrogen. These are considered as alternative resources to traditional fossil-based reducing agents in metallurgical applications.

CASE STUDY – PYROLYSED SECONDARY RAW MATERIALS AS REDUCING AGENT

The shredder light fraction (SLF) is a byproduct of the mechanical treatment of waste electrical and electronic equipment (WEEE), constituting approximately 4.2 per cent of the output materials and containing high concentrations of primary metals like copper, tin, lead, zinc, silver, and gold (Ueberschaar, 2017). The presence of valuable resources in SLF, coupled with rising disposal costs, drives pre-processing companies to seek more efficient approaches for its management. However, direct incorporation into conventional metallurgical processes introduces complexities. These include handling halogens from flame retardants, managing diverse composition, and addressing intricate morphology for charging into the smelters (Diaz, Latacz and Friedrich, 2023).

In this study, SLF was obtained from a local German electronic scrap processing company, consisting of fine particles including dust, rubble, biological fragments, fibres, and plastic sheets. It has relatively low metal content, mainly fine copper wires and electronic components. SLF is carbon-rich (40.6 per cent), with notable oxygen (17 per cent), nitrogen (5.73 per cent), and hydrogen (0.48 per cent). It also contains metals like copper (1 per cent), aluminium (3.34 per cent), silicon (7.51 per cent), and trace elements, resulting in a low lower heating value (LHV) of approximately 18.8 MJ*kg^{-1} (Diaz, 2020). During pyrolysis, SLF transforms into pyrolysis gas (P.Gas), pyrolysis oil and water (P.Oil), and a solid product (PSLF) with metals, oxides, and pyrolysis coke, experiencing a mass loss of about 40 per cent (Diaz, Latacz and Friedrich, 2023).

PSLF can be described as a concentrated version of metals found in SLF, characterised by a black powder containing pyrolytic carbon and oxides. It is highly brittle and contains notable elements, including oxygen (24 per cent), silicon (14.29 per cent), aluminium (6.14 per cent), and copper (3.3 per cent), at higher concentrations than SLF. Additionally, it includes trace amounts of various other elements, including precious metals (PMs). The LHV registered for PSLF was approximately 7 MJ·kg^{-1}. Further details about the research methods, SLF and PSLF characterisation, and the pyrolysis process applied to SLF from WEEE can be found in Diaz, Latacz and Friedrich (2023).

This case study explores an innovative recycling process for printed circuit boards (PCBs), utilising pyrolysis shredder light fraction (PSLF) as an alternative reducing agent to recover copper from oxidised slag. Figure 2 illustrates this process, which is aligned with the principles of a circular economy, converting waste streams–SLF and PCBs–into valuable resources, such as copper and energy.

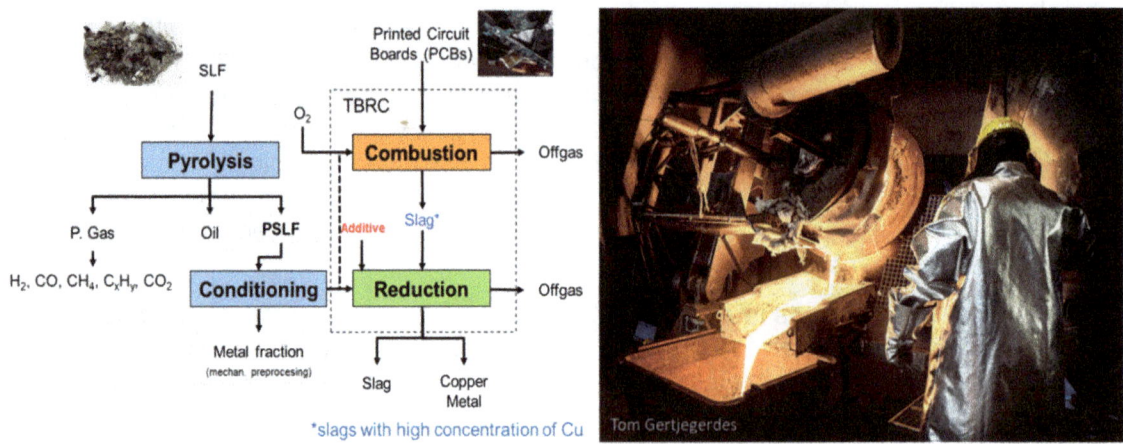

FIG 2 – Sustainable PCB recycling process using PSLF as a reducing agent (Diaz, 2020).

The recycling process commences with two primary inputs: SLF and PCBs. PCBs undergo an autothermal smelting, a combustion in the presence of oxygen that leverages the organic materials within the PCBs as an energy source. This results in the formation of copper-rich slag due to the oxidation of metallic components at high temperatures.

Simultaneously, SLF is processed through pyrolysis, yielding pyrolysis gas, oil, and PSLF. This carbon-rich PSLF, potentially conditioned to remove undesired metals, is then introduced into the smelter as a reducing agent. During the reduction stage, PSLF facilitates the chemical reduction of copper oxide in the slag, producing elemental copper.

The process yields three main outputs: off gas, which is managed to meet environmental standards; residual slag, which may contain non-metallic elements; and the recovered copper metal, now ready for subsequent refining and product development. By leveraging waste-derived materials for metal recovery, this process offers a more sustainable and cost-effective alternative to conventional recycling methods. The main advantage of this approach is that neither for smelting nor for the slag reduction are fossil fuels required.

Experimental work and main results

The experimental work involved the use of a demo scale top blown rotary converter (TBRC) for upscaling. The TBRC located at IME Institute of the RWTH Aachen University in Germany, is a cylindrical reactor with a 240 L volume, capable of processing up to 100 L of molten material. It utilises an Oxyfuel burner that uses natural gas and pure oxygen for heating, providing the advantage of adjustable lambda (λ 0.7–1.3) to control the atmosphere inside the reactor during experiments. The TBRC can rotate at speeds ranging from 0 to 10 rev/min and tilt at angles from 0 to 110°, making it versatile for various applications. It is lined with Cr-magnesite (MgCr) material, suitable for copper-based materials and corrosion-resistant up to 1600°C.

The TBRC includes an off gas cleaning system with two main pathways: hygiene gas and process gas. The system has a maximum suction capacity of 10 000 Nm³/h and utilises fabric filters and electrostatic filters for dust collection. The off gases are transported to a scrubber with a pH of 10 for removal of halogens and heavy metals like As, Cd, Pb, among others, before being released into the environment.

In the trial, 120 kg of Cu-slag was charged into the preheated TBRC at around 800°C. The furnace was heated to 1350°C using the Oxyfuel burner, and the TBRC was set to rotate at 2 rev/min. The reduction experiment occurred in two phases: manual charging and injection of the PSLF. Copper concentration in the slag was monitored throughout the trial.

The stoichiometric amount of carbon needed to reduce copper oxide in the slag was calculated, resulting in the manual charging of PSLF in three periods: 72.5 per cent, 100 per cent, and 145 per cent of the required amount. Rotation speed was increased to 8 rev/min during manual charging to enhance turbulence. After manual charging, the injection of PSLF took place using the same molten slag. The idea was to evaluate the effect of injection and obtain some insights on the kinetics using PSLF as reducing agent.

During the injection phase, the TBRC was placed vertically, rotation speed initially set to 2 rev/min, and PSLF injected at a speed of ~2.75 kg/min using nitrogen as the carrier gas. The first injection period involved injecting 50 per cent of the required PSLF, and subsequent injection periods varied the quantity. After each injection, a gas flushing period occurred. The second injection set-up used a lower λ value to enhance reduction conditions and included an extra stirring period. More details about the experimental set-up can be found in (Diaz, 2020).

The results indicated that manual charging led to poor copper reduction efficiency (7.1 per cent) due to inadequate turbulence and surface burning. On the other hand, injection trials achieved up to 48 per cent efficiency during the first injection period. By optimising PSLF dosification and extending stirring periods, the efficiency increased to 82 per cent, using PSLF equivalent to 93 per cent of the required amount. Besides the outstanding performance, further PSLF charging was hindered by technical problems and clogging of the injector device caused by very small Cu-wires present in PSLF.

Figure 3 illustrated the relationship between Cu reduction efficiency and C:CuO ratio in the system based on the art of charging in the TBRC, showing that process efficiency with manual charging was nearly cero under the tested conditions. Injection technology and submerged gas stirring were identified as crucial factors for improving the utilisation of pyrolysed materials as reducing agents for Cu-slags.

Validation Trials in Demo-scale

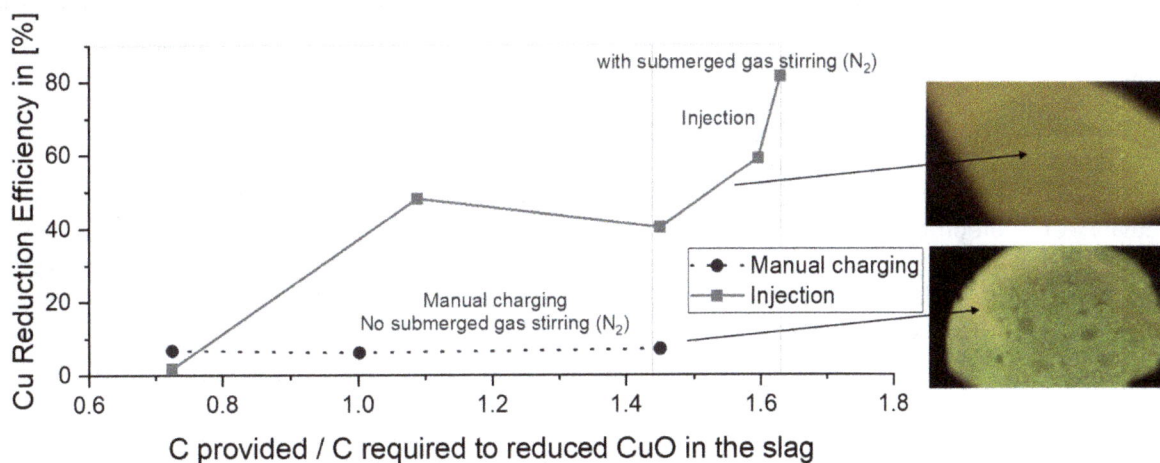

FIG 3 – Cu reduction efficiency versus C:CuO ratio in the system according to the art of charging in the TBRC (Diaz, 2020).

Similar experiments were conducted for other accompanying elements, including Ag, Pb, Zn, Ni, Sn, and Sb. Manual charging resulted in poor reduction efficiencies for these elements, but injection and submerged gas stirring significantly improved their reduction efficiencies. For example, Sb reached 100 per cent reduction, Sn achieved 63 per cent, Ag 100 per cent, Pb 75 per cent, Zn 53 per cent, and Ni 17 per cent.

Thermochemical simulations (FactSage[TM]) identified four key reduction stages when using PSLF as a reducing agent for copper-rich slags: initial copper reduction, metallothermic reduction involving less noble metals, a copper co-reduction phase with metals like Ni, Sn, and Pb, and eventual iron reduction. The limited copper reduction efficiency did not result from chemical limitations but rather from challenges related to melt handling. Future improvements may involve adapting the injector to handle small wires in PSLF or enhancing mechanical conditioning of PSLF to remove some undesired metals.

In the slag system, the addition of PSLF would increase viscosity due to its SiO_2 content, primarily present in glassy structures from WEEE. A suggested slag design included the addition of CaO and

Na_2O to improve viscosity and reduce liquidus temperature. A more detailed publication on the slag design considering PSLF as reducing agent for copper slags is currently under review.

CASE STUDY – BIO-COKE AS REDUCING AGENT

The production of metals, including ignoble ones like chromium, manganese, and silicon, has long relied on submerged arc furnaces, with fossil carbon serving as the primary reducing agent. However, this conventional practice contributes significantly to direct CO_2 emissions, urging the exploration of more sustainable alternatives. Bio-based carbon or Bio-coke has emerged as a promising candidate to replace fossil carbon in metallurgical processes.

Bio-coke, in contrast to traditional metallurgical coke, exhibits distinct proximate analysis characteristics. It is characterised by a lower fixed carbon content and a higher volatile matter content, making it more responsive to temperature changes during high-temperature processes. Additionally, bio-coke boasts a lower ash content and a higher moisture content, which impacts its combustion and reactivity properties (Sommerfeld and Friedrich, 2021).

Accompanying elements in bio-coke also diverge from metallurgical coke, with higher alkaline, chlorine, and phosphorus content, while sulfur content remains notably lower. Furthermore, the ash produced by bio-coke is characterised by increased acidity (Sommerfeld and Friedrich, 2021).

Bio-coke's physical properties further differentiate it, featuring a lower density, higher CO_2 reactivity, and somewhat inferior mechanical stability. More details on general characteristics of bio-coke can be found in the review (Sommerfeld and Friedrich, 2021).

Bio-coke, derived from sources such as coconut, corn, olive, and bamboo, is an emerging carbonaceous material that has been rigorously evaluated in comparison to conventional coke. This exploration has opened doors to its potential applications in metallurgy, particularly in the production of environmentally sustainable ferroalloys (Sommerfeld and Friedrich, 2021).

In the area of metallurgy, bio-coke exhibits promise in several key processes. As it is shown in Figure 4, Its versatility allows for applications in pre-reduction, agglomeration, and direct smelting, each demanding specific attribute such as fixed carbon content, mechanical stability, purity, reactivity, and electrical conductivity. By meeting these varied requirements, bio-coke has the potential to drive the development of greener and more eco-friendly ferroalloy production methods, contributing to the reduction of carbon emissions and the promotion of sustainable practices in the metallurgical industry (Sommerfeld and Friedrich, 2021).

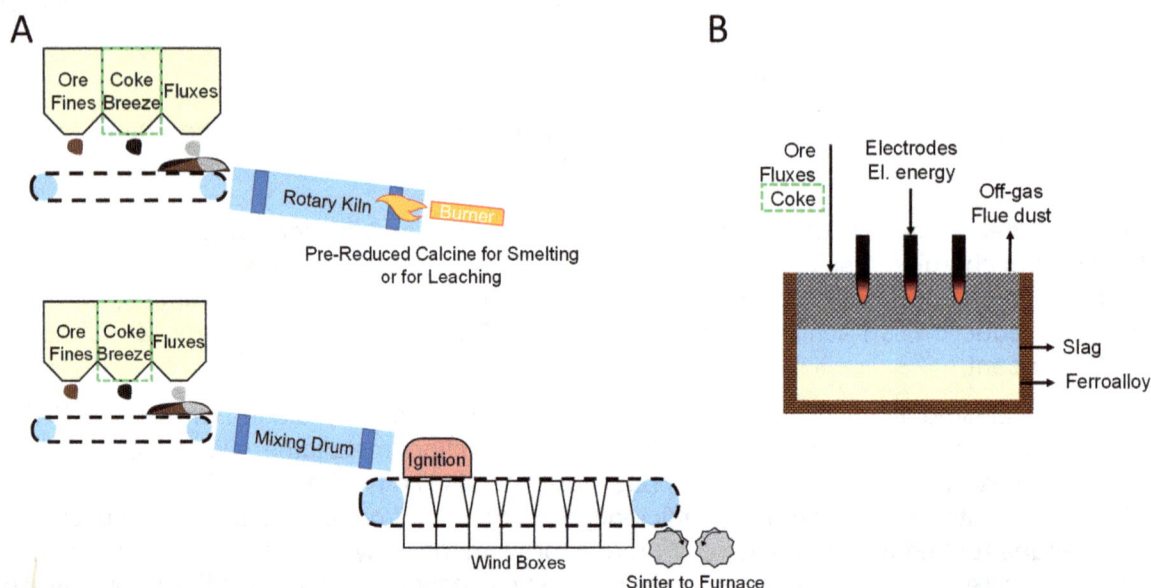

FIG 4 – (a) Prereduction and agglomeration using bio-coke in solid state, (b) bio-carbon for ferroalloy production via submerged electric arc furnaces.

Experimental work and main results

Experiences gathered in this case study are focused on two primary aspects. In the first stage, the study explores the pre-reduction of iron ore containing chromium in solid state. In the second stage, it involves a comparison of melting operations using pre-reduced ore and co-smelting with bio-coal (biomass).

During the initial phase, laboratory-scale experiments were conducted to investigate the pre-reduction process using bio-coke as an alternative reducing agent. A resistance heating furnace was employed for these trials, with each starting with an initial sample mass of 35 g. Different combinations of ore and reductants were explored at varying ratios. Mass loss measurements following each trial assessed chemical reactions, and individual heating of ore and reductants helped isolate their respective mass losses. Samples were charged in 99.7 per cent pure alumina crucibles (50 mm diameter, 75 mm height) and continuously flushed with 5 Nl/min of argon to prevent oxidation. The 16 litre volume furnace chamber accommodated a maximum of six crucibles, ensuring uniform sample temperature. The experiments varied the maximum temperature and duration at that temperature, with a 3.5 hr heating process and uncontrolled cooling. Results are illustrated in Figure 5.

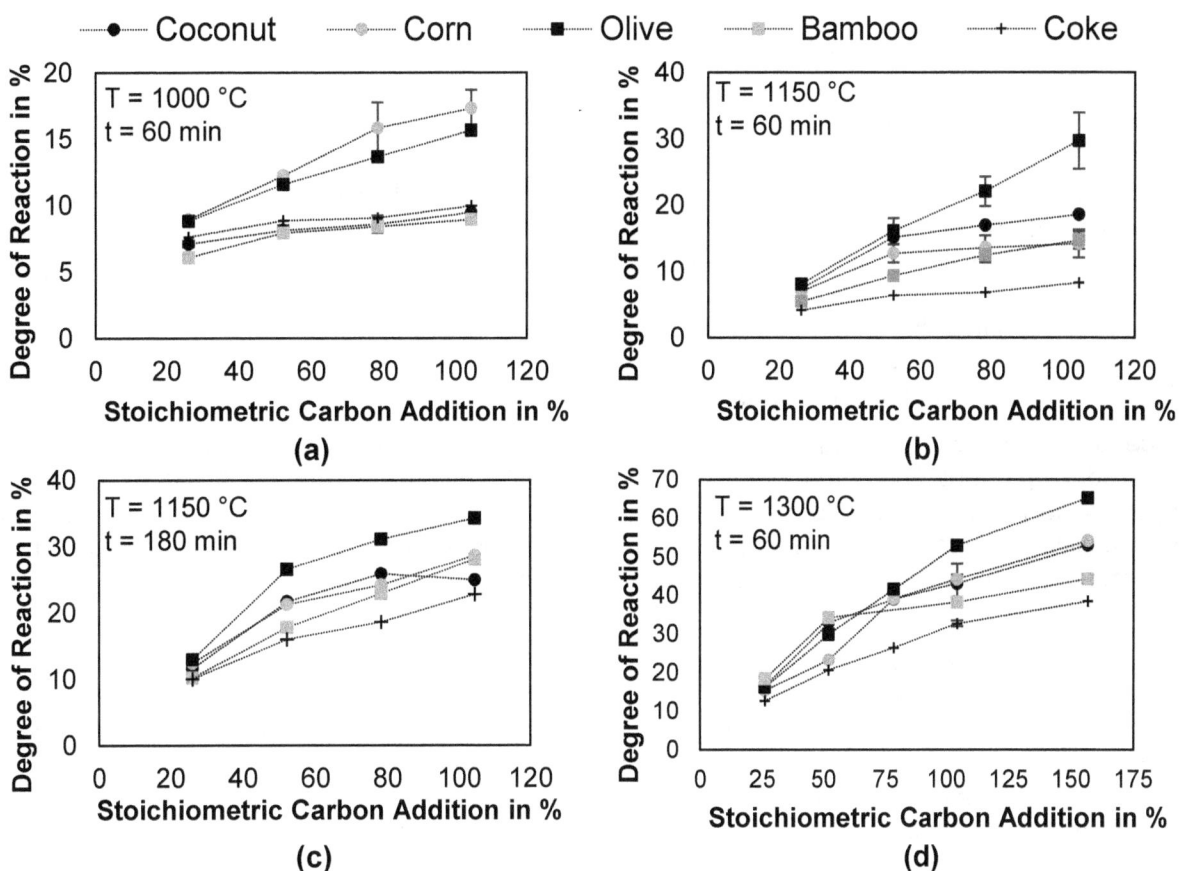

FIG 5 – Degree of reaction dependent on the carbon addition for various reducing agents and temperatures; (a) T = 1000°C, t = 60 min; (b) T = 1150°C, t = 60 min; (c) T = 1150°C, t = 180 min; (d) T = 1300°C, t = 60 min (Sommerfeld and Friedrich, 2022).

Figure 5 provides a comprehensive comparison of various carbonaceous materials as reducing agents for chromite pre-reduction, categorised by operational temperatures. The degree of reaction on the graph represents the portion of carbon involved in reducing the chromite ore.

At the lower temperature of 1000°C, bio-carbons like corn exhibit a moderate degree of reaction after 60 mins, reaching up to 17.4 per cent. Olive, another bio-carbon, demonstrates similar modest reactivity at this temperature. Coke, Coconut and Bamboo performed poorly at this temperature with less than 10 per cent degree of reaction.

Elevating the temperature to 1150°C significantly alters the reaction landscape. Within a 60 min window, all materials, particularly Olive and Coconut, show increased degrees of reaction. Coke as reference stands out with the lowest degree of reaction, indicating low reactivity within this intermediate temperature range and conditions.

Prolonging the duration at 1150°C to 180 mins incrementally raises the degree of reaction for all materials. Although olive's performance improves slightly, it suggests a potential plateau in reactivity at this temperature. Coke improved the degree of reaction, indicating a poor kinetic performance compared to bio-coke at these conditions.

The most striking results are observed at the upper threshold of 1300°C. Here, bio-carbons, while still exhibiting increased reactivity for all materials, are surpassed by Olive, reaching around 60 per cent of the degree of reaction. Coke performs the worst, reaching only 30 per cent of the degree of reaction.

In the second stage of the investigation, various reducing agents were employed for ferrochrome production, involving pre-reduction for different durations and operations without pre-reduction. Notably, both coke and bio-coal (biomass) yielded comparable chromium output during the smelting process. However, the use of bio-coal led to phosphorus content in green ferrochrome exceeding ASTM and DIN standards, whereas lignite coke met these standards. Sulfur content remained consistent regardless of the chosen reducing agent.

Thermochemical simulations were conducted to assess relative energy demand and off-gas production. The base-case scenario involved the use of unreduced material with coke. The results indicated an increase in specific energy demand when bio-based carbon was employed. Nevertheless, pre-reduction effectively mitigated the heightened energy demand associated with bio-based carbon, resulting in reduced energy requirements for the smelting process, thanks to enhanced pre-reduction. A similar trend was observed for off-gas production, with bio-based carbon contributing to increased emissions, but pre-reduction successfully mitigated this effect. It is important to note that the details of this work are currently under review.

CASE STUDY – HYDROGEN AS REDUCING AGENT FOR COPPER SLAGS

The central objective is to produce copper-steel with inherent antibacterial properties using hydrogen as the reducing agent. The primary focus is on efficiently reducing valuable metal oxides, particularly copper, while minimising the formation of undesirable by-products. Environmental and economic sustainability considerations are essential in this study, encompassing evaluations of energy consumption, emissions, and the feasibility of scaling up the process for potential industrial applications.

Research into copper reduction from slags to produce Copper containing steel alloys has primarily utilised carbon carriers like coal and natural gas. Busolic et al (2011) demonstrated a two-stage reduction process, achieving significant copper recovery (up to 51 per cent) from slags initially containing copper oxide and copper sulfide. Gonzalez et al (2005) employed carbothermal reduction, resulting in a ferrous metal phase with high purity. Sarfo et al (2017) and Zhang et al (2018) explored carbothermal and natural gas reduction processes, respectively, obtaining varying copper content in the final slag. Blenau, Stelter and Charitos (2021) investigated graphite injection for improved kinetics, yielding copper-rich metal alloys. These studies underscore the potential for selective copper reduction within iron recovery processes.

Hydrogen-based reduction processes show promise in extracting valuable metals from slags. Qu et al (2020) investigated synthesis gas rich in hydrogen (70 per cent) with copper-sulfur slag, resulting in the formation of a significant copper-rich matte phase. Zhang, Zhang and Zheng (2022) reduced low-grade copper slag using hydrogen in the solid-state, achieving a metal phase with 10.4 wt per cent copper. These studies introduce hydrogen treatment as a viable option for processing low copper slags, offering good insights for the current research.

In Figure 6, the process concept is introduced, aiming to utilise hydrogen as a reducing agent for treating slags generated during sulfur burning. The objective is to eliminate remaining sulfur from the slag and reduce copper oxide to produce a copper-containing steel product (Hovestadt and Friedrich, 2022).

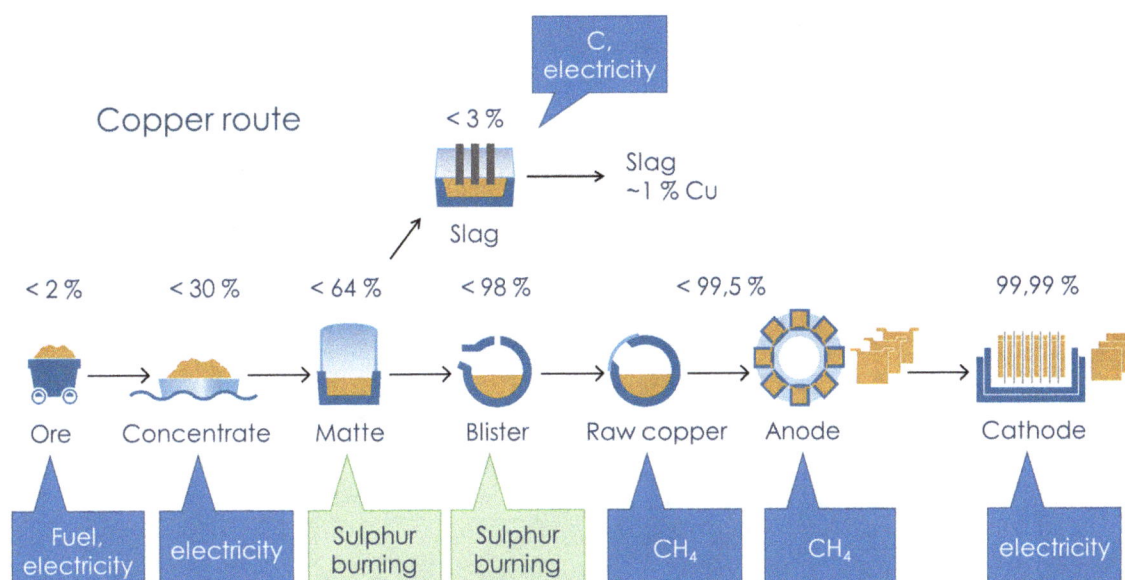

FIG 6 – Hydrogen based-reduction process for Cu-slag cleaning (Hovestadt and Friedrich, 2022).

The project, funded by the European Union under the name HARARE, aims to explore the potential of hydrogen-based reduction processes. It focuses on two residue streams from the metallurgical industry, with the goal of advancing the technology readiness level (TRL) from laboratory scale to potentially extract important technology metals in the future. This project is a collaborative effort led by Sintef AS, involving six European universities and research institutes, along with the assistance of four companies.

In the context of this project, one of the research directions is the reduction of primary copper slags, which is the subject of this report. The objective is to scale up the process from handling 1 kg to up to 3 t of slag melt. The approach involves a two-step reduction process. In the first step, the focus is on achieving a copper-rich metal phase, followed by a second step to obtain a copper-poor iron phase and valuable metal-free slag. The primary emphasis is placed on the selectivity of the first reduction stage.

In their 2022 study, Hovestadt and Friedrich, highlight the importance of safety during the injection of molten materials, emphasizing the need to monitor several precautions and parameters. This includes maintaining a temperature above a 1000°C for direct ignition. The protocol also involves post-combusting hydrogen before initiating the cooling process in a quenching box. Another key safety measure is to avoid cold mixtures containing less than 4 vol per cent hydrogen. Lastly, it is essential to prevent the formation of explosive mixtures outside the melt at temperatures exceeding 560°C.

Experimental work and main results

In this study, the experimental set-ups range from lab-scale to demo-scale. The conditions for both set-ups are presented as follows. However, the main focus of this report is on the demo-scale experiments.

The lab-scale (Figure 7a) equipment includes a resistant heated furnace, refractory lining, ceramic lance, and has a temperature capability of 1300°C. It can process up to 1.5 kg of input material with a volume flow of 2 L/min of hydrogen-reducing gas, equivalent to 4.66 mins per melt volume.

On the other hand, demo-scale (Figure 7b) operations handle 400 to 500 kg of slag, using an oxyfuel burner for heating and injecting 30 Nm³/h of gas, with hydrogen concentrations ranging from 50 to 100 vol per cent. A ceramic shielded steel lance is employed in this larger-scale set-up. The melting unit dimensions and off-gas cleaning system details of the demo-scale plant remain consistent with those presented in the section titled 'Case study – pyrolysed secondary raw materials as a reducing agent'.

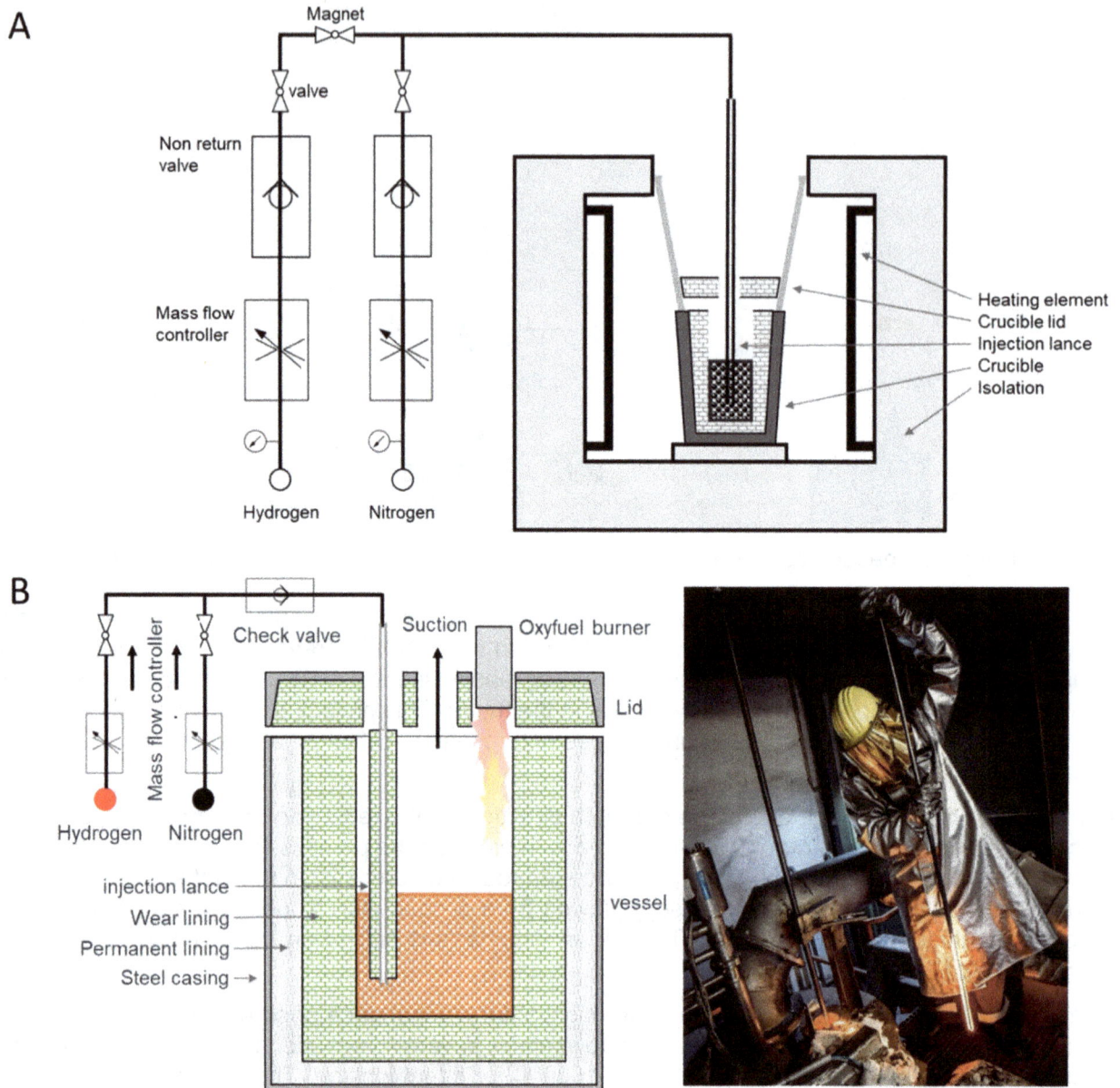

FIG 7 – (a) lab-scale set-up; (b) demo-scale set-up, for H_2 reduction of Cu-slags.

Figure 8 depicts the trends in metal concentration during a two-stage hydrogen reduction process. The results in the left figure indicate a decrease in copper and lead concentrations, both diminishing as the hydrogen volume increases. Copper begins at nearly 1 per cent by weight and falls sharply in the initial stage, then stabilises in the second, suggesting that most copper reduction occurs after 90 L of H_2 per kg of slag. Lead follows a similar pattern, starting below 0.5 per cent by weight and decreasing to almost negligible levels, which implies a significant impact of hydrogen in reducing lead content during these stages.

Regarding zinc, its concentration starts above 4 per cent by weight and exhibits fluctuations with a slight increase as the hydrogen volume rises per cent, likely due to the reduction of other metals, but then decreases to below 1 per cent. However, the second stage shows a clear downward trend in zinc concentration as the hydrogen volume increases. Iron, on the other hand, displays a contrasting behaviour; its concentration increases more steadily in the first stage and even more so in the second, indicating that iron is more reactive to the process as the hydrogen volume rises and copper reduction takes place. Overall, the process demonstrates the differential reactivity of metals during hydrogen reduction, with increasing iron concentrations in the slag, while copper, zinc, and lead are reduced, reflecting their distinct affinities and reaction kinetics with hydrogen.

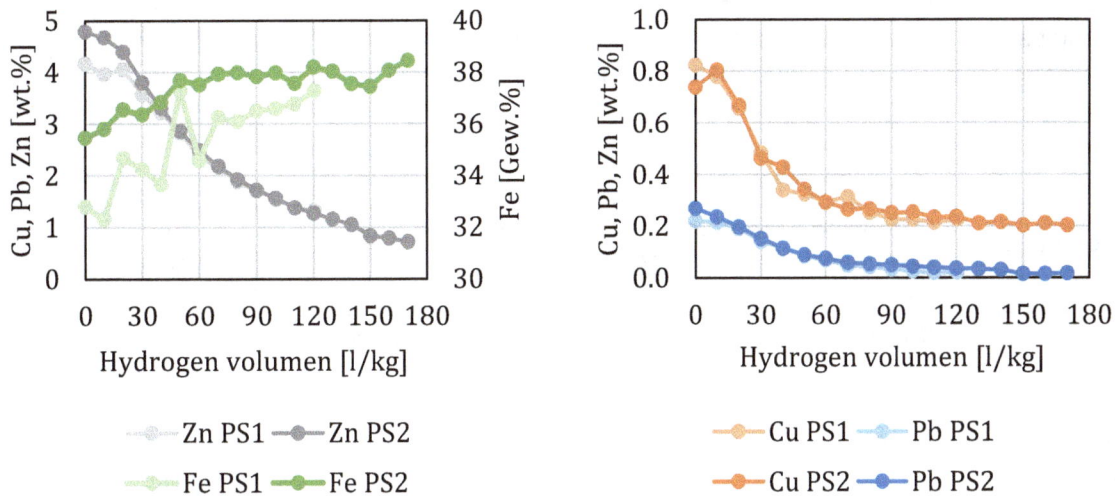

FIG 8 – Composition trends of pilot experiments based on the amount of hydrogen (1300°C, 90 vol H2, 30 Nm³/h and 1350°C, 100 vol H2, 36 Nm³/h).

Table 1 indicates the overall efficiencies and end concentrations of metals in the two-stage hydrogen reduction process. As it can be seen, for copper (Cu), the recovery efficiency in the first stage is 74 per cent, with a final concentration of 0.22 wt per cent. The efficiency slightly increases to 75 per cent in the second stage, with the final concentration remaining at 0.22 wt per cent. This suggests a stable end concentration of copper across both stages.

TABLE 1

Metal recovery efficiencies and final concentrations in a two-stage hydrogen reduction process.

Element	Yield PS1 in% (Process Stage 1)	Final Content PS1 in wt%	Yield PS2 in% (Process Stage 2)	Final Content PS2 in wt%
Cu	74	0.22	75	0.22
Zn	70	1.32	86	0.72
Pb	91	0.02	94	0.02

Zinc (Zn) shows a recovery efficiency of 70 per cent in PS1 with a final concentration of 1.32 wt per cent. There's a notable increase in efficiency to 86 per cent in PS2, accompanied by a decrease in final concentration to 0.72 wt per cent. This indicates a more effective reduction of zinc in the second stage, leading to a lower residue concentration.

Lead (Pb) has a high recovery efficiency of 91 per cent in the first stage, with a very low final concentration of 0.02 wt per cent. In the second stage, the efficiency slightly improves to 94 per cent, maintaining the final concentration at 0.02 wt per cent. The consistently low final content of lead suggests a highly effective reduction process across both stages.

The data reflect the varying responsiveness of different metals to the hydrogen reduction process. The process is especially effective for lead, moderately so for zinc, with the highest reduction occurring in the second stage, and least effective for copper, which maintains a consistent concentration post-reduction after the first stage.

Table 2 provides a comprehensive comparative overview of three alternative reducing agents evaluated in this study: pyrolysed secondary raw materials (Pyrolysed SLF), Bio-coke, and Hydrogen. It highlights their source materials, production processes, carbon footprints, energy efficiency, and metal recovery efficiencies. Pyrolysed SLF is derived from WEEE recycling by-products, Bio-coke from agricultural waste, and Hydrogen typically from water through electrolysis or steam reforming. The table assesses each agent's carbon footprint, energy efficiency, metal recovery efficiency, final metal purity, and process compatibility, revealing distinct advantages for each. The sustainability potential, reached TLR (Technology Readiness Level), economic viability, and technical challenges are discussed, along with their environmental impact in terms of reducing landfill waste, CO_2 emissions, and fossil fuel dependence.

TABLE 2
Comparative of tested alternative reducing agents.

Comparative factors	Pyrolysed secondary raw materials (Pyrolysed SLF)	Bio-coke	Hydrogen
Source material	WEEE recycling by-products	Agricultural waste (corn, olives etc)	Water (via electrolysis/steam reforming)
Production process	Pyrolysis	Pyrolysis, torrefaction, or hydrothermal carbonisation (from biomass)	Electrolysis/Steam Reforming
Carbon footprint	Low to moderate (depends on the process and energy source for pyrolysis)	Low (considering biomass growth absorbs CO_2)	Low to zero (if produced using renewable energy or biomass)
Energy efficiency	Variable (dependent on pyrolysis efficiency, organic and metal content)	High (bio-coke can have high energy content)	High (especially at high temperatures)
Metal recovery efficiency	High for copper and other primary metals	Moderate to high (depending on the metal)	High for iron, variable for others (for Cu < 1 wt per cent)
Final metal purity	High (after metal recovery process)	Variable (affected by ash and other biomass constituents)	High (pure H2 can reduce metals without contaminating)
Process compatibility	Compatible with smelting processes	Suitable for use in furnaces during smelting and potentially in direct reduced iron (DRI) processes	Suitable for both smelting and DRI processes
Sustainability potential	High (turns waste into resource)	High (renewable and carbon-neutral potential)	High (if produced from renewable sources)
Technology readiness level (TLR)	Up to demonstration scale (TLR 5)	Laboratory (TLR 3-4)	Up to demonstration scale (TLR 5)
Economic viability	Moderate (influenced by the value of recovered metals)	Moderate to high (cost-effective if sourced from waste biomass)	High (long-term savings from potential carbon credits and lower emissions)
Technical challenges	Handling halogens, diverse composition, intricate morphology	Lower fixed carbon content, higher volatile matter content, ash composition, high phosphorus content	Infrastructure for hydrogen production and handling
Environmental impact	Reduction in landfill waste, possible emission control challenges	Reduction in direct CO_2 emissions, beneficial use of agricultural waste	Significant potential to reduce CO_2 emissions and fossil fuel dependence

CONCLUSIONS

In conclusion, the search for an optimal approach to minimise CO_2 emissions in the metal production sector underscores the necessity for diverse solutions. As evolving requirements for steel and copper slag metallurgy necessitate alternative carbon sources by 2050, bio-coke and pyrolysed materials from the urban mine sector could play a crucial role. These materials align with the industry's shift towards sustainability, offering pathways to reduce the carbon footprint of the metallurgical industry. However, relying solely on bio-coke for metal production poses technical and logistical challenges, primarily due to limited biomass availability in the required quality, for instance low phosphorous content, in contrast to substantial carbon demand. An integrated approach that combines multiple strategies and materials is required to address this limitation. In this dynamic landscape, the pursuit of sustainable practices remains a primary goal in the metallurgical industry's journey towards a greener and more environmentally friendly future. Transitioning towards sustainable energy resources such as pyrolysed secondary raw materials, bio-coke, and hydrogen presents a promising pathway towards reducing environmental impact and enhancing energy efficiency. Given their varying technological readiness levels and environmental benefits, it is crucial to prioritise further research and development, particularly in advancing production processes and improving economic viability. Fundamental analysis on the exergy distribution of the studied systems through their incorporation in the metal production chain, scaling up technologies to improve the purity of recovered metals in pyrolysis, and enhancing the energy content of bio-coke and pyrolysis gas are key areas that could significantly impact their market readiness and adoption. Moreover, investment in infrastructure to support widespread hydrogen production and utilisation will be essential. Embracing these technologies not only supports waste reduction and resource recovery but also aligns with global efforts to mitigate climate change and foster a sustainable future. Policymakers and industry leaders should consider these factors in their strategic planning to ensure a balanced and effective transition to greener energy alternatives.

ACKNOWLEDGEMENTS

The authors would like to express their gratitude to their technical staff for their support in this publication. This investigation was partially funded by the European Union's Horizon 2020 research and innovation program, under grant agreement number 958307. The work also received support from the collaborative project RessourcenKolleg.NRW, funded by the Ministry of Science of North Rhine-Westphalia.

REFERENCES

Alamgir Ahmad, K, Ahmad, E, Al Mesfer, M K and Nigam, K D, 2023. Bio-coal and bio-coke production from agro residues, *Chemical Engineering Journal*, 473:145340. https://doi.org/10.1016/j.cej.2023.145340

AlHumaidan, F S, Absi Halabi, M, Rana, M S and Vinoba, M, 2023. Blue hydrogen: Current status and future technologies, *Energy Conversion & Management*, 283:116840. https://doi.org/10.1016/j.enconman.2023.116840

Blenau, L, Stelter, M and Charitos, A, 2021. Carbothermic Reduction of Fayalitic Slag with Graphite – Understanding Reaction Kinetics for Pig Iron Production, *Slag Valorisation Symposium 7*, pp 63–67 (Leuven).

Busolic, D, Parada, F, Parra, R, Sanchez, M, Palacios, J and Hino, M, 2011. Recovery of iron from copper flash smelting slags, *Mineral Process Extractive Metall*, 120(1):32–36. https://doi.org/10.1179/037195510X12772935654945

Cavaliere, P, 2016. *Ironmaking and Steelmaking Processes* (Springer International Publishing: Cham).

Diaz, F, 2020. Process concept based on pyrolysis for integration of Shredder light fractions (SLF) in the Recycling of Waste Electrical and Electronic Equipment (WEEE), PhD thesis, RWTH Aachen University, IME publication series, 162 p.

Diaz, F, Latacz, D and Friedrich, B, 2023. Enabling the recycling of metals from the shredder light fraction derived from waste of electrical and electronic equipment via continuous pyrolysis process, *Waste Management (New York, NY)*, 172:335–346. https://doi.org/10.1016/j.wasman.2023.11.001

European Commission, 2021. 'Fit for 55': delivering the EU's 2030 Climate Target on the way to climate neutrality, COM(2021) 550 final, European Economic and Social Committee. Available from: <https://www.eesc.europa.eu/en/our-work/opinions-information-reports/opinions/fit-55-delivering-eus-2030-climate-target-way-climate-neutrality>

Gonzalez, C, Parra, R, Klenovcanova, A, Imris, I and Sanchez, M, 2005. Reduction of Chilean copper slags: a case of waste management project, *Scandinavian Journal of Metallurgy*, 34(2):143–149. https://doi.org/10.1111/j.1600-0692.2005.00740.x

Gorai, B, Jana, R K and Premchand, 2003. Characteristics and utilisation of copper slag—a review, *Resources, Conservation and Recycling*, 39(4):299–313. https://doi.org/10.1016/S0921-3449(02)00171-4

Hovestadt, G and Friedrich, B, 2022. Hydrogen reduction of entrapped metal oxides from fayalitic copper slags, *Conference Copper 2022*, Santiago de Chile, November.

Patisson, F and Mirgaux, O, 2020. Hydrogen Ironmaking: How It Works, *Metals*, 10(7):922. https://doi.org/10.3390/met10070922

Qu, G, Wei, Y, Li, B, Wang, H, Yang, Y and McLean, A, 2020. Distribution of copper and iron components with hydrogen reduction of copper slag, *Journal of Alloys and Compounds*, 824:153910. https://doi.org/10.1016/j.jallcom.2020.153910

Malischek, R, Baylin-Stern, A and McCulloch, S, 2019. Transforming Industry Through CCUS, International Energy Agency. Available from: <https://www.iea.org/reports/transforming-industry-through-ccus>

Roy, S, Datta, A and Rehani, S, 2015. Flotation of copper sulphide from copper smelter slag using multiple collectors and their mixtures, *International Journal of Mineral Processing*, 143:43–49. https://doi.org/10.1016/j.minpro.2015.08.008

Sarfo, P, Das, A, Wyss, G and Young, C, 2017. Recovery of metal values from copper slag and reuse of residual secondary slag, *Waste Management (New York, NY)*, 70:272–281. https://doi.org/10.1016/j.wasman.2017.09.024

Sommerfeld, M and Friedrich, B, 2021. Replacing Fossil Carbon in the Production of Ferroalloys with a Focus on Bio-Based Carbon: A Review, *Minerals*, 11(11):1286. https://doi.org/10.3390/min11111286

Sommerfeld, M and Friedrich, B, 2022. Toward Green Ferroalloys: Replacement of Fossil Reductants in the Pre-reduction Process of Chromite by Bio-Based Alternatives, in *REWAS 2022: Developing Tomorrow's Technical Cycles* (Eds: A Lazou, K Daehn, C Fleuriault, M Gökelma, E Olivetti and C Meskers), vol I, pp 607–619 (Springer International Publishing: Cham).

US Geological Survey 2023. Mineral commodity summaries 2023. Reston, VA, Mineral Commodity Summaries 2023. https://doi.org/10.3133/mcs2023

Ueberschaar, M, 2017. Assessing recycling strategies for critical raw materials in waste electrical and electronic equipment, Dissertation Paper, flieger Verlag.

Vogl, V, Åhman, M and Nilsson, L J, 2018. Assessment of hydrogen direct reduction for fossil-free steelmaking, *Journal of Cleaner Production*, 203:736–745. https://doi.org/10.1016/j.jclepro.2018.08.279

Zhang, B, Zhang, T, Niu, L, Liu, N, Dou, Z and Li, Z, 2018. Moderate Dilution of Copper Slag by Natural Gas, *JOM*, 70(1):47–52. https://doi.org/10.1007/s11837-017-2670-6

Zhang, B, Zhang, T and Zheng, C, 2022. Reduction Kinetics of Copper Slag by H2, *Minerals*, 12(5):548. https://doi.org/10.3390/min12050548

A short review – hydrogen reduction of copper-containing resources

D M Fellicia[1], M I Pownceby[2], S Palanisamy[3], R Z Mukhlis[4] and M A Rhamdhani[5]

1. PhD Student, Swinburne University of Technology, Melbourne Vic 3122.
   Email: dfellicia@swin.edu.au
2. Senior Principal Research Scientist, CSIRO Mineral Resources, Melbourne Vic 3169.
   Email: mark.pownceby@csiro.au
3. Director of Manufacturing Futures Research Platform, Swinburne University of Technology, Melbourne Vic 3122. Email: spalanisamy@swin.edu.au
4. Lecturer, Swinburne University of Technology, Melbourne Vic 3122.
   Email: rmukhlis@swin.edu.au
5. Director Fluid and Process Dynamics, Swinburne University of Technology, Melbourne Vic 3122. Email: arhamdhani@swin.edu.au

ABSTRACT

The trend of global copper production has prospectively increased over time. Based on typical mined ore grades, 1 t of copper ore generates approximately 6–10 kg of copper, which requires much energy, usually in the form of metallurgical coke. Copper production using carbon as a fuel and reductant contributes up to 0.3 per cent to global greenhouse gas (GHG) emissions. As a result, research into decarbonisation processes applicable to reducing low-grade copper sulfide ores, copper oxides, electronic wastes and other alternative complex Cu-rich materials has increased. Alternative fuels for reducing carbon-rich emissions in pyrometallurgical copper processing include methane, ammonia, biomass, solar and wind power, but in recent times increased focus has been on hydrogen as a potential fuel and reducing agent for materials such as oxidised copper scrap, Cu_2O/CuO and Cu-rich slags/e-wastes. From a thermodynamic perspective, hydrogen exhibits a significantly more negative standard Gibbs free energy ($\Delta G°$) than copper oxide making it a suitable reductant and the exothermic thermal effect from reaction between hydrogen and Cu_2O/CuO may be used to control process parameters. These characteristics renders hydrogen an ideal gas for reducing copper oxides and copper-containing slags/e-wastes. This review article assesses previous research on utilising hydrogen for producing and refining copper from primary and secondary feed materials.

INTRODUCTION

Worldwide copper production increased marginally from 24.99 Mt in 2022 to 25.34 Mt in 2023 and is predicted to rise to 26.17 Mt in 2024 (International Copper Study Group (ICSG), 2023). The global copper demand is predicted to rise by 350 per cent in 2050 (Elshkaki et al, 2016). Currently, there are two significant sources of copper able to meet the demand. Natural ores containing primary copper-bearing minerals such as chalcopyrite ($CuFeS_2$) and chalcocite (Cu_2S), as well as less abundant alteration minerals including Cu-rich sulfates, hydroxy-silicates, oxides and carbonates. Secondary resources, including scrap copper, copper alloys and copper-containing resources, such as slag (Schlesinger et al, 2021). The carbon emissions generated by the process heat requirement and the Cu reduction, and ultimately how hydrogen can potentially be incorporated, will depend on the copper resource input used and the process route selected to manufacture the copper.

In conventional primary copper production, for example through a flash smelting route as shown in Figure 1, the copper concentrate is oxidised in a flash furnace to form matte. The matte is further oxidised in a converter to produce blister copper with 99 wt per cent purity, while the slag from smelting is processed in a slag cleaning furnace. The heat for the smelting and converting are supplied mainly by the oxidation reactions themselves. The blister copper is further refined in an anode furnace before going into electrolytic refining. Figure 1 also shows opportunities for decarbonising the process by using hydrogen in specific unit processes. For example, hydrogen can be used as a reducing agent or for generating process heat and can potentially be introduced in the slag cleaning furnace and anode furnace (Schlesinger et al, 2021; Roeben et al, 2021).

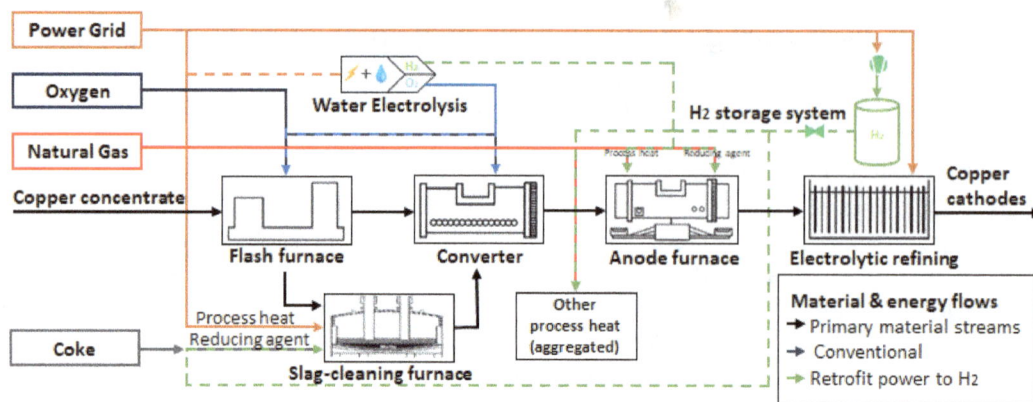

FIG 1 – Flow sheet of primary copper production process, showing possible decarbonisation/hydrogenation approaches (Roeben *et al*, 2021).

For secondary copper resources, such as black copper smelting, the material is either reduced then oxidised (or *vice versa* depending on the input composition and type), before being electro-refined. Figure 2 shows a generic process flow sheet showing where hydrogen can be potentially implemented as a reductant and fuel for heat requirement (eg in a reduction stage). Mairizal *et al* (2023) evaluated the prospects for hydrogen incorporation through a preliminary thermodynamic analysis and found that 82.17 per cent Cu can be obtained in the reduction stage while 94 per cent Cu purity can be achieved from the oxidation stage. In the reduction stage scenario, 59 kg H_2O off-gas is generated with zero CO_2 emissions. In an alternative scenario using copper slag as a feed, a printed circuit board (PCB) as the reductant and H_2 gas as the heat supply, producing 46.6 kg H_2O off-gas and 33.7 kg CO_2 formed.

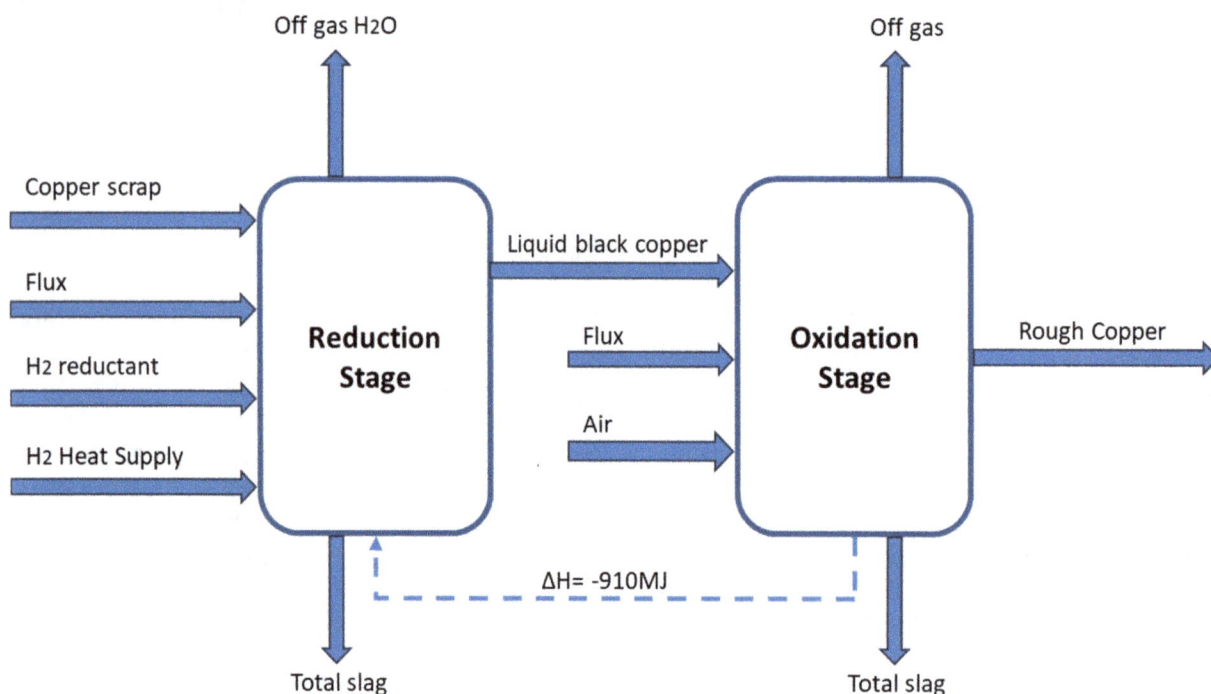

FIG 2 – Process flow sheet for secondary Cu sources using hydrogen as both fuel and reductant (Mairizal *et al*, 2023).

Although some work has commenced, there is a need for a fundamental understanding of the detailed hydrogen reduction mechanism for the different copper sources to optimise its implementation in future processes. This article reviews the use of hydrogen as a reductant for copper-containing materials. The paper starts with a thermodynamic analysis, showing the feasibility of using hydrogen for the critical reactions involved in Cu reduction in the relevant processes. The paper continues with a review and discussion of previous studies that focus on the hydrogen reduction of chalcopyrite, chalcocite, Cu_2O/CuO, and complex copper-containing secondary resources, both in the solid and liquid states. Most of these works were carried out only at a laboratory scale. Nevertheless, they provide the baseline for future industrial applications.

THERMODYNAMICS OF HYDROGEN REDUCTION OF COPPER-CONTAINING RESOURCES

From a thermodynamic perspective, hydrogen reduction reactions have significantly lower standard Gibbs free energies ($\Delta G°$) compared to copper oxidation reactions, as shown in Figure 3. Hence, hydrogen makes an excellent reducing agent for reducing copper-containing ores and slag/e-wastes. Figure 3 also illustrates reaction equilibria for the traditionally used carbon and for plasma hydrogen (H and H^+) which are potentially other hydrogen reducing sources. Equations 1 and 2 show the reactions involved for the reduction of a metal oxide to its lower oxide and/or metal state by molecular hydrogen (Rukini *et al*, 2023):

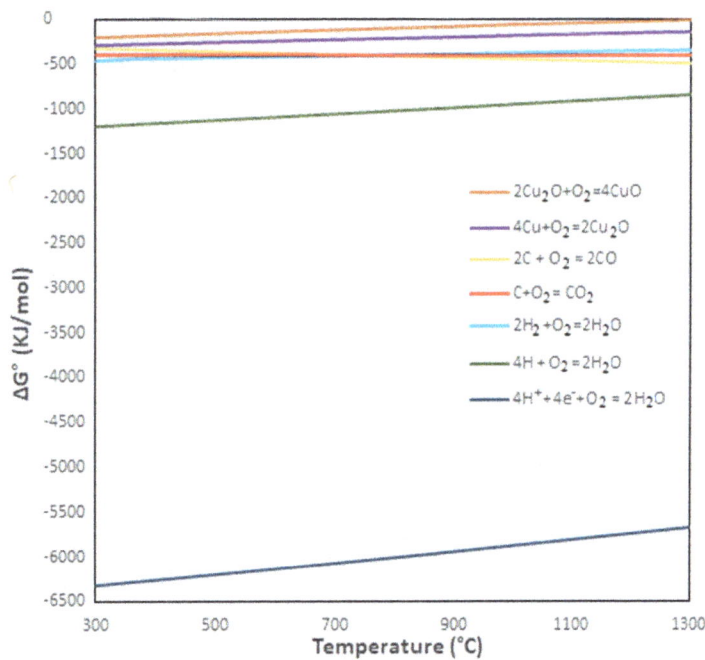

$$MO_x + H_2(g) \rightarrow MO_{x-1} + H_2O(g) \qquad \Delta G = \Delta G° + RT \ln (pH_2O/pH_2) \qquad (1)$$

$$MO_x + xH_2(g) \rightarrow M + xH_2O(g) \qquad \Delta G = \Delta G° + xRT \ln (pH_2O/pH_2) \qquad (2)$$

FIG 3 – Ellingham diagram of Cu reduction reactions and including reactions involving carbon, molecular and plasma hydrogen.

In the case of copper oxides, the reactions and associated free energies for the reduction of Cu_2O and CuO at 250°C are shown in Equations 3 and 4 (Gargul, Małecki and Włodarczyk, 2013):

$$Cu_2O(s) + H_2(g) \rightarrow 2Cu(s) + H_2O(g) \qquad \Delta G° = -87.1727 \text{ kJ} \qquad (3)$$

$$CuO(s) + H_2(g) \rightarrow Cu(s) + H_2O(g) \qquad \Delta G° = -110.8195 \text{ kJ} \qquad (4)$$

The $\Delta G°$ of the reduction reaction for CuO is lower compared to that of Cu_2O, which indicates greater thermodynamic feasibility for the reaction at the same temperature. Kinetically the observed activation energy for reduction of CuO is 14.5 kcal/mol, meanwhile Cu_2O is 27.4 kcal/mol (Kim *et al*, 2003).

Chalcopyrite and chalcocite are the most common ores used for the primary production of copper. As discussed in the previous section, there is possibility for hydrogen reduction of these primary sulfide sources. However, analysis indicates that direct hydrogen reduction of metal sulfides has unfavourable thermodynamics. The equilibrium constant for Equation 5 at 800°C ranges from 2×10^{-3} to 6×10^{-3} for Cu, Ni, Co and Fe sulfides. Shifting the equilibria to the right can boost metal yield by removing hydrogen sulfide (H_2S) immediately after the reaction. This can be achieved by adding a flux with a strong H_2S affinity to the sulfide compound (Equation 6). CaO is a flux that favours the reduction of chalcopyrite by hydrogen at 800°C and this can be achieved through additions of different amounts of CaO as shown in Equation 7–10 (Habashi, Dugdale and Nagamori, 1974; Habashi and Yostos, 1977):

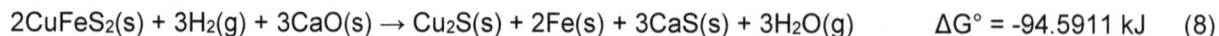

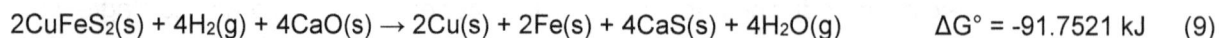

$$MS + H_2 \leftrightarrow M + H_2S \tag{5}$$

$$MS + H_2 + CaO \rightarrow M + CaS + H_2O \tag{6}$$

$$2CuFeS_2(s) + H_2(g) + CaO(s) \rightarrow Cu_2S.2FeS(s) + CaS(s) + H_2O(g) \qquad \Delta G° = -90.0245 \text{ kJ} \tag{7}$$

$$2CuFeS_2(s) + 3H_2(g) + 3CaO(s) \rightarrow Cu_2S(s) + 2Fe(s) + 3CaS(s) + 3H_2O(g) \qquad \Delta G° = -94.5911 \text{ kJ} \tag{8}$$

$$2CuFeS_2(s) + 4H_2(g) + 4CaO(s) \rightarrow 2Cu(s) + 2Fe(s) + 4CaS(s) + 4H_2O(g) \qquad \Delta G° = -91.7521 \text{ kJ} \tag{9}$$

$$2CuFeS_2(s) + H_2(g) + 6CaO(s) \rightarrow 2Cu(s) + 2CaO.Fe_2O_3(s) + 4CaS(s) + H_2O(g) \qquad \Delta G° = -71.0521 \text{ kJ} \tag{10}$$

PREVIOUS STUDIES ON SOLID STATE REDUCTION OF COPPER OXIDES

The kinetics of CuO/Cu_2O reduction in H_2/CO gas mixtures have mainly been studied on submicron-sized powder mixes (Rodriguez et al, 2003; Kim et al, 2003; Jelić, Tomić-Tucaković and Mentus, 2011; Yao et al, 2018), pressed powder pellets (Sabat, Paramguru and Mishra, 2016) and nanoaggregates such nanoparticles, nanowires, rods etc (Pike et al, 2006; Shrestha, Sorensen and Klabunde, 2010). In micro- and nano-scaled assemblies, the building blocks (eg particles, rods, platelets) and their aggregates have broad size and shape distributions, resulting in very different surface morphologies, defect structures and $CuO:Cu_2O$ phase fractions (Jelić, Tomić-Tucaković and Mentus, 2011). Different properties and characteristics of the Cu_2O/CuO powders resulted in contradictive findings regarding the duration of incubation, reaction rates and phase changes during the reduction. There was however, four main findings. Full transformation reduction $CuO \rightarrow Cu_4O_3 \rightarrow Cu_2O \rightarrow Cu$ was observed in the reduction of CuO by hydrogen plasma (Sabat, Paramguru and Mishra, 2016). Meanwhile according to (Rodriguez et al, 2003; Kim et al, 2003; Tyagi, 2018) $CuO \rightarrow Cu$ directly reduced without any intermediate phase of Cu_2O or Cu_4O_3. Sequential reduction of $CuO \rightarrow Cu_2O \rightarrow Cu$ was also reported by some workers (Li, Mayer and Tu, 1992; Pike et al, 2006; Unutulmazsoy et al, 2022). Reduction occurring as a single step of $Cu_2O \rightarrow Cu$ was studied by (Tilliander, Aune and Seetharaman, 2006). A summary of previous studies in solid state reduction of Cu_2O and CuO is presented in Table 1. All these studies were carried out at laboratory scale and the main findings are discussed below for each oxide.

Cu_2O

In secondary sources such as copper slag, the copper oxide form mainly is Cu_2O. Most Cu_2O reduction research has focused on catalysts and has been carried out only at a lab-scale. Reduction of Cu_2O with hydrogen is an autocatalytic process (Hamada, Kudo and Tojo, 1992). Recent investigations on Cu_2O reduction have focused on the kinetics and the mechanism of reduction. In situ time-resolved X-ray diffraction (XRD) involving reducing Cu_2O thin films to metallic copper was studied by Unutulmazsoy et al (2022). High-resolution scanning electron microscopy of the products showed the presence of nano porous copper formation after reduction. Copper film grain size, strain and peak area were all found to be important parameters in real-time reduction kinetics data. Much slower than oxidation, reduction at 300°C was found to take 680s to 1800s to convert Cu_2O to Cu. A single-phase Cu_2O film produced by thermal oxidation of a 300 nm Cu film at 275°C was reduced at 300°C to measure Cu_2O to Cu transformation rates. H_2 exposure causes grain boundaries and porosity at the Cu_2O film surface to behave as short-circuit diffusion routes, trapping hydrogen. Hydrogen diffusion along intergranular areas might cause compressive stress, reminiscent of how atoms diffuse along grain boundaries during the growth of films (Chason et al, 2002; Floro et al, 2001). Another study of the kinetics of reduction Cu_2O by hydrogen gas using thermogravimetric analysis (TGA) at temperatures 300–400°C was conducted by Tilliander, Aune and Seetharaman (2006). This study discovered that the stability of the oxide and the existence of other elements, such as nickel (Ni) or nickel oxide (NiO), did not influence the reduction process. The pace of reduction may be controlled by adjusting the hydrogen flow.

TABLE 1

Summary of previous works on solid state reduction of Cu_2O and CuO using hydrogen.

Author	System and parameters	Key results
Xu et al (2022)	Cu_2O powder - H2, thin film Cu_2O T = 27 - 377°C, 30 min - 23.4 hr pO_2: 5.0×10^{-7} Torr, H_2 pressure: 1.54 - 1.64 Torr	• Activation energy for Cu_2O "29"/Cu (111) reduction by H_2 was 23 kcal/mol, "29" stands for row structure of Cu_2O, meanwhile (111) stands for orientation of the crystal structure. • Cu_2O-"29"Cu and solid Cu were present on the surface before fully turning into Cu (111).
Unutulmazsoy et al (2022)	Cu_2O film - 5%H_2/Ar Cu films: 50, 150 and 300 nm T = 300°C, 30 min	• Reduction of Cu_2O to Cu was identified in 50 nm thick CuO films with an average grain size of 20 nm. • Single-phase Cu_2O and CuO films may be converted to porous Cu, whose size and shape depend on the original Cu-oxide phase's grain size and thickness.
Tyagi (2018)	CuO powder-H_2 pO_2: 76 Torr CuO_{red} T = 160°C - 252°C, 40 min CuO_{red} 0.67 T = 160°C and 200°C, 100 min	• Activation energy of $CuO_{0.67}$ reduction was 10.2 ± 0.7 kcal/mol while CuO was 14.2 kcal/mol. • $CuO_{0.67}$ demonstrated a reduced density and faster reduction rate compared to CuO. • The reduction of $CuO_{0.67}$ occurs at 200°C and ranges from 2% to 85%. In the case of CuO, the reduction ranges from 0.75% to 90% at 150°C.
Tilliander, Aune and Seetharaman (2006)	Cu_2O/NiO/Ni powder-H_2/Ar T = 300 - 400°C, 42 min Heating Rates: 6, 9, 12, 15, 18 K/min Hydrogen flow rate: 0.6 L/min, Mass: 15 mg	• Diffusion through the bed affected the decrease rate using different sample masses. • Presence of Ni/NiO did not affect the reduction kinetics. • Activation energy: isothermal Cu_2O: 92 ± 5 kJ/mol, while non-isothermal Cu_2O: 111 ± 5 kJ/mol.
Jelić, Tomić-Tucaković and Mentus (2011)	CuO powder - H_2 25% H_2 - 99.995% Ar 80 mL/min T = 300 - 450°C Heating rate: 2.5 - 30°C/min	• Both synthesised CuO and commercial CuO with purity >99% were reduced to Cu metal after a 19.7% loss of mass. • No intermediate copper oxidation was found. • Reduced copper particle size was temperature dependent and significantly increased due to sintering at 300–400°C.
Yamukyan, Manukyan and Kharatyan (2009)	CuO powder - H_2 Hydrogen pressure: 0.01 MPa - 2.5 MPa Sample density: 0.42 - 0.57 Combustion temperature: 400 - 600°C Heating rate: 5 - 10 K/s	• Self-propagating diffusion regime can reduce CuO by H_2. • H_2 internal pressure range spans from 0.01 - 2.5 MPa, establishing a surface combustion regime. • The combustion temperature and velocity exhibited constancy at a constant H_2 pressure of 0.075 MPa while operating within a defined density range of 0.42 to 0.57 MPa.
Kim et al (2004)	CuO powder - 5% H_2/95%He T = 150 - 300°C	• CuO reduced to Cu at H_2 flow 15 mL/min while forming intermediate phase Cu_2O at hydrogen flow <1 mL/min. • H_2 flow rate did not influence intermediate phase presence.
Kim et al (2003)	CuO powder - 5% H_2/95%He T = 150 to 300°C, flow rate gas 5 - 15 mL/min H_2 2.5×10^{-6} m^3/s, 150 - 400 min	• Pressure and temperature increase linearly with microwave power and hydrogen flow rate. • Best result for reduction efficiency is 94.5%. • Activation energy of CuO reduction is 14.5 kcal/mol, while Cu_2O is 27.4 kcal/mol.
Rodriguez et al (2003)	CuO powder - 5% H_2/95%He, Gas flow rate 1 - 20 mL/min, 200 - 400°C, pH_2 10^{-4} to 5 Torr	• Activation energy 27.4 kcal/mol. • Cu^{1+} was not a stable intermediate in reducing CuO, resulting in a straight CuO to Cu transition.

CuO

The exothermic reaction between CuO and H_2 to produce metallic Cu and H_2O gas may be divided into induction, autocatalytic and decreasing rate processes. The properties of the initial oxide influence the reduction rates in each step (Tyagi, 2018). Single phase CuO film reduction by 5 per centH_2/Ar at 300°C was evaluated and the phase transformations involving CuO $\rightarrow$ CuO + Cu_2O $\rightarrow$ CuO + Cu_2O + Cu $\rightarrow$ Cu_2O + Cu $\rightarrow$ Cu was observed by in situ XRD (Unutulmazsoy et al, 2022). Three phases, Cu, Cu_2O and CuO, coexist together during the reduction after an incubation time of 1300 sec and with average grain size 30 nm within a 300 nm thick Cu film. Contrary to this, for a 50 nm CuO film with average grain size 20 nm, the formation of Cu metal occurs only after the CuO is completely reduced to Cu_2O (Unutulmazsoy et al, 2022). As oxygen vacancies continue to form on the CuO surface, Cu^+ and Cu^0 coexist. Oxygen vacancies move to the subsurface and create a partly reduced CuO superlattice structure, allowing the Cu_2O phase to nucleate (Sun et al, 2021; Hao et al, 2016; Maimaiti, Nolan and Elliott, 2014). The reduction time for CuO $\rightarrow$ Cu_2O can be reduced by increasing temperature and increasing the partial pressure of hydrogen (Rodriguez et al, 2003; Kim et al, 2003). CuO reduction occurs under two circumstances. First, Cu aggregates form around oxide defects on the surface or bulk in a nonuniform reduction. The reduction mainly occurs at the Cu-CuO surface. Second, most CuO loses oxygen in stages until it becomes metallic copper. Further studies found that when exposed to a steady hydrogen supply with flow rates more than 15 mL/min at temperatures over 200°C, CuO underwent straight reduction to Cu without forming intermediate suboxides (Kim et al, 2003, 2004; Rodriguez et al, 2003).

PREVIOUS STUDIES ON LIQUID STATE REDUCTION OF CU-CONTAINING MATERIALS

Slag

Each ton of copper produced produces 2–3 t of primary copper slag with contains up to 1 per cent copper by weight (Hovestadt and Friedrich, 2023; ICSG, 2023; Gilsbach, 2020) In 2021, copper production generated 57.2 Mt of slag (Jin and Lihua, 2022). From an industrial point of view, the recovered copper content in the slag both from primary smelting and e-waste recycling can produce 350 kt/a copper with 70 per cent recovery rate (Hovestadt and Friedrich, 2023).

A study of primary fayalitic slag reduction using hydrogen was conducted to examine what happened when different flow rates and hydrogen concentrations were added to two slags containing 1 per cent and 2 per cent Cu by weight. Experimental work was conducted at 1300°C with a variation in hydrogen concentration of 15–100 per cent and with a flow rate of around 1–2 L/min. Copper concentration in slag was lowest at 0.3 wt per cent. Turbulence increased the response rate by 40 per cent, suggesting liquid transport was the rate-limiting step (Hovestadt and Friedrich, 2023). Furthermore, the presence of hydrogen gas in slag facilitates the evaporation of zinc and lead. Hydrogen produces equal Zn and Pb levels throughout a broad range of reduction gas concentrations. Thus, increasing hydrogen concentration might significantly reduce the processing time. Zinc concentrations declined linearly over a threshold and decreasingly below 1 wt per cent. Fumigation promoted slag zinc oxide diffusion (Hovestadt and Friedrich, 2023). From thermodynamic point of view, Cu, Zn and Pb oxides were easily reduced, but reducing sulfides was more challenging. The optimal conditions for reducing copper slag were 1450°C, 1.2 alkalinity and 0.225 reducing agent ratio. The newly developed procedure recovered 95.49 per cent copper and iron from slag, as well as 83.54 per cent Pb and 98.30 per cent Zn (Zhang, Zhang and Zheng, 2022). Table 2 represents previous studies of liquid state reduction of copper-containing resources using hydrogen.

Preliminary industrial studies/trials

Current initiatives in Cu metal production by industry seek to use hydrogen to decarbonise copper manufacture. For example, the main Hamburg Cu smelter operated by Aurubis AG, which features two anode furnaces with a capacity of 270 t/batch, has been supplied by hydrogen from September to December 2021. Before this, natural gas treated unwrought copper, emitting much CO_2. Using hydrogen (H_2) as a reducing agent is expected to reduce the Hamburg plant's CO_2 emissions by at least 5000 t/a. Tests using two new anode furnaces showed that the furnaces functioned more effectively and used 30 per cent less natural gas, saving roughly 1.2 t of CO_2 per annum and decarbonised Aurubis output before enough hydrogen was available. Replacing anode furnaces improves metal extraction from metal concentrates and recycling, in addition to climate advantages (Aurubis AG, 2023; Edens and Steindor, 2023).

TABLE 2

Summary of previous works on liquid state reduction of copper containing resources using hydrogen.

Author	System and parameters	Key results
Mairizal et al (2023)	Modelling study, 1200°C - 1400°C pO_2 ranged between 10^{-7} and 10^{-9} atm	• 73% reducing CO_2 emissions using hydrogen as a heat source. • PCBs significantly reduces carbon dioxide (CO_2) emissions from 183.7 kg/h to 123.14 kg/h.
Hovestadt and Friedrich (2023)	Modelling and experimental study Primary slag and Slag mix (≤2%Cu) T = 1300°C, t > 150 min H_2: 15% - 100% and 1 - 2 L/min	• Simulations showed after 0.5 L hydrogen addition, Cu_2O% reduced to 0.54 wt%. • Copper content decreased below the predicted limit of 0.49 wt% across all concentrations. • Lowest slag copper content was 0.31 wt% Cu.
Edens and Steindor (2023)	Industrial trial of 275 t Cu Initial O_2 0.9 - 1.2%, 60 - 97%H_2/N_2, H_2 flow rate 1400 - 2000 m³/h, 150 - 190 min	• Hydrogen decreases O_2 oxygen content with 90% effectiveness. • Low gas H_2 efficiency is produced by unstable operation, such as blockage and leakages, which reduces valid measurements. Uninterrupted operation is predicted to be more efficient.
Hovestadt and Friedrich (2023)	1500 g secondary copper smelter slag T = 1300°C, H_2 flow 0.5 - 2 L/min H_2 25 - 100% H_2 90 L	• Increased hydrogen injection volumes improve fuming and accelerate the reduction process. • Fuming process of H_2 did not follow a clear pattern with different concentrations (25, 50, 75% H_2), but the quantity and turbulence of H_2 (0.5, 1.0, 1.5/L/min H_2) did affect fuming and reduction.
Zhang, Zhang and Zheng (2022)	Experimental Cu slag pellet with 10%H_2 - Ar H_2 flow 4 L/min, pH_2 40% T = 1500°C, 4 hrs, CaO addition	• Activation energy of H_2 copper slag reduction 29.107–36.082 kJ/mol. • Increasing the reducing gas flow rate improves the reduction ratio but after 4 L/min that the reduction process is controlled by internal diffusion, interface chemical reaction control and mixed control.
Huaiwei et al (2021)	Experimental work of copper slag T = 1100°C, t < 1350 sec CO/H_2 Ratios (0 - 6/0 - 4)	• Fastest response rates were seen when the entering gas was hydrogen gas for 300 sec. • Primary cause of the loss of valuable components in the copper slags was the high magnetite content. Activation Energy = 58.8 kJ/mol.
Fasshauer et al (2000)	Industrial trial of 145 t Cu Initial O_2 0.9%, 53 - 81%H_2/Ar, gas flow 200 - 350 m³/h, T = 1250°C	• Optimal hydrogen volume percentage in the H_2/N_2 combination was 60–72% by volume. • Preferable pressure of the reducing gas introduced into the melt at 8 to 12 bar.
Iwamura et al (1991)	40 kg molten copper with O_2 10 ppm H_2 5 - 50%/Ar, T = 1200°C, 10 min	• Increasing volume H_2 > 50% reduce reaction efficiency. • Produced pure Cu with O_2 < 3 ppm by weight.
Fukunaka et al (1991)	5 mm Cu droplets with H_2/Ar, T = 1697°C Initial O_2 0.036 - 1.9 wt%, pH_2 4 kPa Gas flow rate $2×10^{-4}$ m³/s	• Deoxidation dropped droplet weight by a few percent and raised temperature in 30 sec during vaporisation. Deoxidation rises with hydrogen partial pressure and gas flow rate (mixed rates control). • Copper droplets deoxidise faster at high oxygen levels and slower at low oxygen levels.

CONCLUSIONS

Recent investigations have shown significant advances regarding hydrogen's potential in reducing primary and secondary copper resources. The laboratory-scale investigations explicitly focused on the reduction kinetics and reaction mechanisms. More detailed investigations in a simple system (CuO and Cu_2O) and a complex system (copper slag and copper-containing resources) at higher temperatures must be caried out to better understand interaction between phases during the reduction. Variations in the composition of the reductant gas mixture, temperature and reduction time need to be further investigated since these parameters will vary the kinetic driving force. The influence of the H_2/H_2O ratio also needs to be determined to examine its effect on the kinetics and microstructures of the reduction process. Laboratory studies and industrial research indicate that hydrogen reduction can be applied to primary and secondary copper resources, contributing to industrial decarbonisation efforts. However, more research is needed to understand better the parameters impacting hydrogen reduction treatment.

ACKNOWLEDGEMENTS

This article was part of Dian Mughni Fellicia's PhD work, co-funded by Swinburne University of Technology and CSIRO.

REFERENCES

Aurubis AG, 2023. H₂-Ready: Aurubis Investing in Hydrogen-Capable Anode Furnaces in the Hamburg Plant [press release], Aurubis AG. Available from:<https://www.aurubis.com/en/media/press-releases/press-releases-2023/h2-ready-aurubis-investing-in-hydrogen-capable-anode-furnaces-in-the-hamburg-plant> [Accessed: 1 December 2023].

Chason, E, Sheldon, B W, Freund, L B, Floro, J A and Hearne, S, J, 2002. Origin of Compressive Residual Stress in Polycrystalline Thin Films, *Physical Review Letters*, 88:156103. https://doi.org/10.1103/PhysRevLett.88.156103

Edens, T and Steindor, J, 2023. Using Hydrogen as a Reductant in Fire Refining at Aurubis Hamburg's 'Down-town' Smelter, in *Proceedings of the 61st Conference of Metallurgists (COM 2022)*, pp 211–228 (Springer: Cham). https://doi.org/10.1007/978-3-031-17425-4_31

Elshkaki, A, Graedel, T E, Ciacci, L and Reck, B, 2016. Copper demand, supply and associated energy use to 2050, *Global Environmental Change*, 39:305–315. https://doi.org/10.1016/j.gloenvcha.2016.06.006

Fasshauer, K, Steffner, F, Dauterstedt, H-J, Albrech, M and Wernicke, E, 2000. Process for Poling Copper with Hydrogen and Nitrogen, Germany Patent Application 99119000.0.

Floro, J A, Hearne, S J, Hunter, J A, Kotula, P, Chason, E, Seel, S C and Thompson, C V, 2001. The dynamic competition between stress generation and relaxation mechanisms during coalescence of Volmer–Weber thin films, *J Appl Phys*, 89(9):4886–4897. https://doi.org/10.1063/1.1352563

Fukunaka, Y, Tamura, K, Taguchi, N and Asaki, Z, 1991. Deoxidation Rate of Copper Droplet Levitated in Ar-H₂ Gas Stream, *Metallurgical Transactions B*, 22(5):631–639. https://doi.org/10.1007/BF02679018

Gargul, K, Małecki, S and Włodarczyk, M, 2013. Reducing agents in fire refining of copper — analysis of the reduction process and its costs, *Journal: Rudy i Metale Nieżelazne*, 58(12):843–849.

Gilsbach, L, 2020. Copper – Sustainability Information [online], Bundesanstalt für Geowissenschaften und Rohstoffe. Available from: <https://www.bgr.bund.de/EN/Gemeinsames/Produkte/Downloads/Informationen_Nachhaltigkeit/kupfer_en.pdf?__blob=publicationFile&v=4> [Accessed: 01 February 2024].

Habashi, F and Yostos, B I, 1977. Copper from chalcopyrite by direct reduction, *JOM*, 29:11–16. https://doi.org/10.1007/BF03354324

Habashi, F, Dugdale, R and Nagamori, M, 1974. The Recovery of Copper, Iron and Sulfur from Chalcopyrite Concentrate by Reduction (published in German), *Metall (Berlin)*, 28(11):1051–1054.

Hamada, S, Kudo, Y and Tojo, T, 1992. Preparation and Reduction Kinetics of Uniform Particles from Copper(I) Oxides with Hydrogen, *Colloids And Surfaces*, 67(6):45–51. https://doi.org/10.1016/0166-6622(92)80284-9

Hao, G, Zhang, R, Li, J, Wang, B and Zhao, Q, 2016. Insight into the Effect of Surface Structure on H₂ Adsorption and Activation Over Different Cuo (1 1 1) Surfaces: A First-Principal Study, *Computation Materials Science*, 122(9):191–200. https://doi.org/10.1016/j.commatsci.2016.05.023

Hovestadt, G and Friedrich, B, 2023. Fuming process of zinc rich slags by hydrogen injection [presentation], 8th Slag Valorization Symposium, 11 p. https://doi.org/10.13140/RG.2.2.29850.72643

Huaiwei, Z, Bao, L, Chen, Y, Xuan, W and Yuan, Y, 2021. Efficiency improvements of the CO-H2 mixed gas utilization related to the molten copper slag reducing modification, *Process Safety and Environmental Protection*, 146:292–299. https://doi.org/10.1016/j.psep.2020.09.011

International Copper Study Group (ICSG), 2023, 28 April. ICSG Copper Market Forecast 2023/2024 [online]. Available from: <https://icsg.org/download/2023-04-28-press-release-icsg-copper-market-forecast-2023-2024/> [Accessed: 15 January 2024].

Iwamura, T, Koya, T, Sukekawa, I, Hagiwara, H, Asao, H, Shigematsu, T and Sato, H, 1991. Method For Manufacturing Oxygen-Free Copper. US Patent US5037471A.

Jelić, D, Tomić-Tucaković, B and Mentus, S, 2011. A Kinetic Study of Copper (Ii) Oxide Powder Reduction with Hydrogen, Based on Thermogravimetry, *Thermochimica Acta*, 521:211–217, https://doi.org/10.1016/j.tca.2011.04.026

Jin, Q and Lihua, C, 2022. A Review of the Influence of Copper Slag on the Properties of Cement-Based Materials, *Materials*, 15(23):8594. https://doi.org/10.3390/ma15238594

Kim, J Y, Rodriguez, J A, Hanson, J C, Frenkel, A I and Lee, P L, 2003. Reduction of CuO and Cu2O with H2: H embedding and kinetic effects in the formation of suboxides, *Journal of the American Chemical Society*, 125(35):10684–10692. https://doi.org/10.1021/ja0301673

Kim, J, Hanson, J, Frenkel, A, Lee, P and Rodriguez, J, 2004. Reaction of CuO with Hydrogen Studied by Using Synchrotronbased X-Ray Diffraction, *Journal of Physics: Condensed Matter*, 16(33). https://doi.org/10.1088/0953-8984/16/33/008

Li, J, Mayer, J W and Tu, K N, 1992. Nucleation and growth of Cu2O in the reduction of CuO thin films, *Physical Review, B: Condensed Matter*, 45(10):5683–5686. https://doi.org/10.1103/physrevb.45.5683

Maimaiti, Y, Nolan, M and Elliott, S, 2014. Reduction mechanisms of the CuO(111) surface through surface oxygen vacancy formation and hydrogen adsorption, *Physical Chemistry Chemical Physics*, 16(7):3036–3046. https://doi.org/10.1039/C3CP53991A

Mairizal, A Q, Sembada, A Y, Tse, K M, Haque, N and Rhamdhani, M A, 2023. Assessment of Mass and Energy Balance of Waste Printed Circuit Board Recycling through Hydrogen Reduction in Black Copper Smelting Process, *Processes*, 11(5):1506. https://doi.org/10.3390/pr11051506

Pike, J, Chan, S-W, Wang, X and Hanson, J, 2006. Formation of Stable Cu2O from Reduction of CuO Nanoparticles, *Appl Catal A*, 303(2):273–277. https://doi.org/10.1016/j.apcata.2006.02.008

Rodriguez, J A, Kim, J Y, Hanson, J C, Pérez, M and Frenkel, A I, 2003. Reduction of CuO in H2: In Situ Time-Resolved XRD Studies, *Catalysis Letters*, 85:247–254. https://doi.org/10.1023/A:1022110200942

Roeben, F, Schöne, N, Bau, U, Reuter, M, Dahmen, M and Bardow, A, 2021. Decarbonizing Copper Production by Power-to-Hydrogen: A Techno-Economic Analysis, *Journal of Cleaner Production*, 306:127191. https://doi.org/10.1016/j.jclepro.2021.127191

Rukini, A, Rhamdhani, M A, Brooks, G and Bulck, A, 2023. Lead Recovery from PbO Using Hydrogen as a Reducing Agent, *Metallurgical and Materials Transactions B*, 54:996–1016. https://doi.org/10.1007/s11663-023-02745-0

Sabat, K, Paramguru, R and Mishra, B, 2016. Reduction of Copper Oxide by Low-Temperature Hydrogen Plasma, *Plasma Chemistry and Plasma Processing*, 36:1111–1124. https://doi.org/10.1007/s11090-016-9710-9

Schlesinger, M E, Sole, K C, Davenport, W G and Flores, G R F A, 2021. *Extractive Metallurgy of Copper*, 6th edition, pp 1–30 (Elsevier: Amsterdam).

Shrestha, K M, Sorensen, C M and Klabunde, K J, 2010. Synthesis of CuO Nanorods, Reduction of CuO into Cu Nanorods and Diffuse Reflectance Measurements of CuO and Cu Nanomaterials in the Near Infrared Region, *J Phys Chem*, 114:14368–14376.

Sun, X, Wu, D, Zhu, W, Chen, X, Sharma, R, Yang, J C and Zhou, G, 2021. Atomic Origin of the Autocatalytic Reduction of Monoclinic CuO in a Hydrogen Atmosphere, *The Journal of Physical Chemistry Letters*, 12:9547–9556.

Tilliander, U, Aune, R and Seetharaman, S, 2006. Kinetics studies of hydrogen reduction of Cu2O, *Zeitschrift fuer Metallkunde/Materials Research and Advanced Techniques*, 97:72–78. https://doi.org/10.1515/ijmr-2006-0011

Tyagi, D, 2018. Kinetics of Reduction of Copper Oxide, *International Journal for Research in Applied Science and Engineering Technology*, 6:3928–3932. https://doi.org/10.22214/ijraset.2018.4645

Unutulmazsoy, Y, Cancellieri, C, Lin, L and Jeurgens, L P H, 2022. Reduction of Thermally Grown Single-Phase CuO and Cu2O Thin Films by In-Situ Time-Resolved XRD, *Applied Surface Science*, 588:152896. https://doi.org/10.1016/j.apsusc.2022.152896

Xu, F, An, W, Baber, A E, Grinter, D C, Senanayake, S D, White, M G, Liu, P and Stacchiola, D J, 2022. Enhanced Oxide Reduction by Hydrogen at Cuprous Oxide–Copper Interfaces near Ascending Step Edges, *Journal of Physical Chemistry: C*, 126(4):18645–18651. https://doi.org/10.1021/acs.jpcc.2c03719

Yamukyan, M, Manukyan, K and Kharatyan, S. 2009. Copper oxide reduction by hydrogen under the self-propagation reaction mode, *Journal of Alloys and Compounds*, 473:546–549. https://doi.org/10.1016/j.jallcom.2008.06.031

Yao, T, Matsuda, T, Sano, T, Morikawa, C, Ohbuchi, A, Yashiro, H and Hirose, A, 2018. In Situ Study of Reduction Process of CuO Paste and its Effect on Bondability of Cu-to-Cu Joints, *J Electron Mater*, 47:2193–2197.

Zhang, B, Zhang, T and Zheng, C, 2022. Reduction Kinetics of Copper Slag by H2, *Minerals*, 12(5):548. https://doi.org/10.3390/min12050548

Smelting of different hydrogen-reduced bauxite residue-calcite pellets for iron and alumina recovery

M K Kar[1] and J Safarian[2]

1. PhD Researcher, Department of Materials Science and Engineering, NTNU, 7041 Trondheim, Norway. Email: manish.k.kar@ntnu.no
2. Professor, Department of Materials Science and Engineering, NTNU, 7041 Trondheim, Norway. Email: jafar.safarian@ntnu.no

ABSTRACT

Bauxite residue is one of the most important by-products from the alumina industry with no commercial use till now. The utilisation of bauxite residue through pelletising with different lime additions, hydrogen reduction, and smelting has been studied in this work. Three different types of green pellets were made with varying calcite-to-bauxite residue ratio and subsequently sintered at elevated temperatures. The sintered pellets were isothermally reduced at 1000°C under hydrogen gas followed by smelting of the reduced pellets in argon at 1500°C. Smelting of the reduced pellets leads to the recovery of metallic iron and the formation of a leachable calcium aluminate slag. X-ray diffraction (XRD) and scanning electron microscope (SEM) were used to characterise the phases and microstructural analysis. The iron produced in the smelting process has a high purity, above 99.5 wt per cent. While the iron content of the slags is quite low, and the slags shows variations of krotite ($CaAl_2O_4$), gehlenite ($Ca_2Al_2SiO_7$), and perovskite ($CaTiO_3$) phases based on the variation of calcite addition. It was found that the more calcite addition in pelletising yields higher percentage of krotite phase as compared to lower calcite-containing pellets. Employing quantitative XRD analysis, the krotite phase fraction was found approximately 85 wt per cent for higher calcite-added pellet, and around 74 wt per cent for lower calcite-added pellet. FactSage™, ver 8.1 (Thermfact and GGT-Technologies) thermodynamic software was used to evaluate the obtained experimental data, they both revealed that the components of the bauxite residue are mostly distributed into the calcium aluminate slag phase.

INTRODUCTION

As society undergoes continuous modernisation, there is a concurrent rise in the demand for iron metal. However, the quality of traditional iron ore has seen a decline over time. Consequently, iron manufacturers are compelled to explore alternative secondary resources to meet the escalating demand. One such secondary source is bauxite residue, which boasts iron oxide content ranging from 40 to 60 wt per cent (Kar, Önal and Borra, 2023; Kar et al, 2023). Certainly, during the Bayer process, which is employed in the production of alumina from bauxite ore, red mud byproduct is generated. This red mud consists of undigested oxides and other impurities that remain after the extraction of alumina from the bauxite ore. It is a residue with a distinctive reddish colour due to the presence of iron oxide of Fe_2O_3. Managing and finding environmentally sustainable solutions for the disposal of red mud is a significant challenge in the alumina refining industry (Evans, Nordheim and Tsesmelis, 2016; Lazou et al, 2020; Cardenia, Balomenos and Panias, 2019). It is generated in from of slurry and the dewatered form of red mud is known as bauxite residue (Kar and Safarian, 2023). On average, approximately 1.2 tons of bauxite residue is generated per ton of alumina produced. The annual generation of bauxite residue is above 150 million tons which was reported in the year of 2019 (Archambo and Kawatra, 2020). This generated bauxite residue quantity is influenced by various factors, including the processing parameters of the Bayer process and the origin of the bauxite ore. Globally, there is a stockpile of around 4.5 billion tons of bauxite residue (Kar, Önal and Borra, 2023). Numerous utilisation methods and strategies are under exploration to effectively manage and find potential applications for bauxite residue. These applications encompass the production of building materials, decorative tiles, and door filler materials. Additionally, bauxite residue is being investigated for its potential to extract valuable metal elements. Beyond industrial applications, it is being employed for environmental treatment and ecological restoration purposes. This includes its use in water purification, flue gas purification, soil improvement, bauxite residue reclamation, and preparation for environmental restoration materials. The multifaceted potential of bauxite residue utilisation highlights ongoing efforts to transform this byproduct into valuable resources across various domains (Chao *et al*, 2022). Among these, all the utilisation technology, various metal recovery from the bauxite residue will be the most economical. Many works are focusing on iron and alumina recovery from the bauxite residue through the carbothermic route (Ekstrøm *et al*, 2021; Lazou *et al*, 2020).

In earlier studies, iron oxides were reduced to metallic iron utilising carbon as a reductant. However, due to growing concerns about environmental greenhouse gas emissions, using carbon as a reductant has raised issues as it produces greenhouse gases during the reduction process. A more sustainable alternative to carbon is hydrogen, which serves as a reductant for iron oxide, leading to the production of water vapor as a byproduct, in contrast to the carbon monoxide (CO) and carbon dioxide (CO_2) gases generated in carbothermic reduction.

The primary emphasis of this study lies in the examination of phase formation dynamics during the smelting process involving various bauxite residue calcite mixtures that have undergone hydrogen reduction. The investigation encompasses not only the separation of iron from the slag but also delves into the specific types of calcium aluminate phases ($CaAl_2O_4$, krotite) that manifest during the melting of the slag.

MATERIALS AND METHODS

The overall experimental processes of the research work have been presented in the Figure 1. Three different types of bauxite residue calcite pellets have been made and these are named as 1, 2 and 3. The $CaO:Al_2O_3$ (BR) mole ratio was varied 0.85, 1 and 1.15 to examine effect of CaO addition on the formation of calcium aluminate, calcium silicate and calcium titanate phase during hydrogen reduction, and in the slag produced via a further smelting process.

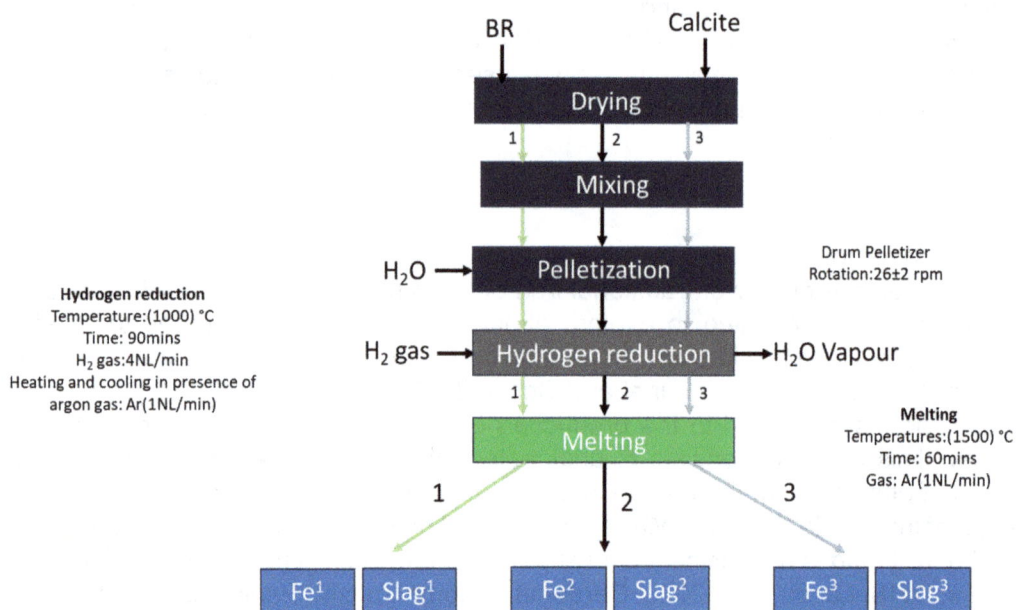

1. Pellet compostion($C_{0.85}A$)
2. Pellet compostion(C_1A)
3. Pellet compostion($C_{1.15}A$)

FIG 1 – Overall flow sheet of the experimental work.

Hydrogen reduction

Three distinct varieties of bauxite residue calcite were produced by employing a drum pelletiser, each characterised by different percentages of calcite. Following the pelletisation process, these pellets underwent sintering at 1150°C for 120 mins to enhance their strength. The sintered pellets were subjected to hydrogen reduction in a TG (Thermogravimetric) furnace. The procedure involved heating the sintered pellets in the presence of argon with a flow rate of 1 Nl/min. Once the target temperature of 1000°C was achieved, the gas flow was switched to hydrogen at a flow rate of 4 Nl/min for 90 mins. Upon completion of the reduction cycle, the reduced pellets were cooled to room temperature in an argon atmosphere. The heating, reduction, and cooling cycle was similar for all types of pellets. The XRD of the three different composition-reduced pellets are shown in Figure 2.

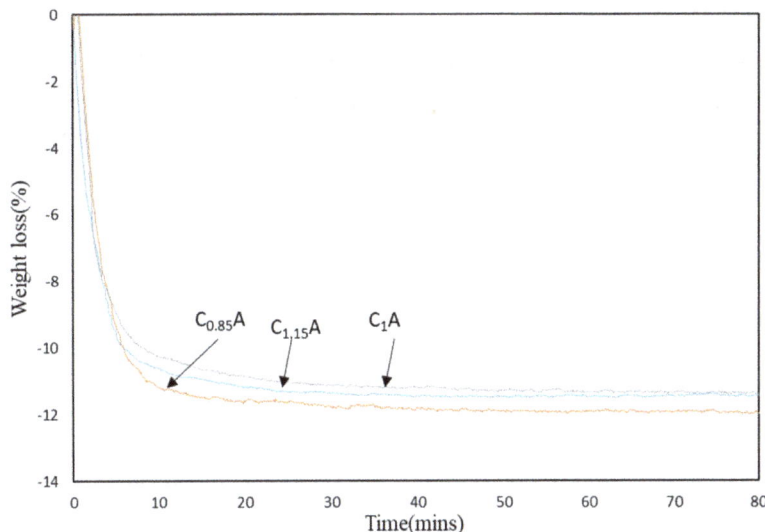

FIG 2 – Weight loss versus Time for different composition bauxite residue calcite pellets (1000°C, 4 Nl/min H_2 flow rate).

Smelting reduced pellets

Following the reduction process of various bauxite residue calcite pellets, smelting was conducted in a vertical tube furnace at 1500°C for 60 mins, maintaining an argon atmosphere. The primary objective of utilising argon during the smelting process was to prevent the oxidation of metallic iron. The complete furnace description and working principle are described here (Rasouli et al, 2022). An Alumina crucible was used to hold the samples.

Characterisation

The phase analysis of slag was conducted through X-ray diffraction (XRD) using CuKα radiation (wavelength λ=1.54 Å) on a Bruker AXS GmbH instrument in Karlsruhe, Germany. The diffractometer scanned within the 15° to 75° 2θ range with a 0.03° step size. Qualitative phase analysis of the raw data employed DIFFRAC.EVA software with the PDF-4+ database (2014, ICDD, Philadelphia, Pennsylvania, USA). Standard samples for XRD were prepared by milling materials in a WC vibratory disk mill (RS 200, RETSCH GmbH, Haan, Germany) for 45 secs at 800 revolutions per minute (rev/min).

Quantitative analysis of phases, including krotite, gehlenite, and perovskite, utilised topas software based on the Rietveld method. To enhance quantitative analysis accuracy, specific parameters in the topas software, such as Zero error, LP (Lorentz–polarization) factor, sample displacement, and background, were adjusted. Microstructural analysis of the slags was conducted using a scanning electron microscope (SEM) (Zeiss Ultra FESEM) equipped with an XFlash® 4010 Detector from the Bruker Corporation for Energy-Dispersive X-Ray Spectroscopy (EDS). Energy dispersive spectroscopy (EDS) was employed to assess the elemental composition in various regions of the slags.

RESULTS AND DISCUSSION

Weight loss during hydrogen reduction

Figure 2 illustrates the weight loss over time for various compositions of bauxite residue calcite sintered pellets. It is evident from the figure that the weight loss is more pronounced for the $C_{0.85}A$ sintered pellets compared to the other two compositions. This higher weight loss can be attributed to the relatively higher fraction of iron oxide within the same mass of sintered pellets. On the other hand, the $C_{1.15}A$ pellets exhibit a higher initial rate of reduction, primarily due to their increased porosity. The elevated porosity of the $C_{1.15}A$ pellets stems from the higher percentage of $CaCO_3$ present, which undergoes decomposition during sintering, resulting in increased porosity. Theoretically, the weight loss should be greater for C_1A pellets compared to $C_{1.15}A$ pellets. However, the figure shows that the difference in weight loss between the two compositions is almost negligible.

Phase analysis of the reduced pellets and obtained slags

As depicted in Figure 3, the primary phases present in the reduced pellets include iron, mayenite, perovskite, larnite, gehlenite, and lime. These phases exhibit a consistent pattern across all the different compositions of bauxite residue-calcite reduced pellets, with variations in intensity observed among the different pellets.

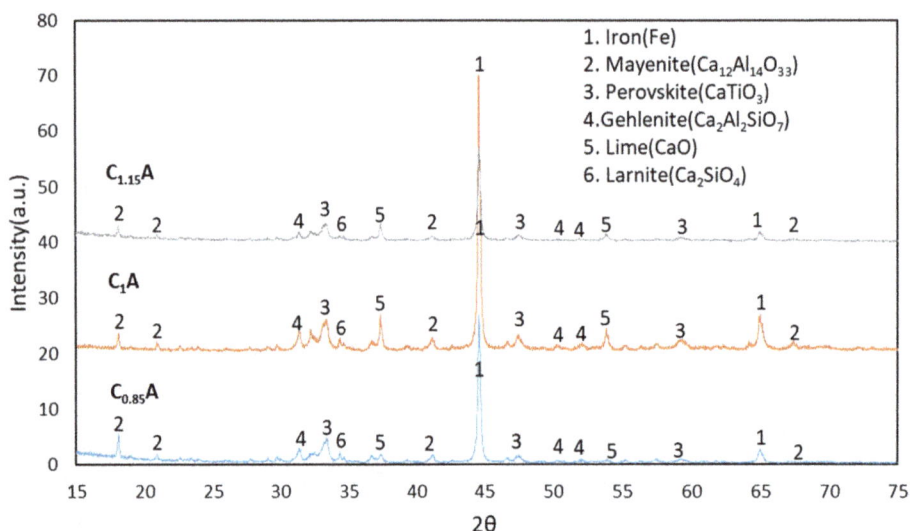

FIG 3 – Phase analysis of different reduced pellets.

Figure 4 illustrates the phases formed in the slags obtained in the smelting of hydrogen-reduced bauxite residue-calcite pellets. Three primary phases, namely krotite ($CaAl_2O_4$), gehlenite ($Ca_2Al_2SiO_7$), and perovskite ($CaTiO_3$), were identified in all smelted slag samples. Krotite emerged as the predominant phase in all the analysed slag samples. From the quantitative XRD analysis, Krotite was high for the higher calcite-added pellets and lower for lower calcite pellets as shown in Table 1.

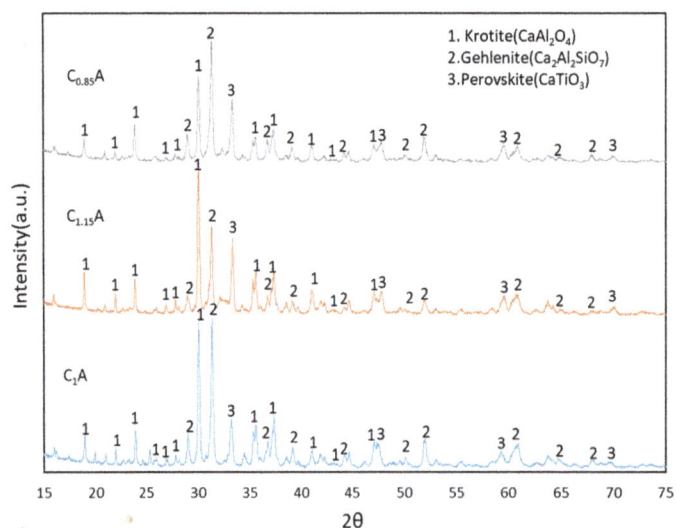

FIG 4 – Phase analysis of different slags (C_1A, $C_{1.15}A$ and $C_{0.85}A$).

TABLE 1

Quantitative XRD analysis of the calcium aluminate phases in slags.

	$C_{0.85}A$	C_1A	$C_{1.15}A$
Gehlenite	17.94	15.41	8.38
Perovskite	10.53	9.89	5.67
Krotite	71.53	74.70	85.95

Microstructure

Figure 5 presents the elemental mapping of slag C_1A. As depicted in the figure, there is a significant overlap of calcium, aluminium, and oxygen in most areas, indicating the presence of krotite. This observation aligns with XRD data, where krotite constitutes a major fraction. In certain regions, the overlap of calcium, aluminium, silicon, and oxygen suggests the potential presence of the gehlenite phase. Additionally, the overlap of calcium, titanium, and oxygen is indicative of the perovskite phase. The presence of iron particles in the slag phase is notable, possibly resulting from incomplete iron separation due to the higher viscosity of the slag. The higher alumina fraction in the slag contributes to increased viscosity.

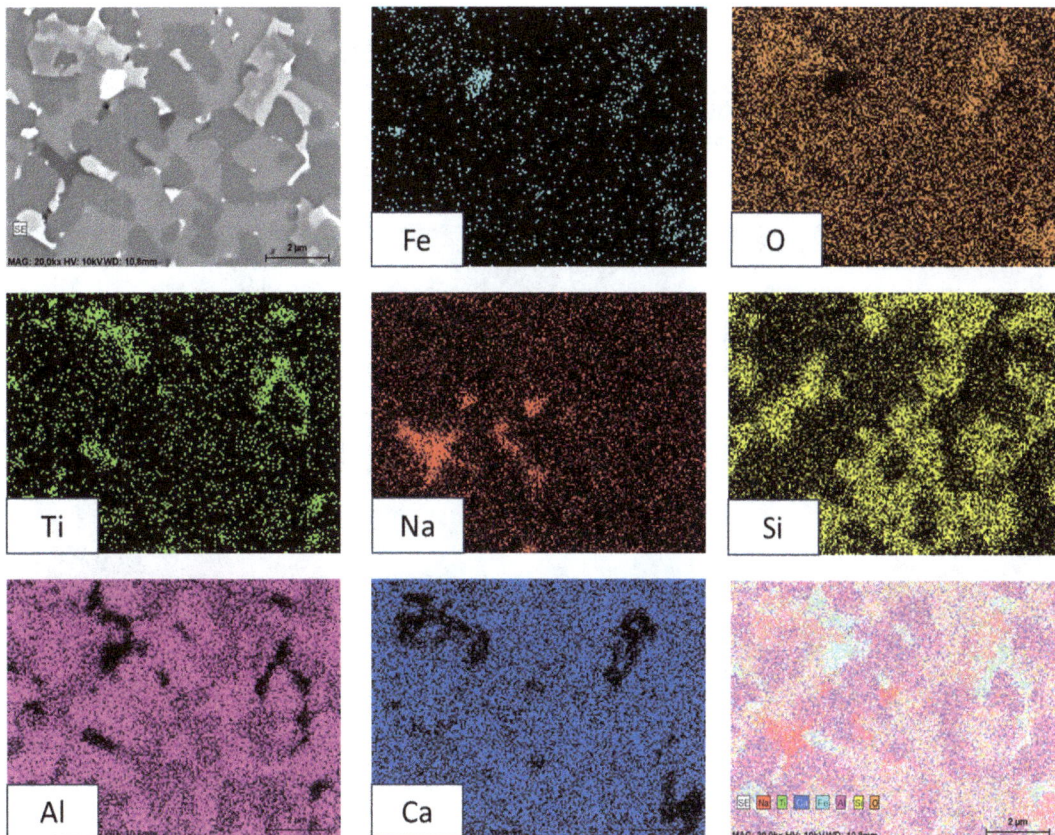

FIG 5 – Elemental mapping of slag C_1A.

Figure 6 displays the elemental analysis of the C_1A slag and metals. On the left side of the figure, the bright areas represent pure iron. On the right side, three distinct phases are observed. The light grey phase consists predominantly of calcium, titanium, and oxygen, with a minor fraction of aluminium. This phase is likely the perovskite phase. The dark phase is composed of calcium, aluminium, and oxygen, with the addition of sodium, indicative of the krotite phase with some sodium present in the lattice. The light dark phase contains calcium, aluminium, oxygen, and silicon, representing the gehlenite phase.

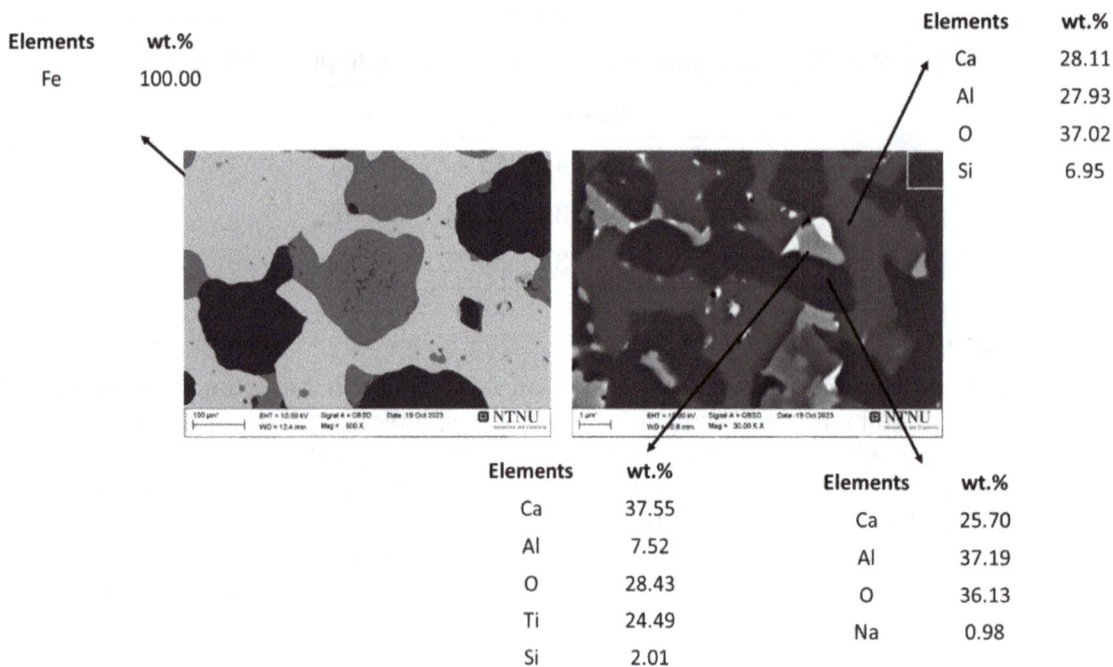

Elements	wt.%
Fe	100.00

Elements	wt.%
Ca	28.11
Al	27.93
O	37.02
Si	6.95

Elements	wt.%
Ca	37.55
Al	7.52
O	28.43
Ti	24.49
Si	2.01

Elements	wt.%
Ca	25.70
Al	37.19
O	36.13
Na	0.98

FIG 6 – Elemental analysis of different phases of slag C_1A.

The elemental mapping of $C_{0.85}A$ and $C_{1.15}A$ are similar (Figures 7 and 8 respectively). However, it has difference in the amount of phases such as krotite, gehlenite, and perovskite, which is shown in the Table 1.

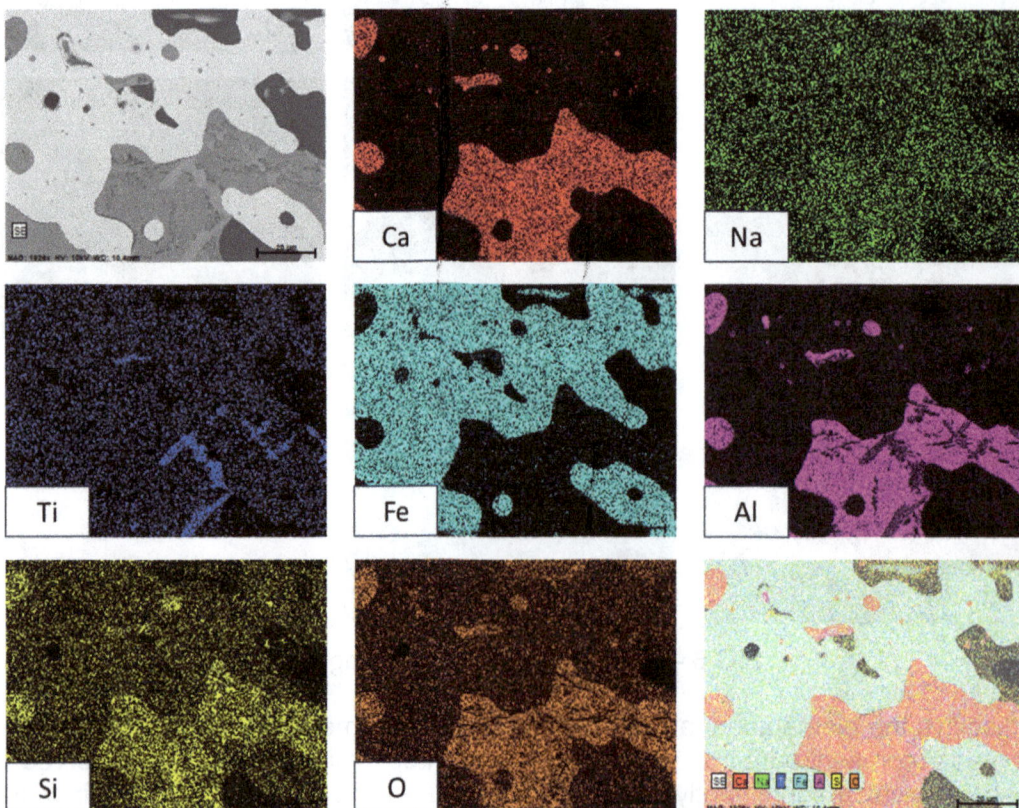

FIG 7 – Elemental mapping of slag $C_{0.85}A$.

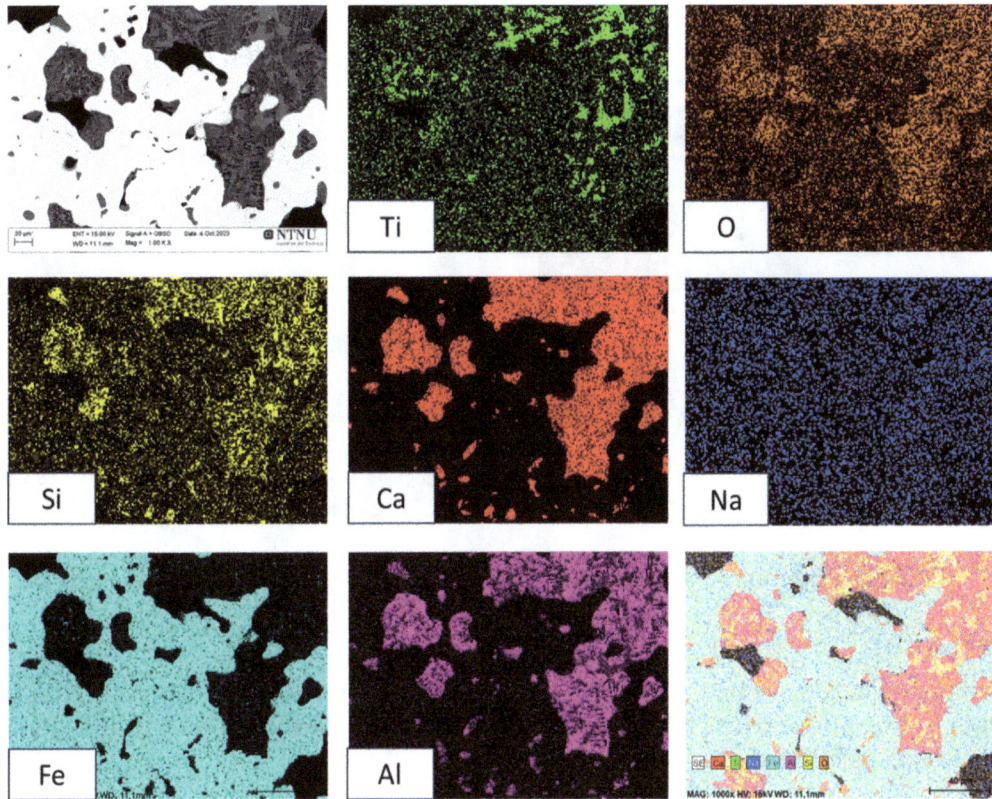

FIG 8 – Elemental mapping of slag $C_{1.15}A$.

The elemental analysis of $C_{0.85}A$ slag of different phases is shown in Figure 9. As shown in the figure white area are the metallic iron with little impurity of calcium, which is likely come from the surroundings signals as the elemental analysis area is smaller. Similar to the previous elemental analysis of C_1A, the light white area is the calcium titanate, and the dark area is the calcium aluminate. As shown in the right image of Figure 9, the shrunk areas are the calcium silicate.

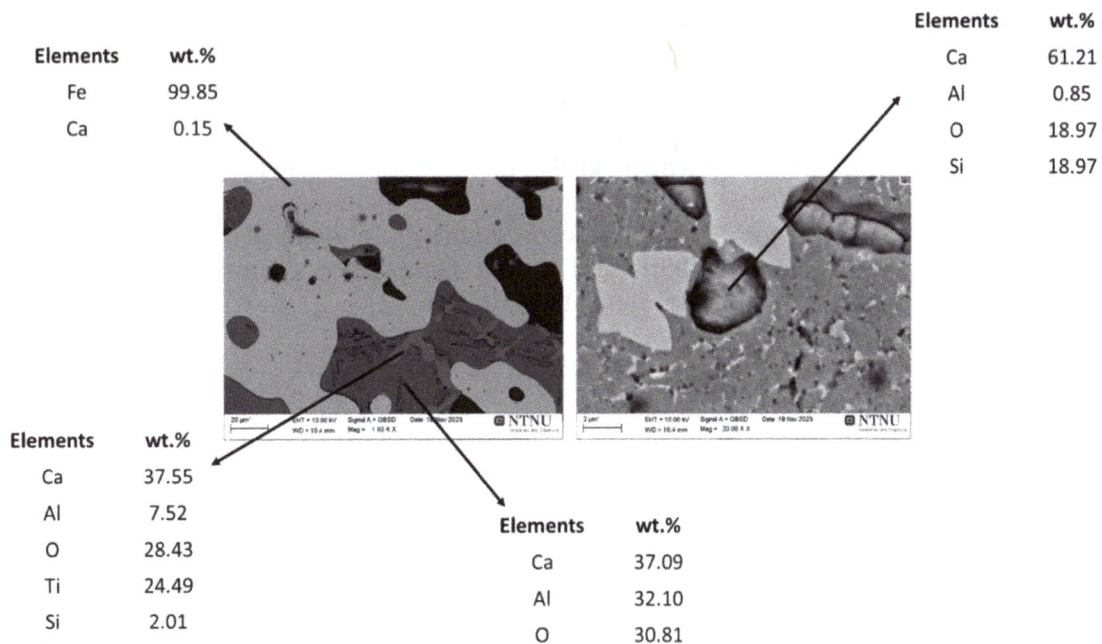

Elements	wt.%
Fe	99.85
Ca	0.15

Elements	wt.%
Ca	61.21
Al	0.85
O	18.97
Si	18.97

Elements	wt.%
Ca	37.55
Al	7.52
O	28.43
Ti	24.49
Si	2.01

Elements	wt.%
Ca	37.09
Al	32.10
O	30.81

FIG 9 – Elemental analysis of different phases of slag $C_{0.85}A$.

Similar to Figure 9, the shrunk areas correspond to calcium silicate with a minor fraction of aluminium. The observed shrunk areas may be attributed to the higher volume shrinkage of calcium silicate compared to other phases, such as calcium alumino silicate, calcium titanate, and calcium aluminate. The elemental analysis of smelted slag of $C_{1.15}A$ is presented in Figure 10.

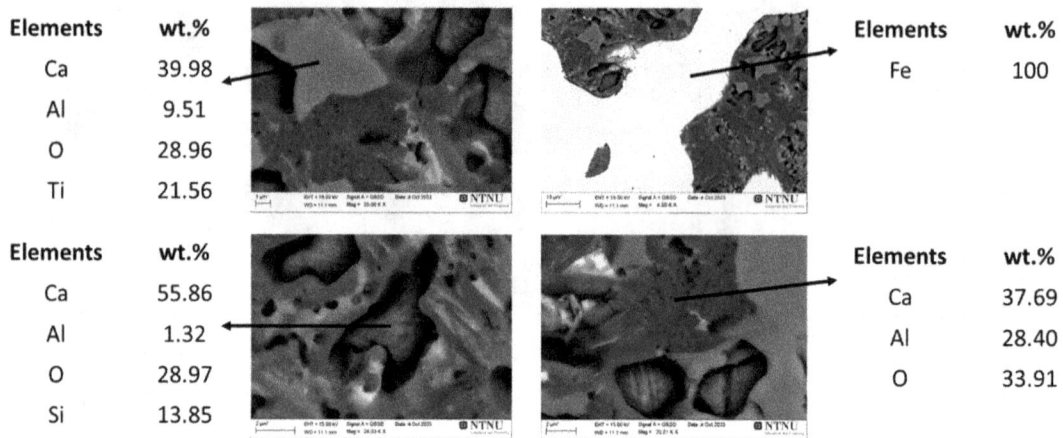

Elements	wt.%
Ca	39.98
Al	9.51
O	28.96
Ti	21.56

Elements	wt.%
Fe	100

Elements	wt.%
Ca	55.86
Al	1.32
O	28.97
Si	13.85

Elements	wt.%
Ca	37.69
Al	28.40
O	33.91

FIG 10 – Elemental analysis of smelted slag of different phases $C_{1.15}A$.

Phase formation

During the slags solidification after the smelting of the reduced pellets of different bauxite residue-calcite pellets, three major phases formed in the slag are krotite, perovskite, and gehlenite. The phases formed are as follows:

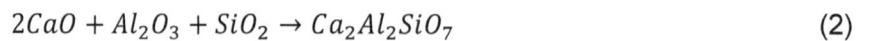

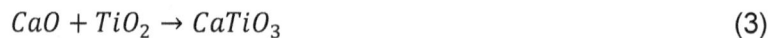

$$CaO + Al_2O_3 \rightarrow CaAl_2O_4 \qquad (1)$$

$$2CaO + Al_2O_3 + SiO_2 \rightarrow Ca_2Al_2SiO_7 \qquad (2)$$

$$CaO + TiO_2 \rightarrow CaTiO_3 \qquad (3)$$

Thermodynamic calculations of slag phases were performed using FactSage™ 8.1 with the FT oxide database. The calculation considered four major oxides from actual measurements in experimental work, namely CaO, Al_2O_3, SiO_2, and TiO_2. Notably, sodium oxide (Na_2O) was excluded from these calculations due to observed sodium losses during hydrogen reduction. The thermodynamic calculations were conducted at a temperature of 1500°C and a pressure of 1 atm, mirroring the experimental conditions.

The Scheil-Gulliver cooling model was selected for thermodynamic calculations, chosen for its faster cooling rate. Figure 11 illustrates that the major phases identified in the calculation include $CaAl_2O_4$, $Ca_3Al_2O_6$, Ca_2SiO_4, and $Ca_3Ti_2O_7$. However, in the experimental analysis, the calcium aluminate phase was observed in krotite form, calcium alumino silicate manifested in gehlenite, and calcium titanate appeared in perovskite form. The observed variation in phases between experimental and thermodynamic calculations may be attributed to the faster cooling rate of the slag and the interactions between high-temperature slag with the crucible.

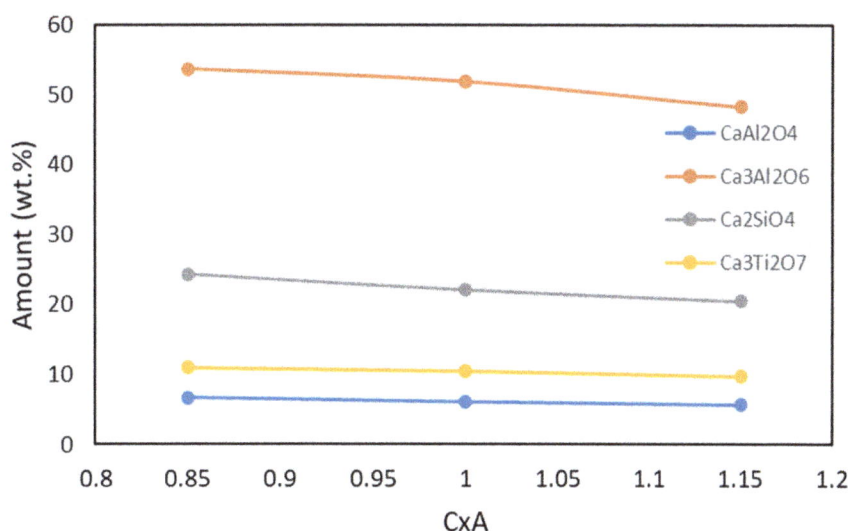

FIG 11 – Phase fraction of different phases calculated by using FactSage™ 8.1.

Gehlenite phase was not found in the thermodynamics calculation may be due to the dissociation of gehlenite to calcium silicate and alumina. The reaction equation is as follows:

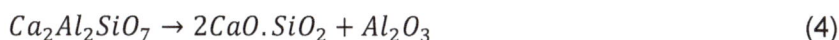

$$Ca_2Al_2SiO_7 \rightarrow 2CaO.SiO_2 + Al_2O_3 \qquad (4)$$

CONCLUSION

The following conclusions from the hydrogen reduction and smelting of different mixtures of bauxite residue and calcite:

- Metallic iron and mayenite dominate as the major phases in the reduced pellets. The phases formed in the reduced pellets from different bauxite residue calcite compositions exhibit similarities with variations in phase fraction.

- Following the smelting of distinct reduced pellets, the primary phases observed in the resulting slag include krotite, gehlenite, and perovskite.

- The leachable calcium aluminate phase, identified as krotite, is present in higher quantities in the pellets with higher calcite content, as revealed by quantitative phase analysis of the slag.

- Thermodynamic calculations indicate that the major phases anticipated in the slag are $CaAl_2O_4$, $Ca_3Al_2O_6$, Ca_2SiO_4, and $Ca_3Ti_2O_7$. Discrepancies between experimental and thermodynamic results may be attributed to the faster cooling rate and interactions during the melting process, including those with the alumina crucible, which could introduce variations in the outcomes.

- The iron purity achieved during the smelting of the reduced pellets exceeds 99.0 wt per cent.

ACKNOWLEDGEMENTS

This study was performed and financially supported by The European Union's Horizon 2020 research and innovation program under grant number of 958307 (HARARE project).

REFERENCES

Archambo, M and Kawatra, S K, 2020. Red mud: Fundamentals and new avenues for utilization, *Mineral processing and extractive metallurgy review*, 24 p.

Cardenia, C, Balomenos, E and Panias, D, 2019. Iron recovery from bauxite residue through reductive roasting and wet magnetic separation, *Journal of Sustainable Metallurgy*, 5(1):9–19.

Chao, X, Zhang, T, Lv, G, Chen, Y, Li, X and Yang, X, 2022. Comprehensive application technology of bauxite residue treatment in the ecological environment: a review, *Bulletin of Environmental Contamination and Toxicology*, 6 p.

Ekstrøm, K E, *et al*, 2021. Recovery of iron and aluminum from bauxite residue by Carbothermic reduction and slag leaching, *Journal of Sustainable Metallurgy*, 7(3):1314–1326.

Evans, K, Nordheim, E and Tsesmelis, K, 2016. Bauxite residue management, *Light metals 2012*, pp 63–66.

Kar, M K and Safarian, J, 2023. Characteristics of Bauxite Residue–Limestone Pellets as Feedstock for Fe and Al_2O_3 Recovery, *Processes*, 11(1):137.

Kar, M K, *et al* 2023. Properties of self-hardened CaO-added bauxite residue pellets and their behavior in hydrogen reduction followed by leaching and magnetic separation for iron and alumina recovery, *International Journal of Hydrogen Energy*.

Kar, M K, Önal, M A R and Borra, C R, 2023. Alumina recovery from bauxite residue: A concise review, *Resources, Conservation and Recycling*, 198:107158.

Lazou, A, *et al*, 2020. High temperature smelting reduction of bauxite residue and a $CaCO_3$-rich bauxite for aluminate slag production and pig iron recovery, in *Proceedings of the 3rd International Bauxite Residue Valorisation and Best Practices Conference, Virtual*, pp 77–83.

Rasouli, A, *et al*, 2022. Magnesiothermic Reduction of Natural Quartz, *Metallurgical and Materials Transactions B*, 53(4):2132–2142.

Effect of mill-scale and calcined dolomite on high Al₂O₃ sinter and its reduction behaviour

S J Kim¹, L Tomas da Rocha², S W Kim³ and S Jung⁴

1. PhD student, Graduate Institute of Ferrous and Eco Materials Technology, Pohang University of Science and Technology, Pohang, North Gyeongsang Province, 37673, South Korea. Email: sjkim96@postech.ac.kr
2. Post-doctorate Researcher, Graduate Institute of Ferrous and Eco Materials Technology, Pohang University of Science and Technology, Pohang, North Gyeongsang Province, 37673, South Korea. Email: leonardoroc@postech.ac.kr
3. Senior Researcher, Ironmaking and FINEX Research Group, Process and Engineering Research Labs., POSCO, Pohang, North Gyeongsang Province, 37763, South Korea. Email: sungwankim@posco.com
4. Professor, Graduate Institute of Ferrous and Eco Materials Technology, Pohang University of Science and Technology, Pohang, North Gyeongsang Province, 37673, South Korea. Email: smjung@postech.ac.kr

ABSTRACT

Al₂O₃ has been regarded as a harmful gangue component in iron ore, especially causing reduction degradation in industrial sintering of iron ore. The current study aimed to utilise low-grade iron ore containing high Al₂O₃ content in sintering process. In order to counteract the adverse effect of Al₂O₃ on reduction behaviour of iron ore sinter, mill-scale and calcined dolomite were chosen as FeO- and MgO-bearing materials. Firstly, mill-scale was added to high Al₂O₃ sinter and addition was optimised as 10 wt per cent, then calcined dolomite was added further. Magnetite and magnesio-ferrite were identified as the major phases in iron ore sinter when adding mill-scale and calcined dolomite. As the dosage of calcined dolomite increased beyond 4 wt per cent of in sinter mix, unreacted MgO particles were observed, which is ascribed to the limitation of MgO dissolution into magnetite matrix. Both mill-scale and calcined dolomite were effective in improving compressive load and Reduction Disintegration Index (RDI) of iron ore sinter although its reduction degree was highly decreased. The addition of both FeO and MgO-bearing materials to high Al₂O₃ sinter has a potential to improve the sinter quality, also contribute to the recycling of industrial by-product. However, the amount should be optimised in views of phase development and deterioration of reducibility.

INTRODUCTION

The price of raw materials being used in steel industry is gradually increasing in recent years (Chai *et al*, 2019; Wang *et al*, 2020). Since several countries with limited mineral resource are largely depending on the import iron ores, ironmaking companies are attempting to substitute low-grade iron ores in their industrial process for high-grade iron ores. This urgent agenda is connected to the numerous studies (Yamaoka *et al*, 1974; Pimenta and Sechadri, 2002; Yang, 2005; Xue *et al*, 2020) in ironmaking research field to clarify the influence of gangue component on characteristics of iron ore sinter in sintering process where low-grade iron ores might be utilised.

Al₂O₃ has been known to be a representative of harmful gangue components affecting iron ore sinter. Several researchers (Yamaoka *et al*, 1974; Xiao *et al*, 2017) reported that Al₂O₃ increased the formation temperature of primary melt during the sintering process. Okazaki *et al* (2003) reported that Al₂O₃ inhibits the penetration of melt with high viscosity and suppresses assimilation between nuclei and adhering fines. This strongly affected the irregular pore distribution and further deteriorated the sinter strength, (Pimenta and Sechadri, 2002; Loo and Leung, 2003) which is inferior to the bed permeability of blast furnace (Umadevi *et al*, 2014). Degradation of sinter matrix is also one of the critical problems when Al atom was entrapped in lattice of hematite and magnetite phase and resulted in rapid propagation of cracks (Sinha and Ramna, 2009).

For counteracting the adverse effect of Al₂O₃ on iron ore sinter, several researchers (Yadav *et al*, 2002; Higuchi *et al*, 2004; Umadevi *et al*, 2012; Umadevi, Mahapatra and Prabhu, 2013) considered the small addition of FeO and MgO. They found that increased FeO is directly connected to the magnetite (Fe_3O_4). Sinter volume was decreased during the reduction of hematite to magnetite, which caused the microstructure of sinter dense (Biswas, 1981). This phenomenon is closely related to the improvement the overall sinter strength as shown by increase of Reduction Disintegration Index (Umadevi *et al*, 2012). Reduction behaviour of MgO in sintering process is similar to that of FeO because ionic radii of both cations are close. Panigrahy, Verstraeten and Dilewijns (1984) explained that MgO favours the formation of magnetite (Fe_3O_4) and Mg^{2+} ions replaced part of Fe^{2+} ions to stabilise the spinel structure. It implies that replacement of Fe^{2+} ions with Mg^{2+} ions can ensure the increase in excess Fe^{2+} content, which might contribute to the improvement of melting behaviour of iron ore sinter (Iwanaga, 1982; Ono-Nakazato, Sugahara and Usui, 2002; Lee *et al*, 2004; Chuang, Hwang and Liu, 2009).

In this study, combined addition of FeO- and MgO-bearing materials to sinter mix with high Al_2O_3 ore was investigated on the basis of reduction behaviour during sintering process. Mill-scale and calcined dolomite were selected to be the source of FeO and MgO, respectively. Mill-scale is FeO-based waste produced during hot-rolling and it is regarded as reusable by-product in both ironmaking and steelmaking process. Dolomite contains both CaO and MgO so melting behaviour control as well as fluxing effect can be expected through addition to high Al_2O_3 sinter. Calcination of dolomite was conducted to exclude the endothermic reaction on reduction process. Mineralogical analyses by X-ray diffraction (XRD) and Scanned electron microscopy (SEM-EDS) were employed to figure out how the combined addition of mill-scale and calcined dolomite interact and affect to the sinter microstructure. Physical properties of sinters were measured by Compression test and Reduction Disintegration Index (RDI) test to assess final sinter quality.

Materials preparation

All raw materials used in the study were supplied by POSCO, an industrial steelmaking company (Table 1). To produce standard sinter mix being used in general ironmaking industrial factory, two kinds of iron ores were prepared: Ore A and Ore B. Ore A is Goethite, FeO(OH)-based ore with high SiO_2 content and Loss on ignition (LOI). Ore B is high-grade hematite-based iron ore, containing low gangue content. Unlike these two iron ores, Ore C, which is typical pisolitic ore, was used as an Al_2O_3 source because it has been classified as a high-Al_2O_3 ore in several ironmaking companies. Dolomite was calcined at 1000°C before experiment.

TABLE 1
Chemical composition of raw materials (wt%).

Raw materials	T.Fe	M.Fe	FeO	Al₂O₃	CaO	SiO₂	MgO	MnO	P₂O₅	LOI
Ore A	57.00	0.07	0.17	1.59	0.06	5.39	0.08	0.02	0.08	10.38
Ore B	65.20	0.07	0.46	1.25	0.07	1.52	0.05	0.22	0.15	3.20
Ore C	57.10	0.06	0.20	3.14	0.04	5.35	0.05	0.03	0.14	8.59
Mill-scale	72.60	2.50	60.40	0.31	0.70	1.00	0.06	0.52	0.05	0.70
Calcined dolomite	-	-	-	-	56.19	-	38.78	-	-	2.53

To make uniform mixture, all samples were crushed under 50 µm and dried in oven for 24 hours. Ore A and Ore B were mixed with the mass ratio of 80 to 20 (Reference 1), which is simple blending of iron ores being typically utilised in actual ironmaking process. Reference 2 was prepared by mixing Ore A, Ore B and Ore C with the mass ratio of 48:12:40. Reference 2 has 2.31 wt per cent Al_2O_3 so that it is possible to compare it Reference 1 (1.76 wt per cent Al_2O_3) in terms of the adverse effect of Al_2O_3. Basicity, defined in this study as CaO/SiO_2, of two mixtures was adjusted to be 1.85 by adding a reagent-grade of CaO (purity 98 per cent) while mill-scale and calcined dolomite were added without considering basicity (Table 2). The sinter mixes were homogenised using a Turbula mixer for three hours and pressed into disk shape for 20 MP by a hydraulic work press machine.

TABLE 2

Calculated chemical composition of sinter mix with addition of mill-scale and calcined dolomite (wt%).

Sample	T.Fe	M.Fe	FeO	Al$_2$O$_3$	CaO	SiO$_2$	MgO	MnO	P$_2$O$_5$	CaO/SiO$_2$
Ref 1	53.97	0.07	0.21	1.40	7.87	4.26	0.07	0.06	0.09	1.85
Ref 2	52.71	0.06	0.19	2.28	8.56	4.63	0.05	0.04	0.11	1.85
Mill5	53.66	0.17	3.06	2.19	8.18	4.45	0.05	0.06	0.11	1.85
Mill10	54.52	0.28	5.67	2.10	7.84	4.30	0.06	0.08	0.11	1.85
Mill15	55.30	0.38	8.05	2.02	7.53	4.15	0.06	0.10	0.10	1.85
Mill10-Dol2	53.54	0.28	5.57	2.06	8.64	4.22	0.70	0.08	0.10	2.05
Mill10-Dol4	52.61	0.27	5.47	2.02	9.41	4.15	1.33	0.08	0.10	2.27
Mill10-Dol6	51.70	0.27	5.37	1.99	10.16	4.07	1.93	0.08	0.10	2.50

Sintering process

A disk-shaped sinter mix was sintered in a vertical tube furnace. That is, it was sintered while heating for seven mins in a furnace at 1280°C by preliminary experiment to optimise the fraction of SFCA phase. SFCA is abbreviation of Silico-Ferrite of Calcium and Aluminium, which is bonding phase in iron ore sinter as complex form of Calcium ferrite (CaO·Fe$_2$O$_3$). This phase is known as desirable for high reducibility and sinter strength of iron ore sinter (Nicol *et al*, 2018). After heating, the sinter was cooled down inside the furnace for 30 mins. Heating was conducted in an atmosphere consisting of CO, CO$_2$ and Ar with the ratio of 2:23:75 to maximise the formation of bonding phases according to Hsieh and Whiteman (1989).

Materials characterisation

All sinters were crushed to 500 μm and analysed by X-ray diffraction (XRD) using wavelength of 1.63Å (Cu-Kα) in the scanning range of 20 to 80°. Then, the sinter was mounted by cold resin and polished up to 1 μm grain size to employ Scanning electron microscopy (SEM). Elemental Dispersive (EDS) X-ray spectrometry was carried out to evaluate the elemental composition of each phase detected by XRD.

To assess the reduction behaviour of sinter, Compression Test and Reduction Disintegration Index (RDI) Test were performed. Compression Test was executed by Instron M5548 machine to characterise the superficial strength of sinter, measuring the maximum compressive load after gradual compression (Zhang *et al*, 2015). RDI Test was performed based on the international standard (ISO 4696-2:2015) Ten disk-shaped sinters were heated at 550°C for 30 mins in an atmosphere of 30 vol per cent CO and 70 vol per cent N$_2$ in a gold image furnace. After heating, the sinters were rotated inside Turbula Mixer at 60 rev/min for 15 mins. RDI$_{-2.8}$ was evaluated based on the weight of sinters retained on the 2.8 mm sieve. Reduction degree was calculated to describe the reducibility of a sinter based on ISO 7215:2015 where the sinter was heated at 900°C for three hours in an atmosphere 30 vol per cent CO and 70 vol per cent N$_2$ gas atmospheres in a gold image furnace.

RESULTS

Effect of adding mill-scale to sinter mix containing high Al$_2$O$_3$ ore on sinter quality

Mill-scale was added to Reference 2 up to 15 wt per cent to evaluate the effect of FeO-bearing materials on sinter mix containing high Al$_2$O$_3$. Three phases were mainly identified by X-ray diffraction (Figure 1): hematite (Fe$_2$O$_3$), magnetite (Fe$_3$O$_4$) and SFCA (Ca$_{2.3}$Mg$_{0.8}$Al$_{1.5}$Fe$_{8.3}$Si$_{1.1}$O$_{20}$). Existence of high Al$_2$O$_3$ ore in the sinter did not show a meaningful difference in case XRD patterns of Reference 1 and Reference 2 were compared. As two sinters showed similar XRD patterns, iron ore sinter manufactured by blending high Al_2O_3 ore in this study is under acceptable range in a view of phase development. This is connected to the previous study that low Al_2O_3 content in the sinter showed positive effect on SFCA formation (Scarlett *et al*, 2004; Webster *et al*, 2012), rather than contribution to the formation of slag phase. Hematite was dominant among the phases of all the sinters while magnetite was drastically increased as 10 wt per cent of mill-scale was added to Reference 2 (Mill10), eventually exceeding the hematite at 15 wt per cent addition (Mill15). Although hematite and magnetite shared the same peak at 33~34° and 62.5~63.5° in the scanning range, the portion of magnetite in the peak intensity is dominant in Mill15. In addition, the fraction of SFCA was barely changed up to Mill10, but it was decreased at Mill15. FeO in mill-scale is helpful to form bonding phases combined with Ca, Si and Al, but excess Fe^{2+} prevents the formation of SFCA because reducing atmosphere is induced during sintering process (Wang *et al*, 2016). At reducing atmosphere, FeO dissolves into Fe_2O_3 to form magnetite. From reason that phase composition with sufficient SFCA and hematite phase is maintained, mill-scale addition can be optimised as 10 wt per cent into high Al_2O_3 sinter.

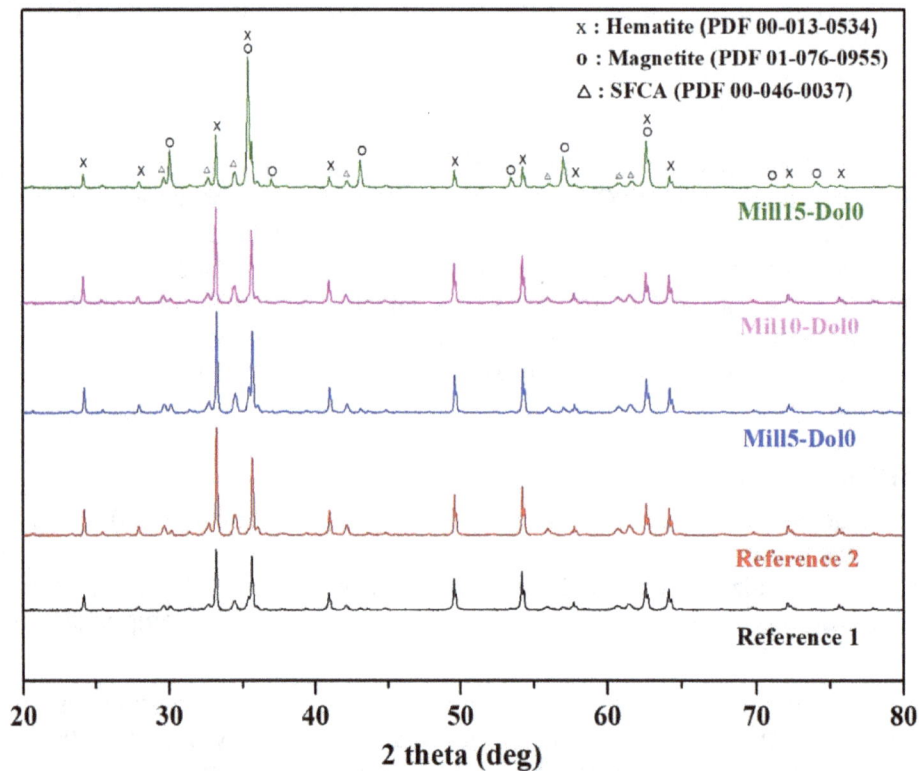

FIG 1 – Phase identification of the sinters to which mill-scale was added by X-ray diffraction.

Figure 2 is the backscattered electron images of the sinters with increasing added amount of mill-scale and EDS results of the selected points at Figure 2 is shown in Table 3. White grains composed of hematite occupied the sinter matrix (Point 1, 3, 5, 7) and grey SFCA were partly observed as the acicular form (Point 4, 6, 8) or columnar form (Point 2, 6, 10). When 5 wt per cent and 10 wt per cent of mill-scale was added, sinters maintained their microstructure and the ratio of hematite and SFCA was consistent from Reference 2. Meanwhile, the fraction of Fe was largely increased at Point 9 in Mill15, indicating that magnetite possessed more area than hematite. Acicular and columnar texture were decomposed as more Fe was diffused into magnetite matrix and the boundaries between magnetite and SFCA were observed to be blurred. It is clearly indicated at Point 10, which portion of Fe in SFCA phase was notably increased compared to that at Point 6 and Point 8. It is not recommended for overall sinter quality, since the irregular morphology of SFCA is detrimental for reducibility and cold strength of iron ore sinter (Ying, Jiang and Xu, 2006; Takayama, Murao and Kimura, 2018).

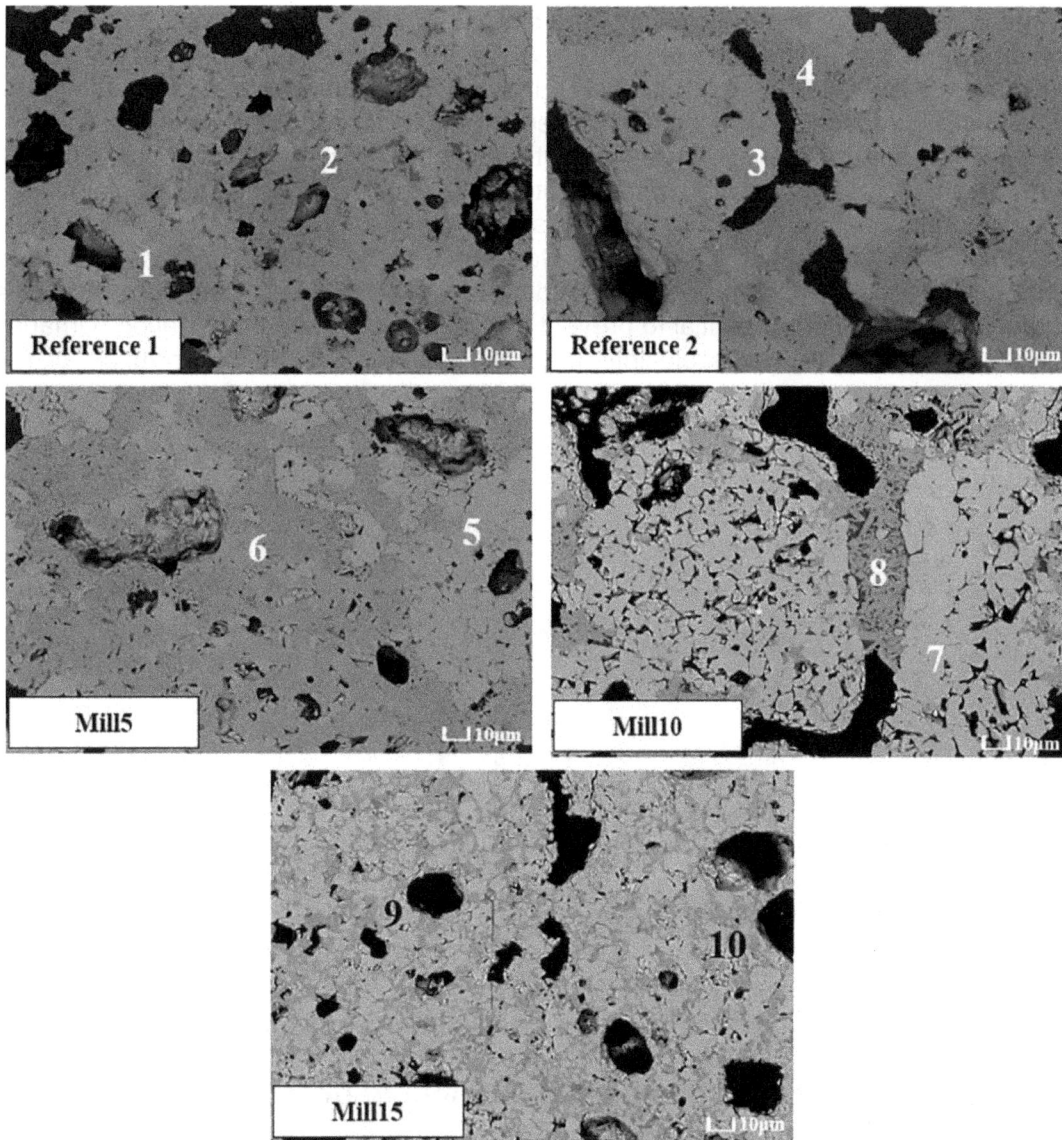

FIG 2 – Morphological features of the sinters to which mill-scale was added.

TABLE 3

Elemental distribution of selected points in the sinters to which mill-scale was added (at%).

Sinters	Pts	Elemental composition						Identified phases
		Fe	Ca	Si	Al	Mg	O	
Reference 1	1	37.15	0.14	0.13	0.49	-	62.09	Fe_2O_3
	2	33.81	4.10	2.69	1.87	-	57.53	SFCA
Reference 2	3	37.93	0.16	0.05	0.55	-	61.23	Fe_2O_3
	4	26.17	7.95	3.29	2.42	-	59.86	SFCA
Mill5	5	34.93	2.29	0.91	0.48	-	61.39	Fe_2O_3
	6	29.74	8.45	3.45	3.15	-	55.26	SFCA
Mill10	7	36.89	0.51	0.31	1.43	-	60.85	Fe_2O_3
	8	31.34	7.75	3.29	2.35	-	55.29	SFCA
Mill15	9	42.28	0.05	0.12	0.38	-	57.16	Fe_3O_4
	10	23.79	8.34	5.26	3.71	-	58.91	SFCA

Figure 3a depicts the reduction behaviour of the sinters as mill-scale was added to the sinter mix. Reduction degree was slightly decreased as high Al_2O_3 ore was blended, which is affected by Al_2O_3 content (Umadevi *et al*, 2009; Sinha *et al*, 2017). In case mill-scale was added to sinters, reduction degree was firstly increased at Mill5, but decreased then. Even it showed lower reducibility than that of Reference 2. It is obvious that FeO acted as a strong factor affecting the decrease in the reduction degree of sinters by increasing portion of magnetite in the sinter. These can be also applied into the change in compressive load and $RDI_{-2.8}$ of the sinters as shown in Figure 3b. For Reference 2, compressive load was decreased while $RDI_{-2.8}$ was increased compared to those of Reference 1. Both trends showed clear improvement after adding mill-scale. Several researchers have already reported that magnetite has a potential to prevent the low-temperature degradation (Umadevi *et al*, 2012; An, 2022), so adding mill-scale into high Al_2O_3 sinter can be the solution of reduction degradation.

(a)   **(b)**

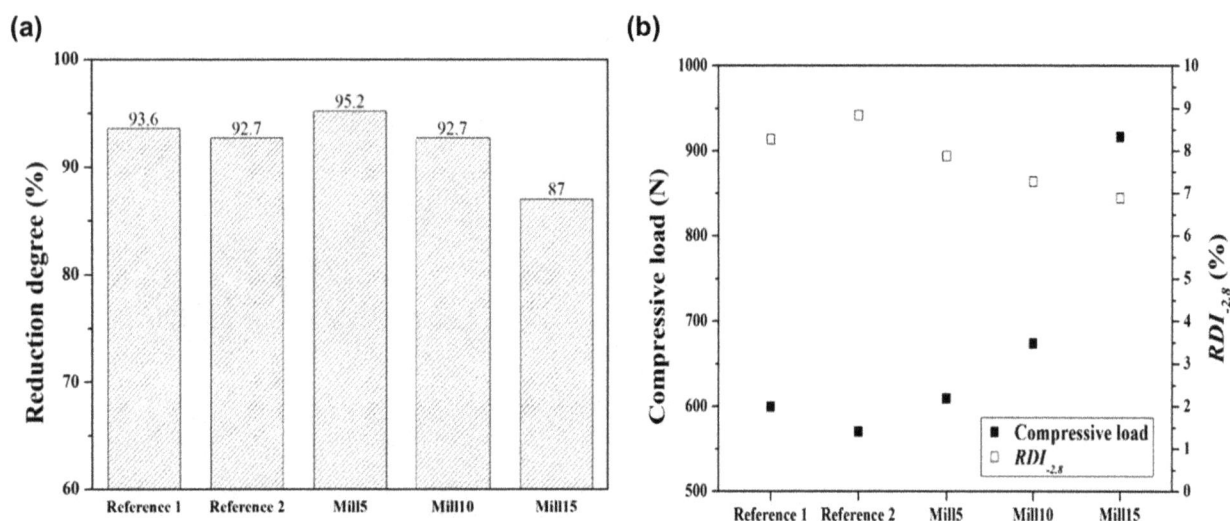

FIG 3 – (a) Change in reduction degree of the sinters to which mill-scale was added; (b) Change in strength and disintegration behaviour of the sinters to which mill-scale was added.

Combined effect of adding mill-scale and calcined dolomite to sinter mix containing high Al_2O_3 ore on sinter quality

Based on the obtained results in the previous section, calcined dolomite was added to the sinter mix (Mill10) up to 6 wt per cent. As shown in Figure 4, three major phases were identified as similar to the case in Figure 2, however, magnesio-ferrite ($MgO \cdot Fe_2O_3$) was mainly observed instead of magnetite at identical peaks. It is required to be considered that several Mg^{2+} ions diffused into magnetite and replaced the Fe^{2+} sites since the cations of Mg^{2+} and Fe^{2+} have the similar ionic radius (Panigrahy, Verstraeten and Dilewijns, 1984). In addition, the fraction of SFCA was maintained at Mill10-Dol2, but it was gradually decreased as more than 4 wt per cent of calcined dolomite was added (Mill10-Dol4). Magnesio-ferrite was increased and became dominant at Mill10-Dol6. The portion of magnesio-ferrite was largest among all other sinters in this study although the total amount of FeO and MgO was lower than that in Mill15. That is, the effect of adding calcined dolomite on the stabilisation of this magnesio-spinel structure was more remarkable than that of adding mill-scale.

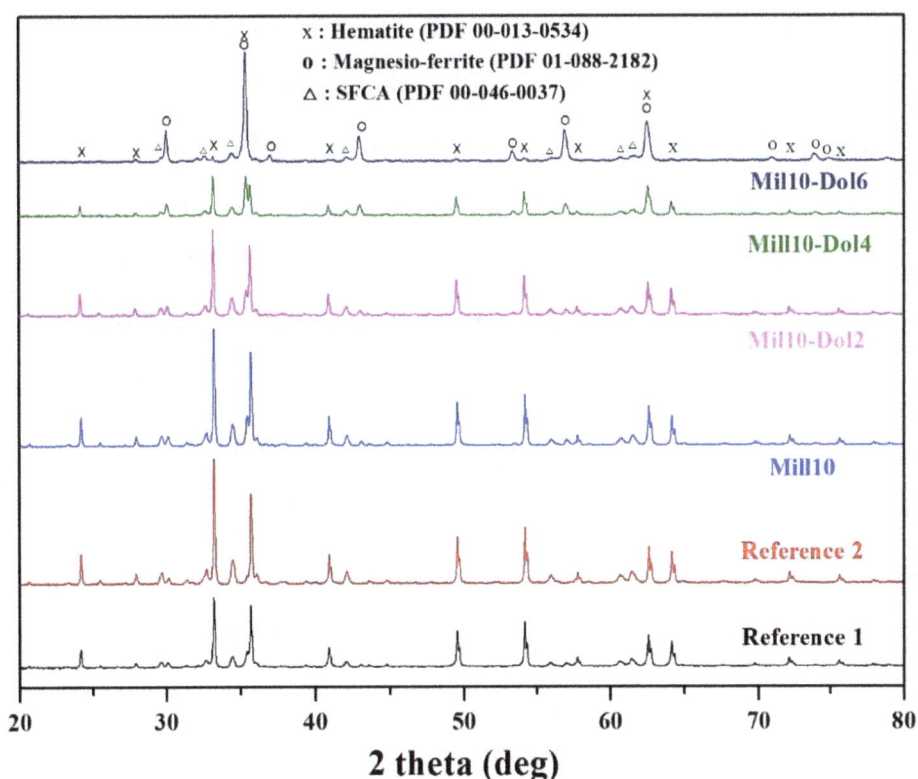

FIG 4 – Phase identification of the sinters to which mill-scale and calcined dolomite were added by X-ray diffraction.

Figure 5 shows the backscattered electron images of the sinters when mill-scale and calcined dolomite were added. Mg was found to be distributed in Fe-enriched regions of all the sinters containing calcined dolomite. EDS results of the selected points at Figure 5 is shown in Table 4. Mill10-Dol2 showed similar morphology to that of the sinter not containing dolomite (Mill10), but residual MgO in hematite matrix was concentrated at Mill10-Dol4 and Mill10-Dol6, which was indicated at Point 8. Residual MgO was surrounded by magnesio-ferrite (Point 9) and calcium silicate (Point 10) while other regions without MgO segregation showed typical microstructure consisting of magnesio-ferrite (Point 5) and SFCA (Point 7). Higuchi, Tanaka and Sato (2007) found similar phenomenon that rim structure consisting of magnetite solid solution was formed by the interaction between MgO and hematite or calcium ferrite, which is similar to the morphology of Mill10-Dol6 in Figure 5. Also, Al_2O_3 preferred to be absorbed into SFCA (Point 2, 4, 7) except Point 10. As addition of calcined dolomite was expanded, more Al_2O_3 became remained in Calcium silicate and caused the increase of slag phase.

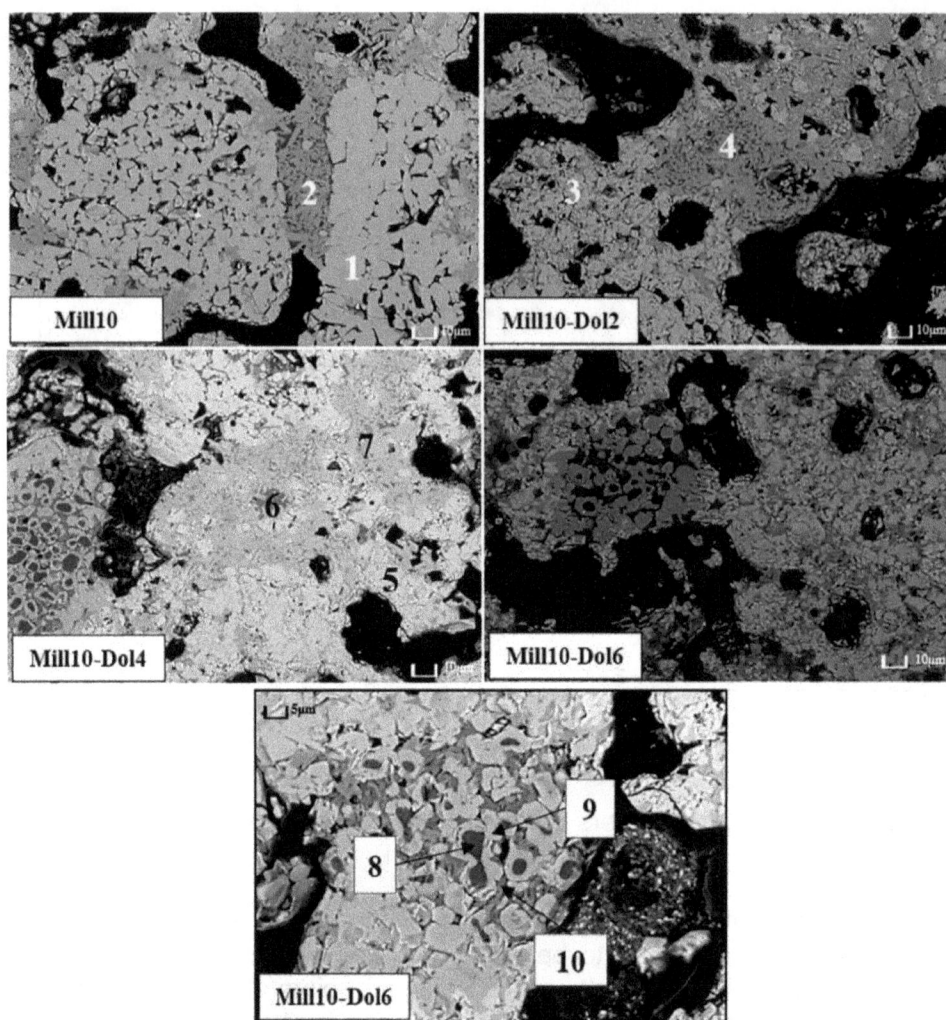

FIG 5 – Morphological features of the sinters to which mill-scale and calcined dolomite were added.

TABLE 4

Elemental distribution of selected points in the sinters to which mill-scale and calcined dolomite were added (at%).

| Sinters | Pts | Elemental composition | | | | | | Identified phases |
		Fe	Ca	Si	Al	Mg	O	
Mill10	1	36.89	0.51	0.31	1.43	-	60.85	Fe_2O_3
	2	31.34	7.75	3.29	2.35	-	55.29	SFCA
Mill10-Dol2	3	37.88	1.32	0.25	0.89	0.94	58.72	Fe_2O_3
	4	23.71	7.77	4.63	2.47	0.25	61.16	SFCA
Mill10-Dol4	5	26.13	0.81	0.19	1.37	12.28	59.21	Fe_2O_3, $MgO \cdot Fe_2O_3$
	6	1.47	21.31	14.88	0.50	0.12	61.73	$2CaO \cdot SiO_2$, $CaO \cdot SiO_2$
	7	25.81	7.43	3.50	2.93	0.54	59.80	SFCA
Mill10-Dol6	8	4.47	0.25	0.19	0.08	44.13	50.88	MgO
	9	28.08	0.72	0.21	2.27	12.09	56.63	$MgO \cdot Fe_2O_3$
	10	3.95	14.84	12.32	3.09	4.03	61.77	$CaO \cdot SiO_2$

Figure 6 show relationship between portion of magnetite and sinter strength of sinters when calcined dolomite was added together with mill-scale. There was no significant difference in reduction degree when calcined dolomite was added 2 wt per cent into Mill10, but it was declined from 4 wt per cent addition. Reduction degree in Mill10-Dol6 was calculated as 69.6 per cent, which is not suitable in industrial sintering process (Umadevi *et al*, 2014). Both compressive load and RDI$_{-2.8}$ were worsened compared to those of Mill10, but it changed to be improved along the calcined dolomite addition. Calcined dolomite seems to be acted as same as mill-scale to improve sinter strength and reduction degradation of high Al$_2$O$_3$ sinter, but further explanation is required to determine the individual role of calcined dolomite separated from that of mill-scale.

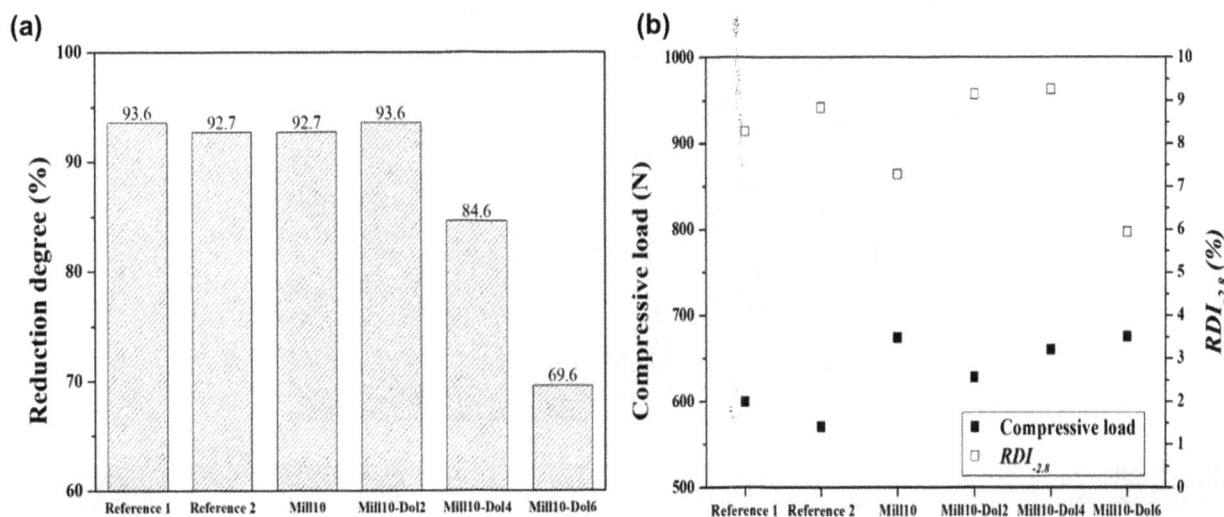

FIG 6 – (a) Change in reduction degree of the sinters to which mill-scale and/or calcined dolomite were added; (b) Change in strength and disintegration behaviour of the sinters to which mill-scale and/or calcined dolomite were added.

DISCUSSION

To clarify the mechanism of phase development of high Al$_2$O$_3$ sinter when mill-scale and calcined dolomite was added, the elemental distribution of Mill10-Dol6 was specified by EDS mapping as shown in Figure 7. The clusters consisting of Mg were surrounded by reddish part comprising Mg and Fe. MgO remained unreacted rather than forming magnesio-ferrite in the sintering process in case the large amount of calcined dolomite was added. It is clear that there is certain limitation in the dissolution of MgO into magnetite matrix. Or the relict MgO with high sintering temperature will block the further assimilation of bonding phases, further act as harmful to overall sinter qualities such as reducibility and toughness (Han *et al*, 2019; Shi *et al*, 2020). In case of Ca and Si, they were locally concentrated around MgO segregation and formed slag phase indicated by blue part. Excessive CaO supplied by calcined dolomite did not react with Fe to form Calcium Ferrite or SFCA. Instead, calcium silicate was formed in microstructure, so this slag phase would affect to the poor reducibility of high Al$_2$O$_3$ sinter.

Physical properties of high Al$_2$O$_3$ sinter with addition of mill-scale and calcined dolomite also can be explained based on phase development after iron ore sintering. Al was found to be homogeneously distributed in whole microstructure, respectively. It is desirable in terms of reduction degradation, since Al was reported to be entrapped in hematite matrix during reduction of hematite to magnetite (Sinha and Ramna, 2009; Umadevi *et al*, 2009) to induce lattice distortion. Since Al$_2$O$_3$ exists in between the lattices of hematite and magnetite and leads rapid crack propagation (Sinha and Ramna, 2009), even distribution of Al in microstructure has an advantage on relaxing this crack propagation by forming dense matrix (Umadevi *et al*, 2012). It is similar to the result of Zhu *et al* (2020), who found the decreased crack density of iron ore sinter to which MgO content was increased. MgO generally decreases the porosity of sinter (Umadevi *et al*, 2014) by stabilising the spinel structure of magnetite so called magnesio-spinel (Panigrahy, Verstraeten and Dilewijns, 1984). Magnesio-spinel phase changes the magnetite matrix dense to resist against external pressure, further contributes to the improvement of compressive strength of iron ore sinter. This effect is dominant in this study, showing the increase of compressive load of iron ore sinter when calcined dolomite was utilised. However, MgO segregation should be considered in detail when controlling the addition of MgO-bearing material into high Al_2O_3 sinter. The portion of the phases where Al_2O_3 can diffuse would be decreased if the portion of MgO clusters is expanded. It implies the acceleration of reduction degradation caused by diffusion of Al into hematite-magnetite matrix. Thus, addition of calcined dolomite in high Al_2O_3 sinter in this study is recommended to be under 4 wt per cent to avoid MgO segregation shown in Figure 7.

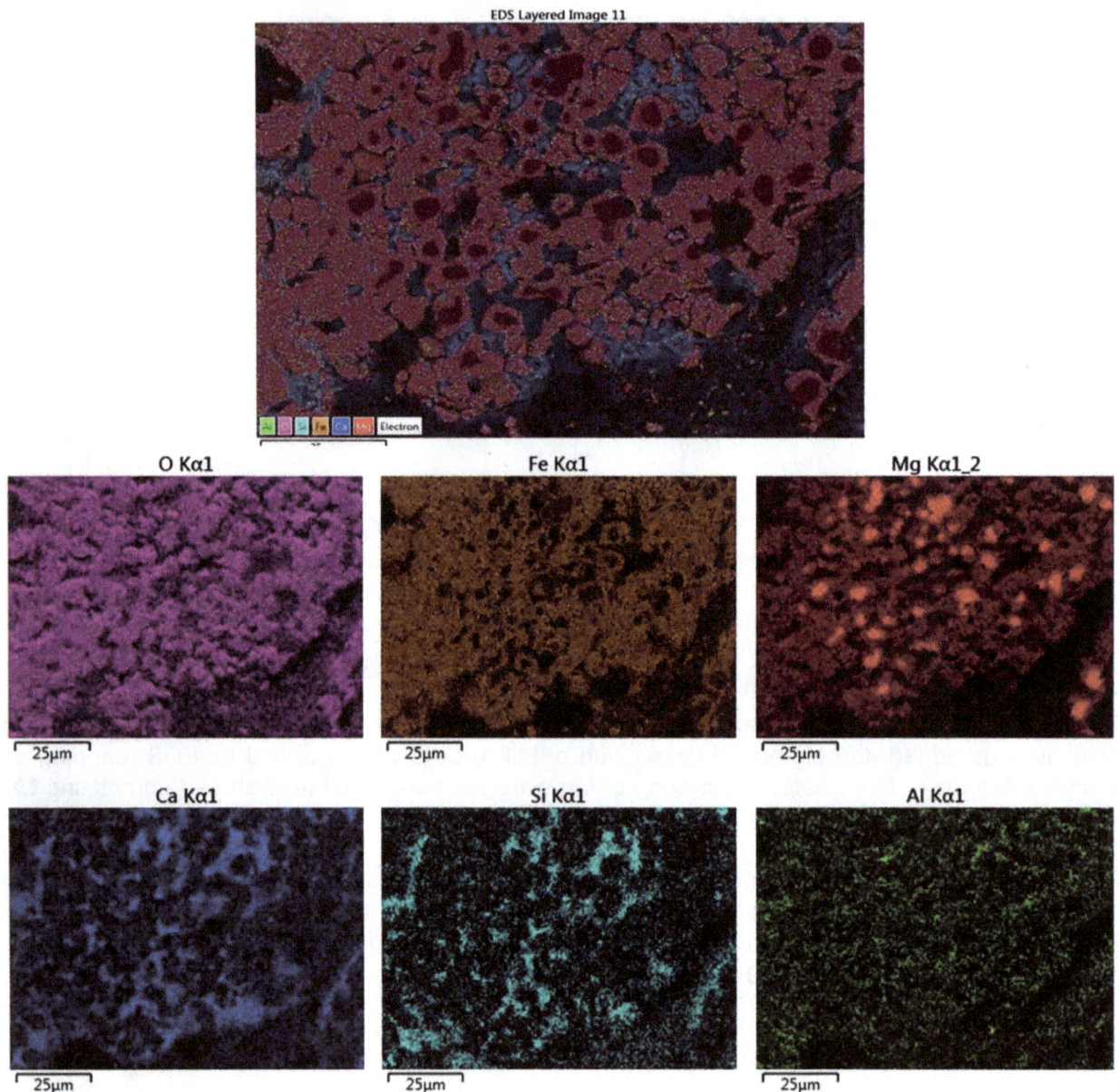

FIG 7 – Elemental distribution of selected area in the sinter to which 10 wt per cent of mill-scale and 6 wt per cent of calcined dolomite were added (Mill10-Dol6) by EDS mapping.

CONCLUSIONS

To clarify the combined effect of adding mill-scale and calcined dolomite to sinter mix containing high Al_2O_3 ore on sinter quality, the reduction behaviour of iron ore sinters was investigated. From the findings, the following conclusions were obtained:

- Magnetite became a dominant phase in the sinter with increasing mill-scale up to 15 wt per cent of mill-scale to sinter mix while magnesio-ferrite was formed instead of magnetite when calcined dolomite was added to sinter mix along with mill-scale. MgO stabilised the magnetite phase, so that magnesio-ferrite in iron ore sinter was remarkably increased while SFCA phase was decreased, respectively.

- Several MgO clusters were observed as a residue with adding more than 4 wt per cent of calcined dolomite to sinter mix. Due to the limitation in the dissolution of MgO into magnetite matrix, MgO remained unreacted in the sintering process, resulting in the suppression of bonding phases including calcium ferrite. In addition, CaO supplied by calcined dolomite formed Calcium silicate as slag phase rather than forming Calcium ferrite or SFCA.

- RDI$_{-2.8}$ and compressive load of the iron ore sinters were improved while their reduction degrees were decreased when adding mill-scale and calcined dolomite. Calcined dolomite can play a same role as mill-scale to stabilise the magnesio-spinel for improving sinter strength and reduction degradation. It has a strong advantage of utilisation of both FeO and MgO-bearing materials for counteracting the adverse effect of Al_2O_3 in high Al_2O_3 sinter, also decreasing FeO amount in final manufactured sinter. However, the total content of additives should be restricted under certain amount to avoid MgO segregation and maintain the overall sinter qualities.

ACKNOWLEDGEMENTS

The authors would like to thank Mr Sang Yeol Lee for help with preparation of experimental procedures.

REFERENCES

An, H, Shen, F, Jiang, X, Zhou, Y, Zheng, H and Gao, Q, 2022. Effects of Fine MgO-Bearing Flux on the Strength of Sinter before and after Low-Temperature Reduction, *ACS Omega*, 7:8686–8696.

Biswas, A, 1981. *Principles of Blast Furnace Ironmaking: Theory and Practice*, pp 105 (Cootha Publishing House: Brisbane).

Chai, Y, Yu, W, Zhang, J, An, S, Peng, J and Wang, Y, 2019. Influencing mechanism of Al_2O_3 on sintered liquid phase of iron ore fines based on thermal and kinetic analysis, *Ironmaking and Steelmaking*, 46:424–430.

Chuang, H, Hwang, W and Liu, S, 2009. Effect of Basicity and FeO Content on the Softening and Melting Temperatures of the $CaO-SiO_2-MgO-Al_2O_3$ Slag System, *Mater Trans*, 50:1448–1456.

Han, H, Shen, F, Jiang, X, Bi, C, Zheng, H and Gao, Q, 2019. Fundamental mechanism of effects of MgO on sinter strength, *J Iron Steel Res Int*, 26:1171–1177.

Higuchi, K, Naito, M, Nakano, M and Takamoto, Y, 2004. Optimization of Chemical Composition and Microstructure of Iron Ore Sinter for Low-temperature Drip of Molten Iron with High Permeability, *ISIJ Int*, 44:2057–2066.

Higuchi, K, Tanaka, T and Sato, T, 2007. Reaction Behaviour of Dolomite Accompanied with Formation of Magnetite Solid Solution in Iron Ore Sintering Process, *ISIJ Int*, 47:669–678.

Hsieh, L and Whiteman, J, 1989. Sintering Conditions for Simulating the Formation of Mineral Phases in Industrial Iron Ore Sinter, *ISIJ Int*, 29:24–32.

International Organization for Standardization (ISO), 2015a. ISO 4696-2:2015 – Iron ores for blast furnace feedstocks – Determination of low-temperature reduction-disintegration indices by static method – Part 2: Reduction with CO and N2, September 2015.

International Organization for Standardization (ISO), 2015b. ISO 7215:2015 – Iron ores for blast furnace feedstocks – Determination of the reducibility by the final degree of reduction index, August 2015.

Iwanaga, Y, 1982. Fundamental Study on the Softening Properties of Sinter by Measuring the Apparent Softening Viscosity, *Tetsu-to-Hagane*, 68:2223–2230.

Lee, Y, Min, D, Jung, S and Yi, S, 2004. Influence of Basicity and FeO Content on Viscosity of Blast Furnace Type Slags Containing FeO, *ISIJ Int*, 44:1283–1290.

Loo, C and Leung, W, 2003. Factors Influencing the Bonding Phase Structure of Iron Ore Sinters, *ISIJ Int*, 43:1393–1402.

Nicol, S, Chen, J, Pownceby, M and Webster, N, 2018. A review of the Chemistry, Structure and Formation Conditions of Silico-Ferrite of Calcium and Aluminum ('SFCA') Phase, *ISIJ Int*, 58:2157–2172.

Okazaki, J, Higuchi, K, Hosotani, Y and Shinagawa, K, 2003. Influence of Iron Ore Characteristics on Penetrating Behaviour of Melt into Ore Layer, *ISIJ Int*, 43:1384–1392.

Ono-Nakazato, H, Sugahara, C and Usui, T, 2002. Effect of Slag Components on Reducibility and Melt Formation of Iron Ore Sinter, *ISIJ Int*, 42:558–560.

Panigrahy, S, Verstraeten, P and Dilewijns, J, 1984. Influence of MgO Addition on Mineralogy of Iron Ore Sinter, *Metal Trans B*, 15:23–32.

Pimenta, H and Sechadri, V, 2002. Characterisation of structure of iron ore sinter and its behaviour during reduction at low temperatures, *Ironmaking and Steelmaking*, 29:169–174.

Scarlett, N, Pownceby, M, Madsen, I and Christensen, A, 2004. Reaction Sequences in the Formation of Silico-Ferrite of Calcium and Aluminum in Iron Ore Sinter, *Metal Mater Trans B*, 35:929–936.

Shi, B, Zhu, D, Pan, J, Liu, X and Li, S, 2020. Combined effect of MgO and basicity varied by different dolomite and burnt lime addition on sintering performance of magnetite concentrates, *Ironmaking and Steelmaking*, 47:567–573.

Sinha, M and Ramna, R, 2009. Effect of Variation of Alumina on the Microhardness of Iron Ore Sinter Phases, *ISIJ Int*, 49:719–721.

Sinha, M, Nistala, S, Chandra, S, Mankhand, T and Ghose, A, 2017. Correlating mechanism properties of sinter phases with their chemistry and its effect on sinter quality, *Ironmaking and Steelmaking*, 44:100–107.

Takayama, T, Murao, R and Kimura, M, 2018. Quantitative Analysis of Mineral Phases in Iron-ore Sinter by the Rietveld Method of X-ray Diffraction Patterns, *ISIJ Int*, 58:1069–1078.

Umadevi, T, Brahmacharyulu, A, Karthik, P, Mahapatra, P, Prabhu, M and Ranjan, M, 2012. Recycling of steel plant mill scale via iron ore sintering plant, *Ironmaking and Steelmaking*, 39:222–227.

Umadevi, T, Brahmacharyulu, A, Sah, R, Mahapatra, P and Prabhu, M, 2014. Optimisation of MgO addition in low and high silica iron ore sinter to improve sinter reducibility at JSW Steel Limited, *Ironmaking and Steelmaking*, 41:270–278.

Umadevi, T, Mahapatra, P and Prabhu, M, 2013. Influence of MgO addition on microstructure and properties of low and high silica iron ore sinter, *Miner Proc Extr Metall*, 122:238–248.

Umadevi, T, Nelson, K, Mahapatra, P, Prabhu, M and Ranjan, M, 2009. Influence of magnesia on iron ore sinter properties and productivity, *Ironmaking and Steelmaking*, 36:515–520.

Wang, Z, Maeda, T, Ohno, K and Kunitomo, K, 2020. Effect of Magnetite on Mineral Phase Formation in Sintering Process, *ISIJ Int*, 60:233–237.

Wang, Z, Pinson, D, Chewm, S, Monaghan, B, Pownceby, M, Webster, N, Rogers, H and Zhang, G, 2016. Effect of Addition of Mill Scale on Sintering of Iron Ores, *Metal Mater Trans B*, 47:2848–2860.

Webster, N, Pownceby, M, Madsen, I and Kimpton, J, 2012. Silico-ferrite of Calcium and Aluminum (SFCA) Iron Ore Sinter Bonding Phases: New Insight into Their Formation During Heating and Cooling, *Metal Mater Trans B*, 43:1344–1357.

Xue, Y, Pan, J, Zhu, D, Guo, Z, Yang, C, Lu, L and Tian, H, 2020. Improving High-Alumina Iron Ores Processing via the Investigation of the Influence of Alumina Concentration and Type on High-Temperature Characteristics, *Minerals*, 10:1–26.

Xiao, Z, Chen, L, Yang, Y, Li, X and Barati, M, 2017. Effect of Coarse-grain and Low-grade Iron Ores on Sinter Properties, *ISIJ Int*, 57:795–804.

Yadav, U, Pandey, B, Das, B and Jena, D, 2002. Influence of magnesia on sintering characteristics of iron ore, *Ironmaking and Steelmaking*, 29:91–95.

Yamaoka, Y, Nagaoka, S, Yamada, Y and Ando, R, 1974. Effect of Gibbsite on Sintering Property and Sinter Quality, *Transaction ISIJ*, 14:185–194.

Yang, L, 2005. Sintering Fundamentals of Magnetite Alone and Blended with Hematite and Hematite/Goethite Ores, *ISIJ Int*, 45:469–476.

Ying, Z, Jiang, M and Xu, L, 2006. Effect of Mineral Composition and Microstructure on Crack Resistance of Sintered Ore, *Journal of Iron and Steel Research, International*, 13:9–12.

Zhang, G, Wu, S, Que, Z, Hou, C and Jiang, Y, 2015. Influencing factor of sinter body strength and its effects on iron ore sintering indexes, *Int J Min Metal Mat*, 22:553–561.

Zhu, D, Xue, Y, Pan, J, Yang, C, Guo, Z, Tian, H, Wang, D and Shi, Y, 2020. An investigation into alumina occurrence impact on SFCA formation and sinter matrix strength, *J Mater Res Tech*, 9:10223–10234.

Slag-metal interfacial reactions in pyrometallurgical processing of industrial wastes for recovery of valuable metals

H J Kim[1], R R Kim[2], H S Park[3] and J H Park[4]

1. Student, Department of Materials Science and Chemical Engineering, Hanyang University, Ansan 15588, South Korea. Email: kimhyunju1158@naver.com
2. Student, Department of Materials Science and Chemical Engineering, Hanyang University, Ansan 15588, South Korea. Email: krr0432@naver.com
3. Senior Researcher, Korea Institute of Geoscience and Mineral Resources (KIGAM), Daejeon 34132, South Korea. Email: hyunsik.park@kigam.re.kr
4. Professor, Department of Materials Science and Chemical Engineering, Hanyang University, Ansan 15588, South Korea. Email: basicity@hanyang.ac.kr

ABSTRACT

This study focuses on the recovery of valuable metals, specifically silver (Ag) and palladium (Pd), from copper-containing sludge (with copper content less than 20 per cent) generated during PCB processing and from spent petrochemical catalysts used in vapour decomposition during oil refining processes. The present results indicate that an increase in the mixing ratio of alumina-based spent catalyst leads to a decrease in the recovery rate of Ag and Pd. The recovery rates for Ag and Pd range from 84 per cent to 96 per cent and 80 per cent to 98 per cent, respectively. This trend is attributed to the increased emulsification of copper droplets containing Ag and Pd in the slag. The emulsification is caused by the decrease in the settling velocity of the copper droplets due to the increased viscosity of the slag as well as by the decrease in the slag-metal interfacial tension due to an increase in sulfur content at higher mixing ratio of spent catalyst.

INTRODUCTION

Wasted printed circuit boards (WPCBs) are components of electronic waste that contain high concentrations of precious metals, such as gold (Au) and silver (Ag), exceeding average ore concentrations by 20 to 30 times (Ebin and Isik, 2016). Similarly, in the petrochemical industry, catalysts rich in precious metals like platinum group metals (PGMs) such as Pt, Pd and Rh exhibit concentrations approximately five times higher than those found in natural resources (Ghalehkhondabi, Fazlali and Daneshpour, 2021). The rise in industrial waste has a negative impact on the environment, but it also presents opportunities for extracting valuable metals and raw materials.

Several studies have been conducted on the recovery of metals from industrial wastes. Kim *et al* (2004) developed smelting processes that extract precious metals, such as Au, Pd and Pt, through mixing PCBs and auto catalysts. Up to 90 per cent of Au, Pd and Pt in the raw materials were concentrated in a Cu-Sn alloy phase. Kwon, Han and Jeong (2005) and Shin *et al* (2008) conducted similar experiments and reported that for the efficient recovery of valuable metals, it is advantageous to use fluxes that can lower the melting point and viscosity of the slag.

Therefore, in the present study, we aim to determine the optimal slag conditions for the recovery of valuable metals through the $CaO-Al_2O_3$ based slag system, which is generated when the spent petrochemical catalyst and copper sludge were employed as raw materials. Several factors (eg basicity, viscosity and interfacial tension) influencing the recovery efficiency of valuable metals were discussed.

EXPERIMENTAL PROCEDURE

The compositions of the raw materials for valuable metal extraction, ie copper sludge and spent petrochemical catalysts, were analysed using an X-ray diffraction (XRD), Inductively Coupled Plasma – Atomic Emission Spectroscopy (ICP-AES) (ACROS, Spectro) and combustion analyser (CS800, ELTRA). The results are presented in Table 1.

TABLE 1

Composition of copper sludge and spent petrochemical catalysts (wt%).

	Al_2O_3	$CaCO_3$	Cu_2O	Fe_2O_3	SnO_2	P_2O_5	SiO_2	MgO	C	S	Ag	Pd
Copper sludge	0.7	29.9	24.7	22.3	6.7	3.5	2.6	1.7	5.6	2.4	0.01	-
Spent catalyst	99.7	-	-	-	-	-	-	-	-	-	-	0.3

When blending the Al_2O_3-based spent catalyst and copper sludge, the mixing ratio of the spent catalyst was varied from 10 per cent to 30 per cent. Here, the mixing condition aims to limit the Al_2O_3 content not more than 30 wt per cent to prevent an abrupt increase in melting point of the slag. Three experiments were carried out using a high frequency induction furnace (Figure 1). A mixture of 100 g raw materials (copper sludge and spent petrochemical catalyst) and 60 g copper powder (to form collecting melt pool) were loaded in a fused magnesia crucible with a graphite heater for efficient heating. The furnace was filled with purified Ar gas controlled by a mass flow controller at a flow rate 500 mL/min. After the equilibration for 60 min at 1773 K, the samples were cooled under an Ar gas atmosphere.

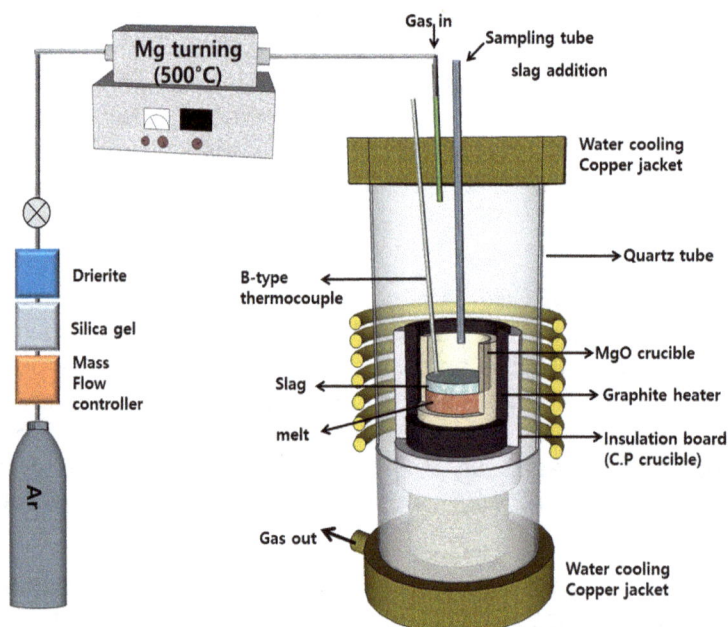

FIG 1 – Schematic diagram of the experimental apparatus.

RESULTS AND DISCUSSION

After experiments, the slag and metal were carefully separated from the crucible. the interface between the metal and slag was clearly defined and allowed a complete separation.

The slag was ground to a fine powder for chemical analysis. The equilibrium composition of the slag was analysed by ICP-AES (OPTIMA 8300; PerkinElmer, Waltham, MA) and combustion analyser (TC-300, LECO). The results of all experiments are represented in Figure 2. Metal droplets and oxides in the slag were observed by using Field Emission – Scanning Electron Microscope (FE-SEM) and Energy Dispersive X-ray Spectroscopy (EDS) (Nova Nano SEM 450, FEI).

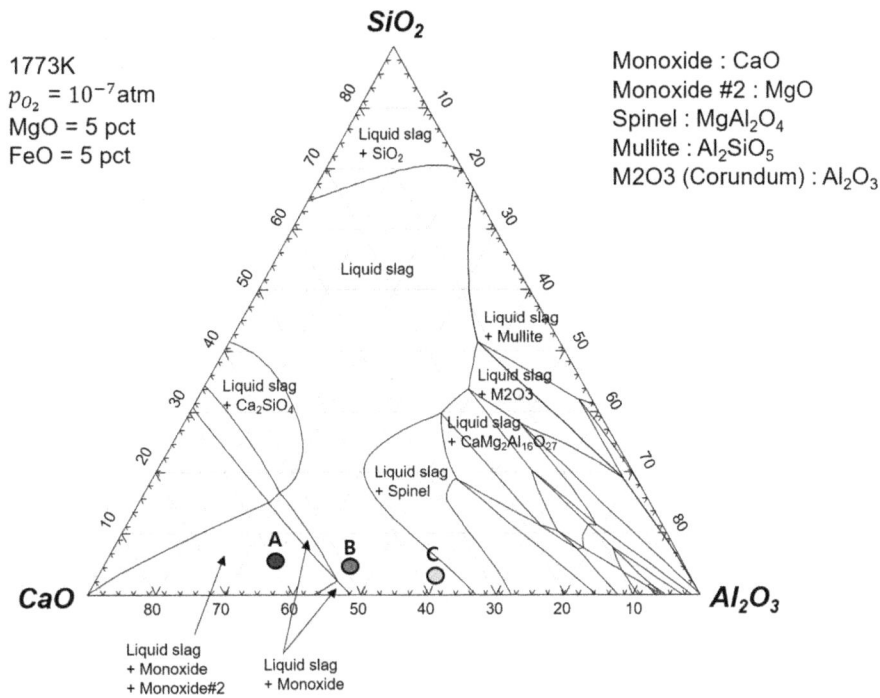

FIG 2 – The 1773 K isotherm in the CaO-SiO$_2$-Al$_2$O$_3$ slag system (wt per cent) (A, B and C represent the final slag compositions with spent catalyst ratios of 10 per cent, 20 per cent and 30 per cent).

As shown in Table 2, the vertical section of metal ingots obtained from the experiments with 20 per cent and 30 per cent spent catalyst, both Cu-rich (yellow) and Fe-rich (grey) phases are present, while the only Cu-rich ingot is produced from the 10 per cent catalyst experiment. However, Ag and Pd have a high solubility in copper phase, representing that they exist in the Cu-rich phase. The recovery rate was calculated using the following formula (Lee *et al*, 2010).

$$\text{Recovery rate of M}(= \text{Ag, Pd}) \ (\%) = \frac{Recovered \ amount \ of \ M \ in \ metal \ ingot \ (g)}{Total \ content \ of \ M \ in \ raw \ materials \ (g)} \times 100 \qquad (1)$$

TABLE 2

Morphology of metal and slag separated after smelting experiments.

	10% spent catalyst	20% spent catalyst	30% spent catalyst
Metal ingot			
Cross-section of ingot			
Slag			

As the alumina content increases, which indicates an increase in the spent catalyst mixing ratio, the rate of recovery of Ag and Pd significantly decreases, as shown in Figure 3a. Figure 3b and 3c illustrate the presence of copper droplets confined to the slag formed during 30 per cent spent catalyst experiment, indicating the incomplete metal separation. Therefore, physical entrainment of copper droplets is the primary cause of copper loss as well as lower recovery rate of Ag and Pd.

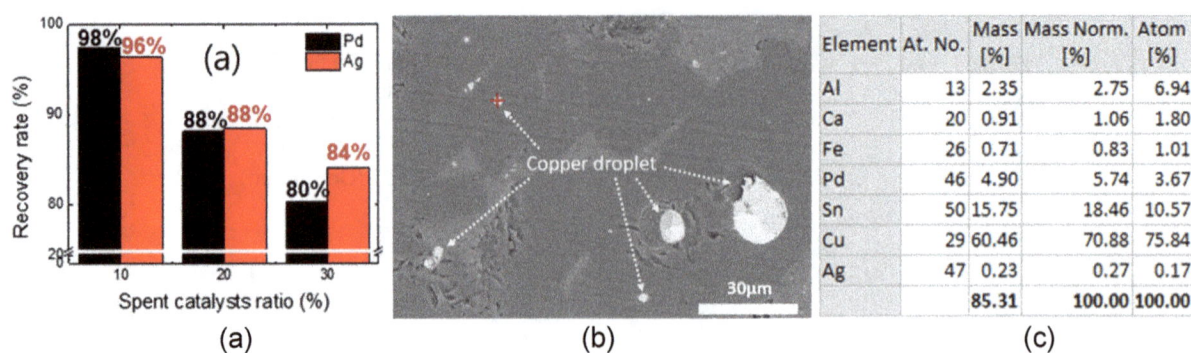

Element	At. No.	Mass [%]	Mass Norm. [%]	Atom [%]
Al	13	2.35	2.75	6.94
Ca	20	0.91	1.06	1.80
Fe	26	0.71	0.83	1.01
Pd	46	4.90	5.74	3.67
Sn	50	15.75	18.46	10.57
Cu	29	60.46	70.88	75.84
Ag	47	0.23	0.27	0.17
		85.31	100.00	100.00

(a) (b) (c)

FIG 3 – (a) Recovery rate of valuable metals as a function of spent catalysts ratio, (b) SEM image and (c) EDS result of copper droplets confined to the slag (30 per cent catalyst experiment).

In the present study, the terminal velocity of copper droplets in the slag was calculated. It was assumed that copper droplets in the slag were spherical in shape. The force balance among gravity, buoyancy and frictional forces on particles was taken into account. Assuming a Stokes regime due to no intentional stirring in the slag, the terminal velocity of metal droplet can be described by Equation 2 (Poirier and Geiger, 1994).

$$v_t = \frac{2}{9} \frac{r^2(\rho_m - \rho_s)g}{\eta} \text{ (m/s)} \tag{2}$$

From Figure 4, a decrease in the C/A ratio of the slag increases the viscosity, which reduces the settling velocity of copper droplets in the slag. As a result, the recovery rate of Ag and Pd decreases. Therefore, it is crucial to precisely control the settling velocity of copper droplets to maximise the recovery of valuable metals.

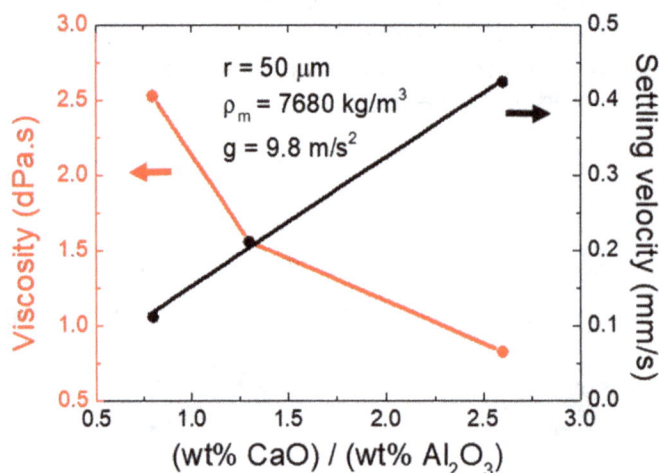

FIG 4 – Relationship between viscosity and settling velocity as a function of the CaO/Al$_2$O$_3$ ratio.

CONCLUSIONS

Loss of valuable metals due to physical entrainment means that copper droplets were not settled down completely but rather dispersed in the slag layer during smelting process. To quantitatively evaluate the physicochemical behaviour of slag-metal interfacial phenomena, the interfacial tension of the slag-metal system and settling velocity of copper droplets was estimated. As the interfacial tension and settling velocity decrease with higher mixing ratio of spent catalyst, the higher the probability of copper droplets in the slag, resulting in a decrease of the recovery rate of valuable metals.

REFERENCES

Ebin, B and Isik, M I, 2016. Pyrometallurgical processes for the recovery of metals from WEEE, *WEEE recycling*, pp 107–137.

Ghalehkhondabi, V, Fazlali, A and Daneshpour, F, 2021. Electrochemical extraction of palladium from spent heterogeneous catalysts of a petrochemical unit using the leaching and flat plate graphite electrodes, *Separation and Purification Technology*, 258:117527.

Kim, B S, Lee, J C, Seo, S P, Park, Y K and Sohn, H Y, 2004. A process for extracting precious metals from spent printed circuit boards and automobile catalysts, *JOM*, 56:55–58.

Kwon, E H, Han, J W and Jeong, J K, 2005. Melting of PCB scrap for the Extraction of Metallic Components, *Korean Journal of Materials Research*, 15(1):31–36.

Lee, B D, Jung, H J, Baek, U H, Hong, S H, Han, J W, Yoo, B D and Lee, J C, 2010. Recovery of Valuable Metals Using Pyrometallurgical Treatment, *Journal of the Korean Society of Mineral and Energy Resources Engineers*, 47(5):628–646.

Poirier, D R and Geiger, G H, 1994. Transport phenomena in materials processing, *The Minerals, Metals and Materials Society*, 5(2.1).

Shin, D Y, Lee, S D, Jeong, H B, You, B D, Han, J W and Jung, J K, 2008. Pyro-metallurgical Treatment of used OA Parts for the Recovery of Valuable Metals, *Resources Recycling*, 17(2):46–54.

Towards integration of pyro- and hydrometallurgical unit operations for efficient recovery of battery metals from waste lithium-ion batteries

L Klemettinen[1], J Biswas[2], A Klemettinen[3], J Zhang[4], H O'Brien[5], J Partinen[6] and A Jokilaakso[7]

1. Staff Scientist, Aalto University, Department of Chemical and Metallurgical Engineering, 00076 Aalto, Finland. Email: lassi.klemettinen@aalto.fi
2. Assistant Professor, IIT Bombay, Powai, Mumbai, Maharashtra 400076, India. Email: biswasj@iitb.ac.in
3. University Teacher, Aalto University, Department of Chemical and Metallurgical Engineering, 00076 Aalto, Finland. Email: anna.klemettinen@aalto.fi
4. Associate Professor, Wuhan University of Science and Technology- The State Key Laboratory of Refractories and Metallurgy, Wuhan 430081, Hubei, PR China. Email: zhangjuhua@wust.edu.cn
5. Senior Researcher, Geological Survey of Finland, 02150 Espoo, Finland. Email: hugh.obrien@gtk.fi
6. Doctoral Researcher, Aalto University, Department of Chemical and Metallurgical Engineering, 00076 Aalto, Finland. Email: jere.partinen@aalto.fi
7. Associate Professor, Aalto University, Department of Chemical and Metallurgical Engineering, 00076 Aalto, Finland. Email: ari.jokilaakso@aalto.fi

ABSTRACT

Waste lithium-ion batteries (LIBs) are important secondary sources of many valuable materials, including Critical Raw Materials (CRMs) defined by the European Union (EU): lithium, cobalt, manganese, and graphite. Additionally, LIBs typically contain nickel and copper, which are classified as Strategic Raw Materials for EU since 2023. In recent years, great effort has been made to develop efficient recycling processes for waste LIBs. Pyrometallurgical processes have been essential in industrial production of metals for many decades. These technologies are relatively mature, with high adaptability for different raw materials. Pyrometallurgical treatment in LIB recycling typically involves the use of smelting processes, in which waste batteries are heated above their melting points and metals are separated through a reduction reaction in the liquid phase. Through this recycling route, cobalt, copper, and nickel can be efficiently recovered in the form of a metal alloy, whereas lithium and manganese are lost in the slag phase.

The goal of this work was to increase the recoveries of valuable battery metals through a combination of hydro- and pyrometallurgical unit operations. First, industrial Li-ion battery scrap underwent a selective sulfation roasting stage, where the aim was to transform $LiCoO_2$ and Mn-oxides into Li, Co and Mn sulfates. After roasting, the battery scrap was leached in distilled water with a solid to liquid ratio of 100 g/L at 60°C and recovered 95 per cent of Li, 61 per cent of Mn and 35 per cent of Co. After leaching, the solid leach residue was mixed with industrial Ni-slag and biochar, followed by reduction at 1350°C in argon atmosphere. The high-temperature smelting experiments were conducted as a function of time (5–60 mins) to investigate the reduction behaviour of battery metals. The results show that Co and Ni from the slag and leach residue can be efficiently recovered in the slag cleaning stage.

INTRODUCTION

Lithium-ion batteries (LIBs) have become extremely important for portable electronics, green energy technologies, electric vehicles, and energy storage systems. Consequently, research related to batteries has drastically increased over last decade (Li *et al*, 2018). Researchers from all over the world are focusing on improving the design of new batteries, developing new battery materials, as well as developing new routes for the recovery valuables from end-of-life LIBs. Waste lithium-ion batteries are important secondary sources of many valuable materials, including Critical Raw Materials (CRMs) defined by the European Union (EU): lithium, cobalt, manganese, and graphite. Additionally, LIBs typically contain nickel and copper, which are classified as Strategic Raw Materials for EU since 2023 (Grohol and Veeh, 2023).

Current developments in the LIB recycling have been described in recently reviewed papers (Makuza *et al*, 2021; Baum *et al*, 2022; Brückner, Frank and Elwert, 2020). Typical recycling processes consist of discharging and mechanical separation followed by pyrometallurgical and/or hydrometallurgical treatment (Neumann *et al*, 2022). To make the battery recycling processes economically more attractive, they can be integrated with already existing primary metal production processes. In our previous studies (Ruismäki *et al*, 2020a; Dańczak *et al*, 2021; Rinne *et al*, 2022), we have investigated the possibility of integrating different battery scrap flotation fractions with pyrometallurgical slag cleaning processes. The slag used in this study came from an industrial nickel flash smelting furnace. It was iron-silicate slag with some magnesia (MgO) and nickel oxide, as well as minor concentrations of Co and Cu (Crundwell *et al*, 2011). As spent lithium-ion batteries typically contain a high concentration of Co and some Ni and Cu, it was proved earlier (Ruismäki *et al*, 2020b) that it is beneficial to mix a battery scrap fraction rich in Co with nickel slag in order to increase the recovery of Co during the slag cleaning process. In the laboratory-scale Ni-slag cleaning process, Ni, Co and Cu from both waste batteries as well as industrial Ni-slag were recovered in metal alloy and matte phases (Ruismäki *et al*, 2020a, 2020b; Dańczak *et al*, 2021). Graphite from waste LIBs was found to be an effective reductant for metal oxides and the reduction reactions appeared to be very fast. No additional reductant was needed in the investigated smelting process. However, lithium and manganese were lost in the slag, therefore the need for an additional pre-treatment stage before the smelting process was identified in order to maximise the valuable metal recoveries. It is not economically viable to recover Li and Mn from huge volumes of base metal smelting slags, which means that their recovery must occur before the smelting process.

The aim of this research was to combine pyro- and hydrometallurgical unit operations for efficient recovery of battery metals Co, Ni, Cu, Li and Mn from Li-ion battery scrap. First, industrial Li-ion battery scrap underwent a sulfation roasting step, where the aim was to transform Li- and Co-containing oxides into Li and Co sulfates. As the formed sulfates were expected to be water soluble, the next step was leaching in water. Water leaching was also selected as it excludes the use of hazardous chemicals and leaching conditions. After water leaching, the solid leaching residue was mixed with industrial nickel slag and smelted in reducing conditions in order to recover Ni, Cu and the remaining Co. Biochar was used as a reductant in the smelting stage, and the experiments were conducted at 1350°C for different times (5–60 mins) to investigate the reduction behaviour of the metals of interest as a function of time. The data provided in this study will be useful for integrating Li-ion battery recycling with already existing industrial-scale unit processes.

EXPERIMENTAL

Sulfation roasting

The main raw material for this study was LCO-rich lithium-ion battery scrap, supplied by AkkuSer Oy (Finland), where the samples were pretreated with a dry technology – two stages of crushing followed by magnetic and mechanical separation (Pudas, Erkkilä and Viljamaa, 2015). A size fraction of <125 µm was separated from the industrial scrap by sieving and employed for the current study. The chemical composition of the battery scrap fraction used in the experiments is presented in Table 1.

TABLE 1

Li-ion battery scrap composition before sulfation roasting.

Element	Al	Co	Cu	Fe	K	Li	Mg	Mn	Na	Ni	P	C	Rest
wt%	1.64	26.45	2.72	0.61	0.05	3.87	0.09	1.67	0.06	2.74	0.45	33	26.65

For each experiment, a batch of 4 g battery scrap was placed in a silica boat in the cold zone of a horizontal tube furnace (Lenton, UK) with a hot zone temperature of 850°C, after which the furnace was sealed and Ar gas flow (500 mL/min) was started. An ejector was used for creating an under-pressure inside the furnace. After approximately 5 mins, the sample was pushed to the hot zone in three stages, keeping a 2 min pause between the stages to avoid crucible cracking, followed by changing the gas mixture to 10 per cent SO_2 – 10 per cent O_2 – Ar gas when the sample reached the hot zone. These conditions were selected based on our previous investigation with LCO-rich black mass (Biswas *et al*, 2023). According to the thermodynamic stability diagram (Biswas *et al*, 2023), it was expected that all lithium and cobalt transform into sulfates, whereas other metals should remain as oxides. After 60 mins of sulfation roasting, the samples were taken out of the furnace following the same three-stage procedure as previously. In total, four roasting experiments were conducted with the same parameters. The roasted battery scrap was analysed using X-ray diffraction technique (XRD, X'Pert Pro MPD, PANanalytical, Netherlands) with a scan rate of 2°/min for 10° to 90° angles using Cu-K$_\alpha$ radiation and HighScore Plus software (version 4.8, PANanalytical).

Water leaching

The sulfation roasted samples were mixed together and a 10 g batch was leached in distilled water for 60 mins with a solid to liquid ratio of 100 g/L at 60°C and with magnetic stirring at 300 rev/min. After leaching, the residue was filtered and the concentrations of metals were analysed by atomic absorption spectroscopy (AAS, Thermo Scientific iCE 3000, USA) after digesting a part of the residue in concentrated aqua-regia (HCl (37 per cent, Merck) and HNO$_3$ (65 per cent, Merck Supelco) at 3:1 molar ratio). AAS was used for the leaching solution analysis as well. The solid residue was also analysed using the X-ray diffraction technique described in the previous section.

Smelting

The smelting experiments were conducted under simulated conditions of an industrial Ni-slag smelting process. Industrial Ni-slag (Table 2) was used in the experiments. The main reductant in the experiments was biochar prepared from black pellets by pyrolysing at 600°C, obtained from University of Oulu (Finland), containing 76.6 wt per cent carbon. The biochar was supplied as pellets, but it was ground in a mortar into fine powder before the experiments in order to increase the reactive surface area. Detailed analysis of the biochar has been presented previously (Attah-Kyei *et al*, 2023).

TABLE 2

Chemical compositions of the Ni-slag and concentrate used in the experiments.

Fe$_3$O$_4$	Fe	Si	Ni	Mg	Ca	Al	Cu	Co	Na	K	Ti	S	Mn
Slag, concentrations in wt%													
20	35.42	10.61	3.41	3.16	0.91	0.84	0.69	0.43	0.35	0.32	0.21	0.16	0.03
Concentrate, concentrations in wt%													
2	29.32	9.25	8.06	2.79	1.05		1.85	0.31				24.41	

Two series of experiments were conducted. In series A, 90 wt per cent of Ni-slag was mixed with 10 wt per cent leaching residue. In series B, 5 wt per cent of the slag was replaced with industrial sulfidic Ni-concentrate (Table 2). The stoichiometric amount of carbon required for complete reduction of NiO to Ni, CuO to Cu, CoO to Co and magnetite (20 wt per cent in slag) to FeO was calculated, assuming the reaction gas to be CO, and 1.25 times the stoichiometric amount was used in both series. For both series, enough reagents for five experiments were weighed and thoroughly mixed in an agate mortar. After mixing, 1.0 g of the mixture was used in each experiment. The weight ratios in the final mixtures were 85.97/9.55/4.48 for slag/leaching residue/biochar in series A and 81.30/9.56/4.78/4.36 for slag/residue/concentrate/biochar in series B, respectively. The small difference in the amount of biochar is due to the concentrate used in series B, which was already sulfidic and therefore did not require reductant.

The experiments were conducted in a vertical tube furnace (Lenton, UK). 1.0 grams of starting mixture was placed in a conical silica crucible and attached to Kanthal A-wire hanging from inside the furnace. The sample was first lifted to the cold zone of the furnace, followed by closing the furnace work tube with a rubber plug. Next the furnace was flushed with Argon (300 mL/min) for 15 mins before lifting the sample to the hot zone at 1350°C and keeping there for 5, 10, 30 or 60 mins. After the set time, the sample was rapidly quenched to ice-water mixture without breaking the inert atmosphere. Before characterisation, the samples were mounted in epoxy, cut in a half by a diamond cutting wheel, mounted in smaller epoxy moulds, ground, and polished with traditional wet metallographic methods. After polishing, the sample surfaces were coated with carbon for microstructural and compositional characterisation.

The sample microstructures and elemental compositions of phases were characterised using a scanning electron microscope (SEM; Mira 3, Tescan, Czech Republic) coupled with an energy dispersive spectrometer (EDS; Thermo Fisher Scientific, USA). The beam current was approximately 10 nA and the acceleration voltage was 15 kV. The elemental analyses were conducted using standards supplied by Astimex, as shown in Table 3.

TABLE 3

EDS standards as well as LA-ICP-MS detection limits (ppmw = parts-per-million by weight).

Element	EDS standard	LA-ICP-MS isotope and detection limit (ppmw)
O	Diopside	
Mg	Magnesium	
Al	Aluminium	
Si	Quartz	
S	Marcasite	
Ca	Fluorite	
Fe	Hematite	
Co	Cobalt	[59]Co: 0.02
Ni	Nickel	[60]Ni: 0.21
Cu	Copper	[65]Cu: 0.04
Mn		[55]Mn: 0.09
Li		[7]Li: 0.04

Trace and minor element concentrations in the slags were below reliable detection with EDS, therefore the slags were analysed using laser ablation-inductively coupled plasma-mass spectrometry (LA-ICP-MS). The laser spot size was set to 40 µm with 50 µm preablation and the laser energy was 40.1 per cent of 5 mJ. The fluence, ie energy delivered per unit area, was 1.25 J/cm² on the sample surface. The laser was operated at 10 Hz frequency and 40 sec of ablation data was collected from each spot. NIST610 glass (Jochum *et al*, 2011) was used as an external standard, [29]Si (from EDS) as the internal standard. NIST612, USGS BHVO-2G and BCR-2G glasses (Jochum *et al*, 2005) were analysed as unknowns. The obtained time-resolved analysis signals were treated with Glitter software (Van Achterberg *et al*, 2001), and the detection limits are shown in Table 3.

RESULTS AND DISCUSSION

Sulfation roasting and water leaching

The XRD pattern of the LCO-rich black mass in Figure 1 suggests that there are two main phases, $LiCoO_2$ and graphite, which originate from the cathode and anode, respectively. The elemental analysis of the black mass in Table 4 also shows 26.6 wt per cent of Co and 3.87 wt per cent Li along with small concentrations of Ni, Mn and Cu, confirming this as LCO-rich black mass. At the first stage of this study, the LIB black mass samples were roasted using 10 per cent SO_2 – 10 per cent O_2 – Ar gas flow at 850°C for 60 mins. The XRD pattern of the sulfation roasted powder in Figure 1 confirms the formation of Li and Co sulfates, while a fraction of Co could be observed to remain as oxide. $LiCoO_2$ peaks were not observed in the roasted powder sample, indicating a high degree of sulfation (and partial transformation into CoO as observed in the XRD graph). The carbon peak also disappeared after sulfation roasting, suggesting full combustion of carbon. At this temperature, it is also expected that the binders, plastics and organic electrolytes are removed during the roasting process.

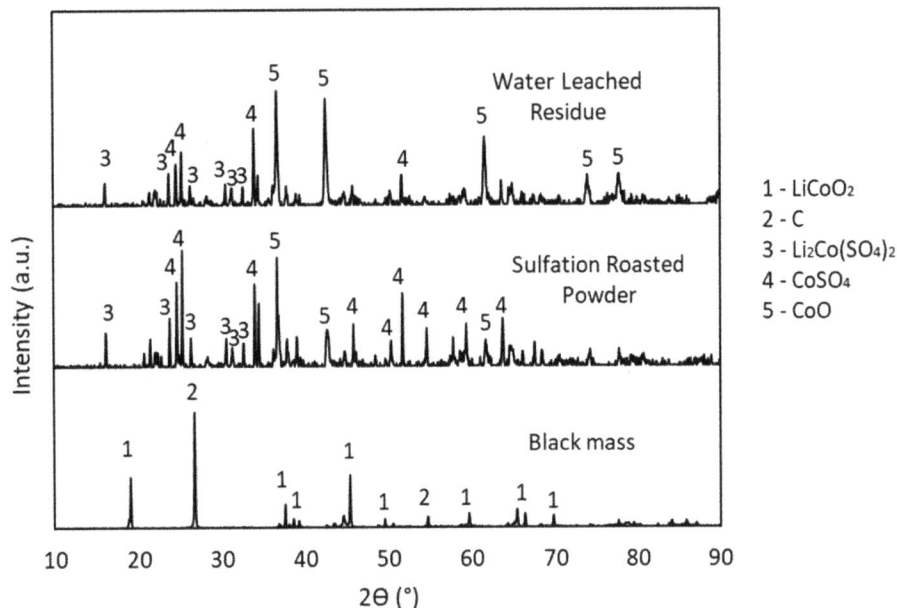

FIG 1 – XRD patterns of black mass, sulfation roasted powder and leach residue after water leaching.

TABLE 4

Elemental composition of black mass and leach residue. The chemical analysis is based on sample digestion in concentrated aqua-regia, followed by atomic absorption spectroscopy.

Element	Co	Ni	Mn	Cu	Fe	Li	Others
black mass, wt%	26.45	2.74	1.67	2.72	0.61	3.87	61.94
leach residue, wt%	43.00	6.10	1.40	6.00	0.20	0.20	43.1

After roasting, the samples were leached with distilled water, where the water-soluble metal sulfates were expected to dissolve, leaving metal oxides in the residue phase. The elemental composition of the leach residue, obtained from the water leaching process, is presented in Table 4. The residue consisted of approximately 43 wt per cent Co, 6 wt per cent Ni, 6 wt per cent Cu and low concentrations of Fe, Li and Mn. The XRD pattern of the water leached residue in Figure 1 shows several strong peaks of CoO along with few peaks of $Li_2Co(SO_4)_2$ and $CoSO_4$. This indicates that majority of Li and a fraction of Co were extracted during the water leaching stage. A longer period of water leaching could have been beneficial for the extraction of remaining sulfates. The composition of the residue shows that it contains valuable metals in high concentrations, and it will be a very good raw material for Ni-slag cleaning furnace, where an alloy containing mostly Ni and Co (as well as Fe) is formed.

The extraction yields of different metals after leaching, presented in Table 5, were calculated based on metal concentrations in water-leached solution and residue digested solution obtained from AAS analyses. Equation 1 was used in the calculations:

$$\% \ Extraction = \frac{m_1}{m_1 + m_2} \cdot 100\% \tag{1}$$

where:

m_1 is the total mass of the specific metal dissolved in water during leaching

m_2 is the total mass of the specific metal remaining in the residue

TABLE 5

Recoveries of metallic elements after sulfation roasting and water leaching.

Element	Co	Ni	Mn	Cu	Fe	Li
Solution after leaching, mg	1130	28	103	12	<1	198
Leach residue, mg	2051.1	291.0	66.8	286.2	9.5	9.5
% Extraction	35.5	8.8	60.7	4.0	<9.5	95.4

Based on the results, approximately 95 per cent of Li, 61 per cent of Mn and 36 per cent of Co were extracted from the black mass in the water leaching stage. According to the thermodynamic stability diagram (Biswas *et al*, 2023), Mn should remain as oxide in the selected roasting conditions, but based on the obtained results, more than half of the Mn-containing oxides were actually transformed to sulfates, as indicated by the 61 per cent extraction efficiency. The surprisingly high Mn extraction during water leaching is beneficial because Mn deports to the slag and cannot be recovered in Ni-slag cleaning stage.

In this research, the focus was mostly on the recovery of metals from the leach residue through Ni-slag cleaning process. The precipitation and purification of metal compounds from the leaching solution was not investigated.

Smelting

The leaching residue was mixed with industrial nickel slag and subjected to a laboratory-scale Ni-slag cleaning process, as described in the Experimental section. SEM images of sample microstructures after smelting in reducing conditions are presented in Figure 2. A typical microstructure consists of the silica crucible (dark grey), glassy slag (medium grey) and matte/metal alloy phase (white). The matte phase had a round shape and was typically found near the bottom of the crucible. In samples without Ni-concentrate (series A, Figure 2a–2b), the matte phase seemed much smaller than in the samples with Ni-concentrate (series B, Figure 2c–2d). Additionally, in the samples without Ni-concentrate, a metal alloy phase was found mostly on the top of the slag layer, which is in line with previous studies on integrating battery scrap recycling with Ni-slag cleaning process (Ruismäki *et al*, 2020a; Dańczak *et al*, 2021).

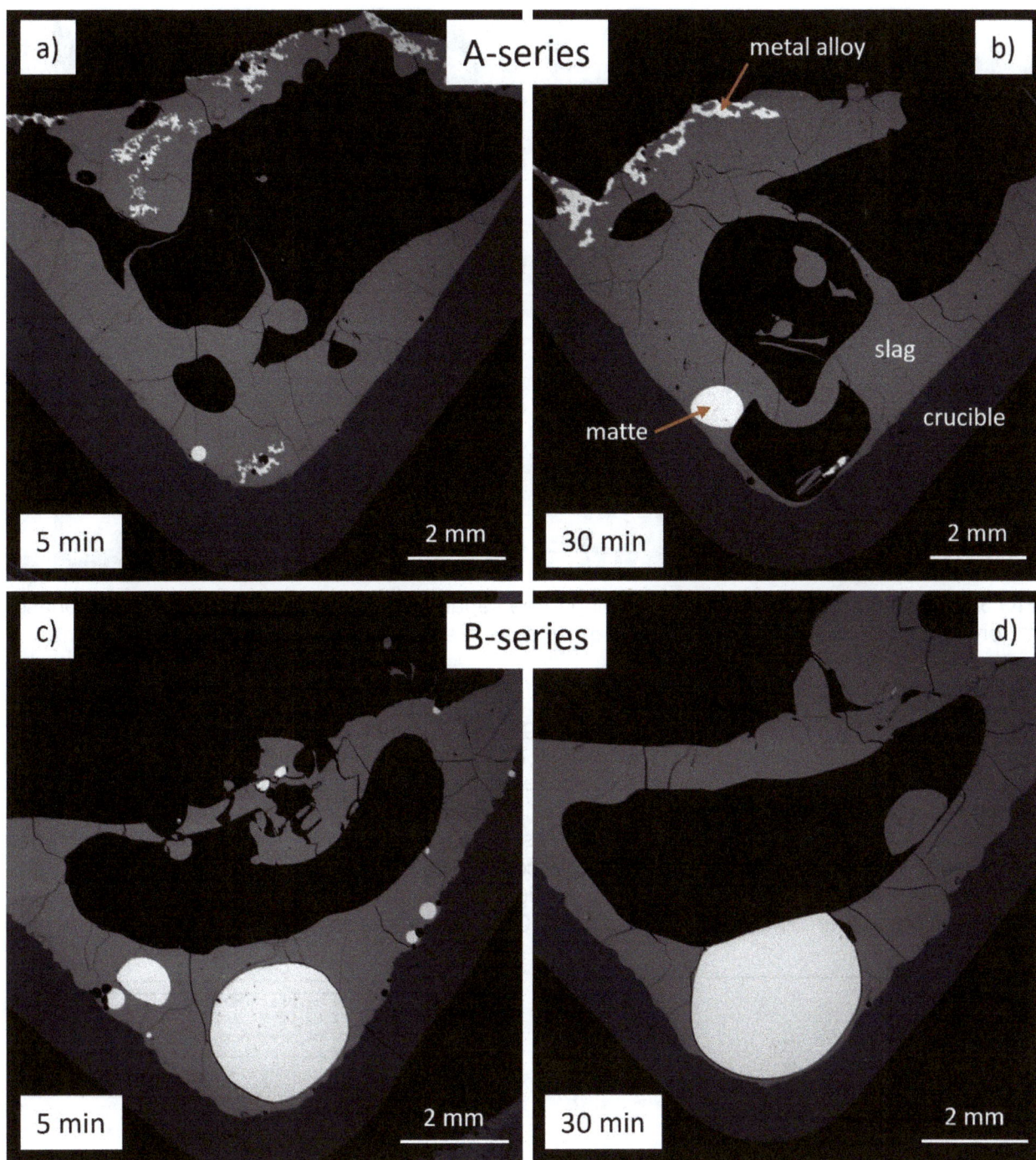

FIG 2 – Microstructures of samples after different reduction times. (a) and (b) are from series A, while (c) and (d) are from series B.

Besides the largest matte phase area close to the bottom of the crucible, some smaller round matte droplets were also found in other places of the samples. In both series, the smaller matte droplets increasingly coalesced with the largest droplet as the reduction time was increased.

A higher magnification SEM-BSE image of a typical microstructure of the matte phase is shown in Figure 3a, and a typical microstructure of the Fe-rich metal alloy present on top of the slag in the samples without Ni-concentrate is shown in Figure 3b. Within the whole matte phase, three different phase areas were distinguished, as marked in Figure 3a: area (1) was metallic Ni-Fe-Co alloy, area (2) was Cu-Ni-Fe sulfide and area (3) was Fe-Ni-Co sulfide. The metal alloy shown in Figure 3b consisted mainly of Fe-Co-Ni alloy, with minor sulfide areas segregated at the grain boundaries.

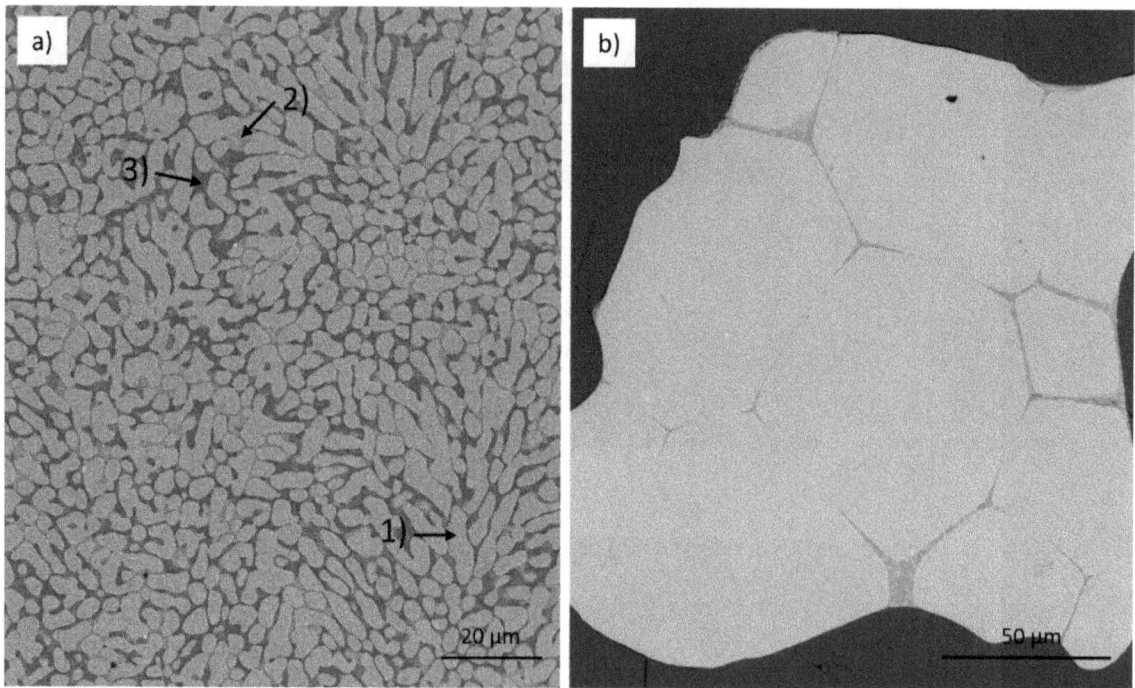

FIG 3 – Microstructure of: (a) matte, showing three separate phase areas; (b) Fe-rich metal phase formed on top of the slag in series A. Both images are from the sample reduced for 10 mins.

Chemical composition of the slag phase

The concentrations of Fe and SiO_2 in the slags are presented in Figure 4, left side. In both experimental series, Fe concentration decreased as the reduction time increased, and at the same time the concentration of SiO_2 increased. An increase in SiO_2 concentration in the slag was caused by iron reduction as well as dissolution of the silica crucible during smelting. The presence of Ni-concentrate in the samples resulted in faster rates of SiO_2 concentration increase and Fe decrease compared to the series without concentrate. The concentrations of MgO, Al_2O_3 and CaO in the slags are shown in Figure 4, right side. There are no significant differences between the two series, and the concentrations remain relatively stable after 10 mins of reduction.

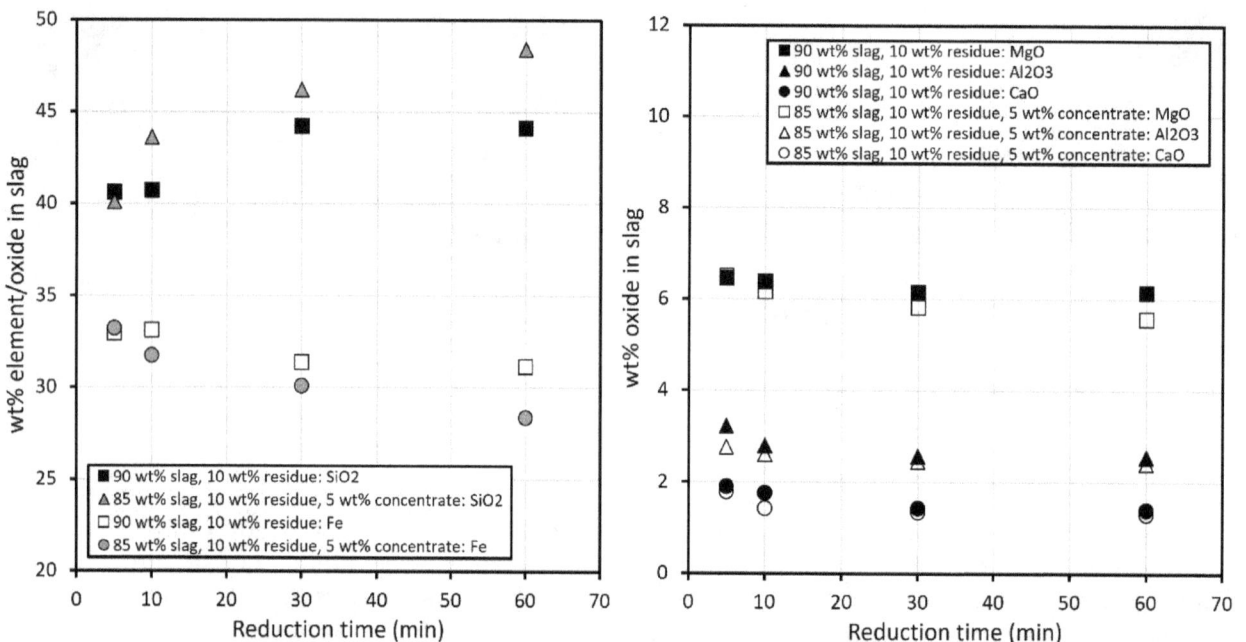

FIG 4 – Concentrations of SiO_2 and Fe (left side) as well as MgO, Al_2O_3 and CaO (right side) in slags after 5–60 min reduction time.

The concentrations of Co, Ni, Cu, Mn and Li in the slag phase are shown in Figure 5. The concentrations of Mn and Li remained relatively constant after 10 mins. This indicates that the slag volume increase that would have been caused by crucible dissolution, resulting in diluted concentrations of stable oxides such as Li_2O and MnO, was quite effectively compensated by iron reduction from the slag.

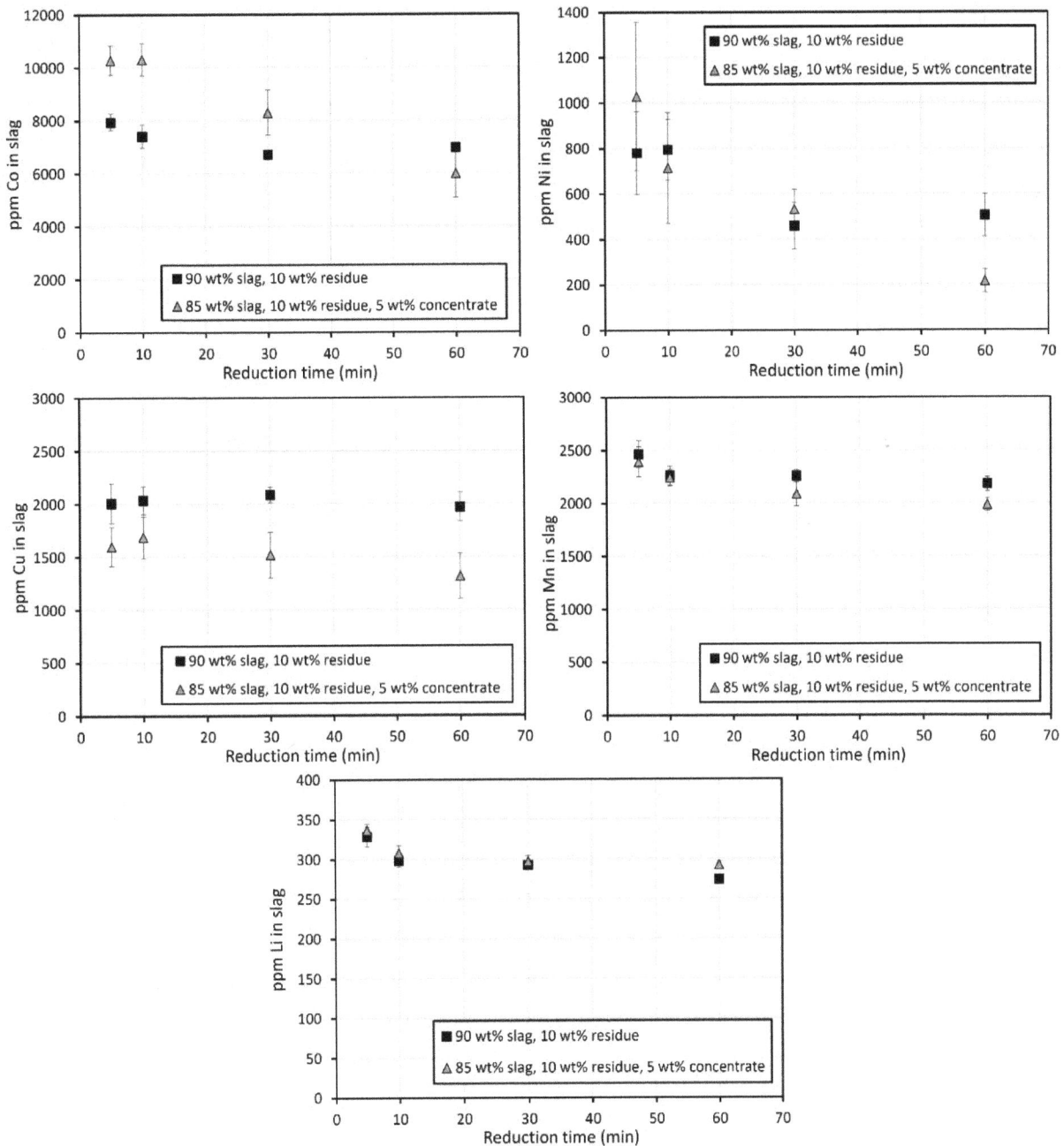

FIG 5 – Concentrations of Co, Ni, Cu, Sn, Pb and Li in slag phase as a function of reduction time at 1350°C.

The concentrations of Co and Ni decreased significantly as the reduction time increased, especially in series B with Ni-concentrate. The concentration decreases in series A stopped after 30 mins of reduction, whereas in series B the decrease continued until 60 mins. At 60 mins, the concentrations of Ni and Co were lower in series B compared to series A. The concentration of Cu in the slag remained surprisingly constant during the entire reduction time range in series A (without Ni-concentrate). In series B (with Ni-concentrate), a decrease in Cu-concentration was observed as the reduction time increased. Mn and Li originated mostly from the leaching residue, and as the residue amount did not change between the experimental series, neither did their concentrations in the slag significantly.

Chemical composition of the matte and metal phases

As presented in Figure 3a, the matte consisted of three different phase areas. The average composition of the matte was calculated according to the method presented by Rinne *et al* (2022). First, the composition of each phase area was analysed using SEM-EDS. Then, several SEM-images were taken from the matte phase of each sample, followed by phase area quantification using ImageJ software and calculation of average composition.

The average concentrations of Fe, Ni, Co, Cu and S in matte after ImageJ calculations are shown in Figure 6. In the series without concentrate, after 10 min reduction, the Ni concentration began to decrease and Fe as well as Co concentrations increased as the reduction time increased. It should be highlighted that even though Ni concentration seems to decrease, it does not mean that the amount of reduced Ni decreased but rather the ratio between Ni and Fe changed as the reduction of iron oxides progressed and the total mass of matte phase increased. The concentrations of Cu and S remained constant after 30 mins of reduction. The elements in the series with concentrate (Figure 6, right side) behaved differently: at all reduction times, the concentration of cobalt was higher than that of nickel, and the iron concentration did not show an increasing trend as a function of reduction time.

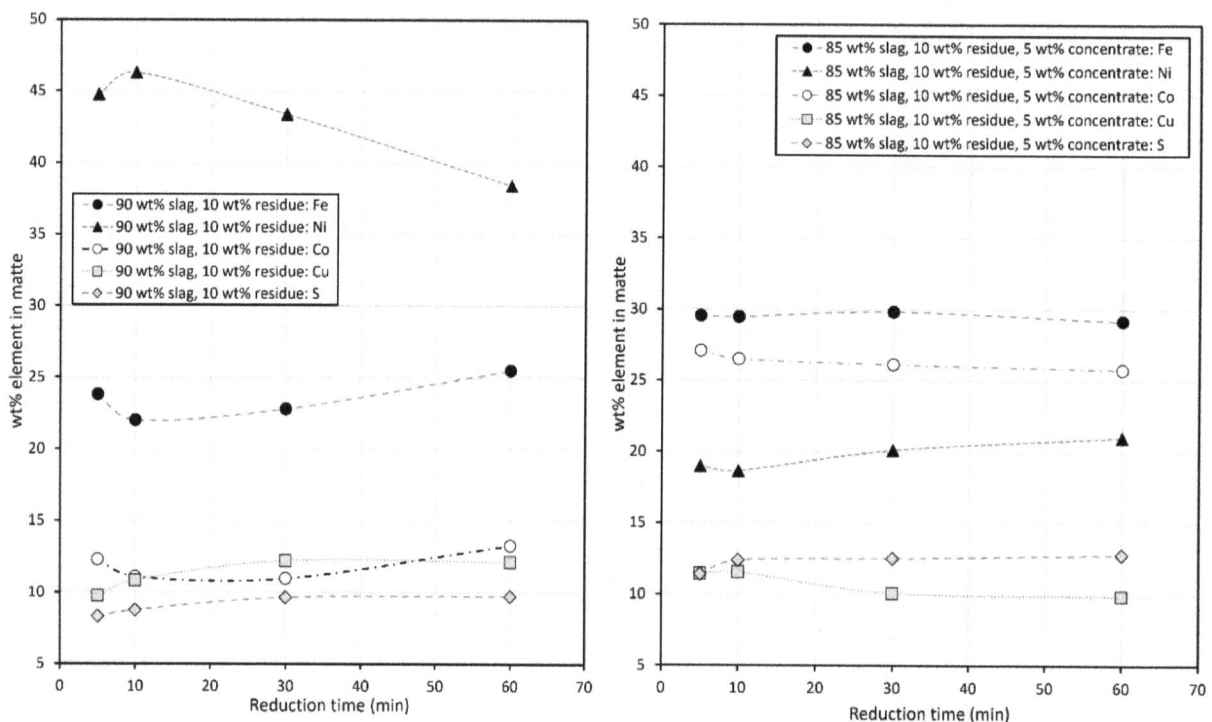

FIG 6 – Concentrations of Fe, Ni, Co, Cu and S in matte as a function of reduction time. Series A on the left side, series B on the right.

The concentrations of iron, nickel, cobalt and copper in the metal phase formed on top of the sample (see Figures 2a–2b and 3b) in series A, without Ni-concentrate, are presented in Figure 7. This phase is rich in iron and cobalt, which explains the lower iron and cobalt concentrations in the matte phase of this series compared to the series with concentrate, where no metal phase was formed on the sample surface. The elemental concentrations in the metal phase were calculated only based on the concentrations in the metallic grains, excluding the small sulfide areas visible in Figure 3b. It should be noted that the formation of such a metal alloy on top of the slag is not desirable from industrial perspective, because it cannot be tapped out of the slag cleaning furnace due to not settling below the slag layer.

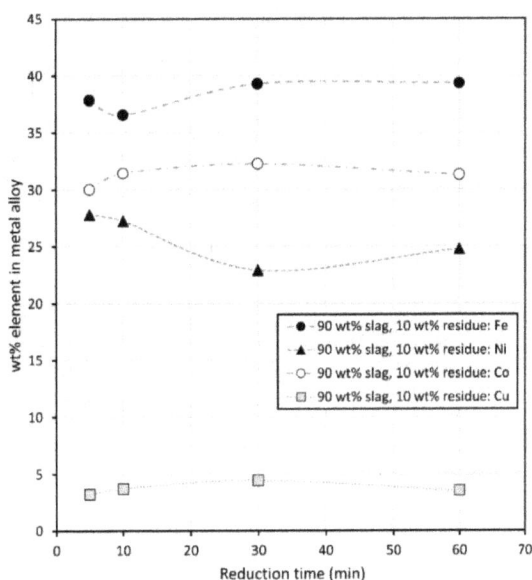

FIG 7 – Concentrations of Fe, Ni, Co and Cu in metal phase of series A.

Distribution coefficients of Co, Ni and Cu between matte and slag

Matte-slag distribution coefficients of Co, Ni and Cu are presented in Figure 8 and were calculated based on Equation 2:

$$L^{m/s}Me = [Me\ wt\%]/(Me\ wt\%) \qquad (2)$$

where:

Me	represents the metal of interest
$Lm^{/s}$	is the distribution coefficient of Me between matte and slag
$[Me\ wt\%]$	is the concentration of Me in matte
$(Me\ wt\%)$	is the concentration of Me in slag

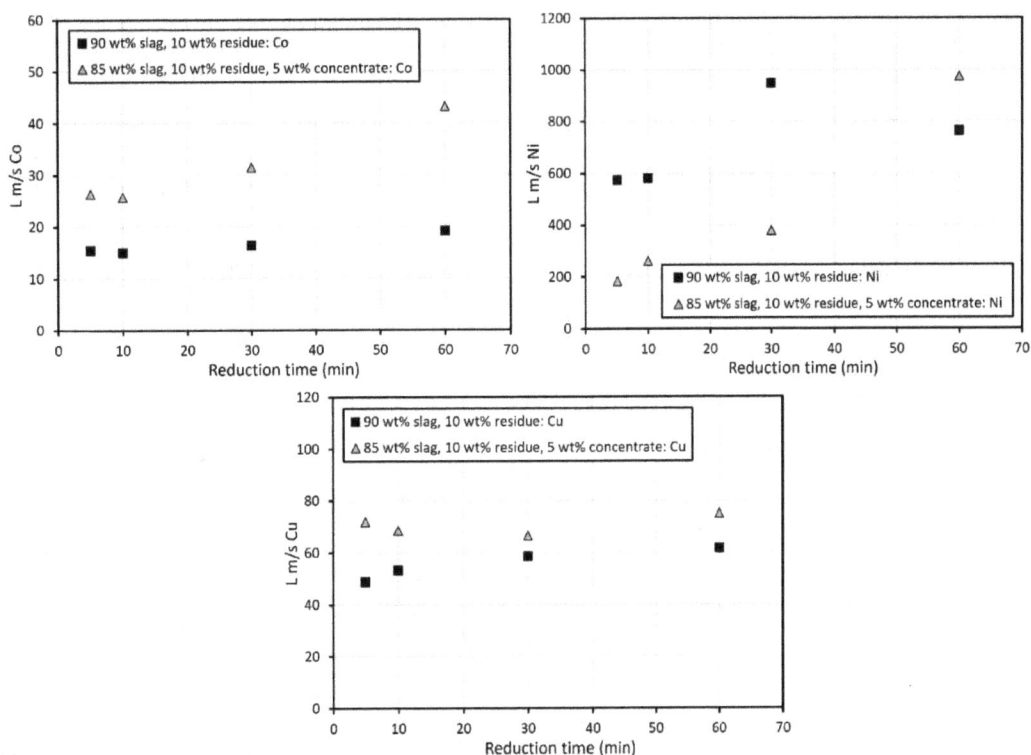

FIG 8 – Matte-slag distribution coefficients of Co, Ni and Cu as a function of reduction time.

The distribution coefficient values of Co and Ni increased significantly as the reduction time increased. The recovery of Co in the matte was significantly increased with concentrate addition (series B) compared to series A, as the distribution coefficient value after 60 mins was approximately 43 compared to 19, respectively. For Ni, the values were systematically higher in the series without concentrate, except for the longest reduction time. In general, the recovery of Ni in matte was approximately 20 times higher compared to Co. The distribution coefficient values for Cu were between 50 and 75, and they did not increase significantly as a function of reduction time. The differences between the two series were relatively small.

The results for Co fit very well with the values obtained previously by Dańczak $et\ al$ (2021) using graphite from battery scrap as a reductant in Ni-slag cleaning. They also reported approximately doubled $Lm^{/s}$ Co values when adding Ni-concentrate, compared to without concentrate. Their $Lm^{/s}$ Ni values after 60 mins were approximately 600, which are slightly on the lower side compared to this work. For Cu, Dańczak $et\ al$ reported higher values than obtained in this study, however the trends as a function of time were the same.

Attah-Kyei $et\ al$ (2023) compared the reduction efficiencies of four different biochars with metallurgical coke in Ni-slag cleaning at 1400°C. The biochar employed in our work was the one which they concluded was the most effective reductant. The distribution coefficient values reported by Attah-Kyei $et\ al$, for Ni after 60 mins were similar as obtained in this study, for Cu they were somewhat lower and for Co they were two to three times higher. A major difference in their work compared to the current study was that only a metal alloy phase was formed and settled at the bottom of the crucible instead of matte phase. Because of this, a direct comparison of distribution coefficient values is not possible.

CONCLUSIONS

Waste lithium-ion batteries are complex mixtures of chemical elements and their compounds. In order to recover as much valuable metals as possible, smart combinations of mechanical and thermal pretreatment steps as well as pyro- and hydrometallurgical processing are needed. This study is based on the concept that battery recycling could be integrated with already existing technologies for primary metals production.

In this study, the mechanically pretreated battery scrap was subjected to sulfation roasting, where lithium and cobalt should transform into water-soluble sulfates according to thermodynamic stability diagrams. Analysis of the roasted battery scrap confirmed the presence of lithium in form of sulfates, whereas cobalt was in form of both sulfate and oxide. The roasted batteries were then used as feed in water leaching, where 95 per cent of Li and 36 per cent of Co were extracted. The low extraction yield of Co is in line with the XRD results, which showed that Co was only partially transformed to sulfate during the roasting step. The extraction yield of Mn reached 61 per cent, which was contradictory to the initial predictions based on the thermodynamic stability of Mn-compounds in the roasting conditions.

The high recovery rate of Li and the relatively good recovery of Mn indicate that the removal of these metals from the battery scrap before further high-temperature treatment is possible, and with further optimisation of the roasting conditions, improved Mn recovery rates are expected. Apart from the roasting parameter optimisation, a second round of water leaching would most likely increase the recoveries of Li, Mn and possibly Co. The low recovery of cobalt in water leaching is actually not an issue, as it is further recovered in smelting process unlike Li and Mn. In our study, the water leaching residue was mixed with industrial Ni-slag and subjected to a laboratory-scale Ni-slag cleaning process conducted at 1350°C, using biochar as a reductant. In the smelting process, the valuable metals Co, Ni and Cu heavily deported to the matte and metal alloy phases formed during reduction. The addition of Ni-concentrate to the smelting mixture had two clear benefits; it increased the deportment of valuables to matte and eliminated the formation of the iron-rich metal alloy on top of the slag. The biochar used in this work, obtained by pyrolysing black pellets at 600°C, was an effective reductant for metal oxides from both battery scrap and industrial Ni-slag.

During sulfation roasting at 850°C in oxygen-containing gas atmosphere, graphite from the battery scrap reacted with oxygen, forming CO_2 emissions and resulting in the loss of graphite, which is also classified as a CRM. From the elemental recovery perspective, it could make sense to separate as much of the graphite as possible before the roasting process, using for example froth flotation. However, this would add another processing stage to the recycling process already containing several stages, leading to negative impacts on exergy efficiency for example. In the future, holistic studies should be conducted regarding the process investigated here as well as other possible processes in terms of economic viability, circularity, life-cycle assessment as well as exergy efficiency and exentropy analysis (Vierunketo *et al*, 2023).

ACKNOWLEDGEMENTS

This research work has been supported by the Business Finland BatCircle 2.0 project (Grant number 43830/31/2020), and the Academy of Finland's RawMatTERS Finland Infrastructure (RAMI) based at Aalto University, GTK and VTT.

REFERENCES

Attah-Kyei, D, Sukhomlinov, D, Tiljander, M, Klemettinen, L, Taskinen, P, Jokilaakso, A and Lindberg, D, 2023. A Crucial Step Towards Carbon Neutrality in Pyrometallurgical Reduction of Nickel Slag, *Journal of Sustainable Metallurgy*, 9:1759–1776.

Baum, Z J, Bird, R E, Yu, X and Ma, J, 2022. Lithium-Ion Battery Recycling–Overview of Techniques and Trends, *ACS Energy Letters*, 7(2):712–719.

Biswas, J, Ulmala, S, Wan, X, Partinen, J, Lundström, M and Jokilaakso, A, 2023. Selective Sulfation Roasting for Cobalt and Lithium Extraction from Industrial LCO-Rich Spent Black Mass, *Metals*, 13(2):358.

Brückner, L, Frank, J and Elwert, T, 2020. Industrial Recycling of Lithium-Ion Batteries – A Critical Review of Metallurgical Process Routes, *Metals*, 10(8):1107.

Crundwell, F K, Moats, M S, Ramachandran, V, Robinson, T G and Davenport, W G, 2011. Chapter 18: Flash Smelting of nickel sulfide concentrates, in *Extractive Metallurgy of Nickel, Cobalt and Platinum-Group Metals*, pp 215–232 (Elsevier: Oxford).

Dańczak, A, Ruismäki, R, Rinne, T, Klemettinen, L, O'Brien, H, Taskinen, P, Jokilaakso, A and Serna-Guerrero, R, 2021. Worth from Waste: Utilizing a Graphite-Rich Fraction from Spent Lithium-Ion Batteries as Alternative Reductant in Nickel Slag Cleaning, *Minerals*, 11:784.

Grohol, M and Veeh, C, 2023. Study on the critical raw materials for the E U, 2023: final report, European Commission, Directorate-General for Internal Market, Industry, Entrepreneurship and SMEs (Publications Office of the European Union). Available from: <https://data.europa.eu/doi/10.2873/725585>

Jochum, K P, Weiss, U, Stoll, B, Kuzmin, D, Yang, Q, Raczek, I, Jacob, D E, Stracke, A, Birbaum, K, Frick, D A, Günther, D and Enzweiler, J, 2011. Determination of Reference Values for NIST SRM 610–617 Glasses Following ISO Guidelines, *Geostandards and Geoanalytical Research*, 35:397–429.

Jochum, K P, Willbold, M, Raczek, I, Stoll, B and Herwig, K, 2005. Chemical Characterisation of the USGS Reference Glasses GSA-1G, GSC-1G, GSD-1G, GSE-1G, BCR-2G, BHVO-2G and BIR-1G Using EPMA, ID-TIMS, ID-ICP-MS and LA-ICP-MS, *Geostandards and Geoanalytical Research*, 29:285–302.

Li, M, Lu, J, Chen, Z and Amine, K, 2018. 30 Years of Lithium-Ion Batteries, *Advanced Materials*, 30(33):1800561.

Makuza, B, Tian, Q, Guo, X, Chattopadhyay, K and Yu, D, 2021. Pyrometallurgical options for recycling spent lithium-ion batteries: A comprehensive review, *Journal of Power Sources*, 491:229622.

Neumann, J, Petranikova, M, Meeus, M, Gamarra, J D, Younesi, R, Winter, M and Nowak, S, 2022. Recycling of Lithium-Ion Batteries—Current State of the Art, Circular Economy and Next Generation Recycling, *Advanced Energy Materials*, 12:2102917.

Pudas, J, Erkkilä, A and Viljamaa, J, 2015. Battery Recycling Method, US patent US8979006B2.

Rinne, T, Klemettinen, A, Klemettinen, L, Ruismäki, R, O'Brien, H, Jokilaakso, A and Serna-Guerrero, R, 2022. Recovering Value from End-of-Life Batteries by Integrating Froth Flotation and Pyrometallurgical Copper-Slag Cleaning, *Metals*, 12:15.

Ruismäki, R, Dańczak, A, Klemettinen, L, Taskinen, P, Lindberg, D and Jokilaakso, A, 2020b. Integrated battery scrap recycling and nickel slag cleaning with methane reduction, *Minerals*, 10:435.

Ruismäki, R, Rinne, T, Dańczak, A, Taskinen, P, Serna-Guerrero, R and Jokilaakso, A, 2020a. Integrating flotation and pyrometallurgy for recovering graphite and valuable metals from battery scrap, *Metals*, 10(5):680.

Van Achterberg, E, Ryan, C G, Jackson, S E and Griffin, W L, 2001. Data reduction software for LA-ICP-MS: Appendix, in *Laser ablation-ICP-Mass Spectrometry in the Earth Sciences: Principles and Applications; Short Course Series* (ed: P J Sylvester), Mineralogical Association of Canada, pp 239–243.

Vierunketo, M, Klemettinen, A, Reuter, M A, Santasalo-Aarnio, A and Serna-Guerrero, R, 2023. A multi-dimensional indicator for material and energy circularity: Proof-of-concept of exentropy in Li-ion battery recycling, *iScience*, 26(11):108237.

Physicochemical properties of steelmaking slags for the mitigation of CO_2 emissions in steel sector

M J Lee[1], J H Heo[2] and J H Park[3]

1. Student, Department of Materials Science and Chemical Engineering, Hanyang University, Ansan 15588, Korea.
2. Senior Researcher, Korea Institute of Geoscience and Mineral Resources (KIGAM), Daejeon 34132, South Korea.
3. Professor, Department of Materials Science and Chemical Engineering, Hanyang University, Ansan 15588, Korea. Email: basicity@hanyang.ac.kr

ABSTRACT

In this paper, the challenging points regarding the high temperature physical chemistry of slags to achieve the improved and stable electric arc furnace (EAF) or electric smelting furnace (ESF) technology on the way to green steel will be reviewed, and the recent experimental and modelling research will be discussed. For example, the initial melting phenomena of hot briquetted iron (HBI) and the slag formation behaviour was observed using a high-frequency induction furnace. Main component of gangue oxides in HBI was SiO_2, Al_2O_3, and CaO in conjunction with unreduced iron oxide. To increase the dephosphorisation efficiency, the distribution ratio of phosphorus between metal and slag was calculated using FactSage™ software, version 8.2 (CRCT ThermFact, Inc., Montreal, Canada) and was compared to the measured results. The optimisation of slag chemistry is required not only for maximum dephosphorisation efficiency with good slag foamability but also for minimum slag volume with less refractory corrosion in EAF process. The slag chemistry is also one of key parameters affecting the operation efficiency in ESF in view of FeO reduction behaviour, viscosity, sulfide capacity, etc.

INTRODUCTION

It is well known that global CO_2 emission from iron and steel sector is *ca.* 7 per cent, which is approximately 1/3 of industrial energy use. Hence, many steel companies are trying to develop the electric arc furnace (EAF) and/or electric smelting furnace (ESF) steelmaking processes instead of blast furnace (BF) and basic oxygen furnace (BOF) integrated routes by employing high amounts of direct-reduced iron (DRI) and/or hot briquetted iron (HBI) to mitigate CO_2 emissions. The DRI/HBI as substitutes for virgin scrap in EAF has been used because DRI/HBI does not have tramp elements. Unfortunately, however, commercially available DRI contains the relatively high levels of phosphorus and gangue oxides, which adversely affects not only the steel properties but also the operation efficiency. Alternatively, integrated steel mills have focused on the ESF process by keeping conventional BOF to produce high-end quality products. The H_2-reduced DRI or HBI are able to be charged in ESF in conjunction with fluxes and carbon sources, producing hot metal and BF type slag. The high-grade iron ores (Fe>68 per cent) are economically used in EAF, whereas low-grade iron ores (Fe<65 per cent) are targeted to be used in ESF (Wimmer, Rosner and Voraberger, 2023).

CHEMISTRY OF DIRECT-REDUCED IRON (DRI)

The composition distributions of gangue oxides including unreduced iron oxide in commercially produced DRI are shown in the CaO-SiO_2-FeO ternary phase diagram (Figure 1). Here, the oxide composition was normalised to ternary system by excluding metallic iron (M.Fe). So, the chemical compositions of oxide phases widely vary, ie 5 to 25 per cent CaO, 10 to 30 per cent SiO_2, and 60 to 80 per cent FeO with minor amounts of MgO and Al_2O_3. By adding fluxing materials such as lime and dolomite, the EAF refining slag can be formed and be working in dephosphorisation reaction after melt-down of the charged raw materials such as DRI, HBI and scrap.

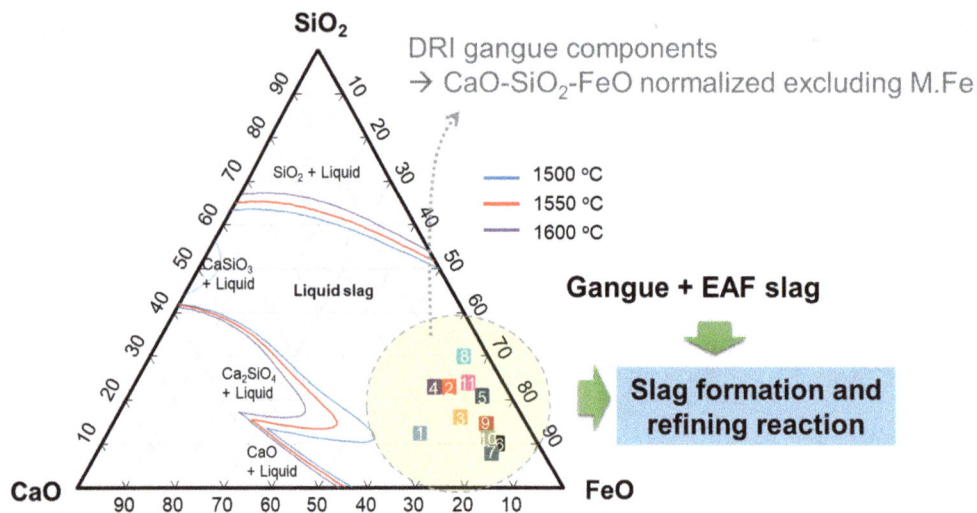

FIG 1 – Compositions of gangue oxides normalised into the CaO-SiO$_2$-FeO ternary phase diagram.

SLAG FORMATION DURING MELTING HOT BRIQUETTED IRON (HBI)

The reaction phenomena for the state of initial melting of HBI during heating are shown in Figure 2. Before melting, HBI samples were cut into small pieces and polished to characterise the gangue oxides and carbon distributions in HBI. A high-frequency induction furnace was used to observe the melting behaviour and the slag-metal reactions. A 400 g HBI was charged in a fused magnesia crucible. The experimental temperature was increased by 10°C/min rate and hold at 1550°C. Partially melted iron briquette, FeO-SiO$_2$ (fayalite) based slag and CO gas evolutions were observed during initial melting. Here, CO gas originated from carbothermic reaction between FeO and carbon in HBI as given in Equation 1.

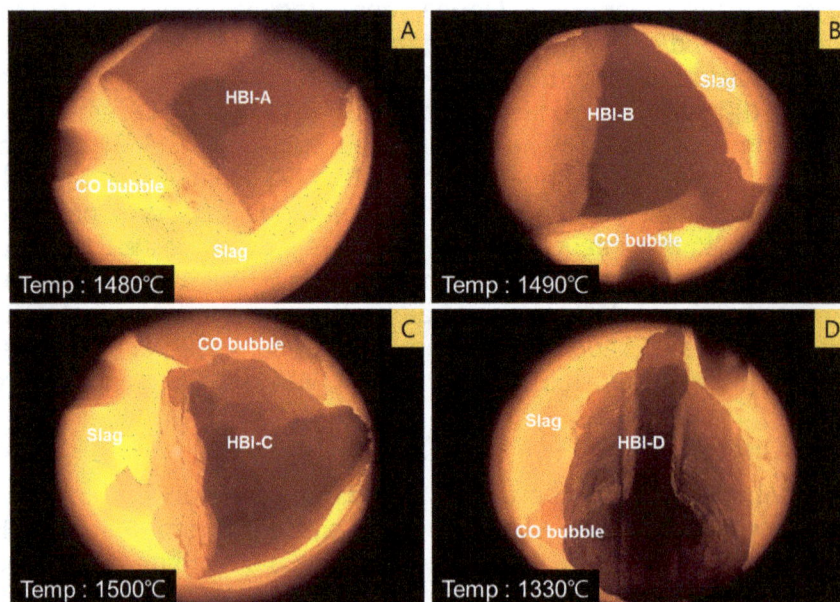

$$FeO(s) + C(s) = Fe(l) + CO(g) \qquad (1)$$

FIG 2 – Snapshot for initial melting of HBI with carbon content of (a) 1.3, (b) 1.0, (c) 1.4 and (d) 2.1 wt per cent.

In Figure 2, it is interesting that the melting initiation temperature decreases from about 1490($\pm$10)°C to 1330($\pm$10)°C by increasing the carbon content in HBI from about 1.2($\pm$0.2) to 2.1($\pm$0.1) wt per cent. It was reported that 1 wt per cent carbon in DRI or HBI can potentially contribute to an increase of productivity (+0.8 to 1.5 per cent) and iron yield (*ca.* +1 per cent) as well as to a decrease of electrode consumption (-0.1 to 1.5 kg/t) and electric power (-15 to 25 kWh/t) in EAF operations (Sunyal, 2015; Hornby, 2021). Also, thin slag layer on the surface of molten steel was produced by the melting of gangue oxides in HBI and the absolute quantity of slag produced by gangue oxides in HBI increased with increasing HBI content. After melting, slag foaming was induced by CO gas evolution from the reaction between FeO in the molten slag with carbon in the molten steel. At this point, the changes in composition and amount of slag are strongly dependent on HBI content. Therefore, it is vital to evaluate the behaviour of phosphorus in molten steel and slag according to HBI mixing ratio during specific reaction stages, such as melting and slag–metal reaction, to thermodynamically understand the dephosphorisation reaction.

EFFECT OF DRI ON PHOSPHATE CAPACITY OF EAF SLAG

Phosphate capacity as a measure of the ability of a slag to absorb phosphorus can be defined as a function of temperature, basicity, and the stability of phosphate ion in the slag as follows (Wagner, 1975).

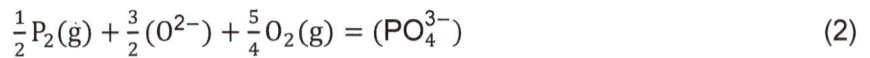

$$\frac{1}{2}P_2(\dot{g}) + \frac{3}{2}(O^{2-}) + \frac{5}{4}O_2(g) = (PO_4^{3-}) \tag{2}$$

$$C_{PO_4^{3-}} = \frac{K_{[2]} \cdot a_{O^{2-}}^{3/2}}{f_{PO_4^{3-}}} = \frac{(\%PO_4^{3-})}{p_{P_2}^{1/2} \cdot p_{O_2}^{4/5}} \tag{3}$$

where $K_{[2]}$ is the equilibrium constant of Equation 2, $a_{O^{2-}}$ is the activity of free O^{2-} ion, $f_{PO_4^{3-}}$ is the activity coefficient of phosphate ion in slag, and p_i is the partial pressure of the gaseous component i. Slag composition is expected to change with DRI content because gangue oxides in DRI dissolve into the slag. In particular, because SiO_2 content in DRI as a gangue oxide is relatively high (~5 wt per cent), slag basicity, which is the driving force of dephosphorisation reaction based on Equation 2, will decrease. Thus, it is vital to confirm the relationship between modified basicity index, eg $\log\{(X_{BO(=CaO+Na_2O+BaO+MnO)}/X_{AO(=SiO_2+B_2O_3)})\}$ and phosphate capacity, $\log C_{PO_4^{3-}}$, as shown in Figure 3a (Heo and Park, 2018). Phosphate capacity, $\log C_{PO_4^{3-}}$ clearly decreased with decreasing basicity and with increasing DRI content, indicating that a higher content of DRI unambiguously reduced dephosphorisation efficiency by contributing SiO_2.

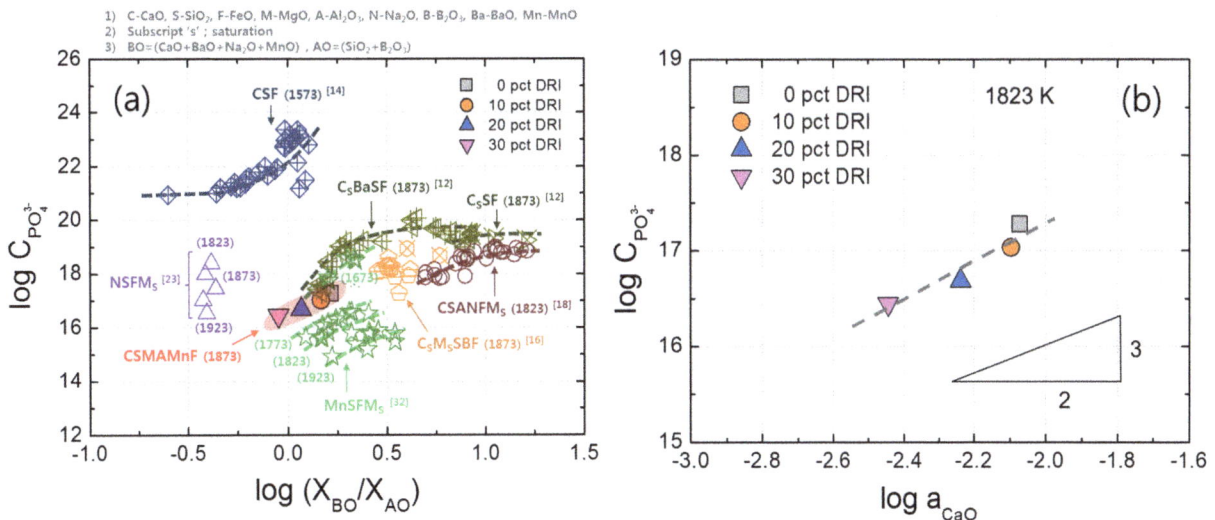

FIG 3 – Phosphate capacity, $\log C_{PO_4^{3-}}$ as a function of (a) modified basicity index, $\log(X_{BO}/X_{AO})$ for the FeO-bearing slags and (b) $\log a_{CaO}$ for the $CaO-SiO_2-FeO-MgO-Al_2O_2-MnO$ slag.

Activity of CaO decreases and that of SiO_2 increases with increasing DRI content. CaO is known to behave as a representative basic component by contributing free oxygen ions (O^{2-}) in the slag as outlined in the following reaction (Sano, 1997).

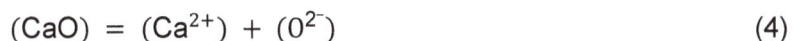

$$(CaO) = (Ca^{2+}) + (O^{2-}) \tag{4}$$

$$a_{O^{2-}} = \frac{K_{[4]} \cdot a_{CaO}}{a_{Ca^{2+}}} \qquad (5)$$

By combining Equations 3 and 5, the correlation between CaO activity and phosphate capacity can be deduced as follows:

$$\log C_{PO_4^{3-}} = \frac{3}{2}\log a_{CaO} - \frac{3}{2}\log a_{Ca^{2+}} - \log f_{PO_4^{3-}} + \text{Const.} \qquad (6)$$

Based on Equation 6, the phosphate capacity and activity of CaO are expected to exhibit a linear relationship with a slope of 1.5 on a logarithmic scale assuming that the composition dependencies of $a_{Ca^{2+}}$ and $f_{PO_4^{3-}}$ are not critical at a given temperature. Here, the activity of CaO was calculated from FactSage™ (ver 8.2) software with FTOXID and FACTPS databases. Figure 3b shows the linear relationship between phosphate capacity and the activity of CaO obtained using least square regression (Heo and Park, 2018). Hence, the activity of CaO can be used as a useful index representing the basicity of slag based on the linear relationship between phosphate capacity and the activity of CaO. It is noteworthy that the phosphate capacity of slag is strongly affected by the activities of CaO and SiO_2 in the slag.

CONTROL OF PHASE EQUILIBRIA FOR GOOD FOAMABILITY OF EAF SLAG

Slag foaming is strongly affected by the slag viscosity, which is dependent on the MgO content as well as basicity. The (Mg,Fe)O monoxide (magnesiowüstite, MW) and Ca_2SiO_4 (dicalcium silicate, C_2S) compounds suspended slags are frequently occurred in commercial EAF plant operation due to the heterogeneous slag composition, temperature fluctuations, and impurities in the fluxing agents. The solid+liquid multiphase slag affects the foamability and dephosphorisation efficiency. For slag composition beyond the MW saturation limit, it is expected that slag viscosity is relatively higher than fully liquid slag due to precipitation of solid compounds according to Einstein-Roscoe equation (Roscoe, 1952; Heo and Park, 2021). A higher viscosity increases the foaming index (Σ), which is equal to the retention or traveling time of gas in the slag, which can be used as a quantitative index of foam stability (Ito and Fruehan, 1989). Hence, it is required to evaluate the effect of basicity on the foaming index. In the present study, the empirical equation of the foaming index of a multicomponent slag system was suggested by considering the thermophysical properties of slags as follows:

$$\Sigma = k\frac{\eta}{\sqrt{\rho \cdot \sigma}} \qquad (7)$$

where k, η, ρ, and σ are a constant, viscosity [Pa·s], density [kg/m^{-3}], and surface tension [N/m], respectively. The constant k was assigned a value of 999 due to the value obtained from a similar slag composition by Kim, Min and Park (2001) The viscosity can be obtained by FactSage™ software. Here, the effect of solid precipitation on slag viscosity is considered by using the Einstein-Roscoe equation. In addition, the density and surface tension are estimated using the polynomial expression by considering pure components, which are functions of temperature. Temperature dependencies of surface tension and density of slag components are available elsewhere (Heo and Park, 2019).

The foaming index of the FeO-bearing slags are compared in Figure 4 including data from Ito and Fruehan (1989) and Jiang and Fruehan (1991) (Heo and Park, 2021). The foaming index of the CaO-SiO_2-FeO-Al_2O_3-MgO-MnO system EAF slag and data from Ito and Fruehan (1989) and Jiang and Fruehan (1991) generally decreases with increasing basicity up to C/S=1.0–1.2, depending on the slag composition, and it rebounds with increasing basicity, which is in excess of the liquidus composition. This behaviour is strongly affected by the slag viscosity based on Equation 7. For slag basicity, C/S<1.2, the viscosity decreases with increasing basicity but beyond 1.2, the viscosity that apparently increases with increasing basicity is affected by the precipitation of solid compounds, contributing to increase the foam stability.

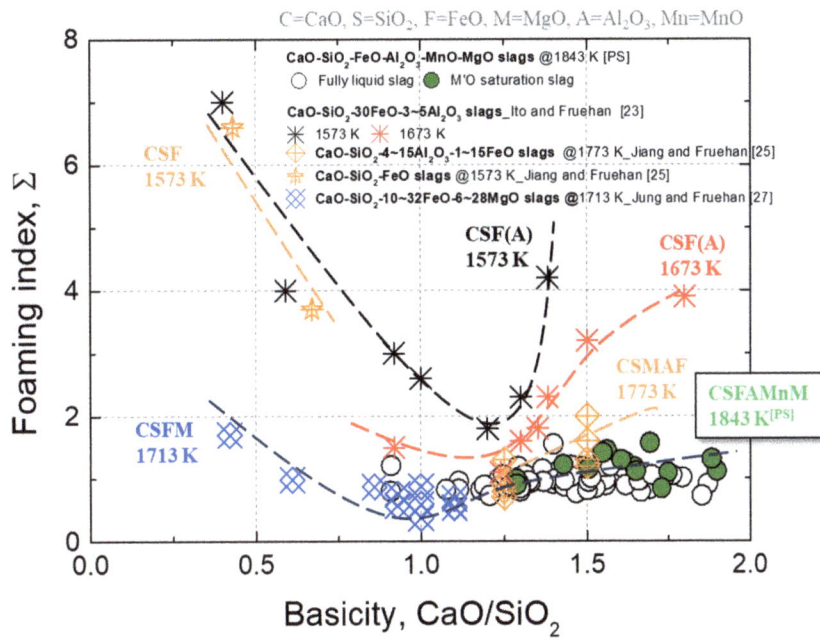

$C=CaO, S=SiO_2, F=FeO, M=MgO, A=Al_2O_3, Mn=MnO$

FIG 4 – Effect of CaO/SiO_2 ratio of slag and temperature on the foaming index (Σ) of FeO-bearing EAF slags.

CONCLUSIONS

The active use of H_2-reduced DRI and/or HBI has been widely increasing to mitigate the CO_2 emissions in steel sector. The contents of unreduced FeO (ie metallisation degree) as well as gangue oxides such as SiO_2, Al_2O_3 etc are significant factor affecting the electric furnace operations efficiency including FeO reduction, carburisation, melting rate of DRI/HBI, dephosphorisation, slag foaming, refractory corrosion etc. These phenomena are expected to be quite different from conventional experiences using NG-reduced DRI/HBI in conjunction with virgin scrap. Consequently, the physicochemical properties of molten slags and fluxes should be further investigated by combining the experimental methodologies and computational modelling.

REFERENCES

Heo, J H and Park, J H, 2018. Effect of direct reduced iron (DRI) on dephosphorization of molten steel by electric arc furnace slag, *Metall Mater Trans B*, 49B:3381–3389.

Heo, J H and Park, J H, 2019. Assessment of physicochemical properties of electrical arc furnace slag and their effects on foamability, *Metall Mater Trans B*, 50B:2959–2968.

Heo, J H and Park, J H, 2021. Effect of slag composition on dephosphorization and foamability in the electric arc furnace steelmaking process: Improvement of plant operation, *Metall Mater Trans B*, 52B:3613–3623.

Hornby, S, 2021. Hydrogen based DRI EAF steelmaking – Fact or fiction?, Proceeding of AISTech2021, Nashville, TN.

Ito, K and Fruehan, R J, 1989. Study on the foaming of CaO-SiO2-FeO slags, *Metall Trans B*, 20B:509–521.

Jiang, R and Fruehan, R J, 1991. Slag foaming in bath smelting, *Metall Trans B*, 22B:481–489.

Kim, H S, Min, D J and Park, J H, 2001. Foaming behavior of CaO-SiO2-FeO-MgOsatd-X (X=Al2O3, MnO, P2O5 and CaF2) slags at high temperatures, *ISIJ Int*, 41:317–324.

Roscoe, R, 1952. The viscosity of suspensions of rigid spheres, *Br J Appl Phys*, 3:267–269.

Sano, N, 1997. *Advanced Physical Chemistry for Process Metallurgy* (Academic Press: New York).

Sunyal, S, 2015. The value of DRI – Using the product for optimum steelmaking, Direct from Midrex, 1Q 2015.

Wagner, C, 1975. The concept of the basicity of slags, *Metall Trans B*, 6B:405–409.

Wimmer, G, Rosner, J and Voraberger, B, 2023. Smelter for processing of low grade DRI, Proceedings of 6th ESTAD, 2023, Düsseldorf, Germany.

Understanding zinc-containing species in basic oxygen steelmaking dust

R J Longbottom[1], D J Pinson[2,5], S J Chew[3,5] and B J Monaghan[4,5]

1. Research Fellow, Faculty of Engineering and Information Science, University of Wollongong, Wollongong NSW 2522. Email: rayl@uow.edu.au
2. Senior Technology and Development Engineer, BlueScope Coke and Ironmaking Technology, Port Kembla NSW 2505. Email: david.pinson@bluescopesteel.com
3. Principal Technology and Development Engineer, BlueScope Coke and Ironmaking Technology, Port Kembla NSW 2505. Email: sheng.chew@bluescopesteel.com
4. Professor, Faculty of Engineering and Information Science, University of Wollongong, Wollongong NSW 2522, Email: monaghan@uow.edu.au
5. ARC Research Hub for Australian Steel Innovation, Wollongong NSW 2522.

ABSTRACT

Recycling steel plant by-products is a critical issue for achieving both environmental and economic sustainability of integrated steel plants. A key steel plant by-product stream is dust from basic oxygen steelmaking (BOS), which can be recycled back to the BOS vessel as iron units and to aid slag formation. The major limitations for its recycling are zinc and the uncertainty in zinc speciation of the dust. This uncertainty affects the amount of dust recycled and how it interacts with the slag during processing. BOS dust in stockpiles can undergo self-sintering reactions that oxidise the iron-bearing components. As this occurs, zinc (or oxide) may combine with iron oxides to form a zinc-iron spinel solid solution phase. Knowing which zinc-bearing species are present in the BOS dust is a key aspect in understanding and developing an efficient recycling process for the dust.

To overcome difficulties associated with phase identification in the BOS dust (both fresh and self-sintered), particularly the magnetite-zinc ferrite spinel solid solution using X-ray diffraction (XRD) / scanning electron microscopy (SEM) / energy dispersive spectroscopy (EDS), high end characterisation tools, transmission electron microscopy (TEM) and Mössbauer spectroscopy, were also used. Such an approach allowed improved phase identification of key zinc- and iron-bearing phases that were not previously resolvable.

The zinc in fresh BOS dust was mostly present as extremely fine separate particles of zinc oxide, attached to the iron particles. These extremely fine particles were <10 nm in size. Additionally, a small amount of zinc ferrite was also identified within the sample (most likely contained within a magnetite-zinc ferrite spinel solid solution).

However, the zinc in the self-sintered BOS dust, was found to be present entirely in a magnetite-zinc ferrite spinel solid solution. While this spinel phase had been previously identified by XRD, it was not clear that it contained zinc. The Mössbauer results provided definitive evidence of zinc within the spinel phase.

INTRODUCTION

On-plant recycling of steel plant by-products will help achieve the environmental and economic sustainability of integrated steel plants, reducing emissions by replacing raw materials and recovering the valuable components within the by-products (Nyirenda, 1991; Ahmed *et al*, 2015). One key by-product in an integrated steel plant is dust from the basic oxygen steelmaking (BOS) process. When oxygen is blown at supersonic velocity to decarburise steel, significant amounts of dust are generated. This dust is made up of very fine particles of steel and slag droplets, as well as flux fines. When the dust is recycled back to the BOS, these slag droplets and flux fines have value and aid slag formation (Longbottom *et al*, 2019a, 2019b).

Unfortunately, there are limitations on the recycling of BOS dust, primarily Zn and the uncertainty in Zn speciation of the dust (Stewart and Barron, 2020; Longbottom *et al*, 2020). Which Zn-bearing phases are present in the BOS dust strongly affects how the BOS dust interacts with the slag during processing. As such, the uncertainty in the Zn speciation in the dust affects the amount that can be recycled to the BOS. Hence, it is important to understand what Zn-bearing phases are present in the BOS dust, whether as metallic Zn, ZnO or in a complex zinc ferrite-magnetite ($ZnFe_2O_4$-Fe_3O_4) spinel solid solution.

Zn in the BOS dust results from the high temperatures (>>1600°C) in the BOS process causing volatilisation of Zn (and other tramp elements) to the gas phase (Nedar, 1996). As the gas phase is cooled in the extraction process, the volatilised Zn can condense as separate particles or onto existing dust particles.

BOS dust is often removed from the off-gas using a wet scrubbing process, resulting in a slurry that is usually dewatered to produce a low moisture filter cake that is stockpiled for future recycling. When stored in stockpiles prior to recycling, BlueScope BOS dust has been found to self-sinter (Longbottom *et al*, 2019a, 2019b, 2020). Using TGA-DSC (thermogravimetric analysis – differential scanning calorimeter) and *in situ* X-ray diffraction (XRD) techniques, the self-sintered dust has been shown to have undergone exothermic oxidation reactions in the stockpiles (Longbottom *et al*, 2019a, 2019b, 2020). As part of these oxidation reactions, ZnO reacted with iron oxides to form the $ZnFe_2O_4$-Fe_3O_4 spinel solid solution. The formation of the $ZnFe_2O_4$ side of the solid solution is shown in Equation (1).

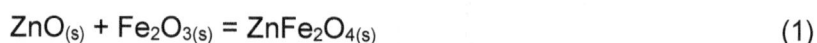

$$ZnO_{(s)} + Fe_2O_{3(s)} = ZnFe_2O_{4(s)} \qquad (1)$$

However, these studies also highlighted a number of issues in the characterisation of the Zn bearing phases. Using scanning electron microscopy (SEM) with energy dispersive spectroscopy (EDS) showed that the BOS dust consisted of very fine iron/iron oxide particles (~200–500 nm) (Longbottom *et al*, 2019a, 2019b). However, it was found that due to the fine particle size, it was difficult to fully resolve the smaller particles and it was not possible to fully resolve the Zn-containing phase/particles by EDS. Similar microstructures containing these very fine particles have been reported in other BOS dust samples (Jabłońska *et al*, 2021; Stewart *et al*, 2022; Wang *et al*, 2021; Hleis *et al*, 2013) and Electric Arc Furnace (EAF) dust samples (Suetens *et al*, 2015; Sofilić *et al*, 2004; Lin *et al*, 2017).

There were also issues with the use of XRD to characterise these samples. It is well known that Fe_3O_4 and $ZnFe_2O_4$ have very similar lattice spacings, resulting in very similar XRD patterns (Stewart and Barron, 2020). In addition, Fe_3O_4 and $ZnFe_2O_4$ form a solid solution, which again is difficult to identify and quantify using XRD. For these reasons, in previous studies a single 'spinel' phase was used to encompass any Fe_3O_4 and $ZnFe_2O_4$ or solid solution (Longbottom *et al*, 2020). Often, these issues can be overcome by substituting a longer wavelength X-ray source, which may also help overcome issues with iron fluorescence with a Cu-Kα source. However, peak broadening from the fine particle size of the BOS dust likely masks any peak shift with composition even with a longer wavelength X-rays.

The aim of the study was to overcome these difficulties in identifying the Zn-bearing phases by characterising the BOS dust using high resolution transmission electron microscopy (TEM) and Mössbauer spectroscopy. TEM microscopy was used to help with the resolving of the very fine particles in the BOS dust samples. Mössbauer spectroscopy was used to help overcome the limitations in XRD in differentiating between Fe_3O_4, $ZnFe_2O_4$ and their solid solution. In this study BOS dust in two conditions was studied, fresh dust collected immediately after the dewatering process, as well as self-sintered dust collected from the stockpiles at BlueScope's Port Kembla steelworks.

EXPERIMENTAL

TEM and Mössbauer spectroscopy were used to characterise BOS dust samples. Fresh and self-sintered samples were both characterised. The fresh dust sample was collected immediately after dewatering, while the self-sintered dust was collected from the stockpiles at BlueScope's Port Kembla steelworks. The composition of the two samples, measured by XRF, is given as simple oxides in Table 1. The speciation of iron (and other phases) is not given by XRF and was initially characterised by XRD phase analysis (see Figure 1) using Cu-kα (λ = 0.154 nm) radiation. It may be seen that iron was mainly present as wüstite and metallic Fe in the fresh BOS dust sample, with a minor amount of 'magnetite'. In the self-sintered sample, iron was mainly present as 'magnetite' with a small amount of hematite. Iron speciation and oxidation state was further investigated as part of this study using Mössbauer spectroscopy.

TABLE 1

XRF analyses of the BOS dust samples. The compositions are given in mass%.

Sample	MgO	Al_2O_3	SiO_2	P_2O_5	SO_3	K_2O	CaO	TiO_2	MnO	Fe_2O_3	ZnO
Fresh	1.8	trace	1.4	trace	trace	trace	3.2	trace	trace	78.2	13.8
Self-sintered	1.5	trace	4.4	trace	trace	trace	3.9	trace	trace	79.8	8.0

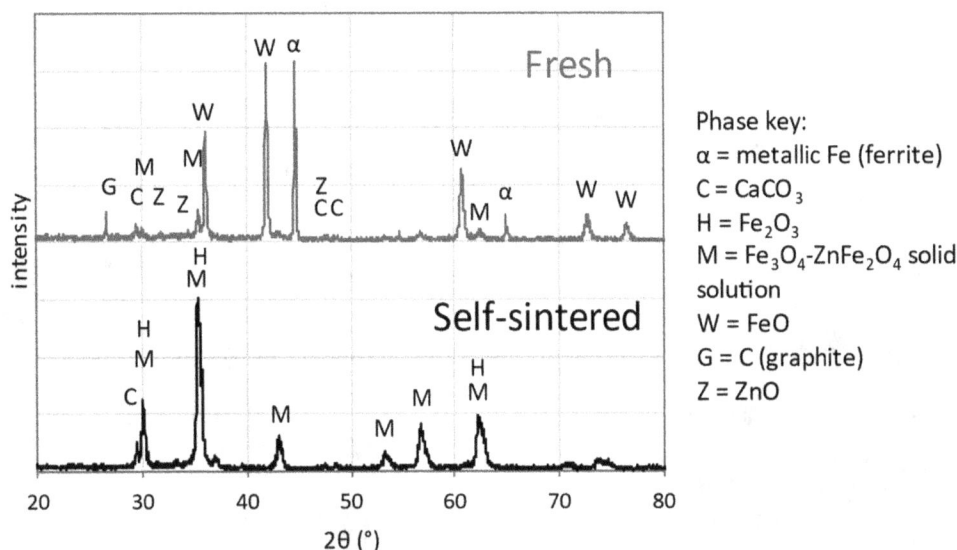

FIG 1 – XRD phase analysis of the BOS dust samples. XRD carried out using Cu-kα radiation (λ = 0.154 nm) at 1 KW power.

TEM characterisation

TEM characterisation was carried out using an aberration corrected JEOL ARM200F equipped with an 80 mm² JEOL Centurio SDD EDS detector. The TEM was equipped with bright field and secondary electron scanning TEM detectors. Each detector measures different characteristics of the samples:

- **Bright field:** image from transmitted beam, related to sample thickness and density.
- **Secondary electron:** equivalent to SEM images, information about surface of particles.

Sample preparation of the fresh BOS dust for TEM was carried out by suspending the sample in ethanol, with the resulting suspension then held for 5 mins in an ultrasonic cleaner to break up clumps. This suspension was then diluted further with more ethanol, which was then drawn into a transfer pipette. Two drops of the suspension were then placed onto a lacy carbon grid.

A ~20 g sample of the self-sintered BOS dust sample for the TEM was ground in a ring grinder for 30 sec to a fine particle size so that the particles would be electron transparent. This powder was then prepared for TEM in a similar manner to that for the fresh BOS dust.

Mössbauer characterisation

^{57}Fe Mössbauer spectra were obtained at room temperature using the standard transmission mode. A ^{57}Co isotope on a Rh foil ($^{57}Co\underline{Rh}$) was used as the γ-ray source. The velocity scale was calibrated using a 6 μm thick α-Fe foil, and all isomer shifts are quoted relative to the α-Fe calibration at source velocity (V) values of 4 and 12 mm/s. The calibration gave a Doppler velocity amplitude of 11.4885 mm/s for the nominal V = 12 mm/s setting and 3.8942 mm/s for the nominal V = 4 mm/s setting.

The collected spectra were fitted using the NORMOS™ software (by Wissenschaftliche Elektronik GMBH) to allow identification of the phases. The sum of expected spectra from different phases were fitted to the measured values. Quantification of the fits to the Mössbauer spectra was conducted without standards. As such, caution should be used when considering the absolute values.

RESULTS

To gain a better understanding of the Zn-bearing phases in BOS dust, fresh and self-sintered samples were characterised by TEM and Mössbauer spectroscopy. As both techniques use small samples (especially in the case of TEM), concerns about sample representation mean that the results (especially with quantification) should be taken as indicative rather than absolute.

Fresh BOS dust

Bright field and secondary electron images of the fresh BOS dust sample are shown in Figure 2. EDS elemental mapping (Figure 3) was used to identify the elements present within the sample, and high magnification images of the lattice fringes (Figure 4) within the sample were used to measure lattice spacing and to aid identification of phases.

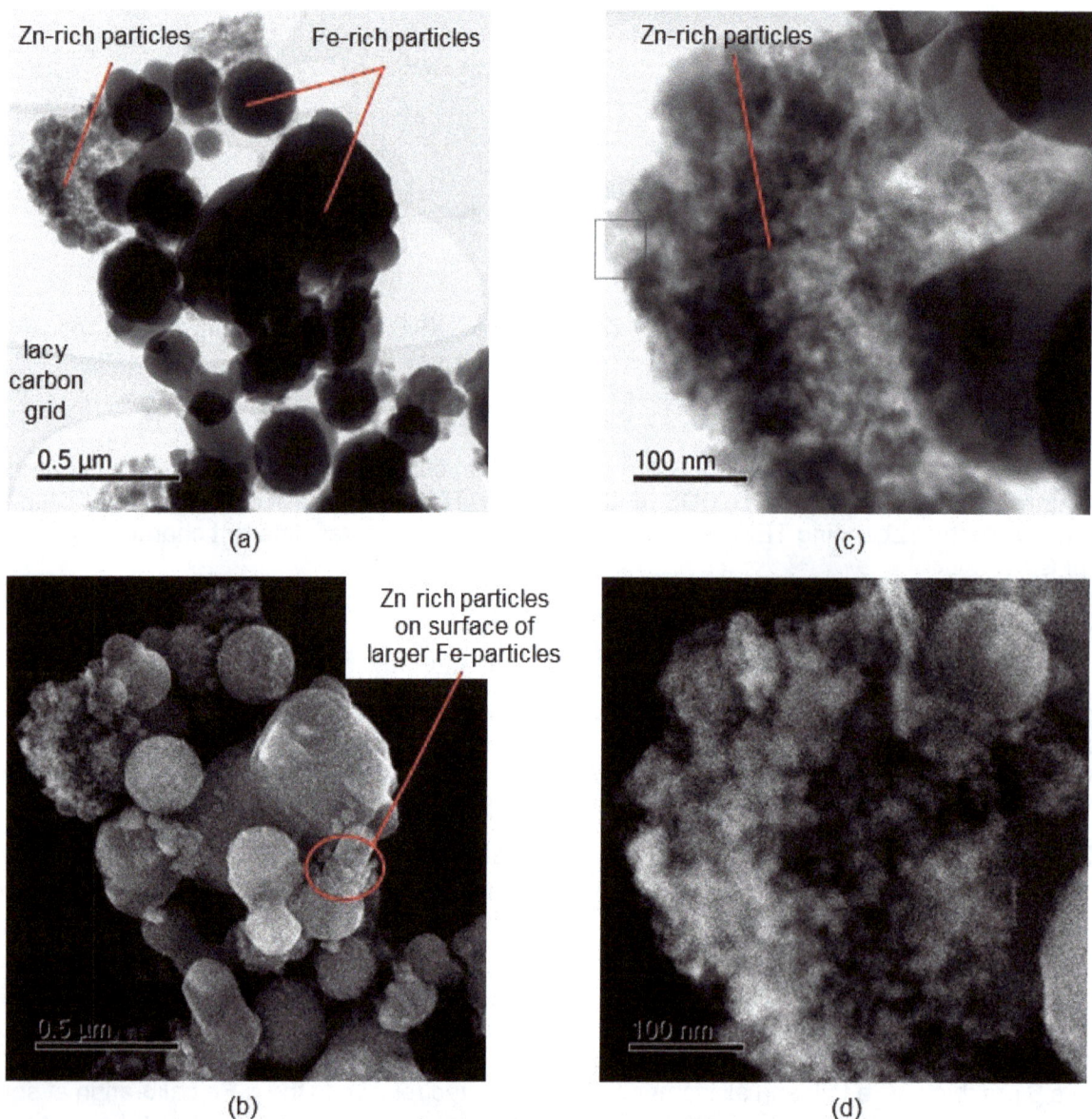

(a)

(b)

(c)

(d)

FIG 2 – TEM micrographs of fresh BOS dust. Lower magnification images: (a) bright field; and (b) secondary electron. Higher magnification images: (c) bright field; and (d) secondary electron.

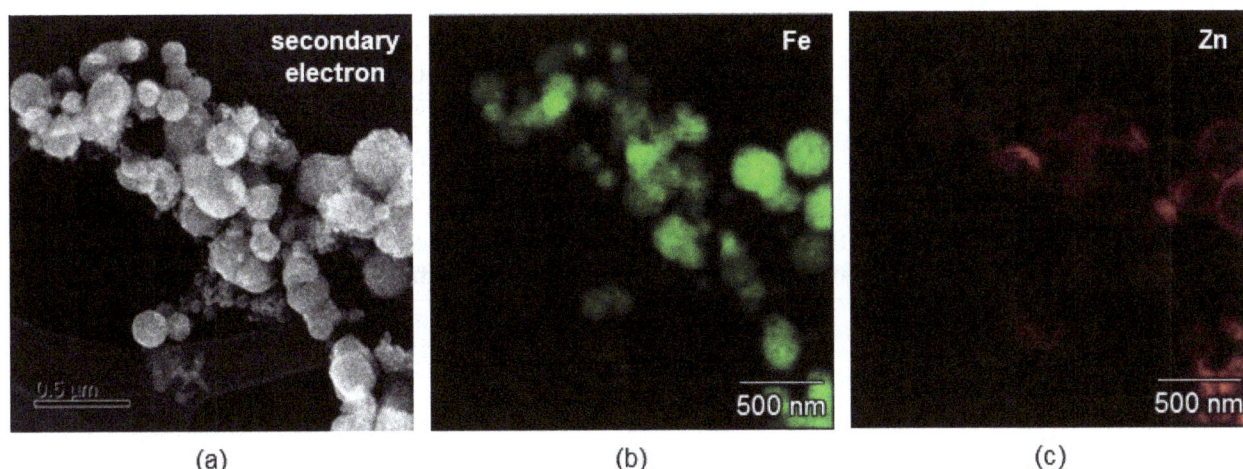

FIG 3 – TEM EDS elemental mapping of a fresh BOS dust sample. (a) Secondary electron image; (b) Fe elemental map; and (c) Zn elemental map.

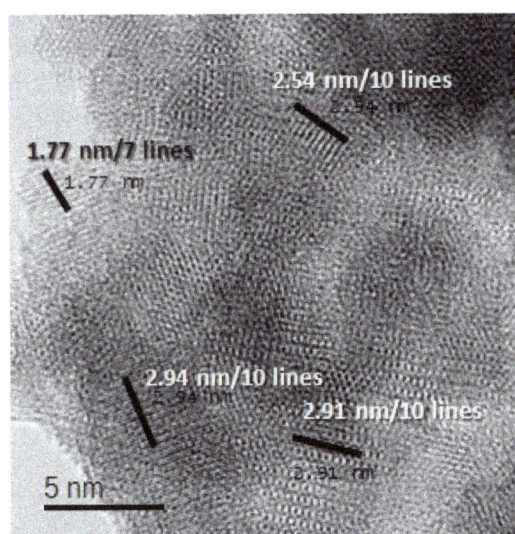

FIG 4 – High magnification bright field TEM image of fresh BOS dust (for the region shown by the rectangle in Figure 2b) showing lattice fringes.

Clumps of particles were observed attached to the carbon grid. Iron was found to be present predominantly as an oxide, but with some metallic particles. The Fe-rich particles were small, with a typical particle size being in the range 200–500 nm. This matches well with the previously reported SEM results (Longbottom *et al*, 2019b).

Zn was mostly found present as extremely fine separate particles, attached to the iron particles. As these particles were extremely small and clustered together, it is difficult to be definitive about the particle size, even from the high magnification TEM images (Figures 2 and 3). Many of the particles appear to be <10 nm in size. These Zn-rich particles were typically much smaller than the Fe-rich particles to which they were attached. In some cases (see Figure 3), the very small Zn-rich particles were found as a layer around metallic Fe particles. However, this behaviour was uncommon, even within the small number of clumps examined, where Zn-rich particles were more often found as clumps of Zn-rich particles attached to both metallic and oxide particles.

Examining the lattice fringes in Figure 4, crystallographic spacing (d-spacing) between the lattice planes was measured (Table 2). Comparing the measured values with the reported values for metallic Zn and ZnO (also given in Table 2), it appears that there is a strong match for hexagonal ZnO, with the measured and published d-spacings matching well. It is possible that there is a limited amount of metallic Zn in the material, with a reasonable match for the (002) reflection for metallic Zn.

TABLE 2

Measured d-spacings from lattice fringes (Figure 4) and Published d-spacings for metallic Zn and ZnO (Downs and Hall-Wallace, 2003; Gražulis *et al*, 2009, 2012, 2015; Merkys *et al*, 2016, 2023; Quirós *et al*, 2018; Vaitkus, Merkys and Gražulis, 2021; Vaitkus *et al*, 2023).

Average measured d-spacing (nm)	Metallic Zn (hexagonal)		ZnO (hexagonal)	
	reflection plane lattice indices	d-spacing (nm)	reflection plane lattice indices	d-spacing (nm)
0.293	(002)	0.247	(100)	0.290
0.253	(100)	0.231	(002)	0.261
–	(101)	0.169	(101)	0.254

Mössbauer spectroscopy results of fresh BOS dust, along with the fitted spectra for metallic iron, wüstite and $ZnFe_2O_4$ are shown in Figure 5. The spectrum of the BOS dust was fitted reasonably well with a non-magnetite, quadrupole-split doublet (likely corresponding to wüstite) between 0–2 mm/s and a magnetite sextet (likely corresponding to metallic iron) spreading out to larger velocity scale values. A third unrestricted non-magnetic phase likely corresponded to $ZnFe_2O_4$. This analysis only considered the iron bearing phases, as only Fe nuclei are detected by Mössbauer spectroscopy using a $^{57}CoRh$ γ-ray source.

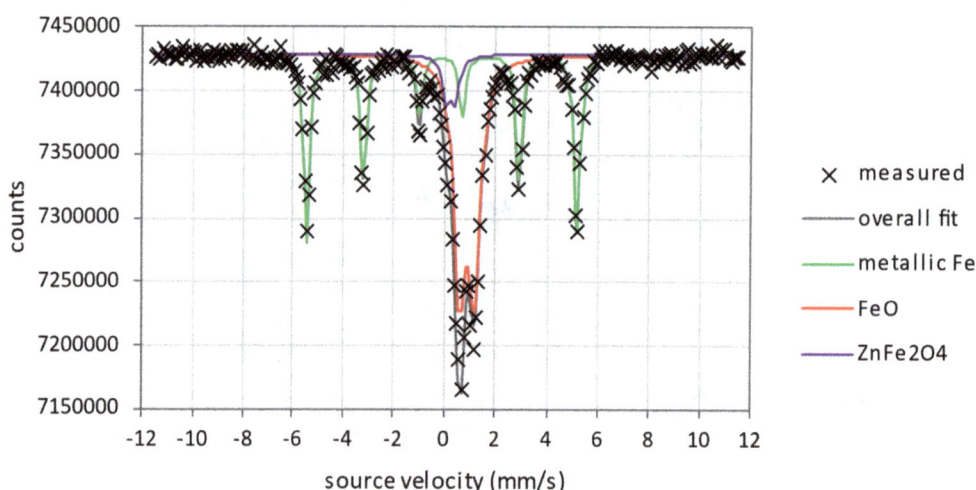

FIG 5 – Mössbauer spectra of fresh BOS dust showing the fitted spectra for metallic Fe, FeO and $ZnFe_2O_4$.

Self-sintered BOS dust

TEM images of a self-sintered BOS dust sample are shown in Figure 6, while EDS elemental maps are shown in Figures 7. Clumps of particles were found attached to the carbon grid (Figure 6). The particles in these clumps were of varying sizes, with relatively few larger particles (>1 µm), with more particles being in the 100–500 nm range. It should be noted that these samples were ground to ensure electron transparency, therefore both the clumps and particle size distributions are artefacts of sample preparation. While there were some very fine particles (<50 nm) attached to the clumps, they appeared quite different to those seen in the fresh BOS dust (Figure 2). At high magnifications (Figure 6) these particles appeared to be strongly crystalline. EDS mapping (Figure 7) showed that most of the particles appear uniform in their Fe and Zn levels, indicating that the particles are likely a Fe_3O_4-$ZnFe_2O_4$ spinel solution.

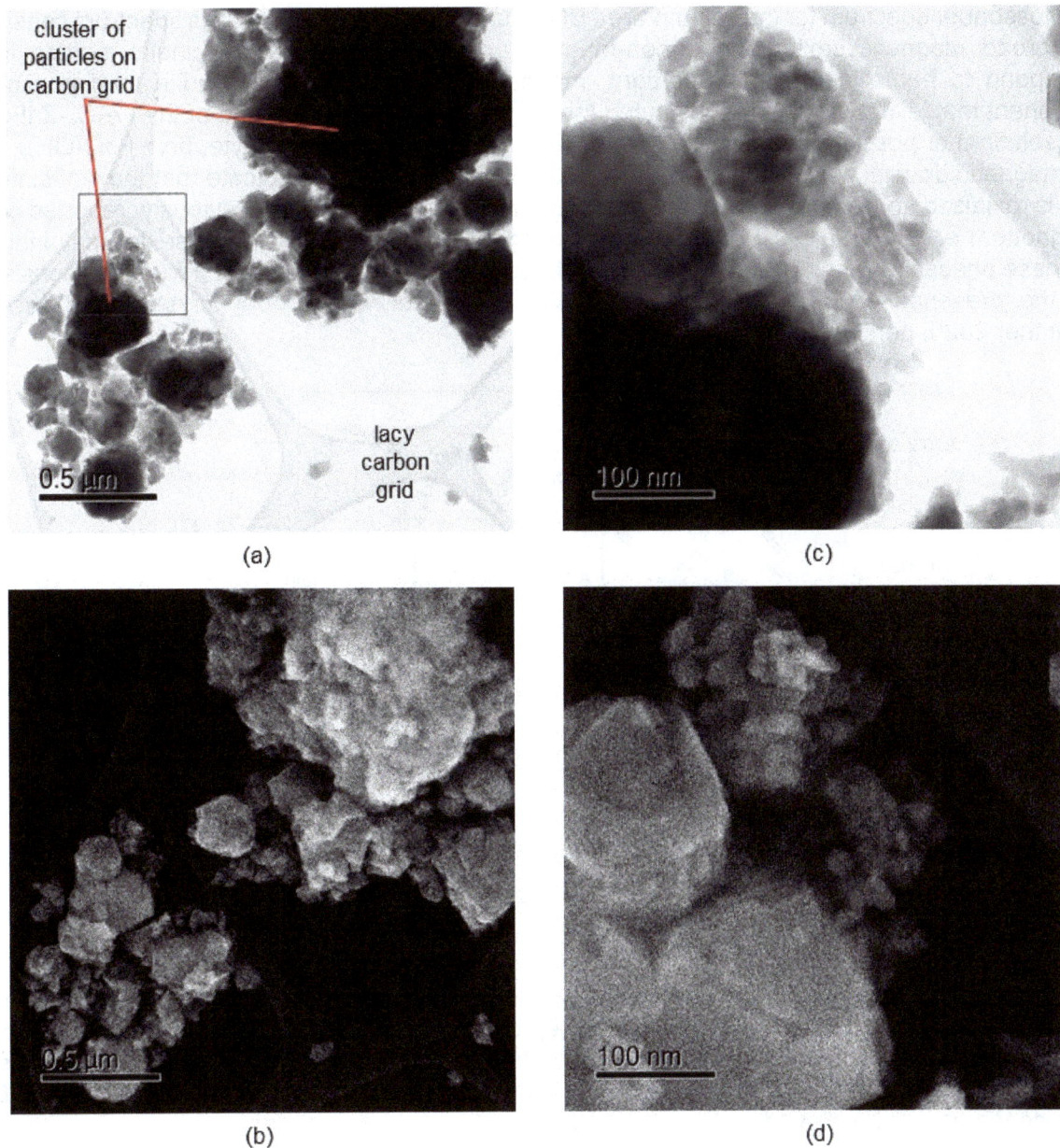

FIG 6 – TEM micrographs of self-sintered BOS dust. Lower magnification images: (a) bright field; and (b) secondary electron. Higher magnification images: (c) bright field; and (d) secondary electron.

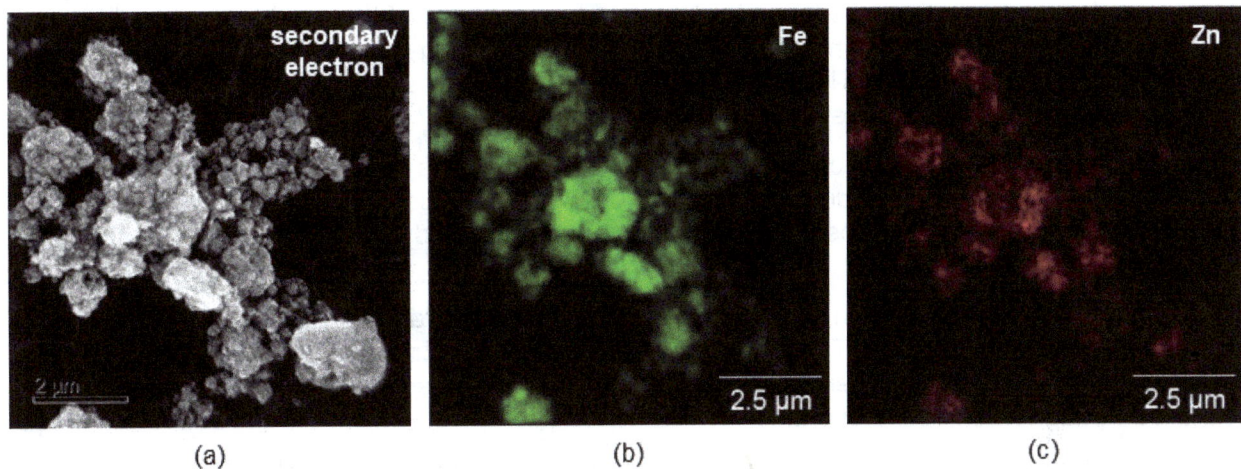

FIG 7 – TEM EDS elemental mapping of a self-sintered BOS dust sample. (a) Secondary electron image; (b) Fe elemental map; and (c) Zn elemental map.

The Mössbauer spectrum for the self-sintered BOS dust is given in Figure 8. This spectrum consisted of a broad magnetic sextet and a non-magnetic doublet. The broad magnetic sextets likely correspond to hematite (major proportion), magnetite (minor proportion) and a third iron oxide component that is likely to be Zn-substituted magnetite (ie the Fe_3O_4-rich end of the Fe_3O_4-$ZnFe_2O_4$ solid solution) or possibly hydrated iron oxides (such as goethite or feroxyhyte, both $FeO(OH)$). The non-magnetic doublet closely matches $ZnFe_2O_4$. The phases present indicate that the self-sintered dust is oxidised in comparison to the fresh sample, consistent with previously reported work (Longbottom *et al*, 2019a, 2019b, 2020). The broadness of the troughs in the sextet may indicate that these phases may have been heavily substituted (most likely with Zn, which would be consistent with the presence of the Fe_3O_4-$ZnFe_2O_4$ solid solution) or were poorly crystalline (Cadogan, November 2020, personal communication).

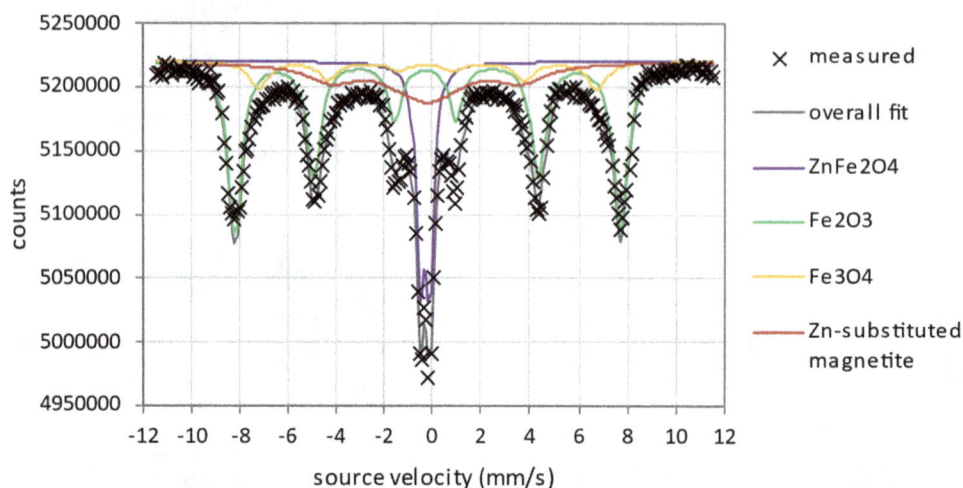

FIG 8 – Mössbauer spectra of self-sintered BOS dust showing the fitted spectra for $ZnFe_2O_4$, Fe_2O_3, Fe_3O_4 and Zn-substituted magnetite.

DISCUSSION

Steel plant by-products such as these BOS dust samples are not commonly examined using these techniques, often considered to be too 'dirty' (eg containing possible volatile components) to be characterised under high vacuum. However, by extensive characterisation of these materials using more routine techniques, such as XRD, X-ray fluorescence (XRF), optical microscopy and SEM, these concerns can be allayed.

It should be noted that there are some limitations to the techniques used. The main issue with both the TEM and Mössbauer characterisation was with the small sample size and whether or not the sample is representative of the bulk material. For this study, there was already a comprehensive characterisation of the bulk material (Longbottom *et al*, 2019a, 2019b, 2020). With this being the case, sample representation was less of an issue. The emphasis of this study was on addressing specific issues with the pre-existing analysis around the Zn-bearing phases. The very fine particulate nature of the fresh BOS dust was identified by SEM (Longbottom *et al*, 2029b), but it was not possible to fully resolve the sample with respect to particle size.

This indicated a higher resolution technique, such as TEM was required. The detailed existing characterisation and understanding of the BOS dust samples, using XRD, SEM-EDS and XRF, allowed the use of TEM and Mössbauer spectroscopy in this study without having to characterise many samples.

Mössbauer spectroscopy used in this study was only able to characterise iron bearing phases due to the γ-ray source (^{57}Co) used. This means that the presence of ZnO could not be tested using this technique. While Zn is Mössbauer active, it is not commonly analysed in this way. $ZnFe_2O_4$ could be identified through its iron component.

The results also indicate that Mössbauer spectroscopy has general applicability in characterisation of metallurgical samples. While Mössbauer spectroscopy is not routinely used in slags, it would be a useful tool, aiding identification of Fe-bearing phases, or using the appropriate radiation source, phases containing any Mössbauer active elements.

Further, slags are almost always characterised at room temperature, often by XRF and/or XRD. The ability of Mössbauer spectroscopy to identify components and associations in less crystalline or amorphous phases, such as glass in quenched slags, offers opportunities to improve the characterisation of these phases within slags, not possible using XRD. When used in conjunction with other techniques (XRD, XRF, SEM-EDS or Wavelength Dispersive Spectrometry (WDS)), Mössbauer spectroscopy will allow a fuller understanding of the slag and the phases that it contains.

The focus of this characterisation of the BOS dust was to better identify the Zn-bearing phases and their form or morphology. There were a few possible Zn-bearing phases in the BOS dust that were examined. These included metallic Zn, ZnO and $ZnFe_2O_4$. The TEM and Mössbauer characterisation given here, in conjunction with existing XRD, XRF and SEM-EDS characterisation (Longbottom et al, 2019a, 2019b), have allowed identification of whether these Zn-bearing phases were present in the BOS dust samples.

Zn

There was little evidence of metallic Zn in either the fresh or self-sintered BOS dust samples. In the fresh BOS dust samples, there is a possibility that the d-spacing measured from the lattice fringes (Table 2) may match that for metallic Zn. However, this measured d-spacing was a better match for ZnO, especially when considered with the other measured d-spacing values (Table 2). Hence, it is unlikely that metallic Zn was present in either the fresh or self-sintered BOS dust samples.

ZnO

The fresh BOS dust samples most likely contained ZnO, as very fine particles surrounding the slightly larger Fe-rich particles, as shown in Figure 2.

One of the motivations for this study was due to a discrepancy between the amount of Zn in the fresh BOS dust (identified through XRF, Table 1) and the amount of Zn that would be contained in the phases identified through XRD, as shown in Figure 1 and in Longbottom et al (2020). The results from the TEM characterisation (Figure 2) give a plausible cause for this discrepancy. While a small amount of ZnO was detected by XRD (Figure 1), the TEM results showed that the ZnO particles were extremely small. Hence, it is likely that at least some of these fine particles will lead to poor diffraction patterns, and as such are likely to report to the amorphous/non-bulk crystalline fraction and unlikely to be identified as ZnO by XRD. This may also be reflected in the low intensities of the ZnO peaks for the fresh BOS dust sample in Figure 1, which may be expected to be higher with 13.8 per cent ZnO in the sample. These very low intensities of the ZnO peaks also made estimation of the particle size from the diffraction pattern difficult.

It is unclear whether these fine ZnO particles had formed by oxidation of metallic Zn particles that had condensed during cooling of the gas, or by oxidation of Zn vapour in the gas phase. What effect these different formation mechanisms would have on the particle size and morphology of the ZnO particles is also unclear.

There was little evidence of ZnO in the self-sintered BOS dust sample.

$ZnFe_2O_4$

In the fresh BOS dust samples, the Mössbauer spectroscopy indicated the presence of $ZnFe_2O_4$. This is consistent with the presence of a spinel phase in the XRD of the fresh BOS dust samples (Figure 1). While it was not possible to determine through XRD whether the spinel was Fe_3O_4, $ZnFe_2O_4$ or a solution of the two, the Mössbauer spectroscopy showed that there was some $ZnFe_2O_4$ in the fresh BOS dust sample, while not detecting Fe_3O_4.

In the self-sintered BOS dust sample, the major Zn-bearing phase was $ZnFe_2O_4$, most likely contained within a Fe_3O_4-$ZnFe_2O_4$ solid solution. Previous characterisation by XRD indicated the presence of spinel with lattice spacing close to that of both Fe_3O_4 and $ZnFe_2O_4$ (Figure 1). This is supported by the TEM characterisation of the sample, which indicated that the sample was largely a single phase (Figure 7). The Mössbauer spectroscopy of the self-sintered sample directly identified

$ZnFe_2O_4$, while also reporting the presence of magnetite with some substitution of Fe with Zn. While the Mössbauer shows the presence of both Fe_3O_4 and $ZnFe_2O_4$, the technique measures iron nuclei, with the response depending on the surrounding atoms in the crystal lattice (Sharma, Klingelhofer and Nishida, 2013). Due to this, a single phase solid solution may be reported as two separate phases. While this seems a drawback with the Mössbauer spectroscopy, it offers a positive identification of the presence of Zn within the spinel phase. Previously this could only be speculated on from the XRD data, due to the similar lattice parameters of Fe_3O_4 and $ZnFe_2O_4$. Thus, it is most likely that the sample contains a Fe_3O_4-$ZnFe_2O_4$ solid solution with differing levels of Zn, consistent with the results from the XRD and TEM characterisation.

The improved understanding of the Zn-bearing phases within the BOS dust will allow better predictions of how it will behave during recycling. Since the Zn in the BOS dust was found to be oxidised, a reductant would be required to volatilise the Zn if extraction or separation of Zn from the dust is necessary. For recycling into the BOS vessel, the Zn-bearing phases in the fresh and self-sintered BOS dust are different, affecting how each would interact with slag. ZnO (high ZnO activity, very small particle size) in the fresh dust will interact differently with the BOS slag than the Fe_3O_4-$ZnFe_2O_4$ solid solution (lower ZnO activity) found in the self-sintered dust.

Comparison with reported steel plant by-products

Other studies have characterised BOS dust samples from steel plants around the world. The current BOS dust samples contain high Zn levels in comparison to other reported compositions for BOS dust (Kelebek, Yörük and David, 2004; Steer et al, 2014; Vereš, Šepelák and Hredzák, 2015; Gargul, Jarosz and Małecki, 2016; Jabłońska et al, 2021; Stewart et al, 2022). Typically, the phases found in the BOS dust samples were a close match to those in the fresh BOS dust sample, with metallic iron, wüstite and smaller amounts of magnetite and/or hematite. The Zn-bearing phases reported in other BOS samples are summarised in Table 3.

TABLE 3

Reported Zn-bearing phases in BOS dust.

Study	Technique(s)	Majority phases	Reported Zn-bearing phases
Kelebek, Yörük and David (2004)	XRD, SEM	FeO, Fe_2O_3	$ZnFe_2O_4$ identified by XRD, not differentiated from Fe_3O_4
Steer et al (2014)	XRD, wet sizing	not reported	ZnO
Vereš, Šepelák and Hredzák (2015)	XRD, Mössbauer	metallic Fe, FeO, Fe_3O_4, Fe_2O_3	ZnO, $ZnFe_2O_4$
Gargul, Jarosz and Małecki (2016)	XRD	metallic Fe, FeO, Fe_3O_4	ZnO
Jabłońska et al (2021)	XRD	Fe_3O_4, ZnO, $ZnFe_2O_4$	ZnO, $ZnFe_2O_4$ identified by XRD, differentiated from Fe_3O_4 using Rietveld fitting
Stewart et al (2022)	Mössbauer, XRD, SEM, XRF, digestion + Atomic absorption spectroscopy (AAS)	metallic Fe, FeO, Fe_3O_4, Fe_2O_3	Small amount of Zn in solution in Fe_3O_4 No $ZnFe_2O_4$ identified

Several studies on BOS dust have used different techniques to identify the presence of $ZnFe_2O_4$ (Table 3). In general, $ZnFe_2O_4$ was identified, although often identified by XRD, which may result in difficulties in differentiation between $ZnFe_2O_4$ and Fe_3O_4 (Stewart and Barron, 2020).

Some studies used Mössbauer spectroscopy to attempt to identify $ZnFe_2O_4$, with some studies reporting $ZnFe_2O_4$ and others not (Vereš, Šepelák and Hredzák, 2015; Stewart and Barron, 2020). The variability between samples was speculated to be related to the specific dust extraction-gas cleaning systems and dust storage for each plant, with samples that contain more oxidised samples having more $ZnFe_2O_4$. These findings correspond well with the phases in the fresh and self-sintered BOS dust samples in the current study, with the more oxidised self-sintered sample containing more $ZnFe_2O_4$ than the fresh sample.

Electric Arc Furnace (EAF) dust often contains more Zn than BOS dust (Suetens *et al*, 2015; Sofilić *et al*, 2004), due to the higher proportion of scrap in the feed. As such, it is useful to consider EAF dust in comparison to the Zn-rich BOS dust samples from the current study. These Zn-rich EAF dust samples were found to contain both ZnO and $ZnFe_2O_4$. Again, these samples compare well with the BOS dust samples in the current study.

CONCLUSIONS

Characterisation of fresh and self-sintered BOS dust by TEM and Mossbauer spectroscopy was carried out to give a better understanding of the Zn-bearing phases within the samples, to improve the understanding of how the BOS dust will behave during recycling.

In the fresh BOS dust sample, TEM showed that Zn was mostly present as extremely fine separate particles of ZnO, attached to the iron particles. These extremely fine particles were <10 nm in size. While most of the Zn was contained within these extremely small particles, Mössbauer spectroscopy identified a small amount of $ZnFe_2O_4$ within the sample (most likely contained within a Fe_3O_4-$ZnFe_2O_4$ spinel solid solution).

The self-sintered BOS dust sample was significantly different. The Zn-bearing phase in this sample was a Fe_3O_4-$ZnFe_2O_4$ solid solution. While this phase had previously been identified by XRD, Mössbauer spectroscopy gave positive identification of Zn within the spinel phase.

The ability to positively identify the phases in complex samples such as these BOS dust indicate that these techniques should have further application in the characterisation of slag, or other metallurgical by-product materials.

ACKNOWLEDGEMENTS

The authors acknowledge use of the facilities and the assistance of David Wexler and David Mitchell at the UOW Electron Microscopy Centre. This research used equipment funded by the Australian Research Council (ARC) Linkage, Infrastructure, Equipment and Facilities (LIEF) grant LE120100104 located at the UOW Electron Microscopy Centre. The authors thank Sean Cadogan (UNSW Canberra) for the Mössbauer analysis measurements and interpretation of results. Funding support from the Australian Research Council Industrial Transformation Research Hubs Scheme (Projects IH130100017; IH200100005) and BlueScope is gratefully acknowledged.

REFERENCES

Ahmed, H M, Persson, A, Ökvist, L S and Björkman, B, 2015. Reduction behaviour of self-reducing blends of in-plant fines in inert atmosphere, *ISIJ International*, 55(10):2082–2089.

Downs, R T and Hall-Wallace, M, 2003. The American Mineralogist Crystal Structure Database, *American Mineralogist*, 88:247–250.

Gargul, K, Jarosz, P and Małecki, S, 2016. Alkaline Leaching of Low Zinc Content Iron-Bearing Sludges, *Arch Metall Mater*, 61(1):43–50.

Gražulis, S, Chateigner, D, Downs, R T, Yokochi, A T, Quiros, M, Lutterotti, L, Manakova, E, Butkus, J, Moeck, P and Le Bail, A, 2009. Crystallography Open Database – an open-access collection of crystal structures, *Journal of Applied Crystallography*, 42:726–729. https://doi.org/10.1107/S0021889809016690

Gražulis, S, Daškevič, A, Merkys, A, Chateigner, D, Lutterotti, L, Quirós, M, Serebryanaya, N R, Moeck, P, Downs, R T and LeBail, A, 2012. Crystallography Open Database (COD): an open-access collection of crystal structures and platform for world-wide collaboration, *Nucleic Acids Research*, 40:D420–D427. https://doi.org/10.1093/nar/gkr900

Gražulis, S, Merkys, A, Vaitkus, A and Okulič-Kazarinas, M, 2015. Computing stoichiometric molecular composition from crystal structures, *Journal of Applied Crystallography*, 48(1):85–91. https://doi.org/10.1107/S1600576714025904

Hleis, D, Fernández-Olmo, I, Ledoux, F, Kfoury, A, Courcot, L, Desmonts, T and Courcot, D, 2013. Chemical profile identification of fugitive and confined particle emissions from an integrated iron and steelmaking plant, *J of Hazardous Materials*, (250–251):246–255.

Jabłońska, M, Rackwał, M, Wawer, M, Kądziołka-Gaweł, M, Teper, E, Krzykawski, T and Smołka-Danielowska, D, 2021. Mineralogical and Chemical Specificity of Dusts Originating from Iron and Non-Ferrous Metallurgy in the Light of Their Magnetic Susceptibility, *Minerals*, 1(2):216.

Kelebek, S, Yörük, S and David, B, 2004. Characterization of basic oxygen furnace dust and zinc removal by acid leaching, *Minerals Engineering*, 17:285–291.

Lin, X, Peng, Z, Yan, J, Li, Z, Hwang, J-Y, Zhang, Y, Li, G and Jiang, T, 2017. Pyrometallurgical recycling of electric arc furnace dust, *J Cleaner Production*, 149:1079–1100.

Longbottom, R J, Monaghan, B J, Pinson, D J and Chew, S J, 2019b. Understanding the Self-Sintering Process of BOS Filter Cake for Improving Its Recyclability, *J Sustainable Metallurgy*, 5:429–441.

Longbottom, R J, Monaghan, B J, Pinson, D J, Webster, N A S and Chew, S J, 2020. *In situ* Phase Analysis during Self-sintering of BOS Filter Cake for Improved Recycling, *ISIJ International*, 60(11):2436–2445.

Longbottom, R J, Monaghan, B J, Zhang, G, Pinson, D J and Chew, S J, 2019a. Self-sintering of BOS Filter Cake for Improved Recyclability, *ISIJ International*, 59(3):432–441.

Merkys, A, Vaitkus, A, Butkus, J, Okulič-Kazarinas, M, Kairys, V and Gražulis, S, 2016. COD::CIF::Parser: an error-correcting CIF parser for the Perl language, *Journal of Applied Crystallography*, 49(1):292–301. https://doi.org/10.1107/S1600576715022396

Merkys, A, Vaitkus, A, Grybauskas, A, Konovalovas, A, Quirós, M and Gražulis, S, 2023. Graph isomorphism-based algorithm for cross-checking chemical and crystallographic descriptions, *Journal of Cheminformatics*, 15. https://doi.org/10.1186/s13321-023-00692-1

Nyirenda, R L, 1991. The processing of steelmaking flue-dust: A review, *Minerals Engineering*, 4(7–11):1003–1025.

Quirós, M, Gražulis, S, Girdzijauskaitė, S, Merkys, A and Vaitkus, A, 2018. Using SMILES strings for the description of chemical connectivity in the Crystallography Open Database, *Journal of Cheminformatics*, 10. https://doi.org/10.1186/s13321-018-0279-6

Sharma, V K, Klingelhofer, G and Nishida, T, 2013. Mössbauer Spectroscopy: Applications in Chemistry, Biology and Nanotechnology, Part V Iron Oxides, pp 349–532 (John Wiley and Sons).

Steer, J, Grainger, C, Griffiths, A, Griffiths, M, Heinrich, T and Hopkins, A, 2014. Characterisation of BOS steelmaking dust and techniques for reducing zinc contamination, *Ironmaking and Steelmaking*, 41(1):61–66.

Stewart, D J C and Barron, A R, 2020. Pyrometallurgical removal of zinc from basic oxygen steelmaking dust – A review of best available technology, *Resources, Conservation and Recycling*, 157:104746.

Stewart, D J C, Scrimshire, A, Thomson, D, Bingham, P A and Barron, A R, 2022. The chemical suitability for recycling of zinc contaminated steelmaking by-product dusts: The case of the UK steel plant, *Resources, Conservation and Recycling Advances*, 14:200073.

Suetens, T, Guo, M, Van Acker, K and Blanpain, B, 2015. Formation of the $ZnFe_2O_4$ phase in an electric arc furnace off-gas treatment system, *J of Hazardous Materials*, (287):180–187.

Vaitkus, A, Merkys, A and Gražulis, S, 2021. Validation of the Crystallography Open Database using the Crystallographic Information Framework, *Journal of Applied Crystallography*, 54(2):661–672. https://doi.org/10.1107/S1600576720016532

Vaitkus, A, Merkys, A, Sander, T, Quirós, M, Thiessen, P A, Bolton, E E and Gražulis, S, 2023. A workflow for deriving chemical entities from crystallographic data and its application to the Crystallography Open Database, *Journal of Cheminformatics*, 15. https://doi.org/10.1186/s13321-023-00780-2

Vereš, J, Šepelák, V and Hredzák, S, 2015. Chemical, mineralogical and morphological characterisation of basic oxygen furnace dust, *Mineral Processing and Extractive Metallurgy, Transactions of the Institutions of Mining and Metallurgy: Section C,* 124(1):1–8.

Wang, J, Zhang, Y, Cui, K, Fu, T, Gao, J, Hussain, S and AlGarni, T S, 2021. Pyrometallurgical recovery of zinc and valuable metals from electric arc furnace dust – A review, *J Cleaner Production*, 298:126788.

Nedar, L, 1996. Dust formation in a BOF converter, *Steel Research*, 67(8):320–327.

Sofilić, T, Rastovčan-Mioč, A, Cerjan-Stefanović, Š, Novosel-Radović, V and Jenko, M, 2004. Characterization of steel mill electric-arc furnace dust, *J of Hazardous Materials*, (B109):59–70.

Effect of oxygen on the interfacial phenomena and metal recovery rate in recycling process

J Park[1], Y-B Kang[2] and J H Park[3]

1. Student, POSTECH, Pohang 37673, South Korea; and formerly Hanyang University, Ansan, South Korea, 15588. Email: jhpark7273@potsech.ac.kr
2. Professor, POSTECH, Pohang 37673, South Korea. Email: ybkang@potsech.ac.kr
3. Professor, Hanyang University, Ansan 15588, South Korea. Email: basicity@hanyang.ac.kr

ABSTRACT

In order to understand the effect of oxygen injection into the molten metal on the valuable metals (VM) recovery and interfacial phenomena, several high-temperature experiments were conducted using a high-frequency induction furnace with a quartz tube as the reaction chamber at 1673 K. Copper-based waste printed circuit board (PCB) that contains some impurities was selected as the end-of-life resource. $CaO\text{-}SiO_2$ based pre-fused slag and oxygen gas were used in present works. The composition change and microstructure of the samples were analysed.

As oxygen was injected into the system, impurities were rapidly removed. After the removal of impurities, continuously injected oxygen lowered interfacial tension between liquid slag and liquid metal. Lowered interfacial tension induced Cu loss into the slag.

INTRODUCTION

Sustainable development and recycling process

Recently, numerous research have been continued for sustainable development. The metal-based scrap recycling process is a representative method. Electronic wastes (E-waste) are usually used as the raw materials in such recycling processes.

In order to recover valuable metals (VM) in E-waste, a batch-type arc furnace is often used (Hagelüken, 2006). Figure 1 shows a comprehensive sequence for a Cu-based scrap recycling process. In the converting stage, oxygen is blown to melt to remove oxidative impurities. Because converting and tapping stage are dynamic, significant amounts of droplets are entrapped in slag phase. In order to recover metal droplets from the slag, additional process such as 'slag cleaning' should be adopted.

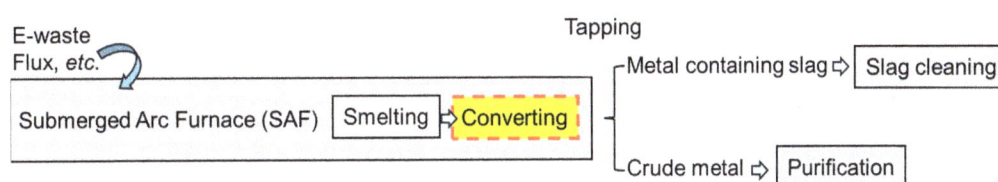

FIG 1 – Recycling process for valuable metals-contained materials (E-waste).

The chemical effect of oxygen blowing

Oxygen can influence on surface tension of the liquid metal. Previous researchers concluded that high oxygen potential reduces the surface tension of liquid metal, and its quantitative description is available (Belton, 1976). However, reports about the effect of oxygen on the quantity of metal loss in molten slag are rare.

The physical effect of oxygen blowing

There have been several research regarding the droplet formation mechanism at the metal/slag interface (Han and Holappa, 2003). It was revealed that the metal droplets were formed by rising bubbles.

Present study focuses on coupled effects of oxygen blowing on Cu loss in order to understand metal loss mechanisms in recycling process.

EXPERIMENTAL PROCEDURE

Furnace information

A high-frequency induction furnace was employed. Metal samples collected from PCB were initially put in a fused magnesia crucible (60 (OD) × 50 (ID) × 120 (H), mm), which was located in a graphite susceptor (80 (OD) × 65 (ID) × 120 (H), mm) for heating of samples. The graphite susceptor was surrounded by a commercial purity alumina board for insulation. A quartz tube (120 (OD) × 114 (ID) × 400 (H), mm) was used as the reaction chamber.

Preparation of the crude metal and the synthetic slag samples

Pre-molten printed circuit board (PCB) was used in the present experiments. Slag samples was using reagent grade powders. Especially, CaO was prepared by calcinating $CaCO_3$ in a muffle furnace at 1273 K for 24 hrs. The mixed powder and was molten at 1873 K for 30 mins in graphite crucible. Collected slag was crushed and calcinated at 1273 K in a muffle furnace for 48 hrs to remove trace C. Information of each sample were also listed in Tables 1 and 2, respectively.

TABLE 1

Mass and chemical composition of the crude metal.

Series ID	Mass (g)	Cu (wt%)	Fe (wt%)	Si (wt%)	O (wt%)
No Slag (B) – 500	500	86.9	12.6	0.3	0.001
CASM (B) – 500	500	88.6	10.9	0.3	0.002
CASM (B) – 300	300	84.7	14.3	0.8	0.001
CSM (B) – 300	300	91.4	8.1	0.4	0.001
CSM (T) – 300	300	86.9	12.5	0.4	0.002

TABLE 2

Mass and chemical composition of the synthetic slag.

Series ID	Mass (g)	CaO (wt%)	Al_2O_3 (wt%)	SiO_2 (wt%)	MgO (wt%)
No Slag (B) – 500	-	-	-	-	-
CASM (B) – 500	130	34.6	19.2	34.6	11.5
CASM (B) – 300	80	34.6	19.2	34.6	11.5
CSM (B) – 300	80	41.2	-	41.2	17.6
CSM (T) – 300	80	41.2	-	41.2	17.6

Procedure for the metal-slag reaction

The reaction was carried out at 1673 K. Temperature of the induction furnace was controlled within ±2 K using a proportional integral differential controller connected to a B-type thermocouple.

When the temperature reached 1673 K and stabilised, molten metal was sampled using a quartz tube and then quenched in ice water. This moment was set to t = 0. The synthetic slag was added to the metal, then oxygen gas (99.99 per cent purity) was blown in metal (bottom blowing, (B), 1 mm above from the crucible bottom) using an alumina lance (10 (OD) × 5 (ID) × 350 (H), mm). In 'CSM (T) – 300', oxygen was injected into surface of the slag. Each sample was sampled periodically using a quartz tube and iron rod, respectively. All of samples were dried using an oven at 323 K for 24 hrs.

Analysis

Metal samples were analysed by combustion method for the O content (TC-300, LECO), by inductively-coupled plasma spectroscopy (ICP-AES, SPECTRO ARCOS). Fe^{2+} content in the slag samples was analysed by titration technique, and Fe^{3+} content was estimated as (%T. Fe)-(% Fe^{2+}) (KS-E-3016). Morphology of the samples was observed using a Field-Emission Scanning Electron Microscope (FE-SEM, S-4800, HITACHI) imaging and Energy-Dispersive X-ray Spectroscopy (EDS, Nova Nano SEM 450, FEI).

RESULTS AND DISCUSSION

Impurity removal and oxygen accumulation

Figure 2 shows the composition evolution of the metal: normalised contents of Si and Fe during oxygen blowing. At the early stage, Si and Fe in the metal rapidly disappeared: the oxygen reacted with Si and Fe by Reaction (1) and (2) (Sigworth and Elliott, 1974). This stage is designated as 'impurity removal regime'. Si was removed faster than Fe, because of the higher affinity of O to Si than to Fe.

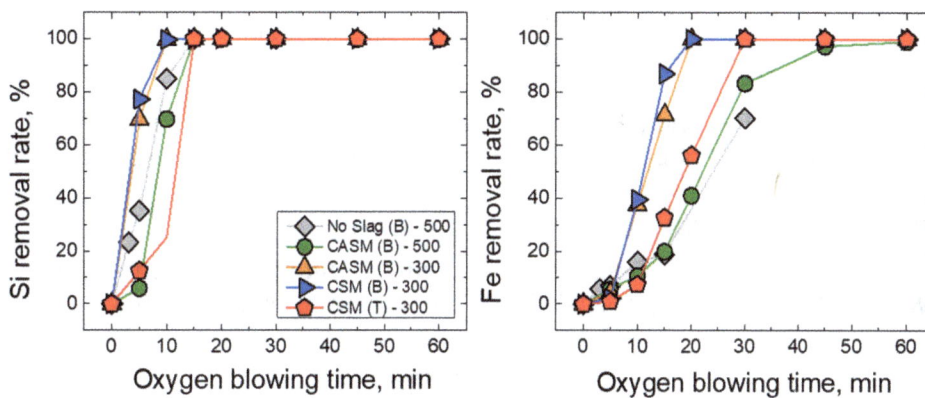

FIG 2 – Si and Fe contents change during the oxygen blowing.

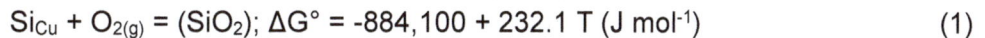

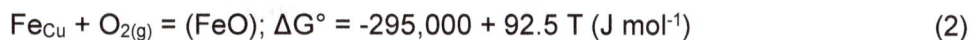

$$Si_{Cu} + O_{2(g)} = (SiO_2); \Delta G° = -884,100 + 232.1\ T\ (J\ mol^{-1}) \tag{1}$$

$$Fe_{Cu} + O_{2(g)} = (FeO); \Delta G° = -295,000 + 92.5\ T\ (J\ mol^{-1}) \tag{2}$$

Figure 3 shows the composition evolution of the slag. Except for 'No slag (B)', all the other cases show an increase in the ratio: oxidising the slag. In particular, 'CSM(T)' shows a considerably higher ratio than the other cases. It has been known that the top blowing can increase the oxygen potential of slag, thereby stabilising Fe_2O_3 relative to FeO. Sano, Lü and Riboud (1997) suggested that stabilising Fe_2O_3 or Fe_3O_4 would increase the melting point and viscosity of slag. Therefore, in case of recycling process which uses Fe-containing sources, top blowing is not recommended.

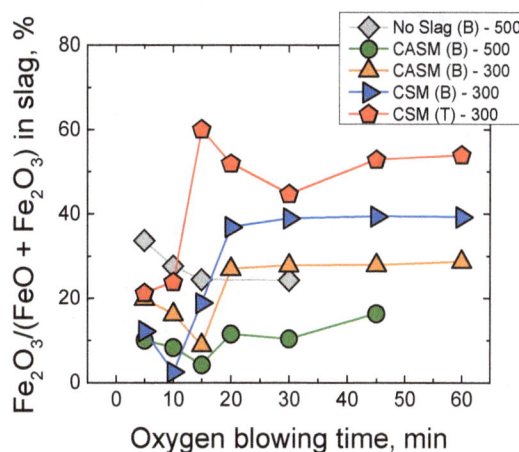

FIG 3 – Fe_2O_3 change in the slag during the oxygen blowing.

As shown in Figure 4, (%T. Cu) increased abruptly during the oxygen blowing. 'No Slag (B) – 500' and 'CASM (B) – 500' were excluded because of lack of related data. The moment of the abrupt increase in (%T. Cu) is more or less close to the moment of termination of the impurity removal (see Figure 2). Therefore, a hypothesis can be developed that excessive oxygen blowing can lead the Cu loss. This moment is considered as a transition from the 'impurity removal regime' to the 'over-blown regime'.

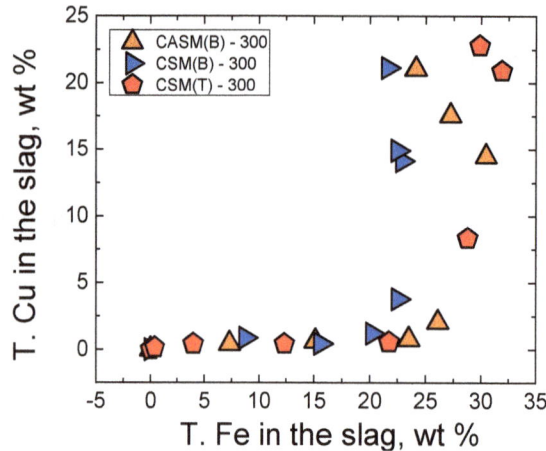

FIG 4 – Changes of (%T. Fe) and (%T. Cu) contents in the slag. Sudden increase of (%T. Cu) is an indicator of the transition point between the impurity removal regime and the over-blown regime.

Surface tension and interfacial tension change

In order to evaluate the physical effect and the chemical effect of oxygen injection, the blowing number (N_B) was evaluated (Subagyo *et al*, 2003) as expressed as Equation 3.

$$N_{\mathrm{B}} = \frac{\rho_{\mathrm{g}} v_{\mathrm{g}}^2}{2\sqrt{g\sigma_{\mathrm{M}}\rho_{\mathrm{M}}}} \tag{3}$$

where ρ_g (kg m^{-3}), v_g (m s^{-1}), σ_M (mN m^{-1}), ρ_M (kg m^{-3}), and g (m s^{-2}) are the density and the velocity of the oxygen gas, the surface tension and the density of liquid Cu, and the gravitational acceleration respectively. Authors concluded that when N_B is over 1, physical turbulence by any gas blowing occurs. As a result, N_B was approximately ~10^{-4}. Therefore, physical effect of oxygen blowing is negligible.

Figure 5 shows the quenched final sample after the experiment of 'CSM (B) – 300'. In all the cases, the two phases stuck close together. It can be assumed that interfacial state is far from other experiment where oxygen is not used.

FIG 5 – Snapshot of slag/metal samples after experiment.

Researchers have investigated the effect of oxygen potential on the surface tension of liquid Cu (alloy). Following Belton (1976), σ_M (mN m^{-1}) can be calculated using Equations 4 and 5:

$$\sigma_{\mathrm{M}} = \sigma_{\mathrm{M}}^{\circ} - RT\Gamma_{\mathrm{sat.}}^{\circ} \cdot \ln\left(1 + K_{\mathrm{O}} \cdot a_{\mathrm{O}}\right) \tag{4}$$

$$\Gamma_{\mathrm{sat.}}^{\circ} = -\frac{1}{RT}\left(\frac{\delta\sigma_{\mathrm{M}}}{\delta \ln a_{\mathrm{O}}}\right)_{\mathrm{constant}} \tag{5}$$

where $\sigma_M^°$ (mN m⁻¹), KO, aO, and $\Gamma_{sat.}^°$ are the surface tension of pure liquid Cu, the absorption coefficient, the activity of O in the liquid Cu alloy and the surface tension absorption at the saturation, respectively. KO was reported to be 40 at 1673 K (Morita and Kasama, 1980). $\Gamma_{sat.}^°$ was reported by others (Monma and Suto, 1961). Surface tension of liquid Cu was calculated using Equations 4 and 5. Surface tension of the slag, σ_S, (mN m⁻¹) was roughly estimated as shown in Equation 6:

$$\sigma_S = \sum X_i \sigma_{i,\,pure}^° \tag{6}$$

where $\sigma_{i,\,pure}^°$ (mN m⁻¹) and X_i are the surface tension and the mole fraction of the pure oxide component i in the slag. Related data are summarised in Mills and Keene's report (1987). Change in σ_S does not appear to be significant compared to the change of σ_M. The interfacial tension, $\sigma_{M/S}$ (mN m⁻¹) was calculated using the equation as shown in Equation 7 (Girifalco and Good, 1957):

$$\sigma_{M/S} = \sigma_M + \sigma_S - 2\emptyset\sqrt{\sigma_M \sigma_S} \tag{7}$$

$$\text{where: } \emptyset = 0.5 + 0.3 X FeO \tag{8}$$

Figure 6 shows relationship between calculated $\sigma_{M/S}$ and (%T. Cu) in the slag. In impurity removal regime, calculation of surface tension of liquid Cu using Belton's equation is not valid. Therefore, points where impurities were not removed completely were excluded.

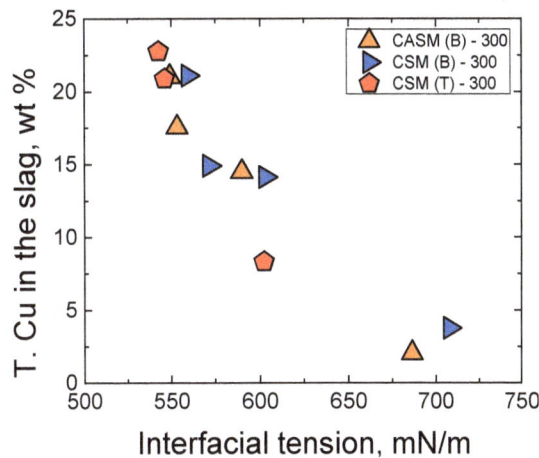

FIG 6 – Relationship between $\sigma_{M/S}$ and Cu loss into slag phase.

After impurity removal, continuously supplied oxygen was then lowered the surface tension of the liquid Cu, thereby lowering the interfacial tension between the liquid Cu and the liquid slag. This was a major reason leading the Cu loss (Park, Lee and Park, 2021).

CONCLUSIONS

The role of the oxygen blowing on the impurity removal and the Cu metal loss was concerned. Entire reaction can be divided with two regimes: the impurity removal regime and the over-blown regime. Key findings of the present study are:

- At the early stage of reaction, oxygen could be used to remove impurities in molten crude Cu metal. Composition of the slag did not show noticeably impact on the impurity removal tendency.

- After the impurity removal, continuously injected oxygen in the liquid Cu lowered surface tension. It was a major reason to lower the interfacial tension between the liquid Cu and the liquid slag. The lower interfacial tension was thought to be a main reason of the Cu metal loss.

REFERENCES

Belton, G R, 1976. Langmuir adsorption, the Gibbs adsorption isotherm and interfacial kinetics in liquid metal systems, *Metallurgical Transactions, A, Physical Metallurgy and Materials Science,* 7(1):35–42. https://doi.org/10.1007/bf02652817

Girifalco, L A and Good, R J, 1957. A theory for the estimation of surface and interfacial energies, Derivation and application to interfacial tension, *The Journal of Physical Chemistry*, 61(7):904–909. https://doi.org/10.1021/j150553a013

Hagelüken, C, 2006. Recycling of Electronic Scrap at Umicore's Integrated Metals Smelter and Refinery, *World of Metallurgy – ERZMETALL*, 59:152–161. Available from: <https://www.researchgate.net/publication/240629115_Recycling_of_Electronic_Scrap_at_Umicores_Integrated_Metals_Smelter_and_Refinery>

Han, Z and Holappa, L, 2003. Mechanisms of Iron Entrainment into Slag due to Rising Gas Bubbles, *ISIJ International*, 43(3):292–297. https://doi.org/10.2355/isijinternational.43.292

Mills, K C and Keene, B J, 1987. Physical properties of BOS slags, *International Materials Reviews*, 32(1):1–120. https://doi.org/10.1179/095066087790150296

Monma, K and Sutô, H, 1961. Effects of dissolved sulphur, oxygen, selenium and tellurium on the surface tension of liquid copper, *Materials Transactions*, 2(3):148–152. https://doi.org/10.2320/matertrans1960.2.148

Morita, Z and Kasama, A, 1980. Effect of slight amount of dissolved oxygen on the surface tension of liquid copper, *Trans Jpn Inst Met*, 21:522–530. https://doi.org/10.2320/matertrans1960.21.522

Park, J, Lee, J and Park, J H, 2021. Effect of oxygen blowing on copper droplet formation and emulsification phenomena in the converting process, *Journal of Sustainable Metallurgy*, 7(3):831–847. https://doi.org/10.1007/s40831-021-00421-8

Sano, N, Lü, W and Riboud, P V, 1997. *Advanced Physical Chemistry for process metallurgy*, Available from: <https://ci.nii.ac.jp/ncid/BA31247737>

Sigworth, G K and Elliott, J F, 1974. The thermodynamics of dilute liquid copper alloys, *Canadian Metallurgical Quarterly*, 13(3):455–461. https://doi.org/10.1179/cmq.1974.13.3.455

Subagyo, S and Brooks, G A, 2003. Generation of Droplets in Slag-Metal Emulsions through Top Gas Blowing, *ISIJ International*, 43(7):983–989. https://doi.org/10.2355/isijinternational.43.983

Measuring circular economy through life cycle assessment – challenges and recommendations based on a study on recycling of Al dross, bottom ash and shavings

E Pastor-Vallés[1], A Vallejo-Olivares[2], G Tranell[3] and J B Pettersen[4]

1. PhD Candidate, Industrial Ecology Programme, NTNU, Trondheim 7034, Norway.
   Email: elisa.p.valles@ntnu.no
2. PhD Candidate, Department of Materials Science, NTNU, Trondheim 7034, Norway.
   Email: alicia.vallejo.olivares@gmail.com
3. Professor, Department of Materials Science, NTNU, Trondheim 7034, Norway.
   Email: gabriella.tranell@ntnu.no
4. Associate Professor, Industrial Ecology Programme, NTNU, Trondheim 7034, Norway.
   Email: johan.berg.pettersen@ntnu.no

ABSTRACT

Metals are essential for the sustainability transition and decarbonisation of society. Yet, it will be paramount to produce them sustainably and minimise the affiliated resource and energy use and the associated emissions. In the circular economy, the metallurgical industry should recycle existing material stocks, and improve its utilisation of wastes, residues and side streams. This increases the complexity of processes, as they become both (often multi-fraction) waste treatment as well as material production processes and brings complexity to the assessment of environmental benefits.

Assessing the environmental impact of technological developments frequently is supported by life cycle assessment (LCA). While the method is well documented, its implementation involves several methodological choices that deserve reasoning and analysis, such as how to define the product when former wastes are turned into new products, the selection of impact methodology when converting emissions to environmental indicators, the definition of system boundaries and co-product allocation and the interpretation of sensitivity and uncertainty in final outcomes.

In this exploratory study, we investigate how the variation of LCA set-up affects the environmental burden of the system. We consider a metallurgical process where a mix of hard-to-recycle aluminium-containing streams is used to produce aluminium cast alloys in a rotary furnace. Re-melting with salt-fluxes allows recovering metals from partly oxidised/contaminated streams, such as dross, bottom ash and industrial shavings, but at the expense of generating significant amounts of salt-slag/salt-cake hazardous waste. The study considers different system alternatives such as landfilling the salt-slag residues versus valourising them into salts, aluminium concentrates, ammonium sulfate and non-metallic-compounds to be used by the metallurgical, construction or chemical industries. Practical recommendations are outlined to facilitate the implementation of LCA in assessing the potential benefits of the circular economy in the metallurgical sector.

INTRODUCTION

Metals are critical for meeting the needs of our society, and we rely on them in multiple applications eg in the energy, transportation, packaging, construction and communications industries. Applied ubiquitously across different sectors, some inherent sustainability implications are associated with their large lifespans and potential for recycling (Norgate and Rankin, 2002), in addition to being critical enablers of some relevant technologies for decarbonisation, such as renewable energy generation, batteries or electronics (World Energy Outlook, 2021).

Facing an unprecedented climate crisis, the demand for major metals might increase up to six-fold over this century as the global population and living standards rise (Watari, Nansai and Nakajima, 2021). Despite the economic and environmental benefits of metal production, sustaining the future provision of metals while considering the ecosystems' limited assimilation and provision capacity remains challenging. The production of metals alone entails resource use (2.8 billion tonnes (Bt) of metal produced each year) (based on information from US Geological Survey, 2023), energy-intensive processes (7–8 per cent of the global energy consumption) (UN Environment Programme, 2017) and emissions of pollutants (among others, approximately 10 per cent of global green house gas emissions) (UN Environment and International Resource Panel, 2019). Some impacts derived from the extraction and processing of these metals are, for instance, climate change, biodiversity loss and particulate matter health impacts (UN Environment and International Resource Panel, 2020).

With increasing environmental pressures to reduce the consumption of resources and generation of emissions in the production of metals, life cycle assessment (LCA) has emerged as a method that quantifies the potential environmental burdens of a product system, identifying hotspots (or the parts of the system where environmental impacts are more significant) and possibilities for improvement. However, even though this method is well documented, its application involves several methodological choices that could significantly influence the results. Reasoning and consideration of these can be critical, especially when dealing with circular economy systems, in which material stocks are recycled and the utilisation of wastes, residues and side streams improved. The complexity of these systems lies in that they become both (often multi-fraction) waste treatment as well as material production processes.

Although LCA is a standard methodology to evaluate the impact through the life cycle of metallurgical systems, there is a lack of specific guidelines for the metallurgical industry that test the influence of methodological choices when implementing circular economy concepts. A study by PE International (2014) and Santero and Hendry (2016) highlights some methodological choices that affect the results of a metallurgical LCA, such as system boundaries, allocation and impact categories. However, their influence has not been tested quantitatively in a case study. In addition, the present article expands on other relevant methodological choices.

Case study

The original model for this study is found in Vallejo Olivares *et al* (2024). The system analysed considers the recycling of hard-to-recycle (partly oxidised or contaminated) aluminium-containing flows via a rotary furnace. This process allows, through re-melting with salt-fluxes and the rotational movement of the furnace, to separate the metal from the non-metallic contaminants (eg oxide layer, carbonaceous residues) and promote its coalescence, while also protecting the molten metal from oxidation during the high-temperature process (Milani and Timelli, 2023).

Even though from an environmental and economic perspective recycling aluminium is considered to be more sustainable than producing it anew (Olivieri, Romani and Neri, 2006; Damgaard, Larsen and Christensen, 2009), the use of salts has a significant downside in the generation of salt slag residues (a mix of salts, non-metallic compounds (NMCs) and entrapped metallic droplets), that are classified as hazardous waste (Environmental Protection Agency, 2002) and pose risks for landfilling (Office of Research and Development, 2015). Another option is to valourise these slags by crushing and dissolving them in water, obtaining both salts and aluminium concentrates that can be fed back into the rotary furnace, as well as ammonium sulfate and NMCs that the chemical and construction industries can benefit from. Various salt-slag recovery treatments are described in the non-ferrous industry's Best Available Techniques (BAT) reference document (Joint Research Centre, 2017).

The LCA methodology is applied to test the influence of relevant methodological choices in the environmental impact for this case study and draw practical recommendations for the use of LCA for the evaluation of metallurgical and circular economy systems for the industry.

LCA methodology

Life cycle assessment is described by the international standards ISO 14040 and ISO 14044 as 'the compilation and evaluation of the inputs, outputs and the potential environmental impacts of a product system throughout its life cycle' (International Standard Organization, 2006a, 2006b).

The evaluation of environmental impacts through LCA is carried out in four iterative steps (Figure 1):

1. *Goal and scope* definition, where the methodological choices of the LCA are defined.

2. *Inventory analysis*, or the calculation of all inputs (resources) and outputs (emissions) that pose a burden in the environment.

3. *Impact assessment*, or the quantification of the environmental impact associated to the flows calculated during the *inventory analysis*.

4. *Interpretation*, in which the outcomes of the previous phases are evaluated in accordance to the *goal and scope*.

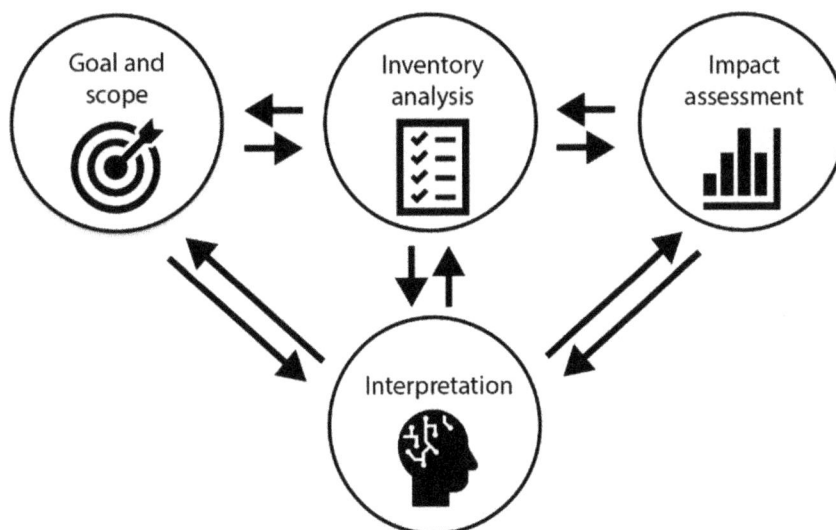

FIG 1 – Phases of an life cycle assessment (LCA).

While methodological choices affect the entire study of an LCA, they are broadly decided during the *goal and scope* phase. However, as LCA is an iterative process, it allows for feedback loops between the different stages and correcting for these when necessary. That could be, for instance, if better information is available or if the application of the study has changed.

The areas identified in the LCA methodology that significantly affect the assessment results are system boundaries; functional unit and allocation procedures; database, impact categories and impact method; and temporal, geographical and data quality considerations. These are thoroughly discussed and evaluated in the next sections.

EVALUATION OF METHODOLOGICAL CHOICES

System boundaries

ISO 14040 states that:

> The system boundary determines which unit processes shall be included within the LCA (...) The deletion of life cycle stages, processes, inputs or outputs is only permitted if it does not significantly change the overall conclusions of the study (International Standard Organization, 2006a).

The term 'cradle-to-grave' involves considering a product's life cycle. However, the metallurgical industry produces many 'cradle-to-gate' studies, ie from the extraction of raw materials to the semi-finished product, excluding its further transport, use and disposal, as these are often uncertain (PE International, 2014). For example, an aluminium sheet can be used in multiple applications (eg in construction, packaging or transport), over which the producer has no control. In contrast, when assessing finished products, such as home appliances, the LCA studies usually take a cradle-to-grave approach.

Since this case study evaluates the production of a semi-finished product (aluminium ingots), a cradle-to-gate approach is preferred. Two different scenarios are considered, as shown in Figure 2:

Scenario 1: Salt slag residue is landfilled.

Scenario 2: Salt slag is processed into recycled salts recirculated in the process and NMCs and ammonium sulfates, which the cementitious and chemical industries can use further.

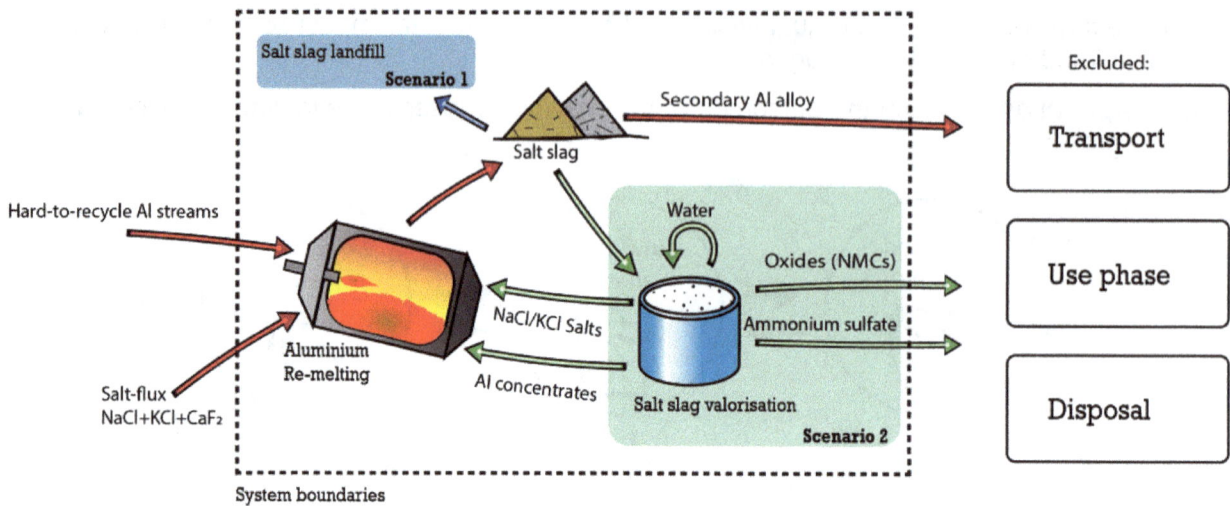

FIG 2 – The system boundaries for this study consider raw material extraction and production but exclude the use and final disposal of products. Modified from Vallejo Olivares *et al* (2024).

The inventory, or inputs and outputs translated to environmental impacts during the LCA, differs depending on the scenario and system boundaries considered and is displayed in Table 1. Note that the quantity of inputs needed when the system boundaries include the recycling of salts (Scenario 2) is generally higher. However, more by-products are also obtained by this process.

Regarding the last negative flow of hard-to-recycle aluminium, it implies that the evaluated process treats 1 tonne of material mix, which would otherwise be considered waste and end up handled by waste management systems. This is based on the assumption that the characteristics of the scrap mix (oxidation, contamination, low aluminium content) make it unsuitable for utilisation by other recycling processes. If treating scrap streams with a higher content of aluminium (eg used beverage containers), this flow should instead be replaced by an input of aluminium secondary material, partly decreasing the benefit of recycling aluminium through the current process. This is because the inherent value of the waste fraction makes it suitable for other recycling processes. There is also a third option, which involves the input of burden-free aluminium, leading to neither positive nor negative impacts. This is called the cut-off approach, in which only the impacts directly caused by a product are considered; in this case, the impacts caused by the collection and treatment of the scrap. The influence of these assumptions is studied later in the section 'Temporal, geographical and data quality considerations'.

To compare the environmental performance of Scenario 1 and 2, the environmental burdens must be distributed among the different co-products through allocation procedures. This will be discussed in the next section, together with the functional unit.

TABLE 1
Life cycle inventory for treating 1 tonne of aluminium-containing waste streams.

Input/output	Unit	Scenario 1: Salt slag landfill	Scenario 2: Salt slag valourisation
Products			
Secondary cast aluminium product	tonne	0.70	0.72
By-product: oxides (NMCs)	tonne	-	0.35
By-product: ammonium sulfate	tonne	-	0.033
Inputs			
Sodium chloride	tonne	0.098	0.01
Potassium chloride	tonne	0.042	0.0046
Calcium fluoride	tonne	0.0028	0.0029
Sulfuric acid	kg	-	10.22
Lime	kg	0.71	0.74
Liquid oxygen	kg	83.84	86.73
Nitrogen	kg	-	0.40
Water	m^3	0.5	1.07
Electricity	kWh	63.88	116
Heat from natural gas	kWh	670.72	917
Diesel	kg	1.04	1.41
Emissions and solid waste			
Carbon dioxide	tonne	0.170	0.233
Sulfur dioxide	tonne	0.21	0.21
Nitrogen oxides	kg	0.10	0.14
Particulates	kg	0.015	0.016
Hydrochloric acid	kg	0.019	0.020
Hydrogen fluoride	kg	0.004	0.004
Heavy metals	kg	0.0006	0.0007
Methane	kg	-	0.028
Dust	tonne	0.02	0.02
Sludge to landfill	tonne	0.53	-

NMC – non-metallic compounds.

Functional unit and allocation procedures

The functional unit (FU) is the reference flow to which all other inputs and outputs are scaled to.

It expresses the function of the system. For example, in a system where the production of a metal is assessed, it can relate to the mass of the product. In multi-output systems, however, selecting one functional unit is not straightforward. In our case study, the main product is the aluminium cast alloy; however, the treatment of the residue fraction is another function provided by this system.

In addition, the studied system produces other by-products: ammonium sulfates and NMCs. The inputs and outputs must be allocated among these co-products, to attribute the corresponding impact to each of them. Allocation is carried out in the following order of preference (International Standard Organization, 2006a):

1. Whenever possible, allocation should be avoided by subdividing the input and output data to each of the co-products, or by expanding the product system (system expansion approach). When two functions are provided in the same system, this last approach is equivalent to crediting the system with the impacts avoided by the alternative production of the secondary function, in the most likely way of producing it (Hauschild, Rosenbaum and Olsen, 2018).

2. Mass allocation – allocating on a physical basis.

3. Economic allocation – allocating on a monetary basis.

The influence of selecting different functional units and allocation methods is tested in this study.

Functional unit

Regarding the functional unit, the results for the global warming impact when considering two functional units: one tonne of hard to recycle mixed scrap to treatment and one tonne of aluminium produced, are displayed in Figure 3. These functional units are tested for the two different scenarios of landfilling the salts (Scenario 1) versus recycling them (Scenario 2).

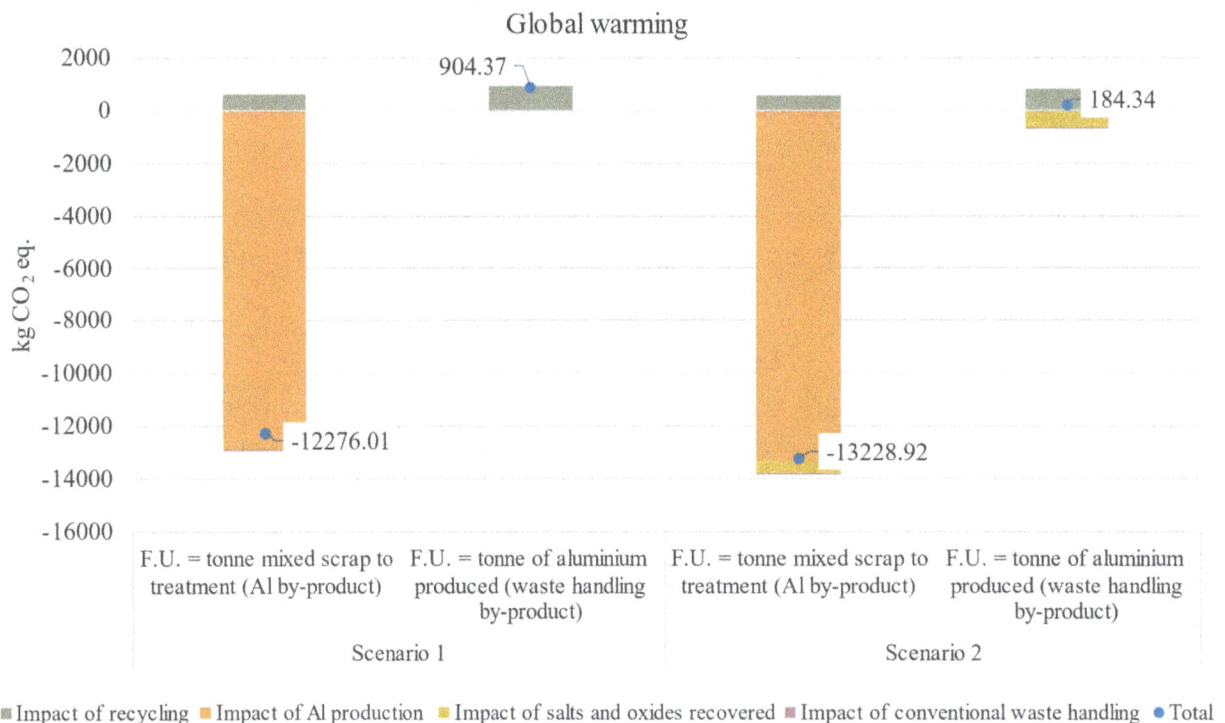

FIG 3 – Global warming impact (evaluated through ReCiPe 2016 Midpoint (H)).

In both scenarios it is observed that, when the FU refers to the treatment of scrap, the impact is negative, meaning an avoided impact on the environment or benefit. Following the system expansion approach, the impacts avoided are more significant than the impacts produced because the impact of the conventional means of producing aluminium (orange bar) outweighs the impact of recycling itself (green bar). Similarly, the impact of producing one tonne of aluminium is still credited with the avoided waste handling (pink bar). However, this is almost inappreciable compared to the impact of producing aluminium. When LCAs are used for comparative assertions, as is the case, the selection of the functional unit must be equivalent for all systems studied. The first FU could be used to compare different waste handling processes for the same waste mix, and the second FU could be used to compare different processes that produce a tonne of cast aluminium.

In addition, and regardless of the functional unit considered, the comparison between Scenario 1 and 2 always results in a higher benefit for the second one: the recovery of salts and oxides contributes to the negative (avoided) impact of the second scenario, being also the process yields higher, with more aluminium produced and lowering the impact per functional unit.

Another important consideration in metallurgical systems when dealing with the functional unit is that, in many cases, we cannot compare the same material of two different qualities. For instance, we cannot develop an LCA comparing high-purity alumina extracted from bauxite through the Bayer process to the lower-purity alumina in the NMC fraction recovered from the salt-slag valourisation because their functions differ. It is also relevant to note that in many applications, mass is not a suitable unit for comparison, especially when comparing two different materials with the same functionality: in the example of producing beverage packaging, the functional unit could be set to eg a container holding half a litre of fluid. If we assess two options, aluminium and glass containers, the mass required to perform this function is not equivalent. In addition, we need to take into account that the lifetime of these containers is not equal; if, for instance, there was a return scheme in place to collect glass containers and refill them again, the same container could be used more than once, which would certainly decrease its impact over its life cycle.

When insufficient data is available, system expansion is not possible and allocation is applied to multi-functional systems. In the next section, allocation methods are discussed.

Allocation method

As explained before, system expansion, such as in Figure 3, is prioritised by the ISO when sufficient data is available.

Whenever system expansion is not possible, the LCA practitioner can apply either mass or economic allocation, where the last is recommended for eg precious metals (PE International, 2014; Santero and Hendry, 2016). The reason behind this is that, in the case of precious metals, these are the ones driving demand and not the other co-products, and it could be unfair to credit the impact on a mass basis, given that, frequently, the less valuable co-products are found in a larger proportion.

In Table 2, it can be observed how the impact of the aluminium produced decreases when by-products are considered. The aluminium cast alloy has the lowest impact when applying mass allocation. Since it is also considered more valuable in the market, this material increases its carbon footprint with economic allocation. When a material is driving demand, one could argue economic allocation is preferable to mass allocation, especially if the difference in mass is substantial.

TABLE 2

Results for three different allocation methods (Scenario 2 – evaluated through ReCiPe 2016 Midpoint (H), for global warming impact).

Product	Mass output (per tonne mixed scrap to treatment)	Unit price	100% allocation to aluminium	Mass allocation	Economic allocation
FU	-	-	Aluminium cast alloy, 1 t	Production of 1 t (per product)	Production of 1 t (per product)
Aluminium cast alloy	0.72 t	$1150	813 kg CO_2 eq per tonne aluminium	533 kg CO_2 eq	696 kg CO_2 eq
Oxides (NMCs)	0.35 t	$375	0	533 kg CO_2 eq	227 kg CO_2 eq
Ammonium sulfate	0.033 t	$300	0	533 kg CO_2 eq	182 kg CO_2 eq

In addition to differences in the allocation method, the results of an LCA will depend on the impact categories and impact methods considered. These are treated in the next section.

Impact categories and impact methods

Previously we have been describing the different methodological choices using global warming and the ReCiPe method (2016) as an example. However, LCA is a methodology intended to study more than one environmental impact, to avoid burden shifting between impact categories, such as eg acidification or human toxicity potentials.

Impact methods are defined as the specific methodologies to convert inventory data into environmental impacts, with different spatial coverage (eg European, global...). Impact methods also look at a wide range of environmental impacts. ReCiPe, the impact method considered in this study, involves the assessment of 13 impact categories (midpoints), which are aggregated in three areas of protection (AoPs) or end points: human health, ecosystems and resources (Huijbregts *et al*, 2017). In the white paper of PE International (2014), later published by Santero and Hendry (2016), some categories such as global warming potential, acidification potential, eutrophication potential, smog potential and ozone depletion potential are recommended to be assessed by metallurgical industry, while others such as resource depletion, toxicity, land use change and water scarcity are considered still too uncertain for the decision-making processes. However, standards and guidelines, the interest of the stakeholders, or the specific case study might broaden the consideration of impact categories.

Another way of considering impact categories is by starting the analysis with areas of protection, and then deciding which environmental impacts to include based on the impacts that have a higher effect on these. An example for the human health impact is shown in Figure 4.

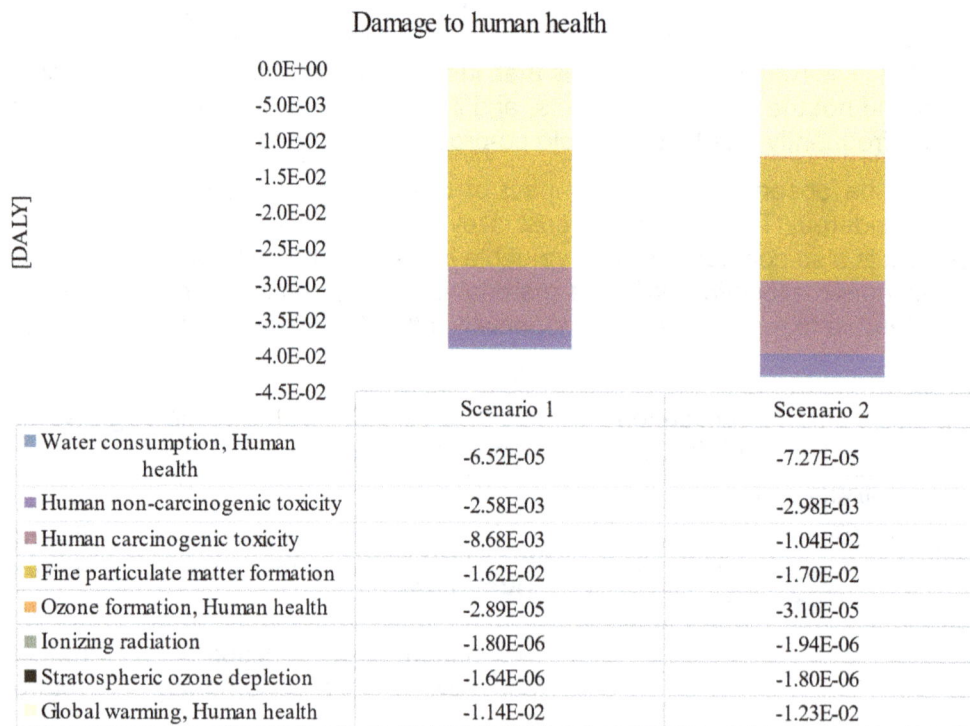

Damage to human health

	Scenario 1	Scenario 2
■ Water consumption, Human health	-6.52E-05	-7.27E-05
■ Human non-carcinogenic toxicity	-2.58E-03	-2.98E-03
■ Human carcinogenic toxicity	-8.68E-03	-1.04E-02
■ Fine particulate matter formation	-1.62E-02	-1.70E-02
■ Ozone formation, Human health	-2.89E-05	-3.10E-05
■ Ionizing radiation	-1.80E-06	-1.94E-06
■ Stratospheric ozone depletion	-1.64E-06	-1.80E-06
Global warming, Human health	-1.14E-02	-1.23E-02

FIG 4 – End point analysis and midpoint contribution, in ReCiPe 2016 (H). Example for human health end point, FU = tonne mixed scrap to treatment.

It is observed that the categories of global warming, fine particulate matter and human carcinogenic toxicity significantly impact human health. Therefore, this is a relevant manner of selecting impact categories for the study. While end points give an overview of the impacts, the decision-maker might prefer midpoints as they are easier to communicate because these show more primary effects. It is important to emphasise that the selection of impact categories should be justified to avoid falling into greenwashing practices by just showing some impact categories that are beneficial to a process or product, especially in comparative assertions.

When midpoint categories have been selected, an in-depth analysis, also called contribution analysis, is developed in Figure 5.

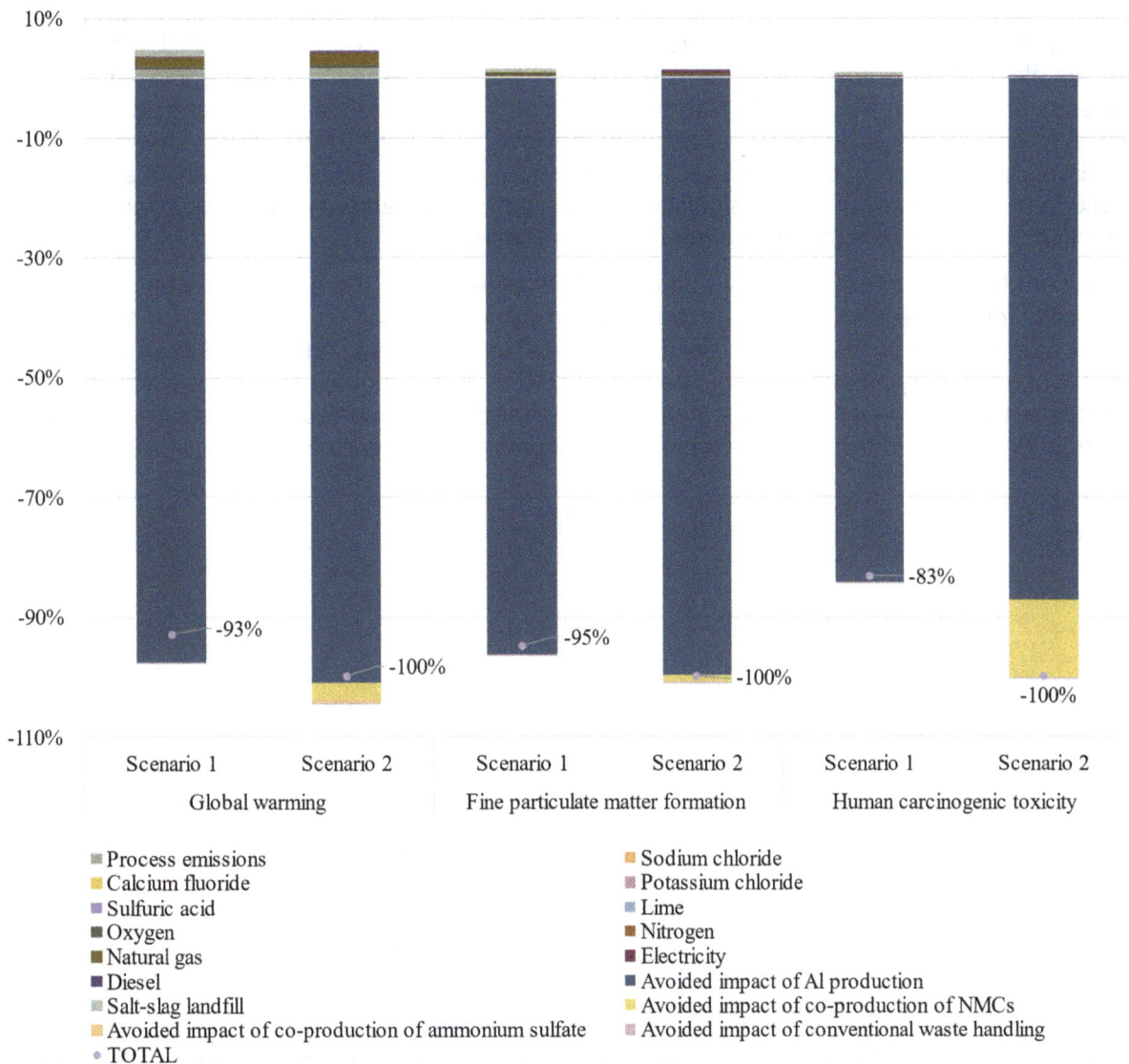

FIG 5 – Contribution analysis (ReCiPe 2016 Midpoint (H)), FU = tonne mixed scrap to treatment. Results are analysed further in Vallejo Olivares *et al* (2024).

Temporal, geographical and data quality considerations

In Figure 5, the avoided impacts of aluminium production stand out in all impact categories studied. Therefore, to truly understand what is driving the environmental impact, the conventional production of aluminium (cast alloy) should be analysed more closely. The initial LCA of this system (Vallejo Olivares *et al*, 2024) considered aluminium from the global market and this flow was chosen in the ecoinvent 3.6 database. Running an analysis of this process in SimaPro, electricity, specifically from China, makes up for more than 44 per cent of this material's global warming environmental impact, evidencing its reliance on the electricity mix of the country of origin. Considering China accounts for almost 90 per cent of coal thermal power, shifting towards producing low-carbon energy sources remains one of the most significant opportunities for reducing its carbon footprint (Saevarsdottir, Magnusson and Kvande, 2021).

In this regard, the geographical scope is highly relevant to determine the impact of the process. As mentioned in the previous example, the electricity source is a dominant parameter in the environmental impact of this and other production systems. Other important parameters could be related to eg the supply chain. In addition, the vulnerability of the ecosystem varies for each location; for instance, regarding water consumption, water scarcity in each territory will affect how sensitive each area is to water use. Spatial-explicit LCA, or regionalised LCA, is a growing field that, over the last decade, has assessed regional impact categories such as air pollution, particulate matter, or land use (Mutel *et al*, 2019).

Temporal data is also relevant in that, for instance, the average electricity mix changes and is predicted to become greener in the future (International Energy Agency, 2023). With a less carbon-intensive electricity mix, the global warming potential from processes that are dependent on energy will also decrease. Investments, the alternative use of waste materials, market considerations, and the vulnerability of the ecosystems, all might be subject to changes in the future. These are studied in different ramifications of LCA (prospective, consequential, dynamic LCAs...), which are not inside the scope of this paper but hold a significant influence in the environmental assessments of metal production and could be analysed further in future research.

Last but not least, data quality should be evaluated regarding eg the age of the data, adequacy of the process, or representativeness. Uncertainty and sensitivity analysis are crucial for assessing the effects of data quality on the LCA results. For instance, as discussed during the 'System boundaries', the input of secondary aluminium could be considered of different qualities, both a material to waste management and an input of secondary aluminium to the process, considering the system expansion approach. Sensitivity to the type of aluminium input is tested in Figure 6.

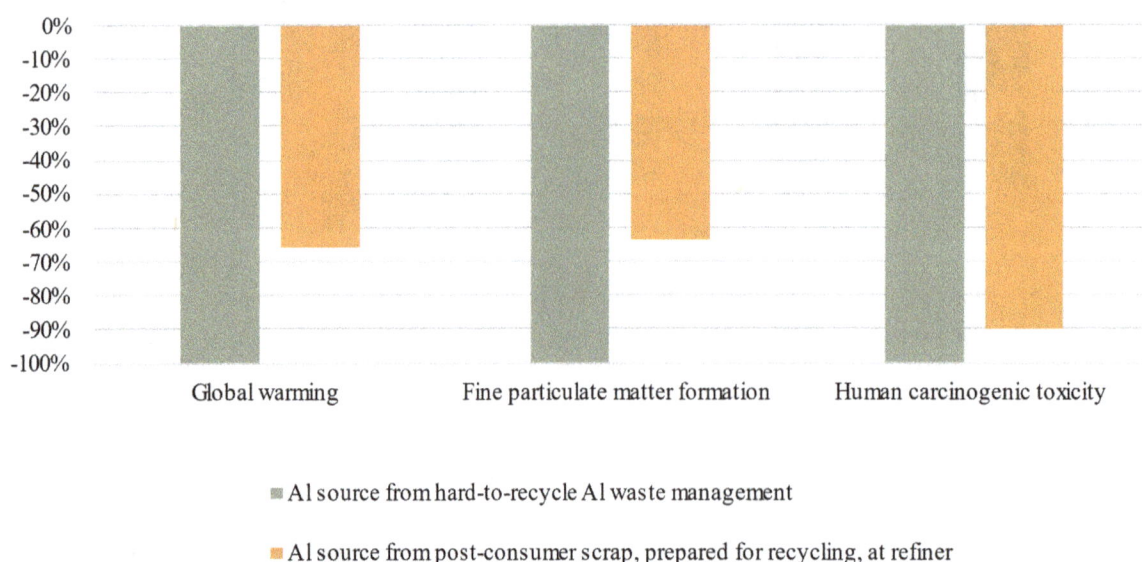

■ Al source from hard-to-recycle Al waste management

■ Al source from post-consumer scrap, prepared for recycling, at refiner

FIG 6 – Effect of different aluminium input considerations (evaluated through ReCiPe 2016 Midpoint (H)).

Even though all impact categories studied still benefited from the recycling of the waste stream, the advantage was smaller when considering an input of post-consumer Al scrap than when considering a flow of hard-to-recycle aluminium-containing streams to waste management. When considering post-consumer Al scrap, it is assumed that this flow is taken from other processes in the technosphere, and the impact of taking out this material from circulation is then added to the process under study because of the system expansion approach. When considering a waste material that is hard to recycle and assumed not currently being used by other processes, then the benefit is higher given that there is no alternative use of this material, highlighting the importance of choosing the correct data and assumptions when developing an LCA assessment of any waste flow.

CONCLUSIONS

The main conclusions and recommendations of this article are summarised in Table 3:

TABLE 3

Relevant methodological choices recommendations summary.

System boundaries	<ul><li>Depends on the object of study:<ul><li>Product oriented → usually cradle-to-grave.</li><li>Material oriented → usually cradle-to-gate.</li></ul></li><li>Possible to use scenario analysis to assess different configurations of the processes included in the system boundaries.</li></ul>
Functional unit	<ul><li>Depends on the function of the system:<ul><li>Single-function, eg mass of product.</li><li>Multifunction: FU accounting for material(s) produced or waste management. A key point is to always take the same functional unit for comparative assertions.</li><li>Always comparing materials that are able to provide the same function and that often means they present similar quality.</li><li>Lifetime can be relevant in the comparison among products.</li></ul></li></ul>
Allocation procedures	<ul><li>Allocation can be done by:<ul><li>System expansion: prioritised, when enough data.</li><li>Mass allocation: generally all other metals, or when the objective is assessing co-products separately.</li><li>Economic allocation: mainly for precious metals, or if a material is driving demand.</li></ul></li></ul>
Impact categories and impact method	<ul><li>Various impact methods, depending on eg spatial coverage.</li><li>Selection of impact categories:<ul><li>Low uncertainty.</li><li>Standards and guidelines.</li><li>Interest of stakeholders.</li><li>Effect on AoPs.</li></ul></li></ul>
Temporal, geographical and data quality considerations	<ul><li>Geographical scope: changes regarding electricity source and supply chain dominant parameters and vulnerability of ecosystems.</li><li>Temporal data: influenced by future-dependant process impacts, market conditions and ecosystems changes.</li><li>Other data quality considerations:<ul><li>Age of data.</li><li>Adequacy to the process.</li><li>Representativeness.</li></ul></li></ul>

ACKNOWLEDGEMENTS

The authors would like to gratefully acknowledge the European Research Council for funding the project SisAl (Grant Agreement #869268) and the Norwegian Research Council for funding the project Alpakka—Circular Aluminium Packaging in Norway (NFR Project #296276).

REFERENCES

Damgaard, A, Larsen, A W and Christensen, T H, 2009. Recycling of metals: accounting of greenhouse gases and global warming contributions, *Waste Manag Res*, 27:773–780. https://doi.org/10.1177/0734242X09346838

Environmental Protection Agency, Ireland, 2002. European Waste Catalogue and Hazardous Waste List [online]. Environmental Protection Agency, Ireland. Available from: <https://www.lehaneenvironmental.com/wp-content/uploads/2016/08/EWC_HWL.pdf>

Hauschild, M Z, Rosenbaum, R K and Olsen, S I (eds), 2018. *Life Cycle Assessment: Theory and Practice* (Springer International Publishing: Cham). https://doi.org/10.1007/978-3-319-56475-3

Huijbregts, M A J, Steinmann, Z J N, Elshout, P M F, Stam, G, Verones, F, Vieira, M, Zijp, M, Hollander, A and van Zelm, R, 2017. ReCiPe2016: a harmonised life cycle impact assessment method at midpoint and endpoint level, *Int J Life Cycle Assess*, 22:138–147. https://doi.org/10.1007/s11367-016-1246-y

International Energy Agency, 2021. The Role of Critical Minerals in Clean Energy Transitions [online], World Energy Outlook report. Available from: <https://www.iea.org/reports/the-role-of-critical-minerals-in-clean-energy-transitions>

International Energy Agency, 2023. Renewables [online]. International Energy Agency (IEA). Available from: <https://www.iea.org/energy-system/renewables> [Accessed: 19 January 2024].

International Standard Organization, 2006a. ISO 14040:2006 – Environmental management – Life cycle assessment – Principles and framework.

International Standard Organization, 2006b. ISO 14044:2006 – Environmental management – Life cycle assessment – Requirements and guidelines.

Joint Research Centre, 2017. Best available techniques (BAT) reference document for the non-ferrous metals industries: Industrial Emissions Directive 2010/75/EU (integrated pollution prevention and control). Publications Office of the European Union, LU.

Milani, V and Timelli, G, 2023. Solid Salt Fluxes for Molten Aluminum Processing – A Review, *Metals*, 13:832. https://doi.org/10.3390/met13050832

Mutel, C, Liao, X, Patouillard, L, Bare, J, Fantke, P, Frischknecht, R, Hauschild, M, Jolliet, O, Maia de Souza, D, Laurent, A, Pfister, S and Verones, F, 2019. Overview and recommendations for regionalized life cycle impact assessment, *The International Journal of Life Cycle Assessment*, 24:856–865. https://doi.org/10.1007/s11367-018-1539-4

Norgate, T and Rankin, W, 2002. The Role of Metals in Sustainable Development, Green Processing 2002 – in Proceedings of the International Conference on the Sustainable Processing of Minerals.

Office of Research and Development, 2015. Secondary Aluminum Processing Waste: Salt Cake Characterization and Reactivity [online]. Available from: <https://cfpub.epa.gov/si/si_public_record_report.cfm?Lab=NRMRL&dirEntryId=311174> [Accessed: 19 January 2024].

Olivieri, G, Romani, A and Neri, P, 2006. Environmental and economic analysis of aluminium recycling through life cycle assessment, *International Journal of Sustainable Development and World Ecology*, 13:269–276. https://doi.org/10.1080/13504500609469678

PE International, 2014. Harmonization of LCA methodologies for metals – a white paper providing guidance for conducting LCAs for metals and metal products. Available from: <https://international-aluminium.org/wp-content/uploads/2021/06/Harmonization-of-LCA-Methodologies-for-Metals.pdf>

Saevarsdottir, G, Magnusson, T and Kvande, H, 2021. Reducing the Carbon Footprint: Primary Production of Aluminum and Silicon with Changing Energy Systems, *Journal of Sustainable Metallurgy*, 7:848–857. https://doi.org/10.1007/s40831-021-00429-0

Santero, N and Hendry, J, 2016. Harmonization of LCA methodologies for the metal and mining industry, *Int J Life Cycle Assess*, 21:1543–1553. https://doi.org/10.1007/s11367-015-1022-4

UN Environment and International Resource Panel, 2019. Global Resources Outlook: Natural Resources for the Future we Want. Available from: <https://wedocs.unep.org/handle/20.500.11822/27517>

UN Environment and International Resource Panel, 2020. Resource Efficiency and Climate Change: Material Efficiency Strategies for a Low-Carbon Future. Available from: <https://wedocs.unep.org/handle/20.500.11822/34351>

UN Environment Programme, 2017. Environmental Risks and Challenges of Anthropogenic Metals Flows and Cycles. Available from: <https://www.unep.org/resources/report/environmental-risks-and-challenges-anthropogenic-metals-flows-and-cycles>

US Geological Survey, 2023. Mineral commodity summaries [online]. Available from: <https://pubs.usgs.gov/publication/mcs2023> [Accessed: 7 December 2023].

Vallejo Olivares, A, Pastor-Vallés, E, Pettersen, J B and Tranell, G, 2024. LCA of recycling aluminium incineration bottom ash, dross and shavings in a rotary furnace and environmental benefits of salt-slag valorisation, *Waste Management*, 182:11–20. https://doi.org/10.1016/j.wasman.2024.04.023

Watari, T, Nansai, K and Nakajima, K, 2021. Major metals demand, supply and environmental impacts to 2100: A critical review, *Resources, Conservation and Recycling*, 164:105107. https://doi.org/10.1016/j.resconrec.2020.105107

A thermodynamic and sustainability assessment of PCB recycling through the secondary Cu smelting process

P Tikare[1], M Mohanasundaram[2], R Kumar[3] and A Kamaraj[4]

1. MTech student, Indian Institute of Technology Hyderabad, Sangareddy Telangana 502285, India. Email: praveentikare@alumni.iith.ac.in
2. Principal Scientist, CSIR National Metallurgical Laboratory, Jamshedpur Jharkhand 831007, India. Email: madan@nmlindia.org
3. Scientist E, Centre for Materials for Electronics Technology, Hyderabad Telangana 500057, India. Email: rajesh@cmet.gov.in
4. Assistant Professor, Department of Materials Science and Metallurgical Engineering, Greenko School of Sustainability, Indian Institute of Technology Hyderabad, Sangareddy Telangana 502285, India. Email: ashokk@msme.iith.ac.in

ABSTRACT

This study presents a comprehensive thermodynamic optimisation and sustainability analysis of the secondary copper smelting process for Waste Printed Circuit Board (WPCB) recycling. The study investigates various scenarios with different proportions of WPCB in the input stream, ranging from 0 per cent to 100 per cent. Through thermodynamic modelling using FactSage™ version 8.1, the optimisation process is conducted in three stages. The results of the thermodynamic modelling indicate that the best-case scenario for WPCB recycling is achieved with a 40 per cent proportion of WPCB in the input stream. This case exhibits a maximum copper recovery of 99.93 per cent while minimising the formation of other waste. In addition to the thermodynamic analyses, environmental factors such as the carbon emission index and resource utilisation efficiency are studied. The 40 per cent WPCB recycling case exhibits a resource utilisation efficiency of 93.055 per cent and a carbon emission rate of 0.133. The present investigation highlights the significant benefits of a 40 per cent WPCB recycling scenario through a secondary copper smelting process in terms of base metal recovery and energy efficiency, thus emphasising its potential for sustainable PCB waste management and resource utilisation.

INTRODUCTION

Pyrometallurgical processes utilise elevated temperatures to extract metals from various sources. These processes encompass a range of techniques such as dressing, incineration, furnace-based smelting, sintering, melting and high-temperature reactions in a gaseous phase. The primary objective is to recover common metals and precious metals from different wastes (Wu et al, 1993). During smelting, e-waste is introduced into a high-temperature furnace, often integrated with a base metal production process such as copper, lead, or zinc smelting. In this process, the base metal acts as a medium for collecting valuable metals such as gold, silver, platinum and palladium (Ghodrat et al, 2017). Over the past few decades, pyrometallurgical methods have been extensively employed to recover metals from diverse waste materials.

Processes like smelting in furnaces, incineration, combustion and pyrolysis are commonly used to recycle electronic waste (e-waste). Numerous studies have documented substantial metal retrieval achieved through pyrometallurgical methods; however, the prominent drawbacks of these processes include their high cost, substantial energy requirements and the emission of hazardous fumes into the environment (Chauhan et al, 2018). It should also be noted that advanced smelters and refineries have proven effective in extracting valuable metals while also isolating hazardous substances. During the pyrometallurgical process, waste electrical and electronic equipment (WEEE) is subjected to high temperatures and combined with fluxes such as copper slag or various salts (eg NaOH) to facilitate the formation of slag. The resulting molten WEEE containing valuable metals is then brought into contact with a molten metal pool, where the valuable metals dissolve and accumulate. This molten metal pool acts as a collector metal (Syed, 2012).

Printed circuit boards (PCBs) play a critical role in electronic and electrical equipment (EEE) and form the backbone of modern electronic and electrical infrastructure (Liu et al, 2016). The PCBs are made up of glass fibre-reinforced epoxy and various metallic materials, including valuable metals (Choubey *et al*, 2015; Bai *et al*, 2016). WPCB represents a significant reserve of recyclable resources (Holgersson *et al*, 2018), with the concentration of metals, especially precious metals, exceeding that of primary minerals, highlighting the economic advantages of recyclability (Cucchiella *et al*, 2016). Moreover, the WPCBs recycling positions them as valuable urban mineral resources (He and Duan, 2017), underscoring their significance in the context of sustainable resource management.

PCBs are crucial components in electronic products, but their recycling presents challenges. PCBs contain precious metals and hazardous elements, requiring proper management to minimise environmental and health risks. These boards consist of materials like phenolic / cellulose paper or epoxy, woven glass fibre and various metals such as copper, tin and lead (Ning *et al*, 2018) Valuable precious metals like gold, nickel, platinum, palladium and silver can be recovered from WPCBs, making them economically attractive for recycling (Lu and Xu, 2016). However, it is essential to recognise that certain chemical elements in circuit boards, such as bromine, antimony, cadmium and lead, exceed their natural crust composition, posing potential environmental and health risks. (Chauhan *et al*, 2018). Aier, Prabhakaran and Kannadasan (2013) utilised NaOH salt during the melting of waste PCB from mobile phones and computers to increase the recovery of copper. Hagelüken (2006) reported that loss of copper during the recycling process will make up for 7 to 42 per cent value loss. ISASMELT process for reductive smelting of e-waste has been extensively studied to understand the major challenges such as control of slag composition, temperature, fume formation due to Zn, Pb and Sn (Stuart *et al*, 2023). Therefore, efficient recycling processes towards maximising the base metal recovery as well as minimising the environmental impact are essential for sustainable e-waste practices. In the present study, thermodynamic simulations are adopted to identify the optimum condition for recycling WPCB towards maximum base metal recovery and minimum environmental impact.

METHODOLOGY

FactSage™ simulation

The present study investigates the material flow while processing copper scrap and WPCBs in the black copper smelting process. The chemical thermodynamic modelling and process flow sheet simulations were employed to generate data for calculations. The thermodynamic calculations were conducted using the FactSage™ thermochemical package (version 8.1), allowing for accurate process modelling. Thermodynamic modelling and calculations offer a comprehensive approach to understanding the behaviour of copper scrap and WPCBs during the black copper smelting process. Varying e-waste and copper scrap proportions (varied from 0 to 100) were utilised as input parameters to simulate the black copper smelting process. Other input material streams are Coal, FCS Slag, enriched air and flux. The proportion of flux used in the simulation is maintained constant, as described in the previous work (Ghodrat *et al*, 2017) for 12 tonnes of charge materials. The detailed composition of various input streams considered in thermodynamic modelling can be found in the annexure (Ghodrat *et al*, 2017). The composition of black copper formed in the different cases is presented in the annexure. It should be noted that there are no solid phases observed in both the output streams (molten black copper and slag) under optimised conditions. By examining different scenarios, the research aimed to gain insights into the material flow during different percentages of e-waste recycling through the secondary Cu smelting process. The results obtained from this study contribute to the knowledge base surrounding the recycling and recovery of valuable metals from electronic waste. The study also informs potential process improvements for sustainable and efficient resource utilisation. Different proportions of the input stream are presented in Table 1.

TABLE 1

List of basic input streams considered for FactSage™ simulation (Ghodrat *et al*, 2017).

WPCB recycling (%)	WPCB (tonne)	Copper scrap (tonne)	FCS slag (tonne)
0	0	12	0.42
10	1.2	10.8	0.42
20	2.4	9.6	0.42
30	3.6	8.4	0.42
40	4.8	7.2	0.42
50	6	6	0.42
60	7.2	4.8	0.42
70	8.4	3.6	0.42
80	9.6	2.4	0.42
90	10.8	1.2	0.42
100	12	0	0.42

Thermodynamic optimisation

A fixed input quantity of 12 tonnes comprising various proportions of e-waste and copper scrap, as detailed in Table 2, was used as input to determine the optimised process parameters. The thermodynamic optimisation has been done stage-wise, as listed:

Stage 1: Identification of coal-air ratio for maximum Cu recovery and minimum residual solid carbon in the product stream.

Stage 2: Identification of optimised process temperature based on maximum liquid slag formation.

Stage 3: Identification of the optimum amount of flux (lime) addition to enhance liquid slag formation and suppress solid slag formation.

In the first stage of thermodynamic optimisation, the coal-air ratio was systematically varied, keeping all other input streams constant. It should be noted that the WPCB in the input stream also contains carbon as a part of plastics. Consequently, the optimal amount of coal addition must be determined for maximum recovery of base metal copper while minimising the formation of solid carbon residue in the output stream.

In the second stage, the process temperature was optimised to enhance the formation of a maximum quantity of liquid slag to facilitate the efficient separation of molten metal and slag. In this stage, the temperature was varied within a specified range, and the temperature at which the weight percentage of liquid slag reached its peak was identified as the optimal temperature.

In the third stage, the optimal amount of flux (lime), was determined to promote the formation of liquid slag in the output stream. The objective was to eliminate the formation of solid slag in the output stream and maximise the quantity of liquid slag, thereby facilitating the separation of metal and solid slag. Lime addition for each case was carefully selected to have minimum slag viscosity, further aiding the separation process.

The distribution coefficient of the element is a measure that compares the amount of element present in black copper (metal phase) to the combined amount of metal found in slag, exhaust gas and metal dust (waste streams). It provides valuable insights into the distribution of valuable metal between these useful products and waste streams.

$$\text{Distribution coefficient} = \frac{\text{weight of element i in black Cu (kg)}}{\text{weight of element i in slag+dust+gas (kg)}} \qquad (1)$$

Metal recovery of element 'i' is defined as a measure that compares the amount of element 'i' present in black copper (metal phase) to the amount of element 'i' found in input streams.

$$\text{Metal recovery} = \frac{\text{weight of element i in black Cu (Kg)}}{\text{weight of element i in input streams (Kg)}} \tag{2}$$

Sustainability assessment

The environmental implications associated with current smelting systems have become increasingly important due to stricter emission targets and escalating energy costs. The present study's primary focus was to examine three indicators from environmental responsibility and sustainability perspectives.

The resource utilisation efficiency is determined by the ratio of the mass of metal/alloy produced and recycled resources to the total input mass, as defined by the following equation:

$$£ = \frac{(M_p + M_r)}{M_{inp}} \times 100 \tag{3}$$

where:

M_p (Kg)   is the mass of the alloy

M_r (Kg)   is recycled mass used

M_{inp} (Kg)   total input stream mass

The carbon dioxide emission ratio is quantified as the quantity of carbon dioxide (Kg) emitted per metric ton of alloy produced, as defined by the following equation:

$$\mu = \frac{M_{CO_2}}{M_{alloy}} \tag{4}$$

where:

M_{CO_2} (Kg)   is the mass of CO_2 and CO emitted

M_{alloy} (Kg)   is the mass of alloy generated

The above two parameters are obtained from the FactSage™ simulation results.

RESULTS AND DISCUSSION

Thermodynamic analysis of the secondary Cu smelting process

In the first stage, thermodynamic optimisation has been done to identify an optimised coal air ratio (CAR) for maximum base metal (copper) recovery and minimum solid carbon residue formation in the output streams. Figure 1 shows the variation of the optimised coal-air ratio for the maximum recovery of base metal and minimum formation of solid carbon residue in the output streams with the WPCB per cent in the input stream. CAR variation with WPCB shows that the coal-air ratio value decreases as the percentage of WPCB increases in the input stream. It depicts the requirement for less amount of coal while increasing the WPCB in the Cu smelter for processing. This is attributed to the presence of carbon in the WPCB. The maximum value is 0.875, which is for 0 per cent e-waste because there will be no plastics. The minimum value is 0.1375, which is for 100 per cent WPCB because it will have the highest number of plastics.

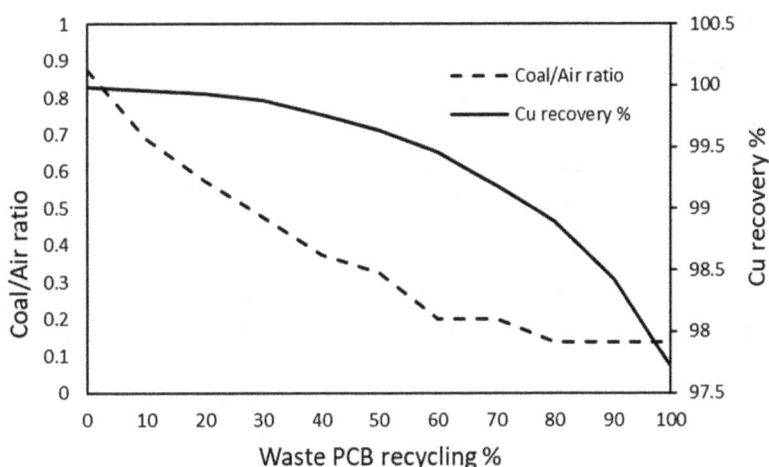

FIG 1 – Variation of optimised CAR for different WPCB recycling per cent.

In the second stage of thermodynamic optimisation, the optimum process temperature towards the formation of maximum liquid slag along with maximum base metal recovery has been identified. The optimised process temperature was determined considering the CAR in the simulation optimised in the previous step. The temperature in the simulation studies has been varied from 1100 to 1600°C to identify the optimum process temperature.

In the final stage, an additional parameter ie flux (lime) addition, was introduced to facilitate liquid slag and minimise solid slag formation. Optimisation of flux addition has been performed to identify the process window for complete liquid slag formation with less viscosity. Introducing lime into the input stream, the intention is to promote the formation of a complete liquid slag with less viscosity. This can significantly improve the overall process performance in terms of refining as well as slag-metal separation. Reducing slag viscosity can facilitate better mass transport, enhancing metal recovery and process efficiency. The simulation experiments have been performed for the optimised CAR and process temperature conditions. Based on the optimisation results, the composition of the slag was identified for each case and is presented in Table 2, along with the optimised process temperature and viscosity of the slag. Slag viscosity has been estimated using the viscosity module of FactSage™ 8.1.

TABLE 2

Optimum CAR, process temperature, slag composition, and viscosity of slag for WPCB recycling in secondary Cu smelter.

WPCB Recycling	CAR	Temperature	Al$_2$O$_3$	SiO$_2$	CaO	MgO	FeO	Viscosity
(%)	-	(°C)			(wt%)			(PaS)
0	0.875	1350	4.73	61.65	27.58	0	5.21	7.723
10	0.687	1550	26.36	34	36.90	1.96	0.008	0.502
20	0.575	1600	40.72	24.48	32.69	1.59	0.001	0.483
30	0.455	1550	38.74	26.53	32.48	1.96	0.001	0.705
40	0.375	1400	19.91	36.45	40.55	2.96	0.008	0.91
50	0.325	1400	21.24	37.09	38.752	2.84	0.008	1.133
60	0.2	1450	24.17	36.40	36.681	2.69	0.005	0.996
70	0.2	1450	23.64	36.30	37.23	2.76	0.005	0.926
80	0.137	1450	22.39	35.69	38.963	2.915	0.004	0.877
90	0.137	1450	23.835	36.647	36.749	2.745	0.004	0.984
100	0.137	1450	23.51	36.548	37.125	2.784	0.003	0.937

Figure 2 illustrates the variation of copper recovery in the final optimised case as a function of the WPCB recycling percentage. The results indicate a marginal decrease in Cu recovery with an increase in the amount of WPCB in the input stream. The maximum recovery of 99.99 per cent, is observed for the 0 per cent WPCB recycling case, indicating optimal conditions for copper recovery in the absence of WPCB. The initial decrease in the Cu recovery is due to more Cu metal dust formation at higher process temperatures (say 1550 to 1600°C). When considering the input stream containing WPCB, the maximum copper recovery achieved is 99.93 per cent for the 40 per cent WPCB recycling case. Conversely, the minimum recovery of 99.7 per cent is recorded for the 100 per cent waste recycling case. The content of Cu in the base metal decreases with the WPCB amount in the input stream. Further, the amount of Cu reported to the dust portion increases with WPCB recycling. This may be due to the influence of other elements that join base metals, such as Fe, Sn and Si, decreasing the activity coefficient of Cu. Considering all these, the 40 per cent WPCB recycling case emerges as the most favourable scenario for achieving maximum copper recovery while recycling WPCB through the secondary copper smelting process.

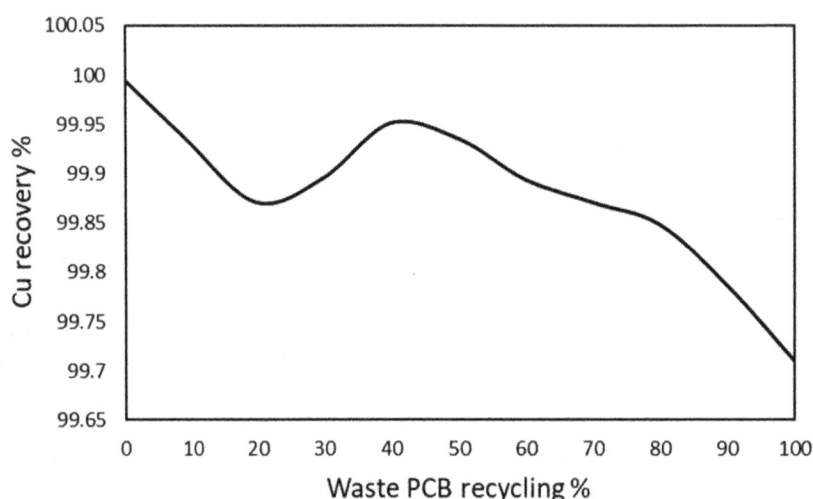

FIG 2 – Recovery of Cu during WPCB recycling through secondary Cu smelting under optimised condition.

The present work aims to recover the maximum precious metals in black copper along with other base metals such as Fe, Si, Sn etc. The discussion pertaining to the formation of fume due to high vapor pressure elements present in the scrap and e-waste in relation with the slag formation will be reported as a part of the future work.

Sustainability analysis

The findings of the environmental analysis are presented and discussed in the following sections. Figure 3a depicts the relationship between resource utilisation efficiency and WPCB recycling percentage. The findings demonstrate a clear trend: resource utilisation efficiency initially rises with increasing WPCB in the input stream till 30 per cent WPCB recycling case. Subsequently, it declines at 40 per cent, gradually increasing till 100 per cent WPCB recycling scenario. Notably, the minimum resource utilisation efficiency is observed when there is no recycling of WPCB (0 per cent recycling case), as it utilises the least amount of recycled mass in the input stream. Conversely, the maximum resource utilisation efficiency occurs at the 100 per cent WPCB recycling case, which utilises the most recycled mass available.

Figure 3b illustrates the relationship between the carbon emission ratio and WPCB recycling percentage. A similar trend has been observed in resource utilisation efficiency. The lowest carbon emission rate is observed when no WPCB is in the input stream. When comparing the cases with WPCB in the input stream, the minimum carbon emission rate occurs at the 40 per cent WPCB recycling case, while the maximum is observed at the 100 per cent WPCB recycling case. This indicates that adopting a recycling rate of 40 per cent WPCB through the secondary Cu smelting process significantly improves the overall environmental and economic sustainability of the process.

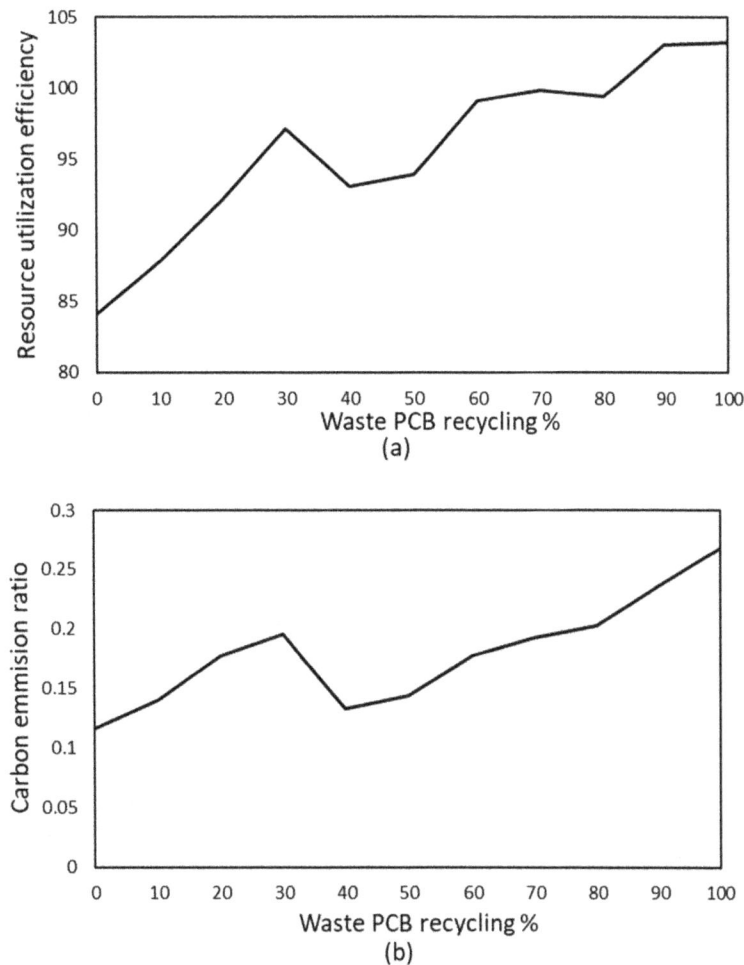

FIG 3 – Sustainability assessment of WPCB recycling through secondary Cu smelting route: (a) resource utilisation efficiency, (b) CO_2 emission rate.

CONCLUSIONS

The salient features of the present investigation are as follows:

- The coal-air ratio decreases with an increase in the WPCB in the input stream. This is attributed to the energy demand required for the process being met from the carbon present in WPCB as plastics.

- The minimum process temperature required to form the maximum amount of liquid slag has been identified for different WPCB recycling rates for the optimised CAR. The optimised process temperature varies from 1350 to 1600°C with extra flux (ie lime) addition to form 100 per cent liquid slag for efficient molten metal-slag separation. However operating at such a high temperature will impact the furnace design and metal and slag containment.

- The 40 per cent WPCB recycling through the secondary copper smelting process under the optimised condition emerges as the efficient scenario for achieving maximum base metal recovery, ie copper.

- It is observed that the resource utilisation efficiency increases with an increase in the WPCB content in the input stream of the process. However, the carbon emission ratio is observed to be minimum (0.133) for 40 per cent of waste PCB recycling cases.

- Adopting a recycling rate of 40 per cent WPCB through the secondary Cu smelting process significantly improves the overall environmental and economic sustainability of the process.

ACKNOWLEDGEMENTS

The authors would like to thank the Director of the Indian Institute of Technology Hyderabad and the Centre for Materials for Electronics Technology (C-MET) Hyderabad for their permission to publish this work. The author acknowledges financial support from the IIT Hyderabad, India, under grant SG/IITH/F282/2022–23/SG-120.

REFERENCES

Aier, A, Prabhakaran, D and Kannadasan, T, 2013. Recovery of Noble Metals from Electronic Waste by Pyrometallurgy Process, *International Journal of Scientific Research,* 2(12):209–212.

Bai, J, Gu, W, Dai, J, Zhang, C, Yuan, W, Deng, M, Luo, X and Wang, J, 2016. The Catalytic Role of Nitrogen-Doped Carbon Nanotubes in Bioleaching Copper from Waste Printed Circuit Boards, *Polish Journal of Environmental Studies,* 25(3):951–957.

Chauhan, G, Jadhao, P R, Pant, K K and Nigam, K D P, 2018. Novel technologies and conventional processes for recovery of metals from waste electrical and electronic equipment: challenges and opportunities – a review, *Journal of Environmental Chemical Engineering,* 6(1):1288–1304.

Choubey, P K, Panda, R, Jha, M K, Lee, J C and Pathak, D D, 2015. Recovery of copper and recycling of acid from the leach liquor of discarded Printed Circuit Boards (PCBs), *Separation and Purification Technology,* 156:269–275.

Cucchiella, F, D'Adamo, I, Rosa, P and Terzi, S, 2016. Automotive printed circuit boards recycling: an economic analysis, *Journal of Cleaner Production,* 121:130–141.

Ghodrat, M, Rhamdhani, M A, Brooks, G, Rashidi, M and Samali, B, 2017. A thermodynamic-based life cycle assessment of precious metal recycling out of waste printed circuit board through secondary copper smelting, *Environmental Development,* 24:36–49.

Hagelüken, C, 2006. Improving metal returns and eco-efficiency in electronics recycling, *IEEE International Symposium on Electronics and the Environment,* San Francisco.

He, J and Duan, C, 2017. Recovery of metallic concentrations from waste printed circuit boards via reverse floatation, *Waste Management,* 60:618–628.

Holgersson, S, Steenari, B M, Björkman, M and Cullbrand, K, 2018. Analysis of the metal content of small-size Waste Electric and Electronic Equipment (WEEE) printed circuit boards – part 1: Internet routers, mobile phones and smartphones, *Resources, Conservation and Recycling,* 133:300–308.

Liu, J, Xu, H, Zhang, L and Liu, C T, 2020. Economic and environmental feasibility of hydrometallurgical process for recycling waste mobile phones, *Waste Management,* 111:41–50.

Lu, Y and Xu, Z, 2016. Precious metals recovery from waste printed circuit boards: A review for current status and perspective, *Resources, Conservation and Recycling,* 113:28–39.

Ning, C, Lin, C S K, Hui, D C W and McKay, G, 2018. Waste printed circuit board (PCB) recycling techniques, *Chemistry and Chemical Technologies in Waste Valorization,* pp 21–56.

Stuart, N, Benjamin, H, Oscar, M and Stuart, N, 2023. Extraction and Recovery of critical metals from electronic waste using ISASMELT Technology, *Processes,* 11(4):1012.

Syed, S, 2012. Recovery of gold from secondary sources — A review, *Hydrometallurgy,* 115:30–51.

Wu, P, Eriksson, G, Pelton, A D and Blander, M, 1993. Prediction of the thermodynamic properties and phase diagrams of silicate systems–evaluation of the $FeO-MgO-SiO_2$ system, *ISIJ International,* 33(1):26–35.

ANNEXURE

The composition of input streams used in the simulation is given below.

Cu scrap

Cu	Cu$_2$O	SnO$_2$	PbO	ZnO	NiO
			(wt%)		
70	7	5	8	5	5

E-waste

E-waste	Element or compound	wt%
Metal	Cu	18.36
	Pb	5.50
	Fe	7.98
	Zn	3.67
	Au	0.091
	Ag	0.18
	Al	4.59
	Sn	3.67
	Ni	1.83
Oxide	Al$_2$O$_3$	5.50
	SiO$_2$	22.03
Plastics	C$_2$H$_3$Cl	1.10
	C$_2$H$_4$	2.19
	CH$_3$NO$_2$	0.66
	C$_7$H$_8$O$_2$	1.53
	C$_{12}$H$_6$Cl$_4$	0.66
	C$_{12}$H$_8$Br$_2$	2.63
	C$_{15}$H$_{16}$O$_2$	13.16
	H$_2$O	4.59

Metallurgical coke

C	H$_2$O	S	Al$_2$O$_3$	FeO
90	5	0.8	2	2.2

FCS slag

FeO	CaO	SiO$_2$
45	17	38

FCS slag

O_2	N_2
54	46

Lime

CaO	MgO	SiO_2
84.32	6.48	9.18

The composition of black copper produced.

Waste PCB recycling (%)	Black copper (wt%)											
	Ag	Al	Au	Cu	Fe	Mg	Ni	Pb	Si	Sn	Zn	Ag
0	0.00	0.00	0.00	81.85	1.23	0.00	4.22	6.87	0.00	4.23	1.60	0.00
10	0.02	0.06	0.01	82.46	2.38	0.00	4.36	4.63	1.02	4.58	0.47	0.02
20	0.04	0.38	0.02	80.86	3.54	0.00	4.40	3.04	2.60	4.87	0.24	0.04
30	0.07	0.35	0.04	78.43	4.83	0.00	4.40	2.99	3.49	5.15	0.23	0.07
40	0.10	0.02	0.05	76.20	6.34	0.00	4.44	5.94	0.64	5.51	0.75	0.10
50	0.14	0.02	0.07	74.05	8.17	0.00	4.52	5.46	0.97	5.97	0.63	0.14
60	0.18	0.06	0.09	71.59	10.38	0.00	4.62	3.64	2.56	6.54	0.33	0.18
70	0.23	0.06	0.12	68.33	13.04	0.00	4.72	3.08	2.91	7.20	0.28	0.23
80	0.29	0.07	0.15	64.31	16.37	0.00	4.86	2.57	3.09	8.04	0.24	0.29
90	0.36	0.06	0.20	58.68	20.47	0.00	4.99	1.79	4.23	9.02	0.18	0.36
100	0.43	0.06	0.25	51.49	25.89	0.00	5.17	1.19	5.00	10.33	0.13	0.43

Melting behaviour investigation of municipal solid waste incineration fly ash samples from different incineration technologies for metal recovery – an integrated experimental and thermodynamic modelling approach

E Soylu[1] and G Tranell[2]

1. PhD candidate, Norwegian University of Science and Technology, 7004.
   Email: ece.soylu@ntnu.no
2. Professor, Norwegian University of Science and Technology, 7004.
   Email: gabriella.tranell@ntnu.no

ABSTRACT

Municipal solid waste incineration fly ash (MSWI FA) is an important waste product that holds considerable potential for valourisation. In addition to major phases such as $CaSO_4$, $CaCO_3$, $NaCl$, KCl, and silicates, these ashes contain significant amounts of valuable elements like copper (Cu), zinc (Zn), lead (Pb), and others, where the specific composition depends on the source of the waste and the incineration process used. This study aims to investigate the melting behaviour of municipal MSWI FA samples from various incineration technologies, including rotary kiln, grate furnace, and circular fluidised bed, as a background for pyro/hydrometallurgical metal extraction. The experimental study was designed to research the effect of salt composition on the melting temperature and phase formations of different ash types, as well as metal migration between phases, using a sessile drop furnace. As a complimentary approach to the experimental study, thermodynamic modelling (FactSage™ ver 8.3, by Thermfact/CRCT and GTT-Technologies) was used to predict the phase formations of different fly ashes using Scheil-Gulliver cooling of molten ash. The observed melting point of the samples varied between 1000–1400°C depending on the ash type, without any trend of salt composition effect on the melting point. Upon solidification, there were three distinct phases observed in the samples: a metallic phase, a crystalline, non-metallic phase with inhomogeneous shape pattern, and an amorphous matrix phase. The findings indicate that the predominant component in the matrix phase was Ca-O-Si, implying the formation of calcium silicate slag. Elemental mapping showed metallic droplets consisting of primarily Fe-P phases, while the crystalline non-metallic phase is concentrated in Ca and S. The furnace atmosphere (Ar versus CO) had no significant impact on the phase formations. Thermodynamic modelling results were in good agreement with the experimental study, except for P-rich metallic phases, showing the formation of non-metallic and complex silicate slag phase formations.

INTRODUCTION

Municipal solid waste incineration fly ash (MSWI FA) is a by-product of the incineration process, which involves the combustion of waste at high temperatures reducing its volume by 85–90 per cent, mass by 60–90 per cent, and organic matter by up to 100 per cent (Zhang *et al*, 2021). Globally, the share of incineration in waste management practices is 11 per cent, where bottom ash and fly ash residue from MSWI account for approximately 30 wt per cent and 1–5 wt per cent of the input waste weight, respectively (Kaza *et al*, 2018; Tian *et al*, 2021). Different technologies such as grate, rotary kiln, and fluidised bed are used for incineration practices. In Europe, approximately 90 per cent of MSW treatment installations are grates, where the share of fluidised beds and rotary kiln are 5 per cent and 2 per cent, respectively (Neuwahl *et al*, 2019). The operation temperature of incinerators varies depending on the feed material and incineration plant, which results in high versatility in the ash composition (Nedkvitne *et al*, 2021). FA contains considerable number of high-value metals such as Zn, Cu, Ni, Co etc along with silicates and varying calcium bearing compounds with a potential of being used in different industries such as construction and glass ceramics (Zhao *et al*, 2021; Fan *et al*, 2022). A study where the chemical composition of approximately 900 different FA samples were analysed showed that 1 kg of fly ash contains 10 g of Zn and 1.2 g of Cu, on average (Nedkvitne *et al*, 2021). Even though the variations are high, and composition of FA depends on many factors such as location, season, incineration technology used etc, FA holds a great potential for recovery of critical metals competing with the ore grades in the concentration of critical metals (Nedkvitne *et al*, 2021; Nedkvitne, Eriksen and Omtvedt, 2023). The current best available technologies (BATs) for utilisation of FA are: 1) stabilisation of heavy metals in FA, 2) utilisation of FA as construction material after stabilisation, and 3) resource recovery in the means of metal and salt extraction via thermal/hydrometallurgical processes (Jadhav and Hocheng, 2012; Becidan, 2018; Wang et al, 2021).

There are various studies on the melting characteristics of FA. One study investigated the effect of the chemical composition on metal separation efficiency during melting (Okada and Tomikawa, 2013). It was observed that decreasing the Cl to Na and K molar ratio in the ash reduced Fe and Cu volatilisation, enhancing metal separation. The studies that investigated the impact of atmosphere on the volatility of metals revealed that volatility of metals was also affected by the amount of liquid slag formed and the temperature. It was concluded that under oxidising conditions and elevated temperatures, elements like As, Bi, Sb, Sn, and Zn tended to be mainly concentrated in condensed phases, whereas, under reducing conditions and high temperatures, these elements were predominantly released into the gas phase, however, Cu and Pb volatility was suppressed under reducing conditions (Lane et al, 2020a, 2020b). Another study suggested a reverse trend for Zn, reducing atmosphere hindering vaporisation of Pb, Zn, Cu, Cr, Co, and Ni due to increased liquid slag formation (Jiao et al, 2022). Other study that investigated the melting characteristic of FA concluded that ash composition was the main parameter affected the melting temperature, whereas atmosphere had minimal impact by a slight shift to higher temperature in oxidising atmosphere. Also, addition of CaO increased the melting point by the formation of new compounds (Li et al, 2007). In one study, it was found that all crystalline structures transformed into molten slag at temperatures exceeding 1300°C (Gao et al, 2021).

Metal extraction from FA via hydrometallurgical methods, has been extensively investigated in the literature (Karlfeldt Fedje et al, 2010; Huang et al, 2011; Tang and Steenari 2016; Jadhav and Hocheng, 2012; Wen et al, 2020; Wang et al, 2021). In terms of metal extraction via pyrometallurgical processes, the main approach has been thermal separation of heavy metal (HM) compounds by taking the advantage of the high volatility of chlorides (Nowak et al, 2012; Yu et al, 2016; Wang et al, 2021). Experimental findings indicated that both Pb and Zn possess considerable volatility, with approximately 80 per cent volatilisation achieved at 900°C even without the addition of chlorination agents (Kurashima et al, 2019). Another pyrometallurgical method investigated for metal extraction from FA is molten salt treatment, in which molten salt served as a metal extraction and separation medium, and requiring relatively low-temperature (600–800°C) processes (Xie et al, 2020). Commonly used reagents include chloride-based salts such as sodium chloride (NaCl) and potassium chloride (KCl), as well as eutectic salt mixtures (eg NaCl-CaCl$_2$).

Even though there are considerable number of technologies available for ash valourisation, the complexity, compositional variations, and relatively low metal concentrations are the biggest limitations for techno-economically robust solutions. This study aims at giving insights into the melting behaviour of different ash types with the prospect of utilising thermal treatment for metal recovery as a part of more holistic FA recovery process. The main focus of the study was on investigating the metallic, non-metallic, and amorphous phase formations upon melting as a basis for potential exploitation of leachability of different crystal structures in varying mediums.

MATERIALS AND METHODS

The fly ash samples investigated in this study were obtained from different incineration plants located in various countries and using different incineration technologies, such as grate furnace (denoted as GF), rotary kiln (denoted as R), and circular fluidised bed (denoted as C). Both raw and salt-washed fly ash samples are used for the experiments, with 'W' denoting the washed samples. The samples were melted without any prior compacting or pelletising steps. The particle size of the raw and washed samples ranges between 37.66–139.52 µm and 30.41–83.38 µm, respectively. The chemical composition of the samples is presented in Table 1. It was determined via X-ray diffraction that the major compounds in fly ash samples were salts (such as KCl, NaCl, and NaSO$_4$), Ca-compounds (including CaO, CaCO$_3$, and CaSO$_4$), and complex alumina silicates.

TABLE 1

ICP-MS results of ash samples (reduced table containing only major and critical elements).

Elements	wt%									
	R	GF1	GF2	GF3	C	RW	GF1W	GF2W	GF3W	CW
Al	1.80	2.10	2.10	3.60	1.20	3.50	7.70	4.20	4.70	10.00
Ba	1.60	0.11	0.07	0.28	0.19	3.30	0.13	0.10	0.33	0.21
Ca	3.30	30.00	28.00	21.00	23.00	6.60	23.00	29.00	24.00	22.00
Cl	8.00	9.90	12.00	3.80	3.80	0.36	0.31	0.36	0.15	0.38
Cu	0.27	0.06	0.05	0.08	0.47	0.49	0.07	0.09	0.08	0.57
Fe	2.80	1.20	0.92	2.10	1.90	6.00	2.40	1.80	2.50	2.60
K	5.20	2.20	2.10	4.30	0.91	2.50	0.56	0.34	0.66	0.45
Mg	0.58	0.78	0.70	0.15	1.50	1.20	1.50	1.50	2.00	2.00
Mn	0.15	0.05	0.05	0.10	0.10	0.30	0.08	0.11	0.16	0.13
Mo	0.10	0.00	0.00	0.00	0.02	0.19	0.00	0.00	0.00	0.00
Na	19.00	2.70	3.40	5.20	1.90	7.20	0.64	0.58	0.79	0.93
Ni	0.10	0.00	0.01	0.01	0.01	0.21	0.01	0.02	0.01	0.02
P	5.90	0.55	0.44	0.75	0.35	11.00	0.71	0.78	0.88	0.47
Pb	0.35	0.21	0.18	0.31	0.20	0.44	0.12	0.20	0.20	0.21
S	8.40	3.60	4.50	11.00	1.60	1.00	2.40	4.00	8.00	1.80
Si	3.60	5.70	3.70	6.60	9.30	6.60	11.00	7.50	8.40	12.00
Ti	1.10	0.54	0.49	1.30	0.24	2.20	1.00	1.00	1.60	1.70
Zn	2.30	0.82	0.82	1.90	0.45	4.50	1.30	1.50	2.30	0.60

In the experimental set-up, a sessile drop furnace was employed to investigate the melting behaviour of fly ash samples. The furnace is composed of a stainless-steel chamber with two windows, featuring a graphite heating element and radiation shield, a pyrometer, and a control C-type thermocouple. Additionally, the furnace is equipped with a high-speed digital CCD camera (Microtron MC 1310, Microtron GmbH, Unterschleissheim, Germany) at 50 frames per sec, allowing real-time observation of the samples during heating (Bao *et al*, 2021; Bublik *et al*, 2021; Canaguier and Tangstad, 2021). To determine the melting point, the ash samples were heated under argon (Ar, 99.9999 per cent purity) atmosphere until the onset of melting was observed in the camera. Before starting the experiments, a calibration experiment with pure iron (Fe) was done to ensure accurate temperature reading. To investigate the salt composition effect on melting temperature, one set of ash samples was washed with water to remove salts and one set of samples are used without any treatment. Based on the melting temperature observations in these experiments, a second part of experiments was conducted at T_m+100°C with a holding time of 1 hr, and under both Ar and carbon monoxide (CO, 99.9993 per cent purity) gas atmospheres. Only washed samples were used in the subsequent experiments. The focus of the second set of experiments was to observe the phase disintegrations and metal migration after melting, and to investigate the effect of a reducing atmosphere (when CO is used) on the system. All experiments were conducted with a consistent gas flow of 0.1 L/min. Alumina was used as the substrate for the experiments, and the sample amount varied between 80–100 mg. The characterisation of ash melts was done using Scanning Electron Microscopy-Energy Dispersive Spectroscopy (SEM-EDS) and Electron Probe Micro-Analysis (EPMA), and samples were epoxy-cast twice to get cross-sectional imaging.

The thermodynamic modelling study was conducted using FactSage™ 8.3 software (Thermfact/ CRCT, Montreal; GTT-Technologies, Aachen), where the Gibbs free energy minimisation is applied. The phase formation upon solidification was modelled in the Equilibrium module selecting FactPS and FTOxide databases (FactSage™). To simplify the system, the chemical composition of ash samples was registered as the main sulfate or carbonate compounds, depending on the sample composition, where the quantitative data was obtained by Rietveld analysis of X-ray diffraction (XRD) scanning, and for metal oxides deriving from elemental composition data provided in Table 1.

RESULTS AND DISCUSSION

Effect of salt composition on the melting temperature

The melting temperatures of the raw and salt washed MSWI FA samples were determined based on the visual observations with a high-speed digital camera during heating. The images are shown in Figure 1, where the first clear images were taken when the temperature reached 900°C. A complete wetting at the melting point was observed for all samples, except sample R. During the heating of sample GF3, gasification of some amount of sample was recorded when the temperature reached close to the melting point, and the evaporation accelerated with further increase in the temperature. On the other hand, sample GF3W did not show any excess gasification, however, bubbling on the melt surface was observable to some extent. This suggests the evaporation of chlorine compounds in the unwashed samples, whereas the removal of most chlorine salts after washing limits the gasification in the washed sample. The results show that the melting temperature of the fly ash samples varies between 1000–1400°C, depending on the ash type and the effect of decreased salt content (Table 2). Salt washing did not have the same effect for all ash types, the melting temperature of samples R and C decreases, whereas the melting point of samples GF1 and GF3 remains constant after salt removal. An increase in melting temperature was observed only in sample GF2. These results were higher than the reported melting temperatures of FA in the literature. According to the previous search results, the melting temperature of MSWI FA varied depending on the specific conditions and composition of the sample. One study found that the melting process of fly ash consisted of three main transitions: dehydration (100–200°C), polymorphic transition (480–670°C), and fusion (1101–1244°C) (Li et al, 2007). Another study reported that the ash fusion temperatures of MSWI FA ranged from 1167°C to 1211°C (Gao et al, 2021).

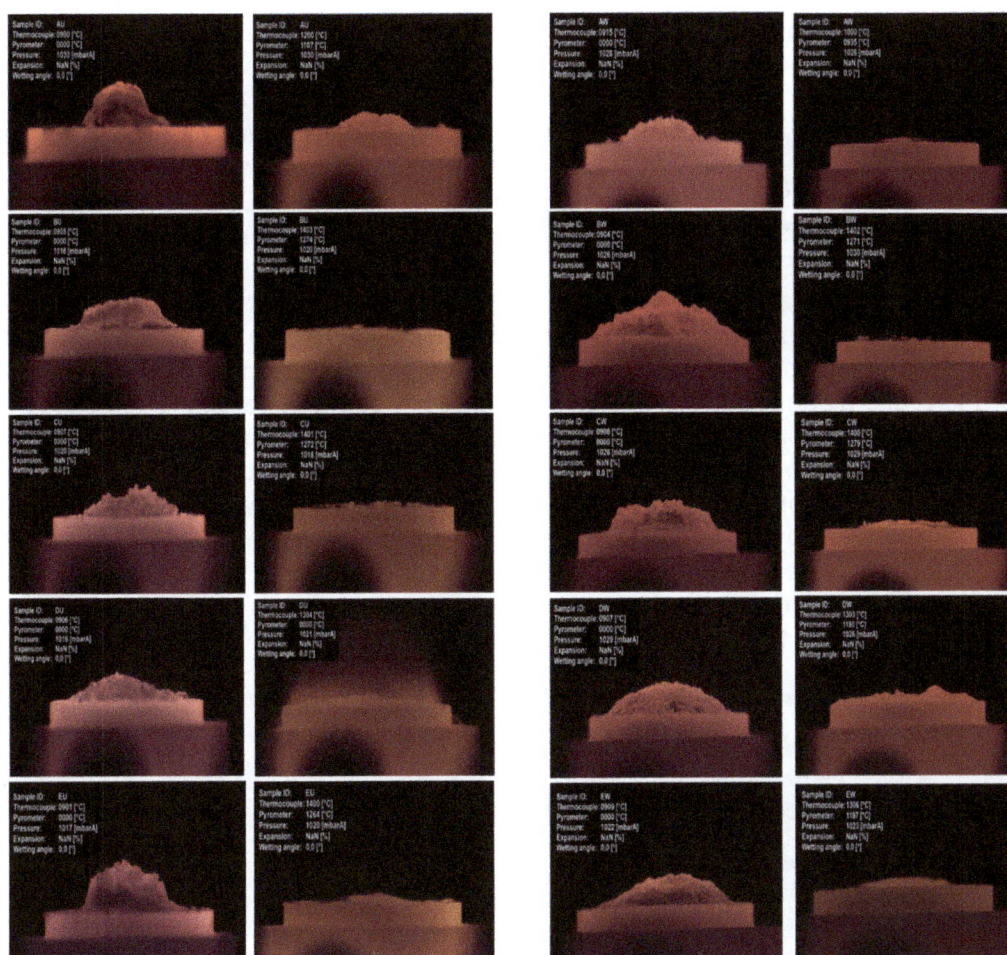

FIG 1 – Images captured during the heating of fly ash samples (Registered sample IDs in the furnace and their corresponding sample denotations: AU=R, AW=RW, BU=GF1, BW=GF1W, CU=GF2, CW=GF2W, DU=GF3, DW=GF3W, EU=C, and EW=CW).

TABLE 2

Approximate melting temperatures of raw and washed ash samples.

Sample	Temp (°C)	Sample	Temp (°C)
R	1200	RW	1000
GF1	1400	GF1W	1400
GF2	1200	GF2W	1400
GF3	1300	GF3W	1300
C	1400	CW	1300

Effect of chemical composition and atmosphere on phase formations and metal migration

To investigate the effect of composition on solidification, three representative samples were chosen based on their major compounds: RW (Na_2SO_4-rich), GF1W ($CaCO_3$-rich), and GF3W ($CaSO_4$-rich). The SEM-EDS results of the melted and solidified ash samples under Ar and CO atmospheres are shown in Figures 2 and 3, respectively. The results suggest that there are three distinct phases upon solidification: a metallic phase, a crystalline, non-metallic phase with inhomogeneous shape pattern, and an amorphous matrix phase. The EDS data points show that metallic droplets mainly contain Fe and phosphorous (P), with some extent of silicon (Si), and other elements such as molybdenum (Mo), nickel (Ni), aluminium (Al), titanium (Ti) and manganese (Mn) found in the metallic phase depending on the initial sample composition. In some samples (Figures 2b and 3a), Cu is found in

Fe-containing metallic phases, suggesting coalescence upon reduction. In Figure 2b, the Cu concentration is high (67.54 per cent) with no P detected in the phase. EDS data points of non-metallic phases reveals that the phase is mainly composed of calcium (Ca) and sulfur (S), suggesting the formation of a CaS phase. It is also observed that non-metallic phases contain some amount of Si, O, Mn, and Al (Figures 2c and 3a–3c), suggesting migration of these elements into the CaS phase. The formation of CaS can be attributed to the reduction of $CaSO_4$ during melting, which also suggests that some extent of reductive conditions provided in the system also under Ar atmosphere that led to formation of CaS, eg in sample GF3W. The sources of reductant, possibly carbon (C), can be the graphite furnace wall in the sessile drop furnace and the elemental C present in the initial composition of the ash. The analysis of carbon is not available for the samples used in this study; however, it is known that most ash samples might contain significant amount of C (5–20 wt per cent) coming from flue gas cleaning operations where activated carbon is added (Geng et al, 2020). In sample RW, S is not found in the Ca-rich non-metallic phase, as opposed to the detected amount of O which is high along with some amount of P, suggesting the formation of $Ca_5(PO_4)$ phase. The results suggest the formation of amorphous Ca-rich $CaO-SiO_2-Al_2O_3$ slag phase as the matrix in the samples where $CaCO_3$ and $CaSO_4$ are the major compounds. In sample RW, however, Ca is not detected in the matrix phase and the results suggest that it is composed of complex $SiO_2-Al_2O_3-K_2O-P_2O5-Na_2O$ slag. In terms of the effect of atmosphere on phase formations, there is no substantial difference between phases that could be linked to furnace atmosphere.

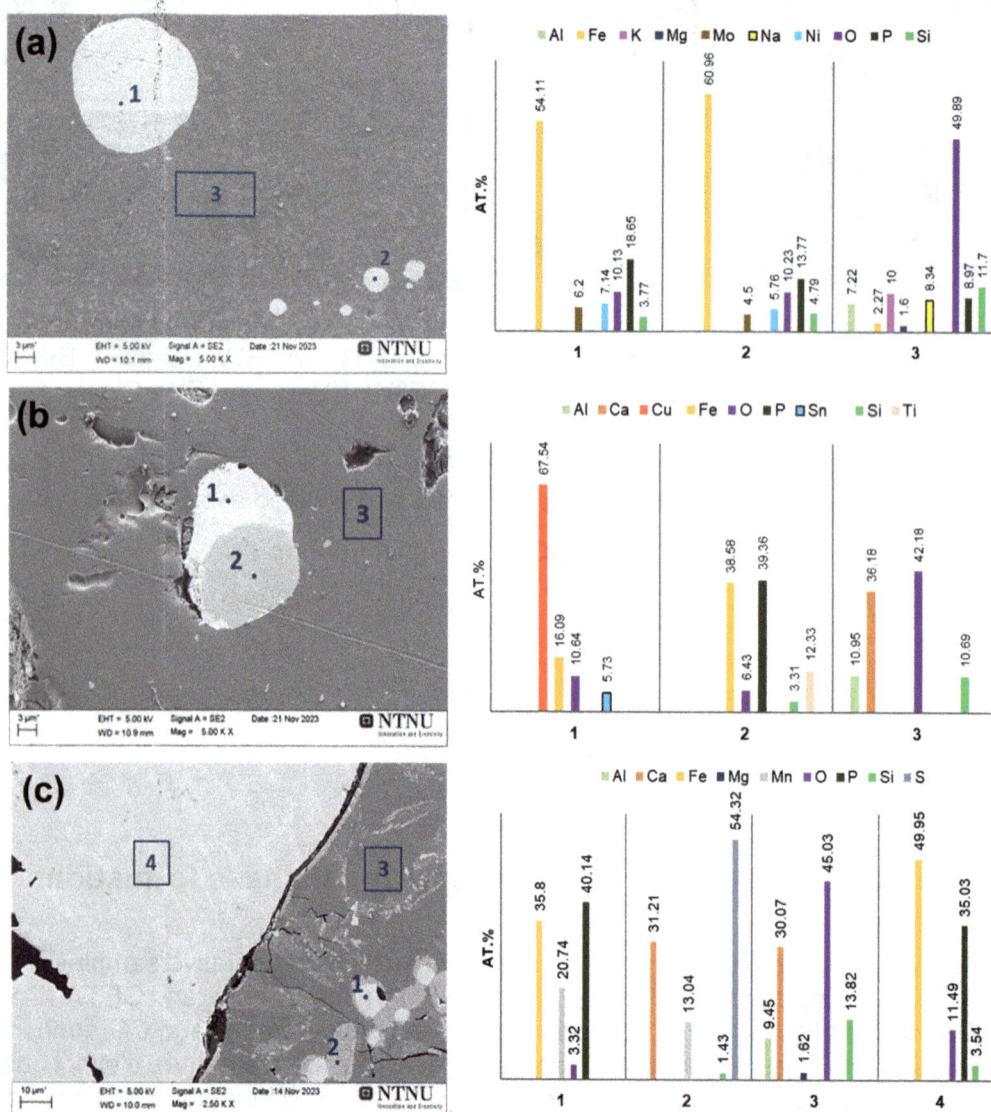

FIG 2 – SEM-EDS results of ash melts under AR atmosphere: (a) RW, (b) GF1W, and (c) GF3W.

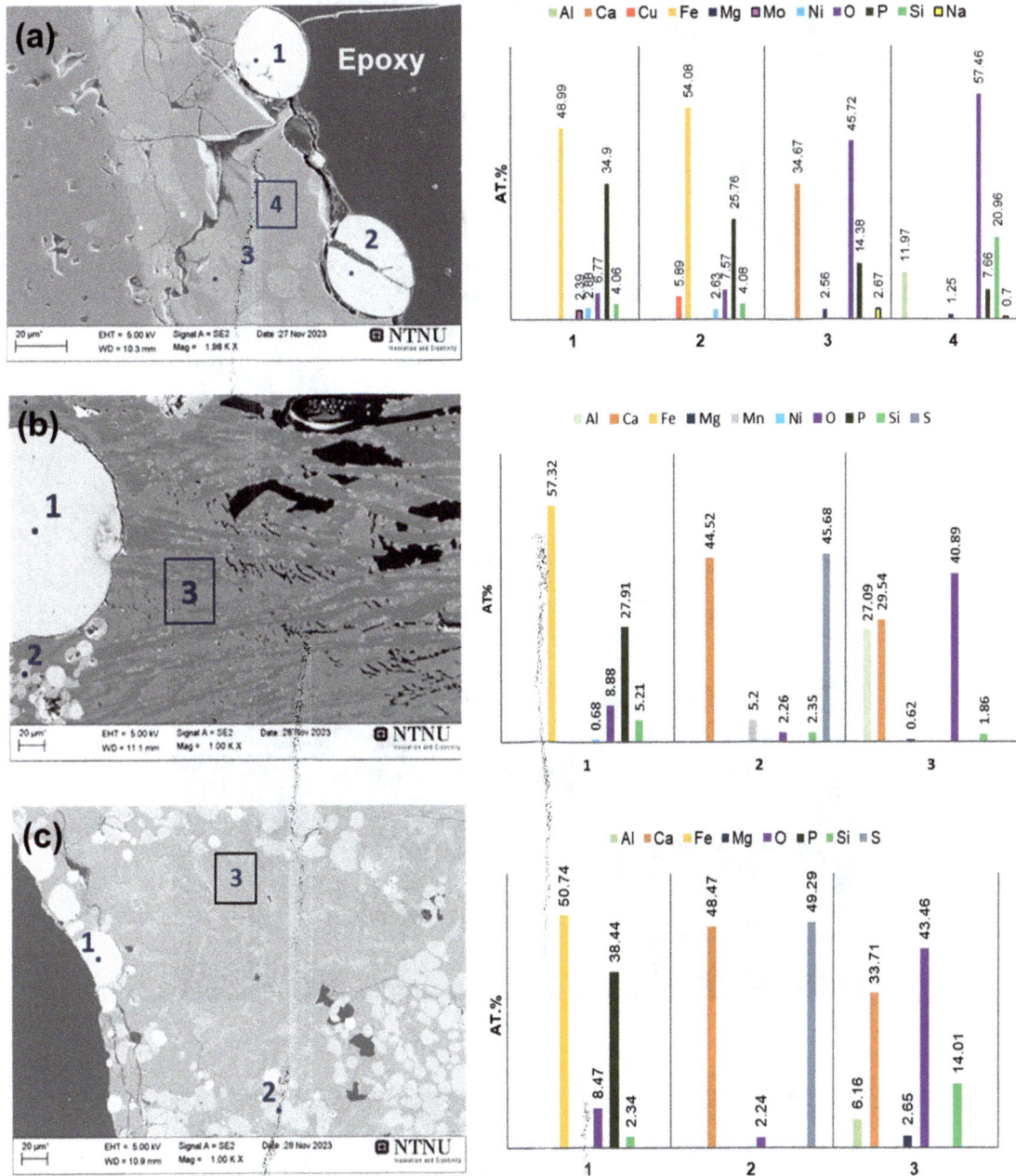

FIG 3 – SEM-EDS results of ash melts under CO atmosphere: (a) RW, (b) GF1W, and (c) GF3W.

Elemental mapping in EPMA correlates well with the EDS data, except for O, where EPMA imaging does not show any O in the Fe-P phase regions as opposed to EDS spectra (Figure 4), as would be expected. It is also observed that metallic droplets composed of two phases, suggesting solidification into different Fe-rich phases, eg Fe_3P and Fe_2P, upon solidification (Okamoto, 1990). The images showed some amount of Cu and Zn in the metallic phase, which indicates reduction followed by coalescence. Differently from EDS results, Zn and Cu are found throughout the sample, concentrating in the metallic phase. The elemental map confirms the formation of non-metallic CaS and Ca-P phases having some amount of Na along with O. The oxygen concentration in the Ca-phase (Figure 4a) is less than P and Ca, which differs from EDS data. Figure 4b suggests regions in the matrix phase where Al_2O_3 and SiO_2 concentrated.

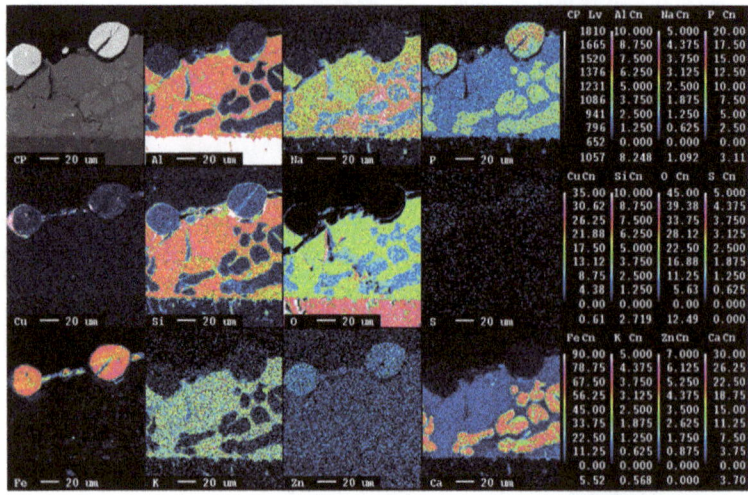

(a)

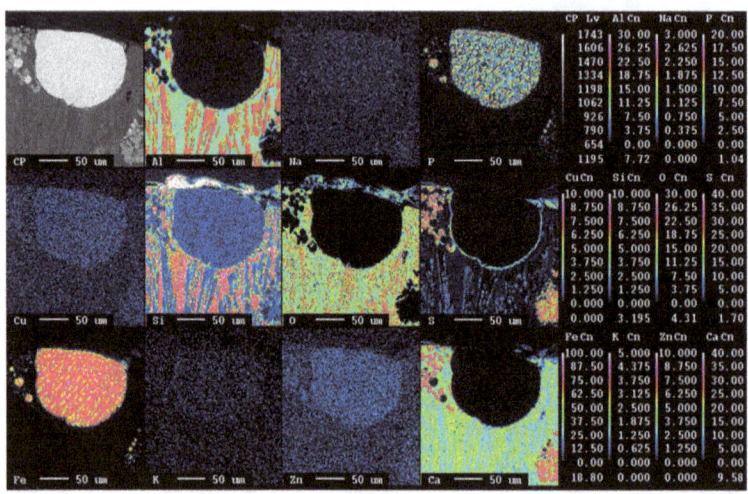

(b)

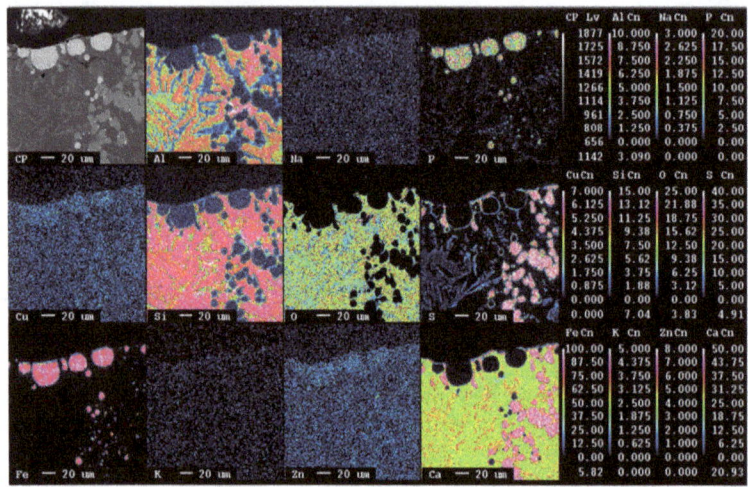

(c)

FIG 4 – EPMA results of selected ash samples melted under CO: (a) RW, (b) GF1W, and (c) GF3W.

FactSage™ modelling

In Figure 5, the calculations of equilibrium phases during heating and cooling of the representative samples under CO atmosphere are shown. Taking into account the composition of the ash systems, initial compositions were used as input to the calculations as shown in Table 3. The melting temperature is interpreted as the temperature where liquid slag formation starts in the heating graphs. The calculated melting temperature of RW and GFW1 is 700°C, and the melting temperature of GF3W is 900°C. These calculated temperatures are lower than the experimental temperature since the melting temperature indicates complete melting of the samples in the experimental observations, whereas, thermodynamic modelling shows the melting initiation temperatures. In the experimental study, high-temperature recording is only possible after 900°C, thus, melting initiation temperatures cannot be observed for low temperatures. The registered complete melting temperatures correspond the temperatures where liquid slag formation has reached its peak, which corresponds to 1100°C for RW, and 1400°C for GF1W and GF3W. In terms of the effect of the dominant compounds on the melting temperatures, it is concluded that high concentration of Na_2SO_4 decreases the melting point, whereas high amounts of Ca-bearing compounds have the opposite effect. The phase formation upon solidification of different ash types was calculated by cooling the liquid slag from 1600°C, using the Scheil-Gulliver functionality in FactSage™. The temperatures for complete solidification were found as 537°C, 642°C, and 546°C for RW, GF1W, and GF3W, respectively. The systems that include $CaSO_4$ shows CaS formation upon cooling, whereas, the calculations suggest the formation of $Na_2Ca_2P_2O_7$ and $Na_2CaP_2O_8$ compounds in RW. In regards to metallic phases, the thermodynamic calculations suggest the presence of FeS and ZnS compounds, starting to form later in the solidification stage. The modelling results on metallic phase formation does not correlate with the experimental observations, where results indicate the solidification of Fe-P compounds. This can be attributed to the well-known limitations in the FactSage™ databases for P containing compounds. The modelling results show the formation of complex silicates ($K_2MgSi_3O_8$, $CaMgSi_2O_6$, $Ca_2FeSi_2O_7$, $Ca_2Al_2SiO_7$, $Ca_7P_2SiO_2O_{16}$ etc) showing a good correlation with the experimental matrix phase observations. The system that has $CaSO_4$ as the major compound shows silicate containing oxide solid solutions (FToxid-BRED, FToxid-bC2SA, and FToxid-Mel_A) along with $Ca_7P_2Si_2O_{16(s)}$ phase:

- [FToxid-Bred] Bredigite OXIDE solution: $Ca_3(Ca,Mg)_4Mg(SiO_4)_4$ – a solid solution originating from $Ca_7Mg(SiO_4)_4$ by substitution of some Ca by Mg.

- [FToxid-bC2SA] OXIDE solution alpha-prime $(Ca,Sr,Ba)_2SiO_4$: Ca_2SiO_4 – Sr_2SiO_4 – Ba_2SiO_4 solution + (Mg_2SiO_4, Fe_2SiO_4, Mn_2SiO_4, Pb_2SiO_4, Zn_2SiO_4, $Ca_3B_2O_6$ in dilute amounts). End-members in pure compound database FToxidBase.cdb: Ca_2SiO_4(S2), Sr_2SiO_4(S2) and Ba_2SiO_4(S1).

- [FToxid-Mel_A] OXIDE solution-melilite: Mineralogical names: Akermanite ($Ca_2MgSi_2O_7$), Iron-akermanite ($Ca_2FeSi_2O_7$), Gehlenite ($Ca_2Al_2SiO_7$), Iron-gehlenite ($Ca_2Fe_2SiO_7$), Hardystonite ($Ca_2ZnSi_2O_7$). End-members in pure compound database FToxidBase.cdb: $Ca_2MgSi_2O_7$, $Ca_2FeSi_2O_7$, $Ca_2Al_2SiO_7$, $Ca_2ZnSi_2O_7$, $Pb_2ZnSi_2O_7$.

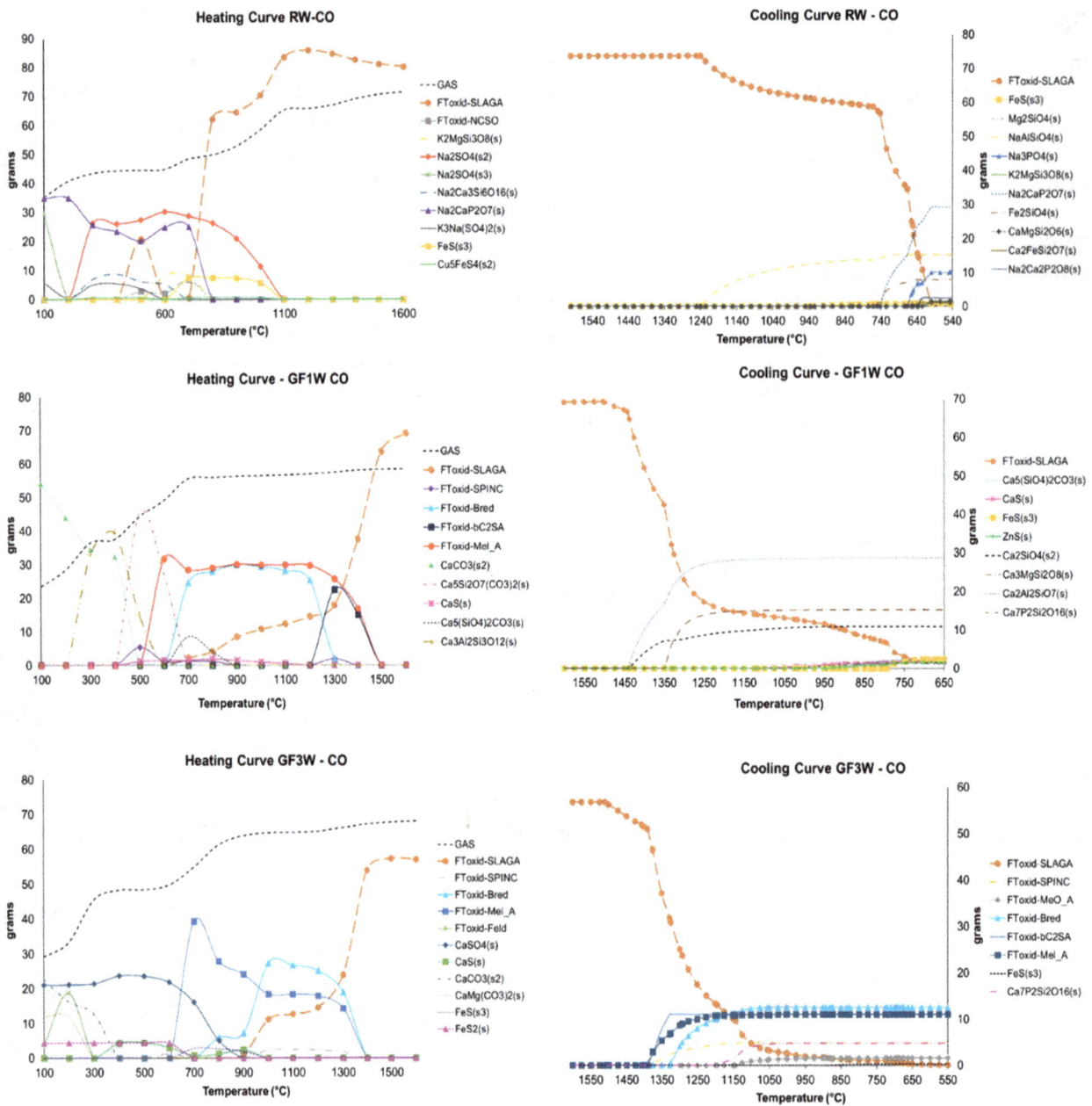

FIG 5 – FactSage™ 8.3 calculations.

TABLE 3

Compositional data used for FactSage™ 8.3 calculations.

RW (wt%)		GF1W (wt%)		GF3W (wt%)	
Na_2SO_4	38.91	$CaCO_3$	55.06	$CaCO_3$	2.09
Al_2O_3	6.17	$CaSO_4$	9.23	$CaSO_4$	64.13
Fe_2O_3	8.00	Al_2O_3	11.13	Al_2O_3	7.49
MgO	1.86	Fe_2O_3	2.62	Fe_2O_3	3.01
MnO	0.18	MgO	1.90	MgO	2.80
SiO_2	13.18	SiO_2	18.00	SiO_2	15.15
P_2O_5	23.52	P_2O_5	1.24	P_2O_5	1.70
K_2O	2.81	K_2O	0.52	K_2O	0.67

CONCLUSION

In conclusion, the findings of this study reveal that, not surprisingly, the most significant effect that has an impact on melting temperature and phase formation upon solidification is the composition of the ash samples. The experimentally observed melting points of the raw and salt-washed samples ranged from 1000–1400°C, showing differences before and after salt-washing, however, without exhibiting a discernible trend. Both experimental findings and thermodynamic modelling indicate that an elevated concentration of Na_2SO_4 lowers the melting point, while the presence of Ca-bearing compounds has the opposite effect. The findings indicate that during the solidification process, three separate phases form: a metallic phase, a non-metallic phase with a crystalline structure and an uneven shape pattern, and a phase characterised by an amorphous matrix. The EDS data show that metallic droplets mainly contain Fe and P, whereas non-metallic phase is mainly composed of Ca-S, suggesting the formation of CaS under reducing conditions. In some samples, low concentration of other metals such as Cu, Ni, Mo etc were detected in the metallic phase, indicating a co-reduction followed by coalescence mechanism. Also, in the CaS phase, the migration of some elements such as Si, O, Mn, and Al into the non-metallic phase is observed. The results indicate the formation of an amorphous Ca-rich CaO-SiO_2-Al_2O_3 slag phase as the predominant matrix in samples characterised by the presence of $CaCO_3$ and $CaSO_4$ as major compounds. Conversely, in sample RW, the absence of Ca in the matrix phase suggests a composition comprising a complex SiO_2-Al_2O_3-K_2O-P_2O_5-Na_2O slag. Regarding the influence of atmospheric conditions on phase formation, no discernible distinctions were observed between phases that could be attributed to variations in the furnace atmosphere. The modelling results were in good correlation showing the formation of complex silicates ($K_2MgSi_3O_8$, $CaMgSi_2O_6$, $Ca_2FeSi_2O_7$, $Ca_2Al_2SiO_7$, $Ca_7P_2SiO_2O_{16}$ etc) as the major components in the solidified system.

REFERENCES

Bao, S, Tangstad, M, Tang, K, Kristian, E, Syvertsen, M, Onsøien, M, Kudyba, A and Bublik, S, 2021. Investigation of Two Immiscible Liquids Wetting at Elevated Temperature: Interaction Between Liquid FeMn Alloy and Liquid Slag, *Metallurgical and Materials Transactions B: Process Metallurgy and Materials Processing Science*, 52(5):2847–2858.

Becidan, M, 2018. Fly ash treatment technologies: An overview of commercial and upcoming technologies for Norway and Scandinavia, report for NOAH AS, SINTEF Energi AS.

Bublik, S, Bao, S, Tangstad, M and Einarsrud, K E, 2021. Interfacial Behaviour in Ferroalloys: The Influence of Sulfur in FeMn and SiMn Systems, *Metallurgical and Materials Transactions B: Process Metallurgy and Materials Processing Science*, 52(6):3624–3645.

Canaguier, V and Tangstad, M, 2021. In situ observation of the carbothermic reduction and foaming of slags in silicomanganese production, *Processes*, 9(11).

Fan, C, Wang, B, Ai, H and Liu, Z, 2022. A comparative study on characteristics and leaching toxicity of fluidized bed and grate furnace MSWI fly ash, *Journal of Environmental Management*, 305.

Gao, J, Wang, T, Zhao, J, Hu, X and Dong, C, 2021. An experimental study on the melting solidification of municipal solid waste incineration fly ash, *Sustainability (Switzerland*, 13(2):1–10.

Geng, C, Liu, J, Wu, S, Jia, Y, Du, B and Yu, S, 2020. Novel method for comprehensive utilization of MSWI fly ash through co-reduction with red mud to prepare crude alloy and cleaned slag, *Journal of Hazardous Materials*, 384.

Huang, K, Inoue, K, Harada, H, Kawakita, H and Ohto, K, 2011. Leaching of heavy metals by citric acid from fly ash generated in municipal waste incineration plants, *Journal of Material Cycles and Waste Management*, 13(2):118–126.

Jadhav, U U and Hocheng, H, 2012. A review of recovery of metals from industrial waste, *Journal of Achievements in Materials and Manufacturing Engineering*, 54(2):154–167.

Jiao, F, Ma, X, Liu, T, Wu, C, Hanxu, L and Dong, Z, 2022. Effect of Atmospheres on Transformation of Heavy Metals during Thermal Treatment of MSWI Fly Ash: By Thermodynamic Equilibrium Calculation 2018, *Molecules*, 27(2):738–754.

Karlfeldt Fedje, K, Ekberg, C, Skarnemark, G and Steenari, B M, 2010. Removal of hazardous metals from MSW fly ash- An evaluation of ash leaching methods, *Journal of Hazardous Materials*, 173(1–3):310–317.

Kaza, S, Yao, L, Bhada-Tata, P and Woerden, F, Van, 2018. *What a Waste 2.0: A Global Snapshot of Solid Waste Management to 2050*, Urban Development Series. https://doi.org/10.1596/978-1-4648-1329-0

Kurashima, K, Matsuda, K, Kumagai, S, Kameda, T, Saito, Y and Yoshioka, T, 2019. A combined kinetic and thermodynamic approach for interpreting the complex interactions during chloride volatilization of heavy metals in municipal solid waste fly ash, *Waste Management*, 87:204–217.

Lane, D J, Jokiniemi, J, Heimonen, M, Peräniemi, S, Kinnunen, N M, Koponen, H, Lähde, A, Karhunen, T, Nivajärvi, T, Shurpali, N and Sippula, O, 2020a. Thermal treatment of municipal solid waste incineration fly ash: Impact of gas atmosphere on the volatility of major, minor and trace elements, *Waste Management*, 114:1–16.

Lane, D J, Hartikainen, A, Sippula, O, Lähde, A, Mesceriakovas, A, Peräniemi, S and Jokiniemi, J, 2020b. Thermal separation of zinc and other valuable elements from municipal solid waste incineration fly ash, *Journal of Cleaner Production*, 253.

Li, R, Wang, L, Yang, T and Raninger, B, 2007. Investigation of MSWI fly ash melting characteristic by DSC-DTA, *Waste Management*, 27(10):1383–1392.

Nedkvitne, E N, Borgan, Ø, Eriksen, D Ø and Rui, H, 2021. Variation in chemical composition of MSWI fly ash and dry scrubber residues, *Waste Management*, 126:623–631.

Nedkvitne, E N, Eriksen, D Ø and Omtvedt, J P, 2023. Grade and Tonnage Comparison of Anthropogenic Raw Materials and Ores for Cu, Zn and Pb Recovery, *Resources*, 12(3). https://doi.org/10.3390/resources12030033.

Neuwahl, F, Cusano, G, Benavides, J G, Holbrook, S and Roudier, S, 2019. Best Available Techniques (BAT) Reference Document for Waste Incineration, report, Industrial Emissions Directive 2010/75/EU (Integrated Pollution Prevention and Control), Joint Research Centre (European Commission) Science for Policy Report.

Nowak, B, Frías Rocha, S, Aschenbrenner, P, Rechberger, H and Winter, F, 2012. Heavy metal removal from MSW fly ash by means of chlorination and thermal treatment: Influence of the chloride type, *Chemical Engineering Journal*, 179:178–185.

Okada, T and Tomikawa, H, 2013. Effects of chemical composition of fly ash on efficiency of metal separation in ash-melting of municipal solid waste, *Waste Management*, 33(3):605–614.

Okamoto, B, 1990. The Fe-P (Iron-Phosphorus) System, *Bulletin of Alloy Phase Diagrams*, 11:404–4012.

Tang, J and Steenari, B M, 2016. Leaching optimization of municipal solid waste incineration ash for resource recovery: A case study of Cu, Zn, Pb and Cd, *Waste Management*, 48:315–322.

Tian, X, Rao, F, Li, C, Ge, W, Lara, N O, Song, S and Xia, L, 2021. Solidification of municipal solid waste incineration fly ash and immobilization of heavy metals using waste glass in alkaline activation system, *Chemosphere*, 283.

Wang, H, Zhu, F, Liu, X, Han, M and Zhang, R, 2021. A mini-review of heavy metal recycling technologies for municipal solid waste incineration fly ash, In *Waste Management and Research,* 39(9):1135–1148.

Wen, T, Zhao, Y, Zhang, T, Xiong, B, Hu, H, Zhang, Q and Song, S, 2020. Selective recovery of heavy metals from wastewater by mechanically activated calcium carbonate: Inspiration from nature, *Chemosphere*, 246.

Xie, K, Hu, H, Xu, S, Chen, T, Haung, Y, Yang, Y, Yan, F and Yao, H, 2020. Fate of heavy metals during molten salts thermal treatment of municipal solid waste incineration fly ashes, *Waste Management*, 103:334–341.

Yu, J, Sun, L, Ma, C, Qiao, Y, Xiang, J, Hu, S and Yao, H, 2016. Mechanism on heavy metals vaporization from municipal solid waste fly ash by MgCl2·6H2O, *Waste Management*, 49:124–130.

Zhang, Y, Wang, L, Chen, L, Ma, B, Zhang, Y, Ni, W and Tsang, D C W, 2021. Treatment of municipal solid waste incineration fly ash: State-of-the-art technologies and future perspectives, *Journal of Hazardous Materials*, 411.

Zhao, S Z, Zhang, X Y, Liu, B, Zhang, J J, Shen, H L and Zhang, S G, 2021. Preparation of glass–ceramics from high-chlorine MSWI fly ash by one-step process, *Rare Metals*, 40(11):3316–3328.

Conversion of hard-to-use wastes to new raw materials for low-energy glass/mineral-wool manufacturing

Z Yan[1], T Htet[2], S Zhang[3] and Z Li[4]

1. Research fellow, Advanced Steel Research Centre WMG, University of Warwick, Coventry CV4 7AL, UK. Email: zhiming.yan@warwick.ac.uk
2. Process Engineer, BUSS Chem Tech., Pratteln CH-4133, Switzerland. Email: theint.htet@buss-ct.com
3. PhD student, College of Materials Science and Engineering, Chongqing University, Chongqing 400044, PR China. Email: xiaoshuo32188@126.com
4. Professor, Advanced Steel Research Centre WMG, University of Warwick, Coventry CV4 7AL, UK. Email: z.li.19@warwick.ac.uk

ABSTRACT

During the pyrometallurgical processes, a substantial quantity of high-temperature ironmaking and steelmaking slags, generated at temperatures ranging from 1400 to 1600°C, is wasted, as the heat contained in these slags is released into the environment during the tapping process without effective utilisation. On the other hand, a significant amount of energy is required to reheat the cold materials for the manufacturing of glass and mineral-wool, often necessitating the addition of high-silica materials to facilitate the production process. In this study, an innovative approach is presented to harness the heat present in molten metallurgical slags to directly produce new raw materials for glass and mineral-wool manufacturing while simultaneously reducing energy consumption and mitigating CO_2 emissions. The theoretical amount of waste glass that can be effectively added in the process has been estimated via thermodynamic calculation. Experimental assessments were conducted to examine the impact of high-silica waste materials on the dissolution, melting and fluidity. Potential recipes for glass and mineral-wool manufacturing were suggested and an assessment of the energy savings, and reduction in CO_2 emissions with manufacturing of glass and mineral-wool using this innovative approach were included. It was determined that the processes could lead to a reduction of over 26 kg of CO_2 emissions per 100 kg of new material used, primarily due to the utilisation of heat from slag and the accelerated smelting process.

INTRODUCTION

The glass and mineral-wool industries are energy intensive foundation sectors, facing significant challenges in reducing energy consumption and CO_2 emissions (Dewick and Miozzo, 2002; Kiss, Manchón and Neij, 2013; Springer and Hasanbeigi, 2017). A significant amount of energy is consumed in raw material preparation and melting, and a great portion of CO_2 emissions is released from carbonate decomposition of raw materials (Deng *et al*, 2023; Schmitz *et al*, 2011). The glass and mineral-wool traditionally produced from cold raw materials, incurs high energy consumption, with heating accounting for over 70 per cent of the total energy use. Now low cost, environmentally friendly, energy-saving and increased-recycling are major driving forces for new raw materials. The glass industry endeavours to maximise the recycled glass content, but a significant amount (100 kTs per annum in the UK) of poor-quality recycled glass is unsuitable to be used in glass furnace due to issues such as contamination and particle size being too fine, in addition to large number of other wastes such as glass fibre (Iacovidou *et al*, 2018). Metallurgical slags such as blast furnace (BF) slag and basic oxygen furnace (BOF) slag offer significant environmental benefits over mined raw materials, such as a higher melting rate and provide a carbonate-free source of calcium (Pan *et al*, 2016).

Molten slag, a high-temperature by-product of the metallurgical industry, holds substantial potential as a valuable resource for the glass and mineral-wool industries. The current focus is largely on investigating BF slag (Oge *et al*, 2019; Piatak, 2018; Piatak, Parsons and Seal, 2015). However, the widespread use of BF slag as a cement raw material has a high utilisation rate. Meanwhile, the considerable volume of steelmaking slag generated by the BOF process and electric arc furnace (EAF) process remains underutilised. This is attributed, in part, to its low acidity (SiO_2/CaO), leading to high melting temperature and strong crystallisation performance that are unfavourable for glass and mineral-wool production. Additionally, the presence of FeO (~25 wt per cent) further hampers product quality. The acidity of metallurgical slag is usually lower than 1.0, while the acidity of mineral-wool is required to be more than 1.2 and even higher for glass (Zhao, Zhang and Cang, 2019). Then high-silica raw materials are needed to adjust the composition of the metallurgical slags and high-silica wastes, such as the waste glass become a potentially available resource.

This research aims to use the waste energy during molten slag tapping to convert various hard-to-use wastes into high value, new raw materials for glass or mineral-wool manufacturing. In this work, the energy utilisation and the effective use of waste glass in new materials generation were calculated. Thermodynamic calculations and experiments were carried out to assess the impact of waste glass on molten mixture properties. Modified mixtures were suggested to meet glass and mineral-wool manufacturing requirements. The study also evaluated energy savings and CO_2 emission reductions achieved through this innovative approach.

EXPERIMENTAL

Materials preparation

The BF slag, BOF slag and low-quality waste glass were sourced from industrial partners. The mineral phases of BOF slag and amorphous phase of BF slag and waste glass were confirmed by X-ray diffraction (XRD) as showed in Figure 1. The chemical composition of slags and glass were analysed by X-ray fluorescence (XRF) as presented in Table 1. The BOF slag exhibits a high content of FeO and a low acidity, contributing to the facile crystallisation of steel slag. The XRD pattern of the BOF slag appears intricate, revealing numerous overlapping peaks indicative of the diverse minerals present in the sample. Iron oxides are prevalent and beyond the presence of FeO, the XRD pattern suggests the likely existence of brownmillerite ($Ca_2(FeAl)_2O_5$) and magnesioferrite ($MgFe_2O_4$).

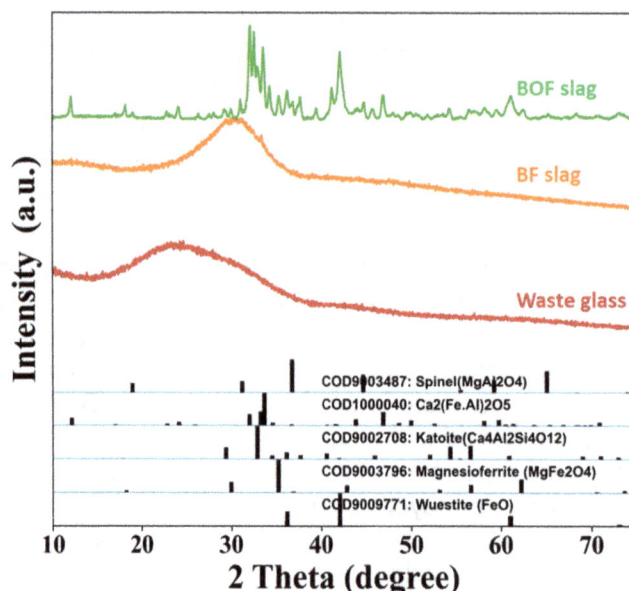

FIG 1 – XRD results of raw materials.

TABLE 1

Composition of BOF slag and waste glass, wt per cent.

wt%	CaO	SiO₂	MgO	Al₂O₃	SO₃	P₂O₅	K₂O	TiO₂	MnO	FeO	Na₂O	Acidity
BF slag	41.10	37.01	7.85	11.24	0.91	/	0.58	0.64	0.37	0.35	/	0.90
BOF slag	40.78	16.28	5.57	4.66	0.32	1.68	/	0.64	2.07	27.55	/	0.40
Waste glass	11.25	72.00	1.15	1.50	0.25	/	0.50	/	/	0.05	13.30	6.40

Thermodynamic calculation

The thermodynamic calculation was conducted by FactSage™ ver 8.2 (by Thermfact and GTT-Technologies) using FactPS and FToxid databases. This includes determining the effect of waste glass addition on the temperature and composition of mixture, the equilibrium phases during slag cooling, the melting behaviour and viscosity of raw materials. When calculating the theoretical addition of waste glass into molten slag during tapping, a key consideration is maintaining the slag-glass mixture in a liquid state. The calculations assumed a BF slag tapping temperature of 1480°C and BOF slag tapping temperature of 1550°C. The oxygen potential keeps at 0.21. Adiabatic calculations were also employed to get the temperature after mixing. The viscosity module was used to calculate the viscosity of the slags and molten mixture at liquid stage, with the aim of evaluating the effect of adding waste glass on the fluidity of the mixture.

Waste glass dissolution

The experiments of dissolving waste glass into BF slag were carried out in a high-temperature laser scanning confocal microscope at 1400°C and 1450°C in Argon, the schematic diagram for device and the temperature program are shown in Figure 2. The dissolution times were 0 sec, 10 sec, 30 sec and 60 sec. The heating and cooling rates were set as quickly as possible at 200°C/min and 300°C/min, respectively. About 100 mg of slag was first melted and then cooled down. A glass particle around 5 mg was placed on the top of the slag and reheated to the target temperature for a certain time. The cooled samples were analysed using a scanning electron microscope (SEM, Zeiss, Sigma) equipped with energy dispersive X-ray spectroscopy (EDS, Oxford, Ultim Extreme).

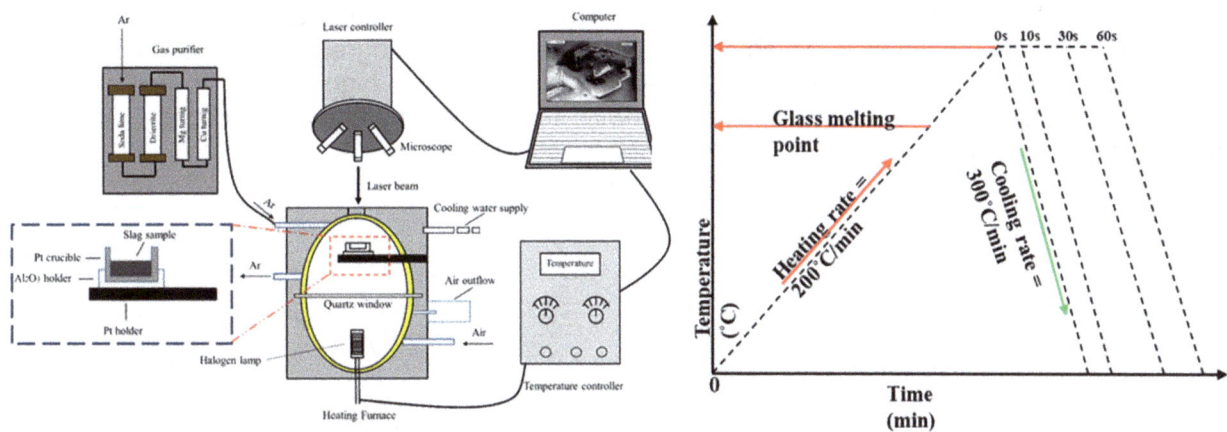

FIG 2 – Waste glass dissolution in BF slag via high-temperature laser scanning confocal microscope.

Melting temperature measurement

Based on the composition range of mineral-wool in the literature, a potential recipe was proposed. A high-temperature vacuum furnace was utilised to determine the melting temperature (T_h) using the Leitz microscope test method. Further information on the equipment and test procedure is available in a previously published paper (He *et al*, 2023). A cylindrical sample (3 mm in diameter and 3 mm in height) was prepared. During this heating phase, the shape of a compressed cylinder of slag powder was continuously monitored every second with a high-resolution optical camera when the chamber temperature over 1100°C. The heating rate was maintained at 6°C/min. The melting temperature is defined as the point at which the height of sample is reduced to half of its original size. Additionally, the fluidity temperature (T_f) is determined as the temperature at which the height of sample is reduced to one-quarter of its initial height.

Mineral-wool preparation

Utilising the proposed recipe, mineral-wool was manufactured through a centrifugal process. The details of the centrifugal apparatus were described in previous paper (He *et al*, 2023, 2020). For the process, a crucible containing 200 g of mixture was heated to approximately 1550°C in the induction furnace and maintained at this temperature for about 30 mins. It was then brought up to a predetermined speed via the rotational drive system. The molten slag was gradually poured from the crucible onto the rotating cup, where it was transformed into wool fibres by centrifugal force. The mineral-wool obtained was observed using a digital optical microscope (OM, Keyence VHX7000).

RESULTS

Thermodynamic calculation

The basic melting characteristics and fluidity of the raw materials can be obtained via thermodynamic calculation, as shown in Figure 3. The liquidus temperature of BF slag is approximately 1375°C. After cooling, the predominant phase formed is Mellite. When the temperature drops to 25°C below the liquidus, the solid phase content exceeds 50 per cent. This indicates a significant increase in the solid fraction at relatively small deviations from the liquidus temperature, reflecting the sensitivity of BF slag composition to temperature changes. The melting point of BOF slag is over 1600°C. Owing to its high basicity and FeO content, the spinel phase is precipitate first. Then there is an extensive precipitation of Dicalcium Silicate (C_2S) due to the high basicity. As the temperature falls, there is a rapid reduction in the quantity of the liquid phase, highlighting the sensitivity of BOF slag to temperature changes as well. The glass starts to soften and transition into a semi-molten state at temperatures above 500°C, with its liquidus temperature being approximately 1150°C. At this stage, the wollastonite phase is the first to precipitate. The viscosity of BF slag in its liquid phase is less than 0.6 Pas, demonstrating relatively low resistance to flow. In contrast, BOF slag exhibits a lowe viscosity even solid phases presents, below 2.0 Pas using the Einstein equation (Yue *et al*, 2018), indicating excellent fluidity. This fluidity is directly influenced by the slag basicity: higher basicity enhances fluidity. This is due to alkaline oxides like CaO, MgO, and FeO disrupting the [SiO_4] network structure in molten slag, which consequently reduces the degree of polymerisation of slag (Yan, Reddy and Lv, 2019; Yan *et al*, 2016). As a result, slag with higher basicity presents lower viscosity in its liquid phase. Conversely, glass, with its significantly high SiO_2 content, exhibits a much higher viscosity.

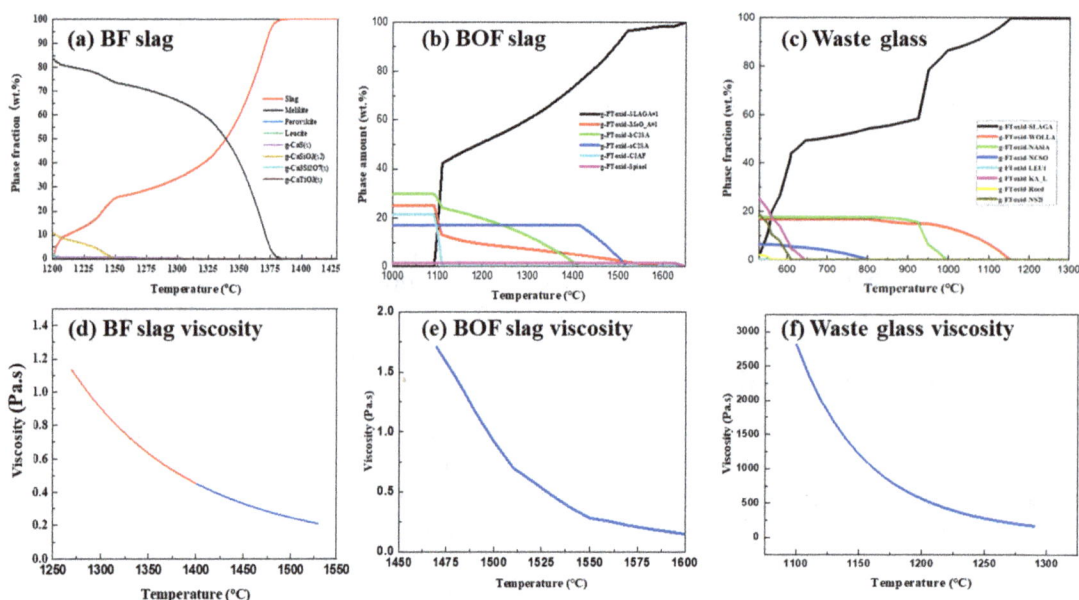

FIG 3 – The equilibrium phases during cooling of raw materials: (a) BF slag, (b) BOF slag, (c) Waste glass, and the viscosity of raw materials obtained via FactSage™: (d) BF slag, (e) BOF slag, (f) Waste glass.

Through adiabatic calculations, the impact of waste glass addition on the temperature of BF slag during tapping was assessed and the melting temperature of these mixtures was also determined. The findings are presented in Figure 4. It should be noted that waste glass can introduce a certain amount of carbon, as it may contain organic oil or plastic. The oxidation of this carbon contributes additional heat. The addition of waste glass results in a gradual decrease in the temperature when waste glass containing 0 per cent carbon and more than 9.4 kg of waste glass led to a temperature less than the liquids (Figure 4a). Conversely, adding more than 50 kg of waste glass with 5 per cent carbon into slag content keeps the mixture in a liquid state (Figure 4b). As shown in Figure 4c, when the temperature is reduced to 1350°C, the viscosity of the mixture with 30 per cent added glass remains below 2 Pa·s, maintaining a good fluidity. For BOF slag, due to the oxidation of FeO and 0 per cent carbon, the slag temperature under adiabatic conditions will exceed 2000°C, as shown in Figure 5. The temperature of the mixture is still higher than the melting temperature when the amount of waste glass added exceeds 50 kg. It is worth noting that heat loss is not considered here. Since the BOF slag possesses a higher basicity compared to the BF slag, its viscosity is more favourable when the same quantity of waste glass is added. Consequently, maintaining fluidity is less challenging with BOF slag.

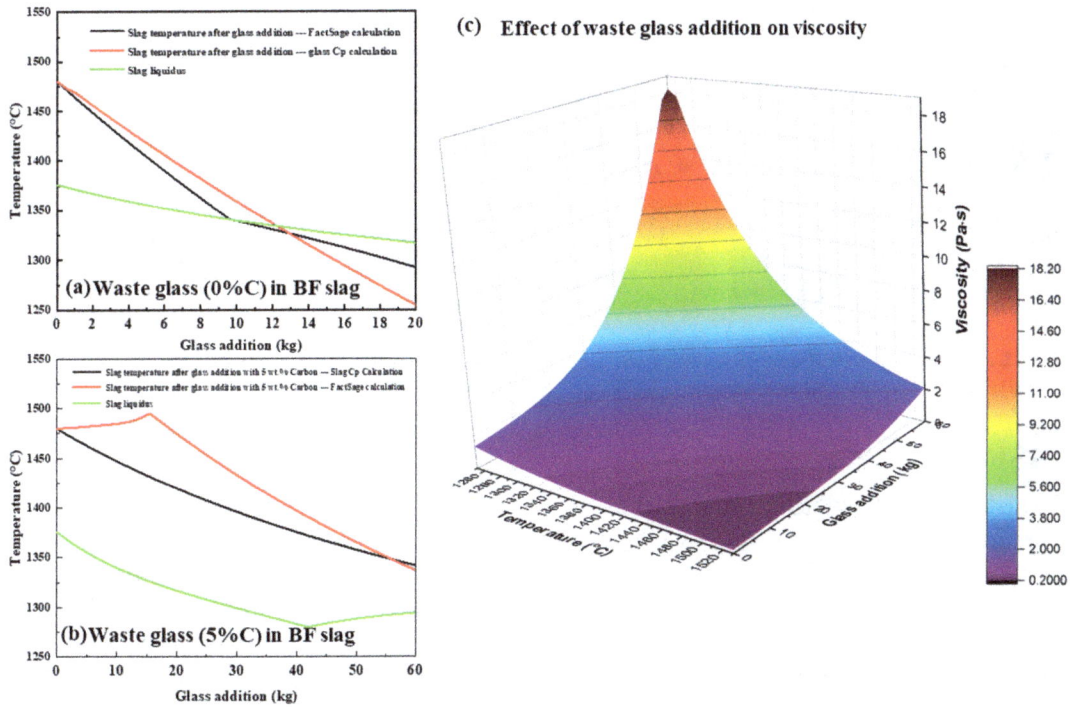

FIG 4 – The theoretical addition of waste glass with different carbon content into BF slag during tapping: (a) 0 per cent Carbon, (b) 5 per cent carbon, and the effect of temperature and waste glass addition on the viscosity.

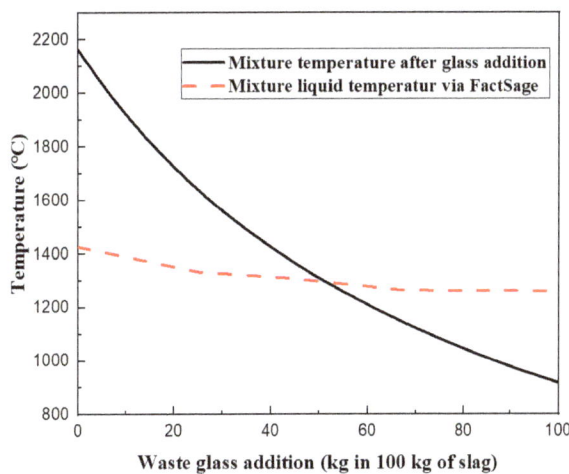

FIG 5 – The theoretical addition of waste glass with 0 per cent carbon into BOF slag during tapping.

Dissolution of waste glass in molten slag

Figure 6 displays the SEM-EDS results illustrating the cross-section microstructure of waste glass dissolution in BF slag at 1400°C and 1450°C for 60 sec. The top surface layer of the slag is identified as glass, characterised by a high concentration of sodium and silicon, as revealed by EDS analysis in Figure 6a. At a temperature of 1400°C, the glass has completely melted, and the liquid glass layer begins to dissolve, but the dissolution is not complete. Remarkably, when the temperature is increased to 1450°C, the glass layer in the BF slag can be completely dissolved in less than 60 sec. Since no apparent stratification is evident and the distribution of sodium is uniform throughout, it indicates a homogenous composition in the slag layer. Due to the high SiO_2 content in waste glass, a lower SiO_2 concentration in the slag leads to a stronger driving force for dissolution, accelerating the dissolution rate. Additionally, SiO_2 dissolution is faster at higher temperatures and lower viscosities. Consequently, at same temperature, waste glass dissolves faster in BOF slag compared to BF slag. The melting and dissolution processes are further enhanced by stirring during the addition of waste glass.

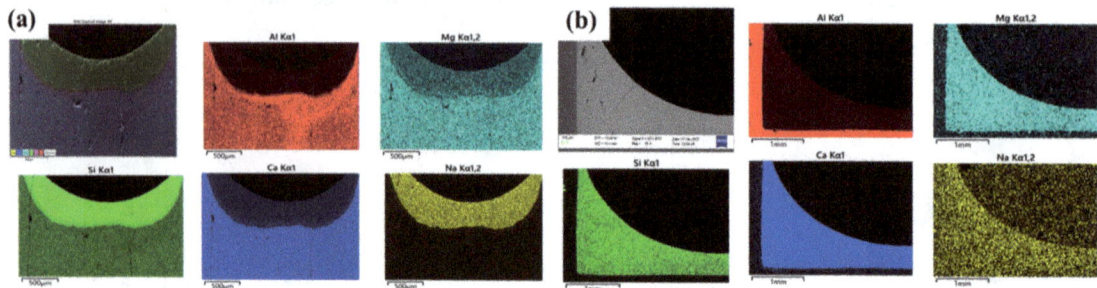

FIG 6 – Dissolution of waste glass in BF slag at different temperature for 60 sec: (a) 1400°C, and (b) 1450°C.

DISCUSSION

Integrating hard-to-use waste glass into the metallurgical slag during the tapping process promises to achieve the multiple purposes of waste utilisation, energy saving and waste heat recovery, as schematically illustrated in Figure 7. This process utilises the sensible heat of metallurgical slag and the heat of reaction from the oxidation of iron and organics, mixing hard-to-use high-silica waste with molten slag to form new raw materials suitable for glass manufacturing or mineral-wool production. The application of the new material after mixing is determined based on the compositional requirements of glass and mineral-wool. According to thermodynamic calculations of melting performance, fluidity, and solubility experiment results, for BF slag, if there is 5 per cent carbon in the waste glass, the addition of waste glass is about 30 per cent which still meets the requirements for slag transportation and tapping. For BOF slag, the addition of waste glass can exceed 50 per cent due to the high tapping temperature, high basicity and high FeO content. Depending on the composition of the new material and the final composition of the expected products, glass or mineral wool is prepared by adding other raw materials. Generally speaking, the new material obtained by BF is suitable for glass manufacturing, while the new material obtained by BOF is suitable for mineral wool. Two common approaches to utilising these materials will be explored below.

FIG 7 – Conversion of wastes to new raw materials for glass and mineral-wool manufacturing.

Based on a typical container glass composition (Shelby, 2020), a potential formula has been proposed as shown in Table 2. This recipe allows for the use of only 20 per cent new raw materials in the manufacturing of container glass. As a result, about 30 per cent of waste glass and 15 per cent BF slag are used in this glass production. By incorporating this new material, there is a notable reduction in lime consumption, which consequently leads to a decrease in CO_2 emissions. The new raw material contains sulfur which is a powerful refining agent and benefits glass manufacturing (Hujova and Vernerova, 2017). Additionally, the new material serves as a replacement for nepheline syenite and feldspar, which are the primary sources of alumina in traditional glassmaking. This substitution not only utilises waste materials but also maintains the necessary chemical composition for high-quality glass production.

TABLE 2

A potential recipe for soda-lime-silica (container glass) using new material, wt per cent.

	Ratio	CaO	SiO_2	MgO	Al_2O_3	SO_3	K_2O	TiO_2	MnO	Fe_2O_3	Na_2O
New material 1: 50 kg waste glass in 100 kg BF slag	20	32.12	47.58	5.84	8.31	0.71	0.56	0.45	0.26	0.28	4.00
SiO_2 sand	46	0.28	98.2	0.03	0.28	0.07	0.01	0	0	0	0.03
Soda ash	16	0	0	0	0	0	0	0	0	0	58.49
Waste glass (good quality)	23	11.25	72	1.15	1.5	0.25	0.5	0	0	0.05	13.3
Lime	2	92.01	1.51	0	0.93	0	0	0	0	0.11	0.00
Proposed glass composition		10.67	70.78	1.30	1.96	0.22	0.22	0.08	0.05	0.06	14.04
Typical container glass composition		10–12	70–73	0.5–2	0–2	0–0.5	0–1	0–0.1	/	0–1	12–15

Referring to the composition ranges of mineral-wool documented in the literature (Yliniemi et al, 2021), a potential recipe was proposed, as shown in Table 3. This formulation permits the incorporation of approximately 60 per cent new raw materials. Consequently, in producing this specific type of mineral-wool, around 20 per cent of the raw materials are derived from waste glass and another 40 per cent from BOF slag. In the laboratory experiments, analytically pure reagents (SiO_2, Al_2O_3 and MgO) were used for replacing silica sand, bauxite and dolime. These were mixed with BOF slag and waste glass to create a mixture with a similar composition to the recipe. Then this mixture was used for both the measurement of the melting temperature and the preparation of mineral-wool and the results are presented in Figure 8. The Lab-made mixture shows a melting temperature at approximately 1160°C and its fluidity temperature is around 1225°C. The diameter of fibres is a crucial determinant of the properties of mineral-wool products, affecting characteristics such as chemical stability and sound absorption (He et al, 2023; Liu et al, 2013). The mineral-wool produced has fibres measuring approximately 10–20 µm in diameter, which are over twice as thick as those found in commercial rock wool products. A high-speed, four-roller centrifuge will be utilised to produce wool fibres and will be investigated in future work.

TABLE 3

TABLE 3

A potential recipe for mineral-wool using new material, wt per cent.

	Ratio	CaO	SiO$_2$	MgO	Al$_2$O$_3$	K$_2$O	TiO$_2$	MnO	Fe$_2$O$_3$	Na$_2$O
New material 2: 50 kg waste glass in 100 kg BOF slag	60	25.8	45.45	3.2	3.48	0.25	0.33	1.06	14.57	4.66
SiO$_2$ sand	15	0.28	98.2	0.03	0.28	0.01	0	0	0	0.03
Bauxite	15	0	18.68	0	56.29	0	2.33	0	1.29	0
Dolime	10	54.75	3.55	40.1	0.33	0.08	0.05	0.03	0.38	0.02
Proposed material composition		21.00	45.16	5.93	10.61	0.16	0.55	0.64	8.97	2.80
Mineral-wool composition from literature		10–25	41–53	6–16	6–14	0–2	0–3.5	0–1.1	0–13	0–6

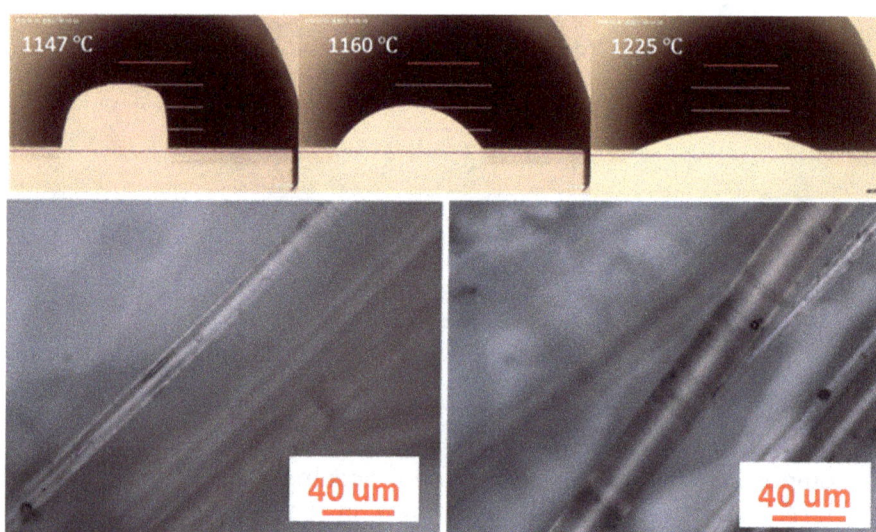

FIG 8 – Height changes of Lab sample (a), and the morphology of obtained mineral-wool (b).

When the raw materials are used directly in their molten state for production, it can significantly enhance the melting rates, reduce the batch-free times and decrease the bubble content in the final product. In this case, the heat from slag could be recovered directly. By directly using the sensible heat and exothermic reactions of slag, there is a potential energy saving of 0.16 GJ per 100 kg of new material with 1350°C used. This translates to a reduction in standard coal consumption by 7 kg and cuts carbon dioxide emissions by 26 kg. When considering the energy used for melting natural raw materials, the reduction in CO$_2$ emissions is further amplified by replacing natural raw material with these new raw materials. This approach not only contributes to lowering greenhouse gas emissions but also promotes the recycling of waste glass, aligning with sustainable waste management practices and environmental conservation efforts.

CONCLUSIONS

- This study introduces an innovative approach for recovering the heat from high-temperature ironmaking and steelmaking slags, which are typically discarded and release heat into the environment during the tapping process. This heat is instead utilised to recycle hard-to-use waste glass in producing raw materials directly for glass and mineral-wool manufacturing.

- Thermodynamic assessments indicate that for BF slag with 5 per cent carbon in waste glass, up to 30 per cent waste glass can be added without affecting transport and tapping requirements. For BOF slag, over 30 per cent waste glass even with 0 per cent carbon content addition is feasible owing to its higher tapping temperature, basicity and FeO content.

- New recipes for container glass and mineral-wool have been suggested, based on the typical products composition ranges. Utilising the sensible heat and reaction heat of slag could reduce coal consumption by 7 kg and CO_2 emissions by 26 kg for every 100 kg of the new material.

ACKNOWLEDGEMENTS

The authors wish to acknowledge the Transforming Foundation Industries Network+ (EPSRC grant EP/V026402/1) for funding this work.

REFERENCES

Deng, W, Backhouse, D J, Kazi, F K, Janani, R, Holcroft, C, Magallanes, M, Marshall, M, Jackson, C M and Bingham, P A, 2023. Alternative raw material research for decarbonization of UK glass manufacture, *International Journal of Applied Glass Science*, 14(3):341–365.

Dewick, P and Miozzo, M, 2002. Sustainable technologies and the innovation–regulation paradox, *Futures*, 34(9–10):823–840.

He, W-C, Luo, M-S, Deng, Y, Qin, Y-L, Zhang, S, Lv, X-W, Zhao, Y, Jiang, C-Z and Pang, Z-D, 2023. Preparation of high acidity coefficient slag wool fiber with blast furnace slag and modifying agents, *J Iron Steel Res Int*, 30(7):1440–1450.

He, W-C, Lv, X-W, Yan, Z-M and Fan, G-Q, 2020. Ferrosilicon alloy granules prepared through centrifugal granulation process, *J Iron Steel Res Int*, 27(11):1–12.

Hujova, M and Vernerova, M, 2017. Influence of fining agents on glass melting: a review, part 1, *Ceramics-Silikaty*, 61(2):119–126.

Iacovidou, E, Hahladakis, J, Deans, I, Velis, C and Purnell, P, 2018. Technical properties of biomass and solid recovered fuel (SRF) co-fired with coal: Impact on multi-dimensional resource recovery value, *Waste Management*, 73:535–545.

Kiss, B, Manchón, C G and Neij, L, 2013. The role of policy instruments in supporting the development of mineral-wool insulation in Germany, Sweden and the United Kingdom, *J Clean Prod*, 48:187–199.

Liu, J, Jiang, M, Wang, Y, Wu, G and Wu, Z, 2013. Tensile behaviors of ECR-glass and high strength glass fibers after NaOH treatment, *Ceramics International*, 39(8):9173–9178.

Oge, M, Ozkan, D, Celik, M B, Gok, M S and Karaoglanli, A C, 2019. An overview of utilization of blast furnace and steelmaking slag in various applications, *Materials Today: Proceedings*, 11:516–525.

Pan, S-Y, Adhikari, R, Chen, Y-H, Li, P and Chiang, P-C, 2016. Integrated and innovative steel slag utilization for iron reclamation, green material production and CO_2 fixation via accelerated carbonation, *J Clean Prod*, 137:617–631.

Piatak, N M, 2018. Environmental characteristics and utilization potential of metallurgical slag, in *Environmental Geochemistry*, pp 487–519 (Elsevier).

Piatak, N M, Parsons, M B and Seal, R R, 2015. Characteristics and environmental aspects of slag: A review, *Applied Geochemistry*, 57:236–266.

Schmitz, A, Kamiński, J, Scalet, B M and Soria, A, 2011. Energy consumption and CO_2 emissions of the European glass industry, *Energy Policy*, 39(1):142–155.

Shelby, J E, 2020. *Introduction to Glass Science and Technology* (Royal Society of Chemistry).

Springer, C and Hasanbeigi, A, 2017. Emerging energy efficiency and carbon dioxide emissions-reduction technologies for the glass industry, Energy Analysis and Environmental Impacts Division, Lawrence Berkeley National Laboratory: Berkeley, CA, USA.

Yan, Z, Lv, X, Liang, D, Zhang, J and Bai, C, 2016. Transition of Blast Furnace Slag from Silicates-Based to Aluminates-Based: Viscosity, *Metall and Materi Trans B*, 48(2):1092–1099.

Yan, Z, Reddy, R G and Lv, X, 2019. Structure Based Viscosity Model for Aluminosilicate Slag, *ISIJ International*, 59(6):1018–1026.

Yliniemi, J, Ramaswamy, R, Luukkonen, T, Laitinen, O, de Sousa, Á N, Huuhtanen, M and Illikainen, M, 2021. Characterization of mineral-wool waste chemical composition, organic resin content and fiber dimensions: Aspects for valorization, *Waste Management*, 131:323–330.

Yue, H, He, Z, Jiang, T, Duan, P and Xue, X, 2018. Rheological evolution of Ti-bearing slag with different volume fractions of TiN, *Metall Materi Trans B*, 49(4):2118–2127.

Zhao, G, Zhang, L and Cang, D, 2019. Fundamental and industrial investigation on preparation of high acidity coefficient steel slag derived slag wool, *Journal of the Ceramic Society of Japan*, 127(3):180–185.

Research on –
Refractory-melt interactions

Radical involved reaction and weak magnetic effect between alumina refractory and high alumina slags

A Huang[1], S H Li[2] and H Z Gu[3]

1. Professor, the State Key Laboratory of Refractories and Metallurgy, Wuhan 430081, China. Email: huangao@wust.edu.cn
2. Doctor, the State Key Laboratory of Refractories and Metallurgy, Wuhan 430081, China. Email: lishenghao@wust.edu.cn
3. Professor, the State Key Laboratory of Refractories and Metallurgy, Wuhan 430081, China. Email: guhuazhi@wust.edu.cn

ABSTRACT

High aluminium steel is a promising advanced steel known for its exception properties and holds significant economic and strategic importance. However, the CaO-Al_2O_3-SiO_2 based slag with high alumina content reacts vigorously with alumina refractories during the smelting process at high temperature, seriously affecting the safe and efficient production of furnaces and resulting in the formation of Al_2O_3-MgO non-metallic inclusions within the molten steel. This is a bottleneck problem for high-quality steel refining, and the key reaction path during the corrosion remains unclear. The dissolution reaction mechanism of alumina refractories in high alumina CaO-Al_2O_3-SiO_2 slags under a weak static magnetic field was investigated in this work, using high temperature laser confocal microscopy and *in situ* radiation spectroscopy techniques. The results indicate that the interaction between the high alumina slags and the alumina refractories is governed by free radical reactions. The dissolution of alumina forms AlO and AlO_2 radicals. The reaction rate increases with the C/S ratio of the slag due to its low polymerisation. The reaction between non-bridging oxygen and AlO_2 free radicals accelerates the generation of AlO and O_2^- radicals. The reaction layer composed of calcium aluminate is formed through the recombination of AlO, Ca^+ and O_2^- radicals. The weak static magnetic field induces the Zeeman splitting of free radical pairs and intersystem crossing occurs, promoting the formation of triplet free radical pairs through the hyper-fine coupling effect. The reaction can be significantly inhibited as the triplet free radical pair does not meet Pauli's incompatibility principle. This discovery perfects the structure theory of molten slags with radicals and provides supplement to the design of the highly slag-resistant refractories, and provides the theoretical basis for the development of external field protection technologies for high-quality steel refining.

INTRODUCTION

High aluminium steel, distinguished by its remarkable strength, plasticity, high strain hardening, and energy absorption capacity, shows broad prospects in applications in national defence, military industries and transportation sectors (Zuazo *et al*, 2014; Zhang *et al*, 2015). The use of CaO-Al_2O_3-SiO_2 and CaO-Al_2O_3 slags are widely used for high aluminium steel refining (He *et al*, 2020; Zhang and Wang, 2002; Zhao *et al*, 2021; Tang *et al*, 2017; Wang, D *et al*, 2012; Wang, W *et al*, 2019), and the high alumina steel can cause the formation of high alumina content slag, which can cause continuous and periodic corrosion patterns at the reaction interface of the slag and the corundum refractory, resulting in rapid dissolution and damage to the refractory (Zou *et al*, 2020a). The periodicity of corrosion patterns is unrelated to the spatiotemporal evolution, but its geometric parameters was highly correlated with the C/S ratio and the dissolution rate of aluminium (Li *et al*, 2021). Free radicals were discovered in the CaO-Al_2O_3-SiO_2 based slag, and the recombination of AlO_4 and SiO_4 tetrahedra in the slag of higher C/S ratio is one of the causes for the formation of radicals. O_2^- and Ca^+ radicals involved in the interface reaction between the corundum refractory and the slag, the Turing pattern corrosion at the interface of alumina ceramic is caused by an increase in O_2^- radical content in the slag (Li *et al*, 2023; Peng *et al*, 2023a, 2023b). This not only affects the safe and efficient operation of furnaces (Dong *et al*, 2022; Jiang *et al*, 2022), but also leads to the formation of non-metallic inclusions in the steel, which is one of the key bottlenecks restricting the development of high-quality steel smelting (Wang, Song and Xue, 2023; Liu *et al*, 2023; Zhou *et al*, 2015; Tsuda *et al*, 1992).

With the development of modern metallurgical technology, electric field and magnetic field have been widely used in metallurgical processes such as electroslag remelting, electromagnetic induction melting, electromagnetic stirring and electromagnetic continuous casting. The physical parameters of slag and the service life of refractories can be affected by external high-intensity electric and magnetic fields. The alternating magnetic field promotes the interface reaction between slag and refractories, accelerating the corrosion and penetration of slag on the refractories (Aneziris *et al*, 2013; Ren *et al*, 2019, 2021), whereas the static magnetic field has the effect of enhancing the corrosion degree of slag on refractory (Huang *et al*, 2018).

An 8.5 mT static magnetic field can induce an 'electromagnetic damping effect' on the slag, leading to a reduction in wettability between the slag and the refractory (Zou *et al*, 2020b). However, the physical effects of the magnetic field, such as electromagnetic damping and polarisation of the reactive group, can only be fully utilised when the magnetic flux density and frequency are high enough to improve the slag resistance of the refractory (Bian *et al*, 2018). Different from the electric field, high intensity magnetic field or alternating magnetic field, it was also found that the weak static magnetic field with a millitesla magnetic flux density has the potential to improve the slag resistance of refractories. The weak static magnetic field of 5 to 10 Gs can influence the chemical reactions between slag and alumina-based refractories, and delaying the occurrence of interface Turing pattern corrosion (Li *et al*, 2022, 2023). Weak static magnetic field has the potential to inhibit the chemical reactions between slag and alumina refractories, offering possibilities for enhancing the slag resistance of the refractory.

However, the interaction between alumina refractories and high alumina slag is rather complex. The lack of information during corrosion at high temperatures limits the exploration of reaction mechanisms (Zhang *et al*, 2021; Ponomar *et al*, 2022). In this work, the dissolution reaction mechanism of alumina refractory in high alumina $CaO-Al_2O_3-SiO_2$ slags under a weak static magnetic field was clarified, and the key reaction path and magnetic field inhibition mechanism were revealed based on high temperature laser confocal microscopy and *in situ* radiation spectroscopy techniques.

EXPERIMENTAL

Al_2O_3 powder (analytical purity, $Al_2O_3 \geq 99$ wt per cent, Shanghai Macklin Biochemical Co., Ltd), CaO powder (analytical purity, $CaO \geq 98$ wt per cent, Shanghai Macklin Biochemical Co., Ltd), and SiO_2 powder (analytical purity, $SiO_2 \geq 98$ wt per cent, Sinopharm Chemical Reagent Co., Ltd) were weighed according to the mass ratios presented in Table 1. The powder were mixed via ball milling at a speed of 60 rev/min for 30 mins. Afterward, the mixed power were pre-melted in a high-frequency induction furnace (Supersonic Frequency Induction Heating Equipment, Hubei Changjiang Precision Manufacturing Technology Co., Ltd). The molten slag was then poured onto a copper plate floating above circulating water to prevent the molten slag from segregating and crystallising. Subsequently, the glassy molten was ground in an agate mortar. The alumina balls (Al_2O_3: 95 wt per cent, SiO_2: 4.5 wt per cent, CaO: 0.5 wt per cent, $\rho = 3.8$ g/cm^3) with a diameter of 1.20 ± 0.05 mm and a mass of 3.45 ± 0.05 mg were selected as alumina refractories.

TABLE 1

Composition of the pre-melted powders (wt%).

No.	CaO	Al$_2$O$_3$	SiO$_2$	C/S ratio
S1	52.00	35.00	13.00	4
S2	54.17	35.00	10.82	5
S3	55.71	35.00	9.29	6

A small amount, 0.2 g, of pre-melted slag was compresses into a platinum crucible measured 8 mm in diameter and 4 mm in height. The crucible was placed on the high temperature laser confocal microscope (HT-CSLM, VL2000DX-SVF18SP, Mikura Seisakusho Co., Ltd., Japan) and heated to 1600°C in an argon atmosphere (the heating ratio referred to Li *et al*, 2022). After soaking at 1600°C for 180 sec, the alumina ball was fed into the slag by using the high-temperature anti-oxidation feeding system (GC-500901, Mikura Seisakusho Co., Ltd.). A pair of magnets as shown in Figure 1, applied a static magnetic field throughout the experiment and it could be adjusted by changing the number of magnets. The flux density of static magnetic in the reaction region was calibrated using a Tesla meter (Changsha Tianheng Measurement and Control Technology Co., Ltd.). The magnetic flux density for each sample is shown in Table 2.

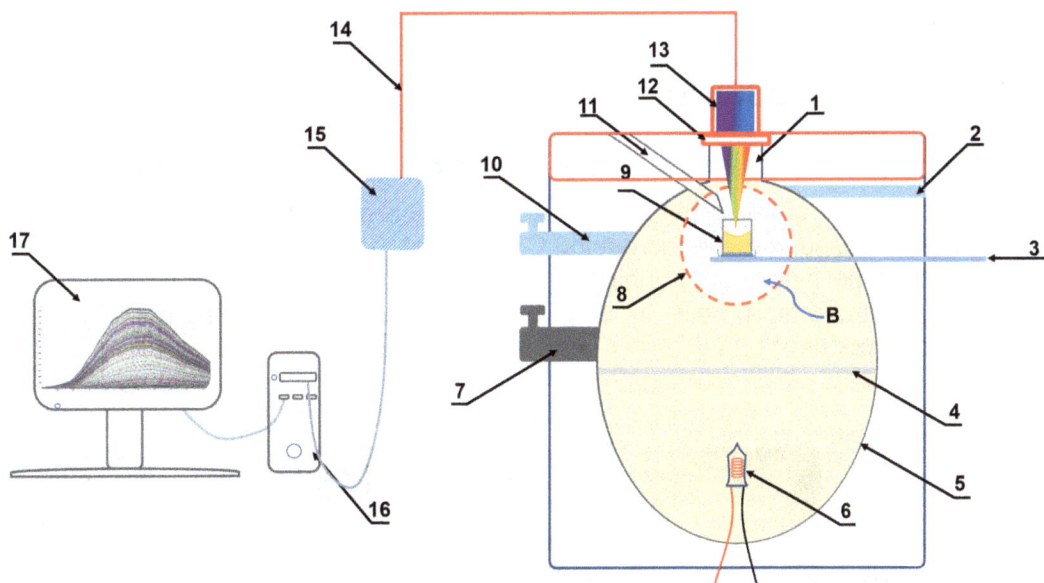

1-Quartz window; 2-Gas outlet valve; 3-Thermocouple with platinum stent; 4-Quartz spacer; 5-Furnace chamber; 6-Halogen lamp
7-Vacuum valve; 8-NdFeB magnet; 9-Pt crucible; 10-Intake valve; 11-oxidation protection feeder; 12-Filter group; 13-Cosine corrector
14- Fibre-optical; 15-Spectrograph; 16-Computing system; 17-Signal output system

FIG 1 – Schematic diagram of the device.

TABLE 2

Magnetic flux density produced in the experimental area.

No.	Magnetic flux density/Gs	
S1	0	15
S2	0	10
S3	0	10

The *in situ* radiation spectroscopic system is composed of a spectrograph (Avenir GmbH, Germany) and a filter group. The radiation passing through the filter group, and the transmitted through a cosine corrector (Avenir GmbH, Germany) and finally analysed by the spectrograph, as shown in Figure 1. The radiation spectrum and the morphology of alumina refractory in dissolution process were recorded by the device.

The quenched samples were longitudinally cut, and polished and gold-plated using vacuum ion sputtering, the microstructure and composition at the interface between slag and alumina balls were analysed by scanning electron microscope (SEM, JSM-6610, JEOL, Tokyo, Japan) and energy dispersive X-Ray spectroscopy (EDX, QUANTAX, Bruker, Berlin, Germany). The synthetic slag was ground into micropowder (0.074 mm), and the superoxide radicals were analysed using electron paramagnetic resonance spectrometer (EPR, A300, Bruker, Saarbrücken, Germany) with DMPO capture agent in a non-light environment.

The average dissolution rate of alumina balls is determined by (Huo *et al*, 2022):

$$\bar{v} = \frac{\Delta W}{\bar{A} \cdot t} \tag{1}$$

$$\bar{A} = \pi D^2 \tag{2}$$

where:

$\triangle W$	is the mass change of the alumina ball, g
$\overline{A}$	is the average superficial area of the alumina ball, cm^2
D	is the diameter of the alumina ball, cm
t	is the dissolution time, sec.

RESULTS AND DISCUSSION

Weak magnetic field effect on the dissolution of alumina balls

The rapid dissolution of alumina balls in the high alumina content slag, refer to Figure 2, and the dissolution rate increased with the increasing C/S ratio of the slag, as shown in Figure 3. However, the dissolution can be significantly inhibited by the external weak static magnetic field, and the dissolution rates of alumina balls in slag S1~S3 experienced a decrease of approximately 20.3 per cent, 27.5 per cent and 35.0 per cent, respectively.

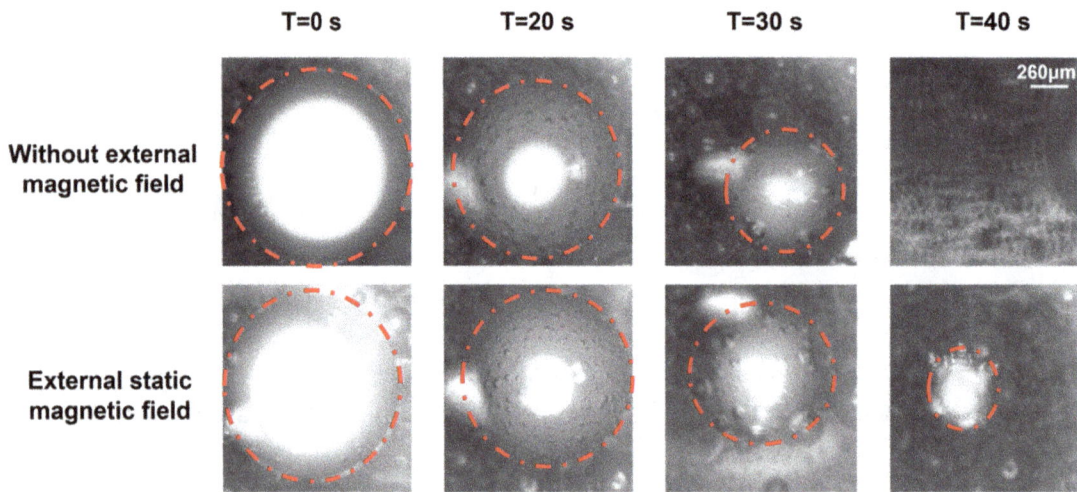

FIG 2 – The recorded picture of the dissolution process of alumina balls in slag at 1600°C.

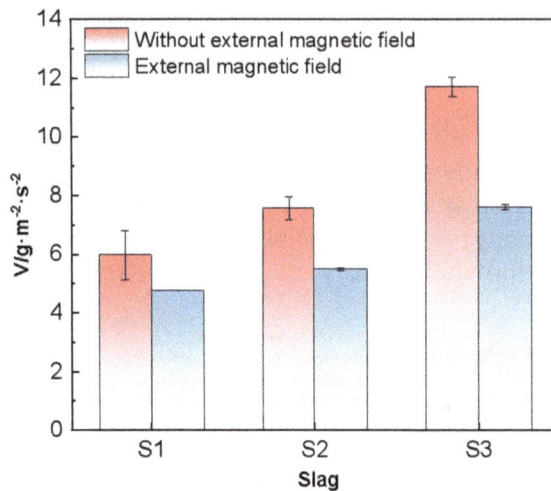

FIG 3 – Dissolution rate of alumina balls in slags.

As an illustration, S3 slag was utilised to investigate, the microstructure of the reaction interface between the alumina balls and the slags during the dissolution was analysed, as shown in Figure 4. A reaction layer composed of calcium aluminate with a thickness of 76.32 μm was observed at the reaction interface as shown in Figure 4 and Table 3, when an external weak static magnetic field was applied the thickness of the layer was reduced to 41.25 μm.

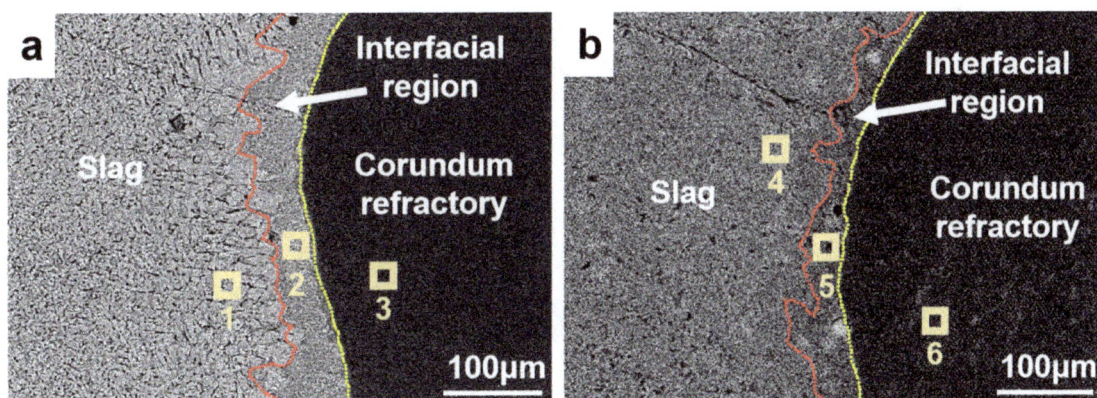

FIG 4 – Microstructure of the reaction interface between the alumina ball and melts at 1600°C in 20 sec, (a) without static magnetic field, (b) with weak static magnetic field.

TABLE 3

Energy dispersive spectroscopy (EDS) analysis results (at %).

Area	Ca	Al	Si	O
Area 1	-	45.76	-	54.24
Area 2	23.35	18.11	2.43	56.12
Area 3	25.87	16.54	2.19	55.40
Area 4	-	45.50	-	54.50
Area 5	24.07	19.29	2.62	54.02
Area 6	28.07	13.38	1.73	56.82

Radical reaction mechanism in the dissolution of alumina balls

The concentration of superoxide radicals in the slags increased with the increasing C/S ratio, as shown by the stronger signals in Figure 5a. Additionally, the concentration of superoxide radicals in the slags gradually increased with the dissolution of alumina balls, as shown in Figure 5b.

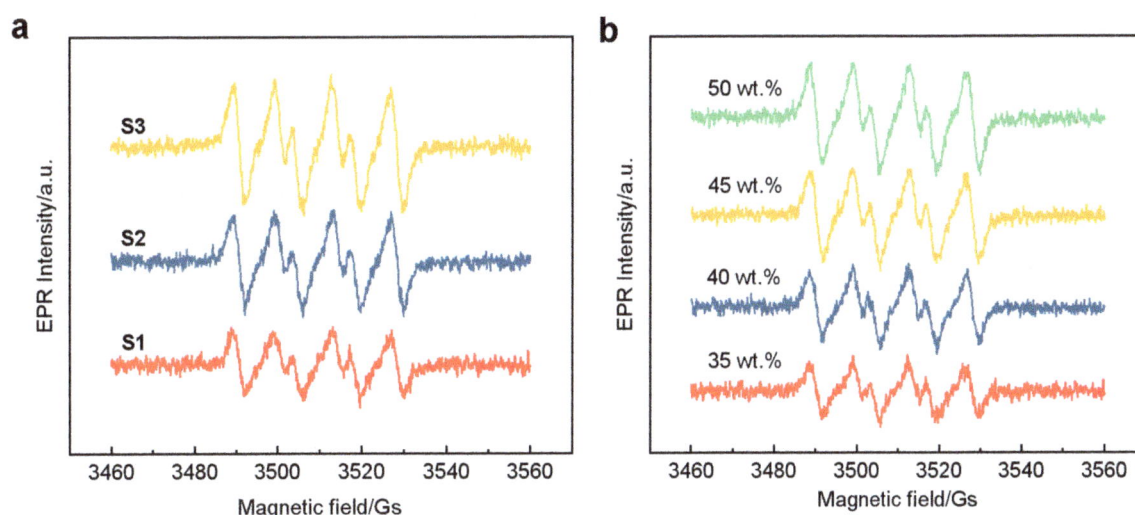

FIG 5 – (a) EPR spectra of S1~S3 slag; (b) EPR spectra of S3 slag with the increase of alumina concentration.

Comparing the radiation spectra before and during the reaction at 1600°C, a signal at 484.21 nm was detected (as shown in Figure 6a), indicating the formation of AlO radical (Varenne *et al*, 2000; Starik *et al*, 2014). The signal intensity gradually increased with the increase in the C/S ratio, indicating a higher formation of AlO radical. The application of an external static magnetic field inhibited the generation of AlO radical as shown in Figure 6b.

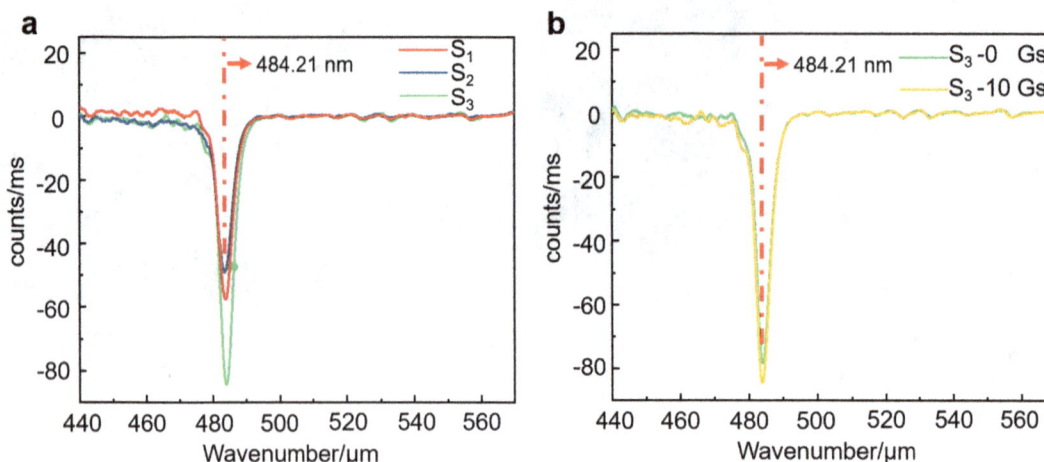

FIG 6 – *In situ* radiation spectrum during the alumina dissolution, (a) in S1~S3 slags, (b) in S3 slag with and without external static magnetic field.

The reaction path between the alumina refractory and the slag is as Equations 3 and 4. The AlO and AlO_2 radicals can be generated by the homolysis of Al_2O_3 in molten slag, the non-bridging oxygen can be attacked by AlO_2 radicals, which accelerates the dissolution of alumina refractory and generates $O_2^{\cdot-}$:

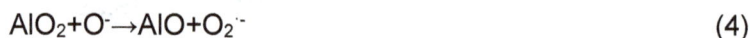

$$Al_2O_3 \rightarrow AlO + AlO_2 \tag{3}$$

$$AlO_2 + O^- \rightarrow AlO + O_2^{\cdot-} \tag{4}$$

Non-bridging oxygen (O^-) in slag plays a crucial role. In slags characterised by a higher (C/S) ratio, a greater concentration of non-bridging oxygen, as indicated by Li *et al* (2023), facilitates Equation 4 and resulting in the rapid dissolution of alumina refractory.

According to our previous investigations (Li *et al*, 2021, 2022, 2023; Peng *et al*, 2023a, 2023b) have indicated that $O_2^{\cdot-}$ and $Ca^{\cdot+}$ are formed spontaneously through the uniform decomposition of CaO_2 in the CaO-Al_2O_3-SiO_2 slag. The reaction layer composed of calcium aluminate can be formed through the recombination of AlO, $Ca^{\cdot+}$ and $O_2^{\cdot-}$ radicals at the reaction interface as shown in Equation 5 and Figure 7.

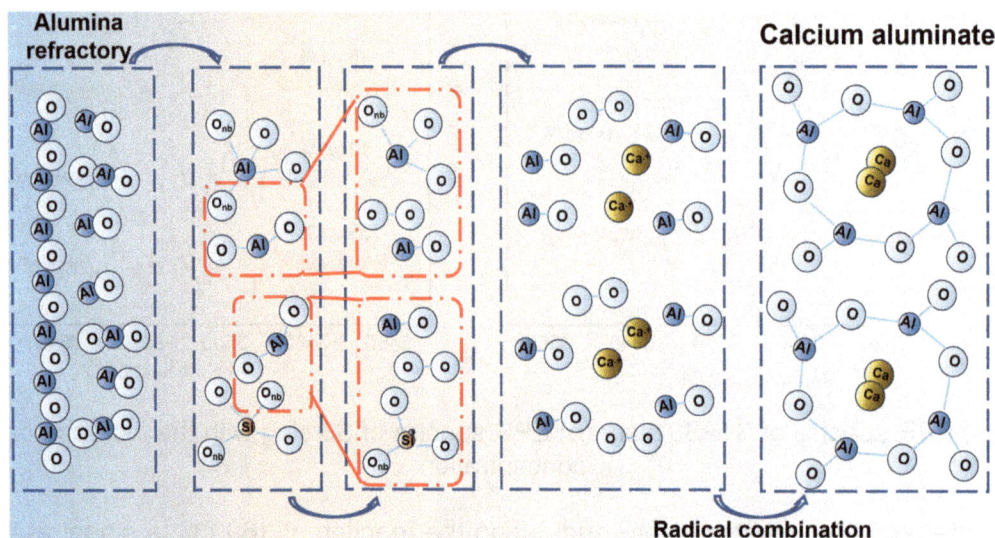

$$2AlO + Ca^{\cdot+} + O_2^{\cdot-} \rightarrow CaAl_2O_4 \tag{5}$$

FIG 7 – Possible radical reaction mechanism at the interface between the alumina refractory and the molten slags.

The energy formula for singlet and triplet state free radical pairs with the external static magnetic field is given by (Zimmt, Doubleday and Turro, 1985):

$$E(S) = \langle S, \chi_N | H_{ex} + H_{mag} | S, \chi_N \rangle = J \tag{6}$$

$$E(T_n) = \langle T_n, \chi_N | H_{ex} + H_{mag} | T_n, \chi_N \rangle = -J + ng\mu_B B + \frac{n}{2}(\sum_i^j a_i m_i + \sum_k^l a_k m_k) \tag{7}$$

where:

a_i is the hyperfine coupling constant of the i atom

m_i is the spin quantum number of the i nucleus

According to Equations 6 and 7, it is clear that the Zeeman splitting (represented by $g\mu_B \cdot B$) of free radical pairs induced by the static magnetic field is the main factor affecting the energy of free radicals in the singlet and triplet states. When an external magnetic field with the appropriate flux density is applied and the Zeeman splitting energy matches the energy difference of the free radical pairs, the intersystem crossing occurs under the hyperfine coupling effect as shown in Figure 8. The transformation of free radical pairs from singlet to triplet states (Sarpoolaky, Zhang and Lee, 2003; Hayashi, 2004) results in a reduction of the bonding probability of free radicals, thereby inhibiting chemical reactions of Equations 4 and 5, and decreasing the interfacial reaction and the dissolution rate of alumina balls in the slags.

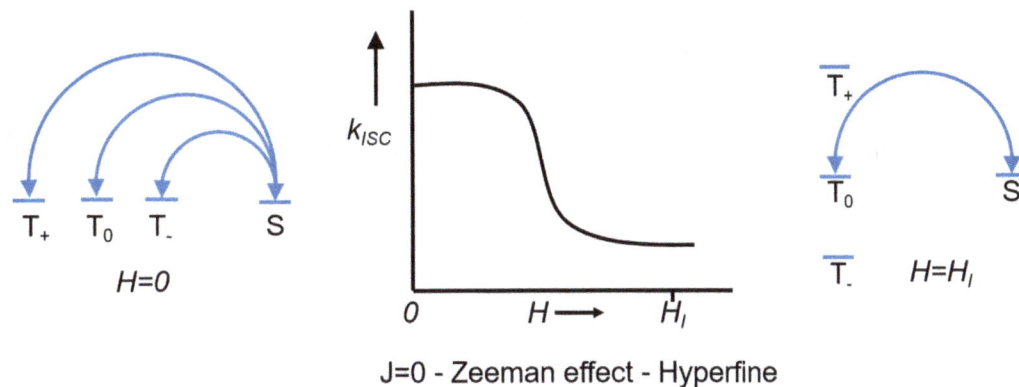

J=0 - Zeeman effect - Hyperfine

FIG 8 – Effect of magnetic field on the radical intersystem crossing (Hayashi, 2004).

CONCLUSIONS

The dissolution mechanisation of the alumina refractory in high alumina content slag was clarified based on high temperature laser confocal microscopy combined with *in situ* radiation spectroscopy techniques. The radical reaction path and the effect of weak static magnetic field during the dissolution of alumina refractory was revealed. The main conclusions are as follows.

- Alumina refractory involved in the intense interfacial radical reaction with high alumina content slag, resulting in the rapid dissolution of refractory and the formation of a calcium aluminate reaction layer on the refractory/slag interface. Free radical splays a crucial role in the process, however when a static magnetic field is applied, the reaction rate decreases significantly and the thickness of the calcium aluminate reaction layer. Taking S3 slag as an example, the dissolution rate of the alumina ball decreased by 35.0 per cent, and the thickness of the interface reaction layer decreased by 46.0 per cent with external the static magnetic field.

- The reaction path between alumina refractory and the slag was revealed. Free radicals including AlO, AlO_2, and O_2^- were generated during the dissolution of alumina refractory at 1600°C. The reaction layer composed of calcium aluminate is formed through the recombination of AlO, Ca^+ and O_2^- radicals at the reaction interface. For the slag with a higher C/S ratio, the non-bridging oxygen was attacked by AlO_2 radicals, thereby accelerating the dissolution rate of alumina refractory.

- The weak static magnetic field induces the Zeeman splitting of free radical pairs and enhances the hyperfine coupling, resulting in intersystem crossing from singlet to triplet states. The radical reactions are inhibited as the triplet free radical pair does not meet Pauli's incompatibility principle. This reduces the interfacial reaction rate between alumina refractories and slags. This work perfects the structure theory of molten slags with radicals and supports the development of external field protection techniques for carbon-free/low-carbon refractories, which is significant for the safe and efficient production of high-quality steel.

ACKNOWLEDGEMENTS

This work was financially supported by the National Natural Science Foundation of China (52272022), Key Program of Natural Science Foundation of Hubei Province (2021CFA071).

REFERENCES

Aneziris, C G, Schroeder, C, Emmel, M, Schmidt, G, Heller, H P and Berek, H, 2013. In situ observation of collision between exogenous and endogenous inclusions on steel melts for active steel filtration, *Metallurgical and Materials Transactions B*, 44:954–968.

Bian, Y, Ding, W, Hu, L, Ma, Z, Cheng, L, Zhang, R, Zhu, X, Tang, X, Dai, J, Bai, J, Sun, Y and Sheng, Z, 2018. Acceleration of Kirkendall effect processes in silicon nanospheres using magnetic fields, *Cryst Eng Comm*, 20(6):710–715.

Dong, J, Gu, H, Huang, A and Xue, Z, 2022. Corrosion behavior of high alumina and low silicon CaO-Al_2O_3 slags on corundum/calcium hexaaluminate based castables, *Journal of Iron and Steel Research,* 34(8):825–833.

Hayashi, H, 2004. *Introduction to dynamic spin chemistry: magnetic field effects on chemical and biochemical reactions*, pp 130–163 (World Scientific Publishing Company: Singapore).

He, S, Chen, Y, Pan, W, Wang, Q, Zhang, X and Wang, Q, 2020. Study on composition control for melting and flowing properties of CaO-Al_2O_3 based mold fluxes with low reactivity, *Journal of Iron and Steel Research*, 32(9):771–778.

Huang, A, Lian, P, Fu, L, Gu, H and Zou, Y, 2018. Modeling and experiment of slag corrosion on the lightweight alumina refractory with static magnetic field facing green metallurgy, *Journal of Mining and Metallurgy, Section B*: Metallurgy, 54(2):143–151.

Huo, Y, Gu, H, Huang, A, Ma, B, Chen, L, Li, G and Li, Y, 2022. Characterization and mechanism of dissolution behavior of Al_2O_3/MgO oxides in molten slags, *Journal of Iron and Steel Research International*, 29(11):1711–1722.

Jiang, X, Huang, A, Gu, H, Fu, L and Xue, Z, 2022. Effect of Al_2O_3 content in CaO-Al_2O_3-SiO_2 slag on corrosion resistance of alumina magnesia castable, *Journal of Iron and Steel Research*, 34(9):991–998.

Li, S, Huang, A, Gu, H, Zeng, F, Zou, Y and Fu, L, 2021. Visual measurement and characterisation of quasi-volcanic corrosion at alumina ceramic-oxides melt-air interface, *Journal of the European Ceramic Society*, 41(16):400–410.

Li, S, Huang, A, Gu, H, Wang, R, Zou, Y and Fu, L, 2022. Corrosion resistance and anti-reaction mechanism of Al_2O_3-based refractory ceramic under weak static magnetic field, *Journal of the American Ceramic Society*, 105(4):2869–2877.

Li, S, Huang, A, Jiang, T, Gu, H, Zeng, F, Wang, X and Zhang, S, 2023. Revealing of rich living radicals in oxide melts via weak magnetic effect on alumina dissolution reaction, *Journal of Molecular Liquids*, 375:121391.

Liu, D, Xue, Z and Song, S, 2023. Effect of cooling rate on non-metallic inclusion formation and precipitation and micro-segregation of Mn and Al in Fe-23Mn-10Al-0.7 C steel, *Journal of Materials Research and Technology*, 24:4967–4979.

Peng, Y, Huang, A, Li, S, Chen, X and Gu, H, 2023b. Radical reaction-induced Turing pattern corrosion of alumina refractory ceramics with CaO-Al_2O_3-SiO_2-MgO slags, *Journal of the European Ceramic Society*, 43(1):166–172.

Peng, Y, Li, S, Huang, A, Gu, H, Zou, Y and Fu, L, 2023a. Effect of slag basicity on turing pattern corrosion of alumina refractory ceramics in the presence of free radicals, *Journal of the American Ceramic Society*, 106(10):6211–6220.

Ponomar, V, Adesanya, E, Ohenoja, K and Illikainen, M, 2022. High-temperature performance of slag-based Fe-rich alkali-activated materials, *Cement and Concrete Research*, 161:106960.

Ren, X, Ma, B, Li, S, Li, H, Liu, G, Zhao, S, Yang, W, Qian, F and Yu, J, 2019. Slag corrosion characteristics of MgO-based refractories under vacuum electromagnetic field, *Journal of the Australian Ceramic Society*, 55:913–920.

Ren, X, Ma, B, Li, S, Li, H, Liu, G, Yang, W, Qian, F, Zhao, S and Yu, J, 2021. Comparison study of slag corrosion resistance of MgO-$MgAl_2O_4$, MgO-CaO and MgO-C refractories under electromagnetic field, *Journal of Iron and Steel Research International*, 28:38–45.

Sarpoolaky, H, Zhang, S and Lee, W, 2003. Corrosion of high alumina and near stoichiometric spinels in iron containing silicate slags, *Journal of European Ceramic Society*, 23(2):293–300.

Starik, A M, Kuleshov, P S, Sharipov, A S and Titova, N S, 2014. Kinetics of ignition and combustion in the Al–CH$_4$–O$_2$ system, *Energy & Fuels*, 28(10):6579–6588.

Tang, H, Wu, G, Wang, Y, Li, J, Lan, P and Zhang, J, 2017. Comparative evaluation investigation of slag corrosion on Al$_2$O$_3$ and MgO-Al$_2$O$_3$ refractories via experiments and thermodynamic simulations, *Ceramics International*, 43(18):16502–16511.

Tsuda, M, Yamaguchi, H, Kaneko, K, Moritani, T and Shinme, K, 1992. Production of ultra super purity ferritic stainless steel by the powder top blowing method under reduced pressure (VOD-PB), *Electric Furnace Conference Proceedings*, 50:259–262.

Varenne, O, Fournier, P G, Fournier, J, Bellaoui, B, Faké, A I, Rostas, J and Taïeb, G, 2000. Internal population distribution of the B state of AlO formed by fast ion beam bombardment or laser ablation of an Al$_2$O$_3$(Al) surface, *Nuclear Instruments and Methods in Physics Research Section B: Beam Interactions with Materials and Atoms*, 171(3):259–276.

Wang, D, Li, X, Wang, H, Mi, Y, Jiang, M and Zhang, Y, 2012. Dissolution rate and mechanism of solid MgO particles in synthetic ladle slags, *Journal of Non-Crystalline Solids*, 358(9):1196–1201.

Wang, J, Song, S and Xue, Z, 2023. Transient evolution of non-metallic inclusions in molten high aluminum and high manganese steel contacting with slag and crucible: experimental investigation and FactSage macros modeling, *Journal of Materials Research and Technology*, 25:2841–2853.

Wang, W, Xue, L, Zhang, T, Zhou, L, Chen, J and Pan, Z, 2019. Thermodynamic corrosion behavior of Al$_2$O$_3$, ZrO$_2$ and MgO refractories in contact with high basicity refining slag, *Ceramics International*, 45(16):20664–20673.

Zhang, D and Wang, H, 2002. Corrosion behavior of refining slags with different basicities to corundum-spinel castable, *Naihuo Cailiao*, 36(4):215–217.

Zhang, J, Wang, C, Jiao, K, Zhang, J, Liu, Z, Ma, H, Fan, X and Guo, Z, 2021. Effect of BaO and MnO on high-temperature properties and structure of blast furnace slag, *Journal of Non-Crystalline Solids*, 571:121066.

Zhang, M, Chen, W, Cheng, Z, Shao, J, Jia, X and Pang, W, 2015. Influence of Physical Property of Silicon Steel Slag with High Alumina on Ladle Slag Buildup, *Bulletin of the Chinese Ceramic Society*, 34(4):1160–1164.

Zhao, X, Zhang, R, Jia, J, Qi, J and Min, Y, 2021. Effect of basicity on the microstructure of molten slag of CaO-SiO$_2$-Al$_2$O$_3$ system, *Journal of Materials and Metallurgy*, 20(3):179–184.

Zhou, M, Yang, S, Jiang, T and Xue, X, 2015. Influence of MgO in form of magnesite on properties and mineralogy of high chromium, vanadium, titanium magnetite sinters, *Ironmaking & Steelmaking*, 42(4):320–320.

Zimmt, M B, Doubleday, J C and Turro, N J, 1985. Magnetic field effect on the intersystem crossing rate constants of biradicals measured by nanosecond transient UV absorption, *Journal of the American Chemical Society*, 107(23):6726–6727.

Zou, Y, Huang, A and Gu, H, 2020a. Novel phenomenon of quasi-volcanic corrosion on the alumina refractory-slag-air interface, *Journal of the American Ceramic Society*, 103(11):6639–6649.

Zou, Y, Huang, A, Wang, R, Fu, L, Gu, H and Li, G, 2020b. Slag corrosion-resistance mechanism of lightweight magnesia-based refractories under a static magnetic field, *Corrosion Science*, 167:108517.

Zuazo, I, Hallstedt, B, Lindahl, B, Selleby, M, Soler, M, Etienne, A, Perlade, A, Hasenpouth, D, Massardier-Jourdan, V, Cazottes, S and Kleber, X, 2014. Low-density steels: complex metallurgy for automotive applications, *JOM*, 66(9):1747–1758.

On the stability of CaS in liquid steel containing alumina or spinel inclusions

S Kumar[1], N N Viswanathan[2] and D Kumar[3]

1. PhD Research Scholar, IIT Bombay, Mumbai, Maharashtra 400076, India.
   Email: sandeep_kumar@iitb.ac.in
2. Sajjan Jindal Steel Chair Professor, IIT Bombay, Mumbai, Maharashtra 400076, India.
   Email: vichu@iitb.ac.in
3. Assistant Professor, IIT Bombay, Mumbai, Maharashtra 400076, India.
   Email: deepook@iitb.ac.in

ABSTRACT

In secondary steelmaking, calcium treatment is practised to modify the solid alumina or spinel inclusions into liquid or partially liquid calcium aluminates. It has been seen that the dissolved calcium reacts with sulfur or oxygen present in the steel to form CaS or CaO type of transient inclusions. Post-calcium treatment these transient inclusions further react with pre-existing alumina or spinel inclusions and modify them. In addition, direct modification of pre-existing inclusions is also reported in the literature.

In the present study, laboratory-scale induction furnace-based experiments are carried out to melt electrolytic iron in an alumina crucible followed by deoxidation with aluminium. The CaS-based composite material, prepared in-house, is added after the aluminium deoxidation to study the stability of CaS at the steelmaking temperature. Multiple steel samples were also taken to track the dissociation of CaS by means of sulfur pick-up in steel. Additionally, inclusion analysis of these steel samples was also performed to track changes in the chemical composition of inclusions in response to CaS addition. It has been seen that the CaS dissociates leading to a gradual pick-up of sulfur in steel. The scanning electron microscope – energy dispersive X-ray spectroscopy (SEM-EDS) analysis reveals that the alumina crucible and floated alumina inclusions are modified into calcium aluminates and the extent of modification depends on the contact time of CaS with the two sources of alumina mentioned earlier. However, the modification of alumina inclusions in the bulk steel did not occur during these experiments.

INTRODUCTION

As the demand for clean steel steadily increases, particularly within the automotive and defence sectors, global steelmakers are not only paying importance to control impurity levels but are also placing a significant emphasis on controlling non-metallic inclusions in the steel. The early works of Hilty and Farrel (1975a, 1975b) on the effect of calcium in aluminium-killed steel for the inclusion modification opened new horizons for controlling non-metallic inclusions in steel. As a result, calcium treatment is a well-established process nowadays at the secondary steelmaking process stage to convert deleterious solid non-metallic inclusions into liquid or partially liquid non-metallic inclusions to enhance castability and improve the mechanical properties of the steel. In aluminium-killed steel, two major types of inclusions are alumina and spinel inclusions. This study primarily focuses on alumina inclusions which can be modified into liquid calcium aluminates with calcium treatment. Verma *et al* (2011a) reported the sequence of reactions after calcium additions as Equations 1–4.

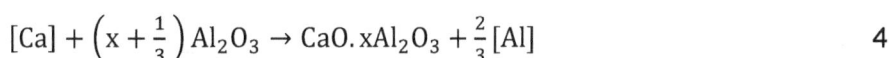

$$Ca(s) \rightarrow Ca(l) \rightarrow Ca(g) \rightarrow [Ca] \qquad 1$$

$$[Ca] + [O] \rightarrow CaO \qquad 2$$

$$[Ca] + [S] \rightarrow CaS \qquad 3$$

$$[Ca] + \left(x + \tfrac{1}{3}\right) Al_2O_3 \rightarrow CaO.xAl_2O_3 + \tfrac{2}{3}[Al] \qquad 4$$

It can be seen from the reactions that the added calcium first melts, vapourises and then dissolves in the liquid steel. Dissolved calcium reacts with dissolved oxygen and sulfur if present in the liquid steel and forms CaO/CaS type of inclusions. The extent of formation of CaS or CaO type of inclusions depends on the dissolved sulfur or dissolved oxygen in the liquid steel. Further, the dissolved calcium reacts with the pre-existing alumina inclusions and converts them into liquid calcium aluminates. The expected alumina modification route was reported by Faulring, Farrel and Hilty (1980) as $Al_2O_3 \rightarrow CA_6 \rightarrow CA_2 \rightarrow CA_x$ (C: CaO, A: Al$_2$O$_3$). This route is reported in almost all the literature for the modification of alumina into calcium aluminates. It should be noted here that the activity of alumina in the calcium aluminates decreases with an increase in CaO content in the inclusion. The above mechanism of modification is often referred to as direct modification of inclusion.

Turkdogan (1988) observed the formation of a CaS layer around the calcium aluminate in high sulfur steels and concluded that the CaS layer hinders the extent of modification of alumina inclusions. (Lu, Irons and Lu, 1994; Higuchi *et al*, 1996; Ye, Jönsson and Lund, 1996; Ren, Zhang and Li, 2014) also reported an increase in CaS or CaO content immediately following calcium treatment, followed by a subsequent decrease, suggesting the formation of CaS/CaO as transitional products. They also suggested that the high sulfur and oxygen can enhance the calcium absorption rate. Verma *et al* (2011b) reported that apart from being CaS or CaO a transient phase they also play a critical role in alumina inclusion modification. They observed that in high sulfur steels, these transient CaS inclusions tend to nucleate on alumina inclusions and destabilise the alumina inclusions and later modify them by the reaction as shown by Equation 5. This modification can be referred to as indirect modification of alumina inclusion and the schematic is shown in Figure 1:

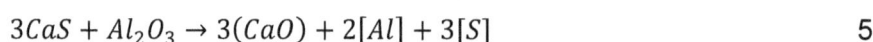

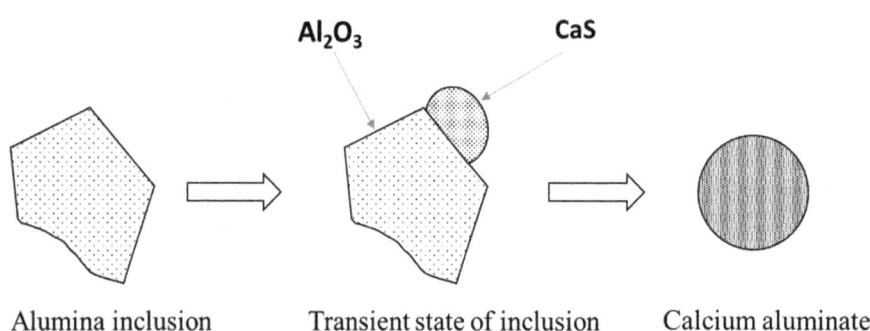

$$3CaS + Al_2O_3 \rightarrow 3(CaO) + 2[Al] + 3[S] \qquad\qquad 5$$

Al₂O₃   **CaS**

Alumina inclusion   Transient state of inclusion   Calcium aluminate

FIG 1 – Schematic of indirect modification of alumina inclusion showing CaS evolution as a transient phase Verma *et al* (2011b).

It should be noted here that the above reaction involves three condensed phases, two solid and one liquid steel phase. It is expected that the reaction will occur via the liquid steel phase. The reaction mechanism can be:

- CaS dissociates into calcium and sulfur.
- Calcium transfer to alumina inclusion through liquid steel and sulfur rejection in liquid steel.
- Alumina inclusion modification to liquid calcium aluminates.

As previously mentioned, the CaS is a transient inclusion. This work focuses on examining the stability of CaS in liquid steel containing alumina inclusions. Thermodynamic calculations were also conducted to explore the conditions determining the stability or dissociation of CaS. The Fe-CaS-based composite material was prepared in-house and its stability was investigated through various experiments involving different durations of experimentation.

EXPERIMENTAL PROCEDURE

The Fe-CaS-based composite material was prepared by melting pure calcium and iron sulfide in an induction furnace under an argon atmosphere (Kumar and Kumar, 2022). The density of the composite material was measured using a helium pycnometer. Microstructural characterisation and elemental composition of the composite lump were carried out using scanning electron microscopy equipped with energy-dispersive X-ray spectroscopy (SEM-EDS, Make and Model: Jeol JSM IT800).

The schematic of the experimental set-up and sampling apparatus is shown in Figure 2. Two sets of high-temperature experiments (No. 1 and No. 2) were performed in the induction furnace by varying the experimentation time. In both experiments, about 800 g of electrolytic iron flakes (containing dissolved oxygen of about 300 ppm and dissolved sulfur of 30 ppm) were kept in the alumina crucible (OD: 57 mm, Height: 87 mm and capacity: 170 mL). The alumina crucible with electrolytic iron was then placed inside the induction furnace chamber. The chamber was evacuated with a rotary vacuum pump followed by continuous backfilling with pure argon (purity: 99.99 per cent) at 1–3/L/min to prevent reoxidation of the liquid steel. The sample was then heated to 1600°C and held at that temperature for homogenisation. Based on the preliminary experiments, a holding time of 15 min was chosen to ensure homogenisation (uniform temperature and composition of bath).

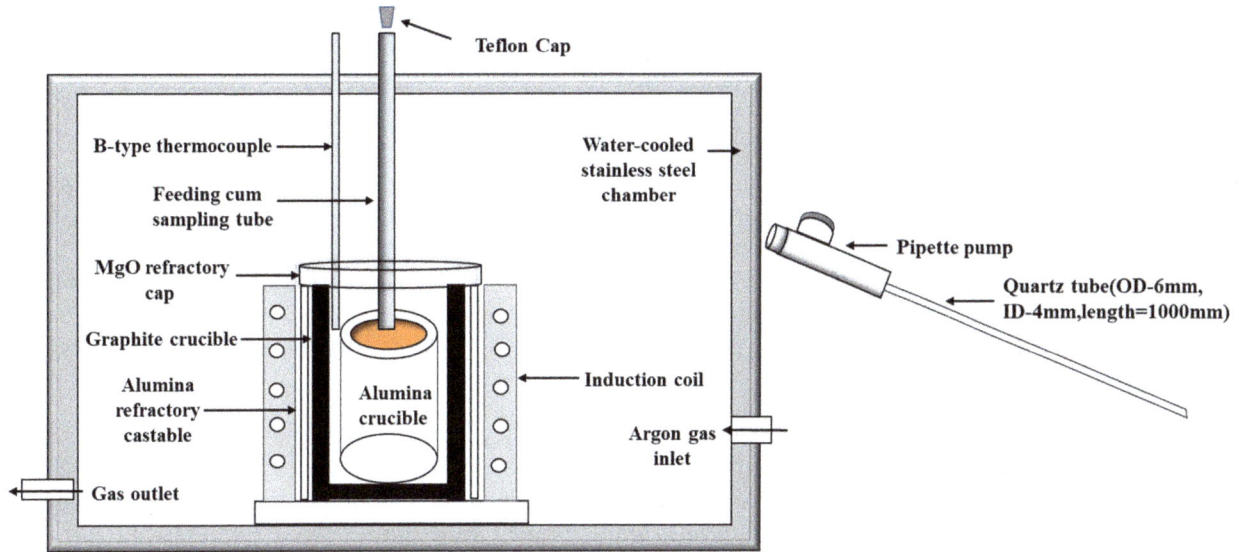

FIG 2 – Schematic of the Induction furnace and sampling apparatus.

After homogenisation, 0.1 wt per cent of pure aluminium chunks (purity >99 per cent) were added through a feeding cum sampling tube to deoxidise the steel melt. The time of Al addition was taken as a reference time and is denoted as t=0. Experimentation time for experiment 1 was 15 min, whereas 60 min for experiment 2. The addition sequence and the sampling timelines for both experiments are shown in Figure 3. After the deoxidation of the steel melt, a pre-determined amount of the Fe-CaS-based composite material powder wrapped in a mild steel foil was dropped into the liquid steel through the feeding tube. Samples from the bulk steel melt were taken at various time intervals by inserting the fused quartz tube (OD: 6 mm, ID: 4 mm and length: 1000 mm) followed by water quenching of the quartz tube containing the steel sample. The sampling cum feeding tube, which was sealed with a Teflon cap during the entire heating cycle to ensure an air-tight chamber; was opened for a very short duration for additions and sampling.

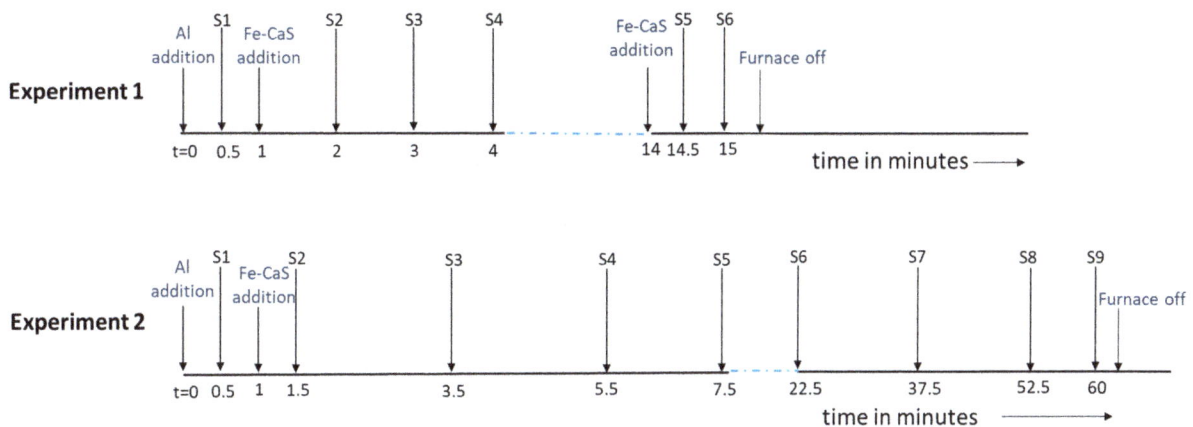

FIG 3 – Sampling timelines and additions for experiments 1 and 2.

The added amount of electrolytic iron, pure aluminium and composite material for both experiments are shown in Table 1.

TABLE 1

Raw material and additions for experiments 1 and 2.

Experiment #	Electrolytic iron (grams)	Aluminium (grams)	Fe-CaS composite (grams)
1	800	0.8	1.44
2	800	0.8	0.72

The bulk steel cylindrical samples aspirated using the quartz tube were then cut into smaller sections approximately 6–12 mm in size. These sections were hot-mounted in the longitudinal direction and subsequently grinded using SiC emery papers of grit sizes of 200, 400, 600, 800, 1000, 1200, 1500 and 2000 respectively. Following this, the samples were mirror polished using 3-micron and 1-micron diamond paste and washed with ethanol. It should be noted that the CaS particles are fairly hydrophilic, so water was not used during the final polishing. Microstructural investigation of these samples was carried out using SEM-EDS for the analysis of inclusion morphology (two-dimensional) and composition. Additionally, the top surface of the solidified steel ingot obtained from experiments no. 1 and 2 was also analysed using SEM-EDS, as floated alumina inclusion is expected at the top portion of the solidified ingot.

Furthermore, the alumina crucible used in both experiments was broken and the inner surface of the alumina crucible, which had been in contact with the solidified steel ingot, was analysed using SEM-EDS. The sulfur content of samples collected at different time intervals (experiment 2) was measured using a LECO CS744 carbon sulfur analyser at a third-party laboratory.

RESULTS AND DISCUSSION

The microstructural features of the composite material are illustrated in Figure 4. Notably, pure CaS microparticles and pure iron particles are embedded within the Fe-CaS network. The sizes of the CaS microparticle vary from 10–100 microns. The average size of CaS particles was determined through ImageJ™ software, ver 1.54g, revealing an average particle size of 50 microns. It should be noted the composite material produced was highly brittle and, thus, difficult to polish. Therefore, the composite material lump was mechanically grinded for SEM analysis, as evidenced by a few scratch marks in the electron image.

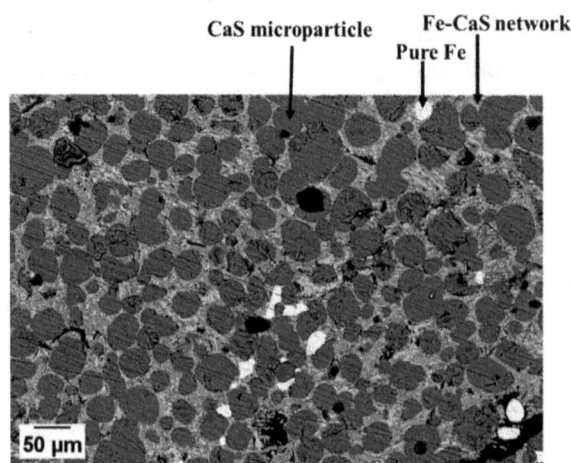

FIG 4 – Microstructure of the Fe-CaS composite material.

The composite material has an average density of 3.95 g/cc, nearly half that of liquid steel, as determined through a helium pycnometer.

In both experiments, the Fe-CaS-based composite material was added one minute after the aluminium deoxidation. The duration of experiment 1 and experiment 2 were 15 mins and 60 mins, respectively. The SEM-EDS analysis of bulk steel samples, obtained through a quartz tube at different time intervals for both experiments to track changes in inclusions composition is shown in

Figure 5. The EDS spectra reveal the irregular morphology and unmodified alumina inclusions within the specified time frame. The CaS particles are expected to have a very short residence time in the liquid steel due to density differences and shallow melt depths of 30–40 mm. It should also be noted here that the CaS particles were not observed in any bulk steel samples during SEM-EDS analysis. The sampling methodology (aspirated through quartz tube) to track changes in inclusion composition has been widely reported in the literature.

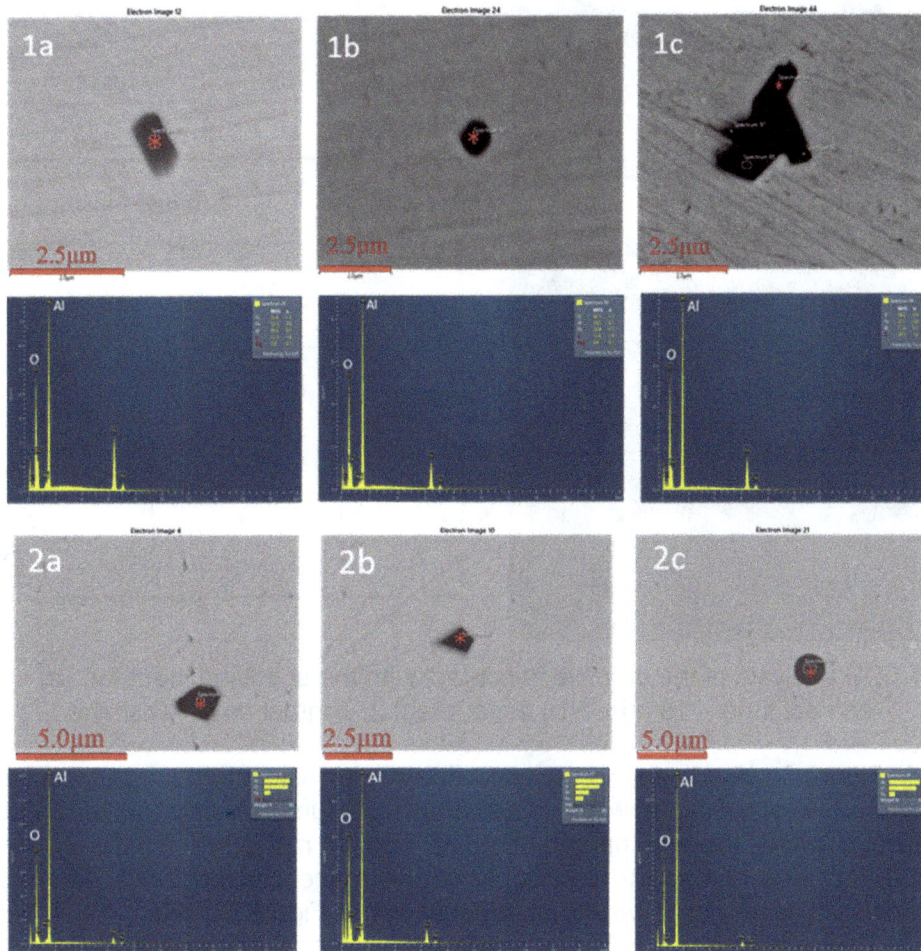

FIG 5 – SEM-EDS spectra of bulk steel samples for experiments 1: (a) 1 min; (b) 2 min; (c) 14 mins; and experiment 2: (a) 0.5 min; (b) 2.5 min; (c) 21.5 mins after composite material addition.

To investigate the dissociation of CaS in the liquid steel the alumina crucible was carefully observed for both experiments. The microstructure and SEM-EDS spectra of the inner surface of the alumina crucible after 15 minutes and 60 minutes of contact time with liquid steel treated by the composite material are shown in Figure 6.

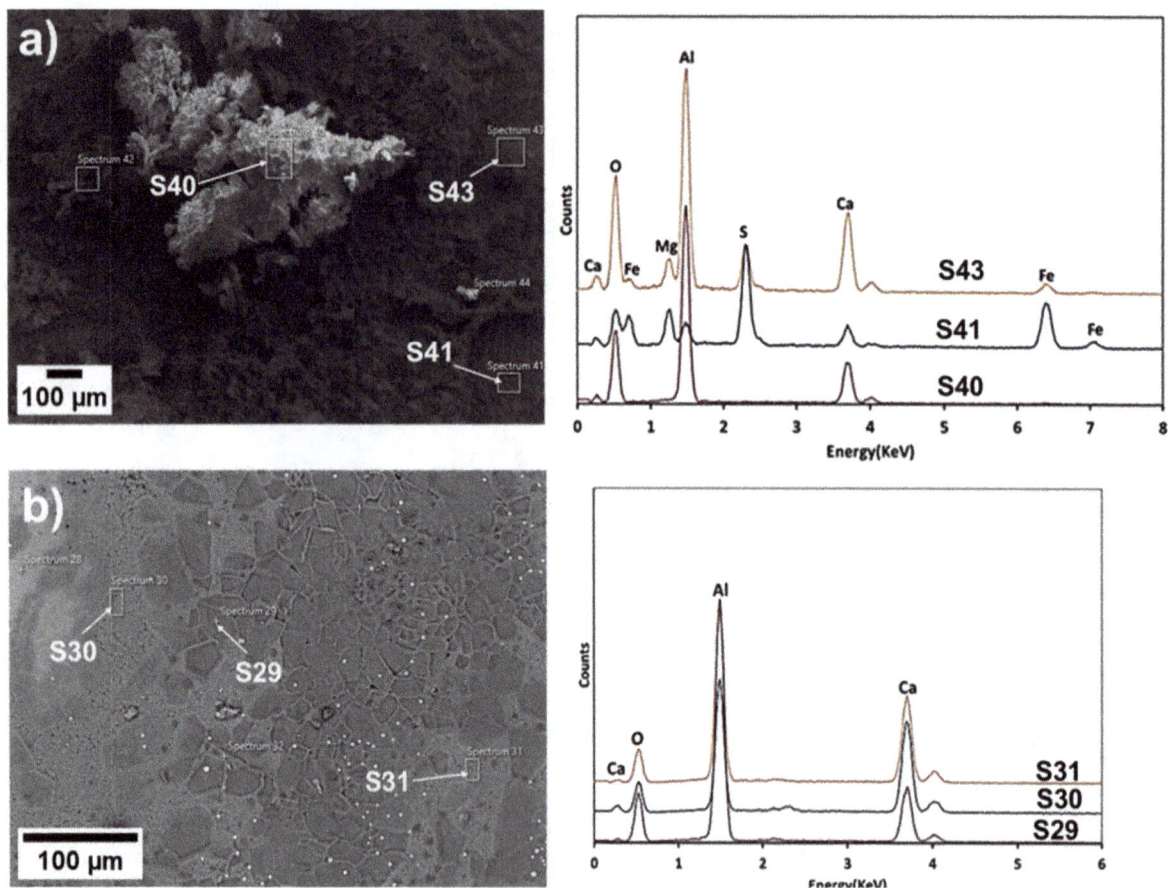

FIG 6 – SEM-EDS analysis of the crucible in contact with the solidified ingot for: (a) experiment 1: contact time – 15 mins; (b) experiment 2: contact time – 60 mins.

Location S40 in Figure 6a clearly shows the presence of a significant amount of CaO and no sulfur. The nearby regions S41 and S42 show a mixture of calcium aluminate and calcium sulfide. However, in experiment no. 2 where the experimentation time was 60 mins (see Figure 6b). A thin layer of liquid calcium aluminates across the crucible's inner surface can be seen. The modification of alumina into calcium aluminates is visible at all locations in Figure 6b. Also, the absence of sulfur peaks at all locations indicates that sulfur must have dissolved in liquid steel. These results indicate that the reaction between CaS and alumina is kinetically driven; that is, the CaS particles in the composite material for experiment no. 1 had insufficient time for complete dissociation and complete modification of alumina, whereas the CaS particles had sufficient time to dissociate and modify alumina crucible for experiment no. 2. To further confirm these findings: (a) the inclusion analysis of the top portion of the solidified ingot, and (b) sulfur pick up in the bulk steel sample at different time intervals for experiment no. 2 have been investigated.

The top surface of the solidified ingot has also been examined to observe the CaS particle flotation and their interaction with floated alumina inclusions. It should be noted here that the top surface also contains floated alumina inclusions after aluminium deoxidation. Upon investigating the solidified steel ingot, the SEM-EDS analysis for experiment no. 1, shown in Figure 7a, reveals partial modification of floated alumina inclusions. The locations S18 and S21 in the SEM micrograph show the floated unmodified alumina inclusions and locations S19 and S20 show the modified alumina inclusions. The SEM-EDS analysis for experiment no. 2 is shown in Figure 7b, which shows that the floated alumina inclusions are modified to calcium aluminates and CaS particles were not observed on the top of the solidified ingot.

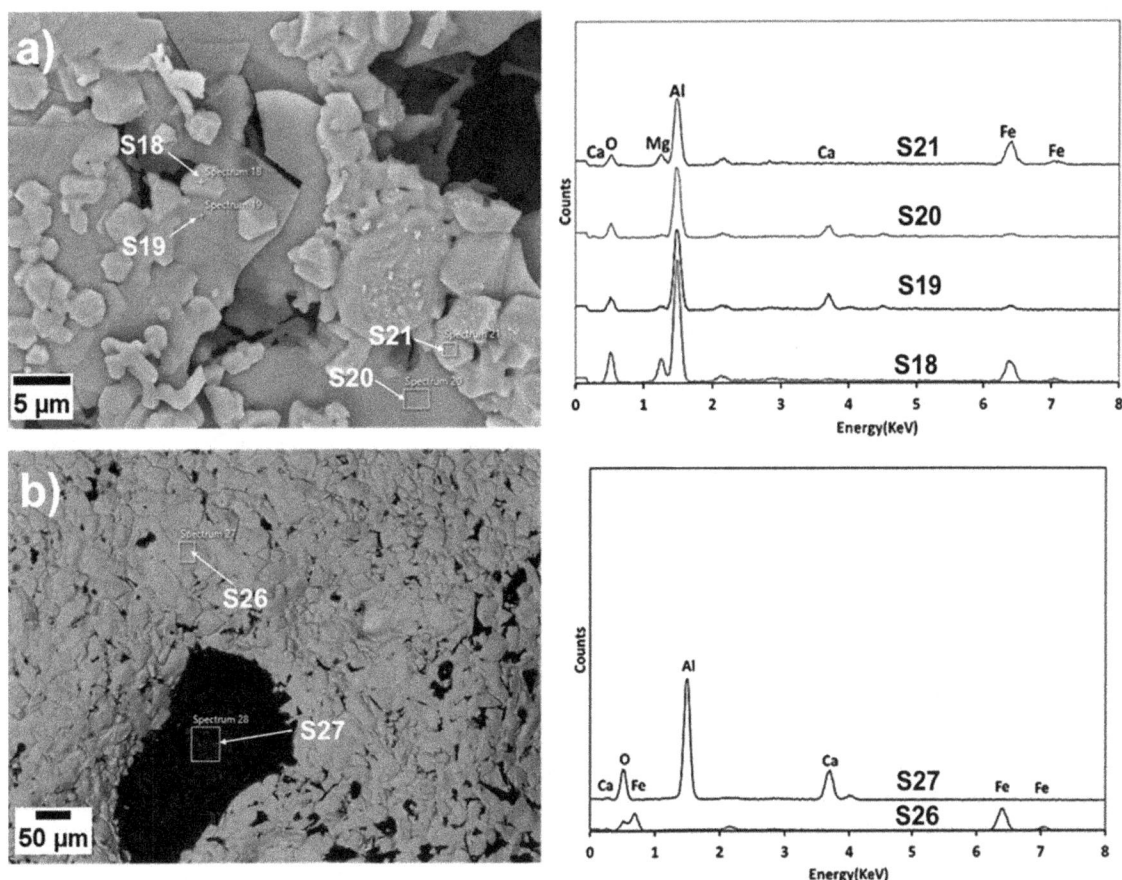

FIG 7 – SEM-EDS analysis of the solidified ingot top: (a) experiment 1; (b) experiment 2.

Furthermore, the steel samples for experiment no. 2 were collected at different time intervals to track the sulfur pick-up in the liquid steel, as shown in Figure 8. The rate of sulfur pick-up is very large at early times, followed by a steady increase in the sulfur concentration. It is evident that the CaS-based composite material undergoes dissociation, facilitating the transfer of calcium to liquid steel for the modification of oxide inclusions and rejecting the sulfur into the liquid steel.

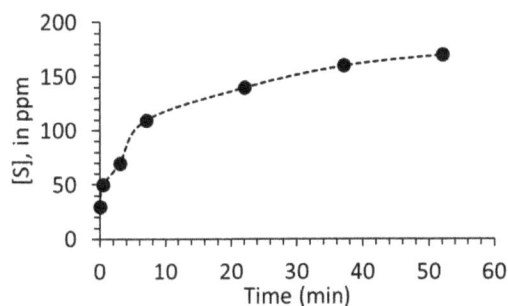

FIG 8 – Sulfur analysis of steel samples at different time intervals for experiment 2.

The thermodynamic stability of CaS in liquid steel was evaluated using the FactSage™ 8.1 thermodynamic software Bale *et al* (2016). The calculations were performed using FactSage™ 8.1 (Equilib module, Databases: FactPS, FactOxid and FTMisc) at 1600°C temperature and 1 atm pressure. The activity of CaS was set to 1 for these calculations. The saturation line representing the equilibrium concentration for S content and total oxygen at which CaS activity is one is shown in Figure 9. The concentration (sulfur, oxygen) region above this line represents CaS is stable, whereas, CaS will dissociate below this saturation line. Both the stability region and dissociation region are depicted in Figure 9.

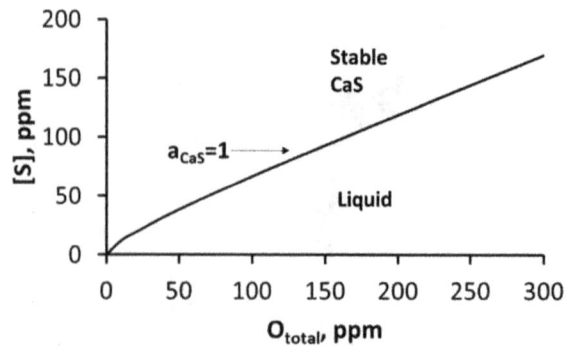

FIG 9 – CaS stability diagram at 1600°C using FactSage™ 8.1.

It can be seen from Figure 9 that a high total oxygen content favours CaS dissociation. The total oxygen in the present system comes from the steel (Fe-Al-Ca-O-S) and the Al_2O_3 crucible interface is expected to have a high oxygen content. Consequently, CaS dissociation is feasible for a wide range of sulfur content, which reasonably aligns with the experimental results. As observed in the experiments, the dissociation of CaS into liquid steel, followed by the transfer of calcium to inclusions and sulfur to melt, depends not on dissolved oxygen but rather on the total oxygen content.

SUMMARY AND CONCLUSIONS:

In the current study, laboratory experiments and thermodynamic calculations were performed to investigate the CaS stability in the liquid steel containing alumina inclusions. Experiments were conducted by adding Fe-CaS-based composite material and the following conclusions can be obtained.

- As expected, since the $CaS-Al_2O_3$ interaction involves three condensed phases, two solid phases and a liquid steel phase the transfer of chemical species occurs via liquid phase.

- The dissociation of CaS, subsequent calcium transfer to alumina inclusions, and sulfur transfer to the melt are influenced by the total oxygen content from both dissolved oxygen in the melt and oxygen present in the inclusions.

- The CaS dissociation is a kinetic-driven process. The dissociation and subsequent chemical species transfer can depend on the mass transfer coefficients of the species in the liquid steel.

ACKNOWLEDGEMENTS

We would like to acknowledge the financial support of the following funding agencies:

- Industrial Research and Consultancy (IRCC), Centre of Excellence in Steel Technology (CoEST) and Department of Metallurgical Engineering and Materials Science, IIT Bombay.

- Science and Engineering Research Board (SERB), Grant number SRG/2020/001495

- Prime Minister's Research Fellowship contingency grant (PMRF, Ministry of Education) Govt. of India.

REFERENCES

Bale, C W, Belisle, E, Chartrand, P, Decterov, S A, Eriksson, G, Gheribi, A E, Hack, K, Jung, I H, Kang, Y B, Melancon, J, Pelton, A D, Petersen, S, Robelin, C, Sangster, J, Spencer, P and Van Ende, M A, 2016. FactSage thermochemical software and databases, 2010–2016, *CALPHAD*, 54:35–53.

Kumar, D and Kumar, S, 2023. A composite for refining of steel and method for preparation thereof, Indian patent application: 202221076794.

Faulring, G M, Farrel, J W and Hilty, D C, 1980. Steel flow through nozzles influence of calcium, *Iron Steelmaker*, 7(2):14–20.

Higuchi, Y, Numata, M, Fukagawa, S and Shinme, K, 1996. Inclusion Modification by Calcium Treatment, *ISIJ International*, 36:S151–S154.

Hilty, D C and Farrel, J W, 1975a. Modification of inclusions by calcium – Part, I, *Iron Steelmaker*, 2(5):17–22.

Hilty, D C and Farrel, J W, 1975b. Modification of inclusions by calcium – Part II, *Iron Steelmaker*, 2(6):20–27.

Lu, D, Irons, G and Lu, W, 1994. Kinetics and mechanism of calcium dissolution and modification of oxide and sulphide inclusion in steel, *Ironmaking and Steelmaking*, 21(5):362–371.

Ren, Y, Zhang, L and Li, S, 2014. Transient evolution of inclusions during calcium modification in line pipe steels, *ISIJ International*, 54(12):2772–2779.

Turkdogan, E T, 1988. *Proceedings of 1st International Calcium Treatment Symposium*, pp 3–13 (The Institute of Metals, London).

Verma, N, Pistorius, P C, Fruehan, R J, Potter, M, Lind, M and Story, S, 2011a. Transient inclusion evolution during modification of alumina inclusions by calcium in liquid steel: Part I background experimental techniques and analysis methods, *Metallurgical and Materials Transactions B: Process Metallurgy and Materials Processing Science*, 42(4):711–719.

Verma, N, Pistorius, P C, Fruehan, R J, Potter, M, Lind, M and Story, S R, 2011b. Transient Inclusion Evolution During Modification of Alumina Inclusions by Calcium in Liquid Steel: Part II, Results and Discussion, *Metallurgical and Materials Transactions B*, 42(4):720–729.

Ye, G, Jönsson, P and Lund, T, 1996. Thermodynamics and kinetics of the modification of Al_2O_3 inclusions, *ISIJ International*, 36.

Corrosion behaviour of ferrite and aluminate refractories in cryolite-aluminium melts

R Z Mukhlis[1] and M A Rhamdhani[2]

1. Lecturer, Fluid and Process Dynamics (FPD) Research Group, Swinburne University of Technology, Hawthorn Vic 3122. Email: rmukhlis@swin.edu.au
2. Professor, Fluid and Process Dynamics (FPD) Research Group, Swinburne University of Technology, Hawthorn Vic 3122. Email: arhamdhani@swin.edu.au

ABSTRACT

Removing the ledge formation in the Hall-Heroult cell during the electrolysis process may significantly reduce the energy requirement of the aluminium production process. In the absence of the ledge, however, the sidewall-material types become heavily restricted since the cryolite is very corrosive at the electrolysis temperatures. This paper study the corrosion behaviour of sintered nickel ferrite, nickel aluminate, and magnesium aluminate refractories (as possible candidates for sidewall material) in molten aluminium and cryolite melts with various alumina content at 980°C. Immersion tests of refractories with various porosity were conducted to assess the suitability of the refractories in the corrosive environment of the electrolysis cells. Microstructural change of the refractory after immersion were analysed using secondary scanning electron microscopy (SEM). Sampling of the melts were taken at 0, 10 min, 30 min, and each hour up to 6 hrs, and subsequently analysed by Inductively Coupled Plasma (ICP) spectroscopy. Based on the projected concentration of the refractories constituents in the melts, it was predicted that the corrosion rate of $NiFe_2O_4$, $NiAl_2O_4$ and $MgAl_2O_4$ in cryolite with 11 wt per cent Al_2O_3 were 461×10^3, 1.03×10^3, and 798×10^3 cm/a, respectively. Much lower corrosion rate of 2.3 cm/a for $NiFe_2O_4$, 2.6 cm/a for $NiAl_2O_4$ and 0.3 cm/a for $MgAl_2O_4$ were predicted if the refractories in contact with aluminium melts. Erosion played an important role in the corrosion as the concentration of the refractory constituents in the melts were beyond its solubility limit. It is therefore crucial to increase the densities of the refractory and improve its wetting characteristics.

INTRODUCTION

Aluminium is produced commercially from alumina through the Hall-Heroult process at around 965°C where the overall chemical reaction produces metallic aluminium and carbon dioxide, as follows:

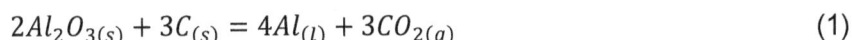

$$2Al_2O_{3(s)} + 3C_{(s)} = 4Al_{(l)} + 3CO_{2(g)} \tag{1}$$

There are two major issues suffered by the process: high cost and great environmental impact. The high cost is the consequence of its high capital cost and its eminent energy requirement. The total specific energy requirement, W_{el} (kWh/kg), of the Hall-Heroult process can be determined through the following relationship by estimating the cell voltage (Thonstad et al, 2001):

$$W_{el} = \frac{UF|V_e|}{3600M_{Al}x_{Al}v_{Al}} = 2.98\frac{U}{M_{Al}} \tag{2}$$

where:

U	is the cell voltage
F	is Faraday constant
$M_{(Al)}$	is the molar weight of Al
$X_{(Al)}$	is the corresponding current efficiency fraction
$|V_e|$ and $v_{(Al)}$	are the stoichiometric numbers of electrons and of product of cathodic reaction, respectively

By considering Reaction 1 and Equation 2, the theoretical energy required to produce 1 kg aluminium at 1000°C and 100 per cent current efficiency is 5.06 kWh. In practice, the energy consumed to produce 1 kg aluminium is ranging from 12 to 15 kWh (Schwarz, 2008; Kvande and Haupin, 2001; Haraldsson and Johansson, 2020). This energy includes the thermodynamic energy needed by the reaction, and additional energy that is required to heat the reactant from room temperature to the operational temperature and to maintain at that temperature. The necessity to have high heat loss through the side of the cell container to form protective frozen ledge is also significantly contributing to the high energy consumption.

The significant environmental impact of the Hall-Heroult process has a strong correlation with its high energy demand, particularly if the source of electricity is coming from fossil fuel. Moreover, the fact that the Hall-Heroult process uses consumable carbon anode and the main reaction of the process produce carbon dioxide gas, induce its effect on global warming even more escalated.

It has been considered since long that the breakthrough technology of inert anode, the anode that theoretically will not be consumed during electrolysis, may revolutionise the Hall-Heroult process (Kvande and Drabloss, 2014). Not only can it alter the aluminium electrolytic process from carbon dioxide generating reaction to oxygen producing reaction (Reaction 3); the implementation of inert anode will eliminate all cost associated with the consumable carbon anode. The capital cost itself has been indicated to be reduced by one-quarter to one-third of the capital cost of the carbon anode potline (Thonstad et al, 2001).

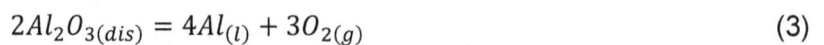

$$2Al_2O_{3(dis)} = 4Al_{(l)} + 3O_{2(g)} \tag{3}$$

The implementation of inert anode to the Hall-Heroult process will also increase its environmental friendliness by eliminating the production of CO, perfluorocarbon gases ie CF_4 and C_2F_6, fluoride and dust emissions during anode change, and the elimination of carbon butts (Kvande and Drabloss, 2014). The inert anode technology, however, needs to be underpinned with the significant advancement in materials technology and the heat flow control (Mukhlis, Rhamdhani and Brooks, 2010).

It is important to be noted that due to the absence of carbon-oxygen reaction that provide energy, the use of inert anode will increase the theoretical energy consumption to 6.54 kWh per kg aluminium production. Lower heat loss, is therefore, required in the cell that used inert anode to be able to have the same amount of energy requirement as the conventional cell.

In the case that heat loss needs to be lowered, the protective frozen ledge will be difficult to be maintained and the sidewall materials will be exposed directly to the very corrosive environments of cryolite bath, the highly reducing molten aluminium zone, as well as the oxidative upper part of the cell due to the existence of air and oxygen at high temperatures. Consequently, the life of the sidewall materials will be significantly reduced. It is obvious that the implementation of inert anode in the cell calls for new ledge-free sidewall materials.

In general, the required properties of the ledge-free sidewalls are very much the same as the required properties of inert anodes except for its electrical conductivity, where in contrast with inert anode, sidewall materials should have low electrical conductivity (Mukhlis, 2015; Yan et al, 2011). Spinel based materials, particularly nickel ferrite ($NiFe_2O_4$), have been proposed as inert anode and ledge free sidewall material for aluminium smelter, mainly due to its low solubility and high corrosion resistance against cryolite and high oxidation resistance (Yan, Pownceby and Brooks, 2007; Mukhlis, Rhamdhani and Brooks, 2010; Longbottom, Nightingale and Monaghan, 2014). While these properties make the spinel suit to be applied in both gas zone and cryolite zone, this is not the case for the application in the aluminium zone. It has been predicted thermodynamically that in equilibrium, liquid aluminium will be contaminated by 29.1 mol per cent Ni and 7.8 mol per cent Fe when 10 mol of aluminium is in contact with 1 mol $NiFe_2O_4$ at 965°C. $NiFe_2O_4$ should therefore be combined with other material that have high resistance against molten aluminium if it is to be used as ledge-free sidewall material.

The current paper study the corrosion behaviour of $NiFe_2O_4$, $NiAl_2O_4$ (nickel aluminate), and $MgAl_2O_4$ (magnesium aluminate) in both aluminium and cryolite melts. $NiAl_2O_4$, and $MgAl_2O_4$ were chosen as both are aluminium contained materials, which hypothetically suit to be applied in aluminium zone, low electrical conductivity as well as having similar crystal structure as $NiFe_2O_4$ for the joining compatibility.

MATERIALS AND METHOD

In the present work, the NiFe$_2$O$_4$, NiAl$_2$O$_4$, MgAl$_2$O$_4$ powders were made from its oxide precursors where the details can be found elsewhere (Mukhlis *et al*, 2011). About 15g of each powder were compacted uniaxially in metal dies at 91 MPa into rectangular bar shape samples, and subsequently sintered under air at 1450°C for 6 hrs for corrosion test. The sintered NiFe$_2$O$_4$, NiAl$_2$O$_4$, and MgAl$_2$O$_4$ sample contained porosity of 12.0 per cent, 42.0 per cent, and 22.4 per cent, respectively, determined by ASTM C20-00 standard (ASTM, 2000). Higher density NiAl$_2$O$_4$ (25.5 per cent) porosity and MgAl$_2$O$_4$ (11.7 per cent porosity) were also tested for comparison.

All surfaces of the sintered bars were grounded up to 800 grits to homogenise the surface roughness and to eliminate the part that may have reacted with the crucible during sintering. Following that, one end of the sintered bar was assembled into a stainless-steel holder that jointed to a stainless-steel rod connected to an overhead electronic stirrer (Figure 1). The assembly were covered with alumina tube to ensure that during experiments, no steel surface is directly exposed to molten bath or molten aluminium and to the atmosphere inside the reactor.

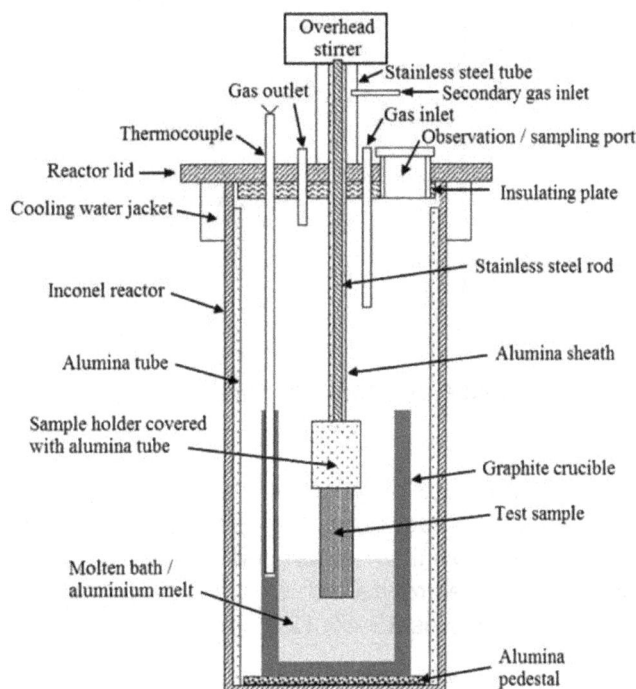

FIG 1 – Schematic diagrams of the stirred finger test apparatus.

The finger test-type apparatus was used to study the corrosion behaviour of the spinel (Figure 1). A pit furnace was used to melt the electrolyte and aluminium. The molten bath or aluminium melt was hold by a graphite crucible with a dimension of 45 mm inner diameter, 65 mm outer diameter and 175 mm high. Type-K thermocouple that was inserted into the hole drilled in the crucible was used to measure the melt temperature. The gastight closure of the reactor enabled the control of the atmosphere.

To study the corrosion behaviour of the spinel materials against Hall-Héroult cell environment, the spinel bar samples were immersed (15 mm deep) in a molten aluminium or in cryolite-based melt with a low cryolite ratio of 2.3 (which is equal to 11 wt per cent excess of AlF$_3$) contained 6 to 11 wt per cent alumina addition. The corrosion tests were conducted for 6 hrs at a rotation speed of 25 rev/min under flowing argon atmosphere of 0.1 L/min. All experiments were carried out at 980 ± 2°C. During the experiment, samples of cryolite-based bath were taken through the observation port by rapid freezing onto a high purity copper rod which was quickly immersed into the bath, while the aluminium sampling was made using a quartz tube. The sampling was made at 0 min (just before the immersion), 10 min, 30 min, 1 hr, then every hour up to 6 hrs. The solidified bath/aluminium samples were then used for elemental analysis using Inductively Coupled Plasma (ICP) spectrometry.

RESULTS AND DISCUSSIONS

Corrosion test of MgAl₂O₄ in cryolite

Figure 2 shows the photographs of $MgAl_2O_4$ sample with 22.4 per cent porosity that were taken before and after corrosion testing in cryolite-based melt with 2.3 CR contained 11 wt per cent alumina at 980°C. No sample left after corrosion testing despite only 15 mm of the sample (out of 34.3 mm length of unsealed sample) immersed in the cryolite. It indicates that the reaction between the melt and the sample took place beyond the surface of the sample that had direct contact with the melt.

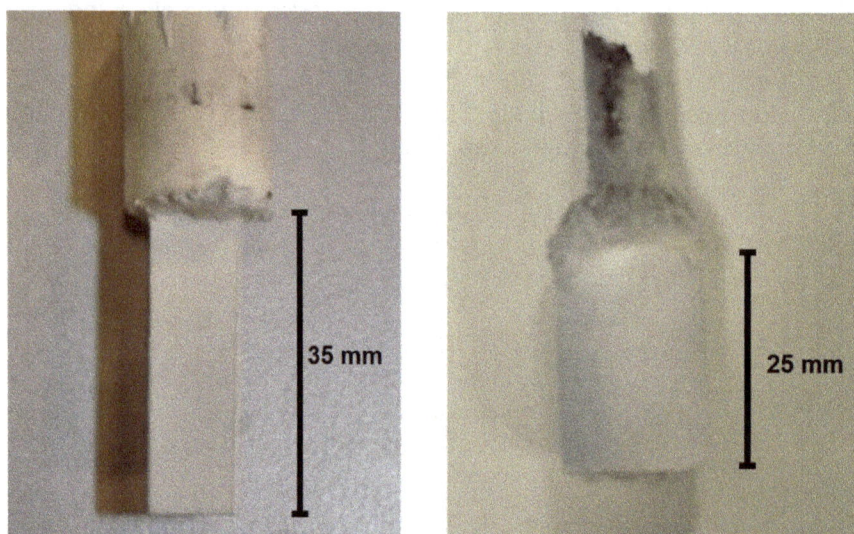

FIG 2 – Sample of $MgAl_2O_4$ with 22.4 per cent porosity: (left) initial sample (b) after corrosion testing.

A liquid in contact with porous substrate may penetrate and progress through the pores of the substrate due to capillarity when the contact angle between the liquid and the substrate is less than 90°. In a separate study conducted by the authors (Mukhlis *et al*, 2014), it was found that the measured apparent contact angle of electrolyte with 0 wt per cent, 6 wt per cent, and 11 wt per cent alumina content on $MgAl_2O_4$ substrate was 19.4°, 12.2°, and 9.3°, respectively.

The photographs of the $MgAl_2O_4$ sample with apparent porosity of 11.75 per cent are shown in Figure 3. Despite there was still some sample left after testing, the denser sample exhibited similar behaviour with the less-dense sample (shown in Figure 2), in which the corrosion proceeded beyond the sample surface that had direct contact with the melt. From 39.7 mm length of the original unsealed sample, only about 10 mm of it remained solid after testing (indicated by 'S' in Figure 3). The rest of the sample were either fully corroded or corroded on its inner part, leaving a hollow structure.

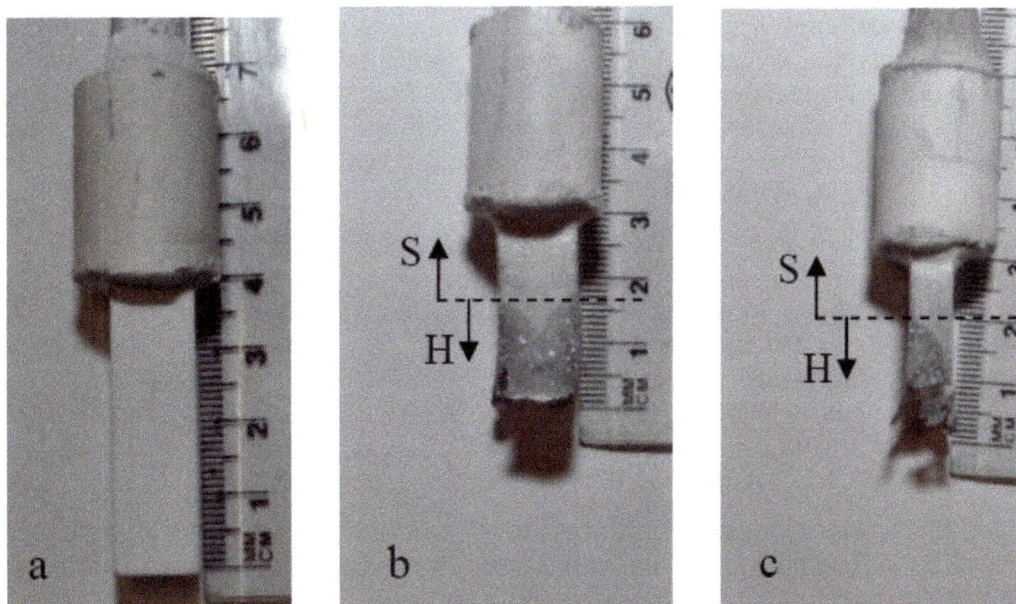

FIG 3 – Photograph of MgAl$_2$O$_4$ sample with apparent porosity of 11.74 per cent reacted with cryolite-based melt (CR = 2.3) contained 11 wt per cent alumina at 980°C: (a) before corrosion testing; (b) after testing (front view); (c) side view. S and H denote solid and hollow, respectively.

The amount of magnesium dissolved in the cryolite-based melt originated from the MgAl$_2$O$_4$ samples over time as per ICP-AES analysis are plotted in Figure 4.

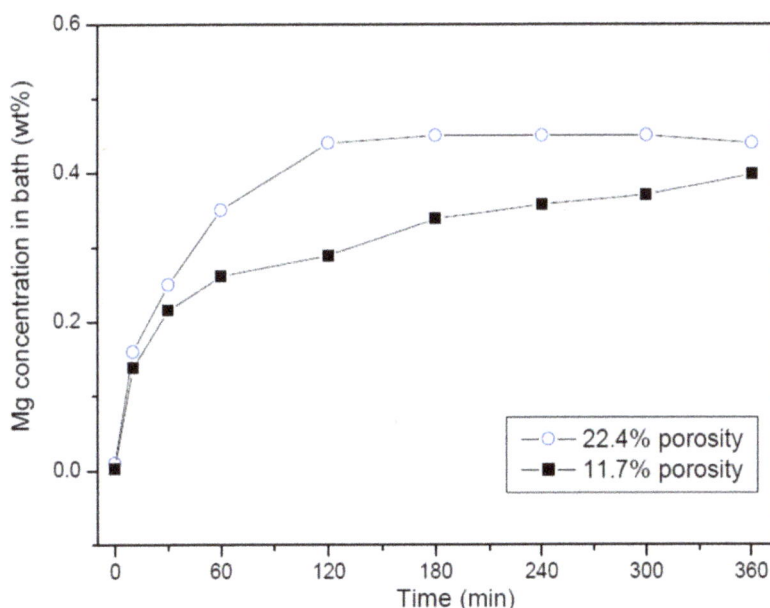

FIG 4 – Elemental analysis of the dissolved magnesium originated from MgAl$_2$O$_4$ with different porosity level in the cryolite-based melt.

It can be clearly seen from the curves plotted in Figure 4 that the amount of magnesium in the melts rose rapidly at the first 10 mins in which the gradient of both curves nearly the same. The concentration of magnesium in the melt increased with time at a reduced rate. As the corrosion proceeds, the surface area of the spinel sample that had direct contact with the melt decreased gradually, which may explain the less steep gradient of the curves over the time. In the case of the experiment with the sample with 22.4 per cent porosity, the increase in the magnesium concentration in the melt ceased at a relatively stable value of 0.44 to 0.45 wt per cent after 120 mins.

Based on the dimension of the unsealed part of the 22.4 per cent porosity sample, it theoretically contains 0.946 g of magnesium since the magnesium concentration in stoichiometric MgAl$_2$O$_4$

sample is 17.08 wt per cent. If the entire unsealed sample is dissolved into 200 g of the cryolite-based melt, the concentration of the magnesium in the melt is expected to be 0.46 wt per cent, which is very close to the steady state value shown in Figure 4. Through the similar dimensional calculation as above, and by also considering the hollow part of the sample, the expected concentration of magnesium in the melt originated from the 11.75 per cent porosity sample is 0.36 wt per cent, which is also close to the maximum concentration of magnesium from the sample shown in Figure 4.

From these calculations, it is concluded that 0.46 wt per cent is not the solubility limit of the $MgAl_2O_4$ in the cryolite-based electrolyte melt with 2.3 CR contained 11 wt per cent alumina. It merely indicated that all available samples have been corroded and all magnesium from the sample has been dissolved as supported by Figure 3. It is expected that if there is still available sample of $MgAl_2O_4$ in contact with the melt, the amount of magnesium in the melt will still increase over time until it reached the solubility limit. As for comparison, the solubility limit of MgO in pure cryolite and in cryolite that contain 5 wt per cent alumina at 1010°C was found to be 11.65 wt per cent and 10.6 wt per cent, respectively (Grjotheim *et al*, 1982).

Corrosion test on the $MgAl_2O_4$ sample with apparent porosity of 22.4 per cent immersed in cryolite-based melt with 2.3 CR contained 6 wt per cent alumina was conducted to study the effect of alumina concentration in the melt to the corrosion behaviour of $MgAl_2O_4$. The concentration of magnesium in the bath over time as per ICP-AES analysis is plotted in Figure 5.

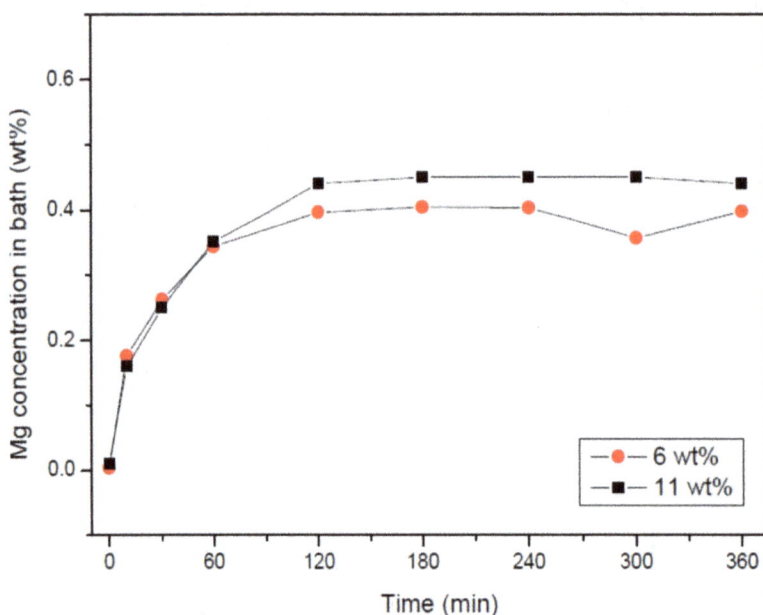

FIG 5 – Elemental analysis of the dissolved magnesium originated from $MgAl_2O_4$ in cryolite-based melts contains different alumina concentration.

As shown in Figure 5, the amount of magnesium in the cryolite-based melt that contain 6 wt per cent alumina increased sharply for the first 30 mins and further increased with a reduced rate up to 60 mins. Up to this period, the gradient of the curve is nearly the same as that for the melt that contain 11 wt per cent alumina. The magnesium concentration in the melt that contain 6 wt per cent alumina reached a plateau at 0.4 wt per cent within 120 mins. Again, this steady value did not indicate the solubility limit, rather indicate that all available $MgAl_2O_4$ sample has been corroded by the melt and all the magnesium has been dissolved into the melt. It should be noted that the Mg concentration at 300 min was slightly lower than expected, however, this most likely due to the variation from the measurement eg slightly smaller sample hence lower amount of available Mg. While overall trend is quite clear, further test at 300 min immersion is planned to validate the results.

All the above results showing a significant corrosion attack of the melt against magnesium aluminate. These results agree with the thermodynamic study that suggest $MgAl_2O_4$ is not suitable to be applied as ledge free sidewall in contact with the cryolite-based electrolyte.

The typical microstructures of the corroded $MgAl_2O_4$ sample are shown in Figure 6, which is associated with Figure 3.

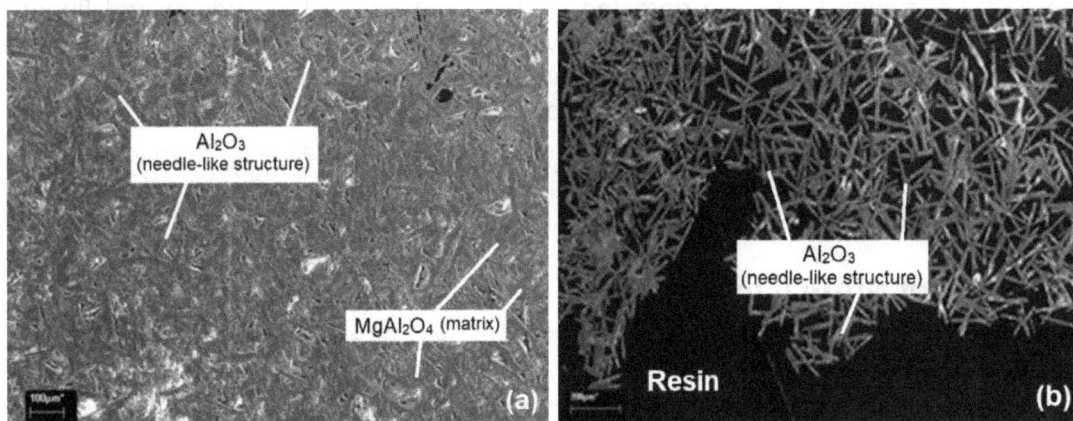

FIG 6 – Microstructure of the $MgAl_2O$ sample that was shown in Figure 3: (a) top section of the solid part; (b) bottom section of the hollow part.

As shown in Figure 6, the microstructures of both solid part and hollow part of the sample contained needle like structures that less than 5 µm width. The only significant difference is that for solid part (Figure 6a), most of the space in-between the needle like structure was filled with solid structures. The energy dispersive X-ray spectroscopy (EDS) point-analysis suggested that the needle-like structure was Al_2O_3, while the solid structures were solidified electrolyte melts. The presence of the solidified cryolite-based melt – in the top section of the solid part of $MgAl_2O_4$ sample – support the previous hypothesis, which suggested that the melt penetrate into $MgAl_2O_4$ beyond the surface that had direct contact with the melt.

In order to investigate whether there are still un-corroded $MgAl_2O_4$ left in the sample, elemental mapping on the top section of solid part of the sample (refer to Figure 3) has been conducted and the result is shown in Figure 7.

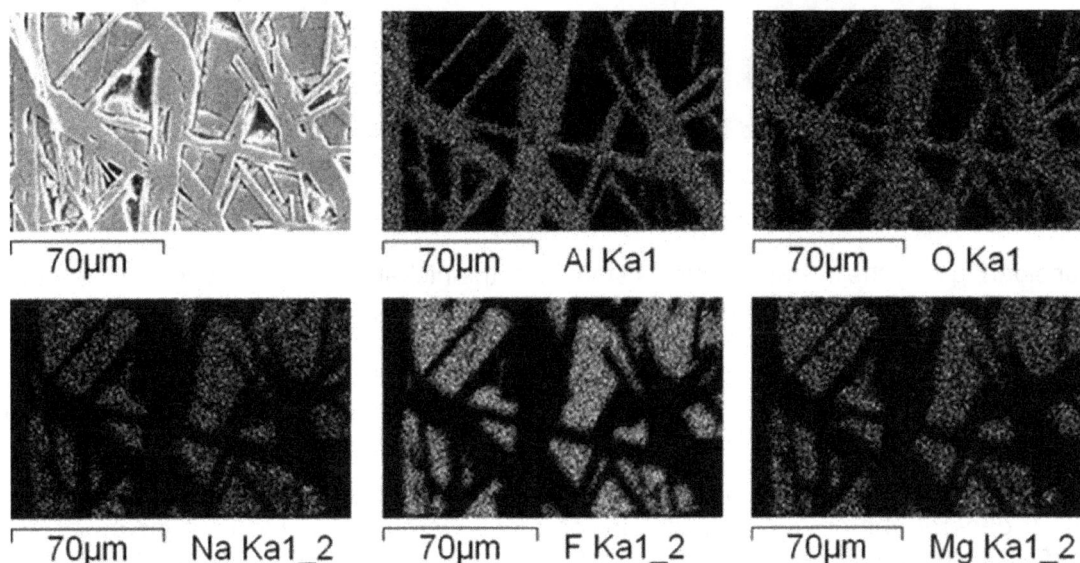

FIG 7 – Elemental mapping on the top section of solid part of $MgAl_2O_4$ sample that shown in Figure 3.

Elemental mapping revealed that there were significant compositional changes on the sample after corrosion testing. Most of the aluminium and oxygen were exist in the needle-like structures. On the other hand, most of magnesium was exist in the solid structure surrounded the needle-like structure. Only insignificant trace of magnesium was detected in the needle-like structure. This indicated that no $MgAl_2O_4$ left in the magnesium aluminate sample after 6 hr immersion in the cryolite-based electrolyte. The concentration of magnesium in the stoichiometric $MgAl_2O_4$ is 17.08 per cent.

Consequently, for a phase to be qualified to be suggested as $MgAl_2O_4$, the trace of magnesium should be substantial at the place where the trace of both aluminium and oxygen are also significant.

Since most of the magnesium were detected on the area where the sodium and fluorine were substantial, it is argued that the magnesium from $MgAl_2O_4$ was dissolved into the cryolite-based melt. It is also reasonable to conclude that the corrosion mechanism of $MgAl_2O_4$ by cryolite-based melt involve reactions that produce Al_2O_3. One of the plausible reactions is the reaction in Equation 4:

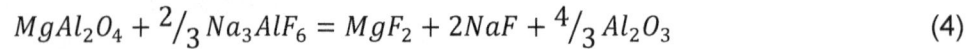

$$MgAl_2O_4 + {}^2/_3 Na_3AlF_6 = MgF_2 + 2NaF + {}^4/_3 Al_2O_3 \qquad (4)$$

This however needs to be studied further.

Corrosion test of NiAl₂O₄ in cryolite

Figure 8 shows the pictures of $NiAl_2O_4$ sample before and after corrosion testing in the 2.3 CR cryolite containing 6 wt per cent Al_2O_3. The $NiAl_2O_4$ samples showed similar behaviour with the $MgAl_2O_4$ samples discussed earlier, in which the part that had direct contact with cryolite – about 15 mm – disintegrated from the sample and went into the melt.

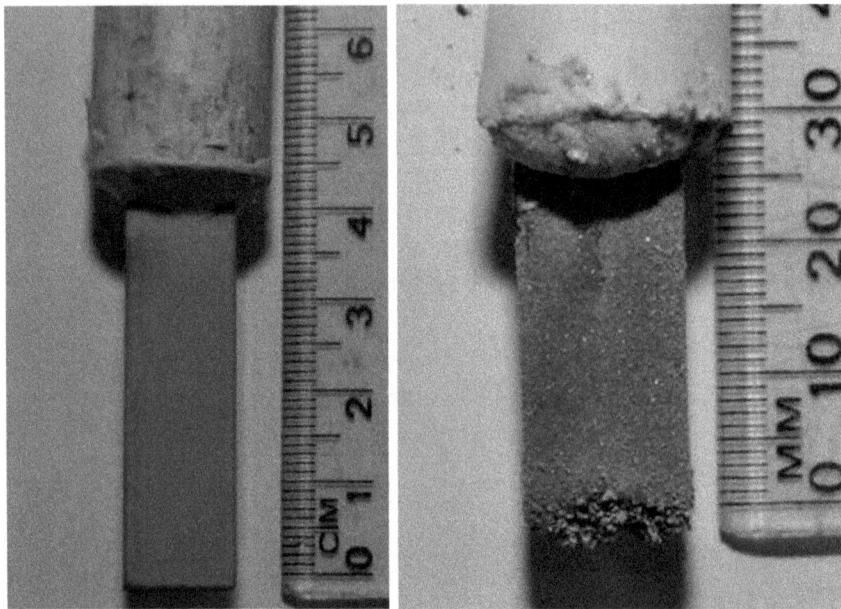

FIG 8 – Typical NiAl₂O₄ samples before (left) and after corrosion test (right). The part of the sample that had direct contact with cryolite goes into the melt.

If one considering only the thermodynamics stability of the nickel aluminate in cyolite melts, the effect of the melts to the sample as shown in Figure 8 was rather unexpected. The Gibbs free energy of the reaction between $NiAl_2O_4$ with cryolite at 980°C is positive (Bale *et al*, 2022). Moreover, predominance area diagram of the nickel solid phase at 980°C shown that $NiAl_2O_4$ was predicted to be a stable phase in the alumina-saturated melts. The $NiAl_2O_4$ samples were therefore expected not to be severely corroded, particularly by the alumina-saturated cryolite melts.

The concentration of nickel from $NiAl_2O_4$ in the cryolite melt over the sampling time is plotted in Figure 9. In general, the amount of nickel in the melts increased as the time increase. The total amount of nickel in the melt containing 6 wt per cent was higher than the nickel in alumina-saturated melt. The rate of the increase in the nickel content was also higher for the melt containing 6 wt per cent alumina.

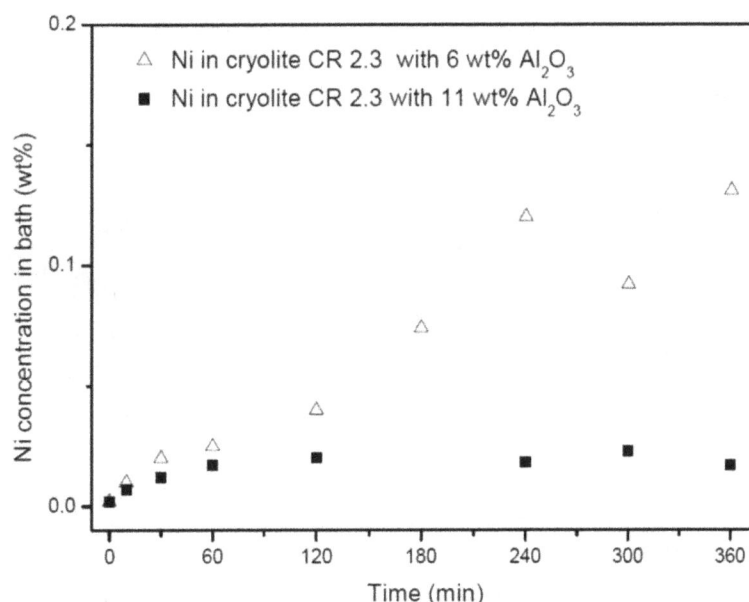

FIG 9 – Concentration of nickel originated from $NiAl_2O_4$ in cryolite-based melts contains different alumina concentration.

According to the previous research conducted by Lorentsen (2000), the saturation concentration, ie solubility limit, of nickel from $NiAl_2O_4$ in cryolite containing 6 wt per cent and 11 wt per cent alumina is 3.8×10^{-2} and 0.86×10^{-2} wt per cent, respectively. Surprisingly, after 6 hrs of experiment, the nickel contents in the melts were much beyond the solubility limit. The concentration of nickel in the melt containing 6 wt per cent and 11 wt per cent alumina was found to be 0.13 and 0.02 wt per cent. Owing to this, it was argued that the corrosion of $NiAl_2O_4$ by cryolite-based melts in current study was not only due to the dissolution of spinel constituents into the melts, but also due to the erosion on the spinel by the melts. The concentrations of nickel in the melt found in current study were suggested as the summation of the dissolved nickel in the melt and the nickel aluminate particles inside the melts due to erosion. By considering the integrity and the density of $NiAl_2O_4$ that were very low, it was reasonable to expect the high erosion level on the sample.

The microstructure of the $NiAl_2O_4$ sample after corrosion testing is shown in Figure 10. The microstructure shown in Figure 10a is associated with the top section of the solid part of the sample closer to sample holder, while Figure 10b is associated with the bottom section of the sample near to the immersed section (refer to Figure 8).

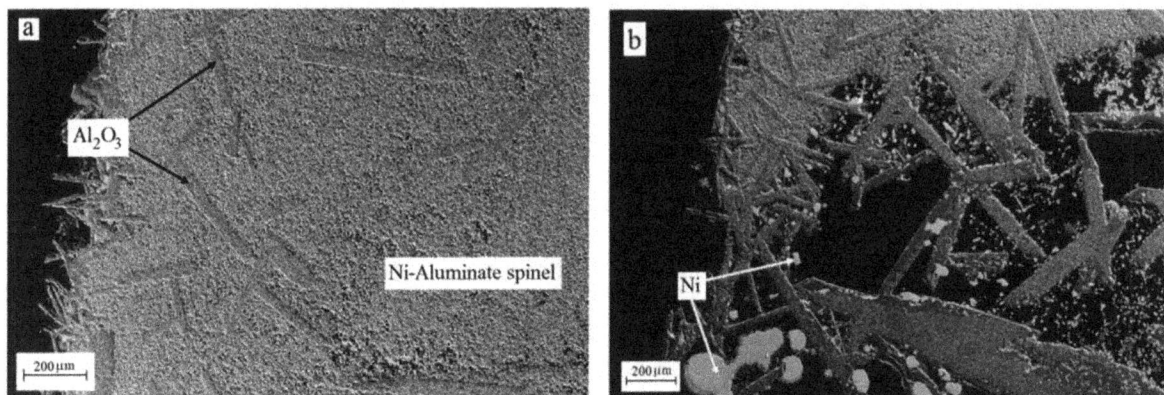

FIG 10 – Microstructure of $NiAl_2O_4$ sample after corrosion testing: (a) section of the sample closer to sample holder; (b) section of the sample near to the immersed section.

Both sections of the corrosion-tested $NiAl_2O_4$ sample in Figure 10 contained needle like-structures which are surrounded by the solid structures. Interestingly, in addition to these microstructures, there was some roundish white microstructure found in the section of the sample near to the immersed section (Figure 10b). EDS point-analysis suggested that the needle like structure was Al_2O_3. The solid structure was suggested as nickel aluminate as it contains Ni, Al and O; while the roundish white microstructure was suggested as metal since nickel was the only element detected in the spot. To further study the microstructure, EDS elemental mapping had been conducted on the zoomed part of the middle section and the bottom section as shown in Figures 11 and 12, respectively.

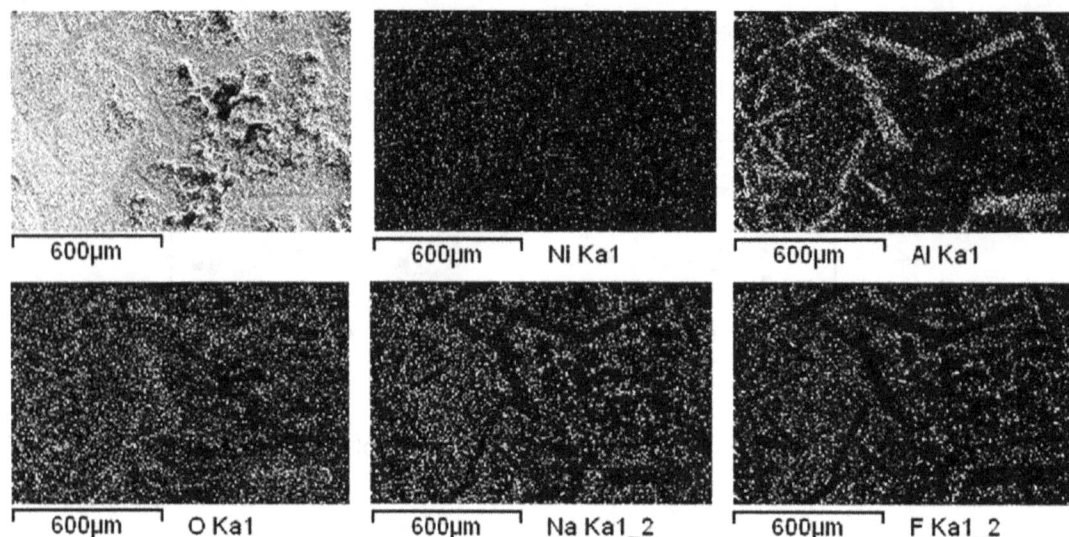

FIG 11 – EDS elemental mapping on the middle section of the corrosion-tested $NiAl_2O_4$ sample.

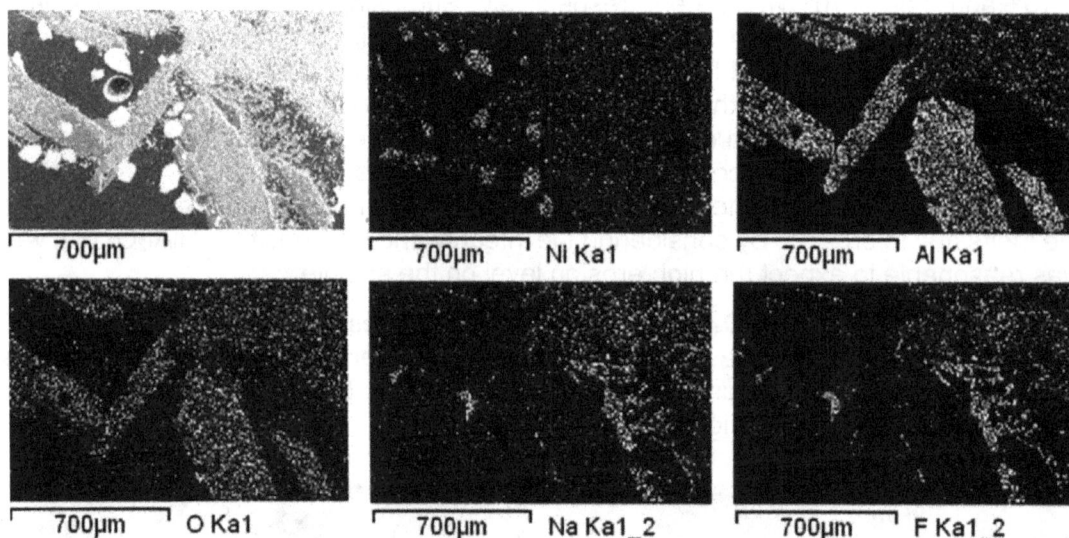

FIG 12 – EDS elemental mapping on the section of the corrosion-tested $NiAl_2O_4$ sample close to the immersed section.

The EDS elemental mapping results confirmed that there were sodium and fluorine elements exist in the solid structure that surrounded the Al_2O_3. This suggests that the cryolite penetrated into the $NiAl_2O_4$ sample beyond the section where the sample had direct contact with the molten bath. However, unlike the $MgAl_2O_4$ where there was no magnesium aluminate left in the sample after corrosion tested, there were still nickel aluminate left in the $NiAl_2O_4$ sample. This support the hypothesis that suggests the corrosion of the $NiAl_2O_4$ sample was not only due to dissolution of the spinel constituent by the melt, but also due to erosion.

Since there were significant amount of Al_2O_3 found in the corrosion-tested $NiAl_2O_4$ sample, the reaction in Equation 5 seems plausible.

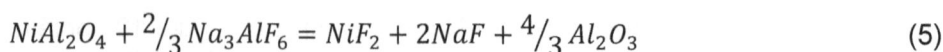

$$NiAl_2O_4 + \frac{2}{3} Na_3AlF_6 = NiF_2 + 2NaF + \frac{4}{3} Al_2O_3 \qquad (5)$$

The presence of metallic nickel in the bottom section of the sample suggested that NiF_2 dissociated into Ni. The explanation on how the nickel was formed need further study.

Corrosion test of NiFe$_2$O$_4$ in cryolite

The pictures of NiFe$_2$O$_4$ sample before and after corrosion testing are shown in Figure 13. The typical sample before test is shown in Figure 13a, while the appearance of the sample after 6 hrs immersion in the cryolite-based melts with CR 2.3 containing 6 wt per cent alumina and 11 wt per cent alumina is shown in Figure 13b and Figure 13c, respectively.

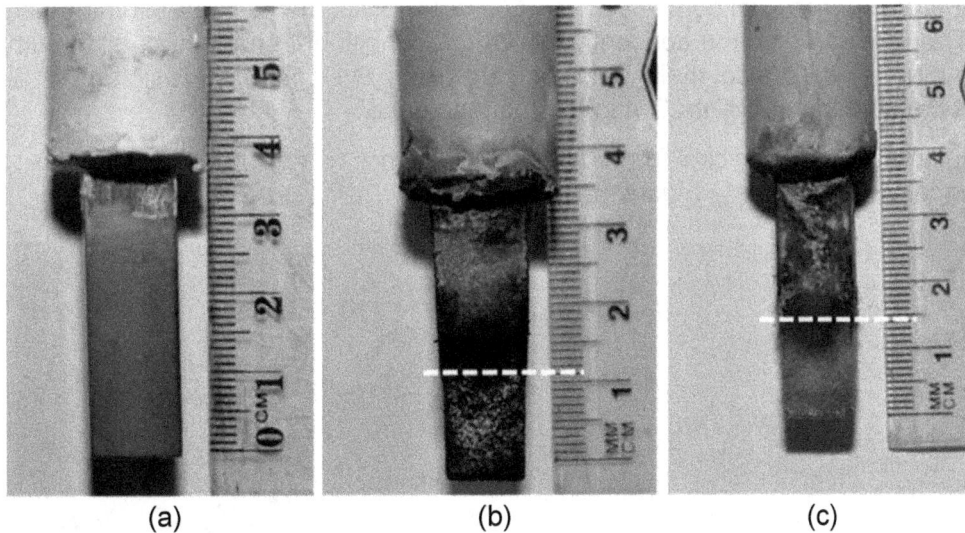

(a)   (b)   (c)

FIG 13 – NiFe$_2$O$_4$ sample for corrosion test in cryolite-based melts: (a) Typical sample before test; (b) after immersion in melt containing 6 wt per cent alumina; (c) after immersion in melt containing 11 wt per cent alumina. The white lines in (b) and (c) indicate the depth of immersion.

On the contrary to the cases of MgAl$_2$O$_4$ and NiAl$_2$O$_4$ samples, most of the immersed sections of the NiFe$_2$O$_4$ samples were still retained. There were indeed noticeable discolorations on the immersed sections. However, the dimension of the section immersed in the melts containing 6 wt per cent alumina ie length, width and thickness was slightly reduced, while the dimension of the section immersed in the melts containing 11 wt per cent alumina showed no change. This indicated that in cryolite-based melts, particularly in the alumina-saturated melt, NiFe$_2$O$_4$ exhibited the lowest corrosion compared to MgAl$_2$O$_4$ and NiAl$_2$O$_4$.

The change in the concentration of nickel ferrite constituents in the melt over time is plotted in Figure 14. In general, the concentration of the constituent increased as the time increase.

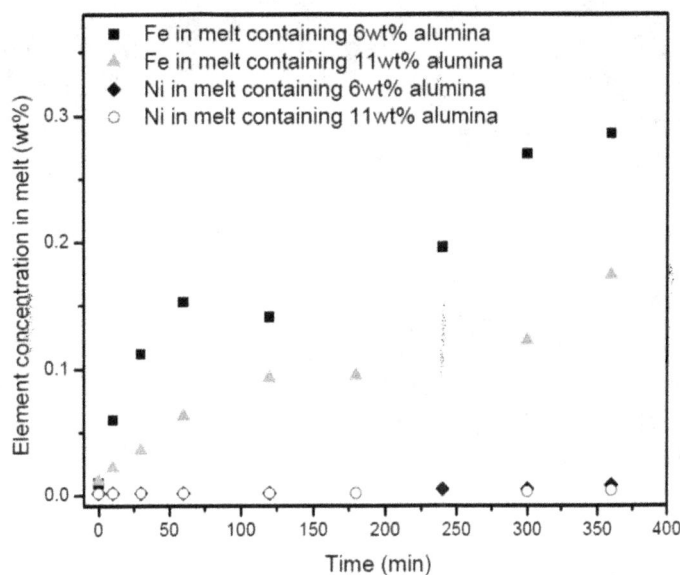

FIG 14 – Concentration of nickel ferrite constituents in cryolite-based melts as a function of time at 980°C.

It can be seen clearly in Figure 14 that the change in the iron concentration was considerably more rapid than the change in the nickel concentration. In both melts containing 6 wt per cent and 11 wt per cent alumina, the final iron concentration after 6 hrs test was also significantly higher than nickel and it did not reflect the stoichiometric ratio of iron to nickel in nickel ferrite. These results indicate preferential dissolution of iron in the melts, which agreed with other' studies (Yan, Pownceby and Brooks, 2007; Olsen and Thonstad, 1999a, 1999b). Compared to the melt containing 6 wt per cent alumina, the total concentration of $NiFe_2O_4$ constituents in the alumina-saturated melt was lower. The rate at which the concentration of the constituent changed over the time was also lower in the alumina-saturated melt. This indicates that the corrosion resistant of nickel ferrite increased as the alumina concentration in the melt increase.

The typical microstructure of nickel ferrite sample after 6 hrs corrosion test and its associated EDS elemental mapping are shown in Figure 15.

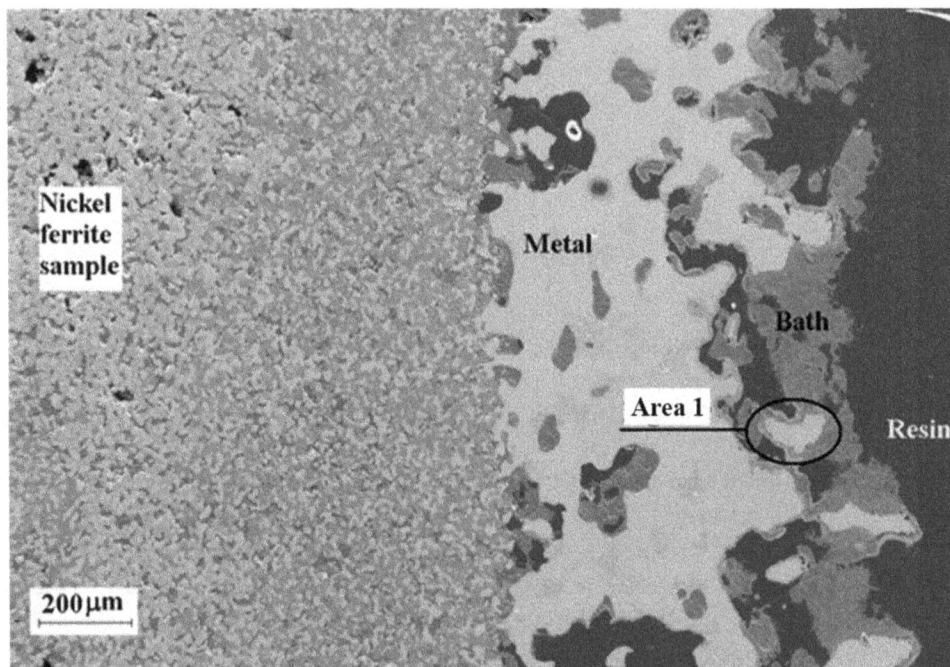

(a)

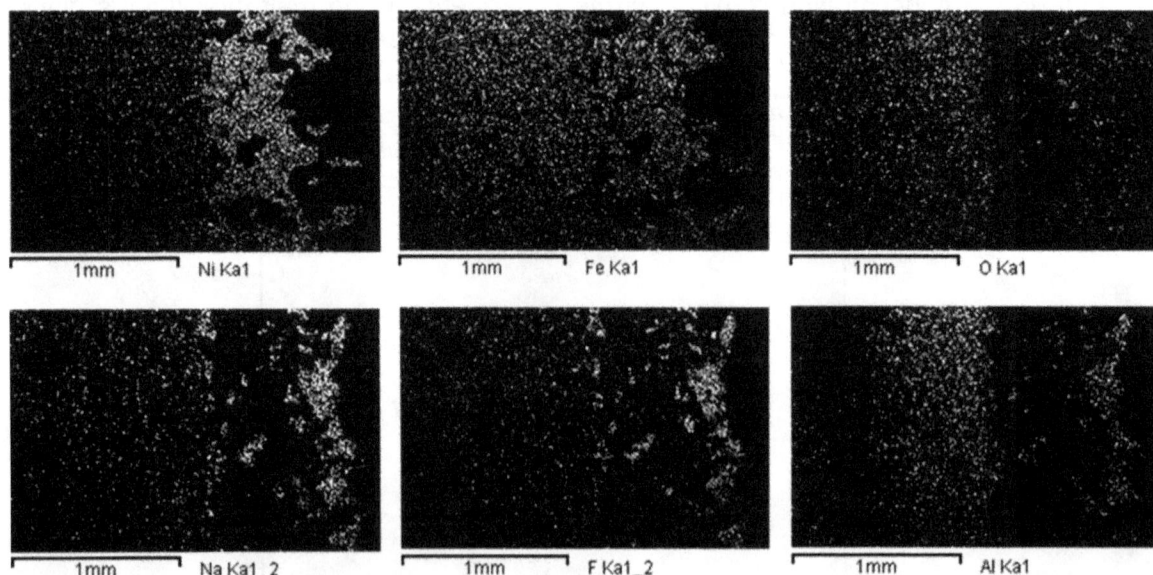

(b)

FIG 15 – (a) Typical microstructure of nickel ferrite sample near the sample-bath interface after 6 hrs corrosion testing in non-saturated cyolite based-melt; (b) EDS elemental mapping confirming the presence of Fe-Ni metallic phase at the sample-bath interface.

Figure 15 shows that there were significant composition changes occurring near the sample-bath interface where Ni-Fe metallic phase was formed. The presence of sodium, fluoride and aluminium in the bulk nickel ferrite sample indicated that the bath penetrated into the sample. A higher magnification of the area designated as 'Area 1' in Figure 15a revealed that there were several different microstructures surrounding the Ni-Fe metallic phases as shown in Figure 16.

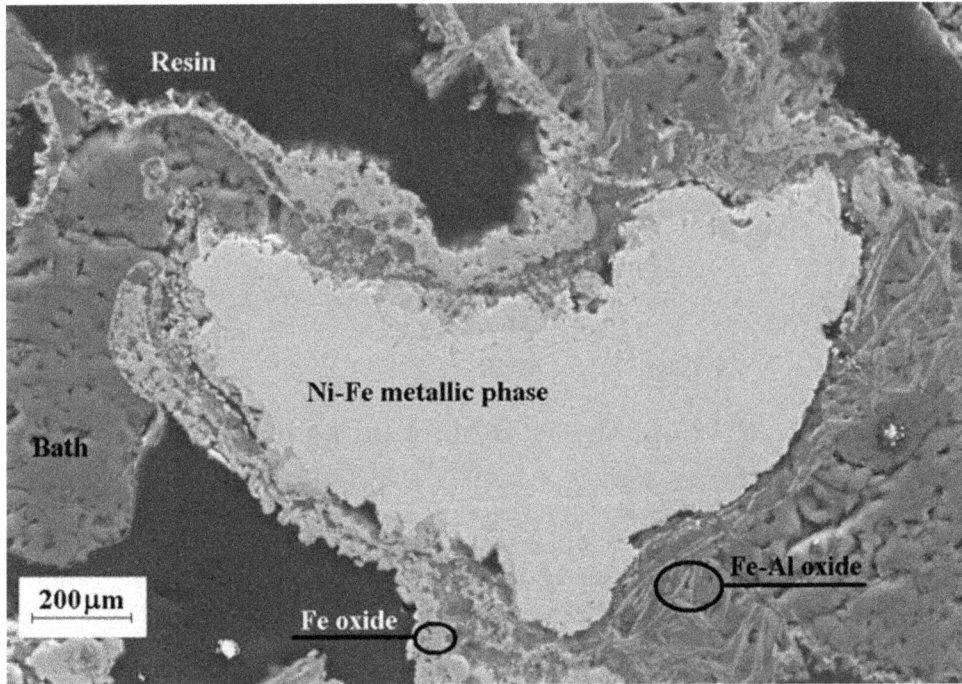

FIG 16 – Higher magnification of the area designated as 'Area 1' in Figure 15a. The phases were suggested by EDS point analysis and elemental mapping.

Dissolution of spinel constituents into cryolite-based melt

The kinetics and the dissolution rate of ferrite and aluminate spinels into cryolite-based melts have been studied elsewhere by the author (Mukhlis, 2015). Plot of $(V/A)(\ln[(c_{sat}-c_0)/(c_{sat}-c)])$ against time time (t) were used to determine the mass transfer coefficient rate (k value) of the process. Here:

V is the volume of electrolyte in m^3

A is the surface area of spinel sample that have direct contact with melt in m^2

c_{sat} is the initial amount of spinel constituent in the melt plus the maximum amount of the spinel constituent that can enter the melt from the sample in wt per cent

$c(t)$ is the time dependent concentration of species in the electrolyte in wt per cent

c_0 is the initial concentration of spinel constituent in melt

k is the mass transfer coefficient of species from spinel to electrolyte in m/s

The same approach, ie plot of $(V/A)(\ln[(c_{sat}-c_0)/(c_{sat}-c)])$ against time (t), using the data shown in Figure 14, was applied in the current study to determine the k value which then used to calculate the corrosion rate. The projected annual corrosion rate for each spinel samples is tabulated in Table 1.

TABLE 1

Projected corrosion rate of spinels against cryolite-based melt with CR 2.3 at 980°C.

Sample	Constituent	Alumina concentration in melt	Projected corrosion rate (cm/a)
MgAl$_2$O$_4$	Mg	6 wt%	644 × 10^3
		11 wt%	798 × 10^3
NiAl$_2$O$_4$	Ni	6 wt%	4.7 × 10^3
		11 wt%	1.03 × 10^3
NiFe$_2$O$_4$	Ni + Fe	6 wt%	947 × 10^3
		11 wt%	461 × 10^3

Projected corrosion rate was calculated based on the assumption that 1 year is equal to 365 days. The value for NiFe$_2$O$_4$ is the combination of values for Ni + Fe. As seen in Table 1, the projected corrosion rates for all spinel in cryolite-based electrolyte were very high.

Corrosion of NiFe$_2$O$_4$, NiAl$_2$O$_4$, MgAl$_2$O$_4$ in molten aluminium

The results of the finger test experiments of spinels in molten aluminium at 980°C through ICP analysis are shown in Figure 17. The concentration of spinel constituents – ie magnesium, nickel, and iron – in aluminium were plotted against the sampling time. The non-zero concentration of the constituents at t = 0 represented the impurity levels of the aluminium used in current study before the immersion of the spinel sample, which agreed with the specification from the manufacturer.

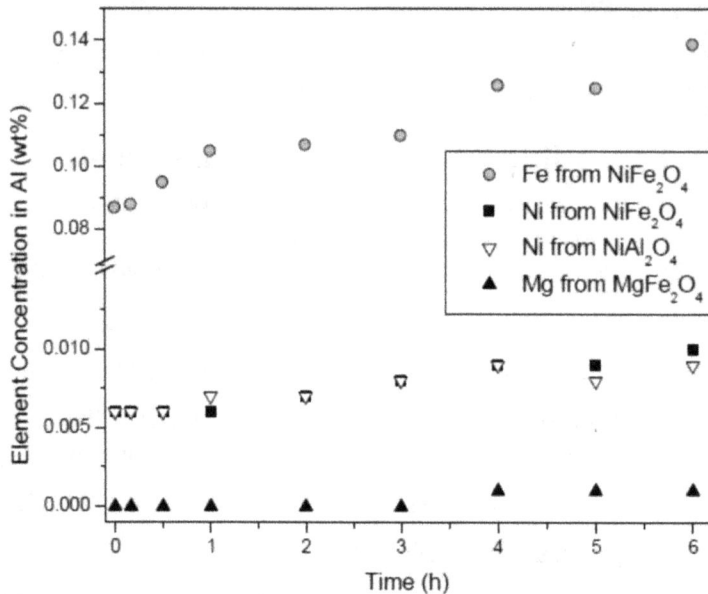

FIG 17 – Element concentrations in molten aluminium from spinel sample.

One can see from Figure 17 that in general, the concentrations of the spinel constituents in aluminium increased with increasing time up to 6 hrs. In agreement with the thermodynamics analysis, MgAl$_2$O$_4$ sample showed the lowest level of contamination to the aluminium at the end of the 6 hr experiments. The amount of nickel contaminations from NiAl$_2$O$_4$ and NiFe$_2$O$_4$ were at the same order of magnitude. The highest contamination in aluminium determined by current study was iron that came from nickel ferrite sample, which reached 0.14 wt per cent after 6 hrs immersion; that was two orders of magnitude higher than the magnesium contamination originated from magnesium aluminate.

Similar to the evaluation on the Spinel-Cryolite system, the rate constant of the corrosion of the spinel due to dissolution of its constituents into molten aluminium (ie mass transfer coefficient) was evaluated from the experimental data points plotted in the coordinates of $(V/A)(\ln[(c_{sat}-c_0)/(c_{sat}-c)])$ against time (t).

In contrast with the spinel-cryolite system, there were no visible changes of dimension on the spinel sample after 6 hr corrosion test in molten aluminium. Therefore, the confidence level of the corrosion rate value of the spinel-aluminium system is higher than that of spinel-cryolite system. The c_{sat} for Mg, Ni and Fe used in current study were 100, 38.2, and 15.5 wt per cent, respectively, which was determined from the associated Mg-Al, Ni-Al, and Fe-Al phase diagrams constructed using Swinburne University of Technology licensed FactSage™ 6.4. FactSage™ is an integrated database computing systems in chemical thermodynamics developed by Thermfact and GTT-Technologies.

Through the identical approach as the calculation for spinel-cryolite melt system, the projected values of the maximum corrosion rate of spinel due to dissolution of its constituent into molten aluminium were calculated to be 0.3, 2.6, and 78.4 cm/a for $MgAl_2O_4$, $NiAl_2O_4$, and $NiFe_2O_4$, respectively. It should be noted that the corrosion rate for $NiFe_2O_4$ is the summation of the dissolution rate of nickel and iron (refer to Table 2).

TABLE 2
Mass transfer coefficient and projected maximum corrosion rate of spinels constituents against molten aluminium.

Sample	Constituent	Element concentration in Al after 6 hr test	Highest possible mass transfer coefficient	Projected maximum corrosion rate
$MgAl_2O_4$	Mg	0.1×10^{-2} wt%	10.0×10^{-11} m/s	0.3 cm/a
$NiAl_2O_4$	Ni	0.9×10^{-2} wt%	8.25×10^{-10} m/s	2.6 cm/a
$NiFe_2O_4$	Ni	1.0×10^{-2} wt%	7.26×10^{-10} m/s	2.3 cm/a
	Fe	13.9×10^{-2} wt%	2.41×10^{-8} m/s	76.1 cm/a

Based on the data tabulated in Table 2, one can conclude that in the molten aluminium environment, $MgAl_2O_4$ showed the best performance as it had the lowest corrosion rate as well as the lowest contamination to the molten aluminium. The data also revealed that $NiFe_2O_4$ will not be suitable to be applied in direct contact with molten aluminium as it will promote substantial contamination to the metal. The projected corrosion rate of $NiFe_2O_4$ was also significantly high, about 50 times higher than the acceptable corrosion rate of the materials applied in the Hall-Héroult cell, which is supposed to be not higher than 1 to 1.5 cm/a (Xiao et al, 1996; Oye and Welch, 1998; Galasiu, Galasiu and Thonstad, 2007).

CONCLUSIONS

The observation on the spinel samples after corrosion testing indicate that $NiFe_2O_4$ has the best corrosion resistance against acidic cryolite-based melt (CR 2.3) compared to the other spinels. After 6 hrs immersion in the melt at 980°C, there were insignificant changes in the dimension of $NiFe_2O_4$ sample from its initial dimension. In contrast, the $MgAl_2O_4$ and $NiAl_2O_4$ samples were corroded heavily. All part of the samples that immersed in the melt disintegrated from the sample bulk and entered the melt. In the case of corrosion test against aluminium, all spinel samples (ie magnesium aluminate, nickel aluminate, and nickel ferrite) showed no dimensional changes macroscopically.

The projected corrosion rates, assuming linear behaviour, of the spinels against the bath were higher than the acceptable corrosion rate of materials applied in the aluminium smelter application. The projected corrosion rates of the spinels against molten aluminium, on the other hand was acceptable except for $NiFe_2O_4$. It was found that the corrosion rate for magnesium aluminate, nickel aluminate, and nickel ferrite against molten aluminium were 0.3, 2.6, and 78.4 cm/a. This result is in agreement with the thermodynamic analysis that suggests $MgAl_2O_4$ is the most stable spinel in aluminium environment compared to the other two spinels.

The nickel in the cryolite-based melt that originated from $NiAl_2O_4$ sample were found to be higher than the solubility limit of the respective element in the given bath composition. This indicate that the corrosion of the spinel not only due to dissolution of the spinel constituent into the bath, but also enhanced with the erosion of the sample. This was supported with the observation on the $NiAl_2O_4$ sample after corrosion test, where all part of the sample immersed in the bath (about 15 mm) was disintegrated from the sample bulk.

The presence of sodium and fluorine in the top part of the samples that far from the immersed section indicate that the cryolite-based melt had a good wettability against spinels; particularly $MgAl_2O_4$ and $NiAl_2O_4$. When the melt wetted the sample, it can readily penetrate into the sample through the sample pores.

The increase of the spinel constituent in the bath over the time for the denser spinel sample was found to be slower than that of high porosity sample. It is then recommended to use a much denser spinel for the aluminium smelter application. The use of denser spinel will reduce the erosion-induced corrosion and the penetration of the bath into the spinel. This however will hinder its industrial application since it will increase the cost and add more complexities to the industry.

Microstructural study on the corroded spinel sample suggested that $MgAl_2O_4$ spinel was unstable in the cryolite-based melt environment as there were no spinel left in the sample even at the section that far from the immersed part of the sample. All magnesium dissolved into the bath, leaving only aluminium oxide in the sample. $NiAl_2O_4$ and $NiFe_2O_4$ appeared to have better resistance to the reaction against cryolite-based melt (kinetically), particularly against alumina-saturated melt, as there were still plenty of spinel left in the sample after 6 hrs immersion. The $NiAl_2O_4$ samples however were susceptible to erosion.

ACKNOWLEDGEMENTS

The first author acknowledges Swinburne University of Technology and Commonwealth Scientific and Industrial Research Organisation (CSIRO) for providing scholarship to conduct the study. The co-supervision of Prof. Geoffrey Brooks, Dr. Kathie McGregor, and Dr. Xiao Yong Yan in the study are also acknowledged.

REFERENCES

ASTM, 2000. ASTM C20-00 Standard test method for apparent porosity, water absorption, apparent specific gravity and bulk density, *ASTM International,* (15).

Bale, C W, Chartrand, P, Degterov, S A, Eriksson, G, Hack, K, Ben Mahfoud, R, Melancon, J, Pelton, A D and Petersen, S, 2002. FactSage thermochemical software and databases, *CALPHAD*, 26(2):189–228.

Galasiu, I, Galasiu, R and Thonstad, J, 2007. *Inert anodes for aluminium electrolysis,* Dusseldorf: Aluminium-Verlag.

Grjotheim, K, Krohn, C, Malinovsky, M, Matiasovsky, K and Thonstad, J, 1982. *Aluminium Electrolysis - Fundamentals of the Hall-Héroult Process,* 2nd edition, Dusseldorf: Aluminium-Verlag.

Haraldsson, J and Johansson, M T, 2020. Effects on primary energy use, greenhouse gas emissions and related costs from improving energy end-use efficiency in the electrolysis in primary aluminium production, *Energy Efficiency,* 13(7):1299–1314.

Kvande, H and Drablos, P A, 2014. The Aluminum Smelting Process and Innovative Alternative Technologies, *Journal of Occupational and Environmental Medicine,* 56:S23–S32.

Kvande, H and Haupin, W, 2001. Inert anodes for Al smelters: Energy balances and environmental impact, *JOM,* 53(5):29–33.

Longbottom, R J, Nightingale, S A and Monaghan, B J, 2014. Thermodynamic considerations of the corrosion of nickel ferrite refractory by Na3AlF6-AlF3-CaF2-Al2O3 bath, *Mineral Processing and Extractive Metallurgy, Trans Inst Min Metall C,* 123(2):93–103.

Lorentsen, O A, 2000. Behaviour of nickel, iron and copper by application of inert anodes in aluminium production, Doctoral Thesis, Trondheim: Department of Electrochemistry, Norwegian University of Science and Technology.

Mukhlis, R Z, 2015. Analysis of Spinel Materials for Aluminium Smelter Applications, Doctoral dissertation, Swinburne University of Technology Melbourne, Australia.

Mukhlis, R Z, Rhamdhani, M A, Brooks, G A and Yan, X Y, 2011. Fabrication of spinel composites for sidewall of Al smelter application, In *European Metallurgical Conference* 2011, pp 865–880.

Mukhlis, R Z, Rhamdhani, M A, Brooks, G and McGregor, K, 2014. Wetting characteristics of cryolite-based melts on spinels substrate, *Light Metals 2014,* pp 609–614.

Mukhlis, R, Rhamdhani, M and Brooks, G, 2010. Sidewall materials for the Hall-Heroult process, *Light Metals*, pp 883–888.

Olsen, E and Thonstad, J, 1999a. Nickel ferrite as inert anodes in aluminium electrolysis: Part I Material fabrication and preliminary testing, *Journal of Applied Electrochemistry*, 29(3):293–300.

Olsen, E and Thonstad, J, 1999b. Nickel ferrite as inert anodes in aluminium electrolysis: Part II material performance and long-term testing, *Journal of Applied Electrochemistry*, 29(3):301–311.

Øye, H A and Welch, B J, 1998. Cathode performance: The influence of design, operations, and operating conditions, *JOM,* 50(2):18–23.

Schwarz, H G, 2008. Technology diffusion in metal industries: driving forces and barriers in the German aluminium smelting sector, *Journal of Cleaner Production,* 16(1):S37–S49.

Thonstad, J, Fellner, P, Haarberg, G M, Hives, J, Kvande, H and Sterten, A, 2001. Aluminum electrolysis, *Fundamentals of the Hall-Heroult Process,* 3rd edition, Dusseldorf: Aluminum-Verlag.

Xiao, H, Hovland, R, Rolseth, S and Thonstad, J, 1996. Studies on the corrosion and the behavior of inert anodes in aluminum electrolysis, *Metallurgical and Materials Transactions B*, 27(2):185–193.

Yan, X Y, Mukhlis, R Z, Rhamdhani, M A and Brooks, G A, 2011. Aluminate spinels as sidewall linings for aluminum smelters, *Light Metals 2011*, pp 1085–1090.

Yan, X Y, Pownceby, M I and Brooks, G, 2007. Corrosion behavior of nickel ferrite-based ceramics for aluminum electrolysis cells, *Light Metals 2007*, pp 909–913.

The interaction between slag and MgO refractory at conditions relevant to nickel laterite ore smelting

Y H Putra[1], Z Zulhan[2], A D Pradana[3], D R Pradana[4] and T Hidayat[5]

1. Junior Business Analyst, PT Bukit Asam, Tanjung Enim, South Sumatera 31716, Indonesia. Email: yudhithp.13@gmail.com; Previously: Researcher, Metallurgical Engineering Research Group, Faculty of Mining and Petroleum Engineering, Bandung Institute of Technology, Bandung, West Java 40132, Indonesia.
2. Professor, Metallurgical Engineering Research Group, Faculty of Mining and Petroleum Engineering, Bandung Institute of Technology, Bandung, West Java 40132, Indonesia. Email: zulfiadi.zulhan@itb.ac.id
3. Business Development Manager, PT Gunbuster Nickel Industry, Jakarta 12190, Indonesia. Email: gika@bharuna.com
4. Operation Supervisor, PT Gunbuster Nickel Industry, Jakarta 12190, Indonesia. Email: dimas@efi.co.id
5. Lecturer, Metallurgical Engineering Research Group, Faculty of Mining and Petroleum Engineering, Bandung Institute of Technology, Bandung, West Java 40132, Indonesia. Email: t.hidayat@itb.ac.id

ABSTRACT

The high temperature processing of nickel laterite ore is currently dominated by the Rotary Kiln-Electric Furnace (RKEF) technology. The ratio of SiO_2/MgO (S/M) in the ore is one of the critical parameters that determines the success of the RKEF process. Incompatible S/M ratios in the ore can cause aggressive chemical interaction between the slag and MgO refractory. The slag-refractory interaction at conditions relevant to nickel laterite ore smelting was investigated in the present study. Synthetic mixtures representing the slag compositions of nickel laterite ore smelting were prepared by mixing MgO, SiO_2, Al_2O_3, CaO, Fe_2O_3, and Fe and heating the mixtures in steel containers at 1300°C under an argon atmosphere for 3 hrs. Pure MgO crucible were used as containment material in the melting process of the mixtures to represent the MgO refractory. The kinetics of the slag-refractory reaction were investigated using samples with selected initial S/M ratios of 2.0 and 3.0 (wt/wt) (with constant CaO = 3 wt per cent, Al_2O_3 = 5 wt per cent, FeO = 10 wt per cent) by performing melting at 1500°C in MgO crucibles for 5, 30, and 120 mins under an argon atmosphere. The effect of the S/M ratio in slag on the slag-refractory interaction was also investigated by performing melting of mixtures with different initial S/M ratios of 1.75, 2.0, 2.5, and 3.0 (wt/wt) in MgO crucibles at 1500°C for 120 mins under an argon atmosphere. After the melting process, each sample was quenched in water, mounted in resin, ground, polished, and coated. The interaction between the slags and MgO refractory was then evaluated by analysing the polished samples using a scanning electron microscope equipped with an energy dispersive spectroscopy (SEM-EDS). Thermodynamic simulations using FactSage™ 8.0 thermochemical software were performed based on the experimental slag compositions, refractory, and conditions to support the evaluation process. The results show that a higher S/M ratio in slag leads to a higher tendency for the refractory component to dissolve into the slag. A higher S/M ratio in slag also leads to an increasing extent of solid-state diffusion of iron oxide into the remaining refractory grains. The dissolution of MgO into the slag is rapid; the MgO concentration in the final slag increased by 2.4 wt per cent and 5.7 wt per cent at initial S/M ratios of 2.0 and 3.0 wt/wt, respectively, within 5 and 30 mins of the melting process.

INTRODUCTION

The Rotary Kiln – Electric Furnace (RKEF) technology is commonly employed for the production of ferronickel (FeNi) or nickel pig iron (NPI), with nickel and iron contents ranging from 10–30 per cent and 70–90 per cent respectively (Crundwell *et al*, 2011). The simplified flow sheet of RKEF process is provided in Figure 1. The RKEF process involves three primary stages in processing nickel ore. The initial stage entails drying the ore in a rotary dryer at 250°C, which serves to decrease the water content from 30 per cent to 20 per cent. Subsequently, the ore undergoes calcination in a rotary kiln at temperatures between 750–900°C, during which remaining water content is removed and the nickel oxide and iron oxide are partially reduced by the reductant. The resulting calcine, which is the product from the rotary kiln, is then smelted in an electric furnace at temperatures ranging from 1450–1550°C in order to form and separate the metal product from the slag. In a few RKEF plants, an additional purification step may be carried out using a ladle furnace to reduce impurity content in the final product.

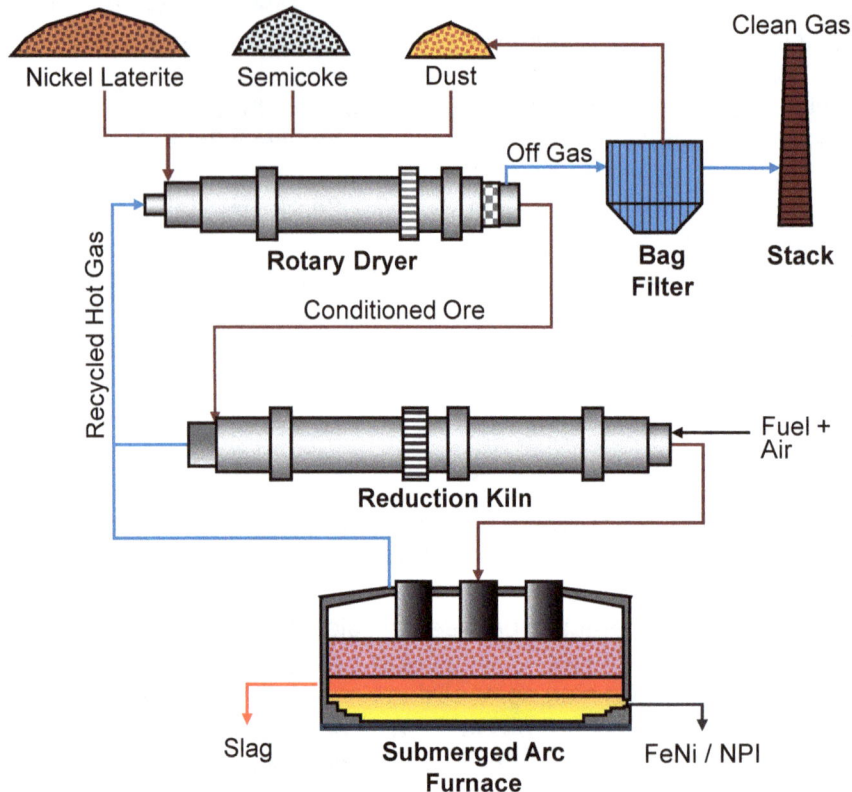

FIG 1 – Simplified flow sheet of RKEF process.

The electric furnace is a widely employed smelting technology in various metal industries. It operates on the principle of resistance heating, where the burden materials such as calcine, slag, and molten metal are heated through the application of electric current (Degel *et al*, 2007). To withstand the high temperatures and harsh conditions within the furnace, refractory materials are used for lining purposes. In the nickel industry, magnesia (MgO) refractory is utilised as a lining for the electric furnace. This choice is due to the high melting point of magnesia, its favourable mechanical properties at elevated temperatures, and its resistance to adverse environments. However, there are a few disadvantages of using MgO refractory, such as it has a very high conductivity, which can result in significant heat loss during the smelting process (Ruh and McDowell, 1962) and it has a relatively high thermal expansion compared to other types of refractories (Gangler, 1950).

The malfunction of refractory material not only leads to a loss of productivity but also poses risks to the surrounding environment (Zhang and Lee, 2013). Consequently, the investigation of the interaction between slag and refractories has garnered significant attention from researchers (Huang, Xue and Wang, 2017; Sagadin *et al*, 2016, 2017, 2018a, 2018b, 2021; Wagner *et al*, 2017) within the nickel industry. Generally, refractory damage can be attributed to three primary factors: chemical attack, thermal pressure, and mechanical pressure. Among these, chemical attack predominantly occurs as a result of the chemical interaction between the refractory and slag. In a number of electric furnaces, water-cooling components are incorporated into the lining structure to absorb heat. This facilitates the formation of a thin freeze lining, which serves to protect the refractories from corrosion caused by the slag (Kotzé, 2002). While in some of electric furnaces, the water-cooling components are not available, hence the furnace integrity relies heavily on the chemical interaction between slag and refractory material. This study aims to investigate the slag and refractory interaction under conditions relevant to ferronickel or NPI smelting as a function of initial SiO$_2$/MgO ratio in slag and melting time. Synthetic slag was used to replicate the industrial slag and a dense pure MgO crucible was used to simulate the actual refractory material. The use of a dense MgO crucible eliminates the effects of physical infiltration of slag due to the heterogeneity and porosity commonly found in the refractory material, thus focusing the investigation mainly on the chemical attack of slag on the refractory material.

EXPERIMENTAL

Preparation and characterisation of synthetic slag

The synthetic slag used in this experiment was produced by mixing high purity reagents, including MgO, SiO$_2$, Al$_2$O$_3$, CaO, Fe$_2$O$_3$, and Fe. These reagents were then blended with specific compositions to achieve constant FeO, Al$_2$O$_3$, and CaO contents of 10 per cent, 5 per cent, and 3 per cent, respectively, in the slag. The chosen compositions were based on typical slag compositions in one of the ferronickel or NPI plants. As for the SiO$_2$ and MgO, their compositions were adjusted so that the final SiO$_2$/MgO or S/M ratios were set around 1.75, 2.0, 2.5, and 3.0. The reagents were mixed and homogenised using a mortar for 10 mins. The resulting mixtures were then placed in cylindrical stainless steel 304 crucibles with dimension of 20 mm in diameter and 60 mm in height. The crucibles were then positioned on an alumina boat to enable heating using a horizontal tube furnace. The synthetic slags were heated in the horizontal tube furnace at a temperature of 1300°C, aimed at pre-reacting the oxides within the synthetic slags before melting them with the MgO crucible. The synthetic slags obtained from the heating in the horizontal tube furnace were then homogenised again using a mortar for 10 mins. The compositions of the synthetic slags with different S/M ratios were checked using scanning electron microscope with energy dispersive spectroscopy (SEM-EDS). The results of the SEM-EDS semi-quantitative analysis for all samples are presented in Table 1.

TABLE 1

Semi-quantitative EDS-measured composition of synthetic slag.

Target S/M	Composition (wt%)					
	SiO$_2$	MgO	FeO	Al$_2$O$_3$	CaO	S/M
1.75	51.9	31.1	8.9	5.4	2.6	1.67
2.0	56.2	27.2	8.6	4.8	2.9	2.07
2.5	60.4	24.2	8.6	4.4	2.4	2.50
3	62.4	21.0	10.1	4.0	2.5	2.97

Melting in vertical tube furnace

There are several experimental methods commonly used to study the slag-refractory interaction, including (Zhang and Lee, 2004):

1. Button or sensible drop test.

2. Dipping, immersion, or finger test.

3. Crucible, cavity, cup, or brick.

4. Induction furnace test.

5. Rotating finger test.

6. Rotating slag test

The interaction between slag and refractory in this study was conducted using the cup test technique. This technique is easier to implement and ensures that the entire slag interacts with the refractory compared to other techniques, such as the finger test. However, this technique also has limitations, such as the inability to study the corrosion effects of fluid movement and the tendency of slag samples to become saturated due to the significantly smaller proportion of slag compared to the refractory. In industrial settings, the proportion of slag should be much larger than that of the refractory. The synthetic slag was weighed at 1.3 g and then placed into the MgO crucible. The MgO crucible with dimension of 16 mm in diameter and 20 mm in height was used. The fusion of synthetic slag with the MgO refractory was carried out using a vertical tube furnace. The sample was suspended using molybdenum wire and held within the hot zone of the furnace under argon atmosphere at 1500°C for 5, 30, and 120 mins. After melting, the samples were rapidly quenched in water.

Product preparation and examination

The quenched samples were dried and embedded in resin to prevent them from breaking during the sample preparation process. The samples were then cut in half at their midsection using a ceramic cutter, remounted in resin, and polished using an automated polishing device. The polished samples obtained from the experiments were examined using an optical microscope (Jiangxi Phoenix L2030A Trinocular Microscope) and semi-quantitatively analysed using scanning electron microscope with energy dispersive spectroscopy (SEM-EDS, JEOL NeoScope JCM-7000). The semi-quantitative SEM-EDS analysis was conducted at the refractory-slag interface at 12 different locations distributed approximately 0.5 mm towards the refractory and 0.5 mm towards the slag, as shown in Figure 2. Various reference materials were used to validate the accuracy of the compositional analysis by the EDS detector. The reference materials used in the present measurement were Basaltic glass (NMNH 113498-1) and Springwater Olivine (USNM 2566) of the Smithsonian Microbeam Standards obtained from the Smithsonian National Museum of Natural History, USA. It was observed that the compositions measured by EDS were within a deviation of less than 1.4 per cent by weight compared to those reported in the reference certificates.

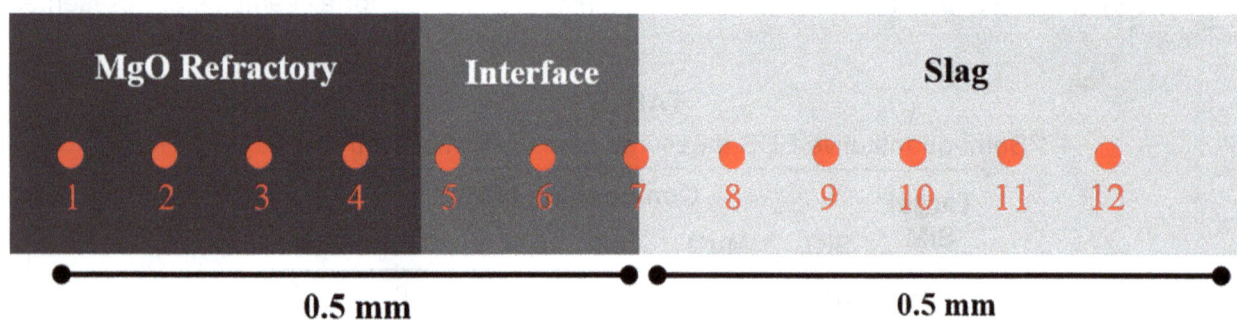

FIG 2 – Illustration of semi-quantitative analysis of composition along refractory and slag interface.

RESULTS AND DISCUSSION

The reaction kinetics of MgO dissolution from MgO refractory into slag

The kinetics of MgO dissolution from the refractory into the slag were investigated at a temperature of 1500°C using synthetic slags with initial SiO_2/MgO (S/M) ratios of 2.00 and 3.00, with melting times varied between 5, 30, and 120 mins. An example of secondary electron image from kinetic experiments using a mixture with an initial S/M ratio of 3.00 reacted at 1500°C can be seen in Figure 3. After melting for 5 mins, the sample shows both slag and several solid particles (see Figure 3a). The solid particles appear to be part of oxides from the initial synthetic slag mixture, which are in the process of dissolving into the liquid slag phase. A fully liquid slag is observed after melting for 30 mins (see Figure 3b). After 120 mins of melting, slag along with olivine is observed (see Figure 3c). It is worth noting that the surface of the MgO refractory appears relatively smooth after melting for 5 and 30 mins compared to after 120 mins of melting. The relatively smooth MgO surface indicates the ongoing dissolution process of solid MgO into the liquid slag.

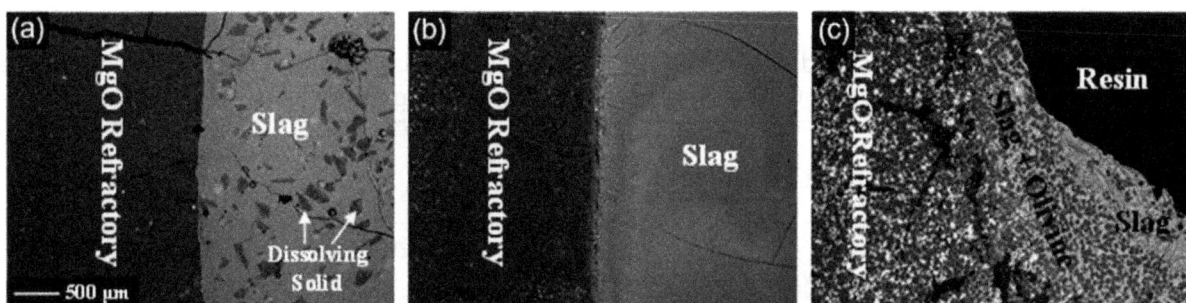

FIG 3 – Examples of secondary electron images from samples melted at 1500°C with an initial SiO_2/MgO ratio in slag of 3.00 for melting times of: (a) 5 mins; (b) 30 mins; and (c) 120 mins.

The semi-quantitative EDS measured compositions of the samples along refractory and slag interface from kinetic experiments using samples with initial S/M ratios of 2.00 and 3.00 are provided in Tables 2 and 3, respectively. The compositions of the dominant components, ie MgO, SiO_2, and FeO, are plotted in Figures 4 and 5. The dissolution of MgO appears to be rapid so that after the reaction continued from 5 mins to 30 mins, the MgO concentration in slag increased significantly, approaching the MgO concentration from melting for 120 mins. On the other hand, SiO_2 concentration in slag continuously decreases due to dilution by the dissolving MgO. There appears to be no clear trend regarding FeO concentration in slag with melting time. In the case of the MgO refractory composition, it appears to remain unchanged in most of experiments, except for the experiment using a sample with an initial S/M ratio in slag of 3.00 and a melting time of 120 mins, which showed a significant increase in FeO concentration within the refractory.

TABLE 2

Semi-quantitative EDS measured composition along refractory-slag interface from samples melted at 1500°C with an initial SiO_2/MgO ratio in slag of 2.00 for melting times between 5 and 120 mins.

Sample		Position											
		1	2	3	4	5	6	7	8	9	10	11	12
t = 5 min	MgO	98.6	98.9	98.2	99.0	98.7	98.6	87.8	30.6	30.3	24.7	24.8	26.9
	SiO_2	0.7	0.6	0.4	0.3	0.3	0.5	0.7	53.5	51.7	56.1	56.3	55.3
	Al_2O_3	0.3	0.3	0.4	0.3	0.5	0.4	2.3	2.7	5.1	5.2	5.6	5.3
	CaO	0.2	0.1	0.7	0.2	0.1	0.1	0.3	1.5	2.5	2.6	2.7	2.6
	FeO	0.2	0.2	0.5	0.1	0.4	0.4	8.9	11.7	10.4	11.4	10.7	10.0
t = 30 min	MgO	98.9	99.0	98.7	98.4	97.9	95.7	82.9	45.3	37.3	30.6	31.8	29.3
	SiO_2	0.4	0.3	0.3	0.7	0.6	0.8	1.1	45.7	48.4	51.0	50.4	52.3
	Al_2O_3	0.5	0.5	0.5	0.4	0.4	0.6	0.3	0.9	3.6	5.2	4.9	5.1
	CaO	0.1	0.0	0.1	0.2	0.1	0.2	0.2	0.4	1.6	2.7	2.1	2.6
	FeO	0.2	0.2	0.5	0.3	1.1	2.7	15.5	7.7	9.1	10.5	10.8	10.8
t = 120 min	MgO	98.6	98.1	98.1	96.4	95.5	92.3	54.3	37.8	34.0	32.4	32.3	31.4
	SiO_2	0.3	0.3	0.3	0.8	0.3	0.5	42.0	50.7	49.1	48.4	49.5	52.0
	Al_2O_3	0.4	0.3	0.5	0.6	0.7	1.1	1.5	3.3	5.5	6.5	6.0	4.5
	CaO	0.0	0.1	0.0	0.0	0.0	0.1	0.1	1.3	2.5	3.4	3.1	3.2
	FeO	0.6	1.2	1.1	2.2	3.4	6.0	2.1	6.9	8.9	9.4	9.1	8.9
		REFRACTORY							**OLIVINE**		**SLAG**		

TABLE 3

Semi-quantitative EDS measured composition along refractory-slag interface from samples melted at 1500°C with an initial SiO_2/MgO ratio in slag of 3.00 for melting times between 5 and 120 mins.

Sample		1	2	3	4	5	6	7	8	9	10	11	12
							Position						
t = 5 min	MgO	98.8	97.6	98.8	98.6	97.6	96.4	85.9	26.5	26.7	26.5	26.0	24.2
	SiO_2	0.6	1.0	0.6	0.6	1.2	1.2	1.7	58.1	59.4	59.8	60.5	59.5
	Al_2O_3	0.5	0.7	0.4	0.6	0.6	0.5	0.8	4.0	3.5	3.8	4.0	4.2
	CaO	0.2	0.3	0.1	0.1	0.2	0.3	1.9	2.8	2.4	2.3	2.4	2.9
	FeO	0.0	0.3	0.2	0.1	0.5	1.6	9.8	8.6	8.0	7.7	7.1	9.2
t = 30 min	MgO	98.5	98.8	98.1	97.9	97.2	94.3	67.6	39.1	30.7	27.4	31.0	29.9
	SiO_2	0.1	0.3	0.8	0.5	0.5	1.2	6.0	50.2	54.8	56.9	54.6	55.7
	Al_2O_3	0.5	0.6	0.6	0.6	0.6	0.6	2.2	1.5	3.4	3.9	4.0	4.2
	CaO	0.0	0.1	0.1	0.1	0.1	0.2	0.2	0.7	2.0	2.4	2.0	2.2
	FeO	0.8	0.3	0.4	0.9	1.6	3.7	24.1	8.5	9.2	9.4	8.4	8.0
t = 120 min	MgO	84.1	84.8	84.7	79.3	65.4	68.6	46.0	41.1	40.1	31.2	30.9	28.9
	SiO_2	3.7	1.0	1.3	1.4	6.5	2.2	32.6	46.0	46.6	50.2	51.4	52.5
	Al_2O_3	3.0	1.0	0.6	1.1	1.6	1.4	4.9	4.2	4.1	6.5	5.1	5.9
	CaO	0.9	0.6	0.5	0.8	3.5	0.8	2.5	2.1	1.8	3.3	2.4	2.6
	FeO	8.4	12.6	13.0	17.4	23.1	27.0	14.0	6.6	7.4	8.8	10.3	10.3
		REFRACTORY						**OLIVINE**		**SLAG**			

The bulk compositions of liquid slag (the compositions of liquid slag at the furthest position from the refractory, ie at point 12) from the kinetic experiments using samples with initial S/M ratios in slag of 2.00 and 3.00 are provided in Figure 6a and 6b, respectively. The increasing MgO and decreasing SiO_2 in the slag can be seen clearly in both figures. Both of these lead to an increasing S/M ratio in the final liquid slag with increasing melting time, as shown in Figure 6c. The dissolution of MgO is rapid, so as the reaction proceeds from 5 mins to 30 mins, the MgO concentration in the slag increases significantly by 2.4 wt per cent and 5.7 wt per cent for initial S/M ratios of 2.0 and 3.0 wt/wt, respectively. However, when the reaction proceeds from 30 mins to 120 mins, the dissolution of MgO is less rapid and approaches completion. The slowdown in the dissolution of MgO is possibly due to the reduced driving force for the MgO dissolution as the liquid approaches MgO saturation condition, which later can lead to the formation of olivine precipitates on the MgO surface.

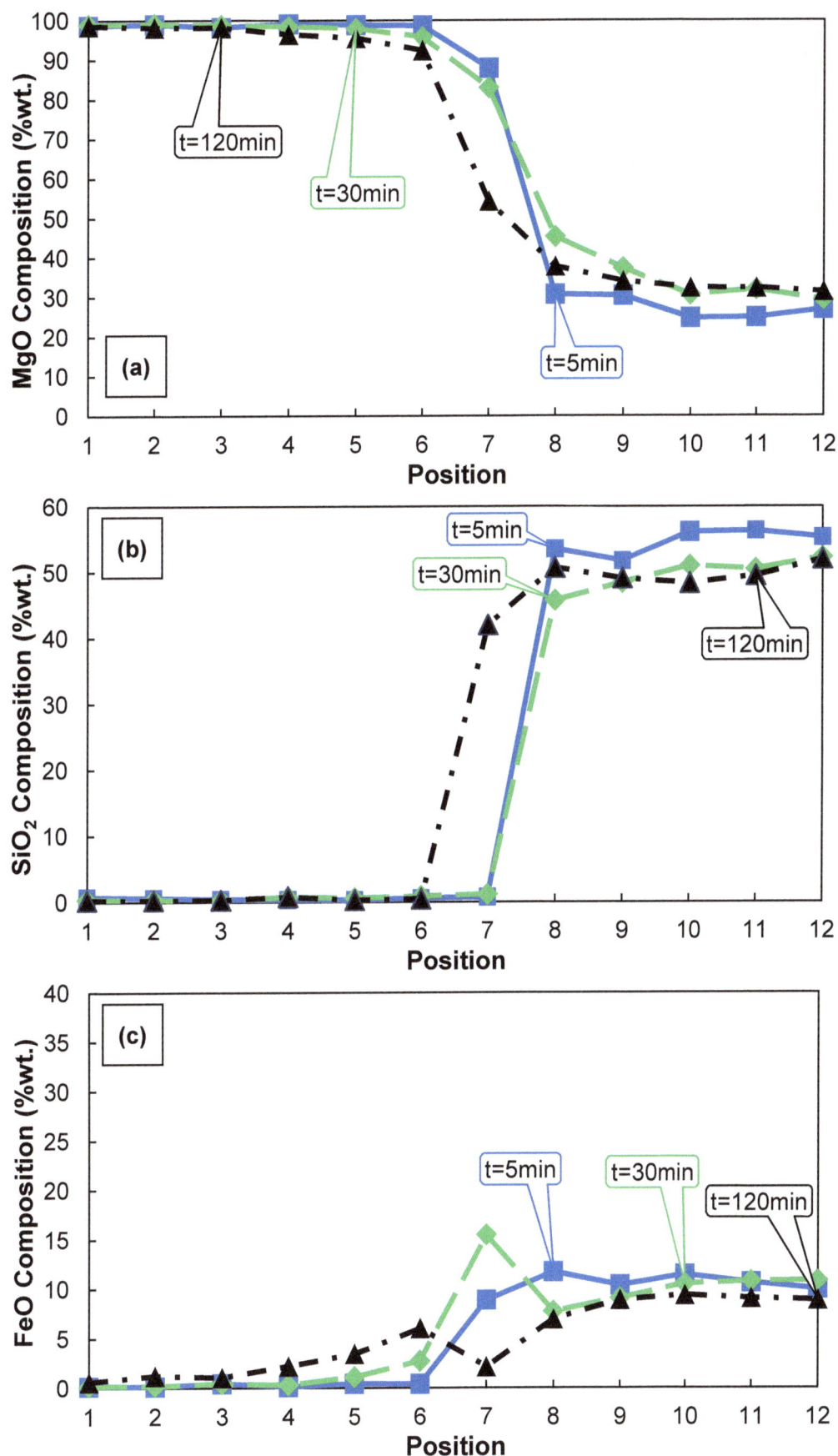

FIG 4 – Semi-quantitative compositional analysis along refractory-slag interface from samples melted at 1500°C with an initial SiO_2/MgO in slag ratio of 2.00 for melting time between 5 and 120 mins: (a) MgO composition; (b) SiO_2 composition; and (c) FeO composition.

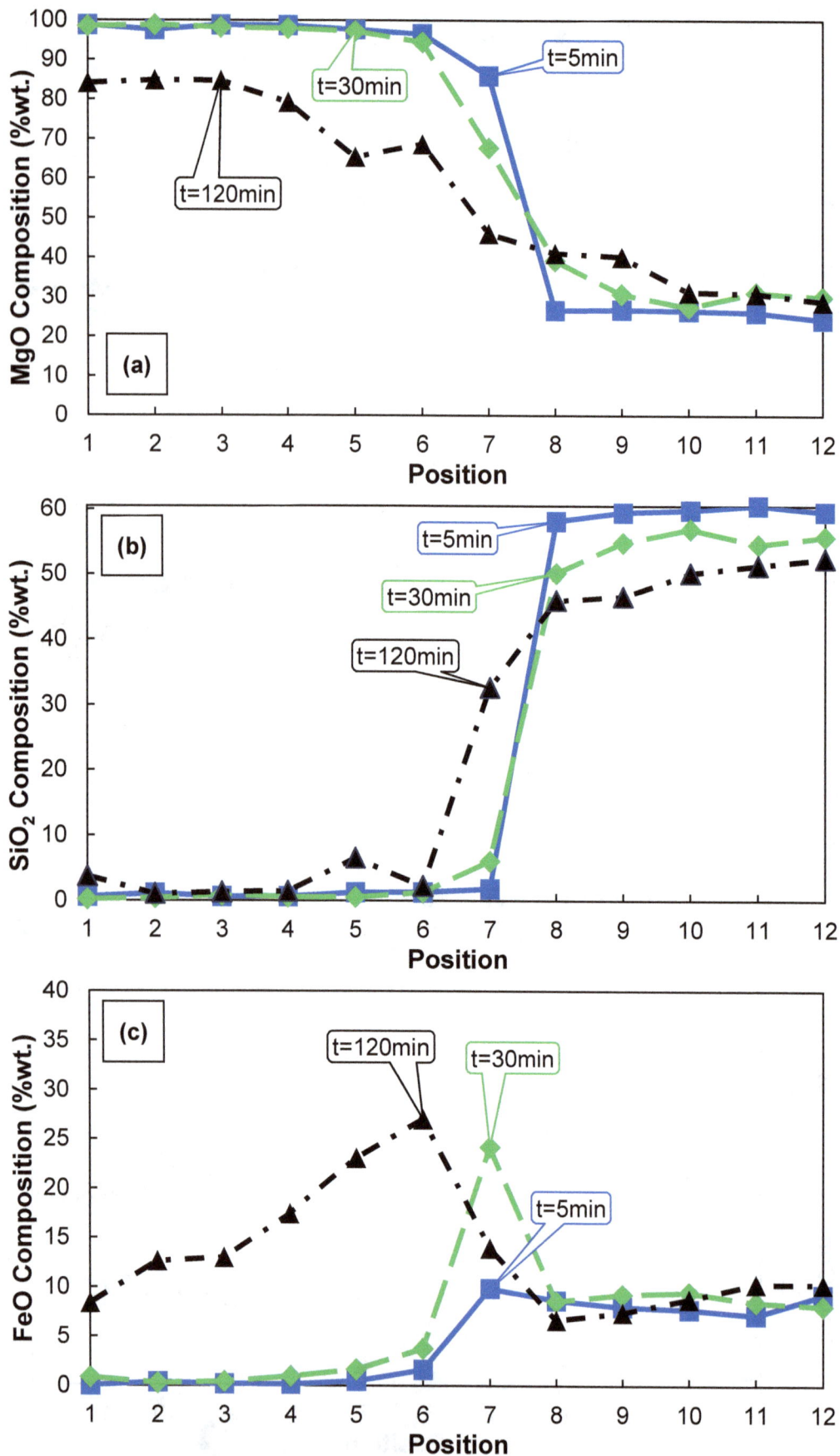

FIG 5 – Semi-quantitative compositional analysis along refractory-slag interface from samples melted at 1500°C with an initial SiO_2/MgO ratio in slag of 3.00 for melting time between 5 and 120 mins: (a) MgO composition; (b) SiO_2 composition; and (c) FeO composition.

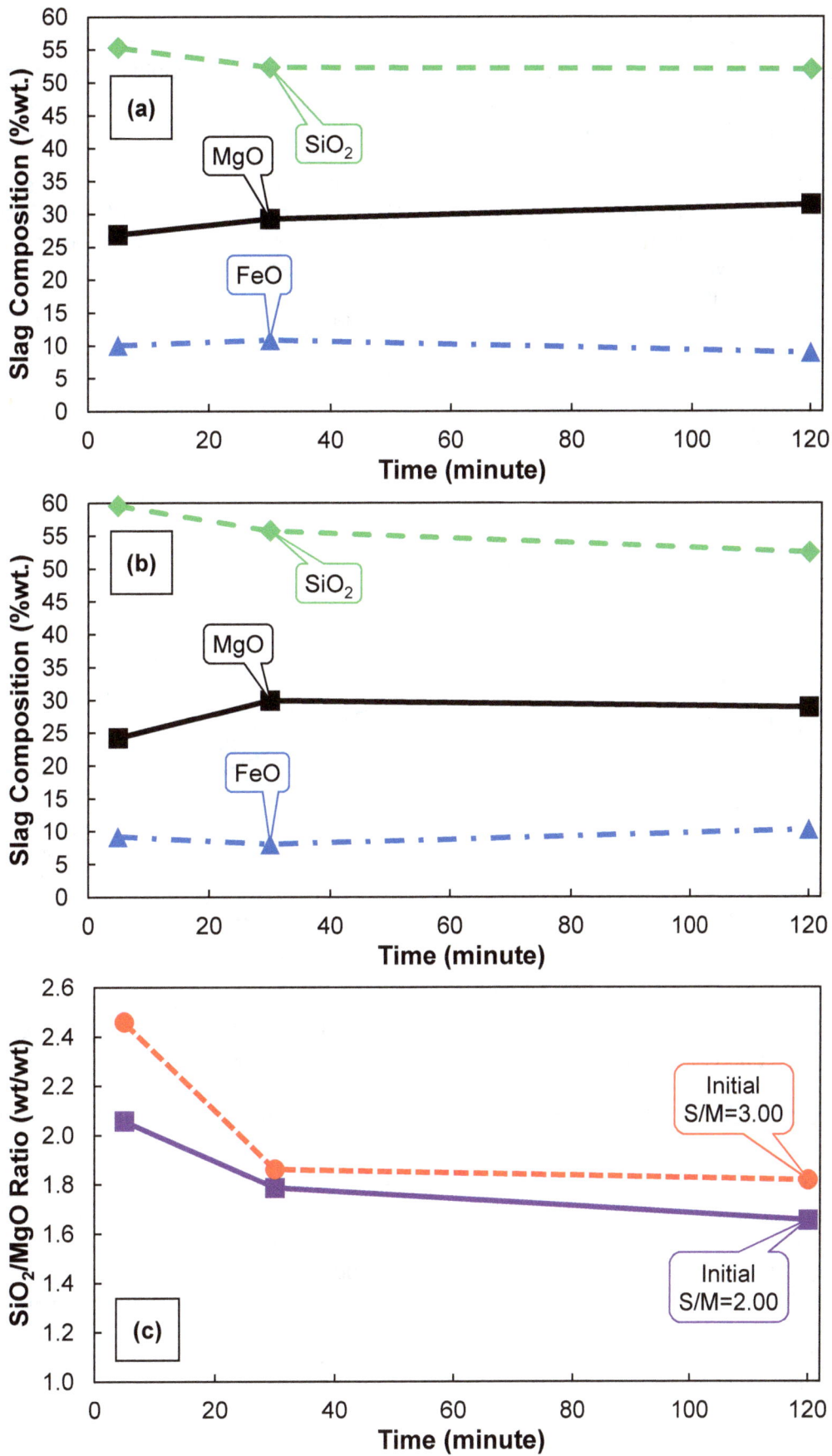

FIG 6 – Bulk composition of liquid slag from semi-quantitative compositional analysis of samples melted at 1500°C for melting time between 5 and 120 mins: (a) Sample having initial SiO_2/MgO ratio in slag of 2.00; (b) Sample having initial SiO_2/MgO ratio in slag of 3.00; and (c) SiO_2/MgO ratio in the final slag.

The pseudo-ternary diagram MgO-FeO-SiO₂, at constant CaO content of 3 wt per cent and Al₂O₃ content of 5 wt per cent, relevant to the slag system in the present study, is shown in Figure 7. The diagram was calculated using FactSage™ 8.0 (Bale *et al*, 2011) with the liquid Fe activity set to 0.9, corresponding to the activity of iron in the NPI metal. The initial slag compositions for samples with initial S/M ratios in slag of 2.00 and 3.00 are plotted in the figure. Additionally, the slag compositions from samples melted at 1500°C with initial S/M ratios in slag of 2.00 and 3.00 for melting times of 5 mins, 30 mins, and 120 mins are also included in the figure. Contacting the mixtures with MgO refractory at 1500°C results in the movement of the liquid slag composition toward the MgO corner of the ternary diagram. Theoretically the movement of the liquid slag should conclude at the liquidus of olivine for 1500°C, where the liquid slag is saturated by MgO and olivine solid is formed. However, the measured liquid slag composition extends beyond the 1500°C liquidus. This discrepancy may arise due to several reasons, including inaccurate liquids prediction by FactSage™ due to an unoptimised database, inaccuracy in temperature set-up during melting experiments, inclusion of olivine solid in the measurement of liquid slag composition, or chemical segregation of the liquid slag during cooling which results in the measured slag composition not accurately representing its actual composition at high temperature.

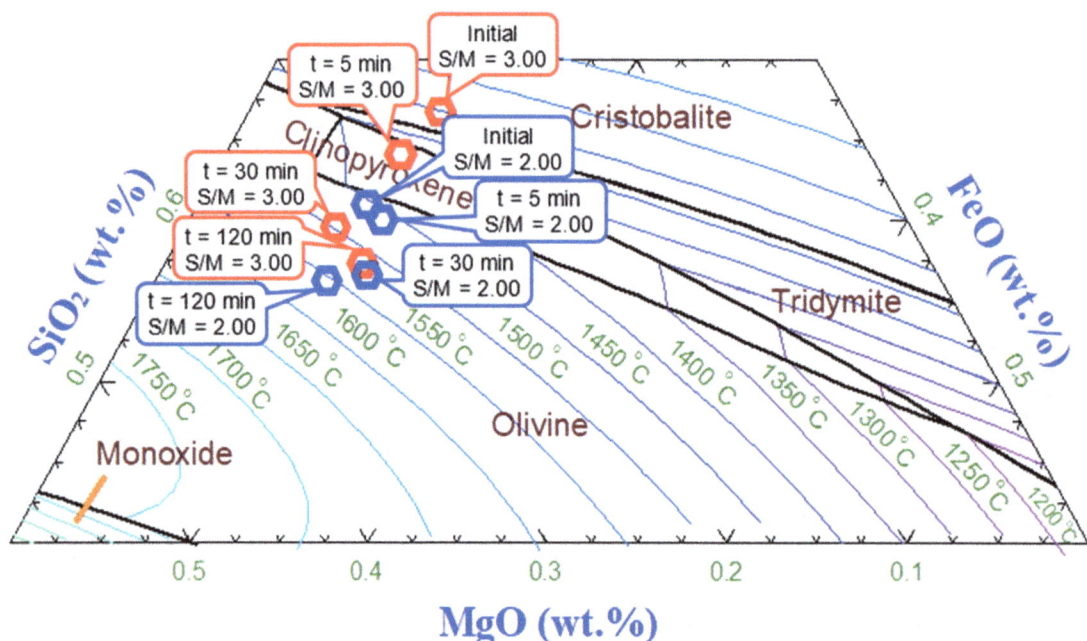

FIG 7 – FactSage™ calculated pseudo-ternary diagram of MgO-FeO-SiO₂ at CaO = 3 wt per cent and Al₂O₃ = 5 wt per cent in equilibrium with liquid Fe (αF_e = 0.9).

The effect of SiO₂/MgO ratio in the initial slag on the interaction between slag and MgO refractory

The effect of initial slag composition on the interaction between slag and MgO refractory was investigated by conducting melting of samples with initial SiO₂/MgO (S/M) ratios in slag of 1.75, 2.00, 2.50, and 3.00 at 1500°C. A melting time of 120 mins was selected to ensure that the interaction between slag and refractory approaches completion, as demonstrated by the melting time variation experiments. Moreover, the 120 mins melting time was chosen since it closely resembles the residence time of melts in the electric furnace in the industrial practice. The secondary electron image and elemental mapping of the melted samples with various initial S/M ratios in slag at 1500°C are provided in Figure 8.

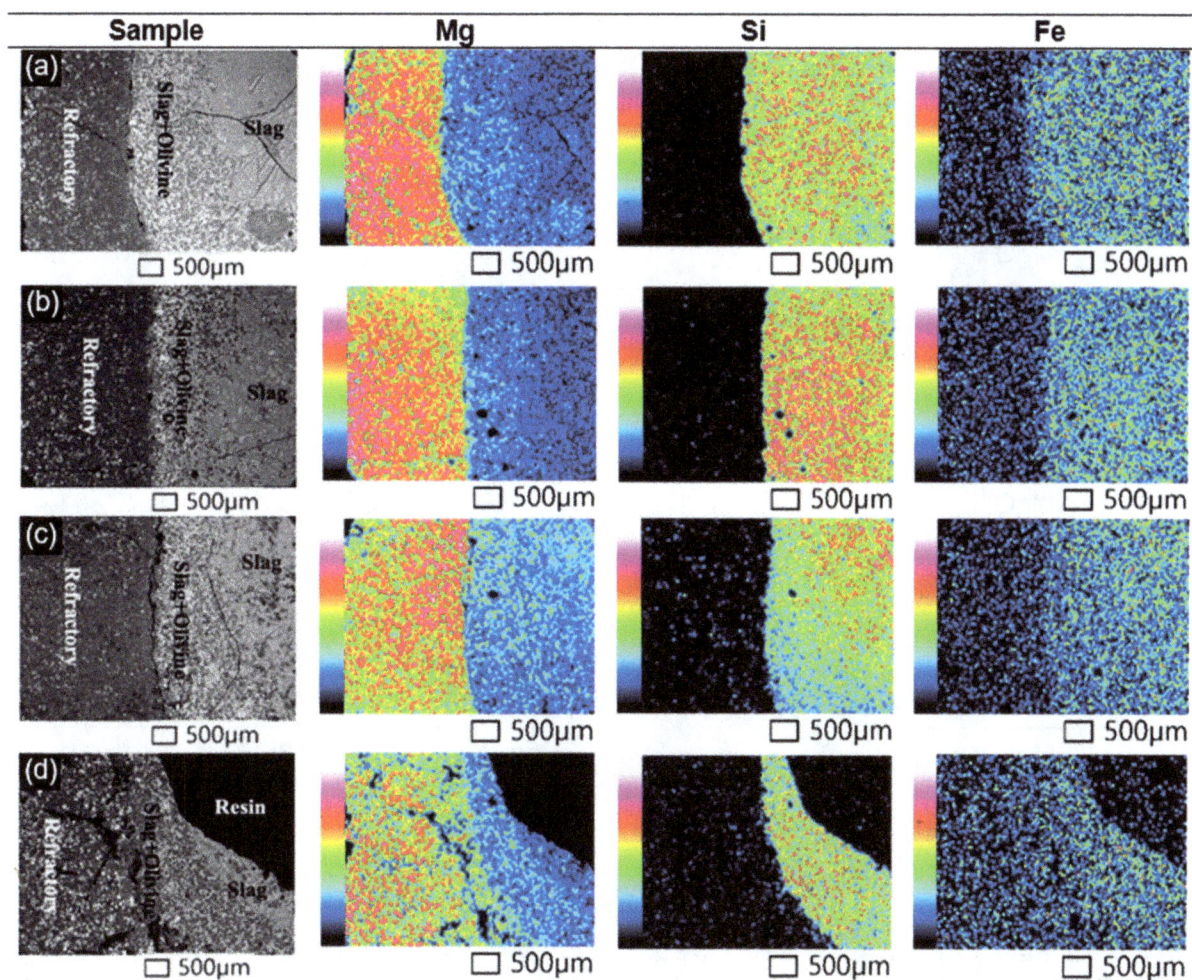

FIG 8 – Secondary electron image and elemental mapping from samples melted at 1500°C for melting time of 120 mins with initial SiO_2/MgO ratios in slag of: (a) 1.75; (b) 2.00; (c) 2.50; and (d) 3.00.

All samples exhibit areas of slag, slag-olivine mixture, and dense MgO. The sample with an initial S/M ratio of 3.00 shows a smaller slag area (see Figure 8d) due to a more extensive reaction between the slag and the MgO refractory. The results of the mapping analysis of Mg and Fe elements in the samples indicate that as the S/M ratio of the slag increases, the interaction between slag and the refractory material becomes more significant. This is evident through the decreasing intensity of Mg and increasing Fe intensity in the crucible. The penetration of FeO into the MgO crucible appears to be more pronounced than that of SiO_2. This phenomenon may be attributed to the ability of FeO to diffuse through and combine with the MgO monoxide solid solution.

The semi-quantitative EDS measured composition of the samples along the refractory and slag interface is provided in Table 4, with the compositions of the dominant components, MgO, SiO_2, and FeO, plotted in Figure 9. The experimental results indicate that the MgO concentration in most of the slags increases after the melting experiments, indicating the dissolution of MgO from the refractory into the slag (see Figure 9a). Conversely, the SiO_2 concentration in the slags decreases after the melting experiments due to dilution by dissolved MgO (see Figure 9b). There is no significant penetration of the MgO solid by the SiO_2. However, the most significant penetration of the MgO solid was by FeO, as clearly observed in the sample with an initial S/M ratio in slag of 3.00, resulting in FeO concentrations in the MgO refractory ranging from 8.4 to 27 wt per cent (see Figure 9c).

TABLE 4

Semi-quantitative EDS measured composition along refractory-slag interface from samples melted at 1500°C for melting time of 120 mins with initial SiO_2/MgO ratios in slag between 1.75 and 3.00.

Sample		Position											
		1	2	3	4	5	6	7	8	9	10	11	12
Initial S/M = 1.75	MgO	98.5	98.4	98.3	97.3	96.0	93.0	90.2	45.1	41.7	37.9	41.5	38.3
	SiO_2	0.2	0.4	0.1	0.2	0.2	0.3	0.6	45.2	47.8	47.2	46.8	46.3
	Al_2O_3	0.5	0.5	0.7	0.5	0.7	1.0	1.9	2.9	3.4	4.2	4.3	3.8
	CaO	0.1	0.1	0.0	0.1	0.1	0.1	0.2	0.9	1.6	2.0	0.9	4.3
	FeO	0.7	0.7	1.0	1.9	3.1	5.6	7.1	5.9	5.5	8.6	6.5	7.4
Initial S/M = 2.00	MgO	98.6	98.1	98.1	96.4	95.5	92.3	54.3	37.8	34.0	32.4	32.3	31.4
	SiO_2	0.3	0.3	0.3	0.8	0.3	0.5	42.0	50.7	49.1	48.4	49.5	52.0
	Al_2O_3	0.4	0.3	0.5	0.6	0.7	1.1	1.5	3.3	5.5	6.5	6.0	4.5
	CaO	0.0	0.1	0.0	0.0	0.0	0.1	0.1	1.3	2.5	3.4	3.1	3.2
	FeO	0.6	1.2	1.1	2.2	3.4	6.0	2.1	6.9	8.9	9.4	9.1	8.9
Initial S/M = 2.50	MgO	98.1	96.2	95.2	93.1	93.7	92.3	49.1	40.5	37.3	27.4	29.8	31.6
	SiO_2	0.4	0.4	0.7	1.0	0.1	0.3	31.3	46.2	49.6	50.0	51.6	53.8
	Al_2O_3	0.5	0.6	0.6	1.0	1.0	1.3	2.7	4.3	4.1	8.5	5.8	4.3
	CaO	0.1	0.4	0.1	0.7	0.0	0.1	2.4	1.9	1.5	3.6	3.0	2.3
	FeO	1.0	2.4	3.4	4.3	5.2	6.0	14.5	7.1	7.5	10.5	9.8	8.1
Initial S/M = 3.00	MgO	84.1	84.8	84.7	79.3	65.4	68.6	46.0	41.1	40.1	31.2	30.9	28.9
	SiO_2	3.7	1.0	1.3	1.4	6.5	2.2	32.6	46.0	46.6	50.2	51.4	52.5
	Al_2O_3	3.0	1.0	0.6	1.1	1.6	1.4	4.9	4.2	4.1	6.5	5.1	5.9
	CaO	0.9	0.6	0.5	0.8	3.5	0.8	2.5	2.1	1.8	3.3	2.4	2.6
	FeO	8.4	12.6	13.0	17.4	23.1	27.0	14.0	6.6	7.4	8.8	10.3	10.3
		REFRACTORY						**OLIVINE**		**SLAG**			

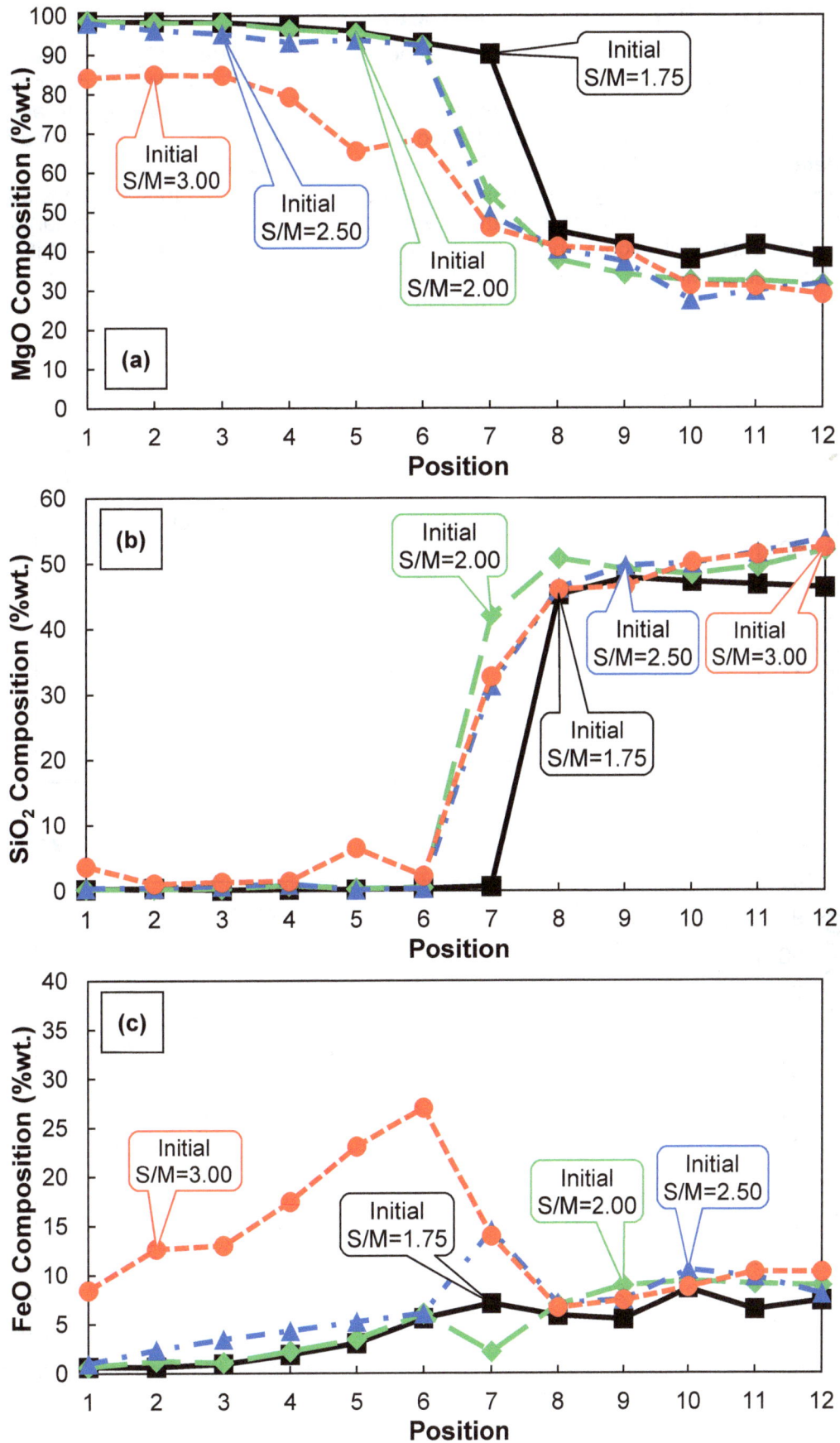

FIG 9 – Semi-quantitative compositional analysis along refractory-slag interface from samples melted at 1500°C for 120 mins with initial SiO$_2$/MgO ratios in slag between 1.75 and 3.00: (a) MgO composition; (b) SiO$_2$ composition; and (c) FeO composition.

The corrosion mechanism of MgO refractory by slag

The phases and their compositions resulting from the interaction between MgO refractory and slag at 1500°C are predicted using FactSage™, as shown in Figure 10. The trends demonstrated by FactSage™ predictions can serve as a guide for understanding the refractory-slag interaction. In the case of sample with an initial SiO_2/MgO (S/M) ratio in slag of 2.00, the slag is entirely liquid and can accommodate MgO up to 29.4 wt per cent. Further addition of MgO leads to the formation of olivine; olivine forms when the ratio of MgO to slag is 4:96. A higher ratio of MgO to slag above 30:70 leads to the stabilisation of monoxide together with traces of olivine. For sample with an initial S/M ratio in slag of 3.00, the slag is fully liquid and becomes saturated with olivine at a ratio of MgO to slag of 10:90. As the ratio of MgO to slag reaches above 33:67, the slag dissipates and is replaced by monoxide with traces of olivine. In general, the phases predicted by FactSage™ are consistent with those observed in actual experiments.

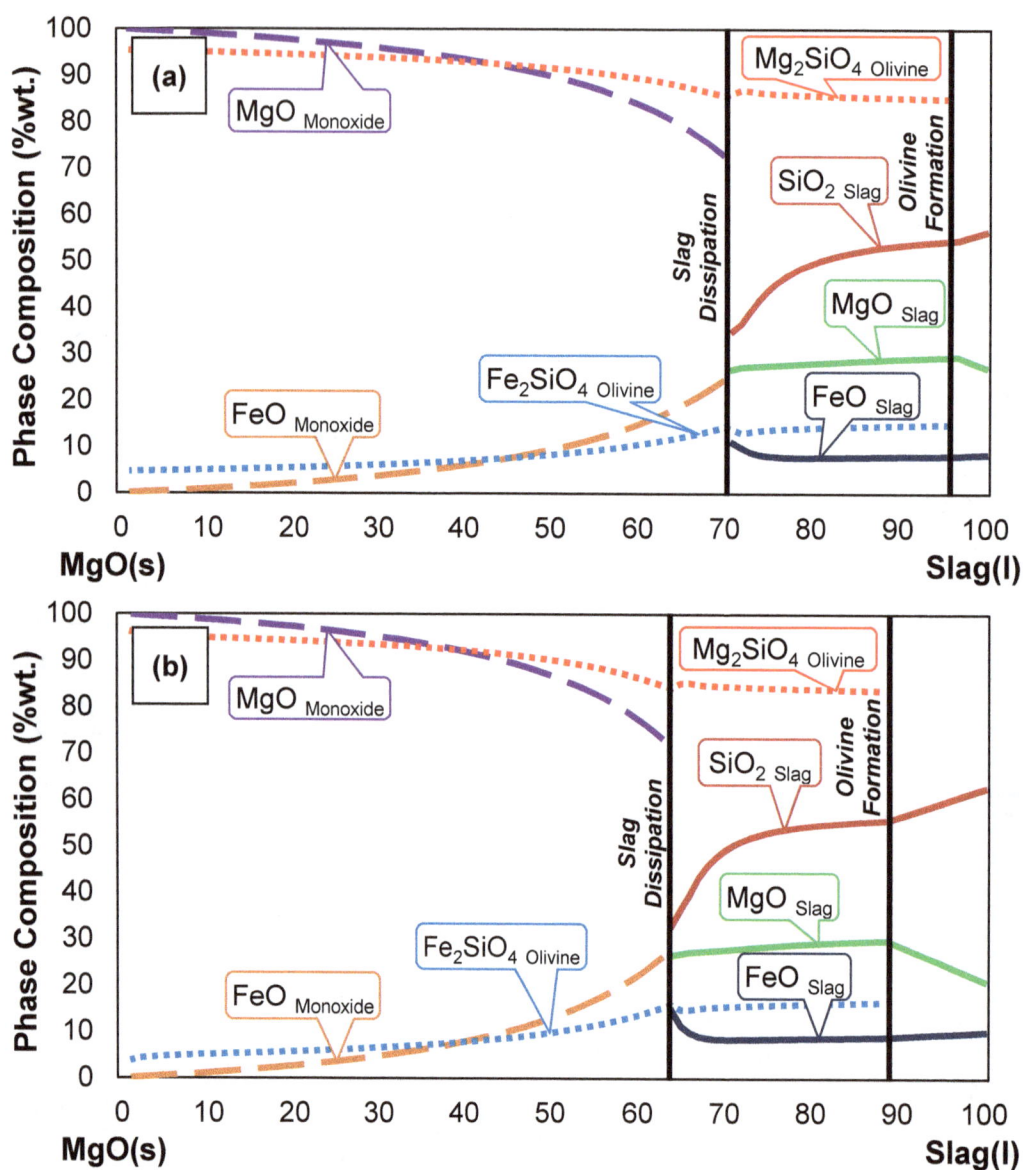

FIG 10 – FactSage™ calculated phase composition from interaction between MgO refractory with slag at 1500°C with initial SiO_2/MgO ratio in slag: (a) 2.00; and (b) 3.00.

The sub-processes involved in refractory corrosion are depicted in Figure 11. Mass transfer of slag components such as FeO and SiO_2 (step-1) to the refractory-slag interface, and MgO (step-6) from the interface can occur due to the concentration gradient between bulk slag and the interface. Contact between FeO from the slag and MgO solid can lead to the incorporation of FeO into the MgO solid solution (step-2); later, this can result in solid-state diffusion of FeO within the MgO refractory (step-3). On the other hand, contact between SiO_2 from the slag and MgO solid (step-4) can lead to the dissolution of MgO into the slag (step-5). Continuous dissolution of MgO into the slag can lead to the saturation of the slag by MgO, which leads to the formation of olivine (step-5b) on the surface of the MgO solid or the formation of olivine crystals within the slag (step-7). The olivine layer protects the MgO refractory from direct contact with SiO_2 but does not hinder the transfer of FeO and MgO via solid-state diffusion through the olivine layer (step-3b and 3c).

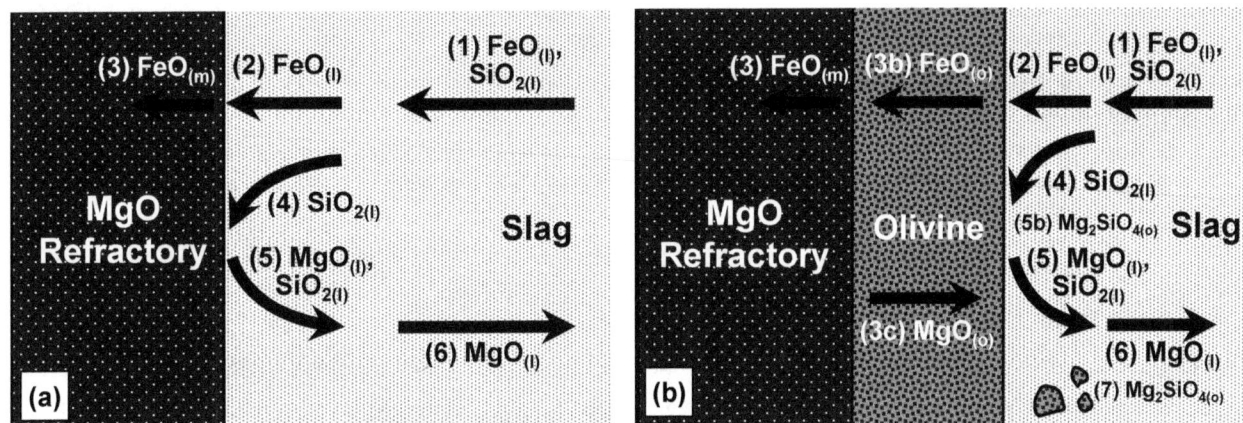

FIG 11 – Mechanism of refractory-slag interaction: (a) without olivine layer; and (b) with olivine layer.

All of the sub-processes in Figure 11 take place simultaneously, and the overall refractory corrosion rate will be controlled by the slowest sub-process. The formation of the olivine layer is expected to minimise the rate of refractory corrosion since it eliminates direct contact/reaction between MgO solid and slag, creating a barrier between the two phases that can only be penetrated by slag components through relatively slow solid-state diffusion. Consequently, the composition of slag in FeNi or NPI smelting must be carefully controlled to aim for the formation of a stable, dense phase that can protect the surface of the MgO refractory and minimise the refractory and slag interaction.

CONCLUSIONS

The interaction between slag and refractory has been examined under conditions relevant to nickel laterite ore smelting. Industrial slag was replicated using synthetic slag, while the actual refractory material was replicated using a dense, pure MgO crucible. The kinetics of the slag-refractory reaction were investigated by conducting experiments at 1500°C for 5, 30, and 120 mins using samples with selected initial S/M ratios in slag of 2.0 and 3.0 (wt/wt) at constant CaO = 3 wt per cent, Al_2O_3 = 5 wt per cent, FeO = 10 wt per cent. The dissolution of MgO into the slag occurs at a rapid rate that within 5 to 30 mins of melting the MgO concentration in slag increases by 2.4 wt per cent and 5.7 wt per cent for initial S/M ratios of 2.0 and 3.0 wt/wt, respectively. The dissolution of MgO is less rapid between 30 and 120 mins, possibly due to a lesser driving force for MgO dissolution as the liquid approaches MgO saturation condition. The slag-refractory interaction was also studied at 1500°C for 120 mins using initial mixtures with different initial SiO_2/MgO (S/M) ratios in slag of 1.75, 2.0, 2.5, and 3.0 (wt/wt). All samples from the 120 mines melting experiments exhibit areas of slag, slag-olivine mixture, and dense MgO. It was found that as the initial S/M ratio of the slag increases, the interaction between the slag and the refractory material becomes more significant, as observed from the increasing dissolution of MgO from the refractory into the slag. A higher S/M ratio in slag also results in a greater extent of solid-state diffusion of FeO into the remaining MgO grains. It is anticipated that the presence of an olivine layer can reduce the rate of corrosion on the refractory material. Therefore, it is crucial to carefully control the slag composition in FeNi or NPI smelting to ensure the formation of a stable and dense protective layer on the surface of the refractory brick.

ACKNOWLEDGEMENTS

The authors would like to acknowledge the support from PT Gunbuster Nickel Industry for providing the Scanning Electron Microscope-Energy Dispersive X-ray Spectroscopy (JCM-7000 NeoScopeTM. JEOL. Tokyo. Japan) used for the analysis of the samples in this work. The authors would also like to acknowledge the support from the Department of Mineral Sciences of the Smithsonian Institution for providing Basaltic Glass and Springwater Olivine reference materials.

REFERENCES

Crundwell, F, Moats, M, Ramachandran, V, Robinson, T and Davenport, W, 2011. *Extractive Metallurgy of Nickel, Cobalt and Platinum Group Metals* (Elsevier: Amsterdam).

Degel, R, Kempken, J, Kunze, J and König, R, 2007. Design of a modern large capacity FeNi smelting plant, in *INFACON XI*, pp 605–620 (The Indian Ferro Alloy Producers' Association).

Gangler, J J, 1950. Some physical properties of eight refractory oxides and carbides, *Journal of the American Ceramic Society*, 33:367–374.

Huang, S, Xue, J and Wang, Z, 2017. Effect of FeO content in laterite nickel slag on the corrosion behaviour of refractory materials, in *8th International Symposium on High-Temperature Metallurgical Processing*, pp 679–688 (The Minerals, Metals and Materials Society).

Kotzé, I J, 2002. Pilot plant production of ferronickel from nickel oxide ores and dusts in a DC arc furnace, *Minerals Engineering*, 15:1017–1022.

Ruh, E and McDowell, J S, 1962. Thermal conductivity of refractory brick, *The American Ceramic Society*, 45:189–195.

Sagadin, C, Luidold, S, Wagner, C and Wenzl, C, 2016. Melting behaviour of ferronickel slags, *JOM*, 68:3022–3028.

Sagadin, C, Luidold, S, Wenzl, C and Wagner, C, 2017. Evaluation of high temperature refractory corrosion by liquid Al_2O_3–Fe_2O_3–MgO–SiO_2, in *8th International Symposium on High-Temperature Metallurgical Processing*, pp 161–168 (The Minerals, Metals and Materials Society).

Sagadin, C, Luidold, S, Wagner, C and Spanring, A, 2018a. High temperature phase formation at the slag/refractory interphase at ferronickel production, in *Extraction 2018*, pp 137–147 (The Minerals, Metals and Materials Society).

Sagadin, C, Luidold, S, Wagner, C, Spanring, A and Kremmer, T, 2018b. Phase reactions between refractory and high-acidic synthetic CaO-ferronickel slag, *JOM*, 70:34–40.

Sagadin, C, Luidold, S, Wagner, C, Pichler, C, Kreuzer, D, Spanring, A, Antrekowitsch, H, Clarke, A and Clarke, K, 2021. Thermodynamic refractory corrosion model for ferronickel manufacturing, *Metallurgical and Materials Transactions B*, 52:1052–1060.

Wagner, C, Wenzl, C, Gregurek, D, Kreuzer, D, Luidold, S and Schnideritsch, H, 2017. Thermodynamic and experimental investigations of high-temperature refractory corrosion by molten slags, *Metallurgical and Materials Transactions B*, 48:119–131.

Zhang, S and Lee, W E, 2013. Use of phase diagrams in studies of refractories corrosion, *International Materials Reviews*, 45:41–58.

Zhang, S and Lee, W E, 2004. Direct and indirect slag corrosion of oxide and oxide-c refractories, in *VII International Conference on Molten Slags Fluxes and Salts*, pp 309–319 (The South African Institute of Mining and Metallurgy).

AUTHOR INDEX

Mu, W	1169, 1259	Petrus, H T B M	495
Muhmood, L	39	Pettersen, J B	1609
Mukhlis, R Z	1533, 1675	Pihlasalo, J	997
Müller, A	1231	Pinson, D	807, 1591
Müller, M	517	Pistorius, P C	1249, 1337
Murakami, T	1387	Polkowski, W	1413
Nahian, M K	895	Pownceby, M I	1445, 1533
Nakashima, K	17	Pradana, A D	1693
Nakayama, A	159	Pradana, D R	1693
Natsui, S	653	Prasad, R	705
Nekhoroshev, E	901, 931, 1029	Preisser, N	1195
Nell, J	343	Putera, A	301, 495
Ni, H W	1259	Putra, Y H	1693
Nicol, S	477, 647	Qi, J	517
Nigam, A	1177	Qin, H	353
Nikolic, S	477, 647	Qin, S	1123
Nogami, H	653	Ranjan, M	101
O'Brien, H	1571	Rapp, D	517
O'Dea, D	625	Reddy, R G	189, 895, 923, 1205
O'Malley, R J	117	Ren, M M	1325
Ohno, K	675	Reuter, M	663
Ostrovski, O	1267	Rhamdhani, M A	301, 495, 569, 1445, 1533, 1675
Pahlevani, F	1495		
Pakhomov, R A	1187	Rich, A	847, 1281
Palai, P	101	Rimal, V	1217
Palanisamy, S	1445, 1533	Ringdalen, E	219
Panda, S K	635	Rodrigues, C M G	839
Pande, M M	201, 251, 413	Roos, Å	1471
Park, G H	423	Safarian, J	361, 759, 1077, 1543
Park, H S	1565	Sahajwalla, V	1495
Park, J	231, 261, 423, 427, 459, 1117, 1169, 1565, 1585, 1603, 1603	Sahoo, P P	101
		Saito, K	375, 505
Park, K S	261	Saito, N	17
Park, S J	427	Samuelsson, C	81
Park, Y-J	487, 915	Sander, T	117
Partinen, J	1571	Sarkar, R	1177
Pastor-Vallés, E	1609	Sarmadi, N	1495
Petersen, J	343	Sartor, G	847
Petersen, S	771	Sasaki, Y	353

www.ingramcontent.com/pod-product-compliance
Lightning Source LLC
Chambersburg PA
CBHW061102210326
41597CB00021B/3958